Mathematics for Elementary School Teachers

6e

TOM BASSAREAR
Keene State College

MEG MOSS
Western Governors University

Australia • Brazil • Japan • Korea • Mexico • Singapore • Spain • United Kingdom • United States

***Mathematics for Elementary School Teachers,* Sixth Edition**

Tom Bassarear, Meg Moss

Vice President, General Manager: Balraj Kalsi

Senior Product Team Manager: Richard Stratton

Content Developer: Erin Brown

Associate Content Developer: Samantha Lugtu

Product Assistant: Jennifer Cordoba

Product Development Manager: Heleny Wong

Marketing Manager: Julie Schuster

Content Project Manager: Jennifer Risden

Art Director: Vernon Boes

Intellectual Property Analyst: Christina Ciaramella

Intellectual Property Project Manager: John Sarantakis

Manufacturing Planner: Rebecca Cross

Production Service: Graphic World, Inc.

Photo Researcher: Lumina Datamatics

Text Researcher: Lumina Datamatics

Copy Editor: Graphic World, Inc.

Illustrator: Graphic World, Inc.

Text Designer: Terri Wright

Cover Designer: Terri Wright

Cover Image: Alex Koloskov Photography/Flickr Open/Getty Images

Interior design images: Magaiza/E+/Getty Images; Robert George Young / Masterfile

Compositor: Graphic World, Inc.

Library of Congress Control Number: 2014943306

ISBN: 978-1-305-07136-0

Cengage Learning
20 Channel Center Street
Boston, MA 02210
USA

Cengage Learning is a leading provider of customized learning solutions with office locations around the globe, including Singapore, the United Kingdom, Australia, Mexico, Brazil, and Japan. Locate your local office at **www.cengage.com/global.**

Cengage Learning products are represented in Canada by Nelson Education, Ltd.

To learn more about Cengage Learning Solutions, visit **www.cengage.com.**

Purchase any of our products at your local college store or at our preferred online store, **www.cengagebrain.com.**

Printed in the United States of America
Print Number: 01 Print Year: 2014

CONTENTS

ABOUT THE AUTHORS

Tom Bassarear

I have been teaching the Mathematics for Elementary School Teachers course for more than 20 years, and in that time, I have learned as much from my students as they have learned from me. This text was inspired by my students and reflects one of the most important things we have taught one another: that building an understanding of mathematics is an active, exploratory process, and ultimately a rewarding, pleasurable one. My own experience with elementary schoolchildren and my two children, Emily and Josh, has convinced me that young children naturally seek to make sense of the world they live in and for a variety of reasons many people slowly lose that curiosity over time. My hope is that this book will engage your curiosity about mathematics once again.

Meg Moss

I am excited and honored to be working with Tom Bassarear on this book. I began teaching Mathematics for Elementary Teachers over 20 years ago. I immediately began seeking advice from others who had taught the course, and volunteered in elementary classrooms to learn more. Teaching these courses has deepened my mathematical understanding as well as my understanding of how people learn math. Helping future elementary school teachers to truly understand mathematics, and see the beauty in mathematics is very rewarding, and I know that all of this will have a major positive impact on their future students. I appreciate you sharing this journey with me!

NEW TO THE SIXTH EDITION!

I am pleased to welcome Meg Moss, Ph.D., to this textbook and to introduce her to you. I have known Meg for about 10 years through conversations and presentations at conferences. I have admired the quality of her work, depth of thought, and commitment to students, so I was delighted when she agreed to join me in continuing this book as I transition toward retirement. While I have been actively involved in the current revision, you will find Meg's footprints throughout the book.

She has done a magnificent job of framing the Common Core State Standards in mathematics (CCSM), which have been adopted by 45 states, in a way that helps readers to see where their future students will learn these concepts and to help them see the importance of such concepts. The CCSM articulates eight mathematical practices (MPs) that replace standards by the National Council of Teachers of Mathematics (NCTM). While I believe that the NCTM standards are more clearly stated and more user-friendly, the eight MPs correlate strongly with NCTM's framework. Meg's genius was not to try to incorporate all the details of the MPs, which would be overwhelming, but to focus on and articulate the big ideas embedded in those practices. She articulates them in the first chapter, connecting them directly to Investigations, and then refers to them in appropriate ways throughout the book.

After many discussions, she constructed a revised and streamlined Chapter 1, which I love. She integrated the number theory concepts—which previously comprised a separate chapter—into the textbook, as those concepts are needed. This connects to research on learning that indicates students are more willing and able to retain ideas if they see how they are connected and if they use them immediately.

With CCSM's emphasis on algebraic thinking, we decided to have a separate Chapter 6 on algebra. Meg did a heroic job of researching best practices in algebra in schools and then organized the material into a coherent framework that addresses the important algebraic ideas articulated by CCSM. She took many Investigations from the fifth edition's algebra section and some from Chapter 1 and has added many of her own.

Meg also wrote Questions to Summarize Big Ideas for the end-of-chapter summaries. These questions help students reflect on what they have learned and articulate major "take-away" ideas from the chapter, ultimately supporting one of the most important ideas of the textbook—this is to OWN knowledge.

In addition, Meg went through every page of the textbook and you will see her work in many places, such as in

- revising text to make points more clear and concise;
- adding extra steps and more concreteness when she felt it would be helpful, especially for students who tend to struggle with those ideas;
- more visual representations including Singaporean bar models; and
- more technology, including references to virtual manipulatives, Geogebra investigations, and several other websites.

I hope you will welcome and appreciate Meg's contributions to the sixth edition as much as I do.

—Tom Bassarear

ANNOTATED CONTENTS

Chapter 1 Foundations for Learning Mathematics

This chapter continues the theme from the fifth edition, but with a new emphasis on the Mathematical Practices of the Common Core State Standards (CCSS). While references to the NCTM standards remain, one of the goals of the revised Chapter 1 is to lay the groundwork for the CCSS so that students can see some of those standards "in action" while they are learning the mathematics throughout the textbook. The explorations in the *Explorations* manual offer diverse types of problems to grapple with that support the strategies used in the rest of the course.

Chapter 2 Fundamental Concepts

Chapter 2 has been shortened, with former Section 2.2 now included in a newly developed Chapter 6 that is devoted to algebraic thinking. Sections 2.1 and 2.3 from the fifth edition remain, with revisions in these sections focused on enhancing discussions of sets and numeration.

Section 2.1 gives students tools that enable them to talk about sets and subsets and to use Venn diagrams when the need arises in other chapters, such as to understand the relationship between different sets of numbers.

Section 2.2 includes the development of children's understanding of numeration and its historical development, both of which students find fascinating. Exploration 2.3 (Alphabitia) is one of the most powerful we have used. Most of our students report this to be the most significant learning and/or turning point in the semester. The exploration unlocks powerful understandings related to numeration, which the text supports by discussing the evolution of numeration systems over time and exploring different bases.

Chapter 3 The Four Fundamental Operations of Arithmetic

Portions of the fifth edition's Chapter 4, Number Theory, are now integrated into Chapter 3, as appropriate. For example, divisibility may now be found in Section 3.4, Understanding Division. Several discussions are now rewritten with more emphasis on place value and visual representations of numbers. The goals of Chapter 3 otherwise are the same. Students see how the concepts of the operations, coupled with an understanding of base ten, enable them to understand how and why procedures that they have performed by memorization for years actually work. In addition to making sense of standard algorithms, we present alternative algorithms in both the text and explorations. Our students have found these algorithms to be both enlightening and fascinating.

Chapter 4 Extending the Number System

The sets of integers, fractions, and decimals represent three historically significant extensions to the set of whole numbers. To enhance the discussion of fractions, Singaporean bar models are used. The concepts of least common multiple and greatest common divisor are integrated into the fraction section when needed for simplifying and for common denominators.

In Exploration 4.5 (Making Manipulatives), students construct fraction manipulatives and then look for rules when ordering fractions, a critically important first step in seeing fractions as more than numerator and denominator. In Exploration 4.19 (Meanings of Operations with Fractions), having students represent problem situations with diagrams requires them to adapt their understanding of the four operations to fraction situations. Having first constructed this concept through exploration, students can approach Investigation 4.2k (Ordering Rational Numbers) with a richer understanding of what it really means to say that one fraction is greater than another.

While Chapters 3 and 4 have been arranged according to the manner in which many instructors prefer this content to appear, it is not fixed. For those instructors who prefer a more "operations-centric" approach to the course, we offer an alternative organization of topics as follows:

Chapter 1 Foundations for Learning Mathematics

Chapter 2 Fundamental Concepts

4.1 Integers

4.2 Fractions and Rational Numbers

3.1 Understanding Addition

3.2 Understanding Subtraction

4.3 Understanding Operations with Fractions (first half addition and subtraction of fractions)

3.3 Understanding Multiplication

3.4 Understanding Division

4.3 Understanding Operations with Fractions (second half, multiplication and division of fractions)

4.4 Beyond Integers and Fractions

Chapter 5 Proportional Reasoning

Chapter 6 Algebraic Thinking, and so on

Chapter 5 Proportional Reasoning

The investigations and explorations in Chapter 5 are conceptually rich and provide many real-life examples so that students can enjoy developing an understanding of multiplicative relationships.

Chapter 6 Algebraic Thinking

In response to requests from reviewers, we have included a new chapter devoted to algebraic thinking. Chapter 6 explores patterns, the concept of a variable, and solving equations and inequalities using different models, including Singaporean bar models. The four sections are arranged under the National Council of Teachers of Mathematics (NCTM) algebra structure of understanding patterns, relations, and functions; representing and analyzing math situations and structures using algebraic symbols; using mathematical models to represent and understand quantitative relationships; and analyzing change in various contexts.

Chapter 7 Uncertainty: Data and Chance

In this chapter, students carefully walk through the stages of defining a question, collecting data, interpreting data, and then presenting data. We are particularly excited that the investigations with the concepts of mean and standard deviation remain successful with students. As a result, students can express these ideas conceptually instead of simply reporting the procedure.

Chapter 8 Geometry as Shape

In Chapter 8, you have the option of introducing geometry through explorations with tangrams, Geoboards, or pentominoes. This more concrete introduction allows students with unpleasant or failing memories of geometry to build confidence and understanding while engaging in rich mathematical explorations.

Chapter 9 Geometry as Measurement

This chapter addresses measurement from a conceptual framework (i.e., identify the attribute, determine a unit, and determine the amount in terms of a unit) and a historical perspective. Both the explorations and investigations get students to make sense of measurement procedures and to grapple with fundamental measurement ideas. Exploration 9.2 (How Tall?) generates many different solution paths and ideas and many discussions about indirect measurement and precision. Exploration 9.5 (What Does π Mean?) has demystified π in the minds of many students and is a wonderful exercise in communication. Exploration 9.11 (Irregular Areas) requires students to apply notions of measuring area to a novel situation. Students will hypothesize many different strategies, some of which are valid and some of which are not. The text looks at the larger notion of measurement, presents the major formulas in a helpful way, and illustrates different problem-solving paths.

Some of the most significant revisions to this chapter have been made to increase conceptual understanding of the concepts of measurement such as perimeter, area, and volume.

Chapter 10 Geometry as Transforming Shapes

The geometric transformations that we explore in Chapter 10 can be some of the most interesting and exciting topics of the course. Quilts and tessellations both spark lots of interest and provoke good mathematical thinking. The text develops concepts and introduces terms that help students to refine understanding that emerges from explorations.

PREFACE

Owning versus Renting

This course is about developing and *retaining* the mathematical knowledge that students will need as beginning mathematics teachers. We prefer to say that we are going to *uncover* the material rather than *cover* the material. The analogy to archaeology is useful. When archaeologists explore a site, they carefully *uncover* the site. As time goes on, they see more and more of the underlying structure. This is exactly what can and should happen in a mathematics course. When this happens, students are more likely to *own* rather than to *rent* the knowledge.

There are three ways in which this textbook supports owning versus renting:

1. Knowledge is constructed.
2. Connections are reinforced.
3. Problems appear in authentic contexts.

1. Constructing Knowledge

When students are given problems, such as appear in Investigations throughout the textbook, that involve them in grappling with important mathematical ideas, they learn those ideas more deeply than if they are simply presented with the concepts via lecture and then are given problems for practice. Additionally, there is a need to shift the focus from students studying mathematics to students doing mathematics. That is, students are looking for patterns, making and testing predictions, making their own representations of a problem, inventing their own language and notation, etc.

Investigation 1.2d (Pigs and Chickens) confronts a common misconception—that there is one right way to solve math problems—by exploring five valid solution paths to the problem. This notion of multiple solution paths is an important part of the book.

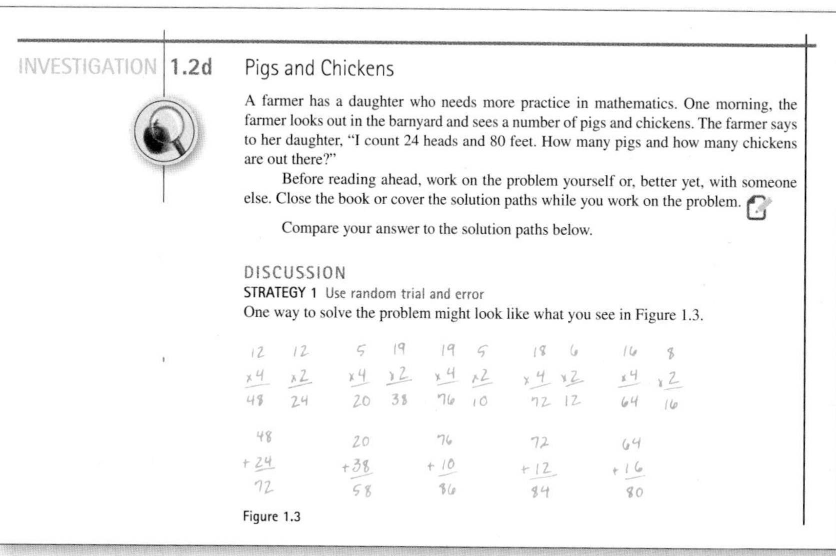

INVESTIGATION 1.2d Pigs and Chickens

A farmer has a daughter who needs more practice in mathematics. One morning, the farmer looks out in the barnyard and sees a number of pigs and chickens. The farmer says to her daughter, "I count 24 heads and 80 feet. How many pigs and how many chickens are out there?"

Before reading ahead, work on the problem yourself or, better yet, with someone else. Close the book or cover the solution paths while you work on the problem.

Compare your answer to the solution paths below.

DISCUSSION

STRATEGY 1 Use random trial and error

One way to solve the problem might look like what you see in Figure 1.3.

Figure 1.3

2. Reinforcing Connections

Understanding can be defined in terms of connections; that is, the extent to which you *understand* a new idea can be seen by the *quality* and *quantity* of connections between that idea and what you already know. There are two ways in which connections are built into the structure of the text.

1. Mathematical connections

Owning mathematical knowledge involves connecting new ideas to ideas previously learned. It also involves truly understanding mathematics, not just memorizing formulas and definitions.

- **CONNECTIONS AMONG CONCEPTS ARE EMPHASIZED**

Investigation 1.1d helps students see how the algebraic formula is closely connected to guess–check–revise. Investigation 1.2i is later connected both to fractions and to remainder. In Chapter 3, the four operations are constantly connected to each other in their development. Then in Chapter 4, the connections between operations with fractions and operations with whole numbers are discussed, as are how decimals are connected to whole numbers and to fractions. In Chapter 5, we look back at some problems done in Chapter 4 and see how they can now be solved more efficiently with the concepts of ratio. In Chapter 10, students see how our work with numbers and shapes is similar.

- **THE *HOW* IS CONNECTED TO *WHY***

In this way, students know not only how the procedure works but also why it works. For example, students understand why we move over when we multiply the second row in whole number multiplication; they realize that "carrying" and "borrowing" essentially equate to trading tens for ones or ones for tens; they understand why we first find a common denominator when adding fractions; and they see that π is how many times you can wrap any diameter around the circle.

2. Connections to children's thinking

In this book you will see a strong focus on children's thinking, for two reasons. First, much work with teachers focuses on the importance of listening to the students' thinking as an essential part of good teaching. If students experience this in a math course, then by the time they start teaching, it is part of how they view teaching. Second, when students see examples of children's thinking and see connections between problems in this course and problems children solve, both the quality and quantity of the students' cognitive effort increase.

3. Authentic Problems

Although most texts have many "real-life" problems, this text differs in how those problems are made and presented.

In Section 6.3, the question of paying a baby-sitter is explored. This situation is often portrayed as a linear function: for example, if the rate is \$10 per hour, $y = 10x$. However, in actuality, it is not a linear function but rather a stepwise function.

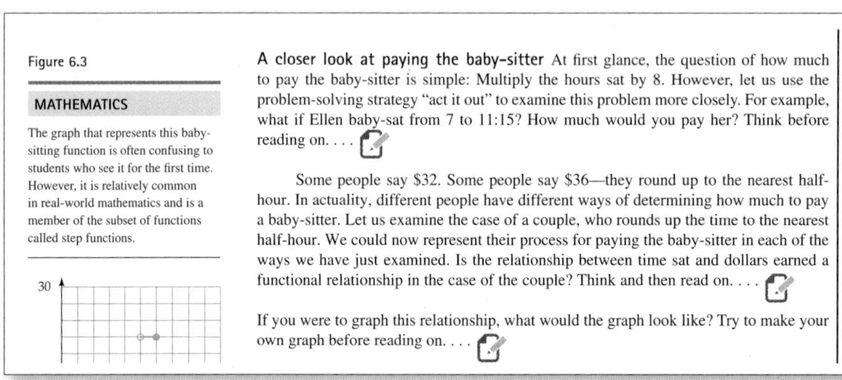

Figure 6.3

MATHEMATICS

The graph that represents this baby-sitting function is often confusing to students who see it for the first time. However, it is relatively common in real-world mathematics and is a member of the subset of functions called step functions.

A closer look at paying the baby-sitter At first glance, the question of how much to pay the baby-sitter is simple: Multiply the hours sat by 8. However, let us use the problem-solving strategy "act it out" to examine this problem more closely. For example, what if Ellen baby-sat from 7 to 11:15? How much would you pay her? Think before reading on. . . .

Some people say \$32. Some people say \$36—they round up to the nearest half-hour. In actuality, different people have different ways of determining how much to pay a baby-sitter. Let us examine the case of a couple, who rounds up the time to the nearest half-hour. We could now represent their process for paying the baby-sitter in each of the ways we have just examined. Is the relationship between time sat and dollars earned a functional relationship in the case of the couple? Think and then read on. . . .

If you were to graph this relationship, what would the graph look like? Try to make your own graph before reading on. . . .

Similarly, when determining the cost of carpeting a room, the solution path is often presented as dividing the area of the room by the cost per square yard; again, this is not how the cost is actually determined.

In this book, you will find many problems—problems we have needed to solve, problems friends have had, problems children have had, problems we have read about—where the content fits with the content of this course.

You will find problems in the text where students are asked to state the assumptions they make in order to solve the problem (e.g., Section 5.2, Exercises 33 to 38). You also will find problems that have the messiness of "real-life" problems, where the problem statement is ambiguous, too little or too much information is given, or the information provided is contradictory.

Problems 33–38 require you to make some assumptions in order to determine an answer. Describe and justify the assumptions you make in determining your answer.

33. Let's say that you read in the newspaper that last year's rate of inflation was 7.2%.

- **a.** If your grocery bill averaged \$325 per month last year, about how much would you expect your grocery bill to be this year?
- **b.** Let's say you received a \$1200 raise, from \$23,400 per year to \$24,600 per year. Did your raise keep you ahead of the game, or are you falling behind?

34. There was a proposal in New Hampshire in 1991 to reduce the definition of "drunk driving" from an alcohol blood content of 0.1 to 0.08. Explain why some might consider this a little drop and others might consider it a big drop. What do you think?

35. Which would you prefer to see on a sale sign at a store: \$10 off or 10% off? Explain your choice.

36. ***Classroom Connection*** Refer to Investigation 5.2b. Jane still doesn't understand the problem. Roberto tries to help her make sense of the problem by saying that the 8% means that if we were to select 100 students at the college, 8 of them would be working full-time. What do you think?

37. Annie has just received a 5% raise from her current wage of \$9.80 per hour.

- **a.** What is her new wage?
- **b.** What would this amount to over a year?
- **c.** What assumptions did you make in order to answer part (b)?
- **d.** What if the raise had been 5.4%?

Features

What do you think? 

What-do-you-think questions appear at the start of each section to help students focus on key ideas or concepts that appear within the sections.

6.1 Understanding Patterns, Relations, and Functions

What do you think?

- How are patterns related to algebraic thinking?
- What are some examples of functions in everyday life?
- What is a reason for developing algebraic thinking in elementary school?

Investigations 

Investigations are the primary means of instruction, uniquely designed to promote active thinking, reasoning, and construction of knowledge. Each investigation presents a problem statement or scenario that students work through, often to uncover a mathematical principle relevant to the content of the section. The "Discussion" that follows the problem statement provides a framework for insightful solution logic.

INVESTIGATION 3.1f Children's Mistakes

The problem below illustrates a common mistake made by many children as they learn to add. Understanding how a child might make that mistake and then going back to look at what lack of knowledge of place value, of the operation, or of properties of that operation contributed to this mistake is useful. What error on the part of the child might have resulted in this wrong answer?

The problem: $38 + 4 = 78$

DISCUSSION

In this case, it is likely that the child lined up the numbers incorrectly:

$$\begin{array}{r} 4 \\ +\ 38 \\ \hline 78 \end{array}$$

Giving other problems where the addends do not all have the same number of places will almost surely result in the wrong answer. For example, given $45 + 3$, this child would likely get the answer 75. Given $234 + 42$, the child would likely get 654. In this case, the child has not "owned" the notion of place value. Probably, part of the difficulty is not knowing expanded form (for example, that 38 means $30 + 8$—that is, 3 tens and 8 ones). An important concept here is that we need to add ones to ones, tens to tens, etc. Base ten blocks provide an excellent visual for this concept as students can literally see why they cannot add 4 ones to 3 tens.

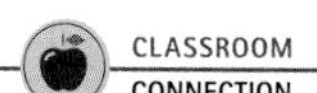

CLASSROOM CONNECTION

A friend of mine, David Sobel, was talking about mathematics with his six-year-old daughter, Tara. David had just shown Tara that $20 + 20 = 40$. Tara thought for a moment and then proudly announced that $50 + 50$ must be 70. When David asked how she had got that answer, she said, "When you add the same numbers that have a zero at the end, you just skip ten!"

Questions in the Text

To encourage active learning outside of the Investigations, questions appear embedded within the text, often accompanied by the icon . These "thinking" questions require students to pause in their reading to reflect or to complete a short exercise before continuing. Answers to these questions can be found in Appendix B in the back of the textbook.

> Translate the following Babylonian numerals into our system. Check your answers in Appendix B.
>
> 1. ▼▼ ◀▼▼ 2. ▼▼▼ ◀◀◀▼▼ / ▼▼ ◀ 3. ▼ ◀▼ ◀◀▼
>
> Translate the following amounts into Babylonian numerals.
>
> **4.** 1202 **5.** 304

Classroom Connections

Connections to the Classroom, denoted with the icon , are found throughout the textbook. The boxed Connections that appear in the margins provide observations, tips, and notes about the elementary/middle-school classroom. Assignments from actual elementary/middle-school books appear throughout as well so that students can see how the material they are learning will directly apply. Connections are also found in the exercises that highlight children's work.

CLASSROOM CONNECTION

This question asking how two things are alike and how they are different is an important teaching structure and one that we will revisit over the course of the book. You may remember it in a common *Sesame Street* feature: Three of These Things Belong Together. For example, they might show a triangle, a square, a hexagon, and a circle. The answer is that the circle doesn't belong because it doesn't have line segments. We will examine this idea of asking how things are alike and how they are different throughout the textbook.

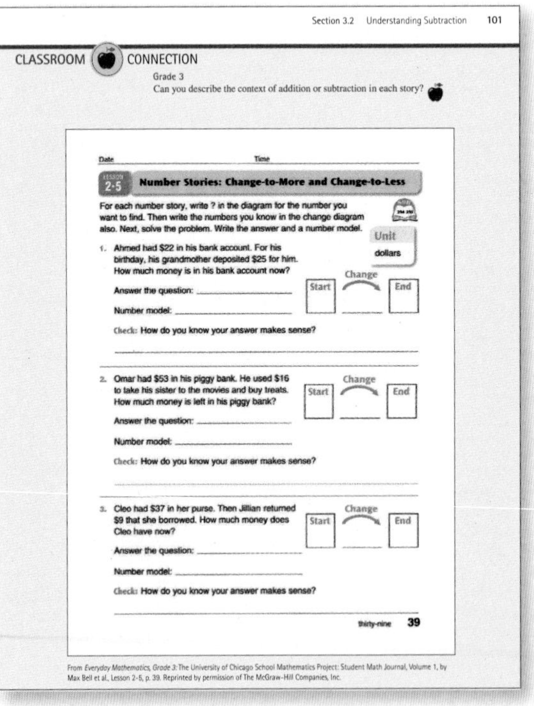

Section 3.2 Understanding Subtraction 101

CLASSROOM CONNECTION
Grade 3
Can you describe the context of addition or subtraction in each story?

Date Time

LESSON 2·5 **Number Stories: Change-to-More and Change-to-Less**

For each number story, write ? in the diagram for the number you want to find. Then write the numbers you know in the change diagram also. Next, solve the problem. Write the answer and a number model.

Unit: dollars

1. Ahmed had $22 in his bank account. For his birthday, his grandmother deposited $25 for him. How much money is in his bank account now?
Answer the question: ______
Number model: ______
Check: How do you know your answer makes sense?

2. Omar had $53 in his piggy bank. He used $16 to take his sister to the movies and buy treats. How much money is left in his piggy bank?
Answer the question: ______
Number model: ______
Check: How do you know your answer makes sense?

3. Cleo had $37 in her purse. Then Jillian returned $9 that she borrowed. How much money does Cleo have now?
Answer the question: ______
Number model: ______
Check: How do you know your answer makes sense?

Start / Change / End

thirty-nine 39

From *Everyday Mathematics*, Grade 3: The University of Chicago School Mathematics Project: Student Math Journal, Volume 1, by Max Bell et al., Lesson 2-5, p. 39. Reprinted by permission of The McGraw-Hill Companies, Inc.

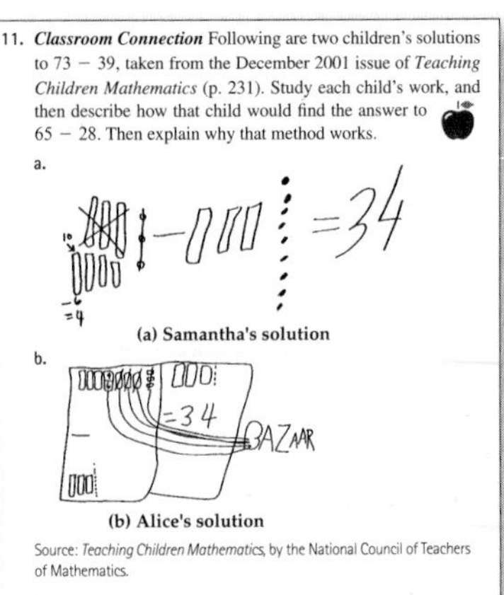

11. ***Classroom Connection*** Following are two children's solutions to 73 − 39, taken from the December 2001 issue of *Teaching Children Mathematics* (p. 231). Study each child's work, and then describe how that child would find the answer to 65 − 28. Then explain why that method works.

a.

(a) Samantha's solution

b.

(b) Alice's solution

Source: *Teaching Children Mathematics*, by the National Council of Teachers of Mathematics.

Margin Notes

To help round out the mathematics education of pre-service teachers, other special margin notes are provided.

▼ ***Outside the Classroom*** boxes highlight applications and uses of mathematical concepts and procedures in the business world, science, and everyday life.

OUTSIDE THE CLASSROOM

Skateboarders and snowboarders use reflex angles to describe some of their moves. What do you think they mean when they talk about doing a frontside 540 or a backside 720?

LANGUAGE

The term *ratio* comes from the Latin verb *ratus,* which means "to think or estimate." Many mathematicians in the sixteenth and seventeenth centuries used the word *proportion* for ratio. Even today you hear the two terms used interchangeably; for example, instructions for making a certain color might say, "Mix the two colors in the following proportion—3 : 2."

◀ ***Language*** boxes include the etymology of selected terms and/or describe nuances of terms.

HISTORY

Much of the mathematical notation we use is actually quite recent. The symbols for addition and subtraction, + and −, first appeared in Germany in the late 1400s. These symbols were first used to indicate sacks that were surplus or minus in weight. In 1631, William Oughtred first used the letter *x* to represent multiplication. Italian merchants introduced the symbol for division (÷) in the 1400s to indicate a half. For example, they wrote 4 ÷ to indicate $4\frac{1}{2}$. The equals sign first appeared in the late 1500s in a book by Robert Recorde.

◀ ***History*** boxes present interesting side notes, relevant to concepts developed in the text.

▼ ***Mathematics*** boxes associate a previously discussed concept or other related math idea with the topic under discussion.

MATHEMATICS

If you did Exploration 2.3, recall how strange the Alphabitian system was to you. What observations made you more comfortable with adding? Can you think of analogous observations that might make the learning of base ten addition facts easier for young children?

Section Exercises

The exercises are designed to give students a deeper sense and awareness of the kinds of problems their future students are expected to solve at various grade levels, as well as to increase their own proficiency with the content. Special subcategories appear toward the end of each set of exercises. *Deepening Your Understanding* exercises go a step beyond, encouraging students to extend their thinking beyond the basics. *From Standardized Assessments* exercises derive from exams such as the NECAP and NAEP to give students a sense of the types of questions found on diverse national exams at various grade levels. Questions are also included from the Smarter Balanced Assessment Consortium which is developing assessments for Common Core State Standards.

DEEPENING YOUR UNDERSTANDING

26. Place the digits 1, 2, 3, 6, 7, and 8 in the boxes to obtain
    a. The greatest difference
    b. The least difference

$$\begin{array}{r} \square\square\square \\ -\square\square\square \\ \hline \end{array}$$

27. Choose among the digits 1, 2, 3, 4, 5, 6, 7, 8, and 9 to make the difference 234. You can use each digit only once. How many different ways can you make 234?

$$\begin{array}{r} \square\square\square \\ -\square\square\square \\ \hline \end{array}$$

28. With three boys on a large scale, it read 170 pounds. When Adam stepped off, the scale read 115 pounds. When both Adam and Ben stepped off, the scale read 65 pounds. What is the weight of each boy?
29. A mule and a horse were carrying some bales of cloth. The mule said to the horse, "If you give me one of your bales, I shall carry as many as you." "If you give me one of yours," replied the horse, "I will be carrying twice as many as you." How many bales was each animal carrying?
30. From each of the following lists, select two numbers whose difference will be closest to the target difference.

Numbers			Target
a. 315	475	764	300
b. 185	372	953	650
c. 382	723	793	350

31. One of my students asked me this question after s the text: "Why do we have addition and multiplic but not subtraction and division tables?" Write yo to her question.

3.2 Exercises

1. Make up a subtraction story problem for each of the following contexts. Briefly *explain* why the story problem is an example of the particular model.
   a. Take-away
   b. Missing addend
   c. Comparison
2. Which model of subtraction best illustrates each of the problems below:
   a. Reena had 25 quilts and sold 10 of them at the show. How many does she have left?
   b. The hall holds 100 people. Currently 65 tickets have been sold. How many more tickets can be sold?
   c. The sixth grade has sold 58 raffle tickets and the fifth grade has sold only 45. How many more has the sixth grade sold?
4. Represent the following problems on a number line. Explain each problem as though you were talking to someone who is not taking this class.
   a. $5 + 4$ b. $8 - 3$
5. a. Explain why the operation of subtraction is not commutative.
   b. Explain why the operation of subtraction is not associative.
6. Determine the following differences mentally. Briefly describe how you obtained your estimate.

a. $\begin{array}{r} 87 \\ -29 \\ \hline \end{array}$ b. $\begin{array}{r} 70 \\ -23 \\ \hline \end{array}$ c. $\begin{array}{r} 82 \\ -34 \\ \hline \end{array}$ d. $\begin{array}{r} 500 \\ -134 \\ \hline \end{array}$ e. 502

f. $\begin{array}{r} 625 \\ -475 \\ \hline \end{array}$ g. $\begin{array}{r} 4000 \\ -555 \\ \hline \end{array}$

Determine these differences mentally by a means other standard algorithm.

FROM STANDARDIZED ASSESSMENTS

NECAP 2006, Grade 5

34. Mrs. Lombardi had 2 hours to prepare for a party. The chart below shows the amount of time she spent completing different tasks.

TIME MRS. LOMBARDI SPENT ON DIFFERENT TASKS

Task	Time
Decorated cake	20 minutes
Made punch	15 minutes
Made sandwiches	50 minutes
Put up balloons	?

How much time did Mrs. Lombardi have to put up the balloons? (1 hour = 60 minutes)

a. 15 minutes b. 25 minutes
c. 35 minutes d. 45 minutes

NECAP 2005, Grade 5

35. The students at Maple Grove School are selling flowers. Their goal is to sell 1500 flowers.
    - On the first day, the students sold 547 flowers.
    - On the second day, the students sold 655 flowers.

    How many flowers must the students sell on the third day to meet their goal?

    a. 298 b. 308 c. 1202 d. 2702

Section Summary

Each section ends with a summary that reviews the main ideas and important concepts discussed.

SUMMARY 3.2

We have now examined addition and subtraction rather carefully. In what ways do you see similarities between the two operations? In what ways do you see differences? Think and then read on. . . . 

One way in which the two processes are alike is illustrated with the part–whole diagram used to describe each operation. These representations help us to see connections between addition and subtraction. In one sense, addition consists of adding two parts to make a whole. In one sense, subtraction consists of having a whole and a part and needing to find the value of the other part.

We see another similarity between the two operations when we watch children develop methods for subtraction; it involves the "missing addend" concept. That is, the problem $c - a$ can be seen as $a + ? = c$.

We saw a related similarity in children's strategies. Just as some children add large numbers by "adding up," some children subtract larger numbers by "subtracting down."

Earlier in this section, subtraction was formally defined as $c - b = a$ if $a + b = c$. The negative numbers strategy that some children invent brings us to another way of defining subtraction, which we will examine further in Chapter 4 when we examine negative numbers. That is, we can define subtraction as adding the inverse: $a - b = a + -b$.

A very important way in which the two operations are different is that the commutative and associative properties hold for addition but not for subtraction.

Looking Back

Each chapter concludes with *Looking Back*—a study tool that brings together all the important points from the chapter. *Looking Back* includes *Questions to Summarize Big Ideas* (NEW!), which ask students to reflect on the main ideas from the chapter; *Chapter Summary*, which lists major take-aways and terminology from the chapter; and *Review Exercises*, which provide an opportunity for students to put concepts from the chapter into practice.

LOOKING BACK on chapter three

QUESTIONS TO SUMMARIZE BIG IDEAS

1. What are some of the different models for addition, subtraction, multiplication, and division?
2. How can you use base ten blocks to model the algorithms for each of the operations?
3. How are these models similar and different in a base other than ten?
4. Which algorithms for the operations are different from what you learned in elementary school?
5. What are the tools for determining divisibility and why do they work?
6. Look back at the Mathematical Practices of the Common Core State Standards. In what ways did you engage in those practices during this chapter?
7. What parts of this chapter are less clear to you at

CHAPTER 3 SUMMARY

1. Many students have said that really understanding base ten and the four operations, was, for them, the beginning of a new attitude toward mathematics. We will continue to examine new and important mathematical ideas throughout this book, but the foundation for much of elementary mathematics has now been laid.
2. Each operation has multiple meanings.
3. Many algorithms have been developed to enable us to compute more efficiently.
4. The standard algorithm for each operation does not connect equally well to each meaning of the operation.
5. Being able to make sense of algorithms requires:
   - The ability to apply base ten and place value concepts
   - The ability to compose and decompose the numbers (for example, to use expanded form)
6. Patterns enable us to understand the operations more deeply.
7. In many real-life problems, the answer depends on knowing how to interpret one's computation.
8. Being able to perform mental math and to estimate requires
   - The ability to apply base ten and place value concepts
   - The ability to compose and decompose the numbers (for example, to use expanded form)
   - The ability to apply properties of the operations, especially the commutative, associative, and distributive properties
9. Numbers in real-life settings are sometimes exact, sometimes rounded, and sometimes estimates.
10. In real-life problem-solving, one needs to know when to find an exact answer and when to find an estimate.
11. Real-life problem solvers need to know whether their estimates are reasonable.
12. People may use rounded numbers rather than exact numbers for a variety of reasons.

BASIC CONCEPTS

Section 3.1: Understanding Addition

Addition terminology:

addition **78** sum **78**
addends **78**

Addition contexts:

discrete **76** continuous, measured **76**

Addition models:

pictorial **76** number line **78**
tables **79**

REVIEW EXERCISES chapter three

1. State the problem that is represented in each case below:

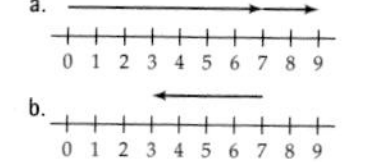

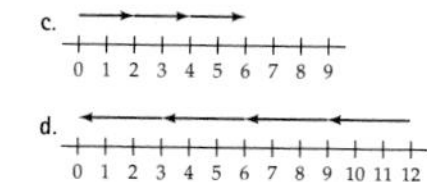

Explorations

The icon that appears throughout the text references additional activities that may be found in the *Explorations Manual*. Explorations present new ideas and concepts for students to engage with "hands on."

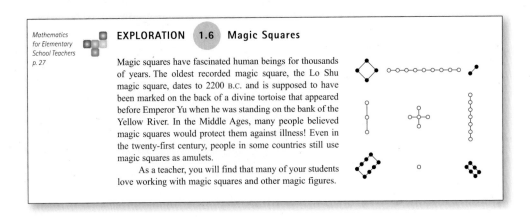

Mathematics for Elementary School Teachers p. 27

EXPLORATION 1.6 Magic Squares

Magic squares have fascinated human beings for thousands of years. The oldest recorded magic square, the Lo Shu magic square, dates to 2200 B.C. and is supposed to have been marked on the back of a divine tortoise that appeared before Emperor Yu when he was standing on the bank of the Yellow River. In the Middle Ages, many people believed magic squares would protect them against illness! Even in the twenty-first century, people in some countries still use magic squares as amulets.

As a teacher, you will find that many of your students love working with magic squares and other magic figures.

SUPPLEMENTS	
FOR THE STUDENT	FOR THE INSTRUCTOR
	Instructor's Edition **(ISBN: 978-1-305-07137-7)** The Instructor's Edition includes answers to all exercises in the text, including those not found in the student edition. (Print)
Student Solutions Manual **(ISBN: 978-1-305-10833-2)** Go beyond the answers—see what it takes to get there and improve your grade! This manual provides worked-out, step-by-step solutions to the odd-numbered problems in the text. This gives you the information you need to truly understand how these problems are solved. (Print)	**Instructor's Manual** The Instructor's Manual provides worked-out solutions to all of the problems in the text. In addition, instructors will find helpful aids such as "Teaching the Course," which shows how to teach in a constructivist manner. "Chapter by Chapter Notes" provide commentary for the *Explorations* manual as well as solutions to exercises that appear in the supplement. This manual can be found on the Instructor Companion Site.
Explorations, Mathematics for Elementary School Teachers, **6e** **(ISBN: 978-1-305-11283-4)** This manual contains open-ended activities for you to practice and apply the knowledge you learn from the main text. When you begin teaching, you can use the activities as models in your own classrooms. (Print)	***Explorations, Mathematics for Elementary School Teachers,*** **6e** **(ISBN: 978-1-305-11283-4)** This manual contains open-ended activities for students to practice and apply the knowledge they learn from the main text. When students begin teaching, they can use the activities as models in their own classrooms. (Print)
Math Manipulatives Kit **(ISBN: 978-1-305-11287-2)** Get hands-on experience when you use the Manipulatives Kit. By using this tool you will see the benefits that will help elementary school students understand mathematical concepts. The kit includes pattern blocks, pentominoes, base ten flats, base ten rods, base ten units, tangrams, and a Geoboard.	**Math Manipulatives Kit** **(ISBN: 978-1-305-11287-2)** These Manipulatives Kits provide preservice teachers with hands-on experience and gives an understanding of why manipulatives are used in the elementary school classroom. The kits include pattern blocks, pentominos, base ten flats, base ten rods, base ten units, Tangrams, and a Geoboard.
Enhanced WebAssign® Instant Access Code: 978-1-285-85803-6 Printed Access Card: 978-1-285-85802-9 Enhanced WebAssign combines exceptional mathematics content with the powerful online homework solutions, WebAssign. Enhanced WebAssign engages students with immediate feedback, rich tutorial content, and an interactive, fully customizable eBook, the Cengage YouBook, which helps students to develop a deeper conceptual understanding of their subject matter.	**Enhanced WebAssign®** Instant Access Code: 978-1-285-85803-6 Printed Access Card: 978-1-285-85802-9 Enhanced WebAssign combines exceptional mathematics content with the powerful online homework solutions, WebAssign. Enhanced WebAssign engages students with immediate feedback, rich tutorial content, and an interactive, fully customizable eBook, the Cengage YouBook, which helps students to develop a deeper conceptual understanding of their subject matter. See www.cengage.com/ewa to learn more.
CengageBrain.com To access additional course materials, visit **www.cengagebrain.com**. At the CengageBrain.com home page, search for the ISBN of your title (see back cover of your book) using the search box at the top of the page. This will take you to the product page where these resources can be found.	**Instructor Companion Site** Everything you need for your course is in one place! This collection of book-specific lecture and classroom tools is available online via www.cengage.com/login. Access and download PowerPoint® images, solutions manual, and more.
	Cengage Learning Testing Powered by Cognero® Instant Access Code: 978-1-305-11304-6 Cognero is a flexible, online system that allows you to author, edit, and manage test bank content; create multiple test versions in an instant; and deliver tests from your LMS, your classroom, or wherever you want. This is available *online* via www.cengage.com/login.

Acknowledgments

We would like to thank the reviewers of this edition:

Marilyn Ahrens, *Missouri Valley College*; Mary Beard, *Kapiolani Community College*; Timothy Comar, *Benedictine University*; Edward DePeau, *Central Connecticut State University*; Sue Ann Jones Dobbyn, *Pellissippi State Community College*; April Hoffmeister, *University of Illinois*; Judy Kasabian, *El Camino College*; Cathy Liebars, *The College of New Jersey*; Kathleen McDaniel, *Buena Vista University*; Ann McCoy, *University of Central Missouri*; Martha Meadows, *Hood College*; Anthony Rickard, *University of Alaska, Fairbanks*; Mark Schwartz, *Southern Maine Community College*; Sonya Sherrod, *Texas Tech University*; Allison Sutton, *Austin Community College*; Osama Taani, *Plymouth State University*; Michael Wismer, *Millersville University*; and Ronald Yates, *College of Southern Nevada*.

In addition, we would like to thank the many reviewers of previous editions noted below for their thoughtful and helpful comments throughout development: Andrew T. Wilson, Austin Peay State University; Anita Goldner, Framingham State College; Art Daniel, Macomb Community College; Bernadette Antkoviak, Harrisburg Area Community College; Beverly Witman, Lorain County Community College; Charles Dietz, College of Southern Maryland; Clare Wagner, University of South Dakota; Deborah Narang, University of Alaska–Anchorage; Dennis Raetzke, Rochester College; Donald A. Buckeye, Eastern Michigan University; Doug Cashing, St. Bonaventure University; Dr. Connie S. Schrock, Emporia State University; Elise Grabner, Slippery Rock University; Elizabeth Cox, Washtenaw Community College; Forrest Coltharp, Pittsburg State University; Fred Ettline, College of Charleston; Gary Goodaker, West Community and Technical College; Gary Van Velsir, Anne Arundel Community College; Glenn Prigge, University of North Dakota; Helen Salzberg, Rhode Island College; Isa S. Jubran, SUNY College at Cortland; J. Normon Wells, Georgia State University; J.B. Harkin, SUNY College at Brockport; James E. Riley, Western Michigan University; Jane Ann McLaughlin, Trenton State College; Jean M. Shaw, University of Mississippi; Jean Simutis, California State University–Hayward; Jeanine Vigerust, New Mexico State University; Jerry Dwyer, University of Tennessee–Knoxville; Jim Brandt, Southern Utah University; John Long, University of Rhode Island; Juan Molina, Austin Community College; Julie J. Belock, Salem State College; Karen Gaines, St. Louis Community College; Karla Karstens, University of Vermont; Kathy C. Nickell, College of DuPage; Larry Feldman, Indiana University of Pennsylvania; Larry Sowder, San Diego State University; Lauri Semarne; Lawrence L. Krajewski, Viterbo College; Lew Romagnano, Metropolitan State College of Denver; Linda Beller, Brevard Community College; Linda Herndon, Benedictine College; Lois Linnan, Clarion University; Lorel Preston, Westminster College; Loren P. Johnson, University of California–Santa Barbara; Lynette King, Gadsden State Community College; Marvin S. Weingarden, Madonna University; Mary Ann Byrne Lee, Mankato State University; Mary J. DeYoung, Hope College; Mary Lou Witherspoon, Austin Peay State University; Mary T. Williams, Francis Marion University; Mary Teagarden, Mesa College; Matt Seeley, Salish Kootenai College; Maureen Dion, San Joaquin Delta Community College; Merle Friel, Humboldt State University; Merriline Smith, California State Polytechnic University; Michael Bowling, Stephens College; Nadine S. Bezuk, San Diego State University; Peter Berney, Yavapai College; Peter Incardone, New Jersey City University; Robert F. Cunningham, Trenton State College; Robert Hanson, Towson State University; Rebecca Wong, West Valley College; Ronald Edwards, Westfield State University; Ronald J. Milne, Gashen College; Sandra Powers, College of Charleston; Stephen P. Smith, Northern Michigan University; Stuart Moskowitz, Humboldt State University; Susan K. Herring, Sonoma State University; Tad Watanabe, Towson State University; Tess Jackson, Winthrop University; Vena Long, University of Missouri at Kansas City; William Haigh, Northern State University.

We offer deep thanks to the people at Cengage and beyond who offered guidance, support, and expertise in ensuring the quality of the resulting product.

These individuals include Richard Stratton, Rita Lombard, Erin Brown, Jennifer Cordoba, Samantha Lugtu, Heleny Wong, Guanglei Zhang, Julie Schuster, Jennifer Risden, Vernon Boes, Chris Waller, Christina Ciaramella, John Sarantakis, Timothy Comar, and Laura Wheel.

We would also like to extend our gratitude to the following instructors who served in an advisory capacity on this edition, offering valuable suggestions and feedback: Edward DePeau, Sue Ann Jones Dobbyn, Cathy Liebars, Allison Sutton, and Michael Wismer.

1 Foundations for Learning Mathematics

Knowing mathematics means being able to use it in purposeful ways. To learn mathematics, students must be engaged in exploring, conjecturing, and thinking rather than only in rote learning of rules and procedures. Mathematics learning is not a spectator sport. When students construct personal knowledge derived from meaningful experiences, they are much more likely to retain and use what they have learned. This fact underlies [the] teacher's new role in providing experiences that help students make sense of mathematics, to view and use it as a tool for reasoning and problem solving.[1]

—National Council of Teachers of Mathematics

SECTION 1.1 Getting Started and Problem Solving

What do you think?

- Respond to the prompt: Mathematics is ___________.
- Describe a few of your experiences learning mathematics as an elementary school student.
- Describe your attitudes toward mathematics and where you think these attitudes come from.
- What attitudes do you have about taking this course?

You are at the beginning of a course where you will re-examine elementary school mathematics to understand these concepts on a much deeper level, and to learn why the mathematical procedures and formulas actually work. On this journey, you will learn several ways to see and think about concepts and procedures that you may have previously simply memorized. This deeper understanding will lead to increased confidence and comfort level with mathematics. Your approach to this course, and to teaching mathematics, depends on the attitudes and beliefs you bring to the classroom; in subtle and not so subtle ways, you may pass these beliefs along when you enter the classroom as a teacher. Reflect on how you answered the questions above. Whatever your feelings about mathematics, consider where these feelings come from. Research suggests that people who have mathematics anxiety can relate it back to a teacher and/or experience in their elementary or middle school years. Think about the best math teacher you have had as well as the worst math teacher you have had. Consider the skills and qualities that each of them had that led to your experience of them. What skills and qualities do you have and need to further develop to become an excellent math teacher?

BELIEFS AND ATTITUDES ABOUT MATHEMATICS

This preliminary exercise is designed with two purposes in mind. First, it will help you examine and reflect on your beliefs and attitudes at the beginning of the course. Second, it will help you see a practical use of mathematics.

Rate your attitudes

Seven pairs of statements concerning attitudes and beliefs about mathematics are given in Table 1.1. Score your beliefs in the following manner:

- If you strongly agree with the statement in column A, record a 1.
- If you agree with the statement in column A more than with the statement in column B, record a 2.
- If you agree with the statement in column B more than with the statement in column A, record a 3.
- If you strongly agree with the statement in column B, record a 4.

Adaptive and maladaptive beliefs

Before we discuss your responses to Table 1.1 let us examine attitudes. I have worked with thousands of students during my time as a teacher, and I know from experience and from reading research that one's beliefs about mathematics can influence how one learns and teaches.

TABLE 1.1

Column A		Column B
1. There will be many problems in this book that I won't be able to solve, even if I try really hard.	1 2 3 4	**1.** I believe that if I try really hard, I can solve virtually every problem in this book.
2. There is only one way to solve most "word" problems.	1 2 3 4	**2.** There is usually more than one way to solve most "word" problems.
3. The best way to learn is to memorize the different kinds of problems—rate problems, mixture problems, coin problems, etc.—and how to solve them.	1 2 3 4	**3.** The best way to learn is to make sure that I understand each step.
4. Some people have mathematical minds and some don't. Nothing they do can *really* make a difference.	1 2 3 4	**4.** Some students may have more aptitude for mathematics than others, but everyone can become competent in mathematics.
5. The teacher's job is to show us how to do problems and then give us similar problems to practice.	1 2 3 4	**5.** The teacher's job is more like that of a coach or guide—to help us develop the problem-solving tools we need.
6. A good test consists of problems that are just like the ones we have done in class.	1 2 3 4	**6.** A good test has problems at a variety of levels of difficulty, including some that are not just like the ones in the book.
7. I don't need to know all the ideas covered in this book because I'm going to teach younger children.	1 2 3 4	**7.** Even teachers of young children need to have a good understanding of the ideas in this book.
	Total ______	

Some students have **adaptive beliefs** that help them approach math with a positive and confident attitude. Some have **maladaptive beliefs** that keep them from thinking of learning as an evolving and enjoyable process.

In Table 1.1, the statements in column A indicate maladaptive beliefs, and the statements in column B indicate the corresponding adaptive beliefs. If you take the arithmetic average, or *mean,* of your scores (by adding up your scores and dividing by 7), you will get a number that we could call your belief index. If your belief index is low and you encounter difficulties in this course, it may be because some of your beliefs are hindering your ability to learn the material. If you find this course frustrating, try to discuss your beliefs with your professor, with someone at a math center (if your college has one), or with a friend who is doing well in the course. Deepening your understanding of mathematics through this course and beyond will help you to have a positive attitude about mathematics. Approach this course with an open mind toward learning and a belief that everyone (including yourself) can understand mathematics. Your future students are depending on you to deepen your understanding of math and to have positive attitudes about it.

OUTSIDE THE CLASSROOM[2]

The pervasiveness of negative attitudes toward mathematics was powerfully illustrated in 1992 when Mattel introduced a new talking Barbie doll that said, "Math is tough." Now this may be true for some people, but having Barbie say it only reinforced that stereotypical perception of mathematics in the United States, especially among females. Mattel was persuaded to change Barbie's statement.

WHAT IS MATHEMATICS?

What is mathematics? Think about this question for a minute and then read on. . . . 

LANGUAGE

Whenever you see the pencil icon, stop and think and briefly write your thoughts before reading on. Students who take the time to think and write after these points (or at least to pause and think) say that it makes a big difference in how much they learn.

You may be surprised to learn that not all mathematicians give the same response to this question. *On the Shoulders of Giants: New Approaches to Numeracy*[3] was written partly to help expand people's views of mathematics beyond the common stereotype of "mathematics is a bunch of formulas and rules for numbers." A group of mathematicians and mathematics educators brainstormed a number of possible themes for that book. In the end, it was agreed that the idea of *pattern* permeates all fields of mathematics. Five mathematicians were asked to write chapters on the following themes:

Dimension. In school, you have studied two- and three-dimensional shapes. Mathematicians have gone far beyond three dimensions for years. Recently, a field of mathematics has opened up the exploration of fractional dimensions. For example, the coastline of Britain (which can be modeled by a long, squiggly line) has been calculated to have a dimension of 1.26.

Quantity. This begins (with children) with the question "how many," for which the counting numbers (1, 2, 3, . . .) are appropriate; it moves in complexity to the question "how much," for which fractions and decimals were invented, and then to questions far more complex, for which other numbers and systems were invented.

Uncertainty. Questions of uncertainty permeate everyday life: How long will I live? What are my chances of getting a job after I graduate? What are the chances that my baby will be "normal"?

Shape. Humans' relationship with shape has a fascinating history—the shape of one's environment (desert, forest, mountain), what shape is best for packaging, the shapes that artists make, and the shapes we manufacture for quilts, clothing, and so on.

Change. We live in a world that is constantly changing. The development of computers enables us better to understand and manage change, whether it be the changing climate, the change in epidemics (such as AIDS), the change in populations (human and animal), or changes in the economy.

Mathematics is far more than titles of courses and chapters in textbooks—whole numbers, fractions, decimals, percents, algebra, geometry, etc. The topics in textbooks represent tools that are needed in order to answer important questions about dimension, quantity, uncertainty, shape, and change.

> The numbers, lines, angles, shapes, dimensions, averages, probabilities, ratios, operations, cycles, and correlations that make up the world of mathematics enable people to make sense of a universe that otherwise might seem to be hopelessly complicated.[5]

MATHEMATICS

Keith Devlin has written several fascinating and readable books on this subject,[4] one of which is a companion to a PBS series entitled *Life by the Numbers* (your college or local library probably has this book). The chapter titles for *Mathematics: The Science of Patterns* are "Counting," "Reasoning and Communicating," "Motion and Change," "Shape, Symmetry and Regularity," and "Position." Devlin discusses (among many other things) how mathematicians helped us to understand why leopards have spots and tigers have stripes, how mathematicians helped American ice skaters learn how to perform triple axel jumps, and how we use mathematics to measure the heights of mountains.

Deborah Meier, an award-winning principal, urged her teachers to think critically and often[6] asked them, "So what?" I encourage this of you, too. There are many wonderful teachers in this country, but there are also many teachers for whom mathematics is just another subject to be "covered." Do you want your students to form this view of mathematics and carry it from your classroom? If not, then you need to ask yourself the "So what?" question frequently, and when you cannot give a satisfactory answer, talk with your instructor or go elsewhere—to the Web, to articles in *Teaching Children Mathematics,* to classrooms—and see for yourself the kinds of mathematical experiences that we want young children to have in school.

Mathematics is both beauty and truth. Two plus two always equals four. The distance around a circle is always a little more than three (actually pi) times the distance across the circle. Throughout this text, we hope you will appreciate more and more of the beautiful truths of mathematics.

USING MATHEMATICS

Let us now turn our attention to how people use mathematics in everyday situations and in work situations. Take a few minutes to jot down some instances in which you have used mathematics in your life and some instances in which you know that mathematics is used in different careers and work situations. Then read on. . . .

People use mathematics for various purposes, for example:

- To persuade a boss that our idea will make money
- To persuade a potential customer that our product will save money
- To predict—tomorrow's weather or who will win the election
- To make a personal decision—whether we can afford to buy a house
- To make a business decision—how much to charge for a new product or whether a new medicine (for example, a cure for AIDS) really works
- To help us understand how the world works (for example, why leopards have spots and tigers have stripes)
- To relax—many people enjoy Sudoku and other math puzzles

Solving problems

In each of these examples, people are using mathematics as a tool for solving problems. To decide whether you can afford a new car, you have to collect data (on insurance, for example), add decimals, and work with percents (such as sales tax and interest on the loan). The mathematics you use will help you solve the problem of whether to buy a car.

Think about this situation involving weather forecasters. In 1994 a hurricane brought severe rains to Georgia. Forecasters predicted that the Flint River would crest at 20 feet above flood level; the river actually crested at 13 feet above flood level, much to the relief of many residents. The forecasters used decimals, volume formulas, and conversions to determine the maximum volume of water that would be flowing. They also used computer models of flooding rivers, and the computer models were based on data collected on previous flooding.

Now that we have discussed mathematics in general, let's focus on the specialized mathematical understanding that teachers need.

MATHEMATICAL KNOWLEDGE FOR TEACHING

INVESTIGATION 1.1a

More Than One Way to Multiply?

First, multiply 49×25 using any method you choose. Then consider how the following students solved the problem.

What method do you think each student is using in each solution below? Do you think each method will work for any two whole numbers?

Student A:	Student B:
49	49
× 25	× 25
245	45
98	200
1225	180
	800
	1225

DISCUSSION

We will look much more deeply into multiplication later in this course. For now, we use this example to illustrate how part of the role of a teacher is to be able to first understand that there are different methods for solving problems, and then be able to understand different strategies, and to know what to do with these different methods to develop deeper understanding. This is the mathematical understanding that is important for elementary teachers, and this deeper understanding is the focus of this course.

Student A used the method that many of us learned in elementary school of 5 × 49 and 20 × 49. Some of us learned to write the second row in the multiplication as 980, some were taught to move over a space as is shown here. Why do we do this? Because we are multiplying 25 × 40, not 25 × 4, we write it as 980, or as 98 with a space in the ones place.

We can use the properties and split the numbers up into their parts to figure this out.

$$49 \times 25 = 49 \times (20 + 5) = 49 \times 20 + 49 \times 5 = 980 + 245 = 245 + 980 = 1225$$

Student B used a method that is sometimes called partial products of 9 × 5 = 45, 40 × 5 = 200, 9 × 20 = 180, and 40 × 20 = 800. Another way to write this, which uses the distributive property more clearly is:

$$49 \times 25 = (40 + 9) \times (20 + 5) = 40 \times 20 + 40 \times 5 + 9 \times 20 + 9 \times 5 =$$
$$600 + 200 + 180 + 45 = 1225$$

The methods of both students are valid. One of the major ideas in this text is that there are multiple ways (often called "solution paths") to get to an answer. There is no ONE "right" way to solve any math problem. This may be different from what you have always thought about mathematics. You may have even had teachers who marked you "wrong" if you did not solve it their way.

CLASSROOM CONNECTION

Many students have told me that before they took this course, they thought there was just *one right way* to do a problem, and so they never looked for patterns but instead looked for formulas or procedures. Once the students started looking for patterns, they found them everywhere, and over time they learned how to use their awareness of patterns more powerfully. We can refer to different ways to solve a problem as different **solution paths**.

Explorations Manual 1.2

As a teacher of elementary school mathematics, a deep understanding of mathematics will enable you to respond to the above type of situation that arises in an elementary school classroom. Simply being able to get the right answer is not sufficient. Teachers need a specialized understanding of mathematics that is flexible, connected, and conceptual. This course will help you develop that.

Teachers use mathematics every day in the classroom, but in different ways than others. In 1986, Lee Shulman used the term "pedagogical content knowledge" to refer to this specialized understanding of mathematics, which includes an understanding of multiple representations and examples, plus an understanding of what ideas may be more difficult for students and why these ideas are more difficult. In 2008, Deborah Loewenberg Ball, Mark Thames, and Geoffrey Phelps developed the framework depicted in Figure 1.1 showing different types of knowledge. This book is focused on the specialized content knowledge, which includes understanding multiple representations and multiple student procedures, and analyzing student errors. It is the type of math knowledge that teachers draw on every day.

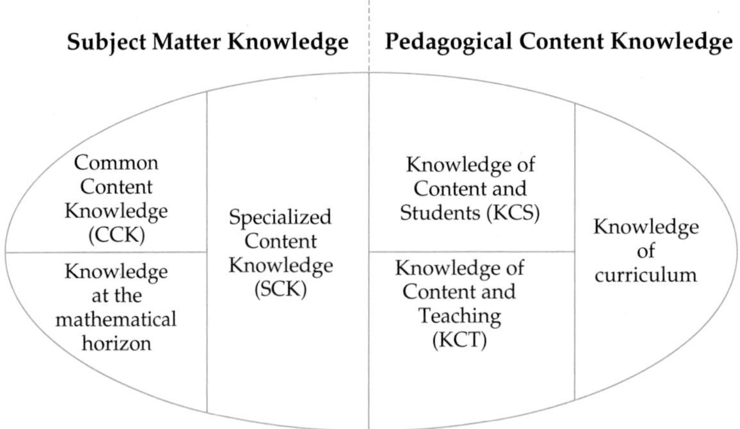

Figure 1.1
Source: Hill, H. C., Ball, D. L., & Schilling, S. G. (2008). Unpacking pedagogical content knowledge: Conceptualizing and measuring teachers' topic-specific knowledge of students. *Journal for Research in Mathematics Education*, 39(4), 372-400.

INVESTIGATION 1.1b Understanding Students' Errors

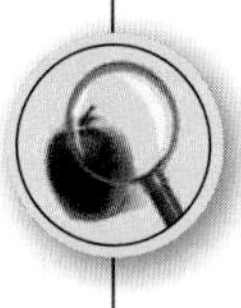

How do you think the answer was produced? What do you think the student was thinking that led to this error?

$$\begin{array}{r} 49 \\ \times\ 25 \\ \hline 245 \\ 98 \\ \hline 343 \end{array}$$

How does this type of scenario draw upon specialized content knowledge?

DISCUSSION

Analyzing student errors draws upon a specialized understanding of mathematics in that the teacher needs to understand the mathematics deeply in order to identify the error and then to help the student to correct the misunderstanding.

This student does not understand that in the second step, we are actually multiplying 20 times 49 so it should be 980, not 98.

While most people could get a correct answer for 49 × 25, teachers need to understand the concept much more deeply. Many of us experienced elementary school mathematics as a series of procedures to memorize (like multiplying 49 × 25). In order to teach true understanding of mathematics, teachers must develop this specialized content knowledge.

LEARNING STYLES

Consider how you best learn, particularly mathematics. Do you prefer visuals? Do you need to experience it by "getting your hands on it"? Do you need to experience it through sound? Do you prefer to talk with others while solving a problem, or think about it alone? There are several theories about how people learn and think differently, including the Kolb learning styles, VAK (Visual Auditory Kinesthetic) learning styles, and multiple intelligences theory. As you go through this course, it may be helpful to think about how you best learn mathematics. There are several online surveys that can help you determine your own learning styles. You can locate free surveys upon searching the Internet for "learning styles surveys"; once you understand your own learning styles more deeply, you can adjust your learning experiences to accommodate them.

LANGUAGE

Many authors define understanding in terms of connections. That is, you truly understand an idea only if it is well-connected to other ideas, and your depth of understanding is connected to how many connections you are making and the quality of those connections.

PROBLEM SOLVING AND TOOLBOXES

A useful metaphor for problem solving is a **toolbox**. Imagine that your car breaks down and is towed to the garage, where a novice mechanic is the first to look at it. The novice will probably try a few standard procedures: Insert the key to see what happens, check the battery connections, look for a loose wire, and so on. If none of those strategies work, the novice mechanic will be stumped and will have to summon the senior mechanic. The senior mechanic may try the same basic procedures and may solve the problem by *interpreting* the results more skillfully. As you go through this course, you will learn new tools and how to use tools you already have more skillfully.

INVESTIGATION 1.1c Real-Life Problem Solving

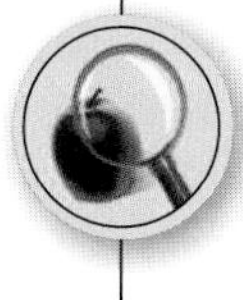

Consider a few problems you have had in your life, and not necessarily math problems.

What steps did you take to solve these problems?

Use this recollection to make a list of general steps that you take to solve a problem. Then read on to learn about a mathematician that made a similar list. Your list will likely be pretty similar to his.

DISCUSSION

There are many kinds of real-life problems that you may have considered here. One that many of us have dealt with is what kind of car to buy. The first step would be to understand what you need. How many seats do you need? What is your budget? Is gas mileage a priority? What models do you like? This part might be called something like "understand the problem." Did you have a step like this in your list? The second step would be to develop a plan. Where will you look? What research will you do? The third step might be to carry out the plan. Research the best models, look around for the best deals, test drive some models and find the one you like. The next step, hopefully before actually buying the one you have your heart set on, might be to reflect on whether it really meets your needs, fits your budget, and so on. Read on to see how this process is the same for solving a math problem.

POLYA'S FOUR STEPS

George Polya developed a framework that breaks down problem solving into four distinguishable steps. In 1945, he outlined these steps in a now-classic book called *How to Solve It.*

When you approach a problem, if you think that you have to come up with an answer immediately and that there is only one "right" way to reach that answer, a solution may seem to be beyond your grasp. But if you break the problem down and thoughtfully approach each *step* of the problem, it generally becomes more manageable. Polya suggests that you first need to make sure you **understand the problem**. Once you understand the problem, you **devise a plan** for solving the problem. Then you **monitor your plan**; you check frequently to see whether it is productive or is going down a dead-end street. Finally, you **look back at your work**. This last step involves more than just checking your computation; for example, it includes making sure that your answer makes sense. For each of these four steps, there are specific strategies that we will explore in this chapter and that you will refine throughout this course.

Owning versus renting

Instead of just listing Polya's strategies, we are going to discover them by putting them into action. You will notice that I often ask you to stop, think, and write some notes. I really

CLASSROOM CONNECTION

A colleague was working through a word problem with her class one day. Suddenly one of the students said, "But you don't need to do all this stuff you are teaching us; you just know the answer." She was stunned, and discovered that many students believe that the difference between a student and a teacher is that the teacher just knows the answer or automatically knows how to get the answer and thus doesn't need such strategies as guess–check–revise, make a table, draw a diagram, look for patterns, etc. The truth is that we do! Virtually all engineers, scientists, businesspeople, carpenters, researchers, and entrepreneurs approach complex problems by using the very tools that are being stressed in this text.

mean it! I have come to distinguish between those students who *own* what they learn and those who simply *rent* what they learn. Many students rent what they have learned just long enough to pass the test. However, within days or weeks of the final exam, it's gone, just like a video that has been returned to the store. One of the important differences between owners and renters is that those who own the knowledge tend to be *active readers*.

Using Polya's four steps

I encourage you to use Polya's four steps (on the inside front cover of this book and *Explorations*) in all of the following ways:

1. Use them as a guide when you get stuck.
2. Don't rent them, buy them. Buying them involves paraphrasing my language and adding new strategies that you and your classmates discover. For example, many of my students have added a step to help reduce anxiety: First take a deep breath and remind yourself to slow down!
3. After you have solved a problem, stop and reflect on the tools you used. Over time, you should find that you are using the tools more skillfully.

Think and then read on . . .

Throughout the book, I will often pose a question and ask you to "think and then read on. . . ." Rather than just look to the next paragraph and see the answer, you will learn more if you immediately cover up the next paragraph . . . stop . . . think . . . write down your thoughts . . . and then read on. The phrase "think and read on . . ." is there to remind you to read the book actively rather than passively. An **active reader** stops and thinks about the material just read and asks questions: Does this make sense? Have I had experiences like this? The active reader does the examples with pen or pencil, rather than just reading the author's description.

WHY EMPHASIZE PROBLEM SOLVING?

Although Polya described his problem-solving strategies back in 1945, it was quite some time before they had a significant impact on the way mathematics was taught.

One of the reasons is that until recently, "problems" were generally defined too narrowly. Many of you learned how to do different kinds of problems—mixture problems, distance problems, percent problems, age problems, coin problems—but never realized that they have many principles in common. There has been too great a focus on single-step problems and routine problems. Consider the examples from the National Assessment of Educational Progress shown in Table 1.2.

To solve the first problem, one only has to remember the procedure for finding an average and then use it:

$$\frac{13 + 10 + 8 + 5 + 3 + 3}{6}$$

TABLE 1.2

Problem	Percent correct Grade 11
1. Here are the ages of six children: 13, 10, 8, 5, 3, 3. What is the average age of these children?	72
2. Edith has an average (mean) score of 80 on five tests. What score does she need on the next test to raise her average to 81?	24

Source: Mary M. Lindquist, ed., Results from the *Fourth Mathematics Assessment of the National Assessment of Educational Progress* (Reston, VA: NCTM, 1989), pp. 30, 32.

TABLE 1.3

Textbook word problems	Real-life problems
1. The problem is given.	**1.** Often, you have to figure out what the problem really is.
2. All the information you need to solve the problem is given.	**2.** You have to determine the information needed to solve the problem.
3. There is always enough information to solve the problem.	**3.** Sometimes you will find that there is not enough information to solve the problem.
4. There is no extraneous information.	**4.** Sometimes there is too much information, and you have to decide what information you need and what you don't.
5. The answer is in the back of the book, or the teacher tells you whether your answer is correct.	**5.** You, or your team, decides whether your answer is valid. Your job may depend on how well you can "check" your answer.
6. There is usually a right or best way to solve the problem.	**6.** There are usually many different ways to solve the problem.

OUTSIDE THE CLASSROOM[2]

Many students still consider the second question to be a "trick" question unless the teacher has explicitly taught them how to solve that kind of problem. However, many employers note that problems that occur in work situations are rarely *just* like the ones in the book. What employers desperately need is more people who can solve the "trick" problems, because, as some may say, "life is a trick problem!"

However, there is no simple formula for solving the second problem. Try to solve it on your own and then read on. . . .

To solve this one, you have to have a better understanding of what an average means. One approach is to see that if her average for 5 tests is 80, then her total score for the 5 tests is 400. If her average for the 6 tests is to be 81, then her total score for the 6 tests must be 486 (that is, 81×6). Because she had a total of 400 points after 5 tests and she needs a total of 486 points after 6 tests, she needs to get an 86 on the sixth test to raise her overall average to 81.

The difference between traditional word problems and many real-life problems

Table 1.3 lists differences between the word problems generally found in textbooks and real-life problems.

When students undertake more authentic problems, they realize that mathematics is more than just memorizing and using formulas, and they begin to value their own thinking.

INVESTIGATION 1.1d Coin Problem

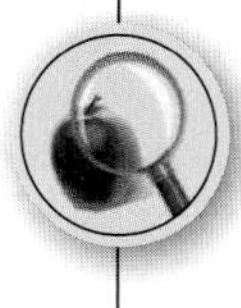

Variations of this problem are often found in elementary school textbooks because it provides an opportunity to move beyond random guess and test.

If 8 coins total 50 cents, what are the coins?

Solve this problem intentionally using and writing out Polya's four steps of problem solving.

Explorations Manual 1.1

DISCUSSION

STEP 1: UNDERSTAND THE PROBLEM

So often students will jump into a problem without stopping to really understand it. Read a problem more than once before attempting to solve. Write down the important information and pay attention to what the question is before starting. Here, you have 8 coins, which might be pennies, nickels, dimes, quarters, or half dollars. All together they equal 50 cents. We need to determine what kind of coins we have.

STEP 2: DEVISE A PLAN

There is more than one strategy to solve any problem. Here we could use a diagram, make a table, use reasoning, or use a bag of coins to help us solve it. Let's consider two strategies of making a diagram and using reasoning.

STEP 3: MONITOR THE PLAN

STRATEGY 1 Use a diagram

We could make 8 circles and begin with all nickels: 8 coins = 40¢. What might be the next step?

With a bit of thinking, we can conclude that each time we substitute a dime for a nickel, the total increases by 5 cents. Thus, we need to trade 2 nickels for 2 dimes, and the answer is 6 nickels and 2 dimes.

STRATEGY 2 Use reasoning

Eight nickels would make 40 cents, and 8 dimes would make 80 cents. Because the 8 coins make 50 cents, your first guess will have more nickels than dimes. Even if the guess is wrong—for example, 5 nickels and 3 dimes make 55 cents—you are almost there.

STEP 4: LOOK BACK AT YOUR WORK

We are almost done but we need to ask questions like: Does my answer make sense? Did I answer the question? Is this the only answer?

We can go back and reread the problem, make sure our solution answers the questions, make sure our answer makes sense, see if we missed any information, and think about alternate ways to get to the solution.

The only other possible solutions are that there might be 1 quarter or 5 pennies. Do you see why? If we make 1 of the coins a quarter, then the other 7 coins must be worth 25 cents. If 5 of those coins are pennies, then we need 2 coins that are worth 20 cents. Aha—2 dimes. Do you see that we could have arrived at the same answer if we had begun with 5 pennies?

In this first section, we have examined the importance of adaptive attitudes and beliefs toward mathematics. We have also broadened the question of what is mathematics and specialized math knowledge needed for teaching. We examined Polya's four steps as the foundation for a toolbox for problem-solving strategies.

1.1 Exercises

1. Write down some personal goals in this course. Keep these in a prominent place in your notebook so that you can refer back to them periodically.
2. Interview a friend, a child, and a parent or grandparent (if possible), asking them the first four questions in the What Do You Think? list at the beginning of this section.
3. Make a list of uses of mathematics in your own life. Ask others to discuss where mathematics is useful in their lives and add these to your list.
4. Read the elementary mathematics curriculum standards for your state/district. For many of you, these can be found at corestandards.org. Write about your impression of them, and how they are similar and different from your current vision of elementary school mathematics.
5. Multiply 86×47 in each of the two valid ways in Investigation 1.1a.
6. Write about your past experiences in math classes. How have these experiences influenced your current beliefs and attitudes about mathematics?

DEEPENING YOUR UNDERSTANDING

Solve the following using Polya's steps for problem solving. See if you can find more than one way to get to the answer.

7. A video store charges \$4 per movie, and the fifth movie is free. How much do you actually pay per movie?

8. At 60 miles per hour, my car's dashboard shows 3000 rpm. This means that the crankshaft, which drives the car, is turning at 3000 revolutions per minute. If a car went 60 miles per hour for 100,000 miles, how many revolutions would the crankshaft have done?

9. Sally works 40 hours a week and makes \$6.85 an hour, but her kids are in child care each day and the day care center charges her \$15 per day. If you deduct her child care expenses, how many dollars per hour does she actually make?

10. A farmer needs to fence a rectangular piece of land. She wants the length of the field to be 80 feet longer than the width. If she has 1080 feet of fencing material, what should be the length and the width of the field?

11. Since its beginning, the U.S. Mint has produced over 288.7 billion pennies.

 a. What if we lined these pennies up? How long would the line be?

 b. The mint currently makes about 30 million pennies a day. How many is this per second? How many is this per year?

12. The record for the longest migration is held by the arctic tern, which flies a round trip that can be as long as 20,000 miles per year, from the Arctic to the Antarctic and back. If the bird flies an average of 25 miles per hour and an average of 12 hours per day, how many days would it take for a one-way flight?

13. A family is planning a three-week vacation for which they will drive across the country. They have a van that gets 18 miles per gallon, and they have a sedan that gets 32 miles per gallon. How much more will they pay for gasoline if they take the van?

 a. First describe the assumptions you need to make in order to solve the problem.

 b. Solve the problem and show your work.

 c. What if the price of gas rose by 40¢ between the planning of the trip and the actual trip? How much more would the gas cost for the trip?

14. This problem was explored in the September 2007 issue of *Teaching Children Mathematics,* pp. 102–106. In September, ruby-throated hummingbirds fly across the Gulf of Mexico to spend the winter on the Yucatan peninsula. The migration takes them 525 miles across the Gulf of Mexico and another 1000 miles farther into Central America. Ruby-throated hummingbirds typically fly about 25 mph. How many hours would it take a hummingbird to make this migration?

15. A hummingbird's wings beat about 60 times per second. How many would this be in a minute? In an hour?

16. a. Using each of the numbers 1–9 exactly once, fill in the blanks below:

 b. How many other solutions can you find?

FROM STANDARDIZED ASSESSMENTS

2006 NECAP, Grade 5

17. Karen used toothpicks to make the two shapes shown below.

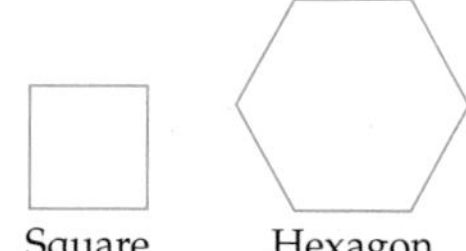

She used a total of 24 toothpicks to make the square. She made the hexagon so that its sides are the same length as the sides of the square. How many toothpicks did Karen use to make the hexagon?

18. Suppose you have 8 coins and you have at least one each of a quarter, a dime, and a penny. What is the least amount of money you could have? (23% of seventh-graders got this correct)

Source: Results of the *Fourth NAEP Mathematics Assessment*, p. 16. U.S. Department of Education, National Center for Education Statistics.

SECTION 1.2 Process, Practice, and Content Standards

What do you think?

- If you were to write a vision for the mathematics the students in elementary school should learn, what would it look like?
- What habits and attitudes does a mathematically proficient student have?

First let's take a brief historical look at the development of math standards, which define what is taught in school. From the "new math" of the 1960s and 1970s to the "back to basics" movement of the 1980s, the pendulum has swung between many ideas of how and what mathematics should be taught. In 1989, the National Council of Teachers of Mathematics (NCTM) published a landmark book called *The Curriculum and Evaluation Standards for School Mathematics.* This was the first document that detailed curriculum standards for elementary, middle, and high school mathematics, and led to individual states creating their own curriculum standards. In

2000, NCTM published a revised version, *Principles and Standards for School Mathematics*, which outlined both content standards and process standards. The content standards are about what math topics should be taught at different grade levels while the process standards are about how students and teachers will engage in the math.

Content standards	Process standards
Standard 1: Number and Operation	Standard 6: Problem Solving
Standard 2: Patterns, Functions, and Algebra	Standard 7: Reasoning and Proof
Standard 3: Geometry and Spatial Sense	Standard 8: Communication
Standard 4: Measurement	Standard 9: Connections
Standard 5: Data Analysis, Statistics, and Probability	Standard 10: Representation

This document also outlines principles of school mathematics that address equity, curriculum, teaching, learning, assessment, and technology.

Based on these NCTM standards, other recommendation documents and international comparisons, the Common Core State Standards (CCSS) were developed in 2010 by educators nationwide to establish clear consistent standards that states could then voluntarily adopt. Currently 45 states plus the District of Columbia have adopted these standards.

Similar to the NCTM content standards, the CCSS defines content standards for which topics should be taught at which grade level. The topics included in this book are found in the CCSS K–8 content standards. The introduction to each section discusses which grade levels look at the content according to the CCSS. If you are in a state that offers a K–5 or K–6 teaching license, your instructor may choose to focus on that content. However, knowing what math is on the horizon for your students is also helpful.

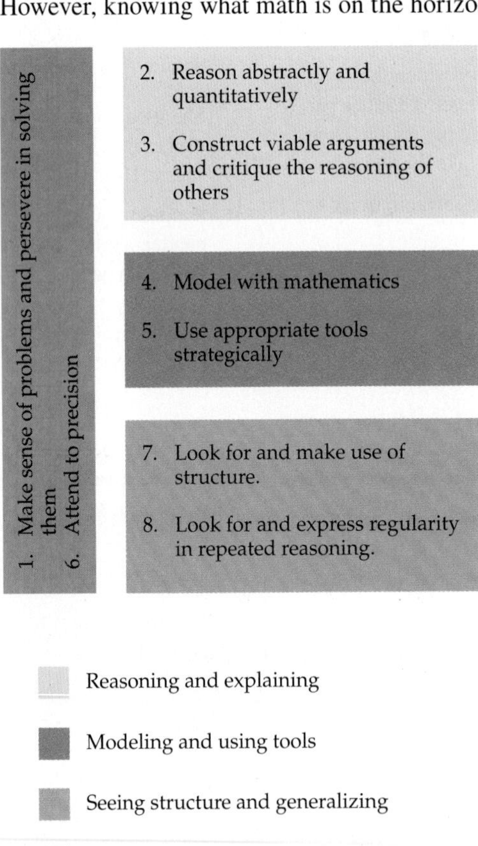

Similar to the NCTM Process Standards, the CCSS describe eight mathematical practices (MPs). These practices are interconnected and although we will look at investigations closely associated with each of the practices below, any worthwhile mathematical activity will involve several of them. These mathematical practices are embedded throughout this course. The following figure helps to organize the higher-order thinking skills of the mathematical practices and provides an interesting grouping that may help us to think about what these practices actually look like in a mathematics classroom.

You will learn more about teaching the curriculum standards as you proceed through your education program. For now, it is important for you to experience thinking about and understanding mathematics through these practices. In the rest of this section we will experience each of these MPs through investigations. Again, all of the practices are quite interrelated and all the investigations connect to more than one. However, in each investigation, we will focus on one to help you to understand these practices better.

OVERARCHING HABITS OF MIND OF A PRODUCTIVE MATHEMATICAL THINKER: MP 1 AND MP 6

As discussed in the previous section, positive attitudes and adaptive beliefs about mathematics are critical for the student and teacher. Perseverance, being able to think outside of the box, and clear communication are important in becoming proficient problem solvers. MP 1 and MP 6 involve these important habits of mind that are necessary to be a productive mathematical thinker.

MP 1: Make sense of problems and persevere in solving them.

- Mathematically proficient students look for a place to get started. Often that is the hardest part—where do I start?
- They think, try something, assess whether it is helpful, and then continue if it was useful or try another plan.
- If they recognize the current problem as similar to one they have solved, they adapt what they used in the similar problem.
- They simplify the problem—making the numbers smaller or simpler.
- If they are heading down a path that is not solving the problem, they become aware of it and try something different.
- They check their solution and strategy and often ask themselves, "Does this make sense?"

Recall in Section 1.1 that we discussed Polya's four steps of problem solving. Those four steps—understand the problem, devise a plan, monitor your plan, and look back at your work—are embedded in this MP. With so many of us exposed to the Internet and the speed and ease of getting quick answers to questions, the art of persevering and "staying the course" is growing increasingly compromised. For students to be successful problem solvers, patience on the part of both the teacher and the student is required to allow the time to grapple with a problem and to be okay with not getting an immediate answer. Having students engaged in sharing solution methods and creating a safe space where mistakes are part of the learning process can help develop the skills that this MP suggests.

INVESTIGATION 1.2a The Nine Dots Problem

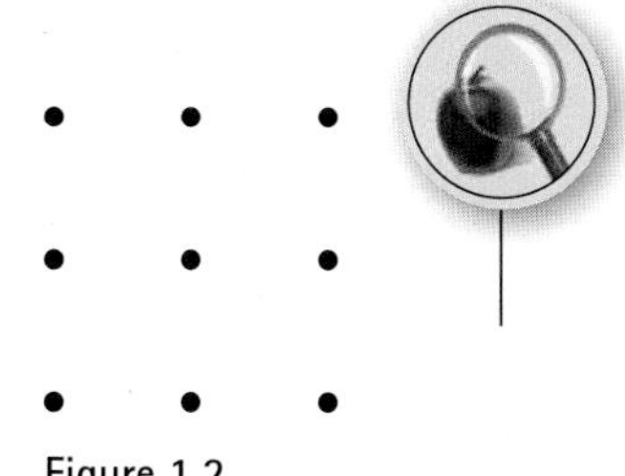

Figure 1.2

Not all problem solving involves computation and formulas, as this investigation shows.

Without lifting your pencil, can you go through all nine dots in Figure 1.2 with only four lines?

DISCUSSION

This is a very famous problem, which some of you may have already encountered because of its moral: This problem is impossible to solve as long as you "stay inside the box." In order to solve the problem, you need to go "outside the box." If you haven't solved the problem yet, try to work with this hint. . . .

The solution to the problem can be seen in Appendix B.

How does this problem relate to MP 1? Being able to "think outside the box" is closely related to good problem solving. In many real-life problems, the solution requires that people think about the problem differently and that they persevere. This relates to MP 1: Make sense of problems and persevere in solving them. In what ways do you see the connection?

INVESTIGATION 1.2b Does Your Answer Make Sense?

This problem has a history that I will share after you solve it. All 261 fifth-graders in a school are going on a field trip, and each bus can carry 36 children and 4 adults. How many buses are needed? Do this problem before reading on. . . .

DISCUSSION

If you divide 261 by 36, you get 7.25. When this problem was given to seventh-graders in one of the National Assessments of Educational Progress, a majority of children gave 7 as the answer. What do you think they did? Do you think they stopped to ask "Does this make sense?" as MP 1 suggests? Many of them had been taught the rules of rounding mechanically, and thus they rounded 7.25 to 7. While they followed the mathematical rule correctly, the real-life application of this computation requires us to round 7.25 up to 8. When solving problems, it is always necessary not only to check your answer but your reasoning. The MP 1 suggests that mathematical thinkers question whether their reasoning as well as their answer makes sense.

MP 6: Attend to precision.

1. Mathematically proficient students are able to communicate their mathematical thinking to others.
2. They are able to articulate clear definitions and the meaning of symbols.
3. They are careful to make sure they use correct units of measure.
4. They calculate accurately and efficiently, and communicate precise answers.

INVESTIGATION 1.2c Precision with Definitions and Symbols

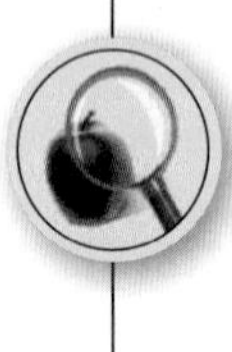

A. Which of the following definitions for even numbers is more precise?

Definition one: An even number has 0, 2, 4, 6, or 8 in the ones place.
Definition two: When an even number is divided by 2, you get a whole number with none left over.

B. Research has shown that when asked to fill in the blank to the following, some students will insert a 5. Why do you think they do this?

$3 + 2 = \_\_\_\_ + 1$

DISCUSSION

A. Although definition one does provide a way to recognize even numbers, it is not very precise for a couple of reasons. For one, a number like 2.4 would be even under the first definition as there is a 2 in the ones place. However, 2.4 is not an even number. This definition also does not tell what an even number really is. Definition two is more precise, because it excludes numbers like 2.4 from meeting the definition and it explains precisely what an even number is.

B. Some students have the misconception that an equals sign is a call to action to perform the operation, instead of realizing that the two sides of the equals sign have to be balanced. These students will see the sign as a call to add 3 + 2 without paying attention to the 1. We will explore this concept further in Chapter 6. For now it is an illustration of how the equals sign may not be as intuitive as you previously believed.

While one aspect of MP 6 is that students are able to perform computations with accuracy (such as finding the correct answer to 35 + 58), MP 6 has greater depth. Students also need to be able to understand precisely what things mean and to be able to consider symbols and units with precision.

REASONING AND EXPLAINING: MP 2 AND MP 3

Mathematical thinking and communication are important aspects of being a mathematically proficient student. As you continue through this course, there will be opportunities for you to analyze your reasoning and communicate your thinking through writing and discussions. Young children naturally make generalizations: They conclude that when the doorbell rings, someone is outside the door; and when they get into the car, they must be strapped in. Elementary school children also regularly make mathematical generalizations: When you add two numbers, you get the same amount when you "add forward as well as backwards" and that if you add two odd numbers together, you get an even number. Reasoning like this and communication of the reasoning are at the heart of these mathematical practices.

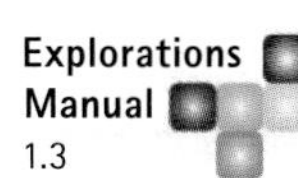

MP 2: Reason abstractly and quantitatively.

- Mathematically proficient students make sense of numbers and their context within a problem.
- They are able to "decontextualize" a problem by representing it with numbers and symbols that abstracts away from the context.
- They are also able to "contextualize" the symbolic manipulations by pausing to go back to the context when needed.

INVESTIGATION 1.2d Pigs and Chickens

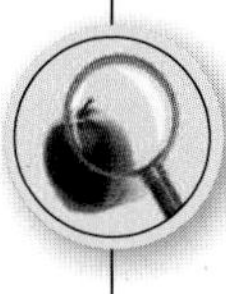

A farmer has a daughter who needs more practice in mathematics. One morning, the farmer looks out in the barnyard and sees a number of pigs and chickens. The farmer says to her daughter, "I count 24 heads and 80 feet. How many pigs and how many chickens are out there?"

Before reading ahead, work on the problem yourself or, better yet, with someone else. Close the book or cover the solution paths while you work on the problem.

Compare your answer to the solution paths below.

DISCUSSION

STRATEGY 1 Use random trial and error

One way to solve the problem might look like what you see in Figure 1.3.

$12 \times 4 = 48$, $12 \times 2 = 24$; $48 + 24 = 72$

$5 \times 4 = 20$, $19 \times 2 = 38$; $20 + 38 = 58$

$19 \times 4 = 76$, $5 \times 2 = 10$; $76 + 10 = 86$

$18 \times 4 = 72$, $6 \times 2 = 12$; $72 + 12 = 84$

$16 \times 4 = 64$, $8 \times 2 = 16$; $64 + 16 = 80$

Figure 1.3

LANGUAGE

The full description of this strategy is "Think–guess–check–think–revise (if necessary), and repeat this process until you get an answer that makes sense." I will refer to this strategy throughout the book simply as guess–check–revise, but I urge you not to let the strategy become mechanical.

The words *trial* and *error* do not sound very friendly. However, this strategy is often very appropriate and can help students to make sense of the problem as the MP 2 suggests. In fact, many advances in technology have been made by engineers and scientists who were guessing with the help of powerful computers using **what-if programs**. A what-if program is a logically structured guessing program. Informed trial and error, which I call **guess–check–revise**, is like a systematic what-if program. Random trial and error, which I call **grope-and-hope**, is what the student who wrote the solution in Figure 1.3 was doing. In this case, the student finally got the right answer. In many cases, though, grope-and-hope does not produce an answer, or if it does produce an answer, it is after many trials.

STRATEGY 2 Use guess–check–revise (with a table)

One major difference between this strategy and grope-and-hope is that we record our guesses (or hypotheses) in a table and look for patterns in that table. Such a strategy is a powerful new tool for many students because a table often reveals patterns. Look at Table 1.4. A key to "seeing" the patterns is to make a fourth column called "Difference." Do you see how this column helps?

TABLE 1.4

	Number of pigs	Number of chickens	Total number of feet	Difference	Thinking process
First guess	10	14	68		We need more feet, so the next guess needs to have more pigs.
Second guess	11	13	70	+2	Increasing the number of pigs by 1 adds 2 feet to the total. What if we add 2 more pigs?
Third guess	13	11	74	+4	Increasing the number of pigs by 2 adds 4 feet to the total. Because we need 6 more feet, let's increase the number of pigs by 3 in the next guess.
Fourth guess	16	8	80	+6	Yes!

The left side of Table 1.4 "decontextualizes" the problem by representing the information numerically. The right side "contextualizes" the problem by returning to the context of the problem to make sense of the numbers.

From the table, we observe that if you add 1 pig (and subtract 1 chicken), you get 2 more feet. Similarly, if you add 2 pigs (and subtract 2 chickens), you get 4 more feet. Do you see why? Think before reading on. . . .

Because pigs have 2 more feet than chickens, each trade (substitute 1 pig for 1 chicken) will produce 2 more legs in the total number of feet. This observation would enable us to solve the problem in the second guess. Do you see how . . . ? After the first guess, we need 12 more feet to get to the desired 80 feet. Because each trade gives us 2 more feet, we need to increase the number of pigs by 6.

It is important to note that the guesses shown in Table 1.4 represent one of many variations of a guess–check–revise strategy.

STRATEGY 3 Make a diagram

Sometimes making a diagram can lead to a solution to a problem. Figure 1.4 shows how one student solved this problem. How do you think she had solved the problem? Write your thoughts before reading on. . . .

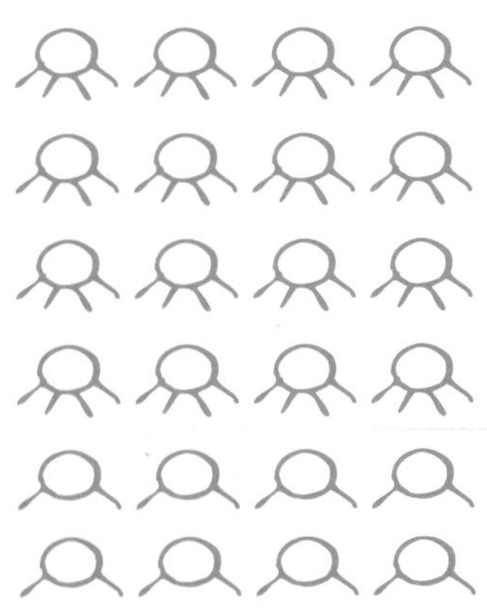

Figure 1.4

She had made 24 chickens, which gave her 48 feet. Then she kept turning chickens into pigs (adding 2 feet each time) until she had 80 feet! I was thrilled because she had represented the problem visually and had used reasoning instead of grope-and-hope. She was embarrassed because she felt she had not done it "mathematically." However, she had engaged in reasoning that MP 2 suggests.

Upon reflection, we realize the enormous potential of this solution path. For example, what if the problem were 82 heads and 192 feet? No way you say? True, it would be tedious to draw 82 heads and then 2 feet underneath each head. This is exactly the power of mathematical thinking—you don't have to do all the drawing. Think about what the diagram tells us, and then see whether you can solve the problem. . . .

If we drew 82 heads and then drew 2 feet below each head, that would tell us how many feet would be used by 82 chickens: 164 feet. Because $192 - 164 = 28$, we need 28 more feet—that is, 14 more pigs. So the answer is 14 pigs and 68 chickens. Check it out!

MATHEMATICS

Most complex mathematical problems are solved first by making a model of the problem—the model makes the problem simpler to see and thus to solve. The model enables us to do things that we could not do in the original situation. In this case, the model enables us to do what is biologically impossible but mathematically possible—we turn chickens into pigs until we get the right answer!

STRATEGY 4 Use algebra

Because the range of abilities present among students taking this course is generally wide, it is likely that some of you fully understand the following algebraic strategy and some of you do not. Let's look at an algebraic solution and then see how it connects to other strategies and to the goals of this course and the MPs.

Go back and review strategy 2. Each guess involved a total of 24 pigs and chickens. Can you explain in words why this is so? Think about this before reading on. . . .

Most students say something like "Because the total number of animals is 24." Therefore, if we say that

p = the number of pigs and c = the number of chickens

then the *number* of pigs plus the *number* of chickens will be 24. Hence, the first equation is

$$p + c = 24$$

CLASSROOM CONNECTION

Many students have told me that before they took this course, they thought there was just *one right way* to do a problem, and so they never looked for patterns but instead looked for formulas or procedures. Once the students started looking for patterns, they found them everywhere, and over time they learned how to use their awareness of patterns more powerfully. We can refer to different ways to do a problem as different **solution paths**.

Many students have difficulty coming up with the second equation. If this applies to you, look back at how we checked our guesses: We multiplied the number of pigs by 4 and the number of chickens by 2 and then added those two numbers to see how close that sum was to 80. In other words, we were doing the following:

$4 \times$ (the guess for number of pigs) $+ \; 2 \times$ (the guess for number of chickens)
$(4 \times p) \qquad + \qquad (2 \times c)$

More conventionally, this would be written as $4p + 2c$.

Using guess–check–revise, we had the right answer when this sum was 80. Thus the second equation is $4p + 2c = 80$.

If solving these two equations, we see that $p = 16$ and $c = 8$.

STRATEGY 5 Visualization by Pictorial Representation

At different places in this book we will use models like the ones used in Singapore, which has some of the most successful math students in the world. A Singaporean colleague, Alice Ho, has shared these methods and also has developed a five-color–coded model that helps communicate mathematical reasoning. Examine the model on the next page and then we will discuss it below.

The green in the first step shows that 4 times the number of pigs plus 2 times the number of chickens together equals 80. Then the blue shows how we can use the fact that the number of pigs plus the number of chickens equals 24, so both of the blue boxes represent 48. This leaves 32 that the red box represents, so the number of pigs must equal 32 divided by 2.

How do each of these strategies relate to MP 2? In each of them, students make sense of quantities and their relationships by considering the numbers of legs, along with the numbers of animals. In the algebra strategy, the context is used to create the equations, but then it is decontextualized as the symbol manipulation occurs, and then it is put back into the context to make sense of the algebraic solution. Reasoning strategies are used in the visualization model, which also helps us to make sense of the numbers within the context of the problem.

STEPS: 1 2 3 4 5

Source: Alice Ho of Math Teach Singapore.

$$p + c = 24$$
$$4p + 2c = 80$$

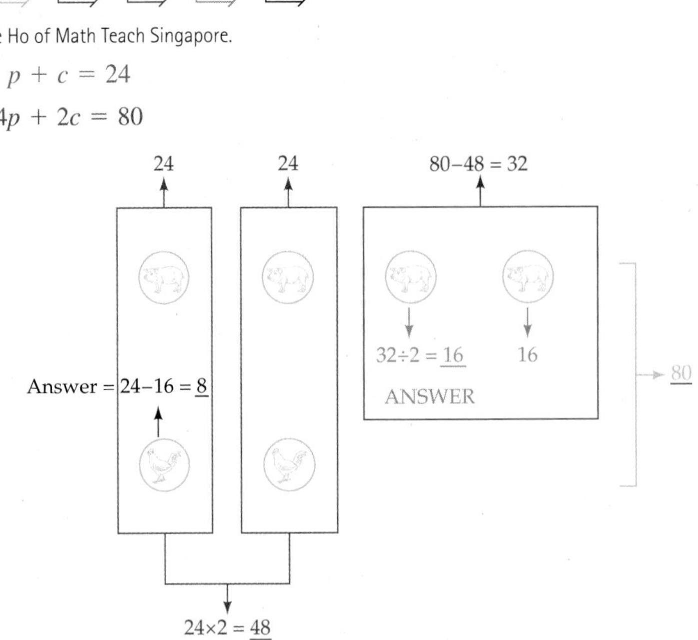

Source: Alice Ho of Math Teach Singapore.

Explorations Manual 1.4

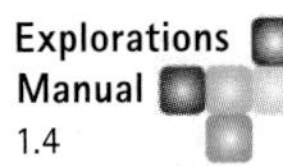

MP 3: Construct viable arguments and critique the reasoning of others.

- Mathematically proficient students are able to use definitions and previous knowledge to communicate their understanding.
- They are able to build a logical progression of their ideas.
- They are able to use counterexamples to make an argument.
- Elementary students can make sense of math and communicate by using objects, drawings, diagrams, or actions.
- They can listen to the reasoning of others and ask useful questions to clarify.

INVESTIGATION 1.2e Why Is the Sum of Two Even Numbers an Even Number?

In the 2009 NCTM Yearbook entitled *Teaching and Learning Proof Across the Grades: A K–16 perspective,* Deborah Schifter describes third-graders working on the question of how to prove that the sum of two even numbers is even. Examine the following responses by the students and think about whether they constitute a proof:

> Paul: I know that the sum is even because my older sister told me it always happens that way.
>
> Zoe: I know it will add to an even number because $4 + 4 = 8$ and $8 + 8 = 16$.
>
> Evan: We really can't know! Because we might not know about an even number and if we add it with 2 it might equal an odd number!
>
> Melody: (Pointing to two sets of cubes she had arranged) This number is in pairs (pointing to the light-colored cubes), and this number is in pairs (pointing to the dark-colored cubes), and when you put them together, it's still in pairs.

DISCUSSION

How are these students engaging in MP 3? Are they communicating their understanding, building a logical progression of ideas, and using drawings to communicate their thinking? Shifter describes four categories of justification common to elementary students (and, I find, with college students too):

appeal to authority (Paul),

inference from instances (Zoe),

assertion that claims about an infinite class cannot be proven (Evan), and

reasoning from representation or context (Melody).

Do you see these categories in the students above?

In the diagrams below, you can see how Melody's assertion closely parallels a more formal proof.

An integer is even if it can be represented as 2 times another integer.	A number is even if it can be broken up into 2 pairs.
If a and b are even numbers, then we can find two integers x and y such that $a = 2x$ and $b = 2y$.	These two numbers are even because they can be broken into pairs. 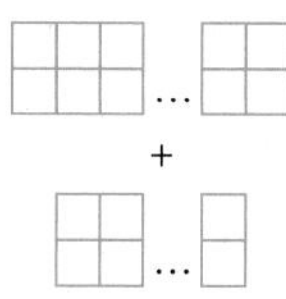
So $a + b = 2x + 2y$; but then $a + b = 2(x + y)$	If you put them together, you still have 2 pairs. 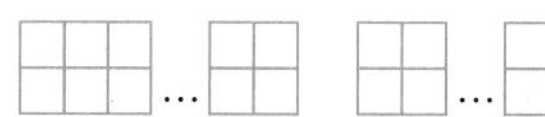
Thus $a + b$ is equal to 2 times an integer, but that is the definition of an even number.	Therefore the sum of the two numbers is also even.

Schifter asserts that young children are capable of making and justifying mathematical generalizations and that making arguments from representations (physical objects, pictures, diagrams, or story contexts) is an effective way to help students develop such reasoning capacity. She proposed three criteria for such representations:

1. The meaning of the operation(s) involved is represented in diagrams, manipulatives, or story contexts.
2. The representation can accommodate a class of examples.
3. The conclusion of the claim follows from the structure of the representation.

Do you see how Melody's argument satisfied these criteria?

1. Her representation modeled two whole numbers.
2. Her language did not say 10 + 16 but rather two whole numbers. That is, her argument did not depend on the actual value of the two numbers (as Zoe's did).
3. When you place the two diagrams together, the resulting amount can also be represented in pairs.

INVESTIGATION 1.2f Darts, Proof, and Communication

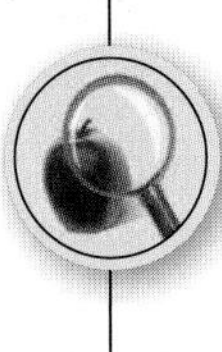

Suppose you have a dart board like the one in Figure 1.5. You throw four darts, all of which land on the dart board. One of the questions I asked fifth-graders was what kinds of scores would be possible and what kinds of scores would be impossible. What do you think?

Explorations Manual 1.4

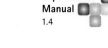

DISCUSSION

After a few minutes, one of the students, Erika, suddenly said, "Only even numbers are possible." I asked her how she came to that conclusion, and she said, "Well I know that an odd plus an odd is even and an odd plus an even is odd. [At this point, she held up four fingers to represent the four darts.] The first two darts are odd and so when you add them, you have an even number. [She joined two of her fingers together to indicate the combined score from two darts.] Now this number (even) plus the next dart (odd) will make an odd number. [She now joined three of her fingers together to indicate the combined score from the first three darts.] Now this number (odd) plus the last dart (odd) will make an even number. So the only possible scores you can get are even numbers."

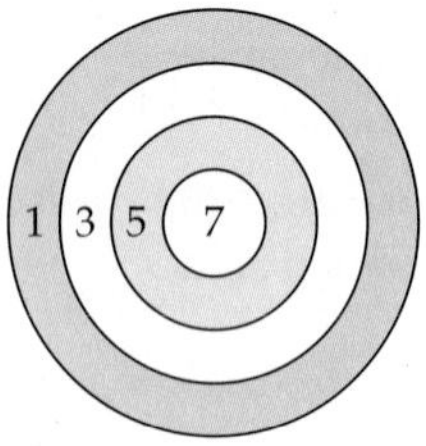

Figure 1.5

We can represent Erika's proof as shown in Figure 1.6.

(odd + odd) + odd + odd
(even + odd) + odd
odd + odd
even

Figure 1.6

Reflect on this problem for a moment along with MP 3. Erika was able to build a logical progression, make sense of the problem, and use actions and objects (her fingers) to communicate her ideas. Because not everyone learns and understands the same way, we have shown another way to represent and communicate the solution above.

We are using this investigation to highlight MP 3, but this is also a great place to consider how the practices work interactively.

INVESTIGATION 1.2g Using Counterexamples

Is the following statement true?

If a number is divisible by 2 and divisible by 6, then it is also divisible by 12. 

DISCUSSION

Here we can look at examples and see if we can find a "counterexample," that is, an example where this statement is not true.

What about 24? Divisible by 2 and by 6 and also 12.

What about 36? Divisible by 2 and by 6 and also 12. Looks good.

What about 30? Divisible by 2 and by 6, but not by 12. So we found a counterexample and therefore the statement is not true.

This illustrates how counterexamples can help us to make the argument that this statement is not true since we found a case where it is not true. There are actually many cases that show it is not true, but finding one is sufficient.

MODELING AND USING TOOLS: MP 4 AND MP 5

These two mathematical practices are particularly related to using math in the workplace and in practical real-life ways. Modeling mathematics problems may involve using tools such as graphs, pictures, concrete materials, and verbal descriptions.

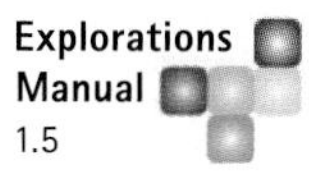

MP 4: Model with mathematics.

- Mathematically proficient students can solve real-life problems, which in elementary school includes being able to write a multiplication equation to solve a problem.
- They are able to identify important information in a real-life problem and analyze relationships using tools.
- They can use models to draw conclusions, make predictions, and reflect on and adjust the effectiveness of the model.

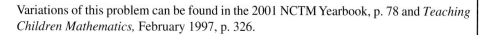

INVESTIGATION 1.2h How Long Will It Take the Frog to Get Out of the Well?

Variations of this problem can be found in the 2001 NCTM Yearbook, p. 78 and *Teaching Children Mathematics,* February 1997, p. 326.

A. A frog is climbing out of a well that is 8 feet deep. The frog can climb 4 feet per hour but then it rests for an hour, during which it slips back 2 feet. How long will it take for the frog to get out of the well?

B. What if the well was 40 feet deep, the frog climbs 6 feet per hour, and it slips back 1 foot while resting? Work on the problem before reading on. . . .

DISCUSSION

A. One of the amazing things about this problem is that, in a class of 25 students, I will often see 10 or more different valid models of the problem. Below are two graph models that both lead to the same answer. First, examine them to see if you understand them. . . .

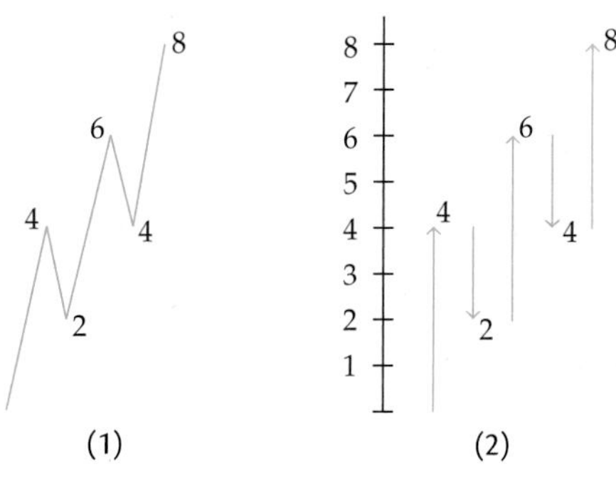

Both models show the frog's progress for each hour and that the frog reaches 8 feet after 5 hours. Let us look more closely: How are models (1) and (2) alike and how are they different?

Alike

1. They have line segments.

Different

1. In the first strategy the line segments are not vertical and in the second they are.

This question asking how two things are alike and how they are different is an important teaching structure and one that we will revisit over the course of the book. You may remember it in a common *Sesame Street* feature: Three of These Things Belong Together. For example, they might show a triangle, a square, a hexagon, and a circle. The answer is that the circle doesn't belong because it doesn't have line segments. We will examine this idea of asking how things are alike and how they are different throughout the textbook.

2. They have numbers: 4, 2, 6, 4, 8.

2. The second strategy has a number line at the left.
3. The second strategy has arrows.

B. As we found in the pigs-and-chickens problem, some models can be "scaled up" and others cannot. Each of the two models shown *could* be used to solve B, but they would be somewhat tedious. In this case, we look for a more efficient model. Using a table to explore the problem is shown below:

Hour	1	2	3	4	5	6	7	8	9	10	11	12	13	14	15
Height	6	5	11	10	16	15	21	20	26	25	31	30	36	35	41

Even this model is a bit tedious. However, if we are always looking out for patterns, we can actually get the answer by making only part of the table.

Hour	2	4	6	8	10	12	14
Height	5	10	15	20	25	30	35

We can see that the numbers when the hours are even are simply multiples of 5. We can then count by 2s to get close to the 40-foot height, or we can see that the height (in even hours) is always $2\frac{1}{2}$ times the number representing the hour. So we can jump to 14 hours when the frog has climbed 35 feet and know that on the 15th hour the frog will get out.

While this problem is not exactly a real-life problem, as MP 4 discusses, it is a great example of how different models can help us to go about solving a problem and how students will come up with different models to think about problems. The next investigation uses math to model a real-life problem.

INVESTIGATION 1.2i How Many Pieces of Wire?

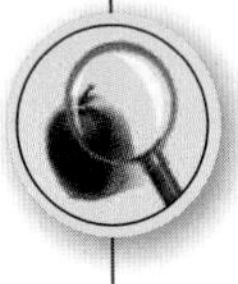

A jewelry artisan is making earring hoops. Each hoop requires a piece of wire that is $3\frac{3}{4}$ inches long. If the wire comes in 50-inch coils, how many $3\frac{3}{4}$-inch pieces can be made from one coil, and how much wire is wasted? Solve this problem on your own and then read on. . . .

But before you do, ask yourself, "Do I understand the problem? What is the important information in this situation, and how can I apply what I know to solve it? Does the problem's wording help me devise a plan for solving it? Once I have a solution, can I check it?"

DISCUSSION

STRATEGY 1 Divide

Some people quickly realize that you can divide "to get the answer." If you use a calculator, it shows 13.333333. . . . If you use fractions, you get $13\frac{1}{3}$ Many people interpret these numbers to mean that you can get 13 pieces and you will have $\frac{1}{3}$ inch wasted. Unfortunately, that is not correct. Can you explain why $\frac{1}{3}$ inch is not the correct answer and what the correct answer is. Then read on. . . .

One of the reasons why math teachers stress the importance of labels is that they illustrate the meaning of what we are doing. The *meaning* of $13\frac{1}{3}$ is 13 whole hoops and $\frac{1}{3}$ of a hoop. That is, what we have left would make $\frac{1}{3}$ of a hoop. Because one whole piece is $3\frac{3}{4}$ inches long, $\frac{1}{3}$ of a piece is $\frac{1}{3}$ of $3\frac{3}{4}$; that is, $1\frac{1}{4}$ inches is wasted.

CLASSROOM  CONNECTION

Grade 2

These are challenging problems for many second-graders. If you had a class of 20 students and $20 to spend, what might you purchase?

Date ______________________ Time ______________________

LESSON 4•5

School Supply Store

21¢ 45¢ 37¢

28¢ 69¢ 76¢ 52¢

You have $1.00 to spend at the School Store.
Use estimation to answer each question.

Can you buy:	Write *yes* or *no*.
1. a notebook and a pen?	________
2. a pen and a pencil?	________
3. a box of crayons and a roll of tape?	________
4. a pencil and a box of crayons?	________
5. 2 rolls of tape?	________
6. a pencil and 2 erasers?	________

7. You want to buy two of the same item.
   List items you could buy two of with $1.00.

   ________ ________

   ________ ________

8. How many pencils could you buy with $1.00? ____

ninety-three **93**

From *Everyday Mathematics, Grade 2:* The University of Chicago School Mathematics Project: Student Math Journal, Volume 1, by Max Bell et al., Lesson 4-5, p. 93. Reprinted by permission of The McGraw-Hill Companies, Inc.

How would we check this answer? Think and then read on. . . . 

One way to check would be to multiply $3\frac{3}{4} \times 13$. This would tell us the length of the 13 whole pieces. If this number plus $1\frac{1}{4}$ equals 50, then our answers are correct. In fact, $3\frac{3}{4} \times 13 = 48\frac{3}{4}$ and $48\frac{3}{4} + 1\frac{1}{4} = 50$.

STRATEGY 2 "Act it out"

Some people understand the problem better when they model the problem with a diagram like that in Figure 1.7.

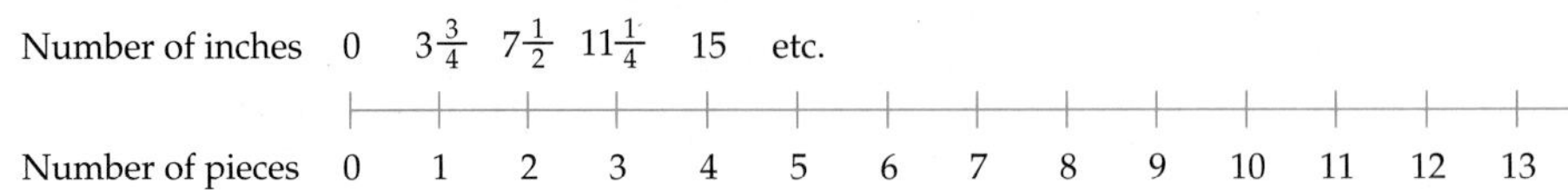

Figure 1.7

STRATEGY 3 Make a table

Yet other people solve the problem by starting with one piece and building up as shown in Table 1.5.

All of these models help us to apply the mathematics to this scenario, as MP 4 suggests. Again, we are focused on MP 4 here, but several MPs come into play here. What others do you see?

TABLE 1.5

Number of pieces	Number of inches	Thinking process
1	$3\frac{3}{4}$	
2	$7\frac{1}{2}$	
3	$11\frac{1}{4}$	
4	15	Using the concept of ratio and proportion, we can reason that if 4 pieces make 15 inches, then 12 pieces would make 45 inches.
12	45	So 1 more piece works.
13	$48\frac{3}{4}$	There are only $1\frac{1}{4}$ inches left, so that is the wasted part.

MP 5: Use appropriate tools strategically.

- Mathematically proficient students can use a variety of tools such as concrete models and technology to find solutions.
- They are able to make good decisions about when to use each of these tools and how to effectively use them.
- They are able to use a variety of tools to investigate and develop their understanding of ideas.

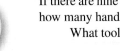

INVESTIGATION 1.2j How Many Handshakes?

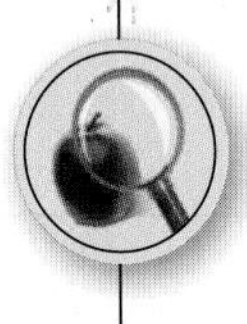

If there are nine people and every person shakes hands once with each person in the room, how many handshakes take place?

What tools could you use to help? 

DISCUSSION

Several tools can be used to answer this. One way may be to have nine people in the class stand up and consider what happens, or you could use beads to represent the people, or a picture. These tools are being used strategically to see that the first person will shake hands with 8 people, the next person will shake hands with 7 other people, until the answer is seen to be $8 + 7 + 6 + 5 + 4 + 3 + 2 + 1$ which equals 36 handshakes in all.

Another tool that can be used is a geometric model and looking at patterns. The following diagram helps us to see that with 2 people there would be 1 handshake, with 3 people there would be 3 handshakes, and so on. Consider these geometric models that illustrate the number of handshakes for 4 people, 5 people, and 6 people.

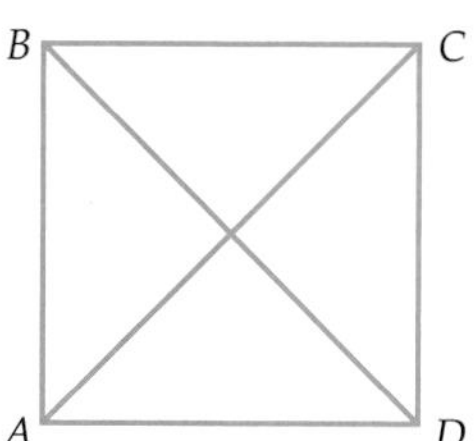

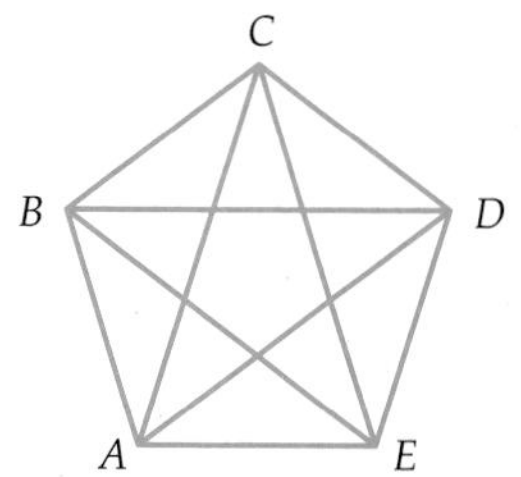

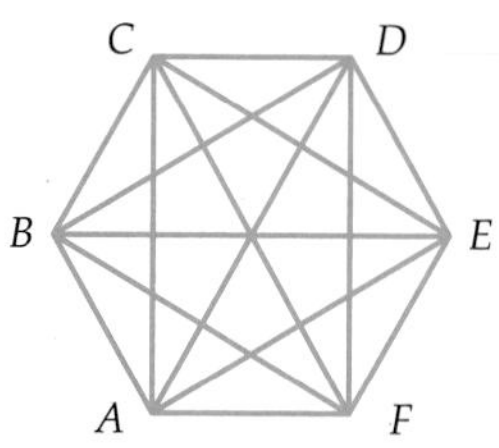

While it is possible to draw geometric models for 7, 8 and 9 people, it might become cumbersome. Therefore a table is another tool that can help us see the pattern.

Number of people	2	3	4	5	6	7	8	9
Total number of handshakes	1	3	6	10	15			

By extending the pattern of adding 2, then 3, then 4, then 5 handshakes each time a person is added, we would continue this pattern by adding 6, then 7, then 8 to get 36 handshakes with 9 people.

Using these tools leads us to a deeper understanding of the problem. Through this investigation, some may intuit a formula for this. Since everyone is shaking hands with everyone else, each of the 9 people will be shaking hands with 8 other people. If we multiply 8 times 9 there is duplication as we are double counting each handshake and so the final answer is $\frac{1}{2}$ of 8 times 9.

INVESTIGATION 1.2k Tools for Defining a Unit Whole

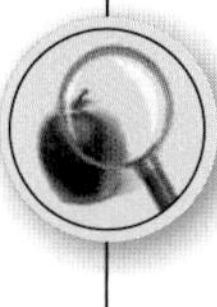

You are teaching a third-grade classroom and are going to draw a picture of $\frac{2}{3}$ on the board. When you ask your students how many circles to draw to illustrate $\frac{2}{3}$, you get two proposals. One student suggests that you start with 1 circle, another student suggests you start with 3 circles. Which student is correct?

DISCUSSION

Each student is using a model that could be used to illustrate $\frac{2}{3}$. If one circle represents the unit whole, then $\frac{2}{3}$ would look like:

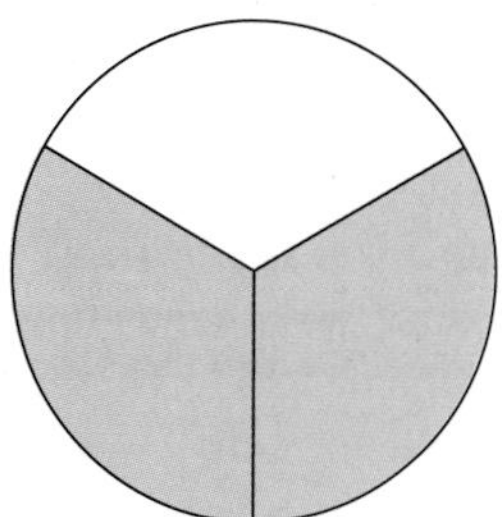

If three circles represent the unit whole, then $\frac{2}{3}$ would look like:

  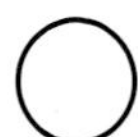

Depending on the context of the problem, which model makes sense to the learner, and the teacher's goals for the discussion in this situation, a strategic decision would be made as to which model to use or to use both models to gain a deeper understanding. Using these pictures strategically illustrates using appropriate tools to solve a problem, as MP 5 suggests.

INVESTIGATION 1.2l Using Base-10 Blocks Strategically

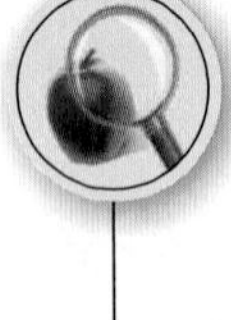

How could you use these base-10 blocks to model 34?

How could you use the base-10 blocks to model 3.4? 

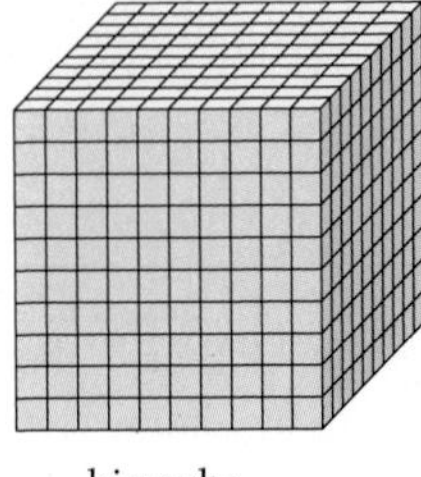
big cube

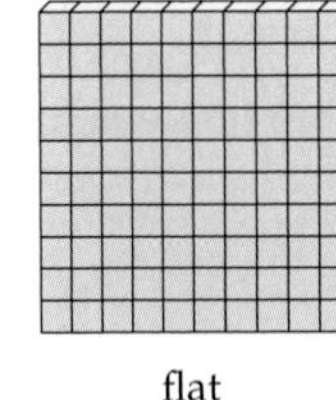
flat

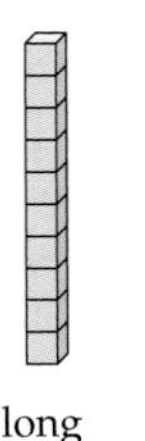
long

small cube

DISCUSSION

These base-10 blocks will be used throughout this course to help us to understand the mathematics more deeply. As we progress through the course, we will see how these tools can be useful. This is a preliminary look at the tools to illustrate this mathematical practice.

As in Investigation 1.2j, strategic choice of unit is important here.

To represent 34, let's define the small cube as "1." Then, 34 would look like:

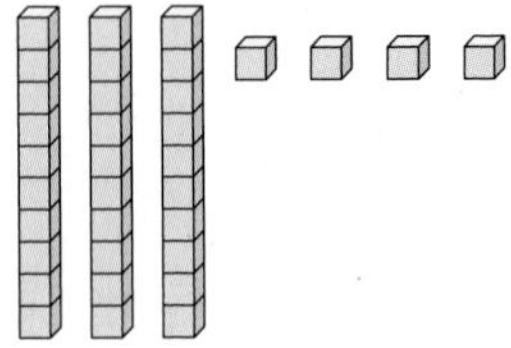

However, when we try to represent 3.4, we find that using the small cube as "1" will not work since there is no way to represent four-tenths. We need to define one of the larger blocks as "1." What if we defined the long one to be "1"? What would 3.4 look like?

Three longs then would equal 3 and each small cube would each equal .1, so we would need 4 of those. Therefore, it would be the same picture as above!

What would 3.4 look like if we defined the big cube as "1"? Then 3 big cubes would model the 3 and 4 flats would equal 4 tenths.

This example illustrates how you as a learner and as a teacher might choose to use these blocks strategically in diverse ways depending on whether you are working with whole numbers or decimals. Mathematical learners (and teachers) must be flexible with how they use tools in order to use them well.

SEEING STRUCTURE AND GENERALIZING: MP 7 AND MP 8

These two practices involve seeing patterns and then using those patterns to generalize ideas. These two practices are very closely intertwined, and even math educators struggle with how they are different. MP 7 is seeing the structure in math to be able to build understanding. MP 8 is more about using those patterns to generalize for solving any problem. Let's take a closer look.

Explorations Manual 1.6

MP 7: Look for and make use of structure.

- Mathematically proficient students are able to find and use patterns or structures.
- They might notice that 5 times 8 has the same answer as 8 times 5 and extend this to discover that the order we multiple numbers in does not affect the answer (commutative property).
- Young children naturally use structure. They may use the word "eated" as past tense for "eat" or "swimmed" instead of "swam" since they have noticed the structure of adding "ed" to a word to make it past tense, such as in words like "played," "talked," and "worked."

The structure in mathematics is far more consistent than in language and can help with developing understanding. For example, if children learn that 3 apples plus 4 apples equals 7 apples, and 3 hundreds plus 4 hundreds equals 7 hundreds, they can use this structure to see that 3 sevenths plus 4 sevenths would naturally equal 7 sevenths $\frac{3}{7} + \frac{4}{7} = \frac{7}{7}$, or that 3 x's plus 4 x's equals 7 x's ($3x + 4x = 7x$). Noticing structure can be useful in being able to do arithmetic quickly without memorizing. For example, 9 can be added easily to any number by realizing that we can add 10 and subtract 1. This can be generalized to adding 39 + 52 (add 40 and subtract 1). Using structure makes memorizing facts less important.

INVESTIGATION 1.2m Using Structure to Do Mental Math

How could you find the answer to the following without finding a common denominator?

$$1\frac{2}{3} - \frac{1}{4} + 2 + \frac{1}{3} - \frac{3}{4}$$

How could you find the answer to the following without doing the pencil-and-paper calculations you traditionally use to multiply and add?

$$25 \times 6 + 25 \times 4$$

DISCUSSION

In the first problem, rearranging the numbers makes this difficult-looking problem quite easy to do in your head. If you look at the parts of this problem, knowing the flexibility to move around, it can be seen as

$$1\frac{2}{3} + \frac{1}{3} - \frac{1}{4} - \frac{3}{4} + 2 = 2 - 1 + 2 = 3$$

In the second problem, if we notice that we can use the distributive property, it becomes $25 \times (6 + 4) = 25 \times (10) = 250$. We can use words to discern this structure: we are adding 6 groups of 25 to 4 groups of 25, so we have 10 groups of 25, which is 250.

Both strategies use the structure of math (the distributive and commutative properties) to simplify these problems.

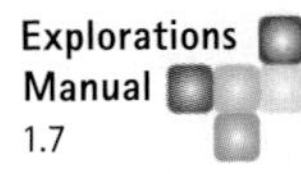

MP 8: Look for and express regularity in repeated reasoning.

- Mathematically proficient students notice when computations are repeated and use this to create methods and shortcuts.
- They continue to look at their process and evaluate their results.
- They develop new methods by generalizing patterns.

Lower elementary students might notice that when counting by 5's, the ones digit is always a 0 or 5. They may notice that adding 0 and multiplying by 1 do not change the number, therefore developing the identity property. Upper elementary students might notice when dividing a number by 10 that the numbers move one place value to the right and then be able to use this whenever dividing by 10.

INVESTIGATION 1.2n Patterns in Multiplying by 11

Let us investigate what happens when we multiply a number by 11. From examining the first three problems in Table 1.6, can you predict the answer to 53 × 11? Make your prediction and then multiply 53 × 11 to check your prediction.

DISCUSSION

TABLE 1.6

The problem	The product
26 × 11	286
35 × 11	385
42 × 11	462
53 × 11	?
62 × 11	?
73 × 11	?
75 × 11	?

If you had trouble, the *algorithm* for multiplying a two-digit number by 11 is to add the two digits together, and that number is the middle digit of the product. Test this out for 62 × 11.

When we get to 73 × 11, the problem becomes more challenging because the sum of the two digits is more than 9. Can you modify your thinking to predict the product of 73 × 11?

First, we know that the digit in the ones place will stay the same. Because 3 + 7 = 10, the digit in the tens place will be a zero and the digit in the hundreds place increases by 1, like "carrying." This is how we get the answer of 803. Using this analysis, predict the product of 75 × 11.

We find that the algorithm works: 7 + 5 = 12. We still have 5 in the ones place, 2 in the tens place, and the hundreds place is now 1 more, 8, giving us the predicted product of 825.

In a classroom this is when we see who has really engaged. Has the problem gotten under their skin? Some people now will ask "what-if" questions: What if we are multiplying by a three-digit number? What if we multiply by 111? These will be left as exercises, if you can wait!

You can make up your own problems with a spreadsheet. The directions below work on Excel with a PC.

First, type 26 in cell A1, then type 11 in cell A2. In cell A3, type =A1 * 11 and 286 will appear.

To do a sequence of problems, like I have done, type this in cell A2: =A1 + 1

In cell B2, type =B1.

Now, highlight cell C1 and drag down 1 space so that C2 is also highlighted. Then press Control D. The answer to 27 × 11 will appear.

Now highlight cells A2, B2, and C2 all together. Drag down as far as you want to highlight more rows. Then press Control D!!!

INVESTIGATION 1.2o Using Patterns to Add Fractions

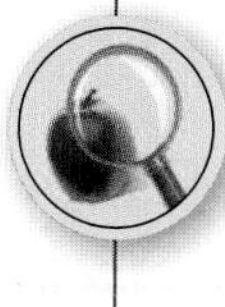

Let's say you don't know how to add fractions with different denominators. Look at the following three examples. Can you see a pattern that would enable you to add other fractions? Think about this before reading on. . . .

$$\frac{1}{3}+\frac{1}{4}=\frac{7}{12}$$
$$\frac{1}{5}+\frac{1}{7}=\frac{12}{35}$$
$$\frac{1}{2}+\frac{1}{5}=\frac{7}{10}$$

DISCUSSION

We can describe the pattern in words by saying that whenever you have two fractions with a 1 in the numerator, the numerator of the sum is found by adding the two denominators, and the denominator of the sum is determined by multiplying the two denominators.

Using this pattern, find the sum of

$$\frac{1}{3}+\frac{1}{41}$$

Using the pattern, we would get $\frac{44}{123}$. Using this pattern makes this much simpler than finding a common denominator.

CLASSROOM CONNECTION

When I wrote this book, one of the criteria that I used when selecting problems for investigations and explorations was that the problems should be extendible. This gets to the "what-if" and "I-wonder" questions that young children naturally ask often. Unfortunately, this frequency seems to decline steadily as we get older. One of the goals in this course is that you resolve to reawaken this curiosity in mathematics. In this case, a "what-if" question is "What if the numerators are not 1 but are the same, as in $\frac{2}{3}+\frac{2}{5}=$?" This will be left as an exercise.

For some students, patterns jump out at them; for others this skill takes more practice. The next page is out of a third-grade textbook provides practice in seeing patterns.

This apparently innate tendency to draw generalizations from specific examples needs careful attention in schools, and is one of the many reasons why elementary teachers need a strong background in elementary mathematics. For example, consider explaining what an isosceles triangle is to young children. If all isosceles triangles shown to them have the base parallel to the bottom of the page, as do the first three triangles below, children will mistakenly not consider the triangle at the far right to be isosceles, although it is.

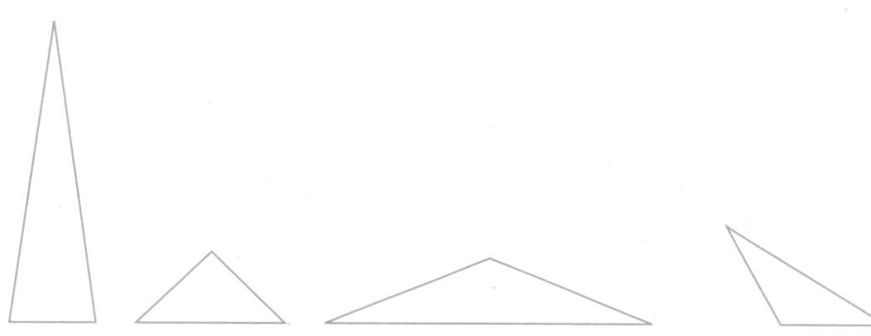

CLASSROOM  CONNECTION

Grade 3

These number sequence problems are challenging for third graders. Do number 6 and then make up a problem that would be challenging for other students in your class.

Date ________ Time ________

LESSON 1·1

Number Sequences

Complete the number sequences.

1. 428, 429, ______, 431, ______, ______, ______, ______, 436, ______, ...

Unit

2. 918, 919, ______, ______, 922, ______, ______, ______, 926, ...

3. ______; 1,416; ______; ______; 1,419; ______; ______; ...

4. ______, 311, ______, ______, 341, ______, ______, ...

5. ______; 4,326; ______; ______; 4,356; ______; ...

Try This

6. 7,628; ______; 7,828; ______; ______; 8,128

one 1

From *Everyday Mathematics, Grade 3*: The University of Chicago School Mathematics Project: Student Math Journal, Volume 1 by Max Bell et al., Lesson 1-1, p.1. Reprinted by permission of The McGraw-Hill Companies, Inc.

Look at the picture below. If we draw 2 points on a circle and connect them, we create 2 regions. If we draw 3 points on a circle and connect each point to every other point, we create 4 regions. If we continue this with 4 points, we create 8 regions. If we continue this with 5 points, we create 16 regions. If we draw 6 points and connect each point to every other point, how many regions will be created?

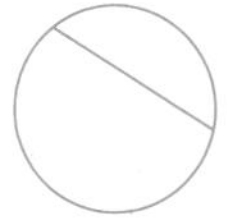  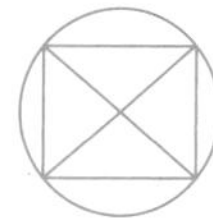  

Most people conclude that the number of regions doubles each time. Alas, this is not so. This pattern breaks down with 6 points. Try as you might (and believe me, mathematicians have tried), no matter where those 6 points are placed, the maximum number of regions formed is 31. This is a classic illustration of the need to be careful with patterns and seeing structure. Sometimes the pattern we see does not hold up.

SUMMARY

In this section we examined the standards movement and where we are today. The NCTM Process Standards were a beginning to the further development of the CCSS Mathematical Practices. The practices are quite interrelated, and any good activity will utilize several. Experiencing mathematics through the eight mathematical practices in these investigations has helped you to not only understand these practices but also to see how the vision of elementary school mathematics is now a deep, connected, reasoning endeavor. I encourage you to continue to experience mathematics in these ways as you proceed through this course and beyond into your teaching career. Deepening our understanding of mathematics is a life-long endeavor. As you continue to deepen your mathematical understanding, and experience mathematics differently in this course, you will continue to gain skills and confidence that will help you to become a successful math student and teacher

1.2 Exercises

1. Read the Common Core State Standards for math for the grade levels you will become certified to teach. To find them, go to corestandards.org and click the link to the mathematics standards. If you live in a state that has not adopted the CCSS, look for the standards in your state. Write a short paragraph describing what you learned.
2. Go to the website of the National Council of Teachers of Mathematics (nctm.org) to read more about this organization and the resources there. Write about what you learn.

DEEPENING YOUR UNDERSTANDING

As you work through the following exercises, consider how you are using the mathematical practices and the problem-solving process.

3. The NCAA men's basketball tournament begins with 68 teams. What is the total number of games played in the tournament?
4. Look at the clock below.

   a. How do this clock and a normal clock differ?
   b. What time is it?
   c. Draw the hour and minute hands in the positions they will occupy when it is 9:45.
   d. Why do you think the direction we call clockwise was selected?

5. When Joe wakes up in the morning the clock is blinking and says 3:30. His watch says 6:45. When did the power go off? What assumptions(s) did you make in order to solve the problem?

6. How many triangles can you find in the figure below?

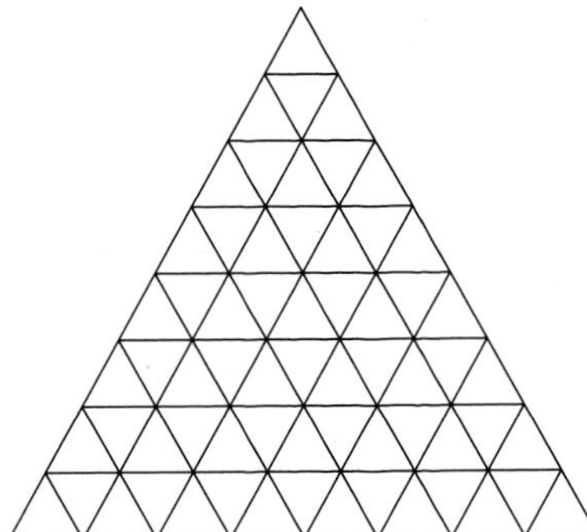

7. ***Classroom Connection*** This problem appeared in *Teaching Children Mathematics* and was discussed by Kim Hartweg. Aunt Katrina just moved into a new house and wants to brighten up the dining room wall by hanging 6 of her favorite decorative plates. Each plate is 8 inches in diameter. Aunt Katrina is having trouble deciding how to hang the plates in a straight line and evenly space them along her 104-inch wall.

   a. Draw a diagram to show the placement and spacing and explain how Aunt Katrina should hang her 6 decorative plates.

   b. What if one plate broke?

   c. What if she wanted to have the two end plates 12 inches from the end of the wall?

   Source: *Teaching Children Mathematics*, December 2004/January 2005, pp. 280–284. National Council of Teachers of Mathematics.

8. ***Classroom Connection*** In "Math Riddles: Helping Children Connect Words and Numbers," Carl Sherrill shows how children can learn many things while creating and solving riddles. Here are a few that my students created.

   a. The number of digits is the same as the number in the ones place. The sum of the digits is 10. All the digits are different. The digits are in increasing order. What is the number?

   b. I have 93 cents and 8 coins. I have the same number of quarters as pennies. Each type of coin is used. What are the coins?

   c. I am thinking of a 4-digit number. The sum of the digits is 27. The largest digit is used twice but not side by side. The smallest digit is in the largest place. The smallest digit is $\frac{1}{3}$ as large as the largest digit. What is the number?

   d. I am thinking of a 4-digit number. The sum of the digits is 21. Starting from the left the third digit is double the first and the fourth is double the second. All the digits are different. Add one more clue to produce a unique solution.

   e. I have 7 coins. There are more nickels than quarters. I have twice as many dimes as nickels. How much money do I have?

   f. Make up your own riddle.

   Source: "Math Riddles: Helping Children Connect Words and Numbers," in *Teaching Children Mathematics*, March 2005, pp. 368–375. National Council of Teachers of Mathematics.

9. At a bicycle store, there were a bunch of bicycles and tricycles. If there were 32 seats and 72 wheels, how many bicycles and how many tricycles were there?

10. A farmer looks out into the barnyard and sees the pigs and the chickens. He says to his daughter, "I count 40 heads and 100 feet. How many pigs and how many chickens are out there?"

11. This problem was explored in the April 2003 issue of *Teaching Children Mathematics,* pp. 457–460: A child was riding her new bike at the playground with other children. Everyone was riding a bicycle or a tricycle. If Jacky counted 12 wheels, not including her own, how many children could be riding and what would they be riding?

12. What if the farmer in Exercise 10 saw 169 heads and 398 feet? How many pigs and how many chickens are out there?

13. A Martian farmer looks out into the barnyard and sees tribbles (which have four legs) and chalkas (which have seven legs). She says to her son, "I count 97 heads and 436 feet. How many tribbles and how many chalkas are out there?"

14. Look back at Investigation 1.2f with the dart board. What if the scores were 1, 5, 10, 25?

    a. List all possible scores from 1 to 100.

    b. Select one score that is more than 4 and less than 100 that is impossible, and prove that it is impossible.

15. Look back at Investigation 1.2f with the dart board. What if the scores were 1, 2, 4, 8?

    a. List all possible scores from 4 to 32.

    b. How many different ways could someone score 12?

    c. Prove that there are no more ways to score 12 than the ones you found in part b.

16. a. A frog is climbing out of a well that is 11 feet deep. The frog can climb 3 feet per hour but then it rests for an hour, during which it slips back 1 foot. How long will it take for the frog to get out of the well?

    b. What if the well was 50 feet deep, the frog climbs 7 feet per hour, and it slips back 3 feet while resting?

    c. What if a caterpillar is climbing out of a glass that is 12 inches high and can climb $1\frac{1}{2}$ inches in an hour but slides and falls back $\frac{1}{4}$ inch during the hour it rests. How long will it take the caterpillar to climb to the top of the jar?

17. a. How many $3\frac{2}{3}$-oz bottles of perfume can be filled from a jug containing 64 oz?

    b. How many ounces of perfume are left over per jug?

    c. Determine a jug size that is a whole number and that would produce no waste.

18. A boatman is to transport a fox, a goose, and a sack of corn across the river. There is room in his boat for only one of the three at a time. Furthermore, if the fox and the goose are left together, the fox will eat the goose. If the goose and the corn are left together, the goose will eat the corn. How can the boatman do the job?

19. The following two jar problems have been around for many years. They are contrived problems to promote reasoning. In

these problems, a jar must be filled *completely*, and emptied *completely*. For example, you cannot say, "Fill the big jar half full and fill the second jar half full" to get 4 gallons.

a. You are given a 5-gallon pail and a 3-gallon pail, both of which are unmarked. You are asked to fetch 4 gallons of water from the well in one trip. How?

b. You have three jugs of capacities 8, 5, and 3 gallons. The largest jug is full of water. The other two are empty. Your task is to redistribute the water so that you wind up with 2 gallons of water in the large jug and 3 gallons of water in each of the other jugs.

20. a. In how many different ways can you make change for 50 cents?

b. In how many different ways can you make change for one dollar without using pennies or half dollars?

21. A special rubber ball is dropped from the top of a wall that is 16 meters high. Each time the ball bounces, it rises half as high as the distance it fell. The ball is caught when it bounces 1 meter high. How many times did the ball bounce?

22. ***Classroom Connection*** Let's say there are 25 students in your class. If each student greeted and shook hands with every other student in the class, how many handshakes would there be?

23. This variation of the handshakes problem is on video 17 of *Teaching Math: A Video Library, K–4,* which many colleges and schools have. "If there are 20 children in a classroom and every child gives every child a valentine, how many valentines are distributed that day?" Does this problem have the same answer as the handshakes problem? Why or why not?

24. ***Classroom Connection*** This problem was posed in the November 2001 "Problem Solvers" section of *Teaching Children Mathematics:* In the Animal Football League, the teams can score only 3-point field goals and 7-point touchdowns; no safeties or extra-credit points are allowed. In one contest, the Anteaters defeated the Bobcats 42 to 37. In how many different ways could each team have arrived at its final score?

a. Do the problem yourself before reading on.

b. Solve the problem using the approach described by this fifth-grader and shown in the table: "I found the different ways that teams could arrive at the score by making a chart and checking each possibility. After subtracting the touchdown points from the total score, the difference would have to be divisible by 3 to be a possible combination."

	Touchdowns	**Field Goals**
1.	7	3
2.	14	6
3.	21	9
4.	28	12
5.	35	15
6.	42	18
7.		21
8.		24
9.		27
10.		30
11.		33
12.		36
13.		39
14.		42

c. What scores are impossible to get?

Source: *Teaching Children Mathematics*, November 2001. National Council of Teachers of Mathematics.

25. Using the work in Investigation 1.2n, predict the products below. If you are not sure, do some problems yourself until you see patterns that enable you to predict the products.

$134 \times 11 = a$
$148 \times 11 = b$
$465 \times 11 = c$

26. Look for patterns in the problems below in order to predict the answers to (a) through (h). In each case, explain your reasoning.

$27 \times 111 = 2997$
$31 \times 111 = 3441$
$42 \times 111 = a$
$53 \times 111 = b$
$68 \times 111 = 7548$
$76 \times 111 = 8436$
$84 \times 111 = c$
$95 \times 111 = d$
$231 \times 111 = 25641$
$314 \times 111 = 34854$
$432 \times 111 = 47952$
$441 \times 111 = e$
$503 \times 111 = f$
$236 \times 111 = 26196$
$526 \times 111 = 58386$
$563 \times 111 = 62493$
$634 \times 111 = g$
$888 \times 111 = h$

27. Describe the pattern in the first four products that enables you to determine the product of 8×1089 without doing any multiplying.

$1 \times 1089 = 1089$
$2 \times 1089 = 2178$
$3 \times 1089 = 3267$
$4 \times 1089 = 4356$
$8 \times 1089 =$

28. Along the lines of Investigation 1.2m, make up a number of fraction addition problems where the two numerators are 2 (such as $\frac{2}{3} + \frac{2}{5}$). Describe the generalization you come up with in words and in mathematical notation.

29. The following pattern, called **Pascal's triangle**, was "discovered" by the mathematician Blaise Pascal (1623–1662). Examine the triangle, play with it, and describe the patterns you see. You may choose to describe patterns with words, in a table or sequence, or by using mathematical notation.

1
1 1
1 2 1
1 3 3 1
1 4 6 4 1
1 5 10 10 5 1
1 6 15 20 15 6 1
1 7 21 35 35 21 7 1
1 8 28 56 70 56 28 8 1

30. Magic squares, triangles, and circles have fascinated human beings for thousands of years. The definition of a magic square is that when you add the numbers in any row, any column, or either diagonal, you get the same number. The square at the left was created by Albrecht Dürer in 1514 in his engraving *Melancholia.* He created a square in which the year appeared in the bottom row of the square! The magic sum for rows, columns, and diagonals is 34. Now take a closer look at this square. What other patterns do you see in this magic square? What observations do you make? Observation will be defined as something interesting, but not repeating—for example, the 16 numbers in the square are consecutive numbers from 1 to 16. Write down your observations and patterns so that someone reading your description could see what you see.

13	3	2	16
8	10	11	5
12	6	7	9
1	15	14	4

31. Find the missing numbers that will turn the grids below into 3×3 magic squares.

a.

9	19	5
17		

b.

14	10	
12		16

c.

8	9	10
18		

32. Find the missing numbers that will turn the grid below into a 4×4 magic square.

	8		19
11		16	14
15	13		
10			7

33. Below is a 5×5 magic square.

11	18	25	2	9
10	12	19	21	3
4	6	13	20	22
23	5	7	14	16
17	24	1	8	15

a. Describe all the patterns you see in this magic square.

b. Describe one of the patterns in words, as though you were talking to someone on the phone.

34. Benjamin Franklin created this magic square. Describe three patterns that you find in this magic square.

52	61	4	13	20	29	36	45
14	3	62	51	46	35	30	19
53	60	5	12	21	28	37	44
11	6	59	54	43	38	27	22
55	58	7	10	23	26	39	42
9	8	57	56	41	40	25	24
50	63	2	15	18	31	34	47
16	1	64	49	48	33	32	17

35. In each problem, briefly explain how you solved the problem by a means other than random trial and error.

a. How many different ways can you make a dollar with 6 coins?

b. How many different ways can you find for 7 coins totaling 95 cents?

c. How many different ways can you find for 21 coins totaling one dollar?

36. Find the missing term in each of the sequences below:

a. 80, 72, __, 56,

b. 5, 10, 16, 23, __, 40

c. 2, 5, 11, 23, __, 95

37. The following magic figure originated in West Africa. Describe the patterns you see in this magic figure.

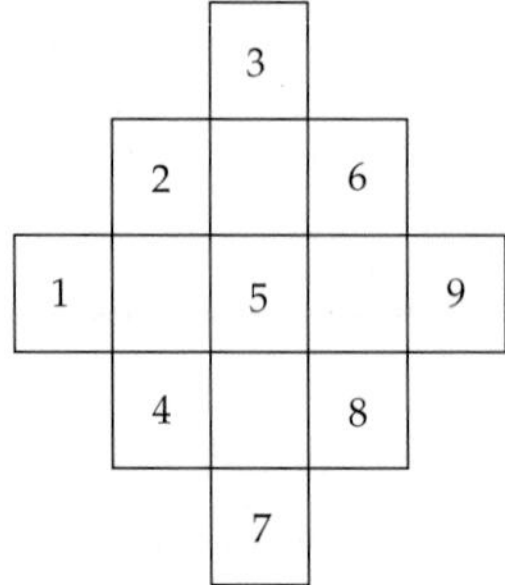

38. Find a pattern in each of the following sets of equations and predict the next equation.

a. $1^2 + 2^2 + 2^2 = 3^2$
$2^2 + 3^2 + 6^2 = 7^2$
$3^2 + 4^2 + 12^2 = 13^2$

b. $1^3 + 2^3 = 3^2$
$1^3 + 2^3 + 3^3 = 6^2$
$1^3 + 2^3 + 3^3 + 4^3 = 10^2$

c. $1 + 2 = 3$
$4 + 5 + 6 = 7 + 8$
$9 + 10 + 11 + 12 = 13 + 14 + 15$

39. One of the most famous patterns in mathematics was discovered by the Italian mathematician Leonardo of Pisa (1170–1250), known to us as Fibonacci. He discovered this sequence, which we today call the Fibonacci sequence, while studying the birth rates of rabbits. He posed the following question: Suppose that a pair of rabbits produce a pair of baby rabbits every month, and that rabbits cannot reproduce until they are two months of age. How many pairs of rabbits will you have after one year (assuming, of course, that no rabbits die)? He found that if he listed the number of pairs he had after each month, he had a very interesting sequence: 1, 1, 2, 3, 5, 8, 13, 21, 34, 55, 89, 144, 233, . . .

a. Add the first seven Fibonacci numbers. What do you notice? Add the first eight Fibonacci numbers. What do you notice? Try to describe this relationship both in words and using notation.

b. Pick three consecutive terms. Multiply the first term and the third term. How does this compare to the square of the middle term? Describe this relationship using notation.

40. Palindromes are numbers whose value is the same backward as forward. For example, 1331 is a palindrome. Most children love explorations with palindromes because they find the patterns fun and the investigations are so mathematically rich. How many palindromes can you find between 100 and 199?

41. Continue the pattern in the three number sentences below, and then describe as many patterns as you can.
$3 \times 3 - 1 = 8$
$5 \times 5 - 1 = 24$
$7 \times 7 - 1 = 48$

42. For each of the triangles below:

(1) Describe as many patterns as you can find in the triangle.

(2) Write the numbers that would be in the next row of the triangle, and explain how you arrived at your answer.

a.
1
2 4
3 6 9
4 8 12 16
5 10 15 20 25

b.
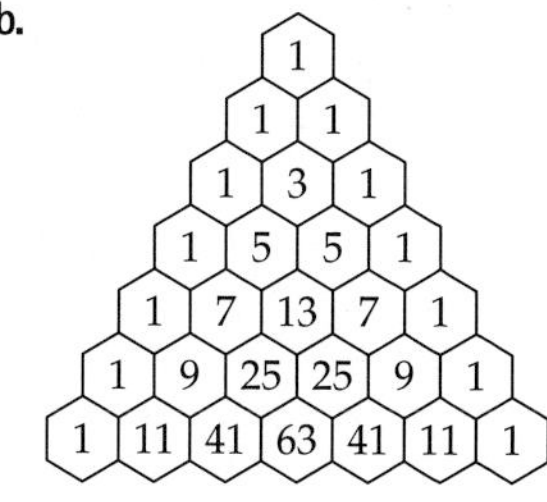

c. d.
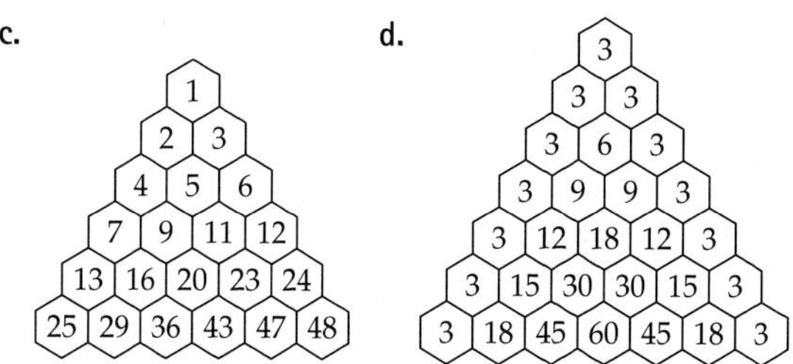

43. Find the sum of the first 25 numbers: 1, 2, 3, 4, . . . 25 by a means other than adding all 25 numbers. That is, look for patterns and different ways to represent the problem.

44. Sheila has a lot of stamps in her drawer and will be mailing a number of 2-oz letters costing 61¢ of postage. How many different ways can she make 61¢ postage using a 44¢ stamp and combinations of 5¢, 3¢, and 2¢ stamps?

45. ***Classroom Connection*** This problem is adapted from the Figure This website, which has 80 rich problems that upper elementary and middle school children enjoy: Is there a combination of 15¢ and 33¢ stamps that can make $1.77?

Source: National Council of Teachers of Mathematics.

FROM STANDARDIZED ASSESSMENTS

2006 NECAP, Grade 4

46. Look at the following pattern.
50, 41, 32, ?, 14, 5
What number is missing in the pattern?

a. 9
b. 23
c. 24
d. 33

2008 NECAP, Grade 4

47. Luis made a pattern on this number chart by circling numbers. He started at 32.

31	(32)	33	34	35	36	37	(38)	39	40
41	42	43	(44)	45	46	47	48	49	(50)
51	52	53	54	55	56	57	58	59	60
61	62	63	64	65	66	67	(68)	69	70

He forgot to circle two numbers *between* 50 and 68 in his pattern. What two numbers did Luis forget to circle?

2005 NECAP, Grade 5

48. If these patterns continue, which pattern will *not* contain the number 100?

a. 2 4, 6, 8, 10, 12, . . .
b. 3, 6, 9, 12, 15, 18, . . .
c. 4, 8, 12, 16, 20, 24, . . .
d. 5, 10, 15, 20, 25, 30, . . .

2008 NECAP, Grade 4

49. Harold is planning a party for 32 people. He has some round tables and some square tables.

- There are 6 chairs at each round table.
- There are 4 chairs at each square table.

What is the *fewest* number of tables Harold needs for 32 people with no chairs left over? Show your work or explain how you know.

2007 NECAP, Grade 6

50. Trisha has a job walking 16 dogs for her neighbors. She can walk one, two, or three dogs on each trip to the park. What is the fewest number of trips Trisha takes to walk all 16 dogs?

a. 16 b. 15 c. 6 d. 5

51. Four pets are competing in a best pet contest. Eighty children vote and the pet with the most votes will win. What is the smallest number of votes that a pet could receive and still win the contest? (28% of seventh-graders got this correct)

Source: From *Results of the Fourth NAEP Mathematics Assessment*, p. 75.

2008 NECAP, Grade 6

52. Hannah is stenciling a border on a wall. She plans to repeat pictures of a tulip, a daisy, a rose, and a lily, in that order, as shown below.

tulip, daisy, rose, lily, tulip, daisy, rose, lily, . . .

The first flower Hannah stencils is a tulip. What will be the 30th flower she stencils?

a. a tulip b. a daisy c. a rose d. a lily

LOOKING BACK on chapter one

QUESTIONS TO SUMMARIZE BIG IDEAS

1. How have your ideas about what math is changed during this chapter?
2. How is the mathematical knowledge needed by elementary school teachers unique?
3. What are the Common Core State Standards in mathematics? Do you live in a state using these standards or does your state have its own?
4. What are the mathematical practices of the Common Core State Standards? If you live in a state that has its own standards, what do they have similar to the mathematical practices?
5. What parts of this chapter are less clear to you at this point? What will you do to clarify those ideas?
6. Look back at the Mathematical Practices of the CCSS. In what ways did you engage in those practices during this chapter?

CHAPTER  SUMMARY

1. One's beliefs about mathematics can powerfully affect one's ability to do mathematics. Many students find that the big ideas of this chapter involve a major shift in the beliefs that they brought into the course.
2. George Polya's description of the four basic steps in solving problems is a useful framework, but it is the starting point, not the ending point, of your using it.
3. Many problems can be solved in a variety of ways. Some of the more common problem-solving strategies are guess–check–revise, make a diagram, develop an equation, make a table, look for patterns, connect the problem to a similar problem, break the problem into smaller pieces and act it out.
4. Solving a problem involves more than simply computation. The number is not the answer, and you are not finished unless the answer makes sense.
5. There is generally more than one way to represent a problem. Thus, it is important to construct a repertoire of representations in order to develop mathematical power and in order to understand other students' representations of problems.
6. Patterns permeate our lives as well as permeating mathematics. Recognizing patterns is the beginning of using them to understand mathematics and solve problems. You should be able to describe, extend, analyze, and create a wide variety of patterns.
7. Reasoning and proof are not just things to be learned by older students; they are also an essential part of elementary school mathematics.
8. Curriculum standards help to provide a coherent and dynamic vision of school mathematics.

9. Communication is an essential aspect of mathematics—being able to communicate with yourself as you learn mathematical ideas and solve problems, and being able to communicate with others. Tools to improve communication include diagrams, vocabulary, and symbols.

10. There are many aspects to being able to develop mathematical connections: making connections between the problem and what is inside your head, connecting new concepts to old concepts, making connections among different concepts, connecting different models for the same concept, connecting conceptual and procedural knowledge, and making connections between school mathematics and "real life" and between mathematics and other disciplines.

BASIC CONCEPTS

Section 1.1: Getting Started and Problem Solving

adaptive beliefs **3**
maladaptive beliefs **3**
problem solving **7**
understand the problem **7**
monitor your plan **7**
math knowledge for teaching **7**
learning styles **7**
rent **8**
Polya's four steps **8**
own **9**
reflect **9**
devise a plan **10**
look back at your work **10**

Section 1.2: Process, Practice, and Content

curriculum standards **11**

Mathematical practices:

MP1: Make sense and persevere **13**
MP2: Reasoning **15**
MP3: Construct arguments **18**
MP4: Model with mathematics **21**
MP5: Use tools strategically **24**
MP6: Attend to precision **14**
MP7: Use structure **27**
MP8: Regularity in repeated reasoning **28**

Singapore Models **17**

REVIEW EXERCISES chapter one

1. A school play charges \$2 for students and \$5 for adults. For the three days of the play, 458 tickets were sold and \$1342 was raised. How many student tickets were sold?

2. A small factory makes three-legged stools and four-legged tables. Last month this factory used 100 legs to build 3 more stools than tables. How many stools did the factory make?

3. How many different ways can you find for 15 coins totaling 92 cents? Briefly explain how you solved this problem by a means other than random trial and error.

4. The Perez family is making a patio in the back yard. They are using patio blocks that measure 12 inches by 8 inches. If the patio is to be 16 feet by 12 feet and each block costs 75 cents, how much will the patio cost?

5. How many different ways can you make 50 cents if you have 1 quarter, 5 dimes, and 10 nickels?

6. If a fence requires a post every 10 feet, how many posts are required for a fence that measures 100 feet by 100 feet?

7. A campus coffee shop charges \$1.30 for a cup of coffee, and the 10th cup is free. How much do you actually pay per cup of coffee?

8. Monarch butterflies flap their wings about 12 times a second in flight. How many times do they flap their wings in 1 hour of flight?

9. A letter was posted that was covered with 10-cent stamps and 5-cent stamps. There were 12 stamps, and the total postage was 70 cents. How many of each stamp were on the letter?

10. Multiply several two-digit numbers by 99. Describe the pattern you see.

11. If the following pattern continues, what letter will be in the 243rd position?

HIPHIPHOORAYHIPHIPHOORAY

HIPHIPHOORAY . . .

12. Look at the 4×4 magic square below. Briefly list at least five observations or patterns you see in this magic square.

13	3	2	16
8	10	11	5
12	6	7	9
1	15	14	4

13. There are so many patterns in Pascal's triangle. This one is often called the hockey stick pattern because when the path is shaded in, it looks like a hockey stick. Two examples of this pattern are shown below.

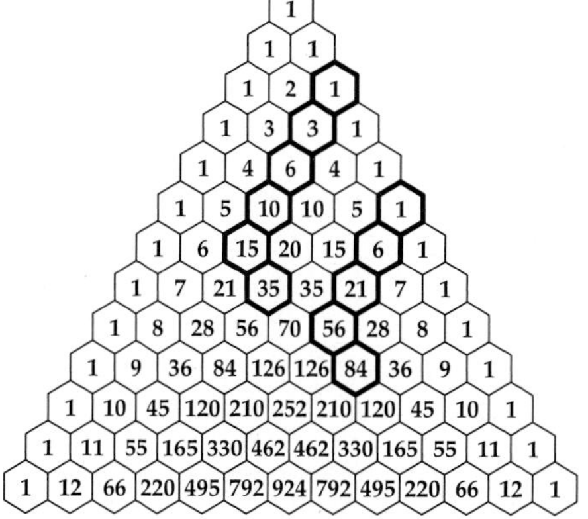

a. What relationship do you see among the numbers along the length of the stick and the number at the end of the stick?

b. Write the numbers of the hockey stick with the largest sum.

c. Write the number of the longest hockey stick.

d. Make up a new hockey stick pattern. Write down the numbers in that hockey stick.

14. Suppose you have a dart board with four concentric circles labeled 1, 2, 4, and 8. You can throw four darts, and the rule is that each dart must score.

a. How many different ways could you get a score of 14?

b. Are there any scores between 4 and 32 that are impossible?

15. Let's say you have a 4-gallon and a 9-gallon pail.

a. How can you measure exactly 1 gallon of water?

b. How can you measure exactly 3 gallons of water?

16. A caterpillar has been trapped in a glass that is 12 inches tall. The caterpillar can climb 3 inches per hour, but then it rests for an hour, during which time it slips back 1 inch. How long will it take for the caterpillar to get out?

17. Sunflower seeds are packed in packages each weighing $2\frac{1}{2}$ ounces. If there is a supply of 67 ounces of sunflower seeds, how many packages of seed can be made? How many ounces of sunflower seeds will be left over?

Fundamental Concepts

2

When we talk about a herd of deer, a bunch of carrots, an army of ants, a band of jackals, or a pack of lies, we are talking about a *set* of objects that have something in common. In mathematics, we use set language in many situations. For example, we talk about the set of even numbers or the set of common factors of 20 and 30. In Section 2.1 we will take a look at sets, visual representations of sets, and relationships between sets. We will investigate sets of numbers to lay the foundation for our following explorations with sets of numbers, beginning with whole numbers and expanding to include real numbers. Although the Common Core State Standards do not specifically include sets in a formal way, sets can be useful in organizing mathematical thinking.

The base ten *numeration system,* which most people use without thinking about it, is one of the greatest inventions in human history. A deeper understanding of our numeration system increases a person's ability to make estimates and to solve more complex problems, to understand connections between whole numbers and decimals, to understand *why* the computation procedures work, and much more. This will be the focus of Section 2.2. The Common Core State Standards have kindergarten students gaining a foundation for place value by using objects and drawings to understand that the numbers 11 to 19 are composed of one ten and a certain number of ones. Developing a deeper understanding of place value continues through the elementary school years and supports understanding the operations as well.

One of the foundations of the NCTM standards and the Common Core State Standards is the notion that mathematics ought to make sense to more people. One part of making sense of mathematics is understanding its vocabulary and symbols. Throughout this book, you will encounter many mathematical words and symbols, some of which will be familiar and some of which will be new to you. It is important for you to realize that these words (e.g., *set, intersection, proportion,* and *parallelogram*) and symbols (e.g., $\cup$, $+$, $\%$, $\neq$, $\frac{a}{b}$) were created to make it easier for people to discuss mathematical ideas. As you encounter new vocabulary and symbols, you might find it helpful to keep the following quotation in mind:

Mathematics is often considered a difficult and mysterious science, because of the numerous symbols which it employs. . . . [T]he technical terms of any profession or trade are incomprehensible to those who have never been trained to use them. But this is not because they are difficult in themselves. On the contrary they have invariably been introduced to make things easy.[1]

SECTION 2.1 Sets

What do you think?

- How do we use and apply set concepts in everyday life?
- Do we have to use circles to make Venn diagrams?

SETS AS A CLASSIFICATION TOOL

CLASSROOM CONNECTION

Learning theorists tell us that a large part of the cognitive development of young children is driven by classification, and classification involves the idea of sets. For example, certain people belong in the set called "family." Most young children divide the human race into two sets: good people and bad people. Children grapple with the defining characteristics of sets; for example, there are many different-looking objects that are called chairs.

Children use set ideas in everyday life as they look for similarities and differences between sets; for example, they want to know why lions are in the cat family and wolves are in the dog family. Children also look for similarities and differences within sets; for example, they look within a set of blocks for the blocks that they can stack and the blocks that they can't.

Whether we realize it or not, we are classifying many times each day, and our lives are shaped by classifications that we and others have made.

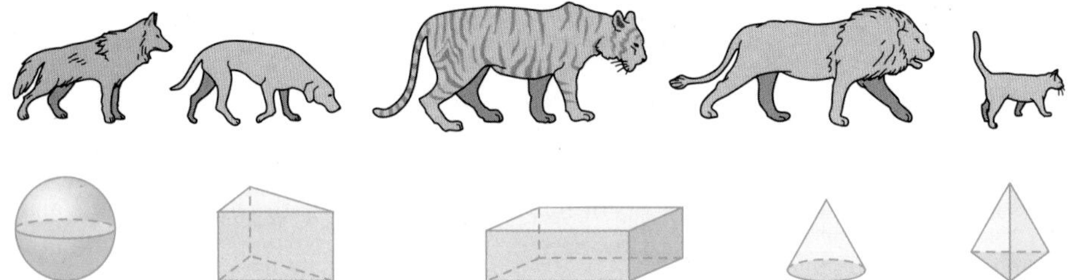

INVESTIGATION 2.1a Classifying Quadrilaterals

Without classifying various objects and ideas, it would not be possible to have mathematics. This introductory investigation will help you to connect set language and concepts to other, more concrete mathematical ideas. Look at the eight shapes below. How can you classify these shapes into two groups so that each group has a common characteristic? Give a name to each group if you can. Work and then read on. . . .

DISCUSSION

There are many possible answers to this question. Let us examine some common ones.

One answer:

Those shapes in which all sides are equal	Those shapes in which not all sides are equal

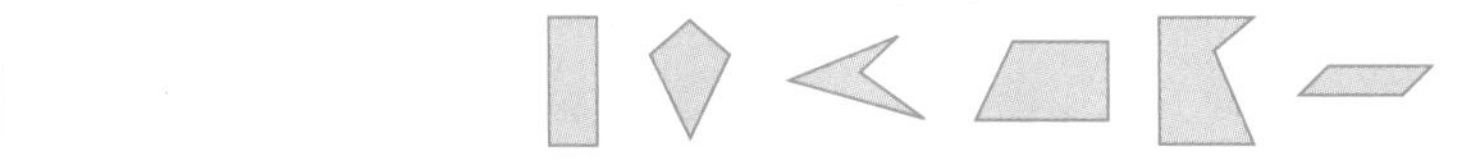

Another answer:

Those shapes with at least one right angle

Those shapes with no right angles

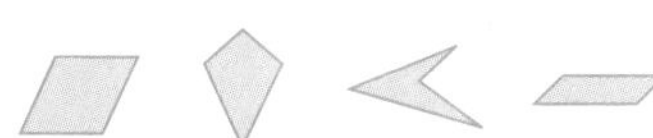

Another answer:

Those shapes in which opposite sides are equal (parallelograms)

Those shapes in which opposite sides are not equal

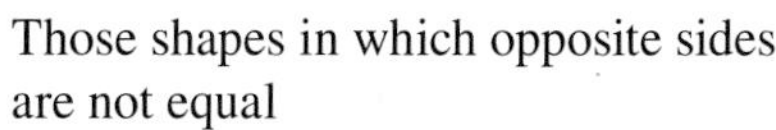

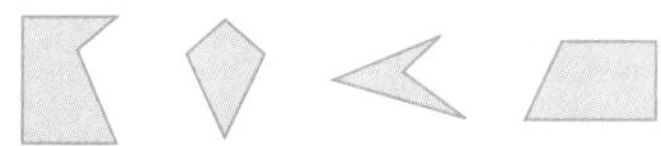

DEFINING SETS

In Investigation 2.1a, we began with a **set**, which is a collection of objects, and classified that set into smaller groups (*subsets*) having certain common features. In general, a **subset** is a set that is part of some other set. (A more precise definition will be given later.) Some subsets have names—for example, the subset of the set of shapes in Investigation 2.1a consisting of parallelograms. Other subsets are well defined but have no names; for example, there is no name for the set of geometric figures with at least one right angle.

We speak of individual objects in a given set as **members** or **elements** of the set. The symbol $\in$ means "is a member of." The symbol $\notin$ means "is not a member of." For example, if E is the set of even numbers, then $4 \in E$ but $3 \notin E$.

DESCRIBING SETS

There are three different ways to describe sets:

1. We can use words.
2. We can make a list.
3. We can use *set-builder notation.*

Words and lists In many cases, one of these representations is simpler or easier than the others. Let us examine some important mathematical sets and how we can describe them.

The first set of numbers that young children learn is called the set of **natural numbers**. We can use **words** to describe this set, or we can describe this set with a **list**.

N is the set of natural numbers or counting numbers.
$N = \{1, 2, 3, \ldots\}$

We use braces to indicate a set. The three dots are referred to as an ellipsis and are used to indicate that the established pattern continues forever.

At some point, children realize that zero is also a number, and this leads to the next set: the set of **whole numbers** (W), which we can describe with words or with a list:

W is the set of natural numbers and zero.
$W = \{0, 1, 2, 3, \ldots\}$

Later, children become aware of negative numbers, so we have the set of **integers** (I):

$I = \{\ldots -3, -2, -1, 0, 1, 2, 3, \ldots\}$

Set-builder notation Another important set is the set of **rational numbers** (Q), which we can describe in words:

Q is the set of all numbers that can be represented as the ratio of two integers as long as the denominator is not zero. Another way to say this is the set of fractions where the numerator and denominator are integers and the denominator is not zero. This includes whole numbers, since any whole number can be written in the form of a fraction—for example, $3 = \frac{3}{1} = \frac{6}{2}$, and, so on.

We cannot represent this set by making a list. Do you understand why? Try listing all the rational numbers between 0 and 1 and it may help you think about why we cannot list them all (there are infinitely many rational numbers between 0 and 1).

In this case, and in many other cases, we describe the set using set-builder notation:

$$Q = \left\{\frac{a}{b} | a \in I \text{ and } b \in I, b \neq 0\right\}$$

This statement is read in English as "Q is the set of all numbers of the form $\frac{a}{b}$ such that a and b are both integers, but b is not equal to zero."

Set-builder notation always takes the form $\{x \mid x$ has a certain property$\}$. Although elementary school students do not use this notation to describe sets, it is helpful occasionally for you to see where your students will go. In high school, when students are working with more complex ideas, this notation makes communication easier.

Let us now apply these different ways of describing sets.

INVESTIGATION 2.1b Describing Sets

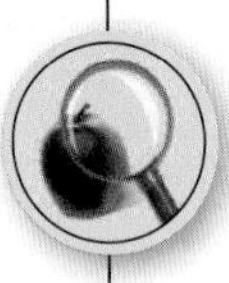

Consider the following set:

$$T = \{10, 20, 30, 40, 50, 60, 70, 80, 90, 100, 110, 120\}$$

Try to describe this set with words and with set-builder notation. What do you see as advantages and disadvantages of each of the three ways to describe this set?

DISCUSSION

Verbal description:

T is the set of all multiples of 10 that are less than or equal to 120.

Set-builder notation:

$$T = \{x | x = 10n, 1 \leq n \leq 12, n \in N\}.$$

Many students initially have difficulty with set-builder notation. Therefore, let us unpack the set-builder description.

This notation tells us that we are looking at all numbers that have the form "$10n$"—that is, multiples of 10. Because we are not talking about *all* multiples of 10, we have to let the reader know which multiples of 10 are in T. The mathematical phrase $1 \leq n \leq 12$ tells us that we are looking for multiples of 10 beginning with $10 \cdot 1$ and ending with $10 \cdot 12$. The last part of the description, "$n \in N$," simply lets us know that n must be a natural number.

Finite and infinite sets If the number of elements in a set is a whole number, that set is said to be **finite**. Some finite sets are small—for example, the set of Nobel Prize winners. Some finite sets are very large—for example, the grains of sand on all the beaches in the world. An **infinite set** has an unlimited number of members.

Consider the following infinite set. Does each way of describing the set make sense?

Verbal description	E is the set of positive even numbers.
List	$E = \{2, 4, 6, 8, \ldots\}$
Set-builder notation	$E = \{x \mid x = 2n, n \in N\}$

DEFINING SUBSETS

The notion of subset is critical because many practical applications of sets involve subsets. What do you think a subset is? Write your current definition of a subset and then read on. . . .

These examples may help you develop a definition:

The set of even numbers is a subset of the set of whole numbers.
The set {1, 2, 3} is a subset of {1, 2, 3, 4, 5}.

One informal definition of a subset is that a set is a subset of another set if all of the elements in the subset are also in the set. We can say this more formally as:

A set X is a **subset** of a set Y if and only if every member of X is also a member of Y.

MATHEMATICS

$\subset$ and $\subseteq$ are like $<$ and $\leq$.

The symbol $\subseteq$ means "is a subset of." Thus we say $X \subseteq Y$. On the other hand, if a set X is not a subset of a set Y, we say $X \not\subseteq Y$.

There is another symbol that we can use when talking about subsets. This symbol ($\subset$) is used when we want to emphasize that the subset is a **proper subset**. A subset X is a proper subset of set Y if, and only if, the two sets are not equal *and* every member of X is also a member of Y. In the case of finite sets, this means that the proper subset has fewer elements than the given set.

OUTSIDE THE CLASSROOM

Databases are closely connected to the idea of sets. In my office, I have a collection of articles on teaching mathematics. Let's say one article is titled "Using Cooperative Learning to Teach Fractions." Physically, I have to store that article in one file folder—"Cooperative Learning," "Fractions," or the folder "Elementary Math Methods." This presents a problem, because I wind up losing articles. I could make copies and have a copy in each folder, but that is costly and wasteful. With a database program, I can enter the article in a file and use the following descriptors: cooperative learning, fractions, and elementary math methods. When I want to review my resources on cooperative learning, the program can tell me where the article is physically stored.

INVESTIGATION 2.1c How Many Subsets?

This investigation opens the idea of families of subsets, shows how subsets might be useful, and provides an opportunity to develop problem-solving tools.

Let's say that you and your friends decide to go out and get a large pizza. Let T represent the set of toppings that this restaurant offers:

$$T = \{\text{onions, sausage, mushrooms, peppers}\}$$

List all the possible different combinations of pizza that you could order, such as a mushroom and onion pizza. Then read on. . . .

DISCUSSION

The text will contain occasional reminders about the process of problem solving. In this investigation, we focus on understanding the problem and looking back. Before you made your list, did you make sure you understood the question? After you make your list and before reading on, take a few moments to look back: How can you check your solution to make sure you didn't miss any combinations? . . .

STRATEGY 1 List all the combinations systematically

Begin with all combinations that involve onion, from the simplest to the most complex, and then do the same for each of the other toppings (see Table 2.1).

There are many patterns within Table 2.1. How many can you see?

TABLE 2.1

Onion	Sausage	Mushroom	Pepper
o	s	m	p
o, s	s, m	m, p	
o, m	s, p		
o, p	s, m, p		
o, s, m			
o, s, p			
o, m, p			
o, s, m, p			

STRATEGY 2 List all the combinations using another system
Begin with all the ways to have one topping, then all the ways to have two toppings, etc. (Table 2.2).

There are many patterns within Table 2.2. How many can you see?

TABLE 2.2

One of the four	Two of the four	Three of the four	All of the choices
o	o, s	o, s, m	o, s, m, p
s	o, m	o, s, p	
m	o, p	o, m, p	
p	s, m	s, m, p	
	s, p		
	m, p		

One of the mistakes that students commonly make in such problems is to overlook a combination. There are patterns in both of the tables above that could help you to find all the combinations. Write down all the patterns you see and then read on. . . .

One pattern (in the "Three of the four" column) is that all of the four toppings are equally represented. Let's say you had somehow omitted the last combination: s, m, p. Looking over the three combinations you had found, you would notice that onion occurred *three* times but that sausage, mushroom, and pepper each occurred only *two* times. This observation itself names the combination you had missed: sausage, mushroom, and pepper.

Mathematically, there are 16 possible subsets of a set containing four elements. However, if you look at either of the tables above, there are only 15 subsets. What is the missing subset?

THE EMPTY SET

The missing subset is plain pizza. How would we represent this subset using set notation?

One way is to put nothing inside the brackets { }. We also use the following symbol to represent a set that is empty: $\varnothing$. We use the terms **empty set** and **null set** interchangeably to mean the set with no elements.

This pizza problem also illustrates two mathematical statements that often baffle students and that will be left as exercises:

- Every set is a subset of itself.
- The empty set is a subset of every set.

EQUAL AND EQUIVALENT SETS

In many situations, the relationship between two sets is important.

Two sets are said to be **equal** if they contain the same elements. For example, $\{1, 2, 3, 4, 5\} = \{5, 4, 3, 2, 1\}$.

Two sets are **equivalent** if they have the same number of elements. More precisely, two sets are equivalent if their elements can be placed in a **one-to-one correspondence**. In such a correspondence, an element of either set is paired with exactly one element in the other set. We use the symbol $\sim$ to designate set equivalence. For example, {United States, Canada, Mexico} $\sim \{1, 2, 3\}$.

HISTORY

Earlier in this century, anthropologists discovered a tribe in New Guinea that did not have numbers for large amounts. For example, tribe members had no way of counting how many warriors went off to war. How do you think the tribe was able to determine how many warriors did not return? The solution involved one-to-one correspondence. Before leaving, each warrior would place a large stone in the center of the village. When he returned, he would take one stone out of the pile. The number of stones remaining told the village how many warriors had not returned.

Equivalence and counting One reason for defining sets in terms of a one-to-one correspondence has to do with the difficulty young children have counting objects accurately. Initially, they do not realize that each number has to match one and only one object, and so they can count a set several times and arrive at a different number each time! At some point, they realize (and you can see it in their pointing) that there must be a one-to-one correspondence between their words and the physical objects. Thus, although our definition of equivalence appears unnecessarily formal to many students, it reflects what children actually go through in learning to count objects accurately.

VENN DIAGRAMS

One way to represent sets is to use **Venn diagrams**, which are named after John Venn, the Englishman who invented these diagrams to illustrate ideas in logic. You have probably seen Venn diagrams used in other contexts, and we used them in Chapter 1.

An elementary teacher explained how she had used the Venn diagram in Figure 2.1 to help her students understand the similarities and differences between butterflies and moths. In this Venn diagram, one region represents the set of moths' characteristics, another region represents the set of butterflies' characteristics, and the overlapping region represents the set of characteristics common to both.

The teacher was using a teaching strategy called concept attainment. You will see the strategy in this book when you are asked how one concept is similar to and different from another, for example, mean and median, square and rectangle.

OPERATIONS ON SETS

MATHEMATICS

In Chapter 3, we will examine the mathematical operations of addition, subtraction, multiplication, and division.

We will use Venn diagrams as we examine three operations on sets: intersection, union, and complement. Just as a doctor's operation consists of something the doctor does to a patient, a mathematical operation consists of something we do to a set of objects.

When we perform operations on sets of objects, it is often useful to refer to the set that consists of all the elements being considered as the **universal set**, or the **universe**, and to symbolize it as U. We represent U in the Venn diagram with a rectangle.

In the following discussions, we will let U be the set of students in a small class:

$$U = \{\text{Amy, Uri, Tia, Eli, Pam, Sue, Tom, Riki}\}$$

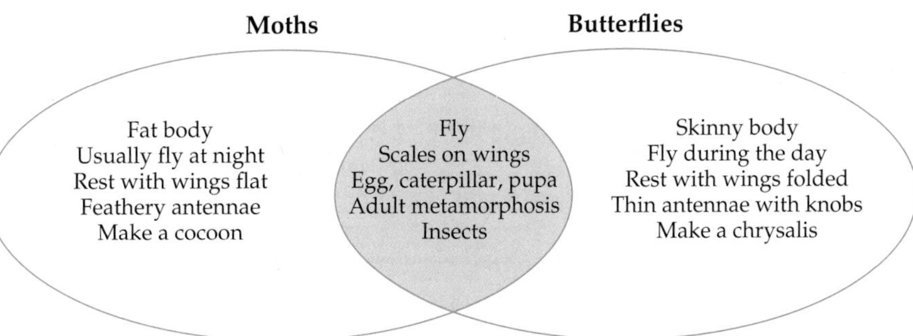

Figure 2.1

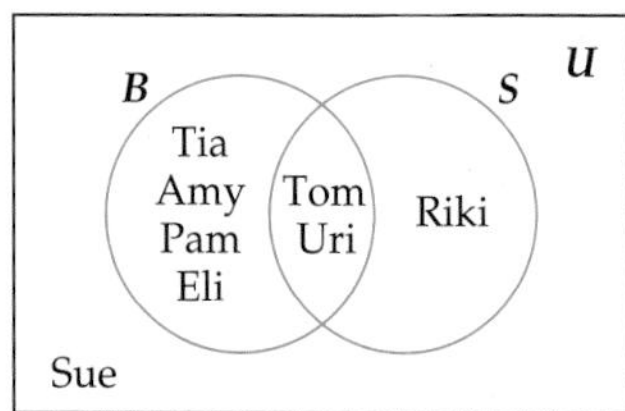

Figure 2.2

We begin with two subsets of U:

$B = \{\text{students who have at least one brother}\}$

$S = \{\text{students who have at least one sister}\}$

Figure 2.2 represents the students in this hypothetical class.

If you were in the class, where would your name appear? . . . 

Intersection We can group this class of eight students into various subsets. How would you describe the subset consisting of Tom and Uri? Think and then read on. . . .

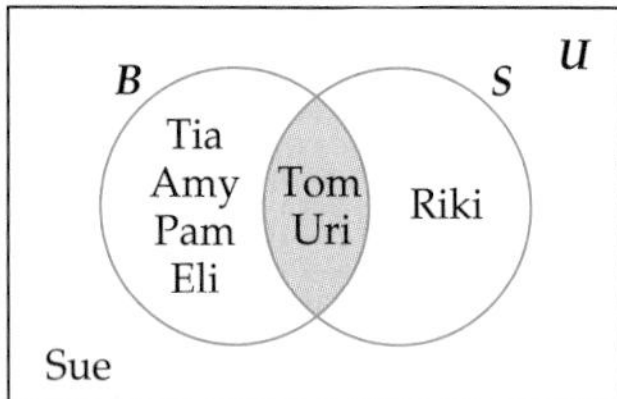

Figure 2.3

One way to describe this subset is "Those students who have at least one brother and at least one sister."

Mathematically, we call this subset the *intersection* of sets B and S. In mathematical language, we say that the **intersection** of two sets B and S consists of the set of all elements common to both B and S.

We represent the intersection of B and S by shading it (Figure 2.3).

Using set-builder notation, we write

$$B \cap S = \{x \mid x \in B \text{ and } x \in S\}$$

The symbol $\cap$ is used to denote "intersection."

In Figure 2.1, the characteristics common to both moths and butterflies are listed in the intersection area. Connecting the concept of intersection to previous notation, we can say Tom $\in B \cap S$, and we can say $(B \cap S) \subset U$. In other words, Tom is an element of the intersection of sets B and S (because Tom is in both sets) and also that the intersection of sets B and S is a proper subset of the set U.

OUTSIDE THE CLASSROOM

Can you see the connection between a mathematical intersection and a highway intersection?

Union Let us examine another subset of the class: {Tia, Amy, Pam, Eli, Tom, Uri, Riki}.

How would you describe this subset in everyday English? Think and then read on. . . .

There are actually several ways to describe this subset:

- Those students who have at least one brother or sister.
- Those students who have at least one sibling.
- Those students who are not an only child.

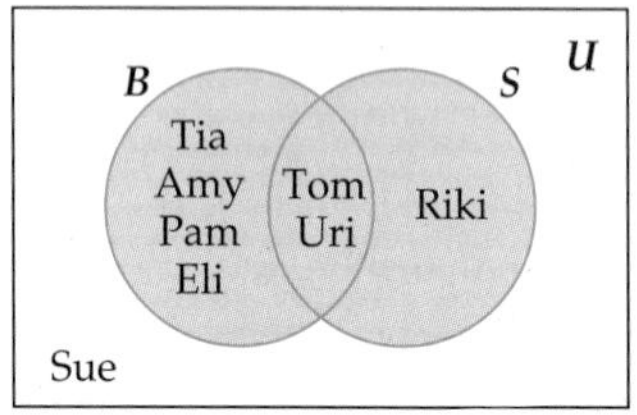

Figure 2.4

Mathematically, we describe this subset as the *union* of sets B and S. In mathematical language, we say that the **union** of two sets B and S consists of the set of all elements that are in set B *or* in set S *or* in both sets B and S.

We represent the union of B and S by shading it (Figure 2.4).

Symbolically, we write

$$B \cup S = \{x \mid x \in B \text{ and/or } x \in S\}$$

The symbol $\cup$ is used to denote "union."

Connecting the concept of union to previous notation, we can say Tom $\in B \cup S$, and we can also say $B \subset (B \cup S)$. In other words, Tom is an element of the union of sets B and S and also the set B is a proper subset of the union of sets B and S.

Complement Let us examine another subset of the class. Look back to Figure 2.2 and consider this subset of the class: {Tia, Amy, Pam, Eli, Sue}. How would you describe this subset in everyday English? Think and then read on. . . .

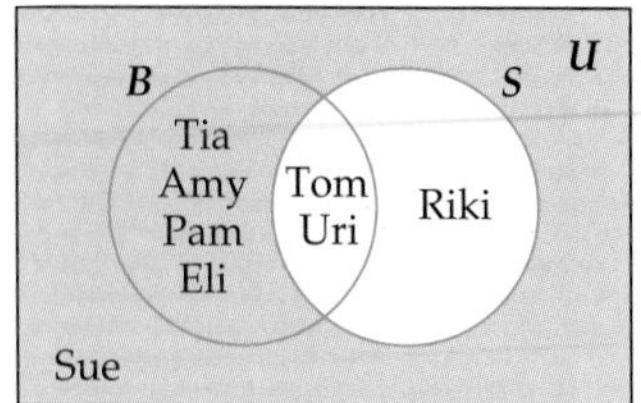

Figure 2.5

One way to describe this subset is "Those students who have no sisters."

We describe this subset as the *complement* of set S. In mathematical language, the **complement** of set S consists of the set of all elements in U that are *not* in S.

We represent the complement of S by shading it (Figure 2.5).

In symbols, we write $\overline{S} = \{x \mid x \notin S\}$. We represent the complement of a set by placing a line over the set's letter.

Some people understand complement better if they think of the complement of S as "not S"—that is, all elements that are not in set S. As another example, if we say the universal set is the set of whole numbers, then the complement of the set of even numbers is the set of odd numbers.

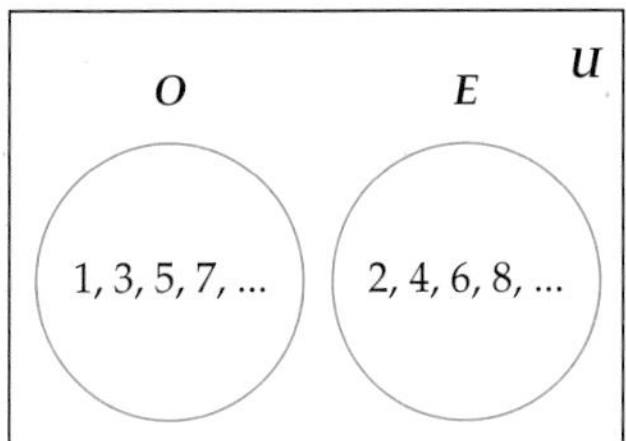

Figure 2.6

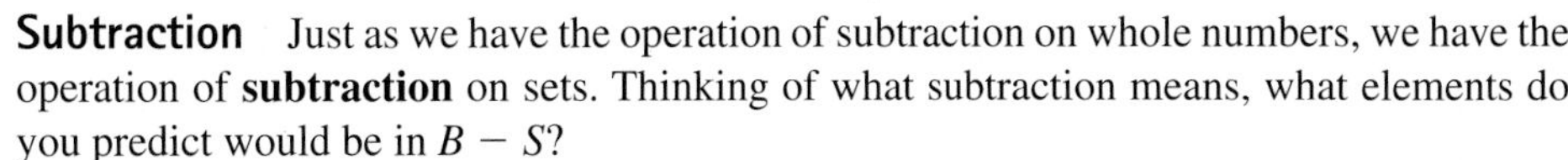

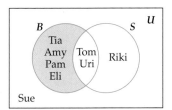

Subtraction Just as we have the operation of subtraction on whole numbers, we have the operation of **subtraction** on sets. Thinking of what subtraction means, what elements do you predict would be in $B - S$?

Verbally, we define set difference as the set of all elements that are in B that are not in S.

We represent $B - S$ by shading it as shown at the left in Figure 2.6.

Symbolically, we write

$$B - S = \{x \mid x \in B \text{ and } x \notin S\}$$

VENN DIAGRAMS AND RELATIONSHIPS BETWEEN SETS

When we are considering two sets, there are three ways in which they might be related.

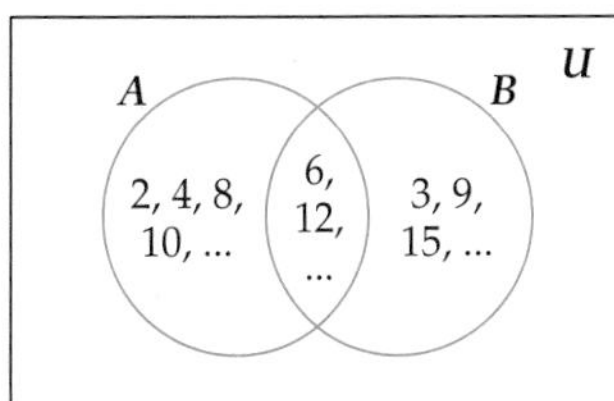

Figure 2.7

(1) They can have nothing in common. In this case, we call them **disjoint sets**. Figure 2.7 illustrates this.

The set of odd and even numbers are disjoint.

$O = \{1, 3, 5, 7, \ldots\}$

$E = \{0, 2, 4, 6, \ldots\}$

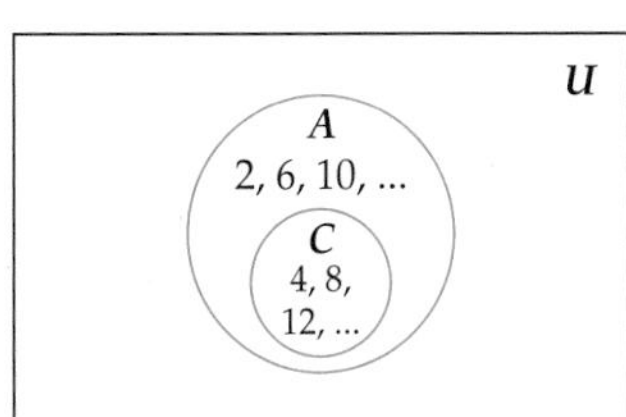

Figure 2.8

(2) They can have some elements in common (Figure 2.8).

Multiples of 2 and 3 have some elements in common.

$A = \{2, 4, 6, 8, 10, 12, \ldots\}$

$B = \{3, 6, 9, 12, 15, \ldots\}$

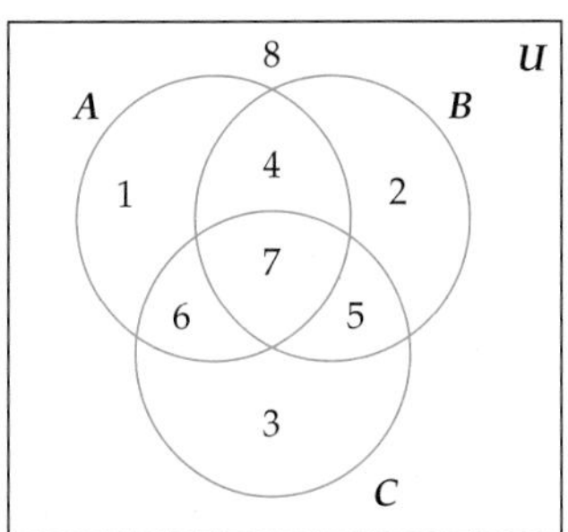

Figure 2.9

(3) One set can be a subset of the other (Figure 2.9).

The multiples of 4 are a subset of the multiples of 2.

$A = \{2, 4, 6, 8, 10, 12, \ldots\}$

$C = \{4, 8, 12, \ldots\}$

When we are considering three sets, there are eight possible regions where elements of the sets could lie. These regions are described in words below and are numbered in Figure 2.10.

Figure 2.10

Elements in region 1 are those elements that are in set A only.

Elements in region 2 are those elements that are in set B only.

Elements in region 3 are those elements that are in set C only.

Elements in region 4 are those elements that are in sets A and B but not C.

Elements in region 5 are those elements that are in sets B and C but not A.

Elements in region 6 are those elements that are in sets A and C but not B.

Elements in region 7 are those elements that are in sets A, B, and C.

Elements in region 8 are those elements that are in none of the 3 sets.

INVESTIGATION 2.1d Translating Among Representations

Consider the following situation:

U = the set of natural numbers from 1 to 30

A = the set of multiples of 2

B = the set of multiples of 3

C = the set of multiples of 5

First, draw a Venn diagram showing these sets and their relationships.
Then, translate the following situations into notation:

A. The numbers that are multiples of 2 and 3

B. The numbers that are multiples of 2 and 3 but not 5

C. The numbers that are only multiples of 2, 3, and 5 

DISCUSSION

The Venn diagram would look like the one in Figure 2.11.

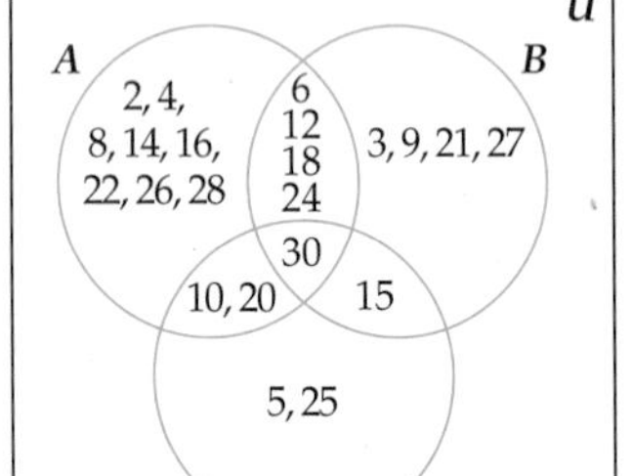

Figure 2.11

A. The numbers that are multiples of 2 and 3 can be represented as $A \cap B$. Here we are looking for what these two sets have in common.

B. The numbers that are multiples of 2 and 3 but not 5 can be represented as $(A \cap B) - C$. Here, we want to take away from $(A \cap B)$ those elements of set C.

C. The numbers that are only multiples of 2, 3, and 5 can be represented as $A \cap B \cap C$. Here, we are looking for numbers that are common to all three sets.

INVESTIGATION 2.1e Finding Information from Venn Diagrams

The owners of Marksalot (a company that makes all sorts of markers) believe that healthy, happy employees are productive employees, so they had a consultant conduct an anonymous survey about employees' exercise habits, their smoking habits, and one particular aspect of their eating habits. Figure 2.12 represents the survey results.

U = the employees of Marksalot

E = employees who exercise regularly

S = employees who do not smoke at all

F = employees who average at least five servings of fruits and vegetables each day

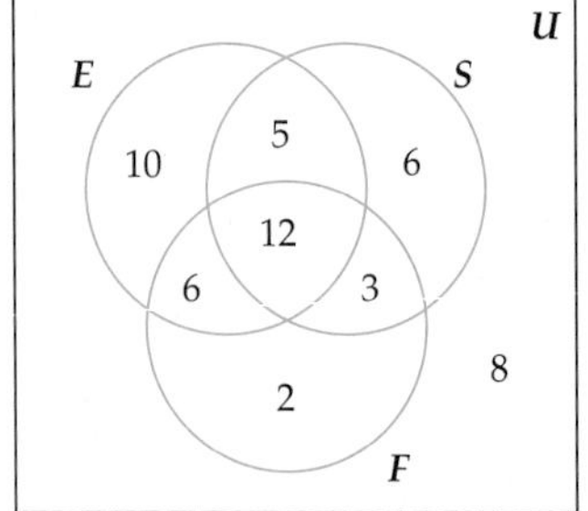

Figure 2.12

Answer the four questions below on your own and then check your answers. . . .

A. How many employees exercise regularly but don't average five daily servings of fruits and vegetables? Represent that subset visually and with symbols.

B. Describe the 8 employees who are outside all three circles (Figure 2.13), first in everyday English and then with symbols.

C. Describe the following subset both in everyday English and visually:

$(E \cap F) \cup (E \cap S)$

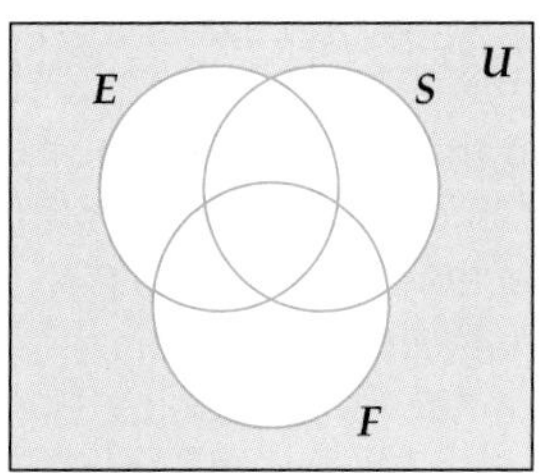

Figure 2.13

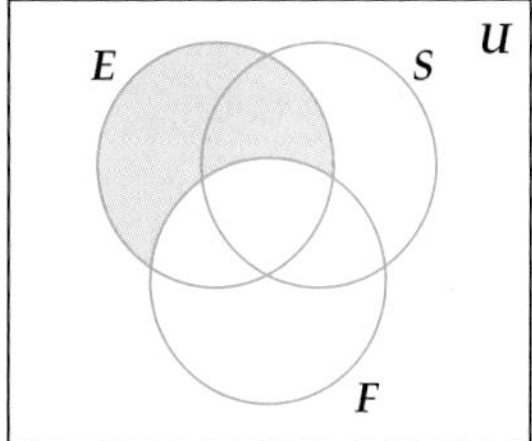

Figure 2.14

DISCUSSION

A. We find that we want to look at the regions that are in E but not in F. There are 15 persons who match the subset "E but not F"—that is, $E \cap \overline{F}$. Visually, we have Figure 2.14.

B. In everyday English, these people do not exercise regularly, they do smoke, and they average less than five servings of fruits and vegetables a day.

In symbols, this set is $\overline{E \cup S \cup F}$. This subset (Figure 2.13) will probably require a disproportionate amount of medical services over their lifetimes!

C. Many students find themselves intimidated by the apparent complexity of this problem.

Many mathematical operations, such as multiplication, are **binary operations**—operations that can be performed on only two elements at a time. If you had trouble with Question C, can you solve it now?

Applying the idea of binary operations, we first find $(E \cap F)$ and $(E \cap S)$. Then we find the union of those two sets (Figure 2.15).

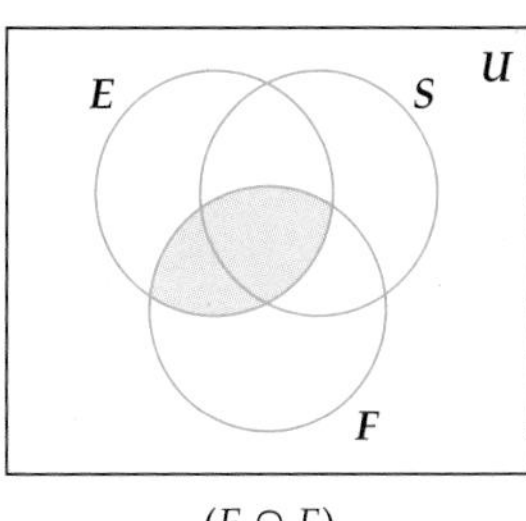

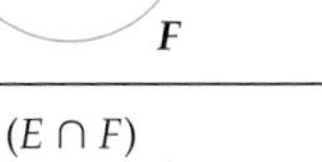
$(E \cap F)$

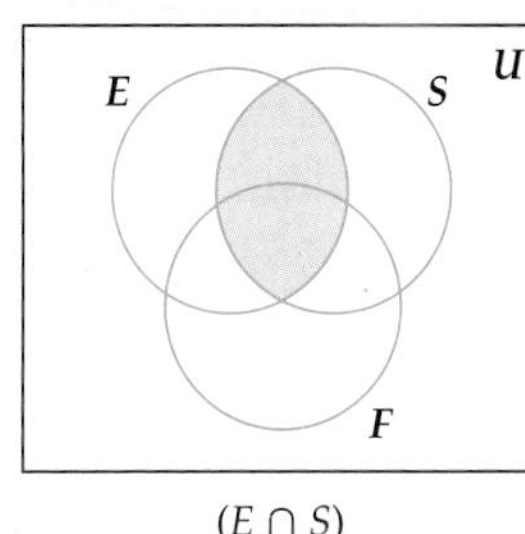

$(E \cap S)$

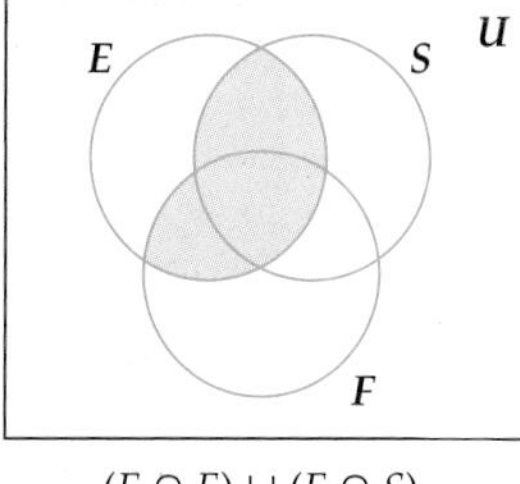

$(E \cap F) \cup (E \cap S)$

Figure 2.15

For some, this type of problem is made easier by numbering each of the eight regions as we did earlier in Figure 2.10. Let's number the regions and translate the expression by replacing the set with the region numbers in each set (circle).

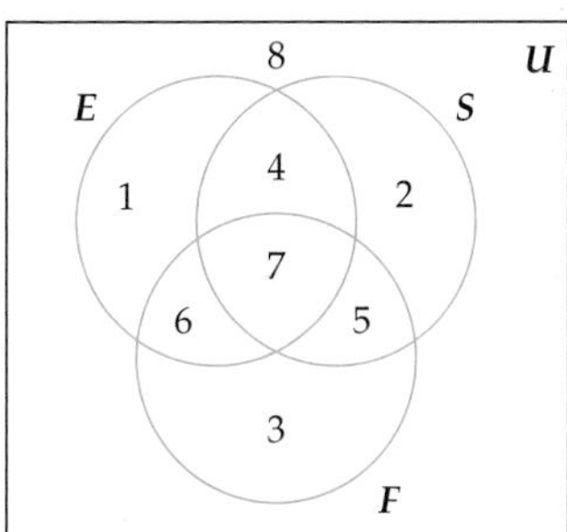

First, replace the letters with the numbers in that set. For example, E will become $\{1, 4, 6, 7\}$ since those are the four regions inside the E circle.

$$E \cap F \cup E \cap S = (\{1, 4, 6, 7\} \cap \{3, 5, 6, 7\}) \cup (\{1, 4, 6, 7\} \cap \{2, 4, 5, 7\})$$

Next, do what is inside the parentheses and we will get

$$\{6,7\} \cup \{4,7\}$$

Then the union of these two sets is $\{4, 6, 7\}$, so these regions are the ones to shade.

One description of this set in everyday English is "Those employees who exercise regularly and who don't smoke and/or who average at least 5 servings of fruits and vegetables each day." Some readers find this description to be awkward. Using the word *subset*, we can describe this group more clearly: We are talking about that subset of regular exercisers who also don't smoke and/or who eat at least 5 servings of fruits and vegetables each day.

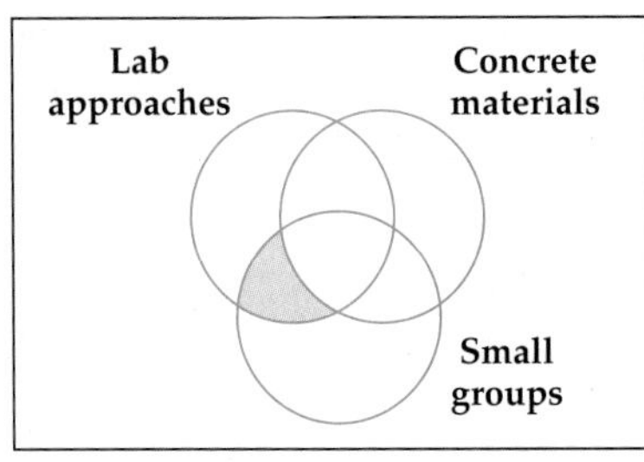

Figure 2.16
Source: Mathematics Resource Project, *Mathematics in Science and Society* (Palo Alto, CA: Creative Publications, 1977).

VENN DIAGRAMS AS A COMMUNICATION TOOL

People other than mathematicians often use Venn diagrams in the same sense that "a picture is worth a thousand words." In the introduction to one of the chapters in *Mathematics in Science and Society*, the authors state: "This section is devoted to three valuable teacher tools: laboratory approaches, concrete materials and small groups of students. All three of these may—and quite often do—appear in the same lesson." The authors use the Venn diagram in Figure 2.16 to illustrate the kind of lesson they would be discussing. Describe, in words, the kind of lesson they were talking about. Then read on. . . .

They would be describing a lesson in which the students would be doing a math lab in small groups but would not be using concrete materials.

SUMMARY 2.1

In this section, we have investigated those concepts from set theory that are most likely to occur in elementary mathematics. Although you are not likely to use these ideas in their formal sense, you will find that you encounter sets and subsets both in elementary school and in everyday life. Most sets in everyday life are finite, but many mathematical sets are infinite. The notion of equivalent sets is one that early childhood educators see young children grapple with as they seek to make sense of relationships. Venn diagrams are used both in elementary school and in other places to describe relationships between various sets. These relationships involve the intersection, union, and complements of sets.

2.1 Exercises

1. Rewrite the following statements using mathematical symbols:
   a. 0 is not an element of the null set.
   b. 3 is not an element of set B.
2. Rewrite the following statements using mathematical symbols:
   a. The set D is not a subset of the set E.
   b. The set A is a subset of the set U.
3. Sets can be described in three ways: verbally, by listing the elements, or with set-builder notation. In each of the statements below, a set has been described verbally. Either describe the set in the other two ways or explain why it would be impossible to do so. Then explain which of the three descriptions you think would be most useful and why.
   a. The set of letters in the word *elementary*.
   b. The set of countries in Europe.
   c. The set of prime numbers less than 100.
   d. The set of fractions between 0 and 1.
   e. The set of students in your class.
4. Let U be the set of all colors and S be the subset {red, orange, yellow, green, blue, violet}.
   For the following questions, place the appropriate symbol in the blank: $\subset$, $\not\subset$, $\in$, $\notin$
   a. S __ U b. red __ U
   c. {magenta} __ U d. {green, blue} __ S
   Which of the following are true? Briefly explain your answer.
   e. $S \subseteq U$ f. red $\subseteq U$
   g. gray $\in S$ h. {green, blue} $\subseteq S$
5. Fill in the most appropriate symbol for each of the following: $\subset$, $\not\subset$, $\in$, $\notin$. *Briefly* justify your choice.
   a. 3 __ {1, 2, 3} b. {3} __ {1, 2, 3}
   c. {1} __ {{1}, {2}, {3}} d. $\{a\}$ __ $\{a, b, c\}$
   e. $\{ab\}$ __ $\{a, b, c, d\}$ f. { } __ {1, 2, 3}
6. a. How many subsets does $A = \{p, i, c, k, l, e\}$ have?
   b. Can you make a generalization about the relationship between the number of elements in a set and the number of subsets? *Hint:* Look at Investigation 2.1c.

DEEPENING YOUR UNDERSTANDING

7. Let $U = \{x \mid x \text{ is an American}\}$
   $F = \{x \mid x \text{ is a female}\}$
   $S = \{x \mid x \text{ is a smoker}\}$
   $P = \{x \mid x \text{ has a health problem}\}$
   a. Represent the following description with a Venn diagram and with symbols: an American nonsmoking healthy female.
   b. Represent $F \cap (S \cup P)$ with a Venn diagram and in everyday English.
   c. Represent $(F \cup P) \cap \overline{S}$ with a Venn diagram and in everyday English.
   d. Convert the information from the first diagram below into everyday English and into symbols.
   e. Convert the information from the second diagram below into everyday English and into symbols.

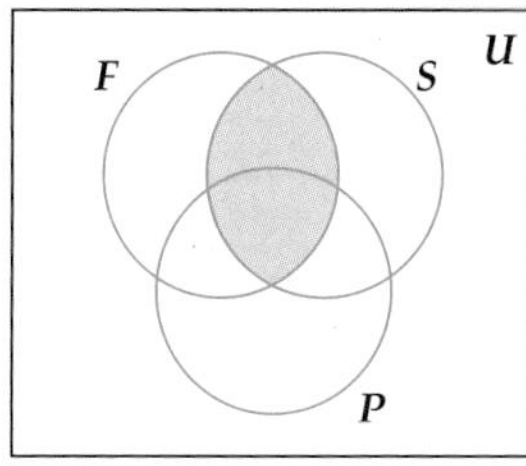

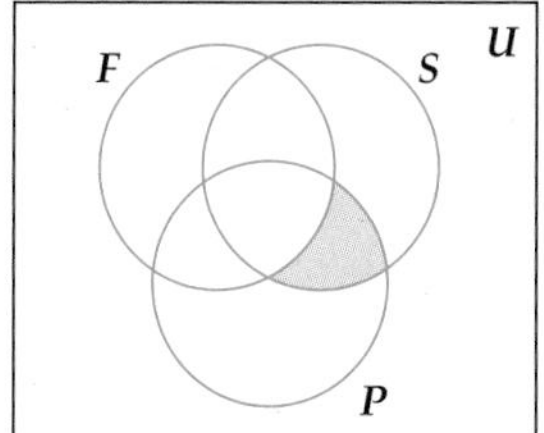

8. In the following Venn diagrams,
   F = the set of students in the film club
   S = the set of students in the science club
   C = the set of students in the computer club
   Describe the following sets in English and in symbols.

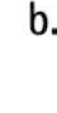
a.

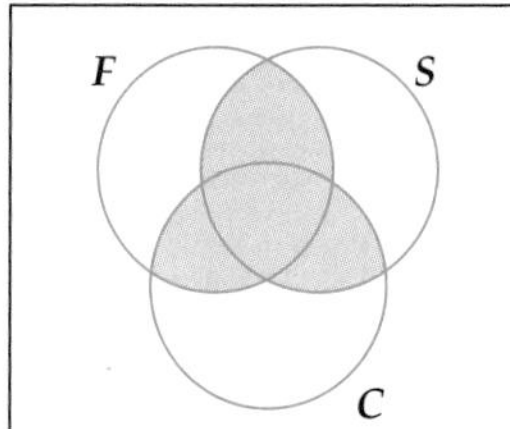

b.

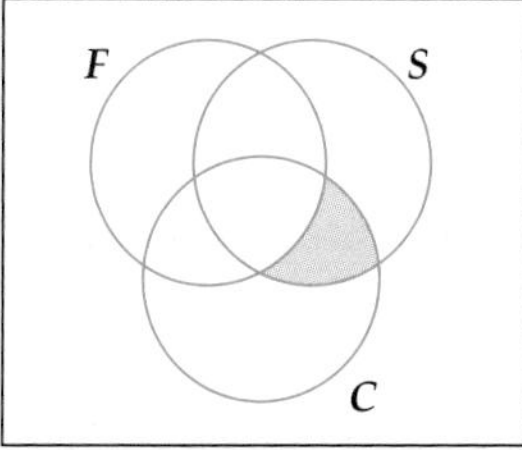

Describe the following subsets in symbolic language and with a Venn diagram.

   c. Those people who are in the science and film clubs but not the computer club
   d. Those people who are in none of the three clubs

9. Use the sets below to answer parts (a) through (f):
   $U = \{1, 2, 3, 4, 5, 6, 7, 8, 9, 10, 11, 12, 13, 14, 15, 16, 17, 18, 19, 20\}$
   A: the numbers in U that divide 12 with no remainder
   B: the numbers in U that divide 15 with no remainder
   C: the numbers in U that divide 20 with no remainder
   a. Make a *clear* Venn diagram showing the sets U, A, B, and C.
   b. Represent the following subset with symbols: $\{7, 8, 9, 11, 13, 14, 16, 17, 18, 19\}$.
   c. Represent the following subset in everyday English and with symbols: $\{2, 4\}$.
   d. Represent the following subset in everyday English and with a diagram: $\overline{A \cap B}$.
   e. Represent the following subset in everyday English and with a diagram: $\overline{A \cup B}$.
   f. Represent the following subset with a diagram and with symbols: Those numbers that divide 12 or 15 with no remainder.

10. An elementary teacher has asked her students to place their names in the region that represents their answers to the question "Which pets live in your home?"
    D = the set of students who have at least one dog
    C = the set of students who have at least one cat
    O = the set of students who have a pet that is not a dog or a cat

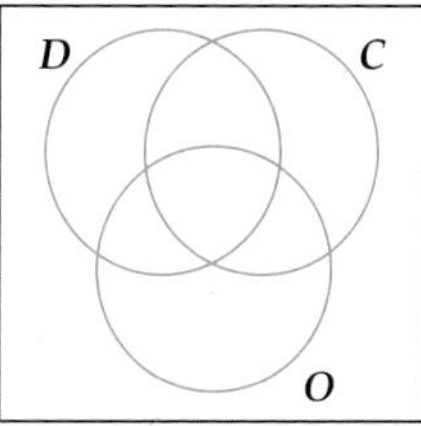

Describe the following sets in everyday English and then with a diagram:

a. $C \cap D$ b. $\overline{D \cup C}$ c. $C \cap O \cap D$

Describe the following sets in symbols and then with a diagram:

d. Students who have dogs but not cats
e. Students who have at least one pet
f. Students who have other pets but neither cats nor dogs

Describe the following sets in everyday English and then with symbols:

g. 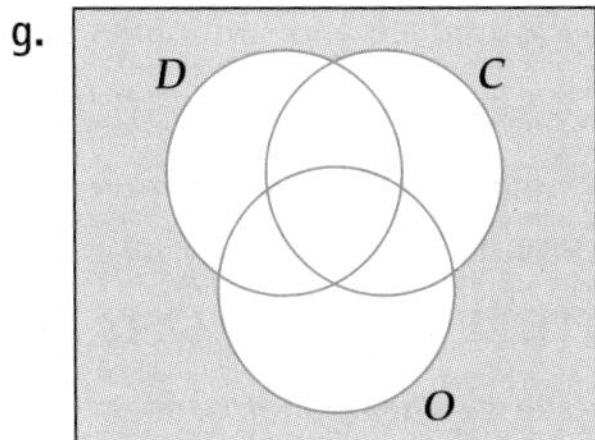

h. 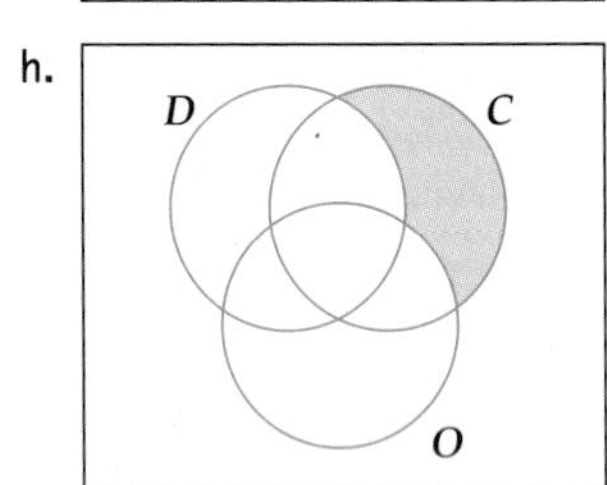

11. Can you apply what you learned in Investigation 2.1c to answer the following question? There are 6 members on the Student Council. A committee consisting of 2 members is to be made. How many different committees are possible?

12. Justify the following statements, as though you were talking to a student who had not read this section.

a. Every set is a subset of itself.

b. The empty set is a subset of every set.

13. Make a Venn diagram with four circles. Use a compass or another device to make good circles. How many distinct regions are there in this diagram? What patterns do you notice in this diagram?

14. Find a Venn diagram from a newspaper, magazine, or elementary school mathematics book. Describe the Venn diagram in words.

15. Why do you think we use circles instead of squares or triangles or other shapes when we make Venn diagrams? For example, is the representation below equivalent to the standard Venn diagram with three overlapping circles? Explain your answer as though you were talking to a fellow student who does not understand.

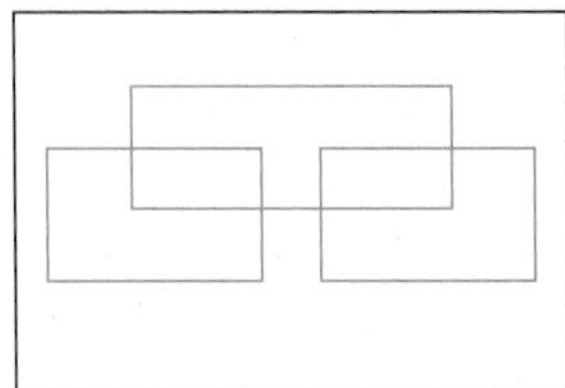

16. Recall the different kinds of sets of numbers described earlier: natural numbers, whole numbers, integers, and rational numbers. Make a Venn diagram to represent how these sets are related.

17. Pollsters often ask people's opinions. Politicians want to know how their position on an issue is viewed by particular constituencies—for example, by young voters, by African Americans, by women, by those who belong to the Sierra Club. Decisions about policies are often made on the basis of this polling information. Let's say that a public opinion survey was conducted to determine how much support there was for the president's policies. People were asked three questions:

Do you support the president's economic policy?

Do you support the president's foreign policy?

Do you support the president's social policy?

Let E, F, and S denote the sets of persons responding yes to the first, second, and third questions, respectively. The results of the survey are shown in the Venn diagram below. The numbers represent the percent of respondents.

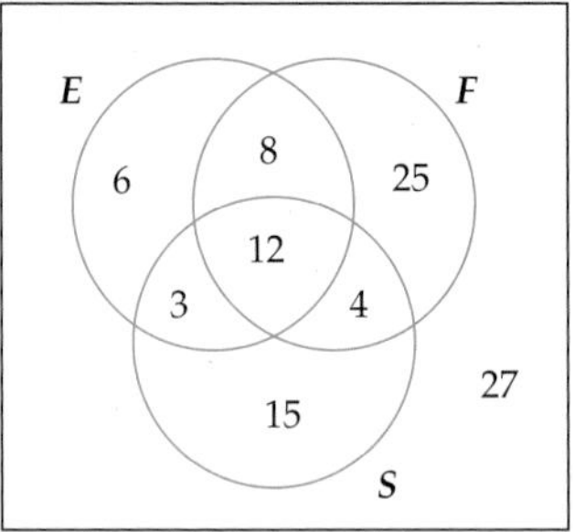

a. What percent agree with his economic policy?

b. What percent agree with just one of his policies?

c. Describe the subset that would be represented by $F \cup (E \cap S)$, either in everyday English or by shading the appropriate portion of the diagram.

d. If the president could make one single region of the Venn diagram larger (that is, make it have more members), which would it be? Why?

18. In a group of 120 students, 75 use craigslist, 65 use ebay, and 20 use neither.

a. How many students know how to use both websites?

b. Are the sets in this problem well defined or not? Justify your response.

19. An advertising firm found that a certain ad that ran on both radio and TV was only heard on the radio by 21 percent of the people and was only seen on TV by 33 percent of the people. Just 10 percent of the population both heard the ad on the radio and saw it on TV.

a. What percent of the people in the area has neither seen nor heard the ad?

b. What percent of the people in the area only heard the ad on radio or only saw the ad on TV?

20. Refer to the discussion about the book *Mathematics in Science and Society* in the last subsection.

a. What would be the focus of the authors' discussion following the Venn diagram below? Describe your answer in everyday English.

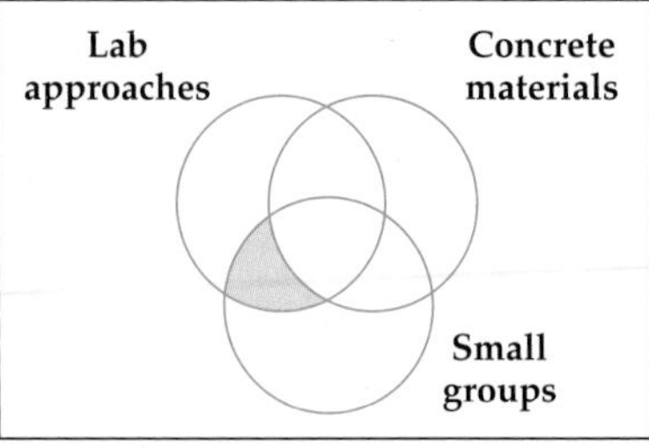

b. The following Venn diagram appeared in the same book.

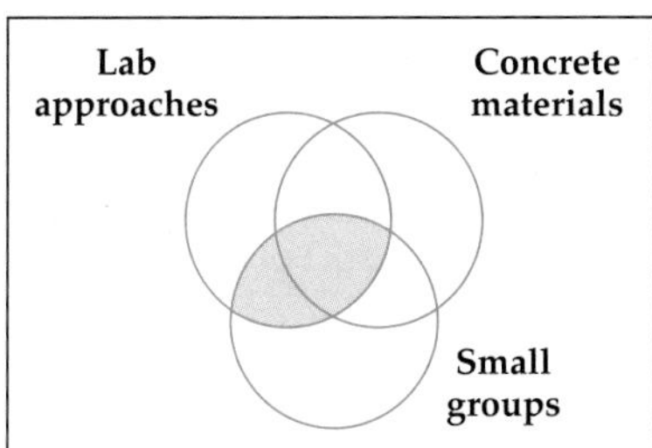

21. Make up a situation for which the following Venn diagram is appropriate.

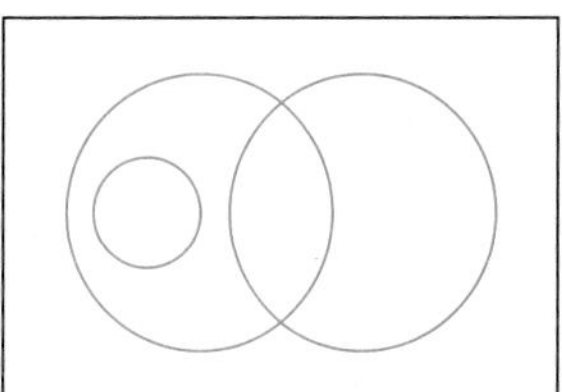

22. Consider the following two subsets of U, the set of all people:

I = the set of intelligent people

S = the set of successful people

a. By yourself, answer the following question: How would you represent these two subsets with a Venn diagram?

b. Discuss your responses with other classmates. What did you learn from this activity? How did the Venn diagram contribute to this learning?

23. Many Internet math activities are on the site www.shodor.org/interactivate/activities/. There are three there with Venn diagrams. When you go to this address, type in "Venn" in the search tool. The three are called "Triple Venn Diagram Shape Sorter," "Venn Diagram Shape Sorter," and "Venn Diagrams." Explore each of these. There is a "help" tab at the top of each activity if you need help figuring them out. Describe each activity and what each one of them helped you to understand more deeply, or at least practice more.

SECTION 2.2 Numeration

What do you think?

- Why and when did humans invent numbers?
- Why do many mathematicians regard the invention of zero as one of the most important developments in the entire history of mathematics?

Explorations Manual 2.3 and 2.4

Did you know that the base ten numeration system you use every day is called the Hindu-Arabic numeration system because it was invented by the Hindus and transmitted to the West by the Arabs? Did you know that this system has been in widespread use in the West for only 500 years?

INVESTIGATION 2.2a What Does "Place Value" Mean?

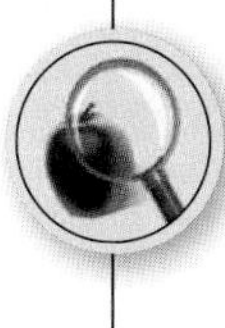

A. Use the numeral 323 to explain what "place value" means and what it means to say we have a "base ten" place value system.

B. What if we had a base five system instead of a base ten system? What would the numeral 323 mean then?

DISCUSSION

We will go deeply into these concepts as we move through this section and the next chapter. This is to get us to start thinking about these concepts.

A. Our system is a base ten system because we group in tens. The value of a symbol is determined by its place. In the number 323 we have the symbol "3" in two different places. The "3" on the left represents 300 while the "3" on the right represents 3 ones.

B. If we used a base five system instead of a base ten system, then we would group in fives instead of tens. We will explore this much more deeply later.

ORIGINS OF NUMBERS AND COUNTING

Did you know that people had to invent counting?

The earliest systems of counting must have been quite simple, probably tallies. The oldest archaeological evidence of such thinking is a wolf bone over 30,000 years old, discovered in the former Czechoslovakia (Figure 2.17). On the bone are 55 notches in two rows, divided into groups of five. We can only guess what the notches represent—how many animals the hunter had killed or how many people there were in the tribe.

Other anthropologists have discovered how shepherds were able to keep track of their sheep without using numbers to count them. Each morning as the sheep left the pen, the shepherds made a notch on a piece of wood or on some other object. In the evening, when the sheep returned, they would again make a notch for each sheep. Looking at the two tallies, they could quickly see whether any sheep were missing. Anthropologists also have discovered several tribes in the twentieth century that did not have any counting systems!

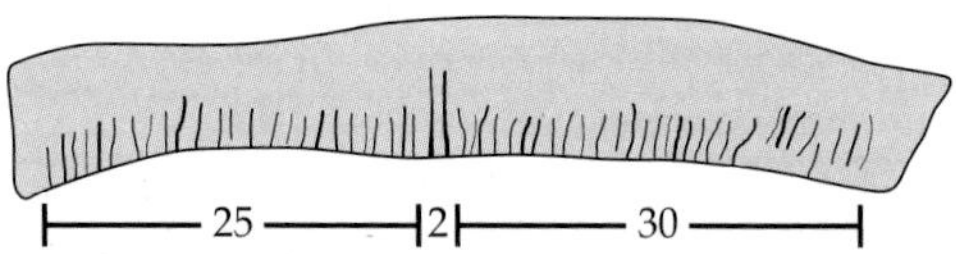

Figure 2.17

The beginnings of what we call civilization were laid when humans made the transition from being hunter-gatherers to being farmers. Archaeologists generally agree that this transition took place almost simultaneously in many parts of the world some 10,000 to 12,000 years ago. It was probably during this transition that the need for more sophisticated numeration systems developed. For example, a tribe need kill only a few animals, but one crop of corn will yield many hundreds of ears of corn.

The invention of numeration systems was not as simple as you might think. The ancient Sumerian words for one, two, and three were the words for man, woman, and many. The Aranda tribe in Australia used the word *ninta* for one and *tara* for two. Their words for three and four were *tara-ma-ninta* and *tara-ma-tara*.

Requirements for counting In order to have a counting system, people first needed to realize that the number of objects is independent of the objects themselves. Look at Figure 2.18. What do you see?

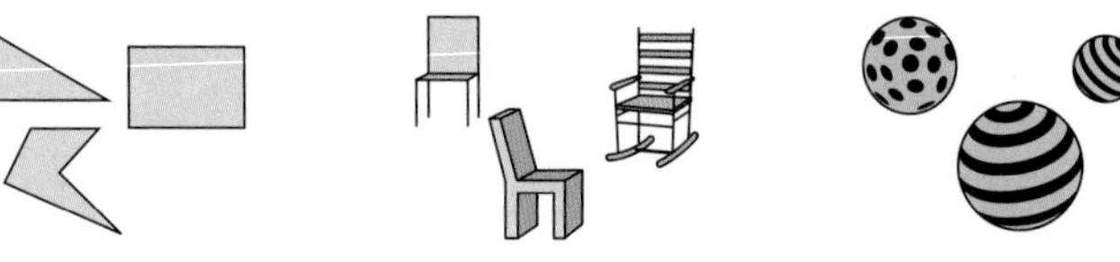

Figure 2.18

LANGUAGE

There is evidence in some of our own words of such origins. For example, we have many different nouns denoting two: a pair, a couple, a brace (of pheasants), a span (of horses), a duet.

There are three objects in each of the sets. However, the number three is an abstraction that represents an amount. Archaeologists have found that people didn't always understand this. For example, the Thimshians, a tribe in British Columbia, had seven sets of words in their language for each number they knew, depending on whether the word referred to (1) animals and flat objects, (2) time and round objects, (3) humans, (4) trees and long objects,

(5) canoes, (6) measures, and (7) miscellaneous objects. Whereas we would say three people, three beavers, three days, and so on, they would use a different word for "three" in each case.

Having a counting system also requires that we recognize a one-to-one correspondence between two equivalent sets: the set of objects we are counting and the set of numbers we are using to count them. Many children miscount objects for some time, either counting too many or too few, because they do not yet realize that they need to say the next number each time they touch the next object. It takes some time for them to realize that each object represents the next number, as shown in Figure 2.19—that is, that there is a one-to-one correspondence between the set of objects and the set of numbers.

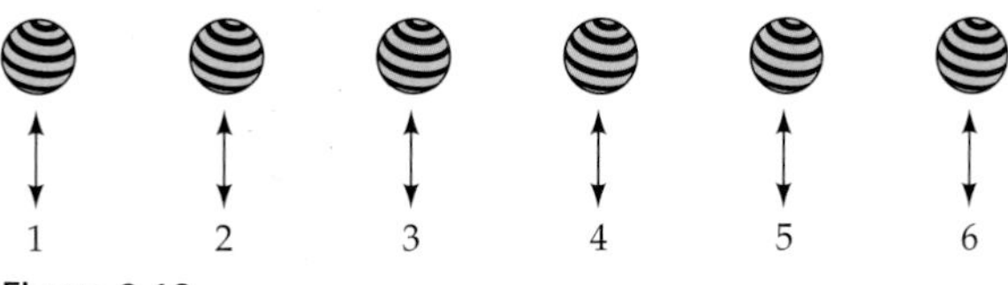

Figure 2.19

There is another aspect of counting that needs to be noted. Most people think of numbers in terms of counting discrete objects. However, this is only one of the contexts in which numbers occur. For example, in Figure 2.20, there are 3 balls, there are 3 ounces of water in the jar, and the length of the line is 3 centimeters. In the first case, the 3 tells us how many objects we have. However, in the two latter cases, the number tells how many of the units we have. In this example, the units are ounces and centimeters. Working with numbers that represent sets of objects is more concrete than working with numbers that represent measures.

Figure 2.20

We distinguish between **number**, which is an abstract idea that represents a quantity, and **numeral**, which refers to the symbol(s) used to designate the quantity. While the number of balls in Figure 2.20 is three, the numeral, or symbol, we would use to represent that number is 3. But the ancient Romans would use the numeral III to represent that number.

PATTERNS IN COUNTING

Explorations Manual 2.4

As humans developed names for amounts larger than the number of fingers on one or two hands, the names for the larger amounts were often combinations of names for smaller amounts.

Can you find the patterns and fill in the blanks for the three systems in Table 2.3? What patterns did you see? Did you find any surprises after you checked your answers in Appendix B?

People who have investigated the development of numeration systems, from prehistoric tallies to the Hindu-Arabic system, have discovered that most of the numeration systems had patterns, both in the symbols and in the words, around the amounts we call five and ten. However, a surprising number of systems also show patterns around 2, 20, and 60. For example, the French word for *eighty*, *quatre-vingts*, literally means "four twenties."

As time went on, people developed increasingly elaborate numeration systems so that they could have words and symbols for larger and larger amounts. We will examine historical numeration systems to not only help us understand the development of place value but also to understand more deeply our own base ten numeration system.

TABLE 2.3

Number	Greenland Eskimos	Aztecs	Luo of Kenya
1	atauseq	ce	achiel
2	machdlug	ome	ariyo
3	pinasut	yey	adek
4	sisasmat	naui	angwen
5	tadlimat	maculli	abich
6	achfineq-atauseq (other hand one)	chica-ce	ab-achiel
7	achfineq-machdlug	chic-ome	ab-ariyo
8	----------------	chicu-ey	----------------
9	achfineq-sisasmat	chic-naui	----------------
10	qulit (first foot)	matlacti	apar
11	achqaneq-atauseq (first foot one)	matlacti-on-ce	apar-achiel
12	----------------	----------------	apar-ariyo
13	----------------	----------------	----------------
15	achfechsaneq (other foot)	caxtulli	----------------
16	----------------	----------------	----------------
20	inuk navdlucho (a man ended)	cem-poualli	piero-ariyo

THE EGYPTIAN NUMERATION SYSTEM

The earliest known written numbers are from about 5000 years ago in Egypt. The Egyptians made their paper from a water plant called papyrus that grew in the marshes. They found that if they cut this plant into thin strips, placed the strips very close together, placed another layer crosswise, and finally let it dry, they could write on the substance that resulted. Our word *paper* derives from their word *papyrus*.

Symbols in the Egyptian system The Egyptians developed a numeration system that combined picture symbols (hieroglyphics) with tally marks to represent amounts. Table 2.4 gives the primary symbols in the Egyptian system.

The Egyptians could represent numerals using combinations of these basic symbols.

Translate the following Egyptian numerals into our system. Check your answers in Appendix B.

1. 𓆼𓍢𓍢𓍢𓎆𓎆𓏺𓏺𓏺 2. 𓆐𓂭𓂭𓂭𓂭𓆼𓆼𓆼𓎆𓎆𓏺𓏺

Translate the following amounts into Egyptian numerals.

3. 1202 4. 304

Working with the Egyptian system Take a few minutes to think about the following questions. Write your thoughts before reading on. . . .

1. What do you notice about the Egyptian system? Do you see any patterns?
2. What similarities do you see between this and the more primitive systems we have discussed?

TABLE 2.4

1,000,000	100,000	10,000	1000	100	10	1
𓁨	𓆐	𓂭	𓆼	𓍢	𓎆	𓏺
Astonished person	Polliwog or burbot fish	Pointing finger	Lotus flower	Scroll	Heelbone	Staff, stroke

3. What limitations or disadvantages do you find in this system?

4. Is this a place value system? Why or why not?

The Egyptian numeration system resembles many earlier counting systems in that it uses tallies and pictures. In this sense, it is called an *additive system*. One of my students called this an "image system" because it is the image that determines the value of the symbol, not the place that it is located. So, it is not a place value system.

Look at the way this system represents the amount 2312. In one sense, the Egyptians saw this amount as $1000 + 1000 + 100 + 100 + 100 + 10 + 1 + 1$ and wrote it as 𓆼𓆼𓍢𓍢𓍢𓎆𓏺𓏺. In an **additive system**, the value of a number is literally the sum of the value of the numerals (symbols).

However, this system represents a powerful advance: The Egyptians created a new digit for every *power of ten*.

Explorations Manual 2.5

They had a digit for the amount 1. To represent amounts between 1 and 10, they simply repeated the digit. For the amount 10, they created a new digit. All amounts between 10 and 100 can now be expressed using combinations of these two digits. For the amount 100, they created a new digit, and so on.

These amounts for which they created digits are called **powers of ten**. Recall, from your work with exponents from algebra, that we can express 10 as 10^1 and 1 as 10^0. Thus we can express the value of each of the Egyptian digits as a power of 10:

1,000,000	100,000	10,000	1000	100	10	1
$10 \cdot 10 \cdot 10 \cdot 10 \cdot 10 \cdot 10$	$10 \cdot 10 \cdot 10 \cdot 10 \cdot 10$	$10 \cdot 10 \cdot 10 \cdot 10$	$10 \cdot 10 \cdot 10$	$10 \cdot 10$	10	1
10^6	10^5	10^4	10^3	10^2	10^1	10^0

The Egyptian system was a remarkable achievement for its time. Egyptian rulers could represent very large numbers. One of the primary limitations of this system was that computation was extremely cumbersome. It was so difficult, in fact, that the few who could compute enjoyed very high status in the society.

THE ROMAN NUMERATION SYSTEM

The Roman system is of historical importance because it was the numeration system used in Europe from the time of the Roman Empire until after the Renaissance. In fact, several remote areas of Europe continued to use it well into the twentieth century. Some film makers still list the copyright year of their films in Roman numerals. Where are some other places where we still use Roman numerals?

Symbols in the Roman system Table 2.5 gives the primary symbols used by early Romans and later Romans.

Translate the following Roman digits into our system using the digits of the Later Roman system. Check your answers in Appendix B.

1. MXI 2. MDXCVII

TABLE 2.5

Amount	Early Roman	Later Roman
1	I	I
5	V or Λ	V
10	X	X
50	↓	L
100	⊙	C
500	ↁ and ɔ	D
1000	ↀ or ∞	M

The X precedes the C in MDXCVII above. Notice how this placement affects the value. We will discuss this aspect of the Roman system later.

Translate these following amounts into Roman numerals.

3. 1102 4. 319

Working with the Roman system Take a few minutes to think about the following questions. Write your thoughts before reading on. . . .

1. What do you notice about the Roman system? Do you see any patterns?
2. What similarities do you see between this system and the Egyptian system?
3. What limitations or disadvantages do you find in this system?
4. Is this a place value system? Why or why not?

Like the Egyptians, the Romans created new digits with each power of 10, that is, 1, 10, 100, 1000, etc. However, the Romans also created new digits at "halfway" amounts—that is, 5, 50, 500, etc. Why do you think they did this? Think and then read on. . . .

This invention reduced some of the repetitiveness that encumbered the Egyptian system. For example, 55 is not XXXXXIIIII but LV.

Basically, the Roman system, like the Egyptian system, was an additive system. However, the Later Roman system introduced a *subtractive* aspect. For example, IV can be seen as "one before five." This invention further reduced the length of many large numbers. However, this is still not a place value system since the value of the symbol itself is not determined by where it is placed.

For example, when writing the amount 444, users of the Later Roman system no longer had to write CCCCXXXXIIII, but rather could write . . . well, what do you think?

Check your answer in Appendix B.

As in the Egyptian system, computation in the Roman system was complicated and cumbersome, and neither system had anything resembling our zero.

THE BABYLONIAN NUMERATION SYSTEM

The Babylonian numeration system is a refinement of a system developed by the Sumerians several thousand years ago. Both the Sumerian and Babylonian empires were located in the region occupied by modern Iraq. The Sumerians did not have papyrus, but clay was abundant. Thus they kept records by writing on clay tablets with a pointed stick called a stylus, just like we use a stylus with today's tablets. Thousands of clay tablets with their writing and numbers have survived to the present time; the earliest of these tablets were written almost 5000 years ago.

HISTORY

We are not sure why the Babylonians chose 60, although many hypotheses have been set forth. One is that because many numbers divide 60 evenly (1, 2, 3, 4, 5, 6, 10, 12, 15, 20, 30, and 60), this choice made calculations with fractions much easier. Their choice of 60 affects our lives today, for it was they who divided a circle into 360 parts and divided an hour into 60 minutes.

Symbols in the Babylonian system Because the Babylonians had to make their numerals by pressing into clay instead of writing on papyrus, their symbols could not be as fancy as the Egyptian symbols. They had only two symbols, an upright wedge that symbolized "one" and a sideways wedge that symbolized "ten." In fact, the Babylonian writing system is called *cuneiform*, which means "wedge-shaped." You can see some pictures of these ancient tablets by searching on the web for "Babylonian mathematics tablets."

Amount	Symbol
1	𒁹
10	𒌋

Amounts could be expressed using combinations of these numerals; for example, 23 was written as 𒌋𒌋𒁹𒁹𒁹.

However, being restricted to two digits creates a problem with large amounts. The Babylonians' solution to this problem was to choose the amount 60 as an important number. Unlike the Egyptians and the Romans, they did not create a new digit for this amount. Rather, they decided that they would have a new *place*. For example, the amount 73 was represented as ▼ ◀▼▼▼. That is, the ▼ at the left represented 60 and the ◀▼▼▼ to the right represented 13. In other words, they saw 73 as 60 + 13.

Similarly, ▼▼▼ ◀▼▼ was seen as six 60s plus 12, or 372.
▼▼▼

We consider the Babylonian system to be a *place value system* because the value of a numeral depends on its place (or position) in the number.

Translate the following Babylonian numerals into our system. Check your answers in Appendix B.

1. ▼▼ ◀▼▼ 2. ▼▼▼ ◀◀◀▼▼ 3. ▼ ◀▼ ◀◀▼
▼▼ ◀

Translate the following amounts into Babylonian numerals.

4. 1202 5. 304

Working with the Babylonian system Take a few minutes to think about the following questions. Write your thoughts before reading on. . . .

1. What do you notice about the Babylonian system? Do you see any patterns?
2. What similarities do you see between this system and the Egyptian and Roman systems?
3. What limitations or disadvantages do you find in the Babylonian system?
4. Is this a place value system? Why or why not?

Place value To represent larger amounts, the Babylonians invented the idea of the value of a digit being a function of its place in the numeral. This is the earliest occurrence of the concept of **place value** in recorded history. With this idea of place value, they could represent any amount using only two digits, ▼ and ◀.

We can understand the value of their system by examining their numerals with expanded notation. Look at the following Babylonian number:

▼▼ ◀◀▼▼▼ ◀◀▼

Because the ◀◀▼ occurs in the first (or rightmost) place, its value is simply the sum of the values of the digits—that is, 10 + 10 + 1 = 21. However, the value of the ◀◀▼▼▼ in the second place is determined by multiplying the face value of the digits by 60—that is, $60 \cdot 23$. The value of the ▼▼ in the third place is determined by multiplying the face value of the digits by 60^2—that is, $60^2 \cdot 2$.

The value of this amount is

$$(60^2 \cdot 2) + (60 \cdot 23) + 21 = 7200 + 1380 + 21 = 8601$$

Thus, in order to understand the Babylonian system, you have to look at the face value of the digits *and* the place of the digits in the numeral. The value of a numeral is no longer determined simply by adding the values of the digits. One must take into account the place of each digit in the numeral.

The Babylonian system is more sophisticated than the Egyptian and Roman systems. However, there were some "glitches" associated with this invention.

What if there were nothing in a place? For example, how could the Babylonians represent the amount 3624? Try this and then read on. . . .

The need for a zero If we represent this amount from the Babylonian perspective, we note that $60^2 = 3600$. Thus the Babylonians saw 3624 as $3600 + 24$. They would use  in the third place to represent 3600, and they would use  in the first place to represent 24, but the second place is empty. Thus, if they wrote , how was the reader to know that this was not $60 + 24 = 84$? Again, try to imagine yourself as a Babylonian. How might you solve this problem? Think and read on. . . .

A Babylonian mathematical table from about 300 B.C. contains a new symbol  that acts like a zero. Using this convention, they could represent 3624 as

The slightly sideways wedges indicate that the second place is empty, and thus we can unambiguously interpret this numeral as representing

$$60^2 + 0 + 24 = 3624$$

This later Babylonian system is thus considered by many scholars to be the first place value system[2] because the value of every symbol depends on its place in the numeral and there is a symbol to designate when a place is empty.

THE MAYAN NUMERATION SYSTEM

One of the most impressive of the ancient numeration systems comes from the Mayans, who lived in the Yucatan Peninsula in Mexico, around the fourth century A.D. Many mathematics historians credit the Mayans as being the first civilization to develop a numeration system with a fully functioning zero. The table below shows their symbols for the amounts 1 through 20. Note that they wrote their numerals vertically.

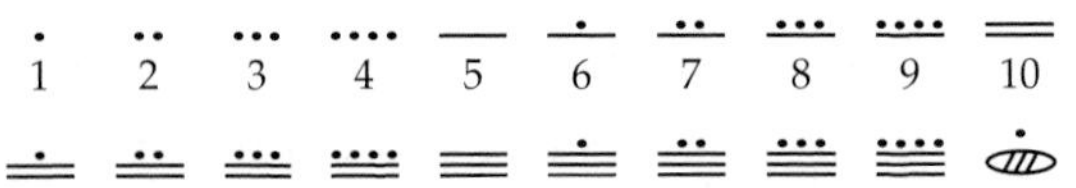

Their numeral for 20 consisted of one dot and their symbol for zero. Thus, their numeral for 20 represents 1 group of 20 and 0 ones, just as our symbol for 10 represents 1 group of 10 and 0 ones.

Theirs was not a pure base twenty system because the value of their third place was not 20×20 but 18×20. The value of each succeeding place was 20 times the value of the previous places. The values of their first five places were 1, 20, 360, 7200, and 14,400.

Use the following questions to reinforce your understanding of the Mayan system. Check your answers in Appendix B.

Translate the following Mayan numerals into our system.

1.  **2.**  **3.** 

Translate the following amounts into Mayan numerals.

4. 123 **5.** 256 **6.** 7300

THE DEVELOPMENT OF BASE TEN: THE HINDU-ARABIC SYSTEM

The numeration system that we use was developed in India around A.D. 600. By A.D. 800, news of this system came to Baghdad, which had been founded in A.D. 762. Leonardo of Pisa traveled throughout the Mediterranean and the Middle East, where he first heard of the new system. In his book *Liber Abaci* (translated as *Book of Computations*), published in 1202, he argued the merits of this new system.

Figure 2.21 traces the development of the 10 digits that make up our numeration system.[4]

HISTORY

In 1229 the City Council of Florence, Italy, passed a law forbidding the use of base ten numbers when entering records of money in account books. Numbers had to be written out.[3]

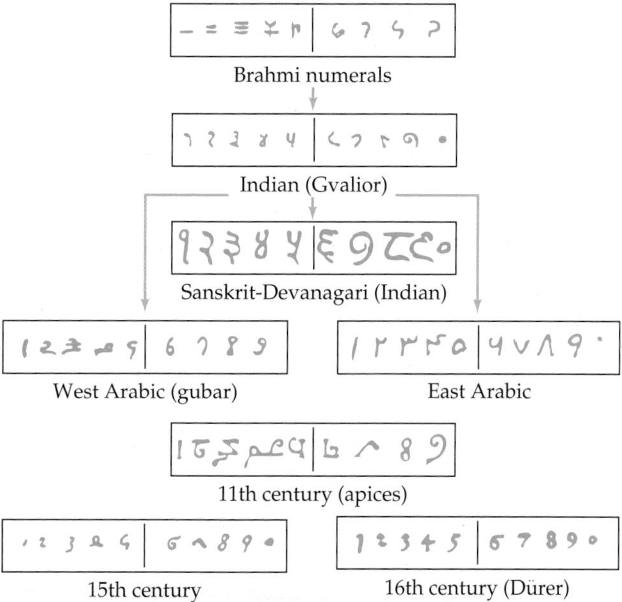

Figure 2.21

Source: Menninger, Karl, *Number Words and Number Symbols: A Cultural History of Numbers*, p. 418, © 1969 Massachusetts Institute of Technology, by permission of The MIT Press.

The development of numeration systems from the most primitive (tally) to the most efficient (base ten) has taken tens of thousands of years. Although the base ten system is the one you grew up with, it is also the most abstract of the systems and possibly the most difficult initially for children. Stop and reflect on what you have learned thus far in your own investigations. Imagine describing this system to a Babylonian, Egyptian, or Roman who has suddenly been transported into our time. Imagine that this person can understand English and has a counting table like Table 2.6 but is still struggling to make sense of this system. How would you describe the essential features of this system to that person? . . .

ADVANTAGES OF BASE TEN

Our **base ten** numeration system has several characteristics that make it so powerful.

No tallies The base ten system has no vestiges of tallies. Any amount can be expressed using only 10 **digits**: 0, 1, 2, 3, 4, 5, 6, 7, 8, and 9. In fact, the word *digit* literally means "finger."

TABLE 2.6

Egyptian	Roman	Babylonian	Hindu-Arabic
ǀ	I	𒁹	1
ǀǀ	II	𒁹𒁹	2
ǀǀǀ	III	𒁹𒁹𒁹	3
ǀǀǀǀ	IV	𒁹𒁹𒁹𒁹	4
ǀǀǀǀǀ	V	𒁹𒁹𒁹𒁹𒁹	5
ǀǀǀǀǀǀ	VI	𒁹𒁹𒁹𒁹𒁹𒁹	6
ǀǀǀǀǀǀǀ	VII	𒁹𒁹𒁹𒁹𒁹𒁹𒁹	7
ǀǀǀǀǀǀǀǀ	VIII	𒁹𒁹𒁹𒁹𒁹𒁹𒁹𒁹	8
ǀǀǀǀǀǀǀǀǀ	IX	𒁹𒁹𒁹𒁹𒁹𒁹𒁹𒁹𒁹	9
∩	X	𒌋	10
∩∩	XX	𒌋𒌋	20
∩∩∩∩∩	L	𒌋𒌋𒌋𒌋𒌋	50
∩∩∩∩∩∩	LX	𒁹	60
𓍢	C	𒁹𒌋𒌋𒌋𒌋	100

Decimal system The base ten system is a **decimal** system, because it is based on groupings (powers) of 10. The value of each successive place to the left is 10 times the value of the previous place:

100,000 10,000 1000 100 10 1

Ten ones make one ten.

Ten tens make one hundred.

Ten hundreds make one thousand.

Ten thousands make ten thousand.

Expanded form When we represent a number by decomposing it into the sum of the values from each place, we are using **expanded form**. There are different variations of expanded form. For example, all of the expressions below emphasize the structure of the numeral, 234—some more simply and some using exponents.

$$\begin{aligned} 234 &= 200 + 30 + 4 \\ &= 2 \cdot 100 + 3 \cdot 10 + 4 \cdot 1 \\ &= 2 \cdot 10^2 + 3 \cdot 10^1 + 4 \cdot 10^0 \end{aligned}$$

Note: $10^1 = 10$ and $10^0 = 1$.

In Chapter 3, we will use expanded form to understand why the procedures we use to add, subtract, multiply, and divide actually work.

The concept of zero "The invention of zero marks one of the most important developments in the whole history of mathematics."[5] This is the feature that moves us beyond the Babylonian system. Recall the Babylonians' attempts to deal with the confusion when a place was empty. It was the genius of some person or persons in ancient India to develop this idea, which made for the most efficient system of representing amounts and also made computation much easier. One of the most difficult aspects of this system is that the symbol 0 has two related meanings: In one sense, it works just like any other digit (it can be seen as the number 0), and at the same time, it also acts as a place holder. It takes young schoolchildren several years to understand this fully and accept it. So, is zero a number? Some will argue that it is "nothing" so it is not a number. Others say that it is just a place holder. However, zero does represent a quantity (and therefore is a number). Let's illustrate it this way. If you were asked how many letter "a"s were in the word "banana," you would answer 3. What if you were asked how many "e"s were in the word "banana"? You would answer 0. Zero is the number of "e"s just as three is the number of "a"s.

HISTORY

A French writer in the 1400s referred to zero as nothing but "a sign which creates confusion and difficulties." Some writers ridiculed this new system, thinking that it would go away once people came to their senses and went back to Roman numerals: "Just as the rag doll wanted to be an eagle, the donkey a lion, and the monkey a queen, the *cifra* [zero] put on airs and pretended to be a digit."[6]

We will use pictures of base 10 blocks to help us understand place value and the operations more deeply. Your instructor may have these in class or you may have them in the manipulative kit. There are also virtual versions available on the National Library of Virtual Manipulatives website (nlvm.usu.edu). (You will find base blocks under the "number and operations" link.)

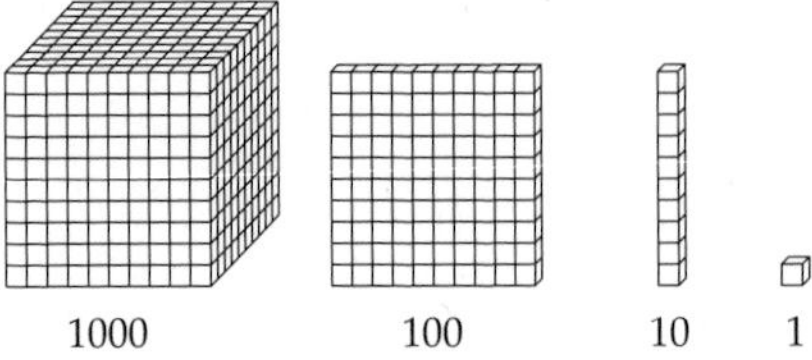

Notice that the ones place is represented by a unit that we will call a "**small cube**." It takes 10 of the small cubes to make one **long**; 10 of the longs to make one **flat**; and 10 flats to make one **large cube**. This visual representation also helps us to geometrically see that the 100 place value is 10^2 because it is a 10 by 10 square and that the 1000 place is 10^3 because it is a 10 by 10 by 10 cube.

What if we had a base five system instead of a base ten system? How would these models change? Instead of grouping in tens, we would trade when there were five. So, five small cubes would equal one long (of five); five longs would equal a flat (which would be a five by five square); five flats would be a cube (which would be a cube five by five by five). We will explore other bases in this course for several reasons. First, it helps us to understand our base ten place value system much more deeply. Second, it helps us to understand how difficult this concept is for young children when they are first learning it. Lastly, there are places where in essence we have to think in other bases. For example, when telling time, we are using a modified base sixty system. Many children, when first learning to tell time, will count time like this: 3:58, 3:59, 3:60, 3:61, . . . 3:98, 3:99, 4:00. Why do you think they do this? They are used to exchanging when they get to 100, but to go up in the digit to the left when they reach 60 is very different. Our coin system of pennies, nickels, and quarters is essentially a base five system as well.

UNITS

There is an old parable that says a journey of 1000 miles begins with a single step. The same can be said for counting. We always begin with 1. However, unlike the phrase "a rose is a rose is a rose," a 1 is not always the same. For example, a 1 in the millions place represents 1 million. This is the power of our numeration system, but it is very abstract.

We will elaborate this idea of unit now because it will recur throughout this book. The first place in our system is called the ones place, the singles place, and the units place. By the time children come to school, they are generally very comfortable with the idea of counting one at a time. Over time they can count higher and higher.

When counting objects, one is our key term. When asked to count a pile of objects, for example, 240 pennies, children will count one at a time. However, if they lose their count, they have to start all over. Some children realize that they can put the pennies into piles of 10. Now if they lose count, they can go back and count by tens, for example, 10, 20, 30, 40, etc. In this case, 10 is a **simple unit**, that is, it is composed of a number of smaller units. Some children can see that 1 pile is also 10 pennies. To be able to hold these two amounts simultaneously is a challenge for young children, and it is an essential milestone along the way.

We have **composite units** everywhere: 100 is equivalent to ten 10s, 1000 is equivalent to ten 100s. In fact, our language shows this: some people will say thirty-four hundred for 3400. We talk about 1 dozen eggs, a case of soda (24 cans), and a pound (16 ounces). When we say that we will need 6 dozen eggs for a pancake breakfast fundraiser, we can see 6 dozen and we also know that this is 72 individual eggs.

Throughout the book, we will stop and take note when we are expanding our understanding of this idea of units.

INVESTIGATION 2.2b What If Our System Was Based on One Hand?

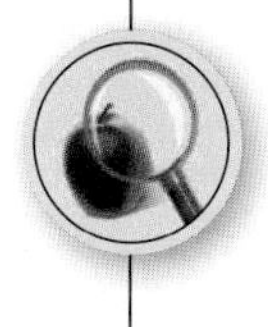

Explorations Manual 2.4

The people who developed base ten decided to base it on two hands. What if they had decided to base it on one hand? That is, what if one-zero had come not after we counted two hands but after we counted one hand? Our counting would look like this: 1, 2, 3, 4, 10, ... The manipulatives (see the figure below) would have the same basic shape as the base ten blocks.

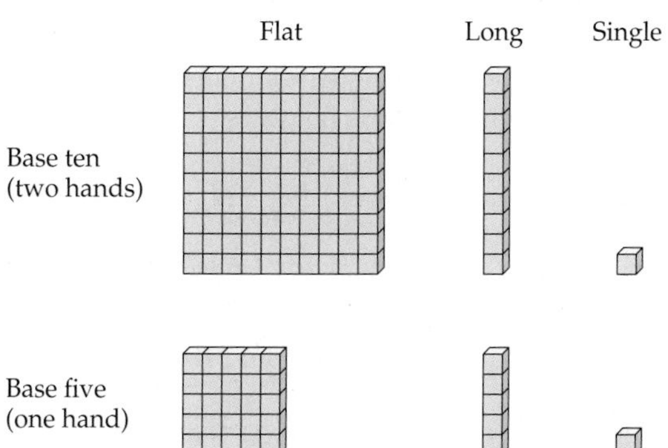

Because the structures of this new base are the same as in the system you grew up with, the counting follows the same rules. The table below shows the beginnings of the new system. If you find yourself struggling, make your own set of manipulatives (cut from graph paper) and represent each number manipulatively: 1 single, 2 singles, 3 singles, 4 singles. . . . On the virtual base blocks on the National Library of Virtual Manipulatives, you can set it to base five and then it groups the singles in fives to make a long. The next number represents one hand and will now be called one-zero because this is what a long will be in this system. The system continues all the way to 44 below. What comes next? Think carefully.

1	2	3	4	10
11	12	13	14	20
21	22	23	24	30
31	32	33	34	40
41	42	43	44	?

DISCUSSION

First of all, counting simply means adding 1 each time, so the next number after 44 is $44 + 1$. In *this* base, a place is full after 4. Thus the ones place is full, and we "move" the five ones into the next place because we can trade five singles into 1 long. However, the longs place is also full once we get one more long. Thus we trade (regroup) 5 longs into 1 flat, and we now have 1 flat, 0 longs, and 0 singles. That is, 100_{five} is the next number after 44_{five}.

If you don't see this, use manipulatives. In the first figure at the left, we see 44 (4 longs and 4 singles) plus 1. In the second figure, we see those five singles have become a new long. In the third figure, we see the five longs have become a flat. We now have 1 flat, 0 longs, and 0 singles—that is, 100_{five}.

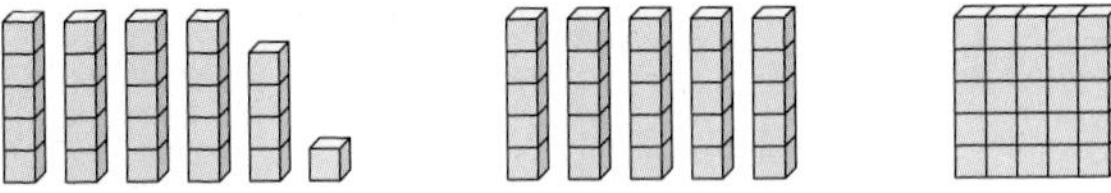

CLASSROOM CONNECTION

Jerome Bruner wrote that children learn mathematical ideas best if they begin at the concrete level and then move at their own rate to the symbolic level. I find this true of most adults. The most basic level is the level of manipulatives. If your instructor does not have base five blocks, you can make your own set by cutting flats, longs, and singles from graph paper. Many students find that actually making and moving the manipulatives helps them grasp the ideas more readily. At some point, many students find that they no longer need the manipulatives but that jumping all the way to using just numbers is abstract. Bruner knew this and articulated a middle level that is a pictorial level. At this level, you don't need the manipulatives but find pictures of them useful. I have come to believe that most adults learn new mathematics concepts more deeply if they first experience these concepts at the concrete level and proceed from the concrete level to the visual level and finally to the symbolic level *at their own pace* and by connecting the representations at each level.

INVESTIGATION 2.2c How Well Do You Understand Base Five?

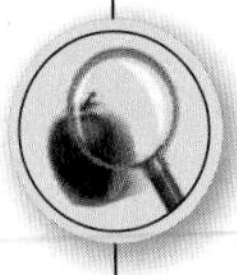

One way to assess your understanding of this new base is to figure out what numbers come after and before given numbers.

A. What number comes after 234_{five}?

B. What number comes after 1024_{five}?

C. What number comes before 210_{five}?

D. What number comes before 3040_{five}?

DISCUSSION

A. At the visual level, one can see that the next number means 2 flats, 3 longs, 4 singles, and 1 more single. Thus we now have 2 flats, 4 longs, and 0 singles—that is, 240_{five}. At the symbolic level, one might reason the first problem like this: After 234_{five}, we have filled the ones place, so the ones place will now be zero, the longs place will have one more, so it will be 4, and the flats place will still be 2.

Flats five^2	Longs fives	Singles ones

→

Flats five^2	Longs fives	Singles ones

B. To determine what number comes after 1024_{five}, let us stop and reflect for a moment. The manipulatives are like training wheels. They are useful to help the ideas develop, but eventually they need to come off, especially as the numbers get larger. If we add to 1024_{five}, the ones place is full—five ones becomes one more long (since five ones = 10_{five}). Now we have 3 longs. No other places are affected, so the next number is 1030_{five}.

C. What number comes before 210? The answer comes more visually from the picture. Do you see how?

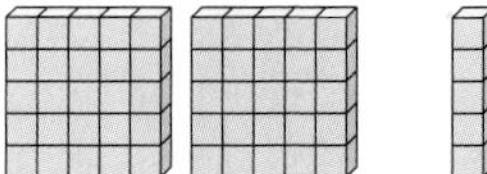

The number before 210_{five} is 1 less than 210_{five}. You can see this by trading (regrouping) the long into five singles and taking one single away so that we now have 2 flats, 0 longs, and 4 singles—that is, 204_{five}. Another way to see it is to cover up one of the singles on the long above and realize that this represents the breaking up of that long into singles.

D. What comes before 3040_{five}? You could represent this as 3 big cubes, 0 flats, 4 longs, and 0 singles. In order to take 1 away from 3040_{five}, you would have to trade 1 long for five singles and take away one single. You now still have 3 big cubes and still have 0 flats, but now you have 3 longs, and 4 singles. Thus the number before 3040_{five} is 3034_{five}.

At a more abstract level, as you internalize the properties of a base system, you simply know that the number before 3040_{five} must have a 4 in the ones place (just as you "know" that the number before 7090 in base ten must end in a 9). Since you are going backwards, you know that you have one less long. You also "know" that the flats and next place are not affected in this case, and thus the answer is 3034_{five}.

Either in explorations or in class, your instructor may have you work with other bases so that your understanding of the fundamental ideas of the base and place value become deeper and deeper. Having explored other bases, let us now revisit the fundamental ideas of base ten. Any base—base two, base five, base ten, base twelve, whatever—will have the following fundamental characteristics.

1. Any base has the same number of symbols as the number of base. In a base ten system, we have ten symbols (0–9) and in a base five system we have five symbols (0–4). With those symbols, we can represent any amount; think of *Toy Story*—"to infinity and beyond!"

2. The value of each place is the base times the previous place. In base ten, the value of the places are ones, tens, ten^2, ten^3, etc. Similarly, in base five the value of the places are ones, fives, five^2, five^3, etc.
3. Each place can contain only one symbol. When a place is full, we "move" to the next place by trading (regrouping) to the next higher place.
4. The value of a digit depends on its place in the numeral.
5. The value of a numeral is determined by multiplying each digit by its place value and then adding these products.
6. Zero represents an empty place and 0 represents an actual amount, with a place on the number line.

INVESTIGATION 2.2d Base Sixteen

Computers don't have fingers to count with; they just have on and off. Thus, computers begin with base two. For a variety of reasons, computers actually compute in base sixteen. Because we do not trade for a long until we have sixteen ones, how do we represent ten, eleven, up to fifteen ones? 10_{sixteen} means 1 long of 16, so we cannot use 10 to mean ten. Thus, we have to make up new digits in base sixteen for the base ten amounts between ten and fifteen. How do you think this problem was solved?

The solution was to use the alphabet. The table below shows the numerals for 1 through 16 in base ten and base sixteen.

Base ten	1	2	3	4	5	6	7	8	9	10	11	12	13	14	15	16
Base sixteen	1	2	3	4	5	6	7	8	9	a	b	c	d	e	f	10

How would you represent the following base ten numerals in base sixteen?

A. 25 B. 100

DISCUSSION

A. If we take 25_{ten} and repackage it in terms of sixteens, we have 16 + 9. Thus, 25 in base ten is equivalent to 19 in base sixteen. Symbolically, we can write $25_{\text{ten}} = 19_{\text{sixteen}}$.

B. If we take 100_{ten} and repackage it in terms of sixteens, we have $100 = 6 \times 16 + 4$. Symbolically, $100_{\text{ten}} = 64_{\text{sixteen}}$.

INVESTIGATION 2.2e Relative Magnitude of Numbers

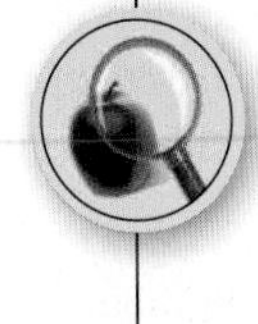

Explorations Manual 2.6

Our modern society deals with large numbers all the time.

Politicians talk about a war costing $100 billion a year.

The federal deficit is more than $17 trillion at the writing of this book.

The closest star to us is about 24,600,000,000,000 miles from earth.

The cleanup from Hurricane Katrina involved the removal of 500 million cubic yards of debris.

More than 25 million people have died of AIDS.

CLASSROOM  CONNECTION

Grade 4

This page is from a fourth-grade math journal. What is your answer to 3d? *Hint:* You will need to answer question 1 first.

Date ____________________ Time ____________________

LESSON 5·8 **How Much Are a Million and a Billion?**

SRB 4

1. How many dots are on the 50-by-40 array page? ________ dots

2. How many dots would be on
   a. 5 pages? ____________ dots
   b. 50 pages? ____________ dots
   c. 500 pages? ____________ dots

3. Each package of paper, or ream, contains 500 sheets. How many dots would be on the paper in
   a. 1 ream? (*Hint:* Look at Problem 2.) ____________ dots
   b. 10 reams? (1 carton) ____________ dots
   c. 100 reams? (10 cartons) ____________ dots
   d. 1,000 reams? (100 cartons) ____________ dots

4. Use digits to write these numbers in the place-value chart below.

   a. 999 thousand b. 1,000 thousand c. 999 million d. 1,000 million

	Billions				Millions				Thousands				Ones		
	100B	10B	1B	,	100M	10M	1M	,	100Th	10Th	1Th	,	100	10	1
a.															
b.															
c.															
d.															

127

From *Everyday Mathematics, Grade 4*: The University of Chicago School Mathematics Project: Student Math Journal, Volume 1 by Max Bell et al., Lesson 5-8, p.127. Reprinted by permission of The McGraw-Hill Companies, Inc.

CLASSROOM CONNECTION

David Schwartz's *How Much Is a Million?*[7] is a classic children's book. My favorite is *Big Numbers: And Pictures That Show Just How Big They Are!*[8] The author begins with one pea and then goes up one place value at a time. When you get to one billion, the peas have completely filled a two-story house and are spilling into the yard.

Let's investigate an example commonly used in elementary classrooms. If a large paper clip is about 2 inches long, how long would a chain of 1 million of those paper clips be?

DISCUSSION

Well, it would be 2 million inches, but can you "feel" 2 million inches? I can't. If we divide 2 million by 12 (the number of inches in a foot), we get 166,667 feet. That's still not a number most people can sense. However, if we divide that number by 5280 (the number of feet in a mile), we get about 32 miles. Most people have a sense of 32 miles.

Now how long would a chain of 1 billion such paper clips be?

It would be about 32,000 miles, because a billion is a thousand million.

How long would a chain of 1 trillion paper clips be?

32,000,000 miles. The distance around the earth is about 24,000 miles. The moon is about $\frac{1}{4}$ million miles from earth, so one round trip would be $\frac{1}{2}$ million and two round trips would be 1 million, so 64 round trips would be 32 million miles.

1 million paper clips	32 miles	A bit longer than a marathon
1 billion paper clips	32,000 miles	More than the distance around the earth
1 trillion paper clips	32,000,000 miles	64 round trips to the moon

Were you surprised at how much bigger a billion is than a million; how much bigger a trillion is than a billion? Most people are. When we are not able to have a reference for thinking about large amounts, they literally swim in our heads. They lose their reality. If we could really sense \$11 trillion, we would clamor to reduce the debt. If we could really sense the number 25 million people living with AIDS, then we would likely take action.

CLASSROOM CONNECTION

Children in the early grades love to show you how high they can count. Once they get to thousands and millions, a common question is "What is the largest number?" When my son asked this question and came to understand there is no biggest number, I told him that a googol, which is 1 with 100 zeros, was the biggest number that had a name. He was so excited with this information that he wowed his friends for weeks! Technically, a much larger number would be a googolplex, which is a 1 with googol zeros, or 10^{googol}, but at age 6 exponents were a much more abstract concept. The name "googolplexian" has been given to the world's largest number with a name, which is a 1 followed by a googolplex of zeros. You can view these numbers written out at googolplexian.com.

SUMMARY 2.2

We have explored different counting systems to give you an appreciation of the significance of base ten and its abstractness—it took humans many thousands of years finally to invent such a powerful numeration system. In the course of working on the explorations, you have come to appreciate the importance of mathematical vocabulary, including the terms *digit* and *place* (ones place, tens place, etc.). You have realized that with the concepts of base and place value, and a symbol to represent "nothing," we can represent any amount using only a few digits (in base ten, we use 0, 1, 2, 3, 4, 5, 6, 7, 8, and 9). You have been introduced, through expanded form, to the tool of decomposing a number into its constituent parts. This notion of breaking an object or idea into its component parts is an essential tool in all scientific disciplines.

In this section, you have probably gained a deeper appreciation of both the power and the abstractness of our base ten numeration system. To use an old phrase, I wish I had a dollar for every student who has said something like "Wow, no wonder it's hard for little kids to learn how to count; I never thought of it [our system] that way before." In Chapter 3, we will learn how the algorithms for addition, subtraction, multiplication, and division really work. By the end of Chapter 3, your appreciation of the Hindu-Arabic system should be even greater!

2.2 Exercises

1. a. Fill in the blanks in these counting systems.
   b. What patterns do you see in these systems?
   c. Describe the "rules" of this system as though you were talking on the phone to a friend who missed the class when this system was discussed.

Base ten	Maya	Luli of Paraguay	South American
1	hun	alapea	tey
2	ca	tamop	cayupa
3	ox	tamlip	toazumba
4	can	lokep	cajesa
5	ho	lokep moile alapea (four and one) or is alapea (one hand)	teente
6	uac	lokep moile tamop	teyente-tey
7	uuc	?	teyente-cayapa
8	uaxac	lokep moile lokep	?
9	bolon	lokep moile lokep alapea	teyente-cajesa
10	lahun	is yaoum (both hands)	caya-ente
11	buluc	is yaoum moile alapea (hands and one)	caya-ente-tey
12	lah-ca	is yaoum moile tamop	?
13	ox-lahun	?	?
15	ho-lahun	?	toazumba-ente
16	?	?	?
20	hunkal	is eln yaoum (hands, feet)	?
21	?	?	?
22	?	?	?

2. Find the base ten equivalent of each of these numerals.
   a. 𓆼𓆼𓆼∩∩∩| b. [illegible]∩||
   c. MDCLXVI d. MDXIX
   e. CIX f. 𒌋𒌋𒁹 𒌋𒌋 𒁹𒁹
   g. 𒁹𒁹 𒌋𒁹𒁹𒁹 h. 𒌋𒌋𒁹𒁹𒁹

3. Represent these base ten amounts in Egyptian, Roman, and Babylonian symbols.
   a. 312 b. 1206 c. 6000
   d. 10,000 e. 123,456

4. Translate the following Mayan numerals into base ten:
   a. [illegible] b. [illegible] c. [illegible]
   Write the following base ten numerals in Mayan:
   d. 245 e. 500 f. 2813

5. I have seen problems like these used in elementary classrooms to help students develop a deeper understanding of the relationship between places.
   a. How many hundreds are in 2600?
   b. How many hundreds are in 54,000?
   c. How many tens are in 250?
   d. How many tens are in 4500?
   e. What is the more common way of saying thirty-four hundred?
   f. What number is equivalent to 345 tens?

6. In the problems below, enter the first number into your calculator. Then describe what number and operation you would enter so that the calculator would display the following number. For example, if you entered 2345 and then the next number were 1845, you would enter "−500 = " on the calculator.
   a. 3456 / 3056 b. 37779 / 30009 c. 123456 / 203456
   d. 30405 / 34445 e. 346734 / 1 f. 445566 / 334455
   g. Make up your own problem.

7. Represent the following base ten numbers by sketching the base ten manipulatives.
   a. 345 b. 2001

8. Represent the following base ten numbers in expanded form.
   a. 345 b. 2001 c. 10,101

9. Rewrite each of the following in standard form.
   a. $4 \cdot 10^3 + 8 \cdot 10^2 + 5 \cdot 10 + 9$
   b. $3 \cdot 10^4 + 2 \cdot 10^2 + 4 \cdot 10$
   c. $7 \cdot 10^5 + 5 \cdot 10^4 + 3$

10. Tell what comes after:
   a. 34_{five} b. 1011_{two} c. 99_{sixteen} d. 7099_{sixteen}
   e. 101_{two} f. 111_{twelve} g. 124_{five} h. 405_{six}

11. Tell what comes before:
   a. 1010_{five} b. 340_{five} c. 100_{sixteen} d. 1110_{two}
   e. 1010_{two} f. 110_{sixteen} g. 120_{four} h. 60_{seven}

DEEPENING YOUR UNDERSTANDING

12. An amusing exercise is to convert English words into Roman values. For example, the English word LID would be worth $50 + 1 + 500 = 551$.

  a. What is the value of MIX?

  b. What is the most valuable English word made up only of letters in the Roman numeration system?

  c. What is the most valuable English word that you can find, if we allow any English word but determine its value by adding only those letters that have values in the Roman system?

13. The symbols developed by the ancient Greeks are visually quite fascinating and remind me of the children's game called hangman. The basic symbols for amounts up to 10,000 are given below.

1	5	10	50	100	500	1,000	5,000	10,000
Ι	𐅃	Δ	𐅄	Η	𐅅	Χ	𐅆	Μ

Translate the following Greek numerals into base ten:

  a. 𐅅ΔΔ𐅃ΙΙ b. Η𐅄ΔΔ𐅃Ι

Translate the following base ten numerals into Greek:

  c. 347 d. 5555

14. An ancient Chinese mathematician named Sun-Tsu who lived in the first century A.D. described the use of calculating rods (made of bamboo) for representing numbers. The digits for 1 to 9 are represented as follows:

1	2	3	4	5	6	7	8	9
𝍠	𝍡	𝍢	𝍣	𝍤	𝍥	𝍦	𝍧	𝍨

However, when the digits from 1 to 9 appear in the tens column, they are represented as follows:

1	2	3	4	5	6	7	8	9
𝍩	𝍪	𝍫	𝍬	𝍭	𝍮	𝍯	𝍰	𝍱

Thereafter, every time you move over one place, you change from one form to the other.

Thus the number 4763 would be represented as 𝍬𝍦𝍮𝍢, and the amount 8888 would be represented as 𝍰𝍧𝍰𝍧.

Translate the following numerals into base ten:

  a. 𝍢𝍪𝍤𝍮 b. 𝍥𝍫𝍥

Represent the following numerals in the Chinese system:

  c. 346 d. 12,345

15. The developers of Braille decided on the notation shown below for numbers in Braille. Each digit has its unique shape of raised dots on a 2 × 3 grid as shown below. The convention is to use a backwards L to let the reader know that a number is coming. The symbol for a comma is shown also.

0	1	2	3	4	5	6	7	8	9
⠚	⠁	⠃	⠉	⠙	⠑	⠋	⠛	⠓	⠊

,
⠂

Here is how the amount 245,608 would appear in Braille:

⠃ ⠙ ⠑ ⠂ ⠋ ⠚ ⠓

Convert these numbers to Braille.

  a. 599 b. 3,603,200

Convert these Braille symbols into base ten symbols.

  c. ⠙ ⠋ ⠚ ⠂ ⠓ ⠑ ⠊

  d. ⠁ ⠉ ⠑ ⠂ ⠃ ⠙ ⠋

16. ***Classroom Connection*** In the October 1993 issue of the *Arithmetic Teacher,* Susan Bohan asked her elementary school class to make changes in the Egyptian system to make it easier to use. Below is the work of several children. In each case, describe which, if any, of the six characteristics of base ten found in this section the refined systems possess. Justify your responses.

  a. In a slight modification of Mary's system, she put dots on each symbol to represent how many of those symbols were represented. For example, 70 was shortened from 𓎆𓎆𓎆𓎆𓎆𓎆𓎆 to 𓎆. Here are two examples of amounts using Mary's system:

  2819 𓆼 𓍢 𓎆 𓏺𓏺𓏺𓏺𓏺𓏺𓏺𓏺𓏺 4240 𓆼 𓍢 𓎆

  b. Three boys, Mark, Jim, and Bob, made up new symbols for 3–9, shown below. They kept | for 1 and || for 2.

  △ → 3 □ → 4 ⛤ → 5 ☹ → 6
  M → 7 J → 8 B → 9

  They then paired these symbols with the Egyptian numerals. For example, 70 was shortened from 𓎆𓎆𓎆𓎆𓎆𓎆𓎆 to J. They decided to have both a vertical and a horizontal aspect to their system—the new symbols (for how many of each amount) were written above the Egyptian symbols instead of next to them. Here are two examples:

  5348 $\frac{⛤}{𓆼}\ \frac{△}{𓍢}\ \frac{□}{𓎆}\ \frac{J}{|}$

  9105 $\frac{B}{𓆼}\ \frac{|}{𓍢}\ \frac{⛤}{|}$

  c. After some discussion of the previous system, someone suggested that the Egyptian symbols weren't necessary anymore as long as everyone knew what each place stood for. The students realized that they did need a symbol for an empty place. They chose Z for 0.

  Source: *Arithmetic Teacher* (now *Teaching Children Mathematics*), October 1993. National Council of Teachers of Mathematics.

17. In some dairy states, a kind of base six is used for shipping milk. Six cartons of milk make 1 box; 6 boxes make 1 crate, 6 crates make 1 flat, and 6 flats make 1 pallet. Thus, if someone said they had filled 2413, that would mean 2 flats, 4 crates, 1 box, and 3 cartons.

  a. How many cartons of milk would this be in base ten?

  b. Which of the six characteristics of base ten (on pages 65–66) does this system possess? Justify your responses.

18. NASA keeps track of time using the following system:

Hours:Minutes:Seconds:Hundredths of a Second

Thus 4 : 2 : 5 : 6 would mean that the program is 4 hours, 2 minutes, 5.06 seconds from launch.

Let's say that a launch was temporarily stopped at 4 : 2 : 5 : 6 and then later stopped at 3:1:7:2. How much time had elapsed between the two stops?

19. ***Classroom Connection*** One common mistake that young children make when counting is illustrated as follows: twenty-six, twenty-seven, twenty-eight, twenty-nine, twenty-ten, twenty-eleven, etc.

Can you describe the nature of the child's mistake? That is, what is the child not understanding or what misconception of counting causes the mistake? Do you see this mistake differently now than you would have before this course? If so, please explain.

20. ***Classroom Connection*** Another common mistake that young children make when counting is illustrated as follows: twenty-six, twenty-seven, twenty-eight, twenty-nine, thirty-one, thirty-two, etc.

Can you describe the nature of the child's mistake? That is, what is the child not understanding or what misconception of counting causes the mistake? Do you see this mistake differently now than you would have before this course? If so, please explain.

21. ***Classroom Connection*** This question came from a fifth-grader: Is 5 the middle number between 0 and 10? What do you think?

22. We have discussed the development of more sophisticated numeration systems in class, culminating in the Hindu-Arabic base ten system that we currently use. Describe the advantages of the Hindu-Arabic system over the Roman system.

23. The first edition of this book was published in 1997, which is in the twentieth century, not the nineteenth century. Can you explain why?

24. When we write large numbers in the United States, why do we insert a comma every three numbers instead of every two or four numbers?

25. We speak of so many children having little or no understanding of place value. What does "place value" mean to you? Define the concept in your own words. This involves giving some meaningful explanation of what "place value" means mathematically.

26. If it were possible to make a line of people standing shoulder to shoulder that went around the world, how many people would it take?

27. a. How long is 1 million seconds? Pick a unit that we can all sense. For example, minutes is not a tangible unit because 17,000 minutes is not an amount that most people can relate to.

b. How long is 1 billion seconds?

28. a. If you laid 1 million dollar bills end to end, how long would that line be?

b. How long would 1 billion dollar bills be?

29. Convert the following numbers into base ten.

a. 41_{five} b. 55_{six} c. 210_{five} d. 2104_{five}
e. 101_{five} f. 1111_{six} g. 303_{four} h. 606_{seven}
i. 1101_{two} j. 10001_{two} k. 99_{sixteen} l. 909_{sixteen}

30. Convert the following numbers from base ten into the designated base.

a. $44_{\text{ten}} = ?_{\text{five}}$ b. $152_{\text{ten}} = ?_{\text{five}}$ c. $92_{\text{ten}} = ?_{\text{two}}$
d. $206_{\text{ten}} = ?_{\text{two}}$ e. $72_{\text{ten}} = ?_{\text{twelve}}$ f $402_{\text{ten}} = ?_{\text{sixteen}}$
g. $44_{\text{ten}} = ?_{\text{six}}$ h. $1252_{\text{ten}} = ?_{\text{six}}$ i. $144_{\text{ten}} = ?_{\text{sixteen}}$
j. $100_{\text{ten}} = ?_{\text{five}}$ k. $99_{\text{ten}} = ?_{\text{sixteen}}$ l. $1052_{\text{ten}} = ?_{\text{five}}$
m. $2{,}500{,}000_{\text{ten}} = ?_{\text{five}}$ n. $2{,}500{,}000_{\text{ten}} = ?_{\text{two}}$

31. In what base does $25 + 25 = 51$?

32. Tell what base makes the following statement true: $23_{\text{ten}} = 25_x$.

33. For what base x is this statement true? $598_{\text{ten}} = 734_x$

34. If $44_x = 28_{\text{ten}}$, how many base x candy bars will fit into a box holding 110_{five} candy bars?

35. Although the base ten long is twice as long as the base five long, the base ten flat is 4 times as large as the base five flat. Why is this so? That is, given that 10 is twice as much as 5, why isn't the base ten flat also twice as big as the base five flat?

36. Explain the following conversion problems as though you were talking to a struggling student.

a. $234_{\text{ten}} = ?_{\text{five}}$ b. $405_{\text{eight}} = ?_{\text{ten}}$

37. Represent the following numbers by sketching what they would look like using manipulatives.

a. 304_{five} b. 1001_{two} c. $23\text{b}_{\text{sixteen}}$

38. a. What is the largest four digit number in base eight?

b. What is the largest three digit number in base sixteen?

39. You are Zirkle, from Zordon, and you have just finished your trip to the solar system. Your planet has managed to develop interplanetary travel without ever developing a sophisticated number system. You have examined three different bases (base two, base five, and base six), and you are preparing your recommendations for your home planet. Summarize the pros and cons of each base, and then tell which base you recommend that your planet adopt. *Note:* Your species has no fingers, only flippers. Therefore, none of the bases has an advantage with respect to your anatomy.

a. Summarize the pros and cons of each base in a table like the one below. *Note:* Entries in the cells need to demonstrate that you understand the advantage or disadvantage/limitation. That is, an entry should state *what* the advantage or disadvantage is and also give a brief *explanation* of the advantage or disadvantage.

Base	+	–
Two		
Five		
Six		

b. Write your final recommendation: an essay of at least one paragraph (more than two sentences). You may refer to your chart, so that you don't have to be redundant.

40. Go to the National Library of Virtual Manipulatives website (http://nlvm.usu.edu), click on Number & Operation, and then on Base Blocks. Change the base to four and use the blocks to count in base four. Write the numeral each time you add another block and trade.

FROM STANDARDIZED ASSESSMENTS

2006 NECAP, Grade 4

41. The chart below describes a number.

Hundreds	Tens	Ones
6	23	5

What is the number?

2005 NECAP, Grade 4

42. Look at the information in the box.

9 tens
5 ones
8 hundreds

What number does the information describe?

a. 589

b. 895

c. 958

d. 985

2008 NECAP, Grade 5

43. What number is 12 tens more than 30,605?

a. 30,617

b. 30,725

c. 31,805

d. 42,605

2007 NECAP, Grade 5

44. Which number has the same value as 6 hundred thousands and 23 hundreds?

a. 6,002,300

b. 623,000

c. 602,300

d. 600,230

45. By how much would 217 be increased if the digit 1 were replaced by a digit 5?

a. 4

b. 40

c. 44

d. 400

(28% of fourth-graders got this correct)

Source: Results of the *Fourth NAEP Mathematics Assessment*, p. 46. U.S. Department of Education, National Center for Education Statistics.

LOOKING BACK on chapter two

QUESTIONS TO SUMMARIZE BIG IDEAS

1. What is a set and how can we describe sets?
2. Describe different relationships between sets.
3. How can Venn diagrams show these relationships and organize information?
4. What are some different number sets and how are they related?
5. What does it mean to say that a numeration system has "place value"?
6. How is a base ten place value system similar and different from a base five place value system?
7. Why are we investigating other bases besides ten?
8. What parts of this chapter are less clear to you at this point? What will you do to clarify those ideas?
9. Look back at the Mathematical Practices of the CCSS. In what ways did you engage in those practices during this chapter?

CHAPTER 2 SUMMARY

1. Many mathematical concepts and problems involve various kinds of sets and subsets of those sets.
2. The primary sets of numbers that children work with in elementary school are the sets of natural numbers, whole numbers, integers, and rational numbers. The relationships among them can be shown with a Venn diagram.
3. When children count, they need to realize that there is a one-to-one relationship between the elements in the set they are counting and the set of natural numbers.
4. Our current (Hindu-Arabic) numeration system evolved over thousands of years, and people from many cultures developed ideas that contributed to our system.
5. Our base ten system has six important characteristics: (1) It has only 10 symbols, with which we can represent any amount. (2) The value of each place is 10 times that of the previous place. (3) Each place can contain only one symbol. (4) The value of a digit depends on its place in the numeral. (5) The value of a numeral is determined by multiplying each digit by its place value and then adding these products. (6) Zero represents both an empty place and an actual amount.

BASIC CONCEPTS

Section 2.1: Sets

set **41** member, element **41**
subset **41** proper subset **43**
universal set, universe **45** Venn diagrams **45**
binary operations **49**

Ways to describe sets:

words, list **41** set-builder notation **42**

Important mathematical sets:

natural numbers (N) **41**
whole numbers (W) **41**
integers (I) **41**
rational numbers (Q) **41**

Kinds of sets:

finite and infinite sets **42**
empty set, null set **44**
disjoint sets **47**

Relationships between sets:

equal **45** equivalent **45**
one-to-one correspondence **45**

Operations on sets:

intersection **46** union **46**
complement **46** subtraction **47**

Section 2.2: Numeration

Origins of numbers and counting **54**

Kinds of numeration systems:

Egyptian **56** Roman **57**
Babylonian **58** Mayan **60**
Hindu-Arabic **60**

Characteristics of numeration systems:

numeral **55** number **55**
additive **57** powers of ten **57**
place value **59** base ten **61**
digit **61** decimal **62**
expanded form **62** base ten manipulatives: small cube, long, flat, large cube **62**
simple unit **63**
composite unit **63**

REVIEW EXERCISES chapter two

1. Describe the set of powers of 10 using:
   a. Set-builder notation
   b. List notation
2. Fill in the most appropriate symbol— $\in$, $\notin$, $\subset$, or $\not\subset$—for each of the following questions.
   a. 3 $\{1, 2, 3\}$
   b. $\{1, 2\}$ $\{1, 2, 3\}$
   c. $\emptyset$ $\{1, 2, 3\}$
   d. $\{3, 4\}$ $\{1, 2, 3\}$
3. Rewrite the following statements using mathematical symbols.
   a. The set D is not a subset of the set E.
   b. 0 is not an element of the null set.
4. Let

   U = the set of whole numbers between 0 and 30

   $A = \{1, 3, 5, 7, 9, 11, 13, 15, 17, 19, 21, 23, 25, 27, 29\}$

   $B = \{0, 5, 10, 15, 20, 25\}$

   a. Represent this system with a Venn diagram.
   b. List the elements of $A \cap B$.
   c. Describe in words the elements of $\overline{A}$.

5. Let

   U = the set of college students at your college

   E = students at your college who are planning to teach elementary school

   C = students who have cars on campus

   S = students who are out-of-state

   a. Draw a Venn diagram to represent this situation: $E \cap (C \cup S)$.

   b. Draw a Venn diagram to represent this situation: Those students at your college who are in-state and who have cars on campus.

6. In a group of 100 people, 70 have a DVD player, 60 have a CD player, and 20 have neither. How many have both?

7. What is the difference between saying that two sets are equal and saying that they are equivalent?

8. Write the following amounts in the Egyptian, (late) Roman, and Babylonian systems.

   a. 47 b. 95 c. 203 d. 3210

9. Indicate what number comes after

   a. 404_{five} b. 1244_{five} c. 1001_{two}

10. Indicate what number comes before

    a. 4320_{five} b. 30040_{five} c. 1100_{two}

11. Convert the number to base ten: $243_{\text{five}} = ?_{\text{ten}}$

12. Would you rather have $\$200_{\text{ten}}$ or $\$1000_{\text{five}}$?

13. Explain why 321_{five} and 321_{six} are not the same.

14. If 10 is twice the value of 5, why doesn't a base ten flat have twice the value of a base five flat?

15. The value of the fifth place in base ten is _____ times the value of the fifth place in base five.

16. Many students say that a major learning from Section 2.2 is that 10 is not just ten, but rather "one-zero." What does "one-zero" mean? Write your answer so that it works for any base.

17. Express the following number in expanded form: 2,068. Note that there are three variations of standard form—any of these is fine here.

18. Describe, in your own words, the primary characteristics that distinguish base ten from earlier, nonbase systems—for example, the Egyptian and Roman systems.

The Four Fundamental Operations of Arithmetic

3

Learning how to add, subtract, multiply, and divide and then applying this knowledge to solve problems makes up the lion's share of elementary school mathematics. The Common Core State Standards introduce addition and subtraction of whole numbers with concrete models, pictures, and strategies for numbers less than 100 in first grade. By fifth grade, students are performing all four operations with whole numbers and decimals. Operations with fractions begin in fourth grade, and with negative numbers in sixth grade.

Our focus here will be understanding that

- The operations have multiple meanings vs. one single meaning—for example, subtraction is more than simply "take away."
- These meanings are often connected—for example, one meaning of multiplication is repeated addition.
- There are many algorithms (standard procedures) for each operation; the one you learned for each operation is simply one of many.
- Knowledge of base ten and place value is vitally important to see how these algorithms work.
- Practice and memorization are important, but practice without understanding is a waste of time.

HISTORY

We cannot overemphasize the fact that most of our everyday mathematics is actually quite recent. During the Middle Ages, many mathematicians considered an operation to be any mathematical technique or procedure that was considered important. It has only been in the last century that mathematicians have linked the concept of operation to the concept of function.[1]

SECTION 3.1 Understanding Addition

What do you think?

- How many words, aside from *combine*, can you think of that describe addition?
- Not having place value, how might the Romans have added, for example, XXXVIII + XXVI?

Children encounter addition even before they have mastered counting. They naturally combine objects and want to know how many. For example, there are 4 people in our family, and 2 guests have come for dinner. How many people will be eating?

Examine the following addition problems and then write your responses to the following questions in your own words.

1. What action words describe what is happening in these and other addition problems? One action word is *combining*. What others can you think of?
2. Other than "they all involve addition," can you think of other ways in which all four problems are alike? In what ways are some of the problems different from each other?

Four addition problems

1. Andy has 3 marbles, and his older sister Bella gives him 5 more. How many does he have now?
2. Keesha and José each drank 6 ounces of orange juice. How much juice did they drink in all?
3. Linnea has 4 feet of yellow ribbon and 3 feet of red ribbon. How many feet of ribbon does she have?
4. Josh has 4 red trucks and 2 blue trucks. How many trucks does he have altogether?

CONTEXTS FOR ADDITION

Problems 1 and 4 are easier for most children, because the child can see the actual marbles and trucks and count 1, 2, 3, 4, 5, 6, 7, 8 marbles and 1, 2, 3, 4, 5, 6 trucks. Problems 2 and 3 are more abstract in that the child cannot actually see 6 ounces or 4 feet. The child might represent the problem with concrete objects, such as 4 buttons for the yellow ribbon and 3 buttons for the red ribbon. These problems represent two basic contexts in which we operate on numbers. Some numbers represent **discrete** amounts, or objects in a set, and some numbers represent **measured** or **continuous** amounts.

A PICTORIAL MODEL FOR ADDITION

One of the themes of this book is the power of "multiple representations," and here we see that if we draw diagrams to represent the four addition problems, they seem on the surface to be quite different (Figure 3.1).

HISTORY

Much of the mathematical notation we use is actually quite recent. The symbols for addition and subtraction, $+$ and $-$, first appeared in Germany in the late 1400s. These symbols were first used to indicate sacks that were surplus or minus in weight. In 1631, William Oughtred first used the letter x to represent multiplication. Italian merchants introduced the symbol for division ($\div$) in the 1400s to indicate a half. For example, they wrote $4 \div$ to indicate $4\frac{1}{2}$. The equals sign first appeared in the late 1500s in a book by Robert Recorde.

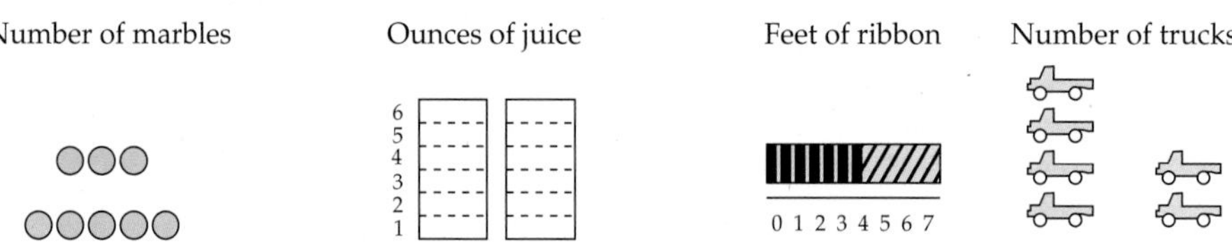

Figure 3.1

In each case, we are joining two sets or we are increasing a set. We can highlight the similarities among the problems with the representations in Figure 3.2.

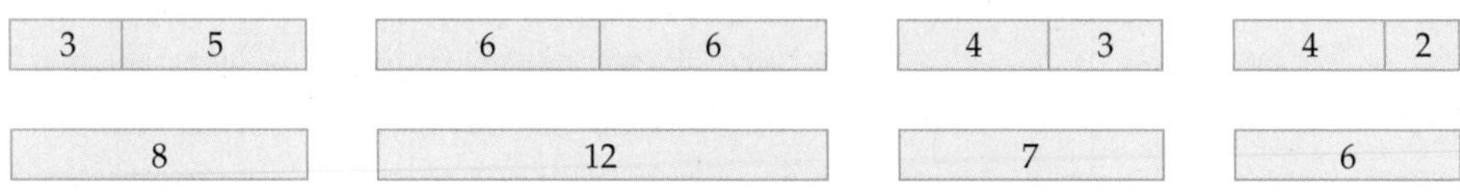

Figure 3.2

In general, we can represent any addition problem $a + b = c$ as shown in Figure 3.3.

Figure 3.3

Do you see any advantages of this general model over having no model at all? If you do, what advantages do you see? Think and then read on. . . .

1. This model captures the way in which *all* addition problems are similar—that is, joining and combining two amounts to make a larger amount.
2. This model is also related to the notion of parts and wholes, an abstraction that is important in the development of whole-number ideas and in understanding other mathematical ideas, like fractions.
3. This model also works well whether the elements to be combined are sets of discrete objects, like the marbles and trucks, or measurements, like the ounces of juice or feet of ribbon.
4. As we examine all four operations, we will see that we can define all four operations in this context. This reveals the essential connectedness of the four operations. When students see this connectedness, they are likely to be more successful with nonroutine and multistep problems. Students in Singapore, who consistently rank high in standardized math tests taken by students in over 100 countries, use similar bar models extensively to understand math. We will continue to explore these bar models in our discussion of fractions in Chapter 4 and of algebra in Chapter 6.

One researcher has stated that "(p)robably the major conceptual achievement of the early school years is the interpretation of numbers in terms of part and whole relationships."[2] It is important to note that understanding part–whole relationships with whole numbers allows numbers to be interpreted simultaneously as positions on the mental number line and as compositions of other numbers. For example, 18 is the number after 17 and before 19, but 18 can also be seen as $10 + 8$, $9 + 9$, $20 - 2$, and so on. Understanding that a number can be composed (put together) and decomposed (broken into parts) is essential for being able to work confidently with the four operations. This notion of composing and decomposing is one of the big ideas of elementary mathematics, and we will come back to it repeatedly throughout the text.

There are several classical stages in children's understanding of addition. At the most basic level, the child counts to determine the sum. For example, in the first problem on p. 98, many young children would answer the question by putting the marbles on the floor and then counting the two groups.

In the next stage of development, the child "counts on." That is, the child begins with the first number and counts however many more the second number represents. A child solving Problem 1 (see Figure 3.4) in this manner would say, "4, 5, 6, 7, 8," probably keeping track of the second number being added (5) with fingers. At the next level, the child realizes that it is possible to begin with the larger number (i.e., $3 + 5 = 5 + 3$) and counts, "6, 7, 8." Finally, the child simply knows that $3 + 5 = 8$.

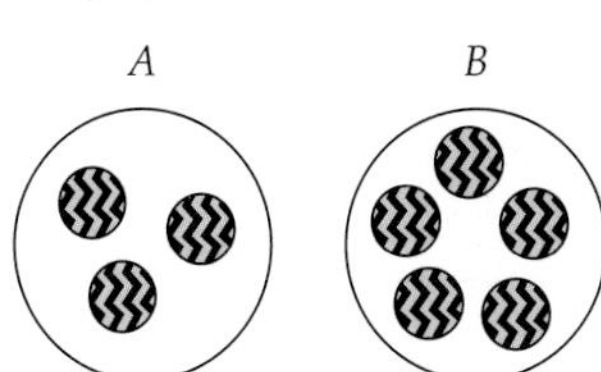

Figure 3.4

Addition is formally defined using set notation:

Let A = the set of Andy's marbles, and let B = the set of Bella's marbles.

Then $n(A) = a$, the number of marbles Andy has, and $n(B) = b$, the number of marbles Bella has.

We then define **addition** in the following way:

> If A and B are disjoint sets, containing a and b elements, respectively, then $a + b = n(A \cup B)$.

In other words, the **sum** of a and b is equal to the number of elements in the union of the two sets. Do you see why the sets have to be disjoint?

The numbers a and b are called **addends**.

REPRESENTING ADDITION WITH NUMBER LINES

The number line plays an important role throughout arithmetic, but research shows that American schoolchildren tend to have trouble understanding and using this model.

Number lines are found on rulers, clocks, graphs, and thermometers.

A **number line** can be constructed by taking a line (not necessarily a straight line) and marking off two points: zero (the origin) and one. The distance from 0 to the point 1 is called the unit segment, and the distance between all consecutive whole numbers is the same. Although number lines are most commonly used to represent length, they may be used to model all kinds of problems. For example, we could use a number line to indicate time, with each unit representing one unit of time—day, minute, year, etc.

How could we have represented the ribbon problem with a number line? Write down your thoughts before reading on. . . .

In the diagram at the left in Figure 3.5, we first draw an arrow (in this case representing the length of the ribbon) 4 units long. We draw another arrow 3 units long and connect the two arrows. The arrow at the top represents the combined length of the two shorter arrows.

Figure 3.5

In the diagram at the right, we start at the point on the line representing the length of the first ribbon and then draw an arrow 3 units long (representing the second ribbon). The location where the arrow ends tells us the combined length of the two ribbons.

Both diagrams represent $4 + 3$ on the number line, although the one on the left more closely resembles the actual laying of the two ribbons end to end.

PROPERTIES OF ADDITION

When young children start to learn about adding, Table 3.1 has been the traditional method of representing the 100 "addition facts" that they have to learn. It can look imposing to some children! However, understanding some properties of addition and base ten can unlock its potential as a learning tool. As you look at the table, what do you observe (insights or patterns) that might make learning the addition facts easier for children?

> **MATHEMATICS**
>
> If you did Exploration 2.3, recall how strange the Alphabitian system was to you. What observations made you more comfortable with adding? Can you think of analogous observations that might make the learning of base ten addition facts easier for young children?

One observation (in children's language) is that "adding zero doesn't change your answer." Mathematicians call this the **identity property of addition**. It is represented in symbols as follows:

$$a + 0 = 0 + a = a$$

Children discover that when you add 1 to any number, you get the next number from counting. In other words, they connect addition and counting.

TABLE 3.1

	0	**1**	**2**	**3**	**4**	**5**	**6**	**7**	**8**	**9**
0	0	1	2	3	4	5	6	7	8	9
1	1	2	3	4	5	6	7	8	9	10
2	2	3	4	5	6	7	8	9	10	11
3	3	4	5	6	7	8	9	10	11	12
4	4	5	6	7	8	9	10	11	12	13
5	5	6	7	8	9	10	11	12	13	14
6	6	7	8	9	10	11	12	13	14	15
7	7	8	9	10	11	12	13	14	15	16
8	8	9	10	11	12	13	14	15	16	17
9	9	10	11	12	13	14	15	16	17	18

We also find that when we add any two numbers, we get the same sum regardless of the order in which we added, which is the **commutative property** of addition.

$$a + b = b + a$$

I have been fortunate to be with a few children when they made this discovery. They were as excited as Edison probably was when he invented the light bulb. Their faces just shone when they shared with me the discovery that you get the same number either way!

There is another discovery, called "bridging with 10," that makes learning addition easier. For example, if you ask a child, "What is 7 + 5?" many children will say something like, "7 + 3 is 10 and 2 more is 12." Many will intuitively decompose 7 + 5 into 7 + 3 + 2.

If we write this as $7 + (3 + 2) = (7 + 3) + 2$ then this leads to a third property, called the **associative property of addition**. As another example, $(37 + 75) + 25 = 37 + (75 + 25)$. Formally, we say

$$(a + b) + c = a + (b + c)$$

For most college students, the closure property evokes one of two reactions: either "so what" or "here we go again, making something obvious look complicated." Let me turn the tables on you. How might this property be related to a question a child might actually ask? Think and read on. . . .

Children will literally ask closure questions: "Will *any* two numbers make a whole number?"

One last property of addition often seems almost trivially obvious; it is called the **closure property of addition**. The closure property states that the sum of any two whole numbers is a unique whole number. There are two parts to this property: (1) uniqueness (the sum will always be the same number) and (2) existence (the sum will always be a whole number).

Not all sets are closed under addition, for example, the set of odd whole numbers. The sum of two odd numbers is not in the set of odd numbers because an odd plus an odd equals an even; thus we say that the set of odd numbers is not closed under addition.

INVESTIGATION 3.1a A Pattern in the Addition Table

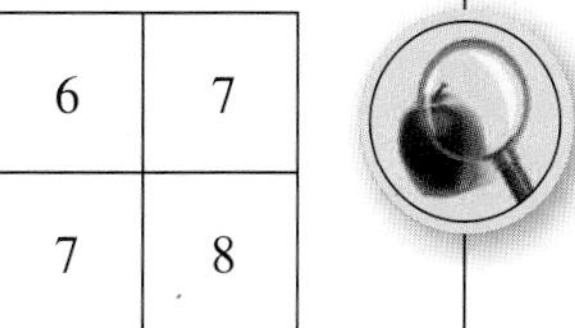

6	7
7	8

As you will come to appreciate by the end of this course, patterns can help make the learning of mathematics easier and more interesting. Mathematical Practice 7 of the CCSS in particular talks about looking for patterns and making use of the structure of math. In the addition table, if you look at any 2×2 **matrix** (that is, a rectangular array of numbers or other symbols), the sums of the numbers in each of the two diagonals are equal. For example, in the matrix to the left, $6 + 8 = 7 + 7$. Can you justify this pattern mathematically? Work on it and then read on. . . .

DISCUSSION

DESCRIPTION 1

At a verbal level, one could justify this pattern by saying that in any 2 × 2 matrix in the table, the two numbers in one diagonal are always identical and the other two numbers are always 1 less and 1 more than this number. Therefore, the sum of the two other numbers will "cancel out" so that you get the "same" sum in either case.

DESCRIPTION 2

We can use some notation to make the description simpler. Noting that the value of each number increases by 1 each time that we move across (or down) the table, we can let x represent the number in the top left corner of the diagonal. Thus, in relation to x, the values of the other three numbers are

x	$x + 1$
$x + 1$	$x + 2$

It is an algebraic exercise to demonstrate that the sum of each diagonal is $2x + 2$. Going from top left to bottom right we have $(x) + (x + 2) = 2x + 2$. The other diagonal is $(x + 1) + (x + 1)$, which also equals $2x + 2$.

DESCRIPTION 3

Yet other students will say that the sums of the diagonals are equal because "it's the same numbers in both cases." What do you think such a student might be seeing? Think before reading on. . . .

Let's say that we are looking at the matrix formed by the intersection of the 2 row and the 3 row and the 4 column and the 5 column. The numbers are 6, 7 and 7, 8. However, if we represent the numbers by their origin, we have

	4	5
2...	2 + 4	2 + 5
3...	3 + 4	3 + 5

The sum of the top-left-to-bottom-right diagonal is (2 + 4) + (3 + 5). However, because of the commutative and associative properties, this sum is equal to (2 + 5) + (3 + 4). In other words, we are indeed using the same numbers!

We can now generalize this cell by saying that the matrix formed by the intersection of the a row and the b row and the c column and the d column is

	c	d
a...	$a + c$	$a + d$
b...	$b + c$	$b + d$

and

$$a + c + b + d = a + d + b + c$$

Our first work with addition in base ten will be doing some addition problems mentally, for a couple of reasons. First, this will require you to think carefully about how your knowledge of place value applies to adding numbers. Second, much of our use of arithmetic does not involve pencil and paper or calculators, but rather mental computation—when estimating or when it is quicker to do a computation or part of a computation in our head than with a pencil or calculators.

CARRYING AND BORROWING

We have noted that we often need to regroup when adding and subtracting. However, using the traditional algorithms we use different words—we use *carry* for addition and *borrow* for subtraction. The following problems illustrate this point.

$$\begin{array}{r} {}^{1}\\ 36 \\ +28 \\ \hline 64 \end{array} \qquad\qquad \begin{array}{r} {}^{5}\\ \not{6}\,{}^{1}4 \\ -36 \\ \hline 28 \end{array}$$

In the addition problem, because $6 + 8 > 9$, we put the 4 in the ones place and "carry" the 1 to the tens place. In the subtraction problem, in order to subtract in each place, we need to "borrow" a 1 from the tens place and move it to the ones place whose value is now 14.

Take a minute to write down what is similar or the same about both processes and what is different. Also, consider what "carrying" and "borrowing" mean in other contexts and whether they are really descriptive of this process. Then read on. . . .

What is the same about both is that (1) an amount is moved from one place to another, and (2) this amount always represents a 10-for-1 exchange (10 ones for 1 ten or 1 ten for 10 ones).

What is different is the direction of the exchange. When adding, we move the 10 from the smaller place to a 1 in the larger place (right to left). When subtracting, we move the 1 from the larger place to a 10 in the smaller place.

Because of the use of these two different words, neither of which is very descriptive of what is actually happening in the process, many children do not realize that these processes are virtually identical. Thus most textbooks now use the words *trading, exchanging,* or *regrouping* instead of the words *carrying* and *borrowing.* This is a very big point, because many, many children are judged to have a learning disability not so much because they are slower or less mathematically inclined but because they find it harder to remember something that doesn't make sense. When we use the same words for the same process, and when children see how closely connected addition and subtraction are, fewer children struggle with subtraction. We encourage you to correct your vocabulary as you proceed through this chapter. It may take time for you to stop using the words *carrying* and *borrowing* since they have been part of your vocabulary since early elementary school. We will use the term *trading* for the rest of the text.

INVESTIGATION 3.1b Mental Addition

Do the following five computations in your head. Briefly note the strategies you used, and try to give names to them.

Note: One mental tool all students have is being able to visualize the standard algorithm in their heads. For example, for the first problem, you could say: "$9 + 7 = 16$, trade for ten, then $5 + 3 = 8 + 1$ traded ten makes 9; the answer is 96." However, because you already know that method, I ask you not to use it here but to try others. There are actually quicker ways to do this problem in your head than using the traditional algorithm. See whether you can discover any of them.

1. $39 + 57$ 2. $68 + 35$ 3. $66 + 19$ 4. $545 + 228$ 5. $186 + 125$

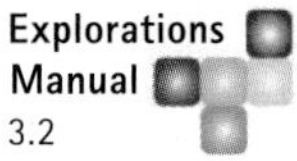

DISCUSSION

Leading digit One strategy that works nicely with most addition problems is **leading digit**. Some people refer to it as front end because we add the "front" of the numbers first. Using leading digit with Problem 1 looks like this:

$$39 + 57 = (30 + 50) + (9 + 7) = 80 + 16 = 96.$$

This strategy can be used with larger numbers too. Try it on Problem 4 and then read on. . . .

$$545 + 228 = 700 + 60 + 13 = 773$$

Compensation Another powerful mental math strategy is called **compensation**. Using compensation with Problem 1 looks like this: $39 + 57 = 40 + 56$.

Do you see how we transformed $39 + 57$ into $40 + 56$?
Which other problems lend themselves to this strategy?
Number 3 could also be solved with this strategy: $66 + 19 = 65 + 20$

Break and bridge We can use the break and bridge strategy in Problem 2 in this way:

$$68 + 35 = (68 + 30) + 5 = 98 + 5 = 103$$

Representing this strategy on a number line makes it easier for some to understand. We break the second number apart (using expanded form) and add one place at a time.

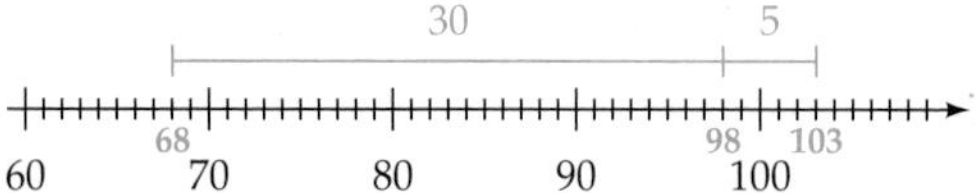

Which other problems lend themselves to this strategy?

Compatible numbers Another powerful strategy is creating compatible numbers. This often involves seeing pairs of digits whose sum is 10. Using compatible numbers with Problem 5 looks like this:

$$186 + 125 = (180 + 120) + (6 + 5) = 300 + 11 = 311$$

This works when we see that $180 + 120 = 300$.

Choosing a strategy Which strategy you use is often a matter of preference. For example, Problem 2 ($68 + 35$) may be done mentally in at least four different ways, each of which is the easiest way for *some* students. One of your goals as a future teacher is to become comfortable with each strategy so that you can support learning for all of your students.

Leading digit:	$68 + 35 = 60 + 30 + 8 + 5$
Compensation:	$68 + 35 = 70 + 33$
Break and bridge:	$68 + 35 = 68 + 30 + 5$
Compatible numbers:	$68 + 35 = 65 + 35 + 3$

Justifying strategies Let's look at the compatible numbers strategy for Problem 2 in detail to connect the mental work with the properties we have discussed.

The action	Justification
$68 + 35 = (65 + 3) + 35$	Substitution
$= 65 + (3 + 35)$	Associative
$= 65 + (35 + 3)$	Commutative
$= (65 + 35) + 3$	Associative
$= 100 + 3$	Addition
$= 103$	Addition

CHILDREN'S STRATEGIES FOR ADDITION

Explorations Manual 3.3

You will explore the teaching of addition in your methods course. Here we will focus on the mathematics underlying common stages in the child's development of computation with addition.

Imagine that you are a child and you haven't yet learned the standard procedure for adding. How might you add 48 + 26?

Add up by 10s

Just as you read earlier that one stage in young children's development of addition is to add up, some children begin multidigit addition in the same way. They add up by 10s from 48. What they say is, "48, 58, 68, 74." Some children will struggle a bit from 68 to 74. They use 10 as a bridge, and say, "48, 58, 68, plus 2 is 70, plus 4 more is 74." Many children will use a number line to explain this strategy.

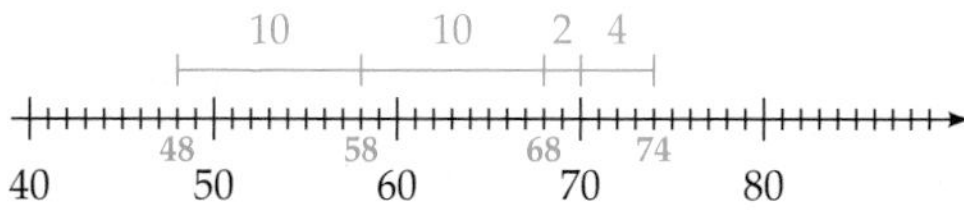

Break the second number apart

Another variation of adding up is to begin with the first number and break the second number into its place value parts. That is, they add the 10s and then the 1s.

$$48 + 20 = 68$$

$$68 + 6 = 74$$

A number line is a useful representation of this strategy.

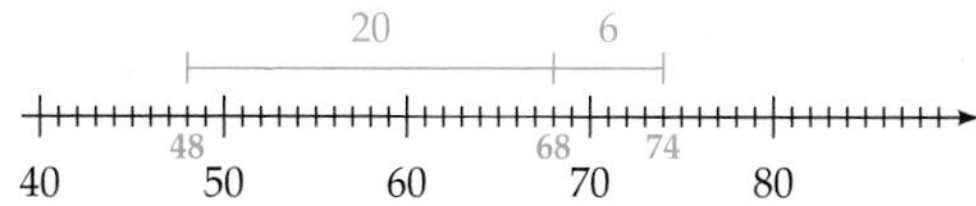

CLASSROOM CONNECTION

Constance Kamii and many others who have explored what is called a constructivist approach to learning arithmetic have found that when children are not just shown how to add, the vast majority of children will actually add from the largest to the smallest place. This is very interesting because we read from left to right. Personally, I have always added (and subtracted) from left to right. It always made more sense to me, and it is faster.

Use partial sums

A very common approach that comes closer to one of the standard algorithms is to add, from left to right. In the beginning, children often write the **partial sums** (which are the sums of each place). The diagram below illustrates this approach with manipulatives. Note that this method is essentially the leading-digit method that many people use when adding mentally.

```
  48
+ 26
----
  60
  14
----
  74
```

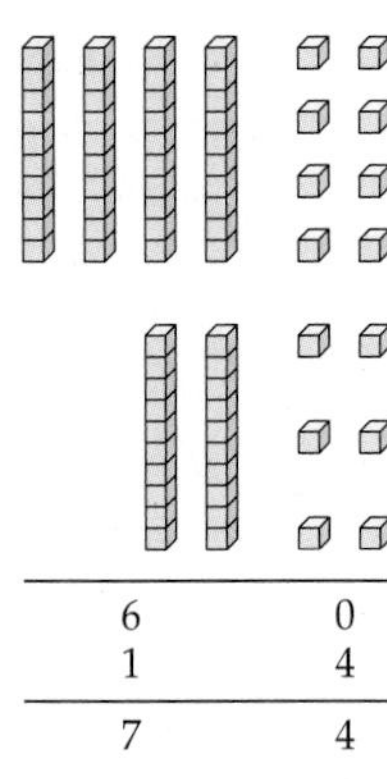

6	0
1	4
7	4

Use money and compensate

Depending on the problem, some children will use other frameworks, often money. For example, because 48 and 26 are close to a half-dollar and a quarter, some children will add 50 and 25 and then compensate, since 48 is 2 less than 50 and 26 is 1 more than 25.

$$50 + 25 - 2 + 1$$

This strategy, though it is not applying base ten in this problem, is actually quite powerful and can be connected to base ten through discussion.

INVESTIGATION 3.1c Children's Strategies for Adding Large Numbers

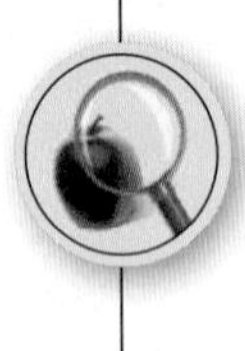

Explorations Manual 3.3

What if the numbers were bigger—for example, 368 + 574? Look back on the approaches described earlier. Can you adapt any of them to this problem? Try to do so before reading on. . . .

DISCUSSION

The "break the second number apart" strategy applies to larger numbers.

$368 + 500 = 868$

$868 + 70 = 938$

$938 + 4 = 942$

The "partial sums" strategy also applies.

$300 + 500 = 800$

$60 + 70 = 130$

$8 + 4 = 12$

$800 + 130 = 930$ and $930 + 12 = 942$

The "partial sums" strategy can be modified to "keep each sum in its proper place," which is shown at the right.

```
  368        368          368
+ 574      + 574        + 574
-----      -----        -----
  800          8     or    12
  130        13           13
   12         12          8
-----      -----        -----
  942        942          942
```

Now that you have done some addition problems mentally and have seen children's approaches, and now that you have examined how place value knowledge enables those methods to work, let us look at one of the standard algorithms for adding.

ALGORITHMS

A major goal of elementary school is to have students become computationally fluent. This means developing efficient algorithms for each operation. An **algorithm** is a single, clearly described method that works in all cases. To master algorithms for all the operations is a long and challenging process, and many of my students remember how difficult certain algorithms were ("borrowing" in subtraction and all the steps of long division), and my recent teaching of fourth grade has reminded me how hard multidigit multiplication is to many students.

The first arithmetic book in Europe was published in Treviso, Italy, in 1478. It described the *new* base ten system and many different procedures for adding, subtracting, multiplying, and dividing. Because base ten was so new to Europe at that time, many different algorithms were in use; over hundreds of years, certain algorithms "won." Thus the algorithms that you learned to add, subtract, multiply, and divide are not *the* algorithms but simply four of many. Furthermore, they are not universally used today. Some of you, in different parts of the country, learned different algorithms, and school children in different parts of the world learn very different algorithms for some of the operations, especially subtraction. So let us examine this notion of algorithm.

HISTORY

An Arabic scholar named Al-Khowarizmi wrote many books on mathematics and astronomy. In A.D. 825, he published a book that explained the use of the Hindu numerals, including the standard procedures for the fundamental operations. The Latin translation of his name is Algorismus. This is where our word *algorithm* comes from.

ADDING BEFORE BASE TEN

Before we begin examining addition in base ten, take a moment to think about how people added before base ten was invented. Imagine that you were a Roman. How might you add these two numbers? As you found in Section 2.2, Romans couldn't just add the VII to the VI and "carry" the III, since it is not a place value system. They would have to combine the symbols and do some rewriting where efficient (such as V + V = X).

```
   XXXVII
+    XLVI
---------
```

There is evidence that the Babylonians did calculations in the sand with pebbles and inscribed the results on clay tablets. In fact, our word *calculate* comes from *calculus,* which means "small stone." This pebble method may have been a precursor to the abacus, which has been used extensively in Asia. The term *carrying* may have also come from carrying the pebbles from one column to the next.

INVESTIGATING ADDITION ALGORITHMS IN BASE TEN

Add 267 + 133 as you normally would. Can you explain the "whys" of each step? Try to do so before reading on. You may or may not wish to use base ten blocks.

In the context of our base ten numeration system, we are combining 2 hundreds, 6 tens, and 7 ones with 1 hundred, 3 tens, and 3 ones. If the student understands the process of combining and regrouping, then this problem is *not* substantially more difficult than one with smaller numbers; it is only longer.

CLASSROOM CONNECTION

The National Assessment of Educational Progress (NAEP) has found that third-graders' success in computation decreases considerably when they move from two-digit problems to three-digit problems. Further research indicates that many students are memorizing rather than understanding the process.

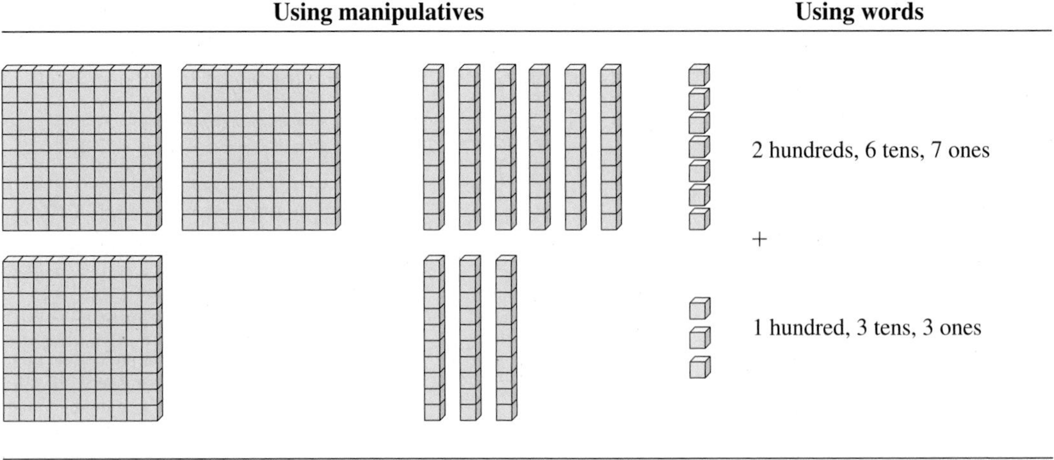

Recalling what addition means, when we literally combine the two sets, we have 3 hundreds, 9 tens, and 10 ones.

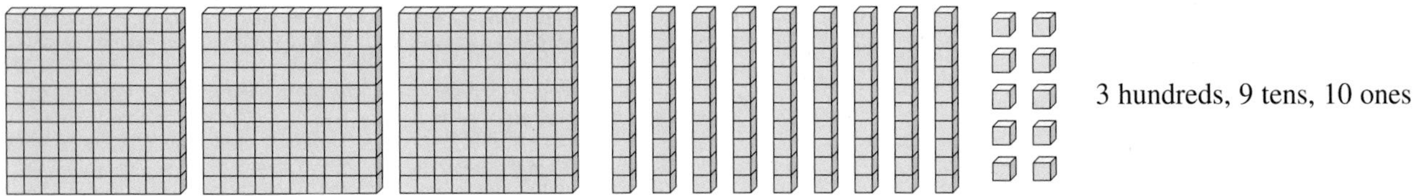

However, this is not a valid answer. Thus we have to regroup. After we "trade" the 10 ones for 1 ten, we have 3 hundreds, 10 tens, and 0 ones.

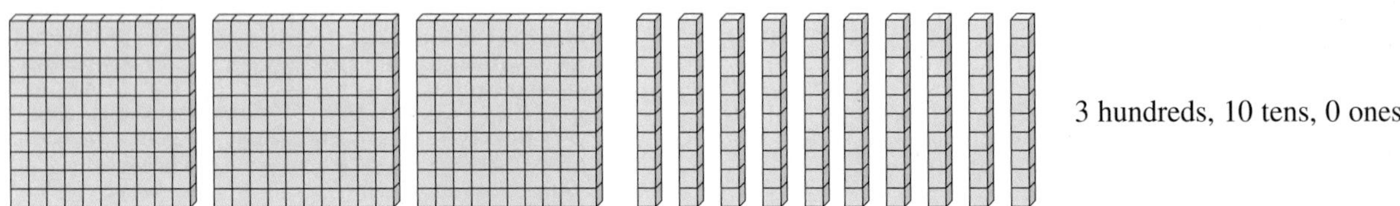

Because we now have 10 tens (once again not a valid representation in base ten), we trade them for another hundred, which gives us 4 hundreds, no tens, and no ones—that is, 400.

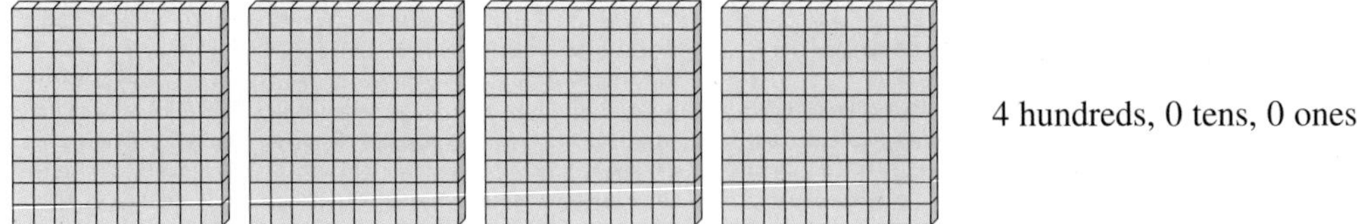

A commonly used algorithm With this manipulative representation, you can now better understand the why of one **common algorithm for addition**.

Step 1 7 + 3 = 10; place the 0 in the ones place and put the 1 above the tens place, because 7 + 3 is equivalent to 1 ten and 0 ones.

$$\begin{array}{r} \scriptstyle 1 \\ 267 \\ \underline{133} \\ 0 \end{array}$$

Step 2 1 + 6 + 3 = 10 (which is really 1 ten plus 6 tens plus 3 tens = 10 tens or 1 hundred); place the 0 in the tens place and put the 1 above the hundreds place (to represent trading 10 tens for 1 hundred).

$$\begin{array}{r} {\scriptstyle 1\,1}\\ 267 \\ \underline{133} \\ 00 \end{array}$$

Step 3 1 + 2 + 1 = 4 (which is really 1 hundred plus 2 hundreds plus 1 hundred = 4 hundreds); place the 4 in the hundreds place. The sum is 400.

$$\begin{array}{r} {\scriptstyle 1\,1}\\ 267 \\ \underline{133} \\ 400 \end{array}$$

Representing the problem in expanded form enables us to prove why it works.

Statement	Justification
267 + 133	
= (2 · 100 + 6 · 10 + 7) + (1 · 100 + 3 · 10 + 3)	Expanded form
= (2 · 100 + 1 · 100) + (6 · 10 + 3 · 10) + (7 + 3)	Commutative and associative properties
= (2 + 1)100 + (6 + 3)10 + (7 + 3)	Distributive property
= 3(100) + 9(10) + 10	Addition
= 3(100) + (9 + 1)10	Distributive property
= 3(100) + 10(10)	Addition
= 3(100) + 100	Multiplication
= (3 + 1)100	Distributive property
= 4(100)	Addition
= 400	Multiplication

Seeing how numbers can be *composed* and *decomposed* makes it possible to understand the algorithm deeply. To understand the addition algorithm, we **decompose** the number using expanded form; we can then see how those different parts can be reconfigured—**composed**—to make our new whole, that is, the sum.

INVESTIGATION 3.1d An Alternative Algorithm

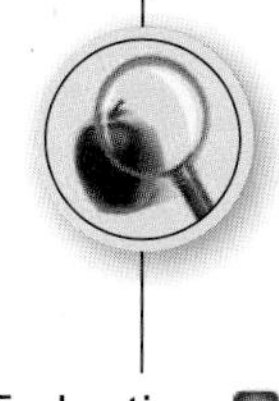

Explorations Manual 3.3

As mentioned earlier, when base ten was invented, many different algorithms for each operation were invented. One algorithm that is a favorite of children is called the **lattice algorithm**. Observe the example below and see if you can figure out how it works and why it works.

```
    6  4  5
 +  7  2  8
  [1/3][0/6][1/3]
1/ 3/ 7/ 3
```

DISCUSSION

First, you write the problem, and below each place draw a square.

Second, draw diagonal lines through each square that extend below the square.

Third, write the result of each partial sum in the box.

HISTORY

The earliest book on arithmetic is the papyrus of Ahmes, created in about 1700 B.C. The book is a compilation of what was known at the time, some of which goes back to 3000 B.C. The title of the papyrus was "Directions for Obtaining the Knowledge of All Dark Things." The first arithmetic book printed in the West was printed in Treviso, Italy, in 1478. The first one in the United States was printed in 1729.[3]

Last, add diagonally. If the sum in any diagonal addition is greater than 10, trade to the next diagonal just as you do with the standard algorithm.

If we represent the problem in terms of the place value of each part of the lattice, we see that the lattice "herds" each digit to the proper place.

```
    6  4  5     O = Ones
 +  7  2  8     T = Tens
[Th/H][H/T][T/O]  H = Hundreds
                 Th = Thousands
```

INVESTIGATION 3.1e Addition in Base Five

Do these addition problems in base five. Visual representations or physical models are encouraged to help you understand.

A. $3_{five} + 2_{five}$ **B.** $21_{five} + 13_{five}$ **C.** $43_{five} + 24_{five}$

DISCUSSION

A. $3_{five} + 2_{five}$

One way to determine the sum, especially if this feels awkward, is to count on from 3 just like children do when learning to add in base ten. When we do this, we get 3_{five}, 4_{five}, 10_{five}.

If we represent this problem with manipulatives, we have

Longs	Singles
	3 singles
	2 singles

Longs	Singles
1 long	

Because in base five, we trade to the next place value when we have five in a place, the five singles become one long with zero singles left over. So, $3_{five} + 2_{five} = 10_{five}$.

B. $21_{five} + 13_{five}$

You might prefer to represent this problem vertically. Since there is no trading, the answer is 34_{five}.

$$\begin{array}{r} 21_{five} \\ +\ 13_{five} \\ \hline 34_{five} \end{array}$$

Longs fives	Singles ones	
2 longs	1 single	21_{five}
1 long	3 singles	13_{five}

C. $43_{five} + 24_{five}$

Since there is trading here, we will solve the problem simultaneously with a manipulative and symbolic representation. It is important to see the connections between these representations. Study the representations below before reading on.

Flats $five^2$	Longs fives	Singles ones	
			$\overset{1}{43_{five}}$ + 24_{five}
			122_{five}

We have to do some trading in this example. In the ones place, we trade five singles for 1 long of five and are left with 2 singles in the ones place. Then we add the 4 longs + 3 longs + 1 long (the one long there from the trading), and because we have more than five we trade five longs for 1 flat and are left with 2 longs. This gives us 1 flat and 2 longs and 2 singles; in other words, 122_{five}. While you will likely not teach adding in other bases in elementary school, this helps us to understand adding in base ten. The process is the same—the only difference is how many we need in one place in order to trade for the next place.

INVESTIGATION 3.1f Children's Mistakes

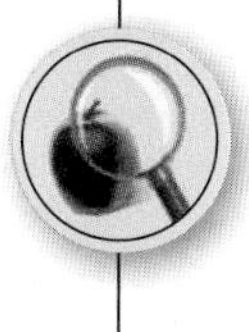

The problem below illustrates a common mistake made by many children as they learn to add. Understanding how a child might make that mistake and then going back to look at what lack of knowledge of place value, of the operation, or of properties of that operation contributed to this mistake is useful. What error on the part of the child might have resulted in this wrong answer?

The problem: 38 + 4 = 78

DISCUSSION

In this case, it is likely that the child lined up the numbers incorrectly:

$$\begin{array}{r} 4 \\ +\ 38 \\ \hline 78 \end{array}$$

Giving other problems where the addends do not all have the same number of places will almost surely result in the wrong answer. For example, given 45 + 3, this child would likely get the answer 75. Given 234 + 42, the child would likely get 654. In this case, the child has not "owned" the notion of place value. Probably, part of the difficulty is not knowing expanded form (for example, that 38 means 30 + 8—that is, 3 tens and 8 ones). An important concept here is that we need to add ones to ones, tens to tens, etc. Base ten blocks provide an excellent visual for this concept as students can literally see why they cannot add 4 ones to 3 tens.

CLASSROOM CONNECTION

A friend of mine, David Sobel, was talking about mathematics with his six-year-old daughter, Tara. David had just shown Tara that 20 + 20 = 40. Tara thought for a moment and then proudly announced that 50 + 50 must be 70. When David asked how she had got that answer, she said, "When you add the same numbers that have a zero at the end, you just skip ten!"

ESTIMATION

Most people now rely on calculators when exact answers are needed. However, estimating skills are still very important in cases where an exact answer is not needed and to check the reasonableness of results obtained on the calculator. Estimation, in turn, requires good mental arithmetic skills, which come from an understanding of the nature of the operations, a firm understanding of place value, and the ability to use various properties.

When are numbers estimates and not exact numbers? Before we examine some methods of estimation, we need to understand when numbers represent estimates and when they represent exact amounts. All of the following numbers are estimates or approximations. What ways can you see to group them according to why they are estimates? For example, the age of a dinosaur bone is an estimate because present dating methods do not enable us to get an exact number; in other words, the exact age is unknown.

CLASSROOM CONNECTION

The following story illustrates the need for number sense. Stopping at a dinosaur skeleton, the museum tour guide said that the skeleton was 65 million and 10 years old. Someone asked how he knew that. The guide replied, "Well it was 65 million years old when I began working here, and I've been here 10 years."

- A certain dinosaur bone is 65 million years old.
- The number of hungry children in the United States is 16 million.
- The area of the Sahara Desert is 3,320,000 square miles.
- The mean July temperature in Tucson, Arizona, is 86 degrees.
- Jane lives 55 miles from the nearest airport.
- My office is 12 feet by 9 feet.

Numbers are estimates when:

1. The exact value is unknown—for example, predictions and numbers that are too large or difficult to determine.
2. The value is not constant—for example, population and barometric air pressure.
3. There are limitations in measurement—for example, the length of a desk and the area of a pond.

Rounding Just as many numbers in everyday life are estimates, many numbers are rounded:

- It took Jackie 10 hours to get from Boston to Buffalo.
- Anna gets 34 miles per gallon with her new car.
- Rosie put 2100 miles on her car last month.
- Fred paid $18,000 for his new car.
- The population of Sacramento, California, is 370,000.

The preceding examples bring another question to mind: Why do we round? Stop for a minute to think of your own response. Then read on. . . .

1. Rounding makes comprehension easier. Which of the following sentences would you prefer to read in a newspaper? Why?

   The school budget for the 34,168 students was $153,167,458.

   The school budget for the 34,000 students was $153 million.

2. Rounding makes computation easier. Let's say you are trying to determine the cost of buying new textbooks for your fifth-grade class. There are 23 students, and the textbooks cost $19.75. If you wanted to get a sense of the cost, you would probably round 19.75 to 20 and multiply 23 by 20.

When do we use estimation and when do we use exact computation? Before you read this section, write down your thoughts about this question. Then read on. . . .

CLASSROOM CONNECTION

Every year in my middle school mathematics methods course, my students go into the schools and ask children questions in order to learn more about what the students know and what they don't know. Every year my students are stunned to find that many of the middle school students have few or no estimation and mental math strategies in their repertoire.

Following are several examples of when people generally estimate:

- Making a budget—cost of college, cost of food per month
- Determining the cost of a trip or vacation—ski trip, camping trip, trip to Europe
- Deciding which to buy—a new car or a used car
- Determining time—how long to get to . . .
- Determining whether we have enough money—being at the grocery store when short on cash
- Deciding how much the tip should be (at a restaurant)
- Determining how long the paper or project will take

Estimation methods As you will find in this chapter, we do not estimate in the same way in all situations. The method(s) we use to estimate depend partly on how close the estimate has to be and on whether we want to over- or underestimate deliberately (for example, underestimating when budgeting can cause serious problems, so budgets are sometimes deliberately overestimated). Finally, people usually prefer certain methods to other methods.

ESTIMATION STRATEGIES FOR ADDITION

Here we will analyze some estimation problems to understand better the application of base ten concepts and mental math strategies.

INVESTIGATION 3.1g What Was the Total Attendance?

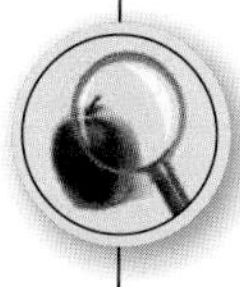

In each of the following estimation problems, first obtain a quick rough estimate, then obtain the best estimate you can, and then read on. . . .

A. Approximately what was the total attendance for the following three football games at Tiger Stadium? 75,145 34,135 55,124

DISCUSSION

Remember that one of the main goals of the investigations is for you to develop a repertoire of strategies.

Leading digit:	7 + 3 + 5 = 15; that is, 150,000 The leading-digit method used alone will always give you an estimate that is lower than the actual sum.
Refined leading digit:	150,000 + (5 + 4 + 5 = 14—that is, 14,000) = 150,000 + 14,000 = 164,000
Rounding:	80,000 + 30,000 + 60,000 = 170,000
Compatible numbers:	75 + 35 + 55 = 110 + 55 = 165, which represents 165,000

B. Approximately what was the total attendance for three baseball games at Wrigley Field in Chicago? Make your own rough estimate and then a refined estimate before reading on. . . .

32,425 31,456 34,234

DISCUSSION

Each of the four methods above could have been used here. You may well have come up with another strategy called **clustering**, because all three numbers are relatively close together. In this case, a very quick, rough estimate would be

$$30{,}000 \times 3 = 90{,}000$$

If we use a refined leading digit strategy, we can get 90,000 + 7000, and looking at the 425, 456, and 234, we can see that this is about 1000. Thus a more refined estimate is about 98,000.

Looking back There are two points to keep in mind when estimating and doing mental mathematics:

1. The method you use is often partially determined by the problem itself. If you want only a rough estimate, you might use leading digit or rounding. If you want a more refined estimate, you might use compatible numbers. If you want to make sure you have enough money, you might round everything up so that the estimated sum is definitely greater than the actual sum.
2. If you have a large repertoire of estimating and mental math techniques in your toolbox, you will be more skillful, and in a class of 20 children, you will see many different strategies.

A REFINED TECHNIQUE FOR ROUNDING

Most textbooks offer this guideline for rounding:

If the appropriate digit is less than 5, round down. Otherwise round up.

For example, if we estimate 35 + 57, we round to 40 + 60 = 100. However, if we have a problem in which many of the numbers end in a 5, this method will give an estimate that is too high.

A more refined set of guidelines for rounding makes the following modification:

1. If the digit to the left of the 5 is odd, round up.
2. If the digit to the left of the 5 is even, round down.

Let us see how this affects the estimate with the preceding problem:

Problem	Standard rounding	Refined rounding
35	40	40
42	40	40
76	80	80
45	50	40
35	40	40
85	90	80
318	340	320

If we use this refined method, our estimate will generally be closer to the actual amount.

INVESTIGATION 3.1h Estimating by Making Compatible Numbers

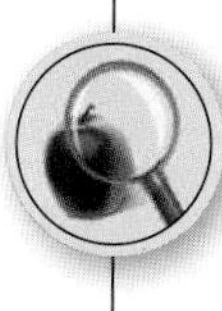

Using compatible numbers is an effective strategy to estimate the following sums. Try this on your own and then check below. . . .

A.	B.
38	23
72	359
89	177
65	675
+ 27	162
	+ 315

DISCUSSION

A. In this case, a quick glance shows us that 38 and 65 will make a sum close to 100, and so will 72 + 27. If we see this, our estimate of 200 + 89 = 289 is quite close to the actual answer of 291.

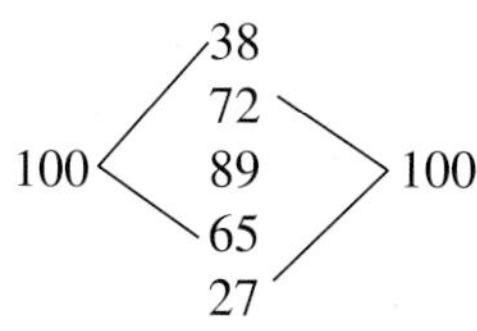

B. Using compatible numbers, we can estimate 200 + 500 + 1000 = 1700.

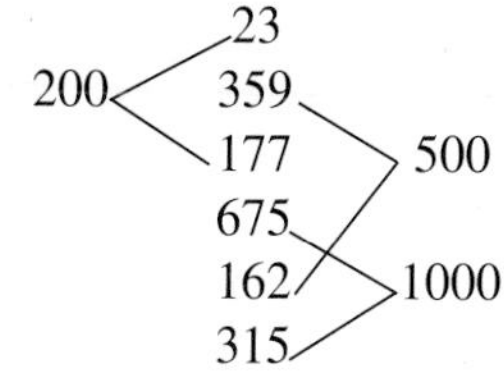

Looking back Before we move on, take a few moments to look back over the strategies for mental addition and estimation that we have explored. Do you find that you prefer some over others? Do you see that some strategies work better with some numbers than with others?

NUMBER SENSE

Number sense involves the ability:

1. To take numbers apart and put them together.
2. To move fluently among different representations.
3. To recognize when one representation is more useful than another.
4. To perform mental computation and estimation flexibly.
5. To determine whether an answer is reasonable.

In each section of this chapter, we will investigate problems that will help to further develop your number sense. It is important to note that number sense is being developed throughout; however, we will use some problems to focus explicitly on number sense.

INVESTIGATION 3.1i Number Sense with Addition

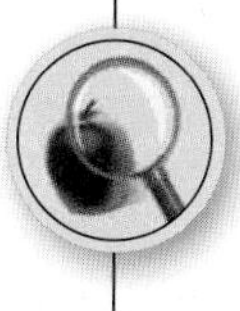

There are generally two ways to proceed. One is to try to explain why it *must* be true. Another is to find a counterexample to demonstrate that the statement is not true.

A. I'm thinking of 3 numbers whose sum is greater than 70.

CLASSROOM  CONNECTION

Grade 2

What kind of strategies do you think children in second grade might use for these problems? 

Date Time

LESSON 10·8 **Ballpark Estimates**

Fill in the unit box. Then, for each problem:

Make a ballpark estimate before you add.

Write a number model for your estimate.

Use your calculator and solve the problem. Write the exact answer in the box.

Compare your estimate to your answer.

Unit

1. Ballpark estimate: ______ 148 + 27	2. Ballpark estimate: ______ 163 + 32	3. Ballpark estimate: ______ 133 + 35
4. Ballpark estimate: ______ 143 + 41	5. Ballpark estimate: ______ 184 + 23	6. Ballpark estimate: ______ 154 + 83

two hundred fifty-five 255

From *Everyday Mathematics, Grade 2:* The University of Chicago School Mathematics Project: Student Math Journal, Volume 2, by Max Bell et al., Lesson 10-8, p. 255. Reprinted by permission of The McGraw-Hill Companies, Inc.

Tell whether each of the following must be true, might be true, can't be true:

1. All three numbers are greater than 20.
2. If two of the numbers are less than 20, the other must be greater than 20.

DISCUSSION

Explorations Manual 3.4

1. Here is a counterexample: 10 + 30 + 40 = 80. The sum is greater than 70 but only two of the three numbers are greater than 20. So, this statement might be true (as in the case of 25 + 30 + 30), but it is not always true as this counterexample proves.
2. Here we can proceed logically. What if the two numbers less than 20 are 19? Then their sum is 38 and the third number must be at least 33. Do you see why? If the two numbers are less than 19, then the third number will have to be more than 33. Therefore, this statement is true.

B. Take no more than 5 to 10 seconds to determine whether this answer is reasonable. Make your determination without doing any pencil-and-paper work.

$$\begin{array}{r} 6{,}563 \\ 4{,}448 \\ +\ 7{,}203 \\ \hline 17{,}214 \end{array}$$

DISCUSSION

If we look at the leading digits, we see that they add up to 17 (meaning 17,000). A very quick look at the rest of the amounts lets us quickly see that the answer must be greater than 17,214, and thus this answer is not reasonable.

C. Place > or < in the circle. Quickly look at the numbers on the left and right of the circle. Determine which symbol is appropriate.

563 + 924 + 723 ○ 842 + 646 + 558

DISCUSSION

The leading-digit method is useful again: 5 + 9 + 7 = 21, whereas 8 + 6 + 5 = 19; therefore, 563 + 924 + 723 > 842 + 646 + 558.

SUMMARY 3.1

Let us review some important ideas from this section and take this time to connect addition to place value explicitly.

Addition means more than just sum or add. Analysis of various story problems show that many words describe addition: *combine, join, extend, increase.* The "combine" description helps us to see that from one perspective, addition can be seen as "fast counting," and indeed most children initially solve addition problems (4 + 3) by combining the two amounts and then counting.

There are two models: discrete and continuous. The discrete is more concrete to children (the concept of 3 balls is easier to "see" than 3 inches).

There are many different methods that can be used to solve addition problems with both small and large numbers

$$8 + 7 = 8 + (2 + 5) = (8 + 2) + 5 = 10 + 5 = 15 \quad \text{(bridging)}$$
$$= (7 + 1) + 7 = 7 + (1 + 7) = 7 + (7 + 1) = (7 + 7) + 1 \quad \text{(doubles)}$$

The properties are not just words but also useful ideas. The commutative and associative properties were both used in the cases above.

To do addition efficiently, we need to connect our place value knowledge of base ten: knowing how to take apart numbers and being able to do so flexibly.

3.1 Exercises

1. Reread the 8 Mathematical Practices of the CCSS in Section 1.2, or on the corestandards.org website and write about relationships between them and what we did in this section.

2. For each number line problem below, identify the computation it models and briefly justify your answer.

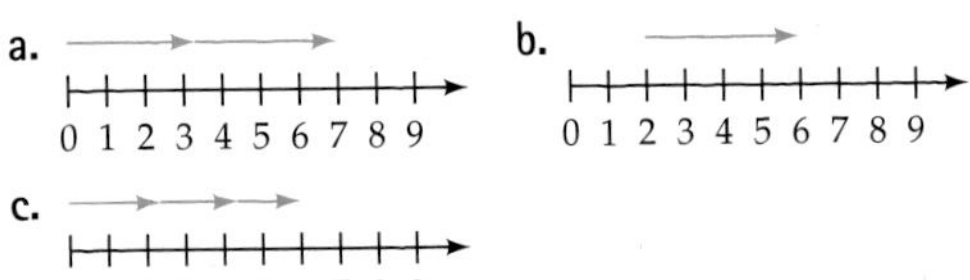

3. Refer to the base ten addition table. If we look at any of the diagonals going in a top-left-to-lower-right direction, the value of each successive term increases by 2 each time. Explain this pattern.

4. In many textbooks, students are encouraged to learn their fact families. This is the fact family for 10: $1 + 9$, $2 + 8$, $3 + 7$, $4 + 6$, $5 + 5$, $6 + 4$, $7 + 3$, $8 + 2$, $9 + 1$. Transform each fact pair to an ordered pair and graph the pairs on graph paper. What do you see?

5. Determine the following sums mentally. Briefly describe how you obtained the sum.

   a. 47 + 55   b. 69 + 77   c. 56 + 19   d. 577 + 126

   e. 735 + 248   f. 75, 36, 187, + 65   g. 24, 53, + 387

   Determine these sums mentally by a means other than the standard add and trade pencil-and-paper algorithm in your head:

   h. $289 + 400$ i. $295 + 394$ j. $186 + 598$

6. A traveling saleswoman is adding up her miles this week. How many miles did she travel? Determine the answer mentally. Briefly describe how you obtained the sum.

   165   345
   78   142
   57

7. This is how a child added $48 + 27$:

   $48 + 2 = 50$
   $50 + 20 = 70$
   $70 + 5 = 75$

   a. Explain why the child's solution works.

   b. Justify the solution path using the commutative and associative properties.

   c. How might the child add $36 + 84$?

   d. How might the child add $268 + 347$?

8. Through illustrations, demonstrate how to solve these problems with manipulatives.

   a. 76 + 47   b. 524 + 268

9. Below is an addition algorithm from an old text. Explain why it works.

```
 36
+48
 14
 7
 84
```

10. How does this addition algorithm work? Imagine someone asking, "How do you know where to put the 1 and the 3 in the 13?"

```
 368
+574
  12
 13
8
```

11. Pretend you are a young child again. Which algorithm do you think would be easiest to learn first: the standard algorithm, the partial sums algorithm, or the lattice algorithm? Explain your reasoning.

12. Finish the following addition problem, which has been done in expanded form.

   345 = 3 hundreds + 4 tens + 5 ones
   + 268 = 2 hundreds + 6 tens + 8 ones

13. ***Classroom Connection*** The following computations represent mistakes that children commonly make in learning to add. In each case, figure out how the child could have gotten the answer and describe the cause of the child's error.

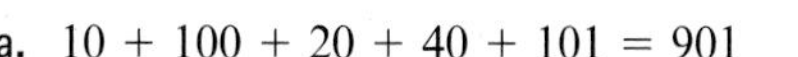

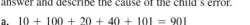

a. 10 + 100 + 20 + 40 + 101 = 901

b. 20 + 2 + 14 + 9 = 36

c. 128 + 133 = 2511

14. ***Classroom Connection*** Identify and explain the errors in each of these computations, all of which have occurred in real classrooms.

a. $\begin{array}{r} 36 \\ +28 \\ \hline 91 \end{array}$ b. $\begin{array}{r} 36 \\ +28 \\ \hline 514 \end{array}$ c. $\begin{array}{r} 365 \\ +287 \\ \hline 742 \end{array}$

15. a. $\begin{array}{r} 1101_{two} \\ +111_{two} \\ \hline \end{array}$ b. $\begin{array}{r} 2004_{five} \\ +2441_{five} \\ \hline \end{array}$ c. $\begin{array}{r} 55_{six} \\ +55_{six} \\ \hline \end{array}$ d. $\begin{array}{r} 992_{sixteen} \\ +265_{sixteen} \\ \hline \end{array}$

16. Solve the following addition problems with the lattice algorithm:

a. $\begin{array}{r} 5568_{ten} \\ +2745_{ten} \\ \hline \end{array}$ b. $\begin{array}{r} 322_{five} \\ +234_{five} \\ \hline \end{array}$ c. $\begin{array}{r} 764_{eight} \\ +215_{eight} \\ \hline \end{array}$

17. Determine the missing numbers that will complete the following addition problem in base six.

$4\square\square4_{six} + 35\square_{six} = \square003_{six}$

18. a. Find the base for which the following statement is true: 35 + 25 = 62.

b. Find the base for which the following statement is true: 135 + 444 = 601.

c. Determine the base(s) in which $31_a = 25_b$.

19. Estimate the following sums. Briefly describe how you obtained your estimate.

a. $\begin{array}{r} 473 \\ 345 \\ +355 \\ \hline \end{array}$ b. $\begin{array}{r} 6963 \\ 3286 \\ +7147 \\ \hline \end{array}$ c. $\begin{array}{r} 88{,}865 \\ 32{,}565 \\ +23{,}784 \\ \hline \end{array}$ d. $\begin{array}{r} 536{,}455 \\ 94{,}352 \\ +659{,}346 \\ \hline \end{array}$

e. $\begin{array}{r} 473 \\ 345 \\ 134 \\ 565 \\ 943 \\ +355 \\ \hline \end{array}$ f. $\begin{array}{r} 69{,}655 \\ 32{,}438 \\ 90{,}432 \\ 24{,}684 \\ 79{,}833 \\ +71{,}347 \\ \hline \end{array}$ g. $\begin{array}{r} 234{,}345{,}343 \\ 325{,}345{,}689 \\ 587{,}247{,}678 \\ 56{,}345{,}865 \\ +\ 234{,}764{,}874 \\ \hline \end{array}$

20. From each of the following lists, select three numbers whose sum will be closest to the target sum. Briefly explain your reasoning—how did you determine those three numbers?

	Numbers				**Target**
a.	37	83	56	74	150
b.	73	32	94	53	200
c.	24	39	47	97	170
d.	115	175	164	153	400
e.	185	372	153	274	650
f.	352	423	439	583	1200

21. Without actually adding the numbers, determine whether to place > or < in the circle. Briefly explain your reasoning.

a. 984 + 62 + 54 ○ 624 + 736

b. 426 + 315 + 408 ○ 378 + 495 + 387

22. Take no more than 5 to 10 seconds to determine if the answer to this addition problem is reasonable. Briefly explain your reasoning.

$\begin{array}{r} 5006 \\ 207 \\ 52 \\ 3264 \\ 921 \\ +\ 386 \\ \hline 8942 \end{array}$

23. Find the missing digits.

$\begin{array}{r} 2\square6 \\ +\square8\square \\ \hline 844 \end{array}$ $\begin{array}{r} 6\square3 \\ 524\square \\ +\ \square65 \\ \hline 6364 \end{array}$ $\begin{array}{r} 5\square\square5 \\ +\ \square64\square \\ \hline \square3423 \end{array}$

DEEPENING YOUR UNDERSTANDING

24. Determine values for a, b, and c that will make the number sentence correct.

$\begin{array}{r} abc \\ +\ bac \\ \hline 788 \end{array}$

25. Place the digits 1, 2, 3, 6, 7, and 8 in the boxes to obtain

$\begin{array}{r} \square\square\square \\ +\ \square\square\square \\ \hline \end{array}$

a. The greatest sum

b. The least sum

26. Choose among the digits 1, 2, 3, 4, 5, 6, 7, 8, and 9 to make the sum equal 500. You may use each digit only once.

$\begin{array}{r} \square\square\square \\ +\ \square\square \\ \hline \end{array}$

27. Can you make this a true statement, using each of the whole numbers 1 through 9 only once?

$\begin{array}{r} \square\square\square \\ \square\square\square \\ +\ \square\square\square \\ \hline 999 \end{array}$

28. Determine the values of N and P that make this a true statement. Explain how you arrived at the answer.

$\begin{array}{r} N \\ N \\ N \\ +\ P \\ \hline PN \end{array}$

FROM STANDARDIZED ASSESSMENTS

NECAP 2007, Grade 4

29. Which number sentence is true?

a. 100 + 85 = 10 + 85

b. 100 + 85 = 150 + 8

c. 100 + 85 = 105 + 80

d. 100 + 85 = 108 + 5

SECTION 3.2 Understanding Subtraction

What do you think?

- What does the term "borrow" mean in elementary school mathematics and what does it mean in other contexts?
- Why do most current elementary texts discourage the term "borrow"?
- What other contexts for subtraction are there besides "take away"?

Assume that you are a young child who has not yet learned subtraction. How might you solve these problems? Purposely "forgetting" what you already know about subtraction will help you to see beyond the mechanics of the operation to its underlying meaning. According to the Common Core State Standards, subtraction of whole numbers is introduced in first grade, and by third grade, students are subtracting three digit numbers.

Subtraction problems in context

A. Write a word problem for $7 - 2$ and draw a model that a young child might use to solve it before they knew subtraction.

B. Draw a model or use physical models for each of the following that a young child might use to solve the problem in the context.

1. Joe had 7 marbles. He lost 2 in a game. How many does he have left?
2. Billy has 2 marbles and Yaka has 7. How many more does Yaka have?
3. At the beginning of the week, a doctor had 7 ounces of insulin. During the week, 2 ounces of insulin were used. How much insulin did the doctor have at the end of the week?
4. You need $7 to go to the movie and you only have $2. How much more do you need?
5. Tom has 7 feet of bubblegum rope and Meg has 2 feet. How much more does Tom have?

Now go back to these five problems and answer the following questions:

- What action words describe what is happening in these five problems?
- How are the models/pictures of the problems similar and different?
- In what ways are the problems different? In what ways are they similar?
- What does subtraction mean? For example, what words come to mind when you think of subtraction?

CONTEXTS FOR SUBTRACTION

In the above problems you likely had three different types of models/pictures, and some of them might have been drawn as a set model, and others as a linear measurement model. Most students when they write a problem and when they think about subtraction, they are thinking about the "**take-away**" context for subtraction. In this context, I have some amount and I take away some amount and my task is to figure out how much is left. Look back over the five problems in the above examples as well as the one you wrote and see which are "take-away." What does the model/visual look like for these problems?

Problems 1 and 3 are take-away problems. It is likely that you wrote a take-away problem as well, as this is the most common one we think of with subtraction. Think back to the previous section where we talked about adding discrete set models (collection of items) and measurement models. We also have that distinction here where Problem 1 is a set model (because our items are sets of marbles) and Problem 3 is a measurement model (think about the markings of a measuring cup). A child would likely solve the first one by representing the marbles (perhaps with actual marbles) and taking 2 away. A child might use a number line to solve Problem 3. Let's look at a take-away set model and a take-away linear model within these contexts.

Joe had 7 marbles. He lost 2 in a game. How many does he have left?

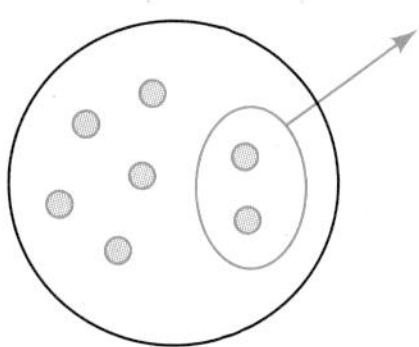

Take-away set model

Joe has 5 left.

At the beginning of the week, the doctor had 7 ounces of insulin. During the week, 2 ounces of insulin were used. How much insulin did the doctor have at the end of the week?

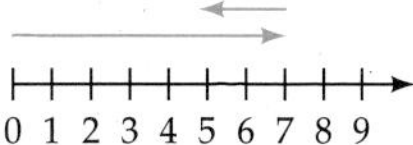

Take-away linear measurement model

Another context for subtraction is when we are comparing two quantities and asking how much bigger one is than the other. Logically, we call this the "**comparison**" context for subtraction. Again, within this context we may have a set model or a linear measurement model. Look back at the problems and decide which ones have you comparing two quantities.

Problem 2 is a set model comparison context since you are comparing numbers of marbles (discrete items in a set). Problem 5 is also a comparison model, but it is better represented on a number line as these are measurements. A child might draw or create the following to figure out the answers to each problem.

Billy has 2 marbles and Yaka has 7. How many more does Yaka have?

Comparison set model

In this context, a child would likely lay out the 7 marbles (or some representation of them) that Yaka has and lay out the 2 that Billy has, and then count how many more Billy has. These models can help us see the difference between comparison and take-away. Compare the two models.

Now, let's model Problem 5 on a number line since it is a linear measurement model.

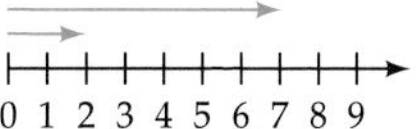

Comparison linear model

Similar to the comparison set model, a child can now count up from 2 to 7 on the number line to answer the question.

The "**missing addend**" context is similar to the "comparison" model but is slightly different. We can think of this one as 2 + ____ = 7 (hence the name of missing addend). In Problem 4, we might think of that as $2 plus how many dollars equals the $7 that I need. A child might answer this by starting with the $2 and adding on dollars until they reach the $7 needed.

Start with $2

Add on $5 until you reach the desired $7

What would a linear measurement model look like in this context? It would be similar to the dollars placed in a line, but the context would be a measurement. Perhaps something like, I have 2 inches of ribbon but I need 7 inches to make a bracelet. How many more inches of ribbon do I need?

0 1 2 3 4 5 6 7 8 9

Here I start with the 2 inches and add on until I get to the 7 inches.

All of the above problems and models are $7 - 2 = 5$, but they look very different. It is important mathematically that we not always think of subtraction as simply take-away, but be flexible to think about it in other contexts.

A PICTORIAL MODEL FOR SUBTRACTION

One way to express the commonality of all five problems just discussed is that in all cases, we have a large amount and two smaller amounts whose sum is equal to the larger amount. Recall the pictorial model for addition: If we invert that diagram, we have a pictorial representation of a general model for subtraction (Figure 3.6). These types of bar models are used extensively in Singapore, which produces some of the best math students in the world. We will use them in other places in the text, like with fractions and algebra.

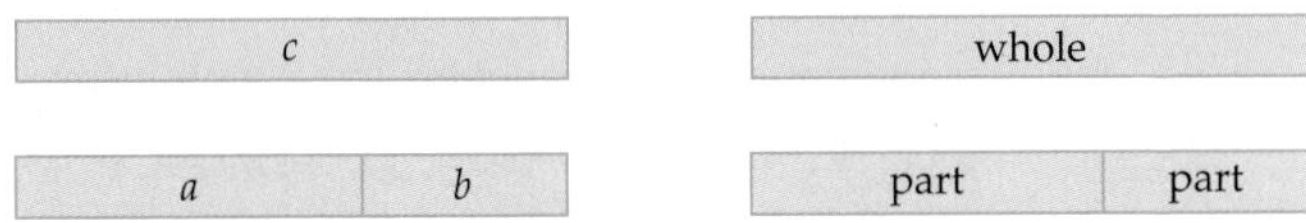

Figure 3.6

It is important to note that *whole* and *part* refer to the numerical values rather than the contexts. For example, in Problem 2, we are comparing two wholes and saying that one whole contains 5 more marbles than the other.

We define **subtraction** formally in the following manner:

$$c - b = a \text{ if } a + b = c$$

That is, the difference between two numbers c and b is a if c is the sum of a and b.

In mathematical language, c is called the **minuend**, b the **subtrahend**, and a the **difference**.

This model also highlights the connections between addition and subtraction problems (Figure 3.7):

$3 + 5 = 8$
$5 + 3 = 8$
$8 - 3 = 5$
$8 - 5 = 3$

8

3 5

Figure 3.7

CLASSROOM CONNECTION

A specialist was called in to diagnose a third-grader who was having difficulties with mathematics. He discovered that the child did not see addition and subtraction as related and was not able to use the commutative property of addition. That is, she saw the four number sentences on the right as four unrelated number sentences. It was no wonder that she was having difficulty remembering her addition and subtraction facts!

CLASSROOM  CONNECTION

Grade 3

Can you describe the context of addition or subtraction in each story? 

Date ______ Time ______

LESSON 2·5

Number Stories: Change-to-More and Change-to-Less

For each number story, write ? in the diagram for the number you want to find. Then write the numbers you know in the change diagram also. Next, solve the problem. Write the answer and a number model.

SRB 254 255

Unit
dollars

1. Ahmed had $22 in his bank account. For his birthday, his grandmother deposited $25 for him. How much money is in his bank account now?

   Start — Change → End

   Answer the question: ______________

   Number model: ______________

   Check: How do you know your answer makes sense?

   ____

2. Omar had $53 in his piggy bank. He used $16 to take his sister to the movies and buy treats. How much money is left in his piggy bank?

   Start — Change → End

   Answer the question: ______________

   Number model: ______________

   Check: How do you know your answer makes sense?

   ____

3. Cleo had $37 in her purse. Then Jillian returned $9 that she borrowed. How much money does Cleo have now?

   Start — Change → End

   Answer the question: ______________

   Number model: ______________

   Check: How do you know your answer makes sense?

   ____

thirty-nine **39**

From *Everyday Mathematics, Grade 3:* The University of Chicago School Mathematics Project: Student Math Journal, Volume 1, by Max Bell et al., Lesson 2-5, p. 39. Reprinted by permission of The McGraw-Hill Companies, Inc.

PROPERTIES OF SUBTRACTION

Think back to the properties of addition—identity, commutativity, associativity, and closure. Do those same properties hold for subtraction? Think and then read on. . . .

Some students think that subtraction has an identity property, because if we subtract zero from a number, its value does not change; that is, $a - 0 = a$. This is true; however, if we reverse the order, the result is not true—that is, $0 - a \neq a$. Therefore, we generally say that the operation of subtraction does not have an identity property.

After examining a few cases, you can see that the operation of subtraction does not possess the commutative property or the associative property. The commutative property (3 − 5 does not equal 5 − 3) is not immediately understood by children. I recall one first-grader arguing that 3 − 5 was 0 because "you can't have less than nothing" and another arguing that it was 2 because "you just turn them around."

Finally, the operation of subtraction is not closed for the set of whole numbers because the difference of two whole numbers can be a negative number, such as 3 − 5 = −2 (which is not a whole number).

Just as we did with addition, we will begin our study of subtraction algorithms with mental subtraction. You know subtraction well, so having to do problems mentally will require you to think about the operation and how you apply your place value knowledge to subtract.

INVESTIGATION 3.2a Mental Subtraction

Do the following computations in your head. Briefly note the strategies you used, and try to give names to them.

Note: One mental tool all students have is being able to visualize the standard algorithm in their heads. For example, for the first problem, you could say, "Cross out the 6, replace it with 5, then add 10 to the 5 in the ones column, and think 15 − 8 = 7 and then 5 − 2 = 3; the answer is 37." However, because you already know that method, I ask you not to use it here but to try others instead.

1. 65 − 28 2. 62 − 29 3. 184 − 125 4. 132 − 36 5. 1000 − 648

CLASSROOM CONNECTION

One of the ways that I assess how well a student understands the various models for subtraction and the relationship between addition and subtraction is by seeing the extent to which a student can develop a repertoire of mental strategies for subtraction.

DISCUSSION

As with addition, there are a variety of subtraction strategies. As you read the discussion below, reflect on the strategies and make sense of them in your mind. Nearly all of the strategies work better with certain kinds of numbers than with others.

Add up In Problem 1, one alternative is to **add up**—that is, to ask how we get from 28 to 65.

One student's actual strategy looks like this:

28 + 30 = 58
58 + 7 = 65
30 + 7 = 37, which is the answer.

The add-up strategy is nicely illustrated with a **number line**.

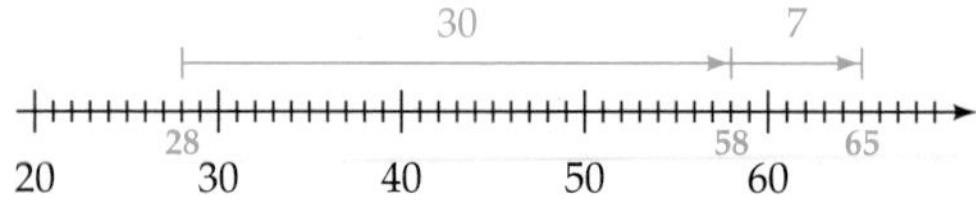

Another variation of adding up looks like this:

28 + 2 = 30
30 + 35 = 65
35 + 2 = 37, which is the answer.

This is illustrated on the number line.

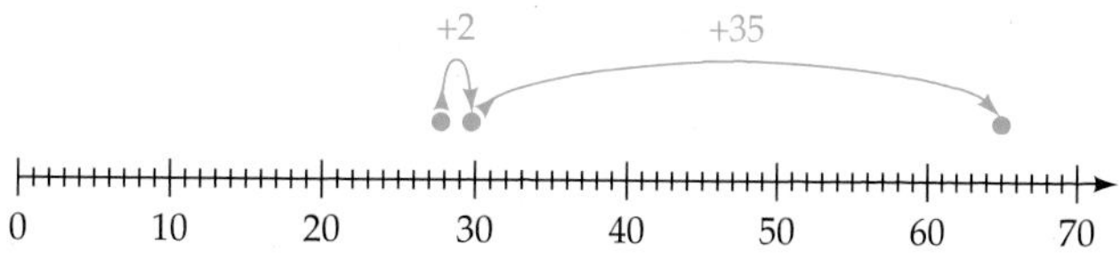

Compensation In Problem 2, we can use **compensation**, which does not work exactly the same way with subtraction as it does with addition.

This is how compensation looks with Problem 2: $62 - 29$
Add 1 to both numbers. We now have $63 - 30$, which is easily solved: 33.

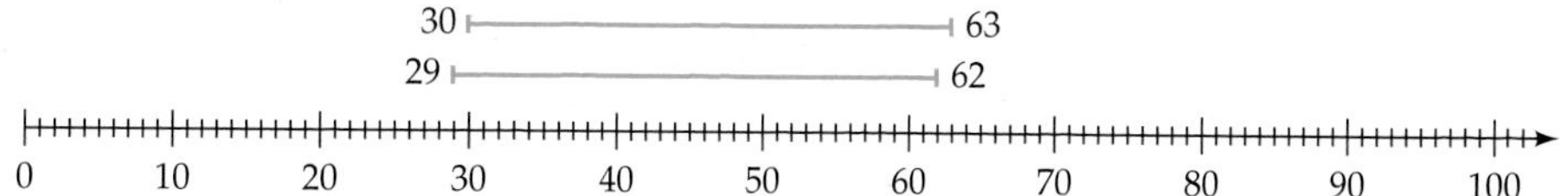

Some students understand the compensation strategy better by using a number line. On a number line, $62 - 29$ can be interpreted as the distance between the two numbers. If we increase both numbers by 1, we have not changed the distance between the numbers.

In Problem 3, we can add up: $184 - 125$
$125 + 60 = 185$
$185 - 1 = 184$, so the answer is 59.

In Problem 4, we can adapt the compensation strategy:

$132 - 36$
$136 - 36 = 100$.

Because we increased 132 by 4, we need to give back the 4, so the answer is $100 - 4 = 96$.

We can also solve Problem 4 by using **compatible numbers** and adding up:

$132 - 36$
$36 + 64 = 100$
$100 + 32 = 132$, so the answer is $64 + 32 = 96$.

In Problem 5, we can use compatible numbers: $1000 - 648$.

Realizing that $48 + 52 = 100$, we can add up: $648 + 52 = 700$
$700 + 300 = 1000$, so the answer is $300 + 52 = 352$.

Problem 5 can also be solved by adding up: $1000 - 648$
$648 + 300 = 948$
$948 + 52 = 1000$, so the answer is $300 + 52 = 352$.

CHILDREN'S STRATEGIES FOR SUBTRACTION

Explorations Manual 3.6

Below are three very different methods that children have invented for subtracting larger numbers. As you explore algorithms that children and adults have invented, you will come to appreciate both the complexity of subtraction (more children struggle with subtraction than addition) and how this process is made easier by connecting addition to subtraction and by understanding how place value works.

Let's explore these strategies with the problem $64 - 27$.

Break the second number apart Essentially, this approach breaks apart the 27 and subtracts in pieces.

$$64 - 7 = 57$$
$$57 - 20 = 37$$

Do you see connection to any addition strategies? Think before reading on. . . . 

This method is similar to adding up, where we begin with one number, break the second number into parts (expanded form), and add one place at a time. Here, we begin with the minuend and then subtract one place at a time.

As with addition, different representations illustrate and illuminate the problem in different ways. On the number line, we move 7 units, which brings us to 57, and then we move 20 units, which brings us to 37. We can also represent the process numerically (below right).

$$\begin{aligned} &64 - 27 \\ &= 64 - (20 + 7) \\ &= 64 - (7 + 20) \\ &= 64 - 7 - 20 \\ &= 57 - 20 \\ &= 37 \end{aligned}$$

MATHEMATICS

As you can see, on many occasions a problem can be interpreted and thus represented in many different ways. Recall the Mathematical Practices of the CCSS, particularly MP 1: Make sense of problems and MP 4: Model with mathematics and reflect on how they apply here.

We can also break both numbers apart into tens and ones. This method works like this: Subtract each place, beginning at the left. Next, combine the two differences: $40 - 3 = 37$.

64	$60 - 20 = 40$
-27	$4 - 7 = -3$
	$40 - 3 = 37$

CLASSROOM CONNECTION

I remember reading in 1986 that a second-grader invented this strategy, and I was skeptical. Then I saw a second-grader invent it in a classroom where I had been working with the teacher. Since then I have read of its invention in many places. Apparently, it is rather common! You can find a discussion of this problem in the October 1991 issue of *The Arithmetic Teacher* (pp. 20–21) and in the February 2003 issue of *Teaching Children Mathematics* (p. 351).

Adding up This method is very common, especially if the context of the problem is not take away but rather comparison or missing addend.

64	$27 + 30 = 57$
-27	$57 + 7 = 64$
	The answer is $30 + 7 = 37$.

When asked to think out loud, the strategy sounds like this: "27 + 30 = 57. 57 + 7 = 64, so the answer is 30 + 7 = 37."

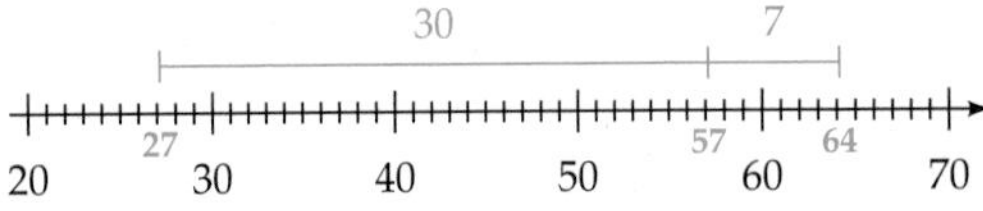

Note that a common variation is to overshoot the number and then come back. For example, "27 + 40 = 67. Now I need to take away 3 to get to 64, so the answer is 40 − 3 = 37."

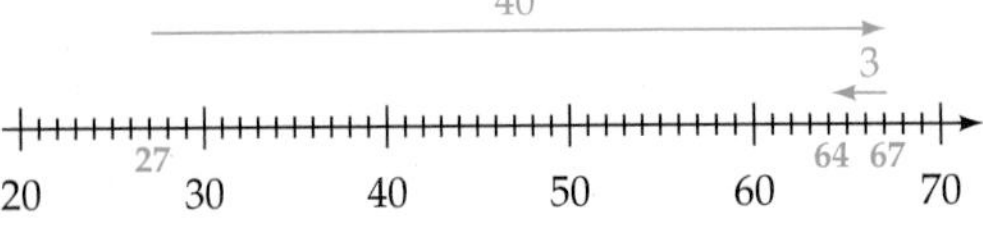

INVESTIGATION 3.2b Children's Strategies for Subtraction with Large Numbers

Explorations Manual 3.6

What if the numbers were bigger, for example 832 − 367? Look back on the approaches described previously. How can you adapt them to this problem? Try to do so before reading on. . . .

DISCUSSION

"Break the second number apart" strategy extends but is now more cumbersome: Begin with the minuend and subtract each place.

$$\begin{array}{r} 832 \\ -367 \\ \hline \end{array} \quad \begin{array}{r} 832 \\ -\ 7 \\ \hline 825 \end{array} \quad \begin{array}{r} 825 \\ -60 \\ \hline 765 \end{array} \quad \begin{array}{r} 765 \\ -300 \\ \hline 465 \end{array}$$

Using negative numbers extends quite well. In fact, many of my students actually find this strategy easier than the algorithm they grew up with.

$$\begin{array}{r} 832 \\ -367 \\ \hline \end{array} \quad \begin{array}{r} 800 \\ -300 \\ \hline 500 \end{array} \quad \begin{array}{r} 30 \\ -60 \\ \hline -30 \end{array} \quad \begin{array}{r} 2 \\ -7 \\ \hline -5 \end{array}$$

$500 - 30 = 470; 470 - 5 = 465$

Adding up also becomes a bit cumbersome, but interestingly, some students feel quite comfortable with it. This is what they write:

$$\begin{array}{r} 832 \\ -367 \\ \hline \end{array}$$

$367 + 3 = 370;$

$370 + 30 = 400;$

$400 + 432 = 832.$

The answer is $3 + 30 + 432 = 465$.

Some children add up from left to right:

$367 + 400 = 767; 767 + 60 = 827; 827 + 5 = 832.$

The answer is $400 + 60 + 5 = 465$.

In the days before computerized cash registers, the cashiers used this method to count back change: If the bill was $3.80 and the customer used a $5 bill, the cashier would hand back two dimes and a dollar, saying twenty cents makes $4 and another dollar makes $5.

HISTORY

"When nothing remains, put down a small circle so that the place be not empty, but the circle must occupy it, so that the number of places will not be diminished when the place is empty and the second be mistaken for the first."[4]

This quote comes from a text written by Al-Khowarizmi in A.D. 825 as he tried to explain the procedure for subtracting. Do you understand what he is saying? What function does the circle serve?

UNDERSTANDING SUBTRACTION IN BASE TEN

Let us now examine some standard and nonstandard algorithms for subtraction. Again, recall the Alphabitian problems; for a child, subtracting 32 − 14 can be as overwhelming and confusing as many students originally found DA − BC in the Alphabitian system from the *Explorations* manual.

Researchers have found that when subtraction problems have zeros, the success rate for most third-graders goes down drastically. However, this need not be so. Let us examine the following problem: 300 − 148. Although there is no single procedure that all students use to solve this problem (unless they are forced to by their teacher), we will examine here how this problem might be solved going from left to right.

Using manipulatives	Using words	Using symbols
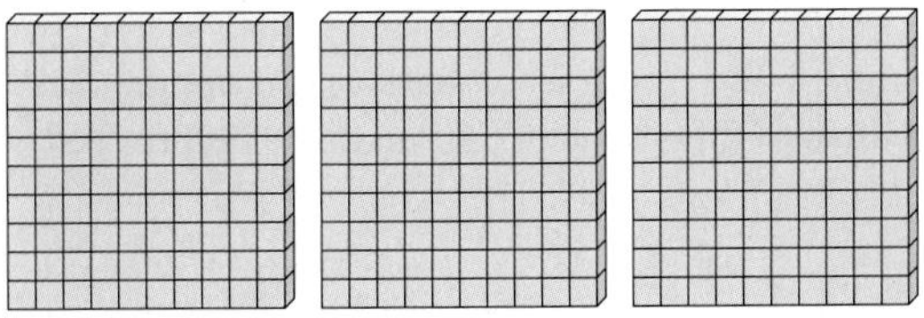	We need to regroup 300 so that we can "take away" 148.	$\begin{array}{r} 300 \\ -148 \\ \hline \end{array}$
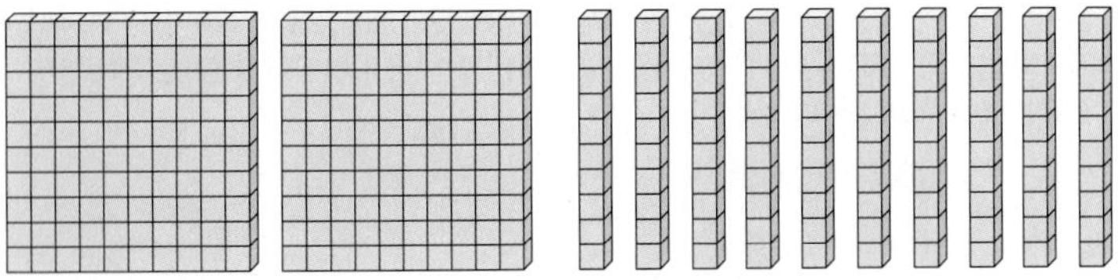	We can trade 1 of the 3 hundreds for 10 tens, giving us 2 hundreds and 10 tens, which has the same value as 3 hundreds.	$\begin{array}{r} \overset{2}{\not{3}}{}^{1}00 \\ -148 \\ \hline \end{array}$
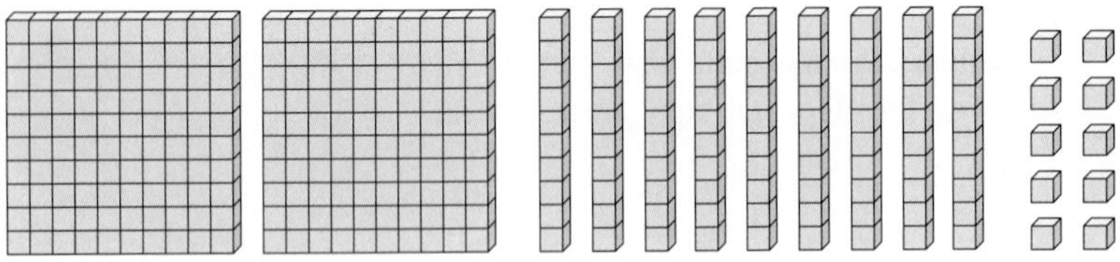	Next, we can trade 1 of the tens for 10 ones. We now have 2 hundreds, 9 tens, and 10 ones, which still has the same value as 3 hundreds.	$\begin{array}{r} \overset{2}{\not{3}}\overset{9}{\not{0}}{}^{1}0 \\ -148 \\ \hline \end{array}$
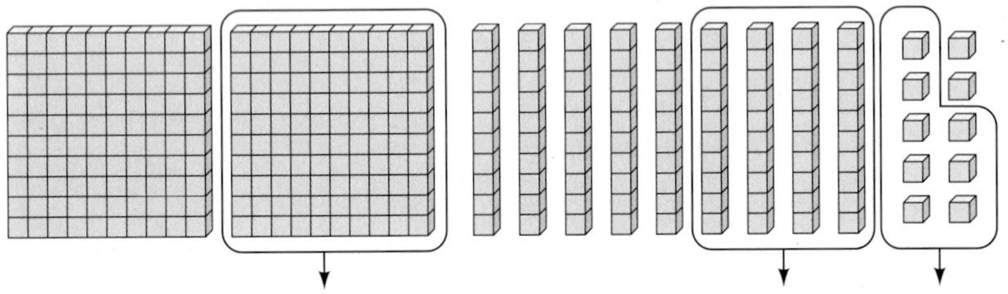	Now we can take away 1 hundred, 4 tens, and 8 ones.	$\begin{array}{r} \overset{2}{\not{3}}\overset{9}{\not{0}}{}^{1}0 \\ -148 \\ \hline 152 \end{array}$

CLASSROOM CONNECTION

It is important to reemphasize that "standard" algorithms are not the "right" ones, or even the "best" ones, but rather the ones that, for various reasons, have become most widespread. Many educators believe that more harm than good is done by forcing all students to learn the "standard" algorithms.

JUSTIFYING THE STANDARD ALGORITHM

It is beyond the scope of this book to prove formally every algorithm, procedure, and theorem. What is essential is that elementary students and teachers understand the algorithms that they use—that they understand the whys of the algorithm. The base ten blocks help us to understand the algorithms, and it is important to make a direct connection between the written algorithms and what is happening with the blocks.

Some students find that representing the problem in expanded form is helpful. A key to understanding this algorithm is to understand the equivalence of 3 hundreds and 2 hundreds, 9 tens, and 10 ones. Although they look different, and the numerals are different (300 versus $29^{1}0$), the amounts are equal. Another key to understanding this algorithm is to know why we needed to do the trading. This is easier to see at the physical level; without trading, we cannot literally "take away" 1 flat, 4 longs, and 8 units.

3 hundreds + 0 tens + 0 ones −1 hundred + 4 tens + 8 ones	→	2 hundreds + 10 tens + 0 ones −1 hundred + 4 tens + 8 ones	→	2 hundreds + 9 tens + 10 ones −1 hundred + 4 tens + 8 ones
Cannot subtract the tens or the ones		Cannot subtract the ones		Subtraction possible

INVESTIGATION 3.2c

Subtracting in Base Five

What if we had a base five system instead of a base ten system? How would the standard algorithm for subtraction look different and how would it be the same? Use models (snap cubes work well if you have them), pictures, or virtual manipulatives to figure out the answer to the following. If you Google "virtual manipulatives," the first site listed is the National Library of Virtual Manipulatives. Under Number & Operations, there are two links that will work for this problem. One is called Base Blocks where you can start with the first number and take away (by throwing in the trash can icon) what you want to subtract. The other is called Base Blocks Subtraction, which is a comparison model where you put the top number in blue and what you want to subtract in red and put one on top of the other to make them disappear. Make sure you change the base to 5! To move blocks from left to right, drag them over and they will break apart in the next column.

$$\begin{array}{r} 321_{\text{five}} \\ -\ 142_{\text{five}} \\ \hline \end{array}$$

DISCUSSION

We will use a "take-away" model, since that is what we did in the previous base ten example. Examine the picture below and see if you can figure it out for yourself; then we will explain.

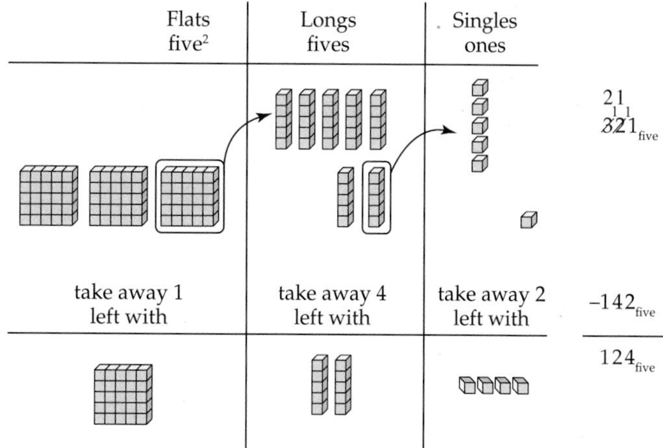

We begin by drawing 3 flats (five by five), 2 longs (five), and 1 single (one). If we start in the ones place as the standard algorithm does, we see that we cannot take away 2 so we need to trade a long for five singles as shown here with the arrow. Now we can take away the 2 singles and are left with 4 singles.

Now we have 1 long and want to take away 4 longs, so we have to trade 1 flat for 5 longs. Then we take away 4 longs and are left with 2 longs. Finally we take away the 1 flat and get our final answer of 124_{five}.

How is this different than in base ten? The only difference is in the trading. Instead of trading ten, we trade five.

Thinking about the algorithms in other bases helps us to understand them better in base ten. You will likely never teach your students to subtract in base five, but it helps you develop a deeper understanding of place value and the algorithms that you memorized in elementary school.

INVESTIGATION 3.2d An Alternative Algorithm

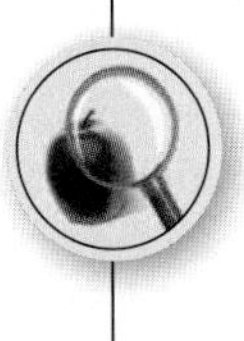

People who grew up in many European countries, Mexico, and others learned a very different subtraction algorithm. Below are two examples of its use and what a person using this algorithm would say.

For the first problem, "You can't take 8 from 4, so you put a 1 next to the 4, cross out the 6, and put a 7. Now subtract: 14 − 8 = 6, 8 − 7 = 1, 9 − 3 = 6."

For the second problem, "You can't take 8 from 3, so you put a 1 next to the 3, cross out the 5, and put a 6. Now, you can't take 6 from 2, so you put a 1 next to the 2, cross out the 1, and put a 2. Now subtract: 13 − 8 = 5, 12 − 6 = 6, 6 − 2 = 4."

Explorations Manual 3.6

$$\begin{array}{r} 984 \\ -368 \\ \hline \end{array} \qquad \begin{array}{r} 98^{1}4 \\ -3\overset{7}{\cancel{6}}8 \\ \hline 616 \end{array} \qquad \begin{array}{r} 623 \\ -158 \\ \hline \end{array} \qquad \begin{array}{r} 6^{1}2^{1}3 \\ -\overset{2}{\cancel{1}}\overset{6}{\cancel{5}}8 \\ \hline 465 \end{array}$$

DISCUSSION

Can you explain why this algorithm works? That is, can you justify it, or prove it? 

The justification for this algorithm is very different from the justification for the previous algorithm, in which the trading didn't change the value of the minuend, but only its representation. Here, however, the value of both the minuend and the subtrahend *do* change: The value of the minuend and the subtrahend are now 994 and 378. Make sure you see this before reading on. Now look at the two number lines below. Can you explain now why this algorithm works?

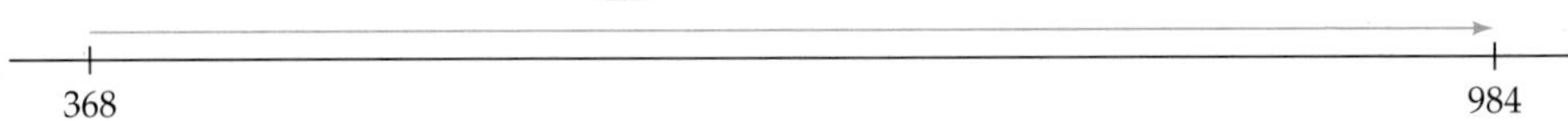

Because the value of both minuend and subtrahend have increased by 10, the difference (distance) between them remains the same. We added ten ones to the 984 and we added one ten to the 368.

The discussion of these two algorithms is a nice place to introduce a framework for examining algorithms that was presented by Hyman Bass in the February 2003 issue of *Teaching Children Mathematics*. Bass proposes that we examine five qualities of an algorithm:

- Accuracy (or reliability). The algorithm should always produce a correct answer.
- Generality. The algorithm applies to all instances of the problem, or class.
- Efficiency (or complexity). This refers to whether the cost (the time, effort, difficulty, or resources) of executing the algorithm is reasonably low compared to the input side of the problem.
- Ease of accurate use (vs. proneness to error). The algorithm can be used reasonably easily and does not lead to a high frequency of error in use.
- Transparency (vs. opacity). What the steps of the algorithm mean mathematically, and why they advance toward the problem solution, is clearly visible.

Let us use this framework to examine the traditional algorithm, the alternative algorithm, and the negative numbers algorithm.

Accuracy. Used correctly, all of these algorithms will always produce a correct answer.

Generality. All the algorithms will work with any subtraction problem.

Efficiency. The traditional algorithm doesn't work exactly the same in all cases—for example, when multiple trading is needed and when there are zeros in the minuend. The alternative algorithm also requires some additional thought if there are 9s in the subtrahend. The negative numbers algorithm works identically in all situations.

Ease of accurate use. In certain situations children are more likely to make errors with the traditional and alternative algorithms. The negative numbers strategy requires stronger computation skills because of the use of negative numbers—for example, $2 - 8 = -6$.

Transparency. The traditional algorithm is more transparent (especially if there is a visual representation like the base ten blocks connected to the written algorithm), which is a primary reason why it supplanted the alternative algorithm in the United States in the twentieth century. It is easier for children to understand the "why" of the traditional (you don't change the value) than the "why" of the alternative (because both increase by the same amount, the difference is still the same). The negative numbers strategy is also transparent.

INVESTIGATION 3.2e Children's Mistakes in Subtraction

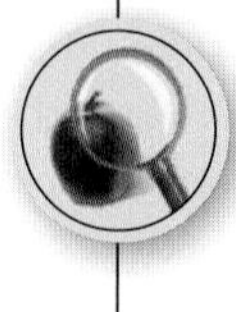

There is a very powerful video in *Integrating Mathematics and Pedagogy to Illustrate Children's Reasoning* that shows a second-grade child not understanding why 70 − 53 is not equal to 23. How do you think she got 23? What misunderstanding(s) of place value and/or subtraction might be causing her problems?

DISCUSSION

$$\begin{array}{r} 70 \\ -23 \\ \hline 53 \end{array}$$

In the video, you hear her say that "0 take away 3 is 3," which is a common misconception among young children. Interestingly, even when the interviewer had the child do the problems with base ten blocks and a hundred chart, even though she got the correct answer using both of those representations, she trusted the answer she got using pencil-and-paper more!

INVESTIGATION 3.2f Rough and Best Estimates with Subtraction

In each of the problems below, first obtain a quick rough estimate. Then obtain the best estimate you can.

A. In 1995, Acstead State College had an enrollment of 4234, whereas Milburn College had an enrollment of 3475. How many more students did Acstead have than Milburn?

$$\begin{array}{r} 4234 \\ -3475 \\ \hline \end{array}$$

B. The attendance at the Yankees game was 73,468, whereas the attendance at the Mets game was 46,743. How much greater was the attendance at the Yankees game?

$$\begin{array}{r} 73{,}468 \\ -46{,}743 \\ \hline \end{array}$$

DISCUSSION

A. You can get a quick estimate by rounding: 4200 − 3500 = 700
To get a more accurate estimate, you could use adding up:

3475 + 800 = 4275, then back off 40 to get 4235. The estimate is 800 − 40 = 760.

B. Leading digit: 7 − 4 = 3, giving an estimate of about 30,000.
Rounding: 70,000 − 50,000 = 20,000.

Rounding and adding up: 47,000 + 3000 makes 50,000. Adding up to 73,000 gives 23,000. Our estimate is 3000 + 23,000 = 26,000.

Looking back Take a few moments to look back over the strategies for mental subtraction and estimation that we have explored. Do you find that you prefer some over others? Can you connect these strategies to different models or contexts of subtraction—for example, take-away, comparison, missing addend, number line? Can you explain which strategies make powerful use of structures of our base ten numeration system—for example, the leading digit strategy? Do you see that some strategies work with some numbers better than with others? Recall Mathematical Practice 7 of the CCSS: Look for and make use of structure. How does that apply here?

INVESTIGATION 3.2g Number Sense with Subtraction

A. Fill in the blanks to make this subtraction problem.

$$\begin{array}{r} \square\square\square\square \\ -\ 4\ 2\ 6\ 3 \\ \hline 3\ 1\ 5\ 9 \end{array}$$

Explorations Manual 3.7

DISCUSSION

At first glance, some students panic. "There is no way I can do this problem!" However, if we stop for a moment and remember the relationship between addition and subtraction, we can see that we can quickly determine the missing number by simply adding 3159 + 4263!

B. Take no more than 5 to 10 seconds to determine whether this answer is reasonable. That is, make your determination, using number sense, without doing any pencil-and-paper work.

$$\begin{array}{r} 6542 \\ -2847 \\ \hline 3125 \end{array}$$

DISCUSSION

For most people the simplest way to proceed here is to add 3125 and 2847 and see if the answer is close to 6542. A quick estimation of the leading two digits shows 3100 + 2900 = 6000, so the answer is not true.

SUMMARY 3.2

We have now examined addition and subtraction rather carefully. In what ways do you see similarities between the two operations? In what ways do you see differences? Think and then read on. . . .

One way in which the two processes are alike is illustrated with the part–whole diagram used to describe each operation. These representations help us to see connections between addition and subtraction. In one sense, addition consists of adding two parts to make a whole. In one sense, subtraction consists of having a whole and a part and needing to find the value of the other part.

We see another similarity between the two operations when we watch children develop methods for subtraction; it involves the "missing addend" concept. That is, the problem $c - a$ can be seen as $a + ? = c$.

We saw a related similarity in children's strategies. Just as some children add large numbers by "adding up," some children subtract larger numbers by "subtracting down."

Earlier in this section, subtraction was formally defined as $c - b = a$ if $a + b = c$. The negative numbers strategy that some children invent brings us to another way of defining subtraction, which we will examine further in Chapter 4 when we examine negative numbers. That is, we can define subtraction as adding the inverse: $a - b = a + -b$.

A very important way in which the two operations are different is that the commutative and associative properties hold for addition but not for subtraction.

3.2 Exercises

1. Make up a subtraction story problem for each of the following contexts. Briefly *explain* why the story problem is an example of the particular model.
   a. Take-away
   b. Missing addend
   c. Comparison
2. Which model of subtraction best illustrates each of the problems below:
   a. Reena had 25 quilts and sold 10 of them at the show. How many does she have left?
   b. The hall holds 100 people. Currently 65 tickets have been sold. How many more tickets can be sold?
   c. The sixth grade has sold 58 raffle tickets and the fifth grade has sold only 45. How many more has the sixth grade sold?
3. For each number line problem below, identify the computation it models and briefly justify your answer.

a.
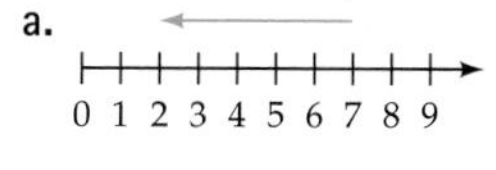

b.

c.
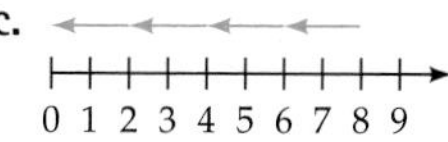

4. Represent the following problems on a number line. Explain each problem as though you were talking to someone who is not taking this class.
   a. $5 + 4$ b. $8 - 3$
5. a. Explain why the operation of subtraction is not commutative.
   b. Explain why the operation of subtraction is not associative.
6. Determine the following differences mentally. Briefly describe how you obtained your estimate.

   a. $\begin{array}{r} 87 \\ -29 \\ \hline \end{array}$ b. $\begin{array}{r} 70 \\ -23 \\ \hline \end{array}$ c. $\begin{array}{r} 82 \\ -34 \\ \hline \end{array}$ d. $\begin{array}{r} 500 \\ -134 \\ \hline \end{array}$ e. $\begin{array}{r} 502 \\ -206 \\ \hline \end{array}$

   f. $\begin{array}{r} 625 \\ -475 \\ \hline \end{array}$ g. $\begin{array}{r} 4000 \\ -555 \\ \hline \end{array}$

   Determine these differences mentally by a means other than the standard algorithm.

   h. $145 - 98$ i. $815 - 399$ j. $562 - 228$
   k. $575 - 197$ l. $1004 - 97$

In Exercises 7–10, determine the exact answer mentally.

7. Peter bought a house for $116,000 and sold it for $145,000. How much profit did he make?
8. I left my house at 3:34 P.M. and arrived at 6:15 P.M. How long did the trip take me?

9. Washington College has 4132 full-time students, whereas Lincoln College has 2824 students. How much larger is Washington than Lincoln?

10. A fund-raiser has a goal of \$55,000. So far, \$34,854 has been collected. How much more is needed?

11. ***Classroom Connection*** Following are two children's solutions to 73 − 39, taken from the December 2001 issue of *Teaching Children Mathematics* (p. 231). Study each child's work, and then describe how that child would find the answer to 65 − 28. Then explain why that method works.

a.

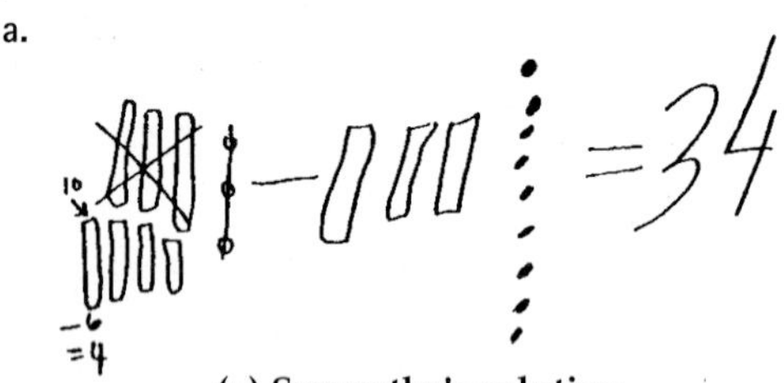

(a) Samantha's solution

b.

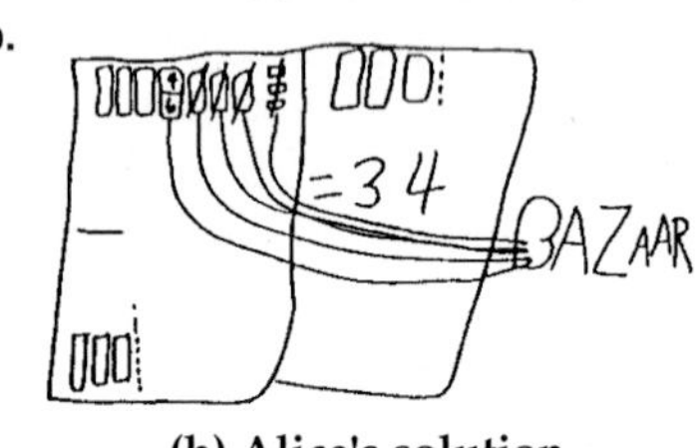

(b) Alice's solution

Source: *Teaching Children Mathematics*, by the National Council of Teachers of Mathematics.

12. A child subtracted 53 − 24 this way:

50 − 20 = 30
30 − 4 = 26
26 + 3 = 29

a. Justify the child's solution path.

b. Explain how the child would solve 72 − 38.

13. ***Classroom Connection*** Below are three methods, described in the February 2003 issue of *Teaching Children Mathematics*, that were invented by different children. In each case, study the child's work. Then solve 904 − 367 using the child's method, and explain why that method works.

a. 872 − 345
872 − 300 = 572
572 − 40 = 532
532 − 5 = 527

b. 872 − 345
345 + 7 = 352
352 + 20 = 372
372 + 500 = 872

c. 872 − 345
872 − 340 = 532
532 − 5 = 527

14. ***Classroom Connection*** As I mentioned earlier, I have always subtracted from left to right. Here is how I solve 837 − 375.

837	8 − 3 = 5, now look to the right: 3 < 7, so put a 4 in hundreds place
− 375	13 − 7 = 6, now look to the right: 7 > 5, put the 6 in the tens place
462	7 − 5 = 2, put the 2 in the ones place

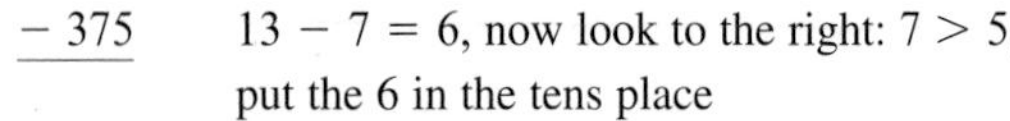

As in Problem 13, solve 904 − 367, this time using "my" method, and explain why that method works.

15. Through illustrations, demonstrate how to solve these problems with manipulatives.

a. 524 − 268   b. 600 − 345

16. A student's work on a subtraction problem is shown here. Explain the process, as though you were talking to a third-grader who is having difficulty with the process.

```
  3 9
  4 0 ¹0
− 1 3 5
  2 6 5
```

17. ***Classroom Connection*** As you might expect, there are many more ways in which children make mistakes in subtracting than in adding. What is the child's mistake in this problem? What aspect of subtraction or place value is the child not applying or applying incorrectly?

a. 76 − 48 = 32	b. 76 − 48 = 22	c. 76 − 48 = 38
d. 70 − 48 = 38	e. 70 − 48 = 32	f. 700 − 482 = 228
g. 364 (with 1 1 written above) − 79 = 395	h. 32 − 29 = 17	

18. Examine the problem at the right.

```
  3 1 1
  4 0 0
− 2 3 6
  1 7 4
```

a. Explain, in mathematical terms, what the student did wrong. Where are the problems in the student's thinking?

b. How would you help the student if the quote below represented her explanation:
"You can't take 6 from 0 and you can't take 3 from 0, so you have to borrow from the four. Cross out 4 and write 3, cross out the zeros and write 10."

19. Estimate the following differences. Briefly describe how you obtained the difference.

a. 4473 − 2355	b. 65,963 − 29,147	c. 73,463 − 28,543
d. 43,433 − 16,328	e. 413,082 − 285,876	f. 320,283 − 184,438
g. 3,133,543 − 1,903,253	h. 71,234,033 − 32,753,962	

20. By estimating, quickly determine if check 238 will bounce. The beginning balance is \$510. Sketch the process by which you estimated.

Check No.	Amount (\$)	Balance
233	143	
234	97	
235	65	
236	212	
237	8	
238	45	

21. This problem requires some nice thinking. Find *a*, *b*, *c*, and *d* that will make the subtraction problem work. Is there more than one solution? Why or why not?

a. $\begin{array}{r} 6ab \\ -c8b \\ \hline 1da \end{array}$ b. $\begin{array}{r} 6xx \\ -1y8 \\ \hline 47y \end{array}$

22. Find the missing digits.

a. $\begin{array}{r} 8\square 6 \\ -\square 6 \square \\ \hline 3\ 1\ 9 \end{array}$ b. $\begin{array}{r} 5\square\square 3 \\ -\ 6\ 4\ 8 \\ \hline 4\ 5\ 6\square \end{array}$ c. $\begin{array}{r} 8\ 2\square\square 6 \\ -2\square 1\ 2\square \\ \hline \square 6\ 2\ 0\ 9 \end{array}$

23. Take no more than 5 to 10 seconds to determine whether this answer is reasonable. Briefly explain your reasoning.

$\begin{array}{r} 8482 \\ -2627 \\ \hline 6865 \end{array}$

24. Find the differences in the given base.

a. $\begin{array}{r} 41_{six} \\ -\ 22_{six} \\ \hline \end{array}$ b. $\begin{array}{r} 401_{six} \\ -\ 222_{six} \\ \hline \end{array}$ c. $\begin{array}{r} 742_{eight} \\ -\ 364_{eight} \\ \hline \end{array}$ d. $\begin{array}{r} 3040_{eight} \\ -\ 507_{eight} \\ \hline \end{array}$

25. Determine the missing numbers that will complete the following subtraction problem in base five.

$3\square 2\square_{five} - 441_{five} = 20\square 4_{five}$

DEEPENING YOUR UNDERSTANDING

26. Place the digits 1, 2, 3, 6, 7, and 8 in the boxes to obtain

$\begin{array}{r} \square\square\square \\ -\square\square\square \\ \hline \end{array}$

a. The greatest difference

b. The least difference

27. Choose among the digits 1, 2, 3, 4, 5, 6, 7, 8, and 9 to make the difference 234. You can use each digit only once. How many different ways can you make 234?

$\begin{array}{r} \square\square\square \\ -\square\square\square \\ \hline \end{array}$

28. With three boys on a large scale, it read 170 pounds. When Adam stepped off, the scale read 115 pounds. When both Adam and Ben stepped off, the scale read 65 pounds. What is the weight of each boy?

29. A mule and a horse were carrying some bales of cloth. The mule said to the horse, "If you give me one of your bales, I shall carry as many as you." "If you give me one of yours," replied the horse, "I will be carrying twice as many as you." How many bales was each animal carrying?

30. From each of the following lists, select two numbers whose difference will be closest to the target difference.

Numbers	Target
a. 315 475 764	300
b. 185 372 953	650
c. 382 723 793	350

31. One of my students asked me this question after she had read the text: "Why do we have addition and multiplication tables, but not subtraction and division tables?" Write your response to her question.

32. Respond to the following as though you were talking to a principal who is familiar with the Common Core and NCTM standards. You want to convince the principal to hire you. The principal has just said that she has only recently realized that the words *borrowing* and *carrying* are no longer used by effective math teachers. Rather, such teachers use alternative terms like *trading* or *regrouping*. The principal asks whether you agree with the change in language and, if so, why.

33. Joanna was driving from Fresno to Bakersfield. When she left Fresno, her odometer read 67,324. Unfortunately, she forgot to check her odometer in Bakersfield. On the way home, she saw a sign that said she was 24 miles from Fresno. At that point, her odometer read 67,564. She now had enough information to determine the distance from Fresno to Bakersfield. How far is it?

FROM STANDARDIZED ASSESSMENTS

NECAP 2006, Grade 5

34. Mrs. Lombardi had 2 hours to prepare for a party. The chart below shows the amount of time she spent completing different tasks.

TIME MRS. LOMBARDI SPENT ON DIFFERENT TASKS

Task	Time
Decorated cake	20 minutes
Made punch	15 minutes
Made sandwiches	50 minutes
Put up balloons	?

How much time did Mrs. Lombardi have to put up the balloons? (1 hour = 60 minutes)

a. 15 minutes b. 25 minutes

c. 35 minutes d. 45 minutes

NECAP 2005, Grade 5

35. The students at Maple Grove School are selling flowers. Their goal is to sell 1500 flowers.

- On the first day, the students sold 547 flowers.
- On the second day, the students sold 655 flowers.

How many flowers must the students sell on the third day to meet their goal?

a. 298 b. 308 c. 1202 d. 2702

NECAP 2007, Grade 4

36. The students in Mr. Hill's class are solving this problem.

> Peter had 10 pennies. Then he found more pennies.
>
> Now Peter has 16 pennies. How many pennies did Peter find?

Three students wrote these number sentences to solve the problem.

$16 - 10 = \square$	$10 - \square = 16$	$10 + \square = 16$
Ella	**Connie**	**Andy**

Who wrote a correct number sentence?

a. Only Connie and Andy b. Only Connie

c. Only Andy and Ella d. Only Ella

37. Kitty is taking a trip on which she plans to drive 300 miles each day. Her trip is 1723 miles long. She has already driven 849 miles. How much farther must she drive?

a. 574 miles

b. 874 miles

c. 1423 miles

d. 2872 miles

Source: Results of the *Seventh NAEP Mathematics Assessment*, p. 152. U.S. Department of Education, National Center for Education Statistics.

38. In a game, Carla and Maria are making subtraction problems using tiles numbered 1 to 5. The player whose subtraction problem gives the largest answer wins the game.

Look at where each girl placed two of her tiles.

Carla	Maria
[1] [ ] [ ]	[ ] [ ] [5]
− [ ] [5] [ ]	− [ ] [ ] [1]

Who will win the game?

Source: Results of the *Seventh NAEP Mathematics Assessment*, p. 318. U.S. Department of Education, National Center for Education Statistics.

SECTION  3.3 Understanding Multiplication

What do you think?

- What other meanings are there for multiplication beyond "repeated addition"?
- Why do we "move over" when we multiply?
- Are there more even or more odd numbers in the multiplication table?

As before, we will first examine the several meanings of multiplication and different models for representing them. Examination of patterns and properties will provide a foundation for understanding how the algorithms work. Then we will examine the connections among the various operations. This, in turn, paves the way for a strong operation sense, which enables you to solve the nonroutine and multistep problems that arise in everyday and work settings.

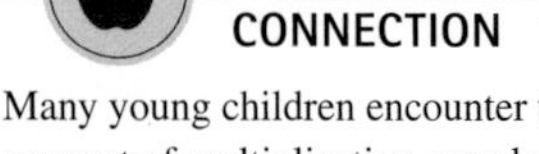

Many young children encounter the concept of multiplication even before school; for example, if the tooth fairy gives me 50¢ for each tooth, how much will I get for all my teeth?

Many elementary methods textbooks group addition and subtraction in one chapter and then multiplication and division in another chapter. This is not simply because children learn addition and subtraction first but also is because multiplication and division are more complex and abstract concepts. According to the Common Core State Standards, multiplication of whole numbers is introduced in third grade and deepened in fourth and fifth grades.

One of the goals of the following examination of multiplication is for you to understand the different meanings of multiplication and why it is useful to know these different meanings.

First write your own word problem for 4 times 3. Assume that you are a young child who has not yet learned multiplication. How might you solve this problem? What picture or model might a child use to help them figure out the answers?

Four multiplication problems

1. One piece of candy costs 4¢. How much would 3 pieces cost?
2. If Jackie walks at 4 mph for 3 hours, how far will she have walked?
3. A carpet measures 4 feet by 3 feet. What is the area of the carpet?
4. Carla has 4 blouses and 3 skirts. How many combinations can she wear? (Assume that all possible combinations go together!)

CONTEXTS FOR MULTIPLICATION

Go back to the four problems and answer the following questions:

- In what ways are the problems different? In what ways are they similar?
- What does multiplication mean? For example, what words come to mind when you think of multiplication?

If we represent the four problems with diagrams, as in Figure 3.8, the differences appear to be tremendous, even though all four problems can be solved as 3 times 4. As in addition and subtraction, we can use discrete set models and linear measurement models to understand multiplication. We also have area models and Cartesian product models in multiplication.

Problem 1 is like the discrete sets we encountered with addition and subtraction. It is also literally **repeated addition**; it is solved by adding 4 + 4 + 4.

Problem 2 involves a certain number of measured units and can be represented with a number line that shows repeated addition of the measures.

Problem 3 involves measures, but it involves repeated addition less than it involves counting the number of new units. This problem illustrates the area context of multiplication.

Problem 4 involves discrete objects, but making sense of it relies not on repeated addition but on putting combinations in an array. In set language, what we did in this problem was to examine two sets and look at all possible ways of pairing the elements of those sets. We will see later in this section that we can represent this context with the *Cartesian product model of multiplication.*

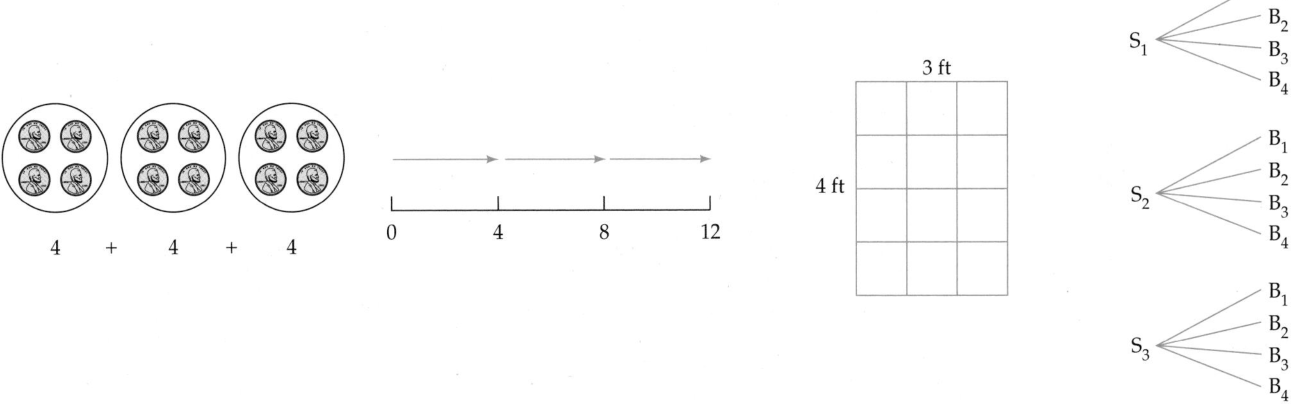

Figure 3.8

A GENERAL MODEL FOR MULTIPLICATION

Can you describe the four problems in such a way that they all have something in common, as we did for addition and subtraction?

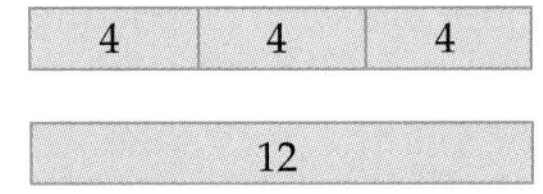

Figure 3.9

Our general model for addition can be extended to multiplication. Figure 3.9 shows one diagram for 3 times 4. Do you see the resemblance between the addition and multiplication models? As with addition and subtraction, the *general model of multiplication* can be cast in part–whole language: The product (the whole) is built from parts that are equal in size or amount.

Traditionally, multiplication is represented as

$$n \text{ times } a = \overbrace{a + a + a + \cdots + a}^{n \text{ times}}$$

In early elementary school, students use "×" to symbolize multiplication, as in $2 \times 4 = 8$. Because in algebra an x is used to denote a variable, students start using a dot or parentheses to denote multiplication later, as in $2 \cdot 4 = 8$ or $(2)(4) = 8$. With technology like calculators and cell phones, we use an asterisk to denote multiplication as in $2 * 4 = 8$. This symbolic notation, especially when it is not consistent, can cause confusion. Let's now consider some related terminology. If $n \cdot a = b$, then n is called the **multiplier**, a is called the **multiplicand**, and b is called the **product**. Furthermore, n and a can be said to be **factors** of b; b is a **multiple** of both n and a.

Using boxes, we can represent a **general model for multiplication** (Figure 3.10):

$$n \text{ times } a = \text{the amount } a \text{ added } n \text{ times}$$

That is, a is the value of whatever is in each box.

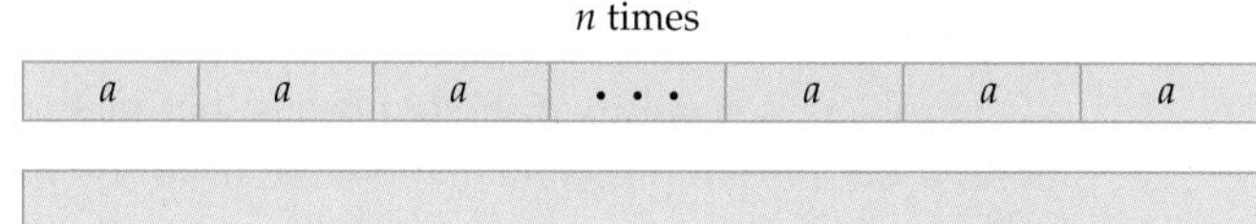

Figure 3.10

The general model shown in Figure 3.10 works well for Problems 1 and 2. However, the last two problems do not fit the model as well. We can force-fit them, but repeated addition is not the essence of those models. Therefore, we look for other models that can represent these contexts better.

The need for this is not readily apparent when we are working with whole numbers. In Chapter 4, we will see that the repeated addition model of multiplication just doesn't work well for $\frac{3}{4}$ times $\frac{2}{3}$; that is, adding three-fourths two-thirds of a time simply doesn't make sense as repeated addition. Let us now examine more closely the area model and the Cartesian product model of multiplication.

AREA MODEL FOR MULTIPLICATION

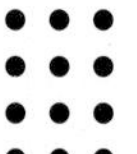

Figure 3.11

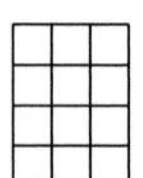

Figure 3.12

The **area model for multiplication** is an important one that we will see throughout the book. Figure 3.11 is a discrete representation for 4 times 3.

If we change each dot to a square, as in Figure 3.12, many students can better see the connection between the repeated addition model and the area model.

As you may recall, the formula for the area of a rectangle is length times width. Thus any multiplication problem $a \cdot b$ can be represented as a rectangle whose length is a units and whose width is b units. We will use this context in the next section when we examine the multiplication algorithm and in Chapter 4 when we examine multiplication of fractions and decimals. This model is a nice connection between operations and geometry.

LANGUAGE

The representations in Figures 3.11 and 3.12 are similar in that they are rectangles. The Common Core State Standards calls Figure 3.11 a rectangular array model and Figure 3.12 an area model.

CARTESIAN PRODUCT MODEL FOR MULTIPLICATION

As we saw in Problem 4, the Cartesian product of any two sets A and B is the set consisting of *all* possible ways of combining elements of the first set with elements of the second set. Using more formal language, we say that the Cartesian product of any two sets A and B consists of all possible ordered pairs such that the first element is from set A and the second element is from set B. In mathematical notation, we write

$$A \times B = \{(a, b) | a \in A \text{ and } b \in B\}$$

In Problem 4, the Cartesian product of the set S of skirts and the set B of blouses is

$$\{(S_1, B_1), (S_1, B_2), (S_1, B_3), (S_1, B_4), (S_2, B_1), (S_2, B_2), (S_2, B_3), (S_2, B_4),$$
$$(S_3, B_1), (S_3, B_2), (S_3, B_3), (S_3, B_4)\}$$

The definition of multiplication as Cartesian product is similar to our definition of addition using set language:

> If $n(A) = a$ represents the number of elements in set A, and if $n(B) = b$ represents the number of elements in set B, then the number of elements in the Cartesian product of sets A and B is equal to the product of a and b.

In notation,

$$a \cdot b = n(A \times B)$$

That is, the value of a times b is equal to the number of elements in the Cartesian product of set A and set B.

MULTIPLICATION WITH NUMBER LINES

Representing multiplication problems on a number line also highlights the context of multiplication as repeated addition. What multiplication problem is represented on the number line below?

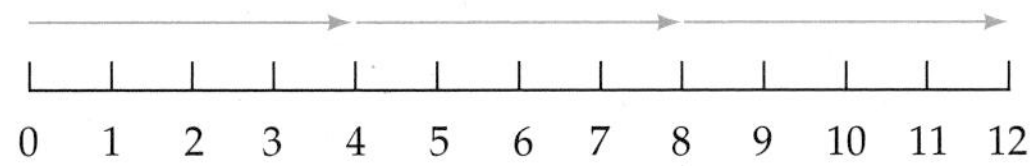

Figure 3.13

We can see the length of 4 represented three times, so this number line represents 3×4.

PROPERTIES OF MULTIPLICATION

MATHEMATICS

Let's say two children have birthday parties. One child has 7 party bags with 8 candies in each, and the other has 8 party bags with 7 candies in each. Numerically, both problems yield 56 candies, but the two problems don't look alike to many children. They are different situations.

Which of the properties that hold for addition also hold for multiplication? How can we use the four models of multiplication discussed earlier in this section to help us see these properties?

When adding, we found that "zero doesn't do anything." However, when we multiply any number by zero, the product is zero. This property of multiplication, which is not intuitively obvious to many children, is known as the **zero property of multiplication**:

$$a \cdot 0 = 0 \cdot a = 0$$

Many children are either intrigued or confused by the fact that when adding, "zero doesn't change your answer," but that when multiplying, "one doesn't change your answer." Stated in everyday language, "when we multiply any number by 1, we get the same number." This is called the **identity property of multiplication**:

$$a \cdot 1 = 1 \cdot a = a$$

Just as with addition, when we multiply any two numbers, we get the same product regardless of the order; that is, $a \cdot b = b \cdot a$. This property is known as the **commutative property of multiplication**. While all the models for multiplication can be used to show the commutative property of multiplication, perhaps the area model is the most intuitive since all you have to do is turn the rectangle sideways. For example, a 3×4 rectangle could be turned sideways to show a 4×3 rectangle, and the two rectangles are the same size.

Just as with addition, grouping doesn't matter when we multiply several numbers. For example, $(5 \cdot 4) \cdot 7 = 5 \cdot (4 \cdot 7) = 140$, although the problem is much easier to do mentally in the first way. This grouping property is known as the **associative property of multiplication**:

$$(a \cdot b) \cdot c = a \cdot (b \cdot c)$$

MATHEMATICS

The distributive property is a powerful one that we will encounter later in this chapter when we learn why the multiplication algorithm works. We will also see it in Chapter 4 when we examine multiplication with fractions and decimals, and it occurs throughout algebra.

The **distributive property** connects multiplication to the operations of addition and subtraction. Consider the following problem given to third-grade students: How many cans of soda are in three 24-can cases?

Some will simply add 24 + 24 + 24.

Others will represent the problem with base ten blocks, as in Figure 3.14. They will count 6 tens and 12 ones, convert to 7 tens and 2 ones, and give the answer of 72.

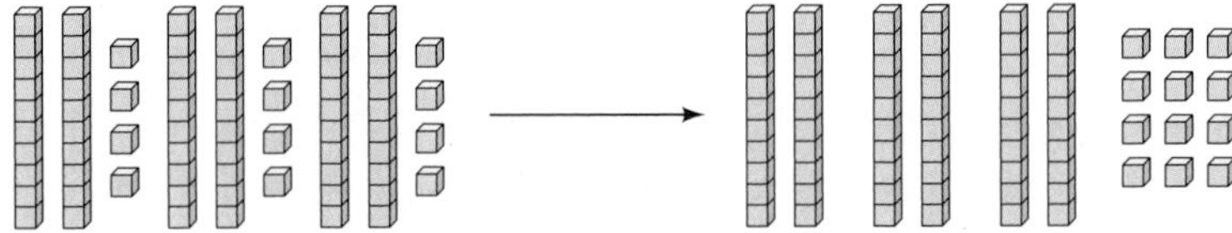

Figure 3.14

In this case, the students are using the **distributive property of multiplication over addition**:

$$a(b + c) = ab + ac$$

They have transformed $3 \cdot 24$ into $3 \cdot (20 + 4) = 3 \cdot 20 + 3 \cdot 4 = 60 + 12 = 72$ (Figure 3.15).

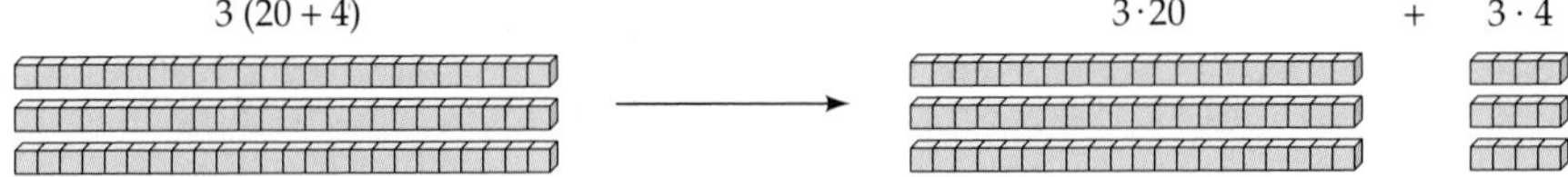

Figure 3.15

Others will solve the problem with money: 24¢ is 1 penny less than a quarter; do this three times and you will have 3 quarters take away 3 pennies—that is, $75 - 3 = 72$. In this case, the students are using the **distributive property of multiplication over subtraction**:

$$a(b - c) = ab - ac$$

The students have transformed $3 \cdot 24$ into $3 \cdot (25 - 1) = 3 \cdot 25 - 3 \cdot 1 = 75 - 3 = 72$.

CHANGES IN UNITS

There is an important difference between multiplication and addition (or subtraction). When we add or subtract two amounts, the units do not change: oranges plus oranges = oranges. However, this is not true with multiplication. Look back to the four multiplication problems presented at the beginning of this section. Do you see this?

- We are multiplying 3 *pieces* by 4 *cents per piece* and getting 12 *cents*.
- We are multiplying 4 *miles per hour* by 3 *hours* and getting 12 *miles*.
- We are multiplying 4 *feet* by 3 *feet* and getting 12 *square feet*.
- We are multiplying 4 *blouses* by 3 *skirts* and getting 12 *outfits*.

This matter of changing units is a crucial one that will develop over the course of the book.

THE MULTIPLICATION TABLE

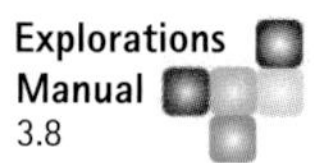
Explorations Manual 3.8

Let us further examine the multiplication table to see how patterns can help children learn the table's 100 "multiplication facts." What patterns do you see in Table 3.2? Write your observations and then read on. . . .

TABLE 3.2

	1	**2**	**3**	**4**	**5**	**6**	**7**	**8**	**9**	**10**
1	1	2	3	4	5	6	7	8	9	10
2	2	4	6	8	10	12	14	16	18	20
3	3	6	9	12	15	18	21	24	27	30
4	4	8	12	16	20	24	28	32	36	40
5	5	10	15	20	25	30	35	40	45	50
6	6	12	18	24	30	36	42	48	54	60
7	7	14	21	28	35	42	49	56	63	70
8	8	16	24	32	40	48	56	64	72	80
9	9	18	27	36	45	54	63	72	81	90
10	10	20	30	40	50	60	70	80	90	100

INVESTIGATION 3.3a A Pattern in the Multiplication Table

A student noted that every odd number in the table is "surrounded" by even numbers but that the reverse is not the case. Why do you think every odd number in the table is surrounded by even numbers? Work on this problem before reading on. . . .

DISCUSSION

A key to unlocking this mystery comes from the following generalizations: The product of two even numbers is always an even number, the product of two odd numbers is always an odd number, and the product of an odd number and an even number is always an even number. If you didn't make much progress on this question before, try to use this information and see whether you can explain this phenomenon now. Then read on. . . .

Let us look first at a specific case and then at a general case. The diagram below shows 35 (from $7 \cdot 5$) and the numbers that surround it in the table.

		6	**7**	**8**
		⋮	⋮	⋮
4	. . .	24	28	32
5	. . .	30	35	40
6	. . .	36	42	48

More generally, we can write

		Even	Odd	Even
		⋮	⋮	⋮
Even	...	Even	Even	Even
Odd	...	Even	Odd	Even
Even	...	Even	Even	Even

Thus we see that each of the eight numbers surrounding an odd number in a multiplication table is either the product of two even numbers or the product of an odd number and an even number. Thus none of them can be an odd number.

Using this pattern I will present two ways in which this knowledge can help with multiplication facts. First, a brief reflection on Investigation 3.3a yields the deduction that 3/4 of the multiplication facts are even numbers; that is, only one in four multiplication facts is an odd number. Second, nowhere in the multiplication table do we have two odd numbers in a row. Thus, for example, if I know that $7 \cdot 7$ is 49, then $8 \cdot 7$ can't be 55 or 57. The generalizations about the products of odd and even numbers would help, for example, a student who wasn't sure whether $7 \cdot 7$ was 48 or 49.

INVESTIGATION 3.3b Mental Multiplication

Do the following computations in your head. Briefly note the strategies you used, and try to give names to them.

1. 64×5 **2.** 16×25 **3.** 15×12 **4.** 849×2 **5.** 60×30

As you read these strategies, once again monitor your own thinking. Do you understand how the strategies work? If not, do you simply need to reread the discussion, or would it be helpful to make up and do some similar problems, or would it be more helpful to practice with a friend?

Explorations Manual 3.9

DISCUSSION

Here are some common strategies. You may have come up with others, as there are many ways to look at each problem. In Problem 1 (64×5), use **multiples of 10** as a reference point:

$$64 \times 5 = \frac{1}{2} \text{ of } 64 \times 10 = \frac{1}{2} \text{ of } 640 = 320$$

In Problem 2 (16×25), the **halve and double** method gives

$$16 \times 25 = 8 \times 50 = 400$$

Another way is to use the fact that $4 \times 25 = 100$ and break up the 16 as

$$16 \times 25 = 4 \times 4 \times 25 = 4 \times 100 = 400$$

Problem 3 is interesting because it lends itself nicely to many different solution paths. It can be solved using the distributive property in two ways.

One way: $15 \times 12 = (15 \times 10) + (15 \times 2) = 150 + 30 = 180$

That is, 15×12 is seen as 12 groups of 15, which break apart into 10 groups of 15 + 2 groups of 15.

Another way: $15 \times 12 = 12 \times 15 = (12 \times 10) + (12 \times 5) = 120 + 60 = 180$

That is, 15×12 is seen as 15 groups of 12, which break apart into 10 groups of 12 and 5 groups of 12.

We can use the halve and double strategy: $15 \times 12 = 30 \times 6 = 180$

In Problem 4 (849×2), try compatible numbers in conjunction with the distributive property:

$849 \times 2 = (850 - 1) \times 2 = 1700 - 2 = 1698$

For Problem 5 (60×30), use multiples of 10:

$60 \times 30 = 6 \times 3 \times 10 \times 10 = 18 \times 100$

What people often think in this case is "18 with two zeros," or 1800.

UNDERSTANDING MULTIPLICATION WITH LARGER NUMBERS

The big jump for children in learning how to multiply is to apply their knowledge of base ten and place value and to let go of strategies that work well for addition. For example, 23×4 can be found in many ways, as shown below. Some ways use base ten; some ways use powerful tools, and some ways do not. Some of these ways, like the first three, help the children to see the close connection between addition and multiplication. The way at the far right illustrates a more powerful, efficient way to determine the answer.

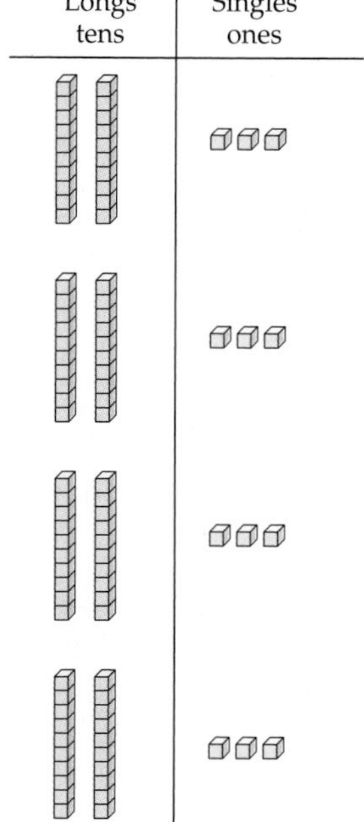

Add	*Add pairs*	*Count the 10s Add the 3s*	*Take apart (1)*	*Take apart (2)*	*Take apart (3)*
1 23 23 23 23 92	23, 23 > 46 23, 23 > 46 46, 46 > 92	23 23 23 23 80 12 92	$25 \times 4 - 2 \times 4$ $100 - 8$	10×4 10×4 3×4 $40 + 40 + 12$	20×4 3×4 $80 + 12$

One starting point for developing an understanding of the standard algorithm is to go beyond the basic multiplication facts, one step at a time. For example, how might you solve 16×7 if you didn't know the algorithm? Do so before reading on. . . .

When this problem is given to children, they solve it in many ways. Children naturally invent the distributive property by applying multiplication as repeated addition to their multiplication facts. For example, 16×7 can be broken into $(8 \times 7) + (8 \times 7)$ or $(10 \times 7) + (6 \times 7)$. We want the children to understand that breaking the numbers apart, using base ten, generally proves more powerful. Thus, this decomposition is ultimately more powerful: $16 \times 7 = (10 \times 7) + (6 \times 7)$.

Toward that end, the problems become larger and larger. For example, the children can construct the answer to 35×14 with the following string, where they begin with what they know and build up:

$3 \times 4 = 12$

$30 \times 4 = 120$ because they have studied the pattern when multiplying by 10

$5 \times 4 = 20$

$35 \times 4 = 140$ because 35×4 is equivalent to four 30s plus four 5s

$35 \times 10 = 350$ multiplying by 10

$35 \times 14 = 490$ because 35×14 is equivalent to four 35s plus ten 35s

These problems, which are called string problems or cluster problems, help to lay the conceptual foundation for understanding the standard algorithm. They also provide good practice in applying the distributive property, which is essential to understanding the algorithm.

Another step in the development is to examine problems in context. For example, we can frame the problem 23×12 as: How much will it cost to blacktop a playground that measures 23 meters by 12 meters? The natural representation of this problem, a rectangle, virtually *requires* the children to move toward the more powerful representation of multiplication as an array. Below you can see various representations of the problem and then some of the solution paths that come out of these representations.

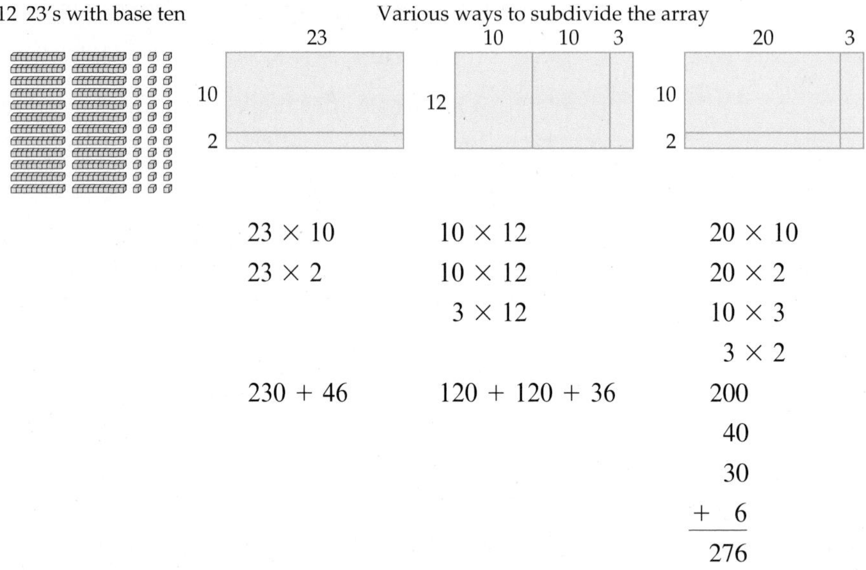

All three rectangles employ the powerful, but often only partially understood, distributive property.

The first rectangle illustrates one way of cutting the rectangle apart. That is, twelve 23s can be broken into ten 23s and two 23s.

The second rectangle illustrates another way of cutting the rectangle apart. That is, twenty-three 12s can be broken into ten 12s plus ten more 12s plus three more 12s.

The third rectangle illustrates a way of cutting the rectangle apart that connects to the standard algorithm. This way can be elicited by asking the children to fill the space with the least number of manipulatives. This can be done by filling the top left section with flats, the bottom right section with singles, and the other two sections with longs.

CLASSROOM CONNECTION

One of the teachers I worked with was so stunned as a result of our work with multiplication that she exclaimed, "I will never tell my students that multiplication is just repeated addition again!"

Investigations like these help teachers to see that although multiplication can be seen as repeated addition, if that is *all* you see multiplication as, then your students can achieve only limited understanding. Teachers also realize that we are still looking at parts and wholes. However, parts and wholes with addition and subtraction are conceptually much simpler. For example, in adding two numbers (say $68 + 43$), almost every way of taking apart the numbers will connect the closest place value ($70 - 2$ or $40 + 3$). However, when we are looking at multidigit multiplication, there are many ways to take apart the numbers, rather than just one or two ways.

Now let us examine the standard multiplication algorithm in a very systematic way.

THE MULTIPLICATION ALGORITHM IN BASE TEN

Explorations Manual 3.12

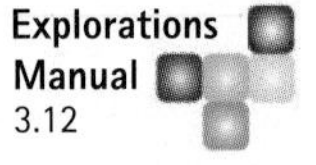

From this section, you now have a better understanding of different models of multiplication and an understanding of the importance of the distributive property. We will now examine the standard multiplication algorithm, which is perhaps the most difficult of the

algorithms to understand at the conceptual level. It rests on an understanding that any multiplication problem can be represented as a rectangle and on an understanding of the distributive property. We will examine this algorithm with 56 · 34.

The rectangular model of the product Without base ten, we are left with an imposing problem! Figure 3.16 shows 34 rows of 56 units. The answer is there, but who wants to count them all? However, if we apply our knowledge of base ten, we can find alternatives to counting each single unit.

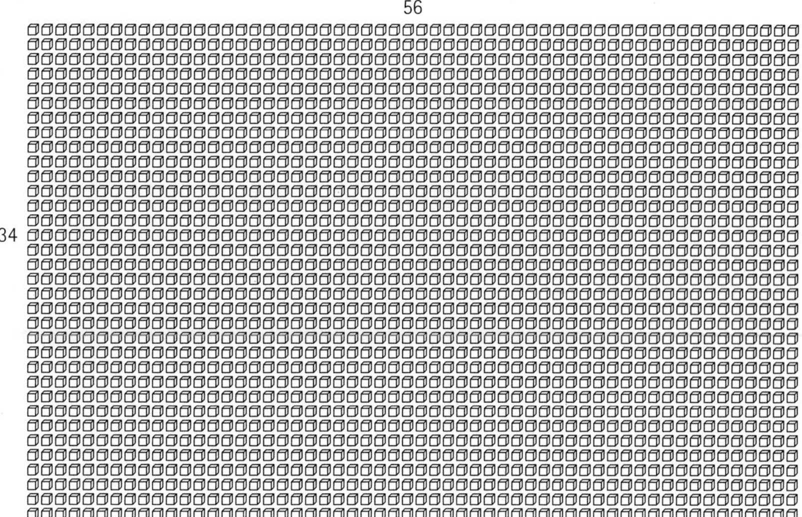

Figure 3.16

Using base ten to model the product One first step is to see what the problem would look like if we represented it using our knowledge of base ten (Figure 3.17). That is, we would have 56 (5 tens and 6 ones) 34 times. This makes our job only slightly easier. Who wants to add 56 thirty-four times?

Once again, our knowledge of base ten can make the counting process less tedious.

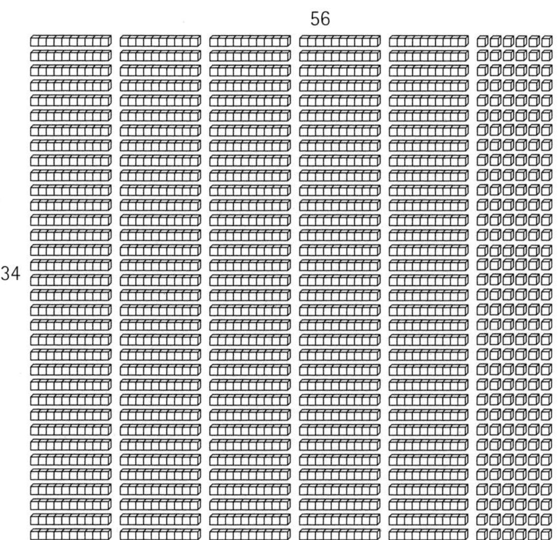

Figure 3.17

Using our knowledge that 10 ones make a ten, and that 10 tens make a hundred, we can turn units into longs and longs into flats (see Figure 3.18). You can do this yourself by drawing a line after every 10 rows of 56s in Figure 3.17. Do you see the connection between these processes and the following diagram? Do you see that this diagram is another representation of 56×34? Do you see connections between this diagram and how you would compute 56×34?

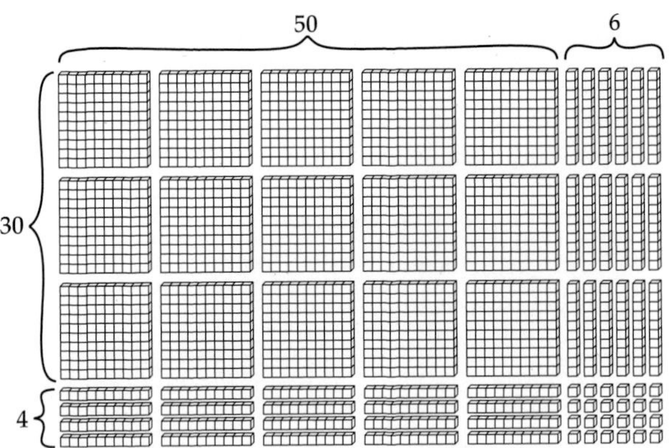

Figure 3.18

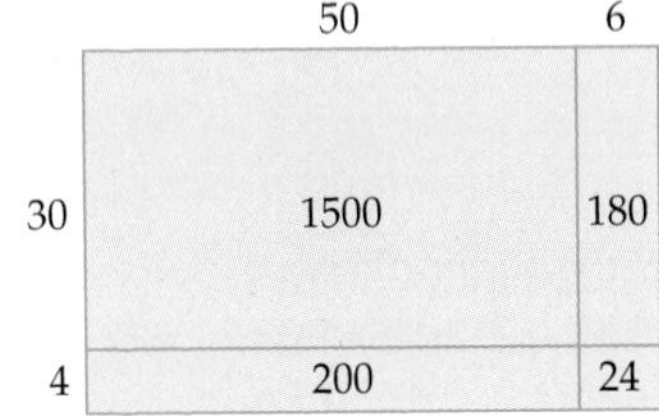

Counting the new groups When we represent the problem using the rectangular model of multiplication, we see that the standard algorithm systematically does the regrouping for us. Let us first look at how we can quickly count the amount in each of the four regions in Figure 3.18. Do this on your own before reading on. . . .

- The units region consists of $4 \cdot 6 = 24$.
- The longs region to the left of the units region consists of $4 \cdot 5 = 20$ longs, which have a value of 200.
- The longs region above the units region consists of $3 \cdot 6 = 18$ longs, which have a value of 180.
- The flats region consists of $3 \cdot 5 = 15$ flats, which have a value of 1500.

Adding these four regions, we have

$$\begin{array}{r} 1500 \\ 180 \\ 200 \\ +\ 24 \\ \hline 1904 \end{array}$$

Thus, when we use the algorithm to determine the product $56 \cdot 34$, we determine four products (in this case, $4 \cdot 6$, $4 \cdot 5$, $3 \cdot 6$, and $3 \cdot 5$), and we know what to do with the products. These products are called **partial products**.

When we represent the problem in expanded form, we can see how the four partial products connect to the algorithm.

$$\begin{array}{r} 50 + \ \ 6 \\ \times\ 30 + \ \ 4 \\ \hline 200 + 24 \\ 1500 + 180 \qquad\quad \\ \hline \end{array} \qquad\qquad \begin{array}{r} 56 \\ \times\ 34 \\ \hline 224 \\ +1680 \\ \hline 1904 \end{array}$$

The standard multiplication algorithm When we move up to larger multiplication problems, the manipulatives and visual representations become more cumbersome. However, if the students understand the process, they can apply that understanding to the larger problems. This understanding is also the foundation for skillful estimation. For example, when we represent 324 times 6 in expanded form, we have three partial products. The value of the partial product 3 · 6 (which represents 300 times 6) is much larger than the value of the other two partial products (120 and 24). Therefore, 300 times 6 (1800) is a quick and reasonable estimate for this problem.

$$\begin{array}{r} 324 \\ \times \quad 6 \\ \hline \end{array} \qquad \begin{array}{r} 300 + 20 + 4 \\ \times \qquad\qquad 6 \\ \hline 1800 + 120 + 24 \end{array}$$

INVESTIGATION 3.3c An Alternative Algorithm

The lattice algorithm that you learned for addition can be modified for multiplication. Observe the example below and see if you can figure out how it works and why it works.

Explorations Manual 3.13

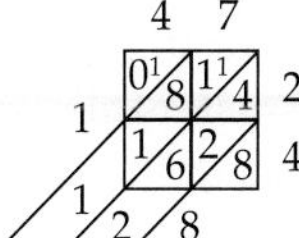

DISCUSSION

It works this way.

First, write the first number horizontally and the second number vertically.

Second, make a rectangle that has the same dimensions as your problem. In this case, since we have a 2-digit by 2-digit problem, we make a 2 × 2 rectangle.

Third, as with lattice addition, we draw diagonal lines through each square that extend below the square.

Fourth, we write the result of each partial product in the appropriate box.

Last, as with lattice addition, add diagonally. If the sum in any diagonal addition is greater than 10, trade to the next place value just as you do with the standard algorithm.

The "why" of this algorithm is very similar to the why of lattice addition. The diagram at the right shows the place value of each digit inside the lattice. As with lattice addition, the diagonals keep each digit with others of the same place value.

4	7	
Th / H	H / T	2
H / T	T / O	4

Th = Thousands
H = Hundreds
T = Tens
O = Ones

INVESTIGATION 3.3d Multiplication in Base Five

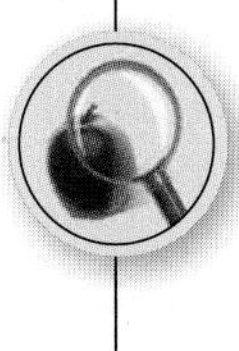

Just as we determined that the base five addition table is one fourth the size of the base ten addition table, so too with the base five multiplication table. I recommend that you take the time to make a base five multiplication table before reading on. It will make your ability to multiply, and later divide, in other bases much more meaningful and thus easier.

Then perform the following computations in base five.

A. $32_{\text{five}} \times 3_{\text{five}}$ **B.** $34_{\text{five}} \times 31_{\text{five}}$

DISCUSSION

You can use your fingers or manipulatives to make the base five multiplication table.

Base five	**1**	**2**	**3**	**4**	**10**
1	1	2	3	4	10
2	2	4	11	13	20
3	3	11	14	22	30
4	4	13	22	31	40
10	10	20	30	40	100

A. $\begin{array}{r} {}^{1} \\ 32_{\text{five}} \\ \times\ 3_{\text{five}} \\ \hline 201_{\text{five}} \end{array}$

Because $2 \times 3 = 11_{\text{five}}$, we place 1 in the singles place, and 1 above the 3 in the longs place.

$3 \times 3 = 14_{\text{five}}$. We add the regrouped 1 from the first partial product to the 14_{five} to get 20_{five}, which is placed appropriately below. Base blocks can help provide meaning to this regrouping as shown below.

Flats five^2	Longs fives	Singles ones	Flats five^2	Longs fives	Singles ones
			2	0	1

B. $\begin{array}{r} 34_{\text{five}} \\ \times\ 21_{\text{five}} \\ \hline 34 \\ 123 \\ \hline 1314_{\text{five}} \end{array}$

In this problem, we have four partial products: 4×1, 3×1, 4×3, and 3×3. The first row in the computation is 34 because there is no trading. In the second row, $4 \times 2 = 13_{\text{five}}$, we place 3 in the longs place and trade for the 1 flat. Last, since $3 \times 2 = 11_{\text{five}}$ we add the traded (from 4×2) to get 12_{five}. We place the 12 appropriately and now we add the two rows to get the final answer.

Some students find lattice multiplication easier especially with different bases. The key is to stay in the base, even when adding along the diagonals.

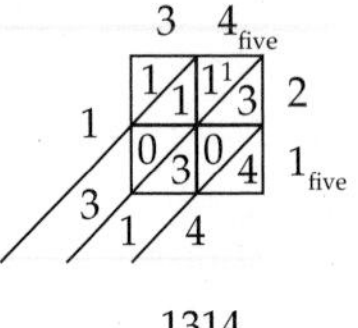

1314_{five}

CLASSROOM  CONNECTION

Grade 4

How might doing the lattice multiplication help fourth-graders to understand the more standard multiplication algorithm better?

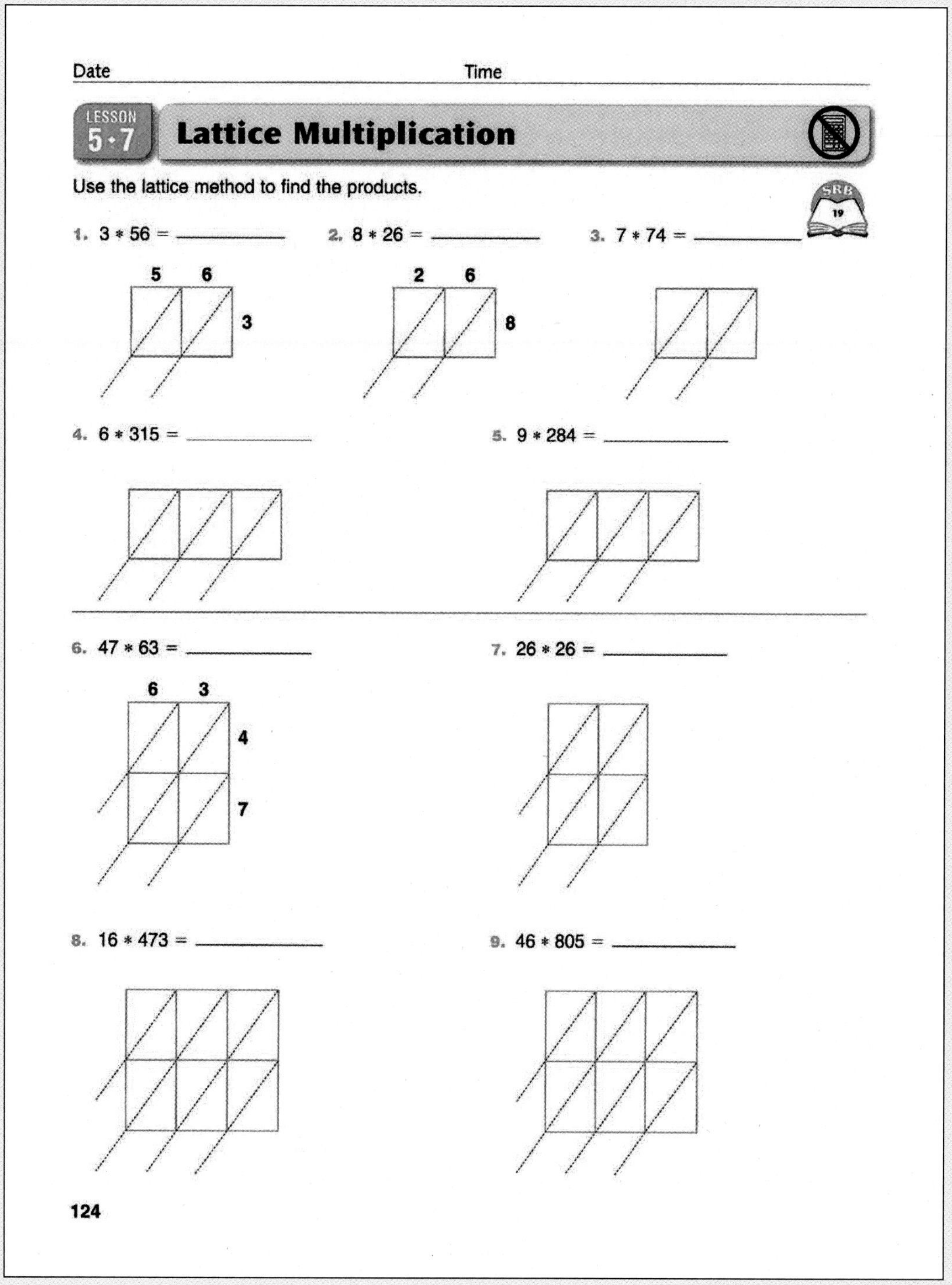

Date ______ Time ______

LESSON 5·7 **Lattice Multiplication**

SRB 19

Use the lattice method to find the products.

1. 3 * 56 = ______
   5 6
   3

2. 8 * 26 = ______
   2 6
   8

3. 7 * 74 = ______

4. 6 * 315 = ______

5. 9 * 284 = ______

6. 47 * 63 = ______
   6 3
   4
   7

7. 26 * 26 = ______

8. 16 * 473 = ______

9. 46 * 805 = ______

124

From *Everyday Mathematics, Grade 4:* The University of Chicago School Mathematics Project: Student Math Journal, Volume 1, by Max Bell et al., Lesson 5-7, p. 124. Reprinted by permission of The McGraw-Hill Companies, Inc.

INVESTIGATION 3.3e Children's Mistakes in Multiplication

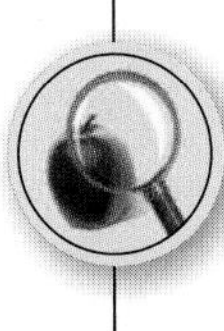

In *Building a System of Tens,* a third-grade child multiplies 49 by 2 and gets 108. How do you think she got 108? What misunderstanding(s) of place value and/or multiplication might be causing her problems?

DISCUSSION

$$\begin{array}{r} 49 \\ \times\ 2 \\ \hline 108 \end{array}$$

In this case, she correctly knew that $9 \times 2 = 18$. She placed the 8 in the ones place below and the traded 1 ten above the 4. It is likely that she generalized the addition algorithm in which you add the traded amount to the addition in the next place. Thus, she added the 1 to the 4 to make 5 and then multiplied 2×5. In multiplication, we need to do the multiplication first (4×2) and then add the traded 1 ten. Base ten blocks can help students see why this is as shown here.

Longs tens | Singles ones

Longs tens | Singles ones

9 8

INVESTIGATION 3.3f Developing Estimation Strategies for Multiplication

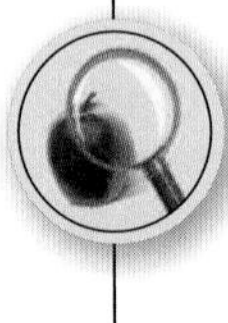

In each of the problems below, first obtain a rough estimate, then obtain the best estimate you can before reading on.

A. Chip rode 78 miles last week. At this rate, how many miles will he ride this year?

DISCUSSION

STRATEGY 1 Find a *lower bound* and an *upper bound* for the answer

Because we want to estimate the product of 78 and 52, if we round both numbers down to the nearest 10, we have $70 \times 50 = 3500$. Similarly, if we round both numbers up to the nearest 10, we have $80 \times 60 = 4800$. The actual product of 78 and 52 lies between these two numbers. The estimate from rounding down is called a **lower bound**, and the estimate from rounding up is called an **upper bound**. Using this technique, we can quickly say that Chip will ride between 3500 and 4800 miles this year.

STRATEGY 2 Round one number up and round the other number down

Thus, 78×52 becomes $80 \times 50 = 4000$.

STRATEGY 3 Use expanded form and estimate the sum of the four partial products

$$\begin{array}{r} 70 + 8 \\ \underline{50 + 2} \\ 140 + 16 \end{array}$$

$$3500 + 400$$

One thought process might go like this: "$70 \times 50 = 3500$, 50×8 is 400, and 70×2 is more than 100. Thus the answer will be 3500 plus more than 500, let's say 4050." Determine the actual answer. How close were the different estimates?

B. Jane plans to start graduate school in September. She figures that she can save \$345 per month for the next 9 months. How much will she have saved? Work on this and then read on. . . .

DISCUSSION

Use the distributive property:

$$345 \times 9 = (345 \times 10) - (345 \times 1) = 3450 - 345 \approx 3100$$

C. There were 47,752 Americans killed or missing in Vietnam. The number of Americans killed or missing in World War II was about 6 times that number. Approximately how many Americans were killed or missing in World War II? Work on this and then read on. . . .

$$\begin{array}{r} 47{,}752 \\ \underline{\times 6} \end{array}$$

DISCUSSION

We can get a rough estimate using rounding:

$$50{,}000 \times 6 = 300{,}000$$

We can get a more refined estimate by rounding and using mental math (double and halve):

$$48 \times 6 = 96 \times 3 = 288, \text{ or } 288{,}000$$

INVESTIGATION 3.3g Using Various Strategies in a Real-life Multiplication Situation

A warehouse has 50 bays (places to stack pallets). Four pallets can be stacked in each bay, each pallet can hold 24 cartons, and each carton holds 12 boxes. How many boxes can the warehouse contain? If each box sells for \$4, what is the value of the merchandise in a full warehouse? Work on this problem yourself before reading on. . . .

DISCUSSION

As with many of our problems, there are several strategies that will lead to a solution.

Bay	
pallet	pallet
pallet	pallet

Figure 3.19

STRATEGY 1 Draw a diagram

The diagram in Figure 3.19 is not "the right diagram" but rather an example of a useful diagram. If you didn't solve the problem or if you didn't solve it with a diagram, take a few moments to look at this diagram. Does it help? How? Does it connect to or stimulate your thinking about what operation(s) might be involved?

STRATEGY 2 Use smaller numbers

What if there were 3 bays, 4 pallets in each bay, 2 cartons in each pallet, and 5 boxes in each carton?

Many researchers have found that if they give two problems that are mathematically identical, but one of which has big or messy numbers, the success rates can be dramatically different. Do the smaller numbers help you to see the problem more clearly so that you can deduce what operation(s) to use?

STRATEGY 3 Use dimensional analysis

Using dimensional analysis, you cancel the larger units to concentrate on the aspect of the problem you want to solve. In this problem, we have 50 bays. However, we don't want to know the amount in terms of bays, we want to know it in terms of a smaller unit—boxes.

We can multiply 50 bays by 4 pallets per bay. Using dimensional analysis, this looks like

$$50 \cancel{\text{bays}} \cdot \frac{4 \text{ pallets}}{\cancel{\text{bay}}} = 200 \text{ pallets}$$

Because we do not want the answer in terms of pallets, we continue to change units:

$$200 \cancel{\text{pallets}} \cdot \frac{24 \text{ cartons}}{\cancel{\text{pallets}}} = 4800 \text{ cartons}$$

Translating to an even smaller unit, we find that

$$4800 \cancel{\text{cartons}} \cdot \frac{12 \text{ boxes}}{\cancel{\text{carton}}} = 57{,}600 \text{ boxes}$$

The value of the merchandise is

$$57{,}600 \cancel{\text{boxes}} \cdot \frac{\$4}{\cancel{\text{box}}} = \$230{,}400$$

A student more confident with using dimensional analysis can do the entire problem in one step:

$$50 \cancel{\text{bays}} \cdot \frac{4 \cancel{\text{pallets}}}{\cancel{\text{bay}}} \cdot \frac{24 \cancel{\text{cartons}}}{\cancel{\text{pallet}}} \cdot \frac{12 \cancel{\text{boxes}}}{\cancel{\text{carton}}} \cdot \frac{\$4}{\cancel{\text{box}}} = \$230{,}400$$

INVESTIGATION 3.3h Number Sense with Multiplication

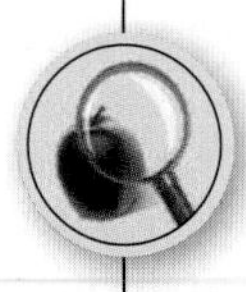

A. Without computing determine whether this number sentence is true or false.

$$68 \times 4 = (34 \times 2) + (34 \times 2)$$

DISCUSSION

If we add 34 groups of 2 and 34 more groups of 2, we get 68 groups of 2, but 68×4 means 68 groups of 4. Therefore, this problem is false. This is also a common mistake

Explorations Manual 3.14

made by children and adults, because both numbers at the right are half of the corresponding numbers at the left, so there is that sense of being equal.

B. This problem is similar to puzzles archaeologists have to solve; the circles represent places where they cannot read the numbers. We have recovered these fragments of this multiplication problem. What was the problem and what is the answer? What knowledge of multiplication can help you?

```
     ○6
  ×  4○
  -----
    32○
   ○○○
  -----
   ○○○4
```

DISCUSSION

This problem requires some analysis. Because the product has a 4 in the ones place, a first step is to ask 6 times what gives a 4 in the ones place? Many people stop with the 4 because 6 × 4 = 24. However, if we are careful, we also see that 6 × 9 = 54. So now what do we do? What number in the tens place of the multiplicand will get us 324? That is 96 × 4, 86 × 4, 76 × 4, etc. None of these work. What about 96 × 9? Way too big. Try smaller 46 × 9? 36 × 9? This works. Therefore, we have now solved part of the puzzle:

```
     36
  ×  49
  -----
    324
   ○○○
  -----
   ○○○4
```

Actually, at this point, it is a simple matter to complete the problem to determine the rest of the missing places, and you can do this on your own.

C. Will the result of 43 × 87 be in the hundreds, thousands, or ten thousands? Explain your choice.

DISCUSSION

If we use our estimation and rounding skills, we quickly see that 40 × 90 = 3600, so the answer is in the thousands.

D. Is this answer reasonable? Why or why not?

```
    84
  × 66
  ----
  4824
```

DISCUSSION

Once we realize that there are four partial products, we can quickly resist the intuitive appeal of 4 × 6 = 24 and 8 × 6 = 48 to see the answer is larger than 4824.

SUMMARY 3.3

In this section, we have seen that multiplication is more than just repeated addition and that the area model of multiplication enables us to understand how the standard algorithm works.

We have examined different contexts and representations for multiplication. Importantly, we have seen how multiplication is similar to addition:

- Properties: closure, associative, commutative
- They both have an identity property
- There are several alternative algorithms

and different from addition:

- Properties: zero property and the distributive property
- When solving real-life problems, the units do not change in addition, but they do in multiplication

We have examined a number of strategies for mental multiplication and for estimating. Finally, we have explored problems to develop your number sense with multiplication.

3.3 Exercises

1. Make up story problems and diagrams to illustrate $5 \cdot 3$ in the four different contexts shown in Figure 3.8.
2. Write a realistic story to represent each problem below.

   a. $12 \cdot 4$ b. $35 \cdot 3$ c. $5 \cdot (3 + 7)$
3. Identify the computation being modeled on the number line below and briefly justify your answer.

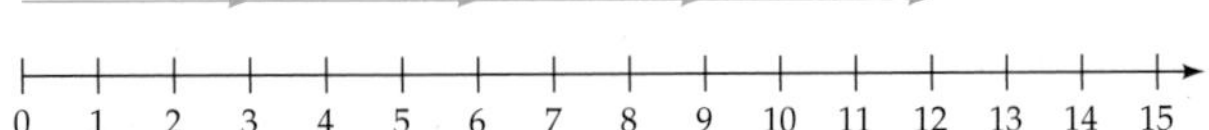

4. Represent 3×5 on a number line. Briefly explain why your number line models this problem.
5. Do you think we could speak of a distributive property of addition over multiplication? How would you determine whether it held or not?
6. a. Someone determined 25×12 by adding (25×10) and (25×2). What property were they using?

   b. Someone determined $8 \times 7 \times 5$ mentally by multiplying $8 \times 5 \times 7$. What properties did they use?
7. For each of the following questions, refer to the base ten multiplication table.

   a. The top-left-to-lower-right diagonal divides the table into three sets: diagonal, upper part, and lower part. What do you notice about the upper part and the lower part? What property is illustrated here?

   b. Look at the top-right-to-lower-left diagonal. Describe the patterns you see. Justify the patterns.

   c. The products of the diagonals of any 2 by 2 matrix are equal. Explain why.

   d. Look at the 9s column in the multiplication table. As we go down the column, the ones digit decreases by 1 each time and the tens digit increases by 1 each time. Can you explain why?
8. Let's say you forgot what $9 \cdot 7$ is. How would you figure out the product, using other multiplication facts? There are many, many different ways!
9. Fingers are a set of manipulatives that most people have, and they can be used to help students remember their multiplication facts for 9. This is how the method works. Suppose you want to know $9 \cdot 7$. Bend down finger number 7. The number of fingers to the left of the bent finger gives you the value of the tens place of the product, and the number of fingers to the right of the bent finger gives you the value of the ones place of the product. Why does this work?

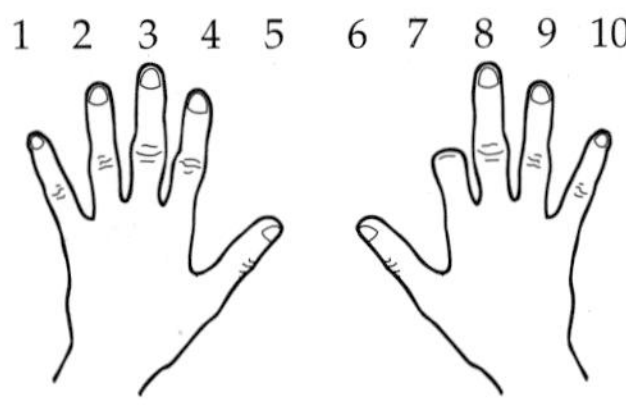

10. Determine the products mentally.

a.	b.	c.	d.	e.	f.
16×16	16×25	66×5	849×2	60×30	450×20

g.	h.	i.	j.	k.
35×12	65×42	736×4	35×16	632×4

In Exercises 11 and 12, determine the answer mentally.

11. A concert hall has 36 seats in a row, and there are 25 rows. How many people can be seated?

12. A camping show rents booth space for \$1.50 per square foot. How much would it cost to rent a 10 foot by 15 foot space?

13. Through illustrations, demonstrate how to solve these problems with manipulatives.

a. $\begin{array}{r} 25 \\ \times 7 \\ \hline \end{array}$ b. $\begin{array}{r} 43 \\ \times 12 \\ \hline \end{array}$ c. $\begin{array}{r} 29 \\ \times 15 \\ \hline \end{array}$

14. Determine the following products in two different ways other than using the standard algorithm or the lattice algorithm.

a. $\begin{array}{r} 42 \\ \times 15 \\ \hline \end{array}$ b. $\begin{array}{r} 41 \\ \times 39 \\ \hline \end{array}$

15. Draw a diagram to represent $34 \cdot 28$. Explain how to obtain the product from the diagram.

16. Look at this multiplication problem, solved using the standard algorithm. In this problem, we multiply $2 \cdot 6$ and trade the 1. What is it that we are trading? What is the mathematical meaning of this trading?

$$\begin{array}{r} 42 \\ \times 36 \\ \hline 252 \\ 126 \\ \hline 1512 \end{array}$$

17. ***Classroom Connection*** Below are three alternative representations, using partial products, of a two-digit by two-digit multiplication problem. These are sometimes called developmental algorithms, because teachers have used them to help children see the four partial products. One child wrote, "I like this (way) better because it allows you to give your brain a rest plus it's faster. . . !" ("Alternative Algorithms," *Teaching Children Mathematics,* April 2001). In each case, explain where the numbers come from, why they are placed where they are, and the value of this particular representation.

a. $\begin{array}{r} 72 \\ \times 34 \\ \hline 8 \\ 280 \\ 60 \\ 2100 \\ \hline 2448 \end{array}$ b. $\begin{array}{r} 72 \\ \times 34 \\ \hline 2100 \\ 280 \\ 60 \\ 8 \\ \hline 2448 \end{array}$ c. $\begin{array}{r} 72 \\ \times 34 \\ \hline 8 \\ 28 \\ 6 \\ 21 \\ \hline 2448 \end{array}$

18. In each of the following, use the information from the computed problem to determine the answer mentally.

$$\begin{array}{r} 587 \\ \times 345 \\ \hline 2935 \\ 23480 \\ 176100 \\ \hline 202{,}515 \end{array}$$

a. $587 \times 40 = ?$

b. $587 \times 45 = ?$

c. $587 \times 305 = ?$

d. $587 \times 300 = ?$

19. a. Write directions for multiplying $46 \cdot 37$ using the lattice algorithm.

b. Explain why this method works.

c. What mathematics do you have to know in order to be able to use this method?

d. Discuss some advantages and disadvantages of this algorithm.

20. Make a base five multiplication table. Use it to determine the following products. Try solving with both lattice and with the standard algorithm.

a. $\begin{array}{r} 43_{\text{five}} \\ \times 12_{\text{five}} \\ \hline \end{array}$ b. $\begin{array}{r} 204_{\text{five}} \\ \times 32_{\text{five}} \\ \hline \end{array}$

21. ***Classroom Connection*** On page 454 in the April 1999 issue of *Teaching Children Mathematics,* a teacher shares a method invented by one of her students to keep track of the regroupings in multidigit multiplication. Explain this child's method. That is, why does she do what she does?

$$\begin{array}{r} 236 \\ \times 57 \\ \hline {}^{2\,4} \\ 1412 \\ {}^{1\,3} \\ 10500 \\ \hline 13452 \end{array}$$

Source: *Teaching Children Mathematics,* by the National Council of Teachers of Mathematics.

22. ***Classroom Connection*** Identify and explain the errors in the problems below, all of which have occurred in real classrooms.

a. $\begin{array}{r} 46 \\ \times 28 \\ \hline 404 \\ 101 \\ \hline 1414 \end{array}$ b. $\begin{array}{r} 46 \\ \times 28 \\ \hline 3248 \\ 812 \\ \hline 4060 \end{array}$ c. $\begin{array}{r} 46 \\ \times 28 \\ \hline 368 \\ 92 \\ \hline 460 \end{array}$ d. $\begin{array}{r} 46 \\ \times 28 \\ \hline 92 \\ 368 \\ \hline 3772 \end{array}$

23. ***Classroom Connection*** A child multiplied 57×4 and got 288. How might the child have done this? What ideas from place value and/or multiplication has the child misapplied?

24. ***Classroom Connection*** When children work to develop methods of multiplying large numbers, the following conceptual error is very common. For example, a child will conclude that 34×26 will be equal to $(30 \times 20) + (4 \times 6)$. Why doesn't this work?

25. Estimate the following as closely as you can.

a. 41×68 b. 56×74 c. 62×83

d. 57×38 e. $34{,}345 \times 48$ f. 417×23

26. Estimate the following multiplication problems by doing the computations entirely in your head within 5 to 10 seconds. (Adapted from *Navigating Through Number and Operations in Grades 3–5* by Natalie N. Duncan el al., Reston, VA: National Council of Teachers of Mathematics, 2007.)

a. 8 erasers in a package and 43 packages

b. 24 crayons in a box and 17 boxes

c. 12 pencils in a package and 29 packages

d. 48 binder clips in a box and 9 boxes

e. 16 pens in a box and 51 boxes

f. 36 exposures per roll of film and 26 rolls of film

27. By estimating, determine which of the following is wrong. Explain your reasoning.

a. $3312 \times 13 = 43{,}056$

b. $23 \times 874 = 30{,}102$

c. $563 \times 86 = 48{,}418$

28. Which two numbers will have a product closest to 2000?

43 34 65 83 111

In Exercises 29 and 30, first determine a rough estimate in 5 to 10 seconds. Then determine your best estimate, which means no pencil and paper and less than 30 seconds or so. (Otherwise, we defeat the purpose of estimating!) Then determine the exact answer.

29. A student lives in Hanover, which is 62 miles from her school. How many miles per semester does she travel if she drives to school twice a week? How much does she spend for gas? Assume a 15-week semester.

30. A fund-raiser has a goal of raising $55,000 in 1 month. After 5 days, the amount that has been raised is $9345. Are they on track?

DEEPENING YOUR UNDERSTANDING

31. What base ten concepts and/or properties are we making use of when we employ the leading-digit method of estimation?

32. Find the missing digits by a means other than random trial and error. Explain your thinking.

a.
```
     7 9
×    □□
   7 1 1
  □□4 0
  □□□□
```

b.

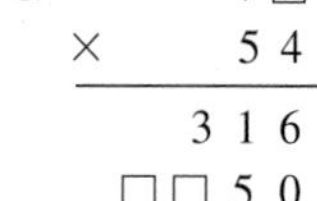

```
     6 □
×    9 7
   □□□
  6 1 2 0
  □□□□
```

c.
```
     7 □
×    5 4
   3 1 6
  □□5 0
  □□□□
```

33. Take no more than 5 to 10 seconds to determine if the answer to each problem is reasonable. Briefly explain your reasoning.

a.
```
   56
× 78
 4798
```

b.
```
   408
× 304
12,032
```

34. Determine what numbers go in the blank spaces and finish the problem in base five.

$$
\begin{array}{r}
2\ 0\ 4\ \square_{\text{five}} \\
\times \qquad \square\ 4_{\text{five}} \\
\hline
\square\ \square\ 3\ 3\ 2_{\text{five}} \\
\square\ 1\ 4\ 1_{\text{five}}\ \ \\
\hline
\square\ \square\ \square\ \square\ \square\ \square_{\text{five}}
\end{array}
$$

35. Find the missing digits by a means other than random trial and error. Explain your thinking.

a.
```
   □2
× 49
 3038
```

b.
```
   9□
× 6□
 6045
```

c.
```
   5□
× □3
 2193
```

d.
```
   8□
× □□
 4611
```

36. If a dripping faucet wastes water at the rate of 75 ounces per day, how many gallons does it waste per year?

37. People at the copy center at Keene State College told me that they used 1200 reams of paper last semester at my college. Departments are charged 3¢ per copy when they use the copy center. If departments use their own copy machines, which they do for small jobs, they are charged 5¢ per copy.

a. How many pages were run off at the copy center?

b. How much did the college pay to the copy center?

c. How much did the college save by having a copy center?

38. Take your pulse for 15 seconds. On the basis of this measurement, how many times would you say it has beaten since you got up this morning? This week? Since you were born?

39. How many Cheerios are in a box? First, define the problem better. Devise a strategy to estimate the number; then devise another strategy. Are your results close? Which strategy do you think is more accurate?

40. Will the product of 347 × 43 be in the thousands or ten thousands? Explain your choice.

41. An article in the January 1995 issue of *Teaching Children Mathematics* (p. 269) reported the following statistic: Every 57 minutes an underage drinker is involved in a traffic fatality.

a. Estimate the number in 1 year.

b. What additional information would you like to have to clarify the matter?

42. Which is a better estimate of 468 × 9: 420 × 10 or 460 × 10?

43. Which of the following might represent the cube of 156?

a. 3,796,416 **b.** 4,251,528

c. 3,944,312 **d.** 3,581,577

44. Select any two-digit number and multiply it by 99. Repeat this process for two or more two-digit numbers. Look for patterns that would enable you to compute the product of any two-digit number and 99 in your head.

a. Describe how you can determine the product of any two-digit number and 99.

b. Describe how you can determine the product of any two-digit number and 999.

45. For certain kinds of two-digit numbers, there is a shortcut to find their product.

$25 \cdot 25 = 625$ $22 \cdot 28 = 616$

$37 \cdot 33 = 1221$ $76 \cdot 74 = 5624$

a. Determine the kinds of numbers for which this shortcut works.

b. Describe the procedure as though to someone on the phone.

c. Explain why it works.

46. ***Classroom Connection*** This problem was described in the October 1993 issue of *The Arithmetic Teacher* (*"Becca's Investigation,"* pp. 78–81). A child, Rebecca, had made an interesting discovery. Two very different-looking problems gave the same product. That is, 36 × 8 and 48 × 6 both yielded 288. She noted that in both cases she had a 48 and a 24, 8 × 6 and 3 × 8 and 4 × 6. She wondered if there were other cases where two different numbers yielded the same product. What do you think? How many pairs can you find where we are multiplying a one-digit by a one-digit number?

Source: *The Arithmetic Teacher*, by the National Council of Teachers of Mathematics.

47. Without doing anything on paper, answer the following question: When we multiply 83 by some number we get a number that ends in 4. What is the number we are multiplying 83 by?

a. What can you tell about the missing multiplier?

b. Can you determine the unknown factor if the product is between 5000 and 6000?

c. Can you determine the unknown factor if the product is less than 2000?

48. Determine the value of a that will make the multiplication problem correct.

$$\begin{array}{r} a6 \\ \times 3a \\ \hline 1564 \end{array}$$

49. Place the digits 0, 1, 3, 5, 7 in the boxes to obtain

$$\begin{array}{r} \square\square\square \\ \times \quad \square\square \\ \hline \end{array}$$

a. The greatest product

b. The least product

50. Let's say you have an inexpensive calculator that has only an eight-digit or nine-digit display, and thus you cannot get the answers to these computations directly from your calculator. Determine the products by a means other than doing them out longhand, assuming that you have only these resources: your knowledge of multiplication and base ten and a calculator with an eight- or nine-digit display. Justify your method; that is, why will it give you the desired product?

a. 999,999,999 · 56

b. 987,654,321 · 9

c. 11,111,111,111 · 45

d. 34,000 · 56,000

e. $(399{,}999)^2$

51. Before the Russian Revolution in 1917, most of the people in Russian villages still used the Roman system and a multiplication algorithm known today as the Russian peasant algorithm. It works by simultaneously doubling one of the numbers and halving the other. Fractions are simply disregarded; for example, half of 19 is $9\frac{1}{2}$—drop the $\frac{1}{2}$. We find the product by first looking in the "Halve" column. Whenever you see an odd number in that column, circle that row. To find the answer, add the circled numbers in the "Double" column. The example below shows how it works.

Problem: 25 · 19.

Double		**Halve**
(25	·	19)
(50	·	9)
100	·	4
200	·	2
(400	·	1)

Circling the appropriate numbers, we add:

$$\begin{array}{r} 25 \\ 50 \\ 400 \\ \hline 475 \end{array}$$

Solve the following problems using the Russian peasant algorithm.

a. 25 · 17 b. 48 · 39 c. 120 · 42

52. Another ancient method of multiplication is shown below with the problem 102 · 96.

Find the average: 99

Take half the difference of the numbers: 3

Look up the square of 99: 9801

Look up the square of 3: 9

Subtract and you have your answer: 9792

a. Use this method to determine the product of 36 and 54.

b. Explain why it works.

53. On first sight, 34 × 23 and $(3x + 4)(2x + 3)$ are very different problems. However, change the representation and the similarities are striking.

a. Describe the similarities between the two computations. *Hint:* Represent 34 × 23 as (30 + 4)(20 + 3).

$$\begin{array}{r} 34 \\ \times 23 \\ \hline 102 \\ 68 \\ \hline \end{array} \qquad \begin{array}{r} 3x + 4 \\ \times\ 2x + 3 \\ \hline 9x + 12 \\ 6x^2 + 8x \\ \hline \end{array}$$

b. Now explain why the famous FOIL algorithm (First, Outer, Inner, Last) works.

54. Take your house number or room number and double it, add 5, multiply by 50, add your age, and subtract 250. What do you get? Can you explain why this trick works?

55. The square on the left is a multiplication magic square. How is a multiplication magic square similar to an addition magic square? How is it different? Can you make another multiplication magic square?

14	39	6	74
111	4	26	21
26	21	111	4
6	74	14	39

56. Estimation game (can be played alone or with a partner—cooperatively or competitively).

a. Write each digit 1 to 9 on 9 small pieces of paper. Mix them up and turn them over.

b. Decide the template for the problem, e.g., two-digit times two-digit numbers or three-digit times one-digit numbers, etc.

c. Turn over the appropriate amount of digits.

d. Estimate the product.

e. Your score is how far your estimate is from the actual product.

57. Target game (can be played alone or with a partner—cooperatively or competitively).

a. Decide a target number, e.g., 800.

b. Decide the template, e.g., two-digit times two-digit numbers.

c. Select the numbers that you think will be closest to the target, e.g., 25 × 28.

d. Your score is how far your estimate is from the actual product.

FROM STANDARDIZED ASSESSMENTS

NECAP 2006, Grade 4

58. Look at this number sentence.

$2 \times 4 = \square \times 8$

What number makes the number sentence true? Show your work or explain how you know.

59. In the multiplication problem below, write the missing number in the box.

$$\begin{array}{r} 2\,3\square \\ \times \quad 8 \\ \hline 1,896 \end{array}$$

Source: *NAEP Mathematics Assessment*, 1992. U.S. Department of Education, National Center for Education Statistics.

SECTION 

Understanding Division

What do you think?

- Do you think our multiplication and division algorithms would work with the Roman or Egyptian numeration system?
- When we do long division, we "bring down" the next number. What does *bring down* mean?

CONTEXTS FOR DIVISION

Assume that you are a young child who has not yet learned division. How might you solve these problems, using manipulatives or diagrams, or using common sense and other mathematical knowledge? Note your thoughts and try writing and modeling your own division problem for 24 divided by 4 before reading on. . . .

Explorations Manual 3.15

Four division problems

1. Applewood Elementary School has just bought 24 Apple computers for its 4 fifth-grade classes. How many computers will each classroom get?
2. Carlos has 24 apples with which to make apple pies. If it takes 4 apples per pie, how many pies can he bake?
3. Melissa, Vanessa, Corissa, and Valerie bought a bolt of cloth that is 24 yards long. If they share it equally, how many yards of cloth does each person get?
4. Jeannie is making popsicles. If she has 24 ounces of juice and each popsicle takes 4 ounces, how many popsicles can she make?

Now go back to these four problems and answer the following questions:

- In what ways are the problems different? In what ways are they similar?
- What picture or model might a child use to solve each of these? How are the pictures/models and strategies similar and different for each?
- What does division mean? For example, what words come to mind when you think of division?

Some of these problems involve discrete objects (Problems 1 and 2), whereas other problems involve measured (continuous) amounts (Problems 3 and 4).

Another difference emerges when we solve these problems as young children do. Left to themselves, most children will solve Problems 1 and 3 in a way that is very different from the way they will solve Problems 2 and 4. Do you see why?

MODELS FOR DIVISION

Let us solve Problems 1 and 2 simultaneously to understand two different models of division.

To solve the first problem, many children will give one computer to each class, dealing them out one by one: one for our class, one for your class, and so on.

Visually, we have

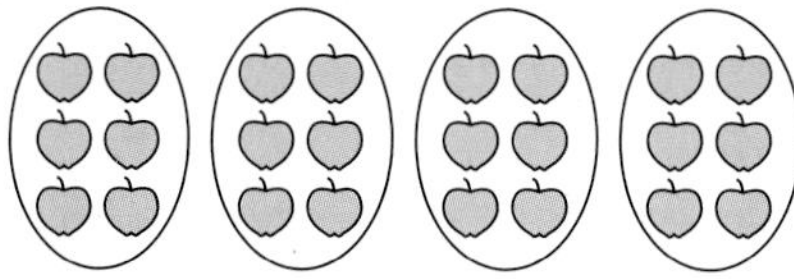

Each classroom gets 6 Apple computers.

This model is called the **partitioning model** because we solve the problem by first setting up the appropriate number of groups (partitions), which we then fill, as in one for you, one for you, etc., until they are gone, just like a child might do if sharing a bag of candies. Because of this, sometimes this is called the sharing model.

To solve the second problem, most children do something very different. They make groups of 4 (each representing one pie) until there are no more apples left.

Visually, we have

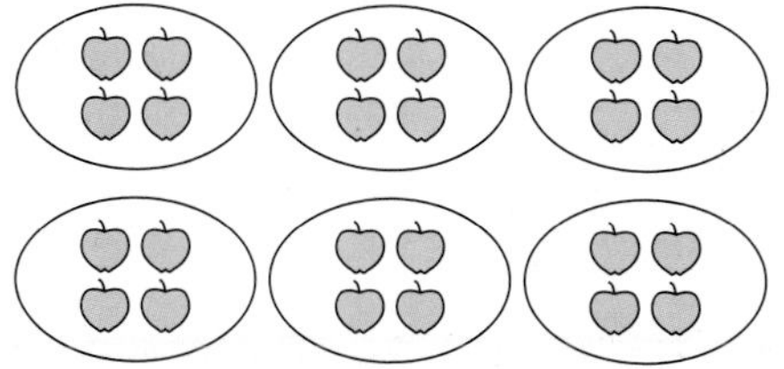

We have enough apples to make 6 pies.

This model is called the **repeated subtraction model** because we repeatedly subtract the given amount until we can do so no more. Some books call this the measurement model of division.

Now let us look at how the models are related. First we need some terminology.

Division terminology In both cases, the problem would be written as

$$24 \div 4 = 6 \text{ or as } 4\overline{)24}^{\,6} \text{ or as } \frac{24}{4} = 6$$

The division of the number b by the nonzero number n is formally defined as

$$b \div n = a \text{ if } a \cdot n = b$$

The number b is called the **dividend**, n is called the **divisor**, and a is called the **quotient**.

Comparing the two division models With this terminology, we can compare the two models in Table 3.3. What patterns or connections between the two models do you notice? Think and then read on. . . .

TABLE 3.3

	Dividend		Divisor	Quotient
Partitioning	24 computers	divided by	4 classes gives	6 computers per class
Repeated subtraction	24 apples	divided by	4 apples per pie gives	6 pies

We can also represent both division contexts with the general part–whole box model we have developed for the other three operations. At the most general level, we have:

TABLE 3.3 *(continued)*

	Dividend	Divisor	Quotient
Partitioning	Whole	Number of parts (groups)	Number in each part (group)
Repeated subtraction	Whole	Number in each part (group)	Number of parts (groups)

Stop for a moment to reflect on these models. How are these two models of division alike, and how are they different?

We can represent the models visually:

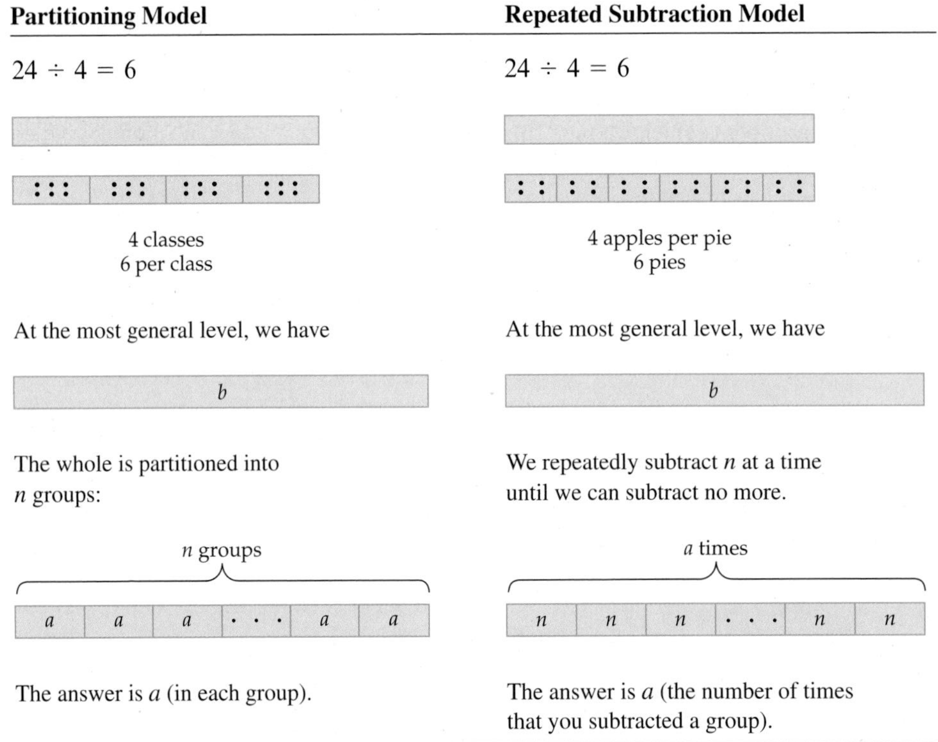

Partitioning Model
$24 \div 4 = 6$
4 classes
6 per class
At the most general level, we have
b
The whole is partitioned into n groups:
n groups
a a a . . . a a
The answer is a (in each group).
Repeated Subtraction Model
$24 \div 4 = 6$
4 apples per pie
6 pies
At the most general level, we have
b
We repeatedly subtract n at a time until we can subtract no more.
a times
n n n . . . n n
The answer is a (the number of times that you subtracted a group).

THE MISSING-FACTOR MODEL OF DIVISION

Just as some children interpret and solve some subtraction problems by finding the missing addend, some children interpret and solve some division problems by finding the missing factor. For example, consider the two problems we have just analyzed. Some children will solve either or both of these problems by asking themselves: 4 times what is equal to 24?

This model for division, called the **missing-factor model** for division, gives us another tool for working with division, one that we will use often in this and future chapters.

REPRESENTING DIVISION WITH NUMBER LINES

What problem is represented in the number line in Figure 3.20? Does the number line fit both models or just one model? Think and then read on. . . .

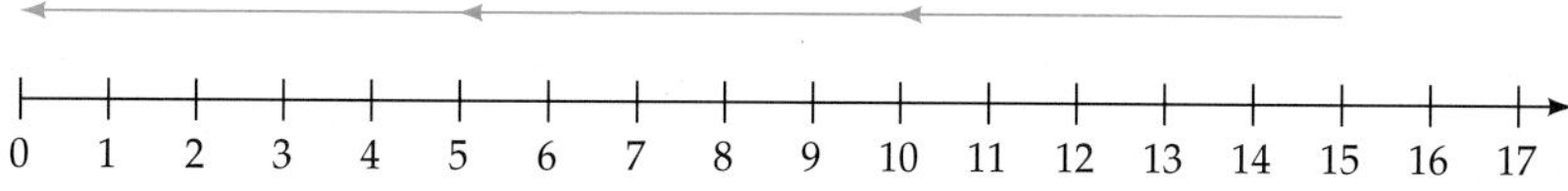

0 1 2 3 4 5 6 7 8 9 10 11 12 13 14 15 16 17

Figure 3.20

If the problem is $15 \div 5 = 3$, then the number line fits the repeated subtraction model. If the problem is $15 \div 3$, then the number line fits the partition model.

PROPERTIES OF DIVISION

Of the properties we have investigated for the other operations, identity, commutativity, associativity, and closure, which do you think hold for division? Think and then read on. . . .

If you have been making connections among the operations, you probably have realized that addition and multiplication are alike in that each operation has an identity and that the commutative and associative properties hold for those operations. You saw that those properties did not hold for subtraction, although subtraction does have a right-identity, i.e., $a - 0 = a$ but $0 - a \neq a$. Therefore, you might be inclined to say the same for division. Division does have a right-identity; that is, $a \div 1 = a$ for all numbers. The commutative and associative properties do not hold for division since $3 \div 4 \neq 4 \div 3$ and $(3 \div 4) \div 5 \neq 3 \div (4 \div 5)$. Similarly, the set of whole numbers is not closed under division since sometimes when we divide a whole number by a whole number, such as $3 \div 4$, it does not equal another whole number.

MATHEMATICS

"*USS Yorktown* dead in water after dividing by zero"
In September 1997 the computer system aboard the *USS Yorktown* shut down the ship. The problem was caused by someone who erroneously entered a zero in the wrong place, which caused the computer to divide by zero, which it cannot do. The computer system crash resulted in shutting down the ship's engines and the ship was dead in the water for three hours.

DIVISION BY ZERO

Our definition for division implies that "you can't divide by zero." Can you explain why? Think about this and then read on. . . .

There are several ways in which we can investigate this problem. We can use inductive reasoning: Make a table and look for patterns. Do you understand Table 3.4? What pattern do you see that can help you to explain why "you can't divide by zero"?

TABLE 3.4

Computation	Dividend	Divisor	Quotient
$5 \div 1$	5	1	5
$5 \div 0.1$	5	0.1	50
$5 \div 0.01$	5	0.01	500
$5 \div 0.001$	5	0.001	5000
$5 \div 0.00000001$	5	0.00000001	?

HISTORY

"Black holes result from God dividing the universe by zero."
Anonymous

As the divisor becomes smaller and smaller, the quotient becomes larger and larger.

We can also approach the problem of dividing by zero by seeing whether either context fits the model. What if we had 24 computers and we wanted to distribute them to zero classrooms? How many computers would each classroom get? That doesn't make any sense. Let's try repeated subtraction. What if we had 24 apples and each pie took zero apples? How many pies could we bake? This doesn't make sense either. From a practical (contextual) level, division by zero just doesn't make sense. Thus we say that division by zero is undefined.

A third approach to this problem is to assume that it is possible; this is called **indirect proof**. If division by zero were possible, then a nonzero x divided by zero would be equal to some number k (that is, $x \div 0 = k$). However, if this is true, then according to our definition of division, $k \cdot 0 = x$, but this is not possible because we know that $k \cdot 0 = 0$.

Many students find this proof to be too abstract. We can use an actual example and begin with 0 divided by a number.

$$\frac{0}{4} = 0 \text{ because } 4 \cdot 0 = 0$$

If $\frac{4}{0}$ is possible, then $\frac{4}{0} = x$, which represents the unknown quotient. But we know that there is no number x that makes $0 \cdot x = 4$.

INVESTIGATION 3.4a Mental Division

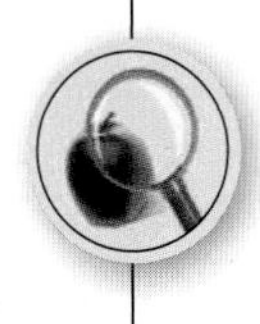

Examine each of the problems below carefully. How might you determine the exact answer, applying what you know about division? Briefly note the strategies you used, and try to give names to them.

1. $20\overline{)6000}$ 2. $400\overline{)20{,}000}$ 3. $8\overline{)152}$ 4. $5\overline{)345}$

Explorations Manual 3.17

DISCUSSION

In Problem 1, we will examine two different ways to use **canceling**.

One way to cancel in Problem 1 is to transform it from $\frac{6000}{20}$ to $\frac{600}{2}$, which is equal to 300.

$$\frac{6000}{20} = \frac{600 \cdot 10}{2 \cdot 10}$$

In this case, we are actually applying the idea of **equivalent fractions**, which we will examine in Chapter 4. That is, we are dividing both numerator and denominator by the same amount, 10.

Another method of canceling transforms the problem from $\frac{6000}{20}$ to $\frac{(60 \cdot 100)}{20}$, which can then be simplified to $3 \cdot 100$ because 20 divides both 60 and 100 without remainder.

$$\frac{6000}{20} = \frac{60 \cdot 100}{20} = \frac{\overset{3}{\cancel{60}} \cdot 100}{\underset{1}{\cancel{20}}} = 300$$

Can you use either of these strategies to solve Problem 2? If so, which?

Problem 3 lends itself to compatible numbers; for example, $160 \div 8 = 20$.

We can use the distributive property to get an exact answer, since

$$\frac{152}{8} = \frac{160 - 8}{8} = 20 - 1 = 19$$

We can also use the distributive property in Problem 4:

$$\frac{345}{5} = \frac{350 - 5}{5} = 70 - 1 = 69$$

In Problem 4, we can also use the idea of equivalent fractions and multiply both numerator and denominator by 2. The resulting fraction, $\frac{690}{10}$, easily simplifies to 69.

DIVISION ALGORITHMS

INVESTIGATION 3.4b Understanding Division Algorithms

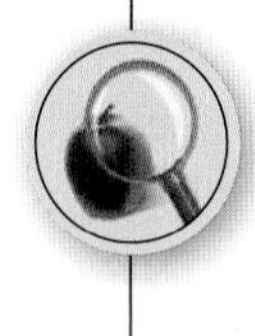

Solve the two division problems as though you didn't know any algorithms. You may use base ten blocks or draw diagrams or use reasoning. Then read on. . . .

Explorations Manual 3.20

A. Warren has 252 guests coming to his wedding. Each table holds 4 guests. How many tables will he need?

B. Mickey has 252 marbles that he wants to distribute equally into 4 piles. How many marbles are in each pile?

DISCUSSION

A. This is a repeated subtraction problem, meaning that we need to put the 252 into groups of 4. This can be done with base ten blocks, as we will show in question B, except here we would put the blocks into groups of 4. However, in this context, the model is not as intuitive. Some students solve this problem by using the missing-factor model: 4 times what makes 252?

Guess	Result	Thinking
50 tables	200 people	Too low, try a bigger number, let's say 60 tables.
60 tables	240 people	We need 3 more tables for the 12 remaining people.
63 tables	252 people	

B. The base ten blocks are more useful for Problem B. In this case, Mickey's problem is to divide (distribute) 252 into four groups equally. Using base ten manipulatives, he can figure out how to do this most efficiently. Because he cannot give 100 to each person, he has to trade the 2 hundreds (flats) for 20 longs. He still has 252, but physically he has 25 longs and 2 units.

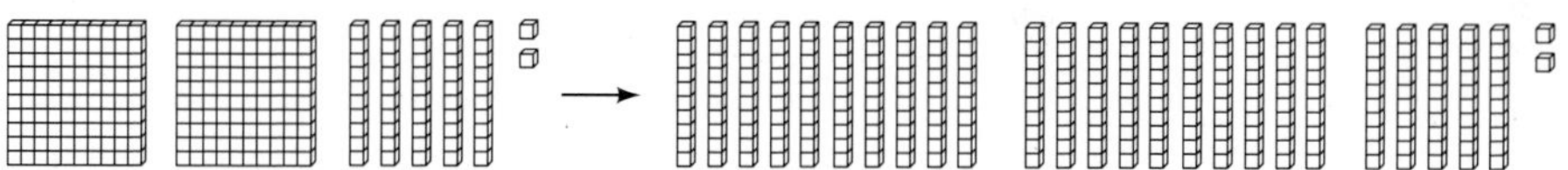

Now he can place 6 longs in each group. The corresponding symbolic step is shown at the left.

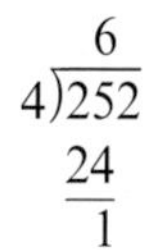

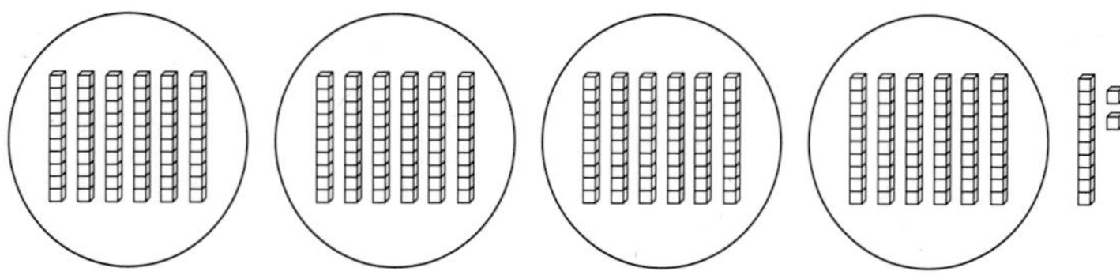

He is left with 1 long and 2 units, representing 12; thus he trades the long for 10 units. This lets him place 3 ones in each group. The corresponding symbolic step in shown at the left.

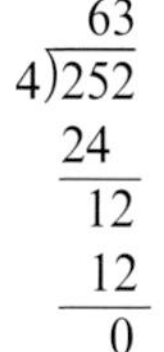

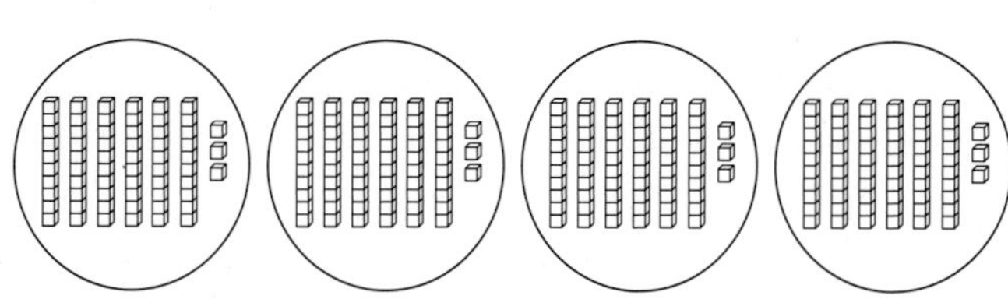

Looking back Write your response to the following questions before reading on. . . .

1. What is happening when we say "4 goes into 25 six times"?
2. What is happening when we multiply and where is this number seen in the model?
3. What is happening when we subtract after we multiply and where is this number seen in the model?
4. What is happening when we bring down the next number?

Let us look at the answers to these questions.

1. When we say divide 25 by 4 we are simply determining the first step in distributing the amount in the dividend into equal groups. For example, in the problem $4\overline{)252}$ modeled above, when we say that 4 "doesn't go into 2," mathematically, this means that when we divide 252 into 4 equal groups, we do not have enough to have 1 hundred in each group, and therefore we have to look at the next place. When we say that 4 goes into 25 six times, mathematically this means that when we distribute 25 tens into 4 groups of equal size, we have enough for 6 tens in each group, and thus we place the 6 over the 5, that is, in the tens place.

   $$\begin{array}{r} 6 \\ 4\overline{)252} \end{array}$$

   In the base ten model above, there are two pieces related to this step. First the 2 hundreds are traded for the 20 tens and put with the 5 tens that are already there. Then the tens are put into 4 groups with 6 tens in each group.

2. When we multiply, we determine how much has been distributed, that is, how much of our original amount has been used up. In the base ten model above, 4 groups of 6 tens have been used up.

   $$\begin{array}{r} 6 \\ 4\overline{)252} \\ \underline{24} \end{array}$$

3. We subtract after we multiply so that we can find out how many tens are left, which in this case is that 1 ten that is not in a group, so we trade it for ones.

   $$\begin{array}{r} 6 \\ 4\overline{)252} \\ \underline{24} \\ 1 \end{array}$$

4. When we bring down the next number, we are combining the ten ones that we just traded the one ten for with the 2 ones that were already there. Now there are 12 ones to be distributed to the 4 groups.

   $$\begin{array}{r} 6 \\ 4\overline{)252} \\ \underline{24} \\ 12 \\ \underline{12} \end{array}$$

HISTORY

The common division algorithm evolved from a method called a *danda,* which is derived from the same root as the word giving. The "giving" part is referred to by most people as the "bringing down" part. That is, when you have finished subtracting, you "bring down" the next digit in the dividend (the number you are dividing). From another perspective, you are giving that number to the remainder.

It is important to emphasize that although our numeration system was invented over a thousand years ago, the division algorithm that you learned in elementary school is only one of many algorithms for division that were invented over this time. The following algorithm is taught in many elementary school textbooks, partly because it is helpful for many children who struggle with the traditional algorithm.

INVESTIGATION 3.4c The Scaffolding Algorithm

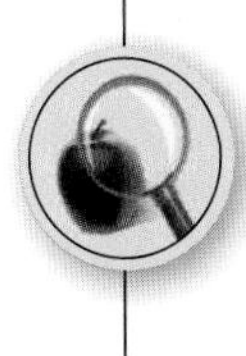

Another algorithm, which has enjoyed favor in different parts of the world over the centuries, is called the **scaffolding algorithm**. Two representations of this algorithm are shown for 4356 ÷ 6 in Figure 3.21.

Explorations Manual 3.19

```
   6
  20
 200
 500
6)4356        6)4356  500
  3000          3000
  1356          1356  200
  1200          1200
   156           156  20
   120           120
    36            36  6
    36            36
```

Figure 3.21

DISCUSSION

```
    699 R 162
 6)4356
   36
    75
    54
    216
     54
    162
```

Figure 3.22

Let us examine how it works. Consider the problem 6)4356. A child with poor multiplication facts will often have trouble with the first step. If the child chooses wrong, such as 6 × 6 = 36, then the whole problem is doomed (Figure 3.22). The alternative approach is to look at the problem from a different vantage point. Instead of asking how many 6s are in 43 or 6 × what is close to 43, we ask the child to look at the problem in context. For example, if a company produces 4356 bottles of soda each day, how many six-packs would this be? Then ask the child to estimate. Let's say the estimate is 500. The child determines how much 500 six-packs would be: 500 × 6 = 3000. This guess is recorded above the problem (see the work shown above), and then what remains after this guess is determined: Subtracting from 4356, we have 1356 left. Now the question is: How many six-packs would this be? Using the knowledge from the previous guess, 200 is a good guess, and 200 × 6 = 1200. Subtracting again, this gives us 156 left. Let's say the child guesses 20; this uses up 120 and leaves 36, at which point the child might guess 6. Note that this example shows only one of many possible solution paths for 4356 ÷ 6 using the scaffolding algorithm.

INVESTIGATION 3.4d Children's Mistakes in Division

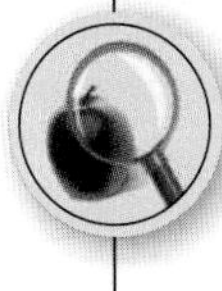

The error below is one that so many of my students have seen and is one reason that many elementary school textbooks present the scaffolding algorithm. What misunderstanding(s) of place value and/or division might be causing the problems?

```
    699 R 162
 6)4356
   36
    75
    54
    216
     54
    162
```

DISCUSSION

In this case, there are likely two problems: weak division facts and not remembering that in each step the remainder must be less than the divisor. When the student had a remainder of 7 in the first step, it was all over!

INVESTIGATION 3.4e Estimates with Division

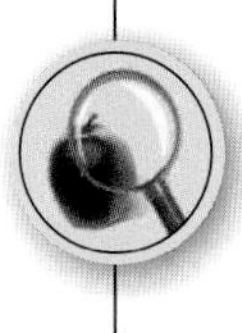

In each of the problems below, obtain an estimate before reading on. . . . 

A. Mr. and Mrs. Smith pay $9100 a year for their son's college. Translate this into a monthly payment that they can put into their budget.

DISCUSSION

Using the **missing-factor model**, ask yourself, "12 times what is closest to 91?" $12 \times 7 = 84$, $12 \times 8 = 96$. Because 91 is just about in the middle (of 84 and 96), it is reasonable to conclude that $12 \times 7\frac{1}{2}$ will be close to 91, and so our estimate is $7\frac{1}{2}$ hundred a month, or, more conventionally, $750 per month.

B. Pierre just bought a new van, and he wants to see what kind of mileage he gets. He filled up with gas after going 489 miles, and the car took 19 gallons of gas. Estimate the mileage.

DISCUSSION

One strategy is to round both the divisor and the dividend up.

$\frac{489}{19}$ will be close to $\frac{500}{20}$, and this is equivalent to $\frac{50}{2}$, and so he is getting about 25 mpg.

This strategy is interesting because it is not identical to the one used in multiplication, where we get more accurate estimates if we round one number up and the other down. In the case of division, we obtain more accurate estimates if we round both numbers up or both numbers down to manageable numbers. Can you explain why?

Hint: It might be helpful to make up some problems and check this out. For example, try $3750 \div 13$. Decrease both: to 3600 and 12, which are compatible numbers. Increase both: to 3900 and 13. Then decrease one and increase the other—for example, 3750 to 4000 and 13 to 10. Compare the results. Make up some other problems. Can you use ideas developed thus far to help you?

Explanation: We can see that rounding 489 up increases the quotient, that is, $\frac{489}{19} < \frac{500}{19}$.

We can see that rounding 19 up decreases the quotient, that is, $\frac{489}{19} > \frac{489}{20}$.

Thus, rounding both numbers up has the effect of canceling their effects, similar to the canceling effect in multiplication when we round one number up and the other down in 59×41.

Rounding both of the numbers down produces the same canceling effect.

C. A large business ran 5432 copies in the last 7 business days. How many copies is it averaging per day?

DISCUSSION

A rough estimate could be obtained by making compatible numbers, increasing the 5432, and solving the problem. For example, $5600 \div 7$ gives an estimate of 800.

CLASSROOM CONNECTION

Grade 4

Do you see how these problems connect with the scaffolding algorithm? 

Date ____________ Time ____________

LESSON 6·2

Solving Division Problems

For Problems 1–6, fill in the multiples-of-10 list if it is helpful. If you prefer to solve the division problems in another way, show your work.

SRB 17

1. José's class baked 64 cookies for the school bake sale. Students put 4 cookies in each bag. How many bags of 4 cookies did they make?

10 [4s] = ________ Number model: ____________

20 [4s] = ________ Answer: ________ bags

30 [4s] = ________

40 [4s] = ________

50 [4s] = ________

2. The community center bought 276 cans of soda for a picnic. How many 6-packs is that?

10 [6s] = ________ Number model: ____________

20 [6s] = ________ Answer: ________ 6-packs

30 [6s] = ________

40 [6s] = ________

50 [6s] = ________

3. Each lunch table at Johnson Elementary School seats 5 people. How many tables are needed to seat 191 people?

10 [5s] = ________ Number model: ____________

20 [5s] = ________ Answer: ________ tables

30 [5s] = ________

40 [5s] = ________

50 [5s] = ________

142

From *Everyday Mathematics, Grade 4:* The University of Chicago School Mathematics Project: Student Math Journal, Volume 1, by Max Bell et al., Lesson 6-2, p. 142. Reprinted by permission of The McGraw-Hill Companies, Inc.

A more refined strategy would be to use multiplication:

$7 \times 700 = 4900$

$7 \times 800 = 5600$

Because 5432 is closer to 5600 than it is to 4900, a rough estimate would be just under 800.

D. Employees put 25¢ in a can for each cup of coffee they drink. Last week the can contained $29.50. How many cups of coffee were drunk?

DISCUSSION

STRATEGY 1 Solve the problem in dollars

That is, 25¢ per cup means 4 cups per dollar.

You can estimate $30 × 4 cups per dollar = 120 cups.

STRATEGY 2 Solve the problem in cents

That is, $29.50 is about $30, which is equivalent to 3000¢.

$$\frac{3000}{25} = \frac{2500 + 500}{25} = 100 + 20 = 120$$

Looking back Take a few moments to look back over the strategies for mental division and estimation that we have explored. Have you added to your repertoire? Do you understand how the distributive property and expanded notation are applied? Do you see that some strategies work with some numbers better than with others? Can you connect these strategies to different models or contexts of division—for example, missing factor or repeated subtraction?

INVESTIGATION 3.4f Number Sense with Division

Explorations Manual 3.18 and 3.21

A. What is the value of the divisor that makes this division problem true?

$$\begin{array}{r} 11 \text{ R } 1 \\ \overline{)463} \end{array}$$

DISCUSSION

Think 11 × ? is less than 463? Because $11 \times 40 = 440$, we can try 11×41, which is 451, and then 11×42, which is 462, which gives a remainder of 1.

B. Take no more than 5 to 10 seconds to determine whether this answer is reasonable. That is, make your determination using number sense, without doing any pencil and paper work.

$$\begin{array}{r} 704 \\ 62\overline{)4268} \end{array}$$

DISCUSSION

Most people find it easier to translate this to a multiplication problem: 704×62. We quickly see that $700 \times 60 = 42{,}000$. Thus, the answer is not reasonable.

C. Determine the three missing digits:

$4\square\square2 \div 8\square = 48$

DISCUSSION

One of the key themes in this book is the idea of multiple representations. Sometimes, one representation sheds more light on a problem than another. If you were stuck, look at the following representation of the problem. Does this help?

$$8\square \cdot 48 = 4\square\square 2$$

What does this tell you now about the ones digit of the first number? Think and then read on. . . .

It must be a 4 or a 9. Do you see why? Test your intuition—which one do you think it is?

Now you can finish the problem.

OPERATION SENSE

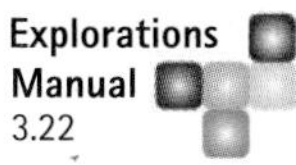
Explorations Manual 3.22

Now that we have examined each of the four operations, let us explore some problems in which we can apply our knowledge. Someone who knows which operation to perform in a problem is said to have good **operation sense**. Operation sense is more than just knowing what operation to use in what situation. Students with good operation sense have the following "knowledge":

- They can see the relationships among the operations—for example, multiplication is the inverse of division.
- They can apply their understanding of the properties of the operations—for example, they can use the distributive property.
- The diagrams that they draw connect the problem to the models of the operations.
- They have a sense of the effects of each operation—for example, the larger the divisor, the smaller the quotient.

DIVISIBILITY

Divisibility is an interesting extension of operation and number sense, so let's look at this concept now. This will help deepen our understanding of place value and of division. We can say that 3 divides 12 because 12 is evenly divisible by 3 (i.e., there is no remainder). We can describe this relationship in other ways as well:

3 is a factor of 12

3 divides 12

12 is divisible by 3

12 is a multiple of 3

Symbolically we write this as $3 \mid 12$.

INVESTIGATION 3.4g Interesting Dates

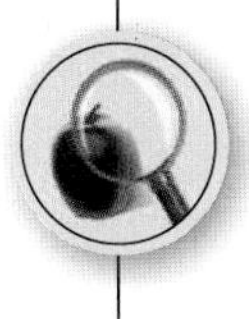

The date March 4, 2012 can be written as 3/4/12, which is mathematically interesting because 3 times 4 = 12. Another such date is June 2, 2012 (6/2/12).

Determine all the dates with this relationship in the year 2016, and in the year 2017.

DISCUSSION

In the year 2016, there will be four dates with this relationship:

1/16/16, 2/8/16, 4/4/16, 8/2/16

What about the year 2017? Here there is only one date with this relationship, 1/17/17.

What is different about the number 16 and the number 17 that the year 2016 has more than the year 2017? Why does the year 2017 only have one such date?

The number 17 is a prime number since its only factors are 1 and 17.

The number 16 is a composite number because it is composed of more than 2 factors.

We define a prime number as a number with exactly 2 factors and a composite number as a number with more than 2 factors. Working with these definitions, is the number 1 a prime or a composite?

The number 1 only has 1 factor (1), so it does not meet either the prime number or the composite number definition, so it is neither.

Over the centuries, many famous mathematicians have been fascinated by prime numbers. Prime numbers are the building blocks of all natural numbers greater than 1. Children are often intrigued by prime numbers as well. The Greeks are believed to have discovered the concept of prime numbers. A famous Greek mathematician, Euclid, was the first to document a proof showing that there is no largest prime number. The set of prime numbers is an infinite set! The largest known prime number at the time of this writing is 17,425,170 digits long! You can read much more about prime numbers at primes.utm.edu.

RECOGNIZING DIVISIBILITY

Exploring divisibility will help us to develop number and operation sense, as well as to develop our understanding of place value. This is also a useful tool when developing examples to use in your classroom. We will explore patterns and visuals that will help us understand divisibility by 5, 10, 2, 4, 8, 3, and 9.

Divisibility by 5

How can we tell by looking at a number whether it is divisible by 5?

Patterns Recognizing whether a number is divisible by 5 is something that many elementary students observe when skip counting by 5.

Let's consider numbers that are divisible by 5 and see the pattern:

5, 10, 15, 20, 25, 30, …

What do all of these numbers have in common? The ones digit is either a 0 or a 5.

Visually Base ten blocks will help us to see all of the divisibility ideas. With 5, why are we only concerned with the ones digit? Why does it not matter how many tens, hundreds, and so on we have in the number? Let's use the number 135 to illustrate.

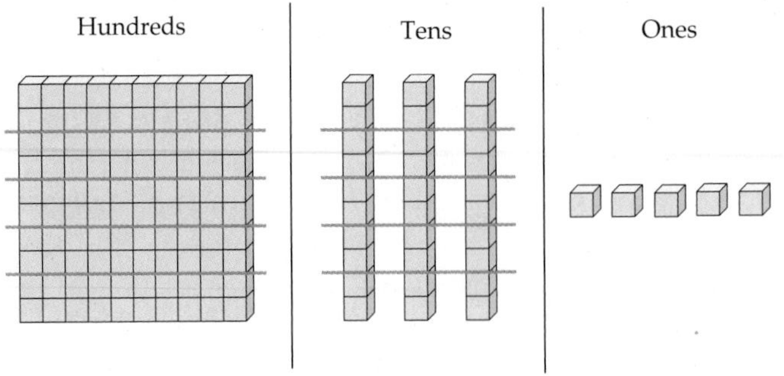

Ten is divisible by 5, so it doesn't matter how many tens there are, we will be able to divide each ten "long" by 5. One hundred is divisible by 5, so it doesn't matter how many hundreds there are, we will be able to divide them by 5. This same reasoning is true for all the place values to the left. This leaves us with only needing to consider the ones place. If there are 5 there, we will be able to divide by 5, or if there are 0 ones, then we will have 0 left over.

Therefore, a natural number is divisible by 5 if the ones digit is either a 0 or 5.

Divisibility by 10

Patterns Just like 5, elementary students will often observe this pattern fairly easily.

The first numbers divisible by 10 are:

10, 20, 30, 40, 50, …

The pattern points to the fact that a number is divisible by 10 if it has a 0 in the ones place.

Visually Similar to 5, the ten long, hundred flat, thousand cube, etc., are all divisible by 10. So, here again we are only concerned with how many are in the ones place. Only if there is a 0 in the ones place will the number be divisible by 10.

Therefore, a natural number is divisible by 10 if there is a 0 in the ones place.

Divisibility by 2

Patterns What do the numbers that are divisible by 2 have in common?

2, 4, 6, 8, 10, 12, 14, 16, 18, 20, 22, …

The pattern we can see here is that all of the numbers have a 0, 2, 4, 6, or 8 in the ones place. In other words, the ones place is an even number, or the ones place is divisible by 2.

Visually Consider how we looked at the divisibility for 5 and 10 visually and consider why with 2 we also are not concerned with any digit other than that in the ones place. Why is that?

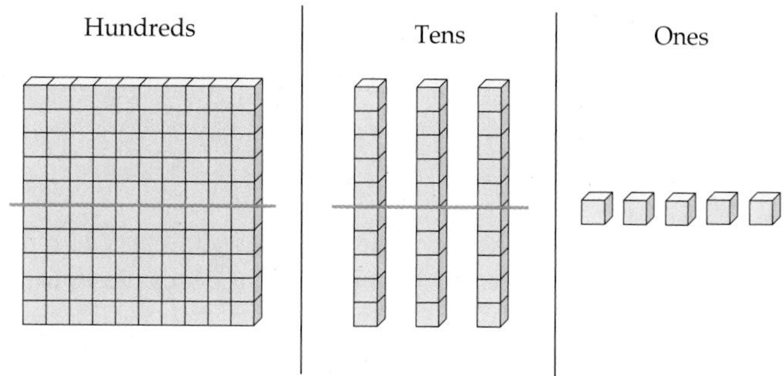

Just like with 5 and 10, 10 is divisible by 2, 100 is divisible by 2, etc., so we only need to look at how many ones we have.

Therefore, a natural number is divisible by 2 if the ones place is divisible by 2.

Divisibility by 4

The pattern of this one does not jump out as easily, so let's consider it visually and connect it to what we did above. Which place values are divisible by 4? Ten is not, but 100 is divisible by 4, and so is 1000, and above. So, it does not matter how many hundreds, thousands, etc., that we have, we will be able to divide those by 4. But since 10 is not divisible by 4, we need to look at both the tens and ones places. Let's use 1232 as an example.

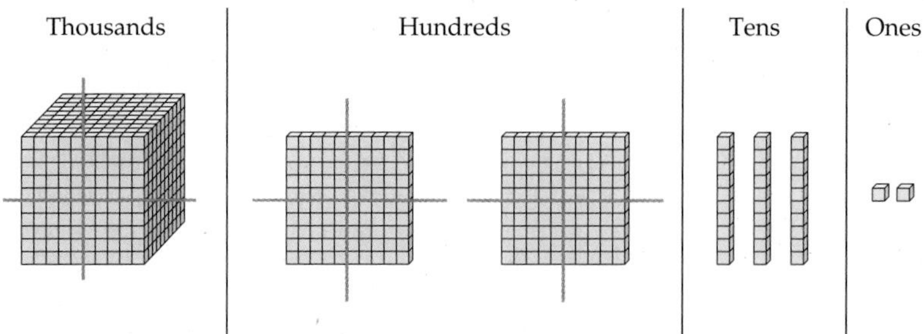

Since 32 is divisible by 4, we know that 1232 is divisible by 4.

Therefore, a natural number is divisible by 4 if the number of cubes in the tens and ones place is divisible by 4.

For example, 1,654,316 is divisible by 4 since 16 is divisible by 4.

Divisibility by 8

How can we apply what was just discussed for 4 to create a divisibility tool for 8? Ten is not divisible by 8, nor is 100. However, 1000 is divisible by 8, and 10,000, etc.

Therefore, a natural number is divisible by 8 if the number of cubes in the number represented by the hundreds, tens, and ones place is divisible by 8.

For example, 1,567,168 is divisible by 8 since 168 is divisible by 8.

Divisibility by 3

What happens when we divide each place value by 3? When we divide a long of 10 by 3, we will have one left over from each long, and when we divide a flat of 100 by 3, we will have one left over from each flat. The same pattern holds true for the thousands, ten thousands, etc. Let's look at 243 as an example.

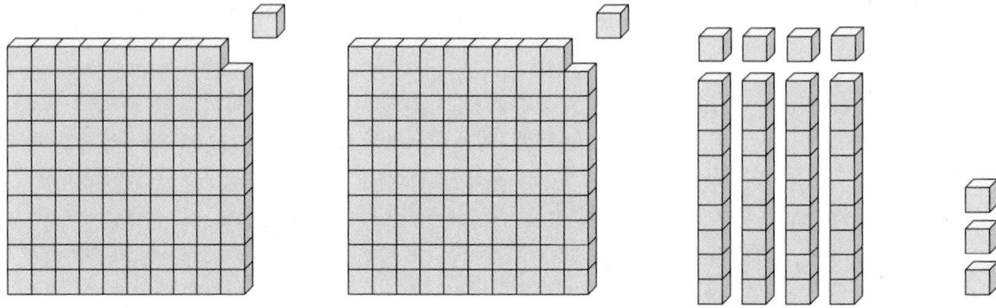

If we take all of these leftovers and put them in a set, we would have 2 left from the hundreds place of 243, 4 left from the tens place, and the 3 ones in the number.

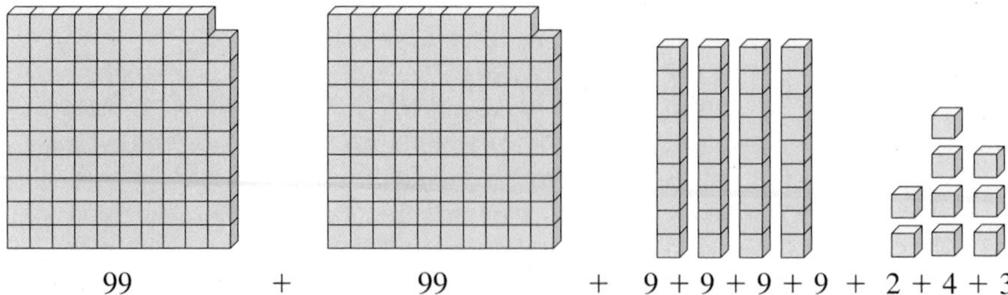

In each place value, the number of leftovers will be however many are in that place value.

Therefore, a natural number is divisible by 3 if the sum of the digits is divisible by 3.

For example, 1,516,113 is divisible by 3 since 18 is the sum of the digits and 18 is divisible by 3.

Divisibility by 9

The number 9 works similarly to 3. Why would that be?

Just like with 3, when each place value is divided by 9, there will be 1 left over for each block in that place value.

Therefore, a natural number is divisible by 9 if the sum of the digits is divisible by 9.

For example, 1,516,113 is also divisible by 9 since the sum of the digits is 18, which is divisible by 9.

Now you have tools for recognizing whether a number is divisible by 5, 10, 2, 4, 8, 3, and 9, and you have seen why these tools work.

INVESTIGATION 3.4h Applying Models to a Real-life Situation

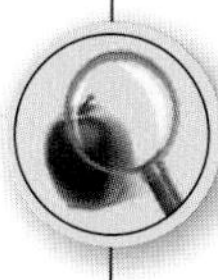

A teacher is going to do a project with the two fifth-grade classes in the school. One class has 21 students, and the other has 15 students. She has decided that all of these students will be divided into 12 groups and that each group will need 72 inches of string. She has a roll of string that is 882 inches long. Does she have enough string, or will she need to buy more? Do this problem yourself without using a calculator before reading on. . . .

DISCUSSION

You can use either multiplication or division to solve this problem. Using multiplication, $72 \cdot 12 = 864$ tells you that you will have enough string for each group, with 18 inches of string to spare. If you use division, the problem can be interpreted either as a partitioning problem or as a repeated subtraction problem. Do you see why?

- If you divide 882 inches by 72 inches per group, this says that you are seeing the problem in terms of repeated subtraction. What does your answer of 12 with a remainder of 18 tell you?
- If you divide 882 inches by 12 groups, this says that you are seeing the problem in terms of partitioning. What does your answer of 73 with a remainder of 6 tell you?

In the repeated subtraction model, the 12 says that if you repeatedly cut 72 inches, you can do this 12 times, and you will have 18 inches of string left over. In the partitioning model, the 73 says that if each group is to have an equal amount of string, you have enough string to make 12 pieces each of which is 73 inches long and you will have 6 inches left over. In either case, you have to understand what your quotient means, and you have to think about what the remainder means.

Although the mathematical answer is that the teacher has enough string, the teacher might actually conclude from her computations that she needs more string. Why is this?

What if the students make mistakes? Because $72 \cdot 12 = 864$, the teacher has only 18 inches of string in reserve. Thus, although the mathematical answer is that the teacher has enough string, the real-life answer would depend on the nature of the project. For example, if the students were to cut the pieces into specified lengths and I anticipated that some groups would mismeasure the lengths, I would not feel confident that I had enough string.

INVESTIGATION 3.4i Operation Sense

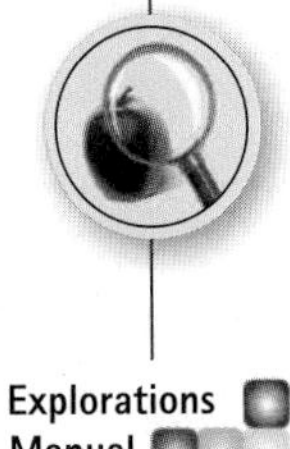

A. Select the correct operations to make the sentence true. You can use an operation more than once.

$$6 \bigcirc 5 \bigcirc 3 = 10$$

Explorations Manual 3.22

DISCUSSION

A few rounds of guess–check–revise generally convince most students that addition and subtraction are not possible. If we explore multiplication and division, trying multiplication 6×5 reveals that $\div 3$ will produce an answer of 10.

B. Make a number sentence that will produce the desired answer once the unknown amounts are known.

Germaine bought ***A*** CDs for ***B*** dollars each. He sold the whole lot for ***C*** dollars. How much profit did he make?

DISCUSSION

If we think about it, we can see the repeated addition in a group of CDs with the same price. In selling, he is taking the CDs away. The tricky part is realizing that, assuming he is making a profit, the C comes first. That is, the number sentence is $C - (A \times B)$.

ORDER OF OPERATIONS

Complex computations that use more than one operation pose a potential problem. To see what I mean, do this problem with pencil and paper and then on your calculator: $3 \cdot 4 - 8 \div 2$. What did you come up with? Stop, think, and then read on. . . .

If you do this on most calculators, you will get 8. However, if you do each operation in the order in which it appears in the problem or on very basic calculators or many cell phone calculators, you will get 2. Why did the calculator give a different answer?

The fact that we can get two different answers from the same problem has led to rules called the **order of operations**. These rules tell you the order in which to perform operations so that each expression can have only one value. Most calculators are programmed to obey the order of operations, though some cell phones are not.

What did the calculator actually do to get 8? Write the process in words and see whether a friend can use your explanation to see how to get 8. Then read on. . . .

In the absence of parentheses, multiply 3 times 4 to get 12, then divide 8 by 2 to get 4. Now subtract 4 from 12 to get 8. You may remember order of operations from school as "Please Excuse My Dear Aunt Sally," which is a mnemonic to remind you of the order in which a computation is done: Start inside the *p*arentheses, then do *m*ultiplication and *d*ivision as they occur from left to right, and finally do *a*ddition and *s*ubtraction as they occur from left to right. (The *e* from "Excuse" refers to exponents, which we will discuss in Chapter 4.)

If we insert parentheses so that more people would understand this problem, it would read as: $(3 \cdot 4) - (8 \div 2)$. How many different answers can you get depending on different placements of parentheses?

SUMMARY 3.4

We have found that multiplication can be represented as repeated addition, as the area enclosed by the multiplier and the multiplicand, or as the number of combinations (Cartesian product). Similarly, division problems can be seen as partitioning, performing repeated subtraction, or finding a missing factor.

There are certain properties that hold for multiplication of whole numbers but not for division: commutativity, associativity, and closure. We also learned of a property that connects operations: the distributive property. This property is a key to understanding how the multiplication algorithm works.

We have also examined the various connections among the four operations, including order of operations and divisibility, and applied our understanding of those connections so that our operation sense becomes more powerful.

What similarities do you see among the operations? What connections do you see among them? Then read on. . . .

Addition and subtraction are inverse operations. Multiplication and division are inverse operations.

One model of multiplication is repeated addition, and one model of division is repeated subtraction. We have a missing-addend model for subtraction, and we have a missing-factor model for division.

The set of whole numbers is closed under the operations of addition and multiplication and is not closed under the operations of subtraction and division. The commutative and associative properties hold for addition and multiplication, but not for subtraction and division. We have identity properties for addition and multiplication; because subtraction and division are not commutative, we have only right-identities for these operations. We can speak of the distributive property of multiplication over addition and of the distributive property of multiplication over subtraction.

With addition and subtraction, the units in each part are always identical; for example, 3 *sheep* plus 5 *sheep* equals 8 *sheep*. However, with multiplication and division, the units are different; for example, 3 *miles per hour* times 4 *hours* yields a product of 12 *miles*.

One of many visual representations of these similarities and connections is shown in the following figure.

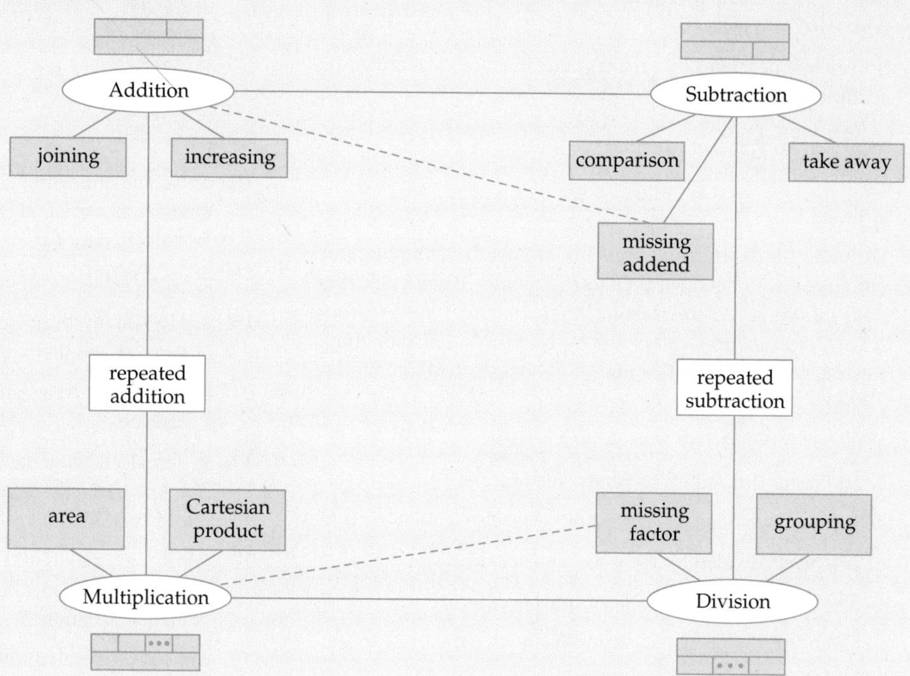

We have seen that the standard algorithms are related to some of the operations' contexts and models. These algorithms rely on our base ten system. Again, it is important to emphasize that these standard algorithms are not the only ones.

3.4 Exercises

1. Four children want to share 168 jelly beans. How many does each get? Which model of division is operative here? Model the problem with a diagram.

2. a. Make up a division story problem for the partitioning model. Make a diagram and explain why this problem is an example of the partitioning model.

   b. Make up a division story problem for the repeated subtraction model. Make a diagram and explain why this problem is an example of the repeated subtraction model.

3. For each number line problem below, identify the computation it models and briefly justify your answer.

   a. 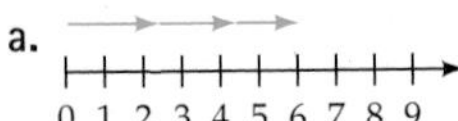
   b. 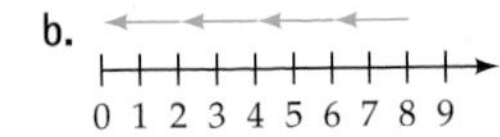

4. Represent the following problems on a number line. Explain each problem as though you were talking to someone who was not taking this class.

   a. $6 \cdot 3$ b. $12 \div 3$

5. Determine the quotients mentally.

   a. $\frac{6000}{20}$ b. $\frac{4800}{60}$ c. $\frac{10{,}000}{200}$

   d. $\frac{9000}{30}$ e. $\frac{3600}{90}$ f. $\frac{20{,}000}{400}$

6. Solve this problem with manipulatives and justify your process: $4\overline{)652}$

7. Solve these problems by a means other than the standard or the scaffolding algorithm.

   a. $641 \div 25$ b. $1842 \div 68$

8. Estimate as closely as you can.

   a. $8\overline{)5432}$ b. $7\overline{)29{,}123}$ c. $20\overline{)342{,}354}$

9. Estimate the following problems, doing the computations entirely in your head within 5 to 10 seconds. Briefly describe your thinking. (Adapted from *Navigating Through Number and Operations in Grades 3–5* by Natalie N. Duncan et al., Reston, VA: National Council of Teachers of Mathematics, 2007.)

   a. 92 beads divided by 6 beads on a bracelet
   b. 500 beads divided by 15 beads on a bracelet
   c. 200 beads divided by 6 beads on a bracelet
   d. 260 beads divided by 20 beads on a necklace
   e. 580 beads divided by 25 beads on a necklace
   f. 850 beads divided by 25 beads in a box
   g. 2400 beads divided by 50 beads in a box

10. Estimate these quotients by breaking the number apart, for example:

$$\frac{168}{14} = \frac{140}{14} + \frac{28}{14}$$

Briefly describe your thinking.

   a. $\frac{256}{4}$ b. $\frac{396}{6}$ c. $\frac{368}{8}$

   d. $\frac{5785}{8}$ e. $\frac{6455}{5}$ f. $\frac{9440}{15}$

In Exercises 11–13, do each of the following. First, determine a rough estimate in 5 to 10 seconds. Then determine your best estimate. ("Best estimate" means no pencil and paper and less than 30 seconds or so. Otherwise, we defeat the purpose of estimating!) Then determine the exact answer.

11. Several years ago, the flat roof over my kitchen was leaking. My friend at the hardware store told me of a new product that would fix my roof. Each can of roof sealant would cover 100 square feet and cost $26. My roof was 23 feet 8 inches long and 15 feet 6 inches wide. How many cans of roof sealant did I need to buy?

12. Mindy rode her bike 3 times last week. At the beginning of the week, the odometer read 302.4, and at the end of the week it read 315.2. At this rate, how many miles will she ride this year?

13. Mr. and Mrs. Smith will pay $8400 this year for their daughter's college tuition. Approximately how much is this per month?

14. ***Classroom Connection*** Identify and explain the errors below, all of which have occurred in real classrooms. 

a.	b.	c.
54	540	79 R 26
$7\overline{)3528}$	$7\overline{)3528}$	$8\overline{)658}$
35	35	56
28	28	98
28	28	72

15. Solve the following division problem and represent the remainder as a fraction in lowest terms. *Note:* The problem must be done in base five.

   a. $4_{\text{five}}\overline{)22202_{\text{five}}}$ b. $13_{\text{five}}\overline{)3203_{\text{five}}}$

16. a. How many buses are needed if each bus can take 25 passengers and there are 334 students?

    b. How many buses are needed if we also place two adults on each bus?

17. A soft drink manufacturer produces 3240 cans in an 8-hour day. Cans are packaged 24 to a case. How many cases are produced each week? each month?

18. A machine makes 400 candies a minute. At this rate, how long will it take this machine to make 2000 boxes, each of which contains 24 candies?

19. Janice has decided to buy a computer for her son. So far, she has $450 in the bank. The computer costs $1234. If she can afford $15 per week, when will she have the money?

20. A factory can make 250 gizmos every hour. If the factory operates 24 hours per day, how many gizmos can they make in one year?

21. Mars is 34 million miles away. Presently, spaceships go about 16,000 miles per hour. How long would it take a spaceship to get to Mars?

22. In 2004 the U.S. federal debt was $8,200,000,000,000. Let's say that the decision was made to reduce the debt by $10 million per day. How long would it take to pay off the debt at this rate?

23. If you divide a number that is larger than 75 by a number that is less than 3, what can you tell about the relative value of the quotient? Explain your choice.

   a. It must be greater than 25.

   b. It must be less than 25.

   c. It will be between 3 and 75.

   d. There is not enough information to conclude any of the above.

24. What is the value of the divisor that makes this division problem?

   32 R 2
)994

25. Take no more than 5 to 10 seconds to determine whether this answer is reasonable.

   67
   68)3976

26. Determine which of these answers are reasonable. You have about 5 seconds to make your determination. That is, make your determination, using number sense, without doing any pencil and paper work or trying to subtract the numbers mentally.

   a. There were 35 students in a class and each student paid $28 for a field trip. The total cost was about $600.

   b. A bookstore chain bought 45,000 copies of a new book and has 652 stores. Each store will have about 70 books.

DEEPENING YOUR UNDERSTANDING

27. Melanie recorded a tape of her guitar playing. It cost her $1200 to record the master tape, and it costs $2.50 to make each copy. If she sells the tapes for $10 each, how many tapes must she sell to break even?

28. Victor read 8 books in 34 weeks. He wanted to figure out about how long he spent reading each book, so he divided. Victor got 4 remainder 2 as an answer and is confused about what the 2 represents. How would you help Victor understand the problem?

29. Wei is having a party. She is expecting 120 people and wants enough juice so that each person can have two glasses. If the juice comes in 32-ounce containers, how many cans of juice should she buy? If each can costs 89 cents, how much will the juice for her party cost?

30. Using a calculator, Ralph multiplied by 10 when he should have divided by 10. The display read 300. What should the correct answer be?

31. Mathematics abounds in children's literature. Consider this age-old children's rhyme:

   As I was going to St. Ives
   I met a man with seven wives.
   Every wife had seven sacks;
   Every sack had seven cats;
   Every cat had seven kits.
   Kits, cats, sacks, and wives:
   How many were going to St. Ives?

32. When someone turns 21, how many days old is that person?

33. Imagine that 275 million Americans each had an Internet account and that an Internet directory was available. If we printed that directory, how many pages would it contain?

34. How many people are listed in the phone book in the town or city in which your college or university is located? First, describe and justify your plan for answering this question.

35. In 1993, a newspaper reported that nearly 3,000,000 crimes occur each year in U.S. schools. What do you think was meant by crimes? Do you believe this figure? Do you think it is too small? too big? Why?

36. A newspaper story said that on any given day, 7 percent of all Americans eat at McDonald's. Is this true or another case of journalistic exaggeration? Write your report.

   One fact: There are approximately 9000 McDonald's restaurants.

37. Julie is working out on a computerized ski machine. She starts at 12:05 and will exercise for 30 minutes. Her goal is to burn 400 calories. At 12:17, the machine says that she has burned 148 calories. At this rate, will she reach her goal?

38. Joe had planned to sell 10 pencils at 20¢ each. However, his little brother broke two of the pencils. If he wants to make the same amount of money, how much should he charge for the pencils now?

39. Orange soda comes in six-packs that are packed four to a carton. To make sure that each of the 247 students in the sixth grade will have a soda for the class picnic, how many cartons should we order? How could you figure this out mentally?

40. Determine which number goes where: 7 90 20

   An adult dolphin is about ___ feet long and has ____ teeth. Dolphins live about ___ years.

41. A student of a teacher with whom I was working made up the following problem: Determine which number goes where: 30 48 1961 20 20,000

   The Berlin wall was set up in ____. The wall was approximately ___ miles long or ____ kilometers long. More than _____ people fled before the wall was erected. The wall was about ____ feet tall.

42. Find the missing digits by estimating. Briefly describe your thinking.

a.	b.	c
□6 53)2438	□9 31)899	□4 82)5248

d.	e.	f.
□7 R 24 83)3925	□8 R 8 69)4010	□4 R 18 43)3200

In Exercises 43–46, determine your answer by estimating.

43. Two college students are riding across the United States on bicycles. They are averaging 112 miles per day.

a. If it is 2987 miles from San Francisco to New York, will they complete the trip in 1 month?

b. Approximately how far would they travel in 30 days if they averaged 167 miles per day?

44. We bought a new van in June 2000. In December 2001, we had logged 18,345 miles. Our plan was to keep the van until 100,000 miles. At the rate at which we were putting miles on the van, approximately when would the van hit 100,000 miles?

45. The Alvarez family has decided that camping will be more fun if they buy the following items. Approximately how much will the family spend?

Lantern	\$35.95
New stove	\$59.99
Four new air mattresses at	\$14.95 each
A screen house to put over the picnic table	\$57.50
A new set of camping pots and pans	\$24.95

46. Sandy wants to get an idea of the cost of lumber for a project. She has determined that she needs a total of about 450 feet of boards, and the cost is 23¢ per foot. Approximately how much will they cost?

47. By estimating, determine which of the following equations is wrong.

a. $75{,}296 \div 16 = 4706$

b. $34{,}272 \div 48 = 714$

c. $38{,}844 \div 498 = 780$

48. Which of the following would be a better estimate of $\frac{225}{9}$: $\frac{250}{10}$ or $\frac{230}{10}$? Explain your answer.

49. A computer company has 3765 copies of a hot-selling program and wants to ship the same number to each of its 18 distributors. Approximately how many programs does each distributor get? Francie estimated by rounding 3765 to 4000 and 18 to 20 and then divided 4000 by 20 in her head; her estimate was 200. However, Nadine says that this way of estimating isn't good because, if you are going to round, you should round one number up and one number down. Nadine rounded 18 up to 20 and 3765 down to 3600, and so her estimate was $3600 \div 20 = 180$. With whom do you agree? Why?

50. In each of the following, use the information from the computed problem to determine the answer mentally.

```
      742
36)26,712
   25,200
    1 510
    1 440
       72
       72
```

a. $36 \times 400 = ?$

b. $742 \times 36 = ?$

c. $25{,}200 \div 36 = ?$

51. A kindergarten teacher has 110_{five} children in her class. She has enough gumdrops so that each child can have 4_{five} gumdrops. On the way to class, she dropped half of the gumdrops, and when she gets to school, she finds that half of the students are absent. How many gumdrops will each child get?

52. In what base is the following problem true?

$$5_x\overline{)3350_x} \quad \text{quotient } 421_x \text{ R } 1$$

53. Determine the numbers in the empty boxes.

a.

a	b	$a + b$	$a \cdot b$
65		131	
	72		86976

b.

a	b	$a + b$	$a \div b$
153			17
		20	19

c.

a	b	$a - b$	$a \cdot b$
		20	1056

d.

a	b	$a \cdot b$	$a \div b$
		20	5
	37		925

54. Find the missing digits.

a. $74 \cdot 8\square = 6\square\square4$

b. $\square\square4 \cdot 64\square = 489{,}346$

c. $18 \cdot (\square\square4 + 29\square) = 10{,}116$

d. $(\square3) \cdot (\square) \cdot (\square7) = 1547$

e. $1530 = \square\square \cdot 48 + 42$

55. Place the digits 2, 4, 7, and 9 in the boxes below to obtain:

$$\square\overline{)\square\square\square}$$

a. The greatest quotient b. The least quotient

56. Write a definition for *divides* in your own words. Then have someone who is not in this course read your definition and explain to you, on the basis of your definition, what *divides* means.

57. Use divisibility rules to answer the following.

a. Which of the following numbers divide 123,456?
3, 4, 6, 8, 9

b. Which of the following numbers divide 2,345,678?
3, 4, 6, 8, 9

58. Test the numbers below for divisibility by 2, 3, 4, 5, 6, 8, 9, and 12.

a. 222,444 b. 213,498 c. 987,987,987

59. a. Is 199,999 divisible by 9?

b. Is 939,393,939 divisible by 8?

c. Is 78,888,888 divisible by 8?

d. Is 26,052 divisible by 13?

60. Find the digits that will make these statements true:

a. 9|35_6 b. 3|5_45 c. 3|54_5 d. 12|468_

61. Find the five-digit number that has the following pattern: If you put a 1 after it, the number is three times as large as it would be if you had put a 1 before it.

62. In the following puzzle, each letter has a value between 0 and 9. Find the value of each symbol so that the following computations are all accurate.

$$\begin{array}{r} A \\ +\,A \\ \hline B \end{array} \quad \begin{array}{r} C \\ +\,C \\ \hline DE \end{array} \quad \begin{array}{r} B \\ -\,E \\ \hline B \end{array} \quad \begin{array}{r} F \\ \times\,D \\ \hline F \end{array} \quad \begin{array}{r} A \\ \times\,A \\ \hline G \end{array} \quad \begin{array}{r} B \\ \times\,F \\ \hline HF \end{array}$$

$$\begin{array}{r} C \\ \times\,A \\ \hline DC \end{array} \quad H\overline{)DB}\ \text{(quotient } J) \quad A\overline{)HD}\ \text{(quotient } K)$$

63. A popular kind of problem in many books is the broken calculator problem.

a. If the 7 key is broken, how would you determine 66 + 77 on your calculator?

b. If the 7 key is broken, how would you determine 23 × 70 on your calculator?

c. If the 5 key is broken, how would you determine 260/5 on your calculator?

d. If the 6 key is broken, how would you determine 26 × 44 on your calculator?

e. The − key is broken. How can you find 834 − 653.

64. In each of the problems below, select the correct operations by a means other than random trial and error. Briefly explain the reasoning behind your guesses.

a. 14(48 ○ 32) = 224

b. (462 ○ 243) ○ 15 = 47

c. 4 ○ (8 ○ 5) = 52

d. 56 ○ 42 ○ 352 = 2000

65. In each of the cases below, the actual numbers have been replaced by variables. Write the number sentence that would correctly determine the answer.

a. The students in the three fifth grade classes saved *A* cans, *B* cans, and *C* cans, respectively. If they receive *D* cents per can, how much will they earn?

b. Emily has just gone to college. Her cell phone will cost *A* dollars per month and her iTunes account is *B* dollars per month. How much money will she spend during the *C* months of the college year?

c. Marissa has *A* dollars in her bank account. She babysat last week for a neighbor for *B* hours at a rate of *C* dollars per hour. She bought a cell phone for *D* dollars and spent *E* dollars on a dress. How much money does she have now?

66. Find an article from a newspaper or magazine that demands from the reader knowledge of numeration, basic operations, or estimation. Describe the mathematical knowledge you used in analyzing the article.

FROM STANDARDIZED ASSESSMENTS

2008 NECAP, Grade 5

67. Mr. Martinez divided 24 students into 8 equal groups. Each group has △ students. Which number sentence is true?

a. 24 = △ − 8 b. 24 = △ + 8

c. 24 = △ ÷ 8 d. 24 = △ × 8

2005 NECAP, Grade 4

68. Anna and Chris are solving this problem.

> There are 4 bags of oranges. There are 6 oranges in each bag. How many oranges are there in all?

Anna writes 6 + 6 + 6 + 6 = □.

Chris writes 6 × 4 = □.

Who has written a correct number sentence?

a. Only Anna b. Only Chris

c. Both Anna and Chris d. Not Anna and not Chris

2005 NECAP, Grade 6

69. Samantha uses 4 round beads and 5 cube beads to make this necklace.

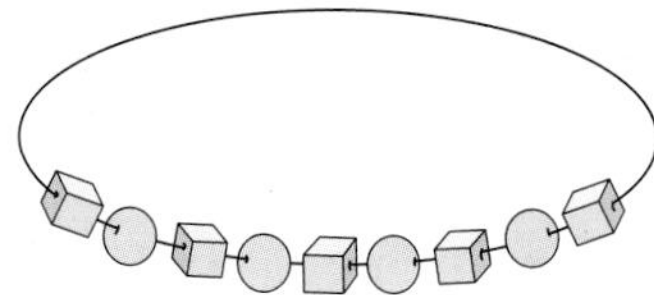

Samantha bought one package of 30 round beads and one package of 24 cube beads. How many of these necklaces can Samantha make?

a. 4 b. 5 c. 6 d. 7

2006 NECAP, Grade 6

70. Reggie received a free movie ticket for every 15 movies he rented at Sky Videos. Reggie rented 88 movies at Sky Videos. What is the total number of free movie tickets Reggie received?

a. 4 b. 5 c. 6 d. 7

71. Anita is making bags of treats for her sister's birthday party. She divides 65 pieces of candy equally among 15 bags so that each bag contains as many pieces as possible. How many pieces will she *have left?*

a. 33 b. 5 c. 4 d. 3 e. 0.33

Source: Results of the *Seventh NAEP Mathematics Assessment*, p. 159. U.S. Department of Education, National Center for Education Statistics.

72. A certain machine produces 300 nails per minute. At this rate, how long will it take the machine to produce enough nails to fill 5 boxes of nails if each box will contain 250 nails?

a. 4 minutes b. 4 minutes 6 seconds

c. 4 minutes 10 seconds d. 4 minutes 50 seconds

e. 5 minutes

Source: Results of the *Seventh NAEP Mathematics Assessment*, p. 159. U.S. Department of Education, National Center for Education Statistics.

73. Mark tried to add the numbers 489 and 263 on his calculator. What is the sum of these numbers? The display on Mark's calculator showed his answer to be 128607? Mark had pressed a wrong key when trying to add. Which wrong key did he press?

Source: Results of the *Seventh NAEP Mathematics Assessment*, p. 152. U.S. Department of Education, National Center for Education Statistics.

74. Lynn had only quarters, dimes, and nickels to buy her lunch. She spent all of the money and received no change. Could she have spent $1.98?

Source: *NAEP Mathematics Assessment*, 1992. U.S. Department of Education, National Center for Education Statistics.

LOOKING BACK on chapter three

QUESTIONS TO SUMMARIZE BIG IDEAS

1. What are some of the different models for addition, subtraction, multiplication, and division?
2. How can you use base ten blocks to model the algorithms for each of the operations?
3. How are these models similar and different in a base other than ten?
4. Which algorithms for the operations are different from what you learned in elementary school?
5. What are the tools for determining divisibility and why do they work?
6. Look back at the Mathematical Practices of the Common Core State Standards. In what ways did you engage in those practices during this chapter?
7. What parts of this chapter are less clear to you at this point? What will you do to clarify those ideas?

CHAPTER 3 SUMMARY

1. Many students have said that really understanding base ten and the four operations, was, for them, the beginning of a new attitude toward mathematics. We will continue to examine new and important mathematical ideas throughout this book, but the foundation for much of elementary mathematics has now been laid.
2. Each operation has multiple meanings.
3. Many algorithms have been developed to enable us to compute more efficiently.
4. The standard algorithm for each operation does not connect equally well to each meaning of the operation.
5. Being able to make sense of algorithms requires:
   - The ability to apply base ten and place value concepts
   - The ability to compose and decompose the numbers (for example, to use expanded form)
6. Patterns enable us to understand the operations more deeply.
7. In many real-life problems, the answer depends on knowing how to interpret one's computation.
8. Being able to perform mental math and to estimate requires
   - The ability to apply base ten and place value concepts
   - The ability to compose and decompose the numbers (for example, to use expanded form)
   - The ability to apply properties of the operations, especially the commutative, associative, and distributive properties
9. Numbers in real-life settings are sometimes exact, sometimes rounded, and sometimes estimates.
10. In real-life problem-solving, one needs to know when to find an exact answer and when to find an estimate.
11. Real-life problem solvers need to know whether their estimates are reasonable.
12. People may use rounded numbers rather than exact numbers for a variety of reasons.

BASIC CONCEPTS

Section 3.1: Understanding Addition

Addition terminology:

addition **78** sum **78**
addends **78**

Addition contexts:

discrete **76** continuous, measured **76**

Addition models:

pictorial **76** number line **78**
tables **79**

REVIEW EXERCISES chapter three

1. State the problem that is represented in each case below:

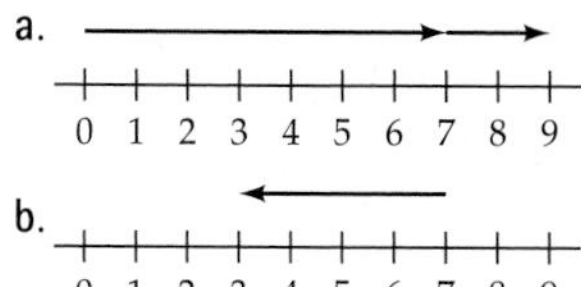

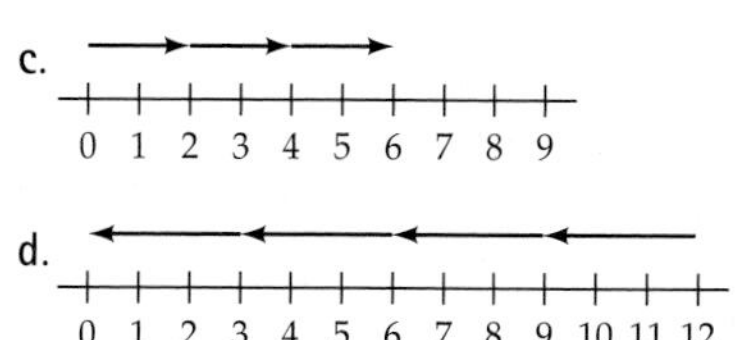

2. Represent the following problem on a number line:
$3(4 + 2)$

3. An alternative algorithm for adding whole numbers is shown below. Make up and solve a few problems using this algorithm. Explain why it works, as though to a parent who has learned only the traditional right-to-left algorithm.

```
  832
+ 549
 1300
   70
   11
 1381
```

4. Place the digits 3, 4, 5, 7, 9 in the boxes to obtain the smallest difference. Briefly explain your reasoning.

$$\begin{array}{r} \square\,\square\,\square \\ -\ \ \square\,\square \\ \hline \end{array}$$

5. The figure below shows one way to find the difference between 800 and 126. Justify this procedure. You're not being asked to explain the procedure (that is, *how* to get the answer of 674) but, rather, to explain *why* this procedure works.

```
  7 9
   ̸1
  8̸0̸0
 −126
  674
```

6. Make up two subtraction story problems, one illustrating the take-away model and one illustrating the comparison or missing-addend model.

7. It has been said that we should not use the terms "borrowing" and "carrying" because they are the same thing. Describe what it is about them that is "the same" mathematically. You need to go deeper than just saying that they both involve regrouping.

8. Using a sheet of graph paper, cut out a piece that measures 37×24 and explain how to get the product *just from the picture* and your understanding of multiplication. Assume you do not know the algorithm, but you do know base ten, place value, and what multiplication means.

9. Why do we move over in the second row when we multiply? Use the example below. Specifically, what is the mathematical reason for moving the 74 over instead of putting it below the 4 and 8?

```
   37
 × 24
  148
  74
  888
```

10. Imagine a parent of a fourth-grader asking you, "Why is 13×25 equal to $(10 \times 25) + (3 \times 25)$?" Write a short paragraph (2–3 sentences) that would help the parent to understand *why*. Note that you are not explaining why we have children break numbers apart. You are simply helping the parent to understand why $13 \times 25 = (10 \times 25) + (3 \times 25)$.

11. Place the digits 1, 2, 4, 5, and 8 in the boxes to obtain the greatest product. Describe your first two guesses and the thinking strategies behind those guesses. Points for this question will be based on the quality of the two guesses *and* on your justification of them. That is, random trial and error gets no points.

$$\begin{array}{r} \square\,\square\,\square \\ \times\ \ \square\,\square \\ \hline \end{array}$$

12. Pete is in fourth grade, and his group has been exploring multiplication. They came up with a method that works, using place value: Multiply the digits in the ones place and then multiply the digits in the tens place, just like we do addition. But it doesn't work. Why not?

```
  34
 ×26
 600
  24
 624
```

13. We can find the sum of 29 and 17 mentally by transforming the problem into $30 + 16$. However, if we transform 29×17 into 30×16, it doesn't work. Why not?

14. Look at the multiplication table in the text. If you take any 2×2 box in the table and multiply the diagonals, the products of the diagonals are equal. Why? Here is one example, shown at the right: $12 \times 20 = 240 = 15 \times 16$.

12	15
16	20

For Exercises 15–19, find the answer using methods other than random trial and error. Briefly explain your solution path—that is, the thinking behind your work.

15. Find a and b if $a + b = 46$ and $a \times b = 480$.

16. How could you figure the remainder of $73{,}932{,}500 \div 97665$? That is, $73{,}932{,}500 \div 97665 = x$ remainder y. Find y.

17. I am thinking of a number that has a remainder of 16 when divided by 75. What is that number? There is more than one answer.

18. Using a calculator, Li multiplied by 5 when he should have divided by 5. The display read 300. What should the correct answer be?

19. Find the missing digits: $(\square 9) \cdot (\square) \cdot (\square 1) = 7821$

20. When we divide, what is going on *mathematically* when we multiply? For example, here we multiply 5 by 8 and later we multiply 5 by 6.

```
    86
 5)432
   40
    32
    30
     2
```

21. Make up two division story problems, one illustrating the partitioning model and one illustrating the repeated-subtraction model.

22. Do the following computations *in your head*, using a strategy other than the standard pencil-and-paper algorithm. Describe the "how" *and* the "why" of the strategy you used in each case. In each case, 2–4 lines should be sufficient.

a. $329 + 567$

b. $34 + 48 + 77 + 66$

c. $91 - 39$

d. $502 - 206$

e. $600 - 246$

f. 35×19

g. 632×4

h. 29×12

i. $40{,}000 \div 80$

23. Explain how you can obtain the answers to these problems easily in your head using properties:

a. $36 + 82 + 64$

b. $13 \times 19 + 7 \times 19$

c. $1592 \div 8$

For Exercises 24–28, obtain your best estimate. Note that "estimate" means two things: (1) all computation is in your head, and (2) what you can do in 10–15 seconds. Then briefly explain how you got your estimate.

24. a. $3684 + 2853 + 6241 + 8312$

b. $44{,}268 - 28{,}843$

c. 468×9

d. $48{,}412 \div 14$

25. Cletha bought a house for $216,250 and sold it for $345,300. How much profit did she make?

26. A student lives in Hanover, which is 62 miles, each way, from school. How many miles per semester does she travel if she drives to school twice a week for 14 weeks?

27. Pierre just bought a new van, and he wants to see what kind of mileage he gets. He filled up with gas after going 489 miles, and the car took 19 gallons of gas. Estimate the mileage.

28. I had planned a spaghetti dinner for my friends. I needed pasta ($1.53), sauce ($2.37), French bread ($1.99), garlic cloves ($.52), parmesan cheese ($3.50), a head of lettuce ($1.19), a pepper ($.89), and salad dressing ($1.20). About how much did this meal cost me?

29. To estimate 638×42, we can use the strategy of rounding one number up and the other down. Explain why 660×40 will give a closer estimate than will 640×40.

30. If a person exercises an average of 3 hours per week from age 18 to age 70, how many days has she or he exercised over this time?

31. The fuel tank in a car holds 20 gallons of gasoline. The car has a fuel efficiency rating of 18 miles per gallon. If the gas tank is full when the family begins, and their destination is 1200 miles away, how many times will they have to stop and fill the gas tank to get to their destination?

32. The copy center at Keene State College told me that it used 1200 reams of paper last semester. It charges departments 3¢ per copy.

a. How many pages were run off at the copy center?

b. How much did the college pay to the copy center?

c. If the departments use their own copy machines, which they do for small jobs, they are charged 5¢ per copy. How much did the college save by having a copy center that semester?

33. Perform the following computations in base five.

a. $\begin{array}{r} 3442_{\text{five}} \\ +\ 2333_{\text{five}} \\ \hline \end{array}$ b. $\begin{array}{r} 4000_{\text{five}} \\ -\ 1234_{\text{five}} \\ \hline \end{array}$ c. $\begin{array}{r} 44_{\text{five}} \\ \times\ 32_{\text{five}} \\ \hline \end{array}$ d. $3_{\text{five}}\overline{)20202_{\text{five}}}$

34. In what base does $13 \times 3 = 42$?

35. Find the base for which the following statement is true: $135 + 444 = 601$

36. Find the base for which the following statement is true: $103_{\text{ten}} = 1213_x$

37. Find the missing digits:

a. $\begin{array}{r} 3\square8\square \\ +\ \square2\square8 \\ \hline 7932 \end{array}$ b. $\begin{array}{r} 6\square\square3 \\ -\ \square28\square \\ \hline 2719 \end{array}$ c. $\begin{array}{r} ab \\ \times\ 3a \\ \hline 1564 \end{array}$

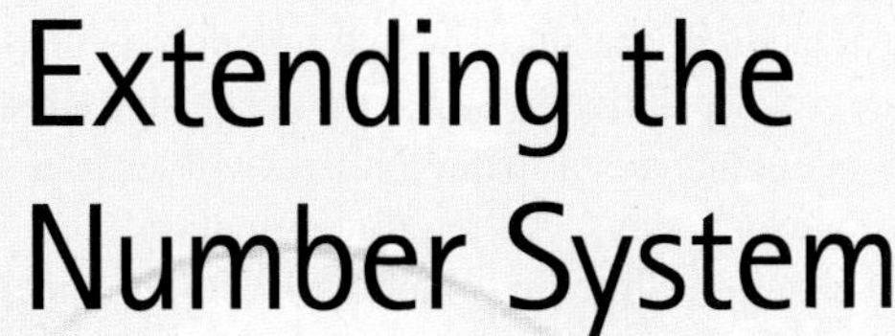

Extending the Number System

4

For many thousands of years, whole numbers were adequate for most people's needs. However, the limitations of whole numbers became more and more problematic as time went on. In this chapter, we will examine four extensions of the set of whole numbers: negative numbers, fractions, decimals, and irrational numbers.[1] We will examine what each of these new sets of numbers means. Knowing what they mean and understanding the four operations will then enable you to make sense of the computation algorithms that we use for integers, fractions, and decimals. As before, knowing why as well as how increases one's mathematical power:

> Teachers [of all grades] should be able to extend the number systems from the whole numbers to fractions and integers, then rationals and real numbers, including a discussion of the extension of the operations, properties, and ordering. Notions of fractions, decimals, percents, ratio, and proportion should be developed through problems with an applied flavor.
>
> (NCTM *Professional Standards*, 1991)

SECTION 4.1 Integers

What do you think?

- Why is it that the sum of two negative numbers is a negative number, but the product of two negative numbers is a positive number?
- How do operations with integers connect with whole-number operations? How do they differ?
- Why is −8 less than −7?

Our first extension of the set of whole numbers is the set of **integers**, which is simply the union of the set of positive integers, the set of **negative integers**, and zero. Although most of people's everyday use of mathematics involves positive numbers, we encounter negative numbers in various ways. Although elementary students may wonder what is to the left of the 0 on a number line and may explore the concept of $2 - 3$ informally, sixth grade is where students learn about negative numbers, according to the Common Core State Standards. One of the major goals of this section is for you to understand how the procedures for computing with positive numbers are related to computing when one or more of the numbers are negative.

INTEGER CONNECTIONS

Before we examine operations with integers, let us take a little time to see how integers connect to the set of whole numbers we have been working with up to now.

- We began our study of numbers with the set of natural numbers, N.

  $N = \{1, 2, 3, 4, \ldots\}$

- With the invention of zero, we have the set of whole numbers, W.

  $W = \{0, 1, 2, 3, 4, \ldots\}$

- With the invention of negative numbers, we have the set of integers, I.

  $I = \{\ldots -4, -3, -2, -1, 0, 1, 2, 3, 4, \ldots\}$

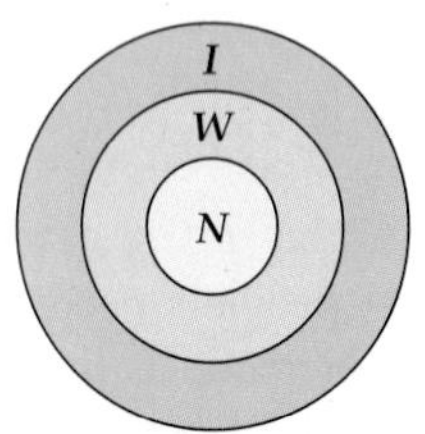

Figure 4.1

Figure 4.1 shows a Venn diagram illustrating the notion of extending our set of numbers as our ancestors invented new kinds of numbers. Each set of numbers contains the previous set. We will extend this diagram as we discuss fractions, decimals, and irrational numbers.

HISTORY

We find the earliest mention of negative numbers and how we might compute with them in the works of Brahmagupta (A.D. 628) in India. Other work is found in the writings of al-Khwarizmi (A.D. 825) in Persia and Chu Shi-Ku (A.D. 1300) in China. Cardan (sixteenth century) accepted negative numbers as solutions of equations. However, he referred to them as "false" numbers.

CLASSROOM CONNECTION

The term *integer* often causes some confusion for many students because the terms *integer* and *whole number* do not have identical meanings in mathematics and in everyday English. In mathematics, *integers* refers to the following set of numbers: $\{\ldots -4, -3, -2, -1, 0, 1, 2, 3, 4, \ldots\}$. In everyday English, the description of this set is not identical to the description in the first sentence of this section. That is, in everyday English, one might say that the integers can be broken down into three sets—negative whole numbers, zero, and positive whole numbers. However, *whole numbers,* in mathematics, refers to the set $\{0, 1, 2, 3, \ldots\}$. Thus saying "negative whole numbers" is like saying "he goed to the store yesterday." My resolution to this dilemma is to use the terms *positive integers* and *negative integers* but not the terms *positive whole numbers* and *negative whole numbers*.

Integers in our world When do we use negative numbers? Stop to reflect on where you have encountered negative numbers before reading on. . . .

People use negative numbers both on and off the job. For instance:

- Businesses use negative numbers to indicate a business deficit, or "negative profit."
- We often use negative numbers when describing change. For example, graphs often have negative numbers.
- We use negative numbers to indicate temperatures below zero.
- Elevations below sea level are often represented with negative numbers.
- A golfer uses negative numbers to indicate a score below par.

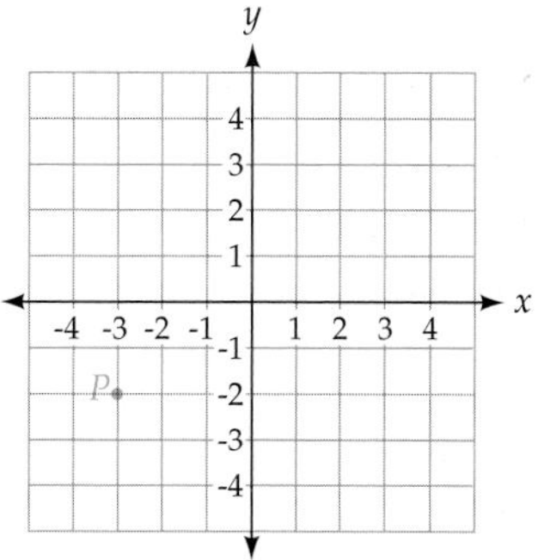

Figure 4.2

- Physicists use negative numbers to indicate negatively charged particles.
- We use a coordinate system to indicate the position of an object in space. For example, the location of point P in Figure 4.2 is $(-3, -2)$.

REPRESENTING INTEGERS

There are many models for representing integers. We will focus on the number line model because it is the model to which most real-life applications connect.

Number lines can be represented horizontally or vertically, as shown in Figure 4.3.

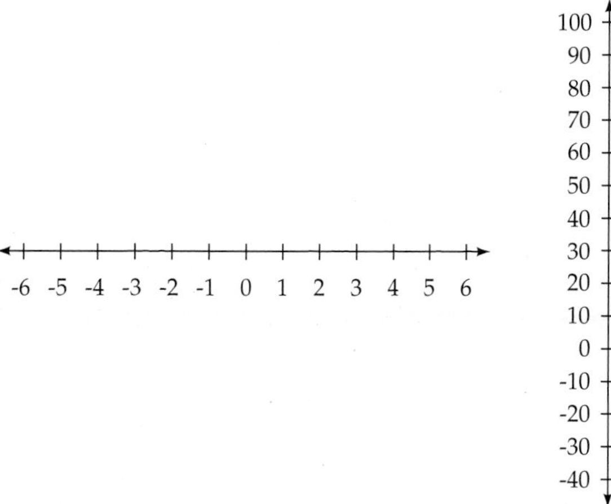

Figure 4.3

Some other models commonly used to represent integers and their applications include black and red chips, positively and negatively charged particles, and money (assets and liabilities). If you did Exploration 4.1, you used black and white dots to represent integers. The discrete model is a representation of integers that makes more sense to some young children than the number line model.

OPERATIONS WITH INTEGERS

In this section, we will examine the four operations with negative numbers. As you recall from Chapter 3, we developed models for each of the operations:

- addition: combine, increase
- subtraction: take away, comparison, missing addend
- multiplication: repeated addition, rectangular array, Cartesian product
- division: partitioning, repeated subtraction, missing factor

As we strive to develop ways to operate *meaningfully* with negative numbers, we will also examine which of these models for the operations work and which do not.

Many textbooks present the students with various algorithms (that is, rules) for adding, subtracting, multiplying, and dividing with negative numbers. The main problem with this approach is that students remember the rules only as long as they use them. I call this rented or Teflon knowledge; after a few months, the rules are gone. If the students do not have a strong understanding of where the rules came from, then they have nothing to show for all that work. Rather than present you with rules, we will investigate how we obtain sums, differences, products, and quotients with negative numbers. If you first solve, or at least attempt to solve, each problem *on your own,* you will find that not only will you be able to compute confidently with negative numbers, but also your problem-solving toolbox will be fuller. Furthermore, even if you forget the rules a year from now, you will be able to use your knowledge to re-create them.

UNDERSTANDING ADDITION WITH INTEGERS

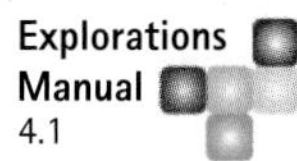

We will use a problem-solving tool to develop algorithms for adding integers: We will examine all possible combinations (cases) and then look for patterns. The four cases are represented below:

	a	b	Example
Case 1: Both numbers are positive.	+	+	$3 + 4$
Case 2: One number is positive, one number is negative, and the magnitude of the negative number is greater than the magnitude of the positive number.	+	−	$3 + (-4)$
Case 3: One number is positive, one number is negative, and the magnitude of the positive number is greater than the magnitude of the negative number.	−	+	$-3 + 4$
Case 4: Both numbers are negative.	−	−	$-3 + (-4)$

Language Sometimes the + and − signs are raised as superscripts when indicating a positive and negative number. We have kept them as regular text here, but we need to be careful of language. For example, we will read the problem $-6 - (-8)$ as "*negative* 6 *minus negative* 8" instead of "minus 6 minus minus 8."

The words *plus* and *minus* will be used to refer to the *operations* of addition and subtraction. The words *positive* and *negative* will be used to refer to the *value* of the number.

For practice, translate each of the equations below into English. Check your translations with the sentences at the right.

$5 - (-4) = 9$	5 minus negative 4 is equal to positive 9.
$32 + (-48) = -16$	32 plus negative 48 is equal to negative 16.

Let us now examine the first two cases.

Case 1 Both numbers are positive. We explored this case in Chapter 3 and drew many models for it like the number line shown in Figure 4.4.

Case 2 One number is positive, one number is negative, and the magnitude of the negative number is greater than the magnitude of the positive number.

Do several other examples and then try to express a rule that would apply to any addition problem in which the magnitude of the negative number is greater than the magnitude of the positive number. Then read on. . . .

We will examine a specific problem in detail and then look at generalizations. How would you represent the problem $3 + (-4)$?

Number line model From our work with number lines in Chapter 3, we learned that the problem $3 + 4$ can be represented as shown at the left in Figure 4.4. That is, we begin at zero and move 3 units to the right, then we move 4 more units to the right; thus we find that $3 + 4 = 7$. If a positive number is represented by an arrow pointing to the right (the positive direction), then a negative number is represented by an arrow pointing to the left (the negative direction). Thus the problem $3 + (-4)$ can be represented as shown at the right in Figure 4.4. That is, we begin at zero and move 3 units to the right and then move 4 units to the left, thus, we find that $3 + (-4) = -1$. As you may have already found, sometimes the sum of a positive number and a negative number is positive (for example, $12 + (-5) = +7$) and sometimes it is negative, as in Figure 4.4.

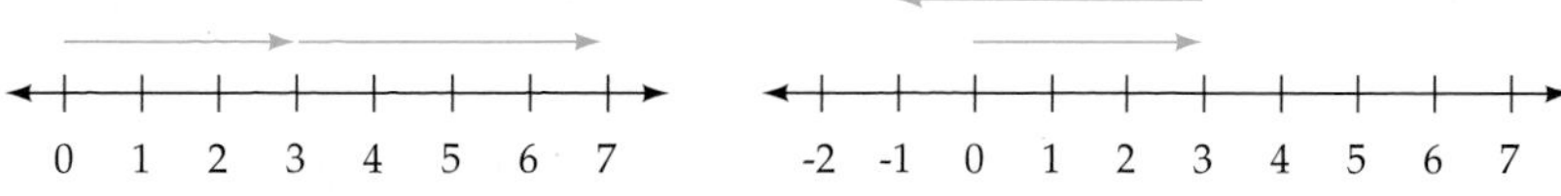

Figure 4.4

If you do a number of problems, you realize that the procedure for adding a positive number and a negative number feels a lot like subtraction. In fact, one description of a general rule is: Disregard the signs and subtract the *smaller* number from the *larger* number; the sign of the sum is the same as the sign of the *larger* number. Look at the number line models again to make sense of this rule visually.

ABSOLUTE VALUE

There is a mathematical way to say, "Disregard the sign in front of the number," and that is to use the term *absolute value*. Let us first define this concept and then use it to state more precisely the rule for adding a positive number and a negative number. This concept first appears in sixth grade in the Common Core State Standards.

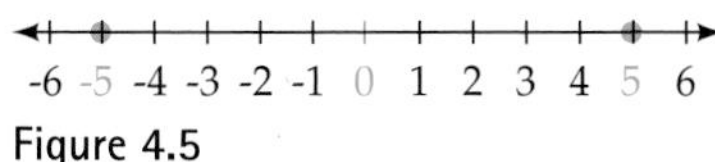

Figure 4.5

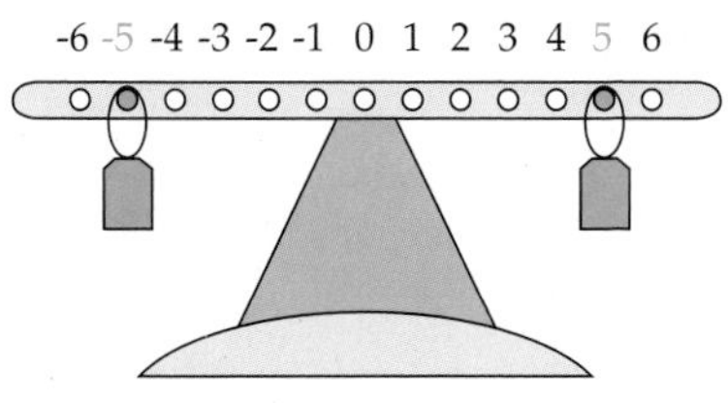

Figure 4.6

One way to illustrate the concept of absolute value is to consider how far the number is from zero (the origin). The **absolute value** tells us the distance of the number from zero. For example, the numbers $+5$ and -5 are both the same distance from zero, so they both have the same absolute value, which is 5 (Figure 4.5).

With notation, we say, $|-5| = 5$. Similarly, $|+5| = 5$.

In English, we say that the absolute value of negative 5 is 5, and the absolute value of positive 5 is also 5.

We can also use a balance scale (a common manipulative used in elementary schools) to illustrate the concept of absolute value. If we place a weight under -5 and an equal weight under $+5$, the scale will balance (Figure 4.6).

There is also language for referring to pairs of numbers whose absolute values are equal: we say they are opposites or negatives of each other. Thus, the **opposite** of $+6$ is -6, and the opposite of -6 is $+6$. Using another meaning of **negative**, we say that the negative of $+6$ is -6 and the negative of -6 is $+6$.

As you can readily see, when we add any integer and its opposite, the result is zero; that is,

$$a + (-a) = 0$$

Thus we can say that every integer has an additive inverse.

In this text, we will use the term **additive inverse** instead of *negative* for two reasons. First, the term *negative* often creates a false impression. For example, if $x = -6$, then $-x = +6$. In other words, when we are working with variables, the value of $-x$ is often positive. Second, when working with fractions, we will develop a similar concept with a similar term: the *multiplicative inverse.*

The concept of absolute value lets us now state more precisely the rule for adding a positive number and a negative number. We first find the difference of the absolute values of the two numbers; the sign of the sum is the sign of the number with the larger absolute value.

THE OTHER TWO CASES OF INTEGER ADDITION

Case 3 One number is positive, one number is negative, and the magnitude of the positive number is greater than the magnitude of the negative number (an example is $4 + (-3)$).

When we look carefully at all integer addition problems that fall in this category, we find that the generalization stated in the preceding paragraph applies to this

MATHEMATICS

With one exception, all the properties of addition, subtraction, multiplication, and division that we developed in Chapter 3 hold *for all numbers*. That exception is the closure property. Whether or not a set is closed under an operation depends on the nature of the set. This concept will be pursued in an exercise.

case too. In other words, the similarity between Case 2 problems $(3 + (-4), 2 + (-8), -7 + 3, -9 + 4)$ and Case 3 problems $(4 + (-3), 7 + (-2), -6 + 9, -1 + 5)$ is that one number is positive and one number is negative. Regardless of which number has the larger magnitude, we can use the same procedure to determine the sum.

When we first examine all the possibilities when adding integers, we have four distinct subsets. However, when we look closely at two of these subsets, we find that their similarity (one positive, one negative) outweighs their difference (which number has the greater absolute value).

Case 4 In Case 4, both numbers are negative. Let us examine a specific problem in detail and then look for generalizations. Figure 4.7 shows a representation of $-3 + (-4)$ on a number line. Do several other examples and then try to express a rule that would apply to any addition problem in which both numbers are negative. Many students find that they can do this more easily if they have a context. For example, if Jeremy's bank account balance is -3 (meaning he is overdrawn by \$3) and he writes a check for another \$4, what will his account balance be? Then read on. . . .

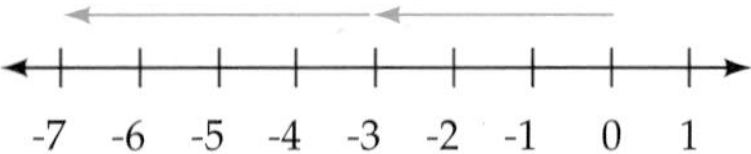

Figure 4.7

With the concept of absolute value, we can state the rule for adding two negative numbers precisely: to find the sum of two negative numbers, we first find the sum of the absolute values of the two numbers and then place a negative sign in front of this sum.

UNDERSTANDING SUBTRACTION WITH INTEGERS

In Exploration 4.2, you examined integer subtraction with the take-away model and the comparison model. Let's now focus on connecting whole-number subtraction to integer subtraction.

INVESTIGATION 4.1a Subtraction with Integers

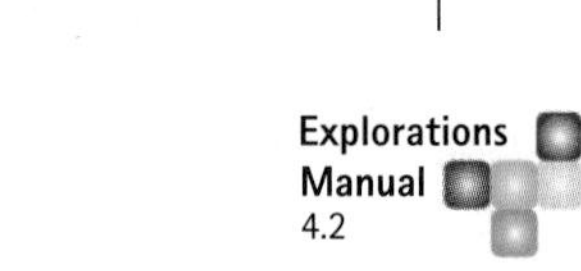

Do the following subtraction problems yourself before reading on. As you work, check to make sure that you are using your understanding of integers rather than just guessing.

1. $14 - (-25) =$ 2. $-5 - 17 =$ 3. $-6 - (-8) =$ 4. $-12 - 5 =$

DISCUSSION

Explorations Manual 4.2

We can define subtraction of negative numbers in terms of addition, which most people understand more easily.

To make the connection stronger, let us examine how we might use what we know about subtraction to determine the answer to $4 - 6$. In one sense, we cannot "take away" 6. However, if we think in terms of a checking account, if we take away 6 from 4, we will have a deficit of 2; that is, we have -2. We could apply our work with subtraction and number lines from Chapter 3: when we subtract one number from another, we move to the left. When we begin at 4 and move 6 units to the left, we end up at negative 2, as shown in Figure 4.8.

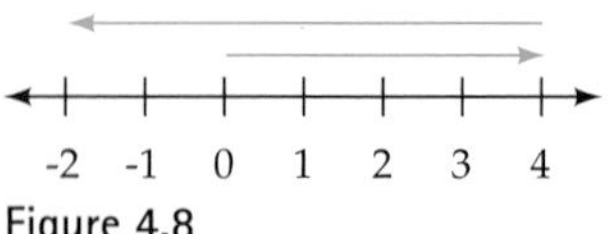

Figure 4.8

Thus we can conclude that $4 - 6 = -2$.

However, this is similar to the addition problem $4 + (-6) = -2$.

We use this knowledge to define subtraction of integers formally in terms of addition:

$$a - b = a + (-b)$$

That is, subtracting is equivalent to adding the additive inverse.

CLASSROOM  CONNECTION

Grade 3

Below are some pretty challenging problems for third-graders! Can you answer each of these questions confidently and quickly?

Date ______________________ Time ______________________

LESSON 9·13

Number Stories with Positive & Negative Numbers

Solve the following problems. Use the thermometer scale, the class number line, or other tools to help.

°F
100
90
80
70
60
50
40
30
20
10
0
−10
−20
−30
−40
−50
−60
−70
−80
−90
−100

1. The largest change in temperature in a single day took place in January 1916 in Browning, Montana. The temperature dropped 100°F that day. The temperature was 44°F when it started dropping.

   How low did it go? ____________

2. The largest temperature rise in 12 hours took place in Granville, North Dakota, on February 21, 1918. The temperature rose 83°F that day. The high temperature was 50°F.

   What was the low temperature? ____________

3. On January 12, 1911, the temperature in Rapid City, South Dakota, fell from 49°F at 6 A.M. to −13°F at 8 A.M.

   By how many degrees did the temperature drop in those 2 hours? ____________

4. The highest temperature ever recorded in Verkhoyansk, Siberia, was 98°F. The lowest temperature ever recorded there was −94°F.

   What is the difference between those two temperatures? ____________

5. Write your own number story using positive and negative numbers.

   ____

   ____

   ____

two hundred thirty-seven **237**

From *Everyday Mathematics, Grade 3:* The University of Chicago School Mathematics Project: Student Math Journal, Volume 2, by Max Bell et al., Lesson 9-13, p. 237. Reprinted by permission of The McGraw-Hill Companies, Inc.

CONNECTING WHOLE-NUMBER SUBTRACTION TO INTEGER SUBTRACTION

In Chapter 3, we examined different models for the four operations and different algorithms. In order for your knowledge of mathematics to be as connected as possible, it is important that you see how we apply these models to each new set of numbers.

Go back to Chapter 3 and examine how we defined subtraction there. In what ways are the two definitions of subtraction the same (that is, equivalent)? In what ways are they different? Why did we define subtraction differently there from the way we define it here? Think and then read on. . . .

Let us compare the two definitions:

Chapter 3: $a - b = c$ if there is a number c such that $c + b = a$
Chapter 5: $a - b = a + (-b)$

If you examine the actual problems involved in subtracting a positive number from a positive number, you find that we are not adding the opposite as much as we are taking away a positive amount or comparing the size of two sets (each with positive values). Therefore, although the definition given in this chapter seems simpler and more practical (to many students), it doesn't connect as well to subtraction with two positive numbers. Therefore, I chose to defer the definition until this chapter so that the work in Chapter 3 would be more focused on the context in which subtraction occurs when all three of the numbers (minuend, subtrahend, and difference) are positive numbers.

CONNECTING SUBTRACTION CONTEXTS TO ALGORITHMS

We now have developed efficient procedures for integer addition and subtraction. However, as you have discovered, in many real-life settings, translating words into a mathematical sentence is not always simple. Therefore, we will examine a few such settings to apply our knowledge.

HISTORY

Our present notation, putting a minus sign in front of a negative number, was a long time in the making. The early Hindus designated negative numbers by circling them—for example, ⑤—whereas the later Hindus and Arabs represented them by placing a dot over the number—for example, $\dot{5}$—and the Chinese used red to represent positive and black to represent negative. Cardan used the symbol *m* to represent a negative number—for example, m5.

Problem 1 Denine realized that she had overdrawn her checking account by $60, and she was fined $15 for a returned check. What is her present balance? Work on this problem and then read on. . . .

If we translate this problem into mathematical language, we find that we need to take away $15 from negative $60. Thus the problem is $-60 - 15$. Using our understanding of subtraction, we translate this subtraction problem into the following addition problem: $-60 + (-15)$. Applying our understanding of integer addition, we have an answer of -75; that is, her present balance is negative $75 (she is $75 in the red).

Problem 2 On one day the high temperature in Nome, Alaska, was -6 degrees. On that same day, the high temperature at the North Pole was -64 degrees. How much warmer was Nome than the North Pole? Work on this problem and then read on. . . .

If we translate this problem to mathematical language, we find that we are using the comparison model of subtraction. Thus the problem is $-6 - (-64)$, which we can translate as $-6 + (+64)$, and the answer is 58; that is, it was 58 degrees warmer in Nome.

A visual representation of this can help us make sense of it as well. Look back at the thermometer on page 169. If we look at -6 and -64 on a thermometer, we can see the difference is 58.

UNDERSTANDING MULTIPLICATION WITH INTEGERS

Explorations Manual 4.3

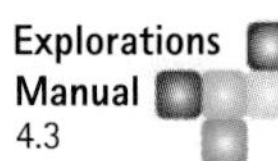

There are very few simple real-world problems in which we multiply and divide integers. Most cases of integer multiplication and division occur in solving equations. Because the

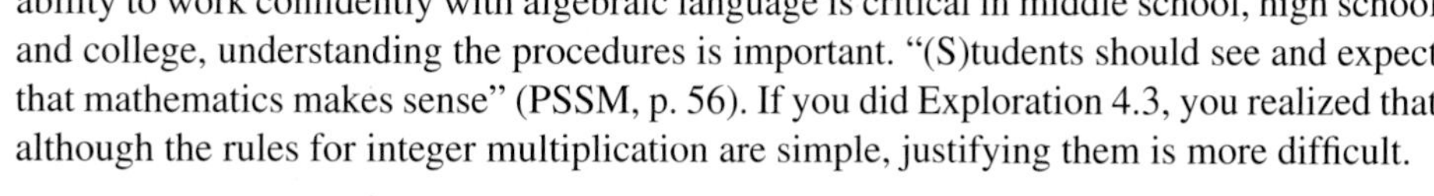

ability to work confidently with algebraic language is critical in middle school, high school and college, understanding the procedures is important. "(S)tudents should see and expect that mathematics makes sense" (PSSM, p. 56). If you did Exploration 4.3, you realized that although the rules for integer multiplication are simple, justifying them is more difficult.

INVESTIGATION 4.1b The Product of a Positive and a Negative Number

Consider the following problem: $5 \cdot (-3)$. Can we apply our understanding of multiplication with positive numbers to determine the product? Think and read on. . . .

DISCUSSION

Applying the repeated-addition model of multiplication, $5 \cdot (-3)$ literally means to add -3 five times, that is, $-3 + (-3) + (-3) + (-3) + (-3)$. We know from integer addition that the sum must be -15. That is, we can deduce that $5 \cdot (-3) = -15$.

What about $-3 \cdot 5$? Think and read on. . . .

Trying to apply repeated addition here is problematical; we must add 5 *negative* 3 *times*. However, the commutative property makes the task easier: $-3 \cdot 5 = 5 \cdot (-3) = -15$.

We can also use patterns to help us understand. What patterns do you see here?

$5 \cdot 3 = 15$

$5 \cdot 2 = 10$

$5 \cdot 1 = 5$

$5 \cdot 0 = 0$

$5 \cdot (-1) = -5$

$5 \cdot (-2) = -10$

$5 \cdot (-3) = -15$

As we decrease what we are multiplying 5 by on the left side, the product on the right is decreasing by 5. Following the pattern, we can see that 5 times $-3 = -15$.

From this discussion, we can recall the rule that you may have memorized in school: the product of a positive number and a negative number is a negative number.

HISTORY

Leonard Euler (1770) produced one of the earliest proofs that $-1 \cdot (-1) = 1$. However, many great mathematicians thought that the idea of a negative number times a negative number was absurd. Many eighteenth-century algebra textbooks rejected multiplication of negative numbers because this was not a problem that they saw as having any real-life relevance.

Two negative numbers What about $-3 \cdot (-5)$? None of the models for positive whole-number multiplication adapt nicely to this problem, and the commutative property does us no good here. However, we can use our knowledge that a positive times a negative is a negative and make use of patterns:

We know that	$4 \cdot (-3) = -12$	
Thus	$3 \cdot (-3) = -9$	
Similarly,	$2 \cdot (-3) = -6$	
And so	$1 \cdot (-3) = -3$	
	$0 \cdot (-3) = 0$	Recall the zero property of multiplication from Chapter 3.
	$-1 \cdot (-3) = ?$	

What does $-1 \cdot (-3)$ *have to* equal if the pattern is to continue? Think and then read on. . . .

As the number line in Figure 4.9 illustrates, in this case, each product is 3 more than the previous product, and 3 more than 0 is $+3$.

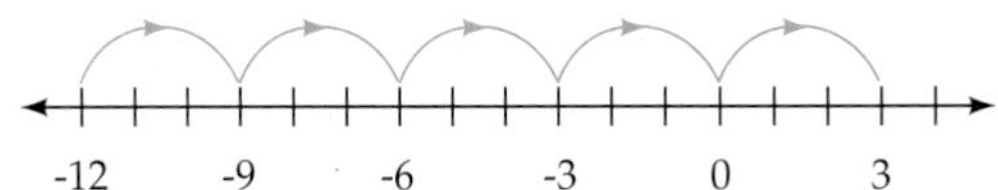

Figure 4.9

From this discussion, we can recall the rule that you may have memorized in school: the product of two negative numbers is a positive number. Seeing where these rules come from not only helps us to make sense of them, it is also more likely that we will remember them if we truly understand them rather than trying to memorize rules with no meaning.

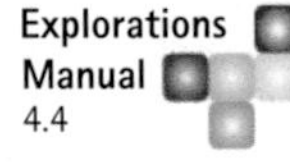

UNDERSTANDING DIVISION WITH INTEGERS

Just as we interpreted integer subtraction by applying our knowledge of the relationship between subtraction and addition, we can now understand integer division by applying our knowledge of the relationship between division and multiplication. From an intuitive perspective, you may sense that the multiplication rules translate quite directly into division:

- The quotient of a positive and a negative number is a negative number.
- The quotient of two negative numbers is a positive number.

We can apply the missing-factor model of division ($y \div n = x$ if $x \cdot n = y$) to verify these rules.

For example, consider the problem $-12 \div 4$. There is little doubt that the quotient is either -3 or $+3$. Many students simply guess, but we can apply this model to help us find the correct answer. Applying our definition of division, we can say that

$$-12 \div 4 = x \quad \text{if} \quad x \cdot 4 = -12$$

What number times 4 is equal to negative 12? We know from multiplication that this number must be -3. Therefore, $-12 \div 4$ must equal -3.

Similarly, consider $-15 \div (-3)$. Applying our definition of division, we can say that

$$-15 \div (-3) = x \quad \text{if} \quad x \cdot (-3) = -15$$

We know from multiplication that this number must be 5. Therefore, $-15 \div (-3)$ must equal 5.

We have learned that the set of integers includes both the set of positive and negative numbers and zero. We have explored two models for representing integers: number lines in the text and dots in Explorations 4.1–4.3. We have adapted our understanding of the four basic operations, learned in Chapter 3, so that the algorithms for adding, subtracting, multiplying, and dividing integers make sense. In the course of exploring these operations, it became important to understand several new terms: *absolute value, opposite, negative,* and *additive inverse.*

Although operations with integers are not generally introduced until middle school, this section is important for several reasons. First, teachers need to have a sense of what lies beyond the concepts their students are learning at the present time. Second, many students have trouble with algebra because there are too many rules to memorize. However, as you have seen in this section, we can apply our understanding of the meaning of the four operations and the properties of the four operations to understand the procedures for operating with integers. Third, you are likely to encounter gifted students. I was visiting a second-grade teacher one day when one of her students invented the procedure for integer subtraction on the spot. The teacher was explaining how to subtract via regrouping with the problem $72 - 58$. She said, "You can't take 8 from 2." He replied, "Yes, you can; you get negative 6; then you take 50 from 70, which is 20, and 20 minus 6 is 14!"

4.1 Exercises

1. Perform the computations:

a. $-356 - (-138)$ b. $\dfrac{-36 + 48}{-6}$

c. $16 - (-3)$ d. $-217 + 139$

e. $-2(3 - 10)$ f. $-3 \cdot 5 + (-6 \div 2)$

g. $16 - (-3)^2$ h. $-6 + (24 \div 3)$

i. $-124 - (-345)$ j. $\dfrac{47 + (-11)}{-6}$

2. Find the missing number:

a. $-65 + \square = 173$ b. $36 - \square = 83$

c. $331 + \square = -86$ d. $-812 + \square = -223$

e. $-342 + \square = -129$ f. $-12{,}348 - \square = -348$

3. Here are the low temperatures (Fahrenheit) for one week in Minneapolis: $-19, -6, -22, 8, -4, 7, 1$. What was the mean (average) low temperature for the week?

4. The Crabby Apple restaurant lost \$2500 in January. If its net worth at the end of the month was $-\$400$, what was its net worth at the beginning of the month?

5. Jacob opens a savings account on January 1 with a deposit of \$250. He has "direct deposit," in which \$25 is deposited every other week. The bank also charges a \$3 monthly processing fee. How much money will he have at the end of the year?

6. Let's say the countdown for a space shuttle launch has begun. At "T minus 27 hours" (that is, 27 hours before launch), a problem occurs. If the technicians have not fixed the problem by T minus 8 hours, the launch will have to be scratched. How much time do the technicians have to correct the problem?

7. John had \$123 in the bank but wrote a check for \$56 and another check for \$86. What is his current balance?

8. The formula for converting from Fahrenheit to Celsius is given by $C = \frac{5(F - 32)}{9}$. If the temperature outside is 23° Fahrenheit, what would the temperature be in Celsius?

DEEPENING YOUR UNDERSTANDING

9. At what temperature would the Fahrenheit and Celsius temperatures be equal?

10. Many Americans do not realize that our calendar is not universal. For example, the copyright date for this textbook is 2016. However, this year would be reported as 5776 in the Jewish calendar. According to the standard calendar, when was the world created, if the Jewish calendar began when the world was created?

11. Show two different ways to determine the following sum:

$$-19 + (-6) + (-22) + 8 + (-4) + 7 + 1$$

12. Have you ever been on an airplane and heard the pilot say that the plane would be a little late because it would be flying into a strong headwind or that even though the plane was taking off a bit late, you would be making up time because you would be flying with a tailwind? This problem asks you to analyze such a situation. You have the following data: A plane flying at its maximum speed can go 200 miles per hour with a tailwind or 160 miles per hour into a headwind.

a. What is the wind speed?

b. What would be the maximum speed of the plane if there were no wind?

13. If you are flying to a city 800 miles away, what will be the difference in flight time between flying into a 40-mph headwind and flying with a 40-mph tailwind if the plane's maximum speed with no wind is 160 mph?

14. The origin of our symbols for + and − can be traced back to the practice of merchants in the Middle Ages, who used the signs p for *piu* (more) and m for *meno* (less) to indicate how much above or below a standard weight each sack was. Thus, if the standard weight was 100 pounds, a sack weighing 96 pounds would have m4 written on it. How much overweight or underweight is the following shipment? Describe at least two different ways you could have solved this problem.

15. In Chapter 1, we investigated magic squares in which all rows, columns, and diagonals had the same sum. The magic square below is a subtraction magic square. Why is this?

15	7	8
5	4	1
10	3	7

a. Make another subtraction magic square.

b. Make a subtraction magic square in which all the numbers are negative numbers.

c. Make a subtraction magic square in which four of the numbers are positive, four of the numbers are negative, and one number is zero.

d. Write instructions explaining to someone how to make a subtraction magic square that works for any number.

16. Below are the low and high temperatures on a winter day in several cities. Make up and answer two questions from the chart.

City	Low	High
Anchorage	−28	−7
Minneapolis	−12	5
Honolulu	68	73
Miami Beach	68	87
Portland	26	46

17. Examine some daily newspapers or weekly magazines to find articles that require some knowledge of negative numbers. Either make up a story problem that might come out of the article or explain how a knowledge of negative numbers is needed in order to understand the article.

18. Consider the positive and negative values of the integers from 1 to 9:

$\pm 1 \quad \pm 2 \quad \pm 3 \quad \pm 4 \quad \pm 5 \quad \pm 6 \quad \pm 7 \quad \pm 8 \quad \pm 9$

The sum of the nine positive integers is $+45$, and the sum of the nine negative integers is -45. How many numbers between -45 and $+45$ can be made by adding and using each number or its opposite exactly once? For example, here is one combination whose sum is -3.

$$1 + 2 + 3 + 4 + 5 + 6 + (-7) + (-8) + (-9) = -3$$

19. Let x and y represent any positive integers, $x \neq y$. For each of the operations below, tell whether the result will be always positive, will be always negative, or might be one or the other. Explain your reasoning.

a. $|x - y|$ b. $x^2 - y^2$

c. $x^2 - xy + y^2$ d. $x^2 + 2xy - y^2$

20. Let x and y represent *any* integers, $x \neq y$. For each of the operations below, tell whether the result will be always positive, will be always negative, or might be one or the other. Explain your reasoning.

a. $|x - y|$ b. $x^2 - y^2$

c. $x^2 - xy + y^2$ d. $x^2 + 2xy - y^2$

21. Sam was told by the doctor that he had to lose about 40 pounds. Six weeks ago he weighed 185 pounds. Below is a weekly record of his progress. How much does he weigh now? Do this problem at least two different ways.

Week	1	2	3	4	5	6
Change	−3	−2	+1	−6	+3	−2

22. The set of integers is closed under which of the four fundamental operations?

FROM STANDARDIZED ASSESSMENTS

2006 NECAP, Grade 6

23. Which picture shows the correct position of $-2°$ and the correct position of $-8°$ on the thermometer?

a. 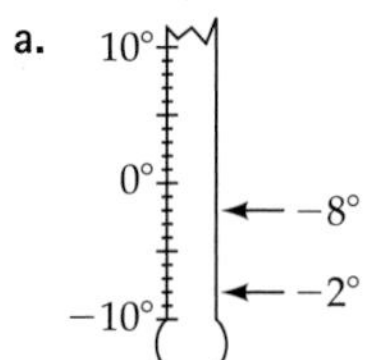

b. 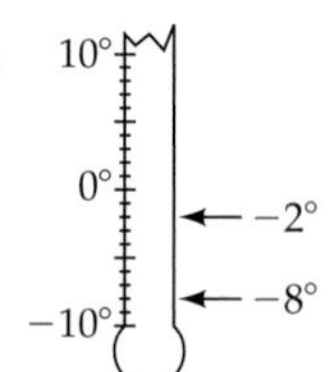

c. 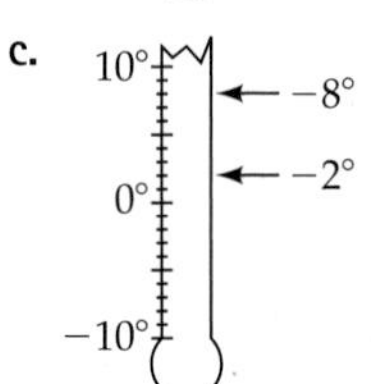

d. 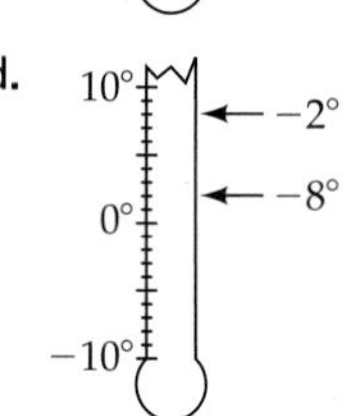

SECTION 4.2 Fractions and Rational Numbers

What do you think?

- How would you define *fraction*?
- What pictures, models, and/or words explain why $\frac{1}{5}$ is less than $\frac{1}{3}$?
- What is the difference between the whole and the unit?

The concept of fractions first appears in the geometry strand in the first grade, according to the Common Core State Standards, where students divide circles and rectangles into equal pieces and use the words "halves," "fourths," and "quarters." This connection between geometry and fractions continues in third grade where this concept is deepened to using the notation of $\frac{1}{2}$ and $\frac{1}{4}$ and more explicitly referring to a fraction of the area of the figure. The third grade is also when the CCSS introduce a separate content strand titled Number and Operations—Fractions, which continues through fifth grade.

Operations with fractions are a relatively recent part of the history of mathematics. Fractions are different from counting numbers and integers in a significant way: *two* numbers are needed to represent one amount! From another perspective, when we move from working with counting numbers to fractions, we are changing the question from *how many* to *how much*. For example, we use counting numbers to count *how many* (such as 200) students go to college. We use fractions to quantify *how much* (such as $\frac{2}{3}$) of a particular graduating class goes to college. A counting number counts the number of units; a fraction tells us how much of a whole there is. This is a very important concept that we will explore more.

CLARIFYING TWO TERMS: FRACTIONS AND RATIONAL NUMBERS

The CCSS start formally with fractions in the third grade with the following:

> Understand a fraction $\frac{1}{b}$ as the quantity formed by 1 part when a whole is partitioned into b equal parts; understand a fraction $\frac{a}{b}$ as the quantity formed by a parts of size $\frac{1}{b}$.
>
> The set of fractions not only includes rational numbers like $\frac{1}{4}$ and $\frac{15}{4}$ as this standard implies, but the set of fractions also includes numbers like $\frac{\pi}{6}$ and $\frac{\sqrt{2}}{3}$.

CLASSROOM CONNECTION

Children encounter the notion of rational numbers and fractions even before they enter school through the notion of sharing between two people. That is, when one whole is divided into two equal parts, each person has one-half.

A **rational number** is a number whose value can be expressed as the quotient or ratio of two *integers* a and b, represented as $\frac{a}{b}$, where $b \neq 0$.

A **fraction** is a number whose value can be expressed as the quotient or ratio of *any two numbers* a and b, represented as $\frac{a}{b}$, where $b \neq 0$. For example, $\frac{\sqrt{2}}{3}$ is a fraction but not a rational number.

In either case, a is called the **numerator** and b is called the **denominator**.

Technically, the set of rational numbers is a subset of the set of fractions, because fractions include amounts like $\frac{\sqrt{2}}{2}$ and $\frac{\pi}{6}$, which are not rational numbers because the numerators are not integers. In elementary school, children work with fractions that are rational numbers, and these will be our primary focus in this chapter. Therefore, we will generally use the term *fraction* just as the CCSS do.

FRACTIONS IN HISTORY

The notion of $\frac{3}{4}$ may make sense to you now, but it is not easy for many children, and it is such an abstraction that it is relatively recent in human history.

The Egyptians expressed all fractions (with the exception of $\frac{2}{3}$) as unit fractions—that is, fractions whose numerator is 1. They used the symbol ⬭, which they placed above a numeral to indicate a fraction. Thus $\frac{1}{12}$ was written as ∩II. The Egyptians' decision to represent fractional amounts using only unit fractions was a consequence of their difficulty with using two numbers to represent a single amount. Notice how the third-grade CCSS above builds fractions from unit fractions as well.

LANGUAGE

The origin of the word *fraction* is also interesting; it is derived from the Latin word *fractio*, which comes from the Latin word *frangere*, meaning "to break." In early American arithmetic books, the term *broken numbers* was often used instead of the word *fraction*. The first known mention of the actual word *fraction* was by Chaucer in 1321.[2]

As we saw earlier, the idea of representing *all* amounts with whole numbers was very appealing to the ancient Greeks, and so they did not even consider the idea of creating numbers that were not whole numbers. Rather, they worked with ratios. For example, instead of saying that $\frac{2}{5}$ of the students at a college are male, they would say that the ratio of males to females is 2 to 3. (We will examine ratios more closely in Chapter 5.)

The Romans also avoided fractions. We live with the effects of one of their ways of avoiding fractions: Rather than dealing with parts of a unit, they created smaller units. Their word for twelfth was *unica*, which is where our words *ounce* and *inch* come from.

Our present method of writing fractions (for example, $\frac{2}{3}$) was probably invented by the Hindus. Brahmagupta (A.D. 628) wrote $\frac{2}{3}$. The bar seems to have been introduced by the Arabs.

INVESTIGATION 4.2a Fraction Contexts with Visual Models: What Does $\frac{3}{4}$ Mean and Look Like?

One of the interesting things about fractions is that we can represent them with different visual models, depending on the context. First draw as many different visual models of $\frac{3}{4}$ as you can. Then read on. . . .

A. Explain how each of the visual models in Figure 4.10 represents $\frac{3}{4}$.

Explorations Manual 4.7

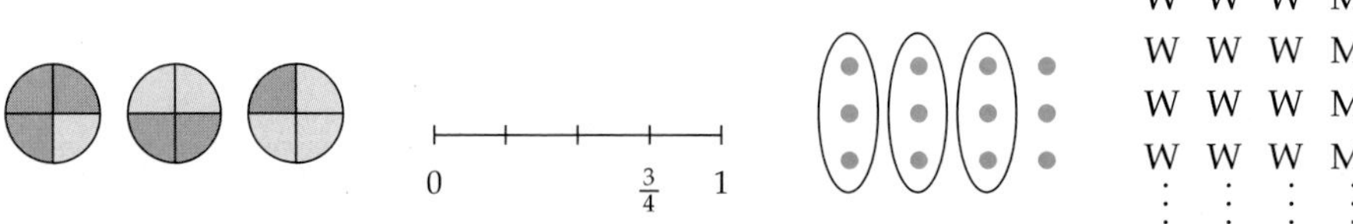

Figure 4.10

B. How are the following contexts for $\frac{3}{4}$ similar and different?

1. Four children want to share 3 pies equally. How much pie does each child get?
2. Joey grew $\frac{3}{4}$ of an inch last month.
3. $\frac{3}{4}$ of a dozen donuts have been eaten.
4. At a college, $\frac{3}{4}$ of the students are women.

C. Which picture from A most closely matches each of the contexts from B?

DISCUSSION

A key idea with fractions is the whole and the unit. We will discuss this throughout the next two sections, but introducing them now will help us to build our understanding. The unit refers to what equals 1. The whole is the given object or total amount.

1. Four children want to share 3 pies equally. How much pie does each child get?

This situation illustrates **fractions as a quotient**. We are dividing 3 pies (the whole) by 4 children. The answer of $\frac{3}{4}$ pie per child is each child's share. In other words, each child gets $\frac{3}{4}$ of a pie (the unit). Here, 1 pie is the context for "1," but the whole, *or total amount we are working with*, is 3 pies. When we are dealing with a fraction $\frac{a}{b}$ in the context of a quotient, an amount a needs to be shared or divided equally into b groups. The first visual model in part A relates to this context. Another way to look at this is in Figure 4.11.

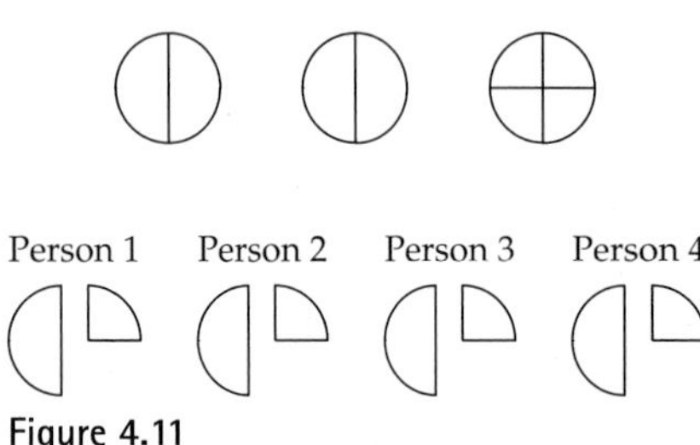

Figure 4.11

This type of visual model is called an **area model** since the size (or area) of the pies is the whole. The pieces also have to be the same size (or area). This is the earliest model of fractions that children encounter, both according to the CCSS and before school with real-life examples like $\frac{1}{2}$ of a sandwich (where the whole is the size of the sandwich). Along with circles, other area models include fraction rectangles, grid paper, and pattern blocks.

2. Joey grew $\frac{3}{4}$ of an inch last month.

This situation illustrates **fractions as a measure**. To measure the appropriate location of $\frac{3}{4}$, we must divide (partition) the unit length (1 inch) into 4 equal lengths. The length of 3 of those equal lengths shows how much Joey grew (Figure 4.12). Here, both the whole and the unit are 1 inch.

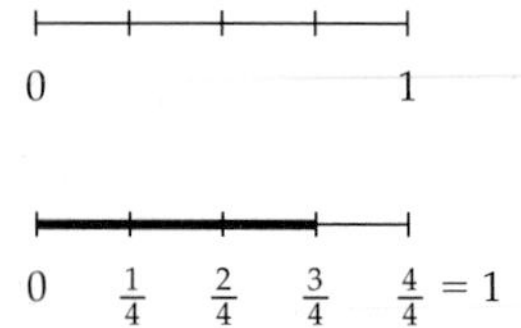

Figure 4.12

This type of visual model is called a **linear model** since the length of the line segment is the whole. The pieces of the line also have to be the same size (or length). Representing fractions on a number line is introduced in third grade, according to the CCSS. Along with number lines, other linear models include Cuisenaire rods and Singapore bar models.

3. $\frac{3}{4}$ of a dozen donuts have been eaten.

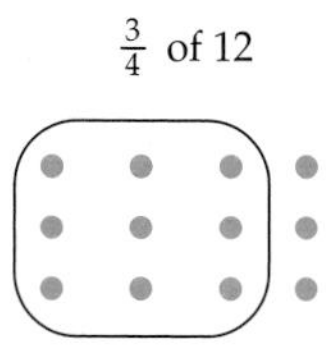

Figure 4.13

This situation illustrates **fractions as an operator** on the set of donuts, because we take 3 out of every 4 donuts in a box of 12 donuts. In this case, the whole is 12 donuts (what we are taking $\frac{3}{4}$ of), and the unit is 1 dozen (because we use "1" to represent a dozen donuts). Taking $\frac{3}{4}$ of the 12 donuts is 9 donuts, which is $\frac{9}{12}$ of the dozen. This model helps us to understand equivalent fractions like $\frac{3}{4} = \frac{9}{12}$, which we will explore more in depth later in this section. Figure 4.13 shows the 9 out of 12 donuts, or $\frac{9}{12}$ of the dozen, which is equivalent to 3 columns out of 4 columns, or $\frac{3}{4}$ of 1 dozen.

This type of visual model is called a **set model** since the set of objects (the 12 donuts) is the whole. While the area and linear models were a size relationship (in other words the areas or lengths were divided into equal-size pieces), the set model is not a size relationship. We can talk about $\frac{3}{4}$ of a set of animals or $\frac{3}{4}$ of a set of shapes. In these examples, the size of the objects is not relevant; the whole is the number of objects, not the size. Any collection of objects (blocks, candies, fruits, etc.) can be used as set models.

4. At a college, $\frac{3}{4}$ of the students are women.

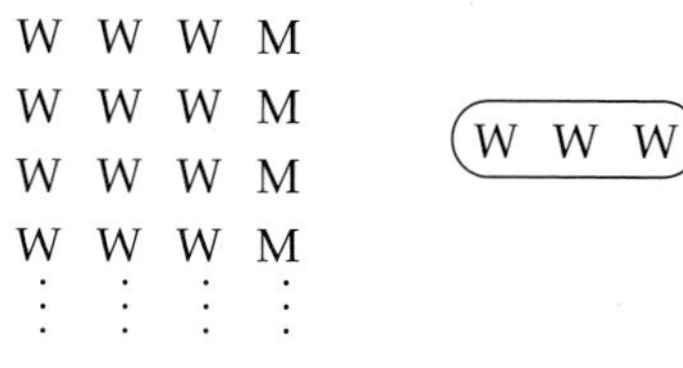

Figure 4.14

This situation illustrates **fractions as a ratio**. We do not know the total number of students in the college, which is the whole, but we do know that if we divided the total number of students into 4 groups with equal numbers of students in each, then the number of women would be 3 of those 4 groups. Sometimes we might see this relationship in ratio notation: the ratio of women to total number of students is 3 : 4. Figure 4.14 shows this relationship visually.

Because the whole is a set of students, this is also a set model (even though in this situation we do not know how many are in that set). The whole is the total number of students.

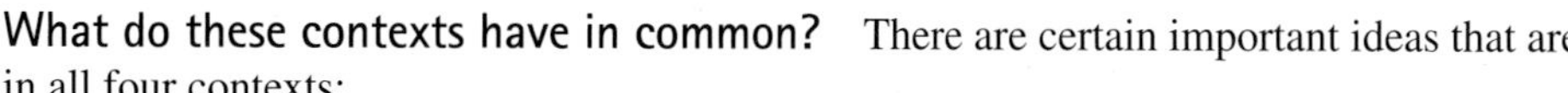

What do these contexts have in common? There are certain important ideas that are in all four contexts:

- Each context can be interpreted in part–whole relationships. In each case a whole has been divided into 4 parts.
- Something is to be partitioned into parts of equal size (value).

  The something can have a value of 1, in which case the unit = the whole.

  The something can have a value $\neq 1$, in which case the unit $\neq$ the whole.
- The numerator and denominator are like codes that tell us about the relative sizes of the parts and the unit, and the code is multiplicative in nature. For example, when we say $\frac{1}{2}$, it is not the difference between the two numbers that contains the key to the value; rather, it is the fact that the value of the denominator is twice the value of the numerator that contains the key. Thus $\frac{1}{2}$ has the same value as $\frac{4}{8}$, not $\frac{7}{8}$.

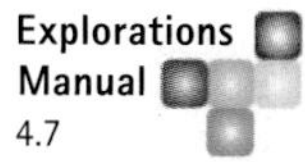

We will explore area models, linear models, and set models throughout our discussion of fractions. These three models have been found useful in helping children to understand fraction concepts and in terms of representing problems visually. There are other models (such as volume), but we will focus on these three main models used in elementary schools. Initially you may feel more comfortable with one type of model, but after learning about them you will become comfortable with all of the models, which will both deepen your understanding and help you to develop models that will be useful to your future students.

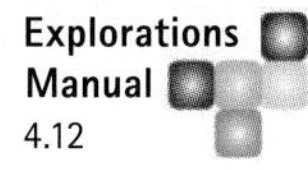
Explorations Manual 4.12

The unit and the whole are not always the same! Equating the unit with the whole is one of the most common misconceptions that people have about working with fractions. Look back on the four problems posed in connection with Figure 4.10 and consider the question: What do we mean by "the whole" and by "the unit"?

The whole is the given object or amount. The unit is that amount to which we give a value of 1: 1 inch, 1 pie, 1 person. In some cases, the whole and the unit are the same. For example, if Lisa gets $\frac{1}{2}$ of a pizza and Liam gets $\frac{1}{3}$ of the pizza, the whole is 1 pizza, and the unit is also 1 pizza. However, if 3 pizzas are divided among 4 people, then the whole is 3 pizzas but the unit is 1 pizza. Understanding the concept of units and wholes is a major key in understanding fractions.

Let us now investigate the context of fractions as division, generally the first context children investigate.

INVESTIGATION 4.2b Wholes and Units: Sharing Brownies

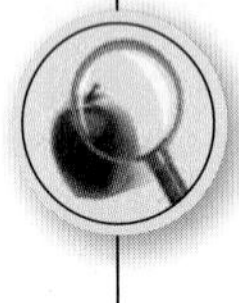

Problems like this one are found in several articles in *Teaching Children Mathematics,* elementary textbooks, and support materials. Five children need to share four brownies. How much does each child get? Solve this using visual and/or physical models.

DISCUSSION

One way to do this is to divide each brownie into fifths and give each child four pieces. However, that is not satisfying to most children because they want bigger pieces. A common solution looks like the figure below. Most children have no trouble saying that each child gets $\frac{1}{2}$ of a brownie and $\frac{1}{4}$ of a brownie, but what is the name for the smallest piece that each child gets?

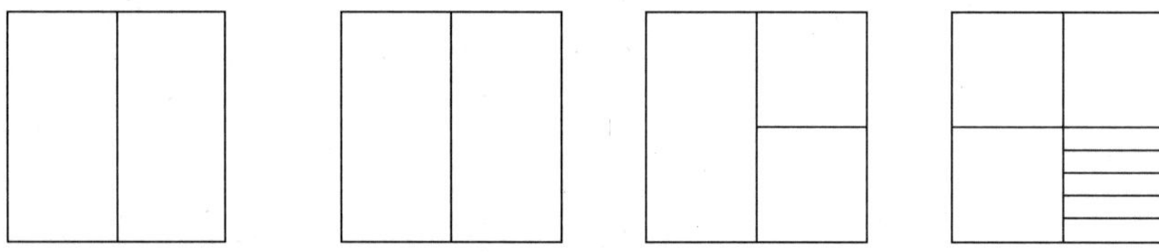

The smallest amount is $\frac{1}{20}$ of a brownie. Children will arrive at this conclusion in different ways. Some will partition the whole brownie so that all pieces are the same size, while others realize that this smallest piece is $\frac{1}{5}$ of $\frac{1}{4}$ of a brownie, and $\frac{1}{5}$ of $\frac{1}{4}$ is $\frac{1}{20}$. When children incorrectly say that the smallest piece is $\frac{1}{5}$, they are seeing that last quarter of a brownie as the whole and that whole has been divided into 5 equal pieces. However, the value of the denominator of a fraction is connected to how many of those pieces it takes to make the unit (that which has a value of 1). In this case, it takes 20 of those small pieces to make 1 brownie, and thus the size of that smallest piece is $\frac{1}{20}$ (of 1 brownie). We use an area model (dividing the area of the rectangles) to represent the brownies.

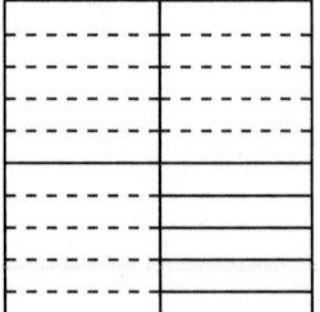

UNITS, WHOLES, AND UNITIZING

Now that we have investigated some important fraction concepts, let us take a step back to examine an important aspect of fractions: unitizing. We first encountered units and

CLASSROOM 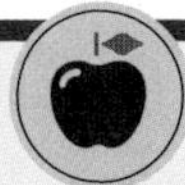 CONNECTION

Grade 2

Draw a line that would divide #5 in half. How could you persuade someone that the line in #9 does divide the shape into halves?

Name Date

Parts of a Whole, Parts of a Group

Daily Practice

Halves and Not Halves

Color the shape if the line shows halves.

NOTE Students determine which shapes are divided into halves by the given line.

SMH **84, 86**

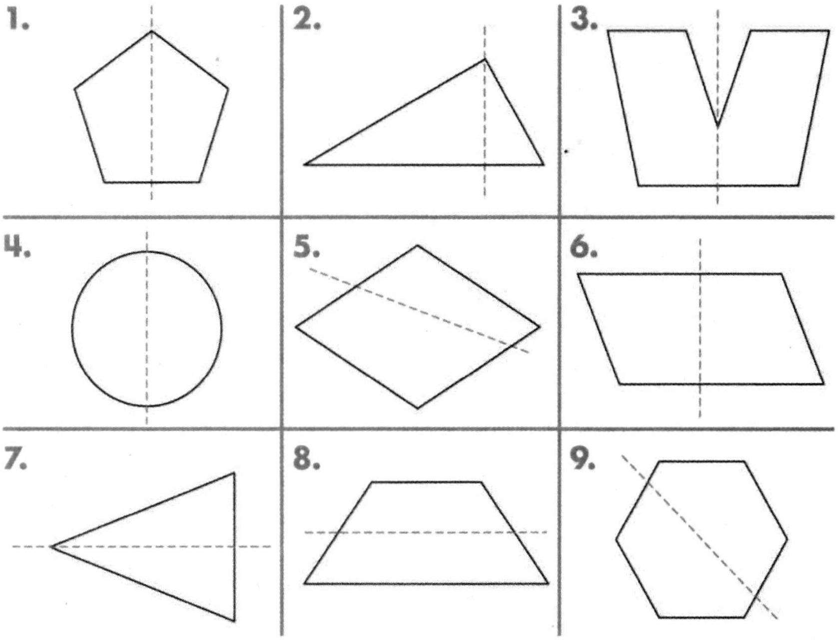

1.

2.

3.

4.

5.

6.

7.

8.

9.

Ongoing Review

10. Which equation describes the groups of dots?

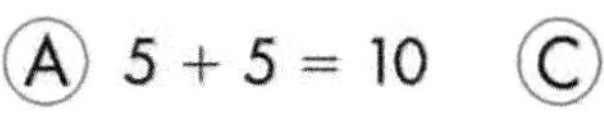

(A) 5 + 5 = 10

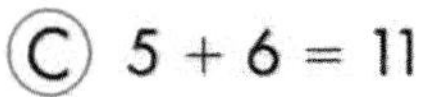

(C) 5 + 6 = 11

(B) 2 + 5 = 7

(D) 4 + 4 = 8

Session 1.2 Unit 7 13

unitizing in Chapter 2, when we saw that we can count by ones and we can also count by tens. In the latter case, ten is now the unit of counting. We need to adapt this use of unitizing to develop flexibility in working with fractions. Susan Lamon's notion that "Unitizing refers to the size of the mental 'bite' in terms of which you think about the unit" is useful.[3]

With whole numbers, counting is simpler. A can of soda is 1 can; a six-pack of soda is 6 cans. We can count by 1 or by 6, e.g., 24 cans or 4 six-packs. With fractions, we are talking about parts of units, and the fraction that our amount represents varies depending on the unit. For example, 6 inches is 6 inches. However, if our unit is a foot, 6 inches is $\frac{1}{2}$ of a foot, but if our unit is a yard, 6 inches is $\frac{1}{6}$ of a yard. Becoming comfortable with unitizing is an important part of developing an understanding of fractions. Below are other examples of this concept.

Hour	4 hours is what part of a day?	4 hours is $\frac{1}{6}$ of a day.
	4 hours is what part of a work week?	4 hours is $\frac{1}{10}$ of a work week.
Nickel	1 nickel is what part of a dime?	1 nickel is $\frac{1}{2}$ of a dime.
	1 nickel is what part of a dollar?	1 nickel is $\frac{1}{20}$ of a dollar.

INVESTIGATION 4.2c Unitizing

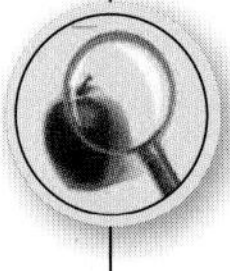

In the figure below, we see several sets of 18 circles. Please do the following problems as children who do not yet have algorithms might do them. For example, to determine $\frac{5}{6}$ of the circles in (a), we need to partition them into 6 equivalent groups. When we do this, $\frac{1}{6}$ = 3 circles, and 5 groups of 3 circles represents $\frac{5}{6}$ of the whole.

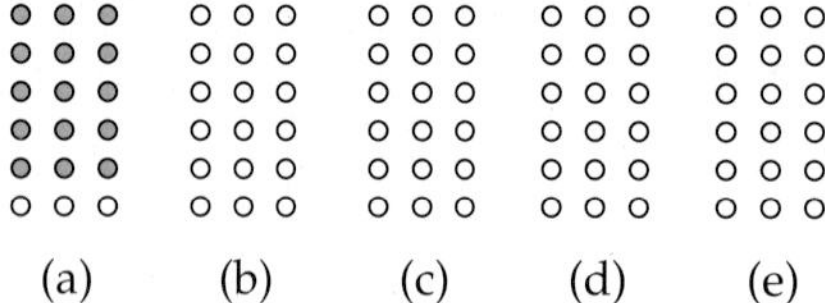

(b) Shade $\frac{2}{3}$ of the circles.

(c) Shade $\frac{4}{9}$ of the circles.

(d) Shade $\frac{7}{18}$ of the circles.

(e) Shade $\frac{5}{12}$ of the circles.

DISCUSSION

The whole in each case is the set of 18 circles (which makes this a set model). The table and models below help show the answers. The second column shows the number of circles to shade.

The third column shows how many are in each group when we partition the set into the number of equal groups the denominator calls for. For example, to make $\frac{5}{6}$, we must partition the circles into 6 equal groups, and we see that $\frac{1}{6}$ is equivalent to 3 circles.

The fourth column shows us the equivalence of the original fraction and our shading. For example, $\frac{5}{6}$ is equivalent to 15 circles, which is equivalent to $\frac{15}{18}$ of the circles. Work on unitizing naturally develops understanding of equivalence, which we will explore very soon. It is important to note that we could get the answers simply by using the algorithm: $\frac{5}{6} = \frac{5 \times 3}{6 \times 3} = \frac{15}{18}$.

Here, our goal is more than getting the answer; it is traveling the terrain that your future students will walk as they develop the ability to work confidently with fractions.

CLASSROOM  CONNECTION

Grade 2

Do you see the connections between these problems and Investigation 4.2c? Did you see anything like this when you were in elementary school? 

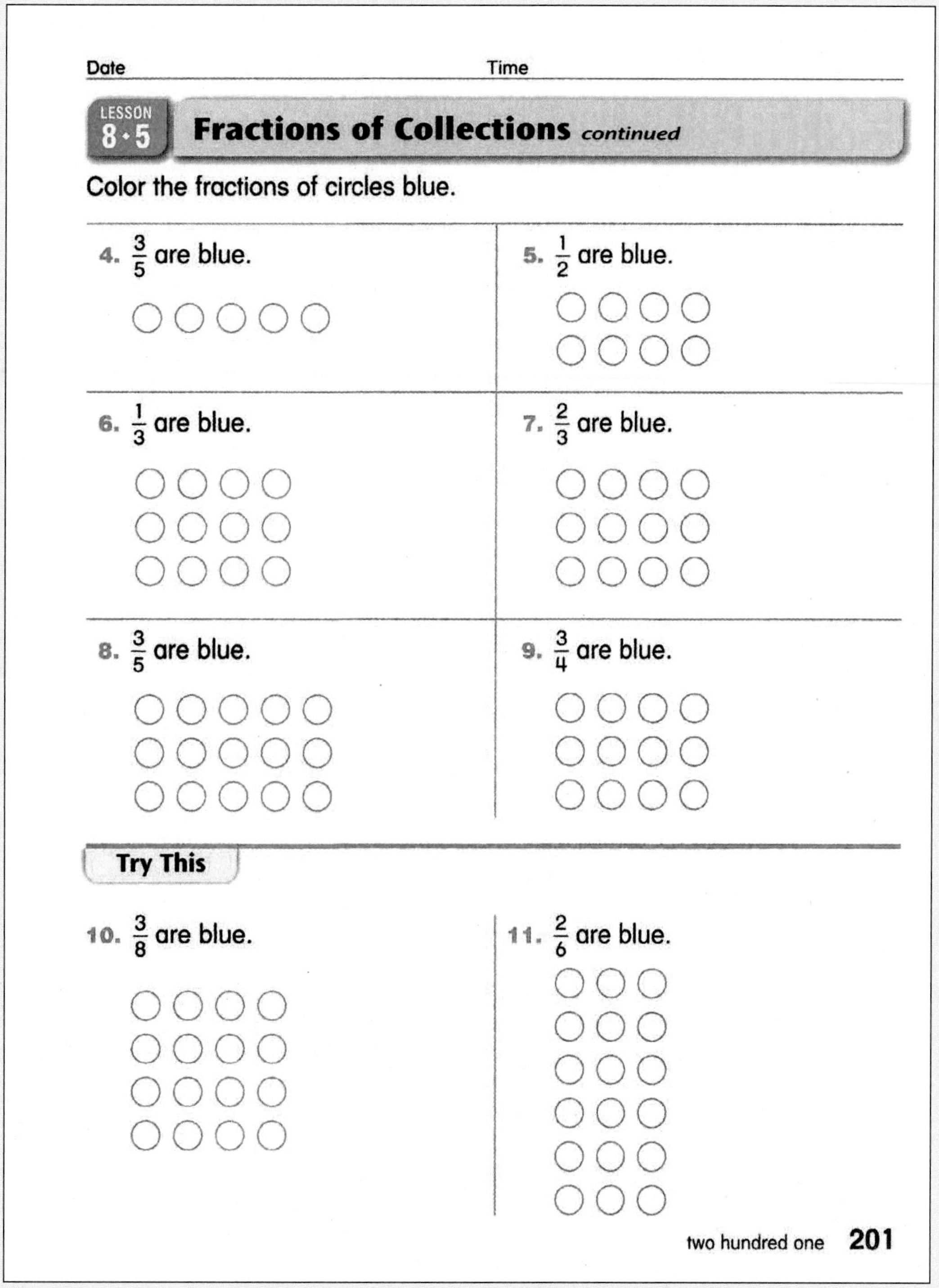

Date Time

LESSON 8·5 **Fractions of Collections** *continued*

Color the fractions of circles blue.

4. $\frac{3}{5}$ are blue.

○○○○○

5. $\frac{1}{2}$ are blue.

○○○○
○○○○

6. $\frac{1}{3}$ are blue.

○○○○
○○○○
○○○○

7. $\frac{2}{3}$ are blue.

○○○○
○○○○
○○○○

8. $\frac{3}{5}$ are blue.

○○○○○
○○○○○
○○○○○

9. $\frac{3}{4}$ are blue.

○○○○
○○○○
○○○○

Try This

10. $\frac{3}{8}$ are blue.

○○○○
○○○○
○○○○
○○○○

11. $\frac{2}{6}$ are blue.

○○○
○○○
○○○
○○○
○○○
○○○

two hundred one **201**

From *Everyday Mathematics, Grade 2:* The University of Chicago School Mathematics Project: Student Math Journal, Volume 2, by Max Bell et al., Lesson 8-5, p. 201. Reprinted by permission of The McGraw-Hill Companies, Inc.

Fraction	Number of circles shaded	Unitizing	Equivalence
$\frac{5}{6}$	15	$\frac{1}{6} = 3$ circles	$\frac{5}{6} = \frac{15}{18}$
$\frac{2}{3}$	12	$\frac{1}{3} = 6$ circles	$\frac{2}{3} = \frac{12}{18}$
$\frac{4}{9}$	8	$\frac{1}{9} = 2$ circles	$\frac{4}{9} = \frac{8}{18}$
$\frac{7}{18}$	7	$\frac{1}{18} = 1$ circle	$\frac{7}{18} = \frac{7}{18}$
$\frac{5}{12}$	$7\frac{1}{2}$	$\frac{1}{12} = 1\frac{1}{2}$ circles	$\frac{5}{12} = \frac{7\frac{1}{2}}{18}$

$\frac{5}{6}$ $\frac{2}{3}$ $\frac{4}{9}$ $\frac{7}{18}$ $\frac{5}{12}$

INVESTIGATION 4.2d Fundraising and Thermometers

An organization has set $300,000 as their fundraising goal. A large sign with a big thermometer sits by the street in front of the building. Figure 4.15 shows their progress after 24 days (on the left) and after 49 days (on the right). Describe the progress of the fundraiser in as many ways as possible. Then read on. . . .

The notion of multiple representations and equivalence are in the foreground here and are essential for a deeper understanding of important mathematical ideas. Answer the following questions, each of which represents different interpretations and representations of the progress.

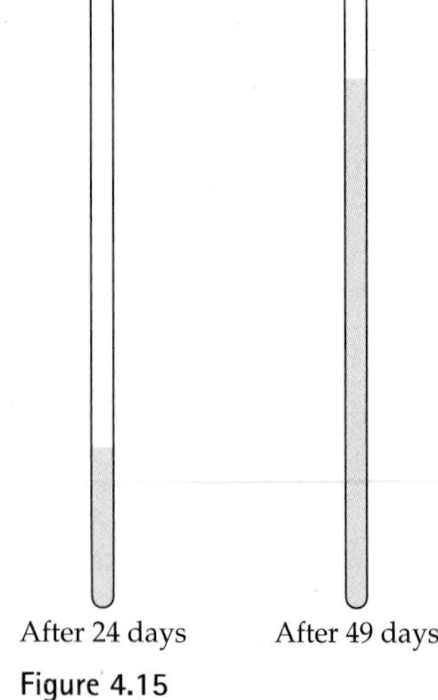

Figure 4.15

A. They are about $\frac{1}{4}$ of the way after 24 days, as shown on the thermometer on the left. Describe how one might arrive at this answer.

B. If the progress of the fundraiser continues at the rate it did during the first 24 days, how long will it take them to reach their goal?

C. Determine a fraction to represent their progress after 49 days.

D. About how many dollars have they raised after 49 days? Can you determine this answer mentally?

E. If progress of the fundraiser continues at the rate it did during the first 49 days, how long will it take them to reach their goal?

DISCUSSION

A. The length of the thermometer (a linear measure) is the whole of this linear model. You could trace the picture on a piece of paper and fold. You could use the amount as a ruler and see how many of these lengths fill the thermometer. In this case, you are using this length as your unit.

B. If they have made $\frac{1}{4}$ of their goal in 24 days, then they would make their goal in $4 \times 24 = 96$ days, or just over 3 months.

The thermometer on the right in Figure 4.15 shows their progress after a total of 49 days.

C. We could use the nonshaded length as our unit. Realizing that it is $\frac{1}{6}$, we would conclude that the organization has reached $\frac{5}{6}$ of their goal. We could also estimate thirds lightly, then fourths, then fifths, etc., until one division seems appropriate.

D. $\frac{5}{6}$ of \$300,000 is \$250,000. Here is one way to determine the amount mentally. First, we can see that $\frac{1}{6}$ of 300,000 is 50,000. Then we multiply 50,000 by 5 to get 250,000.

E. If $\frac{5}{6}$ is equivalent to 49 days, then $\frac{1}{6}$ is equivalent to about 10 days ($\frac{1}{6}$ is $\frac{1}{5}$ of $\frac{5}{6}$, and $\frac{1}{5}$ of 49 is about 10). If $\frac{1}{6}$ is equivalent to about 10 days, then $\frac{6}{6}$ is equivalent to about 60 days.

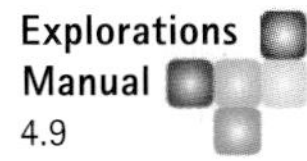

Developing fraction sense One goal of this work and Exploration 4.9 is to develop "fraction sense." Number sense emerges from being aware of the connectedness of, and subtle relationships among, various concepts and procedures. We see this awareness all the time when we watch experts at work: the dancer who can "feel" the subtle rhythms in a new piece of music, the electrician who rewired my house and could tell so much about the wiring problems from so little evidence, the mechanic who can look at a nut and know that he needs a $\frac{5}{16}$-inch wrench.

As you are coming to realize, the set of fractions is in many ways very different from the set of whole numbers. With whole numbers, we are counting how many units, but with fractions we are quantifying how much of a whole (which can vary). If there are 50 people from North Carolina and 50 people from the United States, this is the same number of people in each set. However, $\frac{1}{2}$ of the people from North Carolina is a different quantity of people than $\frac{1}{2}$ of the people from the United States. The relationship is the same, but not the quantity of people. Another important difference between fractions and whole numbers is that the whole numbers are evenly spaced on the number line. This is not true of fractions.

In the next few investigations, you will deepen your understanding of fractions by working with different models (representations) of fractions.

INVESTIGATION 4.2e Partitioning with Number Line Models

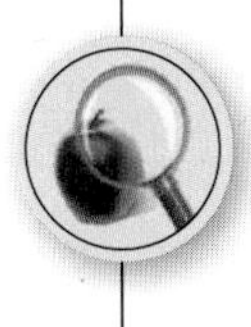

A. Place $\frac{5}{6}$ on this number line.

0 ———————— 2

B. Determine the value of x on the number line.

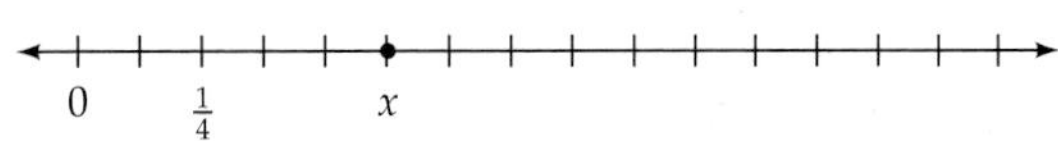

C. Locate $\frac{5}{6}$ on the number line.

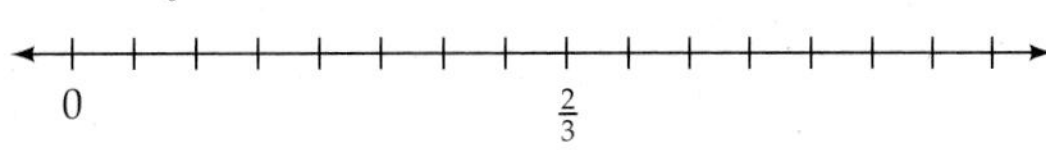

DISCUSSION

A. This question requires you to grapple with the difference between the unit and the whole. In this case, the whole and the unit are not the same. This is important, because the meaning of the denominator *is in relation to the unit.* That is, to find the location of $\frac{5}{6}$, we do not take the whole line and divide it into 6 equal lengths. Rather, we first must determine the unit length and then divide that length into 6 equal lengths. This difference between the whole and the unit cannot be overemphasized.

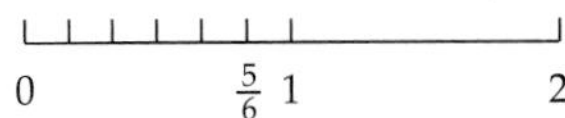

B. We can deduce that if $\frac{1}{4}$ represents two partitions from 0, then 1 partition from 0 would be $\frac{1}{8}$ (which is $\frac{1}{2}$ of $\frac{1}{4}$). Now, we can see that the value of x is $\frac{5}{8}$.

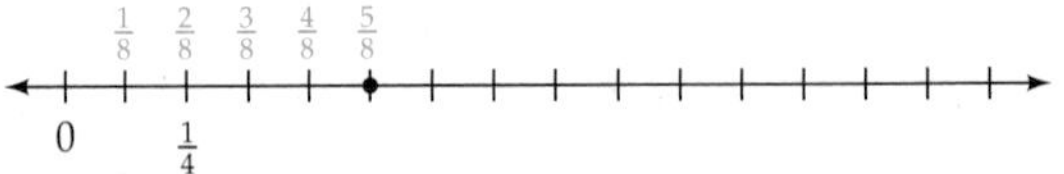

C. In this case, $\frac{2}{3}$ represents 8 partitions to the right of zero. Thus, 4 partitions would have a value of $\frac{1}{3}$, and 2 partitions would have a value of $\frac{1}{6}$. So the location of $\frac{5}{6}$ is $5 \times 2 = 10$ partitions to the right of 0.

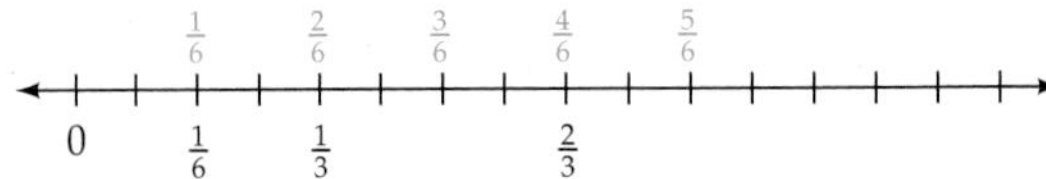

As you will discover when you teach children, it is important to explore concepts with different models. Now we will investigate fractions with area models, which are often encountered in elementary schools with Geoboards, fraction circles or rectangles, and pattern blocks.

INVESTIGATION 4.2f Partitioning with Area Models

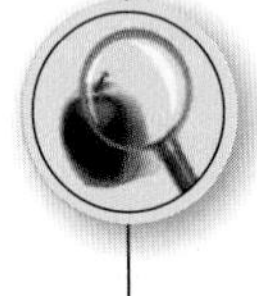

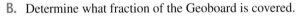

A. If 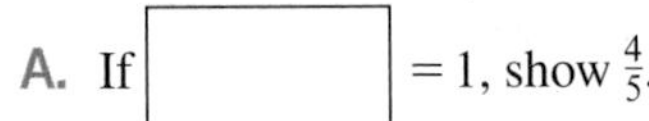 = 1, show $\frac{4}{5}$.

B. Determine what fraction of the Geoboard is covered.

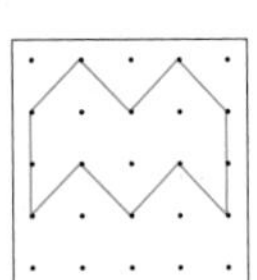

C. If the area of the two hexagons equals 1 unit, what fraction does the area of the two blue parallelograms equal?

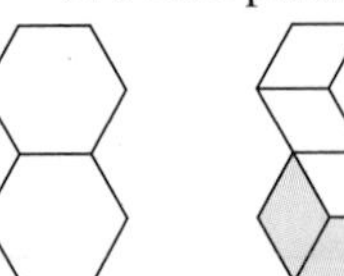

CLASSROOM CONNECTION

The need for fractions is introduced in *Inchworm and a Half* by Elinor Pinczes (Boston: Houghton Mifflin, 2001). An inchworm can measure fruits and vegetables by how many loops it takes her to go from one end to another. When she encounters fractions, smaller worms that are $\frac{1}{2}$, $\frac{1}{3}$, and $\frac{1}{4}$ of her length help her out.

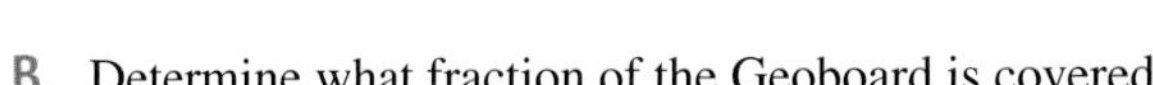

DISCUSSION

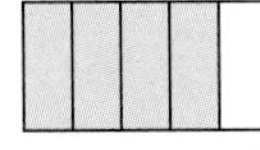

A. This question is relatively straightforward. Divide the rectangle into 5 equal regions, and shade in 4 of them. The area of these 4 regions = $\frac{4}{5}$ of the area of the whole rectangle.

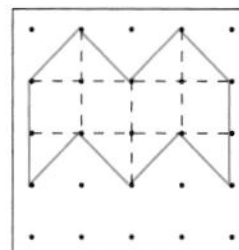

B. In the first case, the value of the entire Geoboard is 16 (unit squares). We can decompose the polygon into squares and triangles whose value is clearly $\frac{1}{2}$ (square). When we count the squares and $\frac{1}{2}$ squares, we have a value of 8, and thus the shape covers $\frac{8}{16}$ or $\frac{1}{2}$ of the area of the Geoboard.

C. The lines help us to see that two of the blue parallelograms cover $\frac{2}{6}$ or $\frac{1}{3}$ of the area of the two hexagons.

Finally, let us investigate fractions with set models.

INVESTIGATION 4.2g Partitioning with Area Models

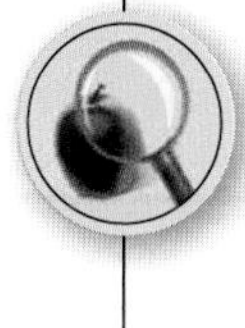

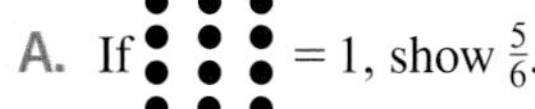

A. If [9 dots] = 1, show $\frac{5}{6}$.

B. If [8 dots] = $\frac{4}{3}$, show 1.

DISCUSSION

A. To make sense of this question, we must partition the dots into 6 equal-size groups. Recall the partitioning model of whole-number division. We then take 5 of those groups to show $\frac{5}{6}$ [Figure 4.16(a)].

(a) (b)

Figure 4.16

B. There are different ways to make sense of this situation and answer the question. For example, we can focus on the numerator, which indicates that we have 4 equal parts. Thus each part—that is, $\frac{1}{3}$ —contains 2 dots. When we then focus on the denominator, we find that 3 of these equal parts represent the unit (that is, 3 of those parts have a value of 1).

On the other hand, we can interpret this statement from a ratio context. For example, if 8 dots have a value of $\frac{4}{3}$, then 4 dots will have a value of $\frac{2}{3}$, and so 2 dots will have a value of $\frac{1}{3}$. Thus, 6 dots will have a value of $\frac{3}{3}$—that is, 1 [Figure 4.16(b)].

INVESTIGATION 4.2h Determining an Appropriate Representation

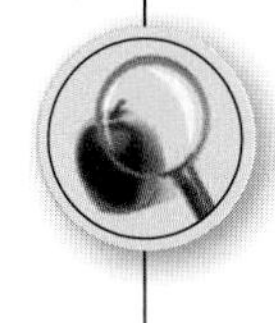

Jose paid \$12 for a box of chocolates that weighed $\frac{3}{4}$ pound. What is the price of 1 pound (at this rate)? Work on this and then read on. . . .

DISCUSSION

An area model is an appropriate representation, because boxes of chocolates are often rectangular in shape. If we look at the box in terms of weight, we have $\frac{3}{4}$ of a pound. If we look at the box in terms of money, it costs \$12. In one sense, we are saying that $\frac{3}{4}$ of a pound is equivalent to \$12.

Thinking of the problem as "we have 3 parts, which are equivalent to \$12," the model helps us see that each part (that is, each quarter-pound) has a value of \$4 (Figure 4.17), and so one pound will cost \$16.

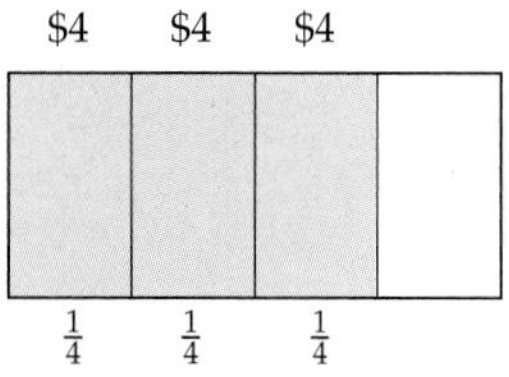

\$4 \$4 \$4

$\frac{1}{4}$ $\frac{1}{4}$ $\frac{1}{4}$

Figure 4.17

CLASSROOM  CONNECTION

Grade 2

Can you explain how this area model connects to each of the contexts for fractions: as quotient, as measure, and as ratio?

Name Date

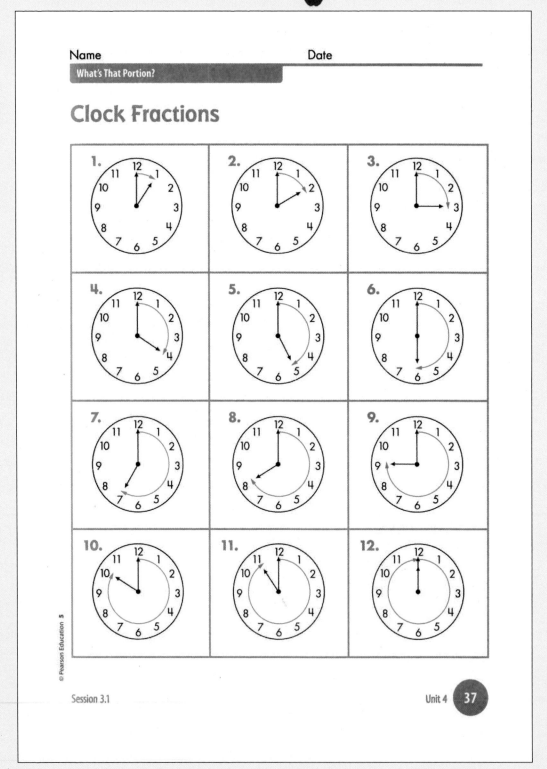

What's That Portion?

Clock Fractions

1.	2.	3.
4.	5.	6.
7.	8.	9.
10.	11.	12.

Session 3.1 Unit 4 37

EQUIVALENT FRACTIONS

Having opened the concept of equivalent fractions in Investigation 4.2c, let us now examine the concept of equivalent fractions.

Equivalence Please address these two questions before reading on. . . . 

- What does *equivalent* mean?
- Where else have you encountered the notion of equivalence in mathematics and in life outside school?

One way of looking at "equivalent" comes from taking apart the actual word: *equivalent*, or equal value. We needed the notion of equivalence when we added and subtracted whole numbers with regrouping; for example, 1 ten and 2 ones is equivalent to 12 ones. We use equivalence with money every day; for example, one quarter is equivalent to 25 pennies.

Two fractions are **equivalent fractions** if they have the same value.

Using models Can you illustrate the equivalence of $\frac{3}{4}$ and $\frac{6}{8}$ using one or more of the fraction models we have discussed: area, length, or set? Do this before reading on. . . .

CLASSROOM CONNECTION

Many teachers use paper folding to introduce the idea of equivalent fractions. For example, take a piece of paper and fold it in half. Then fold it in half again. [*Note:* There are several ways to do this.] Shade 3 of the 4 regions to show $\frac{3}{4}$. Now fold the paper in half again. Now we see that the paper has been folded into 8 regions and that 6 are shaded. Thus $\frac{3}{4} = \frac{6}{8}$: that is, these two fractions represent the same *part* of the whole.

We can use the set model to illustrate equivalence. For example, consider a set of 8 dots [Figure 4.18(a)]. If we take 6 of them, we literally have $\frac{6}{8}$ [Figure 4.18(b)]. However, we can also partition this set of 8 dots into 4 equal groups of 2 dots; if we then take 3 of these 4 equal groups, we have, by definition, taken $\frac{3}{4}$ of the set [Figure 4.18(c)].

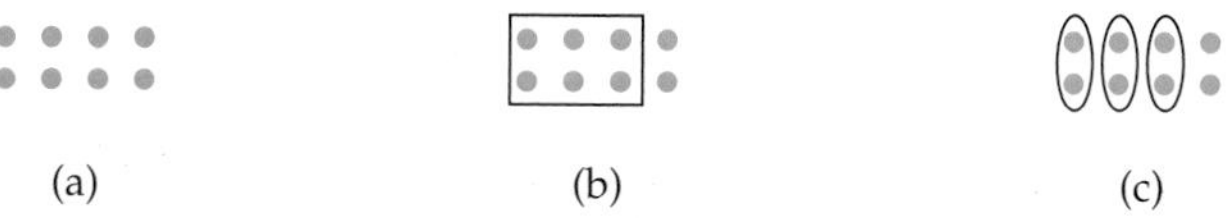

(a) (b) (c)

Figure 4.18

However, we could also have used an area model. For example, consider a rectangle representing 1 [Figure 4.19(a)]. We can partition the rectangle into 4 equal regions and shade in 3 of them to represent $\frac{3}{4}$ [Figure 4.19(b)]. However, if we draw a horizontal line through the middle of the rectangle, the rectangle has now been partitioned into 8 equal regions. Because 6 of those regions have been shaded in, we have, by definition, $\frac{6}{8}$ of the rectangle [Figure 4.19(c)].

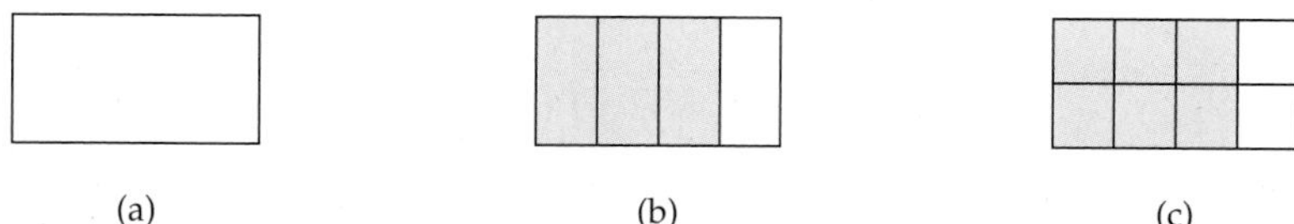

(a) (b) (c)

Figure 4.19

We can also use a linear model, like the bar model below. Here we take a linear bar, divide it into fourths (shown in green), and then subdivide each fourth (shown in blue) to show that $\frac{3}{4} = \frac{6}{8}$.

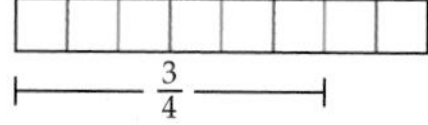

Source: Alice Ho of Math Teach Singapore.

PATTERNS IN EQUIVALENT FRACTIONS

Patterns in a sequence Look at the sequence of equivalent fractions below. What patterns do you notice? What is the next fraction in the sequence? Why?

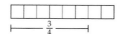

$$\frac{3}{4} = \frac{6}{8} = \frac{9}{12} = \frac{12}{16} \ldots$$

Some of the patterns include:

1. As we move from one fraction to another, the numerator increases by 3 and the denominator increases by 4. In this sense, the $\frac{3}{4}$ acts as an operator.
2. The numerator of each of the fractions is a multiple of 3, and the denominator of each of the fractions is the same multiple of 4.
3. All the denominators are even numbers, and every other numerator is an even number.

We can translate the first statement into notation:

Specific example: $$\frac{3}{4} = \frac{3+3}{4+4} = \frac{6}{8}$$

More general case: $$\frac{a}{b} = \frac{\overbrace{a + a + \cdots + a}^{n \text{ times}}}{\underbrace{b + b + \cdots + b}_{n \text{ times}}} = \frac{an}{bn}$$

We can also translate the second statement into notation:

Specific example: $$\frac{3}{4} = \frac{3 \cdot 2}{4 \cdot 2} = \frac{6}{8}$$

More general case: $$\frac{a}{b} = \frac{a \cdot n}{b \cdot n} = \frac{an}{bn}$$

MATHEMATICS

When exploring equivalent fractions, children have to realize that there are additive and multiplicative relationships. From an additive perspective, $\frac{6}{8}$ is equivalent to $\frac{3}{4}$ because we *add* 3 to the numerator and *add* 4 to the denominator. From a multiplicative perspective, $\frac{6}{8}$ is equivalent to $\frac{3}{4}$ because we have *multiplied* both numerator and denominator by 2.

ADDRESSING A COMMON DIFFICULTY IN LANGUAGE

When trying to explain why $\frac{3}{4}$ and $\frac{6}{8}$ are equivalent, many students use the word *divide*—for example, "We divided the rectangle in half, and now there are 8 equal regions compared to 4 before." This choice of words is interesting and points to a problem that many children have in trying to understand equivalent fractions at a conceptual level. When we look at why fractions are equivalent, we commonly encounter this word *divide*. However, in the procedure for creating equivalent fractions, we *multiply* the top and bottom by the same number! If we physically divide, why do we mathematically multiply? Think about this and then read on. . . .

The answer to this question has to do with the reciprocal relationship between multiplication and division. Figure 4.20(a) shows $\frac{3}{4}$. If we divide each of the regions by 2, we are also multiplying the total number of regions by 2 [Figure 4.20(b)]. We now have 6 out of 8 regions shaded—that is, $\frac{6}{8}$. Starting with $\frac{3}{4}$, we could have divided each region into 3 smaller pieces. By dividing each region into 3 smaller pieces, we are multiplying the number of pieces by 3, and we can name this shaded amount $\frac{9}{12}$ [Figure 4.20(c)].

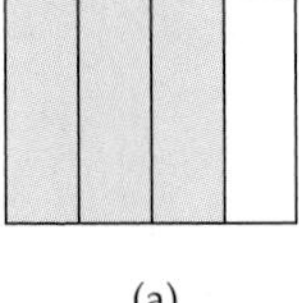
(a)

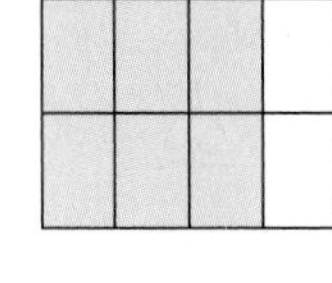
(b)

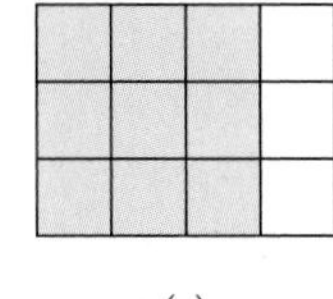
(c)

Figure 4.20

SIMPLEST FORM

A fraction is in **simplest form** if the numerator and the denominator have no common factors (other than 1). The notion of simplifying fractions is clearly connected to the concept

LANGUAGE

Just as we recommend not using the terms *carry* and *borrow* when regrouping during addition and subtraction of whole numbers, we recommend not using the term *reduce* but rather *simplify* for this process.

The word "reduce" means to make smaller. We are not making the value of the fraction smaller, we are just writing the fraction in a simpler way with smaller numbers. When children call this "reducing," it can lead to the misconception that the quantity is being made smaller.

of equivalent fractions, and we will discuss it here. One important connection is that when we are simplifying fractions, we are essentially finding an equivalent fraction in which the numerator and denominator are smaller numbers. For example,

$$\frac{15}{20} = \frac{15 \div 5}{20 \div 5} = \frac{3}{4} \quad \text{or} \quad \frac{15}{20} = \frac{3 \times 5}{4 \times 5} = \frac{3}{4}$$

As has been true for other procedures, there are many strategies for simplifying fractions. Before we discuss them, simplify the following fractions yourself: $\frac{24}{40}$, $\frac{42}{60}$, and $\frac{63}{105}$. Explain your method for each.

One strategy is to divide the numerator and denominator by any common factor, not necessarily the greatest common factor. For example, you might divide both numerator and denominator by 4, which produces $\frac{6}{10}$. A quick glance reveals that this can be simplified further to $\frac{3}{5}$.

$$\frac{24}{40} = \frac{24 \div 4}{40 \div 4} = \frac{6}{10} = \frac{6 \div 2}{10 \div 2} = \frac{3}{5}$$

Another strategy is to determine the prime factorization of each number and then cross out the common factors.

$$24 = 2 \cdot 2 \cdot 2 \cdot 3$$
$$40 = 2 \cdot 2 \cdot 2 \cdot 5$$

Let us examine more closely what happens when we are able to simplify the fraction in one step, as we did earlier to simplify $\frac{15}{20}$. What is the relationship between the divisor and the original numerator and denominator in each of the three fractions given above ($\frac{24}{40}$, $\frac{42}{60}$, and $\frac{63}{105}$)?

$$\frac{24 \div 8}{40 \div 8} \qquad \frac{42 \div 6}{60 \div 6} \qquad \frac{63 \div 21}{105 \div 21}$$

In each case, to simplify the fraction in one step, we divide both the numerator and the denominator by their greatest common factor. Let's explore this concept of greatest common factor a little more before continuing with fractions.

THE GREATEST COMMON FACTOR

Whenever we examine two natural numbers, we can create a set of numbers called their common factors. The greatest of these common factors is called the **greatest common factor** (GCF). We use the notation **GCF(*a*, *b*)** to express the GCF of two natural numbers *a* and *b*.

INVESTIGATION 4.2i Methods for Finding the GCF

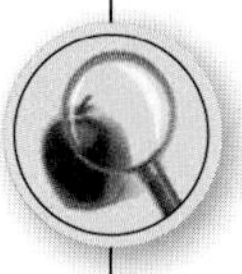

Let us now investigate how we might determine the GCF of two numbers. Rather than give an efficient procedure right away, we will build the foundation of this procedure, much like taking care while constructing the foundation of a house.

Using only the definition of GCF, how would you determine GCF(45, 60)? How does this relate to simplifying the fraction $\frac{45}{60}$? Work on this any way you want. My only recommendation is that you think about the meaning of whatever you do, as opposed to random guess and test. Think and then read on. . . .

DISCUSSION

STRATEGY 1 Use factorization

We could, as we just saw, determine all the factors of each number and then find the largest of the common factors:

Factors of 45 = $\{1, 3, 5, 9, 15, 45\}$
Factors of 60 = $\{1, 2, 3, 4, 5, 6, 10, 12, 15, 20, 30, 60\}$
Common factors = $\{1, 3, 5, 15\}$

We see from this list that 15 is the GCF of 45 and 60. To simplify $\frac{45}{60}$, we could divide both the 45 and 60 by any of these common factors. However, dividing first by the greatest common factor takes the fraction to its simplest form in one step.

STRATEGY 2 Use intuition or number sense

A student who is highly intuitive and has good number sense might just know that 15 divides both these numbers. The fact that 15 divides both numbers simply means that 15 is a common factor. How might you reason that, in fact, 15 is the GCF? Think before reading on. . . .

Let us represent the results of dividing each number by 5 (which we know is not the GCF) and by 15. What do you notice?

$45 = 5 \cdot 9 \qquad 45 = 15 \cdot 3$

$60 = 5 \cdot 12 \qquad 60 = 15 \cdot 4$

When we divide 45 and 60 by 5, we are left with 9 and 12. When we divide 45 and 60 by 15, we are left with 3 and 4. One difference between 9 and 12 and 3 and 4 is that 3 and 4 have no common factors other than 1.

When two numbers have no factors in common other than 1, they are said to be *relatively prime*. Because 3 and 4 are relatively prime, 15 is the GCF of 45 and 60. Do you see why?

LANGUAGE

One of my students came up with an interesting paraphrase of this concept: If two numbers are relatively prime, they are prime to each other.

STRATEGY 3 Repeatedly divide by prime numbers

Another procedure involves an adaptation of the long-division algorithm. The following problem illustrates a systematic application in that we begin with the smallest prime divisor and then move up. That is, we first divide both numbers by 3. At this point, we move up to 5 because 15 and 20 are both divisible by 5. The resulting quotients, 3 and 4, have no factors in common. The GCF of 45 and 60 is the product of their common factors: $3 \cdot 5 = 15$.

$$\begin{array}{r} 3,\ 4 \\ 5\overline{)15,\ 20} \\ 3\overline{)45,\ 60} \end{array}$$

STRATEGY 4 Use prime factorization

This strategy uses a method of breaking a number into its prime factors, called the **prime factorization**. We first determine the prime factorization of each number and then look for common factors. If we look at the prime factorizations of 45 and 60 and circle the factors that the two numbers have in common, we have the following:

$45 = 3 \cdot \boxed{3 \cdot 5}$

$60 = 2 \cdot 2 \cdot \boxed{3 \cdot 5}$

We can further refine this procedure by using exponents:

$45 = 3^2 \cdot 5^1$

$60 = 2^2 \cdot 3^1 \cdot 5^1$

The GCF is determined by examining those factors that both numbers have in common and then taking the *smallest* exponent in each case. The common factors of 45 and 60 are 3 and 5. The smallest exponent of 3 is 1, and the smallest exponent of 5 is 1. Thus $3 \cdot 5$ is the GCF.

Now, we have several ways to find the greatest common factor, which can help us to simplify fractions in one step. When working with a fraction like $\frac{45}{60}$, knowing the greatest common factor helps us to simplify this in one step by dividing by 15 and getting $\frac{3}{4}$.

CLASSROOM CONNECTION

Using Cuisenaire rods, how many different ways can you make a 12 train and an 8 train using the same colors (Figure 4.21)?

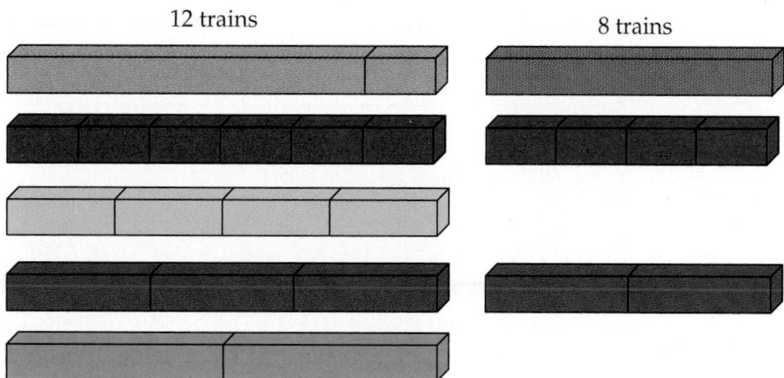

Figure 4.21

Do you see that the answer to this question is the GCF of 8 and 12?

Now that we have explored greatest common factors, which help us to simplify fractions, let's continue our exploration with fractions.

INVESTIGATION 4.2j Sharing Cookies[4]

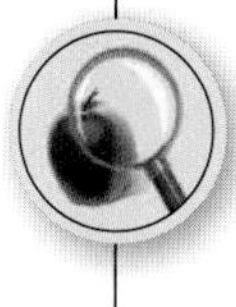

A. Aja came home after school one day and found that her mother had left a plate of cookies. Aja ate $\frac{1}{2}$ of the cookies. When her sister Nolise came home, she ate $\frac{1}{4}$ of the remaining cookies. When their mother came home, there were 3 cookies on the plate. How many did each girl eat?

B. Aja came home after school one day and found that her mother had left a plate of cookies. Aja ate $\frac{1}{2}$ of the cookies. When her sister Nolise came home, she ate $\frac{1}{3}$ of the remaining cookies, and when Clarise came home, she ate $\frac{1}{4}$ of the cookies on the plate. When their mother came home, there were 6 cookies on the plate. How many did each child eat? Work on these problems before reading on . . .

DISCUSSION

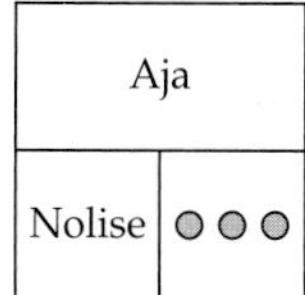

A. The diagram at the left represents the problem and each step. Aja ate $\frac{1}{2}$ of the cookies, and Nolise ate $\frac{1}{2}$ of what was left. If what remains is 3 cookies, that means that 3 cookies represents $\frac{1}{4}$ of what was originally there. Thus, there were 12 cookies, Aja ate 6, and Nolise ate 3.

B. While we could use a similar area model as we used in part A, we will use a bar model to solve this one. First, we choose a bar to represent the entire cookie jar. Then we mark off the $\frac{1}{2}$ that Aja ate. Then we mark off $\frac{1}{3}$ of the remaining to show what Nolise ate. Finally, we mark off the $\frac{1}{4}$ of the remaining that Clarise ate and put the 6 cookies evenly into the remaining three pieces, and we can fill the model from there. The color coding here helps us to communicate the solution.

Aja ate $\frac{1}{2}$ of the cookies. Nolise ate $\frac{1}{3}$ of the remaining cookies, and Clarise ate $\frac{1}{4}$ of the remaining cookies. Then, there were 6 cookies remaining.

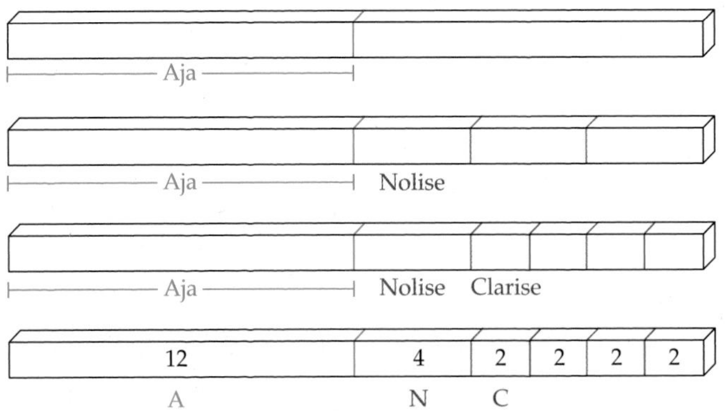

Source: Alice Ho of Math Teach Singapore.

EQUIVALENCE, BENCHMARKS, AND FRACTION SENSE

After children have explored fractions with various manipulatives, a common investigation is to give them pairs or groups of fractions and order them from least to greatest. Children begin by referring to physical models, $\frac{3}{4}$ is clearly greater than $\frac{1}{3}$, and this can be demonstrated with different models.

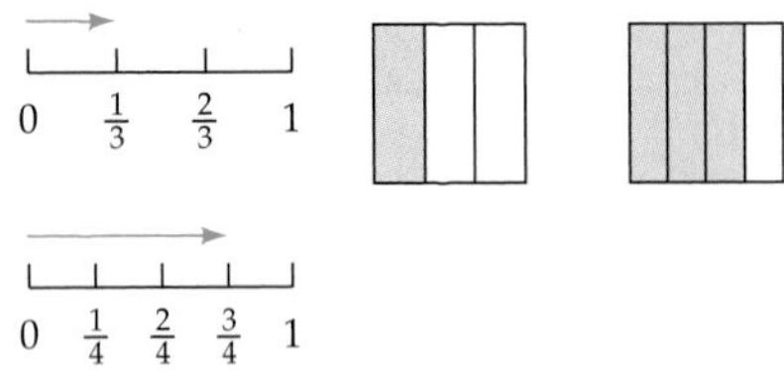

Visuals and manipulatives are powerful tools, but we also want students to develop more tools to navigate through problems and situations involving fractions. One such tool is equivalence. For example, can you explain why $\frac{2}{9} < \frac{1}{4}$ without having to make a picture?

Using equivalence, we can see that $\frac{1}{4} = \frac{2}{8}$. Now how do we know that $\frac{2}{9} < \frac{2}{8}$?

Using our knowledge of the meaning of fractions, we know that 9ths are smaller than 8ths, and since we have the same number of 9ths as 8th, $\frac{2}{9}$ must be less than $\frac{2}{8}$. Children will often use the analogy of pizzas in this case: If you are hungry you'd rather have 2 slices of a pizza that was divided into 8ths than 9ths.

We can use this idea of equivalences to establish benchmarks. Just as powers of 10 serve as benchmarks for whole numbers (1, 10, 100, 1000, 10,000, etc.), unit fractions, $\frac{1}{2}, \frac{1}{3}, \frac{1}{4}$, etc., serve as **benchmarks** to help us keep a sense of the size of fractions. For example, which of these two fractions has a greater value: $\frac{3}{5}$ or $\frac{5}{12}$?

$\frac{3}{5}$ is greater than $\frac{1}{2}$ because $\frac{2\frac{1}{2}}{5} = \frac{1}{2}$ and $\frac{5}{12} < \frac{1}{2}$ because $\frac{6}{12} = \frac{1}{2}$, therefore $\frac{5}{12} < \frac{3}{5}$.

This strategy uses an important mathematical property. The **transitive property** is stated below for any three numbers, a, b, and c.

Transitive property of equality:
If $a = b$ and $b = c$, then $a = c$.

Transitive property of inequality:
If $a < b$ and $b < c$, then $a < c$.

INVESTIGATION 4.2k Ordering Fractions

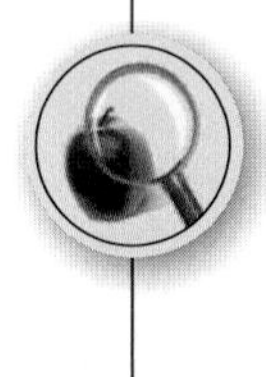

Arrange these fractions from smallest to largest without converting into equivalent fractions, converting into decimals, or drawing diagrams—that is, work by focusing on fraction concepts and reasoning tools. Think and then read on. . . .

$$\frac{3}{4} \quad \frac{2}{5} \quad \frac{5}{6}$$

Explorations Manual 4.10

DISCUSSION

There are a variety of ways in which we can validly answer the question. We will explore several of these ways with the intention of refining or expanding your "fraction sense" toolbox.

Some people start by picking two fractions whose order they know. In this example, let us start that way. We know that $\frac{2}{5} < \frac{3}{4}$. An area diagram would readily show this.

Since one goal of this course is to develop mathematical reasoning, let us examine a tool that goes beyond visual evidence (i.e., "it looks bigger"). This tool uses the fraction $\frac{1}{2}$ as a reference point or benchmark.

We know that $\frac{3}{4}$ is greater than $\frac{1}{2}$ because 3 is more than $\frac{1}{2}$ of 4; similarly, we know that $\frac{2}{5}$ is less than $\frac{1}{2}$ because 2 is less than $\frac{1}{2}$ of 5. Thus we can conclude that $\frac{2}{5} < \frac{3}{4}$.

The next debate concerns $\frac{5}{6}$ and $\frac{3}{4}$. How would you explain which is larger? Think and then read on. . . .

Looking at the two fractions, we see that they are both one piece away from 1. Sixths are smaller than fourths, so we can reason that $\frac{5}{6}$ must be greater than $\frac{3}{4}$ because the distance between $\frac{5}{6}$ and 1 is $\frac{1}{6}$, whereas the distance between $\frac{3}{4}$ and 1 is $\frac{1}{4}$ (Figure 4.22). We can combine $\frac{2}{5} < \frac{3}{4}$ and $\frac{3}{4} < \frac{5}{6}$ to conclude that $\frac{2}{5} < \frac{3}{4} < \frac{5}{6}$.

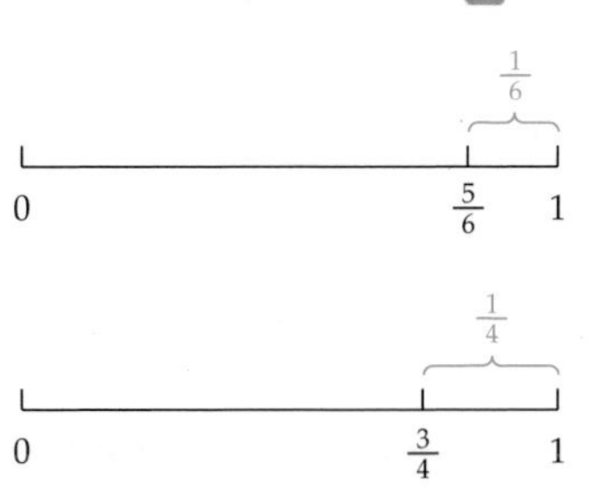

Figure 4.22

MATHEMATICS

One of my favorite examples of students' reasoning with these problems is a student who determined that $\frac{3}{10} > \frac{2}{9}$ by using reasoning by analogy: If a basketball player is 2 for 9 from the free throw line and makes the next free throw, she is now 3 for 10. Thus, $\frac{3}{10} > \frac{2}{9}$.

INVESTIGATION 4.2l Estimating with Fractions

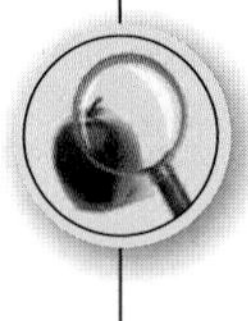

In everyday life, we often see parts of a whole and are more interested in approximations than exact answers. For example, in 2003 the average salary for beginning public school teachers in the United States was $31,073, and the average salary for all public school teachers was $46,597. We can look at these numbers from an additive perspective and say that the beginning teachers' salary is about $15,000 less than the average for all teachers. If we want to turn this into a fraction, we are now reasoning multiplicatively. If we were to do this, we would ask, "Beginning teachers' salaries is about what fraction of the overall average salary?" Can you apply your understanding of fractions (and division!) to find a simple fraction (i.e., not $\frac{31}{47}$) that answers this question?

DISCUSSION

To answer this question accurately, we need to apply three things we have learned:

Fraction can be seen as division.

When estimating division problems we can round both numbers up or round both of them down.

Look for compatible numbers.

$$\frac{31{,}073}{46{,}597} \quad \frac{32{,}000}{48{,}000} \quad \frac{32}{48} \quad \frac{2}{3}$$

Similarly, we could have rounded both down:

$$\frac{31{,}073}{46{,}597} \quad \frac{30{,}000}{45{,}000} \quad \frac{30}{45} \quad \frac{2}{3}$$

THE DENSITY OF THE SET OF FRACTIONS

This activity brings up a question that children sometimes ask: Can we name any point on the number line with a fraction? What do you think?. . .

For example, name a fraction between 0 and 1. If we did this with a whole class, we would get a number of correct responses, although $\frac{1}{2}$ might be the most common.

Can you name another fraction between 0 and $\frac{1}{2}$? How many can you name?

In fact, we can say that between any two fractions, there are an infinite number of fractions. Mathematicians refer to this property by saying that **fractions are dense**. Think of naming any point on a number line, knowing that no matter how close two fractions are, we can find an infinite number of fractions between those two fractions!

SUMMARY

In this section, we have seen that there are four major interpretations of fractions—as measures, as quotients, as operators, and as ratios.

You have learned that a fraction is not simply a number; rather, a fraction expresses a relationship between two quantities. The numerator and denominator can be seen as a code that tells us the relative size of the fraction.

In order to work effectively with fractions as measures, we need a strong understanding of several basic ideas:

- We are dealing with part–whole relationships.
- Something is to be partitioned into parts.
- All of the parts must have the same value or size.
- We need to take care not to confuse the unit and the whole, which are the same in some situations but not in others.
- We can use various models to represent fractions and fraction situations. In this book, we emphasize number line, area, and set models.

We then applied these ideas in order to understand more fully the notions of equivalent fractions and simplifying fractions. A deeper understanding of fractions enables us to sense their relative size; this fraction sense is useful in estimating and problem-solving.

4.2 Exercises

1. Represent $\frac{2}{3}$ in at least five different ways, some numerical and some pictorial.

2. Draw three diagrams to represent each of the following fractions. Justify your diagrams.

	Length	Area	Discrete
$\frac{3}{5}$	0 1		
$1\frac{2}{3}$			
$\frac{9}{4}$			

3. The squares below represent a student's solution to 2 brownies shared by 5 people. Each person gets two pieces. What is the value of each piece? Briefly justify your answer.

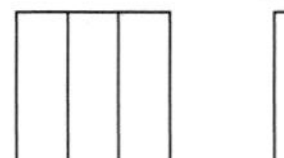

4. Determine the fraction of the thermometer that is shaded.

5. a. 9 inches is what part of a foot?
   b. 9 inches is what part of a yard?
   c. 1 dime is what part of a quarter?
   d. 1 dime is what part of a dollar?
   e. 8 ounces is what part of a quart?
   f. 8 ounces is what part of a gallon?

6. a. Shade $\frac{2}{3}$ of the circles in the first set of circles below.
   b. Shade $\frac{3}{4}$ of the circles in the second set of circles below.
   c. Shade $\frac{5}{6}$ of the circles in the third set of circles below.
   d. Shade $\frac{3}{8}$ of the circles in the fourth set of circles below.

7. a. Name three fractions that are equivalent to $\frac{1}{5}$.
   b. Name three fractions that are equivalent to $\frac{3}{4}$.
   c. Name three fractions that are equivalent to $\frac{2}{3}$.
   d. Name three fractions that are equivalent to $\frac{5}{6}$.

8. Find the value of x.
   a. $\frac{3}{4} = \frac{x}{8}$ b. $\frac{3}{5} = \frac{x}{15}$ c. $\frac{2}{3} = \frac{x}{24}$
   d. $\frac{5}{8} = \frac{x}{32}$ e. $\frac{5}{6} = \frac{x}{24}$ f. $\frac{5}{12} = \frac{x}{60}$

9. Simplify the following fractions.
   a. $\frac{25}{40}$ b. $\frac{32}{48}$ c. $\frac{54}{60}$ d. $\frac{26}{65}$
   e. $\frac{168}{216}$ f. $\frac{84}{132}$ g. $\frac{493}{510}$ h. $\frac{101{,}010}{505{,}050}$

10. Determine what fraction of the figure is shaded.
    a.
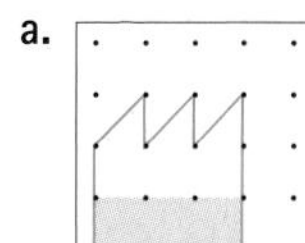
    b.
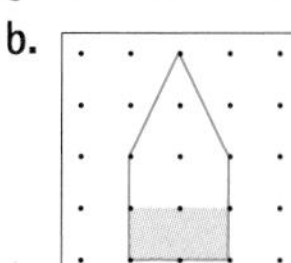

11. ***Classroom Connection*** Below are two problems made by students to show fractional parts of a unit. Determine what fraction of the whole region each region represents. In (c), make up your own problem.

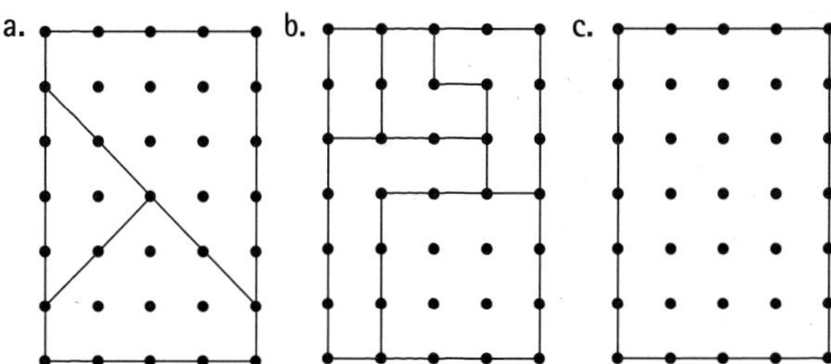

Source: From *Standards-Based School Mathematics Curricula: What Are They? What Do Students Learn?*, edited by Sharon L. Senk, Denise R. Thompson, Erlbaum, 2003, p. 111.

12. Locate $\frac{3}{4}$.
    0 $\frac{1}{2}$

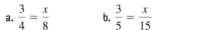

13. Determine the value of x.
    0 $\frac{2}{5}$ x

14. If this set of pattern blocks has a value of 1, shade in $\frac{3}{4}$.

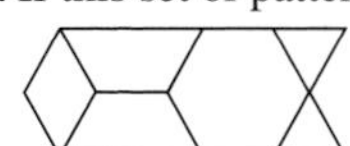

15. In each of the questions below, justify your answer.
    a. If [two-square rectangle] has a value of $\frac{4}{5}$, draw and shade in the amount that would have a value of 1. Justify your answer.
    b. If [9 dots] $= \frac{3}{4}$, show 1.
    c. If [10 dots] has a value of $\frac{5}{8}$, how many dots have a value of 1?

16. Alex came home after school one day and found that his mother had left a plate of cookies. Alex ate $\frac{1}{4}$ of the cookies. When his sister Bernice came home, she ate $\frac{1}{3}$ of the remaining cookies, and when Carla came home, she ate $\frac{1}{2}$ of the cookies on the plate. When their mother came home, there were 2 cookies on the plate. How many did each child eat?

17. Determine which of the fractions below is larger using the kind of reasoning developed in this section, including benchmarks, that is, without converting to decimals, cross multiplying, or using pictures. Justify your reasoning.

a. $\frac{3}{7}$ $\frac{5}{8}$
b. $\frac{5}{6}$ $\frac{9}{10}$
c. $\frac{2}{7}$ $\frac{4}{11}$
d. $\frac{7}{9}$ $\frac{15}{17}$
e. $\frac{2}{9}$ $\frac{5}{16}$
f. $\frac{3}{4}$ $\frac{79}{100}$

18. In 1950 the annual per capita consumption of eggs was 395; that is, the average person ate 395 eggs per year. In 2004 the per capita consumption had dropped to 256. Fill in the blank with a fraction estimate: In 2004 the average person eats ____ as many eggs as the average person did in 1950.

19. In 2006 the average salary for beginning public school teachers was $35,284, and the average salary for all public school teachers was $51,009. The average salary for beginning public school teachers was approximately what fraction of the average salary for all public school teachers?

20. a. Name a fraction between $\frac{2}{3}$ and $\frac{3}{4}$.
b. Name a fraction between $\frac{1}{4}$ and $\frac{1}{5}$.
c. Name a fraction between $\frac{7}{8}$ and 1.
d. Name a fraction between $\frac{1}{4}$ and $\frac{1}{2}$ that has a denominator of 11.

21. The thermometers at the right represent the fraction of the total goal raised in fundraising drives.
a. What fraction is represented in the thermometer (a) at the right?
b. What fraction is represented in the thermometer (b)?
c. If the total is $158,000 and $37,000 has been raised, which of the three thermometers labeled (c) most accurately represents the progress of the fundraising effort?

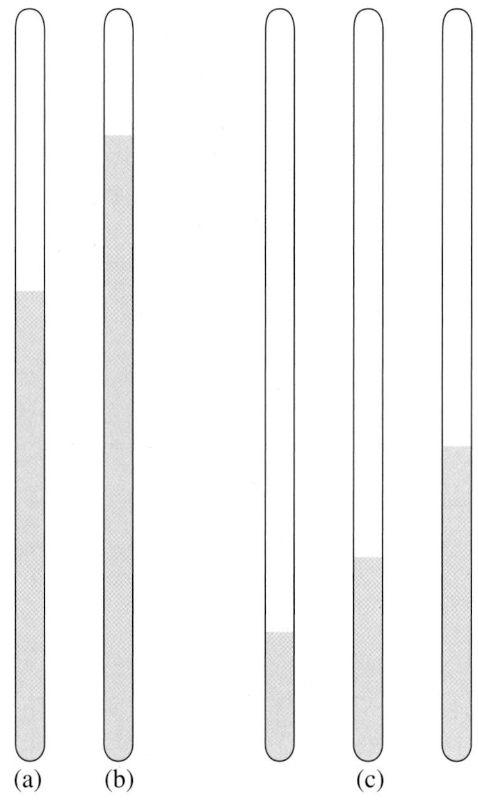

DEEPENING YOUR UNDERSTANDING

22. If each of the hexagons has a value of 1, does each small piece in each hexagon represent $\frac{1}{6}$? Why or why not?

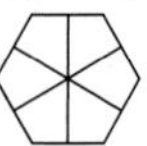

23. Write a fraction to represent the shaded portion. Justify your response.

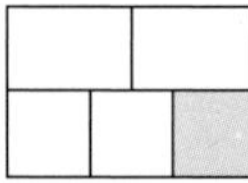

24. Jamila was solving the problem of dividing a rectangular cake into eighths. She drew the diagram at the right. Although this is an unorthodox answer, do you agree with her that it is a mathematically correct solution? If so, explain why. If not, explain what you think is wrong about this.

25. ***Classroom Connection*** I recall a teacher asking for fractions that were equivalent to $\frac{3}{4}$ and one student replying $\frac{1\frac{1}{2}}{2}$. How would you respond to this answer? Justify your response.

26. Determine the value of x.

a.
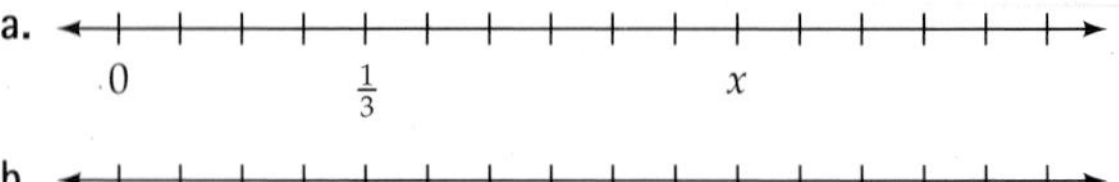

b.
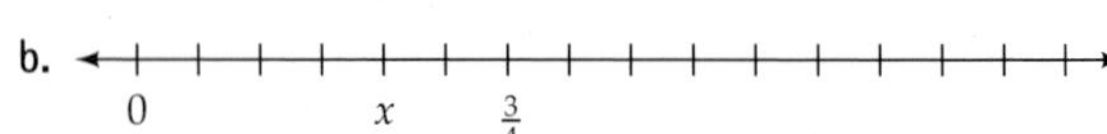

27. a. Locate $\frac{5}{12}$.

0 $\frac{1}{6}$

b. Locate $\frac{5}{6}$.

0 $\frac{1}{4}$

28. a. If [figure] $= 1\frac{1}{2}$, then shade the amount equal to 1.

b. If the figure below has a value of $2\frac{1}{3}$, how many squares has a value of 1?

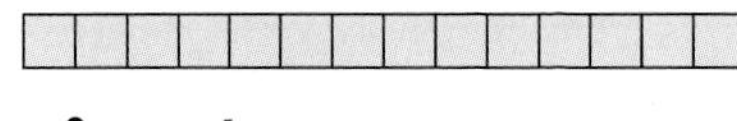

c. If [dots] $= \frac{5}{4}$, show 1.

d. If [figure] $= 1\frac{1}{3}$, then shade in the amount equal to 1. Justify your answer.

29. Below are pictures of gas gauges in a car. What fractions of gas in the tank are remaining? Rather than eyeballing the amount, use partitioning to estimate an actual fraction.

a. **b.**

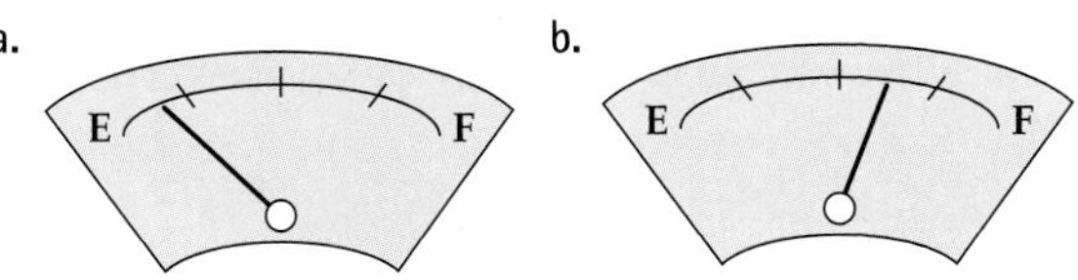

30. Determine what fraction of the figure is shaded.

a.

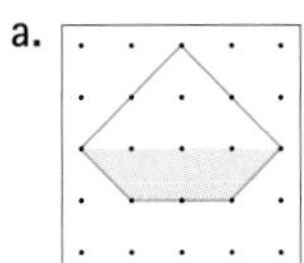

b.

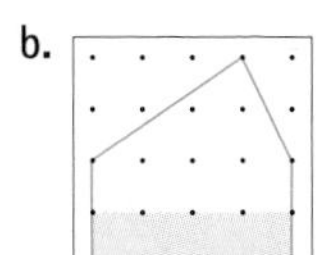

c.

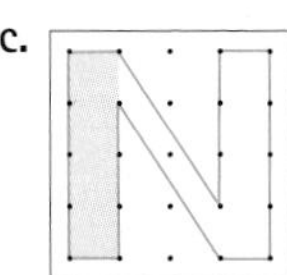

31. Is $\frac{10}{13}$ closer to $\frac{1}{2}$ or to 1? Justify your choice.

32. Let's say that Amy made 15 out of 21 free throws and Jane made 8 out of 12. Which student had a better rate of success? Explain your answer.

33. Arrange the following fractions in order from least in value to greatest in value without converting to decimals or finding the equivalent fractions. Explain your reasoning.

a. $\frac{3}{4}$ $\frac{4}{5}$ $\frac{5}{15}$ **b.** $\frac{6}{7}$ $\frac{5}{11}$ $\frac{2}{3}$

c. $\frac{1}{3}$ $\frac{2}{5}$ $\frac{5}{8}$ $\frac{3}{4}$ $\frac{3}{50}$ **d.** $\frac{2}{5}$ $\frac{5}{6}$ $\frac{4}{7}$ $\frac{7}{8}$ $\frac{3}{10}$

34. Below are the median prices of homes sold in 2009, by region of the country:

South	158,300
Northeast	248,800
Midwest	139,500
West	243,200
U.S.	180,100

The median selling price of a home in the South was approximately what fraction of the median selling price of a home in the West?

35. The data below show the area (in square miles) of several states and some of the Great Lakes:

Lake Superior	31,800
Lake Huron	23,010
All five Great Lakes	94,710
New Hampshire	9,279
California	158,706
Tennessee	42,104
Rhode Island	1,212

For each of the questions below, using mental arithmetic and estimation, find a fraction with a small, "friendly" denominator (2, 3, 4, 5, 6, 8, 10, or 12) that most closely approximates the actual fraction. Justify your reasoning.

a. Lake Huron is approximately what fraction of the size of Lake Superior?

b. New Hampshire is approximately what fraction of the size of Lake Superior?

c. Lake Superior is approximately what fraction of the size of Tennessee?

d. The area of all five Great Lakes is approximately what fraction of the size of California?

36. You have from 10 p.m. to 11:30 p.m. to do a project.

a. At 11, what fraction of the time remains?

b. At 11:20, what fraction of the time remains?

37. You are driving to Memphis, 115 miles away. You have 24 miles to go. Approximately what fraction of the trip is left? Justify your choice of fraction.

38. Phones-R-Us did a survey to determine telephone usage at a college. On the day the survey was done, a total of 5,243 calls were made.

a. If 1,371 calls were made to directory assistance, approximately what fraction of the total calls were made to directory assistance? Justify your choice.

b. The three pizza restaurants that delivered pizza received 737 calls. Approximately what fraction of the total calls were for pizza? Justify your choice.

39. Two elementary school principals are comparing attendance figures for their schools.

School	Total enrollment	Absent
Wheelock	264	32
Fuller	402	58

Which school had the larger fraction of students absent? Justify your answer.

40. In 2014 there were approximately 7,200,000,000 people in the world. Using mental math and estimation, find a fraction to represent the fraction of the world's population that belonged to each of the following religions. Justify your reasoning.

Christianity	2,173,180,000
Islam	1,598,510,000
Hinduism	1,033,080,000
Buddhism	487,540,000
Judaism	13,850,000

41. The music teacher would like to have the same number of girls and of boys in the chorus. She finds that $\frac{5}{8}$ of the chorus are girls but that if she can get 12 more boys, the chorus will have the same number of boys as of girls.
   a. How many students are in the chorus?
   b. If there are 216 children in the school, which unit fraction would you select to represent the fraction of students in the present chorus?

42. Darcy loves tomatoes, but she lives in Minnesota, where the growing season is shorter than in most of the country. Therefore, she begins her tomato plants inside. However, she tells her friend that there were some casualties this spring. One-third of the plants were destroyed by her cat. Furthermore, $\frac{1}{4}$ of the remaining plants were destroyed by a disease. What fraction survived?

43. A few years ago, my son announced that he was $\frac{1}{4}$ of my age. In how many years will he be $\frac{1}{3}$ of my age? $\frac{1}{2}$ of my age?
   a. Find one solution.
   b. How many solutions are there?
   c. Add one piece of information so that there is only one answer.

44. In the following exercise, $0 < a < b < c$. For the following pairs, tell which is larger and explain why, or explain why you cannot be sure which is larger. Try to answer the questions using only reasoning rather than plugging in numbers. You may then substitute actual numbers to check your reasoning.
   a. $\frac{a}{b}$ or $\frac{a}{c}$ b. $\frac{a}{b}$ or $\frac{b}{c}$ c. $\frac{a}{c}$ or $\frac{b}{c}$

45. If $\frac{a}{b} = \frac{3}{4}$, will the value of $\frac{a+x}{b+x}$ be less than, equal to, or greater than $\frac{3}{4}$? Justify your answer by a means other than plugging in values for x.

46. If the numerator and the denominator of a proper fraction are increased by the same amount, is the new fraction greater than, equal to, or less than the original fraction? Justify your answer. Read this problem carefully; it is not a trick problem, but you need to make sure you interpret the wording correctly.

47. Draw a set model, a linear model, and an area model to show that $\frac{6}{12}$ is equivalent to $\frac{1}{2}$.

48. A student in your class says that $3\frac{3}{4}$ is equivalent to $\frac{13}{4}$. She explained that she followed the process of multiplying 3 times 3 and adding the 4 to get 13. What mathematical misunderstanding does this student have? What model could help her understand the correct process?

49. A student says that $\frac{3}{4}$ is equivalent to $\frac{3}{5}$ and uses the following model to explain his reasoning. What mathematical misunderstanding does this student have? What model could help the student understand the correct process?

50. Find the following:
   a. GCF(30, 75) b. GCF(45, 54)
   c. GCF(12, 35) d. GCF(27, 189)
   e. GCF(75, 144) f. GCF(105, 132)
   g. GCF(156, 910) h. GCF(630, 1848)

51. Find the following:
   a. GCF(12, 30, 75) b. GCF(12, 333, 8415)

52. a. Find two numbers such that their GCF = 2.
   b. Find two numbers, both greater than 100, such that their GCF = 2.
   c. Find two numbers such that their GCF = 6.
   d. Find two numbers, both greater than 100, such that their GCF = 6.

FROM STANDARDIZED ASSESSMENTS

2008 NECAP, Grade 6

53. If Shape R represents $\frac{1}{6}$, what number does Shape Z represent? Show your work or explain how you know.

Shape R	Shape Z

2006 NECAP, Grade 7

54. You may use the number line below to answer this question.

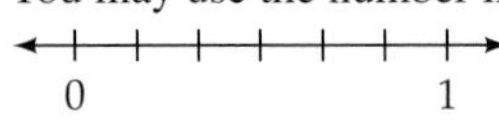

Which fractions are in order from least to greatest?

a. $\frac{1}{2}, \frac{2}{3}, \frac{2}{6}$ b. $\frac{1}{2}, \frac{2}{6}, \frac{2}{3}$ c. $\frac{2}{6}, \frac{2}{3}, \frac{1}{2}$ d. $\frac{2}{6}, \frac{1}{2}, \frac{2}{3}$

55. In which of the following are the three fractions arranged from least to greatest?

a. $\frac{2}{7}, \frac{1}{2}, \frac{5}{9}$ b. $\frac{1}{2}, \frac{2}{7}, \frac{5}{9}$ c. $\frac{1}{2}, \frac{5}{9}, \frac{2}{7}$ d. $\frac{5}{9}, \frac{2}{7}, \frac{1}{2}$

Source: *NAEP Mathematics Assessment*, 2007. U.S. Department of Education, National Center for Education Statistics.

2007 NECAP, Grade 5

56. Which pair of figures has the same fractional area shaded blue?

a. 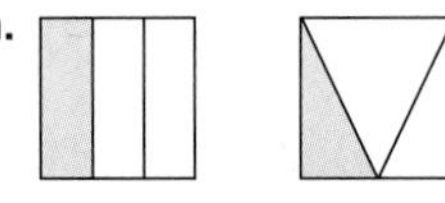 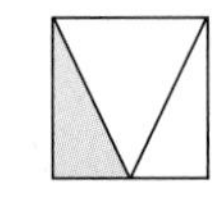

b. 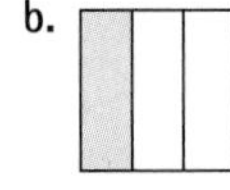 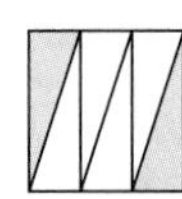

c. 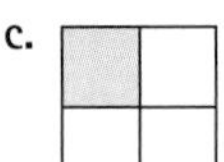 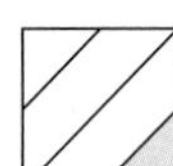

d. 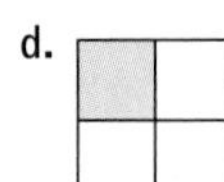 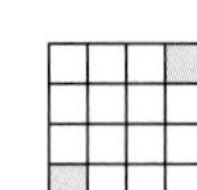

2008 NECAP, Grade 6

57. Nicole measured the height of a tomato plant at the end of each week. Her measurements are labeled on the number line below.

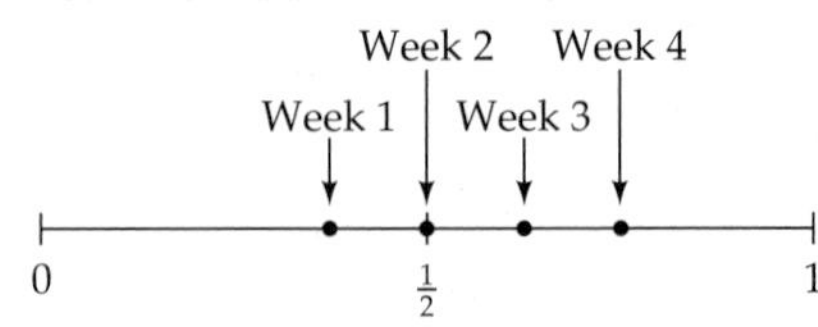

At the end of which week was the height of the tomato plant about $\frac{5}{8}$ yard?

a. Week 1

b. Week 2

c. Week 3

d. Week 4

Source: From *Smarter Balanced Assessment Consortium* (developed for Common Core State Standards).

58. Five friends ordered 3 large sandwiches.

James ate $\frac{3}{4}$ of a sandwich.

Katya ate $\frac{1}{4}$ of a sandwich.

Ramon ate $\frac{3}{4}$ of a sandwich.

Sienna ate $\frac{2}{4}$ of a sandwich.

How much sandwich is left for Oscar?

SECTION

Understanding Operations with Fractions

What do you think?

- Why *do* we need a common denominator to add (or subtract) two fractions?
- Why *don't* we need a common denominator to multiply (or divide) two fractions?
- If whole-number multiplication "makes bigger," why does fraction multiplication sometimes "make smaller"?
- Why isn't $3\frac{1}{4} \cdot 2\frac{1}{3} = 6\frac{1}{12}$?

One of the sentences in the NCTM standards that has burned in my head is that students should "believe that mathematics makes sense." The Common Core State Standards place adding and subtracting fractions with the same denominators and multiplying a fraction times a whole number in fourth grade. In fifth grade this is extended to addition and subtraction of fractions with unlike denominators and multiplying fractions and dividing whole numbers by fractions. There are many algorithms that enable us to manipulate fractions quickly: algorithms for adding, subtracting, multiplying, and dividing, and algorithms for translating improper fractions into mixed numbers and vice versa. It is simply not enough for an elementary teacher to know how to compute. It is crucial that the teacher also know the *whys* behind the *hows* and can use a variety of representations to explain them.

ADDITION OF FRACTIONS

Explorations Manual 4.11

Addition of fractions is one of those situations that mathematics is famous for—the procedure goes against what most people's common sense tells them to do. When we add fractions, we do not add the numerators and the denominators; on the other hand, when we multiply fractions, we *do* multiply the numerators and the denominators.

$$\frac{1}{2} + \frac{1}{3} \neq \frac{2}{5} \quad \text{but} \quad \frac{2}{3} \times \frac{4}{5} = \frac{8}{15}$$

INVESTIGATION  **4.3a**

Using Fraction Models to Understand Addition of Fractions

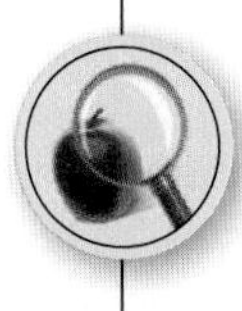

Using the area model, the length model, or the set model, determine the sum of $\frac{1}{2} + \frac{1}{3}$ and then explain *why* we need to find a common denominator in order to add fractions. Then read on. . . .

DISCUSSION

This is an excellent place to demonstrate the role of manipulatives.

Using pattern blocks How might you use pattern blocks to add $\frac{1}{2}$ and $\frac{1}{3}$ (see Figure 4.23)?

Figure 4.23

If we let the hexagon represent 1, then the trapezoid is $\frac{1}{2}$, the parallelogram is $\frac{1}{3}$, and the triangle is $\frac{1}{6}$. Using the notion of addition as combining, we can combine the two and we have $\frac{1}{2} + \frac{1}{3} =$  .

We could be humorous and say that parallelogram + trapezoid = baby carriage! The key question, though, is to determine the value of this amount. The solution lies in realizing that to name an amount with a fraction, we must have equal-size parts. If you are familiar with pattern blocks, you realize that an equivalent representation is to cover this amount with 5 triangles, as in Figure 4.24.

Figure 4.24

Because the value of each triangle is $\frac{1}{6}$, we can now say that $\frac{1}{2} + \frac{1}{3} = \frac{5}{6}$.

This model also helps us see the process for getting the common denominator. Because 3 triangles are the same size as 1 trapezoid, we can see that $\frac{1}{2} = \frac{3}{6}$. How does this model also show that $\frac{1}{3} = \frac{2}{6}$?

Using Cuisenaire rods This is a different question when using Cuisenaire rods, because there is no "natural" choice for a unit. Thus you have to think of a color for which you can represent $\frac{1}{2}$ of that color *and* $\frac{1}{3}$ of that color. One of many different solutions is to choose the dark green rod to have a value of 1. The red rod now has a value of $\frac{1}{3}$, and the green rod has a value of $\frac{1}{2}$ [Figure 4.25(a)]. As with pattern blocks, when we combine these two parts, in order to name the amount we have to find a way to represent this length with equal-size pieces, and 5 white rods have the same length. Because 6 white rods have the same length as the dark green rod, each of the white rods has a value of $\frac{1}{6}$, and therefore 5 of them have a value of $\frac{5}{6}$ [Figure 4.25(b)]. How does this model help us see the common denominator process?

Explorations Manual 4.11

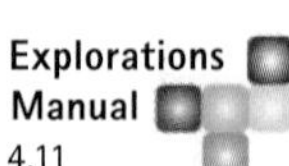

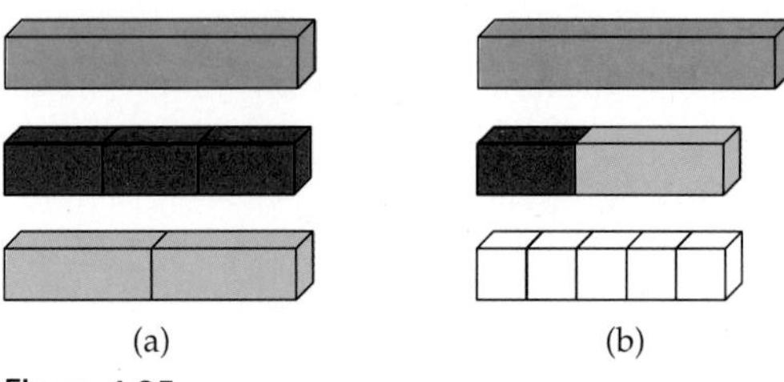

(a) (b)

Figure 4.25

Looking back An important outcome of working with manipulatives is for the students to see that, for example, the yellow pattern block is not 6 but, rather, can be assigned any value, and the value of the other pieces is determined by their relationship to this block. So too with the Cuisenaire rods. We need to define what the unit is in order to know the value of the other pieces.

LEAST COMMON MULTIPLE AND LOWEST COMMON DENOMINATOR

Let us step back for a moment and explore the concept of least common multiple (LCM). What do you think "least common multiple" means and how does it relate to the "lowest common denominator"?

Let us examine this concept word by word, using a specific example. Suppose we wanted to find the least common multiple of 8 and 12. At the most basic level, we can start listing multiples of 8 and 12 until we find the first multiple they have in common.

Multiples of 8 = {8, 16, 24, 32, 40, 48, 56, 64, 72, 80, 88, 96, 104, ...}

Multiples of 12 = {12, 24, 36, 48, 60, 72, 84, 96, 108, 120, 132, ...}

What multiples do the two numbers have in common?

These two numbers have many common multiples: {24, 48, 72, 96, ...}. Because the least of the common multiples is 24, we say that LCM(8, 12) = 24. This helps us to add fractions with denominators of 8 and 12 because the least common multiple, 24, is also the **lowest common denominator**.

STRATEGIES FOR FINDING THE LCM

Now let us examine how we might find the actual LCM of 18 and 40. Again, there are many ways to determine the LCM of two numbers. We will examine several ways.

One way of determining the LCM of 18 and 40 is to construct the LCM by beginning with one of the numbers and applying our understanding of LCM. This process is illustrated in the following discussion:

Reasoning	**The work**
The LCM *must* contain all the factors of 18.	$18 = 2 \cdot 3 \cdot 3$, LCM(18, 40) must contain: $2 \cdot 3 \cdot 3$
Now, in order *also* to be a multiple of 40, the LCM will have to contain all the factors in the prime factorization of 40.	$40 = 2 \cdot 2 \cdot 2 \cdot 5$
Looking now at the factors of 18, what factors of 40 are we missing?	
We need to put two more 2s and one 5 into our prime factorization of LCM(18, 40).	LCM(18, 40) must contain: $\mathbf{2} \cdot \mathbf{2} \cdot 2 \cdot 3 \cdot 3 \cdot \mathbf{5}$ That is, LCM(18, 40) = 360.
Another way of illustrating this process is to note the prime factorizations of 18 and 40 and realize that the least common multiple must contain all the factors in either number with no redundancies. That is, the LCM needs to contain three 2s, two 3s, and one 5.	**2 · 3 · 3** **2 · 2 · 2 · · 5**

USING PRIME FACTORIZATION AND EXPONENTS TO FIND THE LCM

There is a more formal way to find the LCM, and this is connected to one of the ways in which we found the greatest common factor in Section 4.2. This method comes from representing the prime factorization of each number in exponential form:

$$18 = 2 \cdot 3 \cdot 3 = 2^1 \cdot 3^2$$
$$40 = 2 \cdot 2 \cdot 2 \cdot 5 = 2^3 \cdot 5^1$$

When finding the GCF, we took the smaller exponent of all common factors. What do you think we will do when finding the LCM? Think and read on. . . .

In order for a number to be the LCM, it must contain *all* the factors in either number. For example, because the prime factorization of 18 contains a 2, the LCM must contain a 2. However, because the prime factorization of 40 contains three 2s, the LCM must contain three 2s. Thus, when we examine the prime factorization of each number, whenever there is a common factor, we must take the greater exponent.

Using this method, we find that LCM(18, 40) $= 2^3 \cdot 3^2 \cdot 5^1$. Do you see why? 

The only factor that 18 and 40 have in common is 2, and the greatest exponent above 2 is 3 (meaning that $2 \cdot 2 \cdot 2$ is a factor of 40). Therefore, the prime factorization of the LCM must contain 2^3. The noncommon prime factors are 3 and 5, so 3^2 and 5^1 are also placed in the prime factorization of the LCM.

CLASSROOM CONNECTION

Below are two different ways in which the concept of LCM can emerge in elementary school. Teachers may give their students a 100 chart and ask them to draw a circle around multiples of 8 and a square around multiples of 12, as in Table 4.1. When students discover that some numbers have both a circle and a square around them, the teacher can ask, "How would you describe those numbers?" In this manner, the students have constructed the concept of LCM. Then, when the more formal definition is presented, they already have experience that connects to this idea.

TABLE 4.1

1	2	3	4	5	6	7	(8)	9	10
11	[12]	13	14	15	(16)	17	18	19	20
21	22	23	[(24)]	25	26	27	28	29	30
31	(32)	33	34	35	[36]	37	38	39	(40)
41	42	43	44	45	46	47	[(48)]	49	50

Cuisenaire rods can also be used to introduce this concept to children. The teacher might ask the students to select a 4 rod and a 6 rod and then ask, "How many different ways can you make a 4 train and a 6 train that have the same length?" (Figure 4.26). Do you see how the answer to this question connects to the LCM concept?

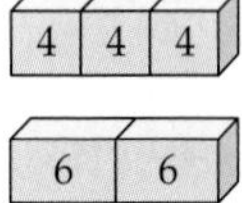

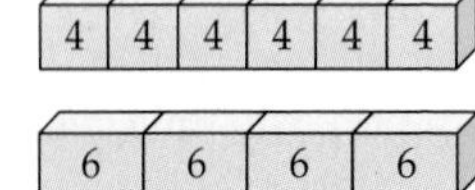

Figure 4.26

Being able to think about multiples with models helps students develop the ability to find common denominators more easily.

CONNECTING CONCEPTS TO A PROCEDURE FOR ADDING FRACTIONS

Let us now connect this work to the general procedure for adding fractions. Look at the addition problem below. Try to explain this procedure meaningfully—that is, to *justify* each step. Try using an area model, a linear model, and a set model to help. Do this before reading on. . . .

$$\frac{2}{3} + \frac{1}{4} = \frac{8}{12} + \frac{3}{12} = \frac{11}{12}$$

A variety of models are helpful in deepening understanding, as well as being able to connect the models to the algorithm. Let's work through each of the three types of models (area, linear, and set) to represent $\frac{2}{3} + \frac{1}{4}$.

Area model

Let's use the area of a circle to model this. Look at these pictures and verify that they represent $\frac{2}{3} + \frac{1}{4}$.

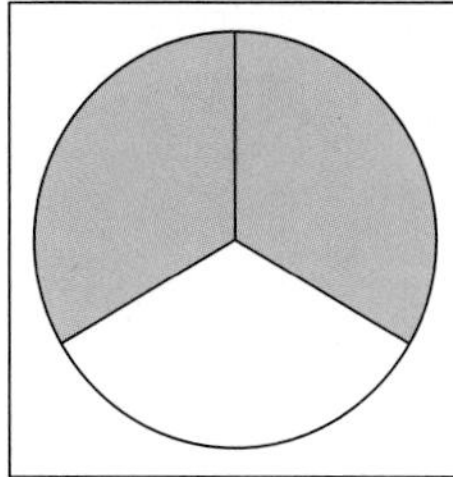
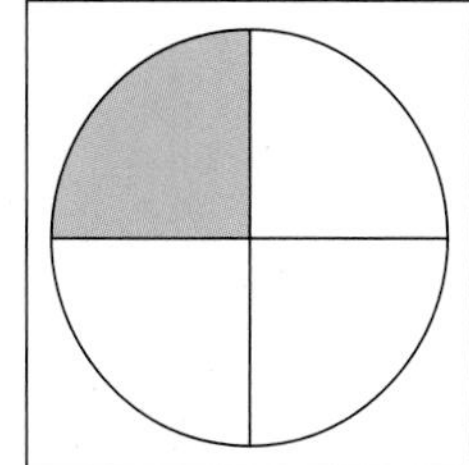

Because the pieces are not the same size, we need to make them the same size. Young students can figure this out by placing pieces on top of the thirds and fourths and learning that the twelfths will fit on both of them. We can also use the fact that we know the LCM(3, 4) = 12.

To turn thirds into twelfths, we cover each of the thirds with four of the twelfths, so $\frac{2}{3} = \frac{8}{12}$. Similarly, $\frac{1}{4} = \frac{3}{12}$ as shown below.

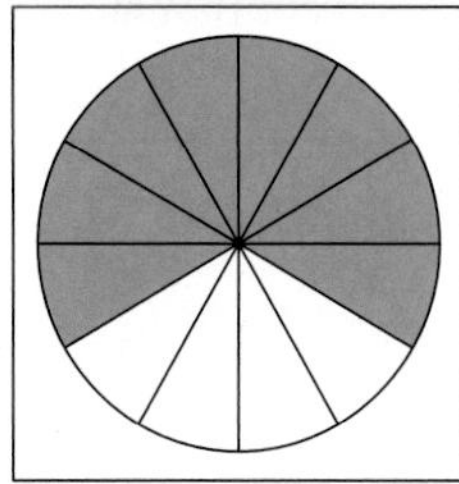
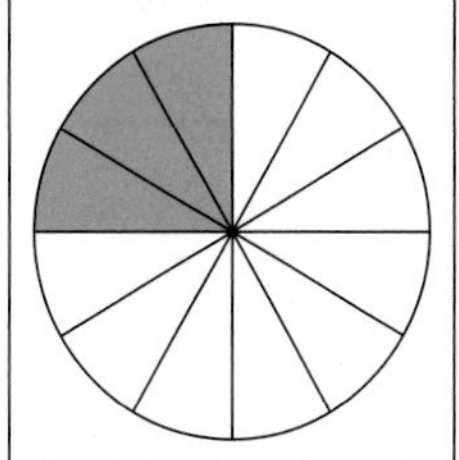

Now we are adding the same size pieces and get $\frac{11}{12}$.

Linear Bar Model

Examine this model and then read on.

We begin with one bar representing the unit, create thirds, and mark two of the thirds. Then we draw a bar of the same size and mark $\frac{1}{4}$. Again, we have to create the same length bars to be able to add them, so we show that $\frac{2}{3} = \frac{8}{12}$ and $\frac{1}{4} = \frac{3}{12}$. The third bar puts these pieces together to show the answer of $\frac{11}{12}$.

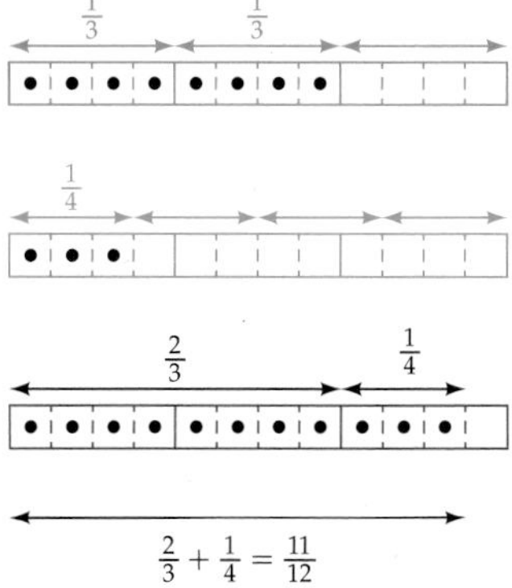

Set Model

Adding fractions with a set model requires that we think ahead to the common denominator to choose how many to put in our set that equals a unit. Since we know that the LCM of 4 and 3 is 12, we know to draw 12 dots in a 3 by 4 array (since the 3 and 4 are the denominators of the respective fractions). We shade two rows of the first array to show $\frac{2}{3}$ and then we shade 1 column of the second array to show $\frac{1}{4}$. Then, $\frac{2}{3}$ and $\frac{1}{4}$ will look like:

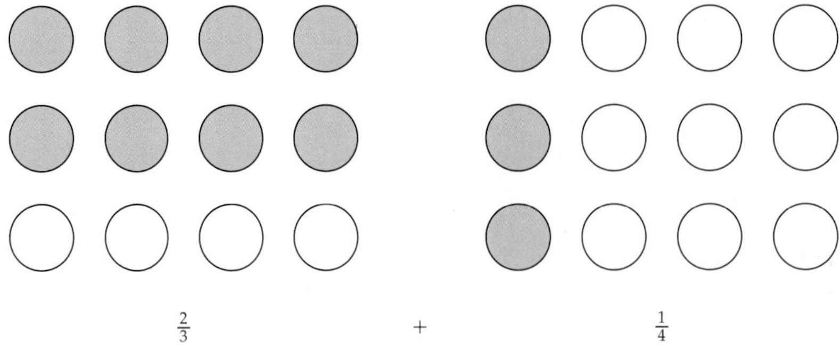

Putting the shaded objects together, we have 11; since it takes 12 to make 1 unit, this is $\frac{11}{12}$.

The procedure is much easier for students to understand and remember when they have these visual models to give meaning to the procedures. The general procedure used in all of these models is that we determine equivalent fractions by finding a common denominator so that all of the parts are the same size. Then we add the pieces.

How is subtraction of fractions different from addition? Look back at each of these models and think about how it would be different if the problem were $\frac{2}{3} - \frac{1}{4}$?

Because we still need to make the pieces the same size, we need a common denominator. Therefore, the process of finding a common denominator is the same for both addition and subtraction.

MIXED NUMBERS AND IMPROPER FRACTIONS

Up to this point, all of the addition problems have involved only **proper fractions**—that is, fractions whose value is between 0 and 1. In many real-life situations, we encounter mixed numbers (for example, $5\frac{1}{4}$) and improper fractions (for example, $\frac{21}{4}$). Write down your own definitions of these two terms and then read on. . . .

An **improper fraction** is one in which the numerator is at least as large as the denominator.

A **mixed number** is a number that has a whole-number component and a proper fraction component.

A National Assessment of Educational Progress contained the following item:

$5\frac{1}{4}$ is the same as:

(a) $5 + \frac{1}{4}$ (b) $5 - \frac{1}{4}$ (c) $5 \times \frac{1}{4}$ (d) $5 - \frac{1}{4}$ (e) I don't know

Only 47 percent of the seventh-graders chose the correct response, (a). Still more startling, an even smaller percentage of eleventh-graders chose the correct response—only 44 percent. This lack of connectedness between concepts and procedures is one of the reasons why students do so much worse on algebraic equations with fractions than on algebraic equations with whole numbers. In this case, the problem is not the algebra so much as it is the arithmetic!

INVESTIGATION 4.3b Connecting Improper Fractions and Mixed Numbers

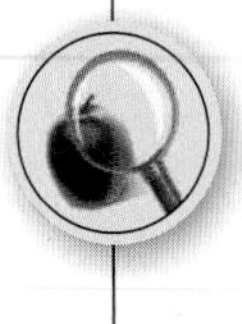

Try drawing an area model, a linear model, and a set model to explain why $3\frac{1}{4} = \frac{13}{4}$.

Most of you remember the procedure for converting a mixed number into an improper fraction. For example, to convert $3\frac{1}{4}$ into an improper fraction, we do the following:

$$3\frac{1}{4} = \frac{3 \cdot 4 + 1}{4}$$

We multiply the *whole number* by the *denominator* and then we add the *numerator,* and then we put this number on top of the *original denominator*.

How would you explain the why of this procedure to, let's say, a fourth-grade student? How can the models you drew help explain this procedure?

DISCUSSION

Examine the area, linear, and set models shown below. How are these similar and different from the ones you drew?

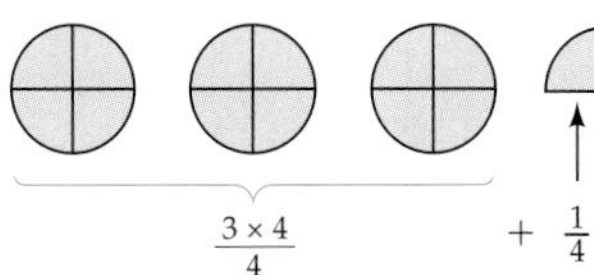

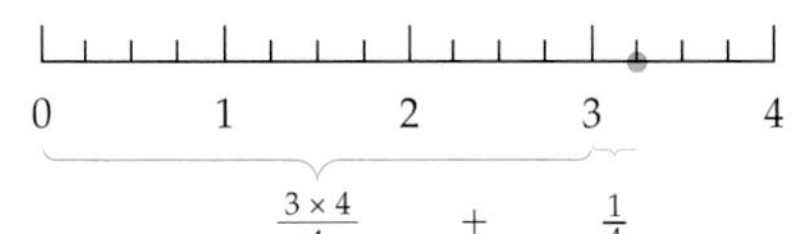

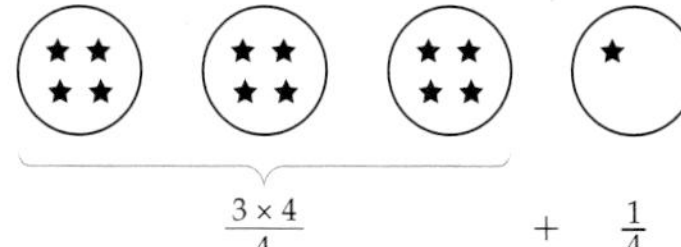

CLASSROOM CONNECTION

Randy Phillipp and Candace Cabral have made some outstanding video clips that are available from Pearson Publishing under the title "IMAP: Integrating Mathematics and Pedagogy to Illustrate Children's Reasoning." In one tape, Rachel, a fifth-grade girl, is struggling to convert $3\frac{3}{8}$ to $\frac{27}{8}$. Her first answer is $1\frac{7}{8}$ because she multiplied $3 \times 3 + 8$. During the discussion with the interviewer, she says, "We did this before but I don't remember it as well because I didn't figure it out for myself. . . . When I figure [it] out it's easier; it stays there because I was the one who brought it there." And that's why this book has an *Explorations* volume and why the investigations ask you to stop and think on your own. When you are the one who "brings it there," then it does "stay there." Rachel so nicely explains the difference between "renting" and "owning"!

In each of the representations, we see that the process is similar to the kinds of regrouping we did with whole numbers. In converting from the mixed number to the improper fraction, we need to convert all the units to fourths because in order for the improper fraction, $\frac{13}{4}$, to have meaning, all the pieces must be the same size—that is, they must be fourths. We can convert each 1 to four fourths and then add all the fourths to get $\frac{13}{4}$. However, because one context for multiplication is repeated addition, we see that we are adding three 4's; hence 3 units times 4 fourths in each unit. When we add these 12 fourths to the 1 fourth we already had, we have a total of 13 fourths. Notice in the set model that we chose to put 4 stars in each set since we were working with fourths.

$$\begin{aligned} 3\frac{1}{4} &= 1 + 1 + 1 + \frac{1}{4} \\ &= \frac{4}{4} + \frac{4}{4} + \frac{4}{4} + \frac{1}{4} \\ &= \frac{3 \cdot 4}{4} + \frac{1}{4} \\ &= \frac{3 \cdot 4 + 1}{4} \\ &= \frac{13}{4} \end{aligned}$$

UNDERSTANDING SUBTRACTION OF FRACTIONS

As discussed earlier, subtraction of fractions involves essentially the same processes as does addition. However, many students encounter difficulty in situations that involve regrouping. Therefore, it is important to examine this process also. Consider the following problem: $23\frac{2}{5} - 17\frac{4}{5}$. First do it on your own, writing down your justification of each step. Then read on. . . .

$$\begin{array}{rcr} 23\frac{2}{5} & = & 22\frac{7}{5} \\ -17\frac{4}{5} & = & 17\frac{4}{5} \\ \hline & & 5\frac{3}{5} \end{array}$$

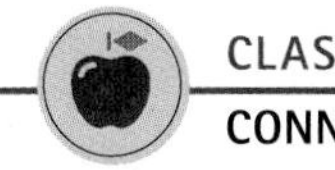

CLASSROOM CONNECTION

Many children and adults do the problem below in the following way:

$$\begin{array}{rcl} 23\frac{2}{5} & = & \frac{117}{5} \\ -17\frac{4}{5} & = & \frac{89}{5} \\ \hline & & \frac{28}{5} = 5\frac{3}{5} \end{array}$$

Some middle school teachers teach this procedure. Their rationale is that so many students stuggle with fraction subtraction that they gave up and taught the students this "foolproof" way, even though it is cumbersome in many cases. The teachers just gave up and focused on the *how*.

Let us discuss the transformation from $23\frac{2}{5}$ to $22\frac{7}{5}$, because this is difficult for many children. How would you explain this to someone who does not understand it?

We must do some renaming because we cannot subtract $\frac{4}{5}$ from $\frac{2}{5}$. With fractions, instead of trading 1 ten for 10 ones (as we did with whole number subtraction), we are now trading 1 unit for five fifths (or six sixths, or ten tenths, depending on the common denominator). Once again, expanded form will help to illustrate the process and justify the equivalence of $23\frac{2}{5}$ and $22\frac{7}{5}$.

$$23\frac{2}{5} = 23 + \frac{2}{5} = (22 + 1) + \frac{2}{5} = 22 + \left(\frac{5}{5} + \frac{2}{5}\right) = 22\frac{7}{5}$$

INVESTIGATION 4.3c Mental Addition and Subtraction with Fractions

Because there are many strategies used in mental addition and subtraction of fractions, we will examine them using a process similar to the one we used when developing mental math strategies with whole numbers.

In each of the following problems, try to find the sum or the difference entirely in your head.

Briefly describe your strategies. Then read the discussion below. 

A. $\begin{array}{r} 2\frac{1}{2} \\ +5\frac{3}{4} \\ \hline \end{array}$ B. $\begin{array}{r} 7\frac{3}{4} \\ 5\frac{1}{8} \\ +2\frac{1}{2} \\ \hline \end{array}$ C. $\begin{array}{r} 7 \\ -2\frac{5}{8} \\ \hline \end{array}$ D. $\begin{array}{r} 9\frac{1}{4} \\ -3\frac{5}{8} \\ \hline \end{array}$

DISCUSSION

Keep the following points in mind for each problem:

- Deciding which strategy is best is often not as helpful as knowing which strategy works best for you. Learning a variety of strategies will help you find strategies that work for you, and that will work for your future students.
- Not all possible useful strategies have been included here, because too many strategies (and 1500-page books) can overwhelm students!

A. $2\frac{1}{2} + 5\frac{3}{4}$

STRATEGY 1 Use the commutative and associative properties

$$2\frac{1}{2} + 5\frac{3}{4} = \left(2 + \frac{1}{2}\right) + \left(5 + \frac{3}{4}\right) = (2 + 5) + \left(\frac{1}{2} + \frac{3}{4}\right) = 7 + \frac{1}{2} + \frac{3}{4}$$

That is, we transform the problem into $7 + \frac{1}{2} + \frac{3}{4}$.

Visualize and decompose $\frac{3}{4}$ into $\frac{1}{2} + \frac{1}{4}$; we then have $7 + \frac{1}{2} + \frac{1}{2} + \frac{1}{4} = 8\frac{1}{4}$.

STRATEGY 2 Add up and break into parts

$$\left(2\frac{1}{2} + 5\right) + \frac{3}{4} = 7\frac{1}{2} + \frac{3}{4} = 7\frac{1}{2} + \left(\frac{1}{2} + \frac{1}{4}\right) = 8 + \frac{1}{4} = 8\frac{1}{4}$$

B. $7\frac{3}{4} + 5\frac{1}{8} + 2\frac{1}{2}$

STRATEGY 1 Look for compatible numbers

$\frac{3}{4}$ and $\frac{1}{2}$ can be added mentally to get $1\frac{1}{4}$; then add $1\frac{1}{4} + \frac{1}{8} = 1\frac{2}{8} + \frac{1}{8} = 1\frac{3}{8}$. Add this to the 14, and we have $15\frac{3}{8}$.

STRATEGY 2 Convert to common denominator

Because the common denominator is relatively small and the conversions to eighths are relatively easy, some students find it easier to add all three at once, thinking something like this: "6 + 1 + 4 = 11, that means 11 eighths, which is 8 eighths (1) + 3 eighths. Add this to the 14 and we have $15\frac{3}{8}$ as the answer."

C. $7 - 2\frac{5}{8}$

STRATEGY 1 Break into parts and subtract

First, $7 - 2 = 5$; now take $\frac{5}{8}$ from 5 to get $4\frac{3}{8}$.

STRATEGY 2 Add up

$2\frac{5}{8} + 4 = 6\frac{5}{8}$. How much more do we need to get to 7? We need $\frac{3}{8}$ more. The answer is $4\frac{3}{8}$.

D. $9\frac{1}{4} - 3\frac{5}{8}$

STRATEGY 1 Add up

$3\frac{5}{8}$ plus 5 equals $8\frac{5}{8}$ plus $\frac{3}{8}$ equals 9 plus $\frac{2}{8} = 9\frac{1}{4}$. What you have to remember to add now is $5 + \frac{3}{8} + \frac{2}{8} = 5\frac{5}{8}$.

A number line helps some students to understand the process better (Figure 4.27).

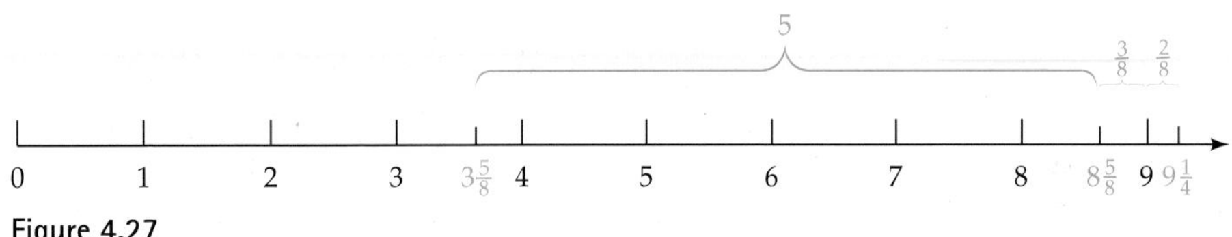

Figure 4.27

INVESTIGATION 4.3d Estimating Sums and Differences with Fractions

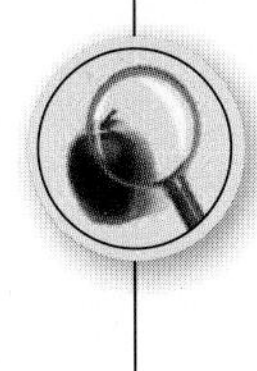

A. In their retirement, the parents of a close friend of mine have created a dollhouse to represent their dream house. They are shopping, and they see some miniature furniture that might fit in one of the bedrooms. The bed is $\frac{7}{8}$ inch wide; the dresser is $1\frac{1}{2}$ inches long, and the desk is $1\frac{3}{4}$ inches long. If the three articles are placed as shown in Figure 4.28, will they fit into the dollhouse bedroom, which is 4 inches wide? Try to add the three lengths in your head, and then read on. . . .

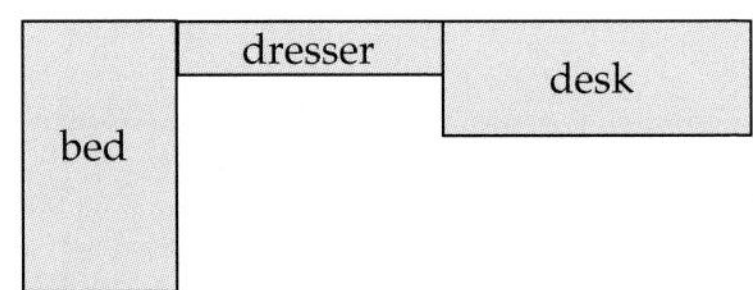

Figure 4.28

DISCUSSION

STRATEGY 1 Make a quick estimate

A quick estimation reveals that it will be close:

$$\frac{7}{8} + 1\frac{1}{2} + 1\frac{3}{4} \approx 1 + 1\frac{1}{2} + 1\frac{1}{2} = 4$$

Thus we need a strategy that will be more precise.

STRATEGY 2 Add the two easier numbers

$$1\frac{1}{2} + 1\frac{3}{4} = 1\frac{1}{2} + \left(1\frac{1}{2} + \frac{1}{4}\right) = 3\frac{1}{4}$$

We can use number sense here; that is, we know that $3\frac{1}{4} + \frac{3}{4} = 4$, and because $\frac{7}{8} > \frac{3}{4}$, we can conclude that the furniture is too big.

B. It rained almost every day this past week. Here are the amounts per day. How much rain fell during the week? First, make a rough estimate. Then make a refined estimate. Then read on. . . .

Monday	$1\frac{3}{4}$ inches	Friday	0 inches
Tuesday	$\frac{1}{2}$ inch	Saturday	$3\frac{3}{4}$ inches
Wednesday	$\frac{5}{8}$ inch	Sunday	$1\frac{5}{16}$ inches
Thursday	$\frac{7}{8}$ inch		

DISCUSSION

STRATEGY 1 Make a rough estimate

One way to get a rough estimate is to round each fraction to the nearest $\frac{1}{2}$ inch and keep a cumulative total:

$$2 + \frac{1}{2} = \mathbf{2\frac{1}{2}};\ 2\frac{1}{2} \text{ plus } \frac{1}{2} = \mathbf{3};\ 3 \text{ plus } 1 = \mathbf{4};\ 4 \text{ plus } 4 = \mathbf{8};\ 8 \text{ plus } 1\frac{1}{2} = \mathbf{9\frac{1}{2}}$$

STRATEGY 2 Look for compatible numbers

One way to get a closer estimate is to look for compatible numbers—numbers whose sum is close to 1 or numbers that you can quickly add mentally.

$$\underbrace{\left(1\frac{3}{4} + 3\frac{3}{4}\right)}_{5\frac{1}{2}} + \frac{1}{2} + \underbrace{\left(\frac{5}{8} + 1\frac{5}{16}\right)}_{\text{about } 2} + \underbrace{\frac{7}{8}}_{\text{about } 1} = \text{about } 9$$

$$5\frac{1}{2} + \frac{1}{2} + \text{about } 2 + \text{about } 1 = \text{about } 9$$

Before we move on, take a few moments to reflect on the mental math and estimation strategies that you used and that were discussed here. What strategies have been adapted from ones we developed with whole numbers, such as compatible numbers?

INVESTIGATION 4.3e Understanding Multiplication of Fractions

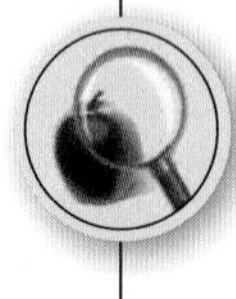

For each of the following problems, represent the situation with a diagram and determine the answer from the diagram. As you are doing this, think back to the models we developed for whole-number multiplication. Which problems connect well to repeated addition? To a rectangular array? Which problems do not connect well? Why don't they?

Which model (area, linear, or set) best represents each problem?

A. Julio runs 4 times a week. His route is $2\frac{1}{4}$ miles long. How many miles does he run each week?

DISCUSSION

This connects nicely to multiplication as repeated addition: 4 times $2\frac{1}{4}$.

We can represent this problem on a number line, as in Figure 4.29.

CLASSROOM  CONNECTION

Grade 5

This activity represents yet another model for exploring fractions. Try these yourself! 

Date ______________________ Time ______________________

LESSON 8·5 Paper-Folding Problems

Record your work for the four fraction problems you solved by paper folding. Sketch the folds and shading. Write an X on the parts that show the answer.

1. $\frac{1}{2}$ of $\frac{1}{2}$ is ____________.

2. $\frac{2}{3}$ of $\frac{1}{2}$ is ____________.

3. $\frac{1}{4}$ of $\frac{2}{3}$ is ____________.

4. $\frac{3}{4}$ of $\frac{1}{2}$ is ____________.

260

From *Everyday Mathematics, Grade 5:* The University of Chicago School Mathematics Project: Student Math Journal, Volume 2, by Max Bell et al., Lesson 8-5, p. 260. Reprinted by permission of The McGraw-Hill Companies, Inc.

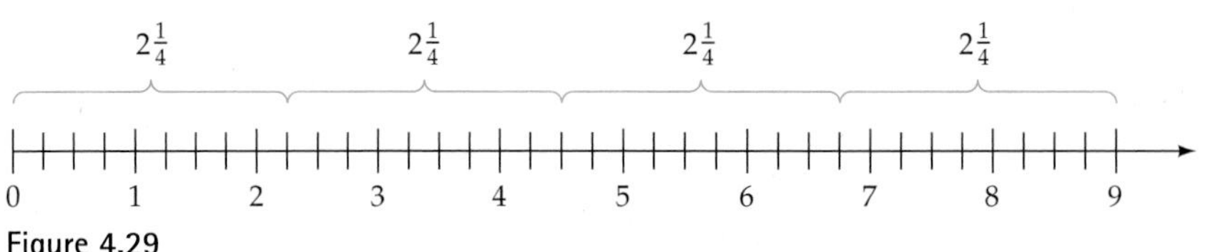

Figure 4.29

B. A group of investors purchased a rectangular parcel of land that is $\frac{3}{4}$ of a mile long and $\frac{2}{3}$ of a mile wide. How many square miles did they buy?

DISCUSSION

This problem is definitely not repeated addition, but it does connect to the area model of multiplication and can be represented nicely with an area model of fractions. Here we can say that we want $\frac{2}{3}$ of $\frac{3}{4}$ (of 1 square mile).

If the large square in Figure 4.30 represents 1 mile by 1 mile, the shaded region represents the area of the land that the investors bought: $\frac{3}{4}$ of a mile by $\frac{2}{3}$ of a mile. But how do we give a name to this amount? Try to do this yourself and then read on. . . .

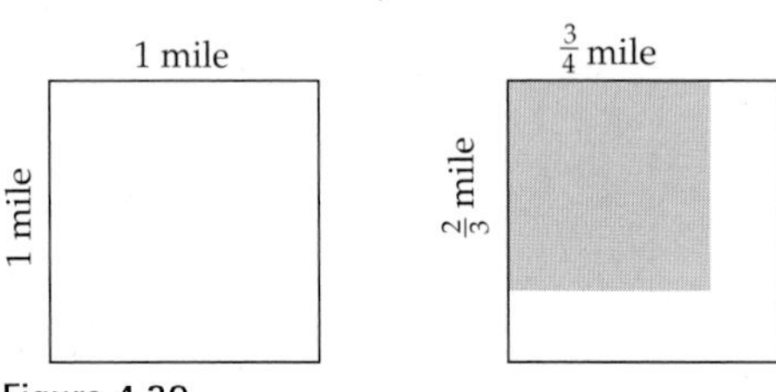

Figure 4.30

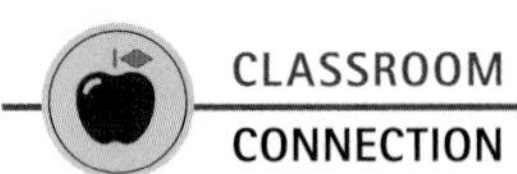

There is another difference between multiplying whole numbers and multiplying fractions that bears noting, for it often baffles children when they encounter fraction computations for the first time. One notion of multiplication that many people (often unconsciously) get from working with whole numbers is that "multiplication makes bigger and division makes smaller," but just the opposite is true when we multiply proper fractions!

Just as we decomposed and recomposed numbers in working with whole numbers, we need to decompose this (shaded) amount in such a way that we can recompose the amount as a set of equal-size pieces.

We can create equal-size pieces (needed to name a fraction) by dividing the square first into fourths (with vertical lines) and then into thirds (with horizontal lines). If we put the two diagrams together and shade in the area enclosed by $\frac{3}{4}$ times $\frac{2}{3}$, we have the figure on the right in Figure 4.31. Can you determine the answer now from the diagram? Can you justify that answer? Think and then read on. . . .

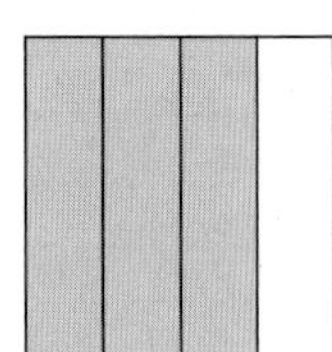
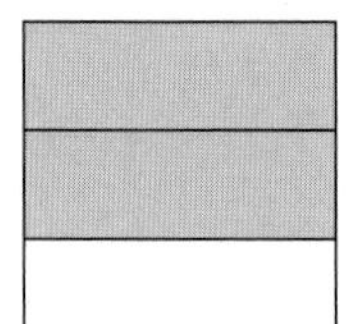
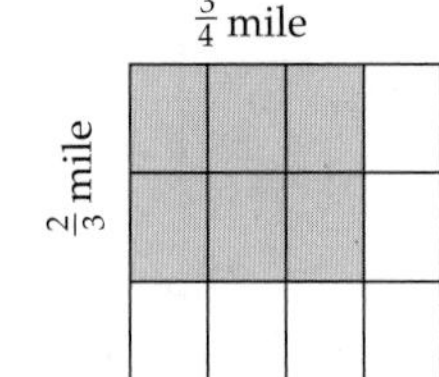

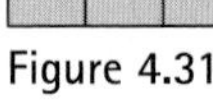
Figure 4.31

Because our unit (1 square mile) has been divided into 12 equal-size regions, each region has a value of $\frac{1}{12}$ square mile. The plot of land covers 6 of these rectangles, so its value is $\frac{6}{12}$ or $\frac{1}{2}$ square mile. This model helps to explain the procedure of "multiply straight across" that is often used for multiplying fractions. The shaded rectangle is 2 by 3 because of the numerators. The entire unit is 3 by 4 because of the denominators.

C. If $\frac{2}{3}$ of the class went on a field trip and there are 24 students in the class, how many students went on the field trip?

LANGUAGE

When students memorize "*of* means multiply," they are memorizing a procedure but not understanding this interpretation of fractions.

DISCUSSION

The operator context can be thought of as a stretching or a shrinking process. In this case, we have a whole (24 students), and we are shrinking that whole—that is, we are considering a part $\left(\frac{2}{3}\right)$ of that whole.

Figure 4.32(a) represents the problem using a set model, with each dot representing one person. Because we have $\frac{2}{3}$ of the people, the $\frac{2}{3}$ tells us that we have to partition this amount into 3 equal groups and then take 2 of those groups. Figure 4.32(b) represents the problem using a bar model. The whole rectangle represents the whole class. We have shaded in $\frac{2}{3}$ of the rectangle. If the whole class represents 24 students, then when we divide the class into 3 equal regions, each region must represent 8 students (that is, $8 + 8 + 8 = 24$ or $3 \cdot 8 = 24$). Both models quickly produce the correct answer of 16 students.

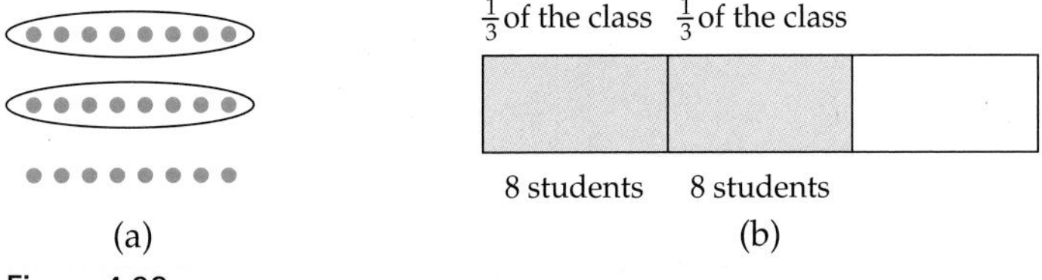

Figure 4.32

CONNECTING MULTIPLICATION OF FRACTIONS TO THE MULTIPLICATION ALGORITHM

As we did with multiplication of whole numbers, we will develop the algorithm with the rectangular model. Let us do so with the problem

$$\frac{3}{4} \cdot 2\frac{1}{2}$$

Let us first represent the problem with a diagram (Figure 4.33). Determine the answer yourself from the diagram before reading on. . . .

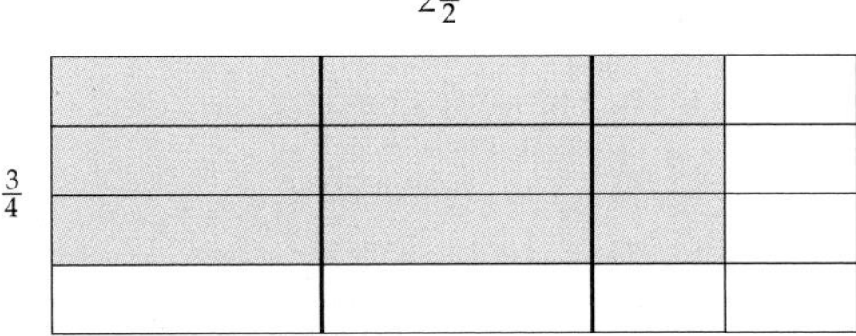

Figure 4.33

We can readily see (from Figure 4.34) that the numerator of our product must be 15 (that is, there are 15 parts). But what is the denominator? (Please resist the temptation to answer this question by doing the algorithm. In my own class, this question is actually a method I use to assess how deeply my students understand the meaning of denominator.) Then read on. . . .

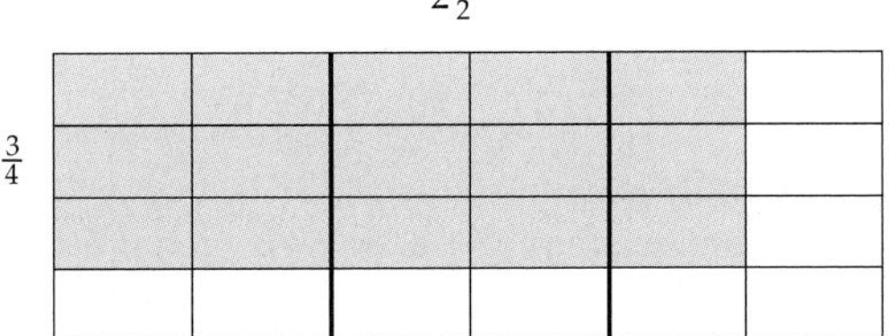

Figure 4.34

I usually get three different responses for the denominator: 24 because there are 24 parts in the "whole" diagram, 15 because there are 5 times 3 parts in the $2\frac{1}{2}$, and 8 because it takes 8 parts to make a 1 by 1 square (the unit).

The correct answer is 8. Do you see why? *Recall that the value of the denominator is determined by its relationship to the unit.* That is, the denominator shows how many pieces it takes to have a value of 1, rather than how many pieces are in the whole.

Therefore, we have determined that

$$\frac{3}{4} \cdot 2\frac{1}{2} = \frac{15}{8} = 1\frac{7}{8}$$

Now let us examine the algorithm and look for connections between the algorithm and the concepts. We must first convert $2\frac{1}{2}$ to an improper fraction, and then we simply find the product of the numerators and the product of the denominators:

$$\frac{3}{4} \cdot 2\frac{1}{2} = \frac{3}{4} \cdot \frac{5}{2} = \frac{3 \cdot 5}{4 \cdot 2} = \frac{15}{8}$$

If we look at the shaded region in Figure 4.34, we have 3 rows, each containing 5 regions; that is, we have 3 times 5 regions. Just as we found that the whole-number multiplication algorithm automatically regrouped for us, so too the fraction multiplication algorithm automatically creates equal-size pieces. Similarly, we have $4 \cdot 2$ pieces in each unit.

JUSTIFYING THE EQUALITY OF EQUIVALENT FRACTIONS

In Section 4.2, we examined equivalent fractions. We can now justify the procedure we used earlier. The process is shown at the left for a specific case and at the right for the general case in the table below.

Specific case	Justification	General case
$\frac{3}{4} = \frac{3}{4} \cdot 1$	Multiplicative identity	$\frac{a}{b} = \frac{a}{b} \cdot 1$
$= \frac{3}{4} \cdot \frac{2}{2}$	Substitution	$= \frac{a}{b} \cdot \frac{c}{c}$
$= \frac{3 \cdot 2}{4 \cdot 2}$	Definition of multiplication of fractions	$= \frac{a \cdot c}{b \cdot c}$
$= \frac{6}{8}$		

INVESTIGATION 4.3f Division of Fractions

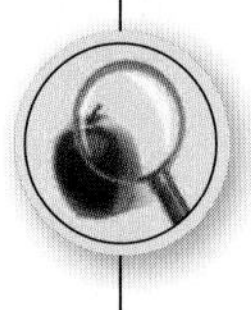

As with most concepts being considered in this course, you probably remember the procedure. This recalls a famous saying of unknown origin: "Ours is not to reason why, just invert and multiply."

$$\frac{3}{4} \div \frac{5}{6} = \frac{3}{4} \times \frac{6}{5} = \frac{18}{20} = \frac{9}{10}$$

Explorations Manual 4.14

Now, though, ours *is* to reason why. Let us begin our investigation of the fraction division algorithm by first examining fraction division problems in context.

Recall the two contexts for division with whole numbers: partitioning and repeated subtraction.

Make up a story for each of the two division problems below.

First, solve the problem using what you know about division and fractions.

HISTORY

The Hindu mathematician Mahavira (A.D. 850) first stated a rule for dividing fractions: "After making the denominator of the divisor its numerator and vice versa, the operations to be conducted then are as in the multiplication [of fractions]."[5]

Then respond to the following questions: Which of your stories were from the partitioning model, and which were from the repeated-subtraction model? If you made up a story connected to one model, could you make up another story for the same problem using the other model?

A. The divisor is a whole number: $3 \div 4$

B. The divisor is a proper fraction: $3 \div \frac{2}{5}$

DISCUSSION

A. Most people make up a story for $3 \div 4$ using the partitioning model, and most of the stories involve sharing.

Let us consider one such story and examine a solution: Basil has 3 pints of ice cream that he wants to share equally with 3 friends—that is, among 4 people. How much is each person's share?

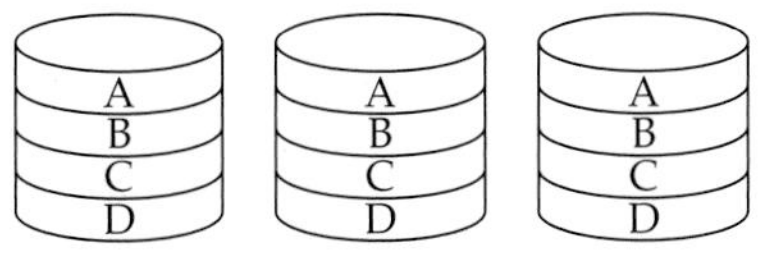

Figure 4.35

Figure 4.35 shows a partitioning (dealing) solution. That is, if we divide each pint into 4 equal parts, each person gets 1 part per pint. Each person's share thus consists of 3 parts, each of which is $\frac{1}{4}$ pint. Therefore, each person's share is $\frac{3}{4}$ of a pint.

Some people literally interpret this problem as a multiplication problem. If we are dividing the ice cream among 4 people, then each person gets $\frac{1}{4}$ of the ice cream. This connection between the two operations produces the following equality: $3 \div 4 = 3 \cdot \frac{1}{4}$, which we just mentioned. We will return to this relationship between multiplication and division shortly. However, first let us examine a story for the other problem.

B. One story for $3 \div \frac{2}{5}$: Jake is stranded in the middle of the desert. He has 3 quarts of water, and he figures that he will drink $\frac{2}{5}$ of a quart each day. How many days' supply does he have?

Solve the problem using only your knowledge of fractions and your knowledge of division. Try drawing a model to help you understand. Then read on. . . .

This is a repeated-subtraction problem because we have an amount (3 quarts) and we are specifying the size of the group $\left(\frac{2}{5}\text{ of a quart}\right)$ to be repeatedly subtracted. The answer will be the number of groups we have.

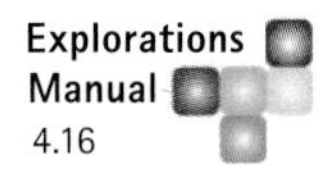

Explorations Manual 4.16

We have chosen to represent this problem with a number line, although an area or set model could also have been used. Each day Jake uses $\frac{2}{5}$ of a quart. We see from Figure 4.36 that he is able to drink $\frac{2}{5}$ of a quart each day for 7 days. However, we have a small problem: We have repeatedly subtracted $\frac{2}{5}$ of a quart, but we have $\frac{1}{5}$ of a quart left over. Does this answer contradict the answer that comes from using the algorithm, which tells us that $3 \div \frac{2}{5} = 7\frac{1}{2}$? How do we reconcile the difference between the answer from the diagram and that from the algorithm? Think and then read on. . . .

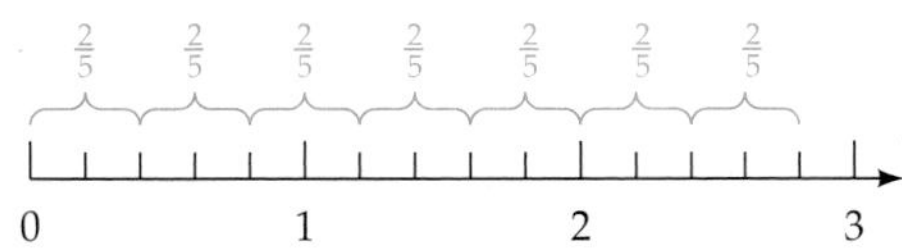

Figure 4.36

MATHEMATICS

In her book, *Knowing and Teaching Elementary Mathematics,* Liping Ma asked a number of U.S. and Chinese elementary school teachers to make up a story problem for $1\frac{3}{4} \div \frac{1}{2}$. I was stunned to read that most of the Chinese teachers' stories involved the partitioning model. Although my conception of this model is not exactly the same as Ma's—who describes it as "finding a number such that $\frac{1}{2}$ of it is $1\frac{3}{4}$" (p. 74), I found her example and her book to be very educational. I recommend the book to you.

It turns out that both answers are correct—they represent different ways of describing the remainder. The $\frac{1}{5}$ tells us that the remainder is $\frac{1}{5}$ of a quart; thus we can say that at the end of 7 days, he will have $\frac{1}{5}$ of a quart left. The $7\frac{1}{2}$ tells us that the remainder is equivalent to $\frac{1}{2}$ day's water. We can think of this problem as how many $\frac{2}{5}$s are in 3? Figure 4.36 shows us that there are $2\frac{1}{2}$ (or $\frac{5}{2}$) two fifths in each 1 quart. Therefore, pictorially when we ask how many two fifths portions are in 3, we can see it is 3 times $\frac{5}{2}$ portions per each quart. In other words, 3 divided by $\frac{2}{5}$ is the same as 3 times $\frac{5}{2}$.

MAKING SENSE OF THE STANDARD DIVISION ALGORITHM

Although using the invert-and-multiply algorithm is rather straightforward, understanding why it works is another matter. We can more formally examine the similarities between how division and multiplication are connected and how subtraction and addition are connected.

We need to introduce a new concept to do so. Just as we discovered in the first section of this chapter that every integer has an additive inverse, every nonzero fraction has a **multiplicative inverse**. These parallel concepts are shown side by side below:

$-a$ is the additive inverse of a	$\frac{1}{a}$ is the *multiplicative inverse* of a, $a \neq 0$
$a + (-a) = 0$ for any integer a	$a \cdot \frac{1}{a} = 1$ for any fraction, $a \neq 0$

Let us now recall the way in which we defined integer subtraction in terms of addition, and then we will define fraction division in terms of fraction multiplication.

$a - b = a + (-b)$	$a \div b = a \cdot \frac{1}{b}$
That is, to subtract b, we add the (additive) inverse of b.	That is, to divide by b, we multiply by the (multiplicative) inverse of b.
In other words, one way to interpret *subtraction* is to *add* the *additive inverse* of the second number.	In other words, one way to interpret *division* is to *multiply* by the *multiplicative inverse* of the second number.

Thus,

$$a \div \frac{x}{y} = a \cdot \frac{y}{x}$$

INVESTIGATION 4.3g Estimating Products and Quotients

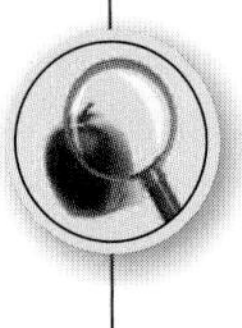

In each of the following problems: First obtain a rough estimate (5 to 10 seconds), and then try to get as close as you can to the actual answer—either a refined estimate or, in some cases, an exact answer (computed mentally).

A. Anastasia walks at a pace of $3\frac{1}{2}$ miles per hour for 2 hours and 15 minutes. How far does she walk during this time?

DISCUSSION

STRATEGY 1 Bound the answer

One kind of estimating involves bounding the answer. For example, the answer will be at least 6. We get this by focusing only on the whole numbers. We call 6 a lower bound.

An upper bound can be obtained by rounding both numbers to the next higher whole number. The upper bound is thus $4 \cdot 3 = 12$.

MATHEMATICS

Remember our discussion of the partial products in multiplication of whole numbers? We can adapt that structure when multiplying fractions.

STRATEGY 2 Get rid of one of the proper fractions (that is, rounding)

Another method of making a rough estimate involves getting rid of one of the proper fractions. We can more easily determine $3\frac{1}{2} \cdot 2 = 7$ or $3 \cdot 2\frac{1}{4} = 6\frac{3}{4}$.

STRATEGY 3 Use partial products

We can estimate and round the four partial products:

$$3\frac{1}{2} \times 2\frac{1}{4} = \left(3 + \frac{1}{2}\right) \times \left(2 + \frac{1}{4}\right) = (3 \times 2) + \left(3 \times \frac{1}{4}\right) + \left(2 \times \frac{1}{2}\right) + \left(\frac{1}{2} \times \frac{1}{4}\right)$$

$$\approx 6 + 1 + 1 + 0 \approx 8$$

B. The Jones family finds that they spend about $\frac{2}{3}$ of their income on food and rent. If their monthly income is \$1635, how much is their monthly food and rent bill?

DISCUSSION

Rough estimate (using compatible numbers):

$$\frac{2}{3} \times 1500 = 1000$$

More refined estimate (using compatible numbers and expanded form):

$$\frac{2}{3} \times 1650 = \frac{2}{3}(1500 + 150) = 1000 + 100 = 1100$$

C. Marvin has 23 yards of cloth with which to make costumes for the play. Each costume requires $3\frac{1}{4}$ yards of material. How many costumes can he make?

DISCUSSION

Rough estimate (using bounding):

$$23 \div 3 = 7\frac{2}{3}$$

$$23 \div 4 = 5\frac{3}{4}$$

The estimate is about 6 or 7 costumes.

Refined estimate (using repeated addition):

$$3\frac{1}{4} \times 2 = 6\frac{1}{2}, \text{ so}$$

$$3\frac{1}{4} \times 4 = 13, \text{ so}$$

$$3\frac{1}{4} \times 8 = 26$$

Because $26 - 3 = 23$, this estimate yields 7 costumes.

D. Shelly has 25 pounds of dog food, and each day she feeds her dog $\frac{3}{4}$ pound. How many days' worth of food does she have?

DISCUSSION

Rough estimate (using bounding):

$$25 \div 1 = 25 \quad \text{and} \quad 25 \div \frac{1}{2} = 50$$

Thus she has enough for more than 25 days and less than 50 days.

Refined estimate:

$$25 \div \frac{3}{4} = 25 \times \frac{4}{3} = \frac{100}{3} = 33\frac{1}{3}$$

which is the actual quotient. Thus, she has 33 days worth of dog food.

APPLYING FRACTION UNDERSTANDINGS TO NONROUTINE PROBLEMS

We have examined carefully the different meanings that fractions can have, and we have examined the four operations in terms of fractions. In the following nonroutine problems, you will need to determine which computation(s) to do and how to interpret the computation(s).

INVESTIGATION 4.3h When Did He Run Out of Gas?

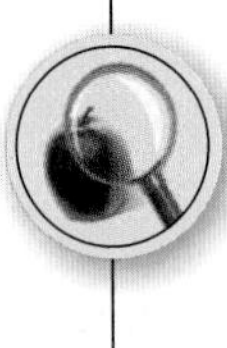

Jeremy started on a trip from Mobile, Alabama, to New Orleans, approximately 120 miles away. His car ran out of gas after he had gone one-third of the second half of the trip. How far is Jeremy from New Orleans? Work on this problem yourself and then read on. . . .

DISCUSSION

This is a case in which a good diagram practically solves the problem by itself (see Figure 4.37).

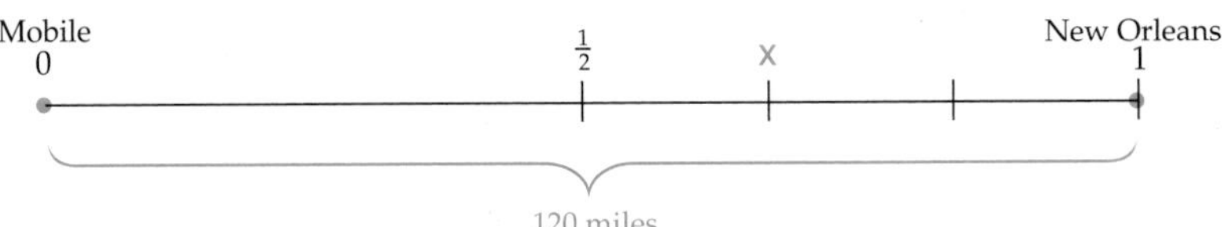

Figure 4.37

If half of the trip is equal to 60 miles, then each of the "thirds of the second half" is equal to 20 miles. X marks the spot of "one-third of the second half of the trip." So Jeremy is still 40 miles from New Orleans.

INVESTIGATION 4.3i They've Lost Their Faculty!

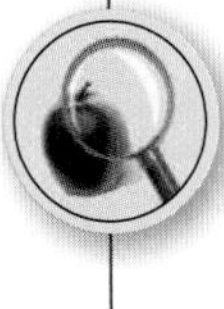

American State College has had to reduce its faculty by $\frac{1}{6}$ because of an economic crisis in the state. The college now has 350 faculty members. A curious reader might ask, "How many did they have before the cut?" Try to solve this problem before reading on. . . .

DISCUSSION

STRATEGY 1 Represent the situation with a diagram

The large rectangle in Figure 4.38 represents the original faculty.

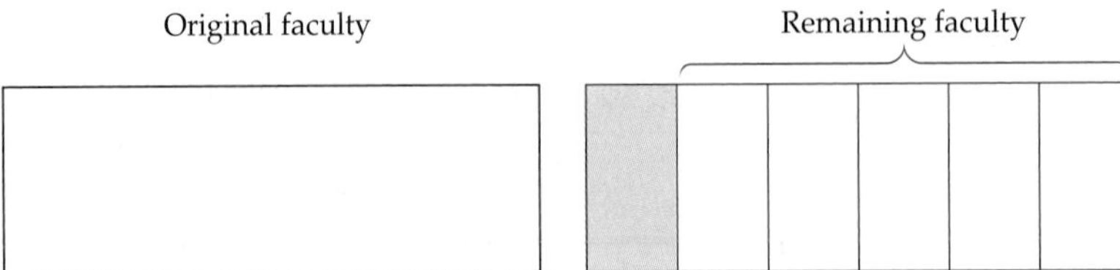

Figure 4.38

If the college lost $\frac{1}{6}$ of its faculty, then we can divide that rectangle into 6 equal pieces and take away 1 piece, which represents $\frac{1}{6}$ of the faculty. Thus the 5 remaining boxes represent the remaining $\frac{5}{6}$ of the faculty.

Because we know that there are 350 faculty members now, the value of those 5 boxes is 350. But if the value of 5 boxes is 350, then the value of 1 box is 70. Now we can answer the problem. Do you see why?

Because the value of the original faculty was represented by 6 boxes, there were $6 \times 70 = 420$ faculty.

STRATEGY 2 Use algebra

Can you find an algebraic solution? Work on it and then read on. . . . 

One algebraic solution comes from letting x = the number of the faculty before the cut. The remaining faculty represents $\frac{5}{6}$ of the original, so 350 is $\frac{5}{6}$ of x; that is, $\frac{5}{6}x = 350$. We solve the equation by multiplying both sides by $\frac{6}{5}$:

$$\left(\frac{6}{5}\right)\frac{5}{6}x = \left(\frac{6}{5}\right)350 \quad \text{and} \quad x = 420$$

SUMMARY 4.3

In this section, we have focused on connecting conceptual knowledge to procedural knowledge—on understanding the whys behind the how of the algorithms for operations with fractions.

We have found that the contexts in which we first studied these four operations in Chapter 3 do not all apply when we are working with fractions; for example, the partitioning model of division does not readily apply when the divisor is not a whole number. In making sense of algorithms, we need to pay attention to the very fraction ideas that we developed in Section 4.2: that the parts in a fraction have the same value, that working with fractions involves parts and wholes, and that we have to pay attention to units and wholes (because they are not always the same). Area, linear, and set models help us make sense of the procedures.

We found that many of the strategies developed in Chapter 3 for mental computation and estimation applied directly to computing with fractions—for example, the commutative and associative properties, compatible numbers, adding up, partial products, and expanded form.

When we applied our knowledge of operations to nonroutine and multistep problems, we found that we needed to think about how we might represent the problem. We also had to make sure that we kept basic fraction ideas in mind—for example, the distinction between the unit and the whole.

4.3 Exercises

1. Perform the following computations.

a. $23\frac{7}{8} + 56\frac{5}{6}$

b. $-3\frac{5}{12} + 1\frac{9}{40}$

c. $7\frac{3}{4} + 8\frac{3}{5} + 7\frac{3}{10}$

d. $213 - 21\frac{5}{16}$

e. $9\frac{1}{4} - 12\frac{5}{6}$

f. $12\frac{1}{3} \times 5\frac{3}{8}$

g. $6\frac{3}{4} \times 14\frac{2}{3}$

h. $36 \times 3\frac{3}{4}$

i. $\frac{3}{10} \times 45$

j. $\frac{3}{4} \div 6$

k. $2\frac{3}{8} \div 6\frac{1}{3}$

2. Represent and solve each of the following problems with area, linear, and/or set models. Explain your solution path.

a. $\frac{2}{3} + \frac{5}{6}$

b. $2\frac{1}{3} + \frac{3}{4}$

c. $\frac{3}{8} - \frac{3}{4}$

d. $\frac{2}{3} \times \frac{1}{2}$

e. $\frac{3}{4} \times 6$

f. $\frac{3}{4} \times 5\frac{1}{2}$

g. $\frac{3}{4} \div 6$

h. $\frac{7}{8} \div \frac{3}{8}$

i. $5\frac{1}{2} \div \frac{3}{4}$

3. ***Classroom Connection*** In each case, describe the error that the student made, then describe the mathematical idea(s) the student did not correctly understand.

a. $\frac{2}{3} + \frac{3}{8} = \frac{5}{24}$

b. $\frac{3}{5} + \frac{2}{5} = \frac{5}{10}$

c. $\frac{1}{5} + \frac{2}{3} = \frac{3}{8} + \frac{10}{8} = \frac{13}{8} = 1\frac{3}{8}$

d. $\begin{array}{r} 7\frac{2}{5} \\ -3\frac{4}{5} \\ \hline 4\frac{2}{5} \end{array}$

e. $8\frac{5}{8} - 2\frac{2}{3} = 6\frac{3}{5}$

f. $\begin{array}{r} 7\frac{1}{8} = 6\frac{11}{8} \\ -3\frac{5}{8} = 3\frac{5}{8} \\ \hline 3\frac{6}{8} = 3\frac{3}{4} \end{array}$

g. $\frac{2}{3} \times \frac{3}{4} = 89$

h. $\frac{2}{3} \times 5 = \frac{10}{15}$

i. $\frac{2}{3} \cdot \frac{3}{8} = \frac{9}{16}$

j. $\frac{8}{15} \div \frac{2}{5} = \frac{4}{3}$

k. $\frac{8}{15} \div \frac{2}{5} = \frac{15}{8} \times \frac{2}{5} = \frac{30}{40} = \frac{3}{4}$

l. $8\frac{1}{8} \div 2\frac{1}{4} = 4\frac{1}{2}$

4. Use a diagram to illustrate why $2\frac{3}{4} = \frac{11}{4}$.

5. a. Convert $3\frac{5}{6}$ to an improper fraction.

b. Convert $5\frac{3}{4}$ to an improper fraction.

6. a. Convert $\frac{18}{5}$ to a mixed number.

b. Convert $\frac{28}{8}$ to a mixed number in simplest form.

7. When we convert $\frac{13}{5}$ to $2\frac{3}{5}$, which model of division do we use? Explain your answer.

8. Use a number line model or a set model to find $\frac{2}{3} + \frac{1}{4}$. Justify each step in the process.

9. Solve the following problems using only the area model.

a. $\frac{2}{3} \times 2\frac{3}{4}$

b. $2\frac{2}{3} \times 3\frac{1}{2}$

10. Determine each of the following mentally, and briefly explain your solution.

a. $\begin{array}{r} 5\frac{3}{4} \\ +7\frac{5}{8} \\ \hline \end{array}$

b. $\begin{array}{r} 8\frac{4}{5} \\ +6\frac{2}{3} \\ \hline \end{array}$

c. $\begin{array}{r} 16\frac{3}{4} \\ 15\frac{1}{2} \\ +7\frac{1}{4} \\ \hline \end{array}$

d. $\begin{array}{r} 26 \\ -6\frac{3}{5} \\ \hline \end{array}$

e. $\begin{array}{r} 7\frac{1}{3} \\ -3\frac{1}{2} \\ \hline \end{array}$

f. $8\frac{2}{3} + 7\frac{5}{6} + 4\frac{2}{3} + 2\frac{2}{3}$

g. $3\frac{2}{3} \cdot 12$

h. $2\frac{1}{2} \cdot 3\frac{1}{4}$

11. For each of the following, determine which of the choices represents the better estimate. Then explain your choice.

a. $5\frac{5}{8} + 4\frac{3}{42}$ Greater than 10 or less than 10?

b. $\frac{7}{8} + \frac{3}{4} + \frac{1}{16}$ Greater than 2 or less than 2?

c. $\frac{1}{4} + 1\frac{9}{10}$ Greater than 2 or less than 2?

d. $8\frac{1}{2} - 2\frac{2}{3}$ Between which two numbers: 5 and $5\frac{1}{2}$ or $5\frac{1}{2}$ and 6?

e. $8\frac{3}{4} \times 2\frac{7}{8}$ Greater or less than 20?

f. $4\frac{7}{8} \div 8\frac{7}{8}$ Greater or less than $\frac{1}{2}$?

12. Javier is $\frac{2}{3}$ as tall as his dad. Javier is 49 inches tall. How tall is his dad?

13. A recent survey concluded that $\frac{3}{4}$ of the teachers in a certain school district had at least 10 years of teaching experience. If there are 152 teachers in the district, approximately how many have at least 10 years of teaching experience?

14. Julianna has $22\frac{1}{2}$ yards of material. If it takes $\frac{3}{4}$ yard of material to make a pair of shorts, can she make enough shorts for 30 children?

15. A fruit juice drink is $\frac{5}{7}$ water and $\frac{1}{6}$ fruit juice (by weight). Other additives (primarily sugar) make up what fraction of its weight?

16. Let's say that a certain antibiotic kills approximately $\frac{1}{2}$ of the germs on the first day, $\frac{1}{2}$ of the remaining germs on the second day, and so on.

a. After 10 days, what fraction of the germs remain?

b. If you had 20 million germs in your body at the beginning, how many of them will be alive at the end of the tenth day?

17. A small office purchases bottled drinking water. The water comes in containers that hold 5 gallons. Two-thirds of the office container was drained last week to put out a fire in the wastebasket.

a. How much water (in ounces) is left in the container?

b. How many 12-ounce glasses can be filled with the water that remains in the container?

DEEPENING YOUR UNDERSTANDING

18. Irene has discovered a rule that works for some addition problems. For example, consider $\frac{1}{3} + \frac{1}{2}$. The rule is this: The numerator of the answer is equal to the sum of the denominators of the two fractions, and the denominator of the answer is equal to the product of the denominators of the two fractions. What do you think of her rule? Describe the set of fractions for which this rule works. Can you modify her rule so that it works for a larger set of problems?

19. ***Classroom Connection*** In the December 2000 issue of *Teaching Children Mathematics,* the authors examined third- and fourth-graders' responses to the question "What is the value of $2\frac{1}{2} + \frac{3}{4}$?"

a. This is the diagram that accompanied one child's explanation. Describe what might have been the child's reasoning behind this diagram. Then describe the mathematical ideas that the child used—properties of addition, expanded form (taking apart and putting numbers back together), meanings of fraction, etc.

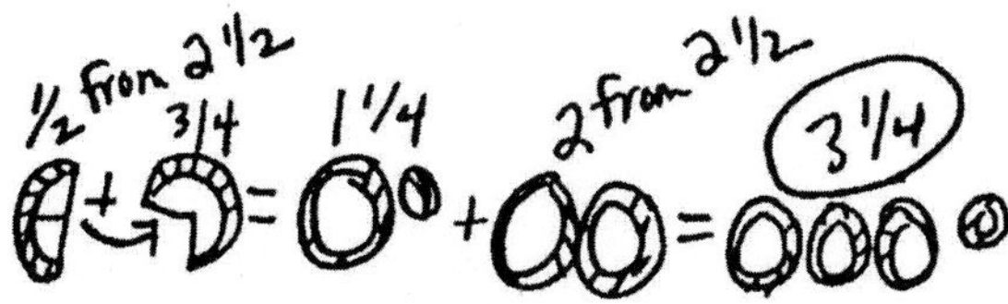

Source: Teaching Children Mathematics, by the National Council of Teachers of Mathematics.

b. Here is another child's solution path: "I took the two off of the two and one half. Then I added the three quarters, and that's two and three quarters. From the one half I took a half away from that. So one quarter plus two and three quarters is three. Then you add the other half of the one half and it's three and one fourth." In this case, write the numbers and/or diagrams that might accompany this verbal description. Then describe the mathematical ideas that the child used—properties of addition, expanded form (taking apart and putting numbers back together), meanings of fraction, etc.

20. ***Classroom Connection*** In the November 1999 issue of *Mathematics Teaching in the Middle School,* the invention of a procedure for subtracting fractions is described. This procedure resembles an invented method for subtraction of whole numbers described in Chapter 3. Here is how Krystal's method works:

	Find LCM	Subtract the whole numbers, and subtract the fractions	Simplify the result
$7\frac{1}{4}$	$7\frac{3}{12}$	$7\frac{3}{12}$	$5 - \frac{1}{12} = 4\frac{11}{12}$
$-2\frac{1}{3}$	$-2\frac{4}{12}$	$-2\frac{4}{12}$	
		$5\frac{-1}{12}$	

a. Make up and solve more problems until you are comfortable with this algorithm.

b. Does it generalize? That is, will it work for all fraction subtraction problems? Justify your response.

c. Justify the algorithm. That is, explain why it works.

d. Compare this algorithm to the standard algorithm.

Source: Mathematics Teaching in the Middle School, by the National Council of Teachers of Mathematics.

21. ***Classroom Connection*** In the February 1998 issue of *Mathematics Teaching in the Middle School,* the authors showed two diagrams made by students to illustrate how they determine the answer to $\frac{1}{2}$ of $\frac{3}{4}$. This is before they learned the algorithm. Provide the answer and justification that come from each diagram.

a.

$\frac{1}{8}$

$\frac{1}{8}$

$\frac{1}{8}$

b.

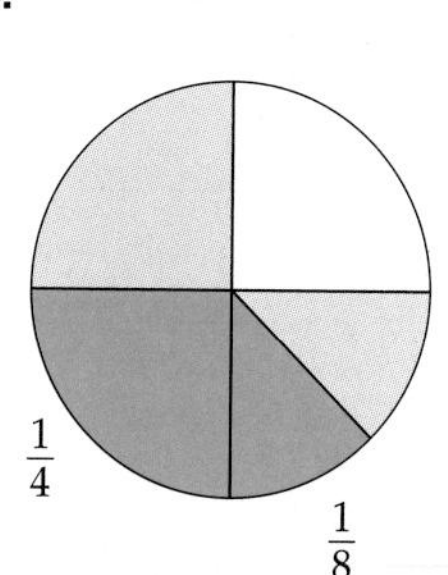

Source: Mathematics Teaching in the Middle School, by the National Council of Teachers of Mathematics.

22. Demonstrate how to obtain the following product two different ways, neither of which can be the standard algorithm: $3\frac{1}{3} \times 2\frac{1}{2}$. Then explain why $6\frac{1}{6}$ is not the correct answer, as though you were talking to a student for whom $6\frac{1}{6}$ is the obvious answer.

23. ***Classroom Connection*** When faced with the problem $\frac{1}{2} \times \frac{1}{4} = \frac{1}{8}$, a child asks, "If multiplication is repeated addition, then what is this?" How might you respond?

24. Without actually multiplying, can you tell which of the following two is greater? Explain your reasoning.

$$26\frac{5}{7} \times 9 \quad \text{or} \quad 26 \times 9\frac{5}{7}$$

25. a. How would you figure $26 \times \frac{11}{12}$ on a calculator without fraction capabilities? Show two ways.

b. How would you figure $26 \div \frac{11}{12}$ on a calculator?

26. Make up a realistic story problem for the following multiplication-of-fractions contexts:

a. A proper fraction times a proper fraction

b. A proper fraction times a whole number

c. A proper fraction times a mixed number

d. A mixed number times a mixed number

27. ***Classroom Connection*** In *Young Mathematicians at Work: Constructing Fractions, Decimals and Percents,* the authors describe a fifth-grade class where the children are exploring multiplication with fractions. The current problem is $15\frac{1}{2} \times 4\frac{1}{2}$. The children do not yet know the standard algorithm and are using their understanding of multiplication, of whole numbers, and of fractions to find the answer. Below are the beginning steps of three solution paths that enabled the children to determine $4\frac{1}{2} \times 15$. In each case, explain how to use this problem to determine $4\frac{1}{2} \times 15$, and then supply additional steps to find the answer to $15\frac{1}{2} \times 4\frac{1}{2}$.

a. One group began with $4\frac{1}{2} \times 60 = 270$.

b. Another group began with $9 \times 30 = 270$.

c. A third group began with $15 \times 36 = 540$.

Source: *Young Mathematicians at Work: Constructing Fractions, Decimals, and Percents* by Catherine Twomey Fosnot and Maarten Dolk.

28. An early division algorithm in India is actually simpler than the current algorithm. It involved converting the fractions to fractions with a common denominator, as we do with addition, and then simply dividing the numerators. For example,

$$\frac{3}{4} \div \frac{5}{6} = \frac{9}{12} \div \frac{10}{12} = \frac{9}{10}$$

Make up some other problems to see whether this algorithm works only with proper fractions or with all fraction problems. Why does it work?

29. ***Classroom Connection*** A student says that she has found a quick way to divide fractions. She demonstrates her method as follows:

$9\frac{1}{4} \div 3\frac{3}{4} = 3\frac{1}{3}$, because $9 \div 3 = 3$ and $\frac{1}{4} \div \frac{3}{4} = \frac{1}{3}$

$20\frac{3}{5} \div 4\frac{3}{5} = 5\frac{1}{1}$, because $20 \div 4 = 5$ and $\frac{3}{5} \div \frac{3}{5} = 1$

Unfortunately, her idea is not correct. Where is the conceptual error?

30. Ginger has $8\frac{1}{3}$ yards of silk cloth from India. She has decided to divide the cloth equally into 3 parts, one for each of her siblings. How much cloth does each person get? *Do this problem without using an algorithm*—that is, using reasoning. You may want to use a diagram, but you are not required to do so.

31. ***Classroom Connection*** In the 1999 NCTM Yearbook, *Developing Mathematical Reasoning in Grades K–12,* Deborah Schifter discusses several solution paths to the following problem that was given to a sixth-grade class: "You are giving a party for your birthday. From the Ice Cream Factory, you order 6 pints of ice cream that they make. If you serve $\frac{3}{4}$ of a pint of ice cream to each guest, how many guests can be served?"

a. First, do the problem yourself.

Below are summaries of different students' solutions. In each case, describe where the number sentence comes from. In other words, supply the why behind the what.

b. $24 \div 3 = 8$.

c. $8 \times \frac{3}{4} = 6$, so the answer is 8.

d. $\frac{3}{4} + \frac{3}{4} + \frac{3}{4} + \frac{3}{4} + \frac{3}{4} + \frac{3}{4} + \frac{3}{4} + \frac{3}{4} = 6$, so the answer is 8.

e. $6 - \frac{3}{4} - \frac{3}{4} - \frac{3}{4} - \frac{3}{4} - \frac{3}{4} - \frac{3}{4} - \frac{3}{4} - \frac{3}{4} = 0$, so the answer is 8.

32. Without doing any computing, order the following from least to greatest. Explain your reasoning.

a. $\frac{4}{11} + \frac{7}{13}$ **b.** $\frac{4}{11} - \frac{7}{13}$ **c.** $\frac{7}{13} \times \frac{4}{11}$ **d.** $\frac{7}{13} \div \frac{4}{11}$

33. The points on the number line are not drawn to scale. Without doing any computation, determine the region in which each answer will lie. Explain your reasoning.

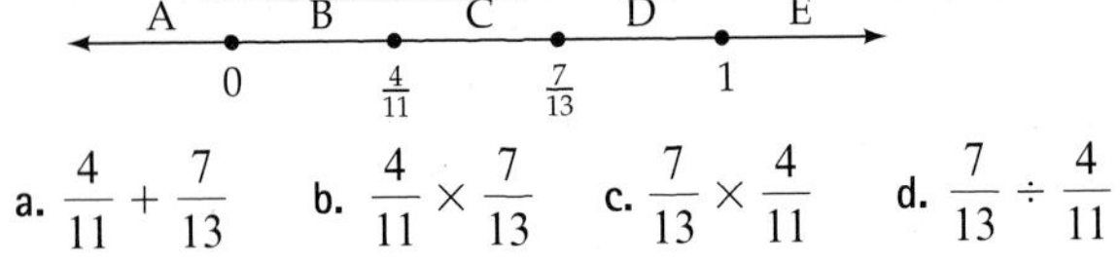

a. $\frac{4}{11} + \frac{7}{13}$ **b.** $\frac{4}{11} \times \frac{7}{13}$ **c.** $\frac{7}{13} \times \frac{4}{11}$ **d.** $\frac{7}{13} \div \frac{4}{11}$

34. What is the weight (in ounces) of a $4\frac{1}{2}$ inch by $7\frac{3}{4}$ inch rectangle of sheet metal if one square inch weighs $\frac{1}{8}$ ounce?

35. Sunflower seeds are sold in packages that weigh $3\frac{1}{4}$ ounces. If there is a supply of 66 ounces of sunflower seeds, how many packages of seed can be made? How many ounces of seeds will be left over?

36. We find from the *Unofficial U.S. Census* that more than 136,800,000 Americans admit to doodling. Of the doodlers, slightly less than $\frac{2}{3}$ said they doodle while talking on the telephone. Approximately how many Americans doodle while talking on the telephone?

37. Betty is stacking boxes at a factory. Each box is $13\frac{3}{4}$ inches high. The ceiling is 16 feet. How many boxes can she stack in one pile?

38. John is making a bookcase. The bookcase is $72\frac{3}{8}$ inches tall and he wants 5 shelves, equally spaced. Tell him where to drill the holes on the side. Each shelf is $\frac{3}{8}$ inch thick. Your solution needs to contain a description that is clear enough so that he can understand it.

39. In a certain town, $\frac{2}{3}$ of the women are married and only $\frac{1}{2}$ of the men are married. What fraction of the community is single? What assumptions do you make in order to solve the problem?

40. An analysis of first-year students at a college revealed that $\frac{1}{4}$ of the first-year women were from homes where both parents were professionals. Of these, $\frac{3}{5}$ were interested in the same profession as one or both of their parents. If this latter group is made up of 18 students, how many first-year women are there?

41. The Petersons have an apple-pressing machine that they use to make apple juice. The first pressing gets about $\frac{1}{3}$ of the total juice in the apples. Each succeeding pressing extracts about $\frac{1}{3}$ of the remaining juice in the apples. How many pressings will it take to get at least $\frac{3}{4}$ of the apples' juice? How many pressings will it take to get at least $\frac{9}{10}$ of the juice?

42. Pete has just made a batch of Rice Krispies marshmallow treats, one of my favorites! The baking pan measures 13″ by 9″. He has decided that each treat will be $1\frac{1}{2}$″ by 1″.

a. First, *estimate* how many treats he can get. Explain your estimate.

b. Now determine exactly how many treats he can get from one pan.

43. ***Classroom Connection*** The following problem appeared in "The Thinking of Students" in the April 2002 issue of *Mathematics Teaching in the Middle School:* "Gordon's Sporting Goods has sold all its wind-breaker jackets except one. It was so ugly that no one wanted to buy it. So Gordon's put the jacket on sale at one-fourth off its original price. The jacket did not sell, so Gordon's marked the sale price down one-third off. Still no one wanted to buy such an ugly jacket, so Gordon's marked the second sale price down one-half off. Finally, customer Ernie Harrison bought the jacket for his cat Wilson to sleep on. Ernie paid only $12 for the world's ugliest jacket. What was the original price of the jacket? What did the jacket look like?"

a. Solve the problem yourself.

b. Read the article and present one of the students' solutions, both the how and the why (that is, the justification of each step).

44. Selecting among the numbers 1 through 9 and repeating none of them, fill in the boxes below to make the sum as close as possible to 1, but not equal to 1.

$$\frac{\square}{\square} + \frac{\square}{\square}$$

45. Selecting among the numbers 1 through 9 and repeating none of them, fill in the boxes below to make the smallest possible difference greater than zero.

$$\frac{\square}{\square} - \frac{\square}{\square}$$

46. Selecting among the numbers 1 through 9 and repeating none of them, make the following equation true.

$$\frac{\square}{\square} \times \frac{\square}{\square} = \frac{1}{2}$$

47. Selecting among the numbers 1 through 9 and repeating none of them, make the largest possible quotient.

$$\frac{\square}{\square} \div \frac{\square}{\square}$$

48. Selecting among the numbers 1 through 9 and repeating none of them, make the quotient closest to $\frac{1}{2}$.

$$\frac{\square}{\square} \div \frac{\square}{\square}$$

49. Do the following problems in base five, as opposed to converting to base ten and solving in base ten.
  a. The Marble Palace has just received a shipment of $4{,}400_{\text{five}}$ marbles in four colors—blue, red, clear, and yellow. One-half of the marbles are blue, and there are half as many red marbles as blue marbles. If $\frac{1}{10_{\text{five}}}$ of the marbles are clear, how many of the marbles are yellow?
  b. The Marble Palace has just received a shipment of 440_{five} marbles in four colors—blue, red, clear, and yellow. One-third of the marbles are blue, and three-quarters of the *remaining* marbles are red. If 20_{five} of the marbles are clear, how many of the marbles are yellow?

50. Do the following division problem and represent the answer as a mixed number in simplified form. Explain how you obtained your answer without resorting to base ten.

$13_{\text{five}}\overline{)2130_{\text{five}}}$

51. A student in your class says that $\frac{2}{3} + \frac{3}{5} = \frac{5}{8}$. What mathematical misunderstanding does this student have? What model could help correct the misunderstanding?

52. Find the following:
  a. LCM(12, 20)
  b. LCM(30, 75)
  c. LCM(44, 66)
  d. LCM(462, 630)

53. Find the following:
  a. LCM(12, 30, 75)
  b. LCM(20, 30, 40)

54. George vacuums the rugs every 18 days, mows the grass every 12 days, and pays the bills every 15 days. Today he did all three. How long will it be before he has another day like today?

55. Ann and Bob are cycling on a track. Ann completes one lap every 12 seconds, and Bob completes one lap every 15 seconds. When will Ann lap Bob, assuming that they started together?

56. A manufacturer sells widgets, each of which is packaged in a box whose dimensions are 4 centimeters by 6 centimeters by 10 centimeters. Design the dimensions of a box to ship widgets. Your box must hold at least 1,000 widgets.

FROM STANDARDIZED ASSESSMENTS

2006 NECAP, Grade 5

57. Beverly uses $\frac{1}{4}$ cup of applesauce in place of every $\frac{1}{3}$ cup of butter in her cookie recipe. How many cups of applesauce will Beverly use in place of 1 cup of butter?

a. $\frac{1}{12}$ b. $\frac{1}{7}$ c. $\frac{2}{4}$ d. $\frac{3}{4}$

2005 NECAP, Grade 6

58. On Saturday, Dora practiced playing her violin 5 times. Each time she practiced for 15 minutes. What is the total number of hours Dora practiced her violin on Saturday?

a. $\frac{5}{15}$ hour b. $\frac{5}{6}$ hour c. $1\frac{1}{4}$ hour d. $2\frac{1}{2}$ hour

2008 NECAP, Grade 5

59. Mr. Bowen used 2 gallons of paint on $\frac{1}{3}$ of his fence. How many more gallons of paint does Mr. Bowen need to finish painting his fence?

2005 NECAP, Grade 6

60. The map below shows the path a boat sailed.

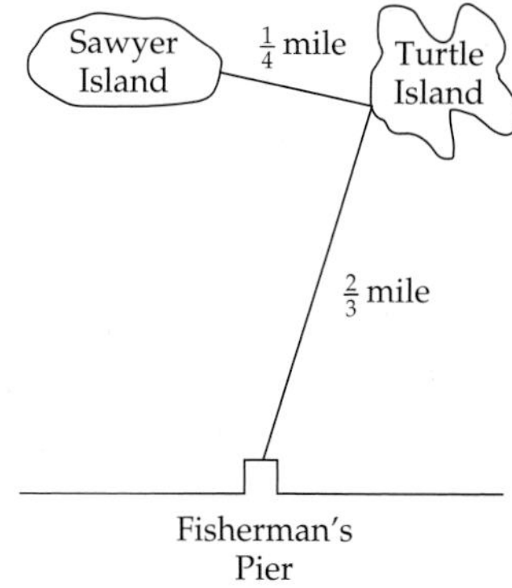

The boat sailed from Sawyer Island to Turtle Island and then to Fisherman's Pier. What is the total distance the boat sailed?

a. $\frac{5}{12}$ mile b. $\frac{3}{7}$ mile c. $\frac{3}{4}$ mile d. $\frac{11}{12}$ mile

61. Which is one way to find $\frac{3}{4}$ of a number?
  a. Divide by 3 and multiply by 4.
  b. Divide by 4 and multiply by 3.
  c. Divide by 3 and divide by 4.
  d. Multiply by $\frac{4}{3}$.

Source: Results of the *Fourth NAEP Mathematics Assessment,* p. 13. U.S. Department of Education, National Center for Education Statistics.

62. Students are running in a relay race. Each team will run a total of 2 miles. Each member of a team will run $\frac{1}{5}$ of a mile. How many students will a team need to complete the race?

Choose the correct number. You may use a number line to help find your answer.

$\frac{2}{5}$ $\frac{5}{2}$ 9 10 20

63. Jared is testing how much weight a bag can hold. He plans to put juice bottles into three bags. He wants each bag to have a total weight within the given range.

Place juice bottles into each bag so that the weight is within the given range.

Leave the bag empty if the given range is not possible using juice bottles.

(Exercises 62 and 63 are samples items from the Smarter Balanced Assessment Consortium, which is developing assessments aligned with the Common Core State Standards.)

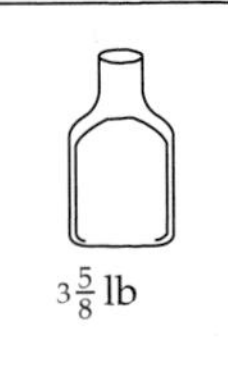

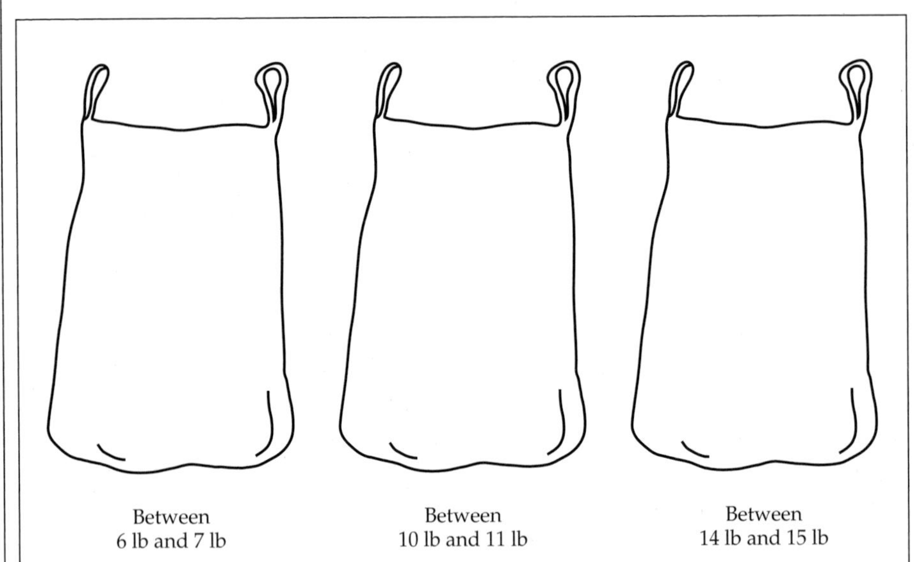

SECTION 4.4 Beyond Integers and Fractions: Decimals, Exponents, and Real Numbers

What do you think?

- How would the structure of a decimal type system be different in a base five system?
- Why is there no "oneths" place?

In this section, we will extend the system of numbers yet again—to decimals and to irrational numbers (that is, numbers that cannot be expressed as the quotient of two integers, such as $\sqrt{2}$ and π). There is a name for the set of numbers that includes both rational and irrational numbers: the set of real numbers (an interesting choice for this set!). There are more sets of numbers beyond the real numbers, called imaginary and complex numbers, but they are not considered until high school. Understanding decimals and decimal operations begins in fifth grade, according to the Common Core State Standards. Irrational numbers are introduced in eighth grade. Whole number exponents are in sixth grade, and integer and rational exponents in eighth grade.

DECIMALS

Technically, the set of decimals is not an extension of the set of fractions. Rather, our use of decimal numbers comes out of an alternative way of representing fractions, using the advantages of our base ten system. In fact, the word *decimal* comes from the Latin word *decem,* which means "ten." Thus decimal numbers are any numbers that are written in terms of our base ten place value. Conventionally, we refer to decimals as fractions whose denominator has been converted into a power of ten.

The history of decimals If you stop to think about it, decimals could not have been invented before our base ten system. Credit for bringing decimals into everyday life goes to Simon Stevin. His book *The Art of Tenths or Decimal Arithmetic* was first published in 1585.

> "We will speak freely of the great utility of this invention . . . for the astronomer knows the difficult multiplication and divisions which proceed from the progression with degrees, minutes, seconds, and thirds. . . . This discovery . . . teaches (to tell much in one word) to compute easily, without fractions, all computations which are encountered in the affairs of human beings. . . ." He went on to say that this work treats of "something so simple that it hardly merits the name of invention."[6]

The use of decimals was spurred primarily by a desire to make computation easier. For example, which computation would be easier to do: $37\frac{3}{4} \times 6\frac{4}{5}$ or 37.75×6.8? You may want to do both. How much faster was the decimal computation?

Even today there is not one universal notation. Some countries use a dot like we do in the United States, but other countries use a comma to represent a decimal. For example:

United States	Continental Europe
54.23	54,23

No decimals? Imagine our world without decimals!

What would life be like if no one had invented decimals—that is, if we expressed amounts using only whole numbers and fractions? Would things just look different (for example, 43 dollars and 26 cents instead of $43.26), or would some things be impossible?

Spend some time writing your answers to these questions before reading on. . . . 

As you may have realized as you thought about these questions, much of our use of decimals involves measurement:

- We measure mass: for example, the package weighs 2.4 ounces.
- We measure distance: for example, Joan lives 3.7 miles from school.
- We measure time: for example, the world record for the 100-meter dash is 9.58 seconds.

Decimals are used in many aspects of daily life to communicate. For example,

- The U.S. birth rate is 14.1 per 1000 population.
- The average life expectancy in the United States is 77.9 years.
- Interest rates are expressed with decimals, such as 7.5 percent.
- The new city budget is $3.48 million.

CONNECTING DECIMALS AND INTEGERS

There are many important similarities and differences between the set of decimals and the set of integers. Before reading on, stop to think about the two sets of numbers. What similarities and differences can you see?

From one perspective, the set of decimals can be seen as an extension of base ten counting numbers.

...	thousands	hundreds	tens	ones	tenths	hundredths	thousandths	...
	10^3	10^2	10^1	10^0	10^{-1}	10^{-2}	10^{-3}	
	1000	100	10	1	0.1	0.01	0.001	

That is, the system of decimals and the system of integers are both base ten place value systems. The value of each digit depends on its place, and the value of each place to the right is $\frac{1}{10}$ the value of the previous place. Investigation 4.4a helps to develop your decimal sense and give you a better "feel" for the relative size of tenths, hundredths, thousandths, and so on. Rather than state the differences, we will let them emerge through investigation.

INVESTIGATION 4.4a Base Ten Blocks and Decimals

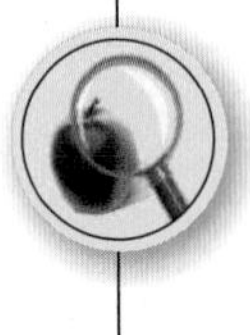

Let's say that a sample of gold weighs 3.6 ounces. How would you use base ten blocks (Figure 4.39) to represent this amount? Work on this and then read on. . . .

Explorations Manual 4.18

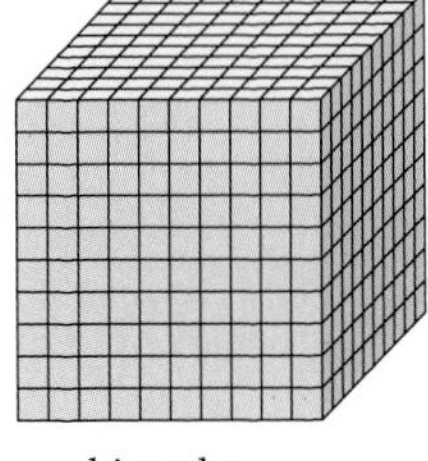

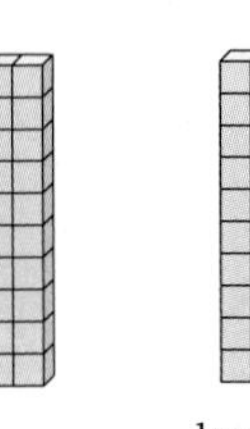

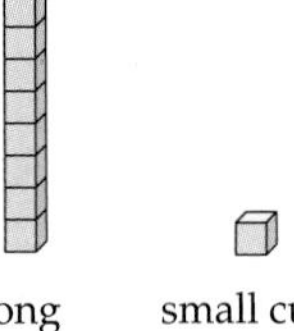

Figure 4.39

DISCUSSION

Any solution requires us to designate one of the pieces as the unit. For example, if we designate the big cube as having a value of 1, then we would represent 3.6 ounces as shown in Figure 4.40. Do you see why?

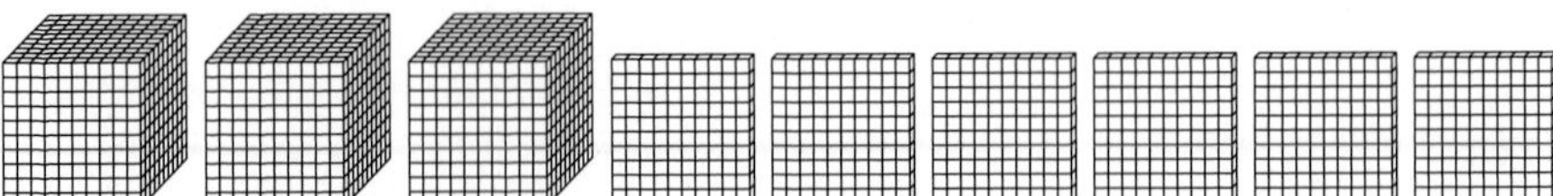

Figure 4.40

However, we could also have selected the flat to be the unit. If we had, then we would represent 3.6 ounces as shown in Figure 4.41.

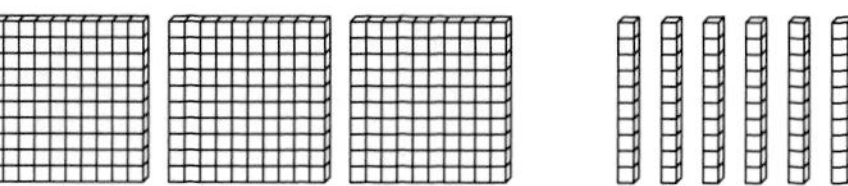

Figure 4.41

EXPANDED FORM

Like whole numbers, decimal numbers can be expressed in expanded form:

$$624.326$$

$$600 + 20 + 4 + \frac{3}{10} + \frac{2}{100} + \frac{6}{1000}$$

$$600 + 20 + 4 + .3 + .02 + .006$$

$$6 \times 10^2 + 2 \times 10^1 + 4 \times 10^0 + 3 \times 10^{-1} + 2 \times 10^{-2} + 6 \times 10^{-3}$$

As with whole numbers, zeros often prove challenging to make sense of. However, the same principles still apply:

$$2.06$$

$$2 + \frac{0}{10} + \frac{6}{100}$$

$$2 + .06$$

$$2 \times 10^0 + 6 \times 10^{-2}$$

ZERO AND DECIMALS

We will examine the role of zero in decimals in the following two investigations.

INVESTIGATION 4.4b When Two Decimals Are Equal

What if we added one or two zeros at the end of 0.2? Would that change its value? That is, do 0.2, 0.20, and 0.200 have the same value? Many sixth- and seventh-graders (and a surprising number of adults) believe that 0.2, 0.20, and 0.200 not only look different but also have different values because of the added zeros. How would you convince such a person that the value is not changed? Work on this and then read on. . . .

MATHEMATICS

The idea of multiple representations connects to our work with whole numbers. For example, just as 24 can be represented by 2 tens and 4 ones or by 24 ones, so too 0.24 can be represented by 2 tenths and 4 hundredths or by 24 hundredths.

DISCUSSION

There are several ways to justify the equality. We will consider two below.

STRATEGY 1 Connect decimals to fractions

- 0.2 is equivalent to the fraction $\frac{2}{10}$.
- 0.20 is equivalent to the fraction $\frac{20}{100}$, which can be shown to be equivalent to $\frac{2}{10}$.
- 0.200 is equivalent to the fraction $\frac{200}{1000}$, which can be shown to be equivalent to $\frac{2}{10}$.

STRATEGY 2 Use manipulatives

In this case, let a square represent a value of 1. Such manipulatives are sold commercially as Decimal Squares. We can divide this square into 10 equal pieces, 100 equal pieces, or 1000 equal pieces, as in Figure 4.42.

Shade in 0.2 of the first square, 0.20 of the second square, and 0.200 of the third square.

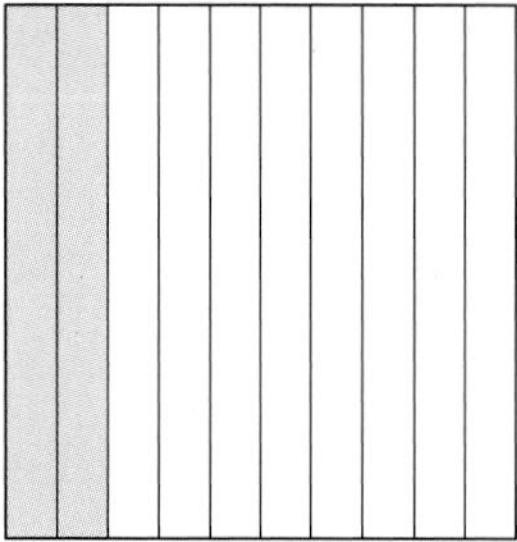
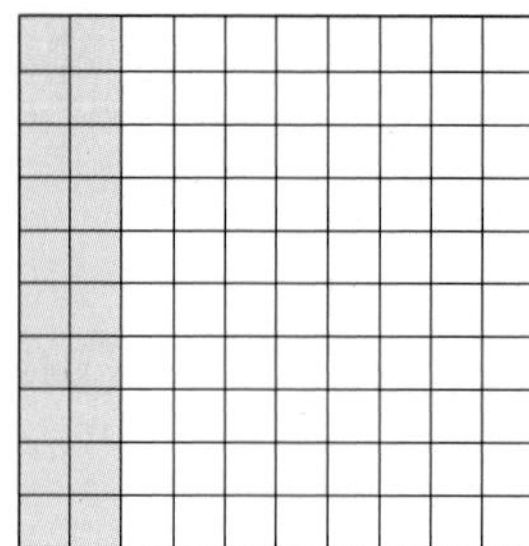
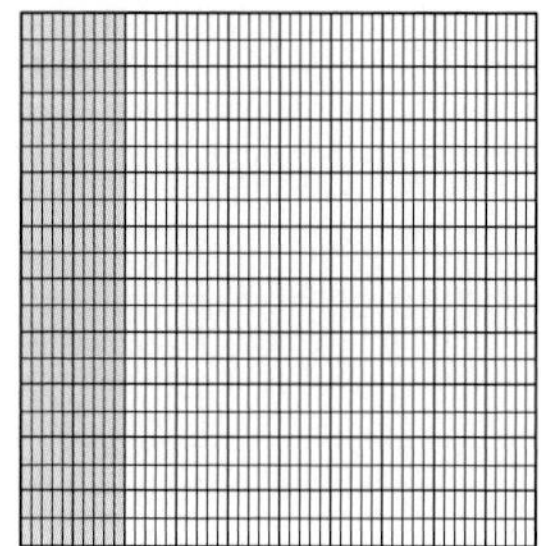

Figure 4.42

We find that in each case, the same area is shaded. Thus, 0.2, 0.20, and 0.200 are all equivalent decimals.

INVESTIGATION 4.4c When Is the Zero Necessary and When Is It Optional?

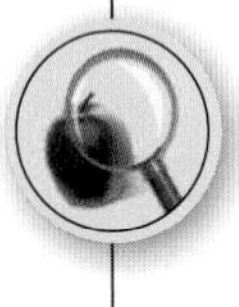

An important difference between decimals and whole numbers has to do with our old friend zero. Examine each number below. In each case, explain whether you think the zero in the number is necessary, optional, or incorrect. If you think it is necessary, explain why. If you think it is optional, explain why the zero doesn't matter. If you think the use of the zero is incorrect, explain why the zero should not be there.

A. 2.08 B. 0.56 C. .507 D. 20.6 E. 3.60

DISCUSSION

A. With 2.08, the zero is necessary: 2 ones, 0 tenths, and 8 hundredths.

B. With 0.56, the use of the zero is optional; it is a convention, like shaking hands with the right hand instead of the left.

C. With .507, the zero is necessary: 5 tenths, 0 hundredths, and 7 thousandths.

D. With 20.6, the zero is necessary: 2 tens, 0 ones, and 6 tenths.

E. In one sense, the zero here is optional. Mathematically, 3.6 = 3.60; you can verify this using the strategies in Investigation 4.4b. In another sense, however, it depends on how the number is being used. For example, if we ask how long the room is, what is the difference between a response of 3.6 meters and a response of 3.60 meters?

A response of 3.6 meters implies that the room has been measured to the nearest tenth of a meter; that is, the length of the room is closer to 3.6 meters than to 3.5 or 3.7 meters.

OUTSIDE THE CLASSROOM

We see examples of pseudo-precision every day. For example, a poll reported that 83.4 percent of people in the Northeast described the winter 1993–1994 as "the worst I have ever experienced." Given that polls generally have an error margin of several percent, this is an example of pseudo-precision. It would have been more valid to report that about 83 percent of the persons polled agreed with that statement.

Some differences between decimals and integers With decimals, when we add a zero at the right end of the number, the value is unchanged; for example, 48.6 = 48.60. As you know, this is not true with integers; for example, 48 ≠ 480. Two other differences between the two systems are worth noting. One has to do with language: With whole numbers, the suffix for each place is *-s*; with decimals, it is *-ths*—for example, the hundreds place versus the hundredths place. Another difference is that there is a ones place but not a oneths place; this lack of symmetry has been noted by more than one youngster. How would you explain it?

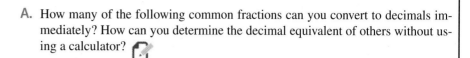

INVESTIGATION 4.4d Connecting Decimals and Fractions

As stated before, each decimal can be translated into a fraction. You may be familiar with common translations between decimals and fractions: $0.5 = \frac{1}{2}$, $0.25 = \frac{1}{4}$, and so on.

Many problem-solving situations require us to convert a fraction into decimal form or vice versa.

A. How many of the following common fractions can you convert to decimals immediately? How can you determine the decimal equivalent of others without using a calculator?

$$\frac{1}{2} \quad \frac{1}{3} \quad \frac{1}{4} \quad \frac{1}{5} \quad \frac{1}{6} \quad \frac{1}{8} \quad \frac{1}{10} \quad \frac{1}{50} \quad \frac{1}{100} \quad \frac{1}{1000}$$

OUTSIDE THE CLASSROOM

Many mathematical concepts are misused in our society. One classic example of misuse of decimals occurs in baseball. Statistics for pitchers include the total number of innings pitched. There are 3 outs per inning, so if the pitcher pitched 7 innings and got 1 out in the eighth inning before being relieved, we would record his total innings pitched as $7\frac{1}{3}$ innings. If he had gotten two outs before being relieved, we would have recorded his total innings pitched as $7\frac{2}{3}$ innings. However, what the newspapers print is 7.1 and 7.2. Explain why this is mathematically incorrect in base ten. Why do you think they do this?

DISCUSSION

$\frac{1}{2}$	$\frac{1}{3}$	$\frac{1}{4}$	$\frac{1}{5}$	$\frac{1}{6}$	$\frac{1}{8}$	$\frac{1}{10}$	$\frac{1}{50}$	$\frac{1}{100}$	$\frac{1}{1000}$
.5	$.\overline{3}$	.25	.2	$.1\overline{6}$	.125	.1	.02	.01	.001

Either by hand or with a calculator, all but $\frac{1}{3}$ and $\frac{1}{6}$ can readily be converted into decimal form. When we convert $\frac{1}{6}$ into a decimal (by dividing 1 by 6), we find that it repeats; that is, the value is 0.1666666. . . . There are two ways to write repeating decimals. We can write $\frac{1}{6}$ in decimal form as 0.166 . . . or as $0.1\overline{6}$. The bar shows the part that repeats. For example, we would write $\frac{348}{999}$ as 0.348348348 . . . or as $0.\overline{348}$.

B. Now let us examine translating decimals into fractions. Translate the decimals 0.1, 0.007, and $0.\overline{18}$ into fractions before reading on. . . .

DISCUSSION

Most people have no trouble with the first two: $\frac{1}{10}$ and $\frac{7}{1000}$. The last decimal is equivalent to $\frac{2}{11}$. Someone (unknown to history) discovered the process that enables us to conclude that there is a unique fraction associated with every repeating decimal. The process works like this:

Let $x = 0.18181818\ldots$

That is, $x = 0.\overline{18}$

Multiply both sides of this equation by 100: $100x = 18.\overline{18}$

Subtract the first equation from the second: $99x = 18$

Solve for x: $x = \frac{18}{99} = \frac{2}{11}$

DECIMALS, FRACTIONS, AND PRECISION

We have used number lines as a model to represent whole numbers and fractions. The number line can also be used to represent decimals, using the analogy of a zoom lens. Suppose you had never heard of decimals and had to determine the length of the wire in Figure 4.43 as accurately as you could. Do this and then read on. . . .

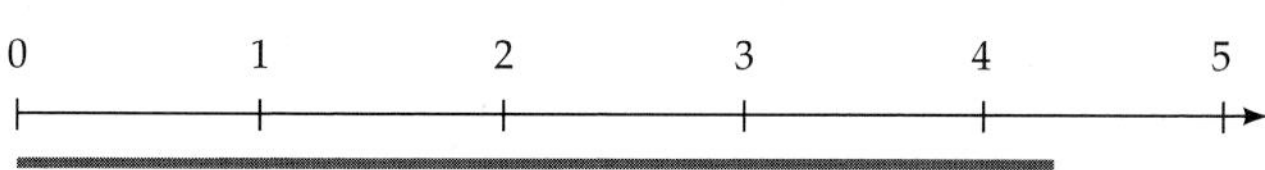

Figure 4.43

MATHEMATICS

Do you see how the number line model connects to the density principle for fractions that we discussed in Section 4.2?

A visual inspection of Figure 4.43 shows that the wire appears to be closer to $4\frac{1}{3}$ inches than to $4\frac{1}{4}$ inches. With decimals, we do not look at the most appropriate unit fraction; instead, we look at the fractional length in terms of tenths, hundredths, thousandths, and so forth. Now imagine using a zoom lens (Figure 4.44). To the nearest tenth of an inch, how long is the wire?

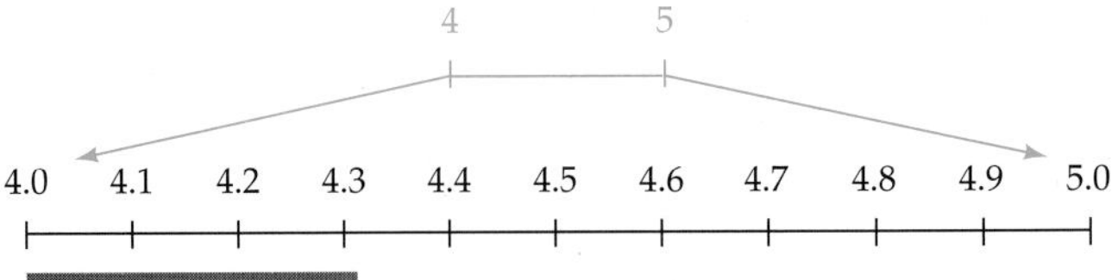

Figure 4.44

Because the length is closer to 4.3 inches than to 4.4 inches, we say that the length is 4.3 inches. If we want more precision, we can keep zooming in, in which case we move more decimal places to the right. Theoretically, we can continue this magnification process indefinitely. Realistically, electron microscopes are able to measure distances of about one ten-millionth of an inch, or 0.0000001 inch. As we will see in Chapter 9, an advantage of the metric measuring system is that when reading a metric ruler, the centimeters are divided into tenths (millimeters).

INVESTIGATION 4.4e Ordering Decimals

Just as young children need time to develop proper ordering relationships with whole numbers (for example, that 30, not 31, comes after 29), older children need time to develop ordering relationships with decimals.

Order the following decimals, from smallest to largest: 0.39, 0.046, and 0.4. Justify your solution.

DISCUSSION

STRATEGY 1 Use equivalent fractions with common denominators

$$0.39 = \frac{39}{100} = \frac{390}{1000}$$

$$0.046 = \frac{46}{1000}$$

$$0.4 = \frac{4}{10} = \frac{400}{1000}$$

Thus the order is 0.046, 0.39, 0.4.

STRATEGY 2 Line up the decimal points and add zeros

$$0.39 = 0.390$$

$$0.046 = 0.046$$

$$0.4 = 0.400$$

STRATEGY 3 Represent them on a number line

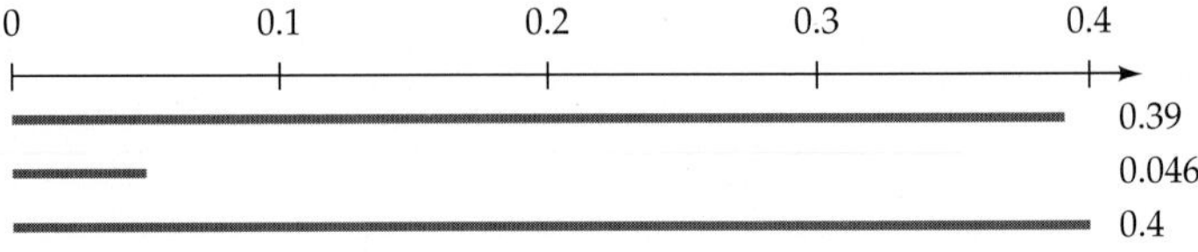

CLASSROOM CONNECTION

Grade 4

Look through these problems for fourth-graders. Are there any that challenge you?

Date ______________________ Time ______________________

LESSON 4·3

Ordering Decimals

1. Write < or >.

SRB 32 33

a. 0.24 _____ 0.18 b. 0.05 _____ 0.1 c. 0.2 _____ 0.35

d. 1.03 _____ 0.30 e. 3.2 _____ 6.59 f. 25.9 _____ 25.72

2. Write your own decimals to make true number sentences.

a. _______ > _______ b. _______ < _______ c. _______ < _______

3. Put these numbers in order from smallest to largest.

a. 0.05, 0.5, 0.55, 5.5 _______ _______ _______ _______
smallest largest

b. 0.99, 0.27, 1.8, 2.01 _______ _______ _______ _______
smallest largest

c. 2.1, 2.01, 20.1, 20.01 _______ _______ _______ _______
smallest largest

d. 0.01, 0.10, 0.11, 0.09 _______ _______ _______ _______
smallest largest

4. Write your own decimals in order from smallest to largest.

_______ _______ _______ _______
smallest largest

5. "What's green inside, white outside, and hops?"
To find the answer, put the numbers in order from smallest to largest.

0.66	1	0.2	1.05	0.90	0.01	0.75	0.35	$\frac{25}{100}$	$\frac{50}{100}$	0.05	0.09	5.5
N	I	O	C	W	A	D	S	G	A	F	R	H

Write your answers in the following table. The first answer is done for you.

0.01												
A												

83

From *Everyday Mathematics, Grade 4:* The University of Chicago School Mathematics Project: Student Math Journal, Volume 1, by Max Bell et al., Lesson 4-3, p. 83. Reprinted by permission of The McGraw-Hill Companies, Inc.

INVESTIGATION 4.4f Rounding with Decimals

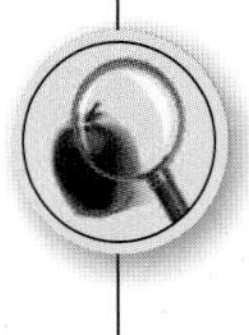

The process of rounding with decimals is very similar to the process of rounding with whole numbers. In fact, if you *really* understand place value, then rounding with decimals is very straightforward.

Round each of the numbers to the nearest hundredth and to the nearest tenth:

A. 3.623 B. 76.199 C. 36.215 D. 2.0368

E. As you work on these, take mental stock of your confidence level. If you do not feel 100 percent confident, what models (discussed earlier) might you use to increase your confidence?

BEYOND THE CLASSROOM

If you stop and think about it, rounding of decimals occurs frequently. For example, the rate I am charged for electricity is 8.82 cents per kilowatt-hour. If someone asked me what the electric rate is, however, I would say about 9 cents per kilowatt-hour. When doing tax forms, the IRS allows people to round some amounts to the nearest dollar. Thus I could record postage expenses as $75 instead of $74.68.

DISCUSSION

A. 3.623 to the nearest hundredth is 3.62, and to the nearest tenth is 3.6.

B. 76.199 to the nearest hundredth is 76.20, and to the nearest tenth is 76.2.

C. 36.215 to the nearest hundredth is 36.22, and to the nearest tenth is 36.2.

D. 2.0368 to the nearest hundredth is 2.04, and to the nearest tenth is 2.0.

E. Some students find the number line or physical models useful in determining how to round. One other common strategy goes like this: Look at 76.199. If we look at only two decimal places, we have 76.19. Therefore, the question can be seen as: Is 76.199 closer to 76.19 or to 76.20?

If we insert zeros so that all three decimals are given in thousandths, we have

76.200

76.199

76.190

Do you see how this strategy helps?

The number 36.215 requires us to deal with a 5 in the relevant place. To the nearest hundredth, this question becomes a decision between 36.21 and 36.22. Remember that with whole numbers, the decision was not to round up whenever we encounter a 5, but rather to round so that the last nonzero digit is an even number. We will apply that rule here. Thus we round 36.215 up to 36.22, but we round 36.245 down to 36.24.

INVESTIGATION 4.4g Decimals and Language

I saw the following line in a newspaper: "The School Board has proposed a budget of $24.06 million." Express the School Board's proposed budget as a whole number. Do this before reading on. . . .

DISCUSSION

I have received many different answers from students in my own class, including the following:

24 million 600 thousand dollars

24 million 60 thousand dollars

24 million 6 hundred dollars

24 million and 6 dollars

24 million dollars and 6 cents

Let us look at several strategies that students have used to determine the correct answer.

LANGUAGE

The language we use when we write checks is connected to place value. How would you write this amount: \$206.45?

It would be incorrect to say "two hundred and six dollars and 45 cents." Can you see why? To reduce confusion, the convention is to use *and* only once—just before giving the cents. Thus the conventional way to write \$206.45 is "two hundred six dollars and 45 cents."

STRATEGY 1 Connect decimals to fractions

Some students who have successfully used this strategy did not know how to represent 24.06 million at first, and so they began with decimal / fraction connections that they did know.

24.5 million	= 24,500,000	because $0.5 = \frac{1}{2}$
24.25 million	= 24,250,000	because $0.25 = \frac{1}{4}$
24.1 million	= 24,100,000	because $0.1 = \frac{1}{10}$ [This is often a hard place for many students to start, but it makes sense now, in light of the first two steps. Does it make sense to you?]
24.06 million	= 24,060,000	because it must be less than 24,100,000

STRATEGY 2 Connect decimals to expanded form

Some students found the solution by analyzing the meaning of 0.06 million.

24.06 million means 24 million and six hundredths of a million.

Now we have to find the value of six hundredths of a million.

Using a calculator or computing by hand yields

$$0.06 \times 1{,}000{,}000 = 60{,}000$$

Thus,

$$24.06 \text{ million} = 24{,}000{,}000 + 60{,}000 = 24{,}060{,}000$$

Once again we come back to the importance of being able to compose and decompose numbers: whole numbers, integers, fractions, and now decimals.

STRATEGY 3 Connect decimals to place value

If your understanding of place value is powerful, you know that the *places* of the digits will not change. Thus, when we represent 24.06 million as a whole number, we simply "add zeros" and we have 24,060,000. The *place* of the 6 does not change!

OPERATIONS WITH DECIMALS: ADDITION AND SUBTRACTION

Explorations Manual 4.19

With the widespread use of calculators, most people no longer do most decimal computations longhand. However, it is crucial to understand why the basic algorithms work because it is often necessary to interpret what the calculator displays.

Few people make mistakes when adding or subtracting decimals if all the numbers have the same number of places, such as 3.24 + 5.56. However, a majority of middle school students have difficulty with a problem like the following: 7.8 + 0.46. Can you explain how to determine the sum and justify each step? You may recall being taught to "line up the decimal places," but why does that rule work?

$$\begin{array}{r} 7.80 \\ +0.46 \\ \hline 8.26 \end{array}$$

We can justify the lining up of decimal places in terms of whole-number computation. When adding whole numbers, we begin with the ones place and move to the left, regrouping as necessary. Adapting this procedure to decimals, we begin with the smallest place—in this case, the hundredths place. We have 6 hundredths in 0.46 and 0 hundredths in 7.8. Moving to the tenths place, we have 12 tenths (regrouped to 1 and 2 tenths). Moving to the ones place, we have 7 plus the regrouped 1, giving us a sum of 8 ones, 2 tenths, and 6 hundredths—that is, 8.26. Just like we have to add ones to ones, tens to tens, and hundreds to hundreds when adding whole numbers, and we have to add the same size pieces when adding fractions, such as twelfths to twelfths, so too with decimals we have to add tenths to tenths and hundredths to hundredths.

MULTIPLICATION WITH DECIMALS

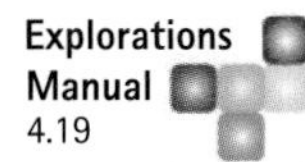

Children often find multiplication with decimals confusing. Let us examine two problems that arise:

- How do we know where to put the decimal point?
- How can you multiply tenths by tenths and get hundredths—for example, $3.2 \times 2.6 = 8.32$?

Let us examine the problem 3.2 meters × 2.6 meters.

Before reading on, please work on the following questions:

- How can you determine the area of the rectangle directly from the diagram shown below?
- What are the connections between the diagram and the standard algorithm for multiplying decimals?

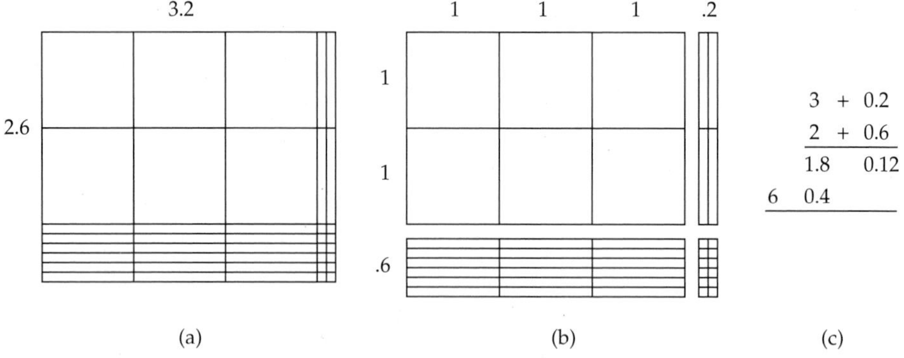

Figure 4.45

In this diagram, we have four distinct regions that correspond to the four partial products [see Figures 4.45(b) and (c)]. The top left region consists of 6 squares (each of which measures 1 meter by 1 meter). Thus the top left region has a value of 6 square meters.

The top right region consists of 4 longs, each of which measures 1 meter by 0.1 meter. Thus each long has a value of 0.1 square meter. The value of the top right region, then, is 0.4 square meter.

The bottom left region consists of $3 \times 6 = 18$ longs and has a value of 1.8 square meters.

The bottom right region consists of 12 little squares, each of which is 0.1 meter by 0.1 meter. Thus the value of each square is 0.1×0.1, or 0.01 square meter. Therefore, the value of this region is 0.12 square meter, because we have $2 \times 6 = 12$ of these little squares.

Adding up the value of the four regions, we have $6 + 0.4 + 1.8 + 0.12 = 8.32$ square meters.

The calculation on the left below shows the sums of the four regions, and the calculation on the right shows the standard multiplication procedure. Do you see the connections?

3.2	3.2
×2.6	×2.6
1.8	192
.12	64
6	8.32
·4	
8.32	

We are still applying the distributive property. Going back to the meaning of whole-number multiplication, the multiplication algorithm simply tells us the sums of the four regions. The beauty of the procedure is more evident when we realize that $3.2 \times .6 = 1.92$ and $3.2 \times 2 = 6.4$ don't "line up." However, when we move 64 over one place, the two partial sums line up just right. We are not generally aware of this lining up; we just remember how many decimal places to move over in the answer column.

DIVISION WITH DECIMALS

MATHEMATICS

This is yet another example of the ideas of multiple representations and equivalence helping us to understand mathematical ideas more deeply!

Most of you know the procedure for dividing with decimals. For example, let us say that you travel 247.9 miles on 7.4 gallons of gas. To determine the miles per gallon, you divide 247.9 by 7.4. If you don't have a calculator, you know to move the decimal points over one place; you also know that the answer is 33.5 because you move the decimal point in the dividend straight up. This procedure is illustrated below. Can you provide a mathematical justification for moving the decimal point? Think and then read on. . . .

$$7.4\overline{)247.9} \longrightarrow 74\overline{)2479.0}$$

```
          33.5
7.4)247.9 ---> 74)2479.0
                  222
                   259
                   222
                    370
                    370
                      0
```

One key is the idea of multiple representations, which has been presented many times. An equivalent representation of this division problem is $\frac{247.9}{7.4}$. This form connects to the idea of equivalent fractions. If we multiply this decimal fraction by $\frac{10}{10}$, we have

$$\frac{247.9}{7.4} \times \frac{10}{10} = \frac{2479}{74}$$

In other words, $\frac{247.9}{7.4}$ has the same value as $\frac{2479}{74}$. By moving the decimal point over, we create an equivalent computation in which the divisor (denominator) is a whole number.

INVESTIGATION 4.4h Decimal Sense: Grocery Store Estimates

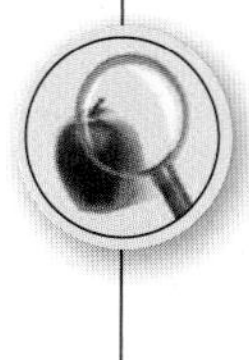

Explorations Manual 4.22

In this and the next investigation, we will examine some estimation problems. We will pay attention to adapting estimation strategies developed with whole numbers and fractions and to using properties developed in Chapter 3 when appropriate.

Let's say that it is Friday afternoon. You drop by the supermarket to get a few items. Suddenly you realize that you didn't go to the bank, and you find that you have only \$11. The milk is \$2.95. The ice cream is \$1.87. Grapes are 89¢ a pound, and you have just over a pound. The bag of potato chips is \$2.69. The loaf of bread is \$3.29. Do you have enough money? How would you estimate this amount in your head? Try it (see the column of numbers at the left) and then read on. . . .

2.95
1.87
.89 a pound
2.69
3.29

DISCUSSION

STRATEGY 1 Round up and use compatible numbers

\$2.95 is almost **\$3**
\$2.70 + 3.30 = **\$6**
\$1.87 + \$.89 ≈ \$1.90 + \$.90 = **\$2.80**
\$3 + \$6 + \$2.80 = **\$11.80**

We know that the actual sum is lower than this estimate. In situations like this, we intentionally overestimate because we don't want to find out at the checkout counter that we don't have enough money.

This next investigation offers several ways to apply ideas from different concepts we have studied.

STRATEGY 2 Use leading digit and compatible numbers

Looking at the "dollars column" (that is, the ones column), we have 4. Next, we look at the "dimes" (tenths) column.

This is what that strategy sounds like:

- 8 (dimes) + 2 (dimes) is one dollar.
- 9 (dimes) + 6 (dimes) is a dollar fifty, so that's two fifty.
- 8 (dimes) more makes three thirty, plus the 8 (dollars). That's eleven thirty plus the cents. You don't have enough money!

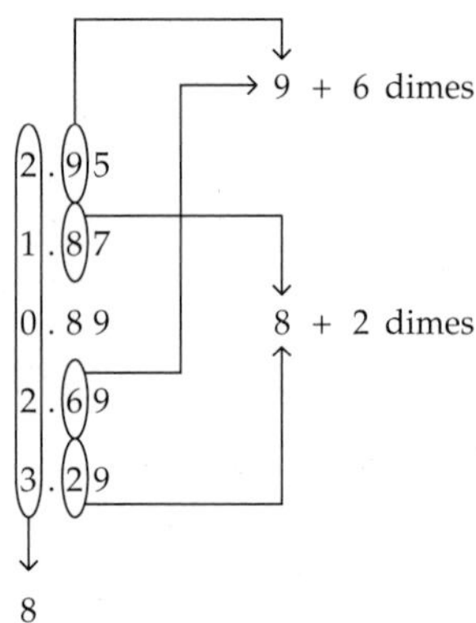

This is just a reminder that there are many ways to estimate. The focus should not be on "getting" the "right" way, but rather on developing ways that make sense to you and that are reasonably sophisticated.

INVESTIGATION 4.4i Decimal Sense: How Much Will the Project Cost?

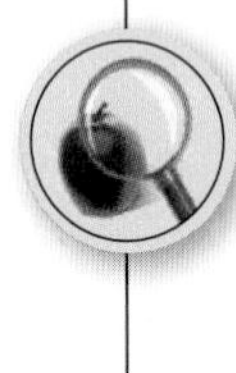

Let's say a contractor needs 308 sheets of plywood for a project. He can get them for \$9.55 each at one store. He could also get them for \$9.39 at another store, but more paperwork would be required. How much money would he save if he went to the second store? Estimate the answer as closely as you can without using a calculator or pencil and paper. Then read on. . . .

Explorations Manual 4.21

DISCUSSION

We could try to estimate 308×9.55 and 308×9.39, but this would be quite tedious. With a bit of reflection, we realize that what we need to know is not the approximate total cost but the difference. Because the difference is 16¢, or \$.16, per sheet, we find that we really have to estimate $308 \times \$.16$ or $308 \times 16¢$.

STRATEGY 1 Converting to fractions

$$0.16 \times 308 \approx \frac{1}{6} \times 300 = \$50$$

STRATEGY 2 Use substitution and the associative property

$$\begin{aligned} 0.16 \times 308 &\approx 0.16 \times (100 \times 3) \\ &= (0.16 \times 100) \times 3 \\ &= 16 \times 3 \\ &= \$48 \end{aligned}$$

CLASSROOM CONNECTION

Consider these examples from the National Assessment of Educational Progress. Students were given the following problem and asked to circle the correct product among five choices.

Select the correct answer to 3.04 × 5.3.

Many of you might say, "Well, 72 percent of all eleventh-graders is not that bad" (Table 4.2). But now look at a very similar problem, which actually should have been easier because the students were asked only to *estimate* the product.

TABLE 4.2

PERCENT CORRECT	
Grade 7	**Grade 11**
57	72

Which of the following is the best estimate of 3.04 × 5.3?

As Table 4.3 shows, barely one out of five seventh-graders was able to see that 16 is the best estimate of 3.04 × 5.3! In fact, more seventh-graders chose 1.6, probably because it was the only answer that had a decimal point. Barely one out of three eleventh-graders recognized that 16 is the best estimate of 3.04 × 5.3!

TABLE 4.3

	PERCENT RESPONDING	
	Grade 7	**Grade 11**
1.6	28	21
16	21	37
160	18	17
1600	22	11
I don't know	9	12

INVESTIGATION 4.4j How Long Will She Run?

A runner is running at 7.6 meters per second. At this rate, about how long will it take her to run 100 meters?

DISCUSSION

This problem is a classic example of how decimals can obscure simple relationships. The problem-solving strategy of using simpler numbers is powerful here. For example, suppose the runner was running at 5 meters per second. How long would it take her to run 100 meters? With these simpler numbers, it is almost immediately apparent that it will take 20 seconds; that is, divide 100 by 5. We can now apply this understanding to this problem and see that it will take her $\frac{100}{7.6}$, or about $\frac{100}{8}$ seconds, which is about 12 or 13 seconds. Over my teaching career, I have seen many students get very intimidated by "ugly" numbers. When you encounter "ugly" numbers, you can use this strategy of thinking it through with simpler numbers.

INVESTIGATION 4.4k Exponents and Bacteria

With the development of negative numbers and decimals, we are now poised to understand better the operation of exponentiation, which is an operation that we have already referred to (for example, in describing place value relationships).

If a kind of bacteria doubles every hour, how many bacteria will there be after 24 hours if we begin with 1 bacterium? Think and then read on. . . .

DISCUSSION

There are several ways to address this problem. For example, we can make a table to represent the growth. The third column simply represents the number of bacteria, using exponents.

HISTORY

In the 1500s, Francois Viete and Michael Stifel introduced symbols for unknowns and powers, for example, x and x^2. Before this time, writing algebraic expressions was very tedious. For example, the expression we write as $x^2 + 2x - 8$ would have been written as Zp 2Rm 8 in the 1500s.[7] Let R represent the unknown amount, Z, the square of the unknown amount, p for plus, and m for minus. Can you imagine solving equations or factoring in the sixteenth century? With this primitive notation, it was very cumbersome.

Does this help you? Using exponents, how would you represent the number of bacteria after 24 hours?

Number of hours	Number of bacteria	Number of bacteria
0	1	
1	2	
2	4	2^2
3	8	2^3
4	16	2^4
5	32	2^5
.	.	.
.	.	.
.	.	.

The number of bacteria after 24 hours will be 2 to the 24th power. Do you see why? The pattern in the table shows that the exponent is the same as the number of hours in each case.

Exponents and multiplication Just as multiplication can be seen as repeated addition, exponentiation can be seen as repeated multiplication, as in Figure 4.46.

$$b^x = \overbrace{b \cdot b \cdot b \cdot b \cdots b \cdot b \cdot b}^{b \text{ occurs } x \text{ times}}$$

Figure 4.46

In mathematical language, b is called the **base** and x is called the **exponent**. We say

$b^2 = b$ **squared**

$b^3 = b$ **cubed**

$b^4 = b$ to the fourth power

$b^x = b$ to the xth power

Computing exponents This then brings us to another problem: How do we determine the value of 2^{24}? What do you think? If you remember anything about exponents, explore alternatives for a while before reading on. If you don't recall anything about exponents, read the following discussion first and then try. . . .

In one sense, we want to know how to make the following computation less tedious:

$$2 \cdot 2$$

One (of several) possibilities is to translate this computation into an equivalent computation, using the associative property of multiplication:

$$(2 \cdot 2 \cdot 2 \cdot 2) \cdot (2 \cdot 2 \cdot 2 \cdot 2) \cdot (2 \cdot 2 \cdot 2 \cdot 2) \cdot (2 \cdot 2 \cdot 2 \cdot 2) \cdot (2 \cdot 2 \cdot 2 \cdot 2) \cdot (2 \cdot 2 \cdot 2 \cdot 2)$$

That is, it is much quicker to press $16 \cdot 16 \cdot 16 \cdot 16 \cdot 16 \cdot 16$ on a calculator than to press 2 twenty-four times.

If you have a scientific calculator, pressing the following entries yields the answer:

2 $\boxed{y^x}$ 24 = 16777216, that is, 16,777,216, and 16 $\boxed{y^x}$ 6 = 16,777,216 also.

Negative exponents When the exponent is 2 or greater, we can interpret it as repeated multiplication. However, what if the exponent is less than 2? For example, what are the meanings and value of the following amounts: 2^1, 2^0, 2^{-1}, 2^{-2}?

One way to understand the meaning and value of such exponents is to look at the connections among all exponents. For example, what is happening to the value of the expression each time we decrease the exponent? Try to verbalize this and then read on. . . .

$$2^4 = 16$$
$$2^3 = 8$$
$$2^2 = 4$$
$$2^1 = ?$$
$$2^0 = ?$$
$$2^{-1} = ?$$

One way to verbalize the pattern is to say that each time the exponent decreases, the value of the expression is divided by 2. It is reasonable to assume that this pattern will continue, and thus it is reasonable to conclude that the value of 2^1 is $4 \div 2 = 2$. This line of reasoning enables us to continue the progression. In this manner, we deduce that the value of 2^0 is $2 \div 2 = 1$.

Continuing this pattern, $2^{-1} = 1 \div 2 = \frac{1}{2}$, and so $2^{-2} = \frac{1}{2} \div 2 = \frac{1}{4}$.

Can you verbalize the pattern now so that you can give the value of 2^{-n}? Work on it and then read on. . . .

Once again, we rely on multiple representations. We can represent 2^{-2} as $\frac{1}{4}$, or we can represent 2^{-2} as $\frac{1}{2^2}$, because $4 = 2^2$. This now leads us to the generalization that

$$2^{-n} = \frac{1}{2^n}$$

INVESTIGATION 4.4I Scientific Notation: How Far Is a Light-Year?

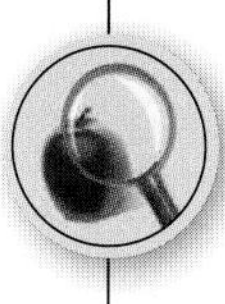

We use exponents to express very small and very large numbers. In many cases, the numbers that we use are so large or so small that computation becomes very cumbersome. For example, the U.S. federal debt in March 2006 was \$8.2 trillion, which is 8,200,000,000,000. The wavelength of red light is 0.00000000000000586 meter. The following situation is a good introduction to the need for an alternative notation, which we call *scientific notation.*

One way in which astronomers address the hugeness of interstellar distance is to measure astronomical distances not in miles or kilometers but rather in a unit called "light-years"—that is, the distance light travels in one year. The speed of light is 186,000 miles per second. Determine the length of a light-year. Work on this yourself before reading on. . . .

DISCUSSION

In this problem, dimensional analysis can greatly simplify the computations.

$$1 \text{ light-year} = \frac{186{,}000 \text{ miles}}{1 \text{ second}} \times \frac{60 \text{ seconds}}{1 \text{ minute}} \times \frac{60 \text{ minutes}}{1 \text{ hour}} \times \frac{24 \text{ hours}}{1 \text{ day}} \times \frac{365.25 \text{ days}}{1 \text{ year}}$$

Do the multiplication now on a calculator.

Most of you will have something like this:

5.8697 12 or 5.8697136E 12

which is the calculator's code for 5.8697×10^{12}.

Let's work with smaller and simpler numbers to ensure that you can understand how scientific notation works. Consider the number 34,000. One way to represent this amount that connects to place value comes from the realization that this number means 34 thousands. In symbols, we can thus say

$$34{,}000 = 34 \times 1000$$

We could also represent 34 as 3.4×10. Substituting 3.4×10 for 34, we have

$$3.4 \times 10 \times 1000$$

We can simplify this as

$$3.4 \times 10{,}000$$

Now we can substitute 10^4 for 10,000, and we have

$$3.4 \times 10^4$$

Formally, we say that a number is in **scientific notation** if it is in the form $a \times 10^b$, where a is a number between 1 and 10, $1 \le a < 10$, and b is an integer.

The reasoning behind restricting a to a number between 1 and 10 is based on convention rather than mathematical structure. This convention makes it easier for us to compare amounts. For example, if we have 6.4×10^7 and 5.6×10^8, we immediately know that the second amount is larger. However, if we have 0.64×10^8 and 56×10^7, it is not obvious that the second amount is larger.

Now let us return to our problem of representing the length of a light-year. Translate this amount into a whole number and then read on. . . .

One strategy is to write the decimal, and then move the decimal point 12 times:

$$5.869700000000$$

When we insert the commas, we have 5,869,700,000,000. How would we say this amount?

We would say 5 trillion, 869 billion, 700 million miles, or we would round further to 5.9 trillion miles.

IRRATIONAL NUMBERS

Thus far, our investigation of numbers has proceeded from counting numbers to whole numbers to integers to fractions and to decimals. There is one more set of numbers that we shall consider in this course. The discovery of this set of numbers is also one of the more dramatic stories in the history of mathematics. Many important contributions to mathematics came from a group of people who called themselves Pythagoreans. The Pythagoreans, a community of mathematicians led by Pythagoras, existed in the 500s B.C. Some interesting things about the Pythagoreans were that vegetarianism was strictly followed and that women and men were treated equally in the Pythagorean community. As you may recall, one of their core beliefs was that all the laws of the universe could be represented using only whole numbers and ratios of whole numbers. Now recall the Pythagorean Theorem (for any right triangle with legs a and b and hypotenuse c, $a^2 + b^2 = c^2$) and my mention that the Greeks were the first people to prove this relationship. Ironically, this proof was part of the undoing of the Pythagoreans. Here is how it happened:

Consider a right triangle in which the length of each of the two sides is one inch, as shown in Figure 4.47.

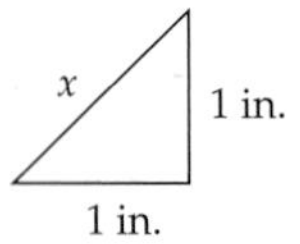

x
1 in.
1 in.

Figure 4.47

If we call the hypotenuse x and apply the Pythagorean theorem, we have

$$1^2 + 1^2 = x^2$$
$$x^2 = 2$$

That is, x is the number that, when multiplied by itself, equals 2.

This problem was perplexing to the Pythagoreans, because try as they might, they could not find a rational number that, when multiplied by itself, came to *exactly* 2.

Using modern notation and base ten arithmetic, we know that we can get *very* close:

$$1.4 \times 1.4 = 1.96$$
$$1.41421 \times 1.41421 = 1.9999899$$

However, there is no rational number (that is, a number that can be expressed as the quotient of two integers) that will produce a product of exactly 2. In modern language, we say $x = \sqrt{2}$.

Finally, one member of the sect proved that there is no rational number that will solve this problem. This discovery was like a child finding out that Santa Claus is not real or an adult finding out that there really are people from other planets. What it meant for the Pythagoreans was that one of their cornerstone beliefs—that all the laws of the universe could be represented as whole numbers and ratios of whole numbers—was not true.

What happened to the person who discovered the proof? Legend has it that he was sent out to sea in a leaky rowboat! This is possibly the source of the saying, "Don't kill the messenger who bears the bad news." It is also interesting to note that the word *irrational,* even in our present time, means "contrary to reason."

INVESTIGATION 4.4m Square Roots

Students in elementary school may come upon situations in which they will encounter square roots that are not rational numbers, so it is important to investigate square roots briefly. Technically, any positive number has two square roots. For example, the square roots of 81 are +9 and −9. Because, in elementary school, square roots are explored only in the context of measurement, and because all lengths are positive, we will focus our explorations on the positive square roots of numbers, also called the **principal square root**.

Before the development of hand-held calculators, schoolchildren had to learn an algorithm for approximating the square root of a number. This algorithm was a standardized procedure that was taught and practiced in much the same way that long division was taught and practiced.

Without using the square root button on your calculator, determine the square root of 800 to the nearest hundredth. As you do so, use the tools that seem most appropriate. Then read on. . . .

DISCUSSION

To determine $\sqrt{800}$, a little estimating and mental math help. For example,

$$30 \times 30 = 900$$

Therefore, a reasonable starting point would be 29. However, if you apply the mental math skills developed in Chapter 3, you realize that 29^2 is still too high. Can you mentally compute 29×29?

We can represent 29×29 as $(30 - 1)(30 - 1)$ and apply the distributive property $900 - 30 - 30 + 1 = 841$. In any case, because we know that $30^2 = 900$ and $29^2 = 841$, a reasonable first guess is 28. If you want to follow the discussion below actively, please cover the table and think about your next guess and your reasoning before reading on. . . .

Guess	Result	Analysis
28	784	Too low. Because 800 is closer to 784 than to 841, the next guess should be closer to 28 than to 29.
28.4	806.56	Not bad! Let's try 28.3.
28.3	800.89	Clearly 28.2 will be too small. Next guess? 28.29? 28.28? Why?
28.29	800.3241	
28.28	799.7584	

Clearly $28.28 < \sqrt{800} < 28.29$. Without getting out your calculator or pencil and paper, which is closer? How did you figure it out? Because 799.7584 is closer to 800 than 800.3241, 28.28 is closer to the square root of 800.

THE REAL-NUMBER SYSTEM

With the set of irrational numbers, we can now extend our system of numbers to the set of **real numbers**, which is defined as the union of the sets of rational and irrational numbers. This is as far as we will go with number systems in this course. The next major expansion, which students encounter in high school, is imaginary and complex numbers—for example, the square root of -1.

Let us look back at the numbers we have examined in this course with respect to how they are related to each other. We began our study of numbers with the set of natural numbers, which is the first set of numbers young children encounter. Our first extension of the number line was to add zero. The union of the set of natural numbers and zero is called the set of whole numbers. We then examined two kinds of numbers that children will encounter in elementary school: integers and rational numbers. Finally, we encountered numbers that cannot be represented as the ratio of two whole numbers—that is, irrational numbers. Figure 4.48 is a visual representation of these different sets of numbers and their relationships.

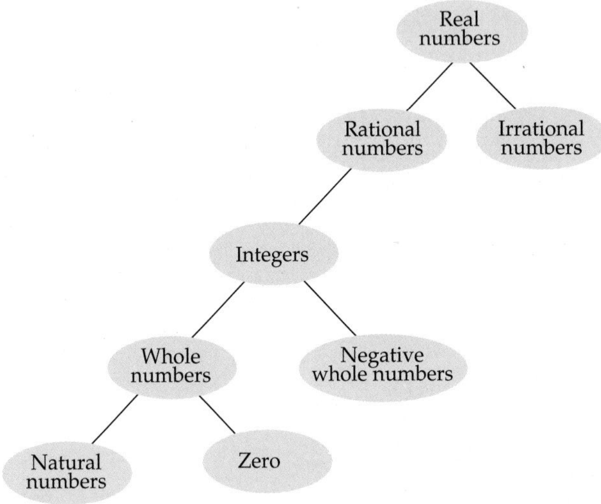

Figure 4.48

PROPERTIES OF REAL NUMBERS

The operations of addition, subtraction, and multiplication are closed for real numbers. That is, for any two real numbers, a and b,

$$a + b \text{ is a real number, } a - b \text{ is a real number, and } a \cdot b \text{ is a real number.}$$

Why do we not say that division is closed for real numbers?

If we were to say that division is closed, we would have to make one qualification. If we say that for any two real numbers a and b, $\frac{a}{b}$ is a real number, we would simply have to note that b cannot equal zero.

Another property that is important for you to know about has to do with the density property of rational numbers, which represents a qualitative difference between the set of rational numbers and the three previous sets. However, as dense as the rational numbers were, we found that the number line was not yet complete, because there are points on the real number line that have no rational name. When we add the set of irrational numbers to the set of rational numbers, our number line is complete. This is known as the **completeness property**.

There is a real-number name for every single point on the number line.

SUMMARY 4.4

To understand decimals, we explored their relationship both to the set of integers and to the set of rational numbers. In doing so, we examined different ways to represent decimals, including base ten blocks and number lines.

When examining connections with decimals, we found that zero presents special issues, just as zero did with integers. When we are representing an amount with a decimal, sometimes the placement of a zero is necessary and sometimes the placement is optional (a convention).

When examining connections with fractions, we found that some decimals terminate ($\frac{1}{2} = 0.5$) but some do not ($\frac{1}{3} = 0.333\ldots = .\overline{3}$).

Understanding decimal computations relied on applying connections with whole numbers and with fractions. When adding and subtracting, we must add digits that represent the same place value, just as we did with whole numbers. To understand why we put the decimal point in a particular place when multiplying, we needed to examine decimal multiplication in the context of an area model, just as we did with multiplication of whole numbers. To justify the movement of the decimal point when we are dividing, however, we made use of the relationship between decimals and fractions.

The operation of exponentiation was introduced as repeated multiplication. Exponentiation enables us to understand a use of decimals that is important in many fields: scientific notation, which is used to represent very large and very small numbers.

Finally, we examined one more extension of the number system, irrational numbers, thus completing our journey through number systems in this course. The set of rational numbers is dense, but the set of real numbers is complete and sufficient for most everyday and business uses of elementary mathematics.

4.4 Exercises

1.

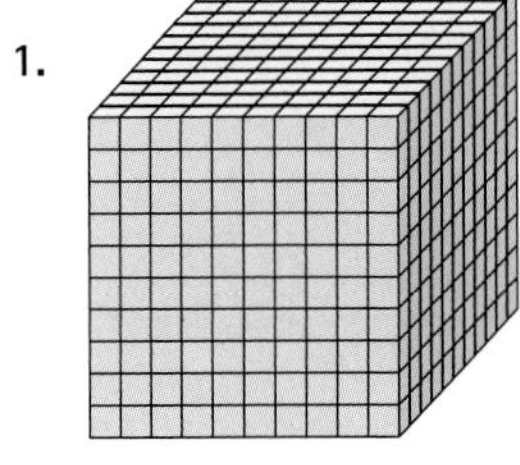
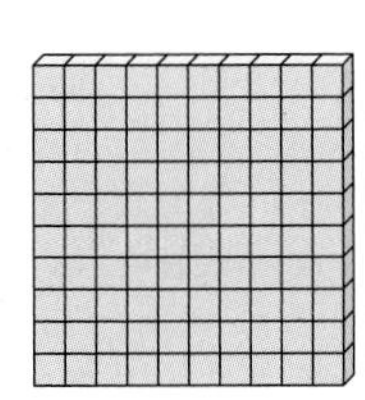
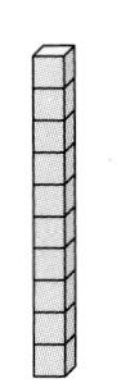

a. Represent 3.4 if the big cube represents 1 unit.

b. Represent 3.4 if the flat represents 1 unit.

c. Represent 2.63 if the big cube represents 1 unit.

d. Represent 2.63 if the flat represents 1 unit.

e. Represent 3.02 if the big cube represents 1 unit.

f. Represent 3.02 if the flat represents 1 unit.

g. What number is represented by the figure on the next page if the flat represents 1 unit?

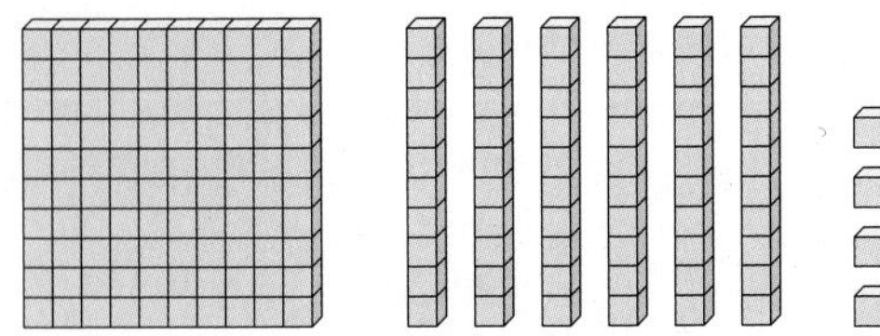

h. What number is represented by the figure above if the big cube represents 1 unit?

i. What number is represented by the figure below if the flat represents 1 unit?

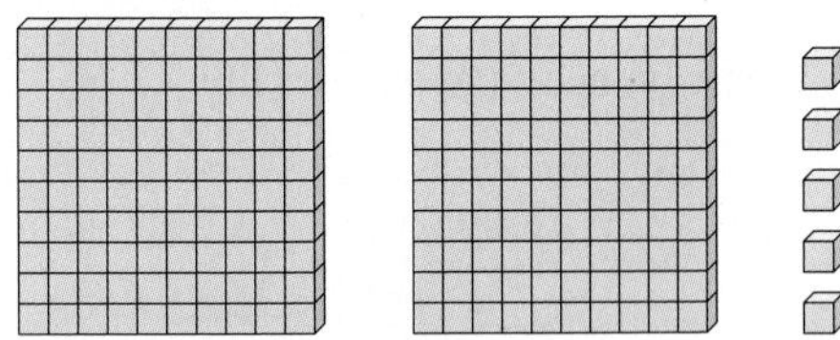

j. What number is represented by the figure above if the big cube represents 1 unit?

2. Express each of the following as directed.

   a. $\frac{3}{40}$ as a decimal b. $\frac{4}{5}$ as a decimal

   c. $\frac{3}{100}$ as a decimal

   d. 0.36 as a fraction in simplest form

   e. 0.005 as a fraction in simplest form

   f. 1.6 as a mixed number

3. Express each decimal in expanded form and as a fraction.

   a. 4.6 b. 0.75 c. 1.234 d. 4.06

4. Using grid paper, represent each of the problems below and explain how to determine the product by interpreting the grid.

   a. 0.3×0.2 b. 1.2×0.4 c. 3.4×2.6

5. Express these decimals in short form using the bar to designate the repeating part. Find the fractional equivalents of each of the decimals below.

   a. .666666666 . . . b. .555555555 . . .

   c. .0909090909 . . . d. .142875142875 . . .

6. In each part, draw a number line and place the given numbers on it.

   a. 3.2, 3.8, 2.9, 3.04 b. 2.43, 2.4325, 2.4396, 2.441

   c. 0.004, 0.00397, 0.00394 d. 0.8, 0.785, 0.81

   e. 0.04, 0.1, 0.039

7. Circle the smaller of the pairs of numbers:

   a. 0.4 0.42 b. 1.03 0.98899

   c. 0.05 0.058 d. 0.302 0.087

   e. 2.72 2.716

8. Which of these numbers is between 0.06 and 0.07?

   a. 0.00065 b. 0.0065 c. 0.065 d. 0.65

9. Order the following decimals from smallest (at the left) to largest (at the right).

   a. 0.56, 0.058, 0.0084, 0.6 b. 1.04, 0.065, 0.0086, 0.9

10. Round the following decimals as directed:

    a. 2.36 to the nearest tenth

    b. 6.04 to the nearest tenth

    c. 2.398 to the nearest hundredth

    d. 5.0006 to the nearest hundredth

    e. 0.9876 to the nearest thousandth

    f. 6.1237 to the nearest thousandth

11. Recall the question about the proposed school board budget in Investigation 4.4g. What if the superintendent recommended adding $380,000 to the proposed budget of $24.06 million? How much would the new budget be, both in whole-number form and in decimal form?

12. Write the following decimal numbers in words.

    a. 32.04 b. 0.00004 c. 508.604

13. Write the following using numbers:

    a. sixty-seven thousandths

    b. four thousand sixty and thirty-four thousandths.

14. Represent each problem with a diagram and then determine the answer from the diagram. Use the grids for *a*, *b*, and *c*. Sketch your own diagram for *d*.

    a. $.2 + .54$ b. $.36 + .05$ c. $.3 \times .5$ d. 3.2×2.4

15. Do the following computations longhand and then check your answers with a calculator.

    a. $3.04 + 5.2$ b. $4.004 + 5.2 + .6 + 7$

    c. $6.2 - 4.35$ d. $2.004 - 0.76$

    e. 4.3×2.6 f. 2.05×8.07

    g. $12 \div .45$ h. $60 \div 2.5$

16. ***Classroom Connection*** In each case, describe the error that the student made, and then indicate what mathematical idea(s) the student did not correctly understand.

    a.
    ```
      .7
     +.5
     .12
    ```
    b.
    ```
       66.9
     - 24.28
       42.68
    ```
    c.
    ```
      32.4
     × .7
     226.8
    ```
    d.
    ```
        3.42
     8)27.4
       24
        3 4
        3 2
          2
    ```

17. Determine the answers mentally and briefly explain your thinking.

    a. $3.4 + .7$ b. $.92 - .28$

    c. $.826 - .799$ d. $.485 - .62$

    e. $.42 - .39$ f. 35×1.2

    g. $8 \times .7$ h. $.25 \times 64$

    i. $.5 \times 62.4$ j. $2.5 \times .2$

    k. $16 \times .125$ l. $88.4 \div 4$

    m. $6.09 \div 10$ n. $45.2 \div 1000$

18. Estimate each of the following; then briefly explain and justify your estimate.

a. 19.4 + 136 + 4.825
b. 23.4 + 24.6 + 5.7 + 34.4
c. 23.34 − 22.56
d. \$10.00 − \$4.34
e. 12.5 × 0.034
f. 2.3 × 6.7
g. 34.2 × 0.007
h. 7.8 ÷ 3.12

DEEPENING YOUR UNDERSTANDING

19. a. Name a decimal between 0.999 and 1.

b. Name a fraction between 3.44 and 3.45.

20. After teaching a lesson on decimals, a sixth-grade teacher asked the students to write out twenty-three hundredths. One student wrote .023. When asked why, the student replied that he had put 23 in the hundredths column! What is the student not understanding?

21. ***Classroom Connection*** Why is there no oneths place?

22. Without doing any computation, determine which is bigger, $(0.8)^3$ or $(0.8)^2$? Briefly explain how you arrived at your decision.

23. a. Find two decimals whose product is 32.76.

b. Find two decimals whose product is 4.76 and one of the decimals has a value less than 1.

c. Find two decimals whose quotient is 0.36.

24. People frequently mispunch the calculator when doing problems.

a. Let's say you were multiplying 24 boards at \$4.50 per board and got \$1080. What is the right answer? What did you punch?

b. Let's say you are adding \$3.45 + \$12.45 + \$16.23 + \$34.45 + \$23.45. You have just entered the last number and hit the = sign, and you realize that you punched \$3.45 for the last number. What could you do with the number that now appears on the calculator rather than starting all over?

c. Let's say you are finding the cost of 12 boards at \$3.45 per board. You have just multiplied 12 × \$3.45 to get \$41.40. Now you decide that you want to buy 14 boards. There are several alternatives to multiplying 14 × \$3.45. Can you find them?

25. Solve each of the following problems. Then explain how you knew whether to multiply or divide and, in the case of division, which number was the divisor and which was the dividend.

a. A box of sand has a volume of 0.8 cubic meter. What is the volume of 0.23 of the box?

b. If a motorcycle gets 40 miles per gallon, how far will it travel on 0.75 gallon?

c. Let's say you make 5 liters of punch, and each cup holds 0.2 liter. How many cups can you make?

d. A box of cookies weighs 0.8 pound. How many boxes can you make with 7.2 pounds of cookies?

e. With 75 roses you can make 5 equal bouquets. How many roses will be in each bouquet?

f. The price of 1 yard of fabric is \$15.00. How much does 0.65 yard cost?

g. A group of 12 friends together bought 5 pounds of cookies. How much does each person get?

h. A rocket travels at a speed of 16 miles per second. How far does it travel in 0.85 second?

i. A piece of elastic can be stretched to 3.3 times its original length. When fully stretched, it is 13.9 meters long. What was its original length?

26. Judy recently took a trip in her car. The trip took 11 hours, including a lunch break of $\frac{1}{2}$ hour. She filled the gasoline tank (11.3 gallons) and noted that the odometer read 38329.8. At the end of the trip, she refilled the tank (12.4 gallons), and the odometer read 38735.4. She paid $\$3.46\frac{9}{10}$ per gallon.

a. How many miles per gallon did the car get on the trip (to the nearest tenth of a mile per gallon)?

b. What was the cost of the gas?

c. What was her average speed?

27. Most gasoline stations price gasoline out to $\frac{9}{10}$ of a cent—that is, $\$3.46\frac{9}{10}$ per gallon rather than just \$3.46 per gallon. How much extra money per day might a gasoline station expect to make from this extra $\frac{9}{10}$ cent per gallon?

28. Henry is currently working at a job that pays \$9.45 per hour, and he regularly works 40 hours per week. Another company has offered to pay him an annual salary of \$24,000. How does this offer compare to his present yearly income? What assumptions did you make in order to solve this problem?

29. Let's say a company receives medicine in 1-gallon (128-ounce) jars and then sells the medicine in vials that hold 1.25 ounces. If we were to ask how many vials can be made from 1 gallon, we would divide 128 by 1.25 and obtain 102.4. What does the 102 mean? What does the .4 mean?

30. A manufacturing plant used 4.2 centimeters of wire on each item. If 3549 cm of wire are used in a day, how many items were manufactured? If the cost of the wire is 3.4¢ per meter, what is the cost per day of the wire?

31. A group of students held a car wash to raise money. They charged \$2.50 for regular cars and \$4.00 for trucks. They washed 150 cars and 68 trucks. If they spent \$15 on soap, sponges, and other supplies, how much profit did they make?

32. The average retail price of a pack of cigarettes was \$5.31 in 2009.

a. If a person smokes 1 pack a day, about how much will it cost for a year?

b. At this rate, how much for 50 years?

33. One day on Sesame Street, a friendly rabbit brought Big Bird a tube of toothpaste that contained one gallon of toothpaste. If regular toothpaste costs \$1.49 for a 2.7-ounce tube, what is the value of the Sesame Street tube, to the nearest cent? [There are 128 ounces in a gallon.]

34. You have decided to open a checking account with Xanadu County Savings and Loan. They offer you two options: (a) pay a flat \$5.00 per month fee or (b) pay a monthly fee of \$2 and 10¢ per check.

a. If you write about 15 checks per month, which option is cheaper?

b. How many checks per month would give you the same fee in both cases?

35. Represent the following numbers in scientific notation:

a. 123,456,789 b. 3,000,000,000,000,000

c. 0.00000000056 d. 0.000000302

36. Perform the following computations using scientific notation:

a. 123,000,000,000 × 34,000,000

b. 123,552,000,000 ÷ 23,400,000

c. 0.0000000034 × 0.0000000045

d. $(0.00043)^3$

37. The national debt at the end of 2009 was just over $12.09 trillion. How could we represent this amount in a way that people could relate to?

38. Jackie is thinking of making extra money by typing papers on her word processor. If she wants to make at least $8 an hour, how much should she charge per page? Assume that she can type 120 words per minute.

39. Let's say the cost of gas just went up by 5¢ per gallon. Approximately how much will this affect the average citizen?

40. Let's say your electric rate is 9.23¢/kWh (kilowatts per hour), and you have 5 different 100-watt lights in the house that run from 5 P.M. until 11 P.M. (1 in the kitchen, 2 in the living room, 1 in the dining room, and 1 in the bedroom). How much do you pay per day for running these lights? (Note: 1000 watts = 1 kilowatt.)

41. I bought a cheap candle for 89¢. It burns at the rate of 4.2 cm per hour. I bought a more expensive candle for $1.59, which burns at the rate of 2.8 cm per hour. Which is the better buy?

a. What additional information would you need in order to answer this question?

b. Using the four numbers supplied in this problem, make a problem that can be answered, and answer it.

42. The width of a certain cell is 3×10^{-6} cm. If we placed 250 of these cells side by side, how long would that line be? Express this length without using scientific notation.

43. a. Explain how you would determine the length of a light-year if you had a calculator that did not "know" scientific notation—that is, that just diplayed an E sign for answers that were too big. Justify your work. What properties that we developed in Chapter 3 did you use?

b. Determine the following product with such a calculator: 14,000 × 3,356,000 × 7,890,000

44. a. Suppose a particular bacterium can divide into two bacteria every 45 minutes. If this process continues for 48 hours, how many bacteria will there be? Express your answer in scientific notation.

b. What if we had started with 100 bacteria instead of 1? Predict the answer without doing any pencil-and-paper or calculator computation. Justify your reasoning.

c. What if we had started with 1 bacterium that doubled every 16 minutes. How many would there be after 48 hours?

Problems 45–50 involve translating between decimal and nondecimal systems.

45. Recall the discussion of reporting of innings pitched in baseball on page 227. Why do you think this is reported in a way that is mathematically inaccurate?

46. Our system of money is not a pure decimal system. Why not? Actually, this question has two parts:

a. What aspects of our money system are not aspects of a pure decimal system?

b. Why don't we use a pure decimal system?

47. Convert 7 minutes, 21 seconds to decimal representation, that is 7.__ minutes.

48. Sam worked 4 days, 2 hours, and 15 minutes last week. Convert this to a decimal, that is, 4.__ days. In this problem, one day is equal to 8 hours.

49. For a time after the French Revolution in 1789, the French had a 10-day week, and each day was divided into 10 hours, each consisting of 100 minutes, each minute of 100 seconds. Thus, noon would have been 5 o'clock and 6 A.M. would have been 2.5 or perhaps two-fifty, since it represents one quarter of a day. Convert the following times into the French day.

a. nine A.M. b. four P.M.

Convert the following times from the French into our time:

c. eight o'clock d. three thirty-three

50. Sherry is a hard-working student who is suffering from fatigue. The doctor says that she is not getting enough sleep. Below are her bedtimes for one work week. What is her average bedtime?

10:30 P.M. 11:15 P.M. 11:45 P.M. 1:15 A.M. 11:00 P.M.

51. Think back to the Alphabitian system that we explored in Chapters 2 and 3. Answer each of the following questions without translating the amounts into base ten. That is, answer them as though you were an Alphabitian and that was the only numeration system you knew.

a. Draw a picture to represent the value of 0.A.

b. Add A.B. + C.D.

52. a. Draw a picture to represent 0.1 in base five.

b. Determine the value of 1.2 + 3.4 in base five.

c. Determine the value of 2.4 × 3.2 in base five.

d. Determine the value of $4\overline{)3044}$ in base five.

FROM STANDARDIZED ASSESSMENTS

2008 NECAP, Grade 6

53. Look at this number line.

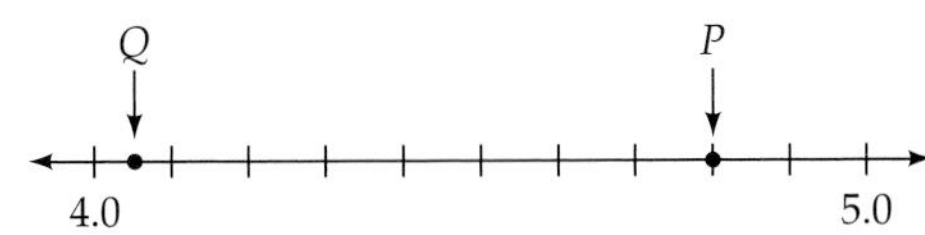

a. What is the value of point *P*?

b. What is the value of point *Q*?

2008 NECAP, Grade 4

54. Libby had a ten-dollar bill. Then she spent $2.85 at a store. How much money does Libby have now?

a. $7.15 b. $7.25 c. $8.85 d. $12.85

55. Of the following, which is closest in value to 0.52?

a. $\frac{1}{50}$ b. $\frac{1}{5}$ c. $\frac{1}{4}$ d. $\frac{1}{3}$ e. $\frac{1}{2}$

Source: Results of the *Seventh NAEP Mathematics Assessment*, p. 251. U.S. Department of Education, National Center for Education Statistics.

56. Jill needs to earn $45.00 for a class trip. She earns $2.00 each day on Mondays, Tuesdays, and Wednesdays, and $3.00 each day on Thursdays, Fridays, and Saturdays. She does not work on Sundays. How many weeks will it take her to earn $45.00?

Source: Results of the *Seventh NAEP Mathematics Assessment*, p. 251. U.S. Department of Education, National Center for Education Statistics.

57. Which of the following numbers is the greatest?

a. 0.36 b. 0.058 c. 0.375 d. 0.4

Source: Results of the *Fourth NAEP Mathematics Assessment*, p. 83. U.S. Department of Education, National Center for Education Statistics.

58. If each grid below has a value of 1, which grid is shaded gray to represent the same decimal as the one marked by P on the number line?

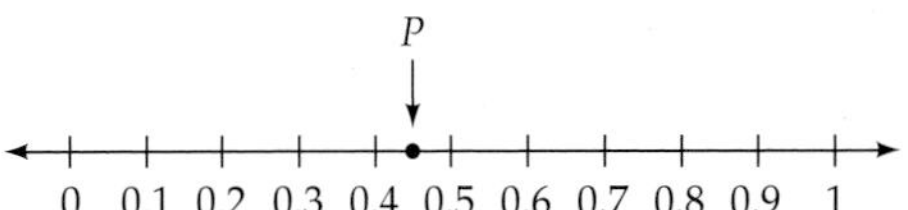

a. 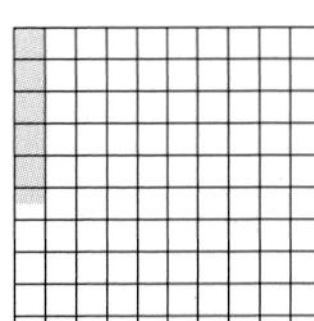

b. 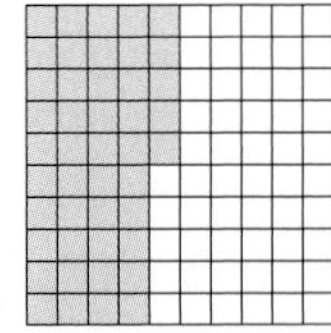

c.

d. 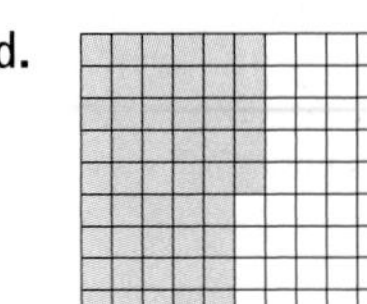

LOOKING BACK on chapter four

QUESTIONS TO SUMMARIZE BIG IDEAS

1. How can we use number line models to represent integer operations?
2. How do the concepts of greatest common factor and least common multiple relate to fractions?
3. How can we use area, linear, and set models to represent fraction operations and fraction equivalence?
4. Why do we have to get a common denominator to add or subtract fractions? Why do we "multiply straight across" when multiplying fractions?
5. How are decimal operations related to whole number operations? Why do we have to "line up" the decimal when we add or subtract decimal numbers?
6. What parts of this chapter are less clear to you at this point? What will you do to clarify those ideas?
7. Look back at the Mathematical Practices of the CCSS. In what ways did you engage in those practices during this chapter?

CHAPTER 4 SUMMARY

1. The set of integers is simply an extension of the set of whole numbers.
2. Operations with positive whole numbers can be adapted to work with integers.
3. A fraction is not simply a number; rather, a fraction expresses a relationship between two quantities. The numerator and denominator can be seen as a code that tells us the relative size of the fraction.
4. A fraction can be interpreted in four ways: as measure, as quotient, as operator, and as ratio.
5. Certain important ideas apply in all the fraction contexts:
   - Something is to be partitioned into parts of equal size (value).
   - The something can have a value of 1, in which case the unit = the whole.
   - The something can have a value $\neq 1$, in which case the unit $\neq$ the whole.
6. The set of decimals has important connections with the set of integers and with the set of fractions.
7. All the sets of numbers that children will study in elementary school are subsets of the set of real numbers.

BASIC CONCEPTS

Section 4.1: Integers

integers **164**
negative integers **164**
absolute value **167**
opposite **167**
negative **167**
additive inverse **167**

Section 4.2: Fractions and Rational Numbers

rational number **175**
fraction **175**
numerator **175**
denominator **175**
fractions as a quotient, as a measure, as an operator, as a ratio **176–177**
models to represent fractions:
area models, length models, set models **176–177**
equivalent fractions **187**
simplest form **188**
greatest common factor **189**
benchmark **192**
transitive property **192**
fractions are dense **194**

Section 4.3: Understanding Operations with Fractions

lowest common denominator **201**
proper fraction **204**
improper fraction **204**
mixed number **204**
multiplicative inverse **214**
estimation and mental arithmetic with fractions **206–210**

Section 4.4: Beyond Integers and Fractions: Decimals, Exponents, and Real Numbers

decimals **223**
base **236**
exponent **236**
squared **236**
cubed **236**
scientific notation **238**
irrational numbers **238**
principal square root **239**
real numbers **240**
properties of real numbers **240**
completeness property **241**

REVIEW EXERCISES chapter four

1. a. What is the additive inverse of 4?
   b. What is the multiplicative inverse of 4?
2. Perform these computations:
   a. $-36 + (-38)$
   b. $16 - (-3)$
   c. $\dfrac{(-26 + 50)}{-4}$
   d. $-3 \cdot 5 + \dfrac{-6}{2}$
3. Find the missing number: $-8(\square + 5) = -24$
4. Express $6 \times (-3)$ as a repeated addition problem.
5. It is conventional to use − to represent "minus" and − to represent "negative." Explain the difference using the example: $-3 - 4 = -7$?
6. Does the shaded part of the following figure below validly represent $\frac{3}{4}$? Why or why not?

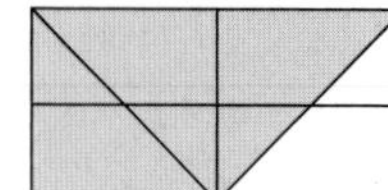

7. Name a fraction between $\frac{5}{6}$ and $\frac{6}{7}$ that is in simplest form. Justify your answer.

8. Name the fraction that tells the value of x on this number line. Explain your reasoning.

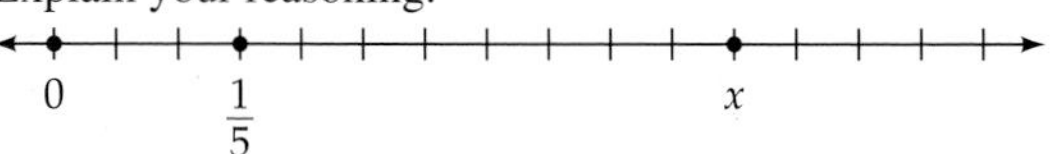

9. If this figure has a value of 1, show $\frac{1}{8}$.

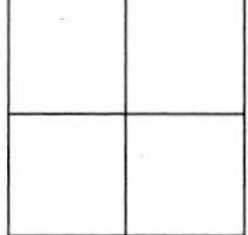

10. If this figure has a value of $1\frac{1}{2}$, show 1.

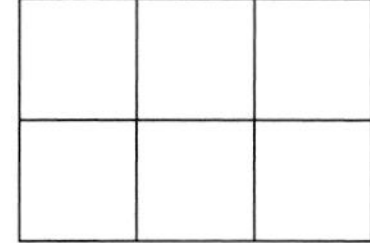

11. If this figure has a value of $\frac{3}{4}$, show $\frac{1}{2}$.

12. For each of the pairs of fractions below, determine which of the fractions is larger without converting the fractions to decimals, finding the LCM, or drawing a diagram to see which is bigger. Briefly explain your reasoning.

a. $\frac{9}{11}, \frac{13}{15}$ b. $\frac{7}{12}, \frac{13}{28}$

13. When we express $\frac{1}{2}$ as a decimal in base ten, we get 0.5. When we express $\frac{1}{2}$ as a decimal in base five, what do we get?

14. What does it mean to say that two fractions are equivalent?

15. Why do two fractions have to have a common denominator in order for us to add them?

16. An approach that children often take to find the product of, say, $3\frac{1}{4}$ and $2\frac{1}{2}$ is to multiply the two whole numbers and then multiply the two fractions; in our example this yields an answer of $6\frac{1}{8}$. Explain why this hypothesis is invalid.

17. Solve the problem $7\frac{3}{4} \times 2\frac{1}{3}$ by a means other than the conventional algorithm of converting to improper fractions and multiplying the two numerators and the two denominators.

18. The distances between the points on the following number line are not drawn to scale. Without determining the actual product, determine the region in which $\frac{8}{15} \times \frac{5}{9}$ lies. Justify your reasoning.

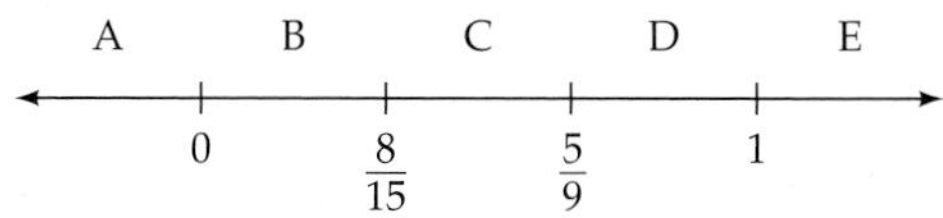

19. Pauline has 35 cups of flour. She makes cakes that require $2\frac{1}{4}$ cups each. If she makes as many such cakes as she has flour for, how much flour will be left over?

20. Without actually doing the problem $4\frac{3}{4} \div 8\frac{7}{8}$, would you estimate that the answer will be greater or less than $\frac{1}{2}$? Explain your reasoning.

21. Make up a realistic story problem for $2\frac{1}{2} \div \frac{3}{4}$ and solve the problem using the meaning of fractions and division (that is, without the algorithm). You need to explain which model of division you are using (partitioning or repeated subtraction), and you need to justify your solution.

22. In each case below, make as good an estimate as you can in your head. Briefly explain how you obtained your estimate.

a. A campus newspaper reported that in a recent all-campus vote, $\frac{2}{3}$ of all the students in a certain college favored the construction of a new recreational center. If 1754 students voted, approximately how many students voted for the new center?

b. A silversmith has a sheet of silver that measures $9\frac{1}{2}$ inches by $6\frac{2}{3}$ inches. About how many square inches of silver does she have?

23. Express the following:

a. $\frac{5}{8}$ as a decimal b. $\frac{4}{1000}$ as a decimal

c. 0.24 as a fraction in simplest terms

24. Name a decimal between 0.4444 and 0.4445.

25. Order the following numbers from smallest to largest:

$\frac{5}{11}$ 0.045 0.454 −0.54

26. Round 23.29 to the nearest tenth. Explain your reasoning.

27. A state budget is $4.306 billion. Express this amount as a whole number.

28. Perform the following computations, using scientific notation.

a. 34,000,000,000 × 56,000,000

b. 582,000 ÷ 23,000,000,000

29. Solve each of the following problems, and explain how you knew whether to multiply or divide.

a. A tank of water has a volume of 0.7 cubic meters. What is the volume of 0.16 of the tank?

b. Rosa is having a party and has bought 8 liters of punch. If each cup holds 0.15 liter, how many cups can she make?

c. A box of candy weighs 0.6 pound. How many boxes can you make with 5.1 pounds of candy?

d. The price of 0.8 ounce of a compound is $4.85. How much does 1 ounce cost?

Proportional Reasoning

5

In one respect, this chapter represents the culmination of the first four chapters. Over the first four chapters, we have explored fundamental mathematical concepts and we have investigated applications of those concepts in real-life settings. In this chapter, we will focus explicitly on the idea of proportional reasoning. *Ratios and Proportional Relationships* is a separate content strand in the Common Core State Standards, with reasoning and problem solving with ratios in sixth grade and with proportional relationships in seventh grade. The NCTM asserts that "the ability to reason proportionally . . . is of such great importance that it merits whatever time and effort must be expended to [ensure] its careful development" (*Curriculum Standards*, p. 82). Let us examine why.

ADDITIVE VERSUS MULTIPLICATIVE COMPARISONS

In this chapter, we will examine two fundamentally different ways in which we can compare amounts and describe changes. For example, let's say we are comparing the cost of two used cars, one of which costs \$10,000 and the other \$15,000.

We can say that the second car costs \$5,000 more than the first car,

$$\$10{,}000 + \$5{,}000 = \$15{,}000$$

or that the second car costs $1\frac{1}{2}$ times as much as the first car,

$$\$15{,}000 = 1\tfrac{1}{2}\,(\$10{,}000)$$

The first description of the relationship is expressed in *additive terms*, while the second is expressed in *multiplicative terms*.

To make the mathematics more visible, let b = the cost of the second car and a = the cost of the first car. The first two representations below are equivalent, and we call them both **additive comparisons**. The last three are equivalent, and we call them **multiplicative comparisons**.

Additive comparisons

$$b = a + 5000 \qquad b - a = 5000$$

Multiplicative comparisons

$$b = 1\frac{1}{2}a \qquad \frac{b}{a} = \frac{3}{2} \qquad b:a = 3:2$$

We tend to use multiplicative comparisons more than additive ones. For example, we hear on television that a paper towel absorbs "50% more liquid than the leading brand" as opposed to "in tests it absorbed 4 more ounces of water than the other brand." Multiplicative comparisons provide a context that helps us to know how much more.

SECTION 5.1 Ratio and Proportion

What do you think?

- Are all ratios fractions? Are all fractions ratios? Why or why not?
- What do we mean by proportional reasoning?
- How do the concepts of ratio and proportion give us tools for making comparisons?

THE UNIT CONCEPT MATURES

A noted educator and scientist once wrote that "it seems odd to refer to a relationship as a quantity."[1] For example, if we say that a "large" drink is 20 ounces, we can see or visualize that amount using our knowledge of measurement. However, when we say that a car gets "35 miles per gallon," the 35 actually expresses a *relationship* between two amounts (miles traveled and gallons consumed). As Schwartz notes, we "refer to [this] relationship as a quantity." However, 35 miles per gallon is much more abstract than 35 ounces or 35 people.

In the primary grades, children mostly work with whole numbers and with the operations of addition and subtraction. As their understanding of mathematical ideas grows, they move from counting physical objects to counting numbers themselves. In both cases, however, the unit is still a single whole entity (for example, 1 ounce or 1 person). With the introduction of multiplication and division, and then the introduction of rational numbers, both operations and numbers become more complex, as you saw in Chapters 3 and 4.

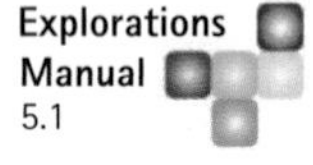

"Underneath all of the surface level changes is a fundamental change with far-reaching ramifications: *a change in the nature of the unit.* Given the difficulty of mastering the concept of unit in whole-number situations, it is not surprising that changes in the nature of the unit in the middle grades bring new cognitive demands and renewed difficulties for students."[2] In the example above, we talked about 35 miles per gallon. In this case, our unit is 1 mile per gallon—a more complex unit than 1 ounce or 1 student. As noted in Chapter 3, one reason why multiplication and division problems are generally more difficult than addition and subtraction problems is that they involve a more complex unit.

Explorations Manual 5.3

RATIOS, RATES, AND PROPORTIONS

A **ratio** is a relationship between two amounts or quantities. It can be expressed in the following equivalent ways: $a:b$, a/b, or $\frac{a}{b}$. For example, a school has a student to teacher ratio of 12 : 1.

When the two amounts in a ratio represent different quantities, we often refer to such ratios as **rates**—for example, miles/gallon or 55 miles/hour.

When we set two ratios equal to each other, we have a **proportion**. For example, 12 students : 1 teacher = 72 students : 6 teachers, or $\frac{55 \text{ miles}}{1 \text{ hour}} = \frac{110 \text{ miles}}{2 \text{ hours}}$.

That is, $a:b = c:d$ if $\frac{a}{b} = \frac{c}{d}$ and $b \neq 0$, $d \neq 0$.

LANGUAGE

The term *ratio* comes from the Latin verb *ratus*, which means "to think or estimate." Many mathematicians in the sixteenth and seventeenth centuries used the word *proportion* for ratio. Even today you hear the two terms used interchangeably; for example, instructions for making a certain color might say, "Mix the two colors in the following proportion—3 : 2."

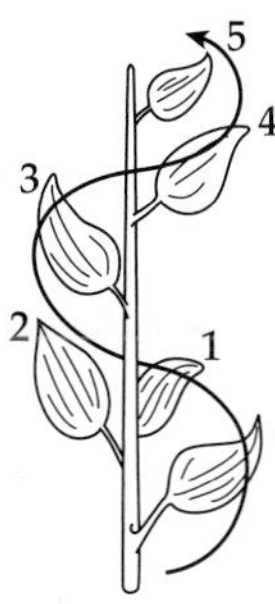

Figure 5.1
From Metamorphosis: A Source Book of Mathematical Discovery by Lorraine Mottershead Copyright © 1977 Pearson Education, Inc., or its affiliates. Used by permission. All Rights Reserved.

RATIOS, RATES, AND PROPORTIONS IN MATHEMATICS AND REAL LIFE

Ratios and rates pervade mathematics. Ratio is inherent in the concept of place value: the ratio of the value of each place to the value of the place to its right is 10 : 1. The concept of similarity involves equal ratios. When we make graphs, we use proportions. When we make scale drawings and scale models, we obey proportions. The essence of probability involves ratios.

Ratios, rates, and proportions show up in all kinds of real-world contexts.

- *Banking:* When I applied for a mortgage, the bank applied a ratio called the 28% rule: If the ratio of fixed monthly payments (mortgage, property tax, car payments, and so forth) to monthly income is more than 28 : 100, the bank is not likely to make the loan.
- *Botany:* If we represent the number of complete turns made around the stem by t and represent the number of leaves between the two points as n, then the fraction $\frac{t}{n}$ is called a divergency constant for that species (Figure 5.1). Table 5.1 gives divergency constants for a number of different trees. The numerators and denominators are all Fibonacci numbers.

TABLE 5.1

Tree	Divergency constant
Elm	$\frac{1}{2}$
Beech, hazel	$\frac{1}{3}$
Apple, oak, apricot, poplar	$\frac{2}{5}$
Pear, weeping willow	$\frac{3}{8}$
Willow, almond, pussy willow	$\frac{5}{13}$

- *Commerce:* Many companies use rates to determine how much to charge for their goods or services. For example, Allied Shipping Company charges $7 for every 100 pounds of goods shipped.
- *Cooking:* If you want to make more than the amount given in the recipe, all the ingredients must be increased in the correct proportion. For example, if the ratio of dry rice to water is 1 cup : 3 cups when cooking rice, and you want to make a double serving, then you would need to double both amounts and use 2 cups rice : 6 cups water.
- *Education:* One of the criteria used to determine the quality of a college is the student : teacher ratio.
- *Sports:* Football announcers talk about the "turnover ratio." If a team has caused a total of 30 fumbles and interceptions and lost the ball 25 times because of fumbles and interceptions, the team is said to have a +5 ratio. Unfortunately, this is not a valid use of the term ratio. Can you explain why not? Would you recommend that these writers and announcers convert the numbers to a ratio or use the same numbers but invent a different term?

In many instances, quantities "should be" proportional but are not. Try to think of a few before reading on. . . .

Airline rates are often not proportional. Longer flights are often cheaper than short flights, and flights between major cities tend to be cheaper than flights between small cities.

The affirmative action debate for the past 30 years has been based on a belief that many people hold: that the ratio of a minority group in society and the ratio of members of that minority group in the work force or the schools should be equal. For example, the ratio of women : men is approximately 1 : 1, but the ratio of female college presidents to male college presidents is not 1 : 1, and so we say that a disproportionate number of college presidents are men.

DELVING FURTHER INTO RATIOS

Look at the following ratio statements. If the statement can be expressed as a fraction statement, do so. If it cannot, try to explain why. Then read on. . . .

1. The ratio of males to females at Mountain State College is 3 : 2.
2. In order to get a stain out of a shirt, Lisa made a mixture of bleach and water in the ratio 3 parts water to 1 part bleach.
3. In the rectangle in Figure 5.2, the ratio of the width to the length is 2 : 3.
4. Fred's car gets 25 miles per gallon.

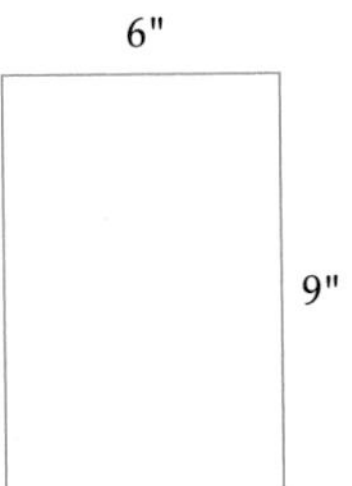

Figure 5.2

Statement 1: We can say that $\frac{3}{5}$ of the students are male and that $\frac{2}{5}$ of the students are female.

Statement 2: The stain mixture Lisa made was a $\frac{1}{4}$ bleach solution. [*Note:* Many people would be more likely to call this a 25% bleach solution.]

Statement 3: There are two ways in which we can translate this statement into fraction language. We can say that the width is $\frac{2}{3}$ as long as the length. Alternatively, we can say that the length is $1\frac{1}{2}$ times as long as the width.

Statement 4: There is no way to translate this statement into fraction language. Do you see why? In Statement 3, the two amounts (parts) had the same unit: length. In this statement, the two amounts are measured with different units: miles and gallons.

Table 5.2 summarizes some of these points.

TABLE 5.2

Statement	Diagram	Whole?	Can be expressed as fraction?
1	Males / Females	The student body	Yes
2	Bleach / Water / Water / Water	The mixture	Yes

TABLE 5.2 *(continued)*

Statement	Diagram	Whole?	Can be expressed as fraction?
3	Rectangle: 6" by 9"	Either number can be seen as the whole	Yes
4	Miles | Gallons: 25 | 1; 50 | 2; 75 | 3	No whole	No

WHEN ARE RATIOS LIKE AND WHEN ARE THEY UNLIKE FRACTIONS?

Ratios are like fractions in that they can be expressed as an ordered pair of numbers representing two sets or two amounts. Ratios are different from fractions in that fractions are only part–whole relationships, whereas ratios can be part–whole relationships, part–part relationships, or whole–whole relationships.

Let us illustrate this difference by returning to the example of the water–bleach mixture.

The ratio of water to bleach is 3 : 1. However, there are actually several possible ratios:

3 water : 1 bleach	1 bleach : 3 water	3 water : 4 total	1 bleach : 4 total
part : part	part : part	part : whole	part : whole

Because a part–whole relationship exists, the water–bleach mixture can be expressed as a fraction. The first two relationships are ratios but not fractions, because they are part–part. The second two are both fractions and ratios because they are part–whole relationships. We can say that $\frac{3}{4}$ of the mixture is water and that $\frac{1}{4}$ of the mixture is bleach.

In the case of the miles per gallon, we cannot make a fraction statement. However, the ratio is 25 miles : 1 gallon, which is a whole–whole relationship even though the two amounts are measures of different units, namely miles and gallons.

Let us now examine these various concepts and ideas that underlie what we call *proportional reasoning*. This first investigation represents a classical problem with a history to it.

INVESTIGATION 5.1a Unit Pricing—Is Bigger Always Cheaper?

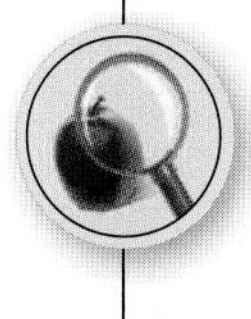

Let's say you want to buy some laundry detergent. The small jug costs \$2.99 for 36 fluid ounces, and the large jug costs \$3.79 for 48 fluid ounces. Which is the better buy?

DISCUSSION

STRATEGY 1 Use fractions

Explorations Manual 5.4

In this case, the additive comparison is not the one that is useful. That is, we get 12 more ounces for 80 more cents. However, we can use this information to make a valid multiplicative comparison.

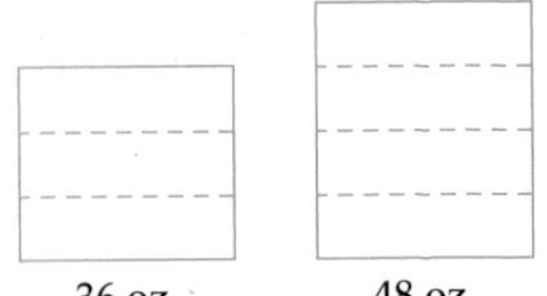

The larger size gives you $\frac{1}{3}$ more detergent (see the figure at the left), so if it costs $\frac{1}{3}$ more, then the value of both sizes will be the same. Does the larger jug cost $\frac{1}{3}$ more? Because $\frac{1}{3}$ of \$2.99 is \$1.00, the two sizes will be the same value if the large size costs \$4. Note that here we are rounding \$2.99 to \$3.00. Because \$3.79 is less than \$4, we conclude that the larger size is a better buy. Why is this?

Perhaps you used this strategy in Exploration 5.4 to determine which box of pancake mix was the better buy. This strategy, used intuitively by many students,[3] is called the **factor-of-change method**. In this case, the factor of change was $\frac{1}{3}$. That is, if we multiply the 36 ounces and the 299 cents by the same factor, we produce an equivalent ratio. The idea of equivalent fractions helps us to understand why this method works.

$$\frac{299 \text{ cents}}{36 \text{ ounces}} \times \frac{1\frac{1}{3}}{1\frac{1}{3}} = \frac{400 \text{ cents}}{48 \text{ ounces}}$$

STRATEGY 2 Solve a proportion

The following proportion illustrates a strategy similar to the previous one. We can let x represent how much a 48-ounce jug of equivalent value would cost:

$$\frac{\$2.99}{36 \text{ ounces}} = \frac{x \text{ dollars}}{48 \text{ ounces}}$$

When we solve for x, we find that $x = \$3.99$, that is, if the 48-ounce jug cost \$3.99, the two jugs would have the same value. Because the 48-ounce jug costs less than \$3.99, it is a better buy.

MATHEMATICS

Many states now require grocery stores to show the unit price below each item. One of the reasons for this law is that many companies realized that people tended to believe that larger packages were cheaper. Even today with unit prices, you can find instances in most stores where the larger amount is not cheaper.

STRATEGY 3 Use ratios and a calculator

We could also compare two ratios.

$$\frac{\$2.99}{36 \text{ ounces}} = 0.0830556 \approx 0.083$$

$$\frac{\$3.79}{48 \text{ ounces}} = 0.0789583 \approx 0.079$$

We have deliberately not placed labels on the two amounts. What do the two decimals mean? Think and read on. . . .

The meaning of 0.083 is \$.083 per ounce, or 8.3 cents per ounce. From a meaningful interpretation of these ratios, we see that the larger size is a better buy because its unit price is less; that is, \$.079 per ounce is less than \$.083 per ounce. Many stores place the price per ounce on the shelf labels to help people identify the best deal.

This strategy, also intuitively used by many students, is known as the **unit-rate method**. That is, we determine the cost of 1 unit (in this case 1 ounce).

STRATEGY 4 Another way to express the ratios?

Maria says she used the previous strategy but "upside down." What do these ratios mean? What do you think of her idea? Is this a variation of the unit-rate method?

$$\frac{36 \text{ ounces}}{\$2.99} = 12.0 \qquad \frac{48 \text{ ounces}}{\$3.79} = 12.7$$

In this case, we have 12.0 ounces per dollar versus 12.7 ounces per dollar. This is a variation of the unit-rate method. In this case, the unit is 1 dollar instead of 1 ounce; that is, the ratios tell us how much detergent we get for a unit of money.

Let's examine another situation, first with smaller numbers, and then we will scale up the problem with larger numbers.

INVESTIGATION 5.1b How Many Trees Will Be Saved?

It is estimated that every 2000 pounds of paper that is recycled saves about 16 trees.

A. If a college recycled 8000 pounds of paper during the semester, how many trees did it save?

B. If a college recycled 13,250 pounds of paper during the semester, how many trees did it save?

C. If a city recycled 450 tons of paper during a year, how many trees did it save?

DISCUSSION

A. Since $4 \times 2000 = 8000$ pounds of paper, the college saved $4 \times 16 = 64$ trees.

B. If we think about the operation we used in the previous problem, we can use the same process. That is, 13,450 is how many times as large as 2000? Since $\frac{13{,}450}{2000} = 6.725$, we can multiply 16 by 6.725, which is 107.6. Thus, we conclude that the college saved about 108 trees.

C. Now the numbers are getting pretty big. If we convert 2000 pounds to 1 ton, we see that 1 ton saves 16 trees. Thus 450 tons saves $450 \times 16 = 7200$ trees.

INVESTIGATION 5.1c How Much Money Will the Trip Cost?

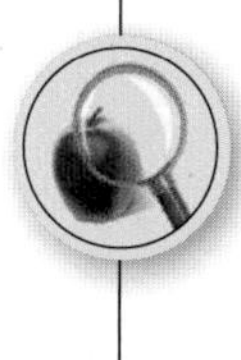

Most real-life uses of ratios and rates involve proportions. For example, let's say you are planning a 2000-mile trip and want to estimate how much money you will spend on gas. How would you do that? Work on this problem on your own and then read on. . . .

DISCUSSION

Explorations Manual 5.5

As you may have realized, the cost of gas for the trip would depend on two variables: how many miles per gallon your car gets and the cost of gasoline. Let's say your car averages 25 miles per gallon and you estimate that gas will cost \$3.49 per gallon on the trip. Now estimate the cost of gas for the trip and then read on. . . .

The rate at which your car uses gas is expressed by the ratio 25 miles : 1 gallon. The following proportion enables us to determine how many gallons of gas you will need at that rate of consumption:

$$\frac{25 \text{ miles}}{1 \text{ gallon}} = \frac{2000 \text{ miles}}{x \text{ gallons}}$$

We find x by solving the proportion for x: $25x = 2000$; therefore $x = 80$ gallons. We can now use another proportion to find the cost of 80 gallons.

$$\frac{\$3.49}{1 \text{ gallon}} = \frac{y}{80 \text{ gallons}}$$

We find y by solving the proportion for y: $y = (\$3.49)(80)$; therefore, $y = \$279$.

We could have solved the entire problem in one step using dimensional analysis. The units for the answer will be dollars. Therefore, we begin with the amount that is represented by dollars: the cost of gasoline. If we multiply the ratio representing the cost of gasoline by the ratio representing the consumption of gasoline, look what happens:

$$\frac{\$3.49}{\text{gallon}} \times \frac{1 \text{ gallon}}{25 \text{ miles}} \times 2000 \text{ miles} = \$279$$

INVESTIGATION 5.1d Reinterpreting Old Problems

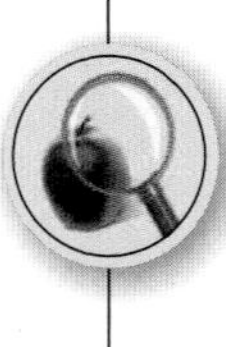

A. In Section 4.2, Investigation 4.2h, Jose paid \$12 for a $\frac{3}{4}$-pound box of chocolates. What is the price of 1 pound (at this rate)?

DISCUSSION

Many students find this problem challenging as a fraction problem but can solve this more easily if they use decimals and proportions:

$$\frac{\$12}{0.75 \text{ pound}} = \frac{x \text{ dollars}}{1 \text{ pound}}$$

B. In Investigation 4.3g, Marvin has 23 yards of cloth to make costumes for the play. Each costume requires $3\frac{1}{4}$ yards of material. How many costumes can he make?

DISCUSSION

In Section 4.3, we solved this problem by interpreting it as division (repeated subtraction). However, it can also be interpreted as a proportion. In this case, 3.25 yards can be seen as the unit—that is, the number of yards needed for 1 costume.

$$\frac{1 \text{ costume}}{3.25 \text{ yards}} = \frac{x \text{ costumes}}{23 \text{ yards}}$$

INVESTIGATION 5.1e Using Estimation with Ratios

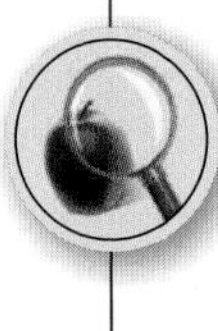

The following investigations consist of problem-solving situations that can be answered with an estimate (rather than an exact answer). Assume that you have no calculator or pencil and paper.

A. It took Rene 24 minutes to go 15 miles from Hingham to Marshfield. What was Rene's average speed for the trip? What is your first impression (5-second estimate)? What is your best estimate? Can you get the "exact" answer? Why is the "exact" answer probably not exact? Work on these questions and then read on. . . .

DISCUSSION

The problem can be represented as the following proportion: 15 miles is to 24 minutes as x miles is to 60 minutes (that is, 1 hour).

$$\frac{15 \text{ miles}}{24 \text{ minutes}} = \frac{x \text{ miles}}{60 \text{ minutes}}$$

STRATEGY 1 Use equivalent fractions

We can see that $\frac{15}{24}$ is just under $\frac{16}{24}$, which is equivalent to $\frac{2}{3}$, and $\frac{2}{3}$ of 60 = 40. Thus a quick estimate (if you connect this question to equivalent fractions) is "slightly under 40 miles per hour."

STRATEGY 2 Use the missing-factor model of division

We also can get a decent estimate by asking: 24 times what is equal to 60? This strategy is related to the compatible-numbers strategy we have used before; it would not have been so straightforward had the number of minutes been, say, 22.

Because 24 times $2\frac{1}{2}$ is equal to 60, we now have to determine 15 times $2\frac{1}{2}$, but that is relatively simple using multiplication as repeated addition: $15 + 15 + 7\frac{1}{2} = 37\frac{1}{2}$, which is the exact answer—well, sort of.

The exact answer of $37\frac{1}{2}$ miles per hour is itself an approximation. Why is this? The 24 minutes and 15 miles are rounded numbers. In this case, the "exact" numbers were 23 minutes and 45 seconds and 15.2 miles. Even if we now computed Rene's "exact" average speed, that is,

$$\frac{15.2}{23.75} = \frac{x}{60}$$

and solved for $x = 38.4$ miles, the term average speed is itself a mental construct, not something that exists in an observable sense. At times Rene was going 55 miles per hour; at times Rene was going 20 miles per hour. We will study the concept of average in more detail in Chapter 7.

B. It is Christmas time, and virtually every store in town is selling wrapping paper. If Zoe doesn't care about looks and just wants the cheapest, which should she buy? Work on this and then read on. . . .

- Store A: 36 square feet for 88¢.
- Store B: 50 square feet for $1.79.
- Store C: 150 square feet for $2.99.

DISCUSSION

If we divide the cost (in pennies) by the square feet, the resulting ratio will tell us the cost per square foot. The ratio that is the smallest number will represent the cheapest paper. Try this on your own and then read on. . . .

In this case, if the prices are not terribly close, you may need only one run-through.

- Store A: $\frac{88}{36}$ = more than 2—that is, more than 2¢ per square foot.
- Store B: $\frac{179}{50}$ = more than 3—that is, more than 3¢ per square foot.
- Store C: $\frac{300}{150}$ = 2¢ per square foot. We can use 300¢ instead of the exact 299¢ because 1¢ makes virtually no difference in this case. There is no need to do more arithmetic here, because store C clearly has the "best" ratio.

C. Ellie bought a new car on August 16. On November 1 the odometer read 2650. Ellie's insurance company gives a discount if she puts less than 10,000 miles per year on the car. If she continues driving at this rate, will she qualify? Work on this and then read on. . . .

DISCUSSION

The time between August 16 and November 1 is very close to $2\frac{1}{2}$ months, so we can represent this problem by the following proportion:

$$\frac{2650 \text{ miles}}{2\frac{1}{2} \text{ months}} = \frac{x \text{ miles}}{12 \text{ months}}$$

We can once again obtain a quick estimate by using the idea of equivalent fractions: What multiple of $2\frac{1}{2}$ will be closest to 12? A bit of mental arithmetic yields the realization that $2\frac{1}{2} \times 5 = 12\frac{1}{2}$. Therefore, the estimated miles per year will be 2650×5. How can you calculate that amount mentally?

Because $2650 \times 10 = 26{,}500$, 2650×5 will be half that amount, or 13,250. At this rate, Ellie will travel about 13,250 miles in just over 12 months and won't qualify for the discount.

Another quick estimate can be obtained by connecting to compatible numbers: 2650 miles is just over 2500 miles, which is $\frac{1}{4}$ of 10,000 miles. Thus, if she had traveled 2500 miles in 3 months, she would be on target to travel 10,000 miles in 12 months. Because she has traveled more than 2500 miles in less than 3 months, her ratio of miles : months is higher than 2500 miles : 3 months.

MORE APPLICATIONS

The next three investigations are not quite as simple and tidy as the previous ones and require a careful application of concepts we have studied thus far.

INVESTIGATION 5.1f Is the School on Target?

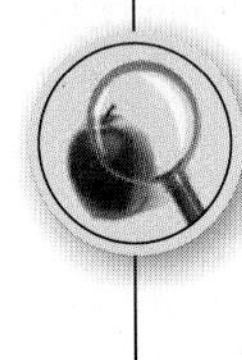

An all-female college first began admitting men three years ago. Its goal for this, the third year, is to have a 2 : 1 ratio of females to males in the incoming freshman class. The admissions department has just reported a problem. When the college sent out letters of acceptance to this year's freshman class, the ratio of females to males was 2 : 1 However, the ratio of females to males in the 1000 students who indicated that they will be coming this fall is 7 : 3 If the college still wants to meet the 2 : 1 goal, how many more males will it need to accept to change the ratio of females to males to 2 : 1? Try to solve this problem on your own and then read on. . . .

Explorations Manual 5.1

DISCUSSION

The desired ratio of females to males is 2 : 1. If the ratio of females to males among the 1000 incoming freshmen is 7 : 3, that means that there are 700 women and 300 men. One way of representing the dilemma is to let x be the number of additional males the college needs to admit.

In other words, if the college admits x more males, the ratio of females to males will be $700 : (300 + x)$, and this ratio needs to be equal to 2 : 1. Because the two ratios are equal, we have the following proportion:

$$\frac{700}{300 + x} = \frac{2}{1}$$

We find that $x = 50$.

This problem could also have been solved using common sense. Targeting a ratio of 2 females to 1 male can be interpreted as wanting the number of males to be one-half the number of females. Because 350 is half of 700, the college needs a total of 350 males—that is, 50 more males than it has presently accepted.

INVESTIGATION 5.1g Finding Information from Maps

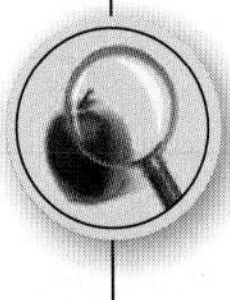

A. Jack and Jill are traveling across the country, but they are not going on the interstate highways, which they find boring, and they do not have a GPS in the car. They just passed Munsonville and their destination is Salyan. Jill asks Jack how far away Salyan is, but Jack does not know. On the map, 1 inch represents 40 miles, and the two towns are about $2\frac{1}{4}$ inches apart. How far until Salyan?

Grade 5

Do you see the connections here both to rates and to functions? 

Date Time

LESSON 10•4

Representing Rates

Complete each table below. Then graph the data and connect the points.

1. a. Andy earns $8 per hour. Rule: Earnings = $8 * number of hours worked

Time (hr) (h)	Earnings ($) ($8 * h$)
1	
2	
3	
	40
7	

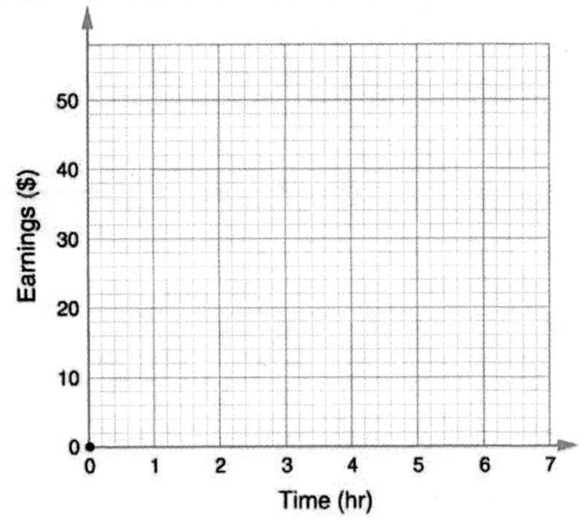

b. Plot a point to show Andy's earnings for $5\frac{1}{2}$ hours. How much would he earn?

____

2. a. Red peppers cost $2.50 per pound. Rule: Cost = $2.50 * number of pounds

Weight (lb) (w)	Cost ($) ($2.50 * w$)
1	
2	
3	
	15.00
12	

30
20
10
0
Cost ($)
0 4 8 12
Weight (lb)

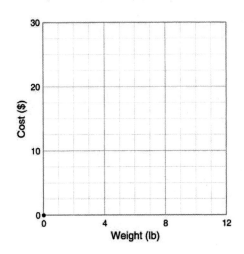

b. Plot a point to show the cost of 8 pounds. How much would 8 pounds of red peppers cost?

____

348

From *Everyday Mathematics, Grade 5:* The University of Chicago School Mathematics Project: Student Math Journal, Volume 2, by Max Bell et al., Lesson 10-4, p. 348. Reprinted by permission of The McGraw-Hill Companies, Inc.

B. Since this is a math course to develop your number sense, let us change the situation to make it more challenging. On this map $\frac{3}{4}$ inch represents 48 miles. If tomorrow's destination, Shangrila, is $4\frac{1}{2}$ inches away, how many miles would it be?

DISCUSSION

A. We can multiply $40 \times 2\frac{1}{4}$ and get 90 miles. The model can help us to see this.

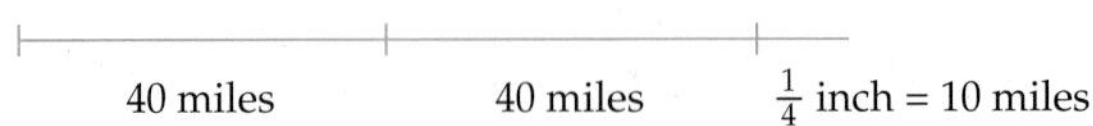

B. There are many ways to solve this problem.

STRATEGY 1 Find the unit distance

If $\frac{3}{4}$ inch represents 48 miles, then $\frac{1}{4}$ inch represents 16 miles.

Do you see why? A common mistake here is to conclude that $\frac{1}{4}$ inch represents 12 miles because students divided by 4. In this case a diagram is helpful to understand why this does not work.

If $\frac{1}{4}$ inch represents 16 miles, then 1 inch represents 64 miles, and the answer is $64 \times 4\frac{1}{2} = 288$ miles.

STRATEGY 2

If $\frac{3}{4}$ inch represents 48 miles,

then $1\frac{1}{2}$ inches represents 96 miles.

and 3 inches represents 192 miles.

so 1 inch represents 64 miles, and the answer is $64 \times 4\frac{1}{2} = 288$ miles.

STRATEGY 3 How many $\frac{3}{4}$ inches are in $4\frac{1}{2}$ miles?

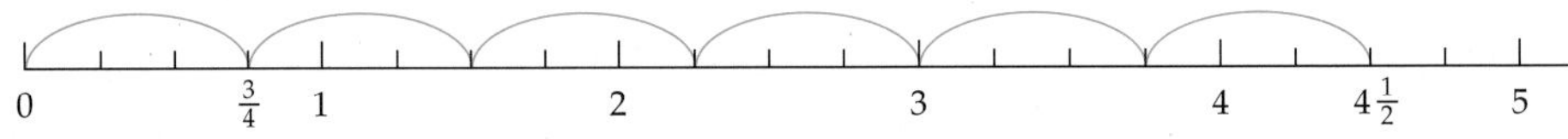

Since 6 sets of $\frac{3}{4}$ inch segments equals $4\frac{1}{2}$ inches, the answer is $48 \times 6 = 288$ miles.

STRATEGY 4 Recognize that this is a proportional situation.

$$\frac{\frac{3}{4}\text{ inch}}{48\text{ miles}} = \frac{4\frac{1}{2}\text{ inches}}{x\text{ miles}} \qquad \frac{3}{4}x = 48 \times 4\frac{1}{2}$$

$$\frac{3}{4}x = 216$$

$$x = 216 \times \frac{4}{3} = 288\text{ miles.}$$

INVESTIGATION 5.1h From Raw Numbers to Rates

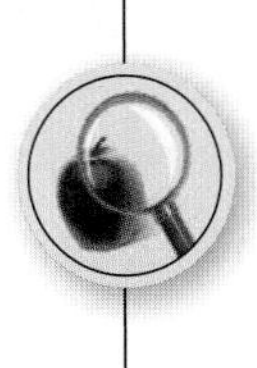

We are bombarded with numbers and data about our world every day. Making sense of this information is not always easy. Sometimes the data are presented in raw numbers and sometimes in rates. Why do you think this is so? Let's examine some actual data to see how rates can help us to "see" data more clearly. In 2013, there were 415 homicides in Chicago and 333 in New York City. What can you conclude from these numbers?

Explorations Manual 5.3

DISCUSSION

Before we examine the numbers, it is important to note that the raw numbers are only one part of the making sense process. For example, I have friends in Chicago and New York City, and they tell me that they do not feel terribly unsafe, but they know when and where they can go in order to stay safe. Here, we will focus on the mathematics of the data. Let's say you were thinking of moving to either city. From a numerical perspective, what can you conclude? For example, would you say that since the two numbers are essentially equal, both cities are basically equally safe in this respect?

Most people will say that because New York City is much larger then, proportionally, it is a safer city than Chicago with respect to homicides. How might we express this conclusion numerically?

The population of Chicago in 2013 was about 2.7 million and the population of New York City was 8.4 million. How can we use these numbers to compare the two cities?

We can find the ratio in each city. What do the numbers mean?

$$\frac{415}{2.83} \approx 147 \qquad \frac{333}{8.4} \approx 40$$

There were 147 murders in Chicago per 1 million people and there were 40 murders in New York City per 1 million people. This calculation has changed the data from raw numbers, where we are essentially comparing apples to oranges, to rates, where we are comparing apples to apples.

In the exercises, you can examine various data to see how rates can help us to interpret data about our lives.

INVESTIGATION 5.1i How Much Does That Extra Light Cost?

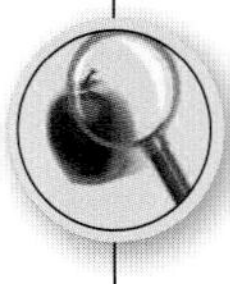

Let's say you are considering the installation of a 300-watt outdoor security light for your home. By how much will your electric bill increase?

DISCUSSION

In order to answer this question, what assumptions do we need to make? What data do we need?

Let us assume that the light will be on an average of 12 hours per day over the course of a year—less than 12 in the summer, more than 12 in the winter. We also need to know how much electricity costs. For this problem, use the rate of 7.85¢/kWh.[4] See whether you can do the problem on your own now, and then read on. . . .

STRATEGY 1 Multiply step by step

- 300 watts = 0.3 kilowatt, so 0.3 kilowatt per hour times 7.85 cents/kWh = 2.355 cents per hour to run the light.
- 2.355 cents per hour times 12 hours per day = 28.26 cents per day to run the light.
- 28.26 cents per day times 365 days per year = 10314.9 cents per year to run the light.

Note: This problem illustrates the need to make sure constantly that each step makes sense. Many of my students give me an answer of $10,314.90. However, the number 10314.9 refers to cents, not dollars. How do you convert cents into dollars?

STRATEGY 2 Use dimensional analysis

By aligning the rates correctly, we can solve the problem in one step:

$$\frac{0.3\ \cancel{\text{kW}}}{\cancel{\text{hour}}} \times \frac{7.85\ \cancel{\text{cents}}}{\cancel{\text{kW}}} \times \frac{12\ \cancel{\text{hours}}}{\cancel{\text{day}}} \times \frac{365\ \cancel{\text{days}}}{\text{year}} \times \frac{\$1}{100\ \cancel{\text{cents}}} = \frac{\$103.149}{\text{year}}$$

Instead of doing each computation separately, what we do on the calculator is simply

.3 (×) 7.85 (×) 12 (×) 365 (/) 100 (=)

What appears on the calculator is 103.149. Dimensional analysis tells us that the meaning of this number is dollars per year.

DIMENSIONAL ANALYSIS EXPLAINED

Dimensional analysis has been discussed in several places in this book. Can you now justify this method?

The justification lies in the concept of equivalent fractions and rates. Usually, we think of finding equivalent fractions by multiplying the numerator and denominator by the same number. However, we can generalize this notion of equal number to equal amount.

When using dimensional analysis, we are treating the units as though they were numbers:

$$\frac{0.3\ \text{kW}}{\text{hour}} \times \frac{7.85\ \text{cents}}{\text{kW}}$$ Our initial problem.

$$= \frac{0.3}{1} \times \frac{\text{kW}}{\text{hour}} \times \frac{7.85}{1} \times \frac{\text{cents}}{\text{kW}}$$ Our work with multiplication of fractions helps us understand that this is an equivalent representation.

$$= \frac{0.3}{1} \times \frac{7.85}{1} \times \frac{\text{kW}}{\text{kW}} \times \frac{\text{cents}}{\text{hour}}$$ The commutative and associative properties tell us that we can change the order.

$$= \frac{2.355}{1} \times \frac{\text{kW}}{\text{kW}} \times \frac{\text{cents}}{\text{hour}}$$ Our understanding of multiplication of fractions tells us that we have the amount 2.355.

$$= \frac{2.355}{1} \times 1 \times \frac{\text{cents}}{\text{hour}}$$ Any amount divided by itself is equal to 1.

$$= \frac{2.355}{1} \times \frac{\text{cents}}{\text{hour}}$$ The identity property of multiplication tells us that when we multiply an amount by 1, the value is unchanged.

$$= \frac{2.355\ \text{cents}}{\text{hour}}$$ Our knowledge of multiplication of fractions tells us that this is an equivalent representation.

Of course, when we use dimensional analysis, we compute and "cancel" units when appropriate.

SUMMARY 5.1

In this section, we have explored the last interpretation of fractions that was introduced in Section 4.2: fraction as ratio. We have seen that not all ratios can be expressed as fractions—only those that are given in part–whole terms or can be translated into part–whole terms. We have seen that rates are a special kind of ratio, one that cannot be expressed in fractional terms. Rates also express a functional relationship between two variables.

In this section, we have made explicit the difference between additive and multiplicative comparisons. Like fractions, ratios and rates signify multiplicative relationships between two amounts. When two ratios are equal (or when two rates are equal), we have a proportion. Thus the concept of a proportion connects to the notion of equivalent fractions, which we studied in Chapter 4. In the next section, we will build the concept of percent upon the concept of proportion.

We have examined several real-life applications of ratios, rates, and proportions. In doing so, we found that it is critical to make sure that the computations make sense. We have given names to two methods that students often invent: the factor-of-change method and the unit-rate method. Dimensional analysis is one tool that helps us to solve proportion problems. Finally, we have seen that many estimation problems involve proportions.

5.1 Exercises

Exercises 1–16 present you with fairly straightforward ratio and proportion situations. In each problem, first estimate the answer and explain how you obtained the estimate. Then determine the actual answer and explain whether the exact answer is, itself, actually exact or an approximation.

1. A quart (32 ounces) container of yogurt contains 920 calories. Approximately how many calories would there be in a 5-ounce serving?

2. a. If the Dow Jones Industrial Average in the stock market began the day at 12,000 and lost 800 points, how much would a proportional fall be if the stock market began the day at 1000?

   b. If the stock market was at 2700 and rose 75 points, how much would a proportional rise be if the stock market began the day at 10,000?

3. An advertisement says that 5 out of 8 dentists recommend the new zigzag toothbrush. If 264 dentists were interviewed, how many recommended the toothbrush?

4. An intravenous solution needs 2 liters (L) of glucose to be mixed with 7 units of blood. How much glucose is needed for 40 units of blood?

5. An employee making $24,000 was given a raise of $1000. All employees were given proportional raises.

   a. How much of a raise would an employee making $18,000 receive?

   b. How much of a raise would an employee making $30,000 receive?

   c. How much of a raise would an employee making $23,450 receive?

6. If $1\frac{3}{4}$ cups of flour are required to make 30 cookies, how many cups of flour (to the nearest $\frac{1}{4}$ cup) are required for 96 cookies?

7. In the summer of 1991, Israel airlifted 14,000 Ethiopian Jews to Israel as immigrants. If the United States were to receive a proportional number of refugees at one time, about how many would be received? Say the population of Israel was approximately 4.4 million and that of the United States was 270 million.

8. a. On one map, $\frac{1}{3}$ inch represents 18 miles. If two cities are $2\frac{1}{2}$ inches apart on the map, what is the actual distance between them?

   b. On another map, 1 inch represents 65 miles. Los Angeles is about 1000 miles from Portland. How many inches apart would Portland and Los Angeles be on this map?

   c. On a map $\frac{2}{3}$ inch represents 84 miles. How far apart are two cities in reality if they are $4\frac{5}{8}$ inches apart on the map?

9. You can use proportions to estimate the height of a tree. John is 6 feet tall, and his shadow is $10\frac{1}{2}$ feet long. How high is a tree whose shadow is 90 feet long?

10. Jane finds that she can read 36 pages of a book in 40 minutes. At this rate, how long, to the nearest minute, will it take her to finish a book 473 pages long?

11. The other day Janet was using the rowing machine in the Fitness Center. Her goal was to row 2200 meters in 10 minutes. At exactly 7 minutes, she saw that she had rowed 1560 meters. Is she going to make her goal if she continues at this rate?

12. Sheila and Dora worked $3\frac{1}{2}$ hours and $4\frac{1}{2}$ hours, respectively, on a programming project. They were paid \$176 for the project. How much did each earn?

13. A photograph is 3 inches high and 5 inches wide. If it is enlarged to be 7 inches high, how wide will it be (to the nearest $\frac{1}{4}$ inch)?

14. The recommended dosage for a particular medication is 6.8 milligrams (mg) per pound of body weight per day, not to exceed 1000 mg per day. For a 125-pound patient, about what would you expect the doctor to recommend for a daily dosage?

15. This problem involves beads for making jewelry.
    a. The beads come in packets of 20, and the packets cost 12 cents each. How much would 80 of these beads cost?
    b. How much would 145 of these beads cost?
    c. The seller offers a discount if you buy more than 500 beads: 20 beads for 9 cents. How much would 965 of these beads cost?

16. What is the average speed in each of these situations?
    a. You travel 11 miles in 16 minutes.
    b. You travel 31 miles in 39 minutes.
    c. You travel 85 miles in 90 minutes.
    d. On a bicycle you travel $12\frac{1}{2}$ miles in 45 minutes.
    e. On a bicycle you travel $3\frac{3}{4}$ miles in $10\frac{1}{2}$ minutes.

17. Make up and solve a real-life problem that lends itself to needing a rough or refined estimate.

18. ***Classroom Connection*** This problem is taken from the August 2006 issue of *Teaching Children Mathematics,* p. 33. During a typical year at the Minnesota State Fair, milk sales total 20,000 gallons.
    a. If the fair runs for 10 days, determine the average number of gallons sold each day.
    b. How many 8-ounce cups per day would this be?
    c. If 1 cup of milk costs \$1.25, about how much money per day is made in milk sales?

19. ***Classroom Connection*** This problem is taken from the September 2006 issue of *Teaching Children Mathematics,* p. 97. To collect the amount of nectar needed to make 1 pound of honey, bees must tap 2 million flowers.
    a. How much honey can they produce if they tap only 300,000 flowers?
    b. How much honey can they produce if they tap 1,750,000 flowers?
    c. How many flowers do they need to tap to make 24.5 pounds of honey?

20. It was reported that Ross Perot spent approximately \$40 million of his own money in the 1992 presidential campaign. His total worth was reported to be approximately \$4 billion. If your total worth were \$50,000 and you spent the same fraction of your worth on an election as Perot did, how much money would you have spent?

21. In a healthy person, the ratio of red blood cells to other blood cells should be about 1 to 5000. Amy just got back a lab report that showed 300 red blood cells out of 230,000 blood cells. Is her red blood cell count low, high, or normal?

DEEPENING YOUR UNDERSTANDING

22. A car travels 60 miles per hour, and a plane travels 15 miles per minute. How far does the car travel while the plane travels 600 miles?

23. On a TV game show, a contestant makes \$700 for every correct answer but loses \$500 for every wrong answer. After answering 24 questions, Sarah broke even. How many questions did she answer correctly?

24. I have seen several elementary textbooks and supplements use *Gulliver's Travels* to explore proportional reasoning. Let's say Gulliver found a button on the ground that fell from the giant's shirt. The diameter of the button was 4 inches. Using measurements from a button on your clothing, how could you estimate the height of the giant, assuming proportionality?

25. In the women's downhill during the Olympics, the difference between first place and second place was 0.04 second. If the two skiers had been racing side by side, what would have been the distance between the two at the finish line? Assume that they were going 60 mph at the end of the race.

26. Convert each of the following figures to miles per hour.
    a. The world record for the men's 100 meters was 9.58 seconds, held by Usain Bolt. What was his average speed over the course of that race?
    b. The world record for the men's 10,000 meters was 26 minutes 17.53 seconds, held by Kenenisa Bekele of Ethiopia. What was his average speed throughout that race?
    c. The world record for the women's 100 meters freestyle swimming was 52.88 seconds, held by Lisbeth Trickett of Australia. What was her average speed over the course of that race?

27. In the *Guinness Book of World Records,* the record for handshaking is 16,615 in 7 hours and 25 minutes. How many is this per minute? The person shook one hand every ___ seconds.

28. At Strawberry State College, the ratio of females to males is 2 : 1. If there are 2400 students, how many are female and how many are male?

29. At a certain college, there are 7 men for every 5 women. If there are 420 more men than women, what is the total enrollment?

30. One measure used in determining the quality of a college is the student to teacher ratio. A lower ratio means smaller classes. Let's say the student to teacher ratio in a college is 16 : 1. If there are 4800 students at the college, how many teachers are there?

31. Yosha is working on a small project that is due in 2 days. She has spent $4\frac{1}{2}$ hours on it and figures that she is about $\frac{3}{4}$ done. How many more hours will she need to spend?

32. Ginger wants to fill her new swimming pool. She has two pumps; the large pump takes 40 minutes to fill the pool, and the small pump takes 60 minutes. How long will it take to fill the pool if both pumps are working?

33. The ratio of physicians to inhabitants of the United States is 1 to 375. The ratio of prison inmates to inhabitants of the United States is 1 to 200. Are there more physicians or prison inmates in the United States? Approximately how many are there of each?

34. Julio makes a lemonade mix every day during the summer.

a. If he mixes more lemonade mix with less water than he did yesterday, will today's lemonade taste stronger or weaker than yesterday's mix, or is there not enough information to tell?

b. If he mixes more lemonade mix with more water than he did yesterday, will today's lemonade taste stronger or weaker than yesterday's mix, or is there not enough information to tell?

35. If the ratio of boys to girls in a class is 3 to 8, will the ratio of boys to girls stay the same, become greater, or become smaller if 2 boys and 2 girls are added to the class? Justify your response.

36. ***Classroom Connection*** This problem is adapted from the article "Proportional Reasoning" in the January 2000 issue of *Mathematics Teaching in the Middle School,* p. 311. Jennifer has a very small picture of her standing next to a large fish she caught while deep sea fishing. Next to Jennifer is a ruler showing the length of the fish. If she puts the photo in a copy machine and blows it up, then puts that enlargement in the machine and enlarges it by the same amount, which set of lines show what could happen to the image of the ruler?

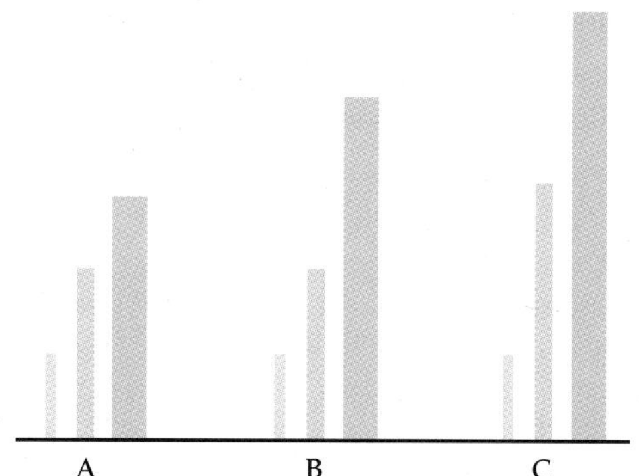

37. Here are the directions from an oatmeal container:

Servings	**1**	**2**	**3**
Water or milk	$\frac{3}{4}$ cup	$1\frac{2}{3}$ cup	$2\frac{1}{3}$ cup
Cereal	$\frac{1}{2}$ cup	1 cup	$1\frac{1}{2}$ cup
Salt (optional)	Dash	$\frac{1}{8}$ tsp	$\frac{1}{8}$ tsp

a. Are the increases in quantities proportional? Why or why not?

b. What amount of each ingredient would you use to make 10 servings? Explain your work.

38. A recipe for chocolate chip cookies follows.

$1\frac{1}{4}$ cups flour
$\frac{1}{2}$ cup sugar
$\frac{1}{2}$ teaspoon salt
$\frac{1}{2}$ cup butter
6 oz chocolate chips
1 teaspoon vanilla extract
1 egg
$\frac{1}{2}$ teaspoon baking powder

The recipe makes 4 dozen cookies.

a. How much of each ingredient would you need if you wanted to make 10 dozen cookies?

b. Unlike in the oatmeal problem, it is essential in this case that certain ingredients be in the "correct proportion." Explain what that phrase means.

39. An advertisement says that 5 out of 8 doctors recommend one brand of aspirin over another. In actuality, 325 doctors were interviewed, and 199 of the doctors recommended the first brand over the other. Is the advertisement accurate? Explain your response.

Exercises 40–44 require you to make some assumptions in order to determine an answer. Describe and justify the assumptions you make in determining your answer.

40. A worker estimates that she spends 75 minutes a day at the copying machine. How many hours would this be per year?

41. Two persons have the same yearly income, but one gets paid every other week, whereas the other gets paid twice a month. Which paycheck is larger?

42. One morning I was listening to National Public Radio, which was conducting its annual pledge drive. The announcer said, "Our goal this morning (from 6 A.M. to 9 A.M.) is to get 40 new pledges. We've received 15 pledges so far, so we're on track." I was puzzled because it was 7:30.

a. Explain why the announcer was inaccurate from a mathematical perspective.

b. Assuming that the announcer was aware that the math wasn't quite right, explain why he might have made that statement.

43. Sonja claims that she was 20 minutes late to work because she got stuck behind a truck the whole way. Is this plausible? If not, explain why. If so, make up some numbers to support your conclusion.

44. Determine the cost of flying from your nearest airport to two different cities. If we consider the fares as rates—that is, cost per mile—are the two rates proportional? If not, why do you think they are not proportional?

45. The following data give the numbers of people in prison in the United States for four different subpopulations.

a. Just looking at these numbers alone, write a sentence or two comparing the different amounts.

b. Now look at the second column, where the raw numbers have been turned into rates. Write a sentence or two comparing the different amounts.

c. Describe how the rates can changes one's perception of people in prison.

	Number	Per 100,000 population
White male	454,300	465
Black male	586,300	3405
White female	39,100	38
Black female	35,000	185

46. When data are presented in either tables or graphs, sometimes we see raw numbers and sometimes the raw numbers have been translated into rates. Unfortunately, we have found that many people do not fully understand rates. Explain the reason why we have rates, as though in response to someone who asks, "Why not just use the raw numbers?"

47. The data below give the numbers of adults living with HIV in several countries.

a. Just looking at these numbers alone, write a sentence or two comparing the different amounts.

b. Now look at the second column, where the raw numbers have been turned into rates—in this case, percentages. Write a sentence or two comparing the different amounts.

c. Describe how the rates can change one's perception of the AIDS epidemic in different parts of the world.

	Adults with HIV	Percent of adult population
South Africa	5,100,000	21.5
India	4,750,000	0.8
Nigeria	3,300,000	5.4
Zimbabwe	1,600,000	24.6
Mozambique	1,200,000	12.2

48. Converting raw numbers to rates.

a. There were 2,777,000 births in the United States in 1900 and 4,266,000 in 2006. Determine the birth rates, per 1000 population, for those two years. The U.S. population in 1900 was 92 million and in 2006 was 299 million.

Source: *2009 Statistical Abstracts of the U.S.*, pp. 63 (birth data) and 7 (population data).

b. There were 1,344,520 reported violent crimes in the United States in 1980 and 1,418,000 in 2006. Compare the crime rate, per 100,000 people, for these two years. The population in 1980 was 227 million.

Source: *2009 Statistical Abstracts of the U.S.*, p. 188.

c. There were 501,886 people in federal or state prisons in the United States in 1980 and 2,258,983 in 2006. Compare the rate, per 100,000 people, for these two years.

Source: 2009 *Statistical Abstracts of the U.S.*, p. 207.

d. There were 54,633 deaths in motor vehicles in 1970 in the United States and 44,700 in 2006. Compare the rate per 100,000 people, for these two years. The population in 1970 was 203 million.

Source: *2009 Statistical Abstracts of the U.S.*, p. 667.

49. Below are infant mortality rates for whites and blacks in the United States. Compare the improvement in the white rate to the improvement in the black rate.

	White	Black
1980	10.9	22.2
2005	5.7	13.7

Source: *2009 Statistical Abstracts of the U.S.*, p. 81.

50. The data below give the numbers of immigrants to the United States in two decades.

Decade	Immigrants	U.S. population at end of decade
1901–1910	8,795,386	91,972,266
2001–2010	10,299,430	305,529,237

a. Which decade had the greater proportion of immigrants coming to the United States?

b. Convert the two immigration numbers to the following common format: x immigrants per 1,000 residents.

c. Do the same as (b), but this time round all four numbers to make the computation easier. How close are the answers to (b) and (c)?

d. How does this relate to the leading-digit strategy of estimation that we learned in Chapter 3?

FROM STANDARDIZED ASSESSMENTS

2008 NECAP, Grade 6

51. One cup of milk has the same amount of calcium as 6 oranges. One orange has the same amount of calcium as 2 eggs. How many eggs have the same amount of calcium as one cup of milk?

a. 3 b. 4 c. 8 d. 12

52. Luis mixed 6 ounces of cherry syrup with 53 ounces of water to make a cherry-flavored drink. Martin mixed 5 ounces of the same cherry syrup with 42 ounces of water. Who made the drink with the stronger cherry flavor?

Source: From the *Seventh NAEP Mathematics Assessment*, p. 81. U.S. Department of Education.

SECTION 5.2 Percents

What do you think?

- What does percent mean?
- To what other mathematical concepts is percent related?
- In what real-life contexts have you encountered percents?

The Common Core State Standards have percents in sixth and seventh grades, under the *Ratios and Proportional Relationships* section. We see percents often in the media, so confidently interpreting percents is an important skill. In this section, you will see how percents rest upon the notion of fractions (in terms of part–whole relationships) and proportions (in terms of equivalent ratios or equivalent fractions).

THE ORIGIN OF PERCENT

The word **percent** literally means "per hundred" and comes directly from the Latin *per centum.*

There are direct and immediate connections among percent, equivalent fractions, and proportions. For example, let's say that a student correctly answered 12 out of 16 questions on a quiz. To determine the student's score as a percentage, we are asking what fraction with a denominator of 100 is equivalent to $\frac{12}{16}$. This question can be stated as a proportion:

$$\frac{12}{16} = \frac{x}{100}$$

That is, what value of x makes $\frac{x}{100}$ and $\frac{12}{16}$ equivalent fractions? Solving for x, we find the student's grade is 75%.

MATHEMATICS

We can also interpret this problem from the perspective of rates. That is, if the student were to continue to get problems correct *at this rate,* how many correct would she or he get out of 100 questions?

USES OF PERCENT

We use percents for a variety of purposes:

- To communicate—for example, we hear on the news that a fire is 30% contained.
- To make sense of situations—a manufacturer may say that the germination rate of a grass seed is 80%.
- To make decisions—should I refinance my house if the interest rate is down to 7.5%?
- To compare—in one high school of 345 there were 12 dropouts, and in another high school of 567 there were 17 dropouts. The first school has a dropout rate of 3.5% and the second school has a dropout rate of 3.0%.

HISTORY

The concept of percent is related to societies' need to compute interest, profit and loss, and taxes. When the Roman emperor Augustus levied a tax on all goods sold at auction . . . the rate was $\frac{1}{100}$ . . . . In the Middle Ages, 100 became a common base for computations. Italian manuscripts of the fifteenth century contained such expressions as "20p100," "x p cento," and "vi p co" to indicate 20 percent, 10 percent, and 6 percent. . . . The first use of a symbol for percent (P°) was used in 1425. By about 1650 the $\frac{o}{-}$, had become $\frac{o}{o}$ so per $\frac{o}{o}$ was often used. Finally the per was dropped, leaving $\frac{o}{o}$ or %.[5]

Let us use this basic understanding of percents to solve a few problems and then look at commonalities and differences among those problems.

INVESTIGATION 5.2a Who's the Better Free-Throw Shooter?

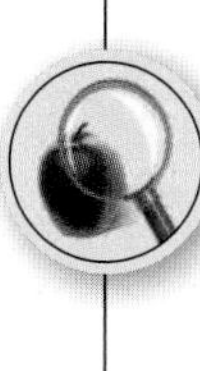

You are the coach of the girls' basketball team at a local middle school. It is the fifth game of the season, the game is tied, and there are only 5 seconds left on the clock. The referee has called a technical foul on the other team. You get to choose any one of your girls to take the shot. Basing your decision only on their free-throw shooting thus far this season, whom would you pick?

- Becky has made 8 of 12 free throws.
- Rachel has made 15 of 20 free throws.

DISCUSSION

STRATEGY 1 Use fractions

Represent the players' free-throw shooting as fractions.

$$\frac{8}{12} = \frac{2}{3} \qquad\qquad \frac{15}{20} = \frac{3}{4}$$

Becky Rachel

Using fraction reasoning and modeling that we discussed in the previous chapter, we know that $\frac{3}{4}$ is greater than $\frac{2}{3}$, so Rachel has the better record.

CLASSROOM CONNECTION

When Investigation 5.2a is given to middle school students, many will pick Becky, saying that she has missed fewer shots. Do you see why? What is their misconception?

STRATEGY 2 Make a proportion

Some students feel more comfortable with proportions. How would you describe why this proportion will tell us the answer?

$$\frac{2}{3} = \frac{x}{20}$$

In this case, x represents how many baskets Rachel must have made in order to have the same ratio as Becky. When we solve for x, we get $x = 13\frac{1}{3}$. Because Rachel has made 15 out of 20, which is better than $13\frac{1}{3}$ out of 20, she is doing better than Becky.

STRATEGY 3 Convert fractions to decimals or percents

$$\frac{8}{12} \approx 0.67 = 67\% \quad \text{whereas} \quad \frac{15}{20} = 0.75 = 75\%$$

In real life, what factors other than the players' free-throw percentages might a coach consider before making this decision?

INVESTIGATION 5.2b Understanding a Newspaper Article

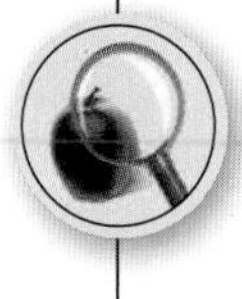

A newspaper story reports that 8% of the 7968 students at Midvale College work full-time. How many students work full-time? First try to estimate the number of students and then determine the exact answer.

DISCUSSION

STRATEGY 1 Use 10% as a benchmark

A very rough estimate: 8% is close to 10%, which is $\frac{1}{10}$, a fraction that we can do mental multiplication with rather simply. If we round 7968 to 8000, then $\frac{1}{10}$ of 8000 is 800.

STRATEGY 2 Use 1% as a benchmark
We could mentally find 1% and use simpler numbers to build up to 8%:

1% of 7968 is about 80.
8% of 8000 is 80 × 8 = 640.

STRATEGY 3 Find a close unit fraction
We could use $\frac{1}{12}$ for 8%. Do you see why? We could think of this as $\frac{1}{12}$ is close to $\frac{8}{100}$.
In this case, we can use the compatible-numbers strategy from Chapter 3.

$$\frac{7968}{12} \approx \frac{8400}{12} = 700$$

When we compare this to $\frac{7200}{12} = 600$, we conclude that the actual number of students working full-time is between 600 and 700.

Converting among percents, decimals, and fractions When we estimate the answers to percent problems, it is helpful to know the basic conversions among percents, fractions, and decimals. Table 5.3 shows some of the more commonly used conversions.

TABLE 5.3

Fraction	Decimal	Percent
1/2	0.5	50%
1/3	0.333. . . or $0.\overline{3}$	$33\frac{1}{3}$%
1/4	0.25	25%
1/5	0.2	20%
1/10	0.1	10%
1/100	0.01	1%

With this in mind, let us examine different ways to determine the "exact" number of students working full-time at Midvale College.

STRATEGY 1 Connect to the meaning of percent
We can represent the ratio of students working full-time to the total number of students as a fraction—that is,

$$\frac{\text{Number of students working full-time}}{\text{Total number of students}} = \frac{8}{100}$$

If we let x represent the number of students working full-time, we have

$$\frac{x}{7968} = \frac{8}{100}$$

Now we can solve for x and see that $x = 637.44$.

STRATEGY 2 Use multiplication procedure

8% of 7968 = 0.08 × 7968 = 637.44

About 637 students work full-time.

An answer of 637.44 or even 637 is too precise; it is an example of what I call pseudo-precision. Why is this? Think before reading on. . . .

When the person in the college determined that 8% of the students work full-time, he or she took the number of students who work full-time and divided by the total number of students, and the number that appeared on the calculator was closer to 0.08 than it was to 0.07 or to 0.09.

CLASSROOM  CONNECTION

Grade 5

Do you see how the previous work with unit fractions and unitizing (Investigation 4.2c) helps the children to make sense of the idea of percents? Which of these problems do you find challenging?

Date ______________________ Time ______________________

LESSON 8·10

Unit Fractions and Unit Percents

Math Message

1. ?

12

If 12 counters are $\frac{1}{5}$ of a set, how many counters are in the set? __________ counters

2. If 15 counters are $\frac{1}{7}$ of a set, how many counters are in the set? __________ counters

3. Complete the diagram in Problem 1 to show your answer.

4. If 31 pages are $\frac{1}{8}$ of a book, how many pages are in the book?
Number model: __________
Answer: __________ pages

5. If 13 marbles are 1% of the marbles in a jar, how many marbles are in the jar?
Number model: __________
Answer: __________ marbles

6. If \$5.43 is 1% of the cost of a TV, how much does the TV cost?
Number model: __________
Answer: __________ dollars

7. If 84 counters are 10% of a set, how many counters are in the set?
Number model: __________
Answer: __________ counters

8. After 80 minutes, Dorothy had read 120 pages of a 300-page book. If she continues reading at the same rate, about how long will it take her to read the entire book?
Number model: __________
Answer: __________ min

9. Eighty-four people attended a school concert. This was 70% of the number expected to attend. How many people were expected to attend?
Number model: __________
Answer: __________ people

280

From *Everyday Mathematics, Grade 5:* The University of Chicago School Mathematics Project: Student Math Journal, Volume 2, by Max Bell et al., Lesson 8-10, p. 280. Reprinted by permission of The McGraw-Hill Companies, Inc.

In other words, the actual number of students who work full-time is between 7.5% of 7968 and 8.5% of 7968. That is, the actual number is between 598 and 677. Do you see why?

We have just found that any number of full-time students between 598 and 677 will round to 8% of 7968. Because we know only that 8% was reported, it would be more accurate to say not that 637 students work full-time but, rather, that the number of students who work full-time is between 598 and 677, or 637 ± 40.

INVESTIGATION 5.2c Buying a House

The Benanders are going to buy a house. Before giving a family a mortgage to buy a home, banks generally require that the total monthly payment (including property taxes) be no more than 28% of the family's gross monthly income. What must the Benanders' monthly income be in order for them to buy a home on which the monthly payment will be $800?

This is not an easy question. Please think about it, grapple with it, apply your understanding of percent and your problem-solving toolbox, and then read on. . . .

DISCUSSION

STRATEGY 1 Use guess–check–revise

One way of paraphrasing the problem is to say that if the bank computes 28% of the Benanders' monthly pay, this number must be at least $800 or the bank will turn them down. In other words, the bank takes their monthly income and multiplies it by 0.28. If the product is greater than $800, they qualify.

For many students, this line of reasoning leads to guess–check–revise, as shown in Table 5.4.

TABLE 5.4

Guess	Computation	Analysis
$3000	$3000 × 0.28 = $840	Too much. Guess less.
$2900	$2900 × 0.28 = $812	Too much. When the guess went down by $100, the payment went down by $28. If the next guess is $50 less, the payment will be $14 less, which will be too low (798). So make a guess of $40 less.
$2860	$2860 × 0.28 = $800.80	

It would take several more trials to get the exact answer, and that is one limitation of guess–check–revise: It is not practical when you need an exact answer. However, when determining eligibility, the bank is not interested in three decimal places. Banks have some flexibility; that is, if you are "close" to 28% and there are other factors in your favor (such as job stability, a promotion due, or an inheritance anticipated), then you are likely to qualify.

STRATEGY 2 Rewrite the problem as an equation

Another set of students paraphrases the problem something like this:

28% of their salary must be at least 800.

Changing the wording to be more "mathematical," we have

28% of what is $800?

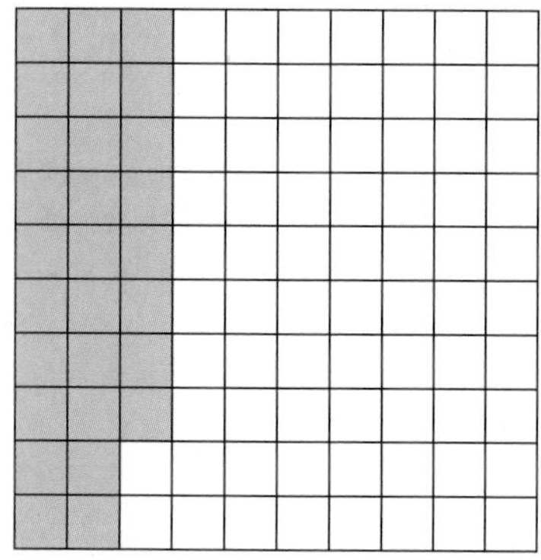

Figure 5.3

The jump to an equation now is not as great a leap to make:

0.28 times x = 800, where x = their monthly salary

That is, if $0.28x = 800$, solve for x, and see that $x = 2857.14$. That is, their monthly income must be at least \$2857.

STRATEGY 3 Use a diagram

We can also represent this problem with a diagram. Even if you "understand" how to get the answer, can you explain the "why" of Figure 5.3? Try to do so before reading on. . . .

In Figure 5.3, the whole box represents their monthly income, and the shaded area represents 28% of their income.

If we look at the problem from a part–whole perspective,

$$\frac{\text{Part}}{\text{Whole}} = \frac{28\text{ boxes}}{100\text{ boxes}} = \frac{\$800}{\text{Total income}}$$

We can also interpret the figure in the following way:

If 28 boxes has a value of 800, then 100 boxes has a value of what?

If 28 boxes	represents a value of	\$800, then
1 box	represents a value of	$\frac{800}{28} = \$28.571428$
100 boxes	represents a value of	\$2857.14

CONNECTIONS BETWEEN PERCENT AND OTHER MATHEMATICAL TOPICS

How would you represent the three previous investigations in terms of part–whole relationships? How would you interpret them in terms of proportions?

One of the keys to seeing the interconnectedness of all three problems is to realize that in percent problems, there are two parts and two wholes.

For example, in Investigation 5.2a, the 12 shots Becky attempted are the whole and 8 are the part she made. When we determine that this part–whole relationship is equivalent to 67%, we are saying that this is equivalent to 67 parts in a whole of 100 [Figure 5.4(a)].

To determine 8% of 7968 in Investigation 5.2b, we need to find the part out of the whole of 7968 that is equivalent to 8 parts out of 100 [Figure 5.4(b)].

In Investigation 5.2c, to determine the Benanders' monthly payment, we ask, 800 is 28% of what whole [Figure 5.4(c)]?

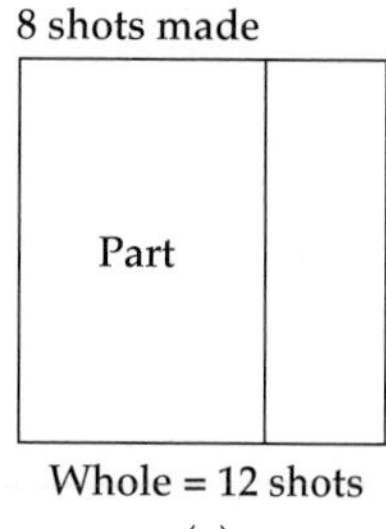

(a)

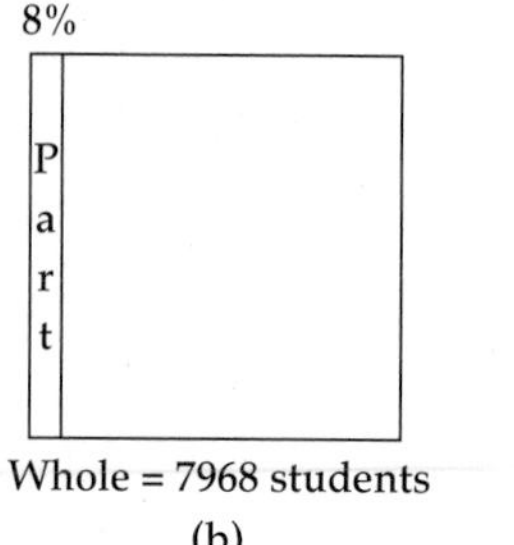

(b)

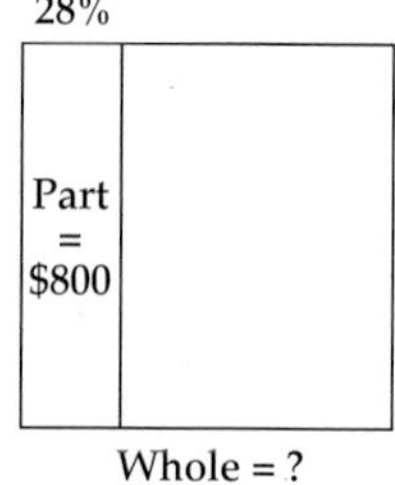

(c)

Figure 5.4

If we now examine these problems in terms of proportions, we can see that the differences among the three problems have to do with what part or whole is missing.

$\frac{\text{Part}}{\text{Whole}}$	$\frac{8}{12} = \frac{x}{100}$	$\frac{x}{7968} = \frac{8}{100}$	$\frac{800}{x} = \frac{28}{100}$

Now that we have investigated some of the basic aspects of percents, the following investigations can serve both as stretching problems and as a self-assessment of how well you "own" the concept (that is, how well you can apply your knowledge to nonroutine problems).

INVESTIGATION 5.2d Sale?

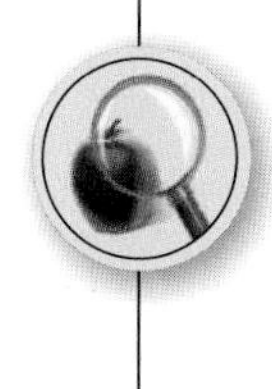

Jorge is excited. He just saw in the newspaper that Showroom Appliances is celebrating 20 years in business by offering 20% off on all merchandise. He just went there yesterday to buy a television set that was regularly priced at \$260. What is the sale price of the television?

DISCUSSION

Explorations Manual 5.6

Probably the most common solution I see to this problem looks like this:

$$0.20 \times \$260 = \$52$$
$$\$260 - \$52 = \$208$$

That is, you determine the discount and then subtract the discount from the regular price. Jorge will pay \$208.

However, Randi says that she solved the problem in one step:

$$0.80 \times \$260 = \$208$$

Do you understand why Randi did this? How would you explain it to someone else? Please reflect on these questions before reading on. . . .

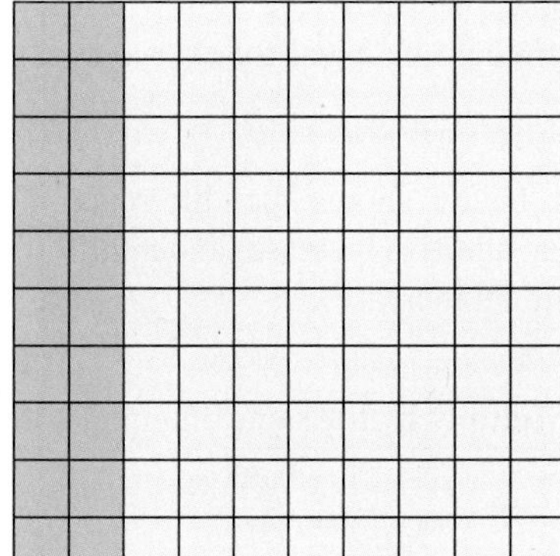

Figure 5.5

One way to get beyond the "how" of this shortcut and into the "why" is to represent this problem with a diagram. In Figure 5.5, if the whole box represents \$260 (the original price), what do the two shaded areas represent (that is, 20 boxes and 80 boxes)? Think before reading on. . . .

The 20-box shaded region represents the **discount**—that is, the amount by which the store will reduce the price, or how much Jorge will save.

The 80 boxes therefore represent the **sale price** of the television. Because it is the sale price that we are looking for, we do not need to find 20% of the price and subtract it from the original price; we can determine the sale price directly.

We can show that the two procedures are mathematically equivalent:

$$260 - 0.20(260) = 260(1 - 0.20) \quad \text{We are using the distributive property.}$$
$$= 260(0.80) \quad \text{Because } 1 - 0.20 = 0.80$$

Thus,

$$260 - 0.20(260) = 260(0.80)$$

PERCENT CHANGE

Many problems involving percents in real life involve change. Such comparisons are called percent increase, percent decrease, percent change, percent error, percent faster or slower, or percent more or less.

We need to be able to understand change in order to interpret intelligently the changes in our society—in employment, the economy, income equality, and other areas. Percent is one of many mathematical tools that can help us better understand and make sense of change.

INVESTIGATION 5.2e What Is a Fair Raise?

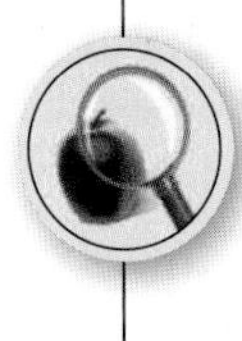

Explorations Manual 5.6

Let's say I am the president of a small company, Bassarear's Bagels, that has done very well in the last year. I have decided that I want to share my good fortune with my hard-working and devoted employees. I announce that I will be giving a \$2-an-hour raise to everyone. Two days later I become aware of some grumbling. A number of employees are complaining that this is not fair. I am stunned. I thought that giving everyone the same raise was the epitome of fairness. Why are some employees grumbling? How would you explain this to me if you were one of my dissatisfied employees? Think and then read on. . . .

DISCUSSION

Some common responses that you may have thought of include:

- Workers with more experience, responsibility, or education should receive "bigger" raises.
- Full-time workers should receive "bigger" raises.
- Hard-working workers should receive "bigger" raises.
- This raise is not fair to the people making more money.

For the concept of percent increase, I want to focus on the last reason. Why would a higher-paid employee feel that my raise was not fair to him or her? Imagine you are such an employee. How might you help me to see your point? Think and read on. . . .

Let's say the janitor is making \$8 an hour and the manager of one of the stores is making \$20 an hour. From a proportional perspective, the janitor is pretty happy because \$2 an hour represents $\frac{1}{4}$ of her salary. The manager, however, is not very happy, because \$2 represents only $\frac{1}{10}$ of her salary. From the proportional perspective, a "fair" raise would be one in which the ratio of raise to present salary would be equal for everyone; that is, all the raises would be proportional.

Additive versus multiplicative increases The chart below shows the difference between *adding* the *same amount* to each person's wage (an additive increase) and *multiplying* each person's wage by the *same amount* (a proportional increase).

Original salary	Add the same amount	Multiply by the same amount
8	$8 + 1 = 9$	$8(1.2) = 9.60$
20	$20 + 1 = 21$	$20(1.2) = 24$

What if I were to be convinced that I should not add the same amount to each person's salary but, rather, should multiply each person's salary by the same amount? How would I announce that to the company in terms of percent? Think and read on. . . .

If we multiply the janitor's salary by 1.2, we are in effect increasing her salary by $\frac{1}{5}$. That is,

$$8(1.2) = 8 + 0.2(8) = 8 + 1.60 = 9.60$$

If we multiply the manager's salary by 1.2, we are in effect increasing her salary by $\frac{1}{5}$. That is,

$$20(1.2) = 20 + 0.2(20) = 20 + 4 = 24$$

Because $\frac{1}{5}$ is equivalent to $\frac{20}{100}$, we say that both employees have received a 20% raise; that is, their salaries are 20% greater than they were before.

In order to understand better the difference between additive and multiplicative changes, look at Figure 5.6, which shows the "before" and "after" salaries using both methods. What do you notice?

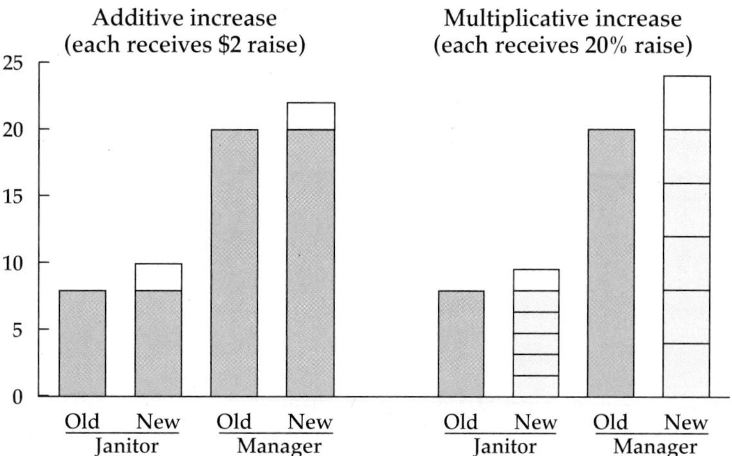

Figure 5.6

In the additive case (at the left in Figure 5.6), what is *equal* is the amount of the raise (represented by the white boxes). That is, both people received a raise of $2 per hour.

In the multiplicative case (at the right in Figure 5.6), what is *equal* is the ratio of the increase to the wage. In both cases, the amount of the raise (represented by the white boxes) is equal to $\frac{1}{5}$ of the wage.

As I hope you are seeing, the idea of "equality" is as complex in mathematics as it is in history and political science. It is also important to know that not everyone would agree that the 20% raise is fair. Many people interpret this scenario as "the rich getting richer" because the difference between the janitor's and the manager's hourly pay was $12 and is now $14.40.

Now let us investigate a problem that is near and dear to the hearts, or at least the pocketbooks, of most students.

INVESTIGATION 5.2f How Much Did the Bookstore Pay for the Textbook?

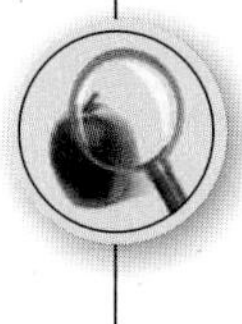

When college bookstores purchase textbooks, they generally sell the books for 20% to 25% more than they paid for them. In other words, their markup is between 20% and 25%. Let's say that you paid $95 for a textbook and your bookstore marked up the price by 25%. How much did the bookstore pay for the textbook? Work on this problem before reading on. . . .

DISCUSSION

STRATEGY 1 Act it out

Let's say you are working in the bookstore. You would take the price of the book (which we don't know at this point) and find 25% of that number, and then you would add that to the price of the book. This sum would be $95. We can solve the problem by using guess–check–revise (Table 5.5) or by forming an equation.

STRATEGY 2 Use guess–check–revise

TABLE 5.5

Guess	Work	Reflection
60	$0.25(60) + 60 = 75$	Too low.
70	$0.25(70) + 70 = 87.50$	Increasing the guess by 10 increases the sum by 12.50.
75	$0.25(75) + 75 = 93.75$	Close.
76	$0.25(76) + 76 = 95$	Done!

STRATEGY 3 Bar model

We can use a bar model to show the cost of the text plus the 25% mark-up. Examine the model and see if you can make sense of it for yourself.

$19	$19	$19	$19	$19

Cost of text + 25%

$95

Since the entire bar represents $95, each of the 5 equal sections in the bar would equal $19 (95 divided by 5). So, the original cost of the book was 4($19) = $76.

STRATEGY 4 Form an equation

If we let x = the price the bookstore paid for the book, many students can "see" the equation emerging by acting out the process. The bookstore employee finds 25% of x and then adds this amount to the amount the bookstore paid for the book; this sum must be $95.

That is,

$$0.25x + x = 95$$

$$1.25x = 95$$

$$x = 76$$

Virtually every day, we see percents described in the context of a variety of situations. One of the difficulties many people have is translating percent ideas into words. The following situation is one of many instances of the need to translate.

PERCENTS LESS THAN 1 OR GREATER THAN 100

In real life, we frequently encounter percents greater than 100% or less than 1%. Researchers and teachers know that when students do not thoroughly understand the concept of percents, their success rate with such problems goes down dramatically.

INVESTIGATION 5.2g The Copying Machine

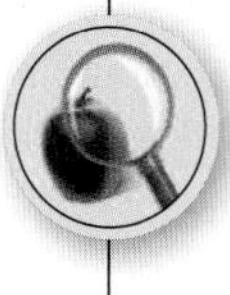

Clark Elementary School has decided to buy a new copying machine. One of the selling points of the new machine is that the manufacturer advertised that the school can expect about 2 paper jams per 1000 copies. After six months, the faculty wanted to see how the copier was actually doing. They had run 42,164 copies and had encountered 96 paper jams. Think about the following questions and then read on. . . .

A. Do you think the ratio of paper jams at Clark is consistent with the advertised paper jam rate? Is the school averaging about 2 paper jams per 1000 copies?

B. Represent the advertised paper jam rate as a ratio, a fraction, a decimal, and a percent.

C. Which one would you use if you were writing the ad for the copier?

D. How do you think the manufacturer determined the figure of 2 paper jams per 1000 copies?

DISCUSSION

A. One way (of several) to answer this question is to set up the following proportion:

$$\frac{96}{42{,}164} = \frac{x}{1000}$$

By setting the ratio $\frac{96}{42{,}164}$ equal to $\frac{x}{1000}$, we are saying the rate of 96 jams per 42,164 copies is equivalent to how many jams per 1000 copies?

Solving for x, we find $x \approx 2.3$, which means that Clark is averaging about 2.3 paper jams per 1000 copies. Because the paper jam rate for Clark is closer to 2 jams per 1000 than to 3, we would conclude that the copier is acting as advertised.

B. All of the following are mathematically equivalent to a paper jam rate of 2 per 1000:

1 in every 500 copies will jam.
$\frac{1}{500}$ of the copies will jam.
0.2% of the copies will jam.
One-fifth of 1% of the copies will jam.

C. As an advertiser, I think I would say that the paper jam rate is one-fifth of 1%. This sounds smaller than 2 copies per 1000. However, I also know that most people will understand 2 copies per 1000 much better than one-fifth of 1%.

D. The manufacturer probably divided the total number of jams by the total number of copies. I would hope that this was done with many different copiers rather than just one copier, and I would hope that these copiers were not all brand new. The ratio that resulted from this calculation was closer to 2 per 1000 than to 1 per 1000 or 3 per 1000.

INVESTIGATION 5.2h 132% Increase?

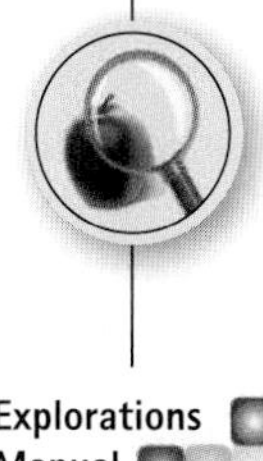

In 2005, I read that "the percent increase since 1975 in births by Cesarean section in the United States is 170." The article went on to say that 640,000 Cesarean sections had been done in 1975. I immediately wondered how many had been done in 2005, and that is the question I want you to answer now. Think and then read on. . . .

Explorations Manual 5.8

A. Before you solve the problem, try to make a rough or refined estimate.

B. Then solve the problem.

DISCUSSION

A. *A rough estimate:* Because a 100% increase is double the original number, the number of Cesarean sections more than doubled, so the answer will be more than 1,280,000 Cesarean births in 2005.

A refined estimate:

An increase of 170% = original amount + 100% + 70% (about $\frac{2}{3}$)
600,000 + 600,000 + 400,000 = 1,600,000 Cesarean births in 2005

B. STRATEGY 1 Break the problem into parts

A 170% increase can be broken into a 100% increase and a 70% increase.
A 100% increase is equivalent to doubling, which gives us 1,280,000.
A 70% increase is 0.7(640,000) = 448,000.
Therefore, a 170% increase brings us to 1,728,000 Cesarean births in 2005.

STRATEGY 2 Connect to a similar, simpler problem
For example, if the problem had said that there had been a 25% increase, many students would have found the answer by multiplying 640,000 by 1.25. Do you see why?

If a 25% increase is determined by multiplying by 1.25, what would we do to determine a 170% increase? Think and then read on. . . .

We would multiply 640,000 by 2.7.

Without an estimate, many students arrive at the wrong answer of 1,088,000 because they multiply 640,000 by 1.7 or they add 70% of 640,000 to 640,000. In this case, even a very rough estimate serves many students nicely: it tells them that the number of Cesarean births in 2005 must be at least 1.2 million. Students who do a very rough estimate and then obtain an actual answer of 1,088,000 see a contradiction between the estimate and the computed answer, and this contradiction causes them to go back and reexamine their work. This is one of the many useful aspects of estimating.

INTEREST

One of the ways in which almost everyone encounters percents is with interest—when you buy a car, when you buy a house, and when you don't pay off your entire credit card balance at the end of the month, you pay interest on the amount that you owe. You also earn interest on money in savings accounts, retirement accounts, and so on.

Most people have a very limited understanding of the consequences of interest. Let's say you buy a house for $225,000 and make a down payment of $25,000. You go to the bank and take out a 30-year mortgage for $200,000, and you are told that your monthly payment will be $1151. Multiply this monthly payment by 360 (there are 360 months in 30 years) to determine how much you actually pay for the house (the $200,000 you borrowed plus all the interest). Then read on. . . .

No, you did not mispunch the calculator. Yes, you really will pay that much for the house! You actually paid $459,360 for the house! Thus, over 30 years, you pay $239,360 of interest!

Let us examine how interest is determined, first in a simple case and then in some instances that are more realistic. Several variables affect how much interest you receive (on an investment) or pay (on a loan): the original amount, called the principal (P), the annual interest rate (r), specified in percent, and the time of the investment/loan (t), specified in years.

Although simple interest—when the principal does not change over the course of the loan—is not common, it is a good starting point for understanding how interest is determined. For example, let's say that Jenna borrows $2000 from a relative and agrees to repay the loan after 1 year at the rate of 6%. At the end of the year, she pays $2000 + 6% of $2000; that is, she pays $2000 + 120 = $2120.

More commonly, interest is compounded; that is, the interest is determined at specified intervals and added to the principal at those times. For example, if Jenna's relative had specified that the loan be compounded semiannually, then to determine how much Jenna would owe at the end of the year, the interest would be determined every 6 months and added to the principal. Thus, after 6 months, she would owe $2000 + 0.03(2000) = $2060. Why is this?

The annual interest rate is 6%, so the semiannual rate is $\frac{1}{2}$ of 6%, or 3%, because 6 months is $\frac{1}{2}$ year. We determine how much she owes after the next 6 months as follows: $2060 + 0.03(2060) = 2060 + $61.80 = $2121.80. In this particular case, a 1-year loan compounded only semiannually, the difference between simple and compound interest is not huge. However, on most deposits and loans, the interest is compounded daily; that is, the annual interest rate is divided by 365.

In the following investigation, we will develop the formula for determining compound interest. In the second investigation, we will examine a very relevant situation.

INVESTIGATION 5.2i Saving for College

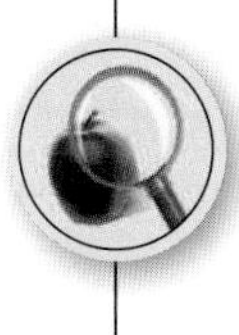

When Emily was born, her grandparents decided to contribute toward her college education by opening a savings account for \$1000. If they add no other money, and if the account earns 6% compounded annually, how much money will there be in the account after 18 years?

DISCUSSION

This is a nonroutine, multistep problem. Therefore, let us go one step at a time.

After 1 year How much money would there be in the account after 1 year? Do this and then read on. . . .

There are several different methods that students will use.

We can find 6% of \$1000 and then add that to 1000; that is, \$1000 + 0.06(\$1000) = \$1060. Or, as we saw earlier in this chapter, we can obtain \$1060 in one step: \$1000(1.06) = \$1060. Do you see why?

This connection between procedures is important because in order to understand this problem, you need to be able to understand the second procedure; the first one is too cumbersome.

After 3 years Now determine how much money would be in the account after 3 years, and then read on. . . .

\$1000(1.06) = \$1060 after 1 year

\$1060(1.06) = \$1123.60 after 2 years

\$1123.60(1.06) = \$1191.02 after 3 years

Let us stop and analyze this strategy. In the first step, we multiplied 1000 by 1.06 to get 1060. In the second step, many students clear the calculator and then multiply 1060 by 1.06. Then, in the third step, they clear the calculator and then multiply 1123.60 by 1.06. If we examine this carefully, we can see a shortcut. See whether you can find it yourself before reading on. . . .

We could do this on the calculator:

1000 [×] 1.06 [×] 1.06 [×] 1.06

Do you see why? If you don't believe this, do it on the calculator to see that it does work. Then try to explain why it works before reading on. . . .

We can represent what we have learned as

$$\$1191.02 = 1000(1.06)(1.06)(1.06)$$

We can now use our knowledge of exponents to say

$$\$1191.02 = 1000(1.06)^3$$

The third, and most efficient, method for finding the amount after 3 years uses this knowledge. How would we enter the last expression in a scientific calculator? Try it before reading on. . . .

Enter 1000 × 1.06 y^x 3 to get the same answer.

Before we move on to solve the original question, this is a good time to understand the basic formula for use with interest. Look at how we determined the value (amount) in the bank after 3 years. Try to represent that procedure with a formula using the interest symbols introduced above, and then read on. . . .

$$\text{The money after 3 years} = (1000)(1.06)^3 = 1000(1 + 0.06)^3$$

That is,

$$A = P(1 + r)^t$$

After 18 years This discovery makes the original problem much less tedious to solve.

STRATEGY 1 Use the y^x button

The most straightforward solution is to use the y^x button on your calculator:

1000 (×) 1.06 (y^x) 18 = \$2854.34

STRATEGY 2 Use the memory key

If your calculator lacks an exponentiation key, you could use the memory key M+ by doing the following:

First, activate the memory by pressing 1.06 and then M+.

Now press 1000 × MR = (you should see 1060, the amount after 1 year).

To get the amount for each succeeding year, you simply press × MR that many times.

That is, you press $1000 \times \overbrace{\text{MR} \times \text{MR} \times \text{MR} \cdots \times \text{MR}}^{\text{18 times}} =$

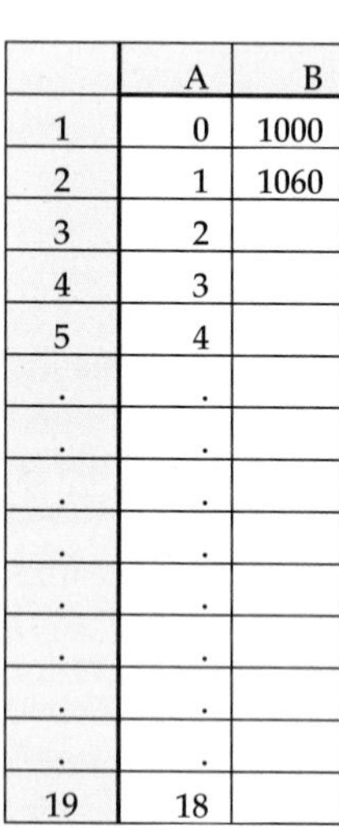

	A	B
1	0	1000
2	1	1060
3	2	
4	3	
5	4	
.	.	
.	.	
.	.	
.	.	
.	.	
.	.	
.	.	
.	.	
19	18	

Figure 5.7

STRATEGY 3 Use a spreadsheet

A spreadsheet (on a computer) can also be used. When you open the spreadsheet, the columns are marked with letters—A, B, C, and so on—and the rows are marked with numbers. In column A, we enter the numbers 0 through 18 (Figure 5.7). In column B, we enter 1000 into the first row, signifying the amount we have at the beginning.

In the B2 cell, we now enter "= B1*1.06"; that is, we tell the computer to multiply the amount in the B1 cell by 1.06. Now we will see 1060 in the B2 cell.

Now, we highlight the B column from row 2 through row 19 and select "Fill down." This command essentially tells the computer to repeat the computation—that is, to multiply the previous amount by 1.06. After we do this, the computer will display the amount at the end of each year!

One advantage of the spreadsheet is that we can play the "what if" game: what if they had started with \$3000, what if the interest rate had been 7%, and so on.

INVESTIGATION 5.2j How Much Does That Credit Card Cost You?

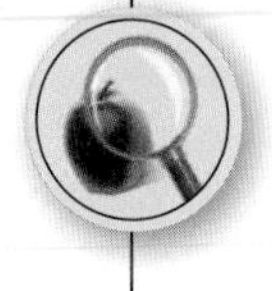

Most credit cards require a minimum monthly payment. If the customer is not able to pay the entire balance (for example, after a vacation or after Christmas), the customer pays at least this minimum amount; then a fee called a finance charge is determined on the basis of the interest rate the company is charging.

Let's say George found himself in just that situation. His VISA bill has come in, and the balance is \$761.34. He decides he can pay \$61.34. If the bank issuing the VISA card determines the finance charge at the annual rate of 18%, called the APR (annual percentage rate), what will his finance charge be? Think and then read on. . . .

DISCUSSION

Because the balance is compounded monthly, the bank will determine George's finance charge by multiplying his unpaid balance ($700) by 0.015. Do you see where the 1.5% comes from?

To find the monthly rate, we divide the yearly rate by 12 (months): 18% ÷ 12 = 1.5%, and 1.5% = 0.015.

Thus George's finance charge will be $700(0.015) = $10.50.

SUMMARY 5.2

In this section, we began with the basic concept of percent, a part expressed as a hundredth, and then we connected this to the concepts of equivalent fractions and proportions. We found, however, that this simple notion of percent needs to be refined when we are dealing with percents greater than 100, with percent increase and decrease, and with nonroutine problems. This need for a richer concept of percent was also underscored in the explorations.

In examining percent problems, you saw applications of percents in several areas, including interest. You also saw the importance of estimating—sometimes because the question calls for an estimate instead of an exact answer, and sometimes because estimation helps you to check your solution or your reasoning.

5.2 Exercises

1. For each of the following questions, estimate first, and briefly explain how you determined your estimate. Then determine the actual answer. Compare your estimate to the actual answer. If your estimate was close, move on; if not, you might want to reread part of the section and/or consult with a friend or the instructor.

   a. What is 4% of 450?
   b. What is 23% of 85?
   c. What is 83% of $1460?
   d. What is 3.2% of 1700?
   e. What is 0.25% of 345?
   f. What is 120% of 200?
   g. 1200 is what percent of 1500?
   h. 30 is what percent of 35?
   i. 56 is what percent of 657?
   j. 1.2 is what percent of 4.6?
   k. 75 is what percent of 400?
   l. 175 is what percent of 120?
   m. What is 3% more than 45?
   n. What is 20% more than 400?
   o. What is 16% more than 6.2?
   p. What is 100% more than 35?
   q. What is 10% less than 36?
   r. What is 3% less than 45?
   s. What is 1.2% less than 1200?
   t. What is 80% less than 800?

In Exercises 2–14, first estimate the answer and briefly describe how you determined the estimate. Then determine the actual answer.

2. Missy Adams's gross pay is $1500 per paycheck. If her total payroll deductions are $340, her take-home pay is what percent of her gross pay?
3. All three second-grade teachers at Uphill Elementary School asked their students how many have pets at home. The response was that 23 out of 65 of the students have pets. What percent of the students have pets?
4. A victim won damages of $2.4 million from a company. If the lawyer's fee was 35%, how much did the lawyer get?
5. If 30% of the patients in a hospital have heart problems and there are 210 patients with heart problems, how many patients are there in the hospital?
6. Last year Mr. Rich paid $27,000 in income taxes because he was taxed at the rate of 38% of his gross income. What was his gross income?

7. a. Between 1990 and 2005, the price of an average house in a town increased from \$42,000 to \$97,000. What is the percent increase (rounded to the nearest whole number) in the average selling price?

   b. What is the percent increase for a house that sells for \$60,000 and then later for \$75,349?

   c. What is the percent increase for a house that sells for \$71,450 and then later for \$102,500?

   d. What is the percent increase for a house that sells for \$62,000 and then later for \$144,000?

8. Two dresses are on sale. The first was selling for \$119 and is being marked down 40%. The second was selling for \$79.99 and is being discounted by 20%. Which dress costs less now?

9. At birth a baby weighed 8 pounds 4 ounces. Two days later, the baby weighed 7 pounds 12 ounces. What percent of its weight had it lost? (This weight loss is normal—part of the adjustment to the living outside the mother's womb.) If a 160-pound adult lost the same proportion of its weight in 2 days, how much weight would the adult have lost?

10. *The Unofficial U.S. Census* reports that 34,177,500 of America's 139,500,000 cars are washed at least once each week and that 6,556,500 cars are never washed. Assuming that these numbers are relatively accurate:

    a. What percent of cars are washed each week?

    b. What percent of cars are never washed?

11. In 2007, 17.6 percent of American children were living in poverty. If there were 72.8 million children, what number were living in poverty? Express the amount both as a decimal and as a whole number.

12. It has been estimated that 0.8% of Americans are homeless. Now 0.8% is a small percentage, but there are approximately 308 million people in the United States, and 308 million is a large number. Approximately how many Americans are homeless?

13. In 2008, 4.8 million of the 11.7 million eligible voters aged 18 to 20 actually voted. What percent voted?

14. In an up-and-coming town, housing prices rose an average of 115% in five years. If a home cost \$142,000 five years ago, how much would that home cost now?

DEEPENING YOUR UNDERSTANDING

15. A store reduced the price of a computer by 20% and sold it for \$1760. How much did the computer originally sell for?

16. Janice Brady borrowed \$4200 to buy a new car. If she pays the loan back in 1 year at 8.2% interest, how much will she actually pay for the car? (Assume simple interest.)

17. A store collected \$53 in sales tax alone in one day. If the sales tax is 5.5%, how much did the store sell that day?

18. John paid \$330 for a new mountain bicycle to sell in his shop. He wants to price it so that he can offer a 10% discount and still make 20% profit. At what price should he mark the bike?

19. A teacher was hired at a salary of \$28,200 and is to receive a raise of 5.5% after her first year and a raise of 4% after her second year. What should her salary be after the 2-year period?

20. Your optimal exercise heart rate for cardiovascular benefits is calculated as follows: Subtract your age from 220 and then find 80% of the difference. Find the optimal heart rate for a 50-year-old and for a 20-year-old. The optimal rate for the 20-year-old is ___ % greater than the optimal rate for a 50-year-old.

21. A city budget is \$21.43 million. The city council voted to increase the budget by 4.4%. What is the new budget, rounded to the nearest \$1000?

22. In the 1980s, the company that built the Seabrook nuclear reactor in New Hampshire went bankrupt. When Northeast Utilities bought the company out, the agreement was that Northeast could raise electricity rates at least 5.5% each year for the next 7 years.

    a. If the Jones family paid an average of \$83.00 per month for electricity last year and the rate rises 5.5% each year, what will they pay after 7 years?

    b. If the Jones family's income is \$80,000 this year and rises at a rate of 5.5% each year, what will be their income after 7 years?

23. Let's say a map has a scale of 1 : 25,000. Two cities on the map are 3 inches apart. How many miles does this represent?

24. Arthur was not able to get his income taxes ready by April 15. The penalty for late returns is assessed at a 20% annual rate of interest. Arthur filed his income taxes on June 20 without having applied for an extension. He owed \$2000 plus the penalty. How much did he owe?

25. Samantha Greene is on the negotiating team for the next teacher's contract. The school board is proposing a 6% increase for each of the next four years. The teachers' counterproposal is 12% this year and 4% for the next three years. Randi says, "What's the difference; in either case, it's 24% over 4 years!" What would you say to Randi?

26. Several years ago, newspapers reported that the average unpaid balance on VISA cards was about \$1500. Let's say that you are an average consumer, that your VISA card charges you at the rate of 18% per year, and that each month for 1 year, you have an unpaid balance of exactly \$1500.

    a. How much will your finance charges add up to for the year?

    b. How much money would you save if you switched to a credit card that charged only an 8% annual interest rate?

27. In 2008, senior high school principals averaged \$97,486 a year. Their counterparts made \$91,334 in junior high and \$85,907 in elementary schools. The average U.S. teacher made \$51,009 in 2008.

    a. Compare the average salaries in percent-greater language; that is, compare all salaries to the \$51,009 figure.

    b. Compare the average salaries in percent-less language; that is, compare all salaries to the \$97,486 figure.

28. Express 0.5% as a ratio in simplest terms. For example, 40% can be expressed as 40 out of 100, and in simplest terms as 2 out of 5.

29. Let's say that tests on a new vaccine show that less than 0.5% of all children have a serious reaction to the drug.

    a. Gerry is not very mathematically literate. Help her to understand mathematically what this means.

    b. What additional information would you like to have in order to help you decide whether or not to have your child take the vaccine?

30. The figures below are death rates for heart disease and cancer (per 100,000 people).

	1960	2006
Heart disease	559.0	200.2
Cancer	193.9	180.7

a. Describe the change in the death rates for heart disease and cancer between 1960 and 2006, using percent language.

b. Compare the 2006 death rates for heart disease and cancer, using percent language.

31. A road sign says, "7% grade." What does that mean?

32. The U.S. Census Bureau reported that there were 46.5 million Americans living below the poverty level in 2012, 15% of the total American population. The U.S. Census Bureau also gives figures for persons living "below 125% of the poverty level." They reported that 19.7 million persons, or 42.4% of those in poverty, were in this category.

a. If the poverty level for a nonfarm family of four in 2012 was $19,157, what would be the cutoff for a family of four "below 125% of the poverty level"?

b. Why do you think they came up with the reference point of "below 125% of the poverty level"?

Problems 33–38 require you to make some assumptions in order to determine an answer. Describe and justify the assumptions you make in determining your answer.

33. Let's say that you read in the newspaper that last year's rate of inflation was 7.2%.

a. If your grocery bill averaged $325 per month last year, about how much would you expect your grocery bill to be this year?

b. Let's say you received a $1200 raise, from $23,400 per year to $24,600 per year. Did your raise keep you ahead of the game, or are you falling behind?

34. There was a proposal in New Hampshire in 1991 to reduce the definition of "drunk driving" from an alcohol blood content of 0.1 to 0.08. Explain why some might consider this a little drop and others might consider it a big drop. What do you think?

35. Which would you prefer to see on a sale sign at a store: $10 off or 10% off? Explain your choice.

36. ***Classroom Connection*** Refer to Investigation 5.2b. Jane still doesn't understand the problem. Roberto tries to help her make sense of the problem by saying that the 8% means that if we were to select 100 students at the college, 8 of them would be working full-time. What do you think?

37. Annie has just received a 5% raise from her current wage of $9.80 per hour.

a. What is her new wage?

b. What would this amount to over a year?

c. What assumptions did you make in order to answer part (b)?

d. What if the raise had been 5.4%?

38. Jack is building an office building. The local building code says that the window area in the building can be no more than 20% of the floor area.

a. If the two-story building will contain 12 offices and have 3000 square feet of floor space, what is the maximum window area allowed?

b. About how many windows would that be?

39. If Jonah puts $25,000 in the bank at 8% interest compounded quarterly (four times a year), how much will his investment be worth in 20 years?

40. If Liam puts $10,000 in the bank at 5% interest compounded quarterly (four times a year), how much will his investment be worth in 5 years?

41. How long will it take $1000 at 6% simple interest to double?

42. A company's sales increased from 2005 to 2006. It is possible to describe this increase either additively or multiplicatively. You will be asked to examine and then compare the two ways.

a. We sold 34,234 more Bender Bobbers in 2006 than we did in 2005. What does that tell you?

b. We sold 25% more Bender Bobbers in 2006 than we did in 2005. What does that tell you?

c. If you were a stockholder in the company, which sentence would be more useful to you in the annual stockholder's report, the first sentence in part (a) or the first sentence in part (b)? Explain your choice.

43. Virtually all sunscreen lotions list the SPF (sun protection factor), which is an indication of how long you will be protected from sunburn when wearing the sunscreen. The amount of time you're protected is proportional to the SPF. If wearing SPF 8 sunscreen will protect your skin for 40 minutes, how long will SPF 30 sunscreen protect you?

44. One day a newspaper reported the following information, gathered from the National Restaurant Association, concerning the percentage of food budgets spent by the average person on eating out in different cities in the United States.

a. Describe how the data might have been obtained.

b. Explain why these data represent an example of pseudo-precision.

	Percentage of food dollars spent eating out
Miami	52.1%
Boston	49.5%
New York	48.9%
San Francisco	48%
Dallas/Ft. Worth	47.6%

45. *The Unofficial U.S. Census* determined the following data concerning the ages of cars on the road.

a. First, estimate the percentage of cars that are 15 years old or older.

b. Determine the "exact" percentage of such cars.

c. How accurate do you think this number is? Justify your choice.

Age of car	Number on road
Less than 1 year	7,812,000
1–5 years	47,569,500
5–10 years	42,687,000
10–15 years	27,760,500
15 years or older	13,671,000

46. In the July 20, 1996, issue of *America*, the author described a school where the violence was out of hand. A new principal tried various measures, "but the most effective change . . . was a school uniform policy that prescribes a white top—a white T-shirt, for instance—with black trousers or skirts. This innovation is said to have helped produce a 100 percent drop in violence at Farragut" (p. 20). Do you believe the number, or is there not enough information to determine whether the actual drop in violence is likely to be as reported? Why or why not?

47. During the Yankees' last weekend series, 46% of those attending Saturday's game were women, and 42% of those attending the double header on Sunday were women. Were there more women in the stadium on Saturday or on Sunday? Explain your answer.

48. Siena saw the following sign in an expensive clothing store: "Any dress in stock—take off 20% or $20 off." What would you advise Siena to do?

49. A manufacturing plant requires lengths of wire that are $3\frac{3}{4}$ inches long.

a. If the wire comes in 48 inch coils, how many $3\frac{3}{4}$ inch segments can be made from one coil, and how much wire is wasted?

b. What is the percent of waste?

c. If the plant spent $40,000 each year on 48 inch coils of wire, how much of that money would be wasted?

d. Determine a length of coil that would result in no waste.

e. If your alternative lengths were 54 inches and 72 inches, would either of these result in less waste than purchasing 48 inch coils?

f. Suppose we did this problem yesterday and Betty was absent. She solves the problem and says that 0.8 inch is wasted in each coil. Explain how you would convince her that 3 inches of wire is wasted.

FROM STANDARDIZED ASSESSMENTS

2007 NECAP, Grade 6

50. This square is shaded gray to represent the part of Gibson's land that is covered with trees.

Gibson's Land

About what percent of Gibson's land is covered with trees?

a. Between 10% and 25%

b. Between 25% and 50%

c. Between 50% and 75%

d. Between 75% and 100%

51. Ken bought a used car for $5,375. He had to pay an additional 15% of the purchase price to cover both sales tax and extra fees. Of the following, which is closest to the *total* amount Ken paid?

a. $806 b. $5,510

c. $5,760 d. $5,940

e. $6,180

Source: From the *1992 NAEP Mathematics Assessment*, p. 252. U.S. Department of Education.

52. Of the following, which is the closest approximation of a 15% tip on a restaurant check of $24.99?

a. $2.50 b. $3.00

c. $3.75 d. $4.50

e. $5.00

Source: From the *Seventh NAEP Mathematics Assessment*, p. 179. U.S. Department of Education.

LOOKING BACK on chapter five

QUESTIONS TO SUMMARIZE BIG IDEAS

1. What is similar and different about a fraction and a ratio?
2. How is the concept of percent related to the concept of ratio and of fraction?
3. What is the difference between a multiplicative comparison and an additive comparison?
4. Where are some real-life places in which we use ratios, proportions, and percents?
5. What parts of this chapter are less clear to you at this point? What will you do to clarify those ideas?
6. Look back at the Mathematical Practices of the CCSS. In what ways did you engage in those practices during this chapter?

CHAPTER 5 SUMMARY

1. When we compare amounts, we can do so in additive or multiplicative ways. Both have value. In many cases, one is more appropriate.
2. Ratios, proportions, and percents involve multiplicative relationships between numbers.
3. Ratios are both like and unlike fractions. Ratios can be expressed in fraction notation, but many ratios involve part-to-part or whole-to-whole relationships. Not all such ratios can be translated into fraction language. Rates never can.
4. Seeing percent only as "what part of 100" is too limiting. For example, a 132% increase expresses a relationship between an original and a new amount.

BASIC CONCEPTS

Section 5.1: Ratio and Proportion

additive comparison **249**

multiplicative comparison **249**

ratio **250**

rates **250**

proportion **250**

unit-rate method **254**

factor-of-change method **254**

Section 5.2: Percents

percent **267**

converting among percents, decimals, and fractions **269**

connections between percent and other mathematical topics **272**

percent change **273**

percents less than 1 or greater than 100 **276**

interest: principal, amount, rate, time **278**

REVIEW EXERCISES chapter five

1. In the fairy tale "Jack and the Beanstalk," the giant's tube of toothpaste holds one gallon. If the cost of Jack's toothpaste is $2.49 for 4.6 ounces, how much does the giant's tube cost if the prices are proportional?
2. An employee making $24,000 was given a raise of $1000. If all employees were given proportional raises, how much of a raise did an employee making $29,460 receive?
3. If the ratio of boys to girls in a class is 3 to 8, will the ratio of boys to girls stay the same, become greater, or become smaller if 2 boys and 2 girls are added to the class?
4. In a school with 560 students, the ratio of students to teachers is 20 to 1. How many new teachers need to be hired to reduce this ratio to 16 to 1?
5. On a map, $\frac{3}{4}$ inch represents 78 miles. How far apart are two cities in real life if they are $5\frac{3}{8}$ inches apart on the map? You can solve this problem by any means except setting up a proportion and solving for x.
6. The admissions department at a college figured that the ratio of females to males in the incoming freshman class is presently 7 : 3. If they have tentatively accepted 1000 students in the freshman class, how many more males would they need to accept to reduce the ratio of females to males to 2 : 1?
7. The following chart lists the U.S. birth rate (per 1000 total population) for the following years. However, there were more babies born in 2000 than in 1930. Explain how this is possible.

Year	Birth rate
1930	21.3
1940	19.4
1950	24.1
1960	23.7
1970	18.4
1980	15.9
1990	16.7
2000	14.1

8. When data are presented in either tables or graphs, sometimes we see raw numbers and sometimes the raw numbers have been translated into rates. Unfortunately, we have found that many people do not fully understand rates. Explain why we have rates, as though to someone who asks, "Why not just use the raw numbers?"
9. A United Way campaign set a goal of $1.6 million. If campaign workers have raised 85% of their goal, how much money do they still need to raise?
10. A mom makes a family favorite juice by mixing 6 cups of grape juice and 9 cups of orange juice. This weekend the family had a party, and mom decided to make a bigger batch. She combined 10 cups of grape juice and 13 cups of orange juice—that is, she increased the original recipe by 4 cups of each. However, it didn't taste quite the same. Explain why it didn't. How much orange juice *should* she have added? Explain your reasoning without using a proportion.

11. A city budget is $21.43 million. The city council voted to increase the budget by 4.4%. What is the new budget, rounded to the nearest $1000?

12. At birth a baby weighed 7 pounds 10 ounces. Three months later the baby weighed 12 pounds 12 ounces. What has been the percent gain in the baby's weight?

13. A pump can fill a pool in 4 hours. After 1 hour another pump is brought in. This pump, by itself, can fill the pool in 2 hours. How much longer will it take for both pumps, working together, to fill the pool?

14. In the *Guinness Book of World Records,* the greatest distance pushing a baby carriage in 24 hours was 342.25 miles. What, to the nearest second, was the average time per mile?

15. A victim won damages of $4.5 million from a company. If the lawyer's fee was 35%, how much did the lawyer get?

16. All the three second-grade teachers ask their students how many have pets. The result is that 43 out of 65 of the students have pets. What percent of the students have pets?

17. In 2008, 308 million of the world's 6.69 billion people lived in the United States. What percent of the world's people lived in the United States?

18. It is estimated that 2% of the students in a school have a hearing impairment. If there are 1500 students in the school, how many does this estimate suggest have a hearing impairment?

19. LaToya's gross pay is $1840 per paycheck. If total payroll deductions are $370, her take-home pay is what percent of her gross pay?

20. A certain item costing $60,000 appears in a $2.6 million budget. What percent of the total budget does this item represent?

21. The other night on my calculator I was doing one of the review problems, finding 5.5% of $156,250. My calculator said 859.375. I knew immediately that I had done something wrong. How did I know? What had I punched to get this answer?

22. An MP3 player was marked down 30% and sold for $87.95. What was its price before the markdown?

23. When is 20% off a better deal than $20 off?

24. Estimate the following and explain your reasoning. Estimate means that you get the answer, in your head, within 20–30 seconds.

a. 16% of 450

b. 123 is approximately what percent of 185?

25. In an up and coming town, housing prices rose an average of 115% in five years. If a home cost $142,000 five years ago, how much would that home cost now?

26. Fernando's salary in 1976 was $8400, and he bought a Honda Civic for $3200. His salary in 2006 was $59,000 and he bought a Honda Civic for $20,400. Has his salary kept pace with the price of cars?

27. The Addams family filled their oil tank just before a surge in oil prices. Their bill was $285 for 202 gallons. The Barnstables were not so lucky. They had to fill their oil tank in October, and their bill was $385 for 160 gallons. The price of oil went up by what percent between the time the Addams family bought oil and the time the Barnstables bought oil?

28. The pollution count in a bay was 72.0 parts per million. This was an increase of 9.2% over the previous year. What, to the nearest tenth of a part per million, was the pollution count in the previous year?

Algebraic Thinking

6

HISTORY

INTERFOTO Personalities / Alamy

Amalie Emmy Noether was born in Germany in 1882. She is likely the most influential mathematician that you have never heard of. Women were not allowed to take university courses, so she sat in on them simply to learn. In 1908 when the University of Erlangen finally permitted women to take courses, she earned a doctorate in mathematics with honors. Her indomitable spirit was in evidence later—she was not allowed to hold paid positions at a university, so she taught for free. She used innovative teaching methods in which she would pose a problem for consideration instead of simply lecturing. Her discoveries in algebra are part of the foundations of modern physics and laid the basis for many of Albert Einstein's discoveries. Her story of determination and resiliency is inspiring.

Many students enter this course wondering why a chapter on algebra is included since many see algebra as an abstract subject not taught in elementary schools. However, algebraic thinking begins in early elementary school when students consider a missing addend approach like $3 + \_\_\_\_ = 5$. This way of thinking leads to solving algebraic equations like $3 + x = 5$. Elementary students also look at patterns as a way to develop algebraic thinking. We will investigate this idea more fully in this chapter.

The Common Core State Standards combine "Operations and Algebraic Thinking" together in a content strand from kindergarten to fifth grade. From grades six to eight, the CCSS have a content strand titled "Expressions and Equations." The NCTM's content standards on algebra will provide a useful structure for this chapter as they organize algebraic content into:

1. Understanding patterns, relations, and functions
2. Representing and analyzing mathematical situations and structures using algebraic symbols
3. Using mathematical models to represent and understand quantitative relationships
4. Analyzing change in various contexts

In elementary grades, the CCSS do not have a separate algebra strand, but rather explicitly connect algebra with operations. There is a trend in U.S. high schools, encouraged by the CCSS and NCTM, to reconnect algebra and geometry rather than to teach them as separate courses. You will see the interconnectedness between algebra and other math content areas in some of the investigations in this chapter.

As an elementary school teacher, you can promote algebraic thinking throughout your teaching. Investigating algebra through this chapter will help develop your own thinking about algebra to prepare you to develop this with your future students.

SECTION 6.1 Understanding Patterns, Relations, and Functions

What do you think?

- How are patterns related to algebraic thinking?
- What are some examples of functions in everyday life?
- What is a reason for developing algebraic thinking in elementary school?

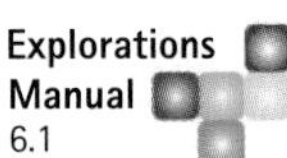

In this section, we will explore patterns, relationships, and functions as we develop algebraic thinking. Recall that the Mathematical Practices (particularly MP 7 and MP 8) of the Common Core State Standards discuss the importance of looking for patterns to help make sense of mathematics. Mathematical Practice 2 says that mathematically proficient students make sense of quantities and relationships. Functions are formally introduced in eighth grade but are examined informally when students explore functional relationships by analyzing patterns and relationships. Let's begin with an investigation where analyzing different patterns can help us to develop an equation that describes a relationship, which is also a functional relationship.

INVESTIGATION 6.1a Patterns Become Equations

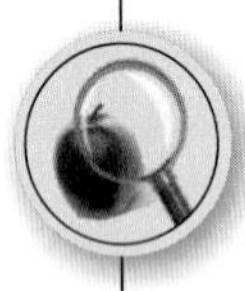

Consider patterns you see in the pictures and as you fill in the table below.

You are designing square patios using brown tiles and white tiles as shown in the picture. How many brown tiles and white tiles will be used in the next largest patio? How many in the 20th patio? Use the patterns to help you write a general rule that can be used for any patio—in other words, the "nth" patio.

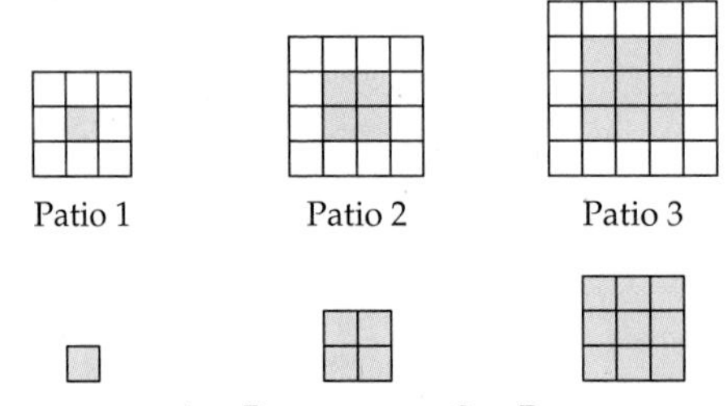

Patio number	Number of brown tiles	Number of white tiles	Total number of brown and white tiles
1	1	8	9
2	4	12	16
3	9	16	25
4			
5			
Pattern in words			
20			
n			

DISCUSSION

What patterns do you see? Does it help you more to look at the table to see patterns or at the pictures?

The brown tiles in the pictures always form squares, so it makes sense that the list of numbers we see for number of brown tiles are all square numbers. This pattern then enables us to represent the number of brown tiles for any patio, so for the fourth patio, the number of brown tiles is 4^2, for the 20th patio, the number of brown tiles is 20^2, and for the nth patio, the number of brown tiles is n^2.

STRATEGY 1

The white tiles may be seen in several ways. Some of you might have seen it as 4 sides which are the length of the inner brown patio plus 4 for the corners. So, for the third patio, there are 4 sets of 3 tiles plus the 4 on the corners. For the 20th then it would be 4 sets of 20 tiles plus the 4 on the corners. Following this pattern, we could say the nth would be 4 sets of n tiles plus the 4 on the corners. Writing this as an algebraic expression would look like $4n + 4$.

Patio number	Number of white tiles
1	$8 = 4(1) + 4$
2	$12 = 4(2) + 4$
3	$16 = 4(3) + 4$
4	$20 = 4(4) + 4$
5	$24 = 4(5) + 4$
Pattern in words	= 4(patio number) + 4
20	$84 = 4(20) + 4$
n	$= 4(n) + 4$

Think back to your algebra classes. When something increases by the same amount each time, what kind of a function is that?

It's a linear function and linear functions have the form $y = mx + b$, where m is the slope and b is the y-intercept. The slope is the constant increase, and since the number of white tiles is increasing by 4 each time, 4 is the slope. This matches the equation $y = 4x + 4$, which is the equation we developed. We will talk more about slope and change in Section 6.4.

STRATEGY 2

Another way to look at this pattern is to see the number of white tiles as the area of the whole patio minus the area of the brown part of the patio. This creates a nice connection between geometry and algebra. The whole patio (white and brown tiles) is a square that has a length that is 2 more than the patio number. For example, the whole patio for number 3 is 5 tiles by 5 tiles, in other words 5^2. The 20th patio would have a total of $(20 + 2)^2$. Following this pattern, the nth patio would be $(n + 2)^2$ for the entire patio. The inside brown patio is n^2 as we have already noted. So, the number of white tiles in this model is the total number of tiles minus the number of brown tiles, or $(n + 2)^2 - n^2$.

Patio number	Total number of tiles	Number of brown tiles	Number of white tiles
1	$3^2 = 9$	$1^2 = 1$	$3^2 - 1^2 = 8$
2	$4^2 = 16$	$2^2 = 4$	$4^2 - 2^2 = 12$
3	$5^2 = 25$	$3^2 = 9$	$5^2 - 3^2 = 16$
4	$6^2 = 36$	$4^2 = 16$	$6^2 - 4^2 = 20$
5	$7^2 = 49$	$5^2 = 25$	$7^2 - 5^2 = 24$
Pattern in words	(patio number + 2)^2	(patio number)^2	Second column minus third column, i.e., total tiles – brown tiles
20	$22 \times 22 = 484$	$20^2 = 400$	$22^2 - 20^2 = 84$
n	$(n + 2)^2$	n^2	$(n + 2)^2 - n^2$

So, Strategy 1 gave us the equation of $4n + 4$ for the number of white tiles, while Strategy 2 gave us the equation of $(n + 2)^2 - n^2$. Both of these equations work, but they look so different. However, with a little expansion we can see that they are the same.

$$(n + 2)^2 - n^2 = (n + 2)(n + 2) - n^2 = n^2 + 2n + 2n + 4 - n^2 = 4n + 4$$

It is interesting how we can use a little symbol manipulation to show that these two expressions that appear to be so different are really the same!

Exploring patterns in elementary school is an important part of developing algebraic reasoning. However, it is important to state that simply noticing patterns alone does not necessarily develop algebraic thinking. Algebraic thinking develops as the students learn to *analyze* various patterns (numerical and visual) in order to make and test predictions and finally to make *generalizations*. The next investigation continues to offer rich possibilities for developing algebraic reasoning by investigating patterns.

INVESTIGATION 6.1b Looking for Generalizations

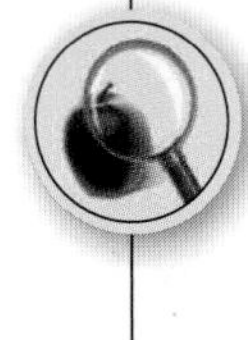

Explorations Manual 6.3

The figure below shows an example of what are called growing patterns in elementary school. How many little squares will it take to make the nth figure in this pattern? Work on this and then read on. . . .

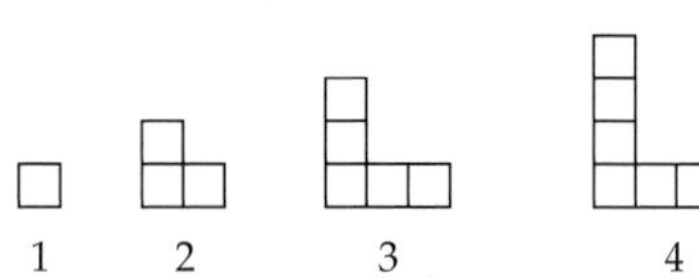

MATHEMATICS

If you connect the table in Strategy 1 to high school mathematics, you remember that when the change is constant, we have a linear function that can be represented in the form $f(x) = mx + b$, where m is the slope (rate of change) and b is the y-intercept. If you make this connection, you see that the slope m is equal to 2, and through quick analysis, you can determine that the equation of the line is $f(x) = 2m - 1$. We will explore this more later in this chapter.

DISCUSSION

As we discussed in the first chapter, there are often different strategies that can be used to solve a problem and that embody different representations of the problem.

STRATEGY 1 Make a table and look for patterns

Figure number	1	2	3	4	5	6	7
Number of squares	1	3	5	7	9	11	13

Looking at the relationship between the corresponding numbers in the first and second sets (input, output), many people will quickly see that the output number is always 1 less than double the input number. Thus the number of squares in the nth L number is simply $2n - 1$. The third row in the table below illustrates this relationship very clearly.

Figure number	1	2	3	4	5	6	n
Number of squares	1	3	5	7	9	11	
Number of squares	1	$2 \cdot 2 - 1$	$2 \cdot 3 - 1$	$2 \cdot 4 - 1$	$2 \cdot 5 - 1$	$2 \cdot 6 - 1$	$2 \cdot n - 1$

STRATEGY 2 Break it apart

Each of the L shapes can be broken down (decomposed) into two "arms" and a base (or "arm connector").

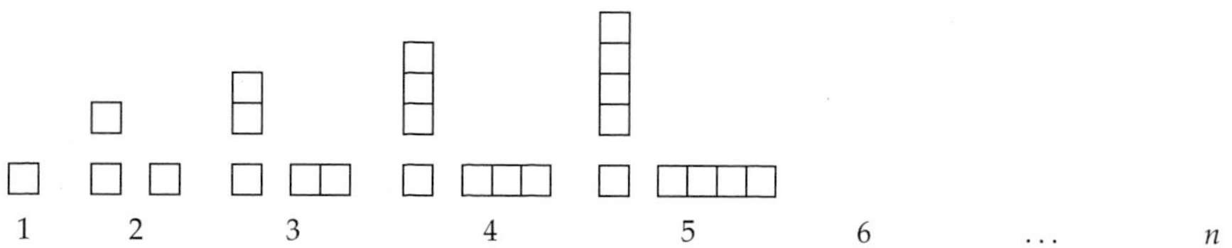

The base is always a single square. In each case, the arms are the same length, and the length is 1 less than the figure number. That is, the length of each of the arms of the 5th figure is 4 (1 less than 5). Thus the nth figure will have two arms, each of whose length is $(n - 1)$, plus one square that represents the base. Thus the number of squares of the nth figure will be equal to $(n - 1) + (n - 1) + 1$, which is mathematically equivalent to $2n - 1$.

STRATEGY 3 Break it apart another way

We could also have broken the shape apart this way: a bottom and a top.

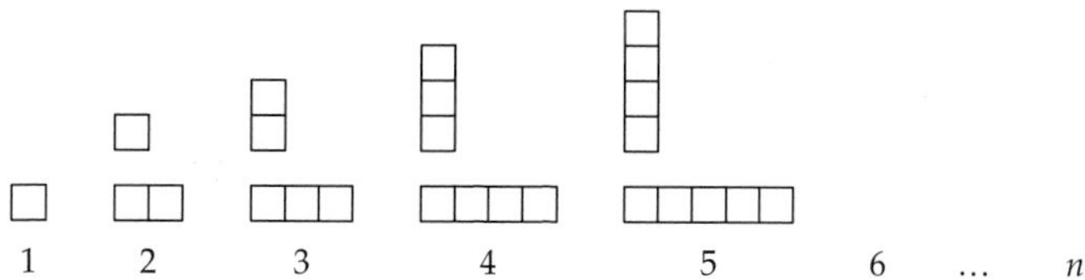

The bottom begins with one square and grows by one each time. The top begins with the second figure and also grows by one each time. The number of squares in the bottom always matches the figure number. That is, the fourth figure has a base of four, the fifth figure has a base of five, and so the nth figure will have a base of n. The number of squares in the top is always one less than the number of squares in the bottom. Thus, the nth figure will have a bottom of n and a top of $(n - 1)$. Adding the two together, we find that the total number of squares is $n + (n - 1)$, or $2n - 1$.

RELATIONSHIPS AMONG QUANTITIES AND FUNCTIONS

In elementary school, children will not investigate functions and other algebraic ideas at a formal level. However, through their investigations, they do need to realize that relationships often exist between different variables (for example, the area of a rectangle varies according to the length and the width of the rectangle); that patterns and relationships can be represented in various ways (with words, symbols, graphs, and pictures); and that being able to communicate clearly the patterns they see is important.

If students engage in these kinds of investigations during elementary and middle school, then they will see many of the concepts that are developed at the formal level in high school as extensions of what they encountered earlier, not as something that is entirely new and foreign.

TABLE 6.1

Hours worked	Dollars earned
1	8
1.5	12
2	16
3	24
4	32
5	40
40	320

HISTORY

The word *function* was introduced by Leibniz, who invented calculus, to denote any quantity connected with a curve. Many famous mathematicians had a conception of functions that would be considered erroneous today. For example, Bernoulli regarded a function as any expression made up of a variable and some constants. The familiar notation $f(x)$ that we use today was introduced by Clairaut and Euler in the 1700s.

CLASSROOM CONNECTION

I love this example from *Teaching Children Mathematics*, February 1997, p 266. The pattern was "add two to the number." This was one student's response: "You say the first number, you don't say the next number, and you say the next number." Experienced teachers know that children will conceptualize and verbalize mathematical situations and problems in many different ways.

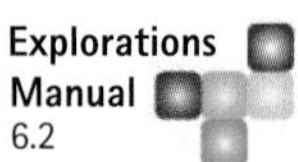
Explorations Manual 6.2

A simple example of a function is the hourly wage. If you know that you are being paid at the rate of $8.00 per hour, you expect to receive $320 for working 40 hours, or $160 for working 20 hours, or $40 for working 5 hours. In other words, there is a clear, consistent relationship between the variable "hours worked" and the variable "dollars earned." We can create a table of values for this function (Table 6.1).

Formally, we define a **function** as a relationship between two sets in which each element of the first set, called the **domain** of the function, is matched with *exactly* one element of the second set, called the **range** of the function. In Exploration 6.3, you found functional relationships between other pairs of variables.

In the wage example, the first column (hours worked) represents the domain, and the second column (dollars earned) represents the range. Technically, what is in these columns is a subset of the domain and a subset of the range, respectively.

Not all relationships between two sets are functions. For example, consider the relationship between the set of positive whole numbers and their square roots. In this case, if we ask, "What is the square root of 9?" the answer is +3 *and* −3. That is, each member of the set of positive whole numbers is matched with two square roots.

When we go outside the classroom to look for instances of functions in situations, we find that their value is often the predictability of the relationship. Consider the following questions:

- If I sell 50 computers, what will be my commission?
- If 75 students sign up for a course, how many sections will be offered?
- What are the consequences when you chew gum in Appleby High School?

If Jack sells 50 computers and receives $1000, whereas Jill sells 50 computers and receives only $700, then there is not a functional relationship between the number of computers sold and the commission.

If one department creates three sections when 75 students preregister and another department creates only two sections, then we say that at that college there is not a functional relationship between the number of students and the number of sections of a course.

If Cindy chews gum in one class and gets detention and chews gum in another class and receives no detention, then there is not a functional relationship between chewing gum and consequences.

Inputs, outputs, and function notation Many elementary textbooks use a game called "What's my rule?" to introduce children to the idea of functions. The students give a number (the *input*), then the teacher performs an operation and gives them the number associated with that input (the *output*). The students then try to guess what the teacher is doing to the inputs. A game of "What's my rule?" is given below (see Figure 6.1).

TEACHER: What's my rule?
STUDENT: 3
TEACHER: 6
STUDENT: 4
TEACHER: 7
STUDENT: I think I know the rule.
TEACHER: What is your rule?
STUDENT: You are adding 3. If I give you 10, you will say 13.
TEACHER: That's right!

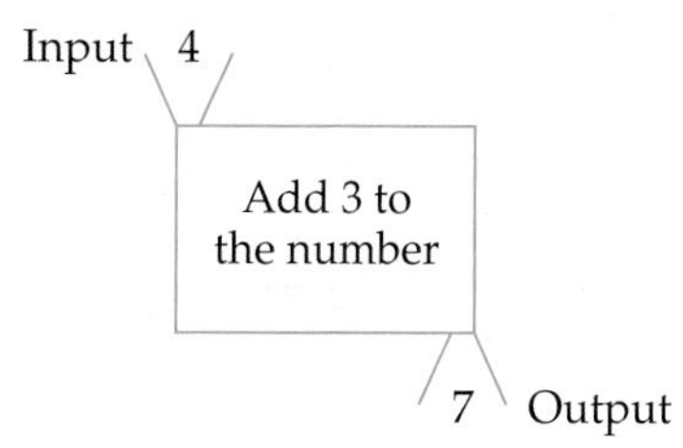

Figure 6.1

Input	Output
1	4
2	7
3	10

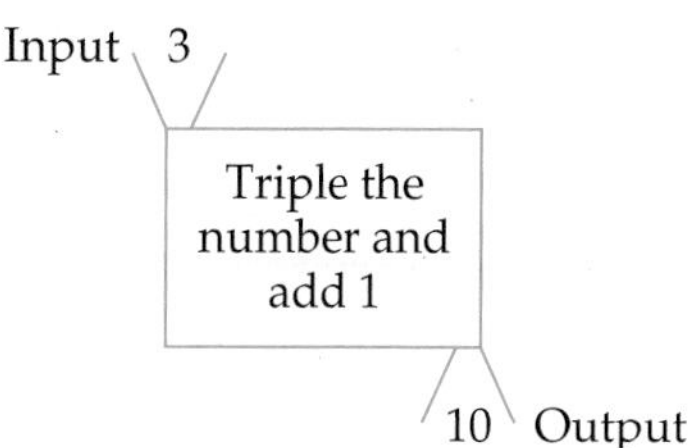

Figure 6.2

The teacher is adding 3. Algebraically, we would say $y = x + 3$. We can also use **function notation** to write the rule for adding 3 to any number: $f(x) = x + 3$. This mathematical sentence is read "f of x equals x plus 3." That is, $f(x)$ means "function of x."

Function notation is more precise because it emphasizes the relationship between the **input**, which is another name for a member of a function's domain, and the **output**, which is the corresponding member of its range. In this particular case of "What's my rule?" the output $f(x)$ is determined by adding 3 to the input x, and so we have

$$f(3) = 3 + 3 = 6$$
$$f(4) = 4 + 3 = 7$$
$$f(10) = 10 + 3 = 13$$

Let's play another round. Look at the inputs and outputs in Figure 6.2. What is the rule? Think and then read on. . . .

Using everyday English, we could say that the rule (or function) is to triple the input and add 1. Using function notation, we could say that $f(x) = 3x + 1$.

Elementary teachers often make a "function machine" to illustrate these situations. Function machines for the two situations just presented can be seen in Figures 6.1 and 6.2.

In this game, the domain and range do not have to be sets of numbers.

Here is one example from Margie Hoey, a second-grade teacher with whom I worked. One student placed some students in one group and some students in another group. The determining factor was whether the student was wearing blue or not. That is, the domain set consisted of the set of students in the class. The range set had two elements: "is wearing blue" and "is not wearing blue." For example,

Input	Output
Anastasia →	Is wearing blue
Benito →	Is wearing blue
Cecilia →	Is wearing blue
Damon →	Is not wearing blue
Emilia →	Is not wearing blue

These types of activities in an elementary classroom can be useful in helping students to understand functional relationships.

SUMMARY 6.1

Through each of the investigations in this section, we examined how patterns can lead to generalized observations about relationships, which then tied into writing the equations and functions that emerged from the patterns. Early elementary students investigate patterns as a way to make sense of relationships, which leads to algebraic and mathematical thinking.

6.1 Exercises

1. Read the algebra parts of the Common Core State Standards (or your state standards if you are not in a CCSS state) and write a paragraph summarizing what you found that was interesting.
2. Explain in your own words how being able to analyze patterns is a form of algebraic thinking.
3. The triangular numbers are 1, 3, 6, 10, ... They are called triangular numbers because they can be drawn as follows:

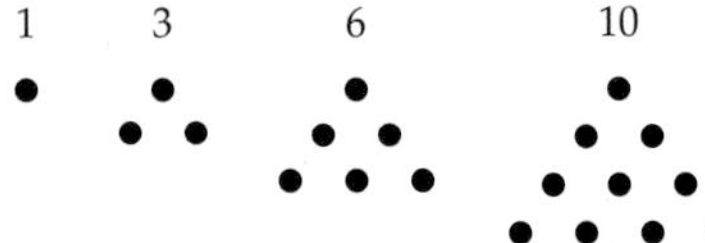

   a. How many dots would be in the fifth triangular number?

   b. How many dots would be in the 12th triangular number?

   c. Write a general expression for the number of the dots in the nth triangular number.

4. The table below shows the process by which a college department determines how many sections of a course to offer.

Number of students	Number of sections
Less than 6	Course not offered this semester
6–29	1
30–59	2
60–89	3
90–119	4

   a. Make a graph to represent this process.

   b. Is the relationship between the number of students and the number of sections a function? Justify your answer.

5. Let's say the entire freshman class of Wannago High School is going on a field trip. The number of buses needed is a function of the number of people going. To keep matters simple, let's say that the school bus company has buses that can hold 36 passengers and that there will be two adults per bus.

   a. How many buses will be needed for 232 students?

   b. Represent this function graphically.

6. Let's say a beautician charges \$20 per haircut. She also determines that overhead not associated with number of haircuts (rent, utilities, equipment) is \$200 per week. Consider the relationship between number of haircuts and profit. Is this a proportional function? That is, does double the number of haircuts mean that profit is doubled?
7. Look at the growing pattern below, in which each figure resembles the letter C. How many squares will it take to make the nth figure in this pattern?

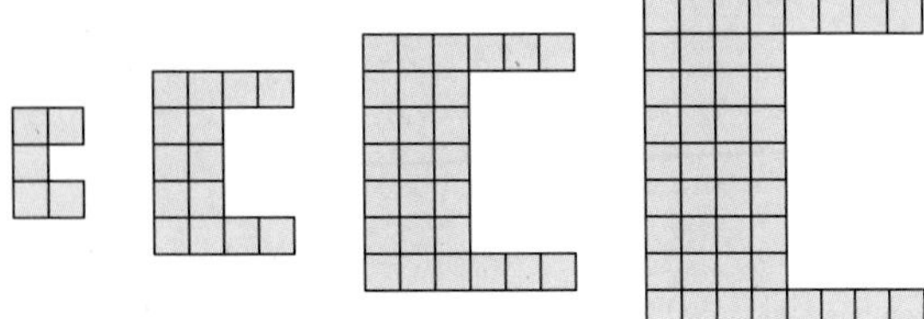

8. Look at the growing pattern below. How many squares will it take to make the nth figure in this pattern?

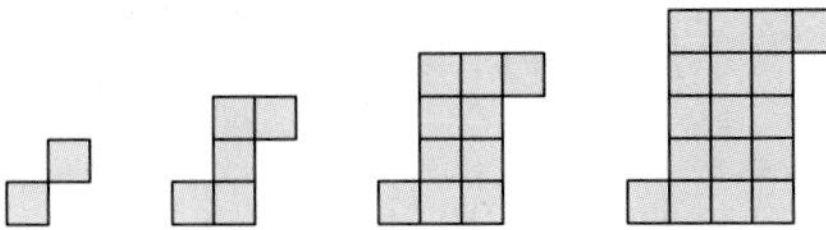

9. Look at the growing pattern below. How many squares will it take to make the nth figure in this pattern?

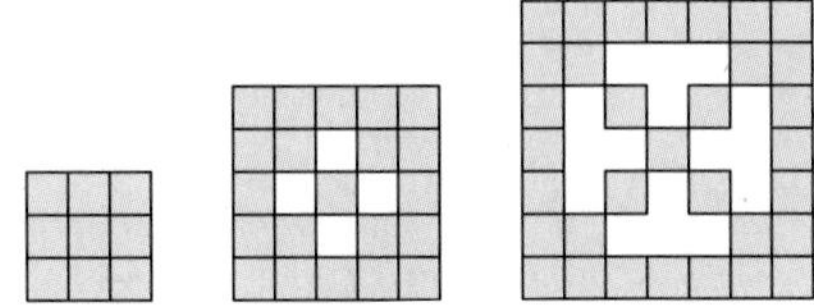

DEEPENING YOUR UNDERSTANDING

10. In Investigation 2.1c, we found that a set containing 4 elements has a total of 16 subsets.

    a. Fill in the blanks in the table that follows.

    b. Describe in words the relationship between the number of elements in a set and the number of subsets. Refer to the table.

    c. Describe the relationship with an equation.

Number of elements in the set	Number of subsets
1	?
2	?
3	?
4	16
5	?

11. There is a relationship between the successive squares of triangular numbers. This relationship can be expressed in everyday English as follows: "Select any triangular number and square it. Then take the previous triangular number and square it. Now take the difference between the two numbers."

    a. Express this relationship in notation.

    b. Is this relationship a function? Justify your response.

12. There is a functional relationship between the frequency of cricket chirps and the temperature. To find the temperature, the rule is to count the number of chirps in one minute, divide by 4, and add 40.

    a. What temperature does 50 chirps per minute correspond to?

    b. Without calculating, determine whether 100 chirps per minute will correspond to twice the temperature of 50 chirps per minute. That is, does chirping at twice the rate mean that the temperature is twice as high? Explain your reasoning.

    c. Translate the rule into an equation.

    d. Draw a graph to represent this relationship.

13. What if we made a pattern by joining pentagons? If the length of each side of the pentagon is 1 unit, then what would be the

perimeter of n pentagons joined together? *Note:* The perimeter is the distance around the outside of the figure.

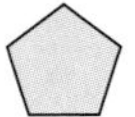
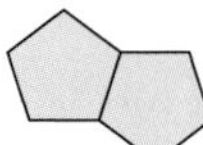
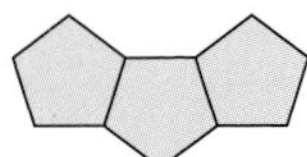
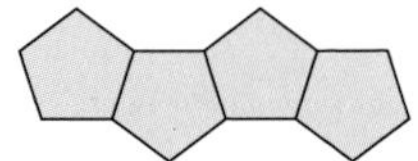

14. Variations of this problem have been found in many elementary schools. Begin with a wooden cube whose dimensions are 1 cm × 1 cm × 1 cm. Glue another cube to this cube. This new rod will be 1 cm longer than the first. Now, let's say we use a rubber stamp with a design, such as a happy face, and that stamp will cover a 1 cm by 1 cm square. Thus it would take 6 stamps to stamp the first cube and 10 stamps for the second figure. How many stamps would it take to cover a rod that was n centimeters in length?

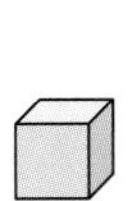
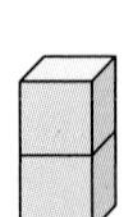

15. a. If the length of each side of each cube is 1 unit, what is the surface area of a tower of n cubes like the one shown at the left below?

    b. What is the surface area of a tower of n cubes like the one shown at the right below?

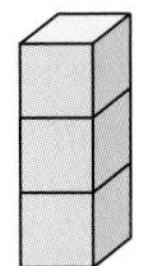
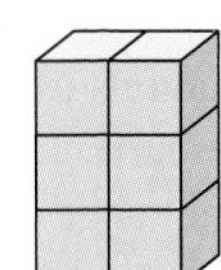

16. ***Classroom Connection*** This problem is taken from the February 1997 issue of *Teaching Children Mathematics.* The class had been studying patterns, and the teacher had assigned the children to create a numerical pattern. Liz made the following problem: Draw the pattern of circles that form a quadrilateral on the top and two triangles on the bottom. The center circle should be a vertex common to all three shapes. Starting at the top and moving down the page, fill in eight counting numbers in order. Any number may be used as the starting number.

    a. Do this for several different sets of numbers. What do you see?

    b. Compute the sum of the 4 numbers that make the quadrilateral, the sum of the numbers that make each triangle. Now add these three sums. Can you prove that the ones place of this sum will always be 4?

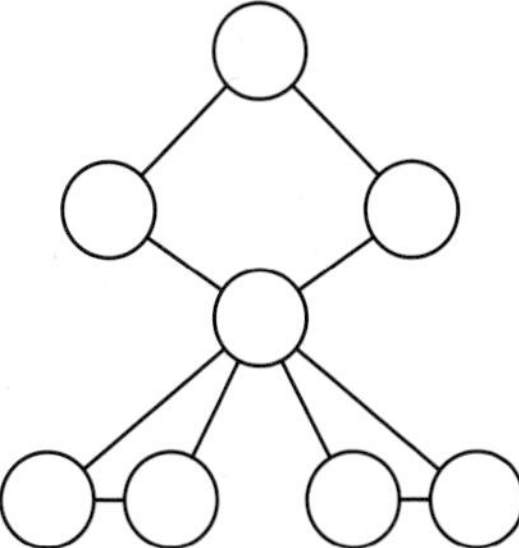

Source: From Teaching Children Mathematics, 1997 by the National Council of Teachers of Mathematics.

17. Variations of this problem have also appeared in many places. Jack and Jill decide to start a rumor that there will be no school on the following Monday.

    a. If each of them tells 1 person, and then everybody tells exactly 1 other person, and so on, how long will it take until all 450 students in the school hear the rumor? Assume that the rate for telling the next person is 1 hour. That is, initially Jack and Jill know; after 1 hour, 2 more people know, ...

    b. What if everybody tells 2 people each hour? How long will it take now?

18. Find a graph from a newspaper or magazine that represents a functional relationship between two variables.

    a. Explain why the relationship is functional.

    b. Describe the relationship in words, as though you were talking on the phone to someone who cannot see the graph.

    c. Make up and answer one question based on data in the graph.

19. Describe a real-life scenario involving a functional relationship between two variables.

20. The first three figures in a pattern of tiles are shown below. The pattern of tiles contains 50 figures.

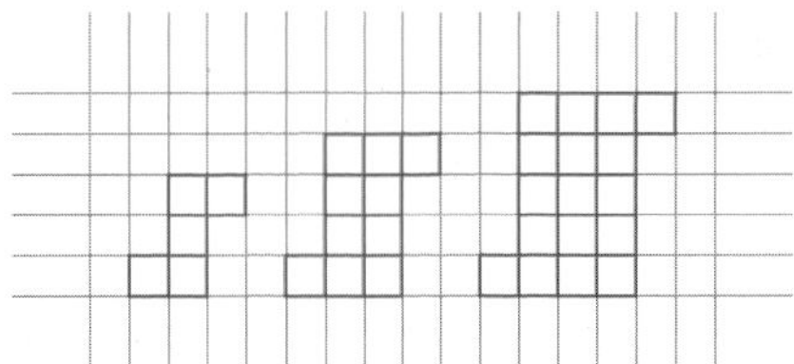

Describe the 20th figure in this pattern, including the total number of tiles it contains and how they are arranged. Then explain the reasoning that you used to determine this information. Write a description that could be used to define any figure in the pattern.

(4 percent of twelfth-graders submitted a satisfactory or extended response)

Source: Results of the *Fourth NAEP Mathematics Assessment,* p. 331. U.S. Department of Education, National Center for Education Statistics.

21. For the following table of numbers:

    a. Describe the pattern in your own words.

    b. Find the value of y when x is 100.

x	y
1	3
2	6
3	9
4	12
5	15
6	18
7	21

    c. Find a general expression for finding the nth term in your own words.

    d. Translate part c into function notation.

22. A plumber charges $40 to come to your house and $30 for every hour of labor.
    a. How much will it cost if he works at your house for 3 hours?
    b. Write an equation for how much it will cost to have him work at your house for x hours.
    c. Is this a function? Why or why not?
23. Find a pattern in each of the following sets of equations and predict the next equation.
    a. $1^2 + 2^2 + 2^2 = 3^2$

       $2^2 + 3^2 + 6^2 = 7^2$

       $3^2 + 4^2 + 12^2 = 13^2$
    b. $1^3 + 2^3 = 3^2$

       $1^3 + 2^3 + 3^3 = 6^2$

       $1^3 + 2^3 + 3^3 + 4^3 = 10^2$
    c. $1 + 2 = 3$

       $4 + 5 + 6 = 7 + 8$

       $9 + 10 + 11 + 12 = 13 + 14 + 15$

SECTION 6.2 Representing and Analyzing Mathematical Situations and Structures Using Algebraic Symbols

What do you think?

- What does the equal sign mean?
- What is a variable?

In the previous section, we began to represent math situations using algebraic symbols as we used the patterns to help us develop the equations. In the current section we will continue this idea of representing mathematical situations using the language of algebra. For many people, algebra is like going into a foreign country with different rules, procedures, and language. Yet, as we began to see in Section 6.1 and will continue to develop in this section, many of the rules and procedures in algebra are actually extensions of rules and procedures in arithmetic. Because of this, the CCSS put *operations and algebraic thinking* into the same category in a content strand through fifth grade. In sixth through eighth grade, the CCSS contains the *Expressions and Equations* domain that includes the use of variables in sixth grade, as well as solving equations and inequalities. The Mathematical Practices also relate here, particularly MP 2 in which students are asked to manipulate symbols, and MP 3, communicating mathematically, which includes understanding symbolic language. The title of this section, which is NCTM language, is similar in nature to these Mathematical Practices of the CCSS, which further shows the relationships between these two sets of standards.

STRUCTURES

Explorations Manual 6.3

All mathematical structures can be described using algebraic notation. Many of these structures are at the heart of the elementary school curriculum:

- Properties of operations—closure, commutative, associative, identity, inverse, distributive, etc.
- Properties of numbers—for example, an "even" number can be divided into two equal halves, whereas an odd number cannot. The algebraic representation of an even number is $2n$, and the algebraic representation of an odd number is $2n + 1$ or $2n - 1$.
- Connections between operations, such as $a - b = a + -b$.

It is exciting to be in a classroom at the moment when a child has discovered that 2 + 9 (which is pretty hard for many first-graders) is "the same as" 9 + 2. This discovery is quickly generalized as "You can switch the numbers when adding." In this case, the student has discovered a very important structure in the system of whole numbers that we call the commutative property of addition. It is stated in its most concise form as $a + b = b + a$ for any whole numbers a and b. In elementary school, coming to understand structures often comes from "playing" with patterns and asking, "What do you see?" If children come to expect mathematics to make sense and look for patterns, the transition from arithmetic to algebra will be so much smoother than it was for most readers of this book.

Is it crucial that second-graders can say and spell "commutative"? I don't think so. Is it crucial that they can apply this concept when adding and that they know it is not true when subtracting? Absolutely. Researchers have found that the mistake shown below is common in second grade. In the following example, the student not only had trouble with place value and "borrowing" but also mistakenly assumed commutativity in subtraction. That is, the child assumed that $2 - 6 = 6 - 2$.

$$\begin{array}{r} 52 \\ -16 \\ \hline 44 \end{array}$$

In high school, students are expected to learn to write formal proofs and to know the properties in formal language. In elementary school, we want the students to continue to ask "why" and "what if" questions (Why doesn't the order matter when you add but it does when you subtract? What if you multiply? What if you are dealing with big numbers? Will it work with fractions too?), and we want them to have an understanding of basic structures and be able to apply this knowledge when solving problems. This type of thinking is referred to in MP 7 of the CCSS, which is titled "Look for and make use of structure."

INVESTIGATION 6.2a Magic Tricks

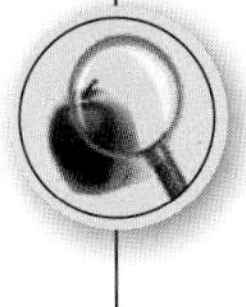

Algebraic notation can even help us explain the structure of magic! Consider this math magic.

1. Pick any number between 1 and 100.
2. Add 4 to that number.
3. Multiply that number by 2.
4. Subtract 2 from that number.
5. Divide by 2.
6. Subtract the original number.
7. Go through these steps with several picked numbers and see what happens.

Abracadabra, your answer is 3!

Magicians never reveal their tricks, but what is the "trick" here? Use the variable "x" to represent the number you pick. Write an expression for the steps given, and see how that shows the answer as 3. Would it only work for numbers between 1 and 100, or all numbers?

DISCUSSION

Using variables, properties, visuals, and symbolic manipulation we can explain the trick.

Steps	Numerical example	Visual representation	Symbolic representation
Pick a number between 1 and 100	5	(Box with the number you picked inside.)	x
Add 4 to that number	$5 + 4 = 9$	(Each oval represents 1.)	$x + 4$
Multiply by 2	$9(2) = 18$		$2(x + 4) = 2x + 8$ Distributive property
Subtract 2	$18 - 2 = 16$		$2x + 8 - 2 = 2x + 6$
Divide by 2	$16/2 = 8$		$\frac{2x + 6}{2} = x + 3$
Subtract the original number	$8 - 5 = 3$	(Remember, the original number was in the box, so we subtract the box.)	$x + 3 - x = 3$

Now you know the magician's trick. So will it work for all numbers, or just between 1 and 100 as the instructions stated? We have shown that it works for x, no matter how many we put in the box, which means we can put any number in for x and we will always get 3.

INVESTIGATION 6.2b Proving Generalizations for Adding Odd and Even Numbers

In Chapter 1, we analyzed why an even number plus an even number equals an even number. Let's expand on that and let algebraic symbols help us.

Since any even number is a multiple of 2, then we can write any even number as 2 times some number, or 2 times any variable we want to use, such as $2n$.

Often it is helpful to think about actual numerical examples to make the letters less abstract. We can also use an area model like we looked at when we learned about multiplication models to help us visualize. Consider the following even numbers and how they can all be written as 2 times some number and how their models will all be rectangles.

$2 = 2(1)$

$6 = 2(3)$

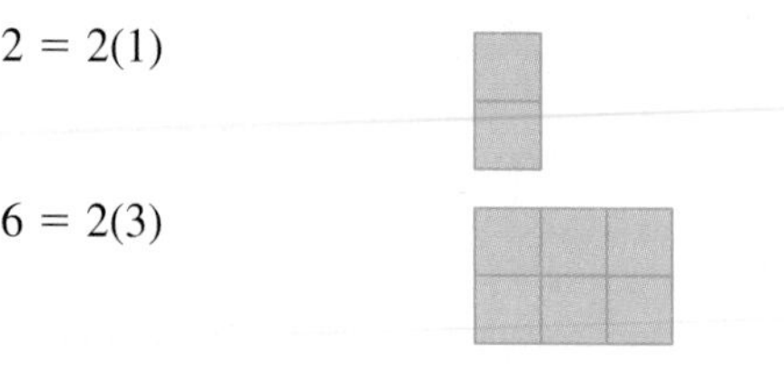

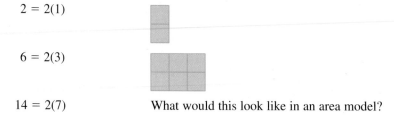

$14 = 2(7)$ What would this look like in an area model?

$122 = 2(61)$ Or this?

Any even number $= 2(n)$ Or this?

Any of these could be represented as rectangles with 2 rows. This visual will help us to generalize.

Odd numbers can be thought of as "one left over" when put into two groups. So, how can we represent an odd number using a variable? What about using the area type model?

To be clear about these representations:

Even $= 2n$ (We represent an even number as 2 times a number.)

Odd $= 2n + 1$ (We represent an odd number as 2 times a number plus 1 left over.)

Use these representations of even and odd numbers, properties, and symbolic manipulation to show that the following are true:

A. even + even = even

B. even + odd = odd

C. odd + odd = even

DISCUSSION

A. An even number is of the form 2 times a number, so $E = 2n$. Because these could be any two even numbers, we cannot represent them both as $2n$. Why? If they were both represented by $2n$, then they would be equal to each other. Therefore, we will represent one of them as $2n$, and the other as $2m$.

So,

even + even $= 2n + 2m = 2(n + m)$ (by distributive property)

Therefore, we have ended up with 2 times some number $(n + m)$, and this proves that it will always be an even answer.

A more visual way to think about this is to say pairs + pairs = pairs, or in the rectangle model a rectangle with 2 rows added to a rectangle with 2 rows will be a rectangle with 2 rows. We do not know how many are in each row, but because they are even, they can be represented as a rectangle with two rows. Elementary students will "prove" this assertion using similar visual models.

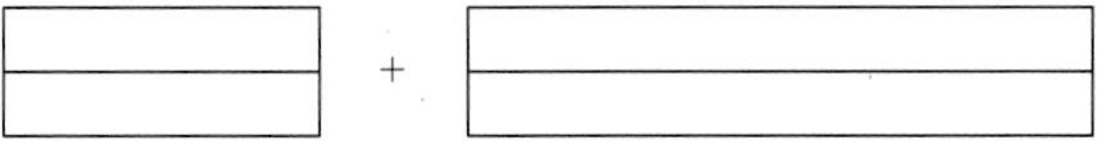

If you did not understand how to do parts B (even + odd = odd) or C (odd + odd = even) of this investigation earlier, go back and try them now that you have seen part A and then read on.

B. First, an odd number can be represented as $2n + 1$ since it is an even number with "one left over" or plus 1.

even + odd $= (2n) + (2m + 1) = 2n + 2m + 1 = 2(n + m) + 1$ (by distributive property)

This answer is of the form 2(number) + 1 (i.e., "one left over"), so it will always be an odd number. Another way to think about this is even + odd = pairs + (pairs + 1), which would equal more pairs + 1. Visually, in the rectangle, the 1 left over would still be there.

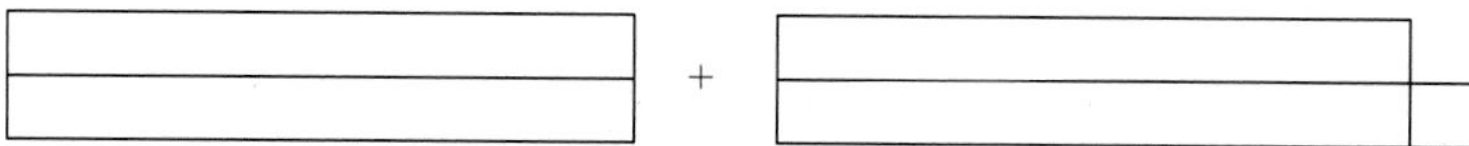

C. Let's represent each odd number as $2n + 1$ and $2m + 1$. Since they are two different odd numbers, we have to use two variables (m and n).

odd + odd $= (2n + 1) + (2m + 1) = 2n + 2m + 2 = 2(n + m + 1)$
(by distributive property)

Here we end up with 2 times a number $(n + m + 1)$, so it will always be even.

Or, if we add (pairs + 1) + (pairs + 1), the +1 in each number would become another pair, so all would be pairs. In the visual representation below, the second number is rotated to show how the "1 left over" from each would pair up to form a rectangle.

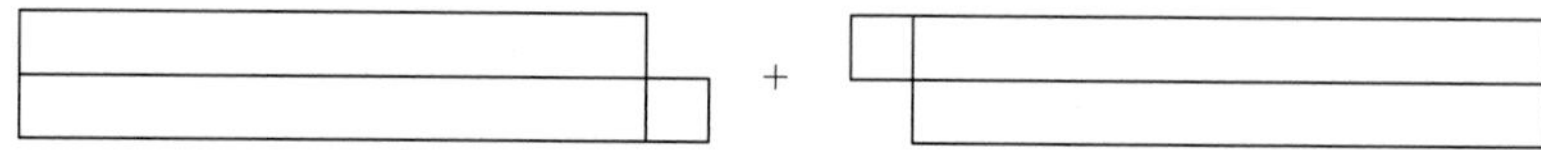

We have used variables and symbols such as the equal sign throughout this text. Let's investigate more deeply what they are and how they are used.

VARIABLES AND SYMBOLS

We use letters and other symbols to communicate algebraically. When students first see letters in a math class, it can seem unusual. Remember, they simply help us to communicate, express ideas, and generalize.

INVESTIGATION 6.2c A Variable by Any Other Name Is Still a Variable

A crucial idea in algebraic thinking is that of the variable. In each of the following examples, the variables have a slightly different meaning. On a separate piece of paper, write down what each of the variables means and then read on. . . .

$$C = \pi d$$

$$5x = 30$$

$$\sin x = \cos x \cdot \tan x$$

$$1 = n \cdot \left(\frac{1}{n}\right)$$

$$y = kx$$

DISCUSSION

The first example is a formula. C and d stand for circumference and diameter, whose values vary according to the circle; however, the value of π (3.14) does not vary.

The second example is generally called an equation. Although x is the variable, in this case its value is 6.

The third example is an example of an identity; it is true no matter what the value of x.

The fourth example represents a property. In this case, n is used as a symbol to represent a property that is true for all numbers (except 0); the formal name of this property is the multiplicative inverse property.

The fifth example represents a family of functions in which the independent variable (x) and the dependent variable (y) are related in a certain way—in this case, a linear relationship.

Table 6.2 illustrates some of the important differences in these examples.

More than one high school teacher has reported a variation of the following response by a student: "If $x = 6$, what did you call it x for?"

HISTORY

Our algebraic notation developed over hundreds of years, and the driving force for inventing this notation was to make it easier to solve problems and communicate ideas. In the 1500s, the expression $x^2 + 2x - 8$ would have been written as Zp 2Rm 8; R for an unknown, Z for zensus (an unknown squared), p for plus, and m for minus.

CLASSROOM CONNECTION

This is a true story. A child comes home from the first day in first grade. His mother asks him what he learned, and he says, "I learned that five plus three equals eight." "That's great," says his mother, at which point the child asks, "Mommy, what does equals mean?"

TABLE 6.2

Example	What it is usually called	What the variables stand for
$C = \pi d$	Formula	C and d are concrete quantities. π is a constant.
$5x = 30$	Equation	x is the unknown.
$\sin x = \cos x \cdot \tan x$	Identity	x is an argument of a function.
$1 = n \cdot (1/n)$	Property	n is a symbol to represent a generalization.
$y = kx$	An equation of a function of direct variation	x is the independent variable, y is the dependent variable, and k is a constant.

CLASSROOM CONNECTION

A first-grade teacher presented this problem to her children: $8 + 4 = \square + 7$, and they all said 12. They simply ignored the 7.[1]

Explorations Manual 6.4

THE EQUAL SIGN AND EQUIVALENCE

The CCSS state that students should understand the meaning of the equal sign in first grade. When people are asked what the equal sign means, as you were asked in the beginning of this section, often the response might be something like, "the equal sign means that two things are equal." Can you see how this circular definition does not really define the equal sign since it requires that we already know what "equals" means? How can we define "equal sign" without using the word "equals"? Try it and then read on.

Many elementary school children think of the equal sign as a call for action as opposed to a relationship between quantities, which can cause confusion later in algebra. They may think of the equal sign as "solve this" or "I did something." This notion is illustrated in the Classroom Connection here. Here, they see the equal sign as a call to add $8 + 4$ instead of as a need to make the same value on both sides of the equal sign. This illustrates the need for students to understand that the equal sign is a relationship between the two sides, that they have the same value.

One way to think of the meaning of the equal sign is simply as "has the same value as." Thus $8 + 4 = 5 + 7$ can be said as, "$8 + 4$ has the same value as $5 + 7$," and 4 quarters = \$1 can be said as, "\$1 has the same value as 4 quarters." This definition helps us to understand equations and formulas better as well. For example, when we say that the area of a triangle equals $\frac{1}{2}$ base times height ($A = \frac{1}{2}bh$), this can be thought of as the area of a triangle has the same value as $\frac{1}{2}$ times its base times its height. When students start solving linear equations like $3x + 4 = 15$, then the idea that $3x + 4$ has the same value as 15 provides a more flexible way of thinking about solving this. In other words, we are looking for a value for x where $3x + 4$ has the same value as 15.

The equal sign can also be thought of as the center of a balance scale. We will use this concept later in this section when we solve equations. There are even commercially available balance scales that can be used to solve equations. Using a balance scale to show that $3 + 5 = 8$ would look something like:

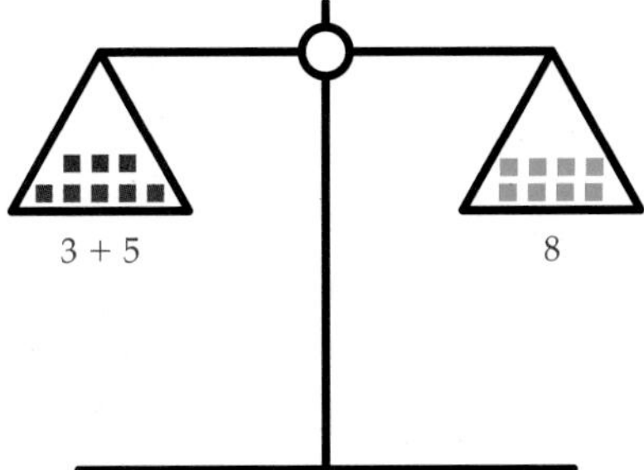

Inequalities

Early in elementary school, students compare quantities to explore the concept of which is more. They may be asked to fill in either > or < in the box given:

3 □ 5

In early middle school, students begin to explore linear inequalities and how to solve them.

The balance scale can help us to think about inequalities. For example, $3 + 5 > 2 + 5$. We could create a visual for this such as:

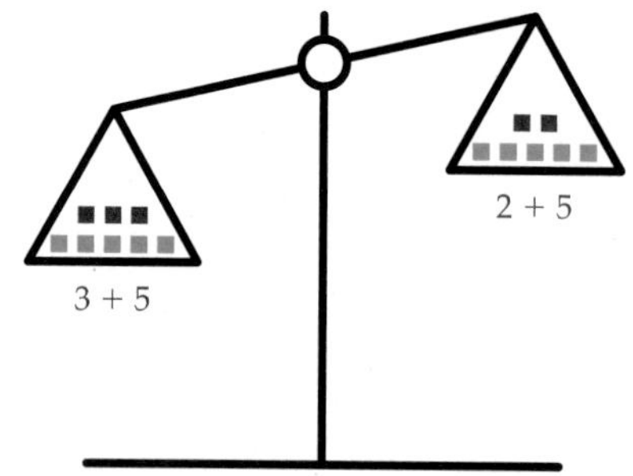

Explorations Manual 6.5

INVESTIGATION 6.2d Does the Balance Remain?

Without performing the calculations, determine whether the statements are true or false and explain how you know. Remember to not do the calculations.

A. It is a fact that $42 + 65 = 107$. Using this fact, is it true or false to say that $42 + 65 + 9 = 107 + 9$? How do you know?

B. Is it true or false that $46 + 54 = 48 + 52$? How can you tell without performing the calculations?

C. It is a fact that $42 + 79 = 121$. What symbol, =,<, or >, goes on the line to make the following statement true? Explain how you can tell without performing the calculations.

$42 + 79 + 3$ _____ $121 + 7$

DISCUSSION

Relational thinking can help us answer these questions.

A. Recalling the balance concept, here we are adding 9 to both sides of the balance. So, if the equality is preserved, the scale will remain balanced. Yes, this is true.

B. Again, picture 46 and 54 blocks on one side of the balance. Can we rearrange them somehow to have 48 and 52 blocks? Yes, if we move 2 blocks from the set of 54 blocks and put them with the 46 blocks, then we would have 48 and 52 without changing the number of blocks. Yes, they are equal.

C. Here we are taking two things that are balanced and adding 3 to the left side and 7 to the right side, which will make them unbalanced and make the right side larger since we are adding more there. Therefore the correct symbol is <.

We will continue thinking about the equal sign and the inequality signs as we move into solving linear equations and inequalities.

SOLVING LINEAR EQUATIONS AND INEQUALITIES

Elementary school students will consider equations early on that help them think about relationships between operations (addition, subtraction, multiplication, and division) and these types of problems lead to algebraic thinking. For example, a student might work on a problem like ____ + 3 = 8. This concept is developed using variables, beginning in the sixth grade according to the CCSS, where they solve equations like $x + 3 = 8$. Let's first consider this using the balance scale model.

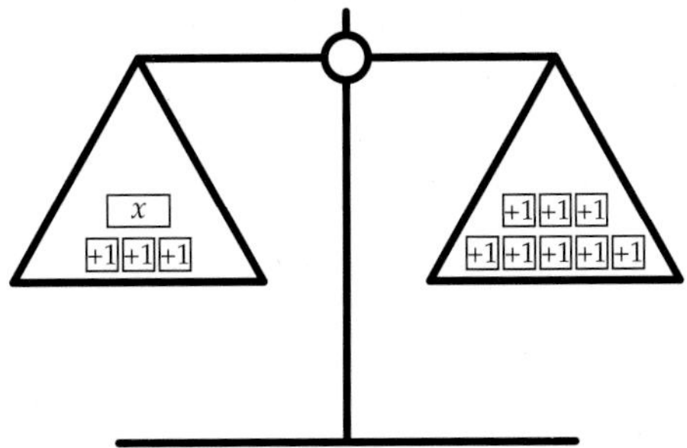

In order to keep the scale balanced, this visual helps us to see that x must equal 5 of the "+1" blocks.

Use symbol manipulation connected to the visual:

$x + 3 = 8$ We need to get the x by itself so we subtract 3 "+1" blocks from each side.

$x + 3 - 3 = 8 - 3$ Can you see how we have to take 3 from each side of the scale (equal sign) to keep the scale balanced?

Therefore, $x = 5$.

Singapore bar models can help us visualize this problem. The concept is the same, although the model is a little different.

+1	+1	+1	+1	+1	+1	+1	+1

x	+1	+1	+1

This model helps us to visualize that x must equal 5 if the whole bar equals 8. Which model makes more sense to you—the balance scale model or Singapore bar model? One of the challenges of being a teacher is that you need to make sense of several models in order to support your students' learning.

Let's consider an example that has subtraction in the original problem rather than addition and solve it both ways.

Consider $x - 4 = 6$. Think about how you might solve this with a balance scale and with Singapore bar models, as well as symbolically.

Balance Scale Model

We need to put an "x" block and 4 negative 1 blocks on one side and 6 positive 1 blocks on the other side to represent this equation.

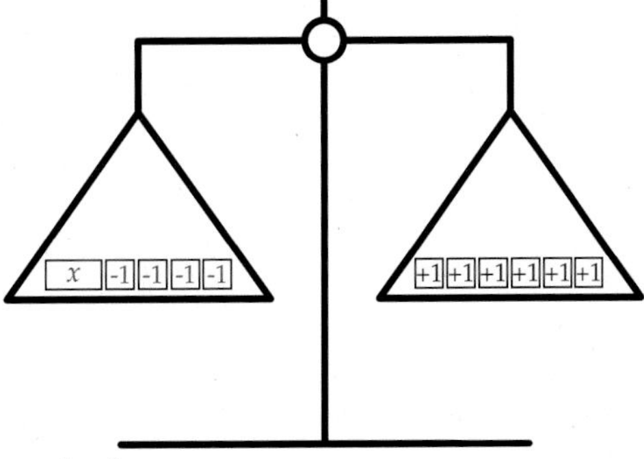

If we add 4 positive 1 blocks to the left side, then the "weight" of the negative 1 blocks would be zeroed out. To keep the scale balanced, this would mean that we also have to add 4 positive 1 blocks to the right side. This would leave x on the left side and 10 positive 1 blocks on the right side, so $x = 10$.

Singapore Bar Model

Because we are subtracting from the x, we let the whole bar represent x and then subtract off the 4; the bar we are left with must equal 6 according to the equation. Therefore, the entire bar, which represents x, would be 10.

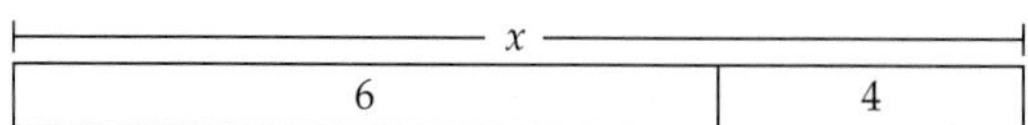

Look back at each of these models and think about the symbolic way we learned to solve this equation.

$$x - 4 = 6$$

We learned to add 4 to both sides:

$$x - 4 + 4 = 6 + 4$$

$$x = 10$$

Look back at both the balance scale model and the Singapore bar model. Can you see how in each of those we added 6 + 4, and how the visual representation makes it less abstract?

INVESTIGATION 6.2e Drawing Models to Solve Equations

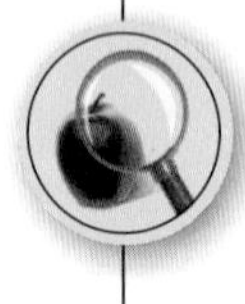

Let's investigate both the balance scale model and the Singapore bar model to solve two-step linear equations.

A. Draw a balance scale model for $3x + 2 = 17$.

B. Draw a Singapore bar model for $3x + 2 = 17$.

C. Compare the two models with each other as well as with the traditional procedure for solving this equation.

DISCUSSION

A. Here we have a balance between the left side showing $3x + 2$ and the right side showing 17.

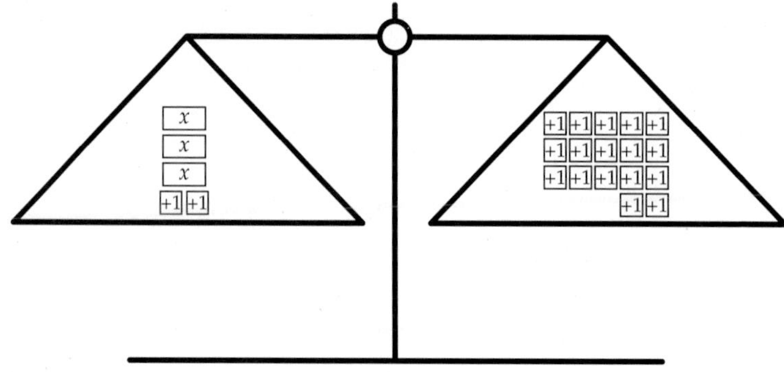

We can remove 2 from each side and keep it balanced. Then it would look like the figure below. What equation does this represent?

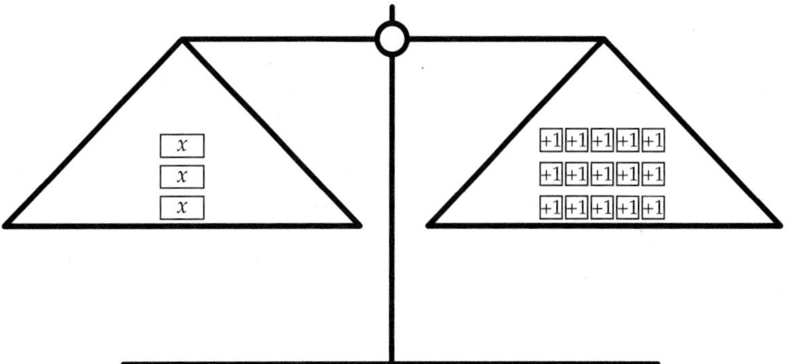

Upon looking at the rows in the picture, we can see that each x must equal 5.

B. The steps here are color coded to help readers understand.

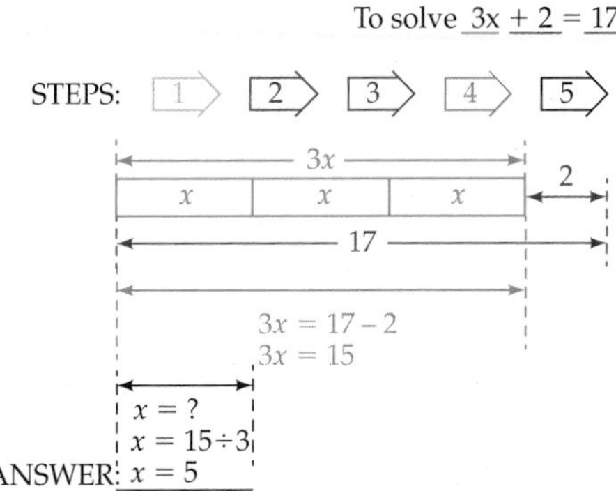

Source: Alice Ho of Math Teach Singapore.

We start with green showing the $3x$. Blue shows us the $+2$, and then red shows that the $3x + 2$ is the same length as 17. From this model we can see that the $3x = 15$, and therefore $x = 5$.

C. In both of these models, we start by representing $3x + 2$ being equal to 17. In each model we subtract 2 from 17 and then divide by 3, just as we do in the pencil and paper process you likely learned in algebra class.

The balance scale model and the Singapore bar model are concrete and visual ways to solve linear equations. When beginning a concept in elementary school, starting with the concrete and moving to visual allows students to move into the abstract with the visuals in their minds as a basis. Let's now turn to solving linear inequalities with visuals.

INVESTIGATION 6.2f Does the Inequality Remain the Same?

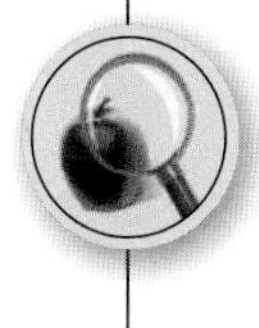

We will begin with the fact that $2 < 5$. Sometimes a number line can be useful in thinking about which number is greater. Because 2 is to the left of 5 on the number line, it is smaller.

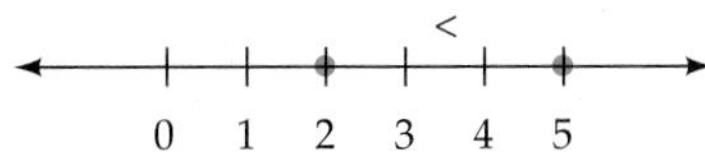

In this investigation we will learn what happens when we perform the same operation on each side of the inequality. In the balance scale models of solving linear equations, we saw that when you do the same thing to both sides, the balance remained intact. Is the same true for inequality, that is, does the relationship (less than or greater than) remain the same?

Let's begin with the fact we showed above that $2 < 5$.

A. What happens when we add 4 to each side? Does the left side remain less than the right side?

B. What happens when we multiply each side by 3?

C. What happens when we multiply each side by -3?

Illustrate each of these on a number line to help provide a visual representation.

DISCUSSION

A. Adding 4 to each side looks like:

$2 + 4 < 5 + 4$

$6 < 9$ which is still true.

On the number line, we are simply moving both numbers to the right four places.

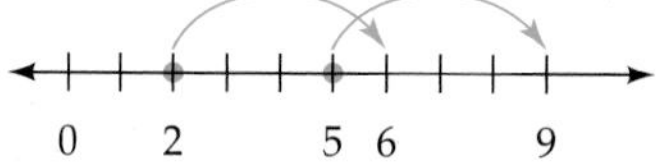

B. Multiplying each side by 3 looks like:

$2(3) < 5(3)$

$6 < 15$, which is still true.

On the number line, we are moving both numbers to the right by multiplying by a positive number.

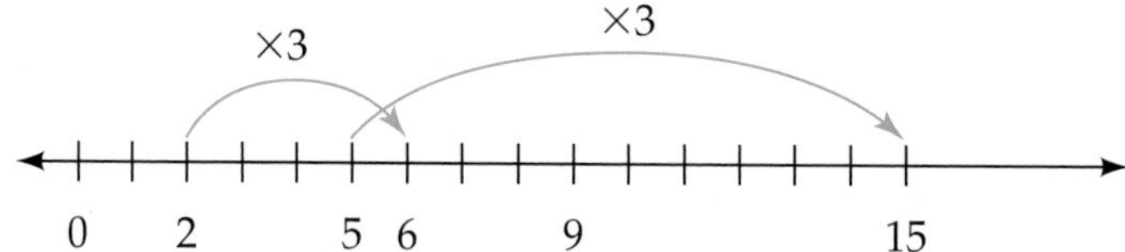

C. Multiplying each side by -3:

$2(-3) < 5(-3)$

But is this still true?

$-6 < -15$

NO! This is no longer true, given that we have to switch around the inequality to make this statement true:

$-6 > -15$

On the number line, we are "flipping" them across the zero mark.

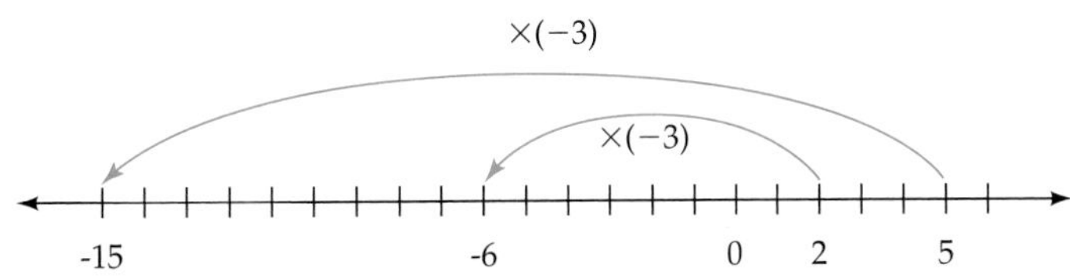

To summarize what happened in this investigation, when does the inequality remain the same and when does it change? From the examples here, we see that when we add a number to both sides, or multiply by a positive number on each side, the inequality relationship stays the same. However, when we multiply both sides by a negative number, the relationship is reversed so we need to "flip" the inequality sign. Subtracting numbers and dividing numbers will be left for you to explore in the exercises.

So, we can use this knowledge to solve linear inequalities.

A. $x + 4 > 5$. Subtract 4 from both sides, and the inequality relationship stays the same, so $x > -1$.

B. $\frac{x}{3} > 5$. Multiply by 3 on each side. Since we are multiplying by a positive number, the inequality relationship stays the same, so $x > 15$.

C. $\frac{x}{-3} < 6$. Multiply each side by -3. Since we are multiplying by a negative number, the relationship gets "flipped." So, $x > -18$.

In this section we used algebraic notation to explain the structure behind magic! We delved deeply into what "equals" really means and looked at balance scale models and Singapore bar models to help us gain a visual and conceptual understanding of solving linear equations. We investigated inequality relationships as we solved them.

6.2 Exercises

1. Do the following steps with a few different number picks. Then use a variable to show why the magic works.

   Pick a number. Multiply it by 3. Then add 6. Divide by 3. Subtract your original number. Your answer is 2.

2. Do the following steps with a few different number picks. Then use a variable to show why the magic works.

   Pick any number. Add 7. Multiply by 2. Subtract 4. Then, divide the result by 2. Subtract the original number you picked from this number. Your answer is 5.

3. Do the following steps with a few different number picks. Then use a variable to show why the magic works.

   Pick any number. Multiply by 2. Add 10. Divide by 2. Subtract 5. Your answer is your original number.

4. Make up your own "trick" where you end up with the same number no matter what number you started with, as in Exercises 1 and 2.

5. Make up your own "trick" where you end up with the number you picked originally as in Exercise 3.

6. A student claims that $3x + 2y = 5xy$. Use a visual representation to show whether this is true or false.

7. Use both models and algebraic representations of even and odd numbers (as we did in Investigation 6.2b) to show that the following are true.

   even times even = even

   odd times even = even

   odd times odd = odd

8. a. Explain in your own words what it means to say two things are equal and what it means to say two things are equivalent.

   b. Define the word "congruent." Look it up if you are unsure of its meaning. Describe how this word relates to the words "equal" and "equivalent."

DEEPENING YOUR UNDERSTANDING

9. Solve each of the following linear equations using a balance scale model.

   a. $x + 4 = 12$

   b. $n - 4 = 6$

   c. $3x + 2 = 14$

   d. $2x - 4 = 6$

10. Solve each of the equations in Exercise 9 with a Singapore bar model.

11. Start with the fact that $2 < 5$.

    a. Does the left side remain less than the right side when we subtract 4 from each side?

    b. How does this idea relate to solving $x + 4 < 12$?

12. Start with the fact that $4 < 6$.

    a. Does the left side remain less than the right side when we divide each side by 2?

    b. How does this idea relate to solving $2x < 12$?

    c. Does the left side remain less than the right side when we divide each side by -2?

    d. How does this idea relate to solving $-2x < 12$?

13. Solve the following linear inequalities.

    a. $-3x < 9$

    b. $5x + 2 > 10$

14. Jack has $700 in a savings account at the beginning of January. He wants to keep at least $200 in the account for a safety net. He withdraws $25 each week for spending money. How many weeks can he withdraw money? Write an inequality for this scenario and then solve it.

15. Go to the website http://www.mathplayground.com/AlgebraEquations.html where there are equations to solve using a balance scale model. Begin with one-step equations and then move on to two-step equations.

16. On another page within the *mathplayground* website, http://www.mathplayground.com/ThinkingBlocks/thinking_blocks_modeling%20_tool.html, you can draw models similar to the Singapore bar models. Go to this website and use the tool to solve some word problems given.

17. Ben lost half of his candies while walking home. His mom gave him 8 more candies. Now he has 22 candies. How many candies did he have to begin with?

    a. Solve this any way you choose.

    b. Write an equation for this, letting x stand for the number of candies Ben had to begin with.

FROM STANDARDIZED ASSESSMENTS

2008 NECAP, Grade 5

18. Look at these number sentences.

    ♡ + 4 = 12

    ☆ ÷ 2 = ♡

    Each heart represents the same number.

    a. What number does each heart represent?

    b. What number does the star represent?

19. If □ represents the number of newspapers that Lee delivers each day, which of the following represents the total number of newspapers that Lee delivers in 5 days?

    a. 5 ÷ □

    b. 5 × □

    c. □ ÷ 5

    d. (□ ÷ □) × 5

    (48% of fourth-graders got this correct)

    Source: Results of the *Fourth NAEP Mathematics Assessment*, p. 249. U.S. Department of Education, National Center for Education Statistics.

20. If k can be replaced by any number, how many different values can the expression $k + 6$ have?

    a. None

    b. One

    c. Six

    d. Seven

    e. Infinitely many

    (72% of eighth-graders got this correct)

    Source: Results of the *Fourth NAEP Mathematics Assessment*, p. 249. U.S. Department of Education, National Center for Education Statistics.

2008 NECAP, Grade 4

21. The scale shown below is balanced.

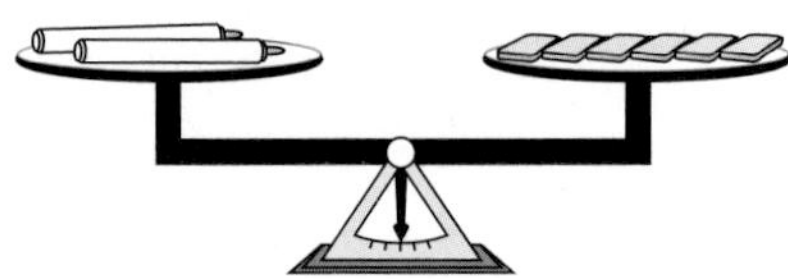

All of the markers have the same weight. All of the erasers have the same weight.

How many erasers balance one marker?

a. 6 b. 4

c. 3 d. 2

22. From the Smarter Balanced Assessment Consortium (one of the new standardized assessments aligned with the Common Core State Standards): Look at each expression. Is it equivalent to $36x + 24y$? Select Yes or No for expressions a through c.

    a. $6(6x + 4y)$

    b. $30(6x + 6y)$

    c. $12(x + 2y + 2x)$

SECTION 6.3 Using Mathematical Models to Represent and Understand Quantitative Relationships

What do you think?

- In what ways have you used mathematical models (pictures, diagrams, manipulatives, graphs, tables, equations, etc.) thus far in this course?
- In what way does mathematics help us model life?

Mathematical models help us to represent and understand quantitative relationships. Models can help us describe situations as well as predict what may happen. Modeling begins in early elementary school when a student might use pictures or blocks to model that $3 + 5 = 8$. Just as mathematical models is a category that NCTM uses for algebra, the Mathematical Practice 4 of the CCSS is titled, "Model with Mathematics." Mathematical models are applied in and can help us understand real-life situations and make sense of them.

Consider what a "model" is. We speak of model airplanes, and architects and engineers build a model of a building or a bridge before constructing the actual object. In one sense, a model is a representation of the original object. There are two important features of many models. First, the model contains many properties of the original object. Second, the model can be manipulated and studied to help us better understand the (usually larger and more complex) object. A mathematical model is a mathematical structure that approximates the features of a situation.

A model, or representation, can take many forms: manipulative, diagram, graph, table, sketch, equation, words, etc. Most problems and most mathematical concepts can be represented in different ways. Throughout this course thus far you have used several ways to model mathematical ideas and to solve problems. Students commonly ask which representation is "best," but there is no "best way." More useful questions are: Does this representation fit the purpose before us? Does it make sense in the context of the problem? Does it help us to solve the problem at hand? Let's investigate a few examples. Math topics are very interrelated so in many of these investigations you will see connections to ideas from previous sections in this chapter. In the next investigation, we will revisit the concept of a function while also modeling a real-life situation.

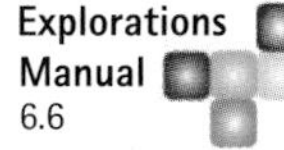

INVESTIGATION 6.3a Baby-sitting

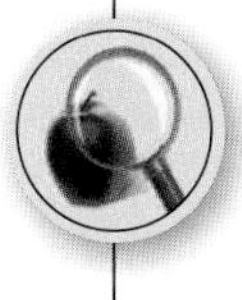

Let us extend our understanding of functions by examining another relationship between the two variables "hours worked" and "dollars earned." Let's say Ellen baby-sits and charges $8 per hour. Describe how you would determine how much to pay Ellen. Determine whether the relationship between the two variables—hours and dollars—is a function. Whether you say it is or is not a function, can you justify your response? Then read on. . . .

DISCUSSION

A common response to describing how to pay Ellen is to multiply the number of hours sat by 8. Let us use this situation to introduce several different ways to represent functions. Then we will examine more closely the relationship between time sat and dollars paid.

For purposes of simplicity, let us first look at different representations of this function using only whole-hour amounts.

TABLE 6.3

h	d
1	8
2	16
3	24
4	32
5	40
.	.
.	.
.	.

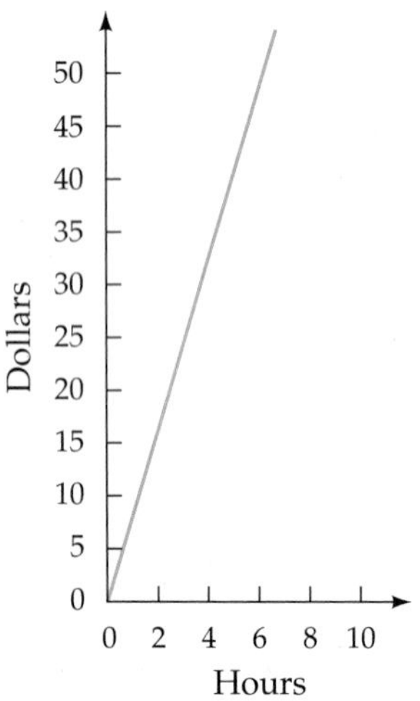

Figure 6.3

Tables We can represent this function with a table in which h represents hours and d represents dollars (Table 6.3). We can then use the table to determine how much money Ellen will make from baby-sitting.

Equations We can represent this function with an equation in which h represents hours and d represents dollars:

$$d = 8h$$

Graphs We can represent this function with a graph (Figure 6.3).

We can refer to the input as the **independent variable** and to the output as the **dependent variable**. In this case, we say that the independent variable is the hours baby-sat and the dependent variable is the dollars earned. That is, the number of dollars a baby-sitter earns is dependent on the number of hours that he or she baby-sits.

Ordered pairs We can represent this function as a set of **ordered pairs** in which the first element of each ordered pair represents hours and the second element represents dollars:

$$B = \{(1, 8), (2, 16), (3, 24), (4, 32), \ldots\}$$

Mappings Finally, we can represent functions with arrow diagrams. Mathematicians often refer to this representation as a **mapping** of one set onto another set.

h		d
1	$\rightarrow$	8
2	$\rightarrow$	16
3	$\rightarrow$	24
4	$\rightarrow$	32

A closer look at paying the baby-sitter At first glance, the question of how much to pay the baby-sitter is simple: Multiply the hours sat by 8. However, let us use the problem-solving strategy "act it out" to examine this problem more closely. For example, what if Ellen baby-sat from 7 to 11:15? How much would you pay her? Think before reading on. . . .

Some people say \$32. Some people say \$36—they round up to the nearest half-hour. In actuality, different people have different ways of determining how much to pay a baby-sitter. Let us examine the case of a couple, who rounds up the time to the nearest half-hour. We could now represent their process for paying the baby-sitter in each of the ways we have just examined. Is the relationship between time sat and dollars earned a functional relationship in the case of the couple? Think and then read on. . . .

If you were to graph this relationship, what would the graph look like? Try to make your own graph before reading on. . . .

MATHEMATICS

The graph that represents this baby-sitting function is often confusing to students who see it for the first time. However, it is relatively common in real-world mathematics and is a member of the subset of functions called step functions.

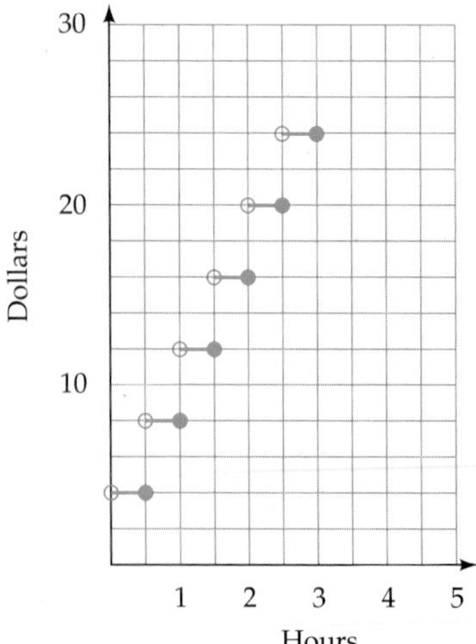

Take a look at the graph on the left and see whether you can make sense of it before reading on.

If you are having trouble making sense of the graph, consider a few examples. Let's say the sitter sits for 2 hours and 50 minutes. Using the couple's process, we round this time to 3 hours and pay Ellen \$24. As you can see, Ellen will receive \$24 for all times between 2 hours and 31 minutes and 3 hours.

In this baby-sitting context, we have seen that the original straight-line graph turns out not to be a useful or accurate model. The step graph is a more accurate representation. This process of examining a situation and then developing a model that accurately represents that situation is called **mathematical modeling**. Constructing and interpreting mathematical models is one of the more important uses of mathematics in the real world.

INVESTIGATION 6.3b Choosing Between Functions

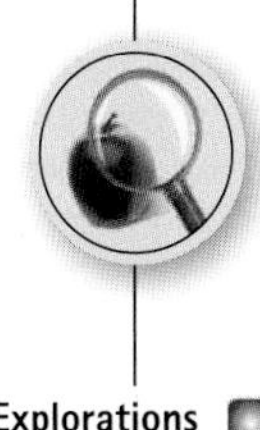

Explorations Manual 6.7

Jackie has been promoted to chief salesperson of Southside Computers. She has two choices as to how she will be paid. She can receive \$50 for every computer system she sells, or she can receive \$250 a week plus \$25 for every computer system she sells. Looking at the past year, she finds that she has averaged 9.3 sales per week, and she figures that she can do a little better this year. What would you recommend? Work on this problem for a bit and then read on. . . .

DISCUSSION

You may want to start by plugging in a few numbers—that is, using guess–check–revise. Do you see that if Jackie sells only a *few* computers, she will be better off with the second plan but that the first plan is better if she sells a *lot* of computers? This realization leads to a specific mathematical question: At what point will the two plans give her the same money? Once she can answer that question, she can base her decision on whether she believes she will sell more or less than that amount. Below we see four very different strategies for solving this subproblem: guess–check–revise, use equations, make a graph, and Singapore bar models.

STRATEGY 1 Guess–check–revise

This strategy is illustrated in Table 6.4.

TABLE 6.4

Sales	Plan A	Plan B	Reflection/Analysis
0	0	250	
2	100	300	Plan A goes up 100; plan B goes up 50. Plan A is still \$200 below plan B. Each time you increase sales by 2 computers, plan A increases \$50 *more* than plan B does.
10	500	500	

What does the last row mean? Think and read on. . . .

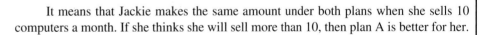

It means that Jackie makes the same amount under both plans when she sells 10 computers a month. If she thinks she will sell more than 10, then plan A is better for her.

STRATEGY 2 Use algebra

Let y = Jackie's weekly salary.
Let x = the number of computer systems she sells.
The equation that represents plan A is

$$y = 50x$$

The equation that represents plan B is

$$y = 250 + 25x$$

Do you understand how both of these equations were constructed?

We have two equations in two unknowns that we can solve:

$$50x = 250 + 25x$$
$$25x = 250$$
$$x = 10$$

STRATEGY 3 Make a graph

Where the two lines cross will show the point at which her earnings under both plans will be equal.

HISTORY

René Descartes (1596–1650) invented the Cartesian coordinate system (named after him) to illustrate such relationships on graphs. Legend has it that he got the idea from watching a fly crawl across a ceiling. He realized that a coordinate system could represent the fly's location with one symbol—that is, the ordered pair.

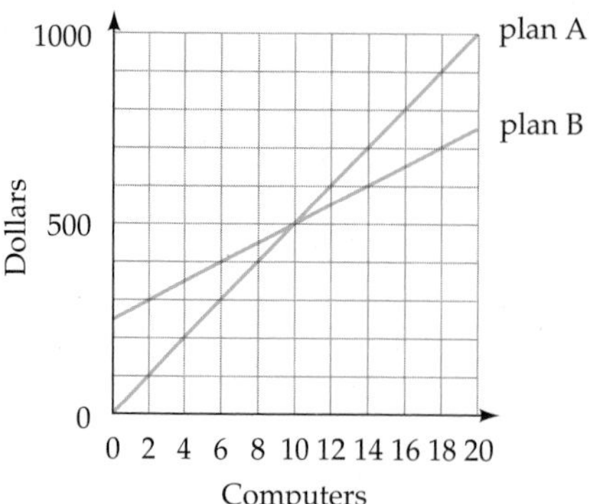

Figure 6.4

The two lines cross at the point (10, 500) (Figure 6.4). In other words, the ordered pair (10, 500) is an element of both graphs.

We choose to use a linear model to represent the situation, with the knowledge that the situation itself is not truly linear in reality. This graph and the equation make it appear that if she sells 2.5 computers, she would make \$125 with plan A, for example. You can plug 2.5 into the equation, or see this on the graph. Of course in reality, she could not sell a fraction of a computer. So, it is important to remember the context of the model, just as we did in the baby-sitting investigation.

STRATEGY 4 Singapore bar models with color-coded communication

Bar models can provide a visual representation for solving these equations. Because we want to know when the earnings for each plan are equal, we need to figure out when the two are the same length as shown. We will use color coding to help us see the model.

Jackie can receive \$50 for every computer she sells, or she can receive \$250 a week plus \$25 for every computer she sells. First we need to find out how many computers sold will make her pay equal, or when the two bars are equal.

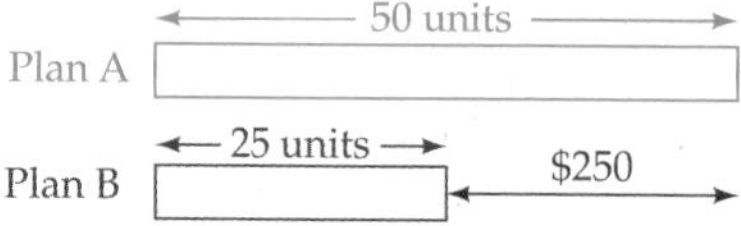

Source: Alice Ho of Math Teach Singapore.

50 units – 25 units = 25 units, which equals \$250. So, 1 unit equals 250 divided by 25 = 10 units. So, her earnings are equal under both plans when she sells 10 computers.

Let's revisit Investigation 6.3b and ask the question a little differently.

For what number of computers sold will plan A be more money than plan B? Try writing an inequality that models this and then read on...

For this, we are asked when the income for plan A is greater than the income for plan B. The expression for the income for plan A equals $50x$ (remember, x represents the number of computers sold) and the income for plan B is represented as $250 + 25x$. Then, if we want to know when plan A is greater than plan B, we could write it as:

$$\text{plan A} > \text{plan B}$$

$$50x > 250 + 25x$$

Solving this algebraically, we can subtract $25x$ from both sides to get

$$25x > 250$$

and then divide both sides by 25 to get

$$x > 10$$

Translate this into words to say that plan A produces more income when the number of computers sold is more than 10. Look back at the graph of this to see how this relates to the picture. Can you see how when x is greater than 10 on the graph, that the income for plan A is higher than for plan B?

SYSTEMS OF SIMULTANEOUS EQUATIONS

Systems of simultaneous equations can be understood more easily when we use visual representations. Strategy 2 of Investigation 6.3b is an example of a system of simultaneous equations. In this we had two equations and two unknowns. Then in the Singapore bar model representation, we had a visual look at the same problem. Let's investigate additional systems of simultaneous equations and use visual tools to help make sense of them. According to the CCSS, these are formally introduced in eighth grade.

INVESTIGATION 6.3c Visualization of Simultaneous Equations

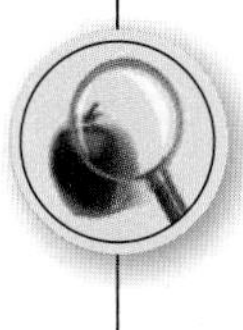

A. 4 apples and 3 cherries cost \$3.40.
3 apples and 2 cherries cost \$2.50.
Find the cost of each apple and each cherry.

B. The weight of Ali and Ben is 44 kg together.
The weight of Ali and Cal is 60 kg together.
The weight of Ben and Cal is 54 kg together.
Find the weight of each of them.

DISCUSSION

We could use several strategies such as guess–check–revise, or algebraic solving of the equations. In Singapore, the students are able to solve these types of problems in elementary school because they use the following visualization tools. The color-coded model, developed by Alice Ho, is also helpful in communicating the solution steps.

A.

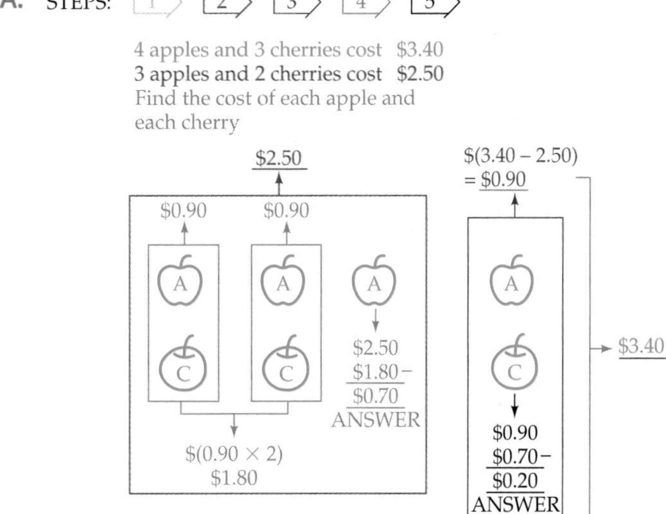

Source: Alice Ho of Math Teach Singapore.

First, we draw 4 apples and 3 cherries that are all worth \$3.40. Then in blue we show that 3 apples plus 2 cherries is \$2.50. Then the 1 apple and 1 cherry in the red box are equal to \$3.40 – \$2.50 = \$0.90. Then in gray we see that each of those apple and cherry combinations must equal \$0.90 each so the 1 apple in the blue box would be the \$2.50 – 2(\$0.90) = \$0.70. Once we know the apple price, then the cherry price is shown in black as \$0.90 – \$0.70 = \$0.20.

B. STEPS: 1 2 3 4 5

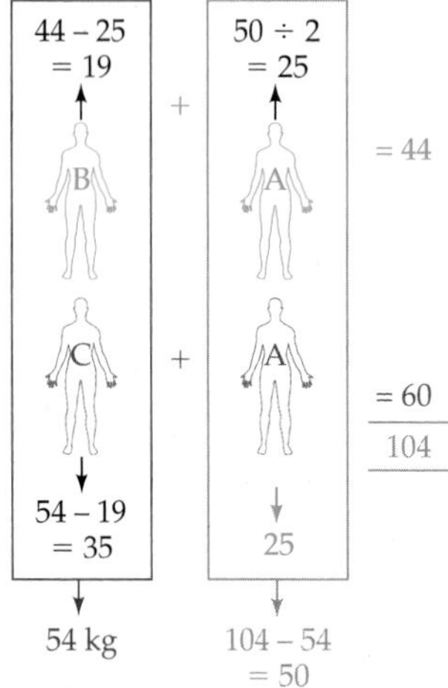

Source: Alice Ho of Math Teach Singapore.

We start in green, which represents that Ali plus Ben = 44 kg. Then in blue we show that Ali plus Cal = 60 kg. In red, going down a column we show that Ben + Cal = 54 kg.

This leaves the two Ali figures equaling 104 – 54 = 50, so each Ali must equal 25. From there we can subtract to get the others.

In equation form, this would look like the following:

$B + A = 44$

$C + A = 60$

$B + C = 54$

With just these equations, the process is not very intuitive, but the pictures greatly help us work through the problem.

SUMMARY 6.3

In this section we continued using visualization strategies to make sense of the abstract concepts. Graphs, diagrams, tables, balance scales, and bar models were all used to model the math situation. Using models can help us to make sense of real-life problems.

6.3 Exercises

1. Let's say you have children, you hire a baby-sitter frequently, and you pay the baby-sitter \$10 per hour. Describe how *you* would pay the baby-sitter, first in words, then with a graph. In order to receive full credit, the reader must be able to pay the baby-sitter the same amount that *you* would pay, based on your descriptions.

2. At the time this book was written, the postage on a letter was determined by the following set of rules:

   The first ounce costs 44¢.

   Each additional ounce, or fraction thereof, costs 17¢.

   a. How much would it cost to mail a letter weighing 6.3 ounces?

   b. Draw a graph of this function.

3. Jack and Jill run a catering business. Their fee for catering banquets is \$150 plus \$4 per person.

   a. Express this relationship with a table.

   b. Express this relationship with an equation.

   c. Express this relationship with a graph.

   d. Jack and Jill catered a banquet for which they charged \$462. How many people attended?

   e. Their goal is to gross \$2000 a month. How many jobs do they need to average in order to make their goal? (There is no single right answer.)

4. My parents used to live in Park Rapids, Minnesota. When I visited them, I flew to Minneapolis and rented a car. It is about 160 miles to Park Rapids. The last time I visited them, I spent three days with them, counting my arrival and departure days. The car rental agency offered two options: \$45 a day plus 30¢ per mile or \$185 for the week. Which do you recommend?

5. Let's say you have a choice between two checking plans. With the first plan, you will be charged \$2 per month plus 15¢ per check. With the second plan, you will be charged \$5 per month regardless of how many checks you write. If you write only a few checks, the first plan is clearly cheaper. Similarly, if you write 200 checks a month, the second plan is clearly cheaper.

   a. Determine the number x that will enable you to make the following statement: If you write fewer than x checks per month, choose the first plan. If you write more than x checks per month, choose the second plan.

   b. Determine the number x using a method different from the one you used in part (a).

   c. Which of the plans are functions: both, only one, or neither? Justify your response.

6. A convenience store advertised two phone cards. Which card would you buy? Why?

   Card 1: 3.9¢ a minute with 37¢ connection charge for each call.

   Card 2: 7.9¢ a minute with no connection charge.

DEEPENING YOUR UNDERSTANDING

7. Archaeologists can estimate a person's height from the length of the femur (thigh bone). The formula for doing so is

   $H = 2.3L + 61.4$

   where all measurements are in centimeters. Let's say an archaeologist has found a femur of a human, and the bone is 45 centimeters long. How tall do we think the person was at the time of death?

8. Have you ever ridden on a bicycle where the saddle was too high or too low? The height of a saddle can be determined by adjusting it until the rider says, "It feels right at this height," but not all people guess right. Bicycle manufacturers have found that for the average person, the saddle height can be determined by the following formula: $h = 1.08i$, where h represents the saddle height and i represents the inseam.

   a. How tall should the saddle height be when you ride a bicycle?

   b. If you have a bicycle, measure the saddle height. If it is different from the number derived by applying the formula above, change it and see whether the height is more comfortable now.

9. I can recall trying to explain to my children that thunder is the sound that lightning makes. It took some time for them to understand this because there is often a delay of many seconds between seeing the lightning and hearing the thunder. Lightning travels at 186,000 miles per second, whereas thunder travels at approximately 750 miles per hour.

   Let's say that you see a flash of lightning and 5 seconds later you hear the thunder. How far away did the lightning strike?

10. Here is another problem that has appeared in many places. Eight adults and two children need to cross a river in a small boat that can hold one adult or one or two children.

    a. How many one-way trips will it take for all of them to cross the river?

    b. How many trips will it take for 2 children and x adults?

    c. How many trips for 8 adults and 3 children?

11. If three books cost \$12, and a book plus a pen plus a notepad equals \$14, and a book plus a notepad equals \$11, find the cost of each item.

12. Find the value of the rectangle and triangle.

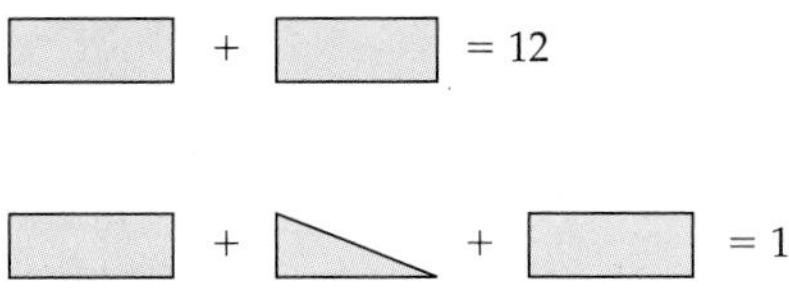

13. Find the value of each figure.

= 15

\+ + = 14

\+ = 9

14. You are deciding between two cell phone plans:

    Plan A: Unlimited text, talk, and data for \$60 per month.

    Plan B: Unlimited text and data and 100 minutes of talk for \$40, plus \$0.20 per minute of talk after 100 minutes.

    a. Draw a graph showing each plan.

    b. At how many minutes of talk are the plans equal?

15. A taxi meter starts at \$2.00 and increases at a rate of \$1 per half mile.

    a. Draw a graph of the cost of the taxi.

    b. How much would a 10-mile taxi ride cost?

16. Suppose another taxi company charges \$1.50 per half mile, with no start cost.

    a. Draw a graph of the cost of this taxi.

    b. How much would a 10-mile taxi ride cost with this company?

    c. At how many miles would the cost of the taxi in Exercise 15 be the same as the company in Exercise 16?

17. Adam and Thomas raised money doing yardwork. They raised \$1200 total. Adam raised \$600 more than Thomas. How much did Thomas raise?

    a. Solve this using Singapore bar models.

    b. Write an algebraic equation for this relationship and solve it.

FROM STANDARDIZED ASSESSMENTS

2005 NECAP, Grade 6

18. Paige rode her bike for one hour and Sally walked for one hour. The graph below shows the number of calories each girl burned.

Calories Burned During Exercise

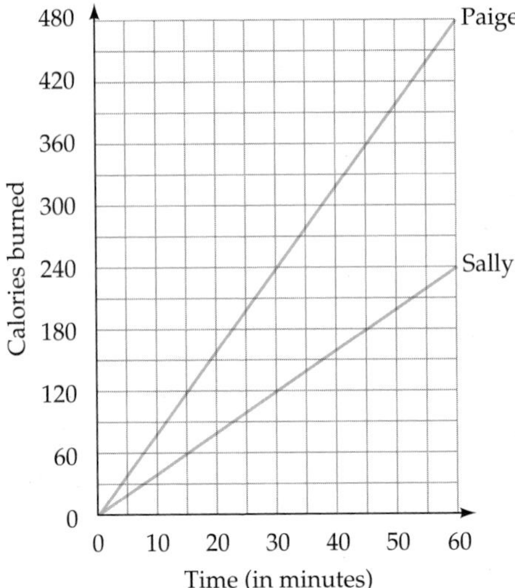

How many more minutes did it take Sally to burn 120 calories than it took Paige?

a. 5

b. 15

c. 25

d. 30

SECTION

Analyzing Change in Various Contexts

What do you think?

- How is change represented on a graph?
- How is change reflected in an equation?
- What types of change are there?

Heraclitus, a Greek philosopher, who lived in Ephesus from 535 B.C. to 475 B.C., is attributed with the famous quote, “The only constant is change.” Many algebraic models answer questions to better understand change. For example, manufacturers test different shapes and materials for cups and then measure the temperature of the coffee to see how fast it cools; scientists measure the effectiveness of a new medicine by measuring how fast it kills the harmful bacteria. Some common elementary school experiments with change include measuring the weight of apple slices over time to measure dehydration and measuring the growth of plants to compare different conditions—more/less sunlight, more/less water, and good/poor soil.

In the next two investigations, we will examine various situations involving change, but we will move away from numbers deliberately. Doing so will make the algebraic ideas come more to the foreground. Slope is formally taught in eighth grade, according to the CCSS.

Explorations Manual 6.8

INVESTIGATION 6.4a Matching Graphs to Situations

Below are descriptions of three runners in a race and three graphs. Match each description to the correct graph and explain your choice. The independent variable is time, and the dependent variable is distance from the start.

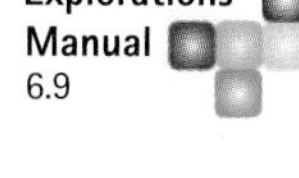
Explorations Manual 6.9

Alex started slowly, then ran a bit faster, and then ran even faster at the end of the race.

Manuel started quickly but then tired and slowed down a bit and then slowed down even more at the end.

Ragib started quickly, stopped to tie his shoe, and then ran even faster than before.

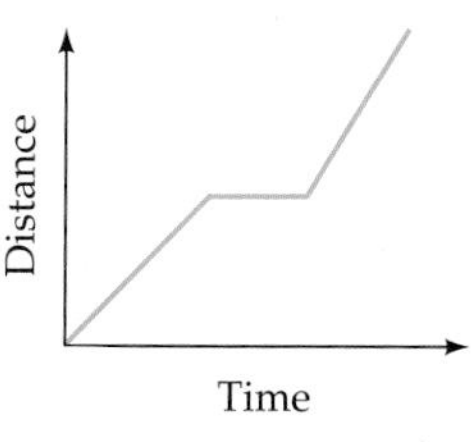

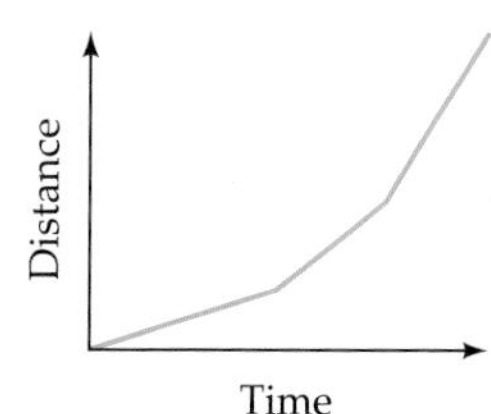

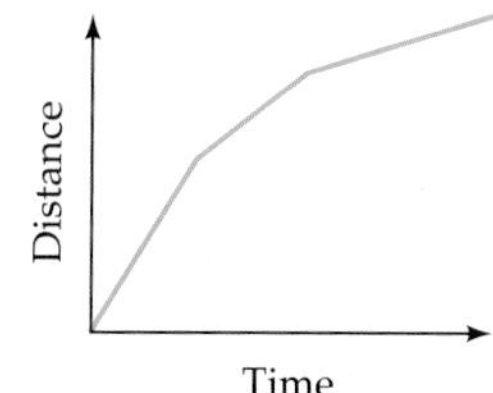

DISCUSSION

A slower speed has a smaller slope than a faster speed. Stopping means that the runner is not going any distance and so is represented by a zero slope (horizontal line). Thus the second graph matches Alex—less steep than the others at the beginning and then steeper and steeper. The third graph matches Manuel—steeper than Alex at the beginning and then less and less steep. The first graph matches Ragib—steep at the beginning, horizontal in the middle to show that he stopped, and then even steeper than before.

INVESTIGATION 6.4b Developing "Graph Sense"

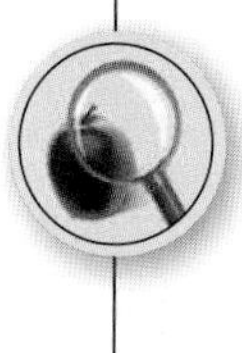

Which of the graphs below best represents the following scenario? A runner is running at a steady rate and then comes to a hill, which causes her to run at a slower rate. Once she reaches the top of the hill, she runs down the hill very fast. Upon reaching the bottom of the hill, she resumes her original pace.

CLASSROOM  CONNECTION

Grade 5

Can you believe that this is from a fifth-grade book? Can you match each event with its graph?

Date ______________________ Time ______________________

LESSON 10•7 Mystery Graphs

Each of the events described below is represented by one of the following graphs.

Time	Time	Time	Time	Time
Graph A	**Graph B**	**Graph C**	**Graph D**	**Graph E**

Match each event with its graph.

1. A frozen dinner is removed from the freezer. It is heated in a microwave oven. Then it is placed on the table.

   Which graph shows the temperature of the dinner at different times? Graph ____________

2. Satya runs water into his bathtub. He steps into the tub, sits down, and bathes. He gets out of the tub and drains the water.

   Which graph shows the height of water in the tub at different times? Graph ____________

3. A baseball is thrown straight up into the air.

   a. Which graph shows the height of the ball—from the time it is thrown until the time it hits the ground? Graph ____________

   b. Which graph shows the speed of the ball at different times? Graph ____________

358

From *Everyday Mathematics, Grade 5:* The University of Chicago School Mathematics Projects: Student Math Journal, Volume 2 by Max Bell et al., Lesson 10-7, p.358. Reprinted by permission of The McGraw-Hill Companies, Inc.

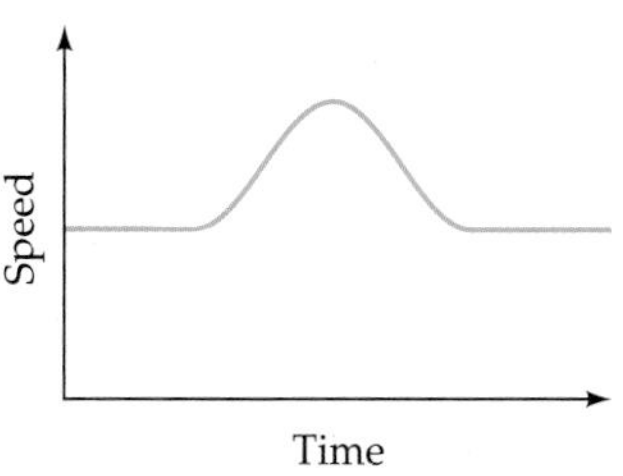

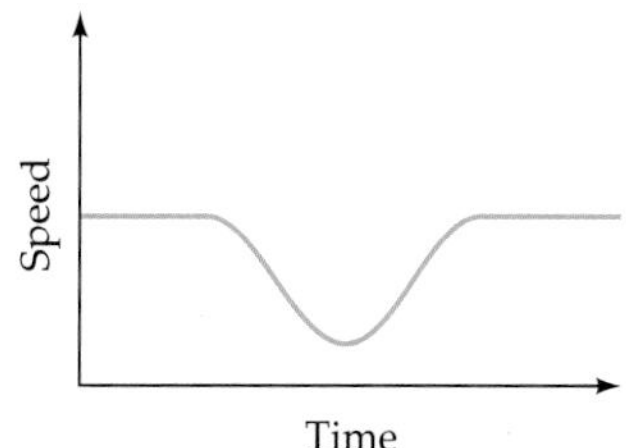

DISCUSSION

Many students pick the first graph because that is what the race looks like—level, then a hill, and then level again. But the graph is a picture not of the layout of the race but, rather, of the person's speed. A steady speed means that the speed is not changing and is thus represented by a horizontal line. When the runner slows down the slope of the graph decreases (i.e., becomes negative); when she speeds up, the slope of the graph increases. Thus the second graph is correct.

SLOPE OF A LINE

We spoke in the previous two investigations about slope as a rate of change. Parts of the graphs were lines, but we also discussed the slope of curves. Actually, a foundation of calculus is studying the slope at a point on a curve. In middle school, students learn about the slope of a line as a measure of steepness, or the rate at which things change. Teaching and learning about change in elementary schools lays a foundation for a more formal look at slope in middle school.

INVESTIGATION 6.4c How Steep Is the Highway?

In mountainous areas, a driver might see a sign such as the following:

imagebroker.net/SuperStock

A. Describe what this means.

B. What if the sign said 12%—how would that be different?

C. Slope is often defined as "rise over run." What does 17% mean in terms of "rise over run"?

D. The triangle in the sign helps us to visualize slope as rise over run. Measure the "rise" and "run" of the triangle in the picture. What percent change is the picture really?

DISCUSSION

A. There are several ways to describe this. One may be, "This highway is really, really steep," which is accurate. The 17% is a measure of how steep. More specifically this means that the highway rises 17 feet for every 100 feet in horizontal distance. This picture is drawn to scale with a 17% slope to illustrate:

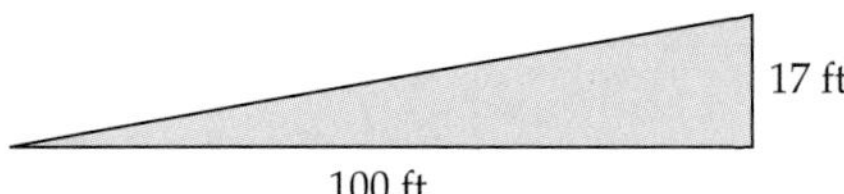

A 12% highway would be less steep as it would rise 12 feet for every horizontal 100 feet. This picture is drawn to scale with a 12% slope to illustrate:

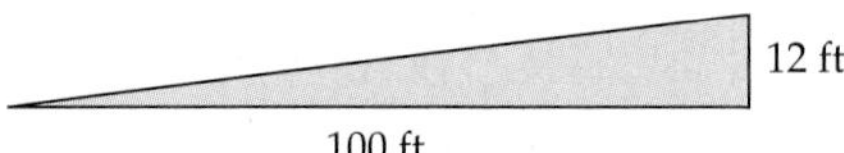

B. In part A, we used the concept of rise over run to talk about and illustrate how 17% means 17/100 as the rise over run or slope.

C. When we measure the triangle on the sign, it actually is a picture of a much steeper triangle. What were your measurements? Divide the rise by the run and your ratio is close to 29%. Now that would really be a steep highway!

INVESTIGATION 6.4d How Many Dots?

Predict how many dots it will take to make the nth figure. Do this yourself and then read on. . . .

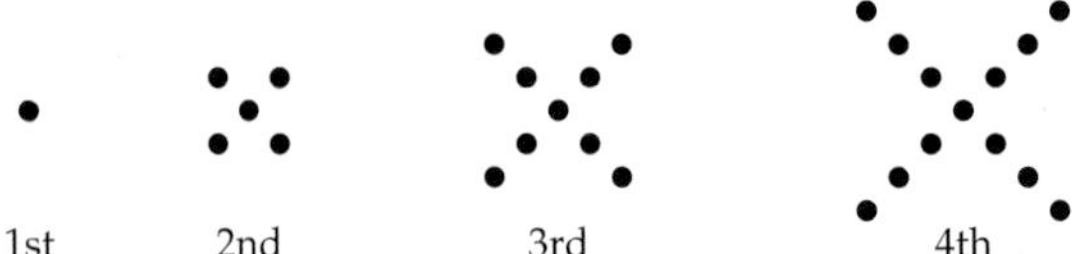

DISCUSSION

While there are several patterns here, and several ways that children may solve it, let's think about it in terms of change. Consider the change between each total number of dots, and use that to determine the equation.

HOW MUCH DOES IT GROW BY?

When looking at the change, the number of dots increases by 4 each time.

Figure number	1	2	3	4	5	6	7	n
Number of dots	1	5	9	13	17	21	25	
Number of dots	1	$4 + 1$	$2 \cdot 4 + 1$	$4 \cdot 4 + 1$	$5 \cdot 4 + 1$	$6 \cdot 4 + 1$	$7 \cdot 4 + 1$	$4(n \cdot 1) + 1$

We used patterns in Section 6.1 to create equations and generalizations. As in this investigation, the pattern of change is often a useful observation. We can use the observation that the pattern grows by 4 each time and modeling the amount this way, such as $4 + 1, 2 \cdot 4 + 1, 3 \cdot 4 + 1$, etc. From this point, we can see that the multiplier of 4 is simply 1 less than the figure number. Thus the nth case will have $(n - 1)\,4 + 1$ dots, which simplifies to $4n - 4 + 1 = 4n - 3$.

Yet another way to use the observation of increasing by 4 each time is to remember that in a straight line, the slope is simply the rate of change. In this case, the rate of change is 4, and so the slope is 4. Thus, the equation is $y = 4x + b$. Now it is a matter of trial and error to determine that $b = -3$. That is, $y = 4x - 3$ is the number of dots in any figure.

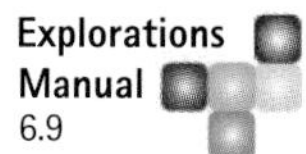
Explorations Manual 6.9

ARITHMETIC AND GEOMETRIC SEQUENCES

A linear change has a slope, which describes the constant change. Another way to look at change is through analyzing sequences of numbers. Let's examine some sequences of numbers and consider how they change from one number to the next to explore this concept further.

INVESTIGATION 6.4e Patterns of Change

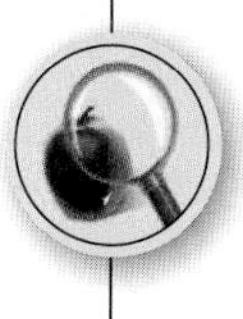

Describe how each of these sequences changes, first in words and then numerically and symbolically.

Sequence 1: 2, 5, 8, 11, 14, . . .
Sequence 2: 3, 6, 12, 24, 48, . . .
Sequence 3: 4, 9, 19, 39, 79, . . .

DISCUSSION

Sequence 1 Let us examine the first sequence: 2, 5, 8, 11, 14, ... It is not terribly difficult to determine that the 6th term will be 17. One way to determine the 20th term is simply to continue the sequence until you get to the 20th term. However, this is tedious. With a bit of analysis, you could realize that you need to add 15 more 3s after the 5th term, and thus, the value of the 20th term is 14 + 45. If you didn't see this, look at the second row of Table 6.5, where we have broken down each term in the sequence so you can see *how it came to be.* Using the fact that multiplication is repeated addition, we can represent the terms more economically. This not only saves time but also begins to reveal the mathematical structure of the sequence.

In the last row of the table, we have shown in bold the number that tells how many 3s. Notice that this number is always 1 less than the number representing the position of the term in the sequence. That is, the 4th term has three 3s, the 5th term has four 3s, etc. Thus, the nth term must have $(n - 1)$ threes.

TABLE 6.5

1st	2nd	3rd	4th	5th	nth
2	5	8	11	14	
2	2 + 3	2 + 3 + 3	2 + 3 + 3 + 3	2 + 3 + 3 + 3 + 3	$2 + \overbrace{3 + 3 + \cdots + 3}^{(n-1)\text{ times}}$
2	2 + 3	$2 + 2 \cdot 3$	$2 + 3 \cdot 3$	$2 + 4 \cdot 3$	$2 + (n - 1) \cdot 3$
2	2 + 3	$2 + \mathbf{2} \cdot 3$	$2 + \mathbf{3} \cdot 3$	$2 + \mathbf{4} \cdot 3$	$2 + (\boldsymbol{n} - \mathbf{1}) \cdot 3$

LANGUAGE

We can describe arithmetic sequences in different ways. For example, we can say that the relationship between each term and the following term is that we add the same number each time. On the other hand, we can say that the difference between each pair of consecutive terms is constant.

Sequences like the one above, where the difference between each pair of consecutive terms is constant, are called **arithmetic sequences**. What is different is the starting number and the common difference. If we represent the starting number by a and the common difference by d, then we can state a rule for finding the nth term of any arithmetic sequence, as shown in the last column of Table 6.6.

TABLE 6.6

1st	2nd	3rd	4th	5th	nth
a	$a + d$	$a + 2 \cdot d$	$a + 3 \cdot d$	$a + 4 \cdot d$	$a + (n - 1) \cdot d$

Sequence 2 Now let's examine the second sequence: 3, 6, 12, 24, 48, ... You probably realized that the two sequences are similar in that there is a pattern and there is a relationship between each term and the next. In this sequence, the relationship is that each term is twice, or double, the preceding term. So the next number in the sequence is 96.

In the second row of Table 6.7, I have broken down each term in the sequence so you can see how it came to be. Using the fact that exponentiation is repeated multiplication, we can represent the terms more economically. Do you understand the *n*th term now?

TABLE 6.7

1st	2nd	3rd	4th	5th	*n*th
3	6	12	24	48	
3	$3 \cdot 2$	$3 \cdot 2 \cdot 2$	$3 \cdot 2 \cdot 2 \cdot 2$	$3 \cdot 2 \cdot 2 \cdot 2 \cdot 2$	$3 \cdot \overbrace{2 \cdot 2 \cdot \cdots \cdot 2}^{(n-1) \text{ times}}$
3	$3 \cdot 2$	$3 \cdot 2^2$	$3 \cdot 2^3$	$3 \cdot 2^4$	$3 \cdot 2^{(n-1)}$

Sequences like the one above, where the relationship between each term and the following term is that they always have the same ratio (in this case 2), are called **geometric sequences**. If we represent the starting number by *a* and the common ratio by *r*, then we can state a rule for finding the *n*th term of any geometric sequence, as shown in the last column of Table 6.8.

TABLE 6.8

1st	2nd	3rd	4th	5th	*n*th
a	$a \cdot r$	$a \cdot r^2$	$a \cdot r^3$	$a \cdot r^4$	$a \cdot r^{(n-1)}$

Sequence 3 Let us examine the third sequence: 4, 9, 19, 39, 79, ... Some students call this a hybrid sequence in that it is not "just like" either of the ones above. That is, the "bad news" is that we can't use either formula just developed. The "good news" is that we can use problem-solving tools and reasoning to determine the *n*th term. Try it yourself before reading on. . . .

One way of describing what repeats is to look at the relationship between each term and the following term. That is, what do you have to do to each term to produce the next term? One way of describing this is that you double each term and then add 1. So the next number in the sequence is 159. We can represent this relationship between each term and the following term as $2n + 1$. This is nice, but it won't help us to determine the value of the 20th term or the *n*th term. Thus we have to look further. Here is where number sense and intuition come into play in mathematics. You might have realized that, after the first term, all of the terms end in 9. What if we wrote out a similar sequence—one in which each term is 1 more than the terms of our sequence. This has been done in the third row of Table 6.9. What do you see?

TABLE 6.9

	1st	2nd	3rd	4th	5th	*n*th
Original sequence	4	9	19	39	79	
Broken down	4	$4 \cdot 2 + 1$	$9 \cdot 2 + 1$	$19 \cdot 2 + 1$	$39 \cdot 2 + 1$	
Similar sequence	5	10	20	40	80	
Broken down	5	$5 \cdot 2$	$5 \cdot 4$	$5 \cdot 8$	$5 \cdot 16$	
Using exponents	5	$5 \cdot 2$	$5 \cdot 2^2$	$5 \cdot 2^3$	$5 \cdot 2^4$	$5 \cdot 2^{(n-1)}$
Original sequence	$5 - 1$	$(5 \cdot 2) - 1$	$(5 \cdot 2^2) - 1$	$(5 \cdot 2^3) - 1$	$(5 \cdot 2^4) - 1$	$5 \cdot 2^{(n-1)} - 1$

By breaking each number apart $(5, 5 \cdot 2, 5 \cdot 2 \cdot 2, 5 \cdot 2 \cdot 2 \cdot 2, 5 \cdot 2 \cdot 2 \cdot 2 \cdot 2)$ or by using exponents (see the fourth and fifth rows of Table 6.9). We can see that the nth term of the original sequence (4, 9, 19, 39, . . .) is simply 1 less; that is, $5 \cdot 2^{(n-1)} - 1$.

Since the only constant is change, we need ways to consider change. Slope is a very common way to think about change, especially in early algebra. Not all change can be described as a linear relationship as seen in the sequences 2 and 3 in the last investigation.

6.4 Exercises

1. Consider the graph below, which shows the depreciation of a copy machine over 5 years.
   a. What is the value of the machine when it is 2 years old?
   b. By how much does the machine decrease in value in 1 year?
   c. Predict when the value of the copy machine will be zero.

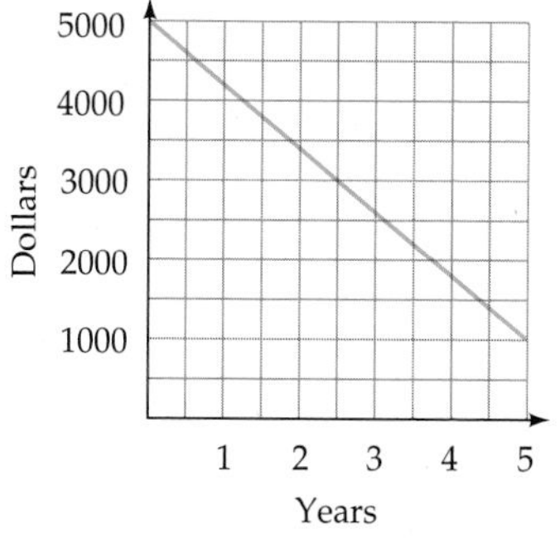

2. Which of the graphs below best matches the situation of a child swinging on a swing, where the independent variable is time, and the dependent variable is height?

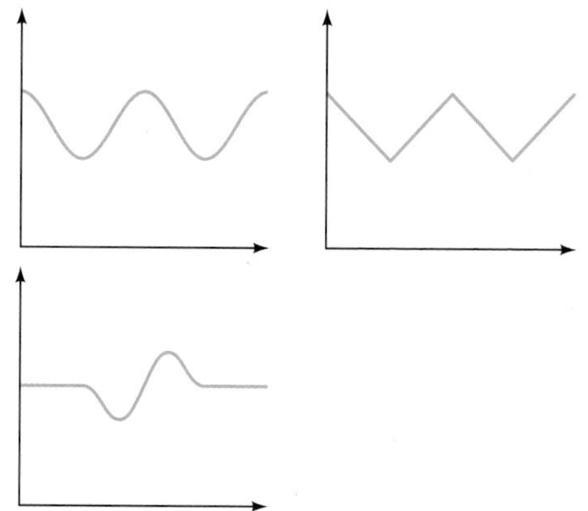

3. Which of the graphs below best matches the situation of a child swinging on a swing, where the independent variable is time, and the dependent variable is speed?

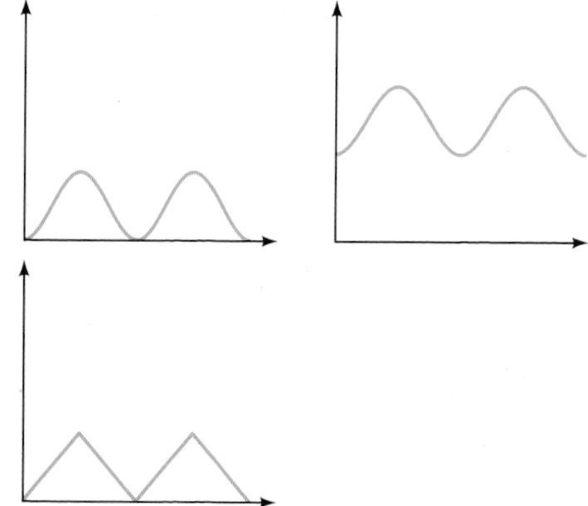

4. A family of four loves popcorn and has made a large batch to eat while they watch a movie. Sketch four graphs to show the amount of popcorn in each person's bowl over time. Dad eats the popcorn steadily. Mom eats steadily but at a slower rate. Emily eats at about the same rate as Dad but refills her bowl when it is half empty. Josh eats slowly and eats only half of his popcorn.
5. Sketch a graph for this situation: Pierre is driving on the highway at a steady rate. He passes a car and then resumes his previous speed. Then he sees a car pulled over for speeding and slows down for a while, before later resuming his original speed.
6. Sketch a graph for the speed of the space shuttle from the time of lift-off until its maximum speed.

7. Below is a graph of someone's drive home from work. Write the story.

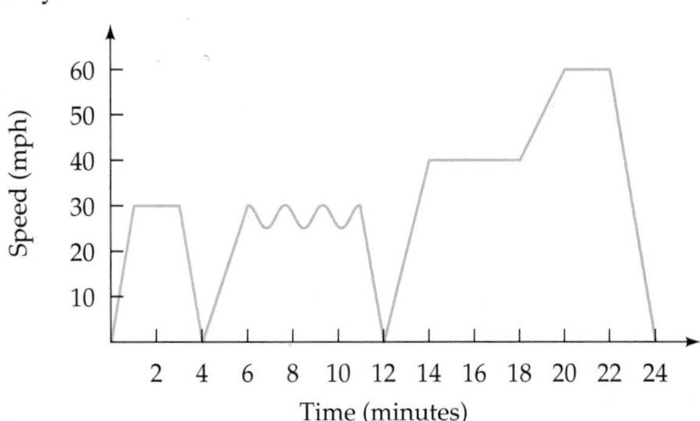

8. Below is a graph of a 400-meter race with three racers, A, B, and C. Write the story.

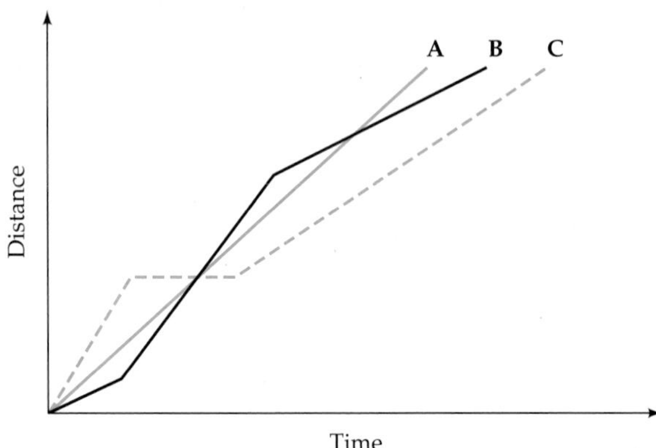

9. Below are several graphs. Write a story for each graph.

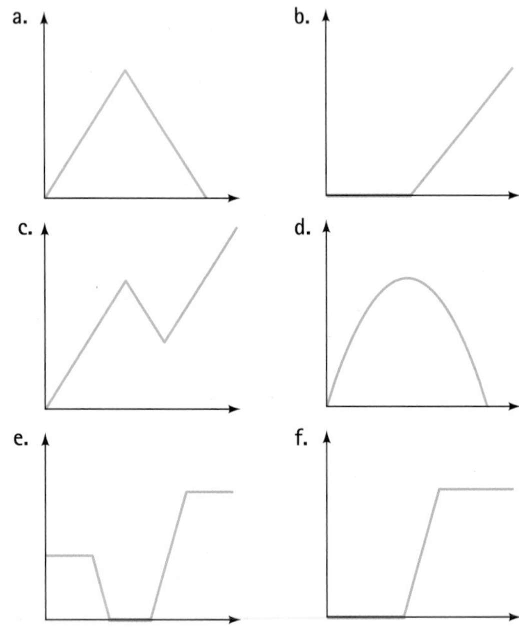

10. In virtually every school, someone raises the U.S. flag every day. Each of the graphs below represents a picture of the flag being raised. Translate each of the graphs into words. Which graph do you think is most realistic? Explain your reasoning.

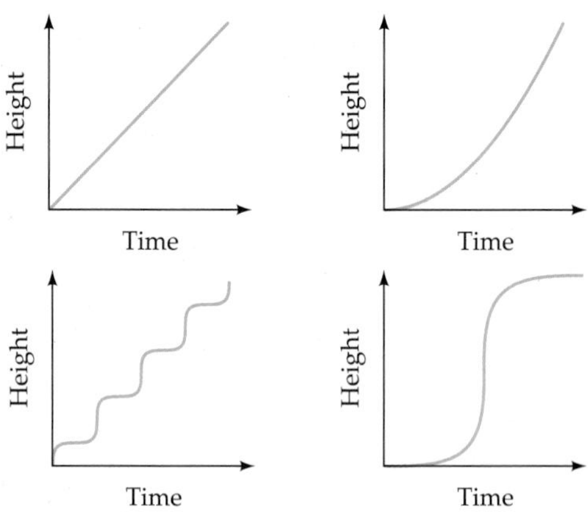

11. The American with Disabilities Act requires a ramp for wheelchair use to have a maximum slope of $\frac{1}{12}$.
    a. Explain what this means.
    b. If the height from the ground to the entry to a building is 6 feet, how far out from the building will the ramp need to start?

12. A road sign says there is an 8% incline. Explain what this means.

13. Which of the following are arithmetic sequences, which are geometric sequences, and which are neither? Describe each sequence and the change in each.
    a. 5, 10, 20, 40, 80, …
    b. 3, 7, 11, 15, 19, 23, …
    c. 1, 4, 9, 16, 25, …

14. Write an arithmetic sequence of numbers. What is the constant change (slope) of your sequence? Write an equation for the *n*th number in the sequence.

15. Write a geometric sequence of numbers. Describe the change.

16. Why would a roof in Alaska need to have a steeper slope than a roof in Florida?

17. A student describes the rate of growth of a plant as follows: "My plant didn't grow at all for 3 days, then it grew slowly for the next 2 days, then it grew faster for the next 6 days, then the growth slowed down again." Sketch a graph that shows the plant growth. Is this a constant rate of change? Explain.

18. In 2013, teachers in a certain state were given a raise of 1%. The rate of inflation that year was 1.2%. Analyze these rates of change. How could teachers make the case that they really didn't get a raise at all?

19. In the first two weeks of 2014, there were roughly 298,000 births and 124,000 deaths worldwide (according to worldometers .info). What do these numbers tell us about the rate of change in world population during these two weeks?

LOOKING BACK on chapter six

QUESTIONS TO SUMMARIZE BIG IDEAS

1. How do patterns relate to algebraic thinking?
2. What types of relationships are functions?
3. What does the equal sign mean?
4. How can balance scale models and Singapore bar models help us to solve algebraic equations?
5. Why do we "flip" the inequality when we divide by -2 to solve $-2x > 6$?
6. Describe different types of change.
7. What is the meaning of the slope of a line?
8. What parts of this chapter are less clear to you at this point? What will you do to clarify those ideas?
9. Look back at the Mathematical Practices of the CCSS. In what ways did you engage in those practices during this chapter?

CHAPTER  SUMMARY

1. Algebraic thinking begins early in elementary school as students see patterns and consider equations like $3 + \_\_\_ = 5$.
2. Some relationships are functions, where we input a number and get one number as the output.
3. Variables can be used in different contexts to help us communicate mathematically.
4. The equal sign can be thought of as the fulcrum of a balance scale, where each side has to be the same.
5. We can use balance scale models and Singapore bar models to make sense of solving linear equations.
6. When solving linear inequalities, we need to flip the inequality when we divide or multiply by a negative number since we are moved to the opposite side of zero on the number line.
7. Graphs, equations, and bar models can help us to model and make sense of real-life problems.
8. Analyzing change is another important piece of algebraic thinking. When there is a constant rate of change, it is the slope and a linear relationship.
9. Arithmetic sequences have a constant rate of change, because they increase by the same amount each time. Geometric sequences are changing exponentially.

BASIC CONCEPTS

Section 6.1: Understanding Patterns, Relations, and Functions

Common Core State Standards **287**

Patterns:

functions **292** — "What's my rule?" **292**

Section 6.2: Representing and Analyzing Mathematical Situations and Structures Using Algebraic Symbols

structures **296** — magic tricks **297**
variables **300**

Symbols:

equal **301** — equivalent **301**
inequality **302** — balance scale models **303**
Singapore bar models **304** — linear equations **304**
linear inequalities **305**

Section 6.3: Using Mathematical Models to Represent and Understand Quantitative Relationships

mathematical models **309** — graphs **310**
systems of simultaneous equations **313**

Section 6.4: Analyzing Change in Various Contexts

change in graphs **317** — slope **319**
arithmetic sequences **321** — geometric sequences **322**

REVIEW EXERCISES chapter six

1. A school play charges \$2 for students and \$5 for adults. For the three days of the play, 458 tickets were sold and \$1342 was raised. How many student tickets were sold?
2. A small factory makes three-legged stools and four-legged tables. Last month this factory used 100 legs to build 3 more stools than tables. How many stools did the factory make?
3. Monarch butterflies flap their wings about 12 times a second in flight. How many times do they flap their wings in one hour of flight?
4. A letter was posted that was covered with 10-cent stamps and 5-cent stamps. There were 12 stamps, and the total postage was 70 cents. How many of each stamp were on the letter?
5. Find the next term, the 20th term, and the nth term in these sequences
   a. 7, 13, 19, 25,
   b. 1, 2, 4, 8, 16,
   c. 2, 8, 26, 80,
   d. 5, 11, 23, 47,
6. What does it mean to say things are equal?
7. A function is described by the formula $f(x) = 3x - 2$.
   a. Draw a table, sketch a graph, or draw an arrow diagram to represent this function.
   b. Find $f(6)$.
8. The function that converts American dollars to Transylvanian rubles is given by the following hookup: M 3, then D 20, then A 13. Use the hookup to convert \$100 into rubles.
9. Determine how many squares it will take to make the nth figure in this growing pattern:

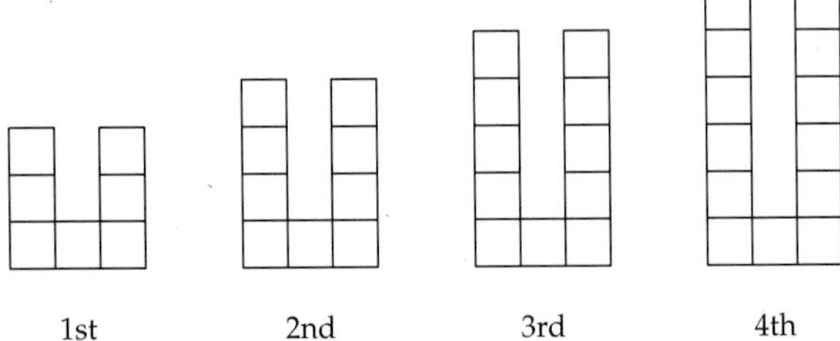

10. Determine how many squares it will take to make the nth figure in this growing pattern:

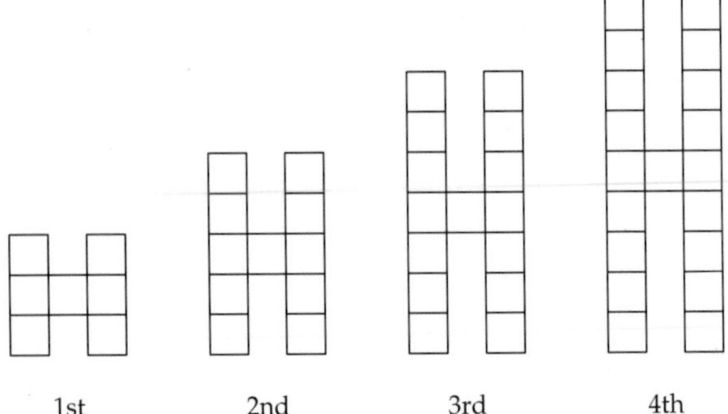

11. The first five rectangular numbers are 2, 6, 12, 20, 30.
    a. What is the 10th rectangular number?
    b. Write a rule for finding the nth rectangular number.
12. The first 6 items in a sequence are: 10, 14, 18, 22, 26, and 30.
    a. What is the 10th number in the sequence?
    b. Write a rule for finding the nth number in this sequence.
13. A cellular phone company offers two plans:
    a. \$30 a month for 300 anytime minutes plus 10 cents a minute for additional minutes
    b. \$40 a month for unlimited anytime minutes

    Fred is not sure which plan he should pick. Tell him the conditions under which each plan would be better for him.
14. Which of the following graphs best matches the situation of the temperature of a cup of coffee that is initially very hot and then sits on a desk? Explain your choice, and also explain why the other choices are not valid representations of the situation.

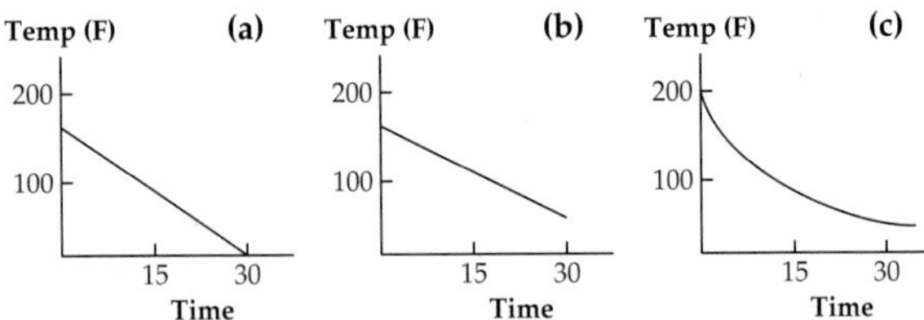

15. Sketch a graph for the following situation. A man walked steadily for awhile and then began climbing a hill. As he walked up the hill, he walked more and more slowly until he reached the top of the hill. He walked more and more quickly as he walked down the hill and then maintained a steady pace once he made it to the bottom of the hill.
16. Write a story for which this graph is an accurate representation. Select and label the independent and dependent variables, for example, time and distance.

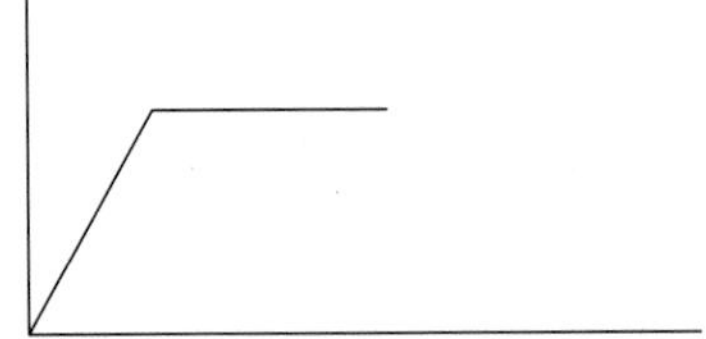

Uncertainty: Data and Chance

SECTION 7.1 The Process of Collecting and Analyzing Data

SECTION 7.2 Going Beyond the Basics

SECTION 7.3 Concepts Related to Chance

SECTION 7.4 Counting and Chance

We have spent a considerable amount of time examining operations with numbers and different ways to represent amounts. We can add, subtract, multiply, and divide amounts; we can represent amounts and relationships between amounts with whole numbers, negative numbers, fractions, decimals, and percents. In this chapter, we will focus on another way in which we use numbers—to understand and deal with uncertainty.

> *In this world, nothing can be said to be certain except death and taxes.*
>
> —Benjamin Franklin

We expend a lot of energy trying to understand the uncertainty and chance in our lives. Let us consider a few examples of uncertainty and chance. What do the examples below have in common? Think and then read on. . . .

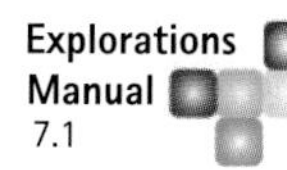

- Newspapers — Which candidate do you prefer? Are you for or against ______?
- Scientists — Are the levels of mercury in fish rising? How well are new AIDS drugs working?
- Educators — What is the best method for predicting success in college? How likely are our graduates to get a teaching job?
- Business — How much should we charge for this product? How many miles should we guarantee the tire for?
- Individuals — Which car should I buy?

As you can see from these examples, we collect data for a variety of reasons: to make predictions, to make decisions, to answer questions, and to better understand aspects of our life.

All of the examples above have two things in common: They all involve **uncertainty**, and they all involve collecting and analyzing data, which increases the chances that the decision or prediction will be a good one despite this uncertainty.

We often refer to that field of mathematics that focuses on collecting and interpreting data as *statistics* and to that branch of mathematics that focuses on determining the chance of something happening as *probability*. However, as you may have discovered from the Chapter 7 explorations, much of our interpretation of data involves chance, and much of our understanding of chance involves collecting and analyzing data. This is one of the reasons why we have chosen to examine these ideas within one chapter rather than in two separate chapters called "Statistics" and "Probability." Furthermore, just as you found that the words *carrying* and *borrowing* mask the essential connectedness of these two procedures, so, too, the terms *statistics* and *probability* tend to mask the essential connectedness of these two fields.

For example, when we say that the average annual snowfall in Buffalo, New York, is 92.2 inches,[1] that does not mean that you should expect 92.2 inches of snow in Buffalo this year (or 92 or 90 inches, for that matter). There are variations in any phenomenon; this number represents an average from data collected over 42 winters. We cannot say with certainty that we will get 92 inches of snow this year. We cannot even say with certainty that Buffalo will get between 82 and 102 inches this year. However, the average gives us an idea of what we might expect. In this sense, it tells us that when we look at the snowfall data for Buffalo, this amount is more common—it is in the middle of the data that have been collected.

Similarly, when we say that the chance of flipping a coin and getting tails is 50%, this does not mean that we will get 5 tails if we flip a coin 10 times. Even with this event, with less *variation* than snowfall in Buffalo, there is variation and a lack of certainty.

DATA INTERPRETATION AND CHANCE IN SOCIETY

There is much evidence that many students understand the fundamental ideas of statistics and probability even less well than they understand other ideas in the elementary school classroom. Not surprisingly, misuse of statistics—by politicians, special interest groups, entrepreneurs, and others—is pervasive. Sometimes the misuse is deliberate—for example, to manipulate decisions and opinion. However, often the misinterpretation of data occurs either out of ignorance or because the problem is very complex.

Toward this end, we find that students need to experience statistics and probability more directly, and so the investigations have been fashioned with that in mind. The NCTM also urges more hands-on work with all aspects of what we call statistical analysis:

> Teachers should have a variety of experiences in the collection, organization, representation, analysis, and interpretation of data. Key statistical concepts for all teachers include measures of central tendency, measures of variation (range, standard deviation, interquartile range, and outliers), and general distributions. Representations of data should include various types of graphs, including bar, line, circle, and pictographs as well as line plots, stem-and-leaf plots, box plots, histograms, and scatter plots.
>
> (From NCTM *Professional Standards*, p. 136, copyright 1991 by the National Council of Teachers of Mathematics.)

In the Common Core State Standards, *Measurement and Data* comprise a content strand in K–5. Analyzing data from a variety of graphs is the focus on data in elementary schools. In the sixth to eighth grades, there is a separate content strand called *Statistics and Probability*, where these topics are explored in more depth.

There are a number of words used in this chapter that are used in everyday life but that have a slightly different or more precise meaning in mathematics—for example, *variation, distribution, uncertainty, population, sampling, reliability, combination,* and *event*. To get too technical is to lose readers; to stay too informal is to cause confusion. Therefore, we have made an attempt to avoid both extremes.

SECTION 7.1 The Process of Collecting and Analyzing Data

What do you think?

- How can the way a set of data is represented affect the way it is interpreted?
- What do we mean by average?
- Why do mathematicians discourage reporting just the average when describing a population?
- Consider a graph that you have recently encountered. Was it accurate? How could you tell?

When we look at this thing we call statistics, we are talking about a process of moving from a question or a problem to gathering and analyzing data that will answer the question or solve the problem. There are four basic components in the process of using statistics to answer a question:

FOUR BASIC COMPONENTS

1. Formulating the Question
   The process begins with a question that can be answered by collecting data, for example, what are the chances that you will get a teaching job after graduating? Then we figure out how to change this question into a "statistical" question that is specific enough that we can collect useful data but yet does not so simplify the original question that the data aren't useful. We are *creating* data as much as we are *collecting* data.

2. Collecting the Data
   We need to decide what data we want to collect, and we need to design a plan for collecting appropriate data. That is, we want the data to be accurate and useful. There are many things we have to think about to ensure that we have accurate data. There are many methods used to collect data: observation, surveys, questionnaires, experiments, interviews, and simulations.

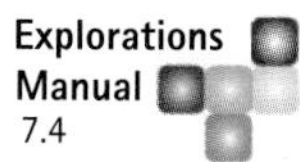

3. Representing and Analyzing the Data
   Raw data is simply a pile of numbers, like all the materials for a building project—nails, boards, shingles, etc. We need to organize the data and how we organize the data depends on what we want to know. We can choose from many kinds of graphs and numerical methods, but there are no hard and fast rules for deciding which ones are most helpful with respect to our question. Each one is more useful for some situations and less useful, or even invalid, in other situations.

4. Interpreting and Presenting Our Results
   We interpret these graphs and numbers and turn them into conclusions. We determine what answers they give us and what answers they do not give us.

We will walk through the four components of this process with several different questions. Our first question involves **categorical data**, that is, data that are not numbers but categories. According to the Common Core State Standards, first-graders "organize, represent, and interpret data with up to three categories." Second-graders draw bar graphs and picture graphs with up to four categories.

INVESTIGATION 7.1a What Is Your Favorite Sport?

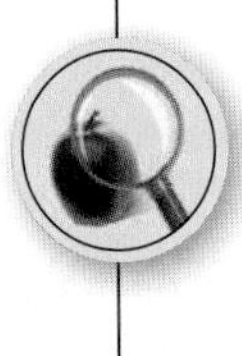

A common introductory activity with children is to have them formulate questions and then collect data with themselves as the population. Let's say this question has been presented: What is your favorite sport? What do we need to do to clarify this question so that when we ask everyone to answer the question, we will have useful data? Think and then read on. . . .

Explorations Manual 7.2

LANGUAGE

The term **population** is used in statistics to refer to the complete set of people or things which we are studying. In Section 7.2, we will investigate how we can make conclusions about a population based on data from a *sample* of that population.

OUTSIDE THE CLASSROOM

Some real-life questions with categorical data include:

- How do unemployment rates for people with less than a high school education, just high school, and college compare?
- How do infant mortality rates for different countries differ?

OUTSIDE THE CLASSROOM

An example of how important wording can be was a poll which concluded that 22% of American adults surveyed thought it was possible that "the Nazi extermination of Jews" never took place. These data were shocking to many people. However, when the pollsters were questioned, it was found that the question was worded this way: "Does it seem possible or does it seem impossible to you that the Nazi extermination of the Jews never happened?" The wording, with a double negative, certainly confused some people. Another poll was done where the question was asked: "Does it seem possible to you that the Nazi extermination of the Jews never happened, or do you feel certain that it happened?" In this poll, only 1% of Americans thought it was possible that the Holocaust never happened, while 8% were unsure.[2]

DISCUSSION

Refining the question Several issues emerge when we consider this question, for example, do you mean favorite sport to play or to watch? If we don't clarify this when we ask the question, people will not be answering the same question. Some will be saying their favorite sport to play and some will be saying their favorite sport to watch.

Let's say we decide to ask: What is your favorite sport to watch? This is still not 100% clear: Favorite to watch on TV or to watch in person? My favorite to watch in person is baseball, but my favorite to watch on TV is football. There are even more questions, what I call "yeah buts"; for example, "yeah, but what is a sport?" Is bowling a sport? Golf? Chess? Poker? In our clarification of the question, it is not so much what is right, but what do we want.

We also need to think about how to ask the question. For example, let's say we decide to ask about favorite sport to watch on TV. Do we want it open-ended, where people can give whatever sport they consider to be their favorite, or do we ask them to select among a list, for example, football, soccer, basketball, and baseball?

As you can see, we can easily get overwhelmed with the many complexities of a seemingly simple question. When this happens, it is helpful to go back to why you are asking the question. For example, if you are just curious as to what people might think of as sports, then you would make it as open-ended as possible. If you were asking this question just before the summer Olympics, you might ask: Which of the summer Olympic sports do you most like to watch? You might also consider having "none" as a legitimate response.

All this thinking in the first step of formulating the question!

Collecting the data With respect to collecting the data, we need to consider what additional data we want. For example, are we interested in investigating differences between boys and girls? If this were a high school poll, would we want to see if there was a difference among freshmen, sophomores, juniors, and seniors?

We also need to think about the process. For example, do we write the questions down and have students write their answer? Do we want to ask them in person? As you might expect, asking in person might change the data. What if the asker is a basketball player? What if you ask three kids at a table and the first two say football?

Representing the data At some point, the question is written, the format determined, and the data collected. Let's say the question asked was: What competitive, team sport do you most enjoy watching?

A common and simple way to display the data is a **frequency table**. This is simply a table showing the number of times (called the *frequency*) that each category occurred.

Sport	Frequency
Football	5
Basketball	3
Baseball	2
Soccer	4
Rugby	1
Lacrosse	2

How might we represent these data? In this case, the two most common representations are a bar graph and a circle graph.

Bar graph A **bar graph** is used to represent situations where the data are categories. The bar graph is pretty straightforward, though there are choices of order. We could arrange them randomly, alphabetically, or by popularity, depending on our preference. For example, we might put football, basketball, and baseball in order if we want to see how many of the class has one of the "big three" as their favorite sport (Figure 7.1).

MATHEMATICS

If you enter these data on a spreadsheet, and the football frequency (5) is in cell B2, then you could enter this equation in cell C2: = B2/17*100. That is, you are telling the computer to take the number in cell B2, divide it by 17 (to get the fraction) and then multiply it by 100, to get the percentage, which is 29. You can then click cell C2 and drag it down to have the computer automatically do the same for the rest of the data. If you have a calculator, you can divide 100 by 17 and then press M+. Then you don't have to do the entire computation each time. You can press 5 × MR = to get the percentage for football.

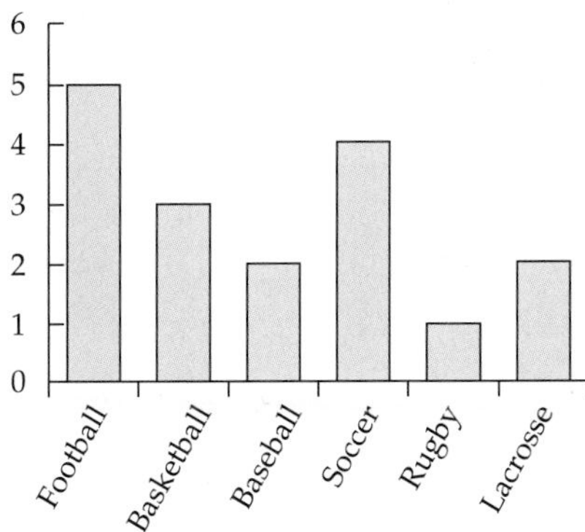

Figure 7.1

Circle graph To draw a **circle graph**, we can turn the numbers into percents and then sketch a circle graph or we can turn them into degrees (fractions of a circle) and make the graph with a protractor. Can you convert the data into percentages and into parts of a circle? (A whole circle is 360 degrees.) Try this before reading on.

	A	B	C	D
1	**Sport**	**Amount**	**Percent**	**Degrees**
2	Football	5	29	106
3	Basketball	3	18	64
4	Baseball	2	12	42
5	Soccer	4	23	85
6	Rugby	1	6	21
7	Lacrosse	2	12	42
	Total	**17**	**100%**	**360°**

To convert to percentages, we first need to know the whole, which is 17 in this case. Thus, football = $\frac{5}{17} \times 100 = 29.41$. Since we are sketching, we can easily round this to 29%.

To find the degrees of the central angle of the sector of the circle, you multiply by 360. For example, the football slice is $\frac{5}{17} \times 360 = 106$ degrees.

As with bar graphs, there are no hard and fast rules for ordering the slices. Again, it depends on what we want to know. If you are sketching the graphs by hand, there are many ways to do this. Let me walk you through one way. First, you can partition the circle into fourths. Thus each quarter circle is 25%. Then, depending on the nature of the data, you can divide each quarter into thirds (so each third is about 8%) or into fourths (so each fourth is about 6%).

We begin with the largest (29%), which is 25% plus 4%, so it makes sense to take the second quarter and break it into thirds (8%) and then divide the first of those thirds in half (4%) [Figure 7.2(a)].

Basketball is next with 18%. We have the 4% slice left plus 8% plus 8% = 20%, so the basketball slice will be not quite the rest of this quarter circle. We can always check with our whole. That is, football plus basketball = 47%, so we need to be just under $\frac{1}{2}$, and we are.

Baseball is 12%. In this case, rather than finding a 12% slice, I find it easier to find where 59% is, because that is how much of the circle I need after these three sports. Since 59% is 50% plus 9%, we can partition the third quarter circle into thirds [Figure 7.2(b)]. We have 50 plus 8, and so our line is just a "bit more" than this.

Next, we look at soccer and see that we will now have 82% of the circle: 75% plus 7. So it makes sense to partition that last $\frac{1}{4}$ circle into fourths, 6, 6, 6, 6 [Figure 7.2(b)]. Soccer is thus about one slice past the 75% mark. We now have three 6% slices—6% for rugby and 12% for lacrosse. The final circle graph is shown in Figure 7.2(c).

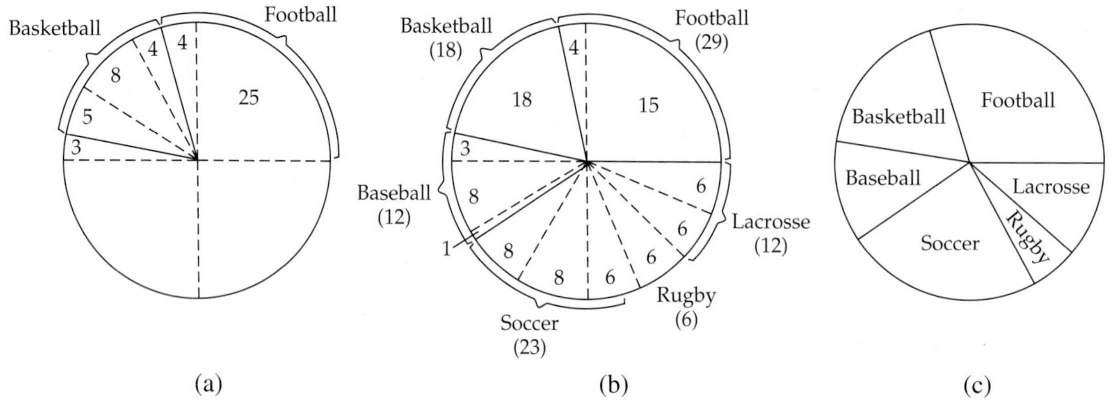

Figure 7.2

Stop for a moment and think about bar graphs and circle graphs. What are the advantages and disadvantages of each?

The bar graph tells you the number in each category (e.g., five people chose football as their favorite sport). However, you cannot tell immediately from a bar graph what percent each category represents, and this is one of the primary advantages of a circle graph. From the circle graph, we can quickly see that just over $\frac{1}{4}$ of the people named football, and we can make statements about combinations of categories, for example, almost $\frac{1}{2}$ of the people named football or basketball. One disadvantage of circle graphs is that they generally do not show the raw data, but rather the percentages, and so we lose the actual data. One cautionary note with circle graphs is that they are valid only when we have a whole for which it makes sense to find parts of, that is, percentages.

Analysis Now, we present our results. What we present depends on our intentions. We can simply present the frequency table and readers can see the number in each sport. The bar graph is a visual representation of the frequency table from which we can make various statements; for example, football and soccer were most popular. The circle graph enables us to make statements about what fraction or percentage of the whole each category or combinations of categories represent; for example, 59% chose the traditional "big 3" of football, basketball, and baseball. Or we could say almost $\frac{1}{2}$ (41%) chose sports that haven't been around as long in the United States.

That's quite a bit of thinking just for one simple question! Let's do another question where the data are numbers.

INVESTIGATION 7.1b How Many Siblings Do You Have?

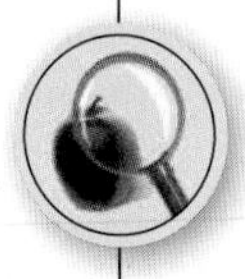

In this case, the response is not a category but rather a number. How would you formulate this question? What are the aspects that need to be addressed so that everyone will be answering the same question; that is, what are the "yeah buts" for this question?

Refining the question There are many considerations. For example: What is a sibling? Does adopted count? What about half-, step-, or foster siblings? What if someone had a

OUTSIDE THE CLASSROOM

Some real-life examples like this investigation would be:

- What is the size of the average American family? How is it different now from 50 years ago?
- How many computers does the average family have?
- What is the average family income?

sibling who died? Do we count that? There are no right answers to these questions. It depends on what you want to find out. If you want to know how many kids live full-time in the house, then you would ask how many siblings live with you all the time. If you wanted to know the number of biological siblings, you would ask for full and half-siblings. If you want to know how many people the child considers to be his or her siblings, you would ask: how many siblings do you have—people that you consider to be your brothers and/or sisters?

Collecting the data The data collection process for this question is fairly straightforward. However, we could expand the question to ask: How many siblings does your father have? Your mother? Of course, that could easily become complicated: what if the parents are divorced? What if the child feels closer to her step-father than to her biological father? *This* is the kind of complexity that is involved in virtually every statistical question that is investigated. Thus, one of the "big ideas" of this unit is for you to see statistics not as cut and dried, black and white, but as complex and inexact.

Representing the data Let's say we collected the data for number of siblings and here are our results.

0 1 4 0 1 2 3 1 7 3 1 3 1 2 1 0 12 1 2

A first step is to organize the data into a frequency table:

Number of siblings	Frequency (how many)
0	3
1	6
2	4
3	2
4	2
7	1
12	1

How might we represent these data, graphically and numerically? Make your own graphs and computations and interpretations before reading on. Really! You learn more by being more active readers.

DISCUSSION

Line plot One plot that is introduced in third grade is the **line plot**. A line plot for our data is shown below. What do you "see" from this graph? That is, what does it tell us with respect to our question that the raw data do not?

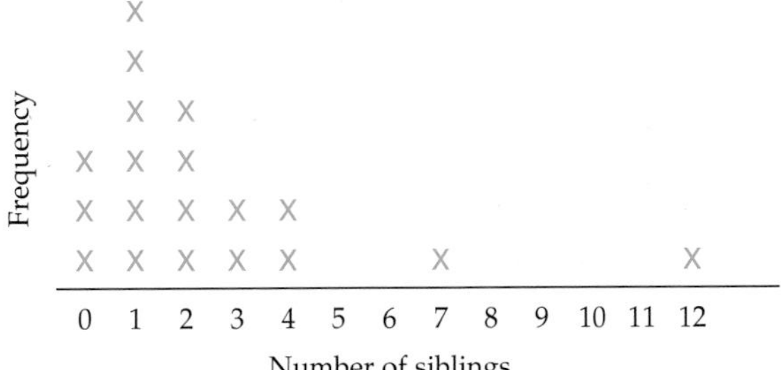

Figure 7.3

LANGUAGE

Some books use the term **dot plot** because they use dots or small circles instead of X's.

There are many ways to verbalize what we can interpret from the line plot. The data range from 0 to 12. We can see a **cluster** of data from 0 to 2. This could be verbalized as "most of the kids have 2 or fewer siblings." We see some **gaps** in the data, and we see some data that lie outside the rest. These are often called **outliers**. We could make quantitative statements about these data, for example, 68% of the students have between 0 and 2 siblings. We could represent this in fractional form also: almost $\frac{2}{3}$ of the children have 0, 1, or 2 siblings.

Histogram A **histogram**, which is introduced in the sixth grade according to the CCSS, can be used to summarize numerical data that are on an interval scale, either discrete or continuous. Look at the histogram below.

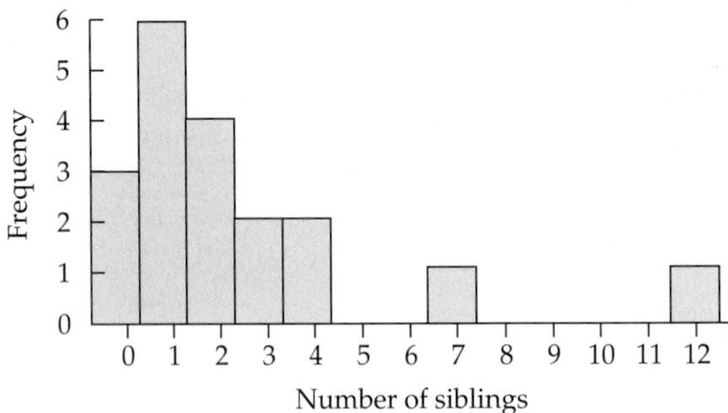

Figure 7.4

In the previous investigation, each bar represented the frequency of each category (that is, each sport). In this case, each bar represents the frequency of each number of siblings.

When some students make a histogram from these data, their histogram looks like the one below. Is this OK? Do you think both the first histogram and the one below are valid, that is, it's a matter of preference which one we use? Or do you think only one is valid? If so, why?

MATHEMATICS

What graph that we have studied most resembles the histogram?

Yes, the line plot. In one sense, the histogram is simply a line plot where we enclose the area taken up by the individual *o*'s or *x*'s with a rectangle.

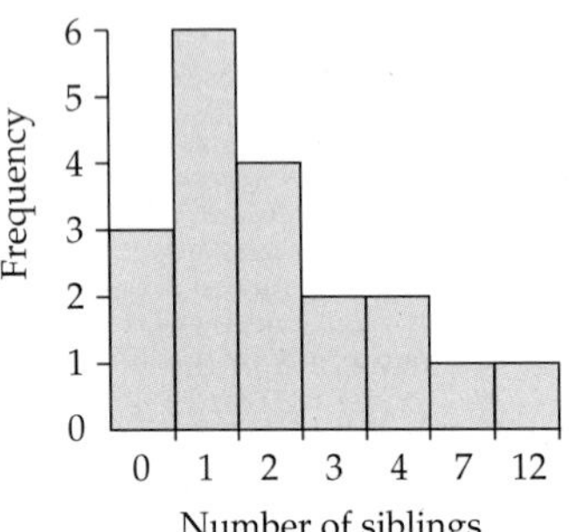

Figure 7.5

The first graph is valid and the second one is not. Do you see why? The second one is not considered valid because it hides or masks the *distribution* of the data. That is, there is a gap between 4 and 7 siblings and a gap between 7 and 12.

Another aspect of the histogram that is confusing for many children and for some of my students is the *y*-axis, which is labeled "Frequency." What does that mean?

It tells the frequency of each amount. That is, 3 students have 0 siblings, 6 students have 1 sibling, etc.

We could also make a *circle graph* from these data. Below are two circle graphs. One is in order of how many siblings. The other is in the order of greatest to least. Which do you prefer? Why?

How many siblings do you have?

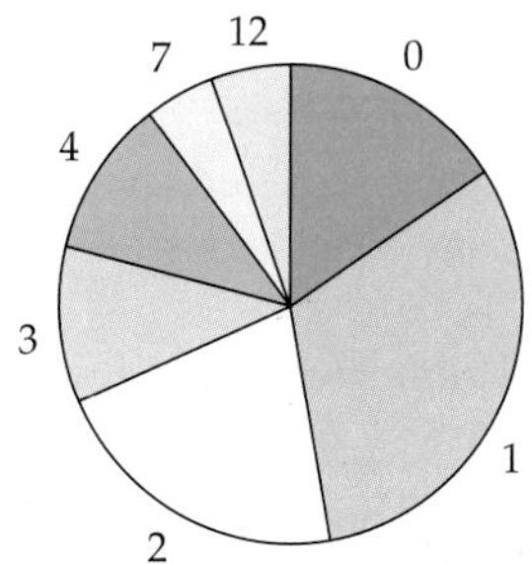

How many siblings do you have?

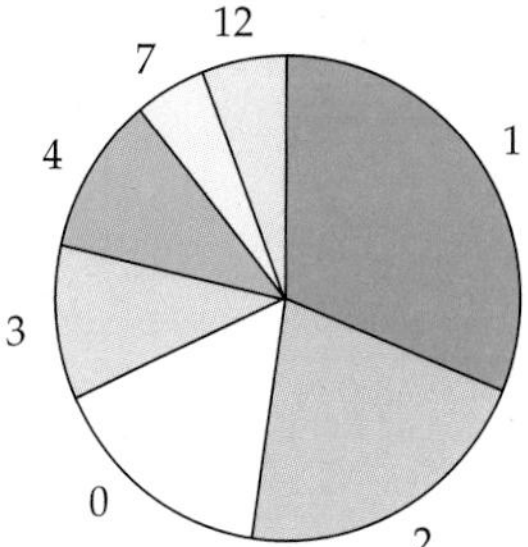

Figure 7.6

Both graphs are accurate; your choice in this case depends on personal preference and other questions you might want to ask. Having the number of siblings in order (like the graph on the left) makes it easier to answer questions such as: What fraction (or percent) of children have 0 or 1 sibling? What fraction (or percent) of children have more than 2 siblings? In contrast, the graph on the right makes it easier to answer a question such as: Which is the most common number of siblings?

MEASURES OF CENTER

With numerical data (compared to categorical data in Investigation 7.1a) we can examine *measures of center* (sometimes called *measures of central tendency*) to summarize the data by finding the middle or average. The measures of center are mean, median, and mode. We begin with how to find each:

Younger children, and even some adults, are bothered when they see the average not as a whole number. In the children's book *The Phantom Tollbooth,* the average family has 2.58 children, and 0.58 of a child tells Milo that being a fraction of a person is actually nice: "Every average family has 2.58 children, so I always have someone to play with" (Norton Juster, *The Phantom Tollbooth* [New York: Random House, 1961], p. 196).

Mean: Add each data value and divide by the number of data values.

Median: Arrange the data values in numerical order. The median is the middle data value. If there are an even number of data, then find the mean of the two closest to the middle. For example, if we have 3, 5, 6, 8, 12, and 15, there is no middle. Since 6 and 8 are closest to the middle, the median is 7.

Mode: The data value that occurs most often.

With our sibling data, we have

$$\begin{aligned}\text{Mean} &= (0 + 0 + 0 + 1 + 1 + 1 + 1 + 1 + 1 + 2 + 2 + 2 + 2 + 3 + 3 + 4 + 4 + 7 + 12)/19 = 2.47\\ &= [(3 \times 0 + (6 \times 1) + (4 \times 2) + (2 \times 3) + (2 \times 4) + 7 + 12]/19 = 2.47\\ \text{Median} &= 2\\ \text{Mode} &= 1\end{aligned}$$

In this case, the mean is 2.47, the median is 2, and the mode is 1. So which one is correct?

Actually, they are all correct. The more useful question is "which one is more useful?" and the answer depends on what we are looking for. If we are looking to answer the question, "what is the most frequent number of siblings?" then the answer is the mode. If we are looking for the middle data value (half above and below), then our answer is the median. If we are looking for what is normally considered to be the "average," then the answer is the mean. You will discover, as we consider more examples and in Exploration 7.2, that each one of these numbers has advantages and disadvantages.

Let's stop for a moment to consider the question: Why do we find the average in the first place?

If you are moving to a new town, you probably want to know the average price of a new home. When you are looking for a job, you might want to know the average starting salary of teachers in the state. In general, the average gives you a sense of the area where most of the data will lie, and it is generally close to the center of the data, which is why these three terms are called **measures of center**. That is, if you look at the neighborhood in which the mean, median, or mode lie, that will generally be where a majority of the data values are found.

Conclusions So what have we discovered about the number of siblings in this group of people? Below are three of many conclusions we could draw from this set of data:

The average number of siblings is either 1, 2, or 2.47, depending on which center we pick.

More than half the class has 2 or fewer siblings.

Two students come from much larger families than the rest of the class.

DEEPENING OUR UNDERSTANDING OF MEASURES OF CENTER

Measures of center are introduced in sixth grade, according to the CCSS. While most students have heard of mean, median, and mode before this course, "measures of center" is a new concept and therefore worth more consideration. Think back on this investigation (and explorations if you did them) and take a moment to note your responses to the following questions and then read on. . . .

What does a measure of center tell us about a set of data?

Why do we determine one of these centers in the first place?

What does it not tell us about a set of data?

Let us use an analogy to help answer these questions. If you saw a snapshot of a classroom, it would give you *some* information about the class. For example, you could determine the number of students, and you would see that the students are not sitting in rows. You might see the teacher standing at the front of the room and you might see a computer projection on the screen. Similarly, the mean, median, or mode gives us a snapshot of a set of data.

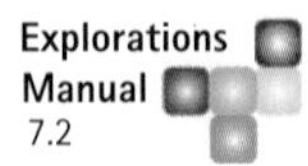

Whichever one is selected is often interpreted as representative of the data or typical of the data. You hear this on television or read it in the newspaper: the typical teenager . . . ; the typical American family. . . . However, there is danger in this characterization. Do you see why?

Let us return to the picture of the classroom to help answer this question. There is much that the snapshot *does not tell* about the classroom, for example, what percent of the time the teacher stands in front of the room or what tools are used (e.g., overhead projector, computer, etc.). Similarly, the mean, median, and mode do not tell us about the *distribution* of the data (the overall shape, clusters, gaps, outliers, the range).

Just as the snapshot of the classroom is an incomplete report of the class, so, too, the mean, median, and mode are an incomplete report of the data. Even worse is if someone generalizes from the picture, for example, concluding that the teacher is always in front of the classroom or that a computer is used most of the time. Similarly, as we will see as we examine more cases, a person who is not statistically literate can make erroneous generalizations from just seeing the mean, median, or mode.

To summarize: Measures of center are simply one of many parts of an analysis of data. At best, they present an incomplete picture. At worst, they can lead to an erroneous sense of the set of data. Thus, a responsible report on a set of data will not give just the mean, median, or mode, but rather more information.

> **MATHEMATICS**
>
> We want to underscore the term *measure* in talking about average. We tend to think of measuring in the sense of weight, distance, and volume. However, when we determine mean, median, and mode, we are essentially giving one measure of the set of data. There are many measures of a room: floor area, perimeter, volume, surface area. So, too, there are many measures of a set of data.

Pros and Cons of Each Measure In some cases, we want to know the mean. If you took five tests in a course, your instructor would generally determine your average score by using the mean—adding up the scores and dividing by 5. The 2003 *Time Almanac*[3] reports that the mean number of people in U.S. households in 2000 was 2.62.

In some cases, we want to know the median. If you determined the height of all the students in your class and ordered the numbers from smallest to greatest, the number in the middle would be the median. The *Statistical Abstract of the United States 2002*[4] reports that the median age in the United States in 1991 was 35.3 years.

In some cases, we want to know the mode. The mode is often used when the characteristic we are studying is not a number. For example, if you were to collect data on what state the students in your class were born in, the mode would be the state that occurred most frequently. We often use the word *typical* with mode. For instance, the typical professional basketball player is over 6 feet.

One of the reasons for determining the average of a set of data is that one number or one phrase can give a quick summary of the data. The mean, median, and mode are all candidates to be considered as a representative of the data. In some cases, the mean, median, and mode are very close, but sometimes they are not. Let us connect this notation of representative of a population to a nonmathematical situation. When I was in the Peace Corps in Nepal, the country director told new volunteers that we needed to think about our behavior because we were representatives of the United States. In one sense, the director was referring to average. Because the people we worked with had had little or no previous contact with Americans, they would probably see whatever we did as typical. If we wore blue jeans or drank alcohol, many people would assume that all (or at least most) Americans wear blue jeans and drink alcohol.

INVESTIGATION 7.1c Going Beyond a Computational Sense of Average

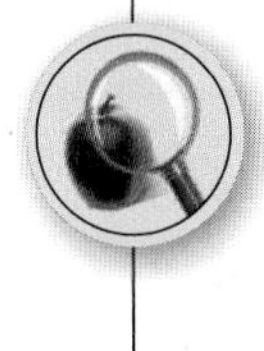

This investigation is designed to help you come to a deeper understanding of the meaning of the mean.

Imagine that five elementary school children were asked how many movies they saw in the past year, and they responded: 7, 2, 9, 8, and 4. First, write down what you think the mean tells you about a set of data. . . .

Explorations Manual 7.3

Figure 7.7 is one physical representation of the data, using pennies to represent each movie seen. Figure 7.8 is a bar graph where the bars are horizontal.

Do not compute the mean. Rather, draw a vertical line across the standard bar graph where, on the basis of your current sense of what the mean is, you "feel" the mean will be. Now, get some pennies and make the "graph" shown in Figure 7.7. If you don't have pennies, other coins or small objects will do.

Now move the pennies so that all the bars are the same length. What did you just learn about the mean? Now read on. . . .

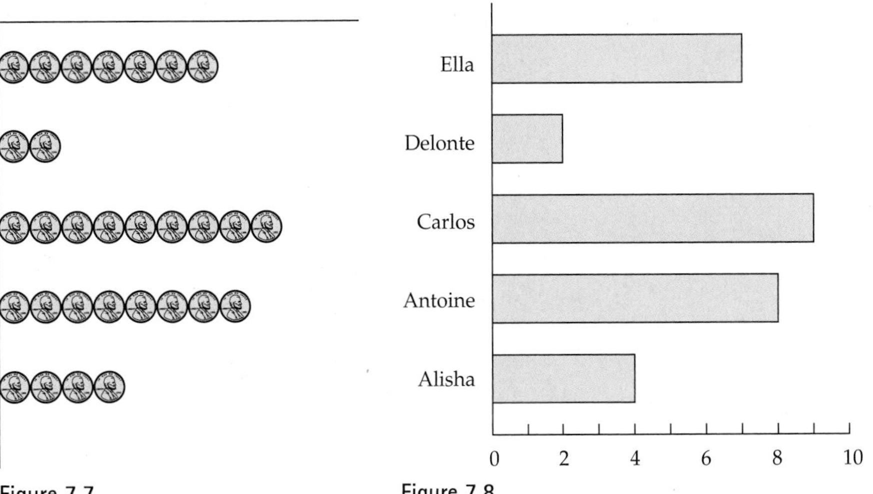

Figure 7.7

Figure 7.8

DISCUSSION

The mean can be viewed as the number you get when all the values are leveled off. In this case, if we "give" values from the larger amounts to the smaller amounts until all the amounts are the same, then the length of the bars and the number of pennies in each row are all the same. This conception of the mean is often referred to as the *fair share conception.*

ANOTHER INTERPRETATION OF MEAN

The mean is also the center of gravity of a set of data; this can also be described as the *balance point* of the data. Think of a seesaw. If two people of the same weight sit the same distance from the center, it balances. If one person sits farther from the center, the seesaw will not balance (Figure 7.9). If you imagine each of the children in this investigation sitting on a seesaw at the number corresponding to the number of movies they saw, the seesaw would balance at 6, the mean. That is, the two persons at 2 and 4 will balance the three persons at 7, 8, and 9 because the 2 and the 4 are farther from the center than 7, 8, and 9 (Figure 7.10).

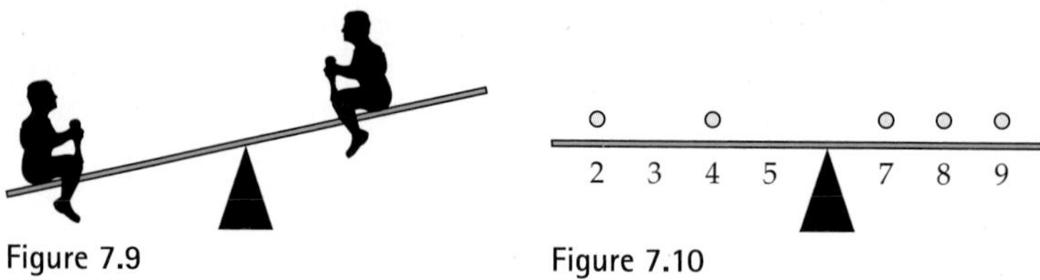

Figure 7.9

Figure 7.10

Let us now move on to another question and another set of data.

INVESTIGATION 7.1d How Many Peanuts Can You Hold in One Hand?

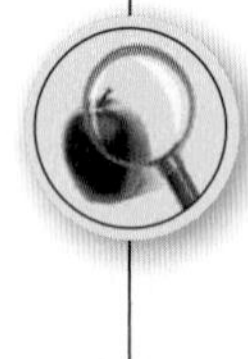

I have done this investigation with my students and with local elementary school children. You might do this also!

Refining the question This question is pretty straightforward. However, the collection process is a bit messy!

Explorations Manual 7.4

OUTSIDE THE CLASSROOM

Similar, real-life questions include:

- How many items (shoes, shirts, gloves) can a person make in one day?
- How long will this battery last?

Collecting the data Most of the leaders in statistics education argue that one of the big ideas of statistics is **variation**. This has to do with the fact that most of the numbers we use when collecting and analyzing data are not the same. Sometimes they are not the same because of the **natural variation** among individuals. Then there is variation that often happens when we take repeated measurements of an individual or an object. In this investigation, when a person grabs a handful of peanuts, they will not get the same number each time. When I first did this investigation, I did it five times and I got 32, 28, 22, 31, and 35 peanuts. This kind of variation is called **measurement variation**. We expect natural variation. However, when collecting data to answer a question, we want to minimize measurement variation. In this case, we have to think about the data collection process so that we minimize the measurement variation. Thus, we have to standardize the procedure for collecting the data.

How might you describe the procedure so that everyone is doing the same thing?

When we tried this, some people scooped the peanuts, that is, they reached in with palm up and open. Then they closed their hand slightly and slowly brought their hand up. Others reached in with their palms down and grabbed. I found that if I groped around, I could feel when I had about as many peanuts as possible. So we had to standardize the procedure. We decided on:

> Reach in with your palm down and grab as many peanuts as you can. You have to raise your hand from the bag within three seconds. Move your hand so that it is above the table. Whatever peanuts fall onto the table will be counted.

We also considered other "yeah buts" that could affect the reliability (consistency) of the data; for example, there were some empty shells and there were double shells and single shells. To reduce this variation, we could empty the peanuts on a table and then put into a bag only those peanuts that consisted of double shells.

Representing the data In this investigation, we will examine the data from my students. In Section 7.2, we will also examine the data from the third-graders in order to answer the question: How many more peanuts can an adult hold than a third-grader? Here are our data.

18, 18, 20, 22, 22, 22, 22, 22, 23, 25, 25, 25, 25, 25, 26, 26, 27, 27, 30, 30, 32, 32, 37

What do you see? What graphs might help us to understand the shape of the data—how the data values are distributed, clusters, gaps, outliers, range, etc.? What measures of center are more useful?

As you are coming to see, the question is not which graphs and centers are right, but which are more useful. In the world beyond the classroom, people often "play with the data." That is, they try different representations to see what light they shed on the question being asked.

Let us consider a line plot as shown below. What does this representation tell us?

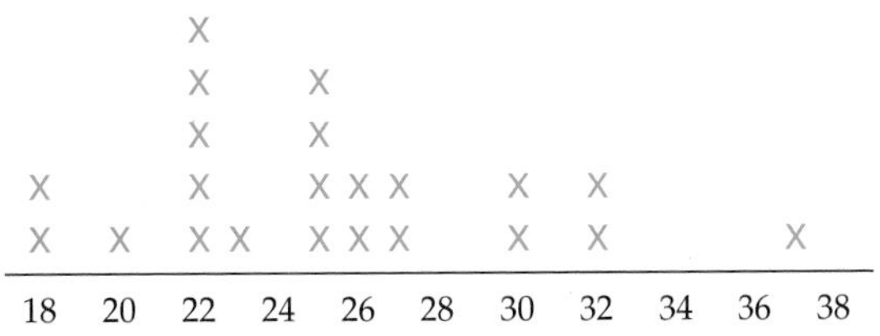

```
          X
          X     X
          X     X
X         X     X X X     X   X
X   X     X X   X X X     X   X         X
--------------------------------------------
18  20    22  24  26  28  30  32  34  36  38
```

Figure 7.11

From the line plot, we can make many statements:

The data range from 18 to 37 peanuts.

The data are pretty well spread out. That is, there is no primary cluster as there was in the previous set. About $\frac{1}{2}$ of the students held between 22 and 26 peanuts.

The mean is 25.3. The median is 25. The mode is 22, well sort of. In this case, the mode is technically 22, but almost as many people held 25. So, 22 is not as "strong" a mode as 1 was for the number of siblings. We will discuss this idea more deeply in the next investigation.

The person who held 37 is a bit of an outlier in the sense that there is a fairly big gap between 37 and the next highest data value, 32.

Just as with the siblings data, we could make a histogram. However, in this case, the data are more spread out. When the data are more spread out, in order to help us interpret the data better, it helps to put the data into intervals. In this case, we can make a **grouped frequency table** for the data. For example, we could do the following:

MATHEMATICS

Remember when we first discussed addition in Chapter 3 and we distinguished between numbers that represent discrete amounts (e.g., peanuts, trucks, etc.) and continuous amounts (e.g., inches, ounces, minutes)? In this case, the data (number of peanuts) are discrete amounts; that is, we can count them.

Number of peanuts	Number of people
15–19	2
20–24	7
25–29	8
30–34	4
35–39	1

CLASSROOM  CONNECTION

Grade 5

The directions for the students are designed to reduce measurement variation. What other ideas can you think of to reduce variation for this activity?

Date ______________________ Time ______________________

LESSON 2·5 Estimating Your Reaction Time

Tear out Activity Sheet 2 from the back of your journal. Cut out the Grab-It Gauge.

It takes two people to perform this experiment. The "Tester" holds the Grab-It Gauge at the top. The "Contestant" gets ready to catch the gauge by placing his or her thumb and index finger at the bottom of the gauge, *without quite touching it.* (*See diagram.*)

When the Contestant is ready, the Tester lets go of the gauge. The Contestant tries to grab it with his or her thumb and index finger as quickly as possible.

The number grasped by the Contestant shows that person's reaction time, to the nearest hundredth of a second. The Contestant then records that reaction time in the data table shown below.

Partners take turns being Tester and Contestant. Each person should perform the experiment 10 times with each hand.

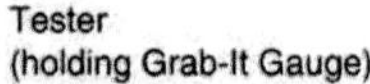

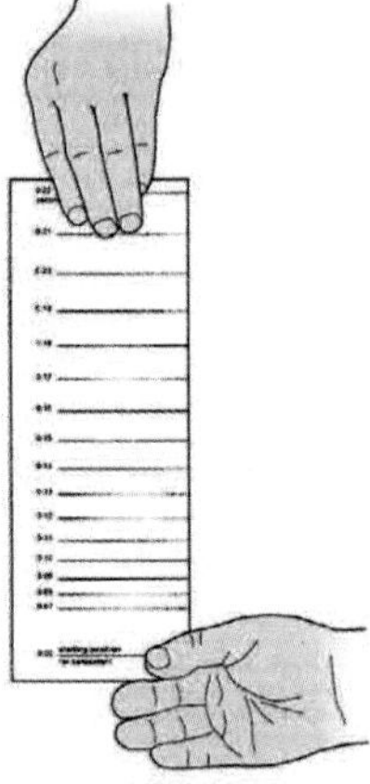

Reaction Time (in seconds)			
Left Hand		Right Hand	
1.	6.	1.	6.
2.	7.	2.	7.
3.	8.	3.	8.
4.	9.	4.	9.
5.	10.	5.	10.

40

From *Everyday Mathematics, Grade 5:* The University of Chicago School Mathematics Project: Student Math Journal, Volume 1, by Max Bell et al., Lesson 2-5, p. 40. Reprinted by permission of The McGraw-Hill Companies, Inc.

We can make a histogram from these data. What do we gain from putting the data into intervals and then making a histogram (Figure 7.12)?

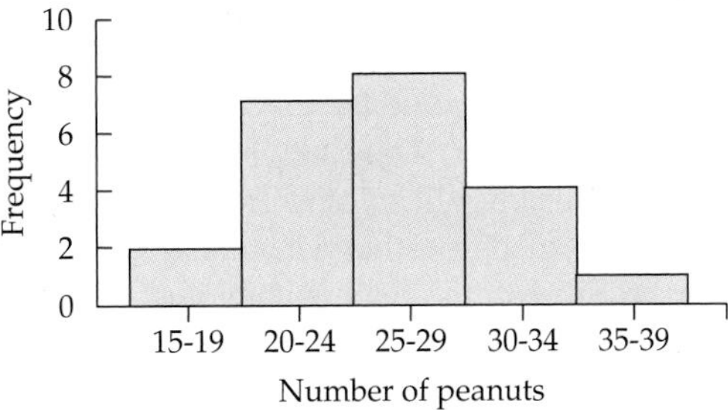

Figure 7.12

In this case, two conclusions we can make from the histogram are that the majority of the data are between 20 and 29 and that there are some below 20 and some above 29.

A question students often ask is: How did you know to put the data in these intervals (e.g., 15–19, 20–24, etc.)? The answer is rather surprising to many students. There is no right answer. You have already seen throughout the first six chapters that mathematics is not as black and white as you thought. Statistics is the least black-and-white field of mathematics.

In the case of intervals, the question is not "what is right?" but again "what is useful?" We generally group data using our base ten system (e.g., 0–9, 10–19, 20–29, etc.). We could have done that with these data, but it wouldn't have been terribly useful:

Number of peanuts	Number of people (frequency)
10–19	2
20–29	15
30–39	6

That is, we didn't need a histogram to see that the vast majority of cases were in the 20s. In many cases, we want to view our data with a *finer grain.* In this case, the finer grain consisted of making the intervals 5 instead of 10. However, we could have picked smaller intervals (e.g., 18–20, 21–23, 24–26). In general, as the interval size increases, the graph gives us less information about the data. We will investigate intervals more deeply in the next investigation.

In the *Explorations,* you have several opportunities to gather your own data. Most students find it more interesting to gather real data than to analyze some made-up set of data. So too with children. *Teaching Children Mathematics* and other publications have had many articles written by elementary teachers about students gathering their own data. On one occasion, fourth-graders had complained that the playground was inadequate and unsafe. They gathered and analyzed data, presented it to the School Board, and experienced great satisfaction when a new playground was built!

Conclusions

So what have we learned about how many peanuts can a person hold in one hand?

We have learned that there is quite a bit of variation: The greatest value is virtually double the smallest value.

The majority of the numbers are clustered between 22 and 26, and both mean and median are in the mid-twenties.

In real life, the process of gathering and analyzing data generally answers some questions and invites new questions. For example, is there a relationship between the size of one's hand and how many peanuts one can hold? (We will investigate this question in the next section.) How do we measure the size of a hand? Did some people get fewer peanuts because they didn't care? What if each person did this 10 times; would they get the same or close to the same number each time? If there is too much variation in our data, we are less confident about being able to make general statements about the data.

INVESTIGATION 7.1e How Long Does It Take Students to Finish the Final Exam?

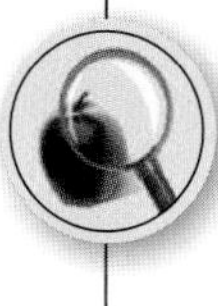

This was a question for which I decided to collect data. I know that a fair exam is one in which there are questions about everything that was addressed during the semester. However, a fair exam would be very long. Therefore, a final exam has a certain inherent degree of unfairness. Thus, having more questions on the final exam increases the chance that the exam will be fair. However, if an exam is too long, students can get stressed. So I decided to gather data on this question. I told my students that they could take as much time on the final as they wished, and I recorded how long each student took to take the exam. The data below are the times, in minutes, that the students took on the exam:

62, 76, 87, 89, 93, 95, 98, 99, 101, 103, 105, 108, 111, 112, 115, 115, 116, 116, 124, 124, 126, 126, 130, 132, 132, 134, 137, 139, 139, 144, 146, 148, 148, 154, 154, 156, 160

What analyses might you do of these data to advise me? Take some time to explore the data, using knowledge that you already have. Summarize what you learned and state your conclusions, and then read on. . . .

DISCUSSION

Examining the spread of the data A line plot (Figure 7.13) gives us a sense of the distribution without losing any data. In this case, the line plot doesn't tell us much beyond what we knew already. The range is defined as the largest value minus the smallest value. Here we have a range of $160 - 62 = 98$. The range is so great that patterns in the data are not apparent.

60 70 80 90 100 110 120 130 140 150 160

Figure 7.13

A **stem-and-leaf plot** helps us to organize the data (see Table 7.1). A stem-and-leaf plot (sometimes simply called a stem plot) display the values in rows. The numbers at the left are the stems and the numbers at the right are the leaves. Consider the two digits at the top row of the stem-and-leaf plot: 6 and 2. The 6 is essentially a code that tells us that all the values on this row are in the 60s. Thus, the 2 next to the 6 represents a data value of 62.

TABLE 7.1

6	2
7	6
8	7 9
9	3 5 8 9
10	1 3 5 8
11	1 2 5 5 6 6
12	4 4 6 6
13	0 2 2 4 7 9 9
14	4 6 8 8
15	4 4 6
16	0

As with the line plot, we don't lose any data (for example, we still know the minimum and maximum). In this case, the stem plot shows us that as we get closer to the middle of the data, the number of students is greater.

As we did with the peanuts data, by selecting an interval size, we can make a grouped frequency table for the data. The intervals are called classes. It is important to note again that there is no one "right" interval size. For example, we could choose an interval size of 10 minutes, in which case we have Table 7.2. This choice produces 11 classes. Alternatively, we could choose an interval size of 20 minutes, in which case we have Table 7.3, which gives us six classes.

MATHEMATICS

What kind of graph does the stem-and-leaf plot remind you of? Rotate the stem-and-leaf plot a quarter turn counterclockwise. Can you see the relationship between this and a grouped frequency histogram?

TABLE 7.2

Interval	Frequency
60–69	1
70–79	1
80–89	2
90–99	4
100–109	4
110–119	6
120–129	4
130–139	7
140–149	4
150–159	3
160–169	1

TABLE 7.3

Interval	Frequency
60–79	2
80–99	6
100–119	10
120–139	11
140–159	7
160–179	1

From these data, we can make a histogram. Examine the two histograms in Figure 7.14. What do you see now that you didn't see before? What do they tell you about the data that is similar? What do they tell you about the data that is different? Then read on. . . .

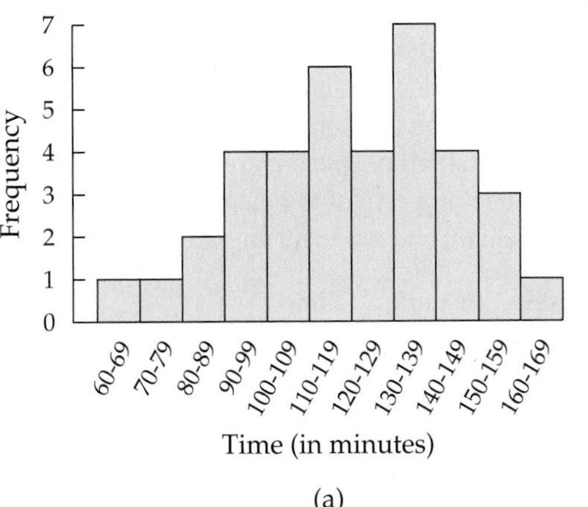

(a)

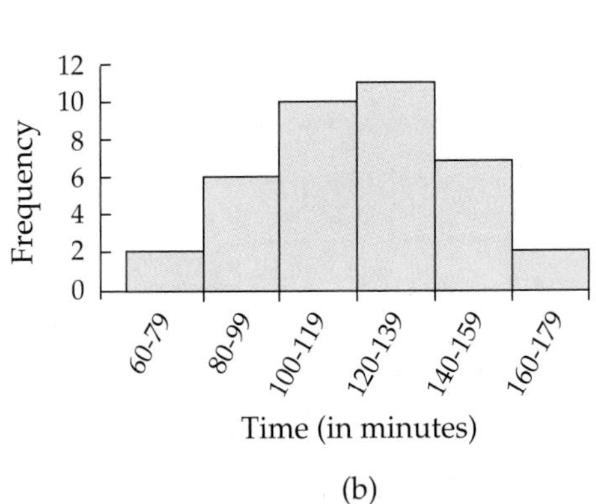

(b)

Figure 7.14

The histogram in Figure 7.14(a) indicates that the majority of the times are between 90 and 150 minutes, and it shows two peak intervals: 110–119 and 130–139. The histogram in Figure 7.14(b) indicates that a majority of the times lie between 100 and 139 minutes.

Note that with the second set of grouped frequencies, we could also make a circle graph for the data. This would more rapidly give a sense of what proportion of the class finished in each of the time intervals. Technically, we could do this with the first set, but a circle graph with 11 slices is a bit much.

Finding the center Estimate the median from the line plot or one of the histograms. Then read on. . . .

From the line plot (Figure 7.13), we can see that there are about as many data values above 120 minutes as below. From Figure 7.14(a), we can see that there are roughly as many data values above the 120–129 group as there are below. In fact, the median is 124 minutes. In this case, the mean is close, 120 minutes.

The strict interpretation of the mode is relatively meaningless in this case. There are several data values that occurred twice, but a frequency of only 2 in a set of 37 hardly makes a number a candidate for typical. Thus, when we make grouped frequency tables, we speak of a **modal class**—that is, the class that occurs most frequently.

A look at Figure 7.14(a) reveals two intervals that stand out: the 110–119 interval and the 130–139 interval. Even though the 130–139 interval has one more data value than the other, both of them stand out from the other intervals, and so we can say that this set of data has two modal classes. In such a case, we say that the distribution is **bimodal**; that is, it has two modes.

We can draw several conclusions from the data and the graphs. When given more than two hours over half the class took the extra time. The "average" time for completion was about two hours, but it is ironic that although the mean and median are both very close to two hours, this is where the actual distribution dipped. I wondered whether that was just a coincidence or whether it was related to the traditional time limit. In any case, because over half the class took longer than two hours, I decided to make a shorter exam.

MEASURES OF CENTER REVISITED

It is crucial that you understand what centers tell us, what they don't tell us, and how they can be misleading. While centers can give us a snapshot of a population, they do not tell us anything about the variation in a set of data, about clusters, gaps, and the range.

While it is said that all people are created equal, it is not true that all three averages are equal. There is one property of the mean that bears explicit attention. The table below shows the mean, median, and mode for Investigations 7.1b, 7.1c, and 7.1d.

Explorations Manual 7.3 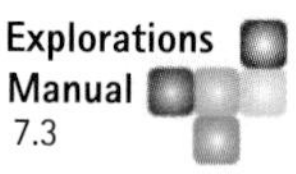

	Siblings	Peanuts	Time
Mean	2.3	25.3	120
Median	1	25	124
Mode	1	22	120s

MATHEMATICS

Here are two quotes about how the mean can be misleading!

Then there is the man who drowned crossing a stream with an average depth of six inches.

—W.I.E. Gates

I abhor averages. I like the individual case. A man may have six meals one day and none the next, making an average of three meals per day, but that is not a good way to live.

—Louis D. Brandeis

In the first case, the mean is significantly different from the median and mode; in the other two cases, all three are similar. One significant way that the mean differs from the other two is that it is affected by extreme values. For example, if the child with 12 siblings had not been in the class, the mean would have been 1.8 instead of 2.3 (a 22% difference); however, the median and mode would still have been 1. Thus, when there is a population with extreme values, it can be misleading to give just the mean.

Table 7.4 summarizes the main reasons for using each measure and some of the disadvantages of each.

TABLE 7.4

	Reasons for using	Disadvantage
Mean	It is often easier to compute with a large set of data.	It can give a distorted sense of the middle if there are extreme data (outliers).
	It is the center of gravity (or balance point) of the data.	It might not make sense. For example, what does a mean shoe size of 6.3 mean?
Median	It is not affected by outliers.	It is not always appropriate—for example, a baseball player's batting average.
	It is the midpoint of the data.	
Mode	It shows the most common datum.	It might be far from the center.
	It is easy to determine from a graph.	There might be no modes or more than one mode.

DISTRIBUTIONS

One of the themes of the course is the power of mathematics to help us make generalizations. When analyzing numerical data, the line plot or histogram will always have a shape, and there are several shapes that represent generalized cases of different patterns of variation in a set of data. Each of these general shapes has names and properties. We will explore them now.

There are many ways in which a set of data can be distributed. According to the Common Core State Standards, students learn about describing data via its distribution in the sixth grade. In this course, we will focus on five **distributions**: uniform, skewed to the right, skewed to the left, bimodal, and normal. The graphs in Figure 7.15 represent idealized (smoothed) versions of these distributions. In real life the data are seldom so smooth. Can you think of a scenario for each of these graphs? Try to do so before reading on. . . .

The **line graphs** shown in Figure 7.15 can be thought of as evolving from histograms (with which most students report being more comfortable). For example, if we collected data on the number of siblings, we would have the histogram shown at the left in Figure 7.16. If we made a line graph from those data, we would have the line graph shown at the right in Figure 7.16. If you look at the idealized graph, you can see that these data are skewed to the right. That is, this graph has the characteristics of a "skewed to the right" distribution.

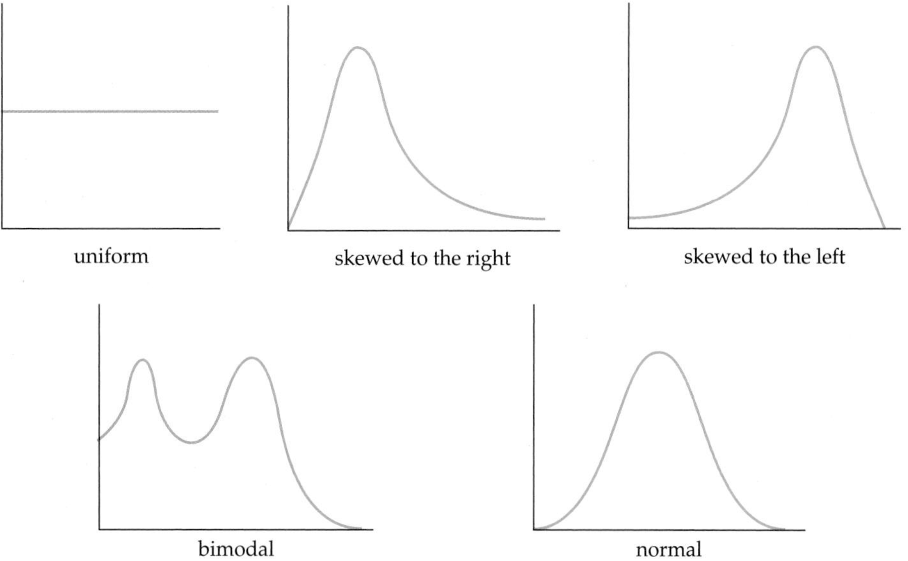

Figure 7.15

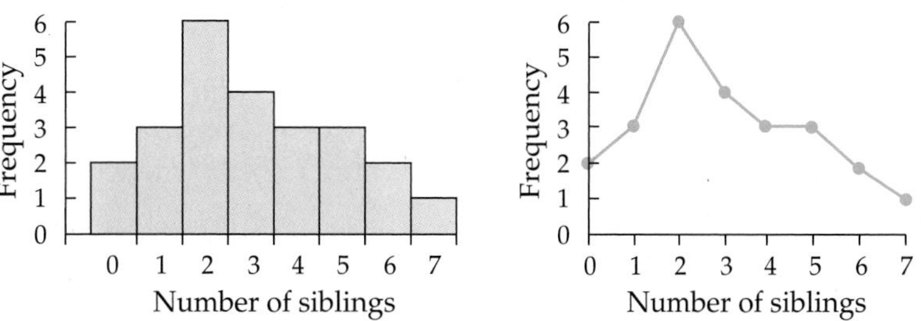

Figure 7.16

Table 7.5 gives one example for each of the five distributions.

TABLE 7.5

Distribution	x-axis (independent variable)	y-axis (dependent variable)
Uniform	Months of the year	Temperature in Hawaii
Skewed (right)	Salaries in a factory	Frequency
Skewed (left)	Class sizes in a high school	Frequency
Bimodal	Time to finish an exam	Frequency
Normal	Heights of people	Frequency

For example, consider the row where "Salaries in a factory" is used to illustrate the distribution "Skewed (right)." Do you see that the graphical representation of this situation will be skewed to the right? Visualize a factory, who works there, what kinds of jobs are involved, and the like before reading on. . . .

If the shape of the graph of "Salaries in a factory" is skewed to the right, that means that the frequency of salaries will peak to the left of the middle, and the graph will slope more sharply to the left than to the right. In other words, there will be people much farther to the right of the center (making much higher salaries) than to the left of the center. From another perspective, the peak of this graph is not in the exact middle of the highest and lowest salaries but is closer to the lowest.

We can make some generalizations about using these terms to describe the center of a set of data:

- If the distribution of the data is skewed, the median will often be more representative than the mean, since the median is not affected by the high salaries of the few top executives.
- If the data are categories rather than numbers (for example, favorite TV show versus age), the mode is used to convey the center of the data. We might say, for example, that the typical American family eats hot dogs on the Fourth of July; this statement indicates that it has been determined that more families eat hot dogs on the Fourth of July than any other food.
- If the distribution is symmetric (for instance, normal) the mean, median, and mode will be close to one another.

Variation comes to play in another way with respect to distributions. Both of the graphs below represent data that are normally distributed. In the former case, the variation is small; in the latter case, the variation is large. We will explore measures of variation in Section 7.2.

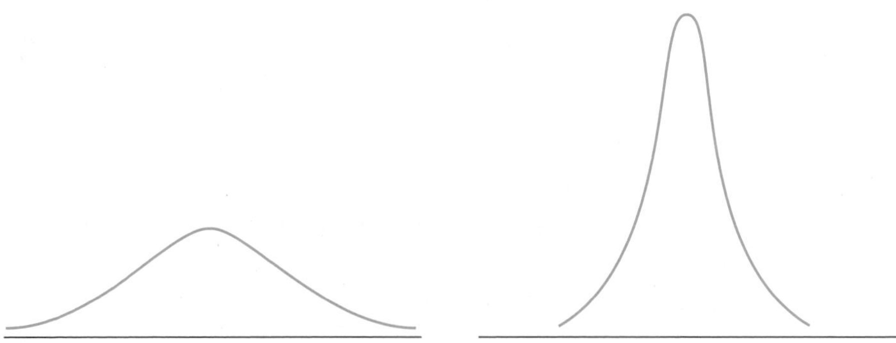

Figure 7.17

EXPLORING DATA WITH LARGER NUMBERS AND DIFFERENT SETTINGS

Up to this point, we have investigated data where the numbers are all relatively small. If you look at a newspaper, you will find data and graphs with large numbers. Now, we will examine how to make and interpret graphs when the numbers are large.

INVESTIGATION 7.1f Internet Access

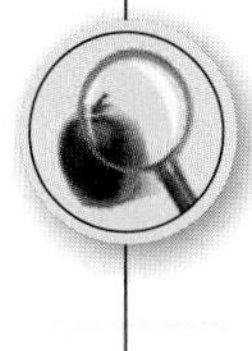

Explorations Manual 7.1

In this first investigation, we will examine two commonly used graphs and some of the problems involved in making them. Take a minute to examine the data in Table 7.6. First, describe the growth of households with Internet access, as mathematically as possible, as though you were talking to someone on the phone. Please be more precise than "Wow, Internet access has increased." Second, describe any questions you have about the data. Then read on. . . .

TABLE 7.6

Year	% of U.S. households with Internet access
2011	76.5
2010	75.9
2009	68.7
2007	61.7
2003	54.7
2001	50.4
2000	41.5
1997	18

Source: United State Census Bureau. Retrieved January 18, 2013 from http://www.census.gov/hhes/computer/.

DISCUSSION

At a basic level, Table 7.6 shows that the percentage of U.S. households with Internet access has increased every year. At a deeper level, we can say that the percentage of U.S. households with Internet access just about doubled between 1997 and 2000 and then about doubled again between 2000 and 2011.

Possible questions about this table include: What has happened since 2011? What about the years before 1997? What percentage were broadband connections?

GRAPHING THE DATA

Let us examine how a graph can help us to understand and interpret the data better. What kind of graph do you think might best describe these data? Make your own graph with graph paper first, before reading on....

Common Graphing Mistakes Below are a bar graph and a line graph showing the data. However, there are two common mistakes in these graphs. Do you see why these graphs are distorted?

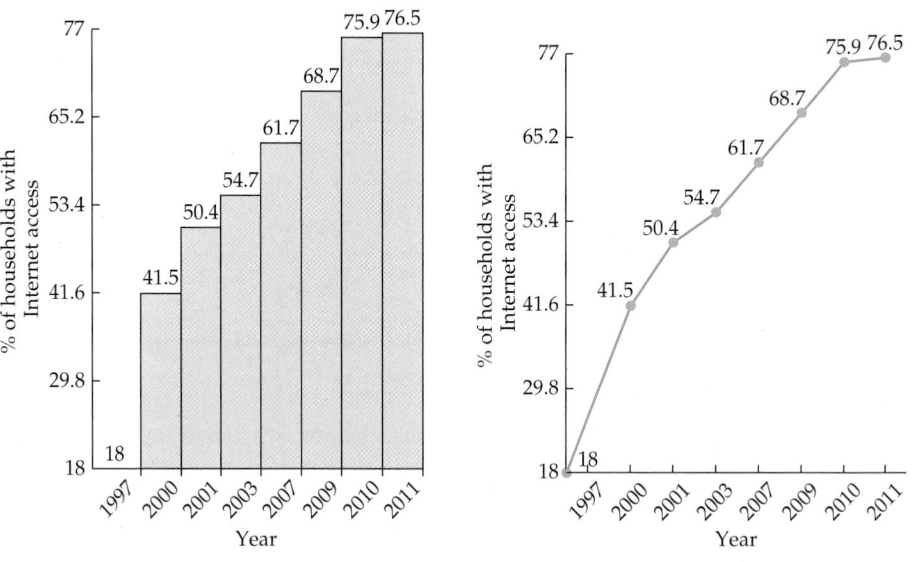

Both graphs skip to 18 on the vertical axis. Sometimes this is done to save space (and is called truncating the graph); however, it gives a distorted view of the data as it does not clearly show the trend. Investigation 7.1g will also explore this concept.

Another mistake is the spacing of years. On the graph, there is one space between 1997 and 2000, which represents 3 years. There is also one space between 2000 and 2001, which represents 1 year. This lack of uniformity can distort the picture the graph shows.

INTERPRETING GRAPHS

In everyday life, we generally don't collect data and graph data as much as we interpret other people's graphs of data that they or yet *other* people have collected. The ability to interpret and to critique graphs is important. As we will find, graphs often distort the data. Sometimes this is intentional on the part of the people making the graph; other times it is unintentional, the result of carelessness or ignorance.

In the next three investigations, we will critically examine graphs made by other people. As you read each graph, think about the following four kinds of questions before you read the discussion.

Conclusions

What conclusion(s) can I draw from the graph?
Do the conclusion(s) that I read seem reasonable?

Construction of the graph

Are the scales and the units clear or are they misleading?
Would another graph be more appropriate? Why or why not?

Reliability/validity

Do I have questions about how the data were obtained that could affect the accuracy of the data?

Further questions

Questions to help you better interpret or understand the data and graph.
Questions that this data set and graph provoke in you.

INVESTIGATION 7.1g Fatal Crashes

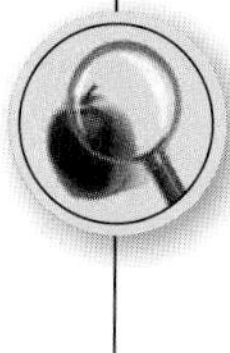

Let us begin with a graph that indicates hopeful news. Examine the graph in Figure 7.18 and answer each of the four kinds of questions before reading on. . . .

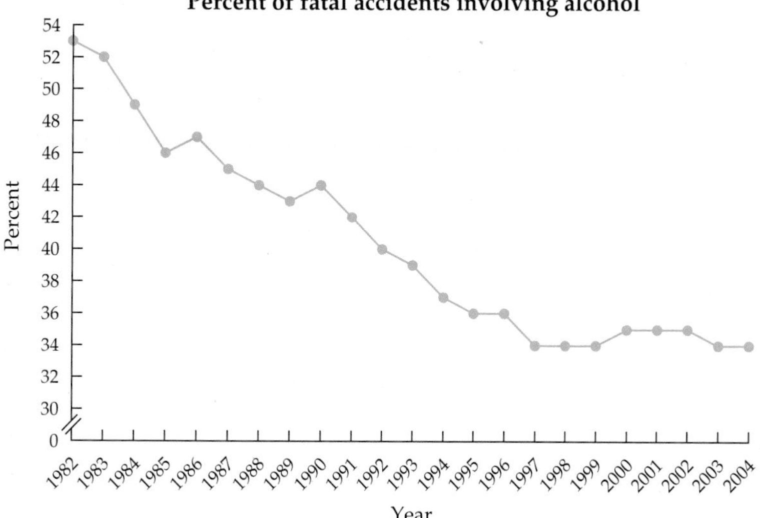

Figure 7.18
Source: NHTSA Fatality Analysis Reporting System (FARS), 2004.

DISCUSSION

One student wrote the following: "The percent of fatal car crashes in which the driver was drunk fell dramatically between 1990 and 1997." What do you think of her summary? Think and then read on. . . .

Actually, there are several problems with the student's summary. First, the data don't seem to be restricted to car crashes. What kinds of "traffic fatalities" do you think count in these data? Think and then read on. . . .

Other kinds of traffic fatalities include crashes between two motor vehicles (trucks, cars, motorcycles) in which one or both drivers were drunk, single-motor-vehicle accidents in which the driver was drunk, and possibly accidents in which a motor vehicle hit a pedestrian or a person on a bicycle.

A second problem with the student's summary has to do with what "drunk" means. What do the graph makers mean by "drunk"? How did the people who recorded the data know that a driver was drunk? How were the data gathered? Think and then read on. . . .

CLASSROOM CONNECTION

In the first edition of this textbook, I had data from a different source. The caption on the graph was "Percent of fatal accidents involving drunken drivers," and the data were from 1982 through 1990. When writing the second edition of the textbook, I could not find an updated set of these data, which had come from an article in *USA Today*. I searched the Web and, finally, I got the current data from the National Highway Traffic Safety Administration. However, the heading was "Percentage of traffic fatalities" and there were three columns: one in which no one involved in the crash had any alcohol in their bloodstream, one in which any person in the crash had a blood alcohol concentration between 0.01 and 0.07, and the third column (which I used) in which the highest blood alcohol concentration of any person in the crash was ≥ 0.08. In the discussion of these data, we are presuming that in virtually all of these cases, at least one of the people was drunk.

It seems reasonable to expect that there are data for every motor vehicle accident in which there was a fatality. However, how did the people who recorded the data determine the number of such accidents in which at least one driver was drunk? Did a sobriety test or a blood test show that the person was drunk? Furthermore, the definition of *drunk* varies from state to state: In some states, a person with a blood alcohol level of 0.08% is considered drunk, whereas in other states the blood alcohol level has to be 0.10%.

Now let us examine the student's use of the word *dramatically* to describe the change in fatalities. Look back at the vertical axis of the graph—what does the jagged line just below 30% mean? Think and then read on. . . .

It means that this is a **truncated graph**—that is, the authors of this graph deleted the 0%–30% interval. Why might someone want to truncate a graph? Think and then read on. . . .

As we began to consider in Investigation 7.1f, sometimes graphs are truncated to save space. However, sometimes they are truncated to distort the data. Let us see how the graph would look if it had not been truncated (Figure 7.19). How does the decline in the percentage of drunk drivers in fatal accidents look now? Think and then read on. . . .

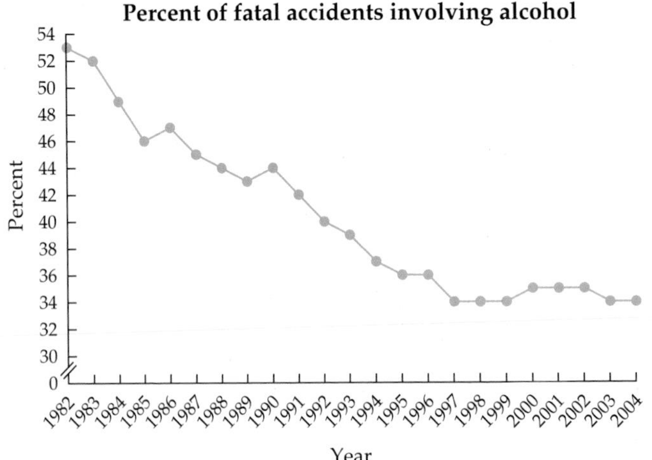

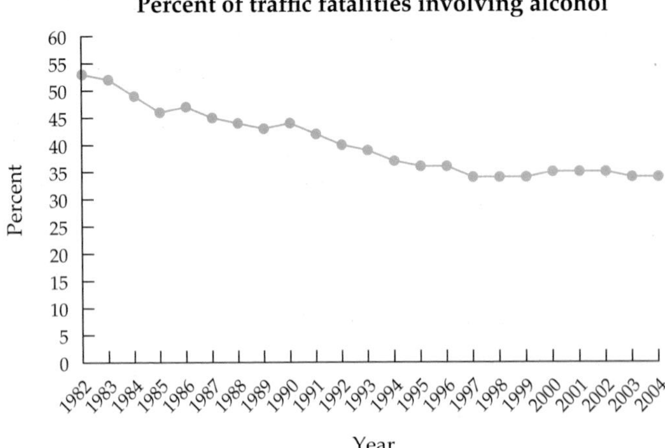

Figure 7.19

As you can see, the decline does not seem so great in the untruncated graph. The actual percent decrease from 53 to 34 is about 36% (that is, $\frac{53-43}{53} \approx 0.36 = 36\%$).

Which graph would you have picked if you were working for the alcohol industry preparing an advertisement showing that drunk driving is on the decline? What if you were a member of SADD (Students Against Drunk Driving)?

INVESTIGATION 7.1h Hitting the Books

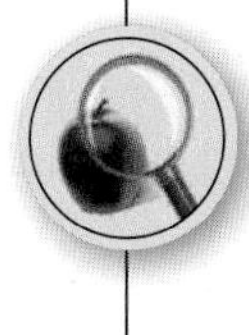

Take a few moments to examine Figure 7.20 and ask yourself questions about conclusions, construction of the graph, and reliability/validity. Write your responses to these questions before reading on. . . .

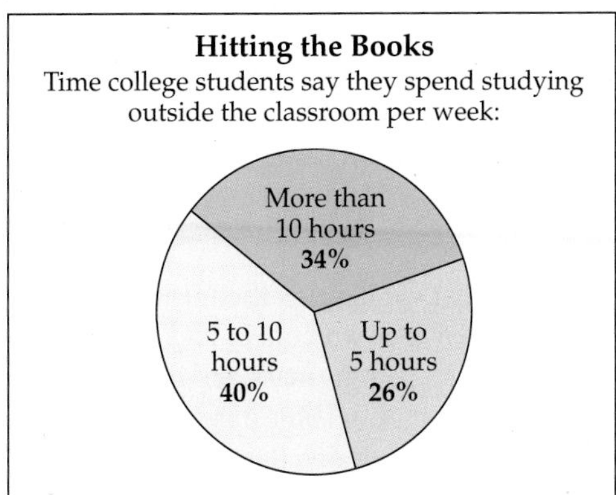

Figure 7.20

DISCUSSION

Describing the graph Critique the following two statements, which represent conclusions that people have made from this graph.

- Barely one-third of college students study more than 10 hours per week.
- Most college students average less than 2 hours a day on homework.

The first conclusion is simply taken straight from the graph: "Barely $\frac{1}{3}$" is consistent with 34%. The second statement represents a valid interpretation of the graph, because less than half of the students spend more than 10 hours a week.

What do the data mean? Now let us look beyond the first impressions to examine what the data mean. If you could meet with the people who collected these data, what questions would you ask? Think and then read on. . . .

Below are three of many possible questions that we might ask. As you read these questions and consider your own questions, ask yourself whether the answers to these questions would make a difference. Then read on. . . .

- How did you get the data (take a survey outside the dining commons, have questionnaires filled out in some classes)?
- What kind of students were included (full-time, a "representative" sample)?
- What did you ask? For example, did you ask, "About how many hours did you study last week?" or "Last week, did you spend less than 5 hours studying, between 5 and 10 hours, or more than 10 hours?"

The way in which we gather data can make a big difference in how the graph appears. For example, if the questions were asked of students coming out of the dining commons, that means that off-campus students were not asked. And if off-campus students tend to study a lot more or less than on-campus students, this could change the data quite a bit. Similarly, if the data included both part-time and full-time students, the number of hours reported would naturally go down. Finally, how we ask the question has an influence on our data. You might want to replicate this study on your own campus and compare the results of asking the question in two different ways.

Choice of graph The makers of this graph chose a circle graph. What other choices would be appropriate, or is the circle graph the "best" choice? What do you think? . . .

Here it is a matter of personal preference. Circle graphs work well when we are looking at parts of wholes. In this case, the whole represents all students, and the makers of the graph have divided the whole into three subsets. However, using a bar graph would also be valid.

SUMMARY 7.1

Let us reflect on what we have learned from examining data gathered by other people.

We have learned that we collect and analyze data for a variety of reasons: to help us make a decision, to help us make predictions, and to help us to understand situations.

We have learned that in some cases, which graph is made is a matter of preference and that each graph has certain characteristics that we might want to use:

- Bar graphs enable us to compare the quantities of the categories.
- Circle graphs enable us to see what part each category is of the whole.
- Line plots enable us to see all data values and aspects of the shape of the data—distribution, clusters, gaps, and outliers.
- Histograms also enable us to see aspects of the shape of the data—distribution and clusters.
- Stem-and-leaf graphs can be used to organize the data and are like histograms turned on one side.
- Line graphs enable us to see change over time.

We have learned that we need to take care that our graph does not distort the data. For example, we need to be careful when choosing the units for our spacing, both horizontally and vertically, in line graphs and bar graphs.

We have also learned that when interpreting others' graphs, we need to think about what the data mean and how the data were gathered, and we need to realize that some graphs are distorted, ether intentionally or out of ignorance or carelessness. At the same time, we have learned that although a truncated graph does indeed distort the data, it sometimes offers advantages. We have learned that whether we are making graphs from data collected by other people or interpreting graphs made by other people, it is important to examine the data and the graphs with respect to fairness, appropriateness, reasonableness, reliability, and validity. "Students should begin to develop a critical attitude toward information presented in the media and learn to ask relevant questions before making judgments based on that information."[5]

Finally, we have learned that statistics is different from the other topics that we have studied in that in many cases, there is not an exact or precise answer, and there is often not a right or best graph to make. Thus, it is even more important that the student have an understanding of the ideas and the whys as opposed to simply knowing how to make the various graphs and how to determine the mean, median, or mode. The following quote nicely sums this:

> *Do not put your faith in what statistics say until you have carefully considered what they do not say.* —William W. Watt

7.1 Exercises

1. Below are several situations for which students might collect data (such as favorite sport or number of siblings). In each case, write the question you would ask so that everyone is answering the same question and the data will be as accurate as possible. Justify your question.
   a. How many times can a third-grader dribble a ball in one minute?
   b. How long can you hop on one foot?
   c. How many concerts have you attended in the past year?
   d. How much time do you study in a week? (A common guideline for how much college students should spend on class work is two hours for every credit. Thus, a student taking 15 credits would be spending about 30 hours a week studying for her courses. Many national surveys report that the actual time students spend is much lower.)
2. A class collected data on how many times in the past 14 days they had exercised. Here are the data:

   10 13 0 14 2 5 7 6 6 10 5 0 14 7 5 6 13 1 6 14

   a. Make a line plot and describe what the line plot tells us. Describe any clusters, gaps, or outliers.
   b. Estimate the mean and briefly describe your thinking.
   c. Determine the mean, median, and mode.
   d. If you were to collect similar data on your campus, describe and justify the actual question you would ask in your survey.
3. Below are the ages of students in a college class at a community college:

   18, 18, 18, 19, 19, 19, 19, 19, 19, 20, 20, 20, 21, 21, 21, 21, 21, 21, 23, 24, 24, 26, 27, 37, 42 43, 46

   a. Make a line plot and describe what the line plot tells us. Describe any clusters, gaps, or outliers.
   b. Estimate the mean and briefly describe your thinking.
   c. Determine the mean, median, and mode.
4. Below are grades in a course and how many students earned each grade:

A	5
B	8
C	10
D	3
F	1

   a. What percent passed the course with a C or better?
   b. Make a bar graph. Write a short summary of the bar graph.
   c. Sketch a circle graph and briefly describe how you did it.
   d. Make an accurate circle graph, using software or a protractor.
   e. Write a short summary of the circle graph.
   f. What are the advantages and disadvantages of representing these data with a bar graph as opposed to a circle graph?
5. My students collected data on how many raisins were in a small box of raisins.

107	107	111	113
117	118	119	119
120	120	120	121
121	121	121	121
121	122	122	122
122	123	124	124
125	125	126	127
127	128	128	128
129	131	132	135
138			

   a. Determine the mean, median, and mode.
   b. What do the mean, median, and mode not tell us?
   c. Make a line plot from these data.
   d. Summarize (2–3 sentences) what the line plot tells us.
   e. What does it not tell us?
   f. Make a grouped frequency table with an interval of 10 for these data.
   g. Make a histogram from these data.
   h. Summarize (1–2 sentences) what the histogram tells us.
   i. Make a grouped frequency table with an interval of 5 for these data.
   j. Make a histogram from these data.
   k. Summarize (2–3 sentences) what the histogram tells us.
   l. How are the two histograms alike and different?
   m. If you were giving a report that described how many raisins are in a box, which graphs and which numbers would you include? Why?
   n. How could there be so much variation in the number of raisins?
6. Here are the distances (in minutes) that students live from school:

5	30	60
5	30	70
5	30	70
5	30	85
10	40	90
10	45	120
20	50	125
20	60	140
25	60	200

   a. Determine the mean, median, and mode.
   b. What do the mean, median, and mode not tell us?
   c. Make a grouped frequency table with an interval of 30 minutes for these data.

d. Make a histogram from these data.

e. Summarize (2–3 sentences) what the histogram tells us.

f. Make a grouped frequency table with an interval of 20 minutes for these data.

g. Make a histogram from these data.

h. Summarize (2–3 sentences) what the histogram tells us.

i. How are the two histograms alike and different?

7. The teachers at Gates Memorial High School got to talking about cars and how long they keep them. Someone wondered about the average age of the teachers' cars. Following is a graph representing the ages, rounded to the nearest year.

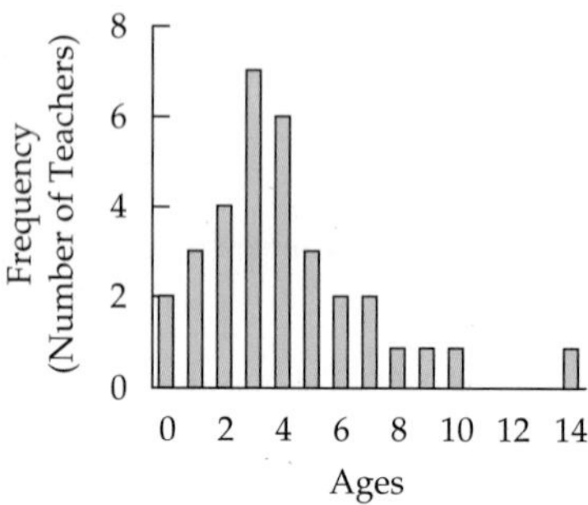

a. Before computing, predict the mean and the median. Explain your reasoning.

b. Determine the mean and the median.

8. In Exploration 7.2, a group in a previous class asked the following question: "What is your favorite sport?" Here are their data:

Basketball	5	Volleyball	4	Soccer	3
Tennis	3	Softball	3	None	2
Skiing	2	Rugby	2	Horseback	1
Gymnastics	1	Swimming	1	Cheerleading	1

a. Select and make an appropriate graph.

b. Justify your choice of graph.

c. Critique the question. If you feel it is fine, explain why; if you feel it has problems, explain why and then suggest a rewording.

d. Determine the mean, median, and mode. If any of these three is inappropriate, explain why.

9. Willie has an average (mean) of 83.4 on the first five exams. What is the least he can score in order to have an average of 85 for all six exams?

10. Lucy took three tests. If her median score was 82, her mean score was 87, and the range was 17, what were her three test scores?

11. The mean of five numbers is 6. If one of the five numbers is removed, the mean becomes 7. What is the value of the number that was removed?

12. Nine people decide to share some candy in an unorthodox way. The most anyone has is nine pieces and one person does have nine pieces. At least one person ends up with no candy (his/her choice). The average (mean) number of pieces is 4. Four is also the median. More people have two pieces of candy than any other number.

a. How many pieces of candy does each person have?

b. Is there more than one solution? Explain why or why not.

13. Here are exam scores for a class of 20 students: 0, 40, 55, 65, 70, 70, 70, 75, 75, 75, 80, 80, 80, 80, 80, 85, 90, 90, 95, 100

a. Determine the mean and median.

b. Remove 2 quiz scores so the median remains the same.

c. Remove 2 quiz scores so the median increases.

d. Remove 2 quiz scores so the mean decreases.

e. Remove 2 quiz scores so the median stays the same and the mean decreases.

14. A student is looking at a prospective college and reads that the average class size at the college is 21 students. Then she goes to that college and has two classes with more than 75 students in her freshman year. How is this possible?

15. The following table shows percentages for daily cigarette use among U.S. eighth-, tenth-, and twelfth-graders.

a. Make a line graph for these data.

b. Describe what you learned or saw from the graph that you didn't see just from the data.

c. Describe the percent change between 1995 and 2007 for each grade.

d. Write a 2–3 sentence summary that might appear in a newspaper article describing changes in cigarette use among children.

	1995	2000	2005	2006	2007
8th	9.3	7.4	4.0	4.0	3.0
10th	16.3	14.0	7.5	7.6	7.2
12th	21.6	20.6	13.6	12.2	12.3

Source: *World Almanac*, 2009, p. 185.

16. The following table records the highest education attainment levels for adults in the United States in 2010. Select and make an appropriate graph for the data below. Justify your choice.

Category	Percent
Not a high school graduate	12.9
High school graduate	31.2
Some college	16.8
Associate's degree	9.1
Bachelor's degree	19.4
Advanced degree	10.5

Source: *Statistical Abstract of the United States 2012*, www.census.gov/compendia/statab.

17. a. Make a line graph for the data shown in the table below.

b. What did you learn or see from the graph that you didn't see just from the data?

c. Write a 2–3 sentence summary for this graph that might appear in a newspaper article describing the increasing numbers of women in these four professions.

FIRST PROFESSIONAL DEGREES EARNED IN SELECTED PROFESSIONS, 1970–2006

Type of degree and sex of recipient	1970	1975	1980	1985	1990	1995	2000	2001	2002	2003	2004	2005	2006
Medicine (M.D.): Degrees conferred,													
Percent to women	8.4	13.1	23.4	30.4	34.2	38.8	42.7	43.3	44.4	45.3	46.4	47.3	48.9
Dentistry (D.D.S. or D.M.D.): Degrees conferred,													
Percent to women	0.9	3.1	13.3	20.7	30.9	36.4	40.1	38.6	38.5	38.9	41.6	43.8	44.5
Law (LL.B. or J.D.): Degrees conferred,													
Percent to women	5.4	15.1	30.2	38.5	42.2	42.6	45.9	47.3	48.0	49.0	49.4	48.7	48.0
Theological (B.D., M.Div., M.H.L.): Degrees conferred,													
Percent to women	2.3	6.8	13.8	18.5	24.8	25.7	29.2	32.0	32.9	34.6	34.2	35.6	33.6

Source: *Statistical Abstract of the U.S.*, 2012, www.census.gov/compendia/statab.

18. a. Sketch a circle graph for the data below.

b. Make a bar graph for the data.

c. Briefly describe the advantages of the circle graph.

d. Briefly decribe the advantages of the bar graph.

VOLUME OF MAIL HANDLED BY THE POST OFFICE

Type	Pieces (billions)
First class	90.3
Second class	10.4
Third class	62.4
Other	1.2

For Exercises 19–23, answer questions a through c and any additional questions that may be asked about that specific graph.

a. *Describe in one major quantitative conclusion that you can make from the graph.*

b. *Critique the data from the perspective of reliability and validity.*

c. *Critique the choice of the graph.*

19. **About 72% of Americans drink coffee. How much they drink:**

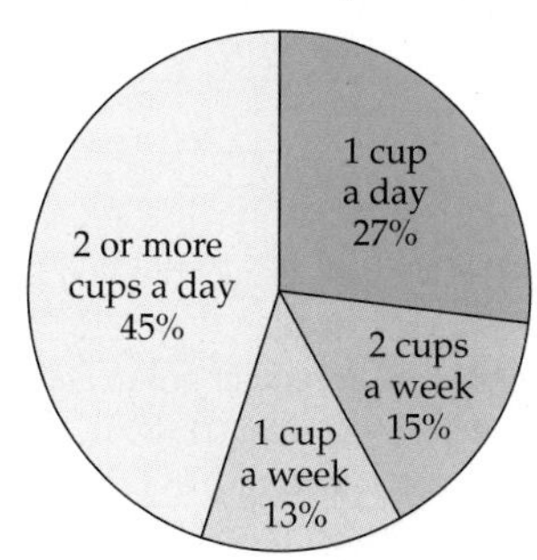

Source: Braun Research for International Delight Coffee House Inspirations/ *USA TODAY* Snapshots, January 8, 2010.

20. **How many pills do seniors take daily?***

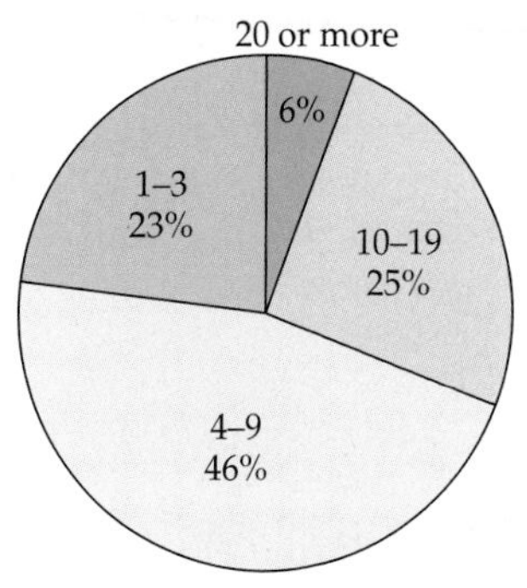

*In 2008, 20% of insured seniors (65+) did not take any medications on a regular basis.

Source: Medio Health Solutions survey/*USA TODAY* Snapshots, January 13, 2010.

21. **Percentage of employers giving a paid day off since the first King holiday in 1986:**

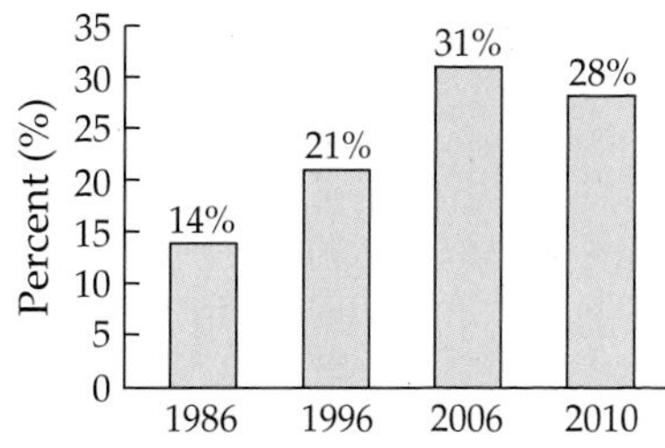

Source: PNA/*USA TODAY* Snapshots, January 18, 2010.

22.

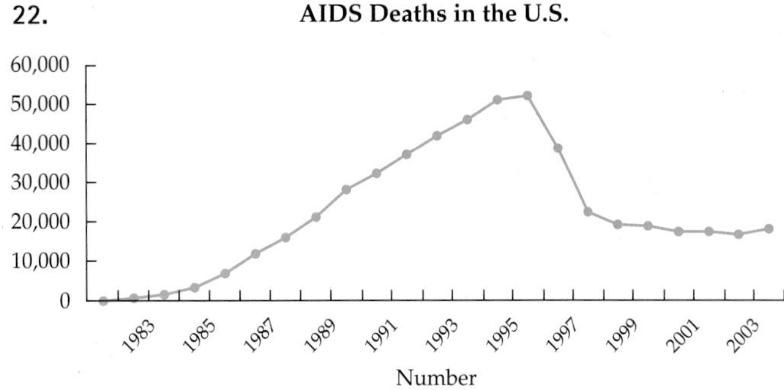

Source: *2006 New York Times Almanac.*

23. The graph below shows the ages of those dying of AIDS.

AIDS Deaths by Age

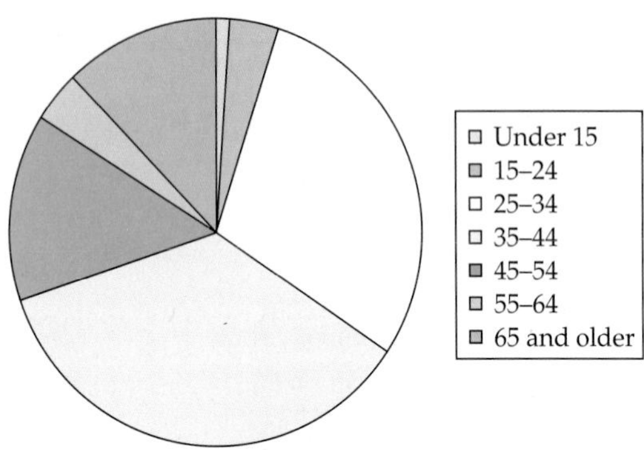

Source: Centers for Disease Control and Prevention, 2004.

DEEPENING YOUR UNDERSTANDING

24. Children from two third grades collected data on the number and kind of pets that they had at home. There were 46 students in the two classes.

Here are the data:

Fish	64
Cats	32
Dogs	28
Birds	12
Hamsters	4
Gerbils	2
Guinea pigs	1
Horses	2
Ferrets	3
Snakes	1

a. Make a bar graph for these data.

b. Determine the mean, median, and mode for these data.

c. Since there are a total of 149 pets and 46 students, could we say that the average student has 3 pets?

d. What questions would you like to ask the students in order to better understand the data?

25. The data below are from Keene State College students in an Introduction to Education course. The data are state of residence.

New Hampshire	26
Massachusetts	6
Connecticut	5
Vermont	3
New York	3
Maine	2
Virginia	1
Rhode Island	1
Pennsylvania	1
New Jersey	1
California	1

a. Some people say the mode is 26, some say the mode is New Hampshire, and some say the mode is 1. Who is right?

b. Some say the mean is 4.6, that is, $\frac{50}{11}$. Some say there is no mean. Who is right?

26. In order to support the case that teachers' salaries are low, a group of high school students has gathered data about the ages of teachers' cars. Write a brief report summarizing what these data say. You decide what graph(s) to make, what number(s) to use, and what words to say.

1994	1	2005	2
1997	1	2006	4
1999	2	2007	4
2001	1	2008	2
2002	3	2009	3
2003	2	2010	1
2004	4		

27. Mr. Arnold asked his students how many brothers and sisters they had. The following graph summarizes what he found.

a. Determine the mean and the median number of siblings.

b. There is something wrong with the graph. Explain not only what is wrong with the graph but also why it is wrong.

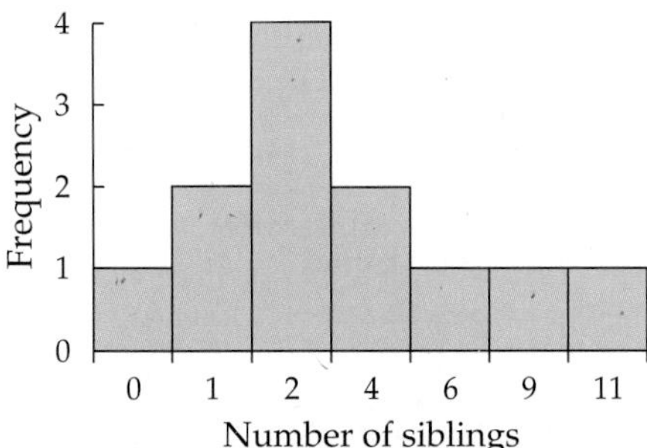

28. Exploration 7.3 asked, "How many drinks did you have in the past week?" Here are the data: 0, 0, 0, 0, 0, 0, 1, 2, 2, 3, 5, 5, 8, 12, 12, 15, 10–15. Imagine you are in that group and have been asked to present your conclusions to the class. In other words, what did you learn about the drinking habits? This includes making a graph—it can be a sketch, as opposed to a polished, precise graph. *Note:* You will need to decide what to do with the 10–15 datum and to justify your decision.

29. Imagine the exam scores of a class of 10 students. Make three very different distributions that all have a mean of 80.

30. Create a set of data containing 11 numbers where the mean is larger than the median. Briefly describe your thinking strategy for this question other than random trial and error.

31. A small company has 19 employees. The mean wage is $10 per hour, but the median is only $7 per hour. What might be causing the two to be different?

32. I recently read an article in which the following sentence appeared: "The average cost of a car is $19,000." How do you think that number was determined? In your answer, you need to state whether you think that figure represents the mean, median, or mode.

33. All averages are not created equal. One of the ways in which they are not equal is how they are affected by outliers. Briefly describe how the mean and median are affected differently by outliers and the implications of this on making conclusions from a set of data.

34. Make up three questions such that in each case one of the centers (mean, median, or mode) would be most appropriate. In each example, all of the three must be able to be determined. ("What is the hair color of children at Riverside Elementary School?" is not acceptable because the mean and median cannot be determined.) Explain why the center you picked is, indeed, the most appropriate.

35. Why do you think many statisticians recommend that when describing a population, one should give at least three pieces of information: some sense of the "average," some sense of the spread, and some sense of how the data are distributed? For example, they would recommend not simply reporting that the average age of teachers in New Hampshire is 43 or that the average adult American goes to church 1.3 times a month (both figures are made up).

36. Describe how the valid changing of the vertical and/or horizontal scales of a graph can change the appearance of a graph.

37. When data are presented, either in tables or in graphs, sometimes we see raw numbers and sometimes the raw numbers have been translated into rates. Unfortunately, many people do not fully understand rates. Explain why we have rates, as though to someone who asks, "Why not just use the raw numbers?"

38. Find or make up a set of data for which:

a. A bar graph would be an appropriate representation

b. A circle graph would be an appropriate representation

c. A line graph would be an appropriate representation

39. Thelma says that she assumes a circle graph is appropriate when the data are percentages. What do you think of Thelma's rule of thumb?

40. a. Make a line graph for the following data on U.S. cellular telephone subscribership.

b. Describe what you learned or saw from the graph that you didn't see just from the data.

c. Based on the graph predict the number for 2017.

d. Write a 2–3 sentence summary that might appear in a newspaper article describing the growth of cell phone ownership.

Year	Number ('000s)	Year	Number ('000s)
1985	340	1999	86,047
1986	682	2000	109,478
1987	1231	2001	128,375
1988	2069	2002	140,767
1989	3509	2003	158,722
1990	5283	2004	182,140
1991	7557	2005	207,897
1992	11,033	2006	233,041
1993	16,009	2007	255,396
1994	24,134	2008	270,334
1995	33,786	2009	290,941
1996	44,043	2010	310,997
1997	55,312	2011	331,595
1998	69,209		

Source: *World Almanac*, 2013, p. 395.

41. The table below shows the estimated number of HIV/AIDS cases in the world, as of 2013.

a. Make a graph for these data.

b. Describe what you learned or saw from the graph that you didn't see just from the data.

c. Write a 2–3 sentence summary that might appear in a newspaper article about HIV/AIDS in the world.

Region	Current cases
Sub-Saharan Africa	23,500,000
South/Southeast Asia	4,200,000
Latin America/Caribbean	1,630,000
East Asia	830,000
Eastern Europe/Central Asia	1,500,000
North America	1,400,000
Western/Central Europe	860,000
North Africa/Middle East	330,000

Source: *World Almanac* 2013 (New York: World Almanac Books), p. 739.

42. Refer to the table below.

a. Write out the actual amount in dollars of the United States sales in 2011.

b. Make two circle graphs from the % of total data.

c. Summarize what you can conclude from the circle graphs.

d. Make two bar graphs, either separate or side-by-side, from the Dollars data.

e. Summarize what you can conclude from the circle graphs.

f. Summarize the advantages and disadvantages of each.

g. If you were to pick one graph, which would you pick? Why?

INTERNATIONAL ARMS SALES
(in millions of current U.S. dollars)

	2004		2011	
	Dollars	**% of total**	**Dollars**	**% of total**
United States	12,368	29.2	66,274	77.7
Russia	8,800	20.8	4,800	5.6
France	2,900	6.9	4,400	5.2
United Kingdom	4,200	9.9	400	0.5
China	1,000	2.4	2,100	2.5
Germany	4,100	9.7	100	0.1
Others	8,900	21.0	7,200	8.4
Total	42,268	100.0	85,274	100.0

Source: The 2013 World Almanac, p. 163

43. The table below shows the U.S. immigration rate by decade, from 1820 to 2000.

U.S. IMMIGRATION RATE BY DECADE, 1820–2000

Period	Total number ('000s)	Rate per 1,000 U.S. pop.
1820–30	152	1.2
1831–40	599	3.9
1841–50	1,713	8.4
1851–60	2,598	9.3
1861–70	2,315	6.4
1871–80	2,812	6.2
1881–90	5,247	9.2
1891–1900	3,688	5.3
1901–10	8,795	10.4
1911–20	5,736	5.7
1921–30	4,107	3.5
1931–40	528	0.4
1941–50	1,035	0.7
1951–60	2,515	1.5
1961–70	3,322	1.7
1971–80	4,493	2.1
1981–90	7,338	2.9
1991–2000	9,095	3.2

Source: www.census.gov and author's calculations.

a. Describe your first impressions from examining these data. Are there any surprises?

b. Make two separate line graphs, one using the total number and one using the rate per 1000 population. Describe the different impressions conveyed by the two graphs. Which graph do you think gives a "better" comparison of immigration to the United States? Why?

c. What did you learn about mathematics from this exercise?

d. What did you learn about immigration from this exercise?

For Exercises 44–48, answer questions a through c and any additional questions that may be asked about that specific graph.

a. *Describe one major quantitative conclusion that you can make from the graph.*

b. *Critique the data from the perspective of reliability and validity.*

c. *Critique the choice of the graph.*

44.

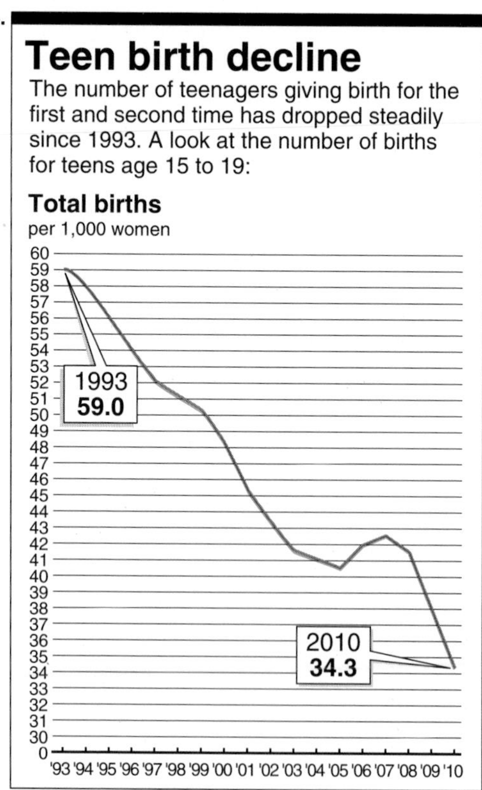

Source: *The 2013 World Almanac*, p. 202.

45. The following graph shows U.S. AIDS diagnoses in 2011 by ethnicity. In addition to answering a through c, answer the following.

d. The numbers of African Americans and Hispanics are disproportionate. Explain what this statement means.

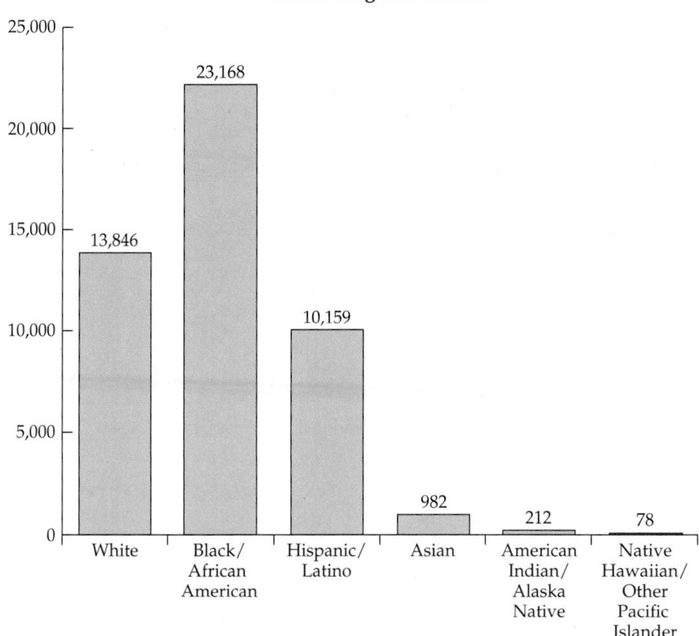

Source: Centers for Disease Control and Prevention. HIV/AIDS Surveillance Report 2011, Volume 23.

The graphs in Exercises 46 and 47 are from the Third International Mathematics and Science Study in which classes in Japan, Germany, and the United States were observed. For more information see http://nces.ed.gov/pubs99/1999074.pdf, p. 46.

46. Teachers were asked to tell the researchers what they wanted students to learn from the lessons that were videotaped. The researchers found that most of the answers fell into one of two categories: *skills,* where the answers focused on students being able to do something, and *thinking,* where the answers focused on students being able to understand something about a mathematical concept or idea.

Percentage of teachers who describe the goal of the video-taped lesson as skills vs. thinking

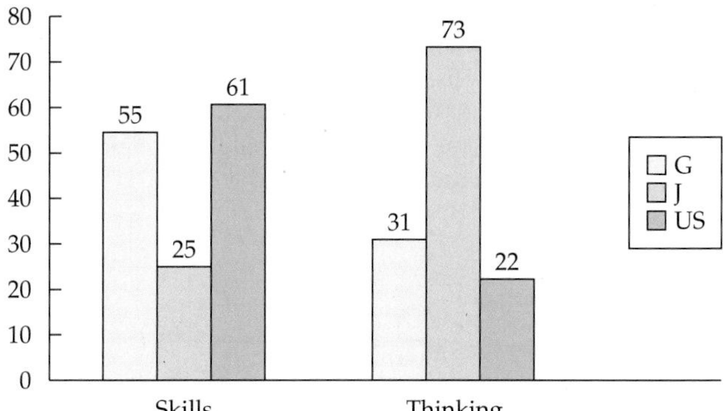

47. In this case, the researchers examined the kind of work that students did during the lesson. They coded three types of student work: practicing routine procedures, applying concepts to new situations, and inventing new solution methods or thinking.

Average percentage of seatwork time in each country spent working on three kinds of tasks

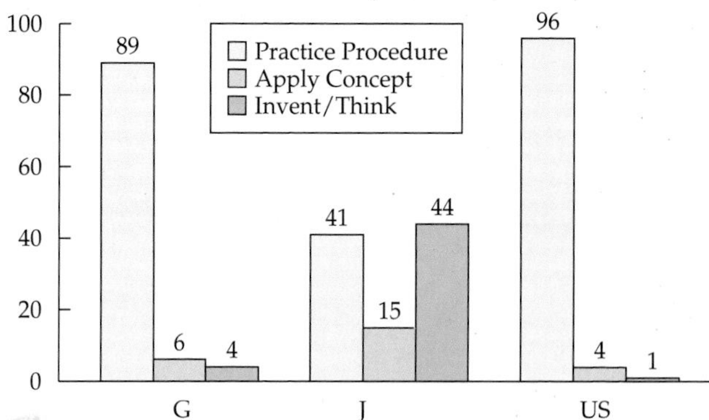

48. **Median Family Income, 1968 to 2007 (2008 dollars)**

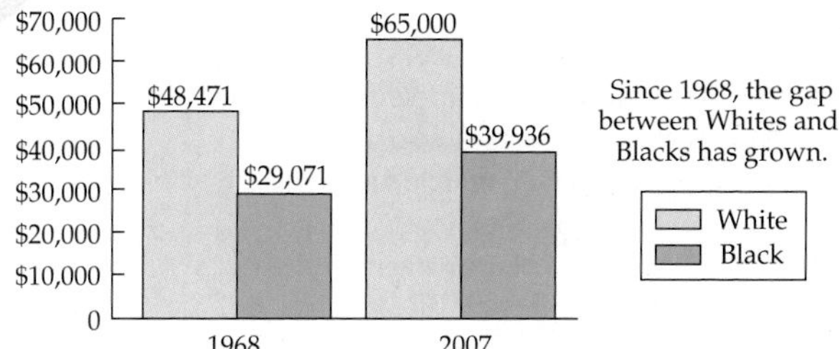

Source: U.S. Census Bureau, Historical Income Tables, Families, Table F-5 <http://www.census.gov/hhes/www/income/data/historical/families/index.html.

49. a. Write a short paragraph that would accompany the following two circle graphs in an article describing changing patterns in immigration.

b. Make a table for these graphs.

c. Discuss the advantages and disadvantages of representing these data in a table as opposed to a graph.

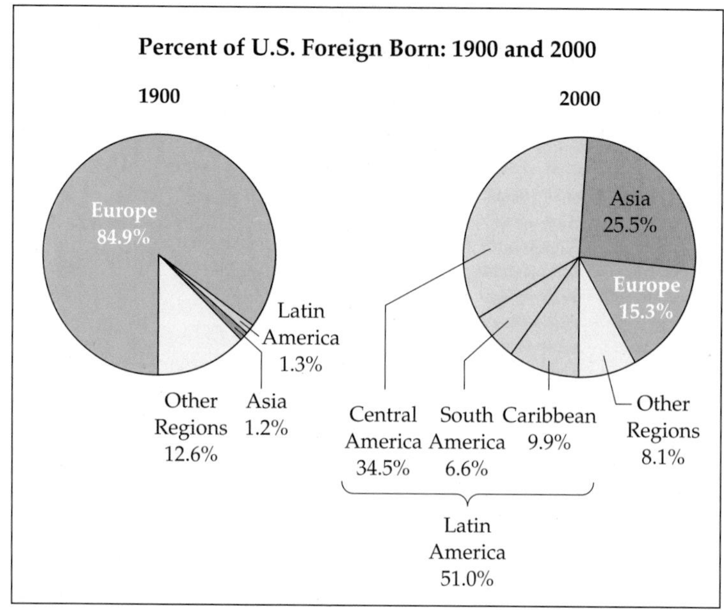

Source: *Time Almanac* 2003, p. 353, from InfoPlease. Reprinted by permission of Pearson Education, Inc.

50. The bar graph shows the percentages of drivers of various ages who were involved in fatal motor vehicle accidents and had blood alcohol levels greater than 0.08. Write a short paragraph describing the differences in age groups.

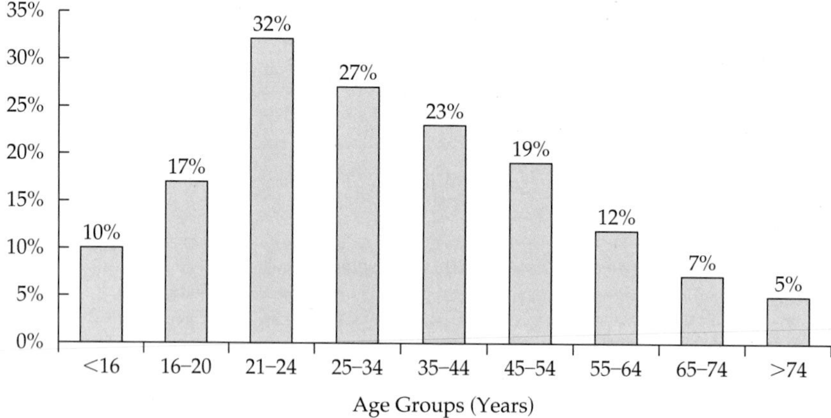

Source: U.S. Department of Transportation, National Highway Traffic Safety Administration, *Traffic Safety Facts*, 2004 (Washington, DC: Author, 2000).

51. We have discussed that there are similarities and connections between different graphs.

a. Describe how a histogram and a line plot are similar.

b. Describe how a line graph and a bar graph are similiar.

c. Find a graph that was not necessary. That is, find a graph where it would have been just as useful simply to supply the data in the paragraph that described the data or in a table. Explain why the graph did not add to your understanding of the situation being reported.

52. a. One important learning described by several students is that they used to take numbers that they saw in tables and graphs as exact. Now they know that in many cases the numbers are not exact after all. Describe this learning more fully.

b. Find a graph that is distorted or misleading. Explain why it is distorted or misleading. Sketch an alternative graph that would not be distorted or misleading.

53. There are many interesting graphing activities at http://www.shodor.org/interactivate/activities/. Explore Bar Graph, Box Plot, Circle Graph, Histogram, and Stem and Leaf Plotter. For each, adjust the scale on the graph and see how that changes the look of the data. Write a paragraph about the activities, which you like best, and what they can teach you.

FROM STANDARDIZED ASSESSMENTS

From 2006 NECAP, Grade 4

54. This line plot shows the number of miles each student in Mr. Miller's class rode his or her bike last week.

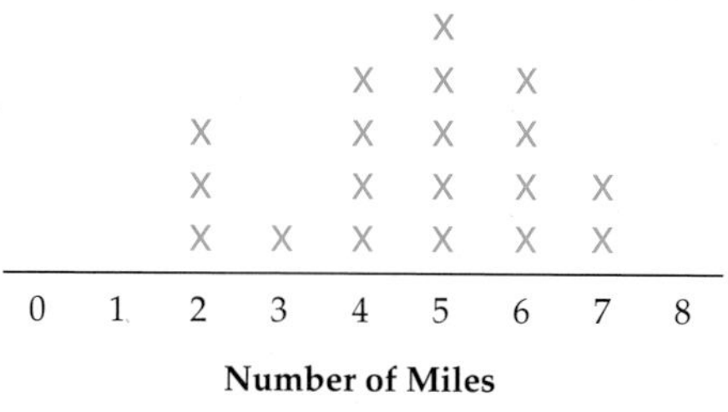

Key
X represents 1 student

What was the greatest number of miles any student in the class rode his or her bike last week?

From 2008 NECAP, Grade 6

55. Look at this graph.

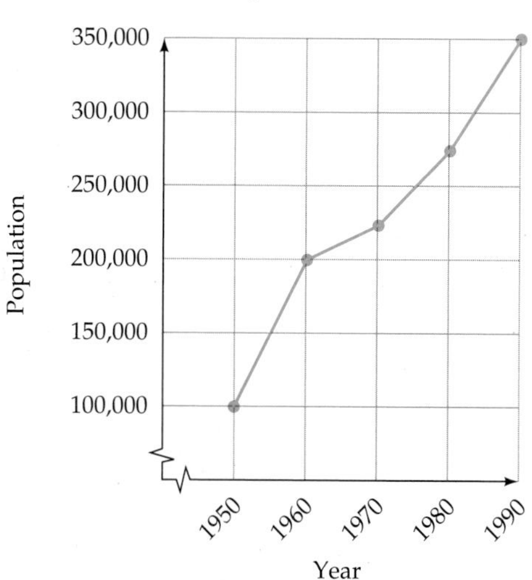

During which time period did Denton's population increase by about 50,000?

From 2005 NECAP, Grade 4

56. Look at this bar graph.

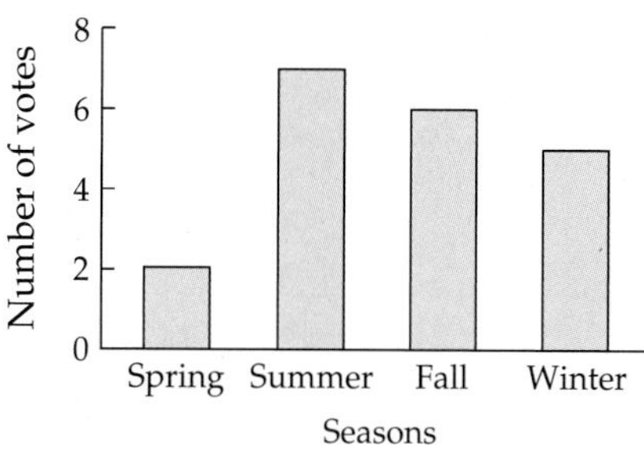

How many more votes did the most favorite season receive than the least favorite season?

From 2007 NECAP, Grade 5

57. Look at this circle graph.

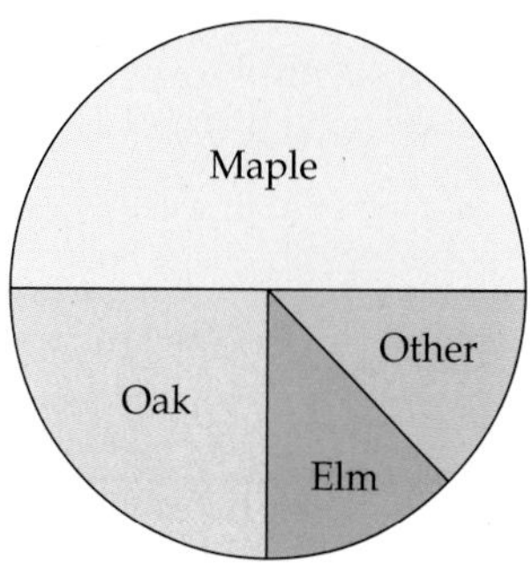

Meadow Park has 12 elm trees. About how many maple trees does Meadow Park have?

a. 24
b. 36
c. 48
d. 96

From 2007 NECAP, Grade 6

58. A yoga teacher collected data on the ages of the people in her class. The median is 36 years old. The mean is 42 years old. The mode is 50 years old. Which of the following statements *must* be true?

a. Most people in the class are 42 years old.
b. The oldest person in the class is 50 years old.
c. Each person in the class is either 36, 42, or 50 years old.
d. At least half of the people in the class are 36 years old or younger.

59. This item is from a recent National Assessment of Educational Progress test (http://nces.ed.gov/nationsreportcard/pdf/demo_booklet/05SQ-grade12-part1.pdf). The table below shows the daily attendance at two movie theaters for 5 days and the mean (average) and the median attendance. Which statistic, the mean or the median, would you use to describe the typical daily attendance for the 5 days at Theater A? Justify your choice.

	Theater A	Theater B
Day 1	100	72
Day 2	87	97
Day 3	90	70
Day 4	10	71
Day 5	91	100
Mean	75.6	82
Median	90	72

SECTION 7.2 Going Beyond the Basics

What do you think?

- How can we compare two sets of data?
- How can we measure how spread out a set of data is?
- How can we see how well two variables are related, like age and height?

In the first section, we examined basic concepts, graphs, and procedures to help us collect and analyze data and to interpret data that was collected by someone else. In this section, we will expand our inquiry. According to the Common Core State Standards, comparing two sets of data is introduced informally in the seventh grade and explored more formally in the eighth grade.

COMPARING TWO SETS OF DATA

In many cases, we collect data because we are comparing two sets of data to one another. For example, we want to compare the test scores of two classes, or we want to compare how well a class performed before and after a lesson. There are two kinds of graphs (boxplots and scatter plots) that help us to understand questions of comparison.

INVESTIGATION 7.2a How Many More Peanuts Can Adults Hold Than Children?

After I had my students collect data about how many peanuts they could hold in one hand, we went to a nearby elementary school and collected these data with third-graders. Below are the data for the children and the data for my students.

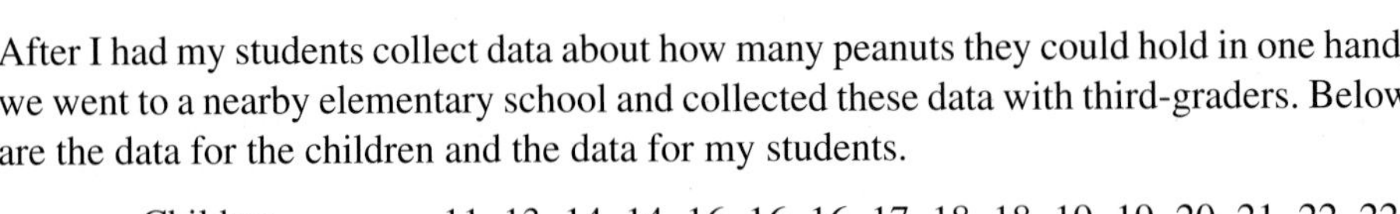

Children	11, 13, 14, 14, 16, 16, 16, 17, 18, 18, 19, 19, 20, 21, 22, 22, 23, 23, 24, 24, 24, 24, 25
College students	18, 18, 20, 22, 22, 22, 22, 22, 23, 25, 25, 25, 25, 25, 26, 26, 27, 27, 30, 30, 32, 32, 37

Explorations Manual 7.9

What kinds of statements can we make about the number of peanuts that children and adults can hold?

DISCUSSION

Just from looking at the data, we can see that the college students could hold more peanuts—the minimum and maximum are greater. Since the numbers of students in each class are identical, we can see that at any point, from lowest to greatest, the amounts held by the college student data are greater.

We could make **back-to-back stem-and-leaf plots**. What additional information about the data can we see from the stem-and-leaf plots below?

Children	Stem	College students
4 4 3 1	1	
9 9 8 8 7 6 6 6	1	8 8
4 4 4 4 3 3 2 2 1 0	2	0 2 2 2 2 2 3
5	2	5 5 5 5 5 6 6 7 7
	3	0 0 2 2
	3	7

Figure 7.21

The stem-and-leaf plots show that the vast majority of the children can hold between 16 and 25 peanuts, and the majority of the adults can hold between 22 and 27 peanuts. This corresponds to the computation of the mean: the adults' average is about 5 to 7 greater than the average for the children.

Look at the line plots of the data. What else can we see or conclude from the line plots below?

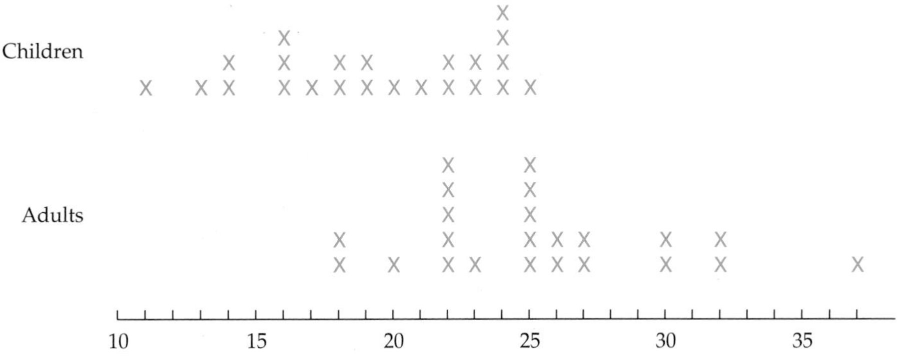

Figure 7.22

We already knew the minimum and maximum. What we get from the line plots is a visual sense of the variation or distribution of the data. We see that the kids are pretty steadily spread out from 11 to 24, no real huge clusters or big gaps or outliers. With the adults, we see a cluster around 22 and another cluster around 25 with some gaps and an outlier. Some students have remarked that if you could pick up the children's data and move it over about 6 or so peanuts, the two sets would be very similar. As with the stem plots, we don't lose any data.

Box-and-whisker graphs According to the Common Core State Standards, in sixth grade students learn about **box-and-whisker graphs**, also known as **boxplots**. Before getting into the details, let us examine the design of this graph, using the children's data. You already know what the median is, and so the idea of dividing a set of data into an upper half and a lower half makes sense. What if we divided each of these halves into half again? In this case, we would have divided the data into quarters, and we could say "this is where the bottom quarter of the data are, the next to the bottom quarter, the next to the top quarter, and the top quarter."

11, 13, 14, 14, 16, 1|6, 16, 17, 18, 18, 19,|19, 20, 21, 22, 22, 2|3, 23, 24, 24, 24, 25

While the "average" is a specific number (which can imply a false sense of exactness about the data), this division of the data into four quarters enables us to specify the middle region. In this case, this division shows us that half of the children held between 16 and 23 peanuts.

Now let's look at the actual procedure for making a boxplot.

1. Find the median of the data. In the case above, there are 22 pieces of data, so the median (19) is between the 11th and the 12th data value.
2. Find the median of the bottom half. In this case, there are 11 pieces of data in the bottom half, so the median of the bottom half (16) is the 6th data value from the bottom. This number is called the **lower quartile**.
3. Find the median of the top half. In this case, there are 11 pieces of data in the bottom half, so the median of the upper half (23) is the 6th data value from the top. This number is called the **upper quartile**.

   We now have five numbers: the minimum, lower quartile, median, upper quartile, and the maximum. In one sense, these five numbers give us a short and sweet summary of the data, more than just the average, but less than a line plot.

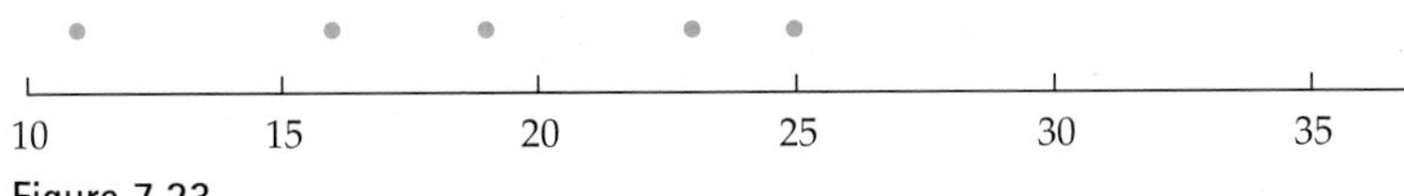

Figure 7.23

4. Draw two line segments (the whiskers) to show the bottom and top quarter of the data. Draw two rectangles (the boxes) to show the two middle quarters of the data.

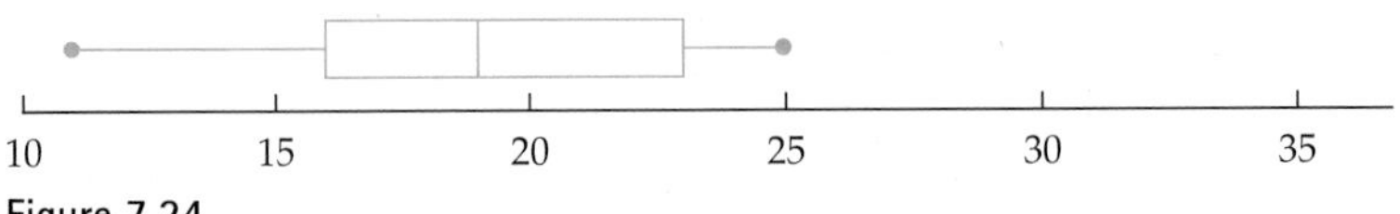

Figure 7.24

Connecting this picture to the previous paragraph, you can see why some authors refer to the boxplot as the **five-number summary**. Now look at the boxplots of the children and college students, one above the other. What can you conclude just from looking at the boxplots?

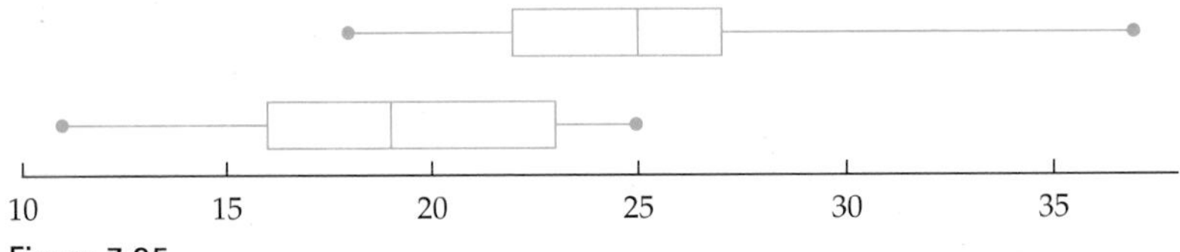

Figure 7.25

	Children	Adults
Range	11–25	18–37
Median	19	25
Bottom quarter	11–16	18–22
Middle half	16–23	22–27
Top quarter	23–25	27–37

Looking at the medians and/or looking at the various regions, it makes sense to say that the average adult can hold about 5 to 7 more peanuts. We can also make other statements. For example, the top children (23–25) are about where the average is for the adults.

We have used the term *outlier* informally thus far. However, how far outside of the rest of the data does a value have to be to officially be considered an outlier? Holding 37 peanuts is a lot compared to the others, but is it big enough to be considered an outlier? To determine whether a data value is formally an outlier, we first find the difference between the upper and lower quartiles. We call this amount the **interquartile range (IQR)**. In one sense, the IQR is like a mini-range—it is the range of the middle half of the data. To determine if a data value is an outlier, we first multiply the interquartile range by 1.5. If a data value is more than 1.5 IQRs lower than the lower quartile or 1.5 IQRs above the upper quartile, we call it an outlier. (It is important to note that not all statisticians use 1.5; even statisticians do not agree on everything!) Apply this procedure to the data on the college students to determine whether there are any outliers, and then read on. . . .

The difference between the upper quartile and the lower quartile is 5 (i.e., $27 - 22$) and $5 \times 1.5 = 7.5$. Thus, to determine the outliers we subtract 7.5 from the lower quartile ($22 - 7.5 = 14.5$) and add 7.5 to the upper quartile ($27 + 7.5 = 34.5$). Thus, any number below 14.5 and greater than 34.5 is an outlier.

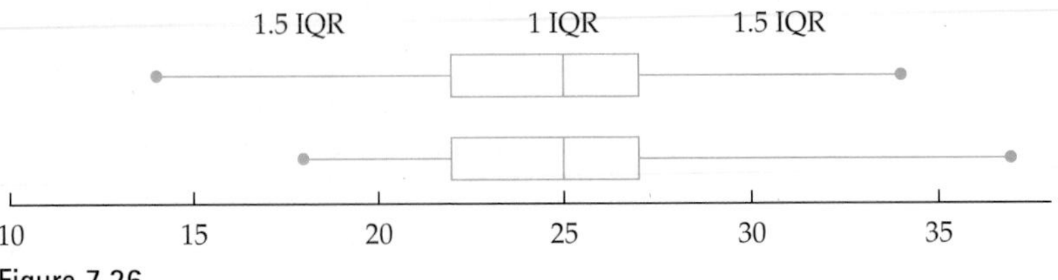

Figure 7.26

So the person who held 37 peanuts is officially an outlier!

Just as we have discussed what the average tells us and doesn't tell us about a set of data, let's look at what the boxplot doesn't tell us: it doesn't show us each individual. Some people would also say that with the boxplot we lose a sense of how the data are distributed, that is, clusters and gaps. Actually, you can get a sense of the distribution with the boxplot, and this has to do with the relative lengths of the boxes and whiskers.

Let's take a set of data where this will be more obvious. Look at the data on the size of classes at a small high school. Both the line plots and boxplots are given. From looking at the two graphs together, can you see what a short box or whisker means and what a long box or whisker means?

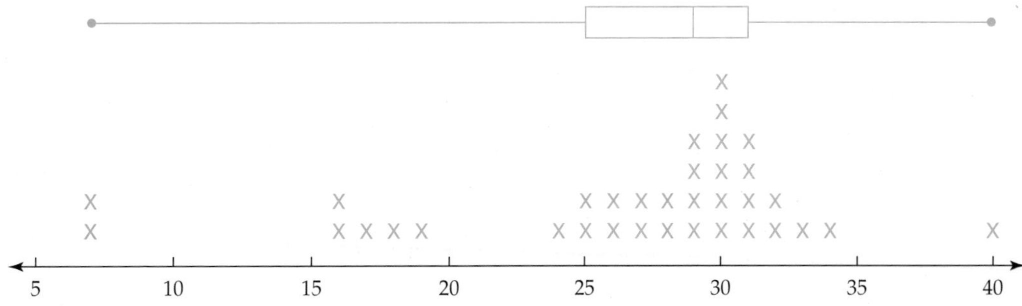

Figure 7.27

A short box or whisker tells us where clusters are, that is, the data are packed into a smaller region. A longer box or whisker tells us that the data are more spread out. Note that a long box or whisker does not necessarily mean there are outliers, because the data might or might not be spread out over this region.

While box and whisker graphs can be useful as a snapshot of one set of data, just like the average is a snapshot, the box and whisker graphs are most often used to give us a snapshot comparison of two sets of data.

INVESTIGATION 7.2b Scores on a Test

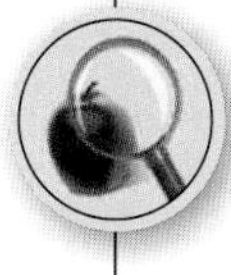

Let's say you have a friend, also a teacher, who has just finished grading his students' midterms. The scores are given below.

77, 96, 58, 100, 66, 76, 88, 73, 94, 75, 76, 84, 91, 74, 87, 92, 67

You ask how his students did, but the teacher does not know much about analyzing data. All he can say is that this is an unusual class. He can see that the scores range from 58 to 100. Take some time to analyze these data. What can you do with the data to help your friend better understand the overall picture of his students' performance? State your conclusions and then read on. . . .

DISCUSSION

After listing the numbers from smallest to largest, we can quickly determine that the median score is 77. Either by observation or from a stem plot, we can also see that grades of C and A dominate the data. Thus grades of C and A are the modal classes for these data. (Recall that when we group the data, we speak of modal classes instead of modes. However, in both cases, the idea is the same: the value or interval that occurs most frequently.)

Look at the histogram and circle graph in Figure 7.28 What do they add to our understanding of the scores? Consider this, and then read on. . . .

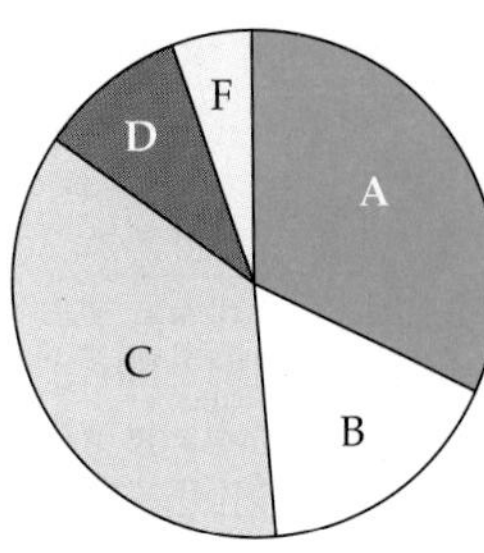

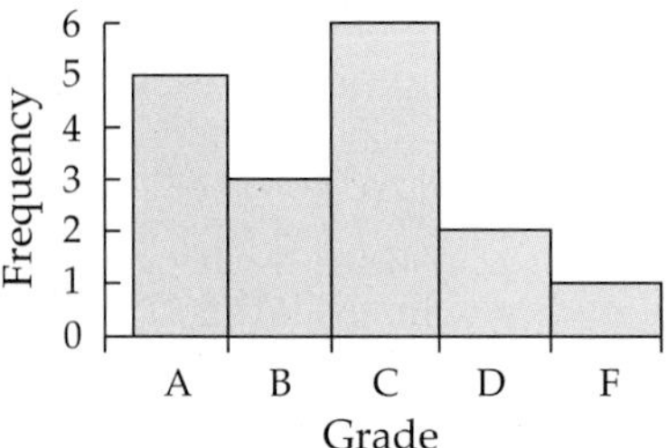

Figure 7.28

From the histogram, we can see that technically the modal class is a grade of C. However, because there is only one more C than A, and because these two intervals stand out, we call this distribution of scores bimodal. From the circle graph, we could frame our analysis optimistically (almost half the class got As and Bs) or pessimistically (more than half the class got C or lower).

What do the line plot and boxplot (Figure 7.29) add to our understanding of the distribution of the scores? Consider this, and then read on. . . .

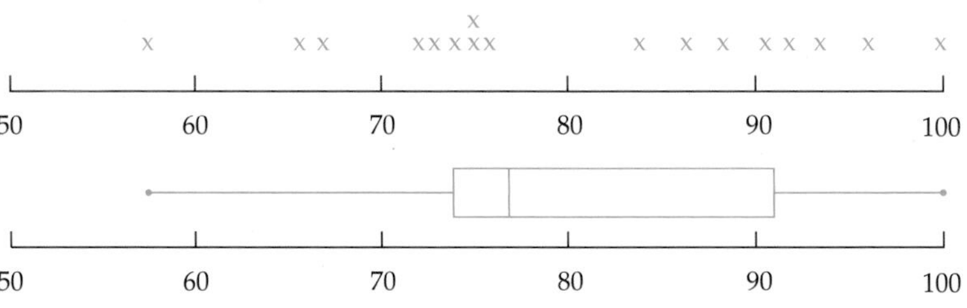

Figure 7.29

The line plot shows two major gaps, one from 58 to 66 and the other from 77 to 84. The line plot also shows a sense of a "top" group and a "bottom" group of the class. From your friend's perspective, this might help him understand why this is an unusual class. If he teaches to the top group, he is likely to lose the bottom group; on the other hand, if he teaches to the bottom group, he is likely to bore the top group.

In this case, because the number of data is small, the boxplot has no advantages over the line plot. From a teaching perspective, an examination of the boxplot is useful, so that you can use it and interpret it when working with larger sets of data. As stated earlier, the boxplot shows the range. Using the idea that the boxplot partitions the data into four roughly equal groups, we can focus on the two boxes and see that about $\frac{1}{2}$ of the students scored between 75 and 90. We can also say that roughly $\frac{3}{4}$ of the class scored above 75. Note that the left-hand box is narrower than the right-hand box. The narrowness of the left-hand box indicates a cluster of scores in that range—that is, the mid-70s.

You will be asked to compare these test scores to another set of test scores in the Exercises.

INVESTIGATION 7.2c Which Battery Do You Buy?

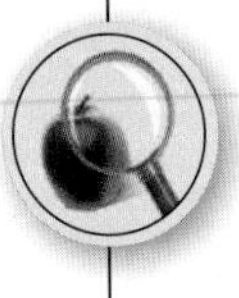

Let's say you were going to buy an expensive, long-life battery for your videocamera. Two companies claim that they have developed batteries that will last an average of 40 hours instead of lasting only a few hours. Both batteries cost the same. What other information would you want to know before making a decision?

Do you think it doesn't matter which you buy? Let's run some numbers. Say you had these data of hours the batteries lasted:

Battery A: 37, 42, 40, 38, 42, 40, 39, 38, 41, 40, 39, 41, 43, 39, 40, 40, 41
Battery B: 30, 48, 44, 36, 42, 36, 40, 34, 46, 38, 40, 40, 42, 44, 38, 40, 32, 50

What analyses of these data might you do to help you make your decision? Take some time to explore these two sets of data, using knowledge you already have. Summarize what you learned and state your conclusions, and then read on. . . .

DISCUSSION

If you did explore the data, you found that for each set, the mean, median, and mode were 40. In this case, the measures of the center tell us nothing about differences between the two sets of data. This occasionally happens with real-life data, and it illustrates a point that statisticians make: Measures of the center of a distribution give an incomplete picture of the data. Let us now discuss some graphs that help us to see how the data are spread out. In this case, because we are comparing two sets of data, we can make a back-to-back stem-and-leaf plot (Table 7.7).

TABLE 7.7

Battery A		Battery B
	5	0
3 2 2 1 1 1 0 0 0 0 0	**4**	0 0 0 0 2 2 4 4 6 8
9 9 9 8 8 7	**3**	0 2 4 6 6 8 8

We see that the data for Battery B are much more spread out (dispersed) than the data for Battery A:

- The range of the data for Battery A is 6 (37 to 43).
- The range of the data for Battery B is 20 (30 to 50).

Here are the line plots and box-and-whisker plots for the batteries. What can we conclude about the differences in the variation from these graphs?

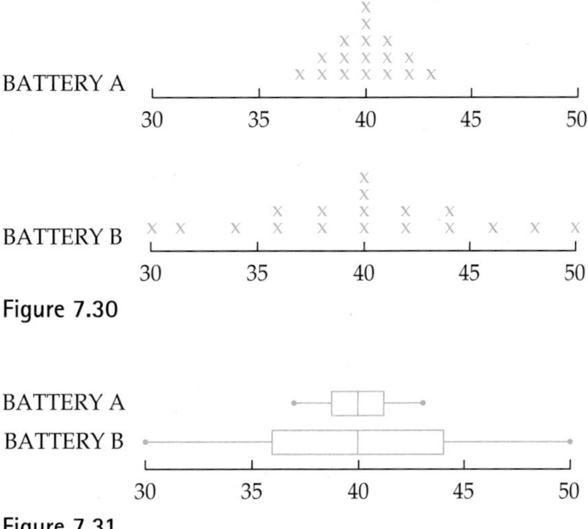

Figure 7.30

Figure 7.31

Both the line plots and the boxplots indicate that the data for Battery A are more clumped together, whereas the data for Battery B are more spread out.

We can make side-by-side histograms. These graphs show that the data for both cases are normally distributed but that the variation for Battery B is much greater; note that the horizontal spacing is different in each histogram.

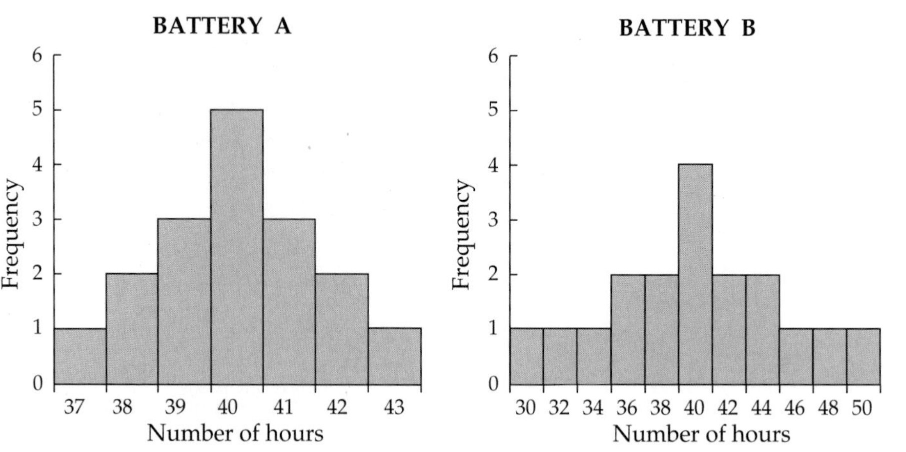

Figure 7.32

From this analysis of the data, most people would likely choose Battery A. Interestingly, their rationale would involve probability language (i.e., with Battery A, the worst battery you are likely to get would last 37 hours, whereas with Battery B, the worst battery you are likely to get would last 30 hours).

MORE TOOLS TO UNDERSTAND VARIATION: MEAN ABSOLUTE DEVIATION AND STANDARD DEVIATION

Thus far we have several tools to describe the variation or spread in a set of data.

Numerically, we have the range and interquartile range. The larger these values, the greater the variation.

Visually, graphs can show clusters, gaps, and outliers.

There are other tools to help us describe variation. Just as average is a number that we can use to give the reader a sense of the typical or middle of a set of data, **standard deviation** and **mean absolute deviation** are numbers that can tell us how spread out the data are. The Common Core State Standards introduce sixth-grade students to the measure of variation called the *mean absolute deviation*. Professional statisticians more commonly use the standard deviation to describe variance. According to the Common Core State Standards, standard deviation is not taught until high school. However, we have this conversation here because often you will see this value when reading standardized test scores or education articles and because it extends the concept of mean absolute deviation.

INVESTIGATION 7.2d Understanding Standard Deviation

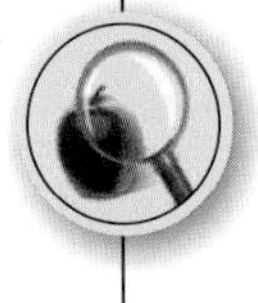

In the previous investigation, we can see that the data for Battery B are more spread out than the data for Battery A. How else might we describe how spread out the data in these two sets are? Think and then read on. . . .

DISCUSSION

One way to do so would be to note how far each data value is from the mean and then take the average (mean) of those numbers. This would give us the mean absolute deviation. Recall that in both cases, the mean life of the batteries was 40 hours. Determine the mean absolute deviation for each of the two classes and then read on. . . .

HISTORY

In the 1830s, a Belgian mathematician, L.A.J. Quetelet, collected measurements of many kinds of data on people, including height, weight, length of arms, and intelligence. He did this for many people. What he found was that in most of these cases, the graphs were normally distributed.

TABLE 7.8

BATTERY A		BATTERY B	
Number of hours	**Distance from the mean**	**Number of hours**	**Distance from the mean**
37	3	30	10
38	2	32	8
38	2	34	6
39	1	36	4
39	1	36	4
39	1	38	2
40	0	38	2
40	0	40	0
40	0	40	0
40	0	40	0
40	0	40	0
41	1	42	2
41	1	42	2
41	1	44	4
42	2	44	4
42	2	46	6
43	3	48	8
		50	10
Total distance from the mean	20	72	
Average distance	20/17=1.2 hours	72/18=4.0 hours	

Table 7.8 shows the computations. If we look at how far each number is from the mean and find the mean of *those* numbers, we can say that for Battery A, the average distance from the mean is 1.2 hours, whereas for Battery B, the average distance is 4.0 hours. Just as the mean, median, and mode are concepts that enable us to give a sense of the middle of the data with a single number, the *mean absolute deviation* enables us to compare the relative spreads of the data with a single number. Because 4.0 is more than 3 times 1.2, if someone were to tell you that the means from both sets of data were equal but that the average distance from the mean for Battery B was 4.0 hours compared with 1.2 hours for Battery A, this number would instantly tell you that the former set of data is more spread out.

What we call the **standard deviation** is quite closely related to the **mean absolute deviation**. For reasons that will be explained very shortly, the computation is slightly more complicated.

Finding the standard deviation To determine the standard deviation, we can use much of the work we did for mean absolute deviation. There are two more steps (Table 7.9).

To determine mean absolute deviation:	*To determine standard deviation:*
Find the distance from the mean of each data value.	Find the distance from the mean of each data value.
	Square each of these numbers.
Add this column of numbers.	Add this column of numbers.
Divide this sum by the number of data values.	Divide this sum by the number of data values.
	Take the square root of this number.

TABLE 7.9

	BATTERY A			BATTERY B		
	Score	**Distance from the mean**	**Square of the distance**	**Score**	**Distance from the mean**	**Square of the distance**
	37	3	9	30	10	100
	38	2	4	32	8	64
	38	2	4	34	6	36
	39	1	1	36	4	16
	39	1	1	36	4	16
	39	1	1	38	2	4
	40	0	0	38	2	4
	40	0	0	40	0	0
	40	0	0	40	0	0
	40	0	0	40	0	0
	40	0	0	40	0	0
	41	1	1	42	2	4
	41	1	1	42	2	4
	41	1	1	44	4	16
	42	2	4	44	4	16
	42	2	4	46	6	36
	43	3	9	48	8	64
				50	10	100
Sum		20	40		72	480
Sum/n		Average distance = 1.2	2.35		Average distance = 4.0	26.7
$\sqrt{\frac{\text{sum squares}}{n}}$			Standard deviation = 1.53			Standard deviation = 5.16

Now that we have examined the meaning of the standard deviation and developed the procedure for determining it, you are more likely to be able to make sense of the algebraic definition of the standard deviation.

If we have a set of data whose values are denoted $x_1, x_2, x_3, \ldots, x_n$, and the mean of these data is represented as $\bar{x}$, then the standard deviation is

$$\text{Standard deviation}^6 = \sqrt{\frac{(x_1 - \bar{x})^2 + (x_2 - \bar{x})^2 + (x_3 - \bar{x})^2 + \cdots + (x_n - \bar{x})^2}{n}}$$

In this case, the standard deviation for the data for Battery A is 1.53, compared to the standard deviation of 5.16 for Battery B. Like the mean absolute deviation, the two standard deviations tell us that the data for Battery B are much more spread out. Whether we use the mean absolute deviation or the standard deviation, this single number can tell us how spread out the data set is.

NORMAL DISTRIBUTION

Let's examine more closely the normal distribution because standard deviation has the most power with data that are normally distributed.

A line graph depicting a set of data that are normally distributed is often referred to as a bell-shaped curve; that is, the frequency values are highest in the middle, and the graph is symmetric. Suppose we have collected data on the heights of all students at a small college. As you have seen, we could make a histogram from these data. The graph at the left in Figure 7.33 shows how the actual histogram might look in this ideal case; that is, the graph is symmetric. If we trace a smooth line over the tops of the bars, we have the line graph at the right in Figure 7.33. When normally distributed data are visually depicted, the writers generally show a line graph. For example, in the case of the heights at the college, the figure on the right is a smoothed-out line graph for the heights. However, in actuality the heights have been measured in whole inches, and therefore the histogram on the left is a truer depiction of the data. Thus, whenever you see a "normal" curve, it will help if you realize that this curve is an idealization of a histogram. For example, we say that SAT scores and IQ scores are normally distributed.

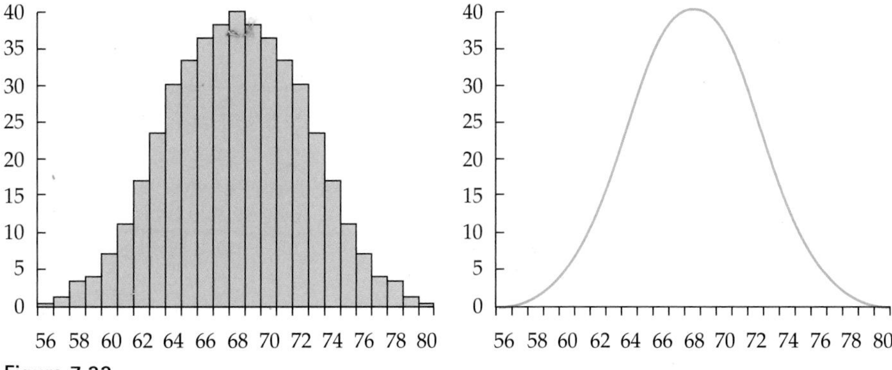

Figure 7.33

Using standard deviation The reason why we use the standard deviation has to do with some rather amazing generalizations that we can make about a set of data that is normally distributed.

When a set of data is normally distributed, we can make the following conclusions (Figure 7.34):

- 68% of the data lie within 1 standard deviation of the mean.
- 95% of the data lie within 2 standard deviations of the mean.
- 99.8% of the data lie within 3 standard deviations of the mean.

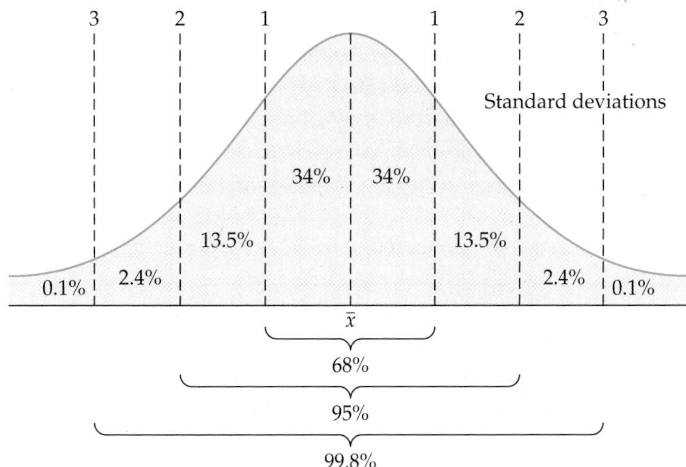

Figure 7.34

Let us apply these conclusions to our data for Battery A. Determine what percent of the data lie within 1 standard deviation of the mean. Then read on. . . .

In this case, the mean was 40 hours and the standard deviation was 1.5 hours. Thus we are looking for how many data points lie between 40 − 1.5 and 40 + 1.5—that is, between 38.5 and 41.5. We find that 11 of the 17 data points lie within this range, and $\frac{11}{17}$ is equivalent to 65%. If we look at the data for Battery B and find how many data points lie within 1 standard deviation of the mean, we are looking for data points that lie between 40 − 5.2 and 40 + 5.2—that is, between 34.8 and 45.2. We find that 12 of the 18 data points lie within this range, and $\frac{12}{18}$ is 67%.

It is important to note that these numbers (68%, 95%, and 99.8%) are theoretical figures that are realized when the number of data is very large and when the data are "perfectly" normally distributed. However, even when the number of data is smaller and the distribution is not perfectly normal, we will generally find close to 68% of the data lying within 1 standard deviation of the mean.

Let us now consider some examples from real life that use the concept of standard deviation.

INVESTIGATION 7.2e Analyzing Standardized Test Scores

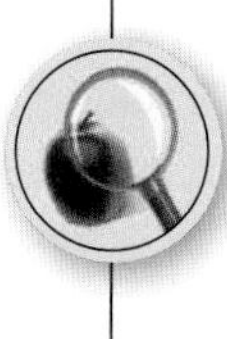

Let's say that a standardized test has a mean of 250 and a standard deviation of 50. Suppose 2000 students took the test, and their scores were normally distributed. Think about the following questions, and then read on. . . .

A. How many students would you expect to score over 350?

B. How many students would you expect to score at least 200?

DISCUSSION

A. Because a score of 350 is 2 standard deviations above the mean, we conclude that 2.5% of the scores will be over 350 [Figure 7.35(a)]. Then we need to compute 2.5% of 2000. We can use a calculator, or we can just as quickly do it mentally: 10% of 2000 is 200, so 5% is 100, and 2.5% is 50.

B. This question is slightly more complex. The answer is that about 1680 students will score at least 200. Figure 7.35(b) illustrates one solution path. If we think of the results of the test in terms of a histogram, we can see that all the shaded bars represent scores over 200. When we superimpose this histogram on the graph showing the percent of scores with respect to standard deviations, we find that this means that 84% of the students scored at least 200.

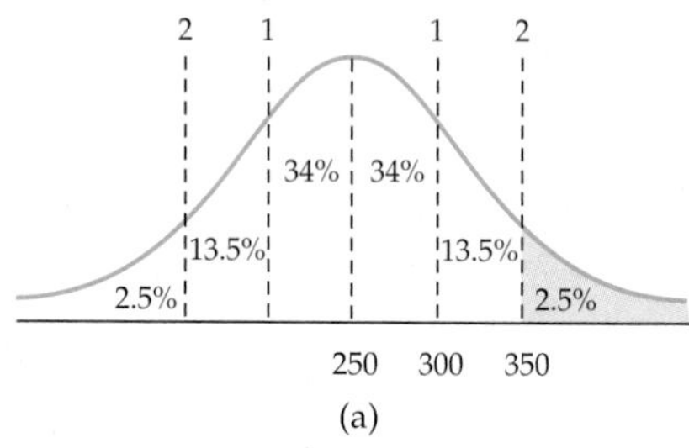

(a)

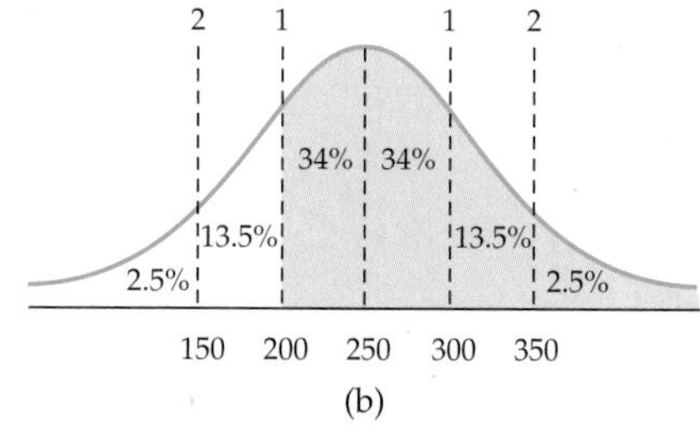

(b)

Figure 7.35

MATHEMATICS

When the SAT was first introduced, it was calibrated so that it had a mean of 500 and a standard deviation of 100. Over the past 20 or so years, the average score has declined considerably. In 1995, the Educational Testing Service "recentered" the test so that the mean would once again be 500. However, this provoked quite a controversy. Why do you think the ETS made the proposal? Why do you think some people objected? Think about these questions before reading on. . . .

The rationale given by the ETS was that if 500 were the "average" score, then the scores would be more meaningful. That is, if a student's score was above 500, that student would know that the score was above average. Many people objected because such a move would make it appear that scores had not gone down.

The concept of standard deviation is used regularly by industries. Let us examine one such example.

INVESTIGATION 7.2f

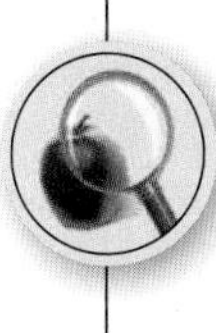

How Long Should the Tire Be Guaranteed?

A tire company has developed a new tire and has tested it extensively. The results of the tests showed that the "average" tire lasted 44,000 miles, that the distribution of wearing was normal, and that the standard deviation was 2500 miles. If the company guarantees that the tire will last 39,000 miles, what percent of the tires are likely to wear out before 39,000 miles and thus be subject to refund?

DISCUSSION

Again, a graph makes the solution to this problem much easier (Figure 7.36). From the graph, we can conclude that 2.5% of the tires are likely to wear out before 39,000 miles.

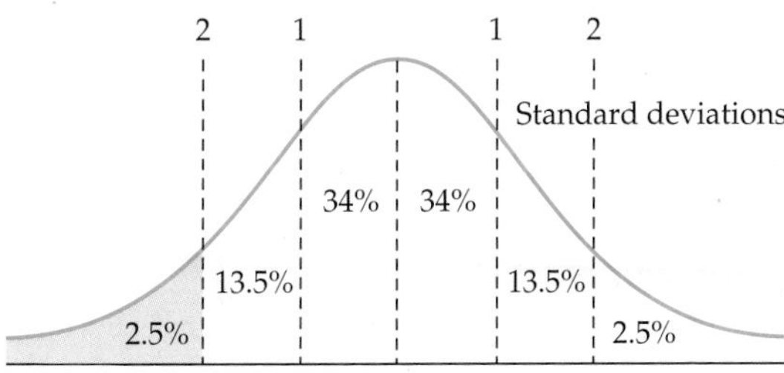

Figure 7.36

What if the company wanted to guarantee 40,000 miles? What percent of the tires are likely to wear out before 40,000 miles? In this case, you cannot get an "exact" answer. What could you do to increase the accuracy of your estimate? This question will be left as an exercise.

SCATTER PLOTS

The last graphical representation of data we will examine is the **scatter plot**, which appears in eighth grade according to the Common Core State Standards. Scatter plots help us see relationships. We have discussed relationships as one of the themes in mathematics throughout the textbook. On many occasions, when we examine two sets of data, we expect a relationship. For example, we expect that older children are generally taller than younger children. On the other hand, sometimes we expect two variables to be related and they are not (e.g., air fares and distances between cities).

Let us begin with a situation where we expect and find a clear and strong relationship. The scatter plot in Figure 7.37 shows the ages and heights of a group of elementary school–aged children. How would you describe the shape of the graph?

OUTSIDE THE CLASSROOM

Given our discussion of truncation in the previous section, some students question why both axes are truncated (i.e., do not start at 0). Make your own scatter plot and begin both axes at 0. What do you notice?

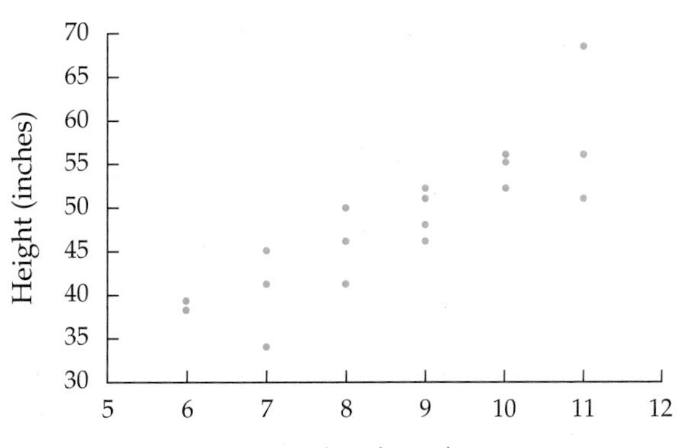

Figure 7.37

LANGUAGE

You can see that there are two outliers from this data set. How would you describe the outliers in everyday English?

We would say that the outlier who is 7 years old is short for his or her age and the outlier who is 11 years old is tall for his or her age.

Some people say that as the age increases the height increases. Others will say that older children are generally, but not always, taller than younger children. Others will say that the data seem to move from the bottom left corner to the top right corner. This wording connects to a relationship you have explored before—linear relationships. The words *linear* and *line* have the same root. Can you "see" a line that best fits the data, that is, a line where most of the points would either be on the line or very close? If you have a pencil, sketch such a line. When virtually all of the points lie on the line or very close, we say that there is a strong linear relationship between the variables. This is useful in science and profitable in business because it means we can predict the value of the dependent variable when we know the value of the independent variable.

Let us continue our investigation of scatter plots with the bungee jump, which interests kids and where the bungee jump operator has to be able to predict the value of the dependent variable (length of the cord) based on the value of the independent variable (weight of the jumper). The children set up a model of a bungee jump—a ruler, a rubber band, a paper clip, and weights that are attached to the paper clip. We know there is a relationship between the weight of a person and how much the bungee cord stretches. The owners of the bungee jump company had better get it right. If the cord stretches more than they predicted, the person dies! Here are the data and the scatter plot for these data (Figure 7.38).

Number of weights	Stretch of the rubber band (cm)
1	0.9
2	1.2
3	1.5
4	1.8
5	2.0
6	2.4
7	3.0
8	3.3
9	3.7
10	4.1
11	4.5
12	5.0
13	5.5
14	5.9
15	6.5
16	6.9
17	7.4
18	8.1
19	8.6

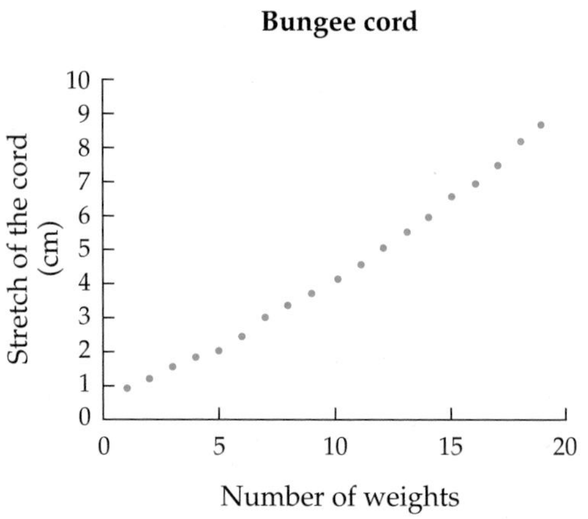

Figure 7.38

MATHEMATICS

You might notice that the line that would "best fit" these data might not be a straight line but rather a curved line. We would need more data and a better model to be more certain. This observation is important because if this is true and the operator takes the average increase for the first 11 weights (about 0.35 cm for each increment in weight) and then uses that to predict how long the cord will be for the 19th weight, the operator would conclude that the length of the cord for that weight would be 7.3 cm when it is actually 8.6 cm! In real life, this would be a catastrophic mistake.

How would you describe the relationship between weight and length?

There is a strong, positive relationship. While the length doesn't increase by the same exact amount each time, the increase is steady and similar. We can draw a straight line that would be a pretty good fit.

Let us consider another example where there is a relationship of a slightly different kind. The College Board has published average SAT scores for states for many years. There is quite a bit of variation and more than just "bragging rights" for states that do well. However, it is not quite so simple, because using those scores alone is like comparing apples and oranges. For example, let's say there are two similar high schools, but in one high school all seniors take the SAT while in the other school, only those going to college take the SAT. Obviously, we would expect the overall average to be higher for the second high school. The participation rates for states vary widely. Therefore, if we are going to compare scores, we had better check this out.

The scatter plot in Figure 7.39 shows the relationship between average SAT math scores for all 50 states in 2004 and the participation rate for the states. What do you see?

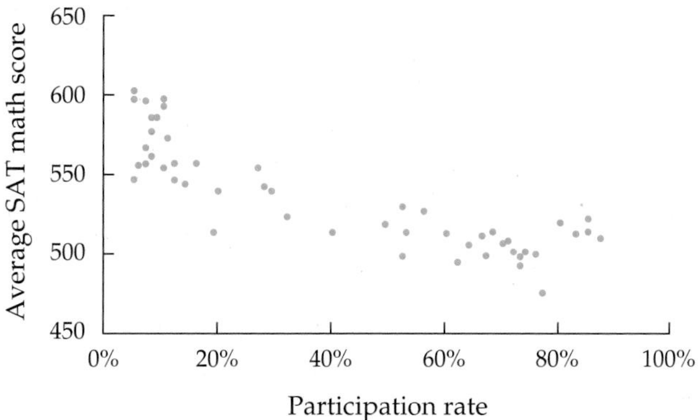

Figure 7.39

In this case, there is "sort of" a linear relationship, but not as strong as the bungee jump or even the ages vs. heights. Yet the data do generally move from the top left to the bottom right. That is, in general, states with lower participation rates have higher scores, and states with higher participation rates have lower scores. When we have data like this, we describe the relationship as a negative relationship. The reason for this is simple and connects to basic algebra: A line that goes up as you go right has a positive slope, and a line that goes down when you go right has a negative slope.

Let's go back to the peanuts data for our last example. It was noticed that there was variation in the number of peanuts that people could hold. Someone hypothesized that people with larger hands should hold more peanuts. We decided to investigate this hypothesis. First, we had to decide how to measure hand size. After some discussion, we decided to measure what is called the hand span: The person stretches their fingers as far apart as possible, and then we measure the distance from the top of the thumb to the top of the last finger. Here are our data and scatter plot (Figure 7.40). What do you think? Do you see a correlation between hand size and number of peanuts held?

Hand size (cm)	Number of peanuts
15.5	25
17	22
17.5	25
18	18
18.5	23
19	22
19	30
19	26
19	32
19.5	29
19.5	22
20	26
20	22
20.5	20
20.5	25
21	32
21	29
21	18
22	22
23	31

Figure 7.40

There really isn't much of a linear relationship here at all. You *could* try to draw a line from bottom left to upper right, but no matter what slope you chose, many points would not be very close to the line. In this case, statisticians say that there is either no relationship or a weak relationship.

Just as average is a number that can give us a "sense" of the center of a set of data and standard deviation is a number that can give us a "sense" of how spread out the data are, there is a number (between −1 and +1) called the **correlation coefficient** that can give us a sense of the extent to which there is a linear relationship. While the formula is beyond the scope of this course, we can say that the closer to 1 the absolute value, the stronger the linear relationship. A larger negative number implies an inverse relationship; a larger positive number implies a positive relationship.

Here are the correlation coefficients for the sets we have examined:

Age vs. height	0.86
Bungee cord weight vs. height	0.99
SAT scores vs. participation rate	−0.85
Hand vs. peanuts	0.20

In this chapter, you have seen the development of three numbers that we can use to measure a set of data.

The number	**What it tells us**
Average	The region where most data are likely to occur
Standard deviation	How spread out the data are
Correlation coefficient	How closely we can predict the relationship (how close to linear)

If you have another area where you are competent (e.g., playing guitar, rock climbing, or skiing), you have experienced that sense of "wow, this is great" when your knowledge has developed to the extent that you can understand and use ideas or tools that just didn't make sense at first. So too with mathematics. It's really a lot more similar to other kinds of knowledge than most people think!

INFERENTIAL STATISTICS

In this section, we have begun to move from descriptive statistics, where we use data and graphs to summarize and describe a set of data, to inferential statistics where we take data from a **sample** (part of the whole population of data), and interpret that data to make predictions. Concepts of sampling are introduced in seventh grade, according to the Common Core State Standards.

The beginnings of inferential statistics are laid when the children begin to realize that all populations may not be the same. That is, the favorite sports of fourth-graders may not be the same as those of eighth-graders; the favorite sports of children in Los Angeles may not be the same as those of children in rural Georgia.

A next step is when they realize that we can't always gather data of the whole population.

One way not to get a representative sample is to get what we call a **convenience sample**. For example, let's say we wanted to determine how many hours a week college students study. It would be pretty convenient to sample students as they came out of a freshman dormitory or the library. But both of those samples would definitely not be representative. Can you explain why?

HISTORY

A classic example of a sampling error occurred in the 1936 presidential election. The *Literary Digest* sent questionnaires to 10 million voters. Of the 2,266,566 responses, 1,293,669 were for Landon and 972,897 were for Roosevelt. That is, more than 57% were for Landon. In fact, Roosevelt received 62% of the vote! What happened?

Analysis of the sampling process showed two flaws. First, the editors did not send the survey to a random sample of potential voters. They had gotten their sample from telephone directories and lists of automobile owners. Second, less than one fourth of the people responded to the survey. These two factors caused their error. Why wasn't the *Literary Digest* sample a random sample?

In those days, a much smaller fraction of the population owned telephones and/or cars. Therefore, the sample was not representative but, rather, was biased toward upper-income voters.

If your sample was taken outside the freshman dormitory, then you would be inferring that the overall distribution of freshmen is the same as for sophomores, juniors, and seniors. If your sample was taken outside the library, then you would be missing those students who never go to the library and under-representing those students who seldom go to the library.

Sometimes we need a **stratified random sample**, a sample in which there are distinct subsets of the population that need to be proportionally represented. For example, if we are surveying students on campus and 25% of the students are married, we might want to make sure that 25% of our sample are married students. In another situation, let's say that 50% of the voters in a district are white, 30% are black, and 10% are Hispanic. In this case, we will want to make sure that our sample contains these proportions also.

MATHEMATICS

The margin of error is usually 1 standard deviation. Thus, if a poll says 61% plus or minus 3%, they are saying that there is a 95% probability that if we were to poll the entire population, the actual percentage would be between 58% and 64%.

As we have said before, statistics does not have that same exactness that most other fields of mathematics have. Thus, the *very nature of sampling* means that our results do not have that same exactness as $2 + 2 = 4$. When people taking polls give their results, they also give the **margin of error**. For example, if a poll states that 61% of the population agree with the president's position on a certain issue, they generally give the margin of error. Let's say the margin of error is 3%. This means that 61% of the sample agreed with the president, and they are pretty certain that the actual percentage who agreed, if you asked everyone, would be within 3% of 61%.

Another concept that needs to be addressed in sampling is the question of **sample size**. That is, how big a sample do you need in order to be confident that the statistics for this sample will be essentially the same as for the entire population? Even though I studied statistics, I must admit to still being surprised that a sample size of only a few thousand can be used to collect data that are then generalized for the entire population of the United States.

We also need to think about **response rate**. For example, let's say a large university decided to see what percentage of students who graduated in May got teaching jobs that September. Let's say they generated a sample of 100 students and sent out questionnaires and e-mails in September, and only 50% of the students responded. Would you feel comfortable making your inferences on those 50 students? How might the 50 nonrespondents be different from the 50 respondents?

Knowing human nature, it is very likely that some of the people didn't respond because they had not gotten jobs and did not want the university to know.

INVESTIGATION 7.2g Comparing Students in Three Countries

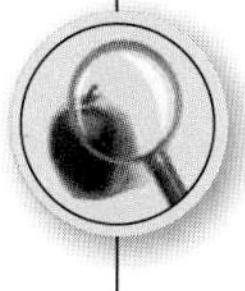

Figure 7.41 is taken from a book titled *Making the Grade in Mathematics: Elementary School Mathematics in the United States, Taiwan, and Japan.* Before reading on, look at the graph. Take some time to understand what information the authors are trying to present. What questions would you ask the authors in order to better understand the data? Then write down what you think they are saying. Then read on. . . .

CLASSROOM CONNECTION

Grade 5

Why do you think a larger sample is more trustworthy than a smaller sample? 

Date ____________ Time ____________

LESSON 6·5

Sampling Candy Colors

1. You and your partner each take 5 pieces of candy from the bowl. Combine your candies, and record your results in the table under Our Sample of 10 Candies.

Candy Color	Our Sample of 10 Candies		Combined Class Sample	
	Count	Percent	Count	Percent

2. Your class will work together to make a sample of 100 candies. Record the counts and percents of the class sample under Combined Class Sample in the table.

3. Finally, your class will count the total number of candies in the bowl and the number of each color.

   a. How well did your sample of 10 candies predict the number of each color in the bowl? ____________

   b. How well did the combined class sample predict the number of each color in the bowl? ____________

   c. Do you think that a larger sample is more trustworthy than a smaller sample? ____________

   Explain your answer. ____________

180

From *Everyday Mathematics, Grade 5:* The University of Chicago School Mathematics Project: Student Math Journal, Volume 1, by Max Bell et al., Lesson 6-5, p. 180. Reprinted by permission of The McGraw-Hill Companies, Inc.

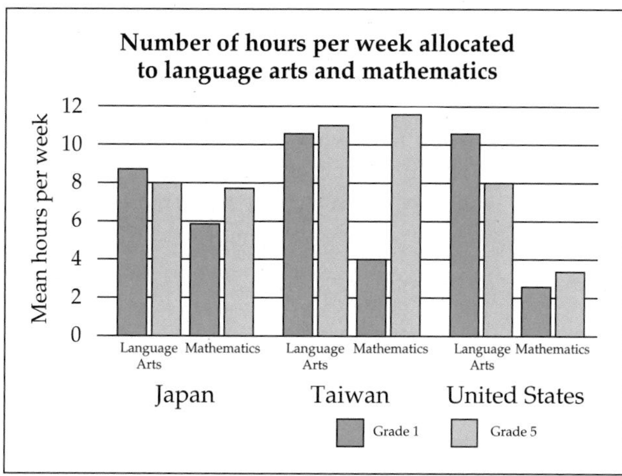

Figure 7.41
Source: Harold W. Stevenson, Max Lummis, Shinying Lee, and James W. Stigler, *Making the Grade in Mathematics: Elementary School Mathematics in the United States, Taiwan, and Japan* (Reston, VA: NCTM, 1990), p. 16. Reprinted with permission from *Making the Grade in Mathematics*, copyright 1990 by the National Council of Teachers of Mathematics. All rights reserved.

DISCUSSION

In their report, the authors state that three cities were selected: Minneapolis (U.S.), Taipei (Taiwan), and Sendai (Japan). They chose "representative samples of ten schools in each city. Within each school, we randomly selected two first-grade and two fifth-grade classrooms. . . . Thus, our report deals with data from 1440 children (240 first-graders and 240 fifth-graders in each of the three cities)."[7] The data were collected in the early 1980s.

Note that there are several cases in which the authors needed to ensure that the samples were representative. All of these choices could affect the results: the choice of cities, the choice of schools, the choice of classrooms, and the choice of students.

For this investigation, we will focus on the differences among the countries in grade 5 mathematics. How would you summarize the differences? Think and then read on. . . .

First, let us make sure we are interpreting the information correctly. The numbers refer to the numbers of hours per week allocated to mathematics. The graph indicates that these are means (as opposed to medians or modes).

Now let us analyze the wording. Does "allocate" mean how much time is scheduled for mathematics, or does it mean how much time is actually spent on mathematics? I was not able to infer from the book exactly how the data for this graph were gathered. My interpretation is that the hours allocated to mathematics were determined by adding up the periods of time scheduled for mathematics in the teacher's plan book.

Now let us examine the hours allocated to mathematics each week in grade 5 and determine how we might convey that information to other people. Looking at the graph, we can conclude that in Japan, the number of hours allocated is about 8; in Taiwan, the number of hours is about 12; and in the United States, the number of hours is slightly over 3, or about $3\frac{1}{2}$ hours.

You have several options for how you write the news item:

1. *Simply report the numbers.* "In the ten Japanese schools, almost 8 hours per week were allocated to math in the fifth grade, compared to almost 12 hours per week in the ten Taiwanese schools and about $3\frac{1}{2}$ hours per week in the ten U.S. schools."

2. *Make additive comparisons.* "The Japanese schools allocate about $4\frac{1}{2}$ hours more per week than the U.S. schools, and the Taiwanese schools allocate about $8\frac{1}{2}$ hours more per week than the U.S. schools."

3. *Make multiplicative comparisons.* "The Japanese schools allocate more than twice as many hours per week as the U.S. schools, and the Taiwanese schools allocate more than three times as many hours per week as the U.S. schools."
4. *Make multiplicative comparisons using percent language.* "The Japanese schools allocate about 130% more time than the U.S. schools, and the Taiwanese schools allocate about 240% more time than the U.S. schools."

WEIGHTED AVERAGE

We will consider one more concept in this section: **weighted averages**. In our work thus far with the mean, we have encountered situations in which we computed the simple mean. However, there are many situations in which we use what we call weighted means. Let us investigate some examples.

INVESTIGATION 7.2h Grade Point Average

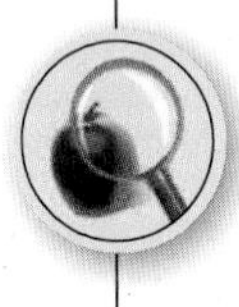

Table 7.10 gives Ed's semester grades. First estimate his grade point average (GPA), and then compute his actual GPA. Recall that A = 4 points, B = 3 points, C = 2 points, and so on. How can you explain and justify your estimation and computations?

TABLE 7.10

Course	Grade	Credits
Mathematics 151	B	3
Philosophy 205	A	3
Computer Science 100	C	3
Biology 102	A	3

DISCUSSION

One way of estimating is to use the concept of balance; recall the seesaw analogy. Where would the balance point be in this case?

C	B	A
x	x	x
		x

This helps us see that it would be between the B and the A. If you place the wedge under the B, the seesaw will not balance there because the two As "weigh" more than the one C. Numerically, we can determine Ed's average by taking the mean of the four grades:

$$\frac{(3 + 4 + 2 + 4)}{4} = 3.25$$

What if the computer course had been only a 1-credit course? What would Ed's GPA be then? Think and read on. . . .

Intuitively, many students realize that his GPA will go up, since the low grade is now worth only 1 credit—that is, it doesn't "count as much" as the others. We can capitalize on this intuitive sense by connecting it to the idea of unit, which we have seen is one of the big ideas of mathematics. That is, we can count the B in mathematics (worth 3 points) three times, the A in philosophy 3 times, the C in computer science only once, and the A in biology three times: (3 + 3 + 3) + (4 + 4 + 4) + 2 + (4 + 4 + 4) = 35. We divide the sum by 10 (i.e., the number of credits Ed is taking), and so we find that his GPA is 3.5 (Table 7.11).

CLASSROOM CONNECTION

There was a time when many (if not most) educators thought that grades should be normally distributed. Thus the grades were curved so that the "appropriate" proportion of students received grades of A, B, C, D, and F. In this course, I have found that the distribution of grades is often bimodal. That is, I often have a large proportion of students who do well (A), a large proportion of students who struggle to master the basic ideas (C), and a smaller proportion of students earning the other grades.

TABLE 7.11

Course	Grade	Numerical equivalent	Credits	Grade points
Mathematics 151	B	3	3	9
Philosophy 205	A	4	3	12
Computer Science 100	C	2	1	2
Biology 102	A	4	3	12
			10	35

We can understand the most efficient procedure for determining GPA by connecting our understanding that multiplication can be seen as repeated addition. We now multiply the numerical equivalent for each grade by the number of credits to get the grade points for each course. Then we add the grade points to get the total number of grade points for the whole semester and divide by the number of credits.

INVESTIGATION 7.2i What Does Amy Need to Bring Her GPA Up to 2.5?

Now let us use this knowledge to solve a problem many college students face: determining how their grades in the present semester will affect their overall GPA. At Keene State College, education majors need to have a 2.5 GPA in order to be able to student-teach. (You might want to check the policies at your college if you haven't already done so.)

Let's say Amy didn't do so well her first three semesters in college (Table 7.12). What GPA will she need this semester (she is taking 16 credits) in order to bring her overall GPA up to 2.5?

TABLE 7.12

	Credits	GPA
Beginning of semester	50	2.35
Spring semester	16	?
Total	66	2.50

DISCUSSION

We can use the balance idea again. The 50 credits Amy has already taken will count more than the 16 credits she is taking this semester. Because the ratio of 50 to 16 is approximately 3 : 1, we can represent the problem by placing three weights at 2.35 and the balance point at 2.5 (Figure 7.42). How far away must this semester's GPA be in order to balance the 2.35?

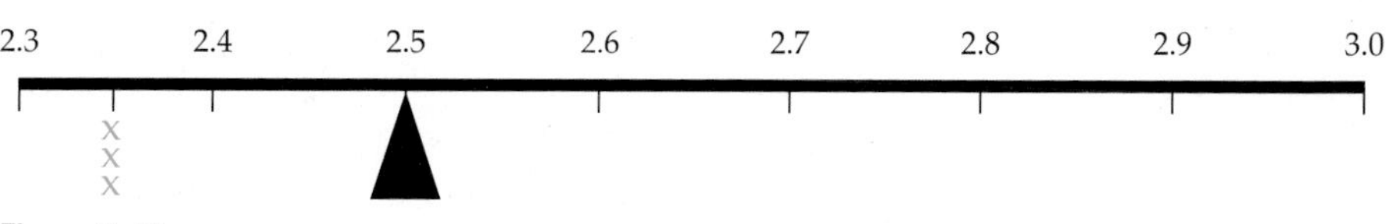

Figure 7.42

If we combine the idea of ratios and number lines with our "balance sense," we realize that the one weight (representing this semester) will have to be three times as far from 2.50 as is 2.35. Therefore, we can estimate that she will need a GPA of about 2.95.

One way to determine the exact answer is to determine how many total grade points she has now (2.35×50) and how many total grade points she needs if she is to have a GPA of 2.5 at the end of the semester (2.5×66). The difference of these numbers will tell us how many grade points her 16 credits this semester have to provide.

That is, $2.35 \times 50 = 117.5$ and $2.5 \times 66 = 165$. We now divide the difference of 47.5 by 16, and we find that she must get a GPA of at least 2.97 to bring her GPA up to at least 2.5. As you can see, the estimate was very close.

SUMMARY 7.2

In this section, we have examined how boxplots can help us to compare two sets of data and how scatter plots can help us to see how closely two variables or sets of data are related.

We have also examined different ways to quantify the dispersion (or spread) of a set of data, from simple ideas like range, clusters, gaps, and outliers, to more complex ideas like interquartile range and standard deviation.

We have learned that the standard deviation is a measure of how spread out the data are. When a set of data are normally distributed, we can make powerful statements about subsets of the data.

We began the transition from descriptive statistics to inferential statistics with the idea of sampling. Taking measures to ensure that the sample is as representative of the total population as possible involves taking measures to ensure that the samples that are obtained are random samples. In cases where there are distinct subgroups within the population, this often involves stratified random samples. We learned that statisticians need to think about the sample size when sampling. When conducting surveys, we also need to consider the response rate.

Finally, we examined the notion of weighed means. We connected the computation of weighted means and simple means, and we connected the concept of both kinds of means—that is, the notion of the balance point of the set of data.

7.2 Exercises

1. A class collected exercise on each person's sitting heart rate:

65 52 74 67 81 63 58 71 62 68 75 68 69 67 59 69

a. Make a line plot and summarize what the line plot tells us about the data.

b. Make a stem-and-leaf plot and summarize what the stem-and-leaf plot tells us about the data.

c. Make a boxplot and summarize what the boxplot tells us about the data.

d. If you were to collect this data on your class, describe the exact procedure you would use to minimize measurement variation.

2. Below is a stem-and-leaf plot for class sizes in education classes at the college.

0	6 8 9
1	6 7 7 8
2	0 0 1 2 4 4 5 5 6 8 9
3	1 1 2 3 3
4	
5	0 1

a. How many students were in the smallest class? the largest class?

b. What is the range?

c. Estimate the mean and describe your reasoning.

d. Determine the mean and median.

e. Determine the mode and modal class. Which do you believe would be more appropriate to use in an article about average class sizes?

3. Below are exam scores from two sections of the same course?

45 64 67 70 72 72 75 76 80 81 82 83 84 88 94 98
62 63 70 71 73 74 74 74 74 76 77 79 81 82 88 96

a. Make two back-to-back stem-and-leaf plots.

b. What conclusions can you make about the two classes from the plots, both similarities and differences?

c. Make two boxplots.

d. What conclusions can you make about the two classes from the boxplots, both similarities and differences?

e. What are the advantages and disadvantages of representing these data with stem-and-leaf plots as opposed to boxplots?

4. Here are data on an exam:

63 63 75 76 77 77 77 85 85 85 89 90 91 95 98

a. First make a stem-and-leaf plot by decades: 60s, 70s, 80s, 90s.

b. Describe three conclusions you can make from this plot.

c. Now make a stem-and-leaf plot by half decades (e.g., the intervals are 60–64, 65–69, etc.).

d. Describe at least one different impression/conclusion that comes from this plot.

5. Suppose the instructor in Investigation 7.2b gave the same test to another class whose scores are given below. Tell the instructor how the students did in each class and how the classes compare.

Class 2: 96, 94, 93, 92, 89, 85, 85, 85, 83, 80, 79, 77, 77, 77, 75, 74, 74, 68, 67, 65, 65

6. ***Classroom Connection*** On page 375 of the March 1999 issue of *Mathematics Teaching in the Middle School,* a class is exploring how data can be analyzed. They are given the following data from a hypothetical survey taken of 30 seventh-graders to find out how many hours of television these children watch in a week. Imagine you are asked to tell the principal what these data tell us about the television habits of seventh-graders at this school. Write that report. Then read the article to see what the children did!

1.5 21 12.5 0 2.5 15 23 19 4 14 8 16
13.5 16.5 6 4.5 9 18 5 10.5 8.5 6 3 9
11.5 3.5 19.5 13 10 9

Source: March 1999 issue of *Mathematics Teaching in the Middle School,* copyright 1999 by the National Council of Teachers of Mathematics.

7. Below are the ages of students in a class in mathematics for elementary teachers:

18, 18, 18, 19, 19, 19, 19, 19, 19, 20, 20, 20, 21, 21, 21, 21, 21, 21, 23, 24, 24, 26, 27, 37, 42, 43, 46

a. Use one or more graphs to show the data.

b. What is the average age? Justify your answer.

c. Describe the spread of the data in as many ways as you can.

8. Below are the annual salaries (in thousands of dollars) of workers in Bates Ball Bearing factory:

12, 12, 13, 13, 13, 16, 16, 16, 16, 18, 18, 18, 21, 21, 21, 26, 26, 28, 31, 51, 51, 75

a. Use one or more graphs to show the data.

b. What is the average salary? Justify your answer.

c. Describe the spread of the data in as many ways as you can.

9. Following are the heights of adults in a college class.

a. Make a bar graph for these data.

b. Determine the mean and standard deviation.

c. What percent of the values lie within 1 standard deviation of the mean?

Height (inches)	Frequency
59	1
60	1
61	3
62	8
63	11
64	13
65	17
66	14
67	11
68	7
69	4
70	2
71	1
72	1

10. Suppose the distribution of heights of a group of 400 children is normal, with a mean of 150 centimeters and a standard deviation of 12 centimeters. About how many of these children are taller than 162 centimeters?

11. The heights of 1500 boys at West High School were measured, and the mean was found to be 66 inches, with a standard deviation of 2.5 inches. If the heights are approximately normally distributed, 95% of the boys are between ____ and ____ inches tall. About how many of the boys are less than 5 feet tall? *Note:* You cannot get an exact answer here. Any answer will involve some estimating.

12. Let's say a student scored 45 on a math achievement test for which the mean was 36 and the standard deviation was 6. Judging on the basis of this performance, what score would you predict for her on a standardized test for which the mean is 100 and the standard deviation 15?

13. A tire company tested a particular model of super radial tire and found the tires to be normally distributed with respect to wear. The "average" (mean) tire wore out at 59,000 miles, and the standard deviation was 2500 miles.

a. If 2000 tires are tested, about how many are likely to wear out before 54,000 miles?

b. What if the company wanted to guarantee 55,000 miles? What percent of the tires are likely to wear out before 55,000 miles? In this case, you cannot get an "exact" answer. What could you do to increase the accuracy of your estimate?

14. Below are data on the lengths of feet (in inches) and heights of a sample of adults. Make a scatter plot for these data. Describe the degree of correlation, that is, positive or negative, and none, weak, or strong. Justify your conclusion.

Student	1	2	3	4	5	6	7	8	9	10	11	12	13
Foot	8.5	9.5	9.5	10	11	11	11	12	12	12.5	13	13.5	14
Height	60	65	66	70	71	72	73	71	72	71.5	70	72	74

15. Below are data about the calories and sodium (salt content) of beef and poultry hot dogs. Analyze the data to see how the distributions of the four sets of data are similar and different.

a. Make two grouped frequency histograms for calories of beef and poultry hot dogs. Summarize what you can conclude from those graphs.

b. Make two box-and-whisker graphs for calories of beef and poultry hot dogs. Summarize what you can conclude from those graphs.

c. Which of the graphs (histogram or box and whisker) were more useful? Explain.

d. Make a scatter plot to examine the relationship between calories and sodium content of the beef hot dogs. Describe the relationship you observe in words.

Beef		Poultry	
Calories	Sodium	Calories	Sodium
186	495	129	430
181	477	132	375
176	425	102	396
149	322	106	383
184	482	94	387
190	587	102	542
158	370	87	359
139	322	99	357
175	479	107	528
148	375	113	513
152	330	135	426
111	300	142	513
148	375	113	513
152	330	135	426
111	300	142	513
141	386	86	358
153	401	143	581
190	645	152	588
157	440	146	522
131	317	144	545
149	319		
135	298		
132	253		

16. The grouped frequency table below shows data on the ages of employees in a company. The company has decided to take a stratified random sample of 40 employees. How many of each age group should they select?

Age range	**Frequency**
20–29	30
30–39	60
40–49	50
50–59	50
60–69	10

17. For each of the pairs of categories, explain whether you would expect a positive correlation, negative correlation, no correlation, or not certain. If any of the categories are ambiguous or not well defined, explain.

a. Height and shoe size

b. Salary and shoe size

c. Year and winning time in the 400-m women's Olympics

d. Numbers of hours studied and test scores

e. Amount of education and annual salary

f. School attendance and grades

g. How often a person watches the news and scores on a current events quiz.

h. Number of times a family eats dinner together each week and a child's grades at school.

In Exercises 18–21, you are given data from various surveys that I have read about. In each case:

a. *State at least two questions you would ask the people who did the survey to satisfy yourself of the validity and reliability of the survey. Briefly explain the reasoning behind your questions.*

b. *Write the actual question that you would ask if you were to conduct a similar survey.*

c. *Make a graph for these data.*

d. *Justify your choice of graph.*

e. *What does the graph add to your understanding of the question? If you feel the graph adds nothing, explain why not.*

18. **HOW PARENTS WAKE UP THEIR CHILDREN**

Response	**Percent**
Call to them	43
Alarm clock	22
Kids wake on their own	16
Other	19

Source: Aunt Jemima survey poll of 400 parents of children under the age of 18.

19. **HOW OFTEN FAMILIES EAT DINNER TOGETHER**

Nights per week	Percent
0	5
1–2	17
3–4	26
5–6	30
7	22

Source: BKG youth survey for Kodak.

20. **HOW MANY FRIENDS PEOPLE SAY THEY HAVE**

Number of friends	Percent
0	1
1	2
2–5	36
6–10	25
11–20	18
More than 20	18

Source: MCI/Louis Harris survey.

21. **AVERAGE NUMBER OF COLDS PER YEAR**

Age group	Number
0–4	5
5–19	3
20–39	2
40 plus	2

Source: University of Michigan.

22. Compute Ed's GPA, to two decimal places, for the semester (A = 4, B = 3, C = 2).

Course	Grade	Credits
Mathematics	B	3
Elementary methods	A	6
Computers in school	C	1
Biology	B	4

23. On a certain exam, Tony corrected 10 papers and found the mean for his group to be 70. Alice corrected the remaining 20 papers and found that the mean for her group was 80. What is the mean of the combined group of students?

DEEPENING YOUR UNDERSTANDING

24. The following table shows the ages of the winners in the Best Actress category at the Academy Awards from 1928 to 2008. Has the age of the winners changed over time? Divide the data into 1928–1967 and 1968–2008.

  a. Make two stem-and-leaf plots.

  b. Summarize what they tell you about the question.

  c. Determine the mean, median, mean absolute deviation, and standard deviation for the two sets.

  d. Summarize what these numbers add to our understanding of the response to the question.

  e. Make and justify at least one additional graph.

  f. How would you answer the question?

Year	Age	Year	Age	Year	Age	Year	Age
1928	37	1948	33	1968	26	1988	26
1929	30	1949	28	1969	35	1989	79
1930	62	1950	38	1970	34	1990	41
1931	32	1951	45	1971	34	1991	28
1932	26	1952	24	1972	26	1992	32
1933	31	1953	26	1973	37	1993	34
1934	27	1954	47	1974	42	1994	44
1935	26	1955	41	1975	41	1995	48
1936	27	1956	27	1976	35	1996	38
1937	30	1957	39	1977	31	1997	33
1938	26	1958	39	1978	41	1998	25
1939	29	1959	28	1979	33	1999	24
1940	24	1960	27	1980	30	2000	32
1941	39	1961	31	1981	74	2001	34
1942	24	1962	37	1982	33	2002	34
1943	29	1963	30	1983	49	2003	28
1944	37	1964	24	1984	38	2004	29
1945	30	1965	34	1985	61	2005	29
1946	34	1966	60	1986	21	2006	60
1947	34	1967	61	1987	41	2007	31
						2008	32

25. The American Federation of Teachers publishes the average salaries of beginning teachers and of all teachers by state. Here is the website where you can get the latest data: http://archive.aft.org/salary/index.htm

  a. Make a boxplot of the data for beginning teachers and all teachers for the latest date available.

  b. Summarize what you can conclude from the boxplots.

  c. Do any additional exercises that your instructor assigns.

26. The Association of Community Organizations for Reform Now (ACORN) presented these data to a Joint Congressional Hearing on discrimination in lending. The data concern the refusal rate for people applying for a mortgage. ACORN contends that these data show a pattern of discrimination.

  a. Make boxplots for the four sets of data.

b. Summarize your conclusions.

Refusal rates of applicants				
Bank	Minority	White	High-income minority	High-income white
Harris Trust	20.9	3.7	21.4	2.2
NCNB Texas	23.23	5.5	8.0	8.0
Crestar	23.1	6.7	11.3	3.6
Mercantile	30.4	9.0	17.3	5.5
First National Bank Commerce	42.7	13.9	38.0	7.6
Texas Commerce	62.2	20.6	33.3	10.3
Comerica	39.5	13.4	33.6	9.4
First of America	38.4	13.2	29.5	7.3
Boatman's National	26.2	9.3	21.7	7.4
First Commercial	55.9	21.0	39.1	15.8
Provident National	49.7	20.1	36.6	15.3
Worthen	44.6	19.1	28.6	10.1
Hibernia National	36.4	16.0	32.9	9.2
Sovron	32.0	16.0	21.0	13.0
Bell Federal	10.6	5.6	5.8	4.2
SEC PAC AZ	34.3	18.4	24.2	14.1
Core States	42.3	23.3	38.3	15.0
Citibank AZ	26.5	15.6	27.3	16.1
MF'ers Hanover	47.2	29.7	41.1	26.8

27. A class of college students collected data about how many siblings they had and how many siblings their mother had.

a. Determine the mean, median, and mode for the two sets.

b. How would you answer this question: How has the average number of siblings changed from the two generations?

c. Can we generalize from these data about how the number of siblings has changed from generation to generation? Why or why not?

d. Make two line plots for these data. Write what you can conclude from analyzing those plots.

e. Make two box-and-whisker plots for these data. Write what you can conclude from analyzing those plots.

f. Determine the mean absolute deviation and standard deviation for each set of data.

g. What additional information would we need in order to make a scatter plot for the data?

h. What would a scatter plot tell us?

i. Write a short report that answers the question "What do we know about the number of siblings we have and our mothers have?" Include at least one graph.

Number of siblings	0	1	2	3	4	5	6	7	8	9	10	11
My students	3	6	4	2	1	1	0	0	0	1	0	0
Their parents	1	2	4	4	1	3	0	1	0	1	0	1

28. Below are the years of experience of teachers in two elementary schools. What do these data tell us about the experience of the teaching staff of the two schools?

a. Determine the mean and median for the two schools.

b. What would you say is the modal class for each school? Justify your answer.

c. What do the mean and median tell us and not tell us about the data?

d. Make a line plot for each school from these data.

e. Summarize (2–3 sentences) what the line plots tells us.

f. Make a grouped frequency table for these data.

g. Make a histogram from these data.

h. Summarize (2–3 sentences) what the histogram tells us.

i. Make a boxplot for these data.

j. Summarize what the boxplot tells us about the data.

k. Compare the spread of the data for the two schools in as many ways as you can.

Years teaching	0	1	2	3	5	6	7	8	10	14	20	25	31
School A	4	3	4	2	1	0	2	0	0	0	1	0	0
School B	1	2	1	2	0	2	1	2	1	1	1	0	1

29. The superintendent and school board have decided to collect data on the years of experience of teachers in the school district, for a variety of reasons—if you have too many veteran teachers, the payroll is higher; if you have too many teachers nearing retirement, it is hard to replace them. Here are the data for a school district concerning the number of years of teaching.

Years teaching	0	1	2	3	4	5	6	7	8	9	10
Teachers	11	8	10	7	6	5	3	6	2	4	4

Years teaching	11	12	15	16	17	18	19	20	21	22	23
Teachers	2	3	2	5	1	3	4	3	5	4	3

Years teaching	24	25	27	28	29	30	32	33	34	35	39
Teachers	5	3	2	4	2	2	2	1	1	1	1

a. Make a grouped frequency bar graph from these data, and justify your choice of intervals.

b. Make one other graph for these data. Explain why you thought that graph would be appropriate.

c. Write a report (maximum of five sentences) that you would give to the school board. Select whatever graph(s) and other information you would want to include.

d. Jacob says "Just give them the average. Why clutter their minds with graphs and all sorts of other information?" How would you respond to Jacob?

30. The graph below is a box-and-whisker graph for the lengths (in miles) of the 18 longest rivers in the world. What information does the graph give us?

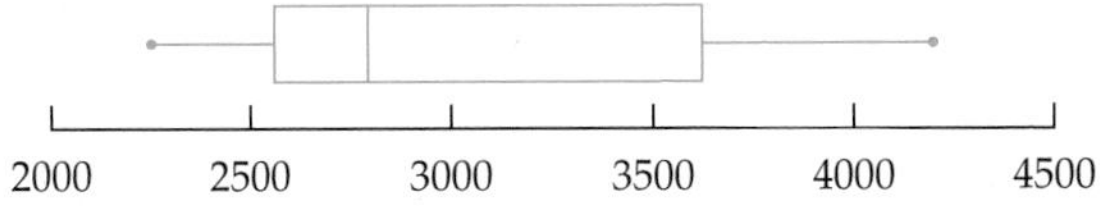

31. Here are two boxplots: the 2008 payroll of the NFL and MLB teams.

a. Describe what this boxplot tells you about the payrolls for each sport.

b. Predict the shape of each distribution from the boxplot. Briefly explain your prediction.

c. Summarize how the payrolls are different.

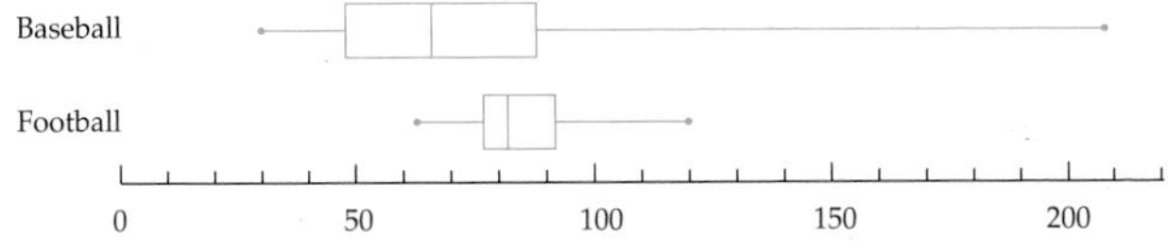

32. Below are data for the average number of tornadoes in the United States from 1950 to 1999 by month.

Jan	Feb	Mar	Apr	May	June	July	Aug	Sept	Oct	Nov	Dec
20	22	54	109	180	171	96	60	41	29	30	17

Consider these possible graphs for the data: bar graph, line graph, circle graph, box and whisker, and line plot.

a. Select *one* of these that would be a good choice and explain why.

b. Select *one* of these that would be a poor choice and explain why.

You do not need to make the graphs, although you can sketch one or two if it would be helpful to you.

33. a. What does a short box or a short whisker mean on a boxplot?

b. What does a long box or a long whisker mean on a boxplot?

c. Sketch a boxplot that would come from a set of data that is normally distributed. Justify your response.

d. Sketch a boxplot that would come from a set of data that is skewed to the right. Justify your response.

34. We have discussed that there are similarities and connections between different graphs.

a. Describe how a line plot and a box-and-whisker graph are similar.

b. Describe how a stem-and-leaf plot and a histogram are similar.

35. A tire company tested another model of tire and found the tires to be normally distributed with respect to wear. The "average" (mean) tire wore out at 46,000 miles, and the standard deviation was 2400 miles. If 2000 tires are tested, about how many are likely to wear out before 45,000 miles? You cannot get an exact answer. I want your best estimate. Please use graph paper to help you estimate.

36. Jack and Jill go to different schools. Jack got an 82 on a test with a mean of 60 and a standard deviation of 15. Jill got a 78 on a test with a mean of 60 and a standard deviation of 9. If both tests are graded on a curve, who did better?

37. Wendy took three standardized aptitude tests: English, mathematics, and general information. The table below gives her score, the mean, and the standard deviation for each test. On which test did she do the "best"? Justify your answer.

Test	Wendy's score	Mean	Standard deviation
English	85	75	7.5
Math	66	55	6
General information	104	94	8

38. We found that in a normally distributed set of data, the mean, median, and mode are virtually identical. However, this is not necessarily true for other distributions. What about data whose distribution is skewed to the right? Can you predict the numerical order of the three averages? For example, will the mean be largest, followed by the median, followed by the mode? Or do you think "it depends"? Justify your answer.

39. Why do you think many statisticians recommend that when describing a population, one should give at least three pieces of information: some sense of the "average," some sense of the spread, and some sense of how the data are distributed? For example, they would recommend not simply reporting that the average age of teachers in New Hampshire is 43 or that the average adult American goes to church 1.3 times a month (both figures are made up).

40. What happens to the mean and the standard deviation of a set of data when the value of each datum is increased by the same amount?

41. A group of students, while collecting data on how many peanuts they could hold in one hand, also collected data on their hand span to see if there was a correlation between the two. Make a scatter plot for these data. What do you think?

Number of peanuts	26	26	27	27	29	31	33	33	35	37	38
Hand span (inches)	6	7.5	6.75	8	6.25	7	6.5	7.25	6.75	6.5	8.25

42. a. Describe some situations where you would expect a scatter plot to show a strong positive correlation (e.g., age of husband and age of wife). Justify your answer.

    b. Describe some situations where you would expect a scatter plot to show a weak positive relationship. Justify your answer.

    c. Describe some situations where you would expect a scatter plot to show a strong negative relationship. Justify your answer.

43. The following scenarios have been taken from the *Instructor's Course Planner* for the textbook *Children.*[8] In each case, do you think the sample is a random sample or a biased sample? Justify your response.

    a. The researchers asked teenage boys about their driving records and habits by going to a movie drive-in, a local bar, the beach, and a baseball park.

    b. Parents at a PTA meeting were interviewed about the quality of the local public school system.

    c. A telephone survey assessed a community's attitudes toward welfare recipients.

    d. Children were asked their opinion of Santa Claus on December 28.

44. Several secretaries in an academic building on a college campus share a copy machine. They believe the copy machine is too slow, and they want the college to buy a newer, high-speed copy machine. Describe what kind of data they should collect and how they should present their data so as to best support their case.

45. In a special triathlon, Betty swam 1 mile in 25 minutes, biked 25 miles in 45 minutes, and then ran 5 miles in 40 minutes. What was her "average" speed for the race?

46. Perry has a 2.53 GPA after 112 credits. This is his last semester. He is taking 15 credits, and he has senioritis.

    a. What is the minimum GPA Perry can get and still graduate with at least a 2.5 GPA?

    b. He is taking five 3-credit courses. Give one scenario that will let him "squeak by."

	Grade
Course 1	
Course 2	
Course 3	
Course 4	
Course 5	

47. Joe has a 3.22 GPA after 46 credits. He says that he has a total of 148.12 grade points. Julie says that's impossible. What do you think?

FROM STANDARDIZED ASSESSMENTS

48. A poll is being taken at Baker Junior High School to determine whether to change the school mascot. Which of the following would be the best place to find a sample of students to interview that would be most representative of the entire student body?

    a. an algebra class

    b. the cafeteria

    c. the guidance office

    d. a French class

    e. the faculty room

    (65% of eighth-graders answered correctly.)

    Source: Results of the Seventh *NAEP Mathematics Assessment,* Grade 8, p. 242. U.S. Department of Education, National Center for Education Statistics.

SECTION 7.3 Concepts Related to Chance

What do you think?

- How can we figure out the probability that something will happen before it actually happens?
- How can we express that probability?

> The study of probability . . . should not focus on developing formulas or computing the likelihood of events pictured in texts. Students should actively explore situations by experimenting and simulating probability models. . . . Students should talk about their ideas and use the results of their experiments to model situations or predict events.
>
> (*Curriculum Standards,* p. 109)

According to the Common Core State Standards, probability is introduced in seventh grade.

Stop for a moment and think of situations in which probability enters into our lives. Then read on. . . .

Explorations Manual 7.13

We make and are affected by probability decisions every day. For example,

- If you decide to have two children, what is the probability that you will have a boy and a girl?
- What is the probability that you will get a teaching job after you graduate?
- If you are independently employed, should you buy health insurance? People's decisions are influenced by their estimate of the probability that they will have a catastrophic illness.
- If you drive over the speed limit or decide not to put money in the parking meter, what is the probability of your getting a ticket?
- What is the probability that everyone who buys an airline ticket will actually take the trip? Airlines regularly sell more tickets than they have seats because they know that the probability of this is very low.

These situations all have in common that their results are unknown. Much of our understanding of probability is based on collecting and analyzing data on occurrences of situations that, individually, are random. For example, because the probability that a 21-year-old male will have an automobile accident is much greater than the probability that a 50-year-old male will have an accident, car insurance rates for 21-year-old males are much higher than those for 50-year-old males.

Our work with data that are normally distributed connects to probability. For example, if a tire company's sample of tires lasted an average of 42,000 miles with a standard deviation of 2000, the company can say that there is an 84% probability that a specific tire will last at least 40,000 miles.

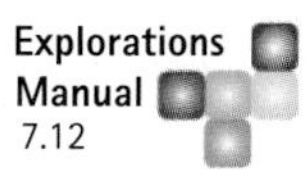

Explorations Manual 7.12

Random phenomena When mathematicians use the word *random,* we are referring to phenomena that are unpredictable individually but that have regular patterns when considered as a group or when considered over the long run. *Random* is not a synonym for *haphazard;* rather, it refers to phenomena for which we cannot make individual predictions. The following example nicely illustrates this point. If we drop a coin from a certain height, we can predict with a great deal of precision the time it will take to reach the ground. However, if we flip that coin, we can give only the probability that it will land heads or tails. We do not say that flipping a coin is haphazard but rather that it is random.[9] When we flip many coins, we find that there are many patterns, as you will see in this section.

In this section, we will investigate both well-defined questions (for example, the probability of tossing 4 heads in a row) and questions that are not as well defined (for example, insurance premiums).

PRELIMINARY TERMS AND CONCEPTS

We hear probability statements all the time. Let's say you turn on the television and hear the weather forecaster say, "There is a 25% chance of rain tomorrow." What does that statement mean? Write your thoughts before reading on. . . .

When we make a probability statement, we are giving a numerical value that represents the degree to which we believe that event will or will not happen.

Probabilities can be represented by percents, fractions, decimals, ratios, and odds. The following five statements are equivalent:

- There is a 25% chance of rain tomorrow.
- The probability of rain tomorrow is $\frac{1}{4}$.

- The probability of rain tomorrow is 0.25.
- There is 1 chance in 4 of rain tomorrow.
- The odds against rain tomorrow are 3 to 1.

Outcomes and events There are two terms whose definition will make our discussion about probabilities much clearer: *outcome* and *event.* Answer the following two questions and then read on. . . .

- If you select a card from a regular deck of playing cards, what is the probability of drawing the queen of spades?
- If you select a card from a regular deck of playing cards, what is the probability of drawing any queen?

The probability of drawing the queen of spades is $\frac{1}{52}$, and the probability of drawing any queen is $\frac{4}{52}$, or $\frac{1}{13}$.

We often use notation as a shorthand to express probabilities:

$$P(Q \text{ of spades}) = \frac{1}{52} \qquad P(Q) = \frac{1}{13}$$

We refer to "queen of spades" as an outcome and "any queen" as an event. Let us examine the difference between the two. When we define a situation (in this case, drawing a card from a deck), each possibility is called an **outcome**. The set of all possible outcomes is called the **sample space**. Thus, each outcome is an element of the sample space. Within the sample space, there are many possible subsets—for example the subset E:

$$E = \{\text{queen of diamonds, queen of spades, queen of hearts, and queen of clubs}\}$$

Any subset of a sample space is called an **event**.

WHEN ALL OUTCOMES ARE EQUALLY LIKELY

In this case, we are examining situations in which each outcome is **equally likely**. In the example of drawing a card out of a deck, we are equally likely to draw the ace of hearts as the king of spades, or any other card. When we are examining a situation in which all outcomes are equally likely, we can determine the probability of an event by computing the following ratio, expressed here both in probability language and in set language. If all outcomes in a sample space are equally likely, then the probability of event E is given by:

$$P(E) = \frac{\text{number of favorable outcomes}}{\text{number of total outcomes}} = \frac{\text{number of elements in } E}{\text{number of elements in } S}$$

WHEN ALL OUTCOMES ARE NOT EQUALLY LIKELY

On many occasions, all outcomes are **not equally likely**. Consider the spinner in Figure 7.43. In this case, there are four outcomes: 1,, 2, 3, 4. However, they are not equally likely. In this case, what is the probability of spinning an odd number?

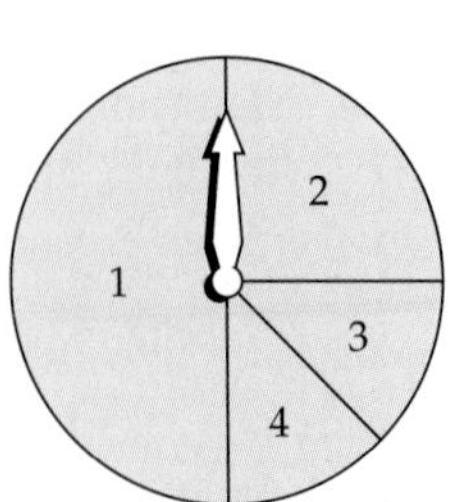

Figure 7.43

$$P(\text{Odd}) = P(1) + P(3) = \frac{1}{2} + \frac{1}{8} = \frac{5}{8}$$

We can generalize from this example to state the probability of an event when all outcomes are not equally likely:

> The probability of an event is equal to the sum of the probabilities of all outcomes in that event.

That is, if event E consists of n different outcomes $O_1, O_2, O_3, \ldots, O_n$, then $P(E) = P(O_1) + P(O_2) + P(O_3) + \cdots + P(O_n)$.

When we speak of the probability of an event happening, at one extreme we have **impossible events**—for example, the probability of a student having a GPA of 4.4. At the other extreme, we have **certain events**—for example, the probability that it will rain this year in Jacksonville, Florida. If an event is impossible, we say that the probability is 0. If it is certain, we say that the probability is 1. Thus, the probability of any event is between 0 and 1.

THEORETICAL AND EXPERIMENTAL PROBABILITIES

When working with probability situations, we distinguish between theoretical probabilities and experimental probabilities. In some situations involving probabilities, we can determine the theoretical probability—for example, the probability of having 5 boys in a row or of rolling doubles 3 times in a row when playing Monopoly. In situations where we can determine the total number of outcomes and we can count the number of outcomes in a specific event, we refer to the probability as the **theoretical probability**. That is, the theoretical probability refers to our expectation, assuming that things turn out ideally. For example, we can say that when we flip a coin, theoretically the probability of its being heads is $\frac{1}{2}$.

There are many real-life situations in which it is either impossible or very expensive to determine theoretical probabilities. In those situations, we determine the experimental probability by collecting and analyzing data. For example, manufacturers determine the probabilities that their products will last a certain time by testing some of the models. For example, if 90% of the sampled tires last at least 40,000 miles, we would say that the experimental probability of a tire's lasting at least 40,000 miles is 90%. In situations in which the probability of an outcome or event has been determined by collecting data and determining the fraction of the time in which the outcome or event actually occurred, we refer to that fraction as the **experimental probability**. Some books and authors use the term **empirical probability** instead.

Let us now investigate some situations to deepen our understanding.

HISTORY

There are some famous historical examples of empirical data concerning flipping a coin. The French naturalist Count Buffon (1707–1788) tossed a coin 4040 times; the results were heads 2048 and tails 1992. Statistician Karl Pearson tossed a coin 24,000 times; his results were 12,012 heads and 11,988 tails. While imprisoned during WWII, South African mathematician John Kerrich tossed a coin 10,000 times; his results were heads 5067 and tails 4933. The percentage of heads in the three cases were 50.69, 50.05, and 50.67.

INVESTIGATION 7.3a Probability of Having 2 Boys and 2 Girls

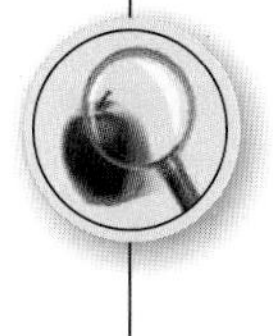

Let's say a couple is planning to begin their family and would like to have 4 children. What is the probability that they will have 2 boys and 2 girls? What do you think? Take some time to work on this question before reading on. . . .

DISCUSSION

Explorations Manual 7.13

Most initial hypotheses range from 20% to 50%. Let us investigate each of these.

Those who answer 50% reason something like this: Each time, there is a 50% chance of getting a boy or a girl, so it makes sense that the "balanced" possibility (same number of boys and girls) will take place 50% of the time.

In the May 1999 issue of *Mathematics Teaching in the Middle School,* Ann Lawrence uses two children's books, *The Giver* and *Twenty-One Balloons,* to engage her students in probability problems. In *Twenty-One Balloons,* a man invites 25 families to move to a remote island. He picks families that have exactly one girl and one boy. Ann began this unit by asking the children how often a family has exactly one girl and one boy.

Those who answer 20% reason something like this: There are 5 possibilities (events)—4G0B (that is, 4 girls and 0 boys), 3G1B, 2G2B, 1G3B, 0G4B. Thus 2G2B is 1 out of 5, which is 20%.

Rather than present the answer just yet (presenting the answer too quickly means that more students are more likely to rent the solution rather than own it), let us do a **simulation**. The process of simulating a situation and then collecting and analyzing a large amount of data is called the Monte Carlo method. It is widely used in science and other fields where a problem is too difficult or complex to be addressed with other methods. The method has been used by mathematicians for some time but was given its current name by Stanislaw Ulam in 1946.

The simulation involves several steps.

Step 1: Make sure we clearly understand the problem.

Step 2: Select a method for modeling the problem.

Step 3: Gather the data.

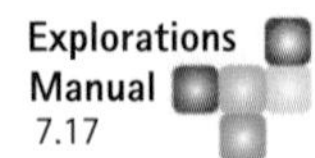

Step 4: Analyze the data, which will result in an approximation of the actual probability.

The method is based on the **Law of Large Numbers**, which states that as we perform an experiment over and over, the experimental probability will converge on a fixed number. For example, if we toss a coin many times, the experimental probability of tails will converge on 0.5. If you have done any sampling, you know that, for example, if you toss a coin 10 times, you are likely to get tails 5 times, but this is not certain to happen. Even if you toss a coin 100 times, you are not certain to get tails 50 times. However, the law states that as you perform the experiment repeatedly, the experimental probability will converge—that is, it will work its way closer and closer to 0.5.

This is what scientists, economists, and mathematicians often do when they are beginning an investigation or when they have different hypotheses: They construct a model of the situation and then run a simulation. How might we run a simulation of this question? Please think for a minute yourself and then read on. . . .

There are many ways to simulate this scenario:

1. We could use a computer program to generate random digits. We could assign 0–4 to represent girl and 5–9 to represent boy.
2. We could roll dice. We could assign 1–3 to represent girl and 4–6 to represent boy.
3. We could write B and G on the same number of slips of paper, put them in a hat, and then draw pieces of paper out one by one. We would need to replace each piece of paper and mix the pieces after each drawing.
4. We could label a spinner as shown in Figure 7.44.

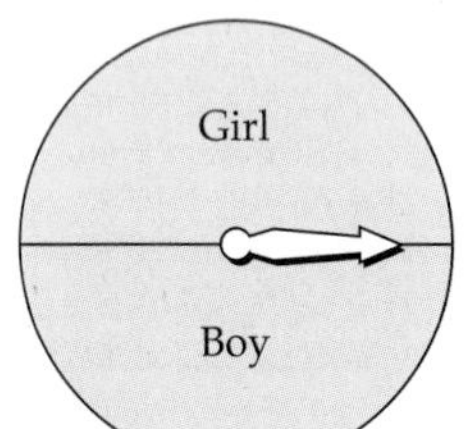

Figure 7.44

When you perform the simulation many times, you will find the experimental probability to be close to 38%.

With the knowledge gleaned from the simulation, let us now explore how the theoretical probability is determined.

STRATEGY 1 Be systematic and look for patterns

There are many ways to be systematic. Let us examine one way.

First, the couple could have all girls or all boys: GGGG and BBBB.

Note that we use letters to represent boy and girl rather than spelling them out. This saves time and makes patterns in the data easier to see.

Next, we could determine the different ways in which the couple could have 3 girls and 1 boy. One way to do this systematically is to see that the boy could be in four different positions: oldest, second, third, youngest. Symbolically, this looks like

B G G G
G B G G
G G B G
G G G B

CLASSROOM CONNECTION

When my daughter and son were small, they would come into our bed when they woke up. One morning, my daughter noted that there was a pattern. She said, "Girl, boy, girl, boy." She then jumped over me and said, "This is another pattern." I was expecting her to say, "Boys on the outside, girls on the inside." What she said was, "Oldest and youngest on the outside."

Do you see a pattern here? How would you describe it?

If you look only at the B, you can see it move diagonally. Do you see that this systematic approach also practically gives us the ways to have 3 boys and a girl?

If you stop and think for a minute, you might realize that because there are four different ways to have 3 girls, there must also be four different ways to have 3 boys. Do you see why? There are many symmetries in probability situations.

Finally, we need to determine how many different ways there are to have 2 boys and 2 girls. Do this first yourself before reading on. . . .

There are a variety of methods you might use to come up with the 6 different ways in which one could have 2 boys and 2 girls. (In probability language, we are saying that the event 2B, 2G consists of 6 outcomes.)

B G B G
G B G B
B G G B
G B B G
B B G G
G G B B

Because 6 of the 16 outcomes have 2 Bs and 2 Gs, the probability we sought is $\frac{6}{16}$, or 37.5%.

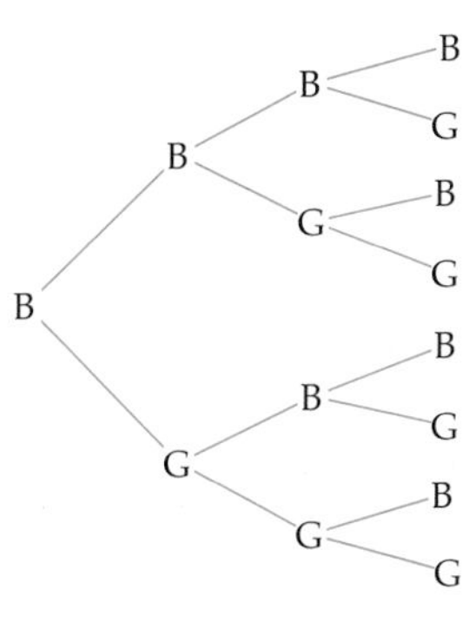

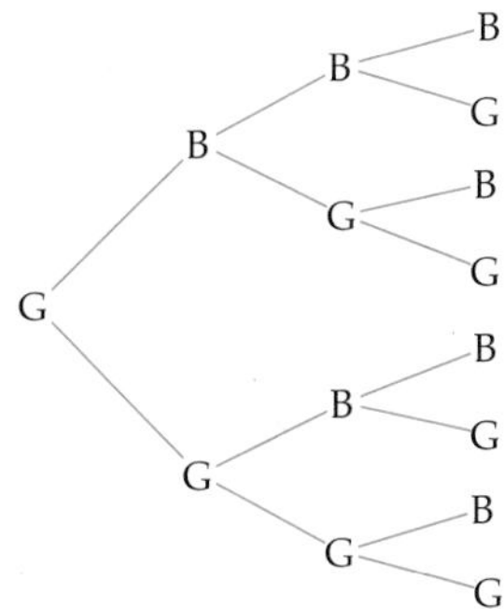

Figure 7.45

STRATEGY 2 Make a tree diagram

Figure 7.45 is a tree diagram for this problem; it is a different way to represent and solve the problem. Can you interpret the tree diagram?

It is called a tree diagram because it branches out. We first begin with B and G to represent the possibilities with one child—a boy or a girl. Then a B and a G branch from the original B and from the G. At this point, we have 4 outcomes: BB, BG, GB, and GG.

Now make a table of the 16 outcomes by using the tree diagram. (The table is not shown—you have to make it!) Write the patterns you see and then read on. . . .

There is a pattern in each of the columns. The left column simply alternates between B and G. The next column alternates between 2 Bs and 2 Gs. The third column alternates between 4 Bs and 4 Gs. The last column consists of 8 Bs and 8 Gs.

Some students see the symmetric nature of the table. Pair each row in the table in the following way: row 1 and row 16, row 2 and row 15, and so on. Another way of seeing this is to fold the table in half and look at the rows that lie on top of each other. Each half is like a mirror image of the other: BBBB and GGGG, BBBG and GGGB, and so on.

Note that this table is not identical to the table that comes from Strategy 1, although they contain the same 16 outcomes. The other table exhibited different patterns. What is important here is that both tables represent a systematic approach to the problem. If a student were to generate the 16 outcomes in a more random fashion, there would probably be fewer patterns and hence a greater probability of missing one or more outcomes.

MATHEMATICS

Blaise Pascal, whom we have encountered before, once said, "Chance favors the prepared mind." A powerful illustration of this occurred one year while my students were investigating this question. Some students noted that since the letters G and B are similar, it was hard to see patterns. One student replaced the Gs with 0 and the Bs with 1. The contrast between 0 and 1 is much greater than that between B and G. The student then listed all 16 outcomes on the board, noting that 0 represented B and 1 represented G. This notation made it easier to discuss patterns and also provoked another discovery. Does Table 7.13 remind you of something you have studied before?

TABLE 7.13

0 0 0 0
0 0 0 1
0 0 1 0
0 0 1 1
0 1 0 0
0 1 0 1
0 1 1 0
0 1 1 1
1 0 0 0
. . .
1 1 1 1

This table is simply the first 16 numbers in base two!

Let us now examine two related questions for this couple, from which new probability concepts will emerge.

INVESTIGATION 7.3b Probability of Having 3 Boys and 2 Girls

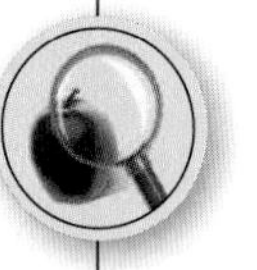

What if the couple were planning to have 5 children? What is the probability of their having 3 boys and 2 girls? Work on this and then read on. . . .

DISCUSSION

There are several different ways to answer this question. You can make a new table, extend the tree diagram, or extend the base two counting table and look for outcomes that show two 0s and three 1s or three 0s and two 1s. However, in any of these cases, obtaining the solution is relatively cumbersome. It becomes downright tedious if we extend the question further. For example, what is the probability of having 4 girls and 4 boys?

Table 7.14 represents another way of solving the problem: by looking at the number of outcomes in different events. I have filled in the first four rows of the table. Take a few minutes to make sense of the table.

TABLE 7.14

NUMBER OF GIRLS							Number of
6	5	4	3	2	1	0	children
					1	1	1
				1	2	1	2
			1	3	3	1	3
		1	4	6	4	1	4
	1					1	5

What this table shows is that if you have 4 children, there is 1 way to have 4 girls, there are 4 ways to have 3 girls (and 1 boy), there are 6 ways to have 2 girls (and 2 boys), there are 4 ways to have 1 girl (and 3 boys), and there is 1 way to have no girls (that is, all boys). Examine the numbers for 1 child, 2 children, and 3 children. If you didn't understand the table before, fill out the table now before reading on. . . .

What patterns do you see? Describe them as though you were talking to someone on the phone. . . .

There are small (local) patterns; for example,

- The bottom-left to top-right diagonal contains only 1s.
- The column representing 0 girls contains only 1s.
- The column representing 1 girl is the counting numbers: 1, 2, 3, 4, 5.

There are also large (global) patterns. For example, there is a connection between each row and the next row. Look at the third row and the fourth row. What do you see? Think and then read on. . . .

Connecting this problem to Pascal's triangle makes our work much less tedious. If a couple has 5 children, what is the probability of 3 girls and 2 boys? Think and then read on. . . .

The cell representing 3 girls and 2 boys contains 10 outcomes, and there are 32 outcomes in all; therefore, the probability of 3 girls and 2 boys is $\frac{10}{32}$, or about 31%. Similarly, the probability of having 2 girls and 3 boys is also $\frac{10}{32}$.

Mutually exclusive events We can simply add the two probabilities together. The justification for this goes back to set theory. The two events (3B, 2G) and (2B, 3G) represent disjoint sets. In probability language, we say that they are **mutually exclusive events**.

If events A and B are mutually exclusive, then $P(A \text{ or } B) = P(A) + P(B)$.

There is another global pattern that comes out of Table 7.14. It arises from adding a new column headed "Total number of outcomes." If you have not yet seen this pattern, add that new column and fill in the numbers. What do you see? Think and then read on. . . .

This column simply can be described most succinctly (in English) as "powers of 2."

INVESTIGATION 7.3c Probability of Having at Least 1 Girl

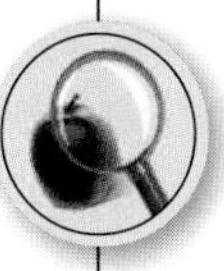

Let us examine another situation. Let's say a couple is planning to have 4 children. What is the probability that they will have *at least* 1 girl? Work on this and then read on. . . .

DISCUSSION

STRATEGY 1 Refer to the table or tree diagram and count

From the tables you created in Investigation 7.3a, count those outcomes in which there is at least 1 girl and divide by 16.

STRATEGY 2 Connect this problem to sets

Let us explore this strategy because it has implications for other probability situations. What if you were asked, "What is the probability of having no girls?" Do you see how the answer to this question enables us to answer the question "What is the probability of having at least 1 girl?"

Do you see why the following formula is true: $P(\text{no girls}) + P(\text{at least 1 girl}) = 1$?

Consider the sample space for having 4 children, and put the outcomes into two subsets:

A = the set of outcomes containing no girls
B = the set of outcomes containing at least one girl

Complementary events In this case, all outcomes go into one subset or the other. Not only are the two subsets disjoint, their union is equal to the whole sample space. Since every outcome is in either set A or set B, that means that $A = \overline{B}$. In other words, these are **complementary sets**. In probability language, we say that these are complementary events. Not surprisingly, if we have two events such that event A and event B are complementary, then $P(A) + P(B) = 1$.

If we recall Investigation 7.3b, we see that complementary events represent a special kind of mutually exclusive event.

In this case, the probability of A (the probability of having no girls) is much easier to determine than the probability of B (the probability of having at least 1 girl), and we can use basic algebra to transform the equation $P(A) + P(B) = 1$ into the equation $P(B) = 1 - P(A)$. We can now determine the probability of having at least 1 girl:

$$P(B) = 1 - P(A) = 1 - \frac{1}{16} = \frac{15}{16}$$

Let us now make use of our understanding of probability to investigate two related problems.

INVESTIGATION 7.3d 50-50 Chance of Passing

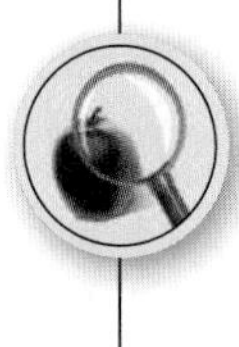

In a Peanuts cartoon, Peppermint Patty is faced with a true-false quiz with 10 questions. She forgot to read the story on which the questions are based, and so she has to guess at every question. Assuming that she has a 50% chance at each question, what is the probability that she will get 70% or better on the quiz? Think about this and then read on. . . .

Explorations Manual 7.12

DISCUSSION

STRATEGY 1 Make a tree diagram

In order to use a tree diagram for this problem, you would need a very big sheet of paper and a lot of time. However, the tree diagram points to another strategy. What is it?

STRATEGY 2 Be systematic and look for patterns

There is only one way she can get all 10 right. Why is this?

Now, how many ways can she get 9 out of 10 right? That is, how many different outcomes are there for getting 9 questions right? There are several ways to determine the number. Work on this question yourself before reading on. . . .

Some students reason like this: "She could get all but the last one (number 10) right, she could get all but number 9 right, she could get all but number 8 right, and so on." What pattern do you see here?

Some students see the systematic nature of this approach better by looking at, and filling out, the diagram below:

RRRRRRRRRW

RRRRRRRRWR

RRRRRRRWRR

etc.

Regardless of how we saw this problem, we find that there are 10 different ways in which she could get 9 out of 10 correct. More formally, we would say that there are 10 outcomes in the event "9 correct answers."

Now does the following row of Pascal's triangle make sense?

1 10 45 120 210 252 210 120 45 10 1

If not, analyze a 4-item test in which she guesses each time: There is 1 way to get all 4 correct, there are 4 ways to get 3 right, there are 6 ways to get 2 right, and so on. Can you use Pascal's triangle now to determine the probability of Patty's getting a score of 70% or greater? Think and then read on. . . .

If you have understood the application of Pascal's triangle to this problem, you have done the computation

$$\frac{(120 + 45 + 10 + 1)}{1024} \approx 0.17$$

Can you express the answer to the problem in a full sentence? Do so and then read on. . . .

There are many valid ways to express the answer. Here are two.

- "Peppermint Patty has a 17% chance of getting at least 70% right."
- "Peppermint Patty has about 1 chance in 6 of getting at least 70% right."

Does the answer surprise you? Is it higher or lower than you expected? Many teachers do not believe in true-false tests because they believe that a lucky student can appear to know more than that student really does. Does this investigation make you feel more or less disposed toward true-false tests? What if there were 20 questions and the student guessed at each one? Will the probability of getting at least 70% be greater or less than on a 10-item quiz?

INVESTIGATION 7.3e What Is the Probability of Rolling a 7?

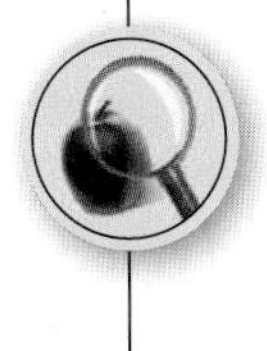

If we roll one die, the probability of each outcome (that is, 1, 2, 3, 4, 5, or 6) is $\frac{1}{6}$ (assuming a fair die). Suppose we roll 2 dice and find the sum of the numbers on the dice. What is the probability of rolling a 7? Work on this problem and then read on. . . .

DISCUSSION

Explorations Manual 7.15

There are several ways to be systematic, and there are several ways to represent this problem. Look at each of the following partially completed representations. How would you finish them?. . . Now, complete the tables.

REPRESENTATION 1

Table 7.15 was created by thinking systematically.

TABLE 7.15

1, 1	2, 1	?, ?
1, 2	2, 2	?, ?
1, 3	?, ?	?, ?
1, 4	?, ?	?, ?
1, 5	?, ?	?, ?
1, 6	?, ?	?, ?

REPRESENTATION 2

Table 7.16 was created by thinking systematically in a slightly different way.

HISTORY

The foundations for probability as a mathematical discipline were laid by the work of Blaise Pascal (1623–1662) and Pierre Fermat (1601–1665) as they responded to several questions asked by a professional gambler, Chevalier de Mere. One of those questions was: Why would you lose money by betting (even money) that you would get at least one double 6 when throwing 2 dice 24 times?

Many people object to students studying gambling theory in school, so we will de-emphasize the gambling connections between dice and probability. This is not hard because dice are also used in many board games.

TABLE 7.16

1, 1	1, 2	?, ?	. . .
2, 1	2, 2		
3, 1	3, ?		
4, 1	?, ?		
5, 1	?, ?		
6, 1	?, ?		

REPRESENTATION 3

Figure 7.46 is yet another tree diagram!

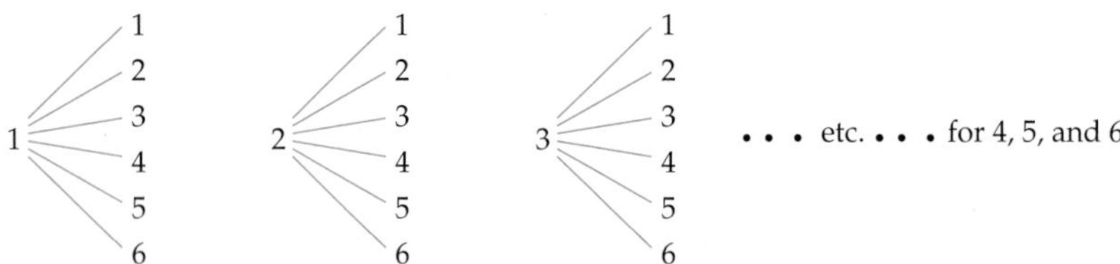

Figure 7.46

In each of these cases, we find that there are 36 distinct possible outcomes. This is a useful time to present formally the **multiplication principle**, which can be stated in outcomes or in probabilities:

Stated in number of outcomes If event A consists of a outcomes and event B consists of b outcomes, then the event "A then B" consists of $a \cdot b$ outcomes. In this case, event A is rolling a die and event B is rolling a die; both consist of 6 outcomes. Thus the event "roll a die and then roll it again" consists of $6 \cdot 6 = 36$ outcomes.

Stated in probabilities If the probability of event A is x and the probability of event B is y, then the probability of the event "A then B" is $x \cdot y$.

We can now count the number of outcomes that produce a 7 and answer the original question. The probability of rolling a 7 is $\frac{6}{36}$, or $\frac{1}{6}$.

What if we wanted to know the probability of *each* event—that is, of rolling a 2, a 3, a 4, and so on? How can you use your work on the previous question to answer this question? What patterns do you see in your previous work that would help you to make the table? Describe the patterns you see in your table and then read on. . . .

One way of describing what we find is to say that each time the sum increases, the number of outcomes increases by 1 until we get to (the event) 7. Then the number of outcomes decreases by 1 (for each event) until we get to 12 (Table 7.17).

TABLE 7.17

Number (event)	Frequency
2	1
3	2
4	3
5	4
6	5
7	6
8	5
9	4
10	3
11	2
12	1

INVESTIGATION 7.3f What Is the Probability of Rolling a 13 with 3 Dice?

Let us extend the investigation now to three dice. Suppose we rolled 3 dice and added the 3 numbers. What is the probability of rolling a 13? Work on this question before reading on. . . .

DISCUSSION

First, we can use the multiplication principle to determine the total number of outcomes from rolling 3 dice—that is, the size of the sample space. There are $6 \cdot 6 \cdot 6 = 216$ possible outcomes!

How many of those outcomes produce a 13? Work on this and then read on. . . . The completed tables are in Appendix B.

As you work, make note of what problem-solving strategies help you do the problem, help to keep you on track, and help you check your solution. Virtually all solutions to this problem require the solver to be systematic. However, there are different ways to be systematic, and you can use many different representations that will help answer this question. Below are the beginnings of two different strategies.

REPRESENTATION 1

Table 7.18 is a table created by thinking systematically.

TABLE 7.18

661	652	643	553	?
616	625	?	?	
166	562			
	526			
	265			
	256			

REPRESENTATION 2

Table 7.19 is another table created by thinking systematically.

TABLE 7.19

661	562	463	?
652	553	?	
643	544		
634	535		
625	?		
	?		

Regardless of the representation used to solve the problem, we find that there are 21 outcomes that produce a sum of 13. Therefore, the probability of rolling a 13 is $\frac{21}{216}$, or slightly under 1 in 10.

An alternative strategy is to construct the entire sample space. The completed tables are shown in Appendix B. If you find all 216 outcomes in the sample space, then you can make use of patterns to count the number of outcomes in the event called "sum of the 3 dice equals 13." Making the table is an exercise in thinking about and looking for patterns that some people find fun and others find tedious. If you are in the latter group, you may look at the ones table in Appendix B. Look for patterns—my students and my children have found hundreds!

INVESTIGATION 7.3g "The Lady or the Tiger"

There is a famous story, written by Frank Stockton, called "The Lady or the Tiger," which I have modified slightly to make an interesting problem. It seems that the king and queen had arranged for their daughter to be married to a prince, but she fell in love with a peasant. When the king discovered this, he ordered that the peasant be thrown into a room full of tigers. However, in response to his daughter's pleas, he agreed to have the peasant walk through a maze to one of two rooms (Figure 7.47). The princess would be waiting in one of the rooms, and the tigers would be in the other room. The princess asked if she could choose the room in which she would wait; the king agreed, for he believed that the chances were equal. If you were the princess, which room would you choose? Think about this and then read on. . . .

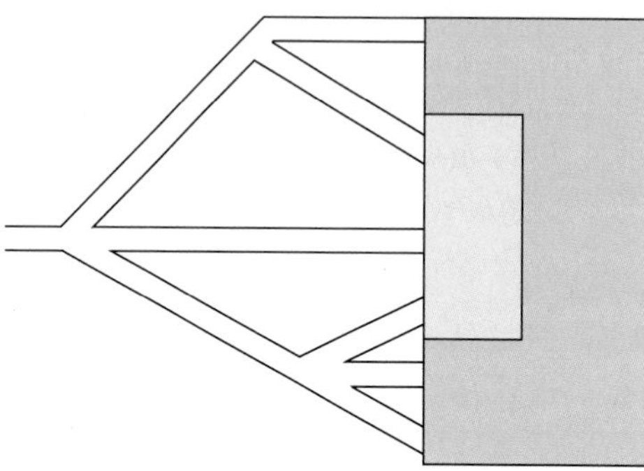

Figure 7.47

DISCUSSION

STRATEGY 1 Solve this as a tree diagram

Can you take your understanding of the tree diagrams used in other problems and apply it to this problem? Try to do so now; determine the probabilities of each event (the princess or the tigers), and then read on. . . .

Figure 7.48 shows 6 different doors (1 through 6) that open into two different rooms (A and B). The probability that the peasant will end up in room A $= \frac{1}{6} + \frac{1}{3} + \frac{1}{9} = \frac{11}{18}$. Do you see why?

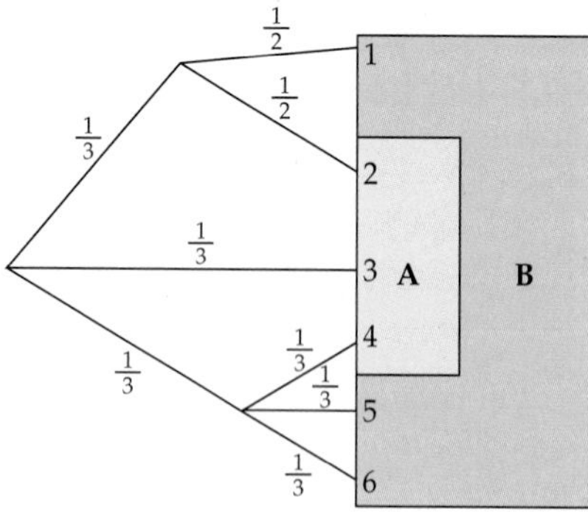

Figure 7.48

We can see why by breaking this problem into parts. We are assuming, of course, that at each point there is an equally likely chance that the peasant will take any path. The probability that, at the first junction, he will take the top path is $\frac{1}{3}$; and the probability that, at the second junction, he will take the top path is $\frac{1}{2}$. Therefore, using the multiplication

Grade 3

What are your answers to these questions from this third-grade lesson? 

Date Time

LESSON 11•5 **Random-Draw Problems**

Each problem involves marbles in a jar. The marbles are blue, white, or striped. A marble is drawn at random (without looking) from the jar. The type of marble is tallied. Then the marble is returned to the jar.

- Read the description of the random draws in each problem.
- Circle the picture of the jar that best matches the description.

1. From 100 random draws, you get:

   a blue marble 62 times.

   a white marble 38 times.

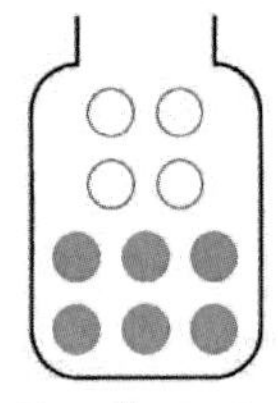

10 marbles in a jar

10 marbles in a jar

2. From 100 random draws, you get:

   a blue marble 23 times.

   a white marble 53 times.

   a striped marble 24 times.

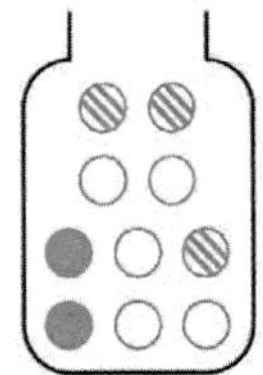

10 marbles in a jar

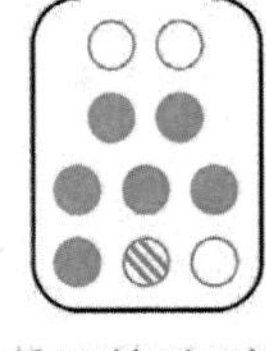

10 marbles in a jar

Try This

3. From 50 random draws, you get:

   a blue marble 30 times.

   a white marble 16 times.

   a striped marble 4 times.

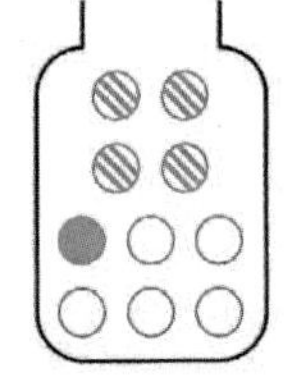

10 marbles in a jar

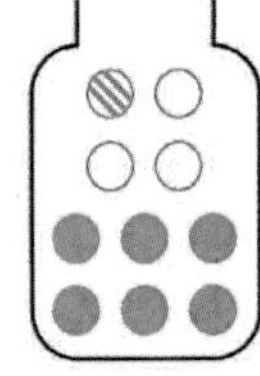

10 marbles in a jar

two hundred seventy-five **275**

From *Everyday Mathematics, Grade 3:* The University of Chicago School Mathematics Project: Student Math Journal, Volume 2, by Max Bell et al., Lesson 11-5, p. 275. Reprinted by permission of The McGraw-Hill Companies, Inc.

principle, the probability that he will end up at door number 1 is $(\frac{1}{3})(\frac{1}{2}) = \frac{1}{6}$. We can determine the probabilities of arriving at the other five doors in a similar manner.

Because A and B are complementary events, we can conclude that the probability that he will end in room B is $1 - \frac{11}{18}$, or $\frac{7}{18}$. Thus the princess should choose to wait in room A.

Top path
Middle path
Lower path

Figure 7.49

A		B
A		
A	B	B

Figure 7.50

STRATEGY 2 Use an area model

At the beginning of the maze, there are three paths. Assuming that each is equally likely to be taken, we can represent them as having the same area (Figure 7.49).

Figure 7.50 represents the outcomes of all of the paths with respect to the probabilities. Do you understand the diagram? How would you explain it?

If the area of the entire rectangle is 1, then the probabilities of A and B are equivalent to their total areas, which will each be fractions of 1.

The fraction of the area of the diagram that is allotted to the three As represents the probability that the peasant will end up at room A—that is, $\frac{1}{6} + \frac{1}{3} + \frac{1}{9} = \frac{11}{18}$.

This next situation is somewhat different from the previous ones. Let us first explore the problem and then examine what is different about it.

INVESTIGATION 7.3h Gumballs

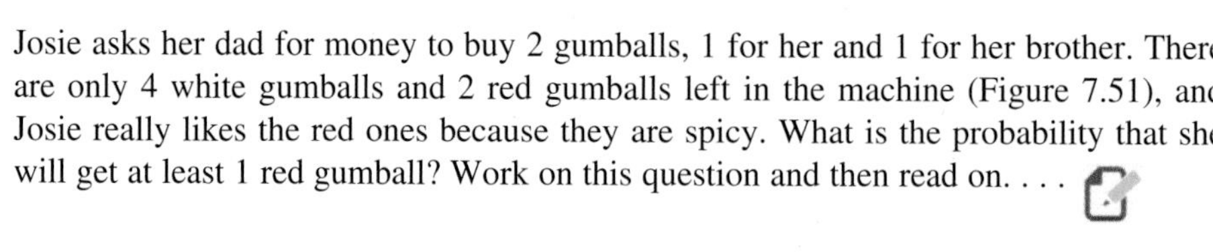

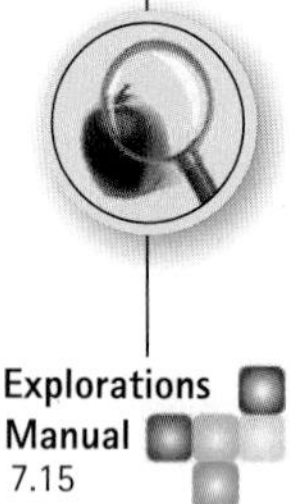

Josie asks her dad for money to buy 2 gumballs, 1 for her and 1 for her brother. There are only 4 white gumballs and 2 red gumballs left in the machine (Figure 7.51), and Josie really likes the red ones because they are spicy. What is the probability that she will get at least 1 red gumball? Work on this question and then read on. . . .

Explorations Manual 7.15

DISCUSSION

STRATEGY 1 Do a simulation

We need to be careful, for there are several ways to go wrong. However, if you write R on two slips of paper and W on four slips of paper and then simulate taking out two gumballs, after a number of trials your experimental probability likely will be close to the theoretical probability. Some students find that they don't really understand the problem at first, but after several trials with the simulation, they suddenly understand the problem more clearly and can then go back and determine the theoretical probability.

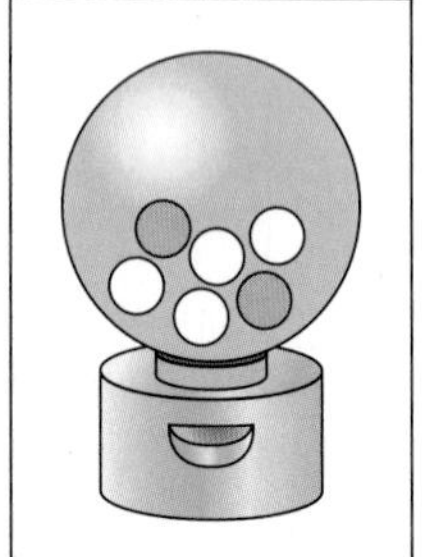

Figure 7.51

STRATEGY 2 Use reasoning

One student claims that there are 4 possible outcomes—RR, RW, WR, WW—and so the probability of getting at least 1 red is $\frac{3}{4}$. What do you think?

The mistake here is similar to the mistake that students often make in the boys-and-girls problem when they reason that there are five possible ways to have 4 children: 4G0B, 3G1B, 2G2B, 1G3B, and 0G4B. This line of reasoning is valid only if each outcome is just as likely to occur as any other.

If we think back to that problem (or to the dice problems) and look at ways to apply what we learned there to this problem, a tree diagram comes to mind. Explore a tree diagram before reading on. . . .

STRATEGY 3 Make a tree diagram

Figure 7.52 shows one way to represent this problem with a tree diagram. There are 6 trees because there are originally 6 gumballs. Each tree shows the outcomes if that gumball comes out first. That is, if a red gumball comes out first, then there are 5 gumballs left, 1 red and 4 white. However, if a white gumball comes out first, then there are 2 reds and 3 whites left.

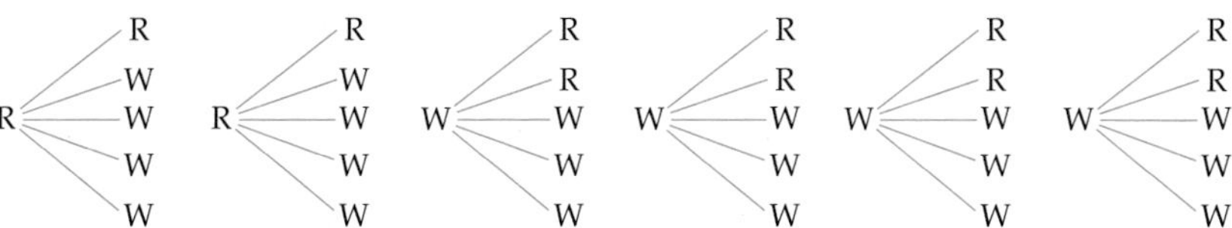

Figure 7.52

From Figure 7.52, we find that there are 30 outcomes and that the event "at least 1 red" occurs 18 times. Thus, the probability of getting at least 1 red gumball is $\frac{18}{30}$, or 0.6.

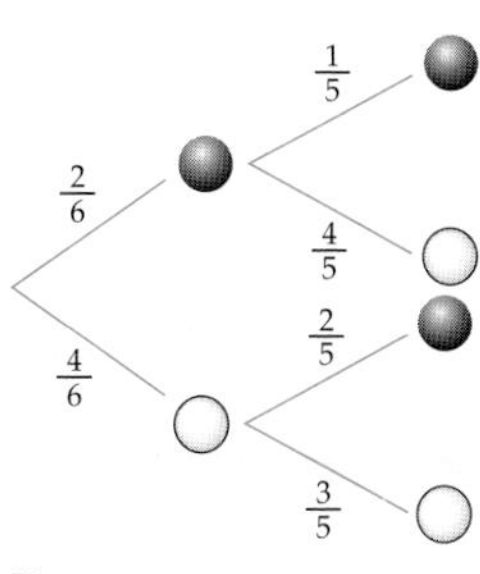

Figure 7.53

STRATEGY 4 Make a more sophisticated tree diagram

A different tree diagram is shown in Figure 7.53. Study it and try to make sense of it. Can you see the solution contained within the diagram?

When we put the money in the machine, the probability of getting a red is $\frac{1}{3}$, and the probability of getting a white is $\frac{2}{3}$. Let's say we get a red the first time. In this case, the probability of getting a red is now $\frac{1}{5}$, and the probability of getting a white is $\frac{4}{5}$. Why is this? Similarly, let's say we get a white the first time. In this case, the probability of getting a red is now $\frac{2}{5}$, and the probability of getting a white is $\frac{3}{5}$.

We can use the multiplication principle to determine the probabilities of each branch of the tree, and we can check our work, because the sum should be 1. Do you see why?

$$P(\text{RR}) = \frac{2}{6}\left(\frac{1}{5}\right) = \frac{2}{30} = \frac{1}{15}$$

$$P(\text{RW}) = \frac{2}{6}\left(\frac{4}{5}\right) = \frac{8}{30} = \frac{4}{15}$$

$$P(\text{WR}) = \frac{4}{6}\left(\frac{2}{5}\right) = \frac{8}{30} = \frac{4}{15}$$

$$P(\text{WW}) = \frac{4}{6}\left(\frac{3}{5}\right) = \frac{12}{30} = \frac{6}{15}$$

These events are mutually exclusive, so we can conclude that the probability of getting at least 1 red is

$$\frac{1}{15} + \frac{4}{15} + \frac{4}{15} = \frac{9}{15} = 0.6$$

Dependent and independent events Now let us return to the question posed at the beginning of the investigation: How is this scenario different from the ones we have explored before? Think and then read on. . . .

With the gumballs, the probability of obtaining a red or white gumball the second time depended on what happened the first time. In the case of having boys and girls or throwing dice, the probability of the second stage of the activity does not depend on what happened first; that is, its probability is independent of the first result.

> Two events are **dependent** if the probability of the second event is affected by the outcome of the first event.
>
> Two events are **independent** if the probability of the second event is not affected at all by the outcome of the first event.

FAIR GAMES

Another application of probability has to do with the idea of fair games and expected value, which has important real-life applications. Let us begin with a simple situation.

INVESTIGATION 7.3i Is This a Fair Game?

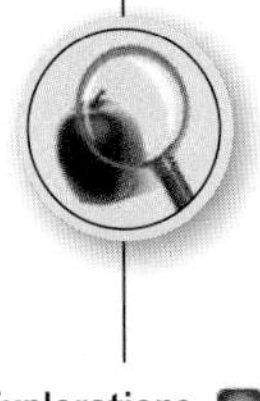

Consider the following game for two players. Each person spins the spinner (Figure 7.54). If the spinner lands on the same animal twice, the first player wins. If it lands on a different animal each time, the second player wins. If you could choose, would you want to be the first or the second player, or does it matter? Think about this and then read on. . . .

Explorations Manual 7.19

DISCUSSION

Can you see similarities between this problem and others we have investigated in this section? If not, look back and try to find connections. Then answer the question and read on. . . .

Figure 7.54

This problem is similar to the questions about the probability of having boys and girls. Do you see why?

Applying these similarities, we find that there are four equally likely outcomes in this game: (1) cat, cat, (2) dog, dog, (3) cat, dog, and (4), dog, cat. Thus, each player has the same theoretical probability of winning.

Using probability language, we say that this is a **fair game**. How would you define *fair game* at this point? Write down your thoughts and then read on. . . .

INVESTIGATION 7.3j What About This Game?

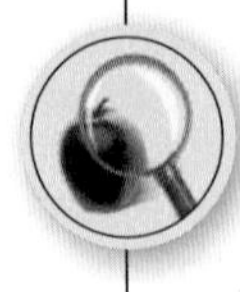

A school is having a carnival to raise funds. In the following game, the player spins the spinner (Figure 7.55) and receives the dollar amount of the number on which the arrow lands. If it costs \$2 to play the game, is this a fair game? What do you think?

DISCUSSION

STRATEGY 1 Make a simulation

You could make a spinner like the one in Figure 7.55 and simulate 100 games. If you made approximately \$200, the game is probably a fair game. If you didn't have a spinner, how else could you simulate this game?

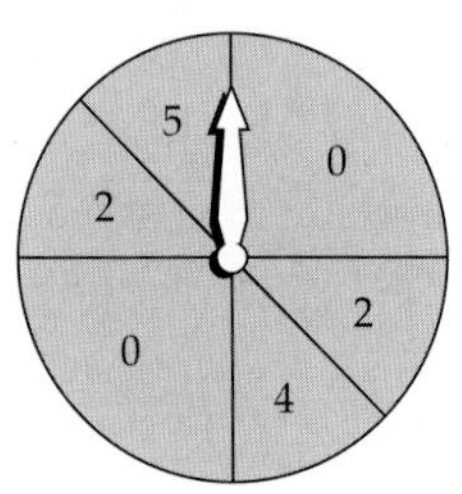

Figure 7.55

First, we need to determine the probability of each outcome. That is, $P(0) = \frac{1}{2}$, $P(2) = \frac{1}{4}$, $P(4) = \frac{1}{8}$, $P(5) = \frac{1}{8}$. So you could make 8 pieces of paper and write 0 on four of them, 2 on two of them, 4 on one of them, and 5 on one of them, and then draw, making sure to put the piece of paper back each time.

STRATEGY 2 Determine the theoretical probabilities

We have spoken of theoretical and experimental probabilities earlier. Can we apply the concept of theoretical probability to theoretical winnings?

We can compute the theoretical winnings per turn by using the concept of weighted average (from Section 7.2):

$$\frac{1}{8}(5) = \frac{1}{8}(4) + \frac{1}{4}(2) + \frac{1}{2}(0) = \frac{5}{8} + \frac{4}{8} + \frac{2}{4} = \frac{13}{8} = \$1.625$$

That is, theoretically, $\frac{1}{8}$ of the time the player will win \$5; $\frac{1}{8}$ of the time \$4; $\frac{1}{4}$ of the time \$2; and $\frac{1}{2}$ of the time nothing. If this solution falls into the category of "it sort of makes sense," what might you do to make it "really make sense"? Think and then read on. . . .

One strategy would be to act it out. For example, let's say the person played the game 80 times. Theoretically, how much money will the player win in 80 turns? (Do you see why I chose 80 turns instead of 50 or 100 turns?)

Theoretically, $\frac{1}{8}$ of the time, or 10 different times (because 10 is $\frac{1}{8}$ of 80), the player will win \$5, and thus the player will win a total of \$50 on those 10 turns. When we apply this strategy to the \$4 and \$2 outcomes, we find that theoretically the player's total winnings will be \$130. Do you see that this is equivalent to the weighted average we obtained above?

If the player wins \$130 after 80 turns, the winnings are an average of \$1.625 per turn. Referring now to the original question, we find that theoretically, the player will lose money playing this game because it costs \$2 to play the game. It is important to note that we are not saying that the player will lose. The player might be very lucky and play 10 games and win \$50. However, playing a large number of games makes losing money more likely than winning it.

Can you revise your original definition of *fair game* so that it works for Investigations 7.3i and 7.3j?

OUTSIDE THE CLASSROOM

Most states now have lotteries. In one lottery scenario, players pick four numbers between 1 and 50. The state then randomly selects four numbers from the sample space—for example, 2, 5, 23, and 47. In another lottery scenario, the players pick four one-digit numbers between 0 and 9. The state then randomly selects one digit at a time, replacing the selected number before drawing the second number. In this case, 1, 1, 2, 2 is a possible winning combination. The first situation is an example of dependent events, and the second is an example of independent events.

EXPECTED VALUE

There is a term for the theoretical winnings per turn: **expected value**. The concept of expected value is an aspect of probability that has many real-life applications. For example, car insurance premiums are based on the insurance company's estimate of the probabilities of different groups of people having accidents. If an experiment (activity) consists of n events, and each event has a specific probability and a specific payoff, we can determine the expected value of that experiment (activity) in the following manner:

Let us denote each event as $E_1, E_2, E_3, \ldots, E_n$.

Let us denote the probability of each event as $P_1, P_2, P_3, \ldots, P_n$.

Let us denote the payoff of each event as $X_1, X_2, X_3, \ldots, X_n$.

The expected value $= P_1 \cdot X_1 + P_2 \cdot X_2 + P_3 \cdot X_3 + \cdots + P_n \cdot X_n$

The concept of expected value is used to define a fair game formally. Before reading on, try to describe whether a game is fair or not using the concept of expected value. . . .

A game is "fair" if the expected value of the game equals the cost of playing the game.

Let us now extend the concept of expected value to a real-life application.

INVESTIGATION 7.3k Insurance Rates

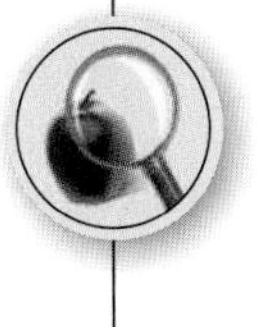

Let's say that an insurance company is determining the rates for car insurance for the next year. The company has determined that for every 10,000 policyholders:

- 1 is likely to have an accident for which the company will have to pay a \$200,000 claim.
- 10 are likely to have an accident for which the company will have to pay a \$100,000 claim.
- 20 are likely to have an accident for which the company will have to pay a \$50,000 claim.
- 100 are likely to have an accident for which the company will have to pay a \$10,000 claim.

- 1000 are likely to have an accident for which the company will have to pay a $1000 claim.
- The remainder of the policyholders will have no claims for the year.

Do you understand how the insurance company will use the concept of expected value in determining its rates? If so, determine the expected value of the claims and then read on. . . .

DISCUSSION

This investigation has many connections to Investigation 7.3j. In order to compute the expected value in the insurance situation, the insurance company wants to get an idea of the "average" payout per customer. This problem is a model of what insurance companies do, not an account of what they actually do.

In order to determine the expected value, the incidence of each event must be converted to probability language. For example, "1 in 10,000 claims will be in the neighborhood of $200,000" translates to "0.0001 probability of a $200,000 claim." Thus,

$$\begin{aligned}\text{Expected payout} &= 0.0001(200{,}000) + 0.001(100{,}000) + 0.002(50{,}000)\\ &\quad + 0.01(10{,}000) + 0.1(1000) = 420\end{aligned}$$

It means that if the accidents occur at the rate and amount that the insurance company is predicting, it will pay out an average of $420 per policy (per year). The actual rate the company charges will be determined by this expected value, the company's overhead (salaries, cost of operations, etc.), and the profit margin.

OUTSIDE THE CLASSROOM

One of my sisters is an insurance agent. As I was consulting with her when writing this material, she told me something that surprised me. I had always thought that insurance companies would ideally like to have as large a percentage of policyholders as possible. However, this is not the case. Can you imagine why an insurance company would not want to have 100% of the policyholders in an area? Think and then read on. . . .

The reason has to do with natural disasters. Several major natural disasters have recently brought even some of the big companies to their knees: the Midwest floods in the summer of 1993, the California earthquake in 1994, and Hurricane Katrina in 2005. Insurance companies had to pay out more than $60 billion to their customers. If even the largest insurance company had most of the homeowners' policies in this region, it would have gone bankrupt. Therefore, insurance companies determine the maximum percentage of total policyholders in any particular area that they want to insure.

As before, some students may better understand the problem by determining the expected payoff for the 10,000 policies.

$$\$200{,}000(1) + \$100{,}000(10) + \$50{,}000(20) + \cdots$$

Consider the connections between these strategies.

SUMMARY 7.3

In this section, we have developed some basic language for exploring and describing probability situations. Probabilities can be expressed as fractions between 0 and 1 and can be expressed using percents, ratios, and odds. We can discuss probabilities using set language, in which case the set of all possibilities is called the sample space. We distinguish between outcomes, which are elements of the sample space, and events, which are subsets of the sample space.

We have learned other distinctions between different kinds of probability situations.

- In some cases, all outcomes are equally likely; in other cases, this is not true.
- Some events are independent and some are dependent.
- In all situations, we can determine the experimental probability of an event or an outcome by collecting data; in some cases, we can determine theoretical probabilities using different problem-solving tools.

We have examined several rules and formulas that have come out of making sense of situations and understanding connections between probability and other areas, such as sets, weighted average, and our understanding of operations.

- Complementary events: $P(A) + P(B) = 1$
- Mutually exclusive events: $P(A \text{ or } B) = P(A) + P(B)$
- The multiplication principle: If $P(A) = x$ and $P(B) = y$, then $P(A \text{ then } B) = x \cdot y$
- Expected value: $P_1 \cdot X_1 + P_2 \cdot X_2 + P_3 \cdot X_3 + \cdots + P_n \cdot X_n$

We have found that there are many different techniques we can use to solve probability problems: tree diagrams, formulas, area models, and simulations. These techniques work better when they are used with other problem-solving tools: making tables, looking for patterns, being systematic, and using reasoning.

Some real-life situations are similar to ones we have studied here—for example, the probabilities associated with genetics, as in Exploration 7.13. Some real-life situations are more complex than the ones we have discussed here, and additional tools are required to solve those problems.

7.3 Exercises

1. In a class, 16 people live on campus and 8 people live off-campus. If we were to draw one student's name from a hat, what is the probability that this student is living off-campus?
2. In a study of blood donors, 225 were classified as group O and 275 had a classification other than group O. What is the approximate probability that a person will have group O blood?
3. A survey found that 132 people brushed their teeth once a day, 368 people brushed twice a day, and 84 people brushed three times a day. If one of the respondents is randomly selected, find the probability of getting someone who brushes their teeth three times a day. What is the probability of selecting someone who brushes two or more times a day?
4. a. Which of the following would you choose if you were picking 1 ball and wanted black? Justify your choice.

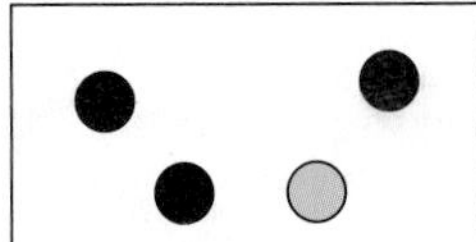
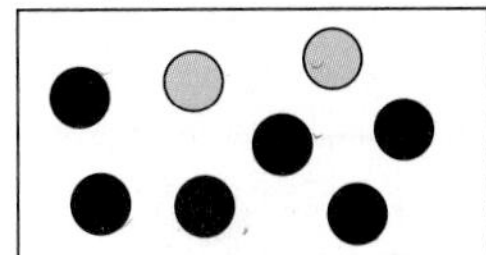

   b. What if you were picking 2 balls (without replacement) and wanted 2 black ones?

5. A bag contains 2 red counters, 6 blue counters, and 12 green counters. What is the probability of drawing
   a. a red counter?
   b. a blue counter?
   c. a green counter?
   d. a red counter or a blue counter?
   e. a green counter or a blue counter?
6. a. If you toss two coins, what is the probability of getting a match (e.g., HH or TT)?
   b. If you toss three coins, what is the probability of all three coins matching?
   c. If you toss five coins, what is the probability of all five coins matching?
7. If a couple has 3 children, what is the probability that
   a. they will be the same sex?
   b. they will have 2 girls and 1 boy?

8. Peppermint Patty is very discouraged about her chances on a 10-item true-false quiz. If she randomly answers each question, what is her probability of getting a grade of at least 50% on a
   a. two-item quiz?
   b. three-item quiz?
   c. four-item quiz?
   d. five-item quiz?
   e. Do you see any patterns in the answers above? Explain.
9. When rolling two dice and adding the numbers, what is the probability of
   a. getting a 2?
   b. getting an even number?
   c. getting a prime number?
   d. rolling doubles?
10. What are the possible results from rolling two dice and subtracting the number showing on the face of one from that showing on the face of the other in such a way that a nonnegative number is obtained? What result is most likely, and what is its probability?
11. Determine the probability of the prince ending up in Room A and Room B in the given scenario.

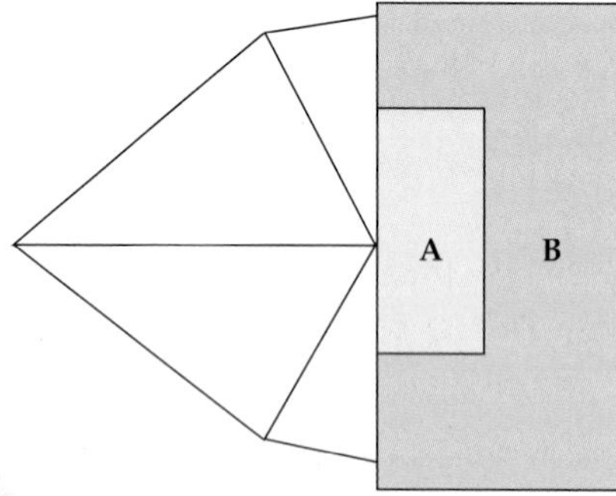

12. A drawer has 4 red socks and 4 blue socks.
   a. If 2 are drawn, what is the probability of a match?
   b. If 3 are drawn, what is the probability of a match?
   c. What is the probability of having all one color after 4 draws?
13. A bag of marbles contains 3 blue, 5 green, and 2 red marbles. If you take out 2 marbles, what is the probability that they are both green?
14. A drawer has 4 red socks and 2 blue socks. What is the probability of a match after picking 2 socks?
15. A gumball machine has 2 red gumballs and 2 green ones. If you buy two gumballs, what is the probability that both will be red?
16. Recall Investigation 7.3h, Gumballs. What if Josie spent a third quarter? Do you think $P(\text{WWR}) = P(\text{WRW}) = P(\text{RWW})$? Why or why not?
17. Consider the following game for two people: A player rolls two standard dice and makes a proper fraction with the two numbers.
   - If the fraction is in simplest form, the person who rolled wins.
   - If the fraction is not in simplest form, the other person wins.
   - If the two numbers are identical, the other person wins.

   Is this a fair game?
18. For each of the three spinners below, determine the probability of getting each number. For each of the spinners, if you did this 100 times, what would the average score be?

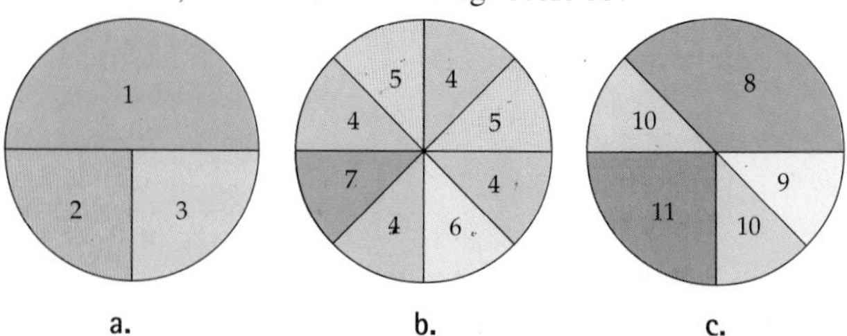

19. This problem is adapted from the Risky Allowance Problem in the April 2006 issue of *Teaching Children Mathematics.* A child can receive $2 per week allowance or spin the spinner. Which would you choose? Over the course of a year, what do you predict would be your average allowance when spinning the spinner? Explain your answer.

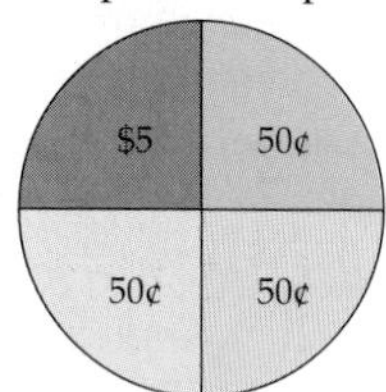

20. If we spin both spinners below, what is the probability of getting different numbers?

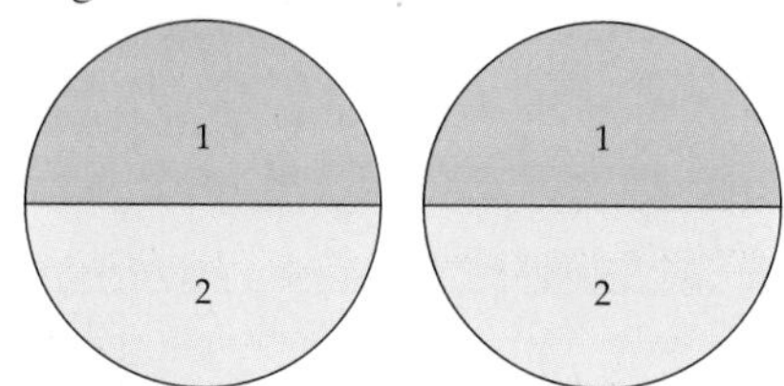

21. If we spin both spinners below, what is the probability of getting different numbers?

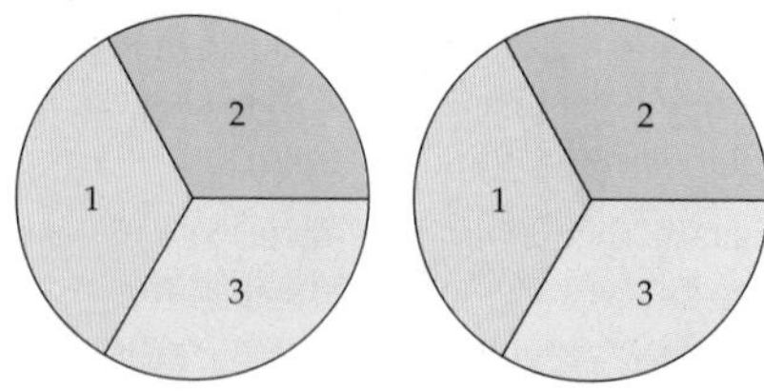

22. At a carnival game, you pay $3 for a ticket and spin the spinner and win the amount shown. Do you play? That is, is it a fair game?

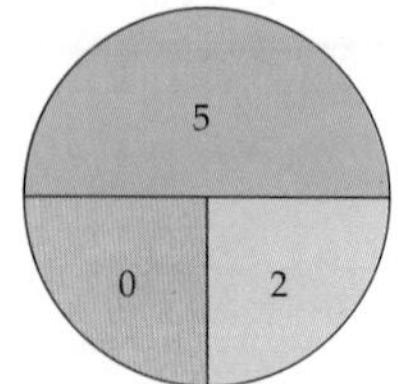

23. Let's say 1000 people play the carnival game in Investigation 7.3j over the course of the carnival. Theoretically, how much money will the school make from this game?

24. Let's say you wanted to make the carnival game in Investigation 7.3j a fair game. Describe one situation that would be a fair game. Explain why that game would be fair.

25. In each of the games below, explain whether or not it is a fair game. If it is not a fair game, modify the game so that it becomes fair.

 a. Two players take turns rolling two dice and adding the numbers. Player A wins if the number is even, and player B wins if the number is odd.

 b. Two players take turns rolling two dice and multiplying the numbers. Player A wins if the number is even, and player B wins if the number is odd.

 c. Two players take turns rolling three dice and adding the numbers. Player A wins if the number is even, and player B wins if the number is odd.

DEEPENING YOUR UNDERSTANDING

26. a. Drawing from a regular deck of playing cards, what is the probability of drawing a face card (jack, queen, or king) 3 times in a row? (Assume that jokers are taken out, and assume no replacement.)

 b. How would the probability change if you assumed replacement?

 c. Marla believes that the answer to the second question is $\frac{3}{13} + \frac{3}{13} + \frac{3}{13}$. Describe how you might convince her that this answer is incorrect.

27. A pinochle deck consists of two copies each of the 9, 10, jack, queen, king, and ace cards of all four suits, for 48 cards per deck.

 a. What is the probability of drawing a face card in a pinochle deck?

 b. What is the probability of drawing a face card two times in a row without replacement?

28. The American Lung Association states that approximately 1 out of every 17 deaths in the United States last year was from lung cancer. If there are 34 students in a class, does that mean that 2 people in this class will die from lung cancer?

29. If it snows, there is a $\frac{1}{3}$ chance that schools will close. The weather person says that there is a $\frac{1}{4}$ chance of snow tomorrow. What is the probability that there will be no school tomorrow?

30. Let's say there is a 15% chance of getting strep if you are exposed to a person with strep. Let's say there is a 50% chance that you were exposed.

 a. What is the probability that you will get strep?

 b. Let's say each person in your (4-person) family had a 50% chance of being exposed. What is the probability that you will all get strep?

31. ***Classroom Connection*** The inspiration for this problem came from a question asked of me by a parent. Let's say the local middle school has five sixth-grade mathematics teachers, two of whom are female, four seventh-grade mathematics teachers, one of whom is female, and four eighth-grade mathematics teachers, one of whom is female. Assuming that the students' placement is random, what is the probability that a student will have at least one female mathematics teacher during the student's three years at the school?

32. Determine the probability of getting 8 heads in a row when flipping a coin 8 times. Give the answer both in decimal form and in ratio form. In decimal form, round the answer to two digits past the string of zeros, and write the decimal you obtained in words. By ratio form, I mean "the chance is 1 in ___."

33. If you flip a coin 10 times, what is the probability of getting at least 4 heads?

34. Suppose that instead of the probability of having a boy or a girl being equal, the probability was $\frac{2}{3}$ girl and $\frac{1}{3}$ boy. If a family had two children, what is the probability that they would have 1 boy and 1 girl?

35. Here are some additional questions inspired by *The Giver* and *Twenty-One Balloons.* Parts (d) and (e) come directly from the article cited in Investigation 7.3a.

 a. In a family of 6 children, what is the probability of having 3 girls and 3 boys?

 b. In a family of 8 children, what is the probability of having 4 girls and 4 boys?

 c. In a family of 10 children, what is the probability of having 5 girls and 5 boys?

 d. "As the number of children in a family increases, what happens to the probability of having the same number of boys and girls, and why does this happen?" (p. 508)

 e. From *The Giver:* "Each class of 50 children in the Community has 25 girls and 25 boys. Assuming that the probability of a woman giving birth to a boy or a girl is the same, how often do you think exactly 25 out of 50 babies will be girls?"[10]

36. Consider a multiple-choice quiz consisting of 5 questions with 3 possibilities per question. What is the total number of possible outcomes? What does "outcome" mean here?

37. Consider a pair of triangular pyramid-shaped dice (that is, dice containing the numbers 1 through 4, rather than 1 through 6 as on regular dice). If you roll two dice at a time, what is the most likely sum and what is the probability of that sum?

38. The following experiment consists of rolling a 12-sided die. The outcome set is {1, 2, 3, 4, 5, 6, 7, 8, 9, 10, 11, 12}. What probability would you assign to each of the following outcomes?

 a. An even number turns up.

 b. A prime number turns up.

 c. A proper divisor of 12 turns up.

 d. What is the most probable event if you roll two such dice and add the numbers? What is its probability?

39. In the board game Monopoly, if you roll three doubles in a row, you go to jail. What is the probability of rolling three doubles in a row? Describe three ways in which you could represent this probability. Then select the one that you think would be most understandable to the general population. For example, if you were writing an article for the daily newspaper about math in society, which representation would you choose and why?

40. When my children were young, they played Junior Monopoly, which uses only one die. One day, Josh rolled a 2, then Emily rolled a 2, then I rolled a 2. Josh and Emily thought this was great fun. What is the probability of rolling three 2s in a row on three tosses of a die?

41. Make 2 six-sided dice so that only even sums are possible and each sum is equally likely. Your solution must be other than the trivial solution of "the same odd number on all 12 faces."

42. If we look at Table 7.24 (Appendix B) and add the number of outcomes in each column, the numbers for the first six columns are 1, 3, 6, 10, 15, 21 which is like the Handshakes Problem in Exploration 1.1. However, the sum of the outcomes in the seventh column is 25 (not the 28 we got with handshakes). Can you explain why we get 25 instead of 28?

43. A game known to most children is rock, paper, scissors. On the count of three, each person displays rock (fist), paper (palm), or scissors (two fingers extended). Rock beats scissors, scissors beats paper, and paper beats rock. Let's say three people play according to these rules. If all three match, player 1 wins, if all three are different, player 2 wins; otherwise, player 3 wins. Is this a fair game? Why or why not? If it is not a fair game, modify the game so that it becomes fair.

44. A popular Hanukkah game is called Dreidel. Each player begins with a number of pieces of gold (actually chocolate in a round, gold-colored wrapper in the shape of a coin). The game consists of spinning a top that has four sides.

If the top lands on the side labeled gimel, the player gets all the gold in the pot.

If the top lands on the side labeled shin, the player has to put one coin in the pot.

If the top lands on the side labeled hay, the player gets half of the gold in the pot.

If the top lands on the side labeled nun, the player gets nothing.

a. What is the probability of getting no gimels after 4 turns?

b. What is the probability of the top landing on each of the four sides on your first 4 turns?

45. There is a carnival game called Chuck-a-Luck. The game works this way:

The player picks a number from 1 to 6. The operator then rolls 3 dice.

If the player's number comes up all 3 times, the player wins $3.

If the player's number comes up 2 times, the player wins $2.

If the player's number comes up 1 time, the player wins $1.

If the player's number doesn't come up, the player pays the operator $1.

If the game costs $1, to play is this a fair game?

46. Let's say a state is considering introducing a new lottery game. The state will charge $1 for a ticket. Each week, one person's name will be selected at random. The amount of the person's prize will be determined in the following manner: Five Ping-Pong balls—four white and one black—will be placed in a hopper. The balls will be selected one at a time until all five balls are in a line. The state expects to sell 250,000 tickets each week.

- If the black ball is first, the person wins $1,000,000.
- If the black ball is second, the person wins $100,000.
- If the black ball is third, the person wins $10,000.
- If the black ball is fourth, the person wins $1000.
- If the black ball is fifth, the person wins $500.

a. Is this a fair game?

b. Do you think this game will be a good money raiser? Why or why not?

47. I bought six tickets for a game where the odds were one in four, and none of the tickets were winners. Does this show that the prizes aren't awarded randomly?

48. Just as insurance companies use the concept of expected value in determining insurance rates, oil companies use expected value in determining whether or not to drill for oil. Let's say that an oil company is considering drilling for oil in a certain area. The company estimates that it will cost an average of $200,000 to drill each oil well in this area. It has determined that there are three kinds of outcomes: a dry well, a "medium" strike that would produce about $750,000, and a big strike that would produce about $2,500,000. The company has estimated that the probability of a medium strike is $\frac{1}{5}$ and the probability of a big strike is $\frac{1}{25}$. Should the company drill for oil? Determine the expected value, and then describe your conclusion as though you were talking to a group of speculators who don't understand expected value.

49. The inspiration for this problem came from one of my Middle School Mathematics Methods students. Let's say you have a bag that contains 10 colored beads. You can pull out one bead at a time and then put it back and then shake up the bag. Let's say you have done this five times and have pulled out a red bead three times, a blue bead, and a green bead. What are the facts and what inferences do you have about the contents of the bag? List and justify your facts and inferences.

50. Go to www.shodor.org/interactivate/activities/Coin/ and simulate tossing a coin 1000 times. Theoretically, how many times should you get heads? Experimentally what happened?

51. Go to http://www.shodor.org/interactivate/activities/Adjustable Spinner/ and simulate the carnival game in Investigation 7.3j and analyze the results.

52. Describe a real-life scenario whose occurrence is a probability and whose probability you would be interested in determining.
    a. Define the scenario.
    b. Describe how you might go about determining the probability.
    c. Describe some aspects of the problem that would influence the actual probability.

    I have sketched one scenario below; this would be the beginning of a good response.

    a. The probability that I will get a job after I graduate.
    b. Interview professors who work with teacher certification and average their guesses; interview principals in this area and average their guesses; contact those students who got certified at KSC last summer and determine how many got jobs.
    c. Probabilities will differ depending on whether I will look only in this area, what my certification is (early childhood, elementary, middle, etc.), whether I am 22 or 32, and so on.

SECTION

Counting and Chance

What do you think?

- What does "counting" mean in the context of probability?
- What is the difference between a combination and a permutation?

Explorations Manual 7.20, 7.21

In many probability problems, simply determining the size of the sample space can be quite tedious. In this section, we will examine some algorithms that can make solving probability problems much easier. While the tree diagrams can be an interesting model of multiplication for elementary students, permutations and combinations are not taught until high school, according to the Common Core State Standards.

INVESTIGATION 7.4a How Many Ways to Take the Picture?

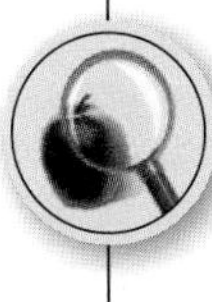

Let's say your college has a chapter of Kappa Delta Pi, the national education society. Let's say this was a new society in your college, and in the first year there were 9 members. Let's also say that the members decided to have a group picture taken and wanted only one row—that is, they wanted everyone in the first row. In how many different ways can they line up for the picture? Work on this and then read on. . . .

DISCUSSION

STRATEGY 1 Make a simpler problem

There are several different strategies that can be successfully applied to this problem. The first one we will discuss is "make a similar, simpler problem and then work up." If you didn't make much progress on your own and you like this method, use it to finish the problem.

Let's see how it works. Let's start with a club with only 3 members:

1 2 3
1 3 2
2 1 3
2 3 1
3 1 2
3 2 1

This gives us a total of 6 different arrangements. Where did the 6 come from?

Most people see 3 groups of 2, and hence 3×2. Some people can jump to the solution of the original problem from here; others need more data from which to generalize.

If you don't yet feel confident that you can connect this to the original problem, try a club with 4 members. If you haven't done this, try it on your own before reading on. . . .

1 2 3 4

1 2 4 3

1 3 2 4

1 3 4 2

1 4 2 3

1 4 3 2

This gives us 6 possibilities in which person number 1 is on the left. Either by reasoning or by working out the other possibilities yourself, you can see that there will be 6 outcomes in which each of the 4 members is on the left. This gives us a total of 24 outcomes.

As before, we have 4 groups, each of which consists of 2 groups of 3. If we put the numbers in order, we have either $4 \times 3 \times 2$ or $2 \times 3 \times 4$.

If you feel that you can jump from here to the solution to the problem, do so. If not, then you can either determine the number of possibilities for a group with 5 members or try another strategy.

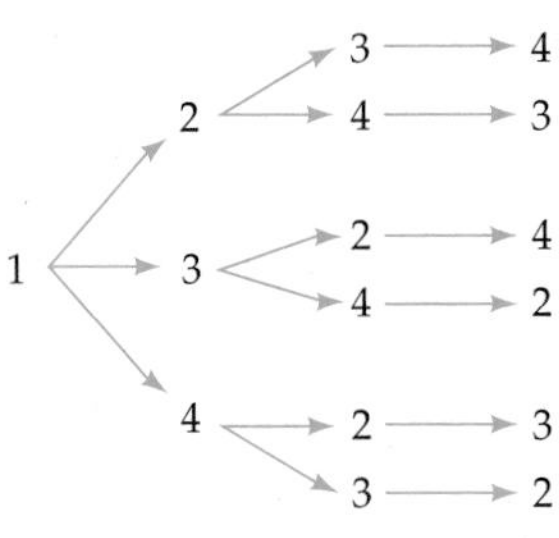

Figure 7.56

STRATEGY 2 Make a tree diagram

In this case, a tree diagram (Figure 7.56) can help us to see the pattern developing in Strategy 1. This figure shows the 6 possibilities with person number 1 on the left.

STRATEGY 3 Connect this problem to something familiar

What about the multiplication principle? Many people understand this strategy better if we simultaneously use the "act it out" strategy.

Let's begin with the first person. We have nine possibilities. Once we have that person set, how many choices do we have for the second position? There are eight possibilities. Do you see why? Do you see how to finish the problem using this line of reasoning? Work on it before reading on. . . .

Using this line of reasoning, the number of possibilities will be 9 times 8 times 7 times 6 and so on, so that the total number of possibilities is

$$9 \times 8 \times 7 \times 6 \times 5 \times 4 \times 3 \times 2$$

Do you see how the first strategy hinted at this solution?

New terminology will make our discussion of this problem and other probability problems easier. How would you describe the computation above to someone—let's say a friend in class who didn't have this book? Do so and then read on. . . .

One way: "Start with 9, then 8, then 7, and keep reducing the number by one till you can't anymore. Now compute the product."

This kind of computation is common in probability and has a name: **factorial**. That is, $9 \times 8 \times 7 \times 6 \times 5 \times 4 \times 3 \times 2$ can be written as 9!

Mathematicians insert a 1 at the end of this product, and define any number n factorial, $n!$, as

$$n! = n \times (n - 1) \times (n - 2) \times \cdots \times 3 \times 2 \times 1$$

INVESTIGATION 7.4b How Many Different Election Outcomes?

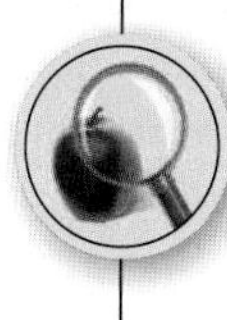

Let's say Kappa Delta Pi has its first election: for president and treasurer. How many possible ways can a president and treasurer be elected from a pool of 9 people? Work on this and then read on. . . .

DISCUSSION

STRATEGY 1 Be systematic, make a table, and look for patterns

TABLE 7.20

1, 2	2, 1	
1, 3	2, 3	
1, 4	2, 4	
1, 5	2, 5	etc.
1, 6	2, 6	
1, 7	2, 7	
1, 8	2, 8	
1, 9	2, 9	

Note that the numbers in Table 7.20 represent possible outcomes. For example, 1, 2 represents the outcome "person 1 as president and person 2 as treasurer." Why is {2, 1} not simply a duplicate of {1, 2}? Why did we skip {2, 2}? When you complete this table, you find 9 columns, each of which has 8 outcomes, giving a total of 72 ways.

STRATEGY 2 Make a tree diagram

Some students find a tree diagram (Figure 7.57) very helpful. Similar to Strategy 1, this diagram shows the possibilities where person number 1 is president. Since there would be 9 of these trees, with 8 branches on each, the total would be $9 \times 8 = 72$.

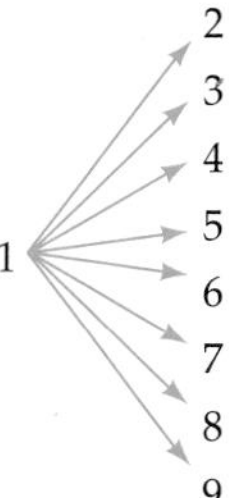

Figure 7.57

STRATEGY 3 Connect to previous problems

We can use the multiplication principle again.

There are 9 different possibilities for president (each member), but only 8 possibilities for treasurer (assuming that no one will be both president and treasurer).

Without too much difficulty, most students can see this as 8 groups of 9, or 9×8, which gives us 72 ways.

Electing 3 officers Now let's see how confident you are about this idea. What if the group decided to have a president, a treasurer, and a secretary? How many different outcomes are there for this scenario? Work on this and then read on. . . .

Each of the three strategies from before still applies (Figure 7.58).

	8 groups						
7 groups	123	132	192	This process repeats 9 times; that is, there is another group of 8 × 7 outcomes that begins with 2, then another group of 8 × 7 outcomes that begins with 3, and so on.	213	231	291
	124	134	193		214	234	293
	125	135	194		215	235	294
	126	136···	195		216	236···	295
	127	137	196		217	237	296
	128	138	197		218	238	297
	129	139	198		219	239	298

Figure 7.58

The total number is $9 \cdot 8 \cdot 7 = 504$.

This question (about president, treasurer, secretary) is more complex than the question in Investigation 7.4a about the picture. Instead of looking at all the possible permutations of a set of specified size, which we did in the picture investigation, we are now asking how many different *permutations* of specified size we can make from a set of a given size. In this situation, the outcome "1 2 3" is considered to be different from the outcome "1 3 2." Do you see why?

Using symbols, we say that our election question seeks the number of different subsets of size r that we can make from a set of size n. We can also refer to this amount as the number of **permutations** of n things taken r at a time. The shorthand for this situation is ${}_nP_r$. (*Note:* Some books use the notation P_r^n.)

In the case of the different possibilities for president and treasurer, we were looking for ${}_9P_2$.

In the case of the different possibilities for president, treasurer, and secretary, we were looking for ${}_9P_3$.

Can you now generalize a formula for the number of permutations of a subset of size r from a set of size n? Try to do so and then read on. . . .

The generalization is

$$ {}_nP_r = n(n-1)\cdots(n-r+1) $$

If this makes sense, great. If not, substitute 9 for n and 2 (and then 3) for r so that it makes sense.

Some authors prefer a different formula, one that helps the learner to see how many terms to include and/or what the last term is. This formula rests on the following observation:

$$ 9 \cdot 8 \cdot 7 = \frac{9!}{6!} $$

(Active readers will make sure that they believe this equivalence!) In this situation, 6 is the number of people who will not be elected. This observation leads to an alternative formula:

$$ {}_nP_r = \frac{n!}{(n-r)!} $$

INVESTIGATION 7.4c How Many Outcomes This Time?

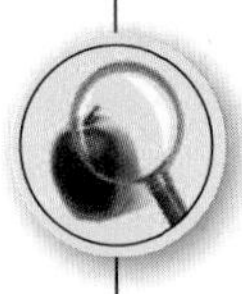

Suppose Kappa Delta Pi decided to elect two members to send to the national convention. How many possible outcomes are there now? Think and then read on. . . .

DISCUSSION

Do you understand why this is not the same as the problem with the president and treasurer? If so, explain why. If not, think and then read on. . . .

Combinations Using the notation of the investigation with the president and treasurer, the outcomes {1, 2} and {2, 1} represent different outcomes. In the first outcome, student 1 is the president and student 2 is the treasurer, whereas in the second outcome, student 2 is the president and student 1 the treasurer. However, delegates {1, 2} and {2, 1} represent the same outcome. Mathematically, we distinguish between permutations and **combinations**.

Permutations In probability situations in which order matters, we speak of **permutations**. When order doesn't matter, we speak of combinations. For example, if outcomes "*abc*" and "*bac*" are considered to be different, we are in a permutation situation; if they are considered to be the same, we are in a combination situation.

Using symbols, we can write the number of combinations of n things taken r at a time as ${}_nC_r$.

Now let us discuss the solution to how many outcomes there are for this question. First, we will be systematic and look for patterns (Table 7.21).

TABLE 7.21

1, 2	2, 3	3, 4	
1, 3	2, 4	3, 5	
1, 4	2, 5	3, 6	
1, 5	2, 6	3, 7	etc.
1, 6	2, 7	3, 8	
1, 7	2, 8	3, 9	
1, 8	2, 9		
1, 9			

Can you finish the problem? Does this problem remind you of a problem or a pattern that we have encountered before? Think before reading on. . . .

If you thought about the handshakes problem (Investigation 1.2j), you are right.

We find that there are 36 different outcomes—that is, $8 + 7 + 6 + 5 + 4 + 3 + 2 + 1$. We would like to be able to generalize this procedure so that we can use it in a variety of situations; for example, what if we had a set of 20 elements (instead of 9) and we wanted all outcomes consisting of 3 elements at a time (instead of 2)? One way to determine the general procedure is to analyze the results of the simpler problems. Thus we will explore where the 36 comes from. Work on this before reading on. . . .

From one perspective, we can see that 36 comes from $8 + 7 + 6 + 5 + 4 + 3 + 2 + 1$. Thus we could generalize that if there were 20 members and we wanted all outcomes of two at a time, there would be $19 + 18 + 17 + \cdots + 3 + 2 + 1$. What if we wanted all possible outcomes of three at a time? In order to develop a procedure that will work in more situations, we have to look more deeply. What other ways do you see to determine the 36? Think before reading on. . . .

Using the algorithm developed in Investigation 1.2j, we can say that

$${}_9C_2 = \frac{9(8)}{2} = 36$$

Now this should look similar to the permutation algorithm:

$${}_9P_2 = 9(8) = 72$$

Why do we divide by 2? Think about this before reading on. . . .

Because each combination is connected to two permutations, we can see that in this case, the number of combinations will be half the number of permutations.

Do you feel able to make a hypothesis for the general combination algorithm, ${}_nC_r$? If so, make it and then read on. . . .

If not, work through the next problem and then try to make a hypothesis.

If 3 people go to the convention What if 3 people could go to the convention? How many different combinations are there? Complete Table 7.22 before reading on. What patterns do you see that could help you to complete the table or help you to determine the total?

In the first set of outcomes, where person number 1 is first, we have 28 total outcomes, and this number is the sum of the sequence $7 + 6 + 5 + 4 + 3 + 2 + 1$. In a similar fashion, the number of outcomes in the second set is equal to $6 + 5 + 4 + 3 + 2 + 1 = 21$. From these patterns, it is reasonable to conclude that we can continue in this manner. Doing so, we find that there are 84 different outcomes for this situation—that is, $28 + 21 + 15 + 10 + 6 + 3 + 1$. In everyday language, we say that there are 84 possible ways to send 3 people to the convention.

Determining a way of connecting this 84 to a formula that we can use is not obvious, but there are connections. What do you think? Work on this before reading on. . . .

There are many possible paths that will enable us to see the general procedure. Let us take one path.

TABLE 7.22

1 2 3	1 3 4	1 4 5	1 5 6	1 6 7	1 7 8	1 8 9
1 2 4	1 3 5	1 4 6	1 5 7	1 6 8	1 7 9	
1 2 5	1 3 6	1 4 7	1 5 8	1 6 9		
1 2 6	1 3 7	1 4 8	1 5 9			
1 2 7	1 3 8	1 4 9				
1 2 8	1 3 9					
1 2 9						
2 3 4	2 4 5	2 5 6	2 6 7	2 7 8	2 8 9	
2 3 5	2 4 6	2 5 7	2 6 8	2 7 9		
2 3 6	2 4 7	2 5 8	2 6 9			
2 3 7	2 4 8	2 5 9				
2 3 8	2 4 9					
2 3 9						
3 4 5	3 5 6	3 6 7	3 7 8	3 8 9		
3 4 6	3 5 7	3 6 8	3 7 9			
3 4 7	3 5 8	3 6 9				
3 4 8	3 5 9					
3 4 9						

First, let us summarize what we know:

We know that ${}_9P_2 = 9 \cdot 8 = 72$, that ${}_9C_2 = 9 \cdot \frac{8}{2}$, and that there is a similarity between these two.

We know that ${}_9P_3 = 9 \cdot 8 \cdot 7 = 504$.

Therefore, we might suspect that ${}_9C_3$ might be something like $\frac{(9 \cdot 8 \cdot 7)}{x}$.

But we know that ${}_9C_3 = 84$, so we can make an equation and solve for x. That is, ${}_9C_3 = 84 = \frac{(9 \cdot 8 \cdot 7)}{x}$.

Solving the equation, we find that $x = 6$. This still prompts the question "Why do we divide by 6?" What do you think? Work on this before reading on. . . .

In this case, each combination outcome is connected to 6 permutation outcomes.

$$\left.\begin{matrix} 1\,2\,3 \\ 1\,3\,2 \\ 2\,3\,1 \\ 2\,1\,3 \\ 3\,1\,2 \\ 3\,2\,1 \end{matrix}\right\} \quad 1\,2\,3$$

Can you make the leap? Do you see a pattern?

$${}_9C_2 = \frac{9 \cdot 8}{2}$$

$${}_9C_3 = \frac{9 \cdot 8 \cdot 7}{6} = \frac{9 \cdot 8 \cdot 7}{1 \cdot 2 \cdot 3}$$

${}_nC_r = ?$ What do you think?

The generalization is:

$${}_nC_r = \frac{n \cdot (n-1) \cdot \cdot \cdot (n-r+1)}{r!} = \frac{{}_nP_r}{r!} = \frac{n!}{(n-r)!r!}$$

INVESTIGATION 7.4d Pick a Card, Any Card!

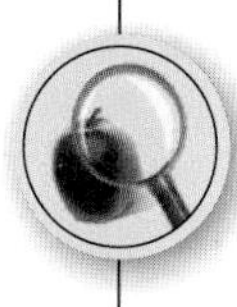

Consider a standard deck of playing cards. Pick two cards. What is the probability of getting two queens? Work on this problem and then read on. . . .

DISCUSSION

This is a combination (not a permutation) problem. Do you see why?

$$_{52}C_2 = \frac{52 \cdot 51}{2} = 1326$$

Does this mean that the probability is $\frac{1}{1326}$? What do you think?

The answer to this question is not $\frac{1}{1326}$. The number 1326 means that there are 1326 different possible combinations. Because we want to know the probability of getting two queens, we also have to determine how many of those 1326 combinations involve two queens. What do you think? Do this now before reading on. . . .

One strategy is to list them systematically—for example, SD, SH, SC, HD, HC, DC. Another strategy is to use the appropriate combination algorithm appropriately. Do you see how $_4C_2$ applies to this part of the problem?

In either case, we find that there are 6 different combinations that involve 2 queens. Thus we find that the probability of two queens is

$$\frac{6}{26 \cdot 51} = \frac{1}{13 \cdot 17} = \frac{1}{221}$$

INVESTIGATION 7.4e So You Think You're Going to Win the Lottery?

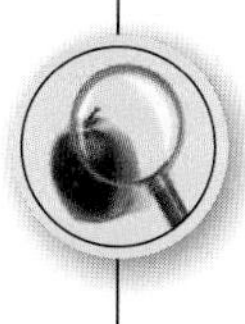

Since 1980, the number of states with lotteries has increased considerably. What factor(s) do you think might have led to the increase in states having lotteries?

During the 1980s, federal aid to states and cities shrank considerably. Many states "solved" this problem by establishing or increasing lotteries and designating the profits from the lottery for expenses such as education. The state lotteries have been called *voluntary taxes*. Can you see why?

Let's begin with a rather straightforward lottery in which players select four digits. The winning number is commonly selected in the following fashion: 10 Ping-Pong balls (each with a digit 0 through 9 written on it) are placed in a hopper. One Ping-Pong ball is selected. That digit now goes in the first place. That Ping-Pong ball is put back in the hopper, and the process is repeated three times.

What is the probability of picking a winning number? Work on this and then read on. . . .

OUTSIDE THE CLASSROOM

There are some facts about state lotteries that are troublesome.

In California, the poor spend 15 times more of their income, as a percentage, on lottery tickets than the rich do. . . . A Detroit study showed that in tough times, the middle and upper classes stopped playing the lotteries—but poor urban dwellers kept right on playing. . . . Other studies show that players are disproportionately minority people with less than average income and education.[11]

DISCUSSION

STRATEGY 1 Connect to the multiplication principle

Each time, there are 10 possibilities, so the sample space is $10 \times 10 \times 10 \times 10$. Because all outcomes are equally likely, the probability of any outcome is $\frac{1}{10{,}000}$.

STRATEGY 2 Act it out

What are some possible winning numbers?

Possibilities include 1234 and 4612—in fact, any number from 0000 to 9999, which is 10,000 possibilities.

SUMMARY 7.4

In this section, we have examined a number of counting situations. That is, we have looked for patterns and order in those situations in order to find easier ways to count the number of outcomes in the sample space. Many children find such counting problems to be interesting. We learned the difference between combinations and permutations, and we examined how the formulas work.

Once again, it is important to emphasize that the algorithms themselves are useless if you do not understand what they mean. It is our hope that over time, more and more students and teachers will strive to make sense of situations in which mathematics applies.

7.4 Exercises

1. Using a regular deck of 52 cards, what is the probability of getting the following card?
   a. the 2 of clubs
   b. a 2
   c. a face card
2. In how many ways can the following students' names be displayed in a row on a poster: Amy, Betty, Carl, Ed, Frank, Gisela?
3. A family of four has decided to have a family picture taken.
   a. How many different ways can they be arranged?
   b. How many arrangements are possible if the two children do not want to be next to each other?
4. You and a friend are at Paul and Elizabeth's Restaurant for a night of fine dining. You can choose among 2 appetizers, 4 main dishes, and 3 desserts.

   Appetizer: soup or salad

   Main dish: fish, chicken, beef, vegetarian

   Dessert: chocolate cake, apple pie, ice cream

   a. How many possible different dinners are there?
   b. If there are 3 choices of salad dressing and 4 choices of ice cream, how many possible dinner combinations are there?
5. A small town has used up their three-digit telephone prefix (357-). They have been granted a new prefix (358-). How many new phone numbers can be granted before they need a new prefix?
6. If you pick a card from a regular deck of cards, what is the probability of getting a face card 3 times in a row? Assume that the face card is put back in after each draw, so that you are selecting from a full deck each time.
7. If you select 2 cards from a deck, what is the probability of getting 2 matching cards (2 kings, 2 fives, and so on)?
8. Find the number of different ways in which 4 flags can be displayed on a flagpole, one above the other, if 10 different flags are available.
9. The German club has 12 members. In how many different ways can a subset of 3 members be selected to go on a field trip?
10. A basketball team has 9 players, 5 of whom are in the starting lineup.
    a. How many different starting lineups are possible?
    b. How many different starting lineups are possible if the star must be in the lineup?
11. Let's say you went out for ice cream. The store had 9 different flavors, and you wanted a triple-decker ice cream cone. How many different triple-decker ice cream cones are possible at this store?
12. Recall Investigation 7.4e. Suppose the Ping-Pong ball was not replaced. Now what is the probability?
13. If you select 5 cards from a deck, what is the probability that you will get 4 of a kind? 3 of a kind? 2 of a kind?
14. Mrs. Olson has 7 brands of cat food and 3 cats, each of which receives 1 can per day. In how many different ways can she serve the cats on any day, assuming that no 2 cats get the same brand?
15. Those students who celebrate Christmas and come from large families may find this problem to be familiar. Let's say that a family of 5 decides that instead of each person buying a present for each other person in the family, all the family members will put their names in a hat. Each person then takes one name out of the hat and buys a present for that person. What is the probability of every family member getting his or her own name?
16. Several years ago I received as a gift a book of different animals. However, the picture of each animal's body had been cut into three pieces, and the pages of the book enabled me to make different combinations of animals. The name of the book was *Por-gua-can* because in the creature shown on the cover, the head was a porcupine's, the torso was an iguana's, and the feet were a pelican's. There were 9 animals. How many different animal combinations are possible?

DEEPENING YOUR UNDERSTANDING

17. This unfortunate event happened to me when my two children were little. Each of them had a friend over for lunch. After lunch, we had popsicles for dessert. There were 9 popsicles in the freezer and 3 flavors: 3 grape, 3 cherry, and 3 orange. What is the probability that each child will get the popsicle that child wants?

18. Let's say a company has 10 members on its management team: 7 men and 3 women. There is a conference in Hawaii, and the company has decided to send 3 people. If the selection of the 3 people to go to Hawaii is determined by lottery, what is the probability that at least 1 woman will be selected?

19. Janine's boss has allowed her to have a flexible schedule. Her boss says she can pick whatever 5 days a week she will work.

    a. How many different work combinations can be made?

    b. How many choices give her consecutive days off?

    c. How many choices give her Wednesday off?

20. Little Caesar's pizza had a commercial on television that said that you could buy one pizza and get another one free, with up to 5 toppings per pizza. The dialogue on the commercial went something like this. The older person said, "5 toppings per pizza; that's 10 different pizzas to choose from." The little guy then said, "No, it's 1,048,576 combinations." What do you think?

21. How many 11-letter "words" can be made using the letters from the word *mathematics*?

22. In how many different ways can the letters of your state be rearranged, with no duplicates?

23. Explain why these two formulas are equivalent.

$$_nP_r = \frac{n!}{(n-r)!} \quad \text{and} \quad _nP_r = n(n-1)\cdots(n-r+1)$$

24. What do you think will be the value of $_nP_n$?

25. Interview a nearby automobile dealership. Ask the people responsible for ordering how they determine which cars to order to have on hand. Compare the selection process for a small dealership that can stock only 5 Plymouth Voyager minivans with that for a large dealership that can stock 50 Plymouth Voyager minivans.

26. ***Classroom Connection*** This question comes from "Figure This," a website for upper elementary and middle school children containing many problems that are particularly interesting and mathematically worthwhile. Here's the question: How many people would have to be in a school before at least two people had to have the same first and last initials?

Source: Reprinted with permission from Figure This! (http://www.figurethis.org), copyright 2006 by the National Council of Teachers of Mathematics.

LOOKING BACK on chapter seven

QUESTIONS TO SUMMARIZE BIG IDEAS

1. What are the different types of graphs discussed in this chapter and when is each one more appropriate?
2. What are the measures of center and the advantages and disadvantages of each?
3. What are the measures of variation and the advantages and disadvantages of each?
4. What are important considerations in sampling?
5. What is the difference between theoretical and experimental probabilities?
6. What does it mean for an event to have a probability of 0? What about a probability of 1?
7. What does it mean for a game to be "fair"?
8. What is the difference between a permutation and a combination?
9. What parts of this chapter are less clear to you at this point? What will you do to clarify those ideas?
10. Look back at the Mathematical Practices of the CCSS. In what ways did you engage in those practices during this chapter?

CHAPTER 7 SUMMARY

1. There are many ways to represent data. In some cases, what graph to use is a matter of preference; in other cases, some graphs are more appropriate than others.
2. Different graphs can give different impressions of a set of data. Graphs can be constructed to give the reader a distorted impression of the data!
3. When examining data that others have gathered, we need to examine the data carefully, thinking about the reliability and validity of the data, and about whether it is fair, appropriate, or distorted.

4. The terms *mean, median,* and *mode* all represent an "average." Each of the three terms has a conceptual base:
   - The mean is the center of gravity of a set of data.
   - The median is the numerical middle of the set of data.
   - The mode is the datum that occurs most often.
5. In some distributions, the mean, median, and mode are equal. In others, they may be very different, and some may not even be appropriate. In those cases, we need to examine which measure is the best representative for "average."
6. Averages are not always highly representative of a population. In most cases, when we are examining a population, we need to have more information about the population, depending on the questions we are asking. We may want to know the center of the data, or we may want to know about the spread of the data, in which case knowing the extreme values, the range, and the standard deviation is important.
7. When a set of data is normally distributed, we can make powerful generalizations and predictions about that set of data.
8. Random and haphazard are not the same thing. Many events that are random can be quantified.
9. When working with probability situations, sometimes we can determine both the theoretical and experimental probability of an event happening. Sometimes we can only determine the experimental probability.
10. Probability formulas and rules are easier to understand when we take the time to be clear about the terms being used, e.g., events vs. outcomes, dependent vs. independent events, complementary events, and permutations and combinations.
11. Care needs to be taken when conducting surveys so that the results are reliable.

BASIC CONCEPTS

Section 7.1: The Process of Collecting and Analyzing Data

Section 7.2: Going Beyond the Basics

Section 7.3: Concepts Related to Chance

Section 7.4: Counting and Chance

REVIEW EXERCISES chapter seven

1. The table shows the number of CDs shipped between 1992 and 2001.
   a. Make a line graph for these data.
   b. Write a brief summary to accompany the graph.

SALES OF RECORDED CDS, BY UNITS SHIPPED (IN MILLIONS) AND VALUE, 1992–2001

	1992	1993	1994	1995	1996	1997	1998	1999	2000	2001	% Change 2000–2001
Units shipped	407.5	495.4	662.1	722.9	778.9	753.1	847.0	938.9	942.5	881.9	−6.4%

Source: Recording Industry Assn. of America, Washington, DC (in millions, net after returns)

2. The table below shows the prevalence of smoking among U.S. adults.
   a. Make a scatter plot for the data for males and females.
   b. Write a brief summary to accompany the graph.

SMOKING PREVALENCE AMONG U.S. ADULTS, 1965–2009

(as a percent of population, 18 years of age and older)

Year	Overall population	Males	Females
1965	41.9%	51.2%	33.7%
1974	37.0	42.8	32.2
1983	31.9	34.8	29.4
1985	29.9	32.2	27.9
1990	25.3	28.0	22.9
1992	26.3	28.1	24.6
1994	25.3	27.6	23.1
1995	24.6	26.5	22.7
1997	24.6	27.1	22.2
1998	24.0	25.9	22.1
1999	23.3	25.2	21.6
2000	23.1	25.2	21.1
2005	20.8	23.4	18.3
2007	19.7	22.0	17.5
2009	20.6	23.4	18.3

Source: Statistical Abstracts of the United States, 2012, www.census.gov/compendia/statab.

3. The table shows the most widely represented nationalities of travelers to the United States in 2008.
   a. Make a bar graph for these data.
   b. Write a brief summary to accompany the graph.

TOP NATIONALITIES OF TRAVELERS TO THE U.S.

2007 rank	Country of residence	2007 total (thousands)
1	Canada	17,761
2	Mexico	14,333
3	United Kingdom	4,498
4	Japan	3,531
5	Germany	1,524
6	France	998
7	South Korea	806
8	Australia	670
9	Brazil	639
10	Italy	634

Source: *The World Almanac and Book of Facts*, 2009, p. 127.

4. The table shows the number of AIDS diagnoses by age at time of diagnosis in 2011.
   a. Make a circle graph for these data.
   b. Write a brief summary to accompany the graph.

Age	Number diagnosed	Percent
13–19	2,293	4.7
20–29	15,538	31.7
30–39	11,494	23.4
40–49	11,317	23.1
50–59	6,263	12.8
60 and older	2,177	4.4
Total	49,082	100.0

Source: Centers for Disease Control and Prevention, *HIV/AIDS Surveillance Report* 2011, Volume 23.

Do the following for Exercises 5–6.

a. *Write a brief summary of the graph.*
b. *Critique the data from the perspective of reliability and validity.*
c. *Critique the choice of the graph (circle, bar, histogram, or line).*
d. *Critique the construction of the graph (scales, categories, and labels).*

5. Examine the graph below. The United States created 21.1% of carbon dioxide emissions but is home to only 5% of the world population. Mathematically, we say that the United States's share is out of proportion. Explain that statement, as though to someone who has no idea what it means.

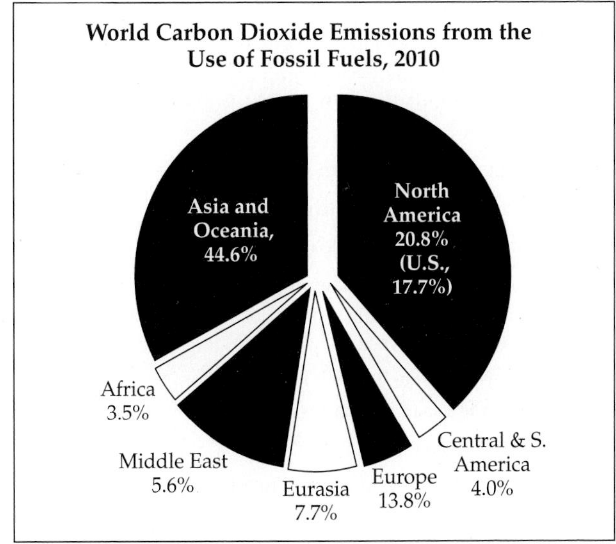

Source: From *The World Almanac and Book of Facts*, 2009. Copyright © World Almanac Education, Inc.

6. The following graph, showing quality of mathematical content in math lessons, is from the TIMSS videotape study of eighth-grade classrooms in Japan, Germany, and the United States, in which classes were observed and videotaped. (Source: http://nces.ed.gov/pubs99/1999074.pdf)

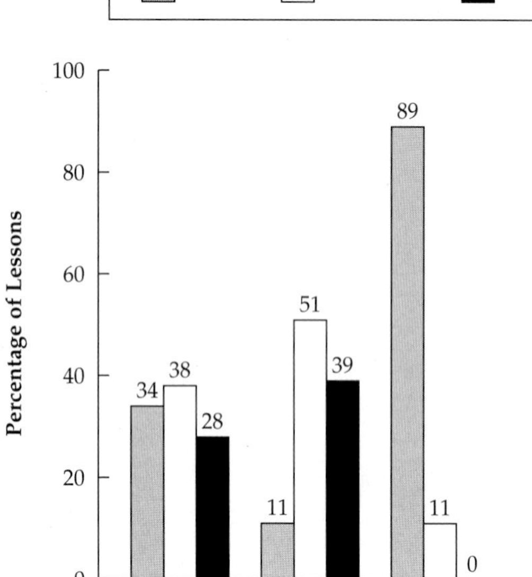

Note: Percentages may not sum to 100 due to rounding.

7. An elementary teacher had his students record the number of drops of water that they could drop onto a penny from an eye-dropper until it spilled: 30, 29, 28, 27, 27, 25, 25, 25, 24, 24, 22, 21, 18, 16, 15.
   a. Use one or more graphs to show these data.
   b. What is the average? Justify your answer.
   c. Describe the spread of the data.

8. A company is testing a new dog food for overweight dogs. They found the following weight loss in a group of 10 dogs after one month. The data are pounds, to the nearest half pound: 0, 1, 2, 2, 3, 3.5, 4, 4, 4.5, 4.5, 5, 6, 6.5, 12.
   a. What is the mean weight loss?
   b. What is the standard deviation?
   c. What questions would you like to ask the company about the study in order to feel more confident about the validity and reliability of the data? Explain the reasoning behind your question(s).

9. Here are the grades on the midterm exam for two of my classes:
   Class 1: 76, 66, 73, 55, 68, 77, 75, 74, 98, 65, 54, 92, 71, 57, 90, 63, 69, 51, 70, 76, 66, 75, 68, 81, 82, 63, 66, 98
   Class 2: 90, 81, 64, 68, 84, 89, 95, 82, 72, 81, 83, 85, 88, 78, 100, 85, 88, 82, 72, 70, 65, 87, 70
   a. Make two box-and-whisker graphs. Write a brief summary that would accompany these graphs in an article.
   b. Make two line plots. Write a brief summary that would accompany these graphs in an article.
   c. Summarize the pros and cons of representing these data with box-and-whisker or line plots.
   d. Compare the average scores in the two classes. Justify your answer.
   e. Describe the spread of the data.
   f. Determine the standard deviations.
   g. What percent of the scores in each class lie within 1 standard deviation of the mean?

10. Willie has an average (mean) of 91.2 on the first 6 exams. What is the least he can score in order to have an average of 90 for all seven exams?

11. Suppose the distribution of heights of a group of 300 children is normal with mean of 160 centimeters and a standard deviation of 15 cm. About how many of these children are less than 145 cm tall?

12. The following graph shows the numbers of siblings reported by a class of students.
   a. How many students are there in the class?
   b. What is the mean?
   c. What is the median?

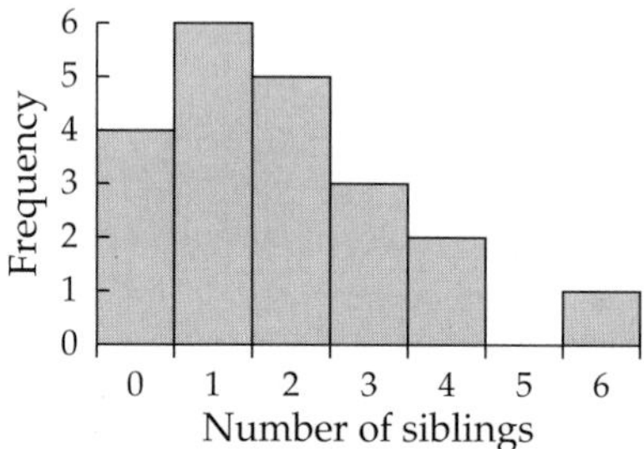

13. a. Describe a situation in which the median would be a more appropriate description of the center of a data set than the mean would.
   b. Describe a situation in which the mean would be a more appropriate description of the center of a data set than the median would.

14. Following are measurements in centimeters of neck, waist, and wrist from a sample of adult females.
    a. Make boxplots for each set of data.
    b. Make two scatter plots: neck:waist and neck:wrist. Describe the degree of correlation, that is, positive or negative, and none, weak, or strong. Justify your conclusion.

Neck	Waist	Wrist
36.2	82.2	17.1
38.5	83	18.2
34	87.9	16.6
37.4	86.4	18.2
34.4	100	17.7
39	94.4	18.8
36.4	90.7	17.7
37.8	88.5	18.8
38.1	82.8	18.2
42.1	88.6	19.2
38.5	83.6	18.5
39.4	90.9	19
38.4	91.6	17.7
39.4	101.8	18.8
40.5	96.4	18.2
36.4	92.8	16.9
38.9	96.4	17.3
42.1	92.8	19.3
38	96.4	18.5
40	97.5	18.2

15. If you randomly draw a circle from each box, what is the probability of your drawing two white circles?

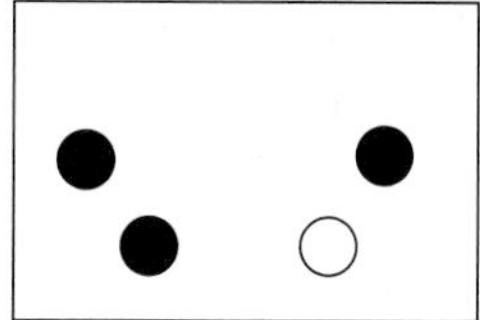

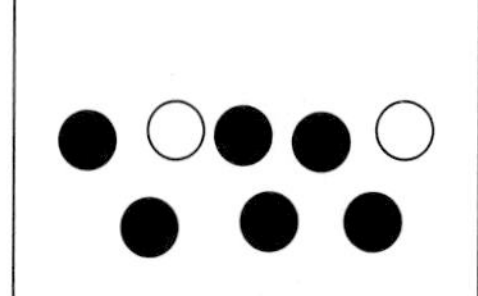

16. A drawer contains 6 red socks and 3 blue socks.
    a. If 2 are drawn, what is probability of a match?
    b. If 3 are drawn, what is probability of a match?
    c. What is probability of all one color after 4 draws?

17. Consider a pair of eight-sided dice—that is, dice that contain the numbers 1–8 as opposed to regular dice, which contain the numbers 1–6. If you roll two dice at a time, what is the most likely sum, and what is the probability of getting that sum?

18. What is the probability of getting two odd numbers in a row if you throw two regular dice and add the two numbers?

19. If the sample space consists of the numbers 10–99, what is the probability of a number drawn randomly containing a zero? A five?

20. Consider the following game: Two dice are rolled, and then the numbers multiplied. If the product is odd, player A gets 1 point. If the product is even, player B gets 1 point. Is this a fair game? Why or why not? If it is not a fair game, change the rules to make it fair.

21. Two people use the spinners below. Player A gets a point if the colors match, player B gets a point if they don't match. Is this a fair game? Why or why not? If it is not a fair game, modify the game so that it becomes fair.

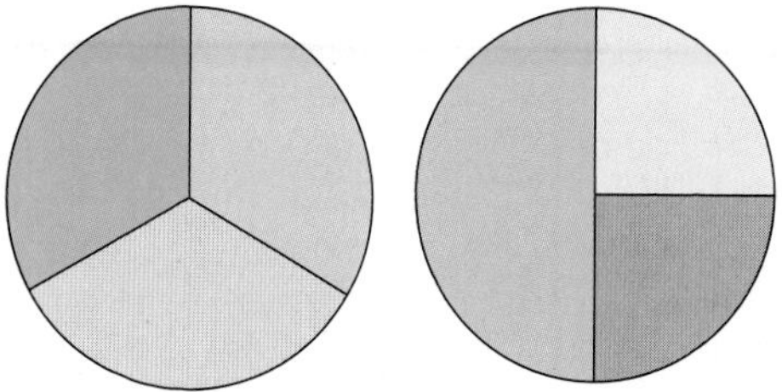

22. In how many ways can the following sports be displayed in a single row on a poster: soccer, basketball, baseball, lacrosse, and field hockey?

23. If you select two cards from a deck, what is the probability of your selecting two matching suits—for example, two clubs?

24. In how many ways can the letters in the word MATH be rearranged?

25. If the digits 1 through 5 are used to make a three-digit code, what is the probability of one of the code numbers beginning with a 5?

26. A committee has 10 members.
    a. How many 2-person subcommittees are possible?
    b. How many 3-person subcommittees are possible?

27. Janine's boss has allowed her to work a flexible schedule. Her boss says she can pick whatever 4 days a week she will work, except Sunday.
    a. How many different work combinations can be made?
    b. How many combinations will give Janine consecutive days off?
    c. How many combinations will give her Saturday off?

28. Let's say a state decided to have a license plate with the following format: one letter followed by three digits followed by one letter—for example, A123B. How many different license plates could be made with this format?

Geometry as Shape

8

Explorations Manual
8.4

From a basic perspective, geometry is the study of shapes, their relationships, and their properties. What is the value of geometry? Who needs to know geometry and who has needed to know geometry? How has understanding of geometry shaped our lives? These are some of the questions that will be addressed in this and the next two chapters.

According to the Common Core State Standards, the study of geometry begins in kindergarten where students identify, describe, analyze, and compare two-dimensional and three-dimensional shapes. Reasoning with shapes and their attributes continues throughout elementary school. In fourth grade, angles and lines are explored, and in fifth grade the coordinate plane (x and y axes) is included in the geometry strand. In sixth through eighth grades the geometry strand continues with area, surface area and volume, transformations, and finally the Pythagorean theorem.

Here are a few examples where people have needed geometry and where this need has produced greater understanding:

- People in ancient Egypt developed methods to determine the boundaries of their land. In fact, the word *geometry* literally means "to measure the earth."
- Explorers needed maps to show where they had been.
- House builders have used many shapes, from Native American teepees, which look like cones but are more like many-sided pyramids, to African houses resembling cylinders, to the more familiar house structures with triangles and quadrilaterals, to geodesic domes with equilateral triangles.
- Surveyors and planners needed to ensure that tunnels or railroad tracks built from both ends would actually meet in the middle.
- Artists wanted to portray convincingly what they saw with their eyes or visualized in their minds.
- Architects and builders needed buildings that would be both beautiful and strong.

In the next three chapters, we will focus on three aspects of geometry: shapes, transformations, and measurement. Look at the pictures in Figure 8.1. What do you see?

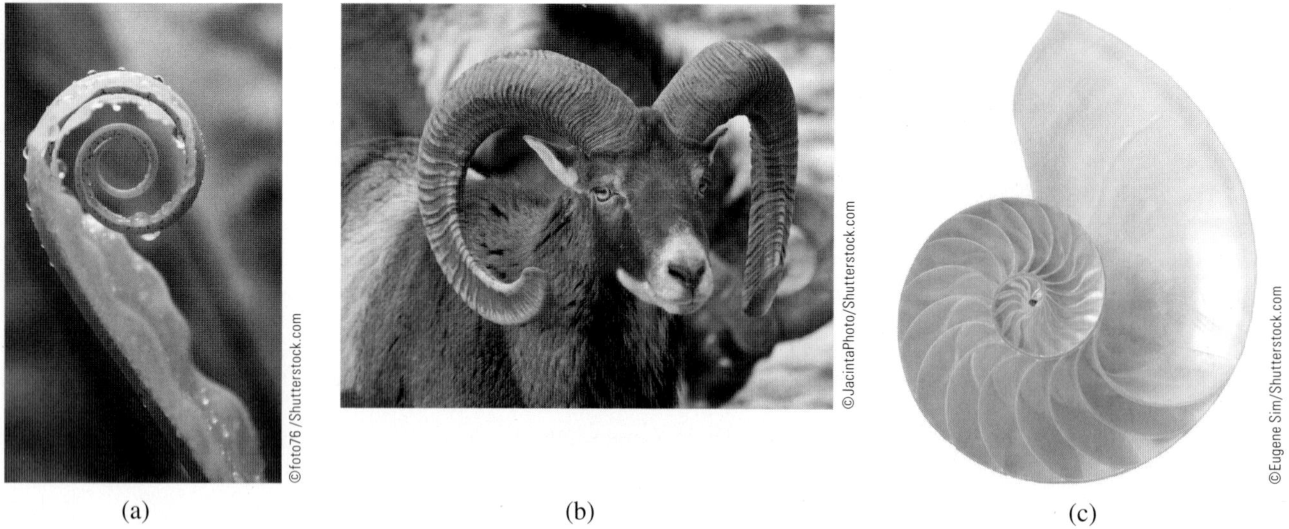

Figure 8.1 Three natural spirals: (*a*) leaves of the sago palm, (*b*) horns of a mountain sheep, (*c*) the chambered nautilus.

Here, a spiral occurs in different parts of the world: in a plant and on two different animals. How is this shape formed? Why is this shape formed in these situations—that is, what purpose or function does this shape serve? Mathematicians, biologists, physicists, and others have asked and answered questions like these. Essential to answering such questions is a deep understanding of shapes, which also involves understanding their measures.

Let us begin our exploration of geometry by addressing the question, "What is the value of geometry?" from three perspectives: fun, function (usefulness), and aesthetics. In each case, you will find that geometric structures (basic ideas) are important.

SECTION 8.1 Basic Ideas and Building Blocks

What do you think?

- Where have you used or seen geometric ideas recently?
- Look around the room or place where you are reading this book. What geometric figures do you see? Do they have a function, or do their shapes serve more aesthetic purposes?

Explorations Manual 8.1, 8.2

WHY GEOMETRY? BECAUSE IT'S FUN!

INVESTIGATION 8.1a Playing Tetris

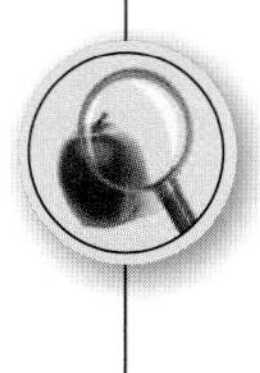

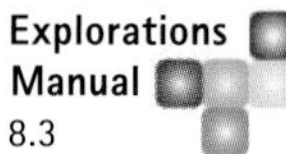
Explorations Manual 8.3

Most students have played or are familiar with the computer game called Tetris. There are many free online versions of the game (such as freetetris.org and tetrisfriends.com). If you are not familiar with the game, go play it and then read on. Let's take a brief look at Tetris and see some of the mathematics that can be developed by playing and analyzing the game.

A shape comes down the screen, and you can move (transform) the shape in several ways: You can move it to the right or left, or you can rotate it 90, 180, or 270 degrees. You get points when an entire row is filled.

In the game that is in progress (see Figure 8.2), there really is no good spot for the shape that is dropping down the screen, even if we turn it. If we could flip it, we would get the shape at the left, but the Tetris game allows only turns and lateral movement, not flips.

Now, many people enjoy playing Tetris, but where is the mathematics in Tetris? Think of questions that could be asked and explored that would result in students learning more mathematics. Then read on. . . .

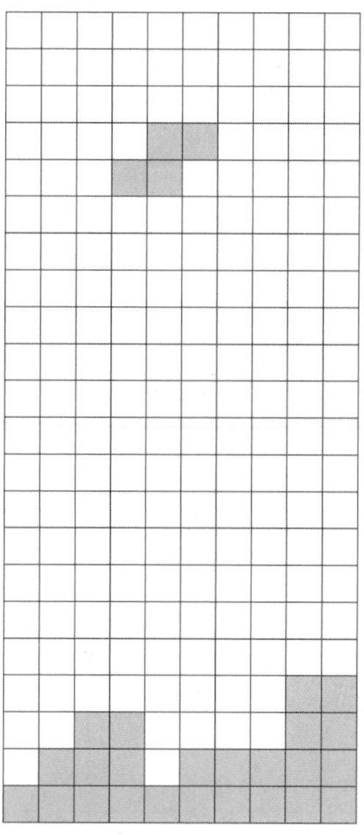
Figure 8.2

DISCUSSION

Below are ten questions that can be explored with Tetris; there are many more. Each of these questions can generate some important mathematics ideas and discussions. I have done many of these with fourth- and fifth-graders also. They love that the game can be played in math class, and they love the related explorations with tetrominoes, pentominoes, and hexominoes. Some of these questions appear in the *Explorations*.

Question	Content Area(s)
1. What is a tetromino (the mathematical name for the Tetris pieces)? If you just say, "four squares put together," then someone not familiar with the game could conclude that either of the shapes at the right is a tetromino. Figure 8.3	Defining terms
2. What are good strategies for playing Tetris? Strategies come both from spatial sense (visualizing what the shape will look like when we slide and/or turn it to get it to the proper position) and probability sense (deciding how long to wait for the "right" piece and when to give up on filling a hole).	Spatial sense Probability
3. Am I getting better? We can measure improvement by recording our scores and comparing early scores to later scores.	Statistics

(continued)

Question	Content Area(s)
4. Are these two pieces "the same"? On the one hand, they have the same shape and size. On the other hand, if the figure at the left starts to come down the screen, you cannot turn it into the figure on the right; so in that sense, they are not the same. Figure 8.4	Congruence Communication-defining terms
5. How many Tetris pieces are there? There are only five different (noncongruent) tetrominoes, but there are quite a few different pentominoes and a lot of different hexominoes!	Problem solving
6. Can we fill in a 5 × 4 grid with all of the Tetris pieces? There are many such puzzle questions that we can ask; some are in *Explorations.*	Spatial sense
7. Can we make bigger versions that are the same shape? For example, using all the Tetris pieces, can we make a bigger version of the L Tetris piece that has the same shape as the L piece?	Similarity
8. Which tetrominoes will cover a surface with no gaps? Will all of them? Will certain combinations?	Tessellation
9. If we wanted to make a set of tetrominoes for each of 24 students, how many sheets of Geoboard Dot Paper would we need?	Measurement Spatial sense
10. How does the computer version work? That is, the computer doesn't really rotate and translate the figures, but rather shades in different pixels on the screen!	Geometry Algebra

As you can see, a lot of mathematics can be developed by looking more closely at a game. There are many popular games that involve geometric thinking: Brick by Brick, Set, Soma cubes, Rubik's Cube, and online games. One of my favorites is Super Collapse.

WHY GEOMETRY? BECAUSE IT HAS PRACTICAL APPLICATIONS

Explorations Manual 8.4

There are many machines, tools, and materials whose shape was determined by geometric properties. Let us explore a few here. The reason why manhole covers are round is explored in Exploration 8.4.

INVESTIGATION 8.1b Different Objects and Their Function

In each of the cases on the next page, there is a reason for the shape of the object. Can you figure it out? Look closely at the pictures and think of the reason for the shapes in each case before reading on. . . .

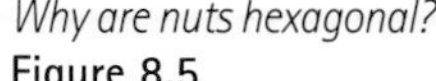

Why are nuts hexagonal?
Figure 8.5

Why is the plug pentagonal?
Figure 8.6

Why are the cells of a honeycomb hexagonal?
Figure 8.7

Why is a soccer ball made with hexagons and pentagons?
Figure 8.8

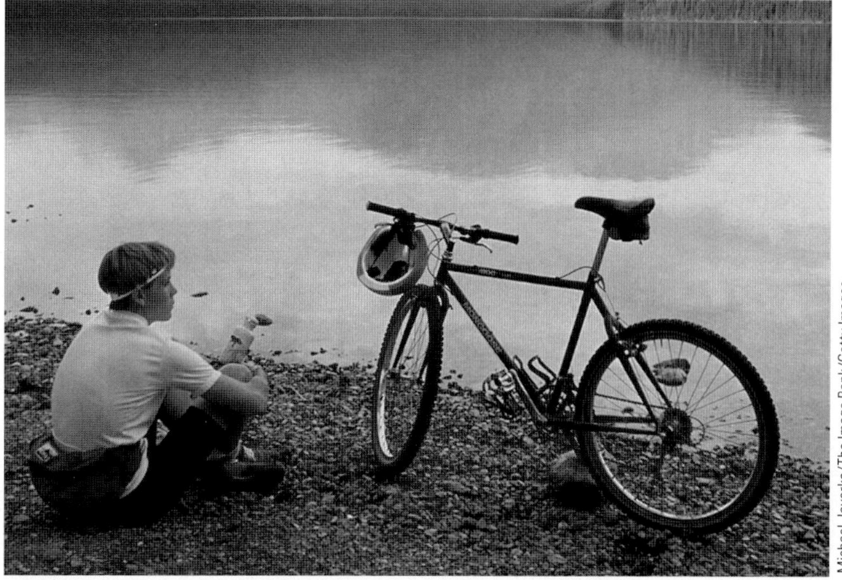

Why are there triangles in a bicycle?
Figure 8.9

DISCUSSION

Most nuts are hexagonal in shape for two reasons. First, opposite sides of hexagons are parallel, and the jaws of the crescent wrench are parallel. Thus you can open the wrench to the appropriate setting, slide it onto the nut, tighten, and then turn. But why not square nuts instead of hexagons? Stop and think before reading on. . . .

For a square nut, you would have to turn the nut 90 degrees (a quarter-turn) before you could slide the wrench off and then slide it back on. In tight places, often it is not possible to turn 90 degrees. With the hexagonal shape, you need only to be able to make a 60-degree rotation.

Why is the fire hydrant plug pentagonal? For the same reason that nuts are hexagonal! If you look at the pentagon, you will see that opposite sides are not parallel. Thus a large pipe wrench, widely available at hardware stores, will not work on a fire hydrant. A special wrench, not sold to the public, is needed.

Why do you see triangles in any bicycle? This question will be explored in Section 8.2.

Why is the soccer ball made of hexagons and pentagons? We will explore questions of covering surfaces with repeating shapes in Chapter 10.

Why is the honeycomb hexagonal? This will be explored in Chapter 9.

WHY GEOMETRY? BECAUSE WE SEE IT IN ART ALL OVER THE WORLD

When archaeologists find artifacts from ancient cultures, we find shapes (squares, pentagons, hexagons, symmetric designs) on pottery, on cloth, and in religious artifacts. Were humans replicating shapes they saw in nature—honeycombs, snowflakes, spirals in many sea animals—or is this love of shapes and designs hard-wired into the human brain like language is? We don't know, but we do know that it is pretty universal.

Let us investigate just a few examples of the value of geometry in art. I remember taking an art history class in college. The professor showed us paintings made before artists fully grasped the geometric connections between the two-dimensional surface of the canvas and the three-dimensional scenes they were representing on that canvas. Figure 8.10 shows the consequences of not understanding the geometry.

Scala/Art Resource, NY

Figure 8.10

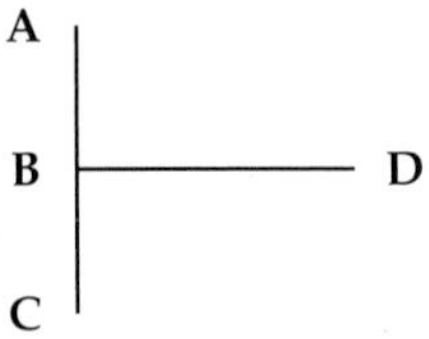

A

B D

C

Figure 8.11

One of the most interesting aspects of art and geometry is seen in optical illusions. Which line in Figure 8.11 is longer, $\overline{AC}$ or $\overline{BD}$? Look and write your answer before reading on. . . .

Most people perceive *DB* as longer than *AC*, but they are the same length. Why does *BD* appear longer? Many theories have been developed. The one that makes the most sense is that *DB* intercepts *AC*. To illustrate this, draw two line segments of equal length that bisect each other (cut each other in half) (Figure 8.12). In these cases, most people will see the two segments as having the same length.

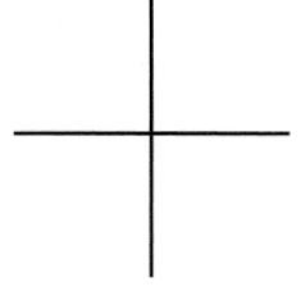

Figure 8.12

Which inner circle in Figure 8.13 is larger? Look at Figure 8.13 and think before reading on. . . .

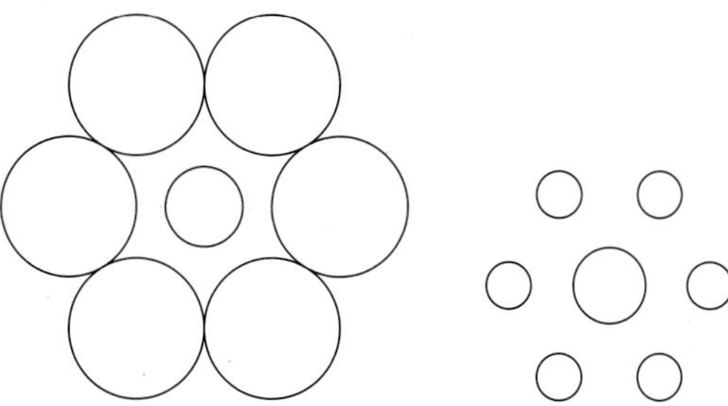

Figure 8.13

The inner circle on the right looks larger to most people because it is larger than its surrounding while the other circle is dwarfed by its surrounding. In fact, the inner circle at the left is slightly larger!

What do you see in the figures below? Think before reading on. . . . 

Figure 8.14

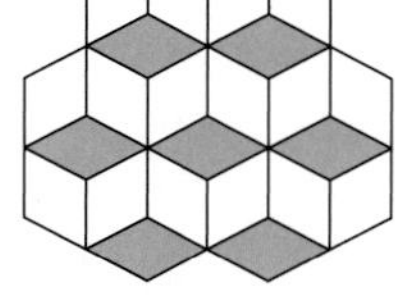

Figure 8.15

Although Figure 8.14 consists only of lines and curves, most people instantly see "Hello." By making the colored parallelograms the top or the bottoms of the boxes in Figure 8.15, you can see either six or seven boxes.

In our limited space, we have examined only a bit of geometry from the perspective of art and aesthetics. When we explore symmetry in the next chapter, we will examine geometry in fabrics and crafts.

GEOMETRY AND SPATIAL THINKING

Douglas Clements and Michael Battista have articulated this notion that we call "spatial sense." Just like "number sense," "spatial sense" is an important goal of school mathematics. However, it is generally left undefined and unarticulated.

One aspect of spatial thinking is **hand-eye coordination**. Most young children are still very much in the process of developing this ability. There are many activities that can help children to develop this ability, such as copying shapes that the teacher or other students have made with pattern blocks, Geoboards (such as those at the end of the *Explorations*), or other manipulatives.

Another important spatial thinking ability is called **figure-ground perception**, the ability to identify a figure against a complex background. One example that most students remember from elementary school is a picture where you have to find objects hidden within the main picture. The activity is far more than just a fun exercise. This ability to discriminate what you are looking for from extraneous lines or information is useful in everyday life and is especially important in many professions.

Another important ability is called **perceptual constancy**, the ability to recognize figures and objects when they are not in their familiar orientation. For example, many students see the triangle on the left in Figure 8.16 as an isosceles triangle but will say that the triangle on the right is not an isosceles triangle. Similarly, other students will see that the triangle on the right is an obtuse triangle but will fail to see that the triangle on the left is also an obtuse triangle. In fact, the two triangles are congruent.

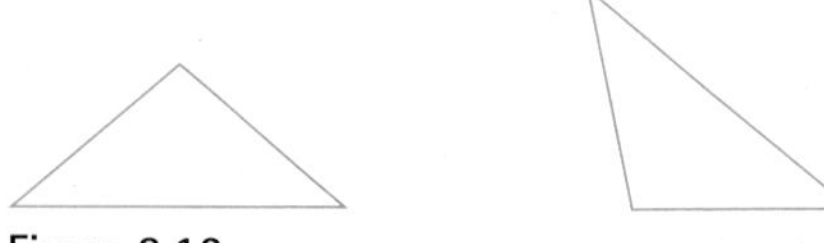

Figure 8.16

Visual discrimination is the ability to find similarities and differences between or among objects. You have probably seen tasks designed to develop this ability on *Sesame Street,* in commercial activity books, and in schoolbooks. For example, five similar objects are shown, and the student is asked to pick the two that are identical.

Visual memory has to do with a person's ability to describe or draw accurately an object that is no longer in view. In Section 8.2, there will be an investigation that asks you to look at an object for only a few seconds and then try to draw that object from memory.

These are but a few of the many kinds of abilities that come under the larger heading of spatial sense and that are important in everyday life and crucial in many occupations. Douglas Clements and Michael Battista state that "much of the thinking required in higher mathematics is spatial in nature,"[1] and they note that Einstein said that he thought not so much in words as in images.

Now that you have seen that there are many uses of geometry and we have examined some aspects of what we want children to learn, let us begin at the beginning, so to speak, and examine the building blocks of geometry and geometric thinking. These ideas will be crucial in the following sections when we look at various shapes and their properties.

Explorations Manual 8.5

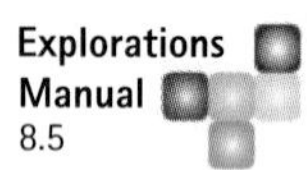

BACK TO THE GREEKS

Although the study of geometry is probably as old as the study of numbers, our modern understanding of geometry began with the Greeks over 2000 years ago when some people felt the need to go beyond merely knowing that certain facts or properties were true to being able to prove why they were true and that they were true everywhere. For example, they wanted to understand the Pythagorean theorem; why the diagonals of a square are the same length and perpendicular, whereas the diagonals of a nonsquare rectangle are the same length but not perpendicular, and so on (Figure 8.17.)

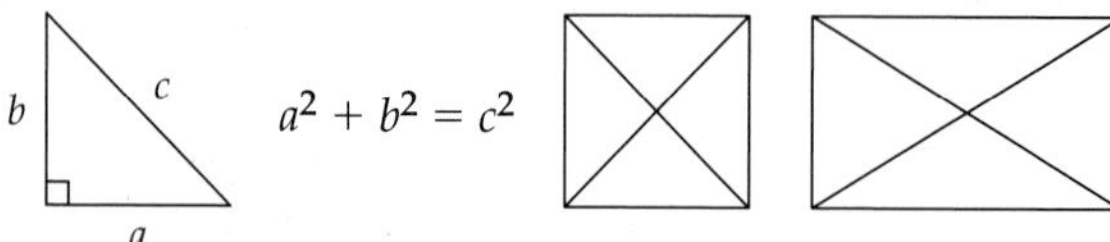

Figure 8.17

This focus on *why,* beyond knowing *that,* represented a huge leap in human history. Even so, modern geometry developed slowly over several centuries. Around 300 B.C. a man named Euclid decided to put together what was known about geometry at that time, but he didn't just record all the scattered information. He organized it systematically, and his book, *The Elements,* ranks as one of the most important and influential books of all time. It has sold almost as many copies as the Bible and has been translated into virtually every language. It was used as the geometry text in high school well into the nineteenth century. So let us take a look at what Euclid did. Along the way, we will also develop the building-block ideas for geometry, and we will look at how children's understanding of these ideas develops too.

As you know from high school, there are many definitions and theorems in geometry. Euclid realized that you can't define everything, and you can't prove everything. That is, you have to have some terms that are undefined, and you have to have some assertions that are simply accepted. If you are not convinced of this, try to define *round* without using the word *circle* or *circular!* Thus certain terms, such as *point, line,* and *plane,* were left undefined, left as common-sense notions.

Similarly, we can't prove everything. Euclid and other Greek mathematicians thought, "What is the least that we can assume as self-evident, and then prove everything else on the basis of those assumptions?" These are known as the five postulates of Euclid:

1. To draw a straight line from any point to any other.
2. To produce a finite straight line continuously in a straight line.
3. To describe a circle with any center and any distance.
4. That all right angles are equal to each other.
5. That, if a straight line falling on two straight lines makes the interior angles on the same side less than two right angles, the straight lines if produced indefinitely, meet on that side on which are the angles less than the two right angles.

HISTORY

Actually, Euclid did try to define these basic terms! Here are his definitions of *point, straight line,* and *plane:*

A point is that which has no part.
A straight line is a line that lies evenly with the points on itself.
A plane surface is a surface that lies evenly with the straight lines on itself.

Most people find the fifth postulate difficult to understand, and many later mathematicians struggled over this one too. A common rewording of this postulate is attributed to an Englishman named John Playfair, who wrote, in 1795, "Given a line and a point not on the line, it is possible to draw exactly one line through the given point parallel to the line."

Whole new fields of geometry were born, some 2000 years after Euclid, when various people played the famous "what if" game with this famous fifth postulate: What if you could draw two lines through the given point parallel to the line? What if you couldn't draw any lines through the given point parallel to the line?

BASIC TERMS AND IDEAS

It is important to note that points, lines, and planes are ideas—mental constructs only. That is, we cannot find a line in our physical world that extends infinitely in two directions. However, we can imagine them, and by imagining them we are able to develop powerful ideas and understandings that are very valuable in many ways. One of the more famous quotations from Albert Einstein is "Imagination is more important than knowledge!"

Before we examine basic geometric definitions and concepts, let us briefly describe three fundamental undefined terms. In everyday life, we refer to the starting point of a race or a point on a map. Geometrically, a **point** has no dimension (length or width), but it does have a location. In everyday life, we refer to lines on a paper and lines of longitude. Geometrically, a **line** is straight and extends infinitely in two directions. We can also think of a line as an infinite collection of points that indicate a straight path. In everyday life, we refer to planes when we talk about floors and countertops. Geometrically, a **plane** is considered to be a flat surface that extends infinitely in all directions.

INVESTIGATION 8.1c Point, Line, and Plane

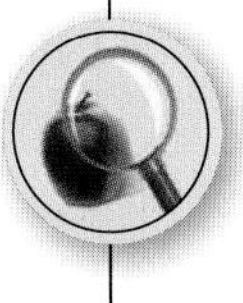

The purpose of this investigation is to have you come to appreciate the kind of thinking the Greeks started. You may find it helpful to use the point of your pencil or pen to represent a *point,* a ruler (or other object with a straight edge) to represent a *line,* and a piece of posterboard to represent a *plane.* By cutting slits in the posterboard, as shown in Figure 8.18, you can more concretely investigate the questions that follow.

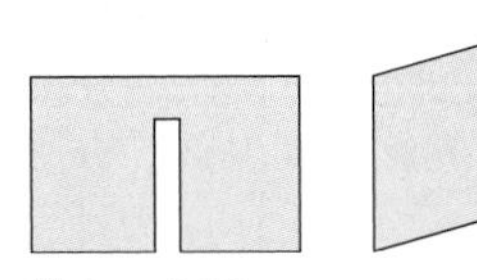

Figure 8.18

POINTS AND LINES

A. Draw a point on a piece of paper. How many different lines can you draw that go through that point?

B. Draw two points on a piece of paper. How many different lines can you draw that go through both points?

C. Draw three points on a piece of paper. How many different lines can you draw that go through all three points?

If you answer any of the questions, "It depends," can you be more specific? Write down your responses before reading on. . . .

DISCUSSION

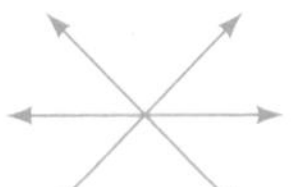

Figure 8.19

A. An infinite number of lines can go through any point. Figure 8.19 shows three of the lines we could draw, but there are infinitely many more.

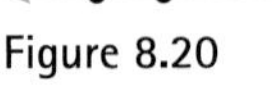

Figure 8.20

B. Only one line can go through any two points. That is, if we draw *any* two points, there will always be one and *only* one line that contains those two points (Figure 8.20).

This powerful, but often only partially understood, conclusion is frequently stated as follows: **Two points determine a line**. Consider the equivalence of the following two statements: "Only one line will go through any two points" and "Two points determine a line."

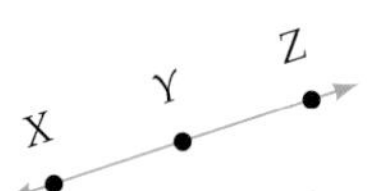

Figure 8.21

C. How many lines we can draw through three points depends on the points. If the three points are on a line, then we can draw only one line (Figure 8.21). However, if the three points are not on a line, then three points will determine three lines (Figure 8.22).

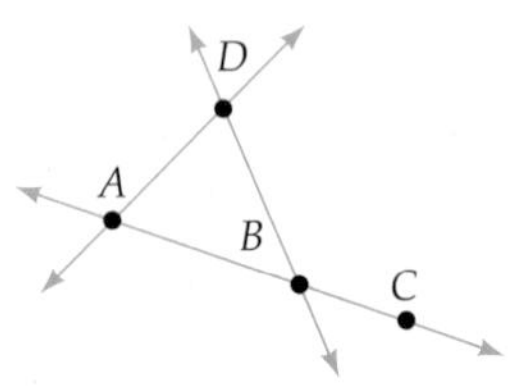

Figure 8.22

This leads us to our first true definition: **Collinear** points are points that lie on the same line. As we have just seen, any two points are collinear.

On the other hand, if three (or more) points do not lie on a single line, then those points are said to be **noncollinear**. In Figure 8.22, points *A*, *B*, and *C* are collinear points, whereas points *A*, *B*, and *D* are noncollinear points.

There is a common misconception about collinear points that needs to be addressed. Draw a point on your paper and place your pencil somewhere above the paper. If the point on the paper is point *A* and the tip of the pencil is point *B* (Figure 8.23), are points *A* and *B* collinear? Think before reading on. . . .

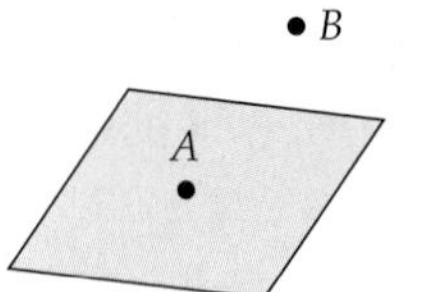

Figure 8.23

Yes; *any* two points are said to be collinear. Even when a line is not already drawn, if it is *possible* to draw such a line, then the points are collinear.

PLANES

A. Now draw a point on a piece of paper. How many different planes can go through that point?

B. Draw two points on a piece of paper. How many different planes can go through both points?

C. Draw three points on a piece of paper. How many different planes can go through all three points?

If you answer any of the questions, "It depends," can you be more specific? Write down your responses before reading on. . . .

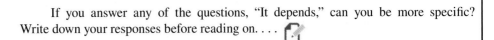

DISCUSSION

A. An infinite number of planes can go through any one point.

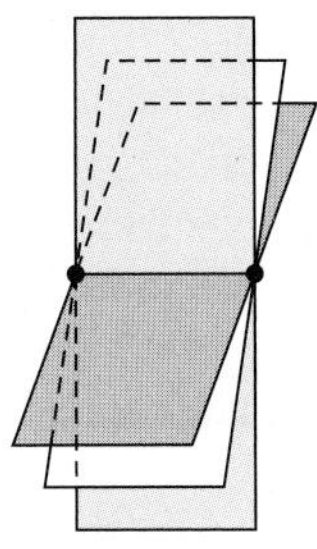

Figure 8.24

B. An infinite number of planes can go through any two points (Figure 8.24).

C. The answer for three points depends on where you put the points. If we draw three noncollinear points, there is only one plane that will contain all three points. Do you see why? Think and then read on. . . .

One of the best ways to demonstrate this requires three people. First, draw one point on a sheet of paper. Next, the first person puts a pencil in the air (tip up). Then the second person puts another pencil in the air (tip up), making sure only that the three points are noncollinear. Now the third person takes a model of a plane (cardboard or a hardcover book). How many different planes contain all three points? Just one!

When two or more points lie on the same plane, they are said to be **coplanar** points. As you have just seen, any two or three points are coplanar, because we can always find a plane that will contain them.

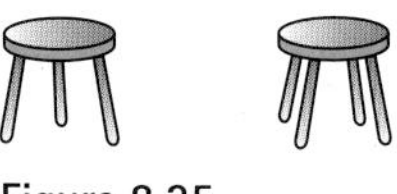

Figure 8.25

There is another way to see this concept concretely. Make a model of a three-legged stool like the one shown in Figure 8.25. You can be creative; for example, use three pencils, some glue, and a piece of cardboard. Put the model on a flat surface. Then make a model of a four-legged stool like the one shown and put it on a flat surface. What do you notice? Think before reading on. . . .

The three-legged stool will never rock. Many four-legged stools will. Why? How would you explain why to a friend who is not in this course?

Because the three points at the bottom of the stool are not collinear, there is only one plane that will contain them. However, the four points at the bottom of the four-legged stool may or may not all lie on the same plane. Thus we have a famous property: Three noncollinear points determine a plane.

LINE SEGMENTS AND RAYS

In most everyday situations, we don't work with lines; instead, we work with line segments and rays.

We define a **line segment** to be a subset of a line that contains two points of the line (which we call the **endpoints**) and all points between those two points.

We define a **ray** as the subset of a line that contains a specific point (the endpoint) and all the points on the line that are on the same side of that point.

Table 8.1 shows the differences among a line, a line segment, and a ray.

TABLE 8.1

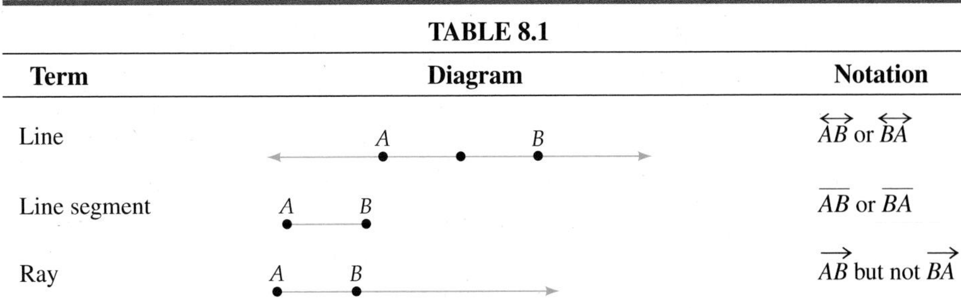

Term	Diagram	Notation
Line	A B (line with arrows at both ends)	$\overleftrightarrow{AB}$ or $\overleftrightarrow{BA}$
Line segment	A B	$\overline{AB}$ or $\overline{BA}$
Ray	A B (arrow to the right)	$\overrightarrow{AB}$ but not $\overrightarrow{BA}$

The notation to represent lines, line segments, and rays is a convention that has evolved. The conventional way to denote a line is by using two points on the line and a double arrow above the two letters.

The conventional way to denote a line segment is to use two letters, representing the endpoints, and a bar above the two letters.

Naming rays requires careful thought, for the order that we name the letters matters. For example, $\overrightarrow{AB}$ and $\overrightarrow{BA}$ are different rays. Why is this? The convention for naming a ray is to use two letters, the first of which is the endpoint of the ray. The second letter can be any other point on the ray and indicates the direction of the ray. For example, in Figure 8.26, we can speak of $\overrightarrow{DA}$ which is different from $\overrightarrow{AD}$. On the other hand, $\overrightarrow{DA}$ and $\overrightarrow{DM}$ are considered to be the same ray. Why is this? If you start at D and go through A, you produce the same ray as if you started at D and go through M. Thus we say $\overrightarrow{DA} \cong \overrightarrow{DM}$, or in other words, these two rays are congruent.

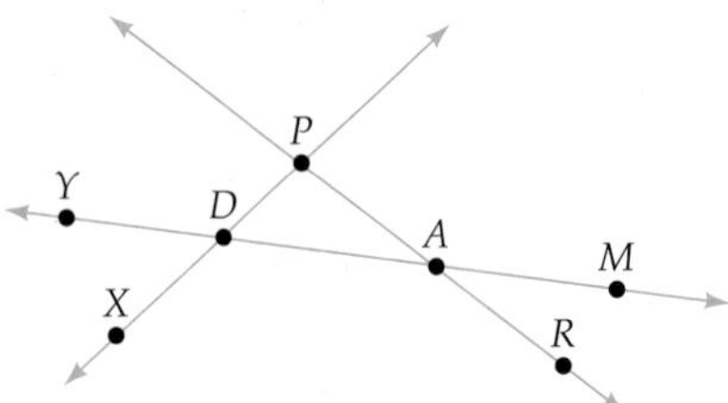

Figure 8.26

RELATIONSHIPS BETWEEN LINES

There are many possible ways in which two or more lines may be related to each other. For example, they may **intersect**, as in Figure 8.27. Two lines are said to intersect if they have exactly one point in common. Describe in your own words the relationship between the lines in Figure 8.28.

Figure 8.27

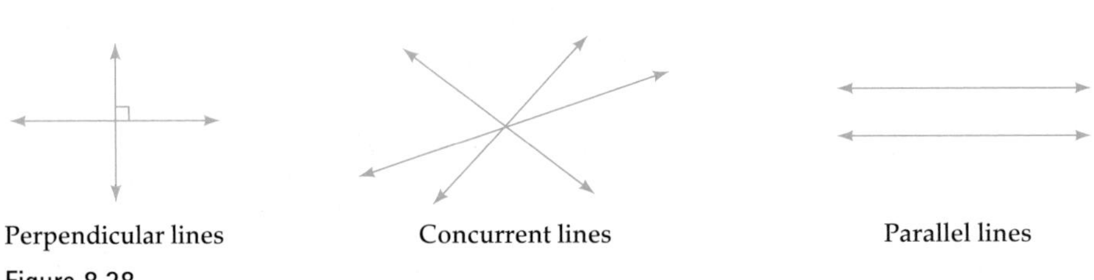

Perpendicular lines Concurrent lines Parallel lines

Figure 8.28

If two intersecting lines form right angles, they are said to be **perpendicular** (Figure 8.28.)

If three or more lines intersect at a single point (that is, have a point in common), they are said to be **concurrent** lines (Figure 8.28.)

As you know, not all lines intersect. If two lines lie in the same plane and never intersect, they are said to be **parallel** lines (Figure 8.28.)

There is a fourth possibility: two lines that do not intersect because they do not lie in the same plane. Such lines are called **skew** lines. For example, in Figure 8.29, $\overleftrightarrow{AD}$ and $\overleftrightarrow{EF}$ are skew lines.

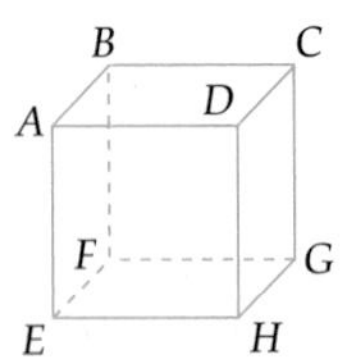

Figure 8.29

OUTSIDE THE CLASSROOM

Some years ago, I was volunteering on a house-building project. It became necessary to cut a piece of lumber $5\frac{1}{2}$ inches wide into three strips of equal width. Easy, huh? So I divided $5\frac{1}{2}$ by 3 and got $1\frac{5}{6}$, but sixths are not on rulers. As I was trying to eyeball $1\frac{5}{6}$, a carpenter saw me and laughed. "Oh, you mathematicians know math theoretically, but not practically. Here's how carpenters do the problem." He laid the ruler on the lumber diagonally so that one end of the ruler was at the beginning and the other was at 6. He drew marks at 2 and 4. Next, he moved up the lumber a ways and repeated the process (Figure 8.30). Then he connected the marks and the problem was solved!

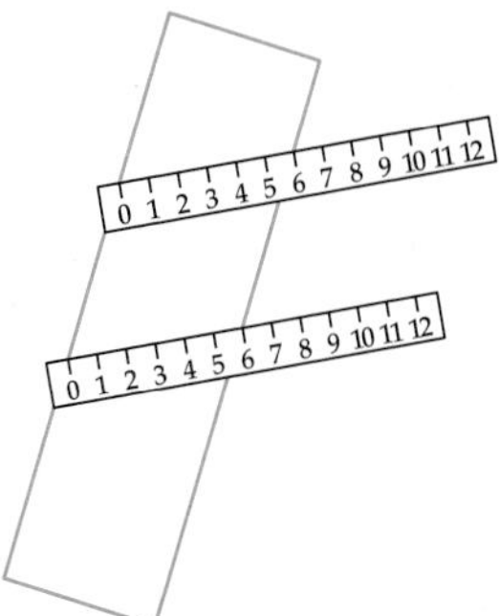

Figure 8.30

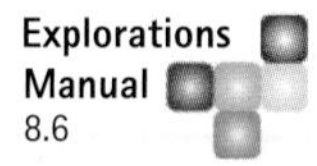
Explorations Manual 8.6

ANGLES

Angles and lines are the focus of the fourth-grade geometry unit, according to the Common Core State Standards. It takes some time for children to understand the ideas about angles fully. In fact, there has been quite a bit of research on how children's understanding of angle develops. Let's begin with your responding to the question "What is an angle?" Write a definition of the term *angle* that you could give to a friend who never took geometry in high school. Think and then read on. . . .

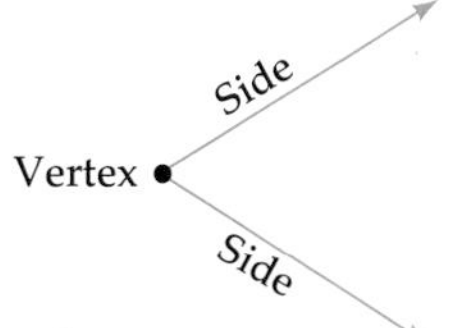

Figure 8.31

A common mathematical definition is that an **angle** consists of the union of two rays that have a common endpoint, which we call the **vertex** of the angle. Each of the rays is called a **side** of the angle (Figure 8.31).

We will argue that many of the cases where we encounter angles do not consist of two rays or even line segments meeting at a common point. We see angles where two flat surfaces meet—for example, the wall and the ceiling; this is called a dihedral (three-dimensional) angle. From another perspective, the definition above, which is considered a static conception of angle, is not easily understandable to children, who more easily comprehend a dynamic conception—that is, angle as movement. Think of situations where you use the term *angle* or *degrees* in the context of movement. . . .

For example, instead of turn right, we say "turn 90 degrees" or "make a 90-degree turn." The saying "Our thinking was 180 degrees apart" mathematically means "in opposite directions." When turning a screw or a valve, we talk about a quarter-turn, which is 90 degrees. We talk about how much motion a damaged joint has: "She can raise her arm only about 45 degrees." In Section 4.2 there is a page from the *Investigations* elementary curriculum that investigates the hands on a clock as fractions, but we can also think of this as the angle between the hands on a clock. At 2:00 the hands of the clock make a 60-degree angle with each other.

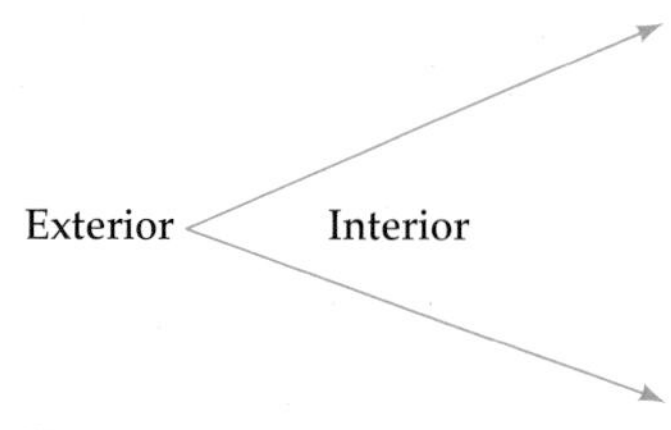

Figure 8.32

While we will use the standard definition in this course, we will occasionally discuss the angles in a dynamic sense when that is appropriate. It is important to note that the two conceptions are not contradictory.

An angle partitions a plane into three disjoint sets: the angle itself, the **interior** of the angle, and the **exterior** of the angle (Figure 8.32).

NAMING ANGLES

We often refer to an angle by using its vertex. For example, we can call the angle below in Figure 8.33 angle A. We can also refer to angles by using a numbering system. In the diagram below in Figure 8.33, we can talk about angle 1, angle 2, angle 3, and angle 4.

Figure 8.33

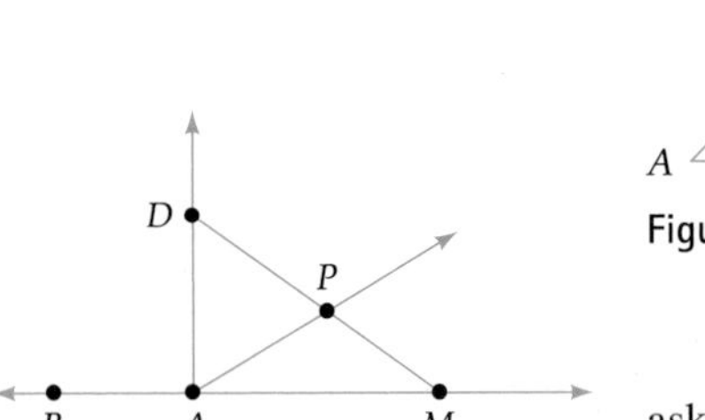

Figure 8.34

Both ways of referring to angles have limitations. Look at Figure 8.34. If someone asks you to look at angle A, what are the possibilities? How might we resolve this dilemma? Think before reading on. . . .

One way out of this dilemma is to use the symbol $\measuredangle$ and three letters to name an angle. The order of the three letters is important. We start with a point on one side of the angle, then give the vertex, and then give a point on the other side of the angle (Figure 8.35). These three points determine an angle in much the same way that three noncollinear points determine a plane. Thus you can see that $\measuredangle DAM$ and $\measuredangle ADM$ are different angles.

Figure 8.35

HISTORY

The Babylonians, for reasons unknown to us, decided (over 3000 years ago) to divide a circle into 360 parts. When the French converted measures of length, mass, and other quantities to the metric system, why didn't they convert measures of a circle? We know that the French attempt to metrify the week (10 days in a week and 10 hours in a day) failed. Perhaps the 24-hour day and 7-day week were just too deeply ingrained, and perhaps this was also true with angles. Actually, 360 has advantages over 100. Consider some common angles—for example, 60° and 45°. If there were 100 degrees in a circle, what would these angles be?

MEASURING ANGLES

We measure angles by how "open" they are, and the tool we use is a **protractor**. The most common unit of measurement for angles is the **degree**.

Although protractors come in a variety of styles, the correct use of all protractors requires the following:

1. The vertex of the angle must lie at the center of the protractor, which is not always the center of the bottom edge of the protractor (as shown in Figure 8.36).

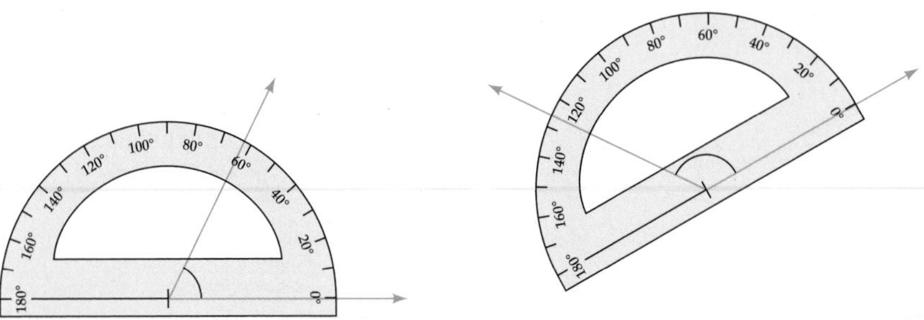

Figure 8.36

CLASSROOM CONNECTION

Grade 5

In this lesson, the children can practice using a protractor, and they can do some data collecting and analyzing! 

Date Time

LESSON 6·3

Finger Measures: Finger Flexibility

The picture shows how to measure the **angle of separation** between your thumb and first (index) finger. This is a measure of finger flexibility.

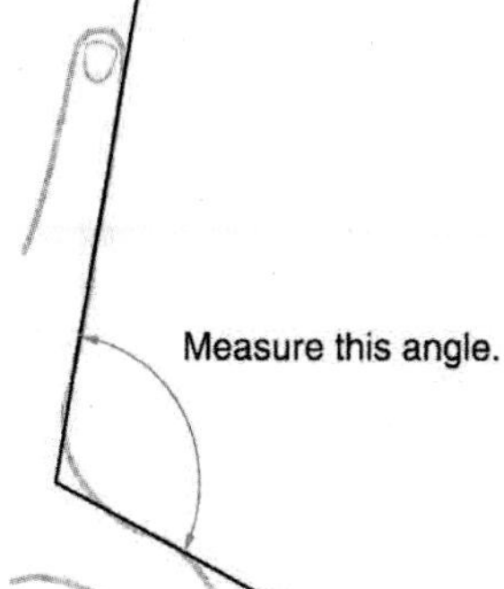

1. Spread your thumb and first finger as far apart as you can. Do this in the air. Don't use your other hand to help. Lower your hand onto a sheet of paper. Trace around your thumb and first finger. With a straightedge, draw two line segments to make a V shape, or angle, that fits the finger opening. Use a protractor to measure the angle between your thumb and first finger. Record the measure of the angle.

   Angle formed by thumb and first finger:

   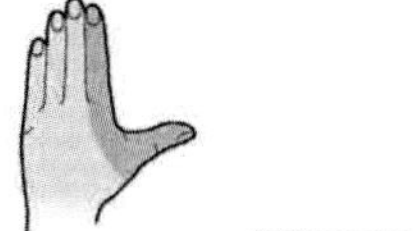

   ______ °

2. In the air, spread your first and second fingers as far apart as possible. On a sheet of paper, trace these fingers, and draw the angle of separation between them. Measure the angle and record its measure.

   Angle formed by first and second fingers:

   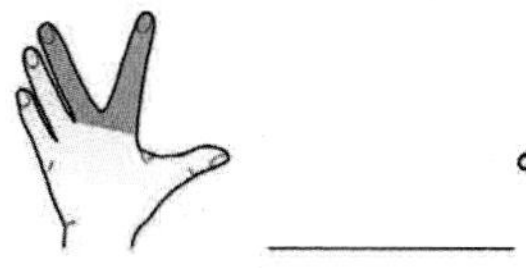

   ______ °

3. Record the class landmarks for both finger-separation angles in the table at the right.

Landmark	Thumb and First	First and Second
Minimum		
Maximum		
Mode(s)		
Median		

174

From *Everyday Mathematics, Grade 5:* The University of Chicago School Mathematics Project: Student Math Journal, Volume 1, by Max Bell et al., Lesson 6-3, p. 174. Reprinted by permission of The McGraw-Hill Companies, Inc.

OUTSIDE THE CLASSROOM

Several years ago, I sprained my ankle badly and had to go for physical therapy. The PT used an instrument called a goniometer, which looked like a protractor, to measure the range of motion of my ankle to use as a baseline to measure my improvement. In the March–April 1997 issue of *Mathematics Teaching in the Middle School,* there is an article about goniometers, and an exercise appears at the end of this section.

2. One ray of the angle must lie directly under the line that goes through a 0 point of the protractor (which is not always labeled).
3. The measure of the angle is read by looking at the number corresponding to where the other ray of the angle crosses the number line that goes around the protractor.

In the right-hand figure, the protractor has been moved in order to measure an angle neither of whose vertices is parallel to the bottom of the page. Measure the two angles, and then check your measurements below or with a friend.

The measure of the angle at the left is 63 degrees, and the measure of the angle at the right is 126 degrees. There are many virtual protractor games online, which will be explored in the exercises.

For the most part, the level of precision of measuring angles in this course will consist of rounding the measure to the nearest whole number. However, an angle can be divided into 60 minutes (denoted by $'$), and a minute can be divided into 60 seconds (denoted by $''$). Can you think of a situation in which this kind of precision would be necessary? For example, an angle of 0 degrees and 1 minute is $\frac{1}{21,600}$ of a circle, too small to show on a piece of paper.

CLASSROOM CONNECTION

When we say that one angle is bigger than the other, we mean that its measure is greater. However, some children will say that angle 2 in Figure 8.37 is bigger than angle 1.

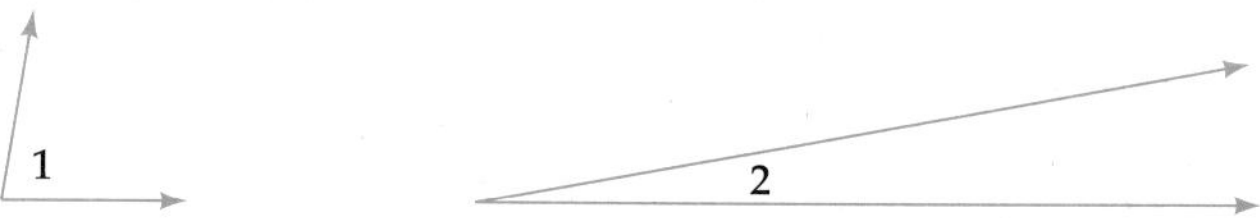

Figure 8.37

INVESTIGATION 8.1d Measuring Angles

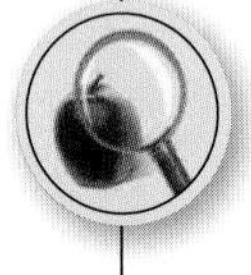

Estimate, and then measure with a protractor, the angles in the figures below.

Clock — Figure 8.38 | Logo — Figure 8.39 | Convex quadrilateral — Figure 8.40

Source: The Octagon logo is a Registered trademark of JPMorgan Chase & Co. and is used here with its expressed permission.

DISCUSSION

When we estimate, we often use benchmarks, just as we did with fractions in Chapter 4. Common benchmarks are 90 degrees, a right angle; 45 degrees ($\frac{1}{2}$ of a right angle), 30 degrees ($\frac{1}{3}$ of a right angle); and 60 degrees ($\frac{2}{3}$ of a right angle).

The angle of the clock at 4 o'clock should be 120 degrees. You can also calculate this, because the angle of the hands at 12 o'clock is 0 degrees and the angle at 6 o'clock is 180 degrees. Because 4 is $\frac{2}{3}$ of 6, we need $\frac{2}{3}$ of 180.

Each of the four trapezoids in the JPMorgan Chase & Co. logo have two 90-degree angles, one 45-degree angle, and one 135-degree angle. As you will see in the next section, we can further break these trapezoids into a rectangle and a right isosceles triangle.

The angles in the convex quadrilateral are 81 degrees, 111 degrees, 84 degrees, and 84 degrees, beginning with the top left angle and moving clockwise.

Let us now examine the kinds of angles that we encounter in mathematics and in everyday life.

LANGUAGE

Some dictionaries define *obtuse angle* as an angle whose measure is greater than 90 degrees. In such cases, straight angles and reflex angles are special kinds of obtuse angles.

CLASSIFYING ANGLES

We have names for different kinds of angles. We will see the importance of this in the next section when we name triangles. We can classify angles with respect to an angle whose measure is one-fourth of a circle—that is, 90 degrees. This is a **right** angle. Just as $\frac{1}{2}$ is a reference fraction, a right angle is a reference angle.

An angle whose measure is less than 90 degrees is called an **acute** angle.

An angle whose measure is greater than 90 degrees but less than 180 degrees is called an **obtuse** angle.

If angles are seen from a static perspective, these terms are often sufficient, but angles can also be seen from the dynamic perspective of *turns.* For example, someone might be told to open a valve "one-quarter of a turn," or 90 degrees. A half turn would be 180 degrees, a full turn would be 360 degrees, and one and a half turns would be 540 degrees. It sometimes makes sense to speak of angles with measures of 180 degrees or greater. As another example, ski jumpers will attempt 540 degree spins in the air, or 720 degrees which is two complete turns.

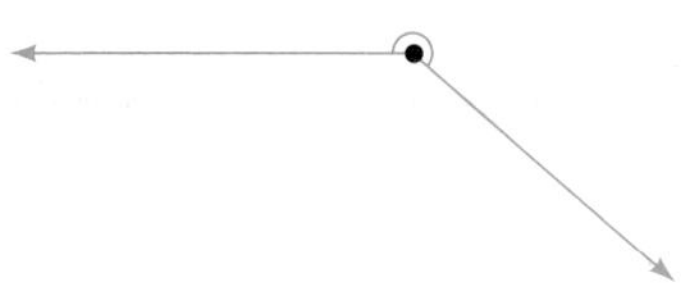

Figure 8.41

An angle whose measure is 180 degrees is called a **straight** angle.

An angle whose measure is greater than 180 degrees but less than 360 degrees is called a **reflex** angle. Determine the measure of the reflex angle in Figure 8.41.

OUTSIDE THE CLASSROOM

Skateboarders and snowboarders use reflex angles to describe some of their moves. What do you think they mean when they talk about doing a frontside 540 or a backside 720?

The measure of the angle is 225 degrees. One way to determine the measure is to extend the horizontal ray to the right and measure the resulting acute angle and then add 180 to the measure of the acute angle. Another way is to measure the obtuse angle and then subtract it from 360. Do you understand both ways?

Figure 8.42 shows one example of each of these angles.

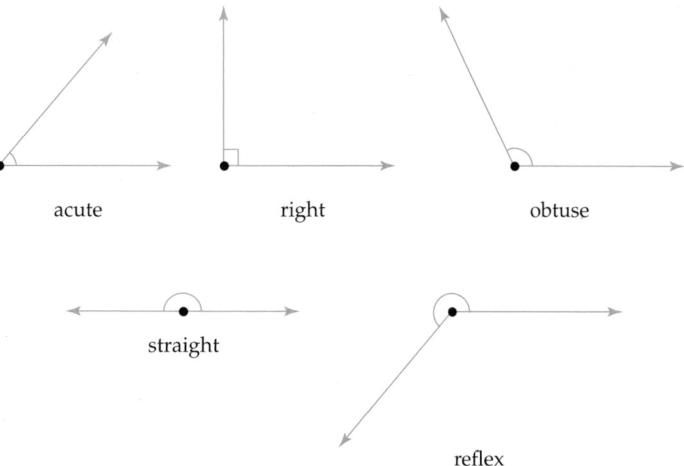

Figure 8.42

Sometimes the relative location of two angles is important. For example, we speak of adjacent angles. What do you think adjacent angles are? Can you draw two adjacent angles? Can you write a definition? If you do, give it to someone who doesn't know the term and see whether that person understands the term in the way that you meant it. Then read on. . . .

Adjacent means "next to," but more specifically, two angles are **adjacent** if they satisfy three conditions:

1. They have the same vertex.
2. They have a common side.
3. Their interiors are disjoint; that is, no point can be in the interior of both angles.

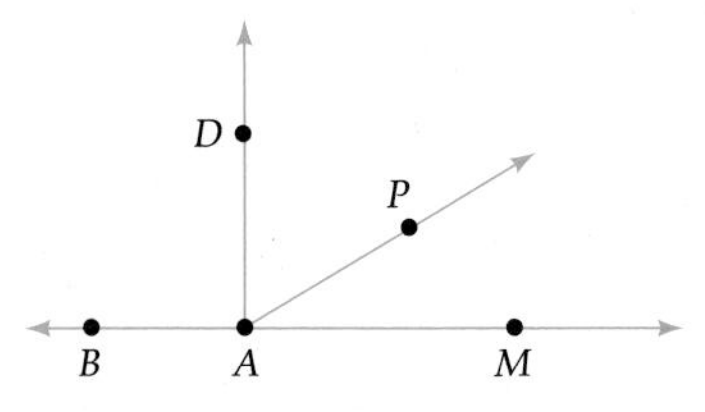

Figure 8.43

Do you see why the third condition is necessary? Can you find two angles in Figure 8.43 that have the same vertex and a common side but are not adjacent? . . . Think about it. Did you discover that $\measuredangle DAM$ and $\measuredangle PAM$ both share the side AM but are not considered adjacent? $\measuredangle DAP$ and $\measuredangle PAM$ are adjacent angles.

We sometimes speak of *complementary* and *supplementary* angles.

If the sum of the measures of two angles is 90 degrees, we call them **complementary** angles.

If the sum of the measures of two angles is 180 degrees, we call them **supplementary** angles.

There is one more kind of angle to examine in this section: vertical angles. Whenever two lines intersect, four angles are formed. The angles that are opposite each other are called **vertical angles**.

LANGUAGE

Many students argue convincingly that their lives in geometry would be much easier if such angles were called opposite angles.

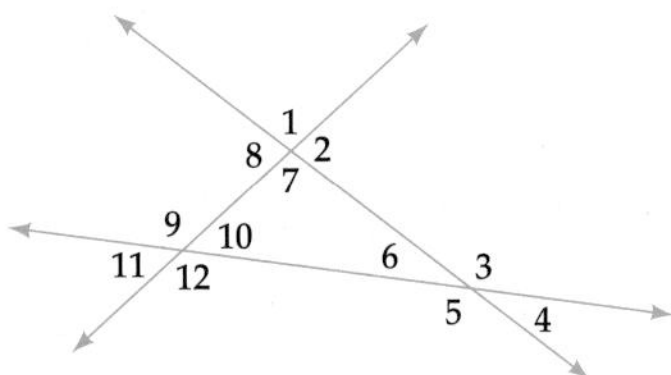

Figure 8.44

That is, $\measuredangle 1$ and $\measuredangle 7$ in Figure 8.44 are vertical angles, and $\measuredangle 2$ and $\measuredangle 8$ are vertical angles. Most people can quickly see that each pair of vertical angles appear to be congruent. Now this brings up one issue (language) and one question (proving they are equal).

Technically, we say either that $\measuredangle 1$ and $\measuredangle 7$ are congruent or that their measures are equal: $\measuredangle 1 \cong \measuredangle 7$ or $m\measuredangle 1 = m\measuredangle 7$.

The notation we use for congruence is $\cong$. *Congruence* means same size and same shape, while *equals* means same size, shape, location, position, that is, the same in every way.

The notation we use for measures is $m\measuredangle$.

We will examine congruence in more detail in Section 8.2.

It certainly seems reasonable that vertical angles are always congruent. Visually, if we draw them on a piece of paper, we can fold one on top of the other to see they match. Although this text is not emphasizing proofs, I believe a few proofs are helpful. How might we prove that the vertical angles *must* be congruent? Think about this before reading on. . . .

A key to the proof is to realize that the members of each pair of adjacent angles in Figure 8.44 are supplementary. For example, $\measuredangle 1$ and $\measuredangle 2$ are supplementary. Do you see why? If you didn't see this, can you now see how we might show that $\measuredangle 2$ and $\measuredangle 8$ must be congruent?

If you see the relationships, the proof is fairly straightforward:

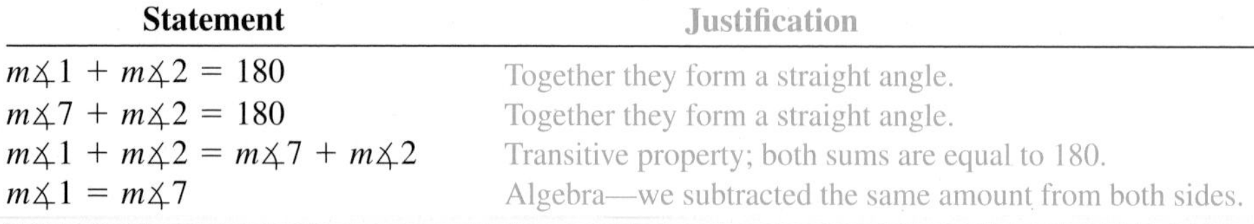

Statement	Justification
$m\measuredangle 1 + m\measuredangle 2 = 180$	Together they form a straight angle.
$m\measuredangle 7 + m\measuredangle 2 = 180$	Together they form a straight angle.
$m\measuredangle 1 + m\measuredangle 2 = m\measuredangle 7 + m\measuredangle 2$	Transitive property; both sums are equal to 180.
$m\measuredangle 1 = m\measuredangle 7$	Algebra—we subtracted the same amount from both sides.

Again, it is this type of thinking that made what the Greeks did over 2000 years ago so different from what had been done before.

SUMMARY 8.1

In this section, we have focused on the building-block language and concepts of Euclidean geometry: point, line, and plane. Line segments can intersect and join together to make plane figures, also called two-dimensional figures. We have examined different subsets of lines (line segments and rays) and different kinds of lines (parallel, perpendicular, concurrent, and skew). Plane figures can intersect to make space figures, also called three-dimensional figures.

We have examined the different kinds of angles that can be formed by intersecting lines. We can classify angles by their measure: acute, right, obtuse, straight, and reflex. We can also classify angles by their relationship to other angles: complementary, supplementary, and vertical.

8.1 Exercises

1. a. What is a tetromino? Imagine defining this term to someone who has no idea what you mean.
   b. Write the definition of tetromino that could have resulted in someone thinking the two figures in Figure 8.3 were tetrominoes. Then fix the definition.
2. This exercise focuses on communication. For each of the following terms, suppose a friend in the class has come to you and told you that the term just doesn't make sense. Describe how you would help that person make sense of the term. Your description must include your own definition of the term in your own words.
   a. Collinear points
   b. Concurrent lines
   c. Adjacent angle
   d. Skew lines
   e. Vertical angles
3. Optical illusions
   a. Which line is longer?

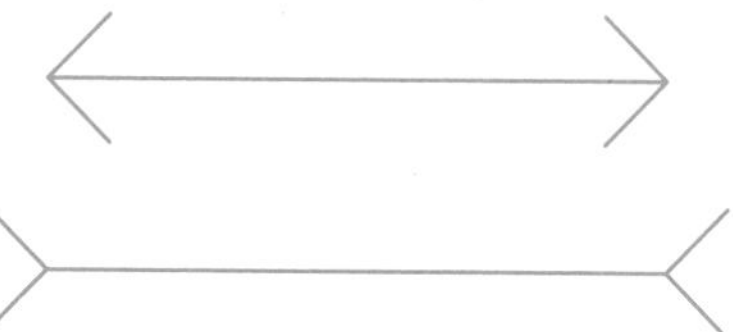

   b. Which circle is larger?

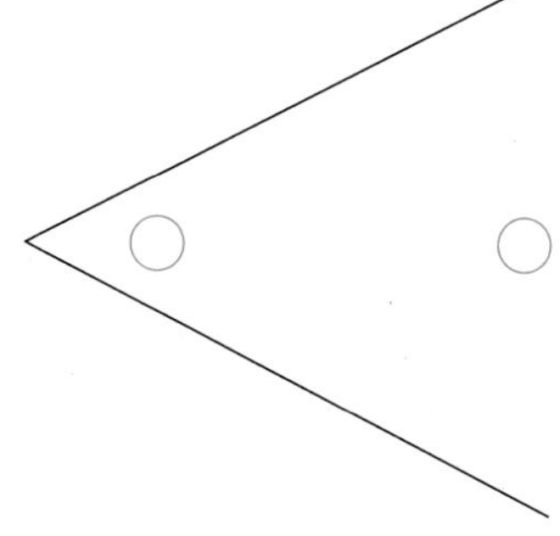

   c. Are the diagonal lines parallel?

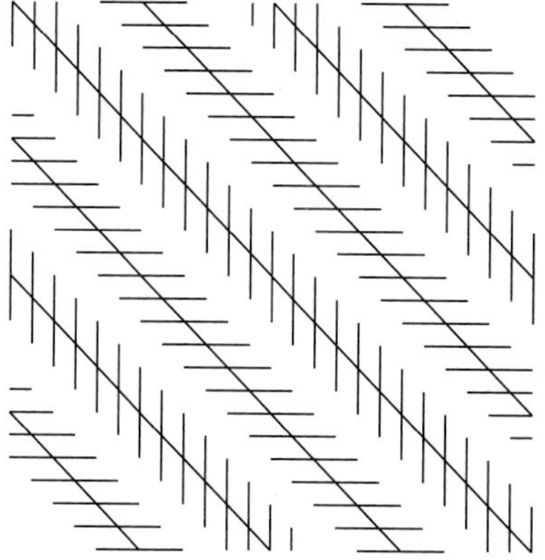

   d. Which segment on the top is a continuation of the segment on the bottom?

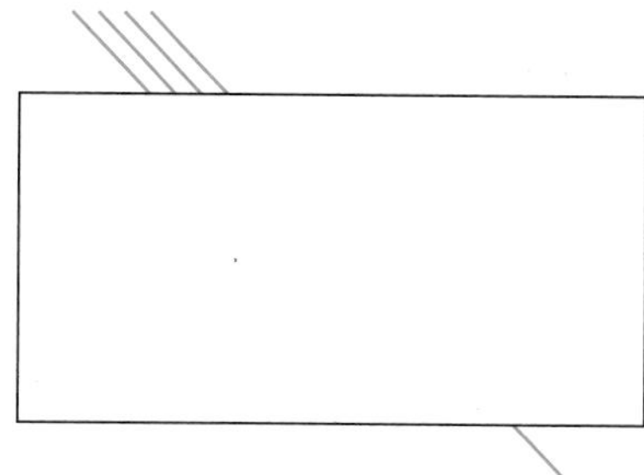

   e. What do you see?

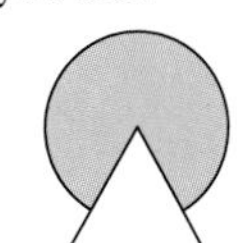

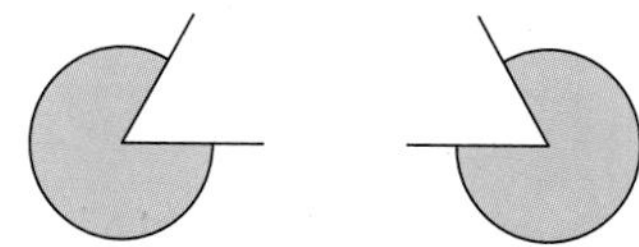

f. Shade in the figure. What do you notice?

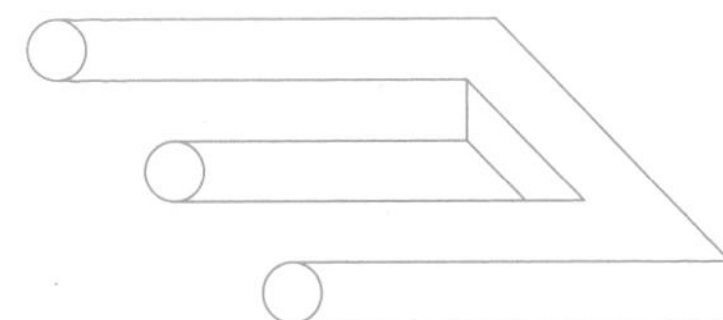

4. Name all the possible different rays that can be formed from the three points below.

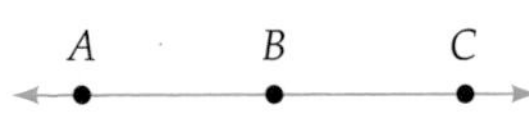

5. How many different rays can be formed from four collinear points?

6. Sketch four lines such that three are concurrent with each other and two are parallel to each other.

7. True or false? If true, briefly explain why. If false, provide a counterexample.
   a. If two distinct lines do not intersect, then they are parallel.
   b. If two lines are parallel, then they lie in the same plane.
   c. If two lines intersect, then they lie in the same plane.
   d. If a line is perpendicular to a plane, then it is perpendicular to all lines in that plane.
   e. If three lines are concurrent, then they are also coplanar.
   f. If two planes intersect, then the intersection is either a point or a line.

8. Refer to the cube pictured below, and use symbols such as $\overline{AB}$ to name the following:

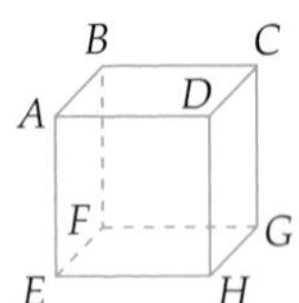

   a. Two parallel line segments
   b. Two line segments that do not lie in the same plane
   c. Two intersecting line segments
   d. Three concurrent line segments that do not lie in a single plane
   e. Two skew line segments
   f. A pair of supplementary angles
   g. A pair of perpendicular line segments
   h. Are points A, B, and H coplanar points? Why or why not?

9. In the figure below, $\overleftrightarrow{AD}$ and $\overleftrightarrow{CF}$ are perpendicular lines.

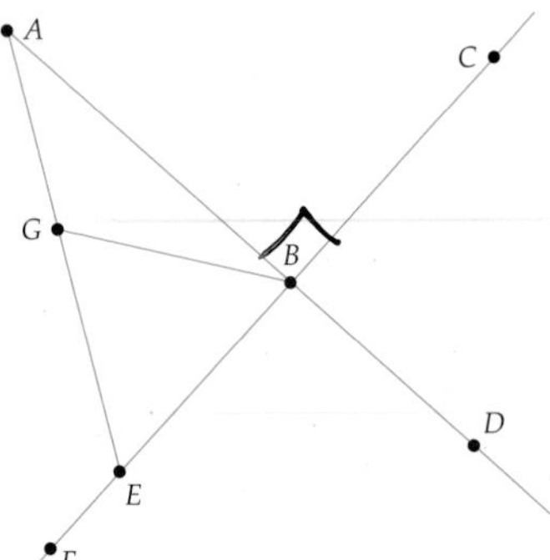

   a. Name two complementary angles.
   b. Name two supplementary angles.
   c. Name two vertical angles.
   d. Name two adjacent angles.

10. Estimate the measure of the following angles. Describe your reasoning process. Then measure the angles with a protractor and determine your percent error.

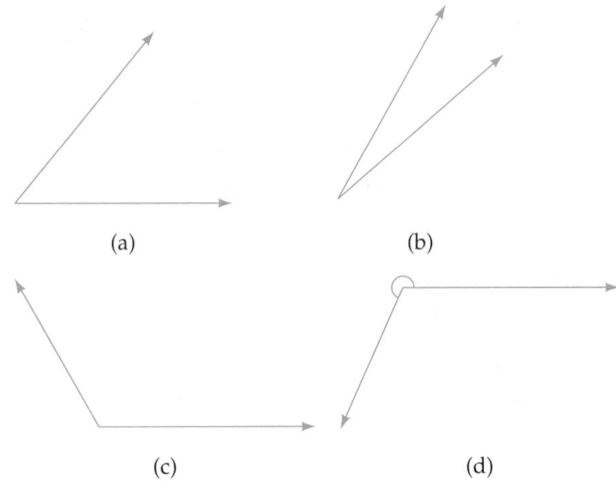

DEEPENING YOUR UNDERSTANDING

11. Clock problems:
    a. How many times a day will the minute hand be directly on top of the hour hand?
    b. What times could it be when the two hands make a 90-degree angle?
    c. What angle do the hands make at 7 o'clock?
    d. What angle do the hands make at 3:30?
    e. What angle do the hands make at 2:06?
    f. Make up and answer a problem like these. Then swap problems with a friend. Have the friend check your work, and you check the friend's work.

12. Make at least five angles. Next, estimate their measure, determine their measure, and determine your percent error. Repeat this process. If you note that the percent error seems to be decreasing, can you articulate any thinking processes that you are developing that are making your estimates more accurate?

13. Using only a straightedge and reasoning, try to make angles with the following measures. Then describe your reasoning process. After making each angle, check your work with a protractor and determine the percent error.

    a. 30° b. 45° c. 150° d. 300° e. 67°

14. a. Make your own goniometer (see p. 440). Briefly describe the thinking process behind your making of the goniometer.
    b. Compare goniometers with other members of your group or class. Which one(s) do you think will produce the most accurate results?
    c. You will use one of the selected goniometers to measure the range of motion of an ankle. Each group separately will

define "range of motion" and then determine a method for determining range of motion. Do that and then measure the range of motion of one of the member's ankles. Each person in the group will make one measurement, but do not say the number of your measurement until all members have measured.

d. Write on the board the three angles determined by your group. Observe the numbers of the other groups. Describe what you see.

e. Repeat the process after the entire class agrees on the same definition and the same measuring technique. Compare data. What is the average range of motion in your class?

f. What did you learn from this exercise?

15. Sketch a pair of angles whose intersection is

a. exactly two points or explain why this is not possible.

b. exactly three points or explain why this is not possible.

c. exactly four points or explain why this is not possible.

16. a. Measure the acute angles in each letter below.

b. In the next step, you will compare your results with other members of your group or class. In which cases do you predict the standard deviation will be greatest? smallest?

c. Compare your measurements to those of other members of your group or class, as directed by your instructor. What did you find?

17. This exercise was taken from the March 2002 issue of *Discover* magazine. Below are the first five letters of the alphabet, designed to fit the vertices of a grid. Below them are the first four letters that were designed on other grids. Draw the "e" that goes with each type style, or font. As stated many times in this book, we want you to go beyond random trial and error and beyond the famous "just do it" commercial slogan. First, take some time and think. One strategy is to try to write the rules that the writer of the letter used. Another strategy is to try to articulate the common characteristics of the four letters.

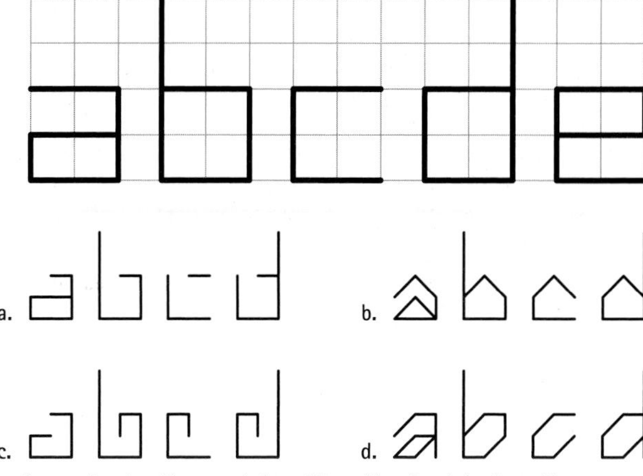

Source: Reprinted by permission of Scott Kim. Puzzle by Scott Kim, scottkim.com.

18. The terms *parallel* and *perpendicular* occur both in algebra (equations of lines) and in geometry (properties of many geometric figures). This exercise requires you to think about these concepts and some of the connections between algebra and geometry.

a. On each of the geoboards below, make a line parallel to the given line. Develop, describe, and justify a rule or procedure for making sure that the two lines are parallel. The rule or procedure should be more precise than "It looks parallel."

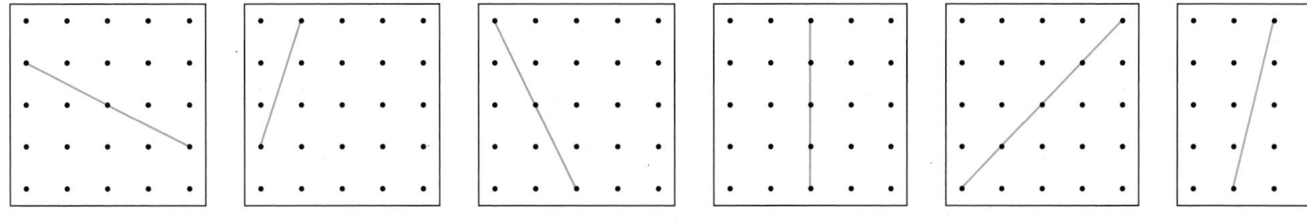

b. On each of the geoboards below, make a line perpendicular to the given line. Develop, describe, and justify a rule or procedure for making sure that the two lines are perpendicular.

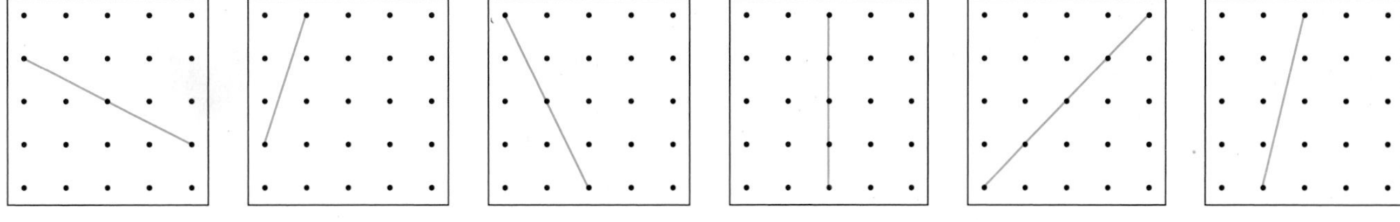

19. In the Classroom Connection on page 440, what mathematical misconception do students have who answer that angle 2 is bigger?

20. There are many geometry-exploration websites. Here are some favorites for exploring angles. For each one, play a few rounds of the activity and write a brief description of the mathematical concepts to which they relate.

    a. On this site you use a protractor to measure angles: www.mathplayground.com/measuringangles.html

    b. On this one, you drag the monkey around the circle to create an angle as close as possible to the angle where "the bananas are located": http://www.oswego.org/ocsd-web/games/bananahunt/bhunt.html

    c. On this site, you use a protractor to measure angles and observe relationships between angles: http://www.amblesideprimary.com/ambleweb/mentalmaths/angleshapes.html

    d. Find a site you like that explores concepts in this section and write a brief description of it.

SECTION 8.2 Two-Dimensional Figures

What do you think?

- In what ways are triangles and quadrilaterals different? In what ways are they alike?
- How are circles and polygons related?
- Can every polygon be broken down into triangles? Why or why not?
- Why do we use two words to name different triangles, but only one word to name different quadrilaterals?

Think of geometric figures that people generally find pleasing, such as those in Figure 8.45. What words would you use to explain why these objects are interesting or appealing? When you look at the various objects and pictures, what similarities do you see between certain objects and shapes—for example, triangles and hexagons?

When we discuss similarities and differences in my class, many geometric terms emerge in the discussion. Some students talk about similar shapes—for example, hexagons

©John Teate /Shutterstock.com

(a)
a quilt pattern

© Poprugin Aleksey/Shutterstock.com

(b)
lamp post

©James Steidl/Shutterstock.com

(c)
snowflakes

Figure 8.45

in honeycombs and snowflakes, squares in some pictures, and triangles in others. Some students observe that many of the shapes are symmetric. In explaining similarities, some students talk about the angles, the length of sides, or the fact that some figures look similar. Many students observe that even the more complex shapes can be seen as being constructed from simpler shapes, such as triangles and quadrilaterals.

In a few moments, you will begin a systematic exploration of geometric shapes. Before you do, let us examine a very important framework for understanding the development of children's geometric thinking. This model was developed by Pierre and Dina van Hiele-Geldorf in the late 1950s and is widely used today. Essentially, the van Hieles found that there are levels, or stages, in the development of a person's understanding of geometry.[2]

Level 1: Reasoning by resemblance

At this level, the person's descriptions of, and reasoning about, shapes is guided by the overall appearance of a shape and by everyday, nonmathematical language. For example, "This is a square because it looks like one." Students at this level may be made aware of the various properties of geometric objects (for example, that a square has four equal sides), but such awareness can be overridden by other factors. For example, if we turn a square on its side, the student may insist that it is no longer a square but now is a diamond.

Level 2: Reasoning by attributes

At this level, the person can go beyond mere appearance and recognize and describe shapes by their attributes. A student at this level, seeing the figure above, can easily classify it as a quadrilateral because it has four sides. However, a student at this level does not regularly look at *relationships* between figures. A student who argues that a figure "is not a rectangle because it is a square" is reasoning at this level.

Level 3: Reasoning by properties

At this level, the student sees the many attributes of shapes and the relationships between and among shapes. A student at this level can see that the square and rhombus have many properties in common, such as opposite sides parallel, all four sides congruent, and diagonals that bisect each other and are perpendicular. This enables the student to understand that a square is simply a rhombus with one additional property—all the angles are right angles.

Level 4: Formal reasoning

Students at this level can understand and appreciate the need to be more systematic in their thinking. When solving a problem or justifying their reasoning, they are able to focus on mathematical structures.

The investigations in the text and the accompanying explorations have been designed to be consistent with this approach. Accordingly, as you are working in this chapter, reflect on your own thinking. Are you looking at the problem only on a vague, general level? Are you fixated on just one attribute? Are you seeing relationships among triangles, among quadrilaterals? As you move from *what* to *why*, are you able to move from solving a problem by random trial and error to being more systematic and careful in your approach?

With this model in mind, let us begin our exploration of shapes with an investigation my students have found to be both fun and powerful. Even more important than knowing the names of all these different shapes will be knowing their properties and the relationships between and among the shapes. It is *this* knowledge that is used by artists, engineers, scientists, and all sorts of other people.

INVESTIGATION 8.2a Recreating Shapes from Memory

For this investigation, you will want to have a pencil and an eraser.

Explorations Manual 8.7

A. Look at Figure 8.46 for about 1 second. Then close the book and draw Figure 8.46 from memory.

Check the picture again for 1 second. If your drawing was incomplete or inaccurate, change your drawing so that it is accurate. Check the picture again for 1 second. Keep doing this until your drawing is complete and accurate.

Now go back and try to describe your thinking processes as you tried to re-create the figure. From an information processing perspective, your eyes did not simply receive the image from the paper; your knowledge of geometry helped determine *how* you saw the picture. What did you hear yourself saying to help you remember the picture? Then read on. . . .

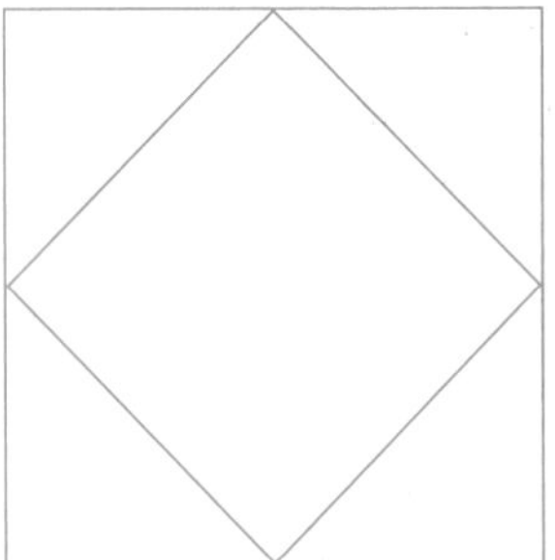

Figure 8.46

DISCUSSION

Some students see a diamond and 4 right triangles. Other students see a large square in which the midpoints of the sides have been connected to make a new square inside the first square. Yet other students see four right triangles that have been connected by "flipping" or rotating them.

B. Now look at Figure 8.47 for several seconds. Then close the book and try to draw it from memory. As before, check the picture again for a few seconds. If your drawing was incomplete or inaccurate, fix it. Keep doing this until your drawing is complete and accurate. Then go back and describe the thinking processes you engaged in as you tried to re-create the figure. . . .

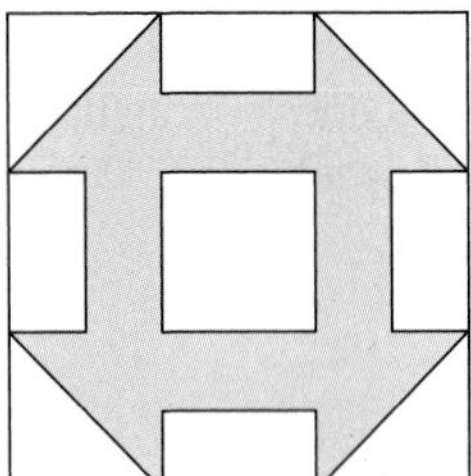

Figure 8.47

DISCUSSION

This figure was more complex. Some students see a whole design and try to remember it.

Some decompose the design into four dark triangles and four rectangles as shown in Figure 8.48.

The figure can also be seen as being composed of 9 squares, which can also be seen in Figure 8.48. The four corner squares have been cut to make congruent right triangles. Each of the other four squares on the border has been cut into two congruent rectangles.

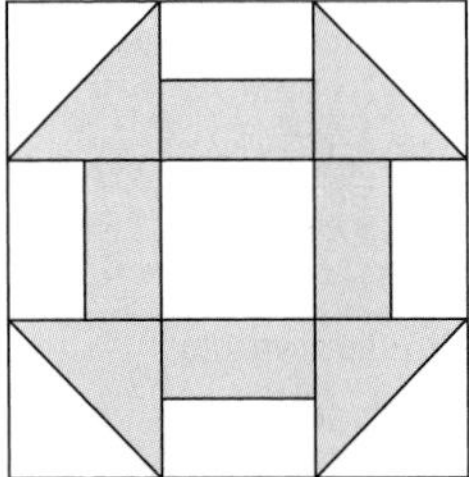

Figure 8.48

Some students re-created this figure by seeing a whole square and then looking at what was cut out (Figure 8.49). That is, they saw that they needed to cut out each corner, and they saw that they needed to cut out a rectangle on the middle of each side. Finally, they remembered to cut out a square in the center.

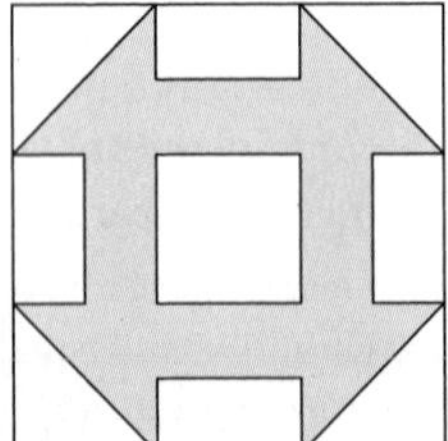

Figure 8.49

There are several implications for teaching from this investigation. How a person re-creates the figure is related to the person's spatial-thinking preferences and abilities. Different people "see" different objects. That is, not everyone sees the figure in the same way. Although there are differences in how people re-create the figure, very few people can re-create the figure without doing some kind of decomposing—that is, without breaking the shape into smaller parts. Being able to do this depends partly on spatial sense and partly on being able to use various geometric ideas (congruent, triangle, square, rectangle) at least at an intuitive level.

OUTSIDE THE CLASSROOM

Figure 8.47 is a famous quilting pattern called the Churn Dash. When people made their own butter, the cream was poured into a pail called a churn. Then the churner rolled a special pole back and forth with his or her hands. At the end of the pole was a wooden piece called a dash, which was shaped like the figure you saw. If you make many copies of this pattern and put them together, you can see what a Churn Dash quilt looks like.

Although some people manage to live happy, productive lives at the lowest van Hiele level, an understanding of basic geometric figures and the relationships among them is often helpful in everyday life (for example, in home repair projects and quilting) and in many occupations.

Before we examine specific kinds of polygons, beginning with triangles, the following investigation will serve to "open your thinking"—to get you to look at polygons not only through the zoom lens, which reveals specific properties and definitions, but also through the wide-angle lens, in which you see *all* the attributes. For example, a person is not just a person. She might be a mother, a sister, a daughter, a scientist, a Democrat, and so on. Should you become a friend of this person, you come to know her many facets. Similarly, if you become a "friend" of shapes, you come to know the many attributes of the shapes with which you are working.

INVESTIGATION 8.2b All the Attributes

Look at the polygons in Figure 8.50. Think of all the attributes, all the characteristics, anything about the two polygons that might be important to state or measure. Use a ruler and protractor to help. Write down your list before reading on. . . .

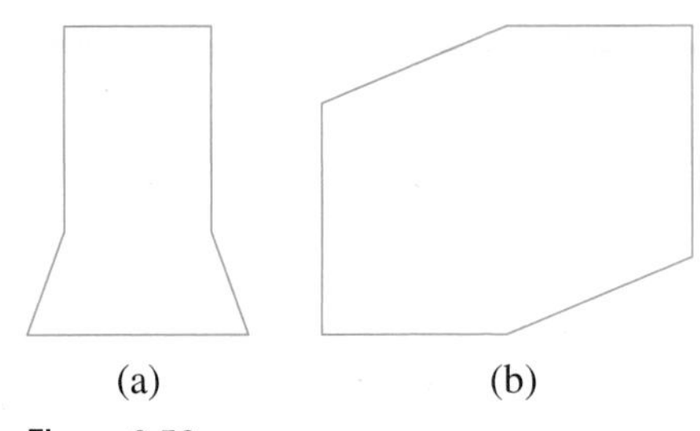

Figure 8.50

BEYOND THE CLASSROOM

Almost everything has multiple attributes. For example, every person has physical attributes (age, height, weight, etc.) and relational attributes (son, sister, mother, etc.). A garage has length, width, height, shape, surface area (if we want to paint it), volume (to consider storage), composition (wood or metal), number of windows, door (automatic or manual), and so on. When interacting with a person, or considering buying a car, or doing something with a garage, we focus on certain attributes. That is exactly what we do in geometry. Which attributes we focus on depend on the problem at hand.

DISCUSSION

As you compare the following lists to yours, read actively. If you made the same observation, did you use the same wording? If not, do you understand the wording here? If you missed one of these attributes, why? Do you understand it now? Are there other attributes that you noticed?

Figure 8.50a	Figure 8.50b
6 sides	6 sides
Top and bottom sides parallel to each other	Top and bottom sides parallel to each other
2 pairs of parallel sides	All 3 pairs of opposite sides parallel
Concave	Convex
2 sets of congruent sides	4 congruent sides; the other pair of sides is also congruent
2 acute angles, 2 right angles, 2 reflex angles	4 obtuse angles, 2 right angles
3 pairs of congruent angles	Opposite angles congruent

A key idea here is to realize that there are lots of attributes and that knowing these attributes and combinations of attributes of a shape helps chemists and physicists to understand the behavior of a molecule or shape; helps builders to know which shapes work together better either in terms of structure and strength or in terms of appearance; and helps artists and designers to make designs that are the most appealing. As you read on, think of multiple attributes and of attributes that different objects have in common.

INVESTIGATION 8.2c Classifying Figures

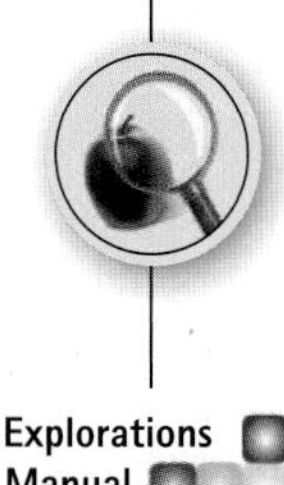

Explorations Manual 8.8

Before we examine and classify important two-dimensional shapes, we first need to investigate the kinds of possible two-dimensional shapes.

As we have done throughout this book, rather than giving you the major classifications, we will engage you in some thinking before presenting them. Look at the 13 shapes in Figure 8.51 and classify them into two or more groups so that each group has a common characteristic. Do this in as many different ways as you can, and then read on. . . .

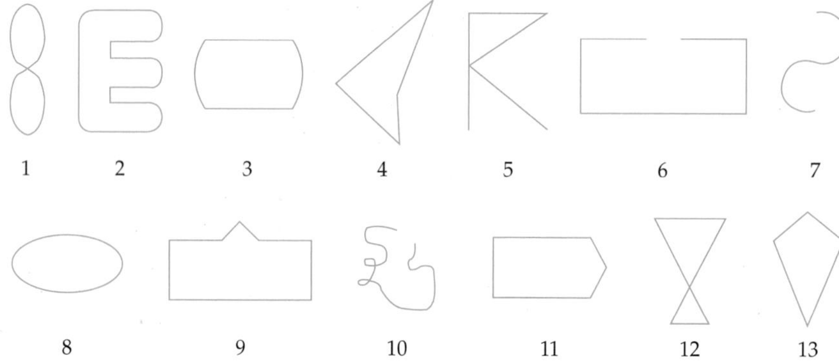

Figure 8.51

DISCUSSION

One way to sort the figures is shown below. How would you describe the figures in set A and the figures in set B? Do this before reading on. . . .

The figures in set A are said to be simple curves. We can describe simple curves in the following way: A figure is a **simple curve** in the plane if we can trace the figure in such a way that we never touch a point more than once. If you look at the figures in set A, you can see that they all have this characteristic; and all the figures in set B have at least one point where the pencil touches twice, no matter how you trace the curve.

Now look at the curves in sets C and D. How would you describe the figures in set C and the figures in set D? Do this before reading on. . . .

Set C

Set D

The figures in set C are said to be closed curves. We can describe closed curves in the following way: A figure is a **closed curve** if we can trace the figure in such a way that our starting point and our ending point are the same. If you look at the figures in set C, you can see that they all have this characteristic; and no matter how you try, you cannot trace the figures in set D with the same starting and ending point.

LANGUAGE

What other words might you use to describe the intersecting and not intersecting subsets? Some students use the phrase "trace over," and others talk about figures that "run over themselves" or "cross themselves." Other students talk about the set of figures that contain two smaller regions within each figure.

CLASSROOM CONNECTION

Using the terms developed in this investigation, how would you classify dot-to-dot pictures?

Now look at the curves in sets E, F, and G. How would you describe the figures in set E, the figures in set F, and the figures in set G? Do this before reading on. . . .

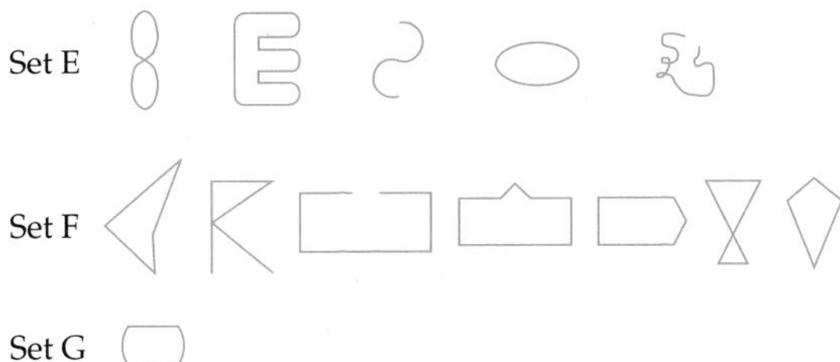

The language used to describe the three sets poses a challenge, for most people describe the figures in set E as consisting only of curvy lines, the figures in set F as consisting only of straight line segments, and the figure in set G as having both curvy and straight line segments. The challenge here is that when mathematicians use the word *curve,* this word encompasses both *curvy* and *straight* line segments—a curve is a set of points that you can trace without lifting your pen or pencil. There is nothing wrong with students' use of the terms curvy and straight. What is important is the realization that we are using the words *curve* and *curvy* in different ways. We do this all the time in everyday English. For example, consider different ways we use the word *hot*: "It sure is a hot day." "I love Thai food because it is hot." "This movie is really hot!"

Most of our investigations of curves will focus on simple closed curves. Looking at the descriptions above, try to define the term *simple closed curve* before reading on. . . .

We will define a **simple closed curve** in the plane as a curve that we can trace without going over any point more than once while beginning and ending at the same point. The set of polygons is one small subset of the set of simple closed curves.

At this point, you might want to do the following activity with another student.

- Draw a simple closed curve.
- Draw a simple open curve.
- Draw a nonsimple closed curve.
- Draw a nonsimple open curve.

Exchange figures with another student. Do you both agree that each of the other's drawings matches the description? If so, move on. If not, take some time to discuss your differences.

HISTORY

Generally, when a theorem is named after a person, it is named after the person who first proved the theorem. In this case, Jordan's proof was found to be incorrect, but the theorem is still named after him!

There is an important mathematical theorem known as the **Jordan Curve Theorem**, after Camille Jordan: Any simple closed curve partitions the plane into three disjoint regions: the curve itself, the interior of the curve, and the exterior of the curve. See the examples in Figure 8.52.

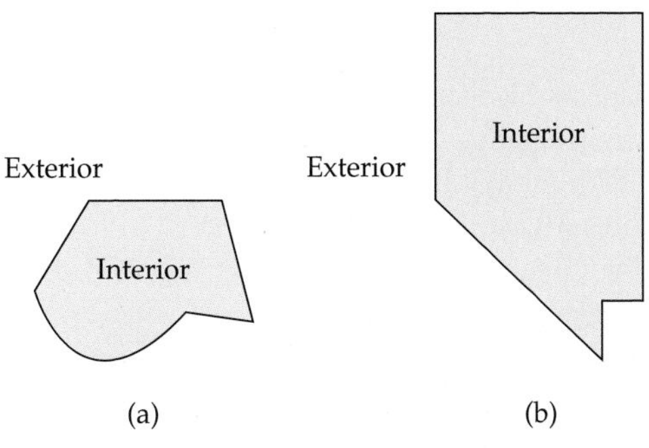

Figure 8.52

In the examples in Figure 8.52, deciding whether a point is inside or outside is easy. However, look at Figure 8.53. Although this figure is a simple closed curve, it is a rather complicated figure, and such complicated shapes are encountered in some fields of science. Is point *A* inside or outside? How would you determine this? Think before reading on. . . .

Figure 8.53

Someone once remarked that mathematicians are among the laziest people on earth because they are always looking for shortcuts and simpler ways to solve problems. Thus you may be wondering whether someone has found an easier way to solve these problems. Look at Figure 8.53 to illustrate the method. Start at a point that is clearly outside the shape and draw a line segment connecting that point to the point you are looking at; it helps if you pick an outside point so that the line segment will cross the curve in as few points as possible. Each time you cross a point, it's like a gate—if you were outside, you are now inside; if you were inside, you are now outside. Thus it is a relatively simple matter to determine that point *A* is inside the curve.

CLASSROOM CONNECTION

The active and curious reader here might realize that this procedure might have applications for solving mazes, and there is a branch of mathematics that does analyze mazes.

POLYGONS

We are now ready to begin our exploration of **polygons**, which can be defined as simple closed curves in the plane that are composed only of line segments. Thus the simple closed curve in Figure 8.52(a) is not a polygon because it has a side that is not a line segment, whereas the simple closed curve in Figure 8.52(b) (which looks like the state of Nevada) is a polygon. On any polygon, the point at which two sides meet is called a **vertex**, the plural of which is **vertices**. The line segments that make up the polygon are called **sides**. Look back at the 13 shapes in Figure 8.51 in Investigation 8.2C. Which of those are polygons? Look at them and decide before reading on. . . .

TABLE 8.2

Number of sides	Name
3 sides	Triangle
4 sides	Quadrilateral
5 sides	Pentagon
6 sides	Hexagon
7 sides	Heptagon
8 sides	Octagon
9 sides	Nonagon
10 sides	Decagon
n sides	*n*-gon

Shapes 4, 9, 11, and 13 are all polygons; the others are not. The word *polygon* has Greek origins: *poly-,* meaning "many," and *-gon,* meaning "angles." You are already familiar with many kinds of polygons. Just as we found in Chapter 2 that the names we give numbers have an interesting history, so do the names we give to polygons. Again, back to the Greeks: These names have Greek prefixes that represent the number of sides (Table 8.2).

Now that we have a good general definition of the term *polygon,* we can spend time examining triangles, quadrilaterals, and a few other specific kinds of polygons.

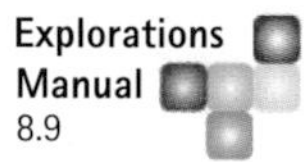
Explorations Manual 8.9

TRIANGLES

Triangles are found in every aspect of our lives—in buildings, in art, in science (see Figures 8.54 and 8.55). They are truly "building-block" shapes. Every bicycle I have seen has triangles. Bridges will always contain triangles. If you look at the skeleton of buildings, and the scaffolding around the building, you will always see triangles. Why? We will use the next investigation to think about this question.

Figure 8.54

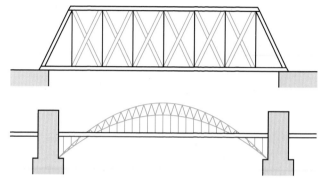

Figure 8.55

INVESTIGATION 8.2d Why Triangles Are So Important

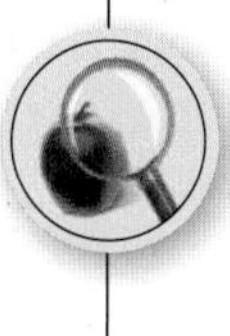

Cut some strips of paper from a file folder or other stiff material. Punch a hole in the ends and use paper fasteners (improvise if you need to; for instance, you can use paper clips). Make one triangle and one quadrilateral, as shown in Figure 8.56. It need not be an equilateral triangle or a square. What do you see? . . .

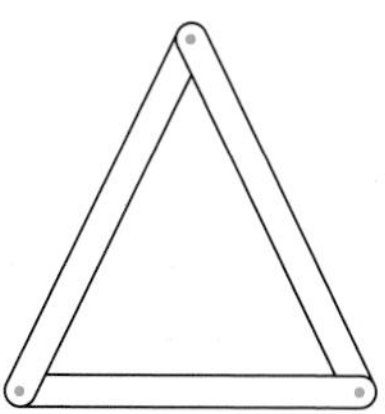
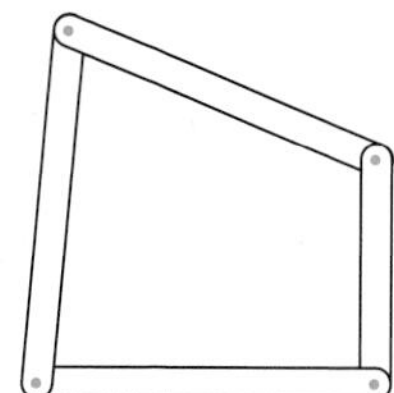

Figure 8.56

DISCUSSION

As you saw, the triangle won't move—we call it a rigid structure. However, the quadrilateral does move; it is not stable. Make another strip and connect two nonadjacent vertices of your quadrilateral. What happens now? It will remain in the shape. If it is a square, it will remain a square; if it is a parallelogram, it will remain a parallelogram. This is because the addition of that diagonal actually created two triangles, which, as you have found, are rigid structures. The next time you walk about campus and about town, look for triangles. You will see them everywhere!

As you already know, there are many kinds of triangles. A crucial goal of the next investigation is for your own understanding of triangles to become more powerful.

INVESTIGATION 8.2e Classifying Triangles

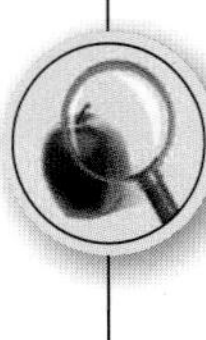

You will find nine triangles in Figure 8.57. Copy them and cut them out, and then separate them into two or more subsets so that the members of each subset share a characteristic in common. Can you come up with a name for each subset? What names might children give to different subsets? Do this in as many different ways as you can before reading on. . . .

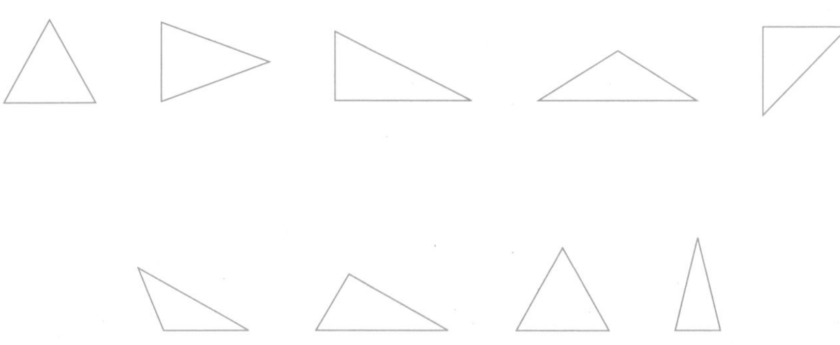

Figure 8.57

DISCUSSION

STRATEGY 1 Consider sides

One way to classify triangles is by the length of their sides: all three sides having equal length, two sides having equal length, no sides having equal length. There are special names for these three kinds of triangles.

- If all three sides have equal length, then we say that the triangle is **equilateral**.
- If at least two sides have equal length, then we say that the triangle is **isosceles**.
- If all three sides have different lengths—that is, no two sides have equal length—then we say that the triangle is **scalene**.

Which of the triangles in Figure 8.57 are scalene? Which are isosceles? Which are equilateral?

STRATEGY 2 Consider angles

We can also classify triangles by the relative size of the angles—that is, whether they are right, acute, or obtuse angles. This leads to three kinds of triangles: right triangles, obtuse triangles, and acute triangles.

- We define a **right triangle** as a triangle that has one right angle.
- We define an **obtuse triangle** as a triangle that has one obtuse angle.
- We define an **acute triangle** as a triangle that has three acute angles.

Many students see a pattern: A right triangle has one right angle, an obtuse triangle has one obtuse angle, yet an acute triangle has three acute angles. What was the pattern? Why doesn't it hold? Think before reading on. . . .

The key to this comes from looking at the triangles from a different perspective: Every right triangle has exactly two acute angles, and every obtuse triangle has exactly two acute angles; thus a triangle having more than two acute angles will be a different kind of triangle. This perspective is represented in Table 8.3.

TABLE 8.3

First angle	Second angle	Third angle	Name of triangle
Acute	Acute	Right	Right triangle
Acute	Acute	Obtuse	Obtuse triangle
Acute	Acute	Acute	Acute triangle

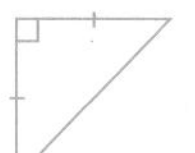

Figure 8.58

STRATEGY 3 Consider angles and sides

This naming of triangles goes even further. What name would you give to the triangle in Figure 8.58? The two marks in the top and left side show that those two sides are congruent.

This triangle is both a right triangle and an isosceles triangle, and thus it is called a right isosceles triangle or an isosceles right triangle. How many possible combinations are there, using both classification systems? Work on this before reading on. . . .

There are many strategies for answering this question. First of all, we find that there are nine possible combinations (Figure 8.59). We can use the idea of Cartesian product to determine all nine. That is, if set S represents triangles classified by side, S = {Equilateral, Isosceles, Scalene}, and set A represents triangles classified by angle, A = {Acute, Right, Obtuse}, then $S \times A$ represents the nine possible combinations.

MATHEMATICS

Did you notice the geometric balance in Figure 8.59, which represents the Cartesian product of the two sets of triangles?

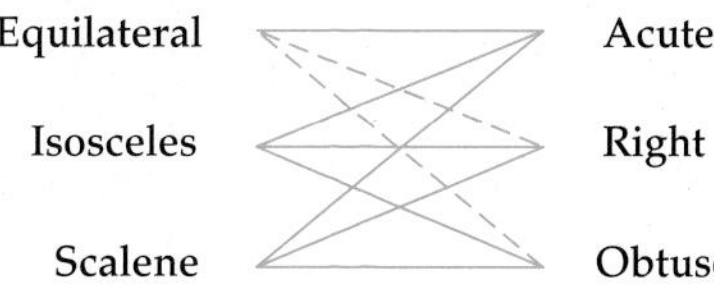

Figure 8.59

However, not all nine combinations are possible triangles. Try drawing each of the nine combinations and decide which are possible, which are not possible, and why.

CLASSROOM CONNECTION

Some students have told me that this investigation is huge for them, because they had never thought about organization and classification with respect to triangles. Triangles just *were!*

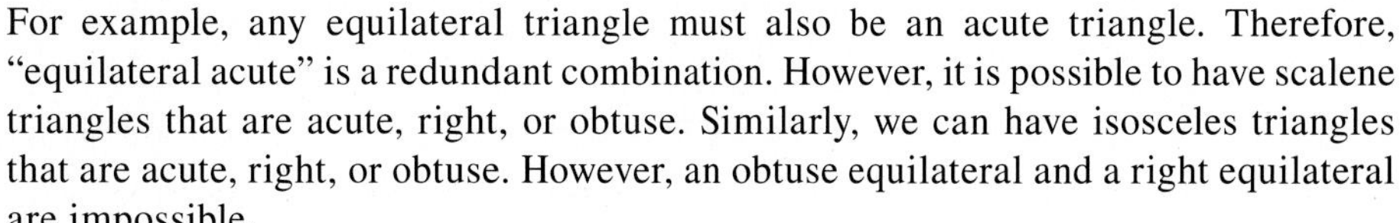

For example, any equilateral triangle must also be an acute triangle. Therefore, "equilateral acute" is a redundant combination. However, it is possible to have scalene triangles that are acute, right, or obtuse. Similarly, we can have isosceles triangles that are acute, right, or obtuse. However, an obtuse equilateral and a right equilateral are impossible.

Name the two triangles in Figure 8.60. Then read on. . . .

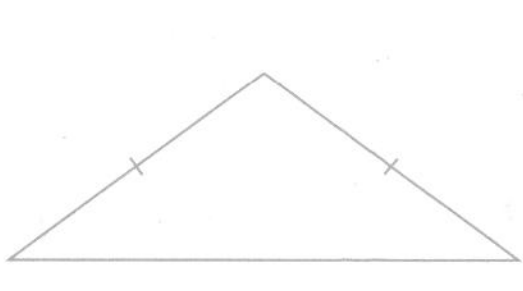

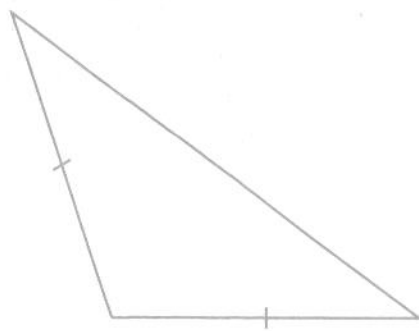

Figure 8.60

Both are obtuse isosceles triangles. The orientation on the left is the standard orientation for isosceles triangles. As stated at the beginning of this chapter, students often see only one aspect of a triangle; for example, they see the triangle at the left as isosceles but not also obtuse, and they see the triangle at the right as obtuse but not also isosceles.

CLASSROOM CONNECTION

Children's development of triangles is fascinating. In several studies, children were given many shapes and asked to identify them. Young children tend to identify the equilateral triangle in the "standard" position (one side parallel to the bottom of the page) as a "true" triangle. They will often reject other triangles because they are too pointy or turned upside down. Recall level 1 in the van Hiele model. One of my favorite examples occurred when a first-grader was given the pattern shown in Figure 8.61 and was asked to continue the pattern.

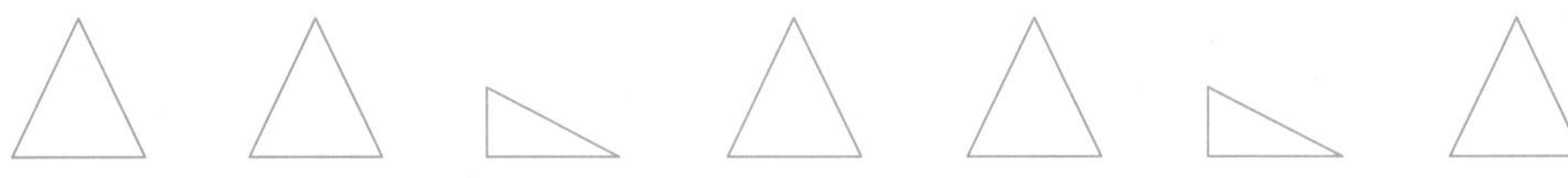

Figure 8.61

After studying the pattern, she said, "Triangle, triangle, wrong triangle, triangle, triangle, wrong triangle, triangle. . . . The next shape is a right triangle!"[3] It is important to note that this child was doing wonderful thinking—she was seeing patterns, *and* she was looking at attributes, and she was at the beginning of her understanding of triangles.

INVESTIGATION 8.2f Triangles and Venn Diagrams

Explorations Manual 8.13

Recall our work with Venn diagrams in Chapter 2. Venn diagrams can help us to understand how concepts are and are not related. Let us take two kinds of triangles: right triangles and isosceles triangles. Draw several right triangles and draw several isosceles triangles. Which of these triangles could be placed in both categories? That is, which are right triangles and also isosceles triangles? Make a Venn diagram that illustrates the relationships between the right and isosceles triangles, and place each triangle in the appropriate region of the diagram.

Now draw several acute triangles and draw several equilateral triangles. Which of these triangles could be placed in both categories? How are these triangles related to each other? Make a Venn diagram that illustrates the relationships between the acute and equilateral triangles, and place each triangle in the appropriate region of the diagram.

DISCUSSION

When we have two groups of objects and look at how they are related, there are three possible relationships. There may be overlap, as some objects are in both groups; as in the diagram on the left in Figure 8.62. They may be disjoint; that is, each object is in one group or the other but cannot be in both, as in the diagram in the middle in Figure 8.62. Finally, one group may be a subset of the other group, as in the diagram to the right. Each of these relationships (remember the van Hiele levels) is represented with a different Venn diagram, as shown in Figure 8.62.

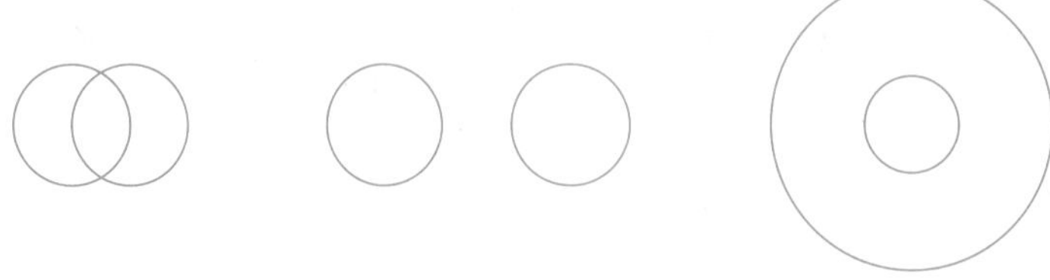

Figure 8.62

Look at your Venn diagrams and think of what you just read. Do you want to change your diagrams? If you couldn't make the Venn diagrams because you just didn't understand the question, can you do so now? Do this before reading on. . . .

Because some triangles are both right and isosceles, those triangles can be placed in the center, showing that they belong to both sets (Figure 8.63). Because equilateral triangles can contain only acute angles, all equilateral triangles are acute triangles. Hence the equilateral triangles are in a ring that is inside the ring that represents acute triangles.

MATHEMATICS

Interestingly, all right isosceles triangles have the same shape. That is, if you found the ratio of lengths of the sides of two right isosceles triangles, converted this to a percent, placed the larger one on a copy machine, and entered that percent in the reduction button, the smaller triangle would come out. We will explore this notion of similarity in Section 10.3.

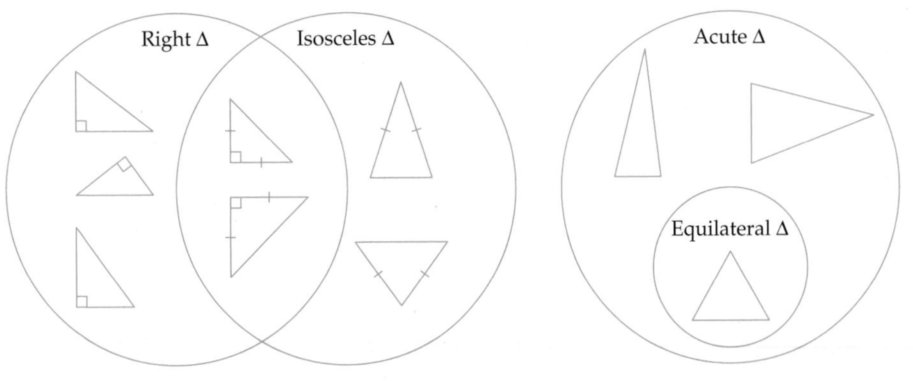

Figure 8.63

SPECIAL LINE SEGMENTS IN TRIANGLES

There are four special lines and line segments that have enjoyed tremendous influence in Euclidean geometry: angle bisector, median, altitude, and perpendicular bisector.

A **median** is a line segment that connects a vertex to the midpoint of the opposite side.

In $\triangle SAT$ (Figure 8.64), $\overline{TR}$ is a median. Hence $\overline{SR} \cong \overline{RA}$.

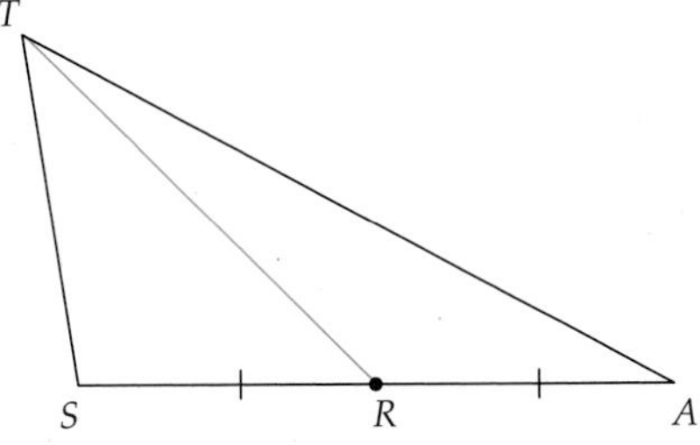

Figure 8.64

An **angle bisector** is a line segment that bisects an angle of a triangle.

In $\triangle ABC$ (Figure 8.65), $\overline{AD}$ is an angle bisector. Hence, $m\measuredangle BAD = m\measuredangle DAC$.

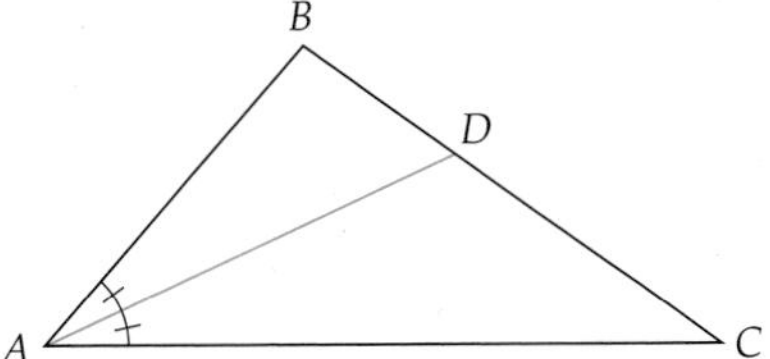

Figure 8.65

A **perpendicular bisector** is a line that goes through the midpoint of a side and is perpendicular to that side. In $\triangle PEN$ (Figure 8.66), $\overleftrightarrow{MX}$ is a perpendicular bisector of side $\overline{PN}$, because M is the midpoint of $\overline{PN}$ and $\overline{MX}$ is perpendicular to $\overline{PN}$.

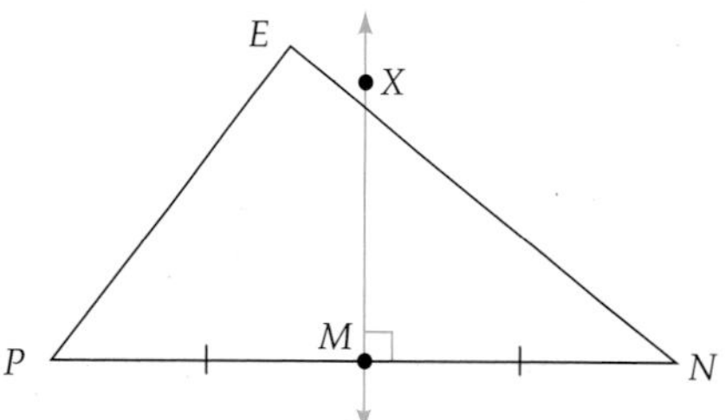

Figure 8.66

An **altitude** is a perpendicular line segment that connects a vertex to a line containing the side opposite that vertex. In some cases, as in $\triangle SAT$ in Figure 8.67, we need to extend the opposite side to construct the altitude.

In $\triangle ABC$ (Figure 8.67), $\overline{BF}$ is an altitude. Hence, $m\measuredangle BFA = m\measuredangle BFC = 90°$.

In $\triangle STA$, $\overline{TP}$ is an altitude. Hence, $m\measuredangle TPA = 90°$.

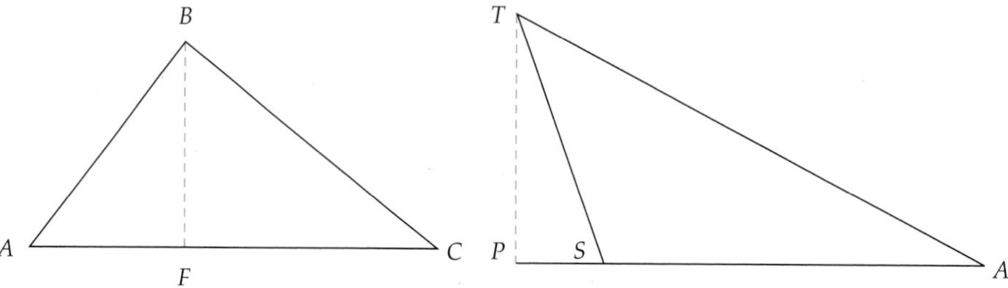

Figure 8.67

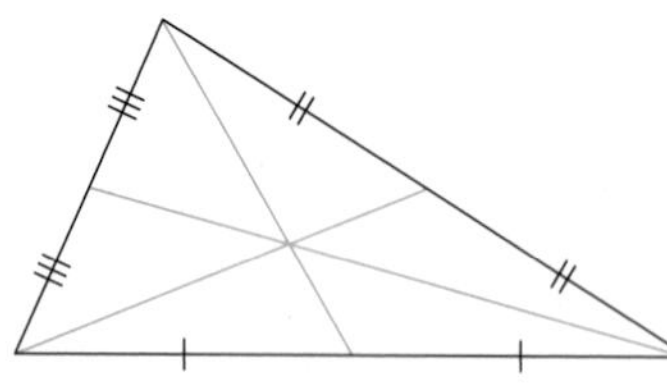

Figure 8.68

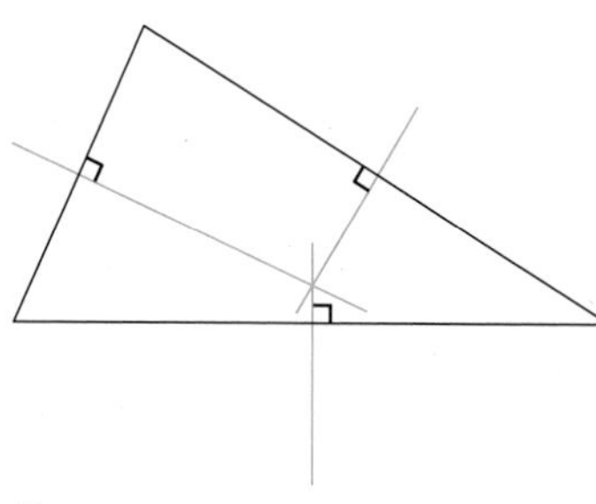

Figure 8.69

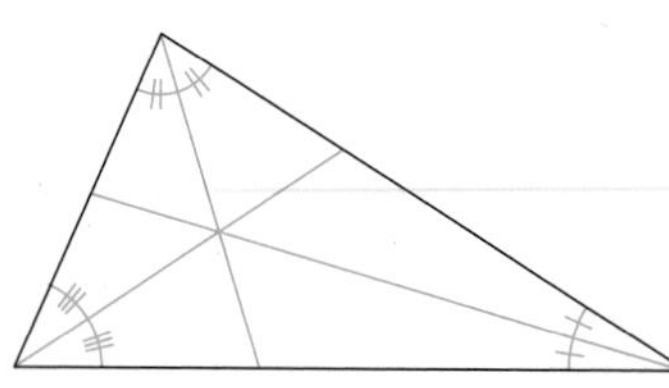

Figure 8.70

Many students have trouble with the idea of an altitude being outside the triangle. This happens when we have an obtuse triangle oriented with one side of the obtuse angle as the base of the triangle. If you are having trouble connecting the definition of *altitude* in these situations, I recommend the following: Trace the triangle and cut it out. Stand it up so that $\overline{SA}$ is on the plane of your desk and T is above that plane. Now draw a line from T that goes "straight down." What do you notice?

If $\triangle STA$ were large enough so that you could stand with your head at point T, the length of line segment $\overline{TP}$ would tell you how tall you were!

The segment with the most practical value is the median. The point where all three medians meet is called the **centroid** and is the center of gravity of the triangle (Figure 8.68). That is, if you found this point, made a copy of the triangle with cardboard, and placed the triangle on a nail at that point, the triangle would balance. Center of gravity is related to balance and is an important concept in the design of many objects—cars, furniture, and art, to mention just a few.

The point where the three perpendicular bisectors meet is called the **circumcenter** (Figure 8.69). It turns out that this point is equidistant from each of the three *vertices* of the triangle. Thus you can draw a circle centered at the circumcenter so that all three vertices of the triangle lie on the circle and the rest of the triangle is inside the circle. We say that the circle **circumscribes** the triangle.

The point where the three angle bisectors meet is called the **incenter** (Figure 8.70). It turns out that this point is equidistant from each of the three *sides* of the triangle. Thus you can draw a circle centered at the incenter so that the circle touches (is tangent to) each of the three sides of the triangle; in this case, the circle is inside the triangle. We say that the circle is **inscribed** in the triangle.

The point where the three altitudes meet is called the **orthocenter** (Figure 8.71). The orthocenter has connections to other geometric ideas. For example, there is a connection between the construction of a parabola and the orthocenter of a triangle.

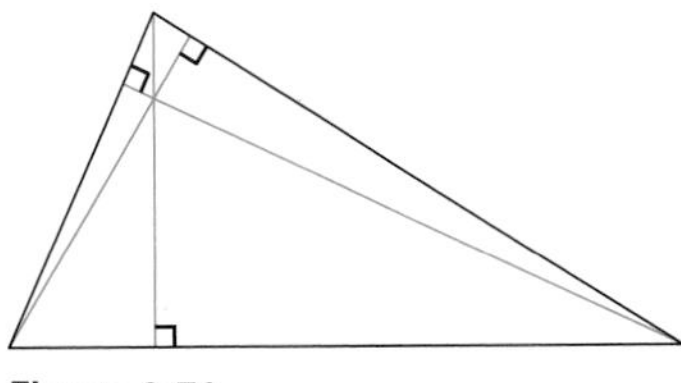

Figure 8.71

There are two amazing things about these lines and line segments. The first is that in any triangle there are three medians, three perpendicular bisectors, three angle bisectors, and three altitudes. In each case, the three line segments will *always* meet at a single point—they are concurrent. That is, the three medians will always meet at one point, the three perpendicular bisectors will always meet at one point, and so on. Exploring this with Geogebra will be left as an exercise.

The second amazing thing is that, except for the case of the equilateral triangle, these points are *not* the same point. In fact, in a scalene triangle, the four points are all different. However (and this is the other cool thing), the centroid, circumcenter, and the orthocenter are always collinear! Figure 8.72 illustrates this fact, which Leonard Euler first discovered. We even call the line containing these three points an *Euler line*.

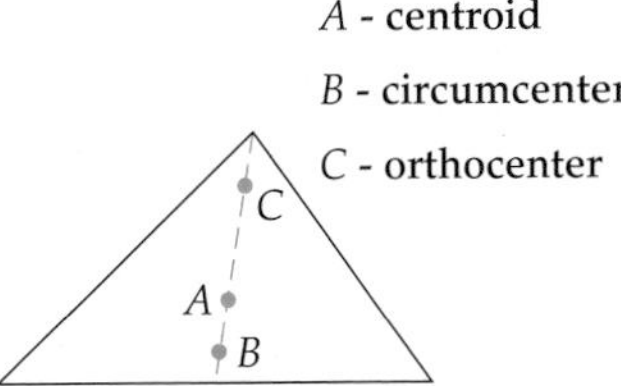

Figure 8.72

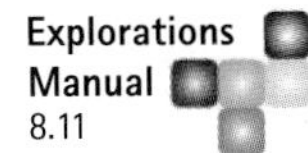

Explorations Manual 8.11

CONGRUENCE

As you have seen from your geometry explorations, questions sometimes arise about whether two figures are "the same" or not. Such observations and questions deal with the idea of congruence.

At an informal level, we can say that two figures are congruent if they have the same shape and size. An informal test of congruence is to see whether you can superimpose one figure on top of the other. This is closely connected to how children initially encounter the concept and is related to the dictionary definition: "coinciding exactly when superimposed."[4] That is, if one figure can be superimposed over another so that it fits perfectly, then the two figures are congruent.

Formally, we say that two polygons are **congruent** if all pairs of corresponding parts are congruent. In other words, in order for us to conclude that two polygons are congruent, two conditions have to be met: (1) each corresponding pair of angles have the same measure, and (2) each corresponding pair of sides have the same length. We use the symbol $\cong$ to denote congruence.

For example, in Figure 8.73, $\triangle CAT$ and $\triangle DOG$ are congruent if $\measuredangle C \cong \measuredangle D$, $\measuredangle A \cong \measuredangle O$, $\measuredangle T \cong \measuredangle G$, $\overline{CA} \cong \overline{DO}$, $\overline{AT} \cong \overline{OG}$, and $\overline{TC} \cong \overline{GD}$.

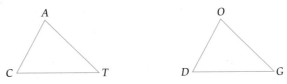

Figure 8.73

The notions of congruent and equal are related concepts. We use the term *congruence* when referring to having the same shape, and we use the term *equal* when referring to having the same numerical value. Thus we do not say that two triangles are equal; we say that they are congruent. Similarly, when we look at line segments and angles of polygons, we speak of congruent line segments and congruent angles. However, when we look at the numerical value of the line segments and angles, we say that the lengths of two line segments are equal and that the measures of two angles are equal.

Congruence is a big idea, both in geometry and beyond the walls of the classroom. Many important properties and relationships come from exploring congruence. The following investigations (and the related explorations) will help move your understanding of congruence to higher van Hiele levels, from the "can fit on top," geometric reasoning by resemblance, to understanding that all corresponding parts are congruent, to geometric reasoning by attributes, to geometric reasoning by properties, which we shall examine now.

OUTSIDE THE CLASSROOM

When do we need congruence in everyday life or in work situations? Take a few minutes to think about this before reading on. . . .

Congruence is important in manufacturing; for example, the success of assembly-line production depends on being able to produce parts that are congruent. Henry Ford changed our world by conceiving of making cars not one at a time but as many sets of congruent parts. For example, the left front fender of a 2003 Dodge Caravan is congruent to the left front fender of any other 2003 Dodge Caravan. One of the differences between a decent quilt and an excellent one is being able to ensure that all the squares are congruent. This is quite difficult when using complex designs. Most of the manipulatives teachers use with schoolchildren (Pattern Blocks, unifix cubes, Cuisenaire rods, and fraction bars) have congruent sets of pieces.

INVESTIGATION 8.2g Congruence with Triangles

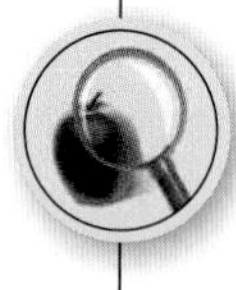

For this you need a protractor and a compass, although you can improvise without them. Do both questions on a blank sheet of paper. First, see how many different triangles you can make that have the following attributes: The base is 50 millimeters (mm). The angle coming from the left side of the base is 30 degrees, and the side coming from the left side of the base is 30 mm. Second, see how many different triangles you can make that have the following attributes: The base is 38 mm. The angle coming from the left side of the base is 30 degrees, and the side coming from the right side of the base is 25 mm. Do this before reading on. . . .

DISCUSSION

There is only one triangle that can be drawn in the first case. However, in the second case, there are two possible triangles. Thus there is not enough information about the triangle to specify exactly one triangle (Figure 8.74).

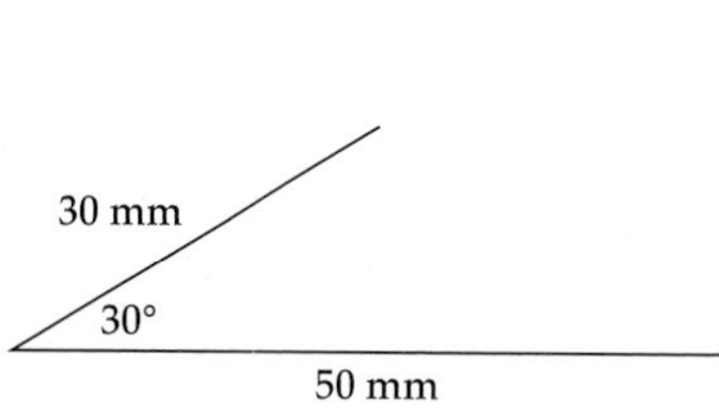

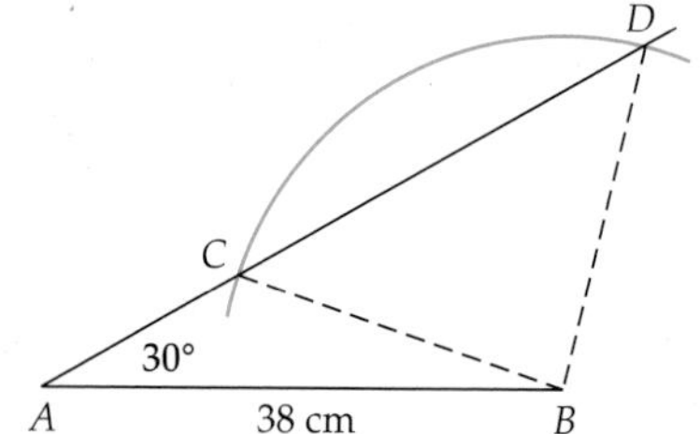

Figure 8.74

In Section 8.1, we said that two points *determine a* line; here, we are looking at what determines a triangle. This notion of *determining* or *specifying* is important in mathematics, both for congruence (When are two figures congruent?) and for definitions (How much do we need to specify to determine a shape?). For example, we can define a rectangle as a quadrilateral with four congruent angles. Now a rectangle has many more properties than four congruent angles. However, mathematicians have discovered that this information—quadrilateral, four congruent angles—is sufficient so that only rectangles can be drawn that meet that criteria. Thus the notion of "determines" is an important one in mathematics. In elementary school we do not get terribly technical, but that is not the same as saying we just have fun and play around. When we ask children to explore well-focused questions, their understanding of shapes *and* relationships between and among shapes *and* their ability to see and apply properties can grow tremendously. When this happens, high school mathematics makes much more sense!

QUADRILATERALS

We found that we could describe different kinds of triangles by looking at their angles or by looking at relationships among their sides. With quadrilaterals, which have one more side, new possibilities for categorization emerge: parallel sides, adjacent vs. opposite sides, relationships between diagonals, and the notion of concave and convex. Thus, how we go about naming and classifying quadrilaterals is not the same as how we name and classify triangles.

In this book, we will define the following kinds of quadrilaterals:

- A **trapezoid** (Figure 8.75) is defined as a quadrilateral with at least one pair of parallel sides.
- A **parallelogram** (Figure 8.76) is defined as a quadrilateral in which both pairs of opposite sides are parallel.
- A **kite** (Figure 8.77) is defined as a quadrilateral in which two pairs of adjacent sides are congruent.
- A **rhombus** (Figure 8.78) is defined as a quadrilateral in which all sides are congruent.
- A **rectangle** (Figure 8.79) is defined as a quadrilateral in which all angles are congruent.
- A **square** (Figure 8.80) is defined as a quadrilateral in which all four sides are congruent and all four angles are congruent.

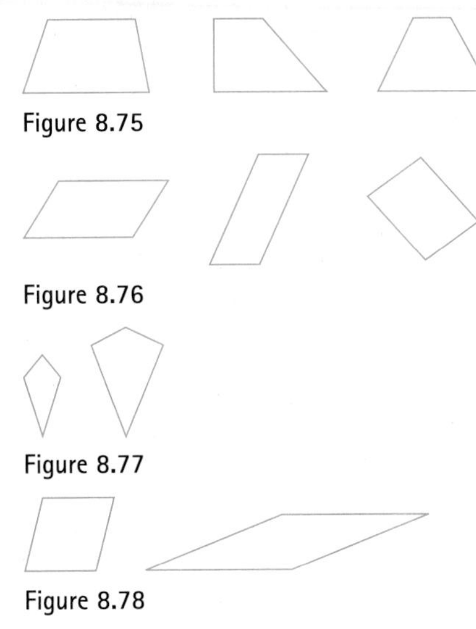

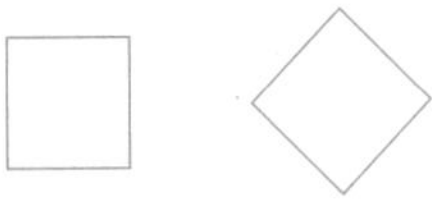

Figure 8.75

Figure 8.76

Figure 8.77

Figure 8.78

Figure 8.79

Figure 8.80

LANGUAGE

Some books define a trapezoid as having exactly one pair of parallel sides. I have spoken with many mathematicians about this particular definition—some prefer the definition used here and some prefer the other one. There is no *right* answer, but rather reasons behind each one. There are two reasons why I use the definition given here. First, I would find it odd to define an isosceles triangle as having at *least* two congruent sides but then a trapezoid as having exactly one set of parallel sides. If we are going to say a trapezoid has exactly one pair of parallel sides, then an isosceles triangle should have exactly two congruent sides. Second, defining *trapezoid* this way enables us to represent relationships among quadrilaterals in a more elegant way.

An active reader may have noted that there are many other possible categories of quadrilaterals. For example, there are many different quadrilaterals that have at least one right angle or exactly three congruent sides. Just as we called the triangle with no sides congruent a scalene triangle, we could call a quadrilateral with no sides congruent a scalene quadrilateral, although this is not commonly done. Following are some examples of scalene quadrilaterals.

OUTSIDE THE CLASSROOM

How many of each kind of quadrilateral can you find in everyday life?

CLASSROOM  CONNECTION

Squares and Rectangles

How many properties can you describe that are true for both rectangles and parallelograms? 

Name Date

Measuring Polygons

Homework

Squares and Rectangles

NOTE Students consider ways in which two types of quadrilaterals, squares and rectangles, are related to each other.

SMH 96–98

1. Write as many statements as you can about this square.

2. Write as many statements as you can about this rectangle.

3. Explain why some statements are on both of your lists.

4. Explain why some statements are on only one of your lists.

12 Unit 5 Session 1.3

Diagonals One characteristic of all polygons with more than three sides is that they have diagonals. The more sides in the polygon, the more diagonals. This term is probably familiar to most readers. However, before reading the definition of *diagonal* below, stop and try to define the term yourself so that it works for all polygons, not just squares and other quadrilaterals. Then read on. . . .

A **diagonal** is a line segment that joins two nonadjacent vertices in a polygon.

Figure 8.81 shows two different diagonals. One of the exercises will ask you to find patterns to determine the number of diagonals in any polygon.

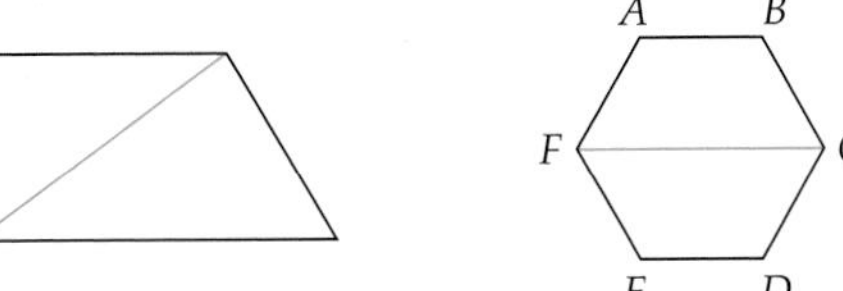

Figure 8.81

LANGUAGE

A few years ago, some students were stumbling over the question of how many diagonals a hexagon had. They contended that $\overline{FC}$ in Figure 8.81 was not a diagonal of the hexagon at the right. Their argument was that it was a horizontal line and that diagonal meant going at a slant. At first, I was stunned. But then I realized that in everyday language, when we use the word diagonal, we do mean going at a slant. Brain research tells us that new information is almost always connected to something we know. The first diagonals we consider are those of a quadrilateral, and those diagonals are almost always shown in a "diagonal" direction.

CONVEX POLYGONS

Another concept that emerges with polygons having four or more sides is the idea of convex. Before reading on, look at the two sets of polygons in Figure 8.82, convex and concave (not convex). Try to write a definition for *convex*. Then read on. . . .

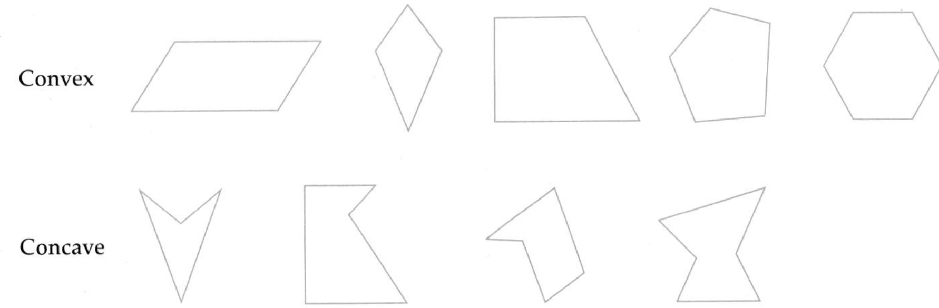

Figure 8.82

LANGUAGE

Your instructor might have had you explore the definition of *convex* in Exploration 8.9. You may be surprised to know that there are multiple ways to view this idea. A common alternative to the definition given at the right is that a polygon is convex if and only if it has no reflex angles. First, make sure you understand the definition. Do you see the equivalence of the two definitions? Do you like one better than the other? Once again, mathematics is not as cut and dried as many people believe!

For many students, defining convex is much like defining balls and strikes in baseball. As one umpire once said, "I knows it when I sees it." Many students focus on the word *concave* and say that the figure has at least one part that is caved in. Such a description is acceptable at level 1 on the van Hiele model, but we need a definition that is not as vague as "caved in."

Examine the following definition to see whether it makes sense. Think before reading on. . . .

A polygon is **convex** if and only if the line segment connecting *any* two points in the polygonal region lies entirely within the region.

If a polygon is not convex, it is called **concave** or nonconvex.

Looking at diagonals is an easy way to test for concave and convex. If any diagonal lies outside the region, then the polygon is concave. In polygon *ABCDE* in Figure 8.83, the diagonal *AD* lies outside the region.

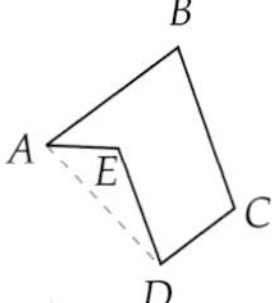

Figure 8.83

INVESTIGATION 8.2h Quadrilaterals and Attributes

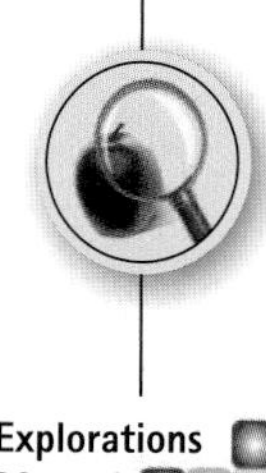

Look at the three quadrilaterals below. Think about their various attributes. Recall Investigation 8.2b. Now answer the following question: Which two of the quadrilaterals in Figure 8.84 are most alike and why? Do your thinking and write your response before reading on. . . .

Explorations Manual 8.12

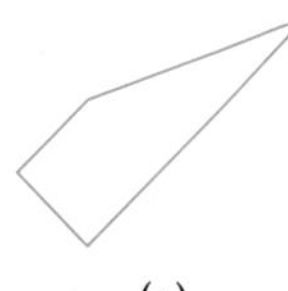

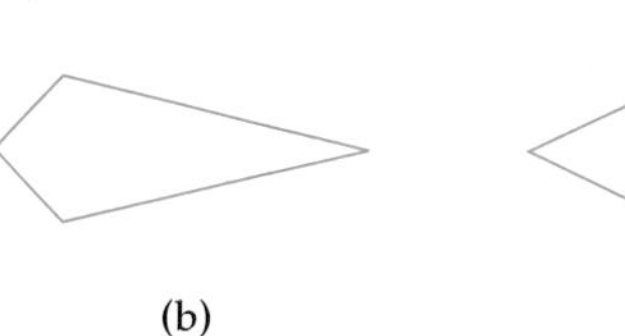

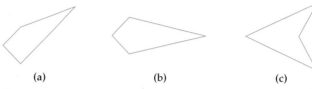

(a) (b) (c)

Figure 8.84

DISCUSSION

It is questions like this that help children, and adults, to move up the van Hiele levels. A good case can be made for different answers. Let us begin by simply listing various attributes.

a.	b.	c.
1 pair of ≅ sides	2 pairs of ≅ sides	2 pairs of ≅ sides
1 pair parallel sides	0 parallel sides	0 parallel sides
2 right angles	1 right angle	0 right angles
1 obtuse angle	2 obtuse angles	0 obtuse angles
1 acute angle	1 acute angle	3 acute angles
0 reflex angles	0 reflex angles	1 reflex angle
convex	convex	concave

This list reflects various attributes: congruence, parallel, angle, and shape (convex and concave). On the one hand, the two figures at the right are both kites and therefore "belong" together under that name. On the other hand, the two figures on the left are both convex and both have right angles, so there is much in common between the two of them also.

One of the most puzzling aspects of how mathematics has generally been taught is how much of it is simply learning and reciting facts and theorems that other people have learned. We ask our students to study mathematics, but we rarely let them do mathematics. If we were to teach art this way, students would learn techniques and be tested on how well they understood those techniques, but they would never get to do art. Most of the various groups advocating change in how mathematics is learned want to present students with problems where solving the problem means not simply applying an algorithm they have learned but, rather, involves what we call mathematical thinking. The investigation that follows has this flavor of doing mathematics.

INVESTIGATION 8.2i Challenges

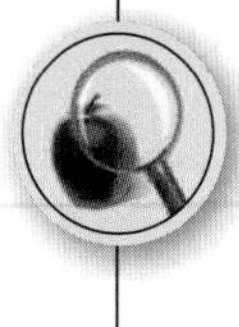

This investigation brings together the notion of attributes and the notion of determinism (e.g., two points determine a line), which are two of the big ideas of geometric thinking. How many different kinds of quadrilaterals can you make that have exactly two adjacent right angles? Play around with this for a while on a piece of paper. Sketch different quadrilaterals that have exactly two adjacent right angles. What do you see? Can you make any conjectures? Can you prove them? Use whatever tools are available. As you do, try to push yourself beyond random trial and error to being more systematic, to thinking "What would happen if," to looking at your solutions to see what they have in common. . . .

DISCUSSION

Figure 8.85 shows three of many possibilities.

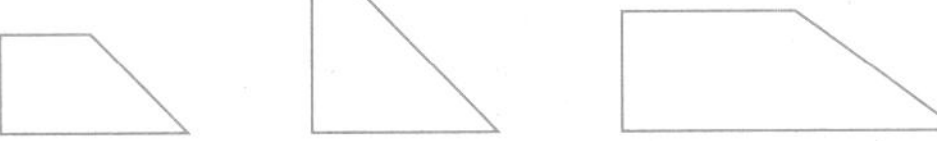

Figure 8.85

What do they all have in common besides two adjacent right angles? Think before reading on. . . .

They all have two parallel sides. That means they are trapezoids. A curious reader might now be asking whether it is possible not to get a trapezoid. What do you think? How might you proceed, rather than just using random trial and error? Think before reading on. . . .

Figure 8.86

If you took high school geometry, you might be remembering a theorem that said something like this: If two lines form supplementary interior angles on the same side of a transversal, then the lines are parallel. When we limit our investigation to two adjacent right angles, we make two interior angles that are also supplementary. Thus we know that the opposite sides of this quadrilateral must be parallel. Hence the condition of adjacent right angles *determines* a trapezoid. Although we can vary the height of the figure and the lengths of the opposite sides, we can get only trapezoids. Figure 8.86 illustrates this.

What if you make a quadrilateral where the two right angles are not adjacent to each other, but opposite each other? Is it possible to have a concave quadrilateral with two right angles? These questions will be left as exercises.

We used Venn diagrams to deepen our understanding of triangles in Investigation 8.2f. We will do so again with quadrilaterals.

INVESTIGATION 8.2j Relationships Among Quadrilaterals

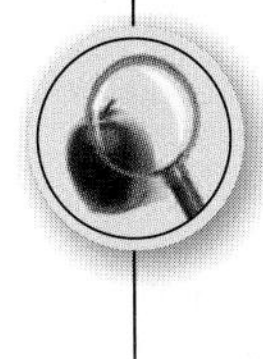

Explorations Manual 8.13

Consider the Venn diagram and set of quadrilaterals shown in Figure 8.87. What attributes do all of the quadrilaterals in the left ring have? What attributes do all of the quadrilaterals in the right ring have? By the nature of Venn diagrams, the quadrilaterals in the middle section have attributes of both right and left.

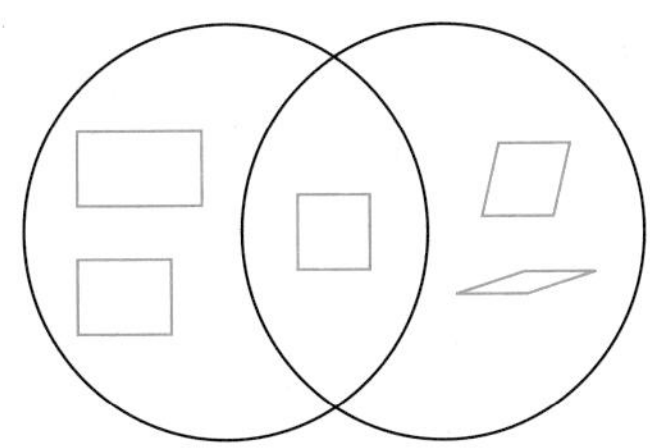

Figure 8.87

CLASSROOM CONNECTION

Classifying is one of the big ideas of mathematics and is found at every level of instruction. It begins with attribute blocks and materials with preschool children. Materials such as buttons can be sorted, using more than one attribute, in many different ways (the shape of the button, the number of holes, the color, and so on). Plastic rings can be used to create Venn diagrams.

DISCUSSION

There are a number of ways to answer the question. Let us begin at the most descriptive level.

In the left ring, all of the shapes have four right angles, they all have opposite sides congruent, and they all have opposite sides parallel. Some students state it slightly differently, saying that the shapes all have two pairs of congruent sides and two pairs of parallel sides.

In the right ring, all the shapes also have opposite sides congruent and opposite sides parallel. Another attribute they possess is that all four sides are congruent. The shape that is in the center, by definition, must have the attributes of both rings—four right angles and four congruent sides.

There is a name for figures that are in both rings—square.

There is a name for figures that are in the left ring—rectangle.

There is a name for figures that are in the right ring—rhombus.

Using set language from Chapter 2, we say that squares are the intersection of rectangles and rhombuses (also called rhombi). Some students will have noticed that all of these shapes are parallelograms. Thus we could actually add another ring encircling all the shapes in Figure 8.87. That is, all rectangles are parallelograms, all rhombuses are parallelograms, and all squares are parallelograms.

RELATIONSHIPS AMONG QUADRILATERALS

It turns out that we can view the set of quadrilaterals in much the same way we view a family tree showing the various ways in which individuals are related to others. Figure 8.88 shows one of many ways to represent this family tree for quadrilaterals. Take a few moments to think about this diagram and to connect it to what you know about these different kinds of quadrilaterals. Write a brief description. Does it make sense? Does it prompt new discoveries in your mind?

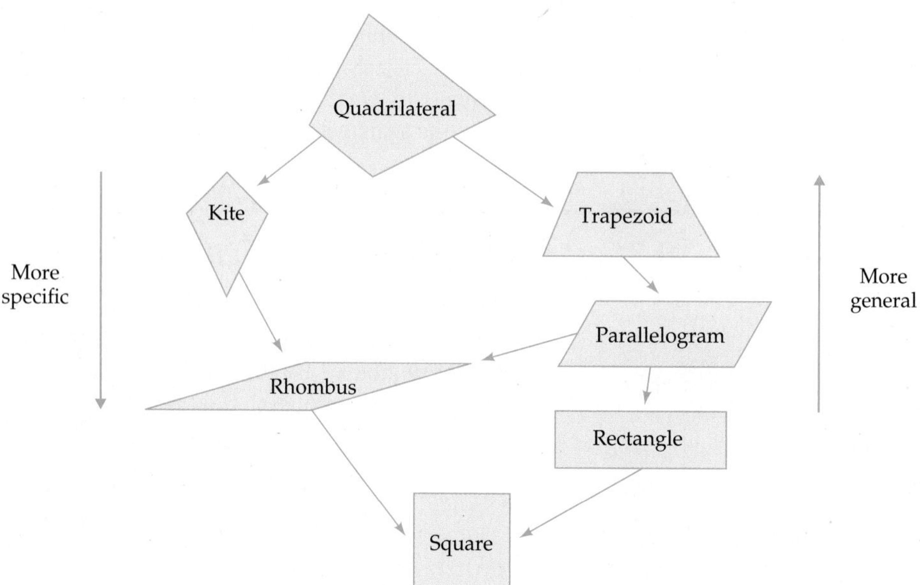

Figure 8.88

> **MATHEMATICS**
>
> Some people are confused with the mathematical statement that a square is a rectangle. However, it is really not that weird. Consider this question: Are humans primates? Are humans mammals? If you have had any biology, the answer is, "Of course!" Same thing. We can go from general (polygon) to a subset (quadrilateral) to a yet smaller subset (rectangle) to an even smaller subset (square). Similarly, we can go from general (animal) to a subset (mammal) to a smaller subset (primate) to a yet smaller subset (human).

One way to interpret this diagram is to say that any figure contains all of the properties and characteristics of the ones above it. The quadrilateral at the top represents those quadrilaterals that have no equal sides, no equal angles, and no parallel sides; this is analogous to the scalene triangle. The kite and the trapezoid represent two constraints that we can make: two pairs of congruent, adjacent sides or one pair of sides parallel. If we take a kite and require all four sides to be congruent, we have a rhombus. If we take a trapezoid and require both pairs of opposite sides to be parallel, we have a parallelogram. If we require the angles in a parallelogram to be right angles, we have a rectangle. If we require all four sides of a parallelogram to be congruent, we have a rhombus. Both the rhombus and the rectangle can be transformed into squares with one modification—requiring the

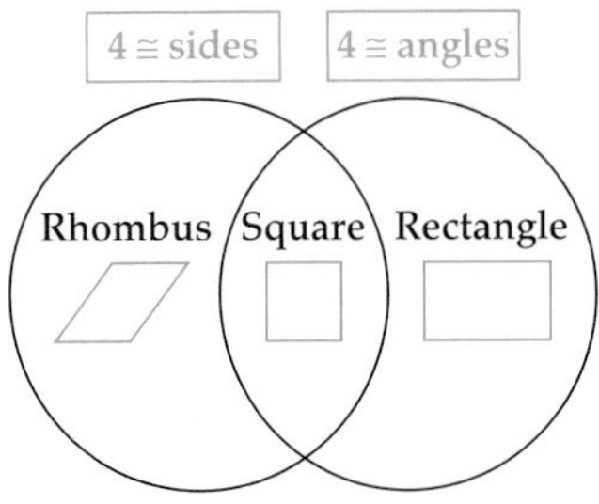

Figure 8.89

rhombus to have right angles or the rectangle to have congruent sides. A key point is to begin to see connections and relationships among figures. Many students find, in this course, that their picture of geometry changes from looking like a list of definitions and properties to looking more like a network with connections among the various figures. This quadrilateral family tree can help students to realize why mathematics teachers say that a square is a rectangle *and* it is a rhombus: It has all the properties of each! In everyday language, we say that a square is a special kind of rhombus and a special kind of rectangle. In mathematical language, we say that the set of squares is a subset of the set of rhombuses and a subset of the set of rectangles. The Venn diagram in Figure 8.89 illustrates this relationship.

OTHER POLYGONS

Although most of the polygons we encounter in everyday life are triangles and quadrilaterals, there are many kinds of polygons with more than four sides. Stop for a moment and think of examples, both natural and human-made objects. Then read on. . . .

All of the figures in Figure 8.90 are polygons.

- The stop sign is an octagon—an eight-sided polygon.
- The common nut has a hexagonal shape—a six-sided polygon.
- The Pentagon in Washington has five sides.

Let us examine a few important aspects of polygons with more than four sides.

First, we distinguish between regular and nonregular polygons. What do you think a regular pentagon or a regular hexagon is? How might we define it? Think about this and write down your thoughts before reading on. . . .

©Frontpage/Shutterstock.com

©Szasz-Fabian Jozsef/Shutterstock.com

©Frontpage/Shutterstock.com

Figure 8.90

A **regular polygon** is one in which all sides have the same length and all interior angles have the same measure.

What do we call a regular quadrilateral? What about a regular triangle?

Is it possible for a regular polygon to be concave?

Think about these questions before reading on. . . . 

Grade 4

What are your answers for Question 3? 

Date ______ Time ______

LESSON 3•1

A Polygon Alphabet

SRB 96 97

Try reading this message:

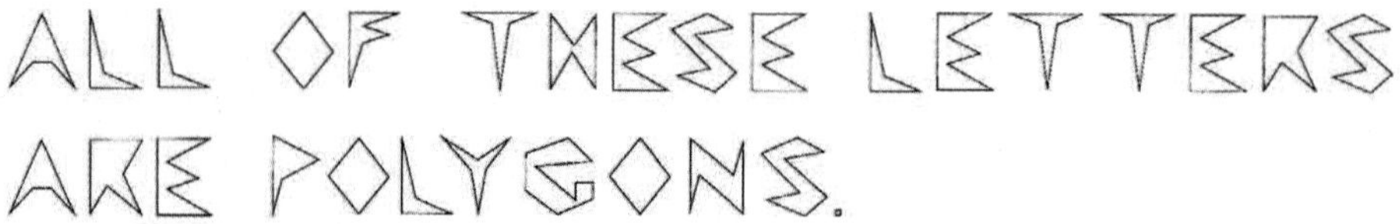

ALL OF THESE LETTERS ARE POLYGONS.

1. Use a straightedge to design a polygon letter for each of the letters shown below. You'll have to simplify, because a polygon can't have any curves, and it can't have any "holes."

   For example, if you look at the letter "P," you see that there is no opening in the upper part. Making it look like this, ⊳, would make it easier to read, but it would not be a polygon.

B	C	D
F	M	X

2. Which of the letters you drew are nonconvex (concave) polygons? ______
   How do you know?
   ______

3. Do any of the letters you drew have special names as polygons? Explain.
   ______
   ______

Try This

4. On a separate sheet of paper, design polygon letters for the rest of the uppercase (capital) letters in the alphabet, the 26 lowercase (small) letters, or the 10 digits (0–9).

54

From *Everyday Mathematics, Grade 4:* The University of Chicago School Mathematics Project: Student Math Journal, Volume 1, by Max Bell et al., Lesson 3-1, p. 54. Reprinted by permission of The McGraw-Hill Companies, Inc.

A regular quadrilateral is called a square. A regular triangle is called an equilateral triangle. A regular polygon cannot be concave.

A critical reader might be wondering whether you can have a polygon—let's say a pentagon—where all the sides have the same length but not all the angles have the same measure. And what about the converse: Can you have a pentagon where all the angles have the same measure but not all the sides have the same length? What do you think? This will be left as an exercise.

INVESTIGATION 8.2k Summing Triangle Angles and Those of Other Polygons

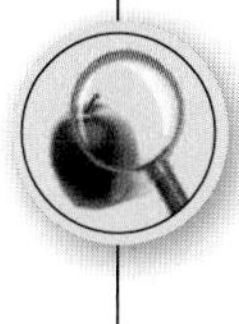

Let's explore an interesting concept with polygons. Just by knowing the number of angles that a polygon has, we can know the answer to finding the sum when we add all of the measures of the angles of that polygon. Let's see how.

We will get started together, and then we will invite you to take some time to see the pattern and develop the formula on your own. You can choose whether to follow using paper and pencil method below, or the Geogebra (free software) method. Initially, use the one that is most aligned with your learning style. We also encourage you to work through both tools, as understanding both will support you in understanding more learning styles of your future students.

A. What is the sum of the degrees of the angles of a triangle? Or does it depend on the type of triangle?

DISCUSSION

STRATEGY 1 Using paper and pencil

Draw one of each of the following types of triangles. Draw them large (i.e., on at least half a sheet of paper) and unlined paper works best.

Acute scalene

Obtuse isosceles

Equilateral (which also means acute)

Right scalene

Take the right scalene triangle and tear the angles off as shown in Figure 8.91.

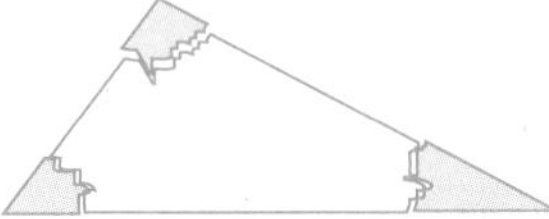

Figure 8.91

Take these three angles and put the three vertex points together, and have each angle adjacent to another angle, as shown in Figure 8.92.

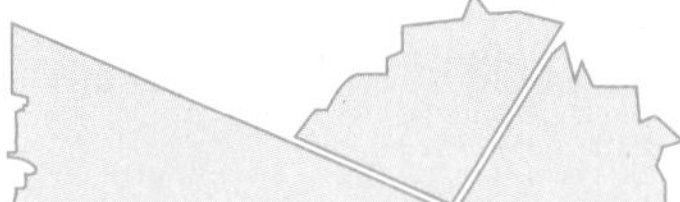

Figure 8.92

Repeat this process with the other three triangles. What happens with each triangle when you put the three angles together this way?

Just as in Figure 8.92, each time you put the three angles of these four very different triangles together like that, they always form a straight line. Because a straight line comprises 180 degrees, the angles of a triangle will always add to 180 degrees.

STRATEGY 2 Using Technology

Geogebra is a free math exploration software. First, go to geogebra.org and download the free app. When you open Geogebra, click on the polygon icon (looks like a triangle, the fifth one from the left) on the tool bar. Draw a triangle of any kind. Then, click on the eighth icon from the left, the angle icon, and choose "angle." Now when you click inside the triangle, it will measure all of the angles. Verify that they add to 180 degrees. This is a great place to practice your estimation and mental math skills. Move a vertex of the triangle to create each of the types of triangles listed previously, and add the angles. Repeat this several times.

B. What is the sum of the degrees of the angles in other polygons, such as a quadrilateral or pentagon?

DISCUSSION

With the knowledge that the sum of the measures of the angles of a triangle is always 180 degrees, we can determine the sum of the measures of the angles of any polygon.

The paper and pencil strategy and the Geogebra strategy are similar here. Some people prefer to draw figures on paper, some on the computer. Geogebra does have the added benefit of measuring the angles so you can support your conjecture.

Draw a quadrilateral, pentagon, and hexagon either with pencil, or on Geogebra. To do this on Geogebra, you choose the polygon icon again (the one that looks like a triangle). Now, our task is to turn these into triangles, without creating new angles.

In the quadrilateral on the left, new angles are created in the center. Drawing all the diagonals possible from a single vertex is the way to avoid creating new angles. In the quadrilateral on the right, the angles of the two triangles would add to the same as the angles of the quadrilateral.

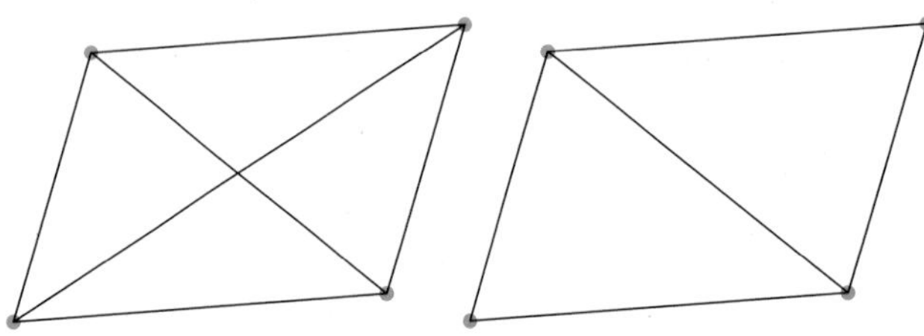

1. With your quadrilateral, draw a diagonal from one vertex to the opposite one. How many triangles do you now have? What is the sum of the angles?
2. With your pentagon, pick one vertex and draw all the diagonals possible from that one vertex (should be two). How many triangles are there? What is the sum of the angles?
3. Now, do the same with the hexagon. What is the sum? If you are doing this on Geogebra, measure and add the angles to check. What is your conjecture of how we find the sum of the angles of any polygon?

Here is the picture of the pentagon and hexagon.

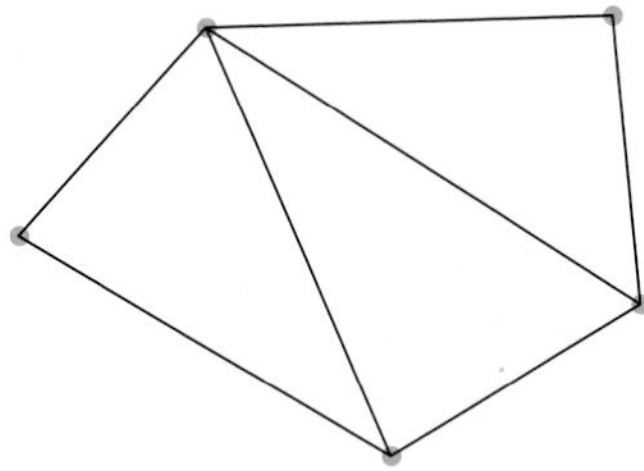

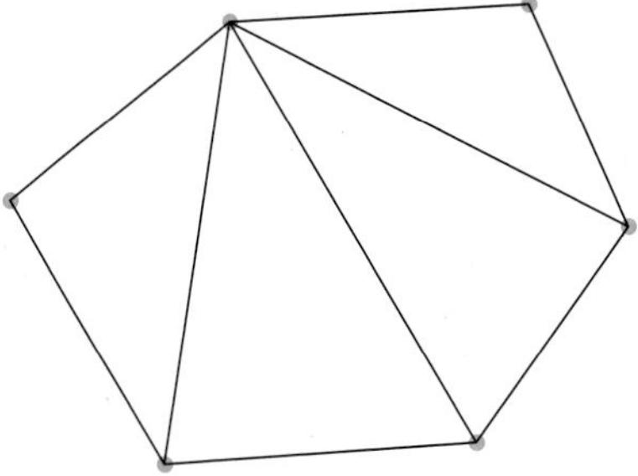

Appearing below is a table showing what happens. What patterns do you see?

Name of Polygon	Number of Angles	Number of Triangles Created	Sum of Angles
Triangle	3	1	1(180) = 180
Quadrilateral	4	2	2(180) = 360
Pentagon	5	3	3(180) = 540
Hexagon	6	4	4(180) = 720
Heptagon	7	5	5(180) = 900
Octagon	8	6	6(180) = 1080
Nonagon	9	7	7(180) = 1260
Decagon	10	8	8(180) = 1440
n-gon	n	$(n - 2)$	$(n - 2)\ 180$

In each case, there are two less triangles than there are number of angles in the polygon. The general form of this, with n sides, is $n - 2$. Since each triangle has a total of 180 degrees, we multiply the number of triangles by 180.

CURVED FIGURES

There is one more class of two-dimensional geometric figures that we need to discuss: those figures that are composed of curves that are not line segments.

How many words do you know that describe such shapes? Think and then read on. . . .

There are many such geometric figures—for example, circle, semicircle, spiral, parabola, ellipse, hyperbola, and crescent (Figure 8.93).

Figure 8.93

In this course, we will focus on the simplest of all curved geometric figures: the circle. Stop for a moment to think about circles. How would you define a circle? Try to do so before reading on. . . .

A **circle** is the set of points in a plane that are all the same distance from a given point, the center.

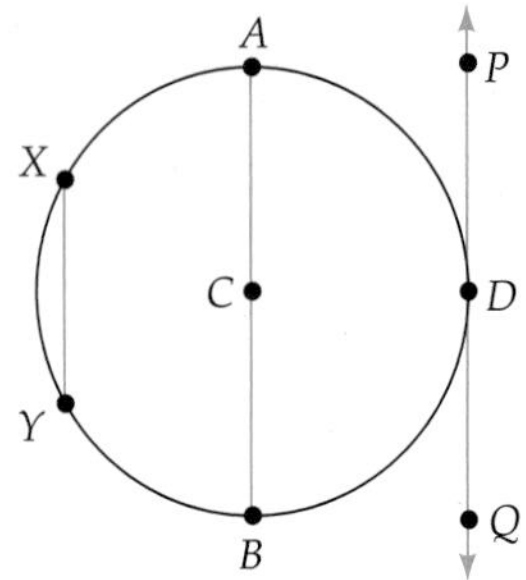

Figure 8.94

In Figure 8.94, C is the *center* of the circle.

The line segment $\overline{CA}$ is called a *radius*, the plural of which is *radii*.

The line segment $\overline{AB}$ is called a *diameter*.

The line segment $\overline{XY}$ is called a *chord*.

The line $\overleftrightarrow{PQ}$, which intersects the circle only at point D, is called a *tangent*.

An *arc* is any part of a circle. We use two letters to denote an arc if the arc is less than half of the circle. However, for larger arcs, we use three letters. Do you see why?

We do this to distinguish between the arc at the left ($\overset{\frown}{AD}$) in Figure 8.95 and the arc at the right ($\overset{\frown}{ABD}$) because both of them have points A and D as endpoints.

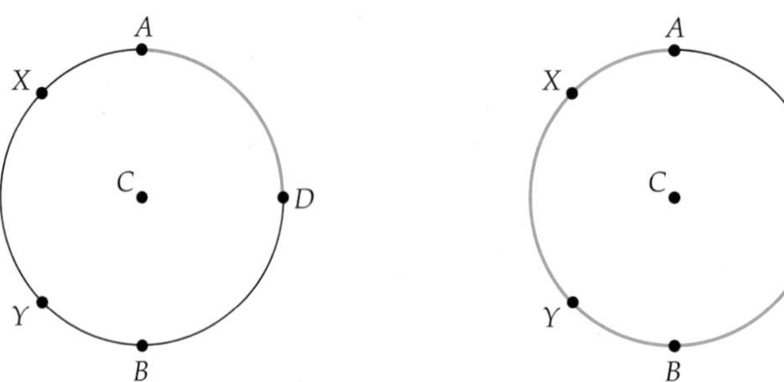

Figure 8.95

On the basis of these descriptions and your previous experience with circles, try to write a definition for each of these terms. Then compare your definitions with the ones that follow. . . .

A **radius** of a circle is any line segment with one endpoint on the circle and another endpoint at the center.

A **diameter** of a circle is any line segment with endpoints on the circle and also goes through the center of the circle.

A **chord** is any line segment with endpoints on the circle. Thus, a diameter is also a chord.

A **tangent** line intersects a circle at exactly one point.

An **arc** is a subset of a circle—that is, a connected part of a circle.

COORDINATE GEOMETRY

One of the themes of this book is multiplicity—multiple solution paths to most problems, multiple connections between and among ideas, and multiple ways of representing many ideas. One of the most powerful ways to illustrate the value of multiplicity is through coordinate geometry. Up to now, all of our exploration has been without coordinates—that is, we have simply used sketches of geometric figures. Let us look at what is added when we place the figures in a coordinate plane.

HISTORY

For many hundreds of years, algebra and geometry were developed for the most part separately. It was in the 1600s that René Descartes began the mathematical work of connecting geometry to algebra.

A brief review of the Cartesian coordinate system

Any point on a plane can be represented by an ordered pair. The first number represents the point's horizontal distance from the center of the coordinate plane, which is called the **origin** and is denoted by the ordered pair (0, 0). The second number represents the point's vertical distance from the origin. At some time during the development of mathematics, mathematicians adopted the convention that right is positive, left is negative, up is positive, and down is negative.

As we noted at the beginning of this chapter, a powerful contribution of the Greeks was to move us from the *how* to the *why*. It turns out that some geometric proofs that are very difficult in "normal" representation are actually quite simple with coordinate geometry. Before we get to some proofs, let us do one investigation to review your skills with the coordinate system.

INVESTIGATION 8.2l What Are My Coordinates?

First, let's play a game. I'm thinking of a rectangle. Three of its coordinates are (3, 5), (7, 5), and (3, 10). What is the fourth coordinate? Do this yourself before reading on. . . .

DISCUSSION

Using their basic understanding of the properties of rectangles and their intuitive, visual understanding of the coordinate plane (Figure 8.96), most people will deduce that the fourth coordinate is (7, 10).

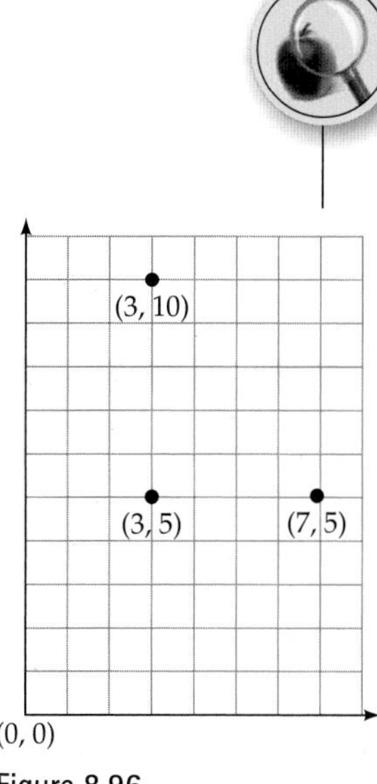

Figure 8.96

One pattern that emerges here is that if you look at the coordinates of two consecutive vertices of the rectangle, either their x values or y values will be equal, e.g., (3, 5) and (7, 5).

We need to develop one more idea before we jump more deeply into coordinate geometry, and that is to learn how to find the distance between two points on the coordinate plane. If you took geometry in high school, you probably memorized this formula. However, if you read this carefully, you will find that it doesn't have to be memorized.

INVESTIGATION 8.2m Understanding the Distance Formula

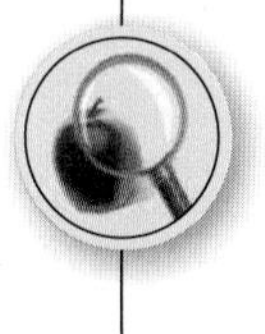

Let us consider two random points on the plane: $Y = (x_1, y_1)$ and $S = (x_2, y_2)$. If we consider the line segment connecting these two points to be the hypotenuse of a right triangle, we can draw the two sides of the triangle (Figure 8.97). What must be the coordinates of the third vertex, E? Remember the discussions in this chapter about determinism—in this case, both the x and y values of the third vertex are determined. Think before reading on. . . .

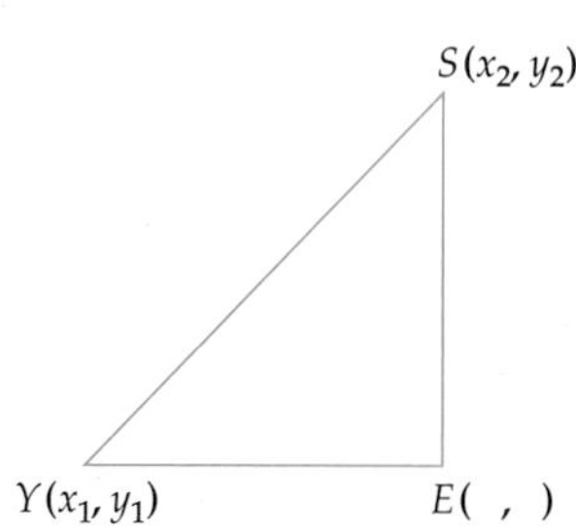

Figure 8.97

DISCUSSION

The coordinates of that third vertex are (x_2, y_1). Do you see why? If not, recall that this vertex is the same horizontal distance from the origin as the top vertex (S); thus it has the same x value. Similarly, it is the same height above the origin as the left vertex (Y); thus it has the same y value.

Now we can apply the Pythagorean Theorem: $c = \sqrt{a^2 + b^2}$. This formula simply says that in any right triangle, the length of the hypotenuse "c" is equal to the square root of the sum of the squares of the other two sides, "a" and "b".

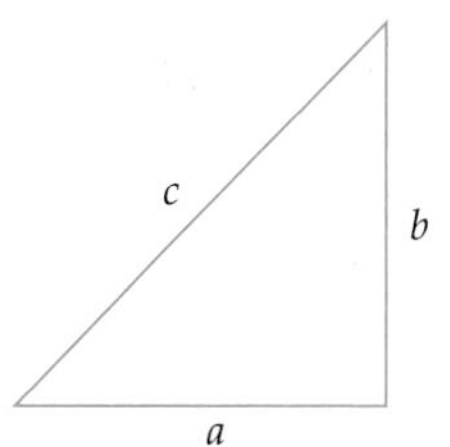

Figure 8.98

If this seems a bit overwhelming, look at Figure 8.98 and connect it to Figure 8.97. That is, in order to find the distance from Y to S, we have to find the distance from Y to E

and the distance from E to S. Then we will square those distances, add them, and take the square root.

But the distance from Y to E is easy because it's a horizontal line segment. It is just the difference between the x values—that is, $(x_2 - x_1)$. Similarly, the distance from E to S is the difference between the y values—that is, $(y_2 - y_1)$.

Thus the distance from Y to S, substituting our distances into the Pythagorean theorem, must be $\sqrt{(x_2 - x_1)^2 + (y_2 - y_1)^2}$.

Phrased in English, if we square the x distance and the y distance, add them, and take the square root, we have the distance between any two points.

Now we are ready for one of the proofs. Recall that a parallelogram is defined as a quadrilateral in which opposite sides are parallel. Using only this knowledge, we can *prove* that the opposite sides of a parallelogram must be congruent.

INVESTIGATION 8.2n The Opposite Sides of a Parallelogram Are Congruent

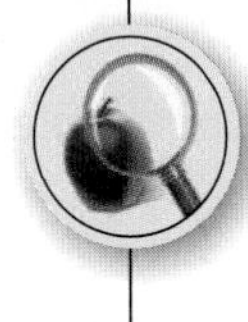

Let us begin by sketching a parallelogram on the coordinate plane (Figure 8.99). Since the opposite sides are parallel and since, from an algebraic perspective, parallel lines have the same slope, we know that the slopes of the opposite sides are equal. Thus, if we place one vertex at the origin and the next vertex at $(c, 0)$, we have one side lying on the x axis. Because we know that $\overline{PL}$ and $\overline{NA}$ must be parallel, we can draw $\overline{NA}$ parallel to the x axis.

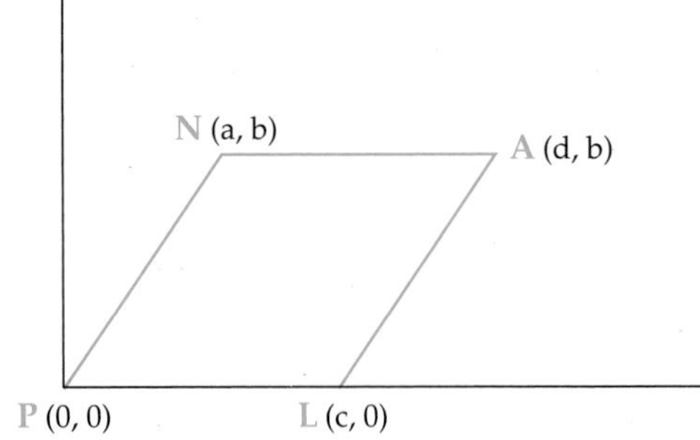

Figure 8.99

DISCUSSION

Thus, if we let N be represented as the point (a, b), we know that the y coordinate of A must also be b. Do you see why?

If not, recall that slope is the ratio of rise to run. That is, $\frac{y_2 - y_1}{x_2 - x_1}$.

We know that the slope of $\overline{PL}$ is 0. Thus the slope of $\overline{NA}$ must also be zero. If both N and A have the same y coordinate, b, then the slope is $\frac{b - b}{d - a} = 0$.

Because $PLAN$ is a parallelogram, the other two sides must also be parallel—that is, they must have the same slope. Let us first determine the slope of $\overline{AL}$ and $\overline{PN}$.

$$\text{The slope of } \overline{PN} \text{ is } \frac{b - 0}{a - 0} = \frac{b}{a}.$$

$$\text{The slope of } \overline{AL} \text{ is } \frac{b - 0}{d - c} = \frac{b}{d - c}.$$

What does this tell us about the relationship among a, d, and c? 

It means that since the two lines have to have the same slope and thus be parallel, $a = d - c$, which is equivalent to $d = a + c$.

We can now substitute $(a + c)$ for d and have the new representation of the coordinates of the points of the parallelogram (Figure 8.100).

N (a, b)
A (a + c, b)
P (0, 0)
L (c, 0)

Figure 8.100

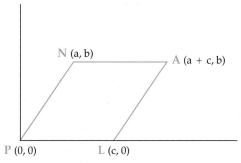

Now we can prove that the lengths of opposite sides are equal. Let's begin with the easier case, the horizontal sides. The distance from P to L is $c - 0$; that is, it is c. The distance from N to A is $(a + c) - a = c$.

What about the distance from P to N and from L to A? Using the distance formula from Investigation 8.2m, we can find the length of PN—that is, the distance between P and N.

$$PN = \sqrt{(a - 0)^2 + (b - 0)^2} = \sqrt{a^2 + b^2}$$

Now, we find the distance from L to A.

$$LA = \sqrt{((a + c) - c)^2 + (b - 0)^2} = \sqrt{a^2 + b^2}$$

Because the two distances are the same, the lengths of the two line segments are congruent. Therefore, we have proved that the opposite sides of a parallelogram are the same length.

INVESTIGATION 8.2o Midpoints of Any Quadrilateral

Now let us investigate one of my favorite theorems in geometry. I want you to see it first; then we will prove it. On a blank piece of paper, draw a quadrilateral, any quadrilateral. Find the midpoints of each side. I recommend using the metric side of your ruler. If you don't have a ruler, you can still find the midpoints. Do you see how? . . . Yes, you can fold the paper to find the midpoints of each side. Now make a new quadrilateral by connecting the four midpoints consecutively. What do you see?

Do another one. What you just observed will always happen. That is, in all cases, you will wind up with a parallelogram. Now, let's prove that.

DISCUSSION

Before we can do the proof, we simply need to know how to find the midpoint of a line segment. In this case, the theorem makes intuitive sense, and I will simply present it. The midpoint of the line segment connecting points (x_1, y_1) and (x_2, y_2) shown in Figure 8.101, is $\left(\frac{x_1 + x_2}{2}, \frac{y_1 + y_2}{2}\right)$. That is, it is the average (mean) of the x values and of the y values.

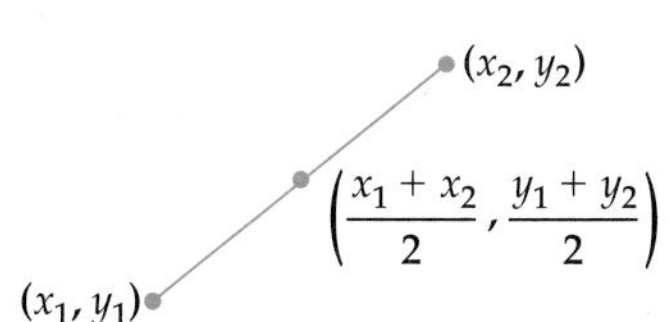

Figure 8.101

We will prove the theorem for all quadrilaterals. As we saw from the quadrilateral hierarchy, once we prove a theorem for one quadrilateral, it is true for all quadrilaterals below that quadrilateral. Because we are dealing with any quadrilateral, we cannot assume any properties—congruence or parallel. Thus we will set one vertex at the origin (0, 0) and one vertex on the x axis $(a, 0)$. Then the other vertices are at arbitrary points (b, c) and (d, e) (Figure 8.102).

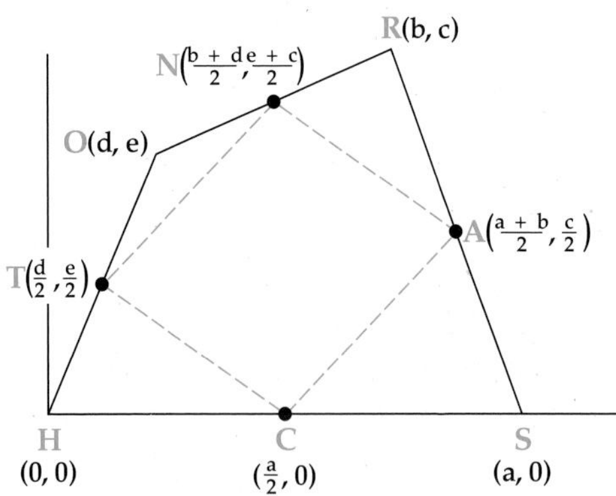

Figure 8.102

We will label the original quadrilateral *HORS* and the quadrilateral formed by the midpoints *CANT*.

Using the midpoint formulas, we determine the coordinates of the midpoints and then connect them. We now need to show that the opposite sides must be parallel—that is, that they have the same slope.

Let us begin with $\overline{CA}$ and $\overline{TN}$.

$$\text{Slope of } \overline{CA} = \frac{\frac{c}{2} - 0}{\frac{a+b}{2} - \frac{a}{2}}$$

A little algebra makes this next step much easier. If we multiply the top and bottom of this expression by 2, we don't change the value.

Doing this, we have

$$\text{Slope of } \overline{CA} = \frac{c - 0}{(a+b) - a} = \frac{c}{b}$$

Now for $\overline{TN}$.

$$\text{Slope of } \overline{TN} = \frac{\frac{e+c}{2} - \frac{e}{2}}{\frac{b+d}{2} - \frac{d}{2}}$$

If we multiply the top and bottom of this expression by 2, we have

$$\text{Slope of } \overline{TN} = \frac{(e+c) - e}{(d+b) - d} = \frac{c}{b}$$

Thus we have shown that the slope of $\overline{CA} = \frac{c}{b}$ and that the slope of $\overline{TN} = \frac{c}{b}$. That is, these two line segments are parallel.

By a similar means, we can show that the slope of $\overline{CT}$ and the slope of $\overline{AN}$ are equal.

Thus we have proved that if you take the midpoints of *any* quadrilateral and connect them in turn, you will always get a parallelogram. To prove that by other means is much more tedious and difficult.

SUMMARY 8.2

While many of the terms in this section are review for many students, it is important to realize that learning geometry is much more than memorizing language. In this section, we have emphasized relationships among figures, for example, the different kinds of triangles and the different kinds of quadrilaterals *and how they are related to each other.* We have examined the many terms that can be used to help us as we solve problems and communicate what we see—congruent sides and angles, diagonals, polygons, etc. We went beyond triangles and quadrilaterals to other kinds of polygons and nonpolygons that are found in everyday life. Finally, we took a brief tour of coordiate geometry and saw how some geometric ideas and theorems can be more easily proved from a different representation system.

8.2 Exercises

1. Look at each figure below for just 2 to 3 seconds and draw it from memory as you did in Investigation 8.2a. Describe your thinking process.

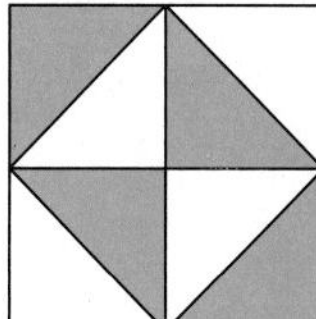
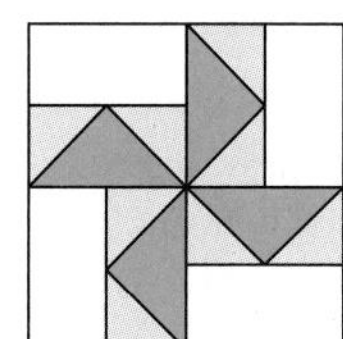

2. Write down all the attributes of each of the figures.

a. 
b. 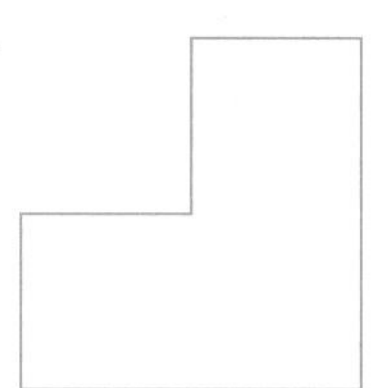
c. 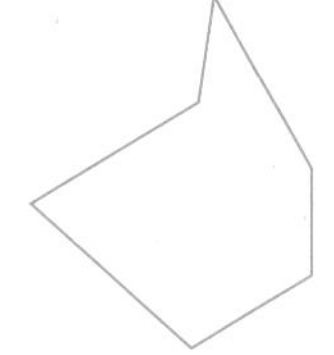

3. In each case below, which two figures are most alike? Explain your reasoning.

a. 
b. 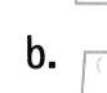

c. 
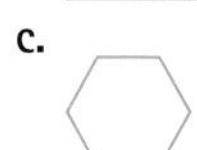

4. Name at least six different polygons that you can see in this shape. Trace and number each shape.

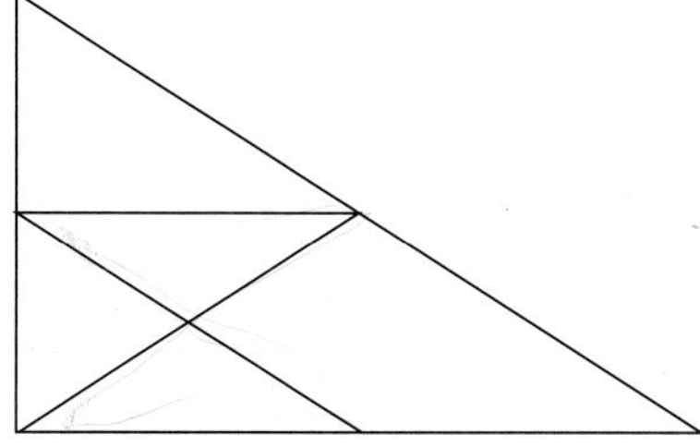

5. ***Classroom Connection*** This question was posted in *Teaching Children Mathematics.* The children's responses can be found in the April 2004 issue on pp. 403–406: "Tiffany is creating different shapes on her geoboard. . . . [S]he shares her shape with her friend Leticia [who] notices that she can see triangles and rectangles in the design."

a. How many different triangles can you find in the design?

b. How many different rectangles can you find?

c. What other polygons do you see in the design?

d. Create your own design. What shapes can you see? What other questions would you ask a friend about your design?

Source: Reprinted with permission from *Teaching Children Mathematics,* April 2004, pp. 403–406, copyright 2004 by the National Council of Teachers of Mathematics.

e. (My extension) What is the polygon with the most number of sides that you can find in this design?

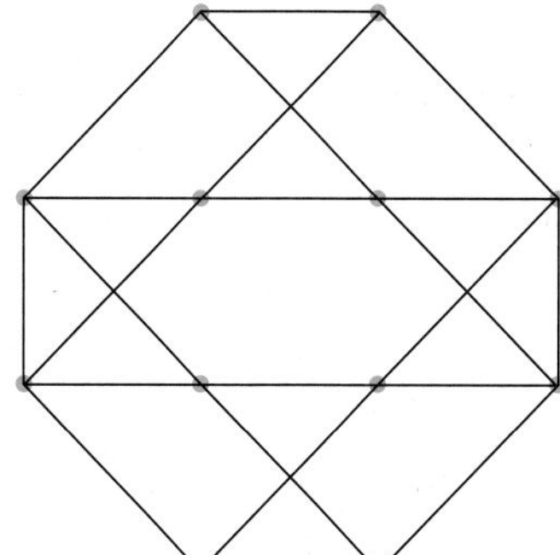

6. Describe all the geometric shapes you see in the quilt designs below:

a.

b.

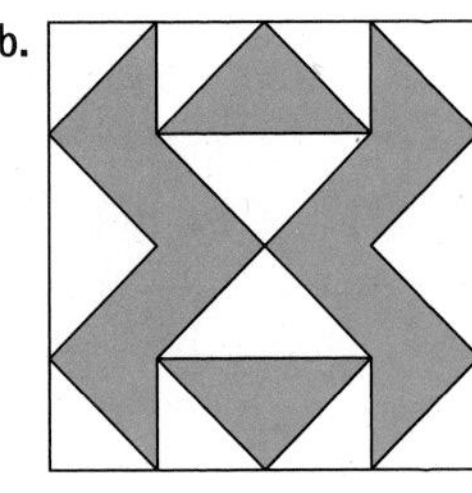

Fool's Puzzle

c.

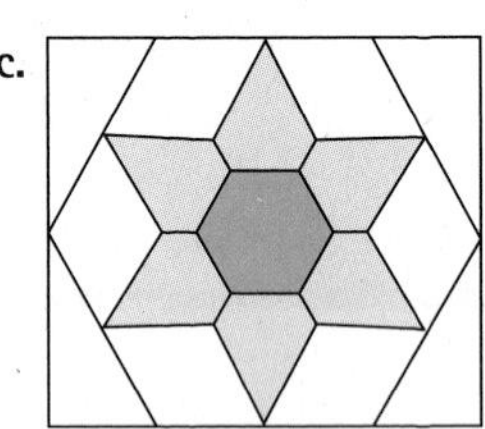

Texas Star

Source: Jinny Beyer, *The Quilter's Album of Blocks & Borders* (Delaplane, VA: EPM Publications, Inc., 1986), p. 117. Reprinted with permission.

7. Try to make each figure on Geoboard Dot Paper *and* on Isometric Dot Paper. If you can make the figure, do so and explain why your figure is an example of the specified triangle or quadrilateral. If you cannot make it, explain why you think it is impossible to make the figure on that type of grid. Your reasoning needs to be based on properties and attributes (see the van Hiele discussion), as opposed to, for example, "It's a right triangle because it looks like a right triangle."
   - a. acute scalene triangle
   - b. right isosceles triangle
   - c. obtuse isosceles triangle
   - d. equilateral triangle
   - e. trapezoid
   - f. kite
   - g. parallelogram
   - h. rectangle
   - i. rhombus
   - j. square
   - k. square with no sides parallel to the sides of the paper.

8. Fill in the table below.

Shape	Number of diagonals
Quadrilateral	2
Pentagon	
Hexagon	
Heptagon	
Octagon	
n-gon	

9. For each figure below, write "polygon" or "not a polygon." If it is a polygon, also write "convex" or "concave."

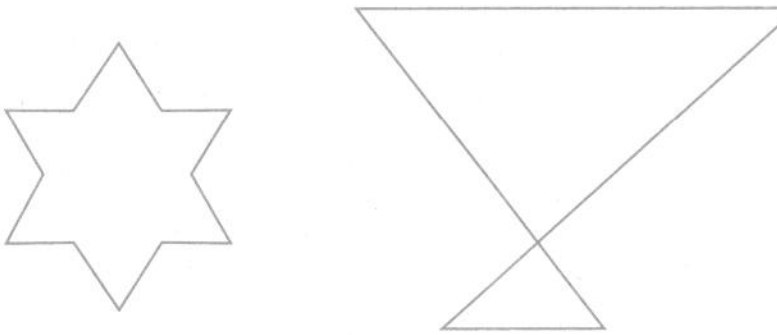

a. b. c.

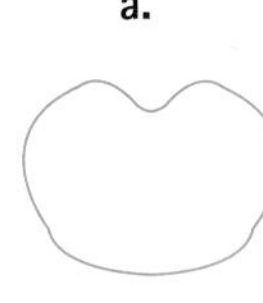

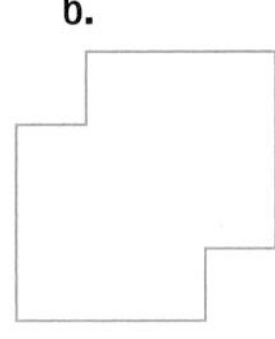

d. e.

10. a. Does this figure appear to be a kite? Why or why not?

b. Does this figure appear to be a rectangle? Why or why not?

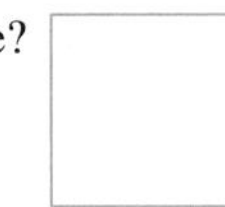

c. Does this figure appear to be an isosceles triangle? Why or why not?

11. Describe all quadrilaterals that have these characteristics. If there is more than one, say so.
   - a. A quadrilateral with opposite sides parallel
   - b. A quadrilateral with 4 right angles
   - c. A quadrilateral with all sides congruent
   - d. A quadrilateral in which the diagonals bisect each other
   - e. A quadrilateral in which the diagonals are congruent
   - f. A quadrilateral in which adjacent angles are congruent
   - g. A quadrilateral in which opposite angles are congruent
   - h. A quadrilateral in which no sides are parallel
   - i. A quadrilateral with 4 congruent sides and 2 distinct pairs of congruent angles
   - j. A quadrilateral with 4 congruent angles and 2 distinct pairs of congruent sides

12. Draw each of the following or briefly explain why such a figure is impossible.
   - a. An isosceles trapezoid
   - b. A concave quadrilateral
   - c. A curve that is simple and closed but not convex
   - d. A nonsimple closed curve
   - e. A concave equilateral hexagon

f. A concave pentagon having three collinear vertices

g. A pentagon that has 3 right angles and 1 acute angle

13. Use a Venn diagram to represent the relationship between:

a. scalene and obtuse triangles.

b. equilateral and isosceles triangles.

c. parallelograms and rectangles.

d. rectangles, rhombi, and squares.

14. Write in the labels for each set in the problems below. Justify your choice. Add at least one new figure to one of the regions.

a.

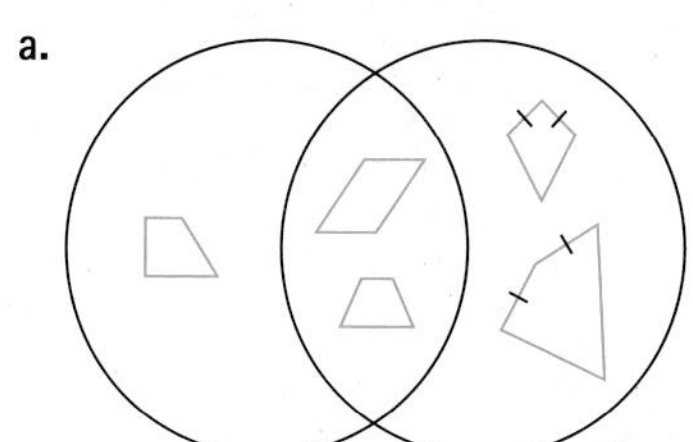

b.

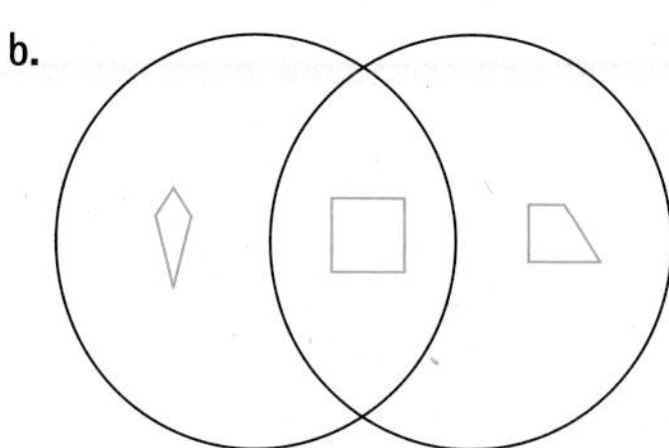

15. Find the following points on a sheet of graph paper:
$A\ (3, 5)$ $B\ (5, 5)$ $C\ (5, -5)$ $D\ (-2, -5)$
$E\ (-2, 5)$ $F\ (0, 5)$ $G\ (0, -3)$ $H\ (3, -3)$

Now connect the points in order. What do you see?

16. If a polygon is regular, how can we use the sum of the interior angles formula that we developed in Investigation 8.2k to find the measure of each angle?

17. Find the midpoints of the line segments connecting these pairs of points:

a. (0, 4) and (4, 10) b. (3, 4) and (7, 12)

18. a. Three vertices of a kite are (6, 8), (9, 11), and (12, 8). What are the coordinates of the fourth vertex?

b. The two vertices that form the non-congruent side of an isosceles triangle are (−5, 3) and (2, 3). What are the coordinates of the other vertex?

c. The coordinates of the endpoints of the hypotenuse of a right triangle are (7, 5) and (3, 1). Find the other vertex. There are two possible solutions.

d. Three vertices of a parallelogram are (0, 0) (4, 0), and (0, 6). Find the fourth vertex. There are three possible solutions.

e. A rectangle is oriented so that its sides form vertical and horizontal lines. Two coordinates of a rectangle are (−3, −2) and (7, 6). Do you have enough information to determine the other two coordinates? If so, find them. If not, explain why not and describe what information you would need (for example, a third coordinate, the length of one side, the relative location of one of the points).

f. Make up a problem similar to the ones above and solve it.

19. We can play a variation of a child's game called "What am I?" that I will call "Where am I?"

a. I am a square. The intersection of my two diagonals lies at the point (3, 3), and the length of each of my sides is 6. My sides form horizontal and vertical lines. Where am I?

b. I am an isosceles triangle. The midpoint of my non-congruent side is the point (7, −2). This side is horizontal and my vertex opposite this side is at the point (7, −9). Oh, I almost forgot to tell you. I am upside down. Where am I?

c. I am a right isosceles triangle. I have an area of 50 square units. The coordinates of the vertex at which the two congruent sides meet is (0, 0) and my sides form horizontal and vertical lines. Do you have enough information to determine the other two coordinates? If so, find them. If not, explain why not and describe what information you would need.

d. Make up a problem similar to the ones above and solve it.

20. Determine the coordinates of the vertices of the following quilt pattern if the bottom left-hand corner is the origin and the length of each side of the whole square is 9. The small square in the center has an area of 9, and each of the four rectangles has an area of 4.5.

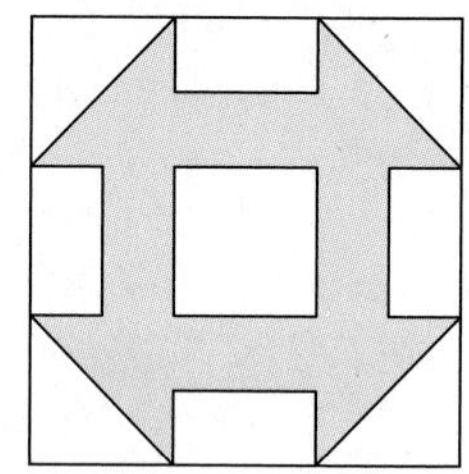

DEEPENING YOUR UNDERSTANDING

21. The definition of a regular polygon states that all sides have the same length and all interior angles have the same measure. Why is the second part of the definition necessary? That is, why can't we just say that a polygon is a regular polygon if all the sides are the same length?

22. How many different quadrilaterals can you make that have at least one pair of adjacent congruent sides? Sketch and label your figures. For example, you can make many different trapezoids, but they are all trapezoids. See the figures below.

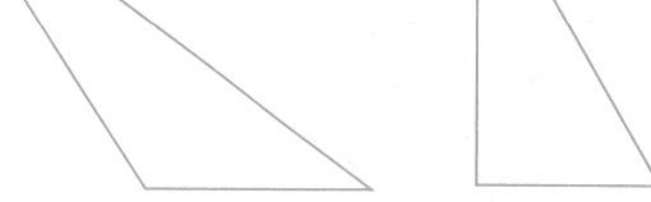

23. Write directions for making the following figures. Following your directions, the reader should be able to make the same figure.

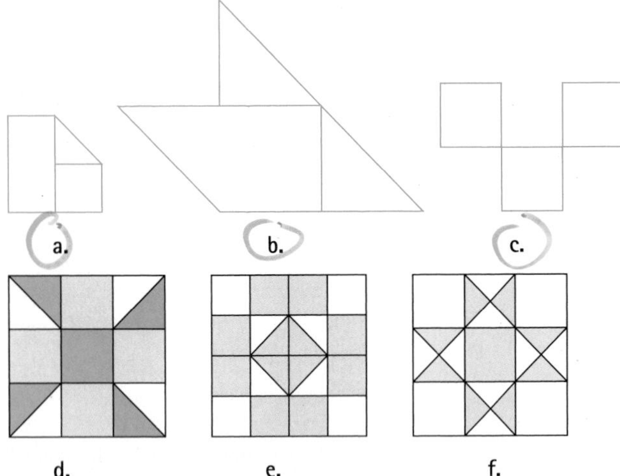

24. In each of the following cases, determine whether the two figures are congruent only by using your mind. That is, you cannot trace one figure and see whether it can be superimposed on the other figure. Describe your reasoning—that is, how you arrived at your conclusion.

a. Below are two parallelograms made on a Geoboard.

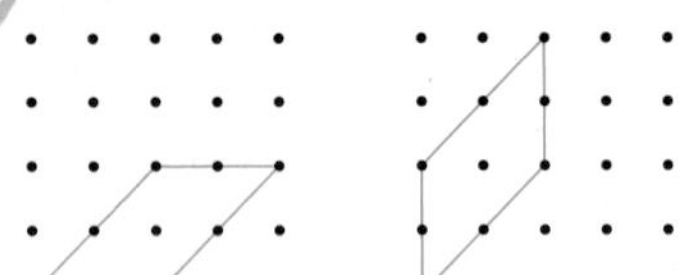

b. Below are two figures made with tangram pieces.

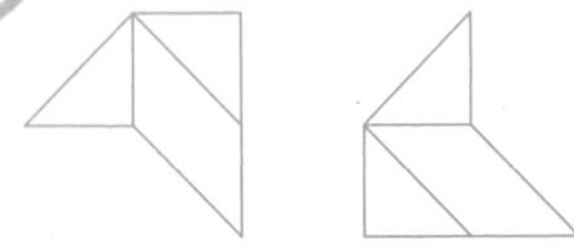

c. Below are two pairs of hexominoes.

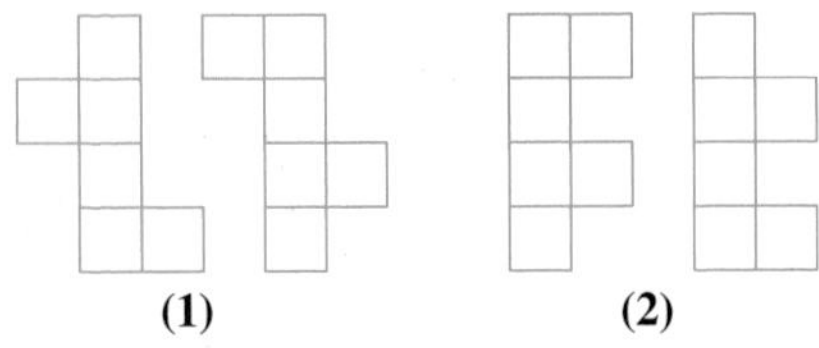

25. In each case, explain and justify your answer—that is, indicate why you think there are none, just one, or many.

a. How many different hexagons can you draw that have all sides equal but not all angles equal?

b. How many different hexagons can you draw that have exactly 2 right angles?

c. How many different hexagons can you draw that have exactly 3 right angles?

d. How many different hexagons can you draw that have exactly 4 right angles?

e. How many different hexagons can you draw that have exactly 5 right angles?

f. Can you make a trapezoid with no obtuse angles?

g. How many different kinds of quadrilaterals can you make that have exactly two opposite right angles?

h. Can you make a concave quadrilateral with exactly 2 right angles?

i. Can you make a concave pentagon with exactly 2 right angles?

26. Show that the family tree for quadrilaterals also holds for the characteristics of the diagonals.

27. In this section, we found that we could name some quadrilaterals in terms of their "ancestors." For example, we could say that a rhombus is a parallelogram with four congruent sides. What other quadrilaterals could we describe in terms of their ancestors?

28. Draw a triangle. Find the midpoints of each side. Connect those points. What is the relationship between the new triangle and the original triangle? Prove it.

29. Take a blank piece of paper. Fold it in half and then fold it in half again. Draw a scalene triangle and then cut it out. You should now have four congruent triangles. For each of the following questions, explain your reasoning.

a. Put these triangles together to make a large triangle.

b. Is there only one way or are there different ways?

c. How could you prove that the new triangle is actually a triangle, as opposed to a figure that is almost a triangle?

d. Does the new large triangle have anything in common with the original triangle?

30. What is the relationship between the number of sides in a polygon and the total number of diagonals in that polygon?

31. Prove that the sum of the interior angles of any polygon is equal to $180(n - 2)$, where n represents the number of sides in the polygon.

32. Describe all possible combinations of angles in a quadrilateral (for example, acute, acute, acute, obtuse). Briefly summarize methods you used other than random trial and error.

33. Trace this circle following #34 onto a blank sheet of paper. Describe as many ways as you can for finding the center of the circle. In each case, explain why the method works.

34. Trace the circle onto a blank sheet of paper. After cutting out the circle and finding the center, fold the paper so that the top point of the circle just touches the center of the circle. Fold the circle again so that another point on the circle just touches the center *and* the two folds meet at a point on the circle. The two folds will be congruent. Fold the circle one more time so that your three folds will all be congruent. What kind of triangle has been created inside the circle? Prove that you will get this kind of triangle by making these three folds.

35. A 3RIT is a figure made from 3 right isosceles triangles. On the left are two examples of RITs, and on the right are two figures that are not RITs.

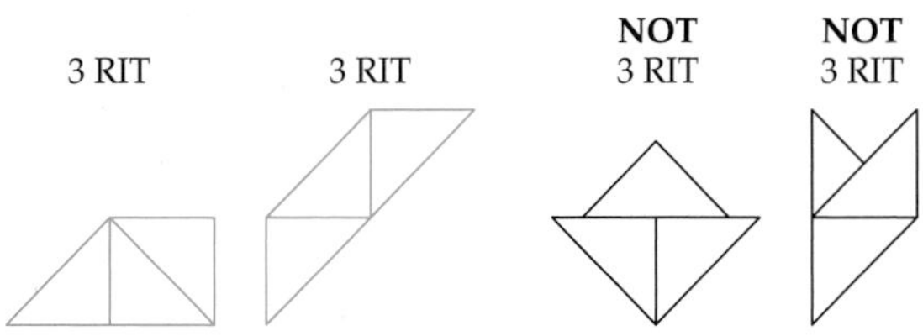

a. Write a definition of 3RIT.

b. Find as many different 3RITs as you can. Sketch (or tape, or glue) them on a separate piece of paper.

c. Are the two 3RITs at the right different or not? Explain why you believe they are (or are not) different.

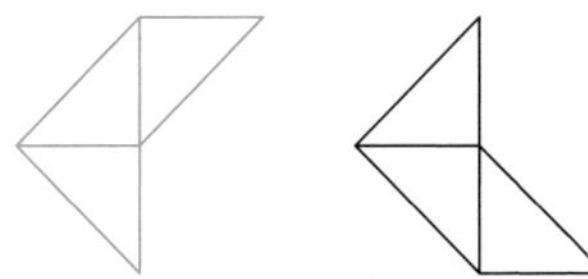

d. Find as many 4RITs as you can.

36. Make as many different hexiamonds as you can. A hexiamond is made by joining six equilateral triangles—when a side is connected, a whole side connects to a whole side. Briefly (in two or three sentences) describe your method(s) for generating as many shapes as possible. Describe any method(s) other than random trial and error; there are many different ways to be systematic. Cut your hexiamonds out and tape them to a piece of paper. (There are more than 9 and fewer than 16.)

37. Determine the coordinates of the vertices of tangram pieces if the bottom left-hand corner is at the origin and if the length of each side of the square is 8.

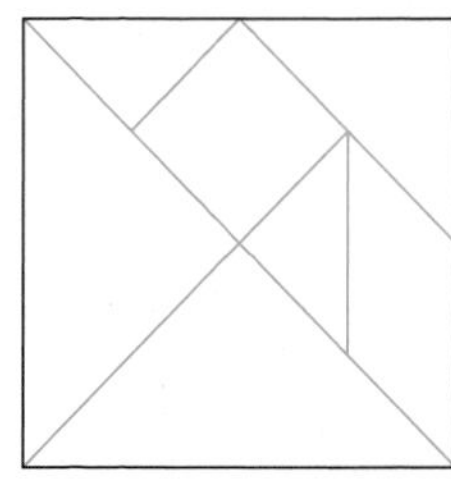

38. If we set a side of a triangle on the x axis, we can specify the coordinates of any triangle as follows: $A(0, 0)$, $B(r, m)$, and $C(e, 0)$. However, if we have an isosceles triangle, there is a relationship between some of these coordinates that enables us to be more specific, just as we were in Investigation 8.2n with a parallelogram. If $\overline{BA}$ and $\overline{BC}$ are the congruent sides, there is a relationship between e and r that enables us to specify e in terms of r. What is the relationship?

39. a. Prove that if we connect the midpoints of the two congruent sides of an isosceles triangle, the length of the line segment connecting those two points is one-half the length of the base.

b. Is this true just for isosceles triangles or for all triangles? Support your conclusion.

40. Using Geogebra, draw a triangle.

a. Using the midpoint tool and line segment tool, draw the three medians. Click and drag a vertex of the triangle to see how it affects the centroid (the intersection point of the medians).

b. Using the angle bisector tool in Geogebra, draw the three angle bisectors of the triangle. Click and drag a vertex of the triangle to see how it affects the incenter (the intersection point of the angle bisectors).

c. Using the midpoint tool and the perpendicular line tool, draw the three perpendicular bisectors of the triangle. Click and drag a vertex of the triangle to see how it affects the circumcenter (the intersection point of perpendicular bisectors).

41. a. Show that the diagonals of a square have the same length.

b. You may recall from high school algebra that two lines are perpendicular if the product of their slopes is equal to -1. Use this knowledge to show that the diagonals of a square are perpendicular.

FROM STANDARDIZED ASSESSMENTS

42. In what ways are the two figures in the grid below alike? In what ways are they different?

Source: Results of the *Seventh NAEP Mathematics Assessment.* U.S. Department of Education, National Center for Education Statistics.

43.

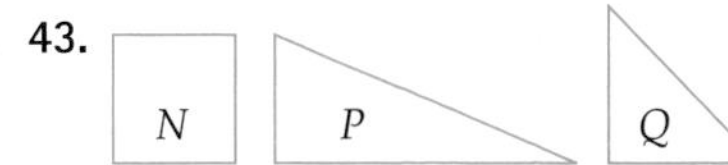

Laura was asked to choose 1 of the 3 shapes N, P, and Q that is different from the other 2. Laura chose shape N. Explain how shape N is different from shapes P and Q.

Source: Results of the *Seventh NAEP Mathematics Assessment.* U.S. Department of Education, National Center for Education Statistics.

2007 NECAP, Grade 4

44. Which statement is true?

a. One line segment can divide a triangle into two smaller triangles.

b. One line segment can divide a square into two smaller squares.

c. One line segment can divide a circle into two smaller circles.

d. One line segment can divide a hexagon into two smaller hexagons.

2006 NECAP, Grade 6

45. Marcy is drawing a *right triangle* by placing stars on the grid below and connecting them with line segments.

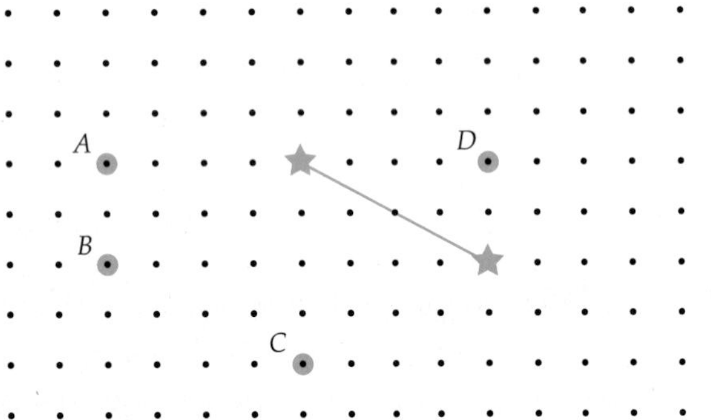

Which point could be the location of the third star of Marcy's right triangle?

a. Point *A** **b.** Point *B*

c. Point *C* **d.** Point *D*

2008 NECAP, Grade 6

46. Which shape is possible?

a. A rhombus with 4 acute angles

b. A parallelogram with 4 angles of equal measure

c. A rhombus with sides that measure 4 cm, 4 cm, 8 cm, and 8 cm

d. A parallelogram with sides that measure 2 cm, 4 cm, 6 cm, and 8 cm

2007 NECAP, Grade 5

47. Look at these shapes.

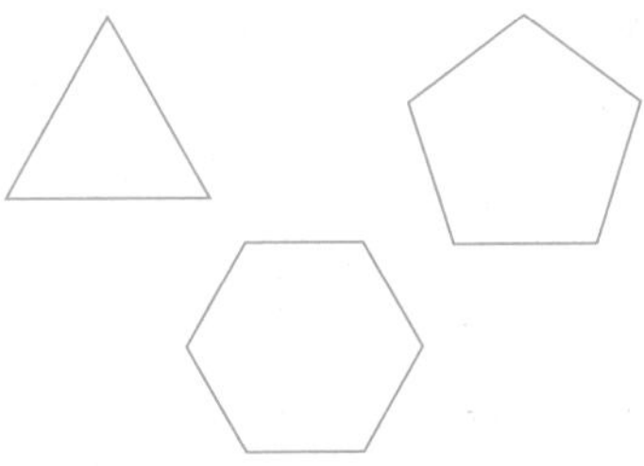

a. Use mathematical language to write *one* way all three shapes are different.

b. Use mathematical language to write *one* way all three shapes are alike.

SECTION 8.3 Three-Dimensional Figures

What do you think?

- What do pyramids and cones have in common?
- How are properties of two-dimensional objects and three-dimensional objects related?
- How can you represent three-dimensional objects in two-dimensional space?

According to the Common Core State Standards, students in kindergarten learn to identify shapes as two-dimensional or three-dimensional. Almost everything we interact with is three-dimensional: people and pets, buildings, our rooms, our cars, what we see when we look around. Look at the six pictures on the next two pages. What do you see? Think, jot some notes, and then read on. . . .

Four of the pictures are of objects made by humans, and two are natural—one organic and one inorganic. The snail (in its genetic makeup) and the crystal (in its molecular makeup) have the "blueprints" for the ultimate shape of the shell and of the crystal. Both the pyramid and the Parthenon are over 2000 years old, and yet the geometry that was used to make them is relevant to building today. The soccer ball and the design on the roof of the mosque represent the solution to two different questions about connections between two dimensions and three dimensions. In the former case, someone discovered that a specific combination of two polygons (a regular pentagon and a regular hexagon) will produce a nearly perfect ball (sphere). In the latter case, the designers had to figure out how to take a two-dimensional tessellation design and "fold" it around the roof of the shrine so that it would "work" in three dimensions.

Just as we discovered patterns and relationships among many two-dimensional objects, there are many patterns and relationships among three-dimensional figures, also called space figures. Understanding the geometry of human-made objects helps us to make them work better and, in the case of objects such as bridges, overpasses, and airplanes, to make them work more safely. Geometry also helps us to understand natural phenomena better—for example, why certain animals have the shapes they have. An understanding of shapes has many applications in science. For example, many carcinogens are virtually identical in size and shape to other compounds, and thus they fool the body into thinking they are not harmful. The silicon chip has the same structure as the diamond, except that there are silicon atoms instead of carbon atoms at these positions.[5] With respect to aesthetics, geometry helps us to understand why some shapes are so appealing to people and to understand patterns within those shapes (Figure 8.103).

(a)
Pyramids at Giza

(b)
the Parthenon

(c)
nautilus shell

(d)
pyrite crystal

Figure 8.103

(e) *soccer ball*

(f) *mosques*

Figure 8.103 *(continued)*

In this section, we will begin simple and build up. Let's do a "What do you see?" investigation here, as we did in Section 8.2. As you get better at seeing the various attributes of a solid object, relationships between those attributes, and relationships to other similarly shaped objects, your appreciation of the object grows, too.

HISTORY

The massive pyramid at Gizeh is one of the seven wonders of the ancient world. Even today we are not certain how it was constructed. Many of the blocks weighed more than 10 tons! The pyramid was originally covered with white marble and must have dazzled like a mirror in the desert. It was so well built that some of the edges fit together so well that a razor blade cannot be inserted into the space between two blocks.

INVESTIGATION 8.3a What Do You See?

Examine a cube carefully (Figure 8.104). What do you see? Write down all the attributes you can think of before reading on. Next, look at the "box" (Figure 8.105). Which of the attributes of a cube does the box possess? What different attributes does it have? . . .

Explorations Manual 8.14

DISCUSSION

Cube

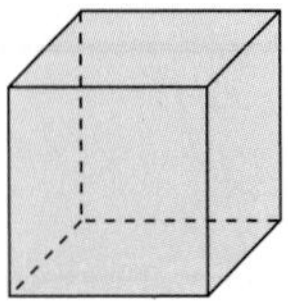

Figure 8.104

Box

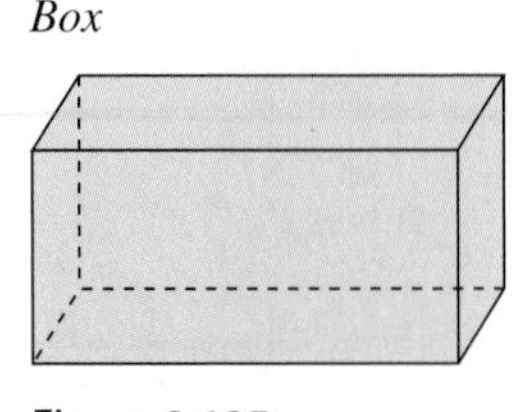

Figure 8.105

A cube has 6 faces, though you might have called them sides.	This box also has 6 faces.
All of the faces are squares, which also means right angles, parallel sides, and so on.	Some of the faces are squares and some are rectangles. As you discovered in the last section, our definitions enable us to say that all of the faces are rectangles.
All of the faces are congruent.	Not all of the faces are congruent, but some are. The opposite faces are congruent. We have names for opposite faces: front-back, side-side, top-bottom.
There are 12 edges on the cube.	The box also has 12 edges.
There are 8 vertices.	The box also has 8 vertices.

This investigation serves as what is called an advance organizer of this section. That is, it got you to grapple with many of the important ideas that we will examine in more detail. Our first order of business toward that end is to learn some of the language we will use with three-dimensional objects. Rather than just presenting you with the new terms, we will do another investigation in which you will be asked to think about what language and concepts from our work with polygons will make sense with our work with polyhedra, and where we need new terms or where the addition of one dimension "changes" things. For example, the addition of one side resulted in a way of naming quadrilaterals different from the way we named triangles.

INVESTIGATION 8.3b Connecting Polygons to Polyhedra

Explorations Manual 8.14

Just as we examined families (subsets) of triangles and quadrilaterals, we will now investigate families of three-dimensional geometric figures (Figure 8.106).

Let us explore the connection between polygons and *polyhedra,* which will be loosely defined (for now) as three-dimensional figures made up of polygons.

The second column of Table 8.5 describes several attributes of polygons. Which of these attributes do you think hold for polyhedra or can be modified to describe different kinds of polyhedra? Fill in as much of the third column as you can. The questions below are given to help you focus on the connections. Then read on. . . .

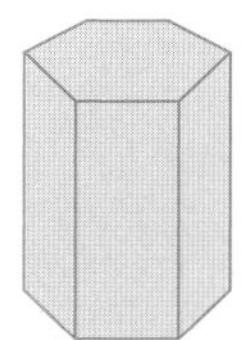
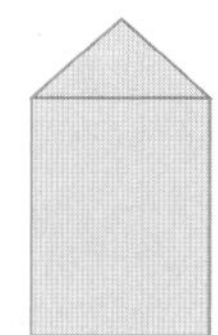
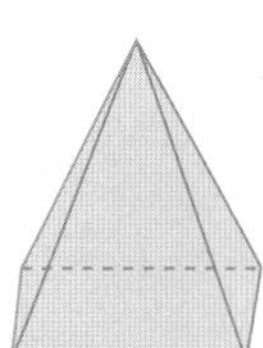
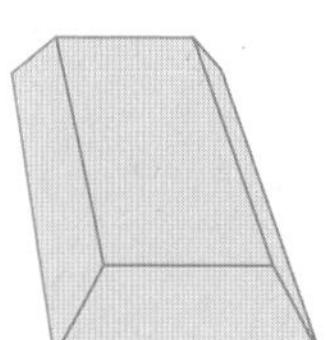
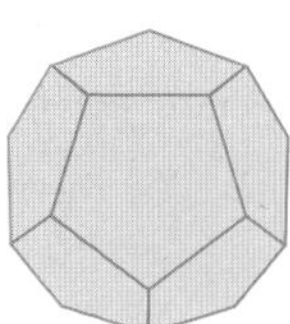

Figure 8.106

- If a polygon is defined as a simple closed curve, is there an analogous definition for a polyhedron?
- All polygons have vertices, line segments, and angles. Do these terms work for describing component parts of polyhedra? Do we also need new terms?
- Can we classify polyhedra by the number of sides?
- If there are regular polyhedra, how might they be defined and how many are there?
- If there are convex polyhedra, how might they be defined?

CLASSROOM CONNECTION

Children view solids (three-dimensional shapes—3D) as entities instead of seeing the parts of the solid as a collection of related shapes.

TABLE 8.5

	Polygons (two-dimensional)	Polyhedra (three-dimensional)
What they are	Simple closed curves	
Definition	Union of line segments	
Component parts	Vertices	
	Line segments	
	Angles	
	Other?	
Classification	By number of sides	
	Regular vs. not regular	
	Convex vs. concave	

DISCUSSION

In Section 8.2, we began with simple closed curves that partitioned a plane into three disjoint sets: the curve, inside, and outside.

Though we will not rigorously define simple closed surfaces, we can say that they partition space (three dimensions) into three disjoint sets: the surface itself, inside, and outside (Figure 8.107).

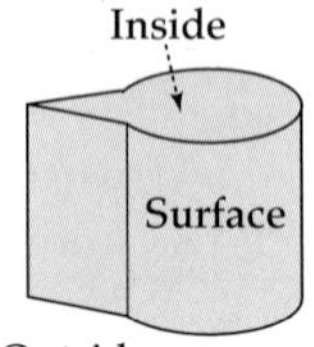

Figure 8.107

We will use the term **space figure** to describe any three-dimensional object.

We will use the term **polyhedron** (the plural is **polyhedra**) to describe those simple closed surfaces that are composed of polygonal regions.

We will use the term **solid** to describe the union of any space figure and its interior.

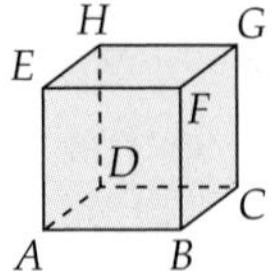

Figure 8.108

Component parts Just as the component parts of polygons have special names, so do those of polyhedra.

Each of the separate polygonal regions of a polyhedron is called a **face**; for example, square *ABFE* is a face of the cube in Figure 8.108.

The sides of each of the faces are called **edges**; for example, $\overline{AB}$ is an edge of the cube in Figure 8.108.

The **vertices** of the polyhedron are simply the vertices of the polygonal regions that form the polyhedron; for example, *E* and *F* are vertices of the cube in Figure 8.108.

Convex and concave Just as polygons can be convex or concave, so can polyhedra. Before reading the definition of a convex polyhedron, think back to the definition of a convex polygon and see whether you can modify that definition for three-dimensional objects. Then read on. . . .

A polyhedron is **convex** if and only if any line segment connecting two points of the polyhedron is either on the surface or in the interior of the polyhedron (Figure 8.109).

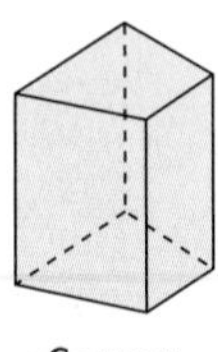
Convex

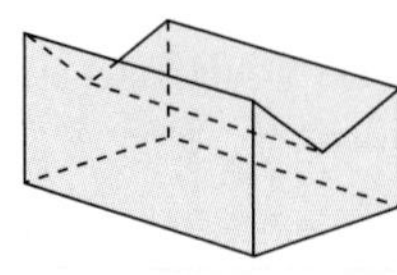
Concave

Figure 8.109

LANGUAGE

Some mathematics dictionaries and textbooks define *edge* as the side of a polyhedron. Some focus on the term *line segment.* Yet others define edge as the intersection of two faces. On one mid-term exam, I asked my students to define *edge.* Interestingly, all three of these perspectives and interpretations appeared in the students' definitions!

INVESTIGATION 8.3c Features of Three-Dimensional Objects

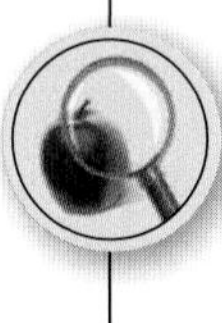

Look at the picture of a cube and a ramp , or triangular prism (Figure 8.110). In what ways are they "the same"? That is, what characteristics do they have in common that not all three-dimensional objects have? In what ways are they different? Do this before reading on. . . .

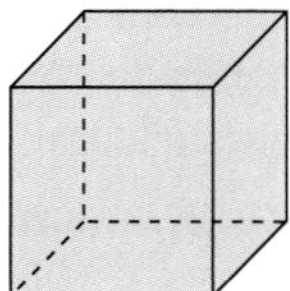
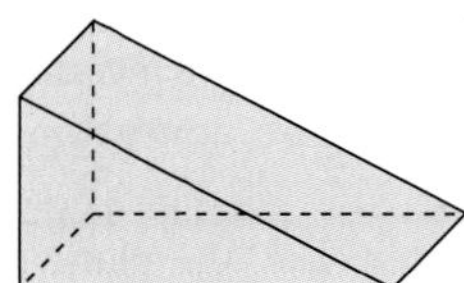

Figure 8.110

DISCUSSION

Some of the things they have in common:

All the faces (sides) are polygons.

In both cases, at least some of the sides are quadrilaterals.

At least one pair of sides are congruent and parallel to each other. In the ramp, the two triangles on the side are parallel and congruent.

Some of the differences between them:

In the cube, there are an even number of faces, and opposite faces are congruent.

In the ramp, only the triangle faces are opposite. The other three faces are noncongruent rectangles.

The numbers of faces, edges, and vertices are different.

The cube: 6, 12, 8
The ramp: 5, 9, 6

However, the relationship between the numbers of faces, edges, and vertices is the same.

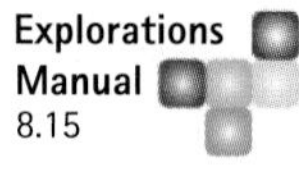

FAMILIES OF POLYHEDRA: PRISMS

Now let us investigate some of the families of polyhedra. Take a few minutes to examine the figures in Figure 8.111. How are these figures alike? How are they different? Write your thoughts before reading on. . . .

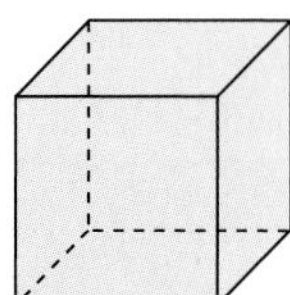
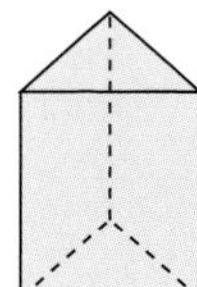
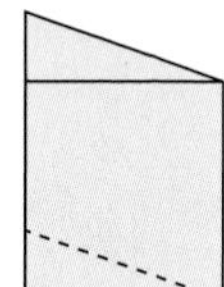
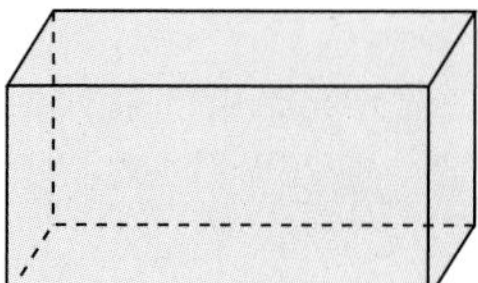

Figure 8.111

All of these figures have at least two faces that are parallel; some students would say that the top and bottom faces are parallel. And the faces are all polygons. All of these figures are called prisms. We use the word **prism** to describe all polyhedra that have two parallel faces called **bases** that are congruent polygons. It is a convention to call the other faces of prisms **lateral faces**. Even if we turn these prisms on their sides, the bases are still the parallel polygons, even though they are now on the side. When we move into volume in Chapter 9, it will be important to think of these bases as the parallel polygons, not necessarily "top and bottom."

What one shape can be used to describe the lateral faces of *all* prisms? In other words, all lateral faces of all prisms are __________. Think and read on. . . .

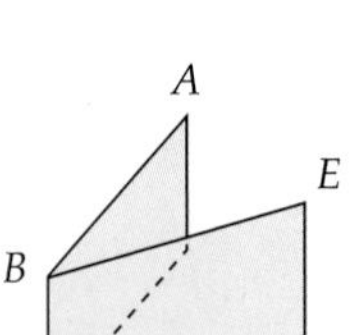

Figure 8.112

In all prisms, the lateral faces are parallelograms. In some cases, all of the lateral faces are rectangles. How would you describe the differences between those prisms whose lateral faces are nonrectangular parallelograms and those whose lateral faces are rectangles?

In the latter case, the plane of the base $\triangle ABE$ and the plane of the lateral faces are perpendicular (Figure 8.112). We could also say that the *dihedral angle* formed by either base and any lateral face is a right angle. (A **dihedral angle** is simply a three-dimensional angle—that is, an angle whose vertex is a line and whose sides are planes.)

Thus we can define a **right prism** as a prism in which the lateral faces are rectangles. Alternatively, we could define a right prism as a prism in which the angle formed by either base and any lateral face is a right dihedral angle.

A prism that is not a right prism is an **oblique prism** (Figure 8.113).

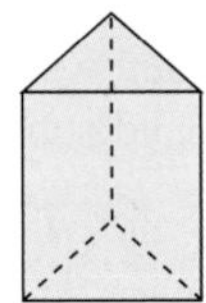
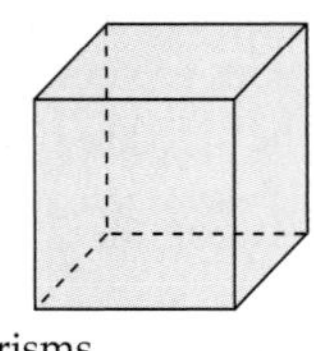
Right prisms

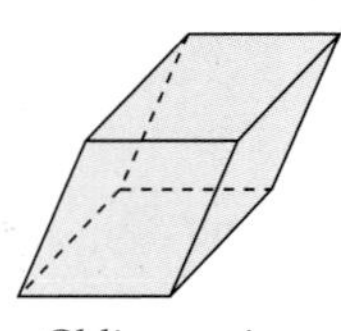
Oblique prism

Figure 8.113

LANGUAGE

As you can see from this presentation of prisms, the cube and ramp in Investigation 8.3c are both prisms.

Long before they study formal geometry, many children know the names for two special kinds of prisms.

Although this is not a term mathematicians use, what we call a **box** is actually a prism in which all six faces are rectangles. If all six faces are squares, we call the figure a **cube** (Figure 8.114).

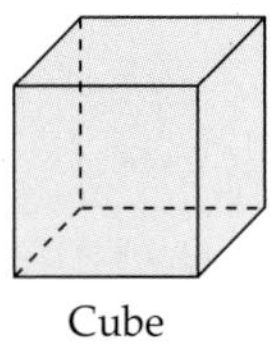
Cube

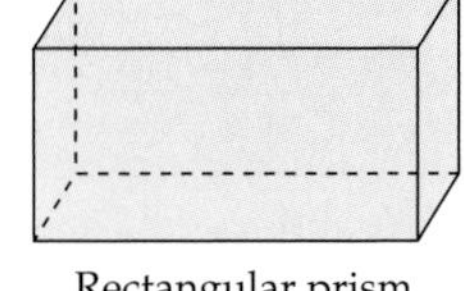
Rectangular prism

Figure 8.114

PYRAMIDS

Let us consider now another family of polyhedra. You may recognize the polyhedra in Figure 8.115 as pyramids. How might we define that term? Make your own definition and then read on. . . .

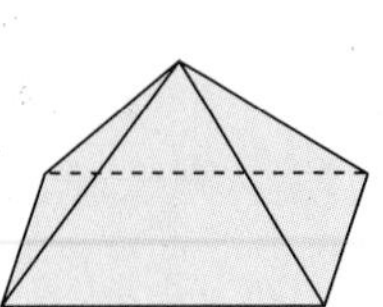
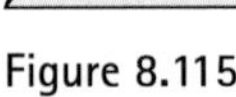
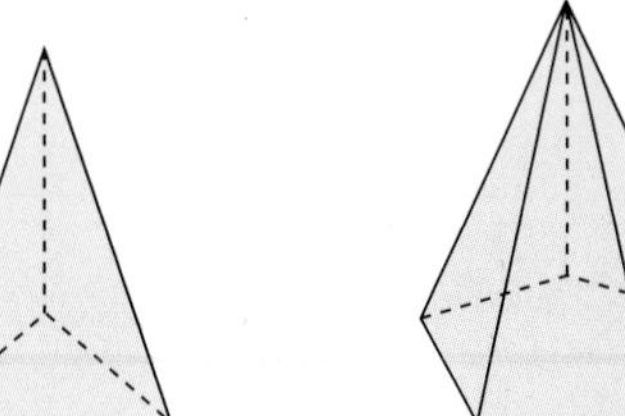

Figure 8.115

We use the word **pyramid** to describe those polyhedra whose base is a polygon and whose faces are triangles that have a common vertex. That common vertex is called the **apex** of the pyramid.

CLASSROOM CONNECTION

Grade 4

Compare your answers to these questions to the answers of other students in your class.

Date ____________________ Time ____________________

LESSON 11•2 **Modeling a Rectangular Prism** SRB 101 102

After you construct a rectangular prism with straws and twist-ties, answer the questions below.

vertices
edges
faces

1. How many faces does your rectangular prism have? ________ face(s)

2. How many of these faces are formed by rectangles? ________ face(s)

3. How many of these faces are formed by squares? ________ face(s)

4. Pick one of the faces. How many other faces are parallel to it? ________ face(s)

5. How many edges does your rectangular prism have? ________ edge(s)

6. Pick an edge. How many other edges are parallel to it? ________ edge(s)

7. How many vertices does your rectangular prism have? ________ vertices

8. Write T (true) or F (false) for each of the following statements about the rectangular prism you made. Then write one true statement and one false statement of your own.

   a. ________ It has no curved surfaces.

   b. ________ All of the edges are parallel.

   c. ________ All of the faces are polygons.

   d. ________ All of the faces are congruent.

   e. True ______________________________

   f. False ______________________________

290

From *Everyday Mathematics, Grade 4:* The University of Chicago School Mathematics Project: Student Math Journal, Volume 2, by Max Bell et al., Lesson 11-2, p. 290. Reprinted by permission of The McGraw-Hill Companies, Inc.

An alternative way to think of a pyramid is to start with any polygon and a point above the plane of the polygon. Now connect that point to each vertex of the polygon.

Most of the pyramids you have seen have square bases. However, the base can be any polygon. A pyramid is named according to its base: triangular pyramid, square pyramid, and so on.

As you have seen from our work with two-dimensional objects, the question of examining how objects are alike and how they differ is an important part of the learning process. Remember that it begins in preschool, where the teacher might give the children an assortment of buttons and have the children put them in piles so that each button belongs in one pile. The next investigation involves looking at similarities and differences.

INVESTIGATION 8.3d Prisms and Pyramids

Look at the set of prisms in Figure 8.111 and the set of pyramids in Figure 8.115. Note that these are just some examples of prisms and pyramids. What attributes do all prisms and pyramids have in common? Write your thoughts before reading on. . . .

DISCUSSION

In all prisms and all pyramids:

There are bases, although prisms have two and pyramids have one.

There are faces, edges, and vertices.

The bases and faces are polygons.

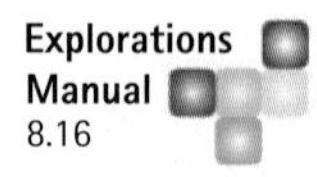

REGULAR POLYHEDRA

In Section 8.2, we discussed regular polygons. How might we define a regular polyhedron? Try to do so before reading on. . . .

One of the ways we classified polygons was by the number of sides: triangles, quadrilaterals, pentagons, hexagons, and so on. We could speak of a regular hexagon and a nonregular hexagon. However, that doesn't work with polyhedra. Do you see why?

We define a **regular polyhedron** as a convex polyhedron in which the faces are congruent regular polygons and in which the numbers of edges that meet at each vertex are the same.

Which of the prisms and pyramids we have discussed so far do you think might be regular polyhedra? Think before reading on. . . .

A cube is a regular polyhedron. A triangular pyramid composed of equilateral triangles is a regular polyhedron and has a special name, **tetrahedron**. The origin of the name is Greek: *tetra* ("four") and *hedron* ("face").

A fact that surprises many people is that there are not a large number of regular polyhedra. In fact, there are only five regular polyhedra: the tetrahedron, the cube, the **octahedron** (with 8 triangular faces), the **dodecahedron** (with 12 pentagonal faces), and the **icosahedron** (with 20 triangular faces) (Figure 8.116). The solids made from the regular polyhedra are called *Platonic solids* after the Greek philosopher Plato.

MATHEMATICS

The Five Regular Polyhedra

Tetrahedron
3 edges meet at each vertex

Octahedron
4 edges meet at each vertex

Cube
3 edges meet at each vertex

Icosahedron
5 edges meet at each vertex

Dodecahedron
3 edges meet at each vertex

HISTORY

Over 2000 years ago (long before we knew about atoms), many Greek philosopher-scientists believed that there were four basic elements from which all things arose: earth, air, fire, and water. Some of the Greeks believed that the smallest particle of earth had the form of a cube, the smallest particle of air had the form of an octahedron, the smallest particle of fire had the form of a tetrahedron, and the smallest particle of water had the form of an icosahedron. The dodecahedron was associated with the universe, probably because it was the last solid discovered, although it has been speculated that it is associated with the universe because it has 12 faces and there are 12 signs in the zodiac.[6]

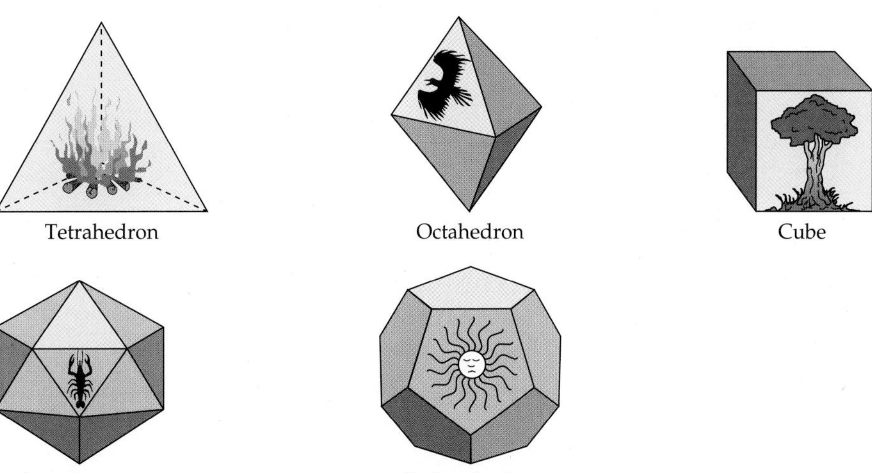

Source: The five regular solids drawn by Johannes Kepler in *Harmonices Mundi, Book II*, 1619.

Figure 8.116

OUTSIDE THE CLASSROOM

These regular polyhedra occur in nature:

- Crystals of salt and of pyrite are formed in the shape of a cube.
- Crystals of chrome alum are formed in the shape of a tetrahedron.
- Crystals of pyrite have been found in the shape of an octahedron.
- Skeletons of microscopic sea animals have been found in the shape of a dodecahedron and in the shape of an icosahedron (Figure 8.117). I didn't make this up![7]

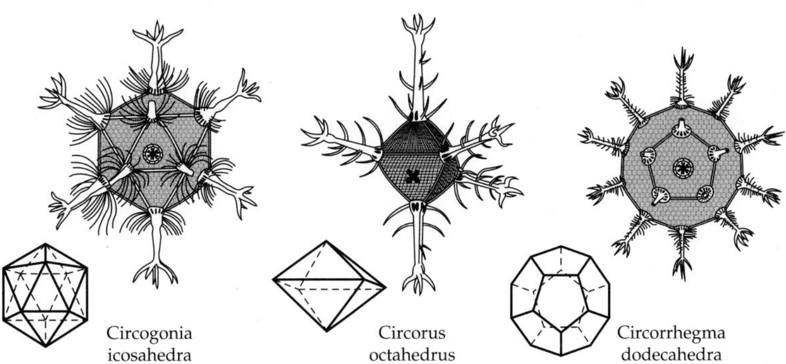

Figure 8.117

RELATIONSHIPS AMONG POLYHEDRA

A special relationship among polyhedra that we will consider here was discovered by Leonard Euler and bears his name. If you did Exploration 8.15, you discovered this on your own. If not, you will see it here. It is included both because it is a famous formula and because it is one that children can discover, with some guidance, and it is another example of the rich interconnectedness that permeates mathematics.

Any polyhedron has a certain number of faces, vertices, and edges. For example, count the number of faces, vertices, and edges of the cube on page 487, the square pyramid

Explorations Manual 8.15

on page 485, and the truncated pyramid (the polyhedron to the right of the square pyramid) on page 485. Then read on. . . .

The cube has	6 faces, 8 vertices, and 12 edges.
The pyramid has	5 faces, 5 vertices, and 8 edges.
The truncated pyramid has	6 faces, 8 vertices, and 12 edges.

There is a relationship among the number of faces (F), vertices (V), and edges (E) in any polyhedron, and knowing this relationship enables us to construct a formula that connects the number of faces, vertices, and edges. Can you guess it from these three examples?

What Euler discovered is that the sum of the number of vertices and faces is always two more than the number of edges, and this is true for all polyhedra. In symbols, we write:

$$V + F = E + 2$$

CONNECTING TWO-DIMENSIONAL REPRESENTATIONS TO THREE-DIMENSIONAL OBJECTS

There are many ways in which the two-dimensional and three-dimensional worlds connect. All buildings, from small sheds to large skyscrapers, are designed before they are built. To enable the architects and the engineers to communicate, blueprints are designed and studied. So that the electricians, plumbers, and other members of the building team will know where to place the appropriate wires and fixtures, other kinds of drawings are used. Each of these drawings requires someone to think about the object in three dimensions and then represent that information two dimensionally, although computer simulation is changing the nature of these representations.

And this is only one example. Archaeologists work with the three dimensions of the excavation site and the two-dimensional representations of the "dig." Painters need a thorough understanding of geometry so that their paintings (on a two-dimensional surface) will look like the three-dimensional objects or landscapes that they are representing. In this section, we will examine several ways in which the two-dimensional and three-dimensional worlds connect: simple (**isometric**) drawings, **cross sections**, and **nets**.

First, we will focus on simple buildings, the kind that can be made with cubes. One powerful investigation is for children to build block buildings and then give directions for making the buildings.

INVESTIGATION 8.3e Different Views of a Building

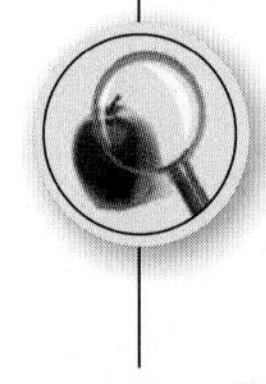

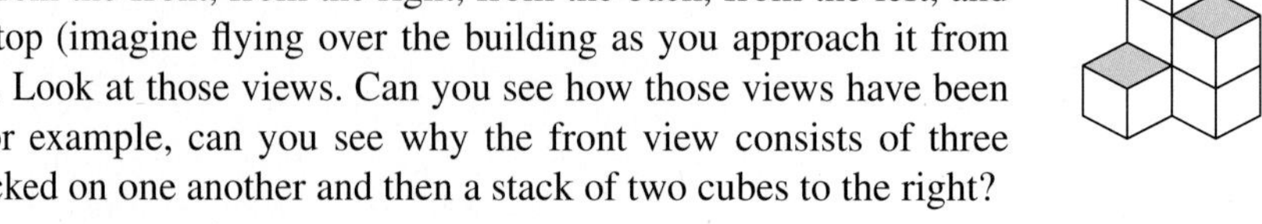

Look at the building at the right. Following are the profile views of the building from the front, from the right, from the back, from the left, and from the top (imagine flying over the building as you approach it from the front). Look at those views. Can you see how those views have been made? For example, can you see why the front view consists of three cubes stacked on one another and then a stack of two cubes to the right?

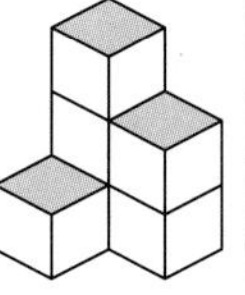

Explorations Manual 8.17

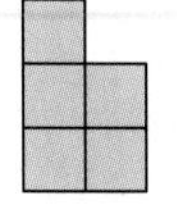
Front

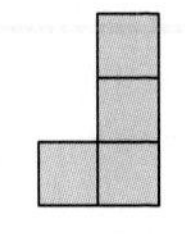
Right

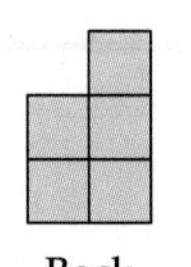
Back

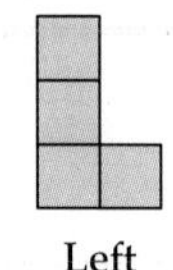
Left

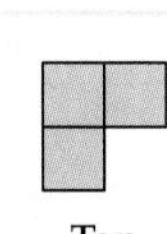
Top

Look again at those views. What do you see? The question here: Do we need all five views in order to make the figure? Why or why not? Think about this question before reading on. . . .

DISCUSSION

As you may have noticed, the right and left views are mirror images of each other. Similarly, the front and back views are mirror images of each other. Do you think this is true just in this case, in some cases, or all cases?

It turns out that it will be true in all cases. Thus we can cut out two pieces of information. For the sake of convention, we will denote the front and right-side views. What about the top? Is that really necessary? For example, if you were given only the front and right-side views of the building above, could you make the building? Think and read on. . . .

There is another building that has the same front and right-side views as the one pictured above. It is shown below. However, its top view is different. Both the building and its top view are shown below. Thus the front, right, and top views are all necessary. A curious reader might be wondering whether the top, front, and right views will be sufficient in all cases. That is a great question and will be left as an exercise.

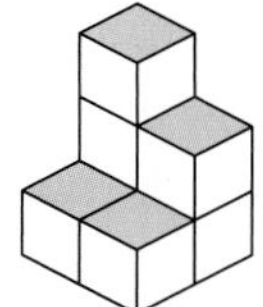

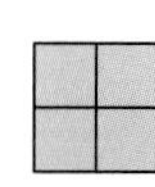

Top view

OUTSIDE THE CLASSROOM

After finishing this section in my course, a student showed me a card from the game Mindtrap, in which the players are given two views of an object and have to sketch the object. We both found the question difficult, and we disagreed with the answer!

Another way to represent our three-dimensional block buildings would be to sketch them as was done above. Before proceeding further, cover the rest of this page and look at the block building shown above. Sketch it on a blank sheet of paper. Then read on. . . .

Some of you remember being told to make a cube by first drawing two overlapping squares. If you try to use Geoboard Dot Paper, you won't draw very good models. But it turns out that Isometric Dot Paper will enable us to draw pictures of buildings quite nicely.

INVESTIGATION 8.3f Isometric Drawings

As we noted in Section 8.1, paintings became much more realistic when artists learned perspective during the Renaissance. I'm not sure when isometric drawings were developed, but they offer us a tool to sketch simple polyhedra. First, look at the Isometric Dot Paper in Figure 8.118. What do you see?

The dots are not in a rectangular array. Rather, the dots in each row are staggered so that when you connect dots, you form equilateral triangles. Believe it or not, this equilateral triangular array is an efficient way to draw objects whose angles are right angles! Now go back and try to sketch the block building from the previous investigation on the isometric grid below. Then read on. . . .

Figure 8.118

DISCUSSION

Let us begin at the beginning. Figure 8.119(a) below shows 1 cube. Figure 8.119(b) shows a column of 2 cubes, and Figure 8.119(c), show a column of 2 cubes next to a column of 3 cubes. Finally, the earlier block figure illustrates how to put columns of cubes side by side, especially when there is a blank spot in the building.

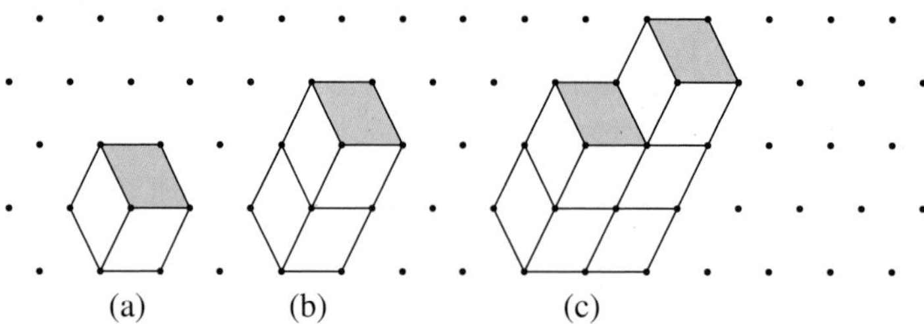

Figure 8.119

Now that we have examined buildings from different views and isometric drawings, let us examine another connection between two- and three-dimensional objects. A cross section of a solid is what the exposed face would look like if we sliced through the solid. Because there are many ways that we might slice through a solid, the shape of the cross section will depend on the nature of the slice. Let's examine a few.

INVESTIGATION 8.3g Cross Sections

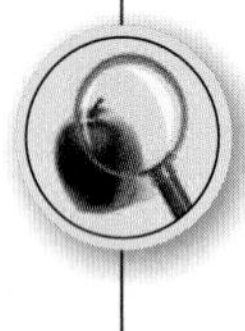

Many textbooks introduce the notion of cross sections to children through silhouettes and shadows.

This investigation strengthens your spatial visualization skills in three dimensions. Figures 8.120 through 8.122 below illustrate three different cross sections of a cube. In each case, describe the shape of the cross section. One way to visualize the task is to imagine that the cubes are made of clay and someone is cutting slices of the cube with piano wire. The piano wire is indicated by a line segment, and the two rays represent the handles that the cutter holds onto when slicing through the cube.

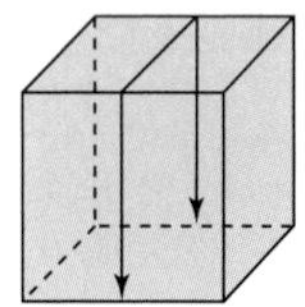

Figure 8.120

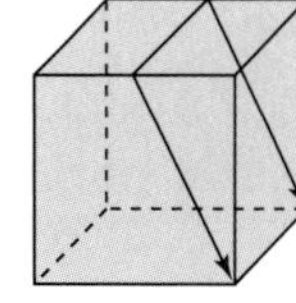

Figure 8.121

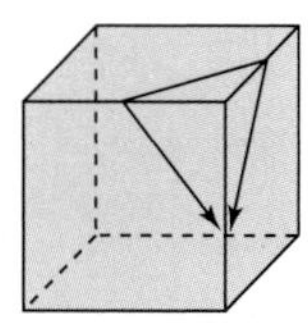

Figure 8.122

DISCUSSION

In the first case, the cross section is a square. In the second case, the cross section is a rectangle. In the third case, the cross section is a triangle. There are different ways of slicing that will result in different rectangles and triangles. Slicing three-dimensional figures and describing the two-dimensional figure that results is in the seventh grade according to the Common Core State Standards.

Nets

One last connection between two- and three-dimensional figures that we will explore here is nets. Nets are explored in sixth grade, according to the Common Core State Standards. A net is a two-dimensional representation of a three-dimensional object, in which:

1. Every face of the object is represented.
2. If you cut out the net and fold along the edges, it will fold up into the three-dimensional object.

The figure at the left in Figure 8.123 is a net of a cube, whereas the figure at the right is not. If you fold the first figure up, you will get a cube. Visualizing this can be a challenge, so you are encouraged to trace the figures, cut them out, and see if you can fold them. There are many printable nets online (google geometry nets) that are fun and good practice. Another way is to make use of the properties of a cube. I have labeled the faces: Bo, T, F, Ba, S, and S for bottom, top, front, back, and sides. Does that help? In the second case, if you cut out the figure, it won't fold up.

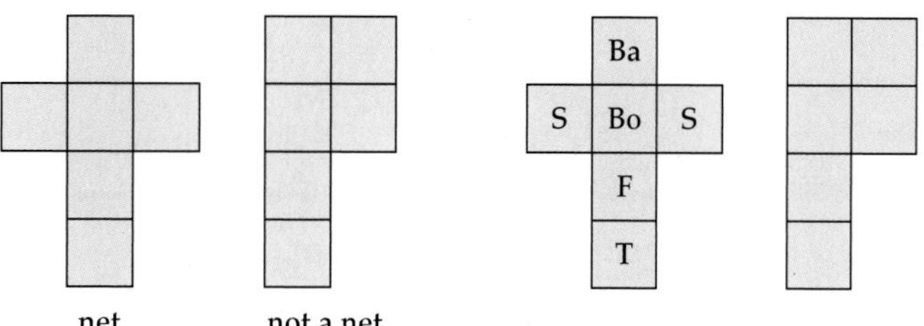

Figure 8.123

INVESTIGATION 8.3h Nets

One of the nets that people experience regularly (especially if they recycle) is a flattened cereal box. One net for a standard cereal box is shown in Figure 8.124. What do you notice about this net? This includes, but is not restricted to, the question "What attributes and characteristics do you see?"

Explorations Manual 8.19

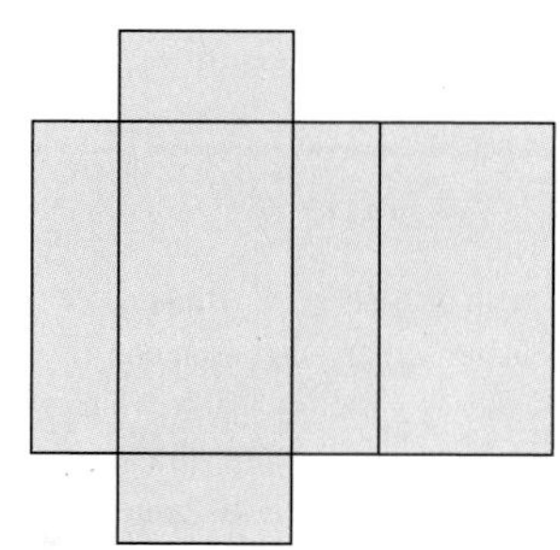

Figure 8.124

DISCUSSION

One of the key aspects of "really" understanding nets is to see certain attributes of nets and then to connect that information. One important observation is that the net has six faces. If you recall, cubes have six faces. The cereal box is a rectangular prism, and thus it has many of the attributes of a cube. Another observation is there are three pairs of congruent faces.

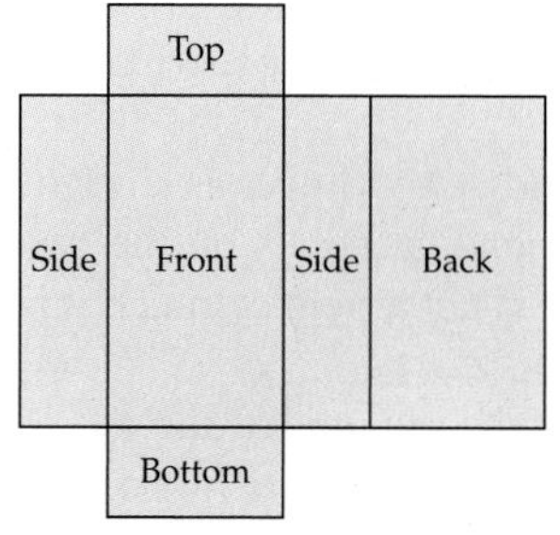

Figure 8.125

If we think of a cereal box, we think of front, back, sides, top, and bottom. If we label those faces on our net (Figure 8.125), this leads to another observation: Two congruent faces are never side-by-side. Do you see why?

A good way to deepen your understanding of the connection between the three-dimensional and two-dimensional worlds is to sketch several other nets for the cereal box. Try this yourself before reading on. . . .

One thing that makes this task easier is to realize that each face of the box is connected to three other faces. Thus we can take our original net and slide the bottom underneath the back, as shown in Figure 8.126(a)—it still folds up. In the original net, the top and bottom were connected to the front. However, on the actual box, they are also connected to the sides. Figure 8.126(b) represents that connection. Finally, we can move the back so that it is connected to the bottom, as shown in Figure 8.126(c). My students have worked on this problem and have found many, many nets!

MATHEMATICS

If you did Exploration 8.19, you found that one category of hexominoes that fold up to a cube consists of those with four cubes in a column and one cube on either side of the column. Because cubes and boxes are both rectangular prisms, some of the properties of a cube also hold true for the box. In this case, that categorization of hexominoes holds. Look at the three nets for the cereal box above. Do you see "four faces in a row and then one face on either side?"

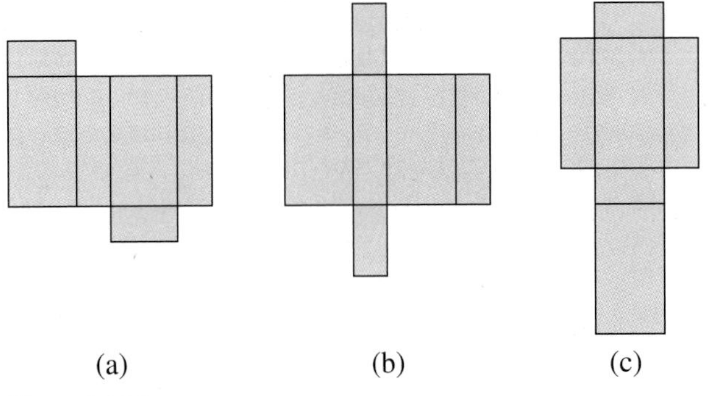

Figure 8.126

CYLINDERS, CONES, SPHERES

The polyhedra we have defined thus far have all been simple, closed surfaces in which all the faces are polygons. There are three other kinds of three-dimensional objects that are commonly found and that elementary schoolchildren study. *Cylinders, cones,* and *spheres* are related to polyhedra we have studied. Before we examine these three, stop for a moment and consider which polyhedra are related to cylinders, which to cones, and which to spheres. Then read on. . . .

MATHEMATICS

Technically, the base of a cylinder can be any simple, closed curve. In everyday life, cylinders usually have circular bases.

Think of a prism with more and more faces (see the prism at the left in Figure 8.127). At some point, a prism with a lot of faces begins to look more like a cylinder than like a prism. From one perspective, we can think of a cylinder as a prism in which the bases are circles. Technically, this is not true, because the bases of prisms are polygons, and a circle is not a polygon.

MATHEMATICS

We can also define a cylinder by using set language (remember Chapter 2?) and say that a cylinder is constructed by starting with a simple, closed curve and a line segment that has one endpoint on the curve and the other endpoint not on the curve. A cylinder consists of the union of the simple closed curve, the union of all line segments congruent to and parallel to the given line segment, and the curve formed by connecting the other endpoints of each of the line segments.

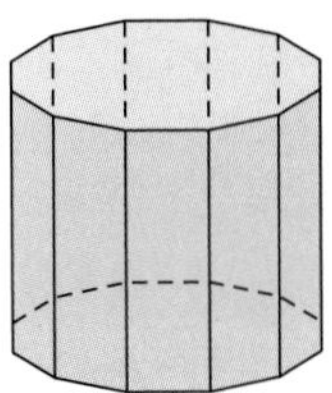
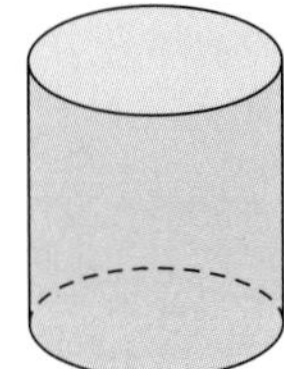
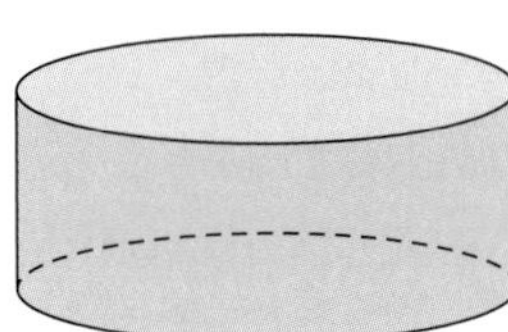

Figure 8.127

Thus we will describe a **cylinder** more formally as a closed solid that has two parallel and congruent circular bases connected by a curved surface.

Earlier, we talked about right prisms and right pyramids. A cylinder is a **right cylinder** if and only if the line segments joining two corresponding points on the two bases are perpendicular to the planes of the bases. If a cylinder is not a right cylinder, it is called an **oblique cylinder**.

Now think of a pyramid with more and more sides (see the pyramid at the left in Figure 8.128). At some point, a pyramid with a lot of sides begins to look more like a cone than like a pyramid. From one perspective, we can think of a cone as a pyramid in which the base is a circle. Technically, this is not true, because the base of a pyramid is a polygon, and a circle is not a polygon.

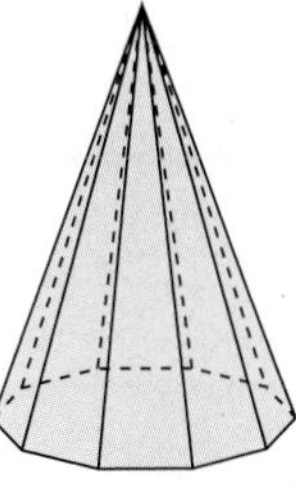
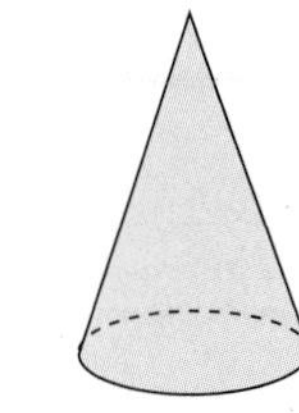
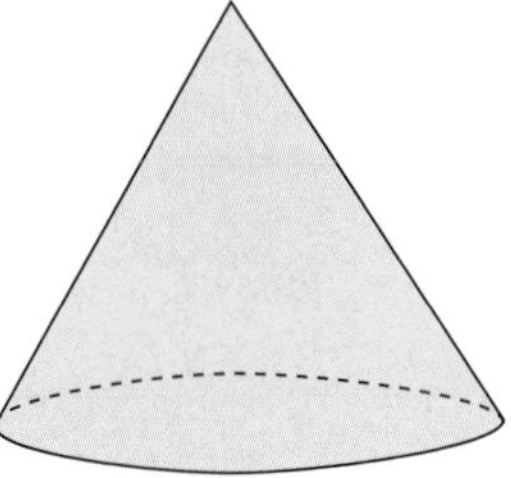

Figure 8.128

CLASSROOM CONNECTION

When children are asked to describe a cone, they say things like "A triangle with a flat bottom," "A round triangle," "A large circle with smaller and smaller circles on top until it reaches a point," "A cylinder, triangle, and a circle in one."[8]

As we did with cylinders, we can define the term *cone* by using set language and say that a cone is constructed by starting with a circle and a point not on the circle. A cone consists of the union of the circle and the union of all line segments connecting that point and the circle.

Thus we will describe a **cone** more formally as a simple, closed surface whose base is a simple, closed curve and its interior, and whose lateral surface slopes up to a vertex that we call the **apex**.

If the apex of the cone lies directly above the center of the base, then we call it a **right cone**. If a cone is not a right cone, it is called an **oblique cone**. In everyday life, we generally experience only right cones and right cylinders. Therefore, in this book, we will use the terms *cones* and *cylinders* unless referring to oblique cones or cylinders.

A sphere is conceptually related to a circle. Can you apply the earlier definition of a circle to define the term *sphere*? Try to do so before reading on. . . .

A **sphere** is the set of points in space equidistant from a given point, which is called the **center**.

How would you define the radius and diameter of a sphere? Try to do so before reading on. . . .

Any line segment joining the center of the sphere to a point on the surface is called a *radius*.

Any line segment whose endpoints lie on the surface of the sphere and that contains the center is called a *diameter*.

SUMMARY 8.3

We want to make explicit that the introduction to three-dimensional geometry was done by connecting your understanding of two-dimensional figures to three-dimensional figures. There is much research on learning that shows that when new ideas are connected to existing ideas, the knowledge is stored in more powerful ways and is more likely to be retained. Thus, the terms used to describe three-dimensional figures—face, vertex, edge—can be retained better. We learned about many kinds of three-dimensional figures. An important aspect of polyhedra is that there are families of polyhedra (just as there were with polygons) including prisms and pyramids. We saw how prisms are similar to cylinders and pyramids to cones. We found that just as there are regular polygons, so too there are regular polyhedra. Finally, we did some investigations with three-dimensional figures that are commonly found in elementary school curricula: block buildings, cross sections, and nets.

8.3 Exercises

1. This problem is a variation of one given to children called "What's My Shape?" There is a polyhedron in a bag.
   a. First clue: It has 6 faces. What might it be?
   b. Second clue: None of the faces are parallel. What might it be?
2. This problem is a variation of one given to children called "What's My Shape?" There is a solid figure in a bag.
   a. First clue: It rolls. What might it be?
   b. Second clue: Two faces are flat. What might it be?
3. Given the tetrahedron at the right, name the following:
   a. A face
   b. A vertex
   c. An edge

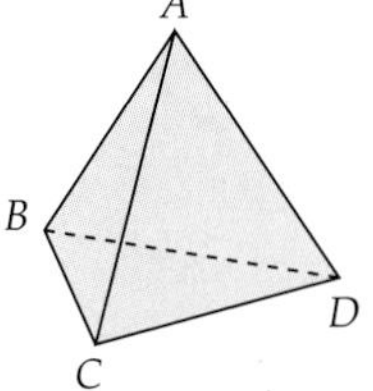

4. Identify the numbers of vertices, edges, and faces of the following figures.

a.

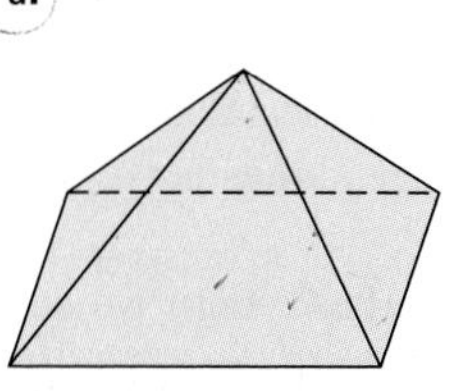

b.

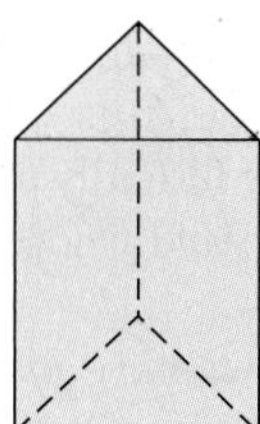

c.

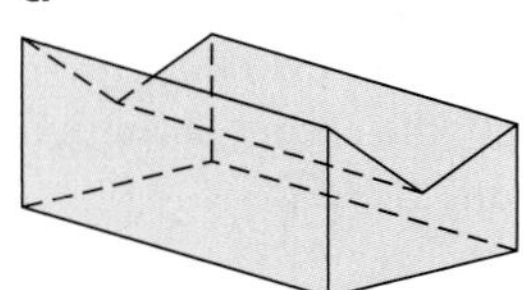

5. Name the figures below.

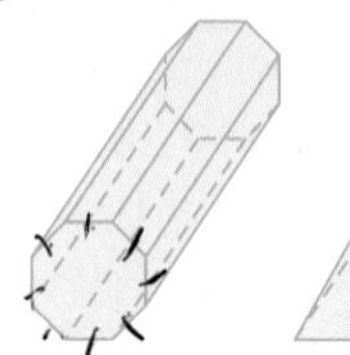
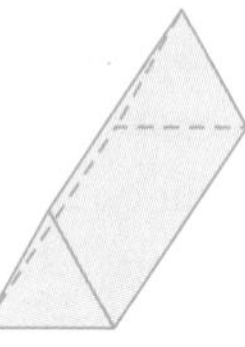
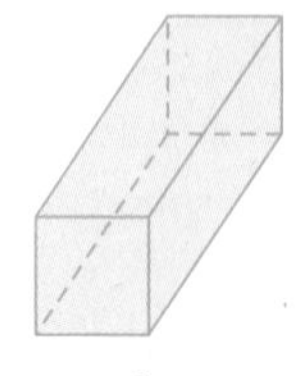
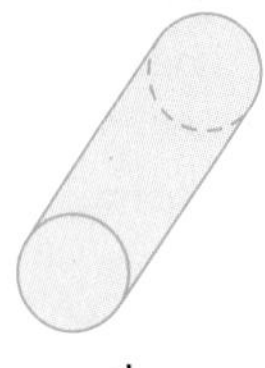

a. b. c. d.

6. Which of the polyhedra below are convex?

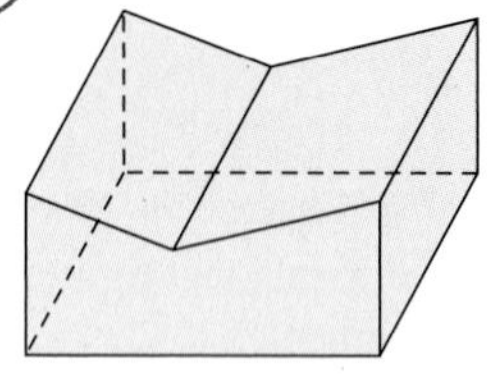
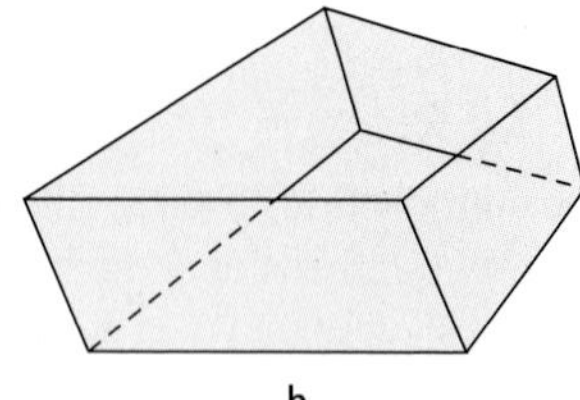

a. b.

7. a. What attributes do all cylinders and all prisms have in common that not all polyhedra have?

b. What attributes do all prisms have that only prisms have?

c. What attributes do all cylinders and cones have that not all three-dimensional figures have?

8. a. Write directions for making each of the following block buildings any way you want.

b. Write directions for the same block buildings, using a different method.

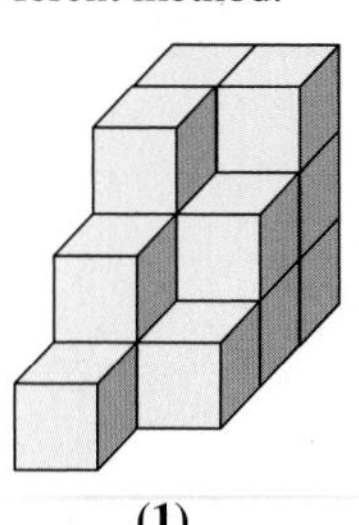

(1)

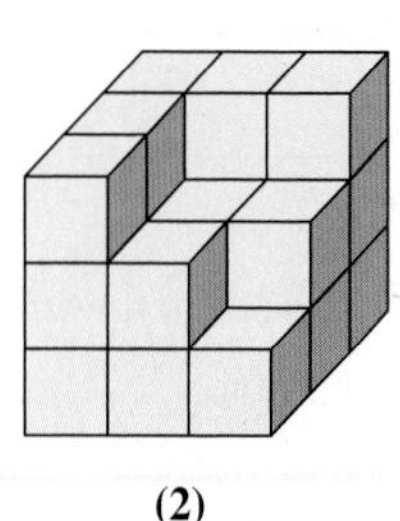

(2)

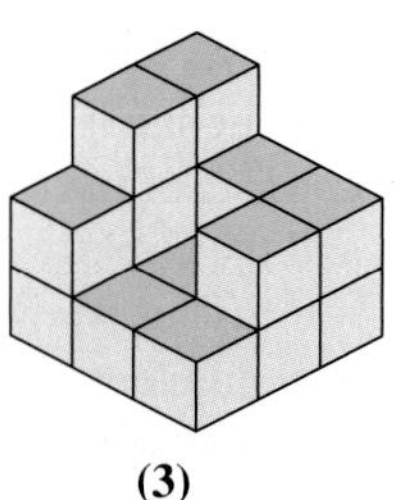

(3)

9. Sketch the figures below on Isometric Dot Paper.

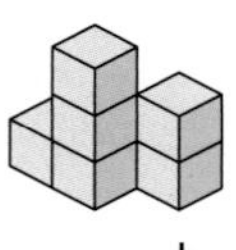

a. b.

10. Sketch the front, side, and top views of the buildings below.

Front Right Top

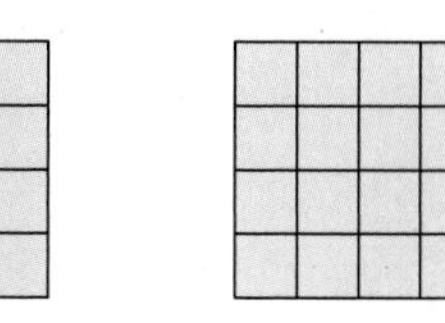

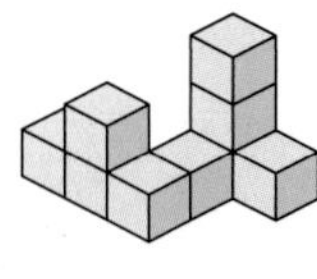

a.

b.

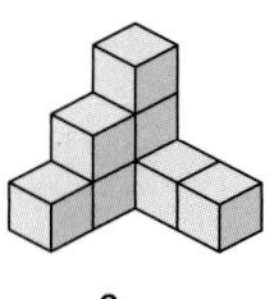

c.

11. In each of the figures below, identify the cross section formed when the plane indicated by the bold line intersects the figure. Note that some people visualize this better if they imagine the figure is made of clay, the line segment is the piano wire, and the two rays are the handles that the cutter holds onto.

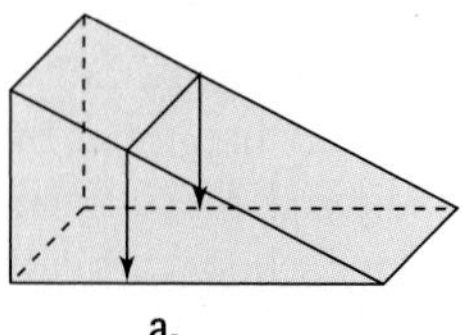

a.

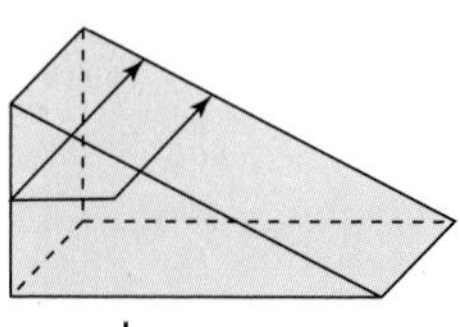

b.

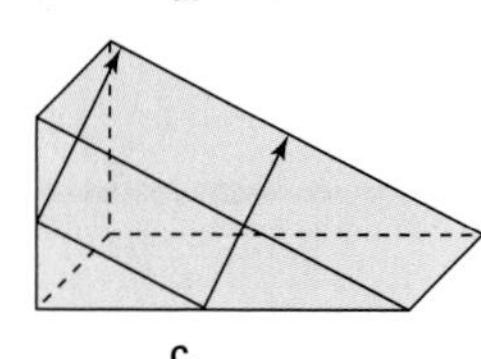

c.

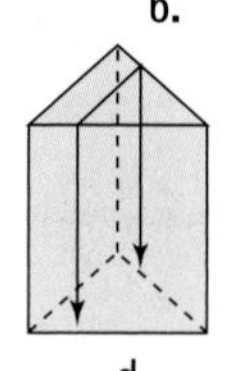

d.

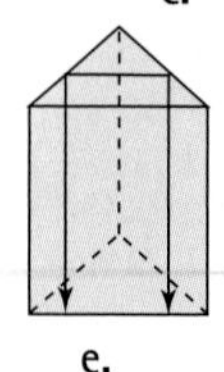

e.

12. Below are three nets for a cereal box. They count as one family because all it takes is a simple transformation—in this case a translation (slide)—to change one into another. Draw three more nets for a cereal box that are all in different families. Each net needs to have 6 whole faces; that is, do not cut one face into two or more pieces—otherwise, we have an almost infinite number of possibilities.

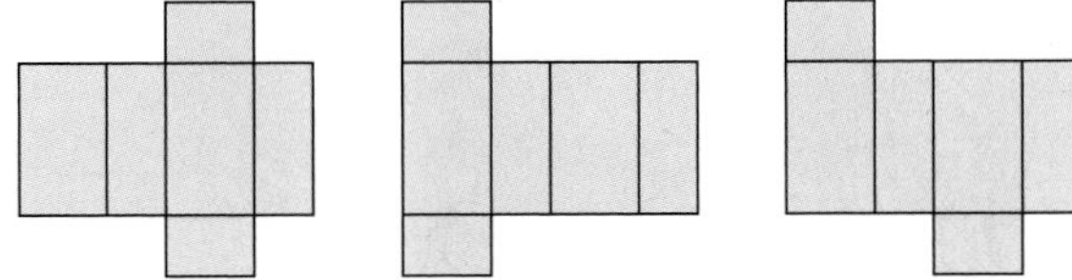

DEEPENING YOUR UNDERSTANDING

13. Consider a prism whose base is a regular n-gon—that is, a regular polygon with n sides. How many vertices would such a prism have? How many faces? How many edges? You may want to start with a triangular prism, square prism, pentagonal prism, and so on, and look for patterns.

14. Consider a pyramid whose base is a regular n-gon—that is, a regular polygon with n sides. How many vertices would such a pyramid have? How many faces? How many edges?
   a. Describe the relationship between the numbers of vertices of an n-gon prism and of an n-gon pyramid.
   b. Describe the relationship between the numbers of faces of an n-gon prism and of an n-gon pyramid.
   c. Describe the relationship between the numbers of edges of an n-gon prism and of an n-gon pyramid.

15. Draw a nonconvex rectangular or pentagonal prism.

16. We defined a regular polyhedron as a convex polyhedron in which the faces are congruent regular polygons and in which the numbers of edges that meet at each vertex are the same. Carlos says that instead of saying that the numbers of edges that meet at each vertex are the same, we could have said that the numbers of faces that meet at each vertex are the same. What do you think? Support your choice.

17. Can you make a pyramid in which the triangular faces are not all congruent?

18. Write a definition of *diagonal* for polyhedra.

19. There is a relationship between the number of diagonals of a prism and the number of vertices of a prism. That is, triangular prisms have a certain number of diagonals, square prisms have a certain number of diagonals, and so on. Determine this relationship so that you can answer the following question: How many diagonals does a prism with v vertices have? [Your instructor may or may not give you hints for this problem. If not, I suggest looking at the 4 Steps for Problem Solving on the inside front cover of this book.]

20. At the center of every tissue of toilet paper is a cardboard cylinder. Find and examine one of these cylinders. You can see a curved line running along the face of the cylinder.
   a. If you cut the cylinder along this line, what would the unfolded shape look like? Predict the shape and explain your reasoning.
   b. Why do you think these cylinders are manufactured this way instead of having a vertical cut?

21. Is there one geometric shape that describes *all* the sides of (right) pyramids? If there is, name it and justify your answer. If there is more than one shape, describe the shapes and justify your response.

22. Using Polyomino Grid Paper, make as many nets as you can for the triangular prism shown below. There are 5 faces in the figure—the 2 triangular bases and 3 rectangular lateral faces. Do not cut any of the faces—this would create many, many possible nets. That is, each net will consist of 5 distinct faces, joined together. The three lateral faces are all 1×3 rectangles, and the two bases are equilateral triangles. There are fewer than 10 possible nets.

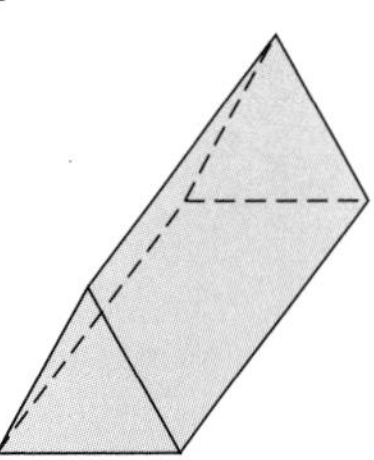

23. How many different heptominoes (made from 7 squares) are there that have 5 squares in a column and 1 square attached to opposite sides of the column? One is sketched below. Templates are given to make it easier to draw others—you will not necessarily need all of them. Describe your method, other than random trial and error. How confident are you that you have drawn all of them? Why?

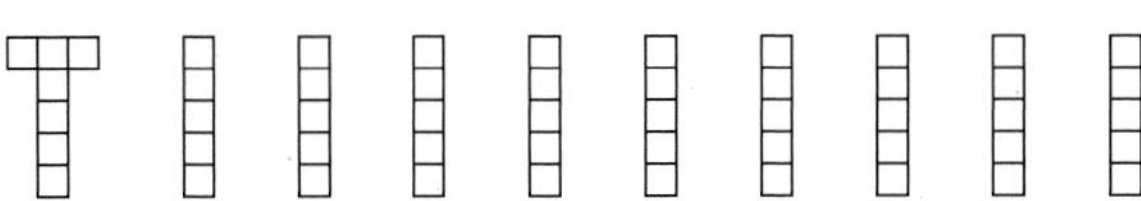

24. Below are drawings of two polyhedra. Draw the top, front, and right side views for each polyhedron.

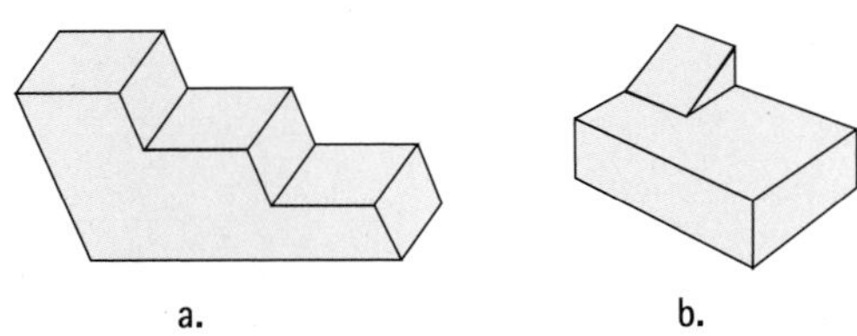

a. b.

25. Which of the nets below is a possible net for a cereal box?

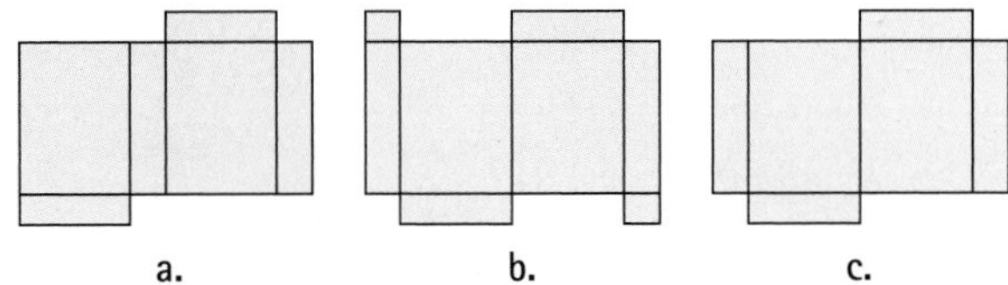

a. b. c.

26. Which of the nets below is a possible net for an oatmeal box (which has the shape of a cylinder)?

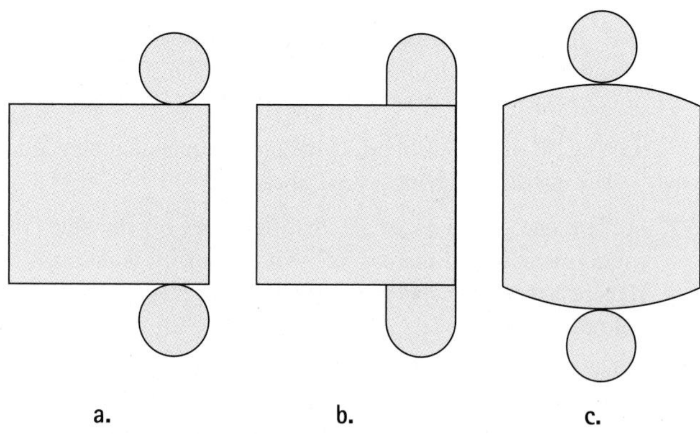

27. How would you make the octahedral die shown below?

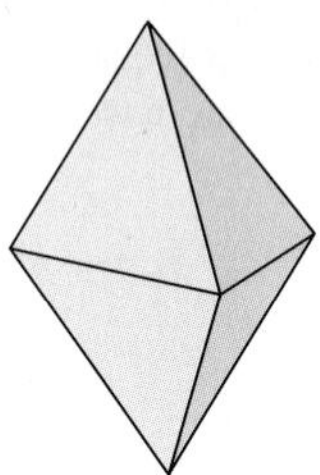

28. Most doors are rectangles. Occasionally, however, you see a door whose shape is a nonrectangular polygon. In *Lord of the Rings,* the door to Frodo's house was round. Explain why we don't see round doors very often!

FROM STANDARDIZED ASSESSMENTS

2005 NECAP, Grade 6

29. Look at these figures.

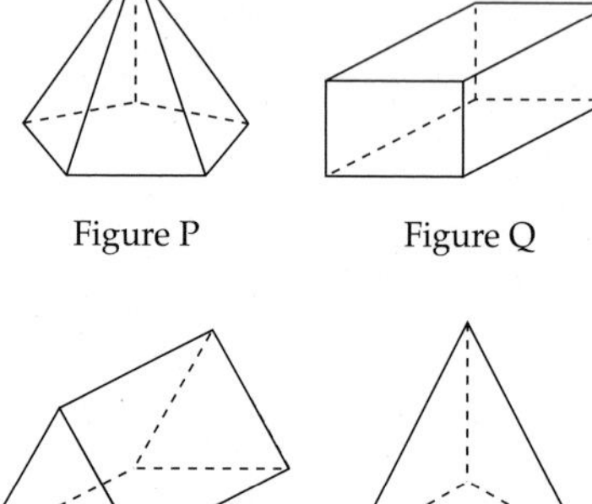

Which two figures have the same number of faces?

a. Figure P and Figure Q
b. Figure S and Figure R
c. Figure P and Figure R
d. Figure S and Figure Q

2008 NECAP, Grade 5

30. Look at this prism.

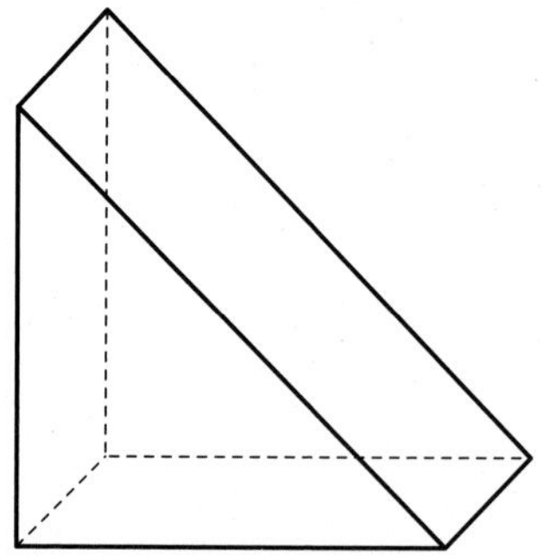

How many *rectangular* faces does the prism have?

a. 2 **b.** 3
c. 5 **d.** 6

2007 NECAP, Grade 6

31. These shapes are the 5 faces of a three-dimensional figure.

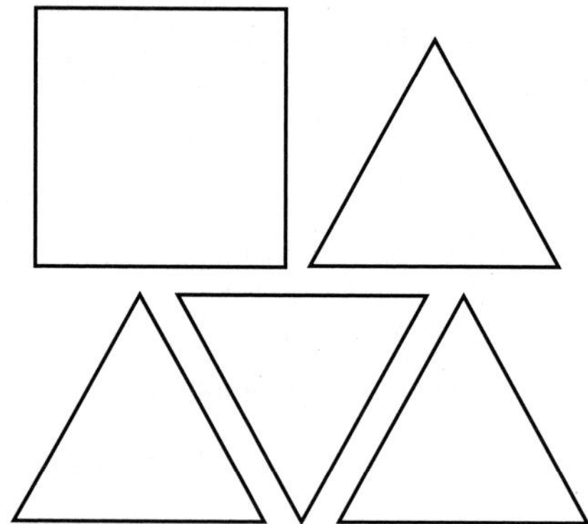

What is the three-dimensional figure?

a. cube **b.** cone
c. prism **d.** pyramid

2006 NECAP, Grade 5

32. Jack and Diane each picked a mystery solid from the ones shown below.

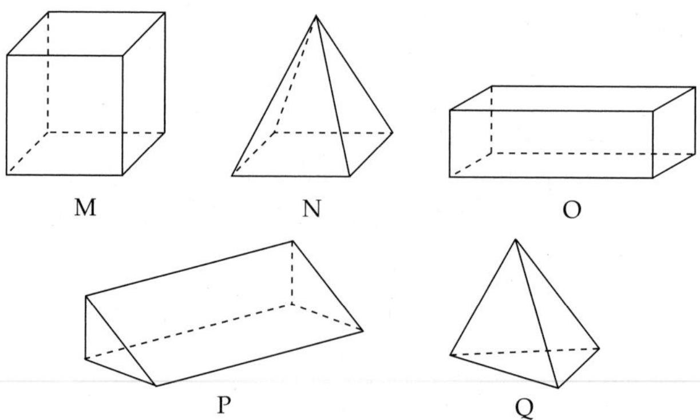

a. Here are the clues to Jack's mystery solid.
Clue 1: They mystery solid is a prism.
Clue 2: The mystery solid has 5 faces.
Which solid is Jack's mystery solid?

b. Here are the clues to Diane's mystery solid.

Clue 1: The mystery solid is a prism.

Clue 2: All of its faces are the same shape.

Which solid is Diane's mystery solid?

2007 NECAP, Grade 6

33. Juan made cones and cylinders out of paper. He cut a total of 10 circles for the bases. Juan made 4 cones. How many cylinders did he make?

a. 1 b. 2

c. 3 d. 6

34.

	B		
A		X	*D*
	C		

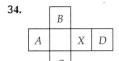

The squares in the figure above represent the faces of a cube that has been cut along some edges and flattened. When the original cube was resting on face *X*, which face was on top?

a. *A* b. *B*

c. *C* d. *D*

Source: Results of the *Seventh NAEP Mathematics Assessment.* U.S. Department of Education, National Center for Education Statistics.

LOOKING BACK on chapter eight

QUESTIONS TO SUMMARIZE BIG IDEAS

1. What are some uses of geometry?
2. How do we measure and name angles?
3. What are the van Hiele levels of geometric understanding?
4. What are ways we can classify polygons?
5. How would you respond to a child who says that a square is not a rectangle?
6. What are some relationships among different types of quadrilaterals?
7. How do we find the sum of the interior angles of any polygon? Explain the formula.
8. How do we classify three-dimensional figures?
9. What are some ways to connect two-dimensional figures with three-dimensional figures?
10. What parts of this chapter are less clear to you at this point? What will you do to clarify those ideas?
11. Look back at the Mathematical Practices of the CCSS. In what ways did you engage in those practices during this chapter?

CHAPTER  SUMMARY

1. Important spatial thinking abilities include eye-motor coordination, figure-ground perception, perceptual constancy, visual discrimination, and visual memory.
2. The van Hiele levels of geometric thinking help us to see that the level of our geometric thinking determines what we see (for example, the churn dash investigation) and how powerfully geometric ideas can be used.
3. Our current knowledge of geometry took thousands of years to develop and is the result of intuitive thinking, inductive thinking, and deductive thinking. The Greeks were the first people to develop mathematical systems, that is, a coherent system of mathematical ideas.
4. Our understanding of geometry has been built carefully on a foundation of axioms, undefined terms, definitions, and theorems. The building blocks of this knowledge are the ideas about points, lines, planes, and space.
5. Shapes are found everywhere, and virtually all shapes have functional and/or aesthetic value.
6. Each geometric shape represents the common characteristics of a set of objects. For example, the set of objects we call prisms can look very different to a novice, but they are all prisms because they have two bases that are polygons, parallel, and congruent, and they have faces that are parallelograms.

7. Every shape has multiple attributes, which means that we can classify any set of shapes in multiple ways. Recognizing and understanding these attributes helps us to understand the shape more deeply and leads to practical applications.
8. Classifying leads us to deeper understanding of mathematical structure which leads to greater mathematical power, for example, that the sum of the angles of any polygon is equal to $180(n - 2)$.
9. Looking for and recognizing relationships within and between shapes also leads to understanding of mathematical structure; for example, the many relationships among quadrilaterals (which we will further develop in Chapter 9), and seeing the relationships between prisms and cylinders.
10. Coordinate geometry is a useful tool for understanding attributes of shapes and relationships among shapes.
11. There are rich connections between two-dimensional and three-dimensional shapes. This knowledge has many practical uses and applications—drawings, representations of buildings, and nets.

BASIC CONCEPTS

Section 8.1: Basic Concepts of Geometry

hand-eye coordination **431**
figure-ground perception **431**
perceptual constancy **431**
visual discrimination **432**
visual memory **432**
Point, line, and *plane* are undefined terms. **433**
collinear **434** noncollinear **434**
coplanar **435**
Sets of points may be coplanar or not; lines may be coplanar or not. **435**
$\longleftrightarrow$ line **433** — line segment **435**
• endpoint **435** $\longrightarrow$ ray **435**
intersect **436**
Relationships between lines: perpendicular, parallel, concurrent, skew **436**
Angles have vertices and sides. **437**
Angles have an interior and an exterior. **437**

Naming angles:
with one letter, one number, three letters **438**

Measuring angles:
protractor **438** degree **438**

Kinds of angles:
right **441** acute **441**
obtuse **441** straight **441**
reflex **441** adjacent **442**
complementary **442** supplementary **442**
vertical **442**

Section 8.2: Two-Dimensional Figures

Kinds of lines and curves:
simple curves **450** closed curves **450**
simple closed curve **451** Jordan curve theorem **451**
polygons **452** vertex (vertices) **452**
sides **452**

Classifying triangles:
By length of sides: equilateral, isosceles, scalene **454**
By size of angle: right triangle, obtuse triangle, acute triangle **454**

Special line segments in triangles:
median **457** angle bisector **457**
perpendicular bisector **457** altitude **458**

Congruence:
congruent **459**

Quadrilaterals:
trapezoid **461** parallelogram **461**
kite **461** rhombus **461**
rectangle **461** square **461**

Polygons:
diagonal **463** convex or concave **463**
regular polygon **463**

Properties:
The sum of the measures of the angles of a convex polygon = $180(n - 2)$ degrees. **471**

Circles:
circle **471** radius **472**
diameter **472** chord **472**
tangent **472** arc **472**
origin **472**

Section 8.3: Three-Dimensional Figures

space figure **486** polyhedron (polyhedra) **486**
solids **486**

Parts:
face **486** edges **486**
vertices **486**

Classifying Polyhedra:
convex **486** concave **486**

Polyhedra:
prism **487** base **488**
lateral face **488** dihedral angle **488**
right prism **488** oblique prism **488**
cube **488** box **488**

Pyramid:
pyramid **488** apex **488**

Regular polyhedra:
regular polyhedron **490** tetrahedron **490**
octahedron **490** dodecahedron **490**
icosahedron **490**

REVIEW EXERCISES chapter eight

1. Without using a protractor, draw an angle that is approximately 30 degrees. Explain your reasoning. Now measure the angle. Repeat the process for a 120-degree angle.
2. In the figure at the right, $\overleftrightarrow{AE}$ and $\overleftrightarrow{CF}$ are perpendicular lines.
   a. Name two complementary angles.
   b. Name two supplementary angles.
   c. Name two vertical angles.
   d. Name two adjacent angles.
3. True or false? If true, briefly explain why. If false, provide a counterexample.
   a. If three distinct lines intersect, then they are coplanar.
   b. If two lines do not intersect, then they are parallel.
4. Why is it necessary to start with undefined terms in geometry?
5. A teacher defined *triangle* as a shape made by three line segments. By that definition, the shape at the right is a triangle. Fix the definition so that it works.

6. How many rectangles can you make that will fit on a 3 × 3 Geoboard and none are congruent?
7. How could you convince someone that the sum of the angles of any quadrilateral is 360 degrees?
8. What attributes do all rectangles have in common?
9. For each of the following, draw the figure or explain why it is impossible.
   a. A triangle that is isosceles and obtuse.
   b. A quadrilateral that has exactly one set of parallel sides and two right angles.
   c. A nonregular hexagon that has all sides congruent.
   d. A pentagon with three right angles and two sets of parallel sides.
   e. A concave pentagon.
10. How many right angles can a hexagon have?
11. Write down all the attributes of each of the following figures. Then write down the attributes they have in common.

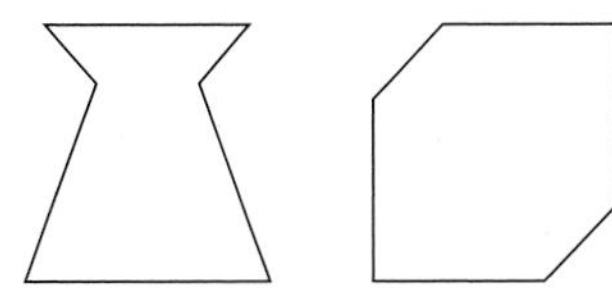

12. Name six different polygons that you can see in this figure in such a way that the reader can easily find the polygons that you have found.

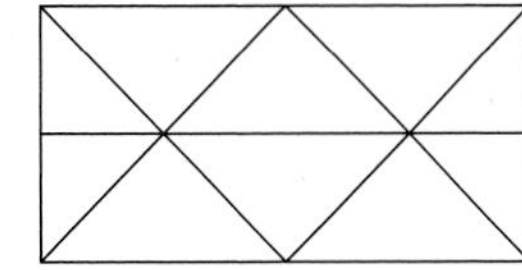

13. The four figures at the left below are kiwis. The four figures in the middle group are not-kiwis.

Kiwis

a. 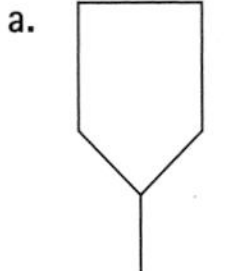
b. 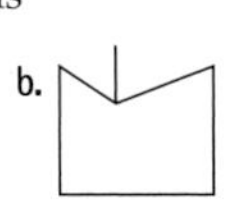
c.
d. 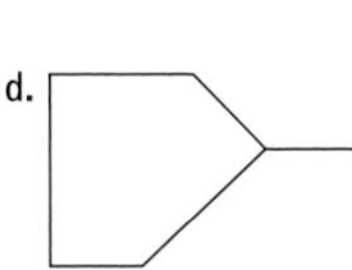

Not-kiwis

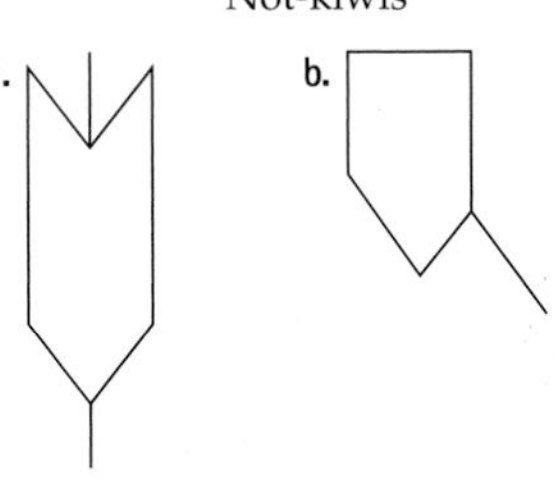

Kiwis or not-kiwis

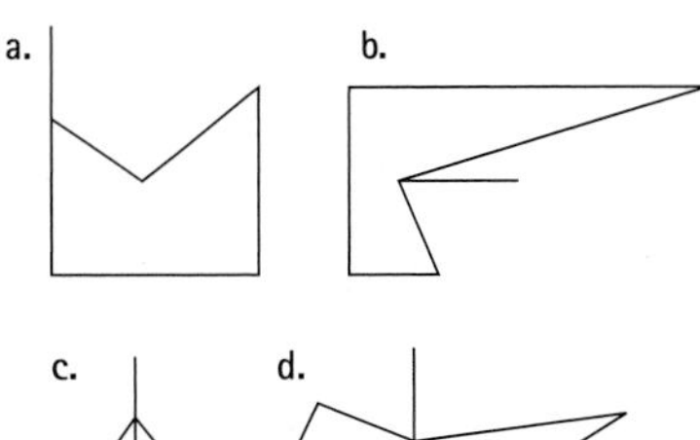

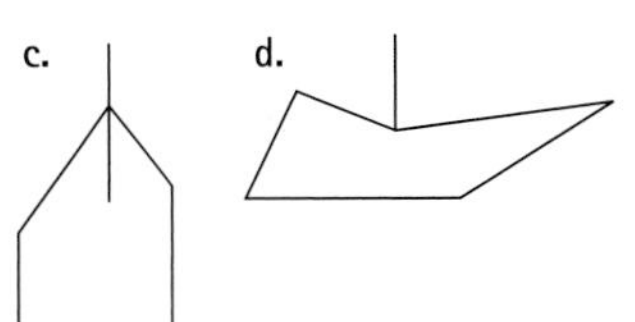

   a. Judging on the basis of the four kiwis and the four not-kiwis, list the attributes that all kiwis possess.

b. Which of the four figures in the group at the right are kiwis? Justify your answers.

14. Write directions for making the figure below. Following your directions, the reader should be able to make the same figure.

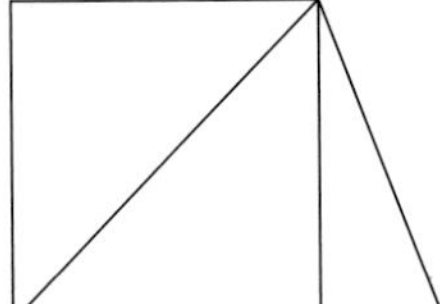

15. Is the Venn diagram below a valid representation of the relationship between parallelograms and rectangles? If yes, explain why. If not, explain why not.

Parallelograms Rectangles

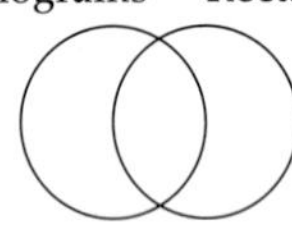

16. Draw a Venn diagram to represent the relationship between regular polygons and convex polygons.

17. Find the midpoint of the line segments connecting these pairs of points: (0, 3) and (5, 12).

18. Three vertices of a square are (5, 0) (5, 6), and (8, 3). What are the coordinates of the fourth vertex?

19. I am a parallelogram. Two of my sides are parallel to the bottom of the paper. The endpoints of one side are at (2, 3) and (8, 3). The slope of the line segments that form my two sides is 0.75, and the length of each side is 5 units. What are my other two vertices?

20. Identify the number of vertices, edges, and faces of the accompanying figure (*right*).

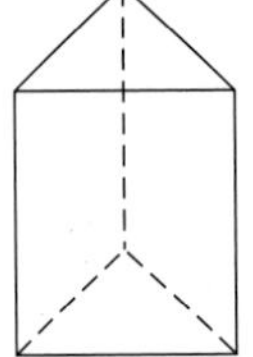

21. What attributes do all cones and pyramids have in common that not all three-dimensional figures have?

22. a. Write directions for making the block building at the right any way you want.

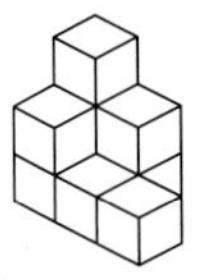

b. Write directions using a different method.

23. Sketch the figure at the right on Isometric Dot Paper.

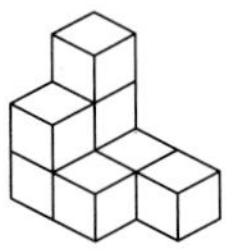

24. Sketch the front, side, and top view of the building above.

25. Following are the top view, the front view, and the right-side view of a building created only by cubes.

a. Without making the building, predict how many cubes it will have. Explain your prediction.

b. Predict the left-side view. Explain your prediction.

c. Sketch the building on Isometric Dot Paper.

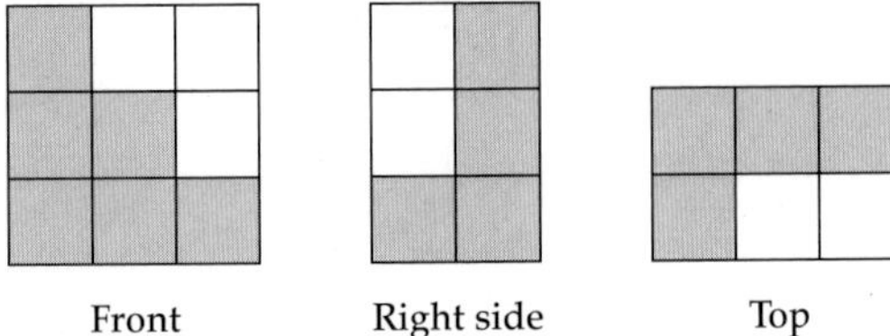

26. a. Draw a net for a prism whose base is a right isosceles triangle.

b. Draw a different net for the same prism.

27. Write a definition of *diagonal* for polyhedra.

28. Describe the polygon formed by the following cross sections.

a. You cut a cross section that is perpendicular to the base of a square pyramid but does not pass through the apex of the pyramid.

b. You cut a cross section that is perpendicular to the base of a circular cylinder.

29. Below is a drawing of a polyhedron. Draw the top, front, and side views for this polyhedron.

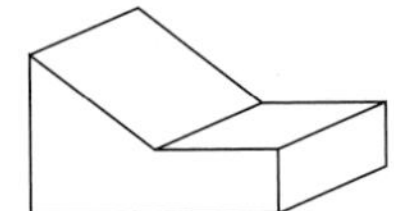

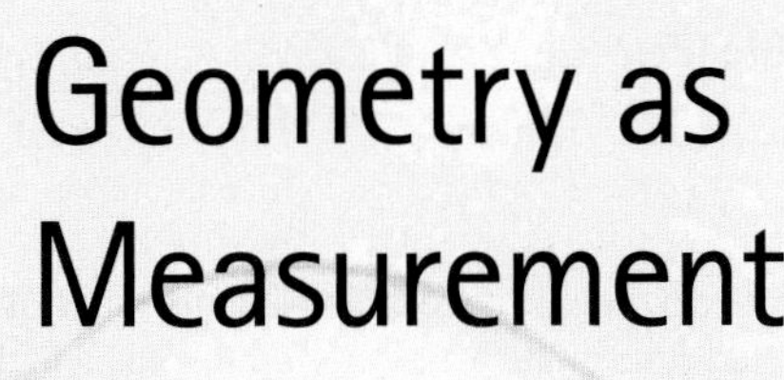

Geometry as Measurement

What do you think of when you hear the term *measurement*? In what ways do you use measurement, and in what ways are you measured?

We measure and are measured all the time.

- How much time will it take?
- How far is it from here to there?
- How much paint do we need to paint the house?

Explorations Manual 9.1, 9.2, 9.3

Long before they start school, children are fascinated by measurement, and some of their measurement questions leave parents confused, too:

The child asks, "How long until we get there?" The parent says, "One hour." The child asks, "Is that long?"

Or the parent divides the candy bar into two pieces (measures), and one child says, "His half is bigger than mine."

By the end of this chapter, you will come to understand that measurement is far more than just an examination of the quantities we use in everyday life, such as length, area, volume, and time.

In this chapter, we will investigate several important aspects of measurement.

Big Idea	Example/Elaboration
1. Most objects have multiple attributes that are measureable.	Consider a garage that is not attached to a house. We might want to know its perimeter, for a fence; or the surface area of the walls, to paint it; or its volume, to know how much it will hold.
2. There is a process that we use to measure all attributes.	This process is the same whether we are measuring distance or density.
3. The units that we use are arbitrary as opposed to absolute.	All units were invented by people.
4. Virtually all measurements are approximations.	There is no "exact" height of a building. How we report the height depends on the precision that is needed: 12 m, 12.2 m, 12.21 m, 12.214 m, etc.

(continued)

Big Idea	Example/Elaboration
5. Precision and accuracy are not the same thing.	A measurement can be very precise but still be inaccurate, or it can be accurate and not be very precise.
6. Different attributes of an object are generally not related in simple ways.	Doubling the area of a garden does not necessarily double the length of the fence around the garden.

SECTION 9.1 Systems of Measurement

What do you think?

- Why is measurement important?
- When do we need our measurements to be precise, and when is it acceptable to have just a "rough" measurement?
- How do we measure heights of mountains?

Visualize yourself measuring something—for example, how far it is to the beach, how much paint you need to paint a house, how much a package weighs. We might want several measures of something. Consider a room. What measurements would we need if we wanted to stencil the top of the walls around the room (the perimeter of the room)? What measurements would we need if we wanted to paint the room (the surface area of the room)? What measurements would we need if we wanted to fill the room with boxes (the volume of the room)? Think of the process of measuring in each of these situations. What is common to all of them? Think and then read on. . . .

In each case, we must

1. Identify the *attribute* we want to measure.
2. Select or determine the *unit* we will use to tell us "how much."
3. Determine the *amount* we have in terms of the units we have chosen.

Keep these three aspects in mind as you read the text and do the investigations.

The Common Core State Standards link measurement with data in grades K–5, since data we analyze is often a measurement of something. In kindergarten, students begin by describing measurable attributes and comparing sizes of objects. By second grade, students are using standard units (inches, meters, etc.) to measure as well as working with units of time.

DEVELOPMENT OF MEASUREMENT SYSTEMS

Measuring is probably one of human beings' earliest forms of mathematical activity. In the following pages, we will examine the development of the units and systems of measurement that people commonly use in everyday life.

Time Many thousands of years ago, people developed ways to determine the number of days in a year. Why do you think they wanted to know this?

There are many reasons, such as wanting to know when to celebrate certain rituals, when to hunt, or when to plant. Early divisions of the day were probably marked by three significant times: sunrise, midday, and sunset. Later, midnight became significant too. We say that 1 day = 24 hours, 1 hour = 60 minutes, and 1 minute = 60 seconds. However, this has not always been so. Where do you think 24 and 60 came from? Think and then read on. . . .

Over 3000 years ago, the Babylonians divided the day into 12 hours and the night into 12 hours. However, the length of an hour depended on the time of year; that is, in the winter, a day hour was shorter than a night hour. At some point, the Greeks decided to divide the entire day into 24 equal parts, but it was not until around 1330 B.C. that the hour was standardized—that is, that an hour in January and an hour in June were equal, and an hour in Stockholm and an hour in Venice were equal.

LANGUAGE

The origins of our words for the days of the week can be traced to the Babylonians and correspond to the seven "planets" they had identified: Sun, Moon, Mars, Mercury, Jupiter, Venus, and Saturn. Which ones do you recognize? What might have happened to the names of the other days?
Hint: Look up the Latin names of the days of the week.

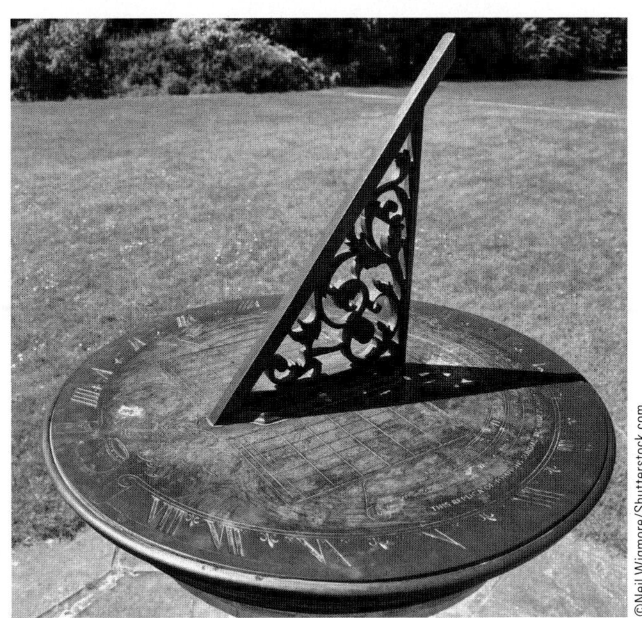

©Neil Wigmore/Shutterstock.com

For centuries, people used the sundial to measure the passage of time. The earliest known sundial comes from Egypt and dates from about 1500 B.C. The sundial, of course, has several limitations: It is not useful on cloudy days or at night, and it is not useful for small amounts of time. An early attempt to address this limitation was the water clock. The first water clock in Egypt dates from about 1400 B.C., and the first evidence of an hourglass is from ancient Rome in about 300 B.C. Columbus used a sand clock to measure time on his voyage.[1] It was not until the 1700s that the mechanical watch (with springs) was both accurate enough and inexpensive enough to be used outside of scientific experiments. The present definition of a second is the duration of 9,192,631,770 periods of the cesium-133 atom![2] Believe it or not, there are machines that can count these periods. The ability to measure accurately time intervals of one-billionth of a second is crucial in fields like particle physics.

Length The history of the various units that different peoples have selected to measure length (Figure 9.1) is quite fascinating.

The earliest recorded linear unit of measurement was the cubit, the distance from the point of the elbow to the outstretched tip of the middle finger. The word *cubit* is derived from the Latin *cubitum*, meaning "elbow."

It should not be surprising that the foot was one of the units of measurement in ancient times. The Roman writer Plutarch stated that the foot was based on the actual length of Hercules' foot. It is likely that the French foot was originally the actual measurement of the length of King Charlemagne's foot.[3] In the tenth century, King Edgar I decreed that the yard would be the distance from the tip of his nose to the tip of the middle finger of his outstretched arm. The reason behind the decree was a desire to regulate trade in textiles, so that 1 yard of cloth would be approximately the same in all parts of England.

LANGUAGE

In the language of King Henry I (1496), the word for yard literally meant "stick." Thus our term *yard-stick* is redundant. This is similar to speaking of the river that flows along the border between Texas and Mexico as the Rio Grande River. In Spanish, Rio Grande literally means "big river."

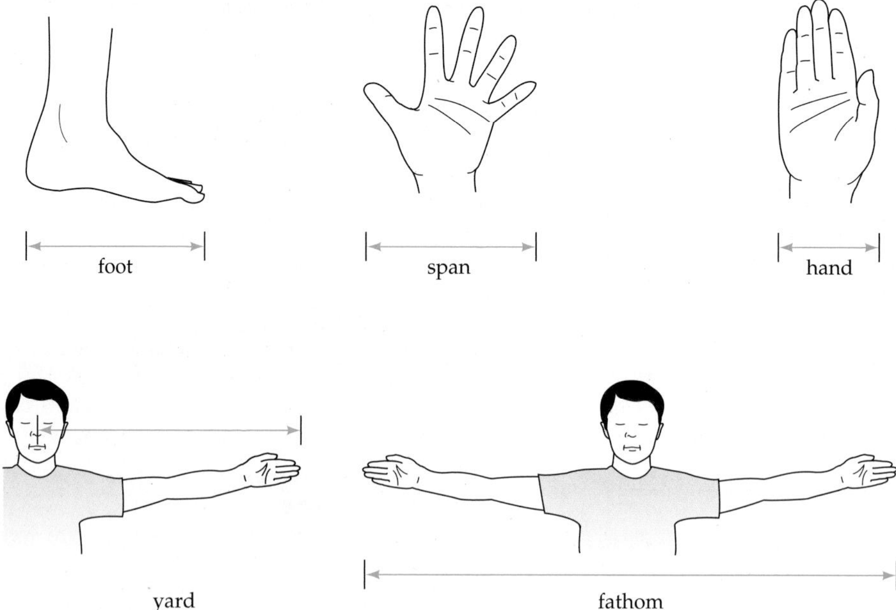

Figure 9.1

CLASSROOM CONNECTION

Virtually all mathematics textbooks now have young children first measure lengths with nonstandard units. For example, they might ask students to find the length of a desk in terms of paper clips. Why do you think teachers are being encouraged to start with nonstandard units?

There are several reasons. First, it helps to keep the focus of the students' thinking on the attribute that is being measured and learned. If you did Exploration 9.2, you may have seen this concept in action. For example, it is much simpler (conceptually) to line up and count paper clips than it is to line up and count units on a ruler. Second, the answers can be more understandable to the children—for example, the length of the soccer field in terms of footsteps or the number of body lengths of one student in the class. An important concept that comes out of measuring with number of body lengths, or footsteps, is that the number of body lengths needed to measure depends on the length of the body being used. For example, it would take more body lengths of an elementary school student (the smaller unit) to measure a distance than it would take if the body length of LeBron James (the larger unit) were used. This later translates to converting from inches to feet and knowing we would have a smaller number of feet than if we measured an item in inches.

CLASSROOM CONNECTION

Eunice Hendrix-Martin uses the children's book *Biggest, Strongest, Fastest* to spark interest in measurement. In order to get a sense of the size of large animals, the students determine the length of the animals in relation to the students' height. For example, a blue whale is about as long as 24 students. First the students had to determine the average height of a student in the class, and then they had to use proportional reasoning to determine the length of the whale in standard units. The article describing this activity appears in the April 1997 issue of *Teaching Children Mathematics,* pp. 426–430.

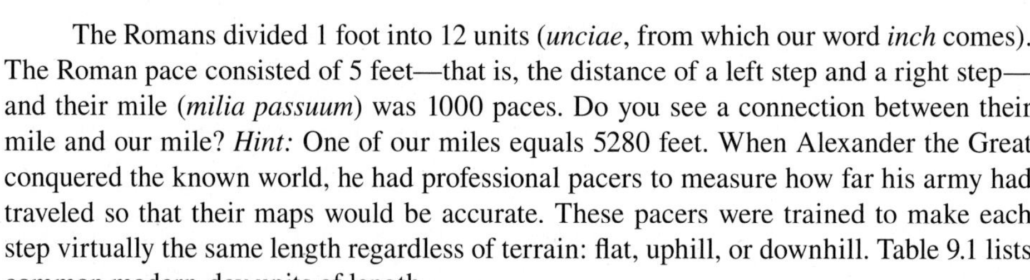

The Romans divided 1 foot into 12 units (*unciae*, from which our word *inch* comes). The Roman pace consisted of 5 feet—that is, the distance of a left step and a right step—and their mile (*milia passuum*) was 1000 paces. Do you see a connection between their mile and our mile? *Hint:* One of our miles equals 5280 feet. When Alexander the Great conquered the known world, he had professional pacers to measure how far his army had traveled so that their maps would be accurate. These pacers were trained to make each step virtually the same length regardless of terrain: flat, uphill, or downhill. Table 9.1 lists common modern-day units of length.

TABLE 9.1

Units	Abbreviation	Relationship to other unit(s)
1 inch	in	
1 foot	ft	12 inches
1 yard	yd	3 feet = 36 inches
1 mile	mi	1760 yards = 5280 feet

Weight Two ancient units for weight were the barleycorn and the seed of the carob plant (from which comes the word *carat* for the measure used to weigh gold and diamonds).

Grade 5

This activity is similar to Exploration 9.1 in the *Explorations* manual. 

Date ______________________ Time ______________________

LESSON 2·1

Estimation Challenge

Sometimes you will be asked to solve a problem for which it is difficult or even impossible to find an exact answer. Your job will be to make your best estimate and then defend it. We call this kind of problem an Estimation Challenge.

Estimation challenges can be difficult, and they take time to solve. Usually, you will work with a partner or as part of a small group.

Estimation Challenge Problem

Imagine that you are living in a time when there are no cars, trains, or planes. You do not own a horse, a boat, or any other means of transportation.

You plan to travel to ______________________ . You will have to walk there.
(location given by your teacher)

Information needed to solve the problem.

Definition of a *step*. 1 mile = 5,280 feet

Length of a step

1. About how many miles is it from your school to your destination?

   About ______________________ miles

2. a. About how many footsteps will you have to take to get from your school to your destination?

   About ______________________ footsteps

   b. What did you do to estimate the number of footsteps you would take?

   ____

   ____

   ____

29

From *Everyday Mathematics, Grade 5:* The University of Chicago School Mathematics Project: Student Math Journal, Volume 1, by Max Bell et al., Lesson 2-1, p. 29. Reprinted by permission of The McGraw-Hill Companies, Inc.

Over the centuries, three different systems of weights evolved: the avoirdupois system for everyday use, the troy system to weigh precious metals and gems, and the apothecaries' system for very small amounts. However, conversion was confusing. For example, in the troy system, there were 12 ounces in a pound, whereas in the avoirdupois system, there were 16 ounces in a pound!

Volume and capacity Over time, two related systems of measure for what we call volume evolved, depending on whether the thing being measured was dry or wet. For dry volume, some basic units were the ounce, pint, quart, peck, and bushel. For liquid volume, there were a host of units. Thomas Jefferson almost certainly memorized the following ditty, which was commonly used in the eighteenth century to teach schoolchildren the units:

> Two mouthfuls are a jigger; two jiggers are a jack; two jacks are a jill; two jills are a cup; two cups are a pint; two pints are a quart; two quarts are a pottle; two pottles are a gallon; two gallons are a pail; two pails are a peck; two pecks are a bushel; two bushels are a strike; two strikes are a coomb; two coombs are a cask; two casks are a barrel; two barrels are a hogshead; two hogsheads are a pipe; two pipes are a tun—and there my story is done![4]

HISTORY

The familiar nursery rhyme Jack and Jill was written as a protest against King Charles of England's taxation of jack(pots) of liquor. King Charles not only lost his crown, he lost his head! (He was beheaded.)

Which measures do you recognize? Table 9.2 lists common modern-day units of volume.

TABLE 9.2

Units	Abbreviation	Relationship to other unit(s)
1 teaspoon	tsp	
1 tablespoon	tbsp	3 teaspoons
1 liquid ounce	oz	2 tablespoons
1 cup	c	8 liquid ounces
1 pint	pt	2 cups or 16 ounces
1 quart	qt	2 pints or 32 ounces
1 gallon	gal	4 quarts or 128 ounces

Temperature In 1714, a German instrument maker named Gabriel Fahrenheit made the first mercury thermometer. He designated the lowest temperature he could create in the laboratory as 0° and the normal temperature of the body as 98°. On his scale, the freezing point of water is 32° and the boiling point of water is 212°.

Units of measure have evolved through the years, and some of the ones mentioned so far we still use today. Now let's turn our attention to understanding the metric system, which is a base ten system (like our numeration system) and is used by most of the world.

HISTORY

In 1785, Congress passed Thomas Jefferson's proposal for a decimal system of currency. On July 4, 1790, Thomas Jefferson presented to Congress a report on weights and measures that included two proposals: the first a complete transformation of our system of weights and measures based on powers of 10, and the second a plan for a partial transformation. George Washington supported Jefferson's efforts. However, the Congress decided to adopt neither plan.

THE METRIC SYSTEM

The idea of a system of measures based on powers of 10, which is what the **metric system** is, was first proposed in 1670 by Gabriel Mouton, from Lyons, France. Over the next 100 years, many scientists and nonscientists made various proposals for a uniform system of measurement. Before the development of the metric system, there were literally hundreds of systems of measurement in Europe alone. What is particularly important is that they were not uniform, so a bushel in one location was not the same as a bushel in another location. These differences were often used by the rich to exploit the poor, who obviously resented this. One of the first acts of the French government after the French Revolution in 1789 was to develop a uniform system of measurement so that the rich could not cheat the poor. In 1793, the French Academy of Sciences proposed a new metric system for *all* units of measurement.

The United States has had an interesting history with the metric system. In 1866, because of trade with Europe, Congress passed an act saying that the use of the metric system was allowed. In 1975 Congress passed the Metric Conversion Act as an attempt to get the United States to adopt the metric system, but this was met with much resistance from consumers. In 1988, because of global trading, Congress passed the Omnibus Trade and Competitiveness Act, which designated the metric system as the preferred system in trade and commerce. In spite of all this, the metric system is still not used completely in the United States. (Our system is called the **U.S. customary system.**) However, metric units are finding their way into everyday life—for example, some road markers now give distances in kilometers as well as in miles, and most soda is now sold by the liter. Furthermore, international travel is so common that many Americans encounter the metric system in their travels, and most industries that compete in international markets have gone metric, since most of the rest of the world uses the metric system. The Common Core State Standards expect students to understand both metric and the U.S. customary system beginning in second grade, although we no longer have to convert between the two systems.

There is one feature of the metric system that it is helpful to know before we examine the different units. Once the French had determined the unit, whether it was meter for length, liter for volume, or gram for mass, they used Greek prefixes for *multiples* of this unit and Latin prefixes for *fractions* of this unit.

Several of these prefixes show up in familiar words; for example, a millennium is 1000 years, a century is 100 years, and a decade is 10 years.

Table 9.3 shows the metric prefixes and their relationship to the basic metric units. Note that in everyday life, we rarely use *deci-*, *deca-*, or *hecto-*.

MATHEMATICS

In 1999, NASA lost a $125 million Mars orbiter because one team working on the project was using U.S. Customary units of measurement while another team was using metric units.

TABLE 9.3

Prefix	Milli	Centi	Deci	Deca	Hecto	Kilo
Relationship to the basic unit	$\frac{1}{1000}$	$\frac{1}{100}$	$\frac{1}{10}$	10	100	1000
Example	milliliter	centimeter				kilogram

We also use metric prefixes when we describe the memory on our computers: 1 kilobyte = 1000 bytes; 1 megabyte = 1 million bytes; 1 gigabyte = 1 billion bytes; and 1 terabyte = 1 trillion bytes. Imagine someday talking about 1 yottabyte which is 1 septillion (that is 24 zeroes!) bytes. Also as we begin to study smaller units, we use the metric system prefixes micro- (1 millionth), nano- (1 billionth), and pico- (1 trillionth). Nanotechnology, for example, is an emerging field focused on the study of extremely small things such as atoms.

METRIC LENGTH

The standard unit of length in the metric system is the **meter**, which was defined (most likely after much debate) as one ten-millionth of the length of the line that starts at the equator and goes to the North Pole through Barcelona, Paris, and Dunkirk. In other words, this distance was decreed to be 10 million meters, and the meter was the length that was one ten-millionth of that distance. How they decided upon 10 million and how they knew the exact distance to the North Pole is an interesting story in itself—it took 7 years to accomplish! The most commonly used metric units of length are, from largest to smallest, the *kilometer, meter, centimeter,* and *millimeter.* Although units like dekameter are not used very often, they are shown in Table 9.4 to help us see the entire base ten system.

TABLE 9.4		
Units	**Abbreviation**	**Relationship to meter**
1 millimeter	mm	$\frac{1}{1000}$ of a meter
1 centimeter	cm	$\frac{1}{100}$ of a meter
1 decimeter	dm	$\frac{1}{10}$ of a meter
1 meter	m	1 of meter
1 dekameter	dam	10 meters
1 hectometer	hm	100 meters
1 kilometer	km	1000 meters

In Table 9.4, the relationships are based on the meter. How many millimeters in a centimeter? How many centimeters in a decimeter? Think and then read on. . . .

The answer to both of these questions is 10. Just like in our base ten number system, we trade 10 of the smaller unit to 1 of the next unit.

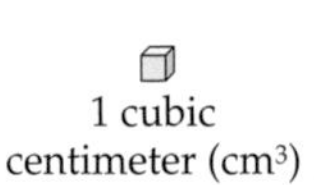
1 cubic centimeter (cm^3)

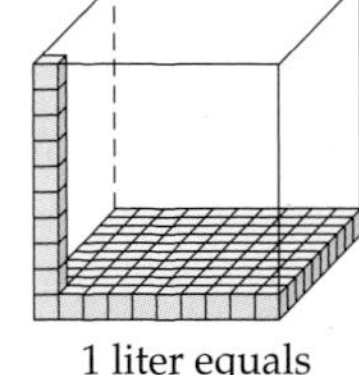
1 liter equals

Figure 9.2

METRIC VOLUME

One of the advantages to the metric system, unlike the U.S. customary system, is that it is defined so that units of length are related to units of volume and mass. The standard metric unit for volume is the **liter** (Table 9.5). The liter is defined to be the volume of a cube whose edges are 10 centimeters. Thus 1 liter is equivalent to 1000 cubic centimeters, abbreviated as cm^3 (Figure 9.2), since it would take 1000 of the small cubes to fill up the large one. A nice visual of this can be seen with the 1000 cube of the base ten blocks, which is the size of 1 liter, and also is made up of 1000 of the small cubes, which are each 1 cubic centimeter because they are 1 cm by 1 cm by 1 cm.

Since 1 liter = 1000 milliliters, and 1 liter = 1000 cm^3, then 1000 mL = 1000 cm^3, leading to 1 mL = 1 cm^3. Therefore, we have the equivalence between length and volume measures: 1 cm^3 = 1 mL. Most canned goods give the amount in milliliters. If a can contains 240 mL, how many liters is that? Do this before reading on. . . .

Because there are 1000 mL in a liter, we must divide by 1000. Therefore, 240 mL = 0.24 liter, about one-fourth of a liter.

TABLE 9.5		
Units	**Abbreviations**	**Relationship to liter**
1 milliliter	mL	$\frac{1}{1000}$ of a liter
1 centiliter	cL	$\frac{1}{100}$ of a liter
1 deciliter	dL	$\frac{1}{10}$ of a liter
1 liter	L	1 liter
1 dekaliter	daL	10 liters
1 hectoliter	hL	100 liters
1 kiloliter	kL	1000 liters

The commonly used units in Table 9.5 are liter and milliliter. Even though the other units are not used very often, we show them in this table to help us see the base ten relationship.

METRIC MASS

The creators of the metric system not only worked to connect length and volume measures but also connected volume and mass. They defined a **kilogram** to be the mass of 1 liter of pure water at 4°C.

The standard of mass in the metric system is the **gram** (Table 9.6). Technically, the terms *mass* and *weight* refer to different attributes. **Weight** is the force that gravity exerts on an object, and **mass** is the amount of matter that makes up the object. If you went to the moon, you would weigh about one-sixth as much as you do on Earth, but your mass would be the same. Because our planet is much larger than the moon, the force of gravity on your body is much greater on Earth. You may recall hearing that because the force of gravity is less on the moon, we could jump six times as high on the moon as we can on Earth. Grams are used for both mass and weight, and these measures are the same as long as we stay on Earth.

TABLE 9.6

Units	Abbreviations	Relationship to gram
1 milligram	mg	$\frac{1}{1000}$ of a gram
1 centigram	cg	$\frac{1}{100}$ of a gram
1 decigram	dg	$\frac{1}{10}$ of a gram
1 gram	g	1 gram
1 dekagram	dag	10 grams
1 hectogram	hg	100 grams
1 kilogram	kg	1000 grams

Some metric units of mass—kilogram, gram, and microgram—are encountered more often than others. Once again, we are including the less commonly used units in this table to help us see the base ten relationship more clearly. Where do you see metric mass in everyday life in the United States? Think before reading on. . . .

The net weight on most canned goods is generally given in both ounces and grams. This is one of the few places where one might see metric units for mass used in the United States at present. Some scales also have weights in metric units.

METRIC TEMPERATURE

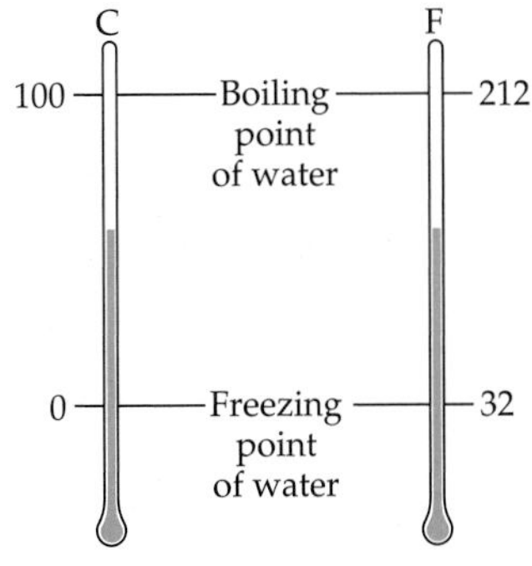

Figure 9.3

In 1742, a Swedish astronomer named Anders Celsius proposed a modification in the units of measurement that the Fahrenheit system used. He proposed that the reference points be the freezing point of water (0°) and the boiling point of water (100°) (Figure 9.3). This *Celsius* system was also called the *centigrade* system (that is, "100 grades").

TIME AND ANGLES

As mentioned earlier, the French Academy of Sciences recommended that *all* known measures be based on powers of 10. Thus they recommended changing the calendar so that there would be 10 months in a year, 10 days in a week, and 10 hours in a day. They also proposed that there be 400 degrees in a circle, which would mean that a right angle had 100 degrees. As you know, not all of these proposals survived! When Napoleon came to power, he repealed the law making metric measure compulsory. Although the French continued to use metric units for length, volume, mass, and weight, they (happily) threw out the metric system for time and returned to the more familiar 7-day week, 24-hour day, 60-minute hour, and 60-second minute.

BECOMING COMFORTABLE WITH METRIC MEASUREMENTS

What if you heard that someone is 182 centimeters tall and weighs 80 kilograms? Most Americans can easily visualize 6 feet and 176 pounds, the U.S. customary equivalents of the metric measures given in the first sentence, but are less comfortable with measurements given in metric units. We will address this problem by providing some reference measures.

The metric system is a base ten system, just like our numerals. One goal is to be able to think in metric terms, to be able to estimate the length of something in metric units directly, instead of converting from feet to meters. Rather than converting back and forth from U.S. Customary to metric, let's use some benchmarks to help us be more comfortable with thinking in metric. Converting between U.S. Customary units and metric units is no longer taught in elementary schools according to the CCSS.

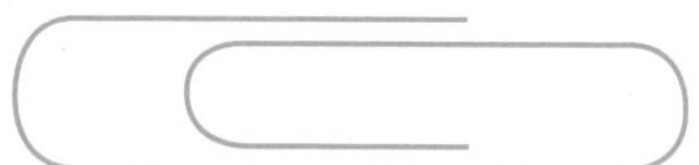

A jumbo paperclip can help serve as a metric benchmark. The width of the paperclip is about 1 centimeter, the thickness of its wire is about 1 millimeter and its mass is about 1 gram. Other useful benchmarks are given in Table 9.7.

TABLE 9.7

Item	Approximate measurement in metrics
Width of a doorway	1 meter
Height of an average man	177 centimeters or 1.77 meters
Mass of hardback textbook	1 kilogram
Mass of a thin adult	60 kilograms
Temperature on a warm day	27° Celsius
Temperature on a snowy day	−4° Celsius
Coffee cup	250 milliliters or 0.25 liters

INVESTIGATION 9.1a Developing Metric Sense

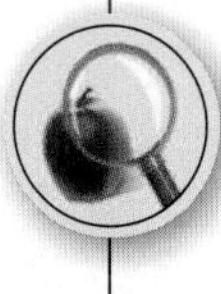

Try to do these problems yourself. Then read on. . . . 

A. Insert the decimal point in the proper place.

- The diameter of a penny is 19 centimeters.
- The length of a page of notebook paper is 279 centimeters.
- The common adult height of an elephant is about 39 meters.

B. Determine which metric unit represents the weight and volume as described.

- The weight of a box of cereal is about 425 grams or 425 kilograms?
- The volume of a cup of coffee is 250 liters or 250 milliliters?

DISCUSSION

A.
- The diameter of a penny is 1.9 centimeters.
- The length of a page of notebook paper is 27.9 centimeters.
- The common adult height of an elephant is about 3.9 meters.

B. Our first suggestion is to notice these measurements next time you are in the grocery store. A box of cereal weights about 425 grams, or about half a kilogram. Picture a liter bottle of water or soda to help with this. A coffee cup is about one-fourth of a liter bottle, not 250 times one, so the answer is 250 milliliters.

INVESTIGATION 9.1b Converting Among Units in the Metric System

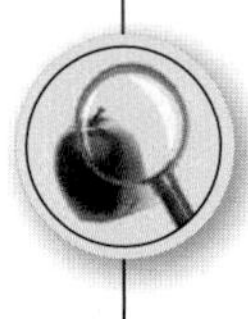

A. A path is measured as 1.5 km. How many meters is this?

B. A recipe calls for 250 mL. How many liters is this?

C. A sample weighs 700 mg. How many grams is this?

D. Change 235 mm to centimeters.

DISCUSSION

A. Because there are 1000 meters in 1 meter, we multiply 1.5×1000 to get 1500 meters. We can also use dimensional analysis.

$$1.5 \cancel{\text{km}} \times \frac{1000 \text{ m}}{1 \cancel{\text{km}}} = 1500 \text{ m}$$

B. Because there are 1000 milliliters in 1 liter, we divide 250 by 1000 to get 0.25 liter.

$$250 \cancel{\text{mL}} \times \frac{1 \text{ liter}}{1000 \cancel{\text{mL}}} = 0.25 \text{ liter}$$

C. Because there are 1000 milligrams in 1 gram, we divide 700 by 1000 to get 0.7 gram.

$$700 \cancel{\text{mg}} \times \frac{1 \text{ gram}}{1000 \cancel{\text{mg}}} = 0.7 \text{ gram}$$

D. Because there are 10 millimeters in 1 cm, we divide 235 by 10 to get 23.5 centimeters.

$$235 \cancel{\text{mm}} \times \frac{1 \text{ cm}}{10 \cancel{\text{mm}}} = 23.5 \text{ cm}$$

Hint: You can see this on your ruler, which has centimeter and millimeter gradations.

An important concept in all of these is knowing when to multiply and when to divide. When we are going from a larger unit to a smaller unit (like cm to mm), we would need a larger number of the smaller unit to measure the item so we need to multiply. The opposite is true if we are going from a smaller unit to a bigger unit, so we would divide.

MEASUREMENT OF OTHER QUANTITIES

The quantities that we have examined in this section—length, volume, mass, weight, time, and temperature—represent only a small subset of the quantities measured in our world. What other things do we measure? Why do we need to measure them? What are some of the units used to measure those things? What instruments are used? Think about things that we measure, and make a table like Table 9.8 for those things you can think of. Do this before reading on. . . .

TABLE 9.8

What do we measure?	Why?	Units	Instruments
Noise	To study noise levels that can cause hearing damage	Decibels	Audiometer

The following is only a partial list: speed (miles per hour), fuel consumption (miles per gallon), light intensity (candela), electric current (amperes), efficiency (production per unit of time), density (mass per unit volume), and infant mortality (deaths per 1000 live births).

PRECISION

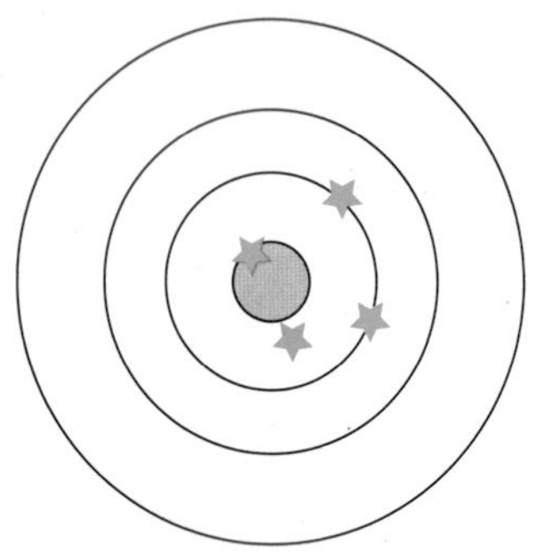
Accurate, but not precise

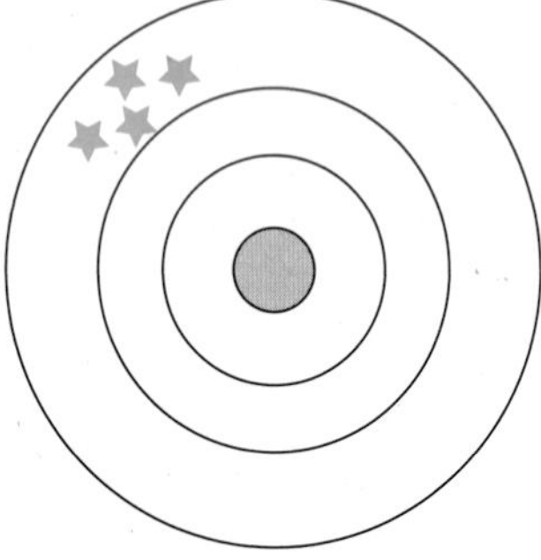
Precise, but not accurate

There is another aspect of measurement that requires some attention related to MP 6 of the CCSS, "Attend to Precision." Virtually all measurements are approximations. For example, if you ask for the dimensions of a room, you might be told that it is 14 feet by 11 feet. If you ask for more precision—let's say you are buying a rug—you might be told that it is 14 feet 2 inches by 10 feet 11 inches.

The precision of our measurement depends on the reason why we are measuring. Sometimes we want to measure amounts with much precision, and sometimes less precision is quite acceptable. For example, when selling cloth, clerks generally measure the length roughly and then add an extra couple of inches. However, if you go to a candy store or buy meat, the amount is generally weighed to the nearest tenth of an ounce, and cylinders in cars need to be accurate to the nearest thousandth of an inch. See also Explorations 9.1, 9.2, and 9.3.

Although many people use the words **precise** and **accurate** synonymously, the two concepts are not identical. A measurement can be very precise (for example, 34.628 meters) but be inaccurate (if the measurer made a mistake). Similarly, a measurement can be not very precise (for example, $16\frac{1}{2}$ feet) but be very accurate (that is, the actual distance is closer to $16\frac{1}{2}$ feet than to 16 feet or 17 feet). As illustrated at left, a dartboard can provide us with a nice visual of the difference as well.

In this section, we have explored what it means to measure an object, and we have realized that many objects have several attributes that can be measured—for example, weight, surface area, and volume. We have also examined where many of the units came from, both U.S. customary and metric. We have noted that many, if not most, of the numbers we use represent measurements. We have learned that most measurements are approximations and that there is a difference between precision and accuracy.

9.1 Exercises

1. Think about each object below and describe all of the measurable attributes of that object. Next to each property, briefly describe why someone might want to measure that attribute.

   a. A pond
   b. A garage
   c. A house
   d. A carpet
   e. A banana
   f. A tree

2. Select, from the following units, the unit that you would use to express each of the attributes below. Briefly explain your choice.

   Length: millimeters, centimeters, meters, kilometers

   Volume: milliliters, liters

   Mass: milligrams, grams, kilograms

   a. Your height
   b. Your weight
   c. The length of a butterfly
   d. How much fluid there is in a drinking cup
   e. The mass of a fingernail
   f. The volume of a thimble

3. Fill in the blanks:

   a. 500 m = ____ km
   b. 4.5 cm = ____ mm
   c. 670 mL = ____ L
   d. 3.6 L = ____ mL
   e. 450 g = ____ kg
   f. 35 kg = ____ g
   g. 24 mg = ____ g

DEEPENING YOUR UNDERSTANDING

4. There are about 6 billion people on Earth. If they all lined up and held hands, how long a line would they form?

5. The lengths of most older U.S. swimming pools are multiples of yards, generally 25 yards. However, newer pools are 25 or 50 meters long. Suppose an American swimmer is practicing for a 400-meter race. Eight lengths of the pool is 400 yards, which is "close" to 400 meters. How close is it? That is, how much shorter is 400 yards than 400 meters? First predict the amount, using only mental math. Explain how you got your prediction and how you did the operations in your head. Then determine the amount (to the nearest yard).

6. Alan and Bill are going to have a 400-meter race, but Alan gets a head start because he is slower. Their best times of the season are 82 seconds and 75 seconds, respectively. How much (distance) of a head start should Alan get?

7. Let's say your estimate of a length was 100 feet and the actual length was 105 feet. Clearly this would not be "as big" a difference as that between an estimate of 10 feet and an actual length of 15 feet, although both estimates were "off" by 5 feet. Compare the accuracy of the estimates in such a way that we can see from the language that the first estimate was much closer than the other.

8. How many millimeters are in a centimeter? How many milligrams are in a centigram? How many milliliters are in a centiliter?

9. What is the relationship between length, mass, and volume measurements in the metric system?

10. a. How many days would it take you to spend $1 million if you spent $3 per minute?

    b. How high would a stack of 1 million pennies be?

    c. How high would a stack of 1 million quarters be?

    d. How high would a stack of pennies worth 1 million dollars be?

    e. How high would a stack of quarters worth 1 million dollars be?

11. ***Classroom Connection*** This problem comes from "Figure This!" Every year many birds migrate. The record holder is the Arctic tern, which flies from the Arctic to the Antarctic and back. If the one-way distance is 9000 miles and these birds can fly for 12 hours a day at an average speed of 25 miles per hour, how many days does it take them to make the trip?

    Source: Reprinted with permission from Figure This! (http://www.figurethis.org), copyright 2006 by the National Council of Teachers of Mathematics.

12. Wagon trains in the 1800s took from 4 to 6 weeks to travel from Missouri to California. Although the direct distance is less, the actual mileage for the wagons was 2000 miles. About what average distance did they travel per day?

13. The Biblical Goliath was said to be 6 cubits and 1 span tall (1 Samuel 17:4). How tall do you think he was in feet and inches? Explain both your reasoning and your calculations.

14. Suppose you were King Henry I (1496), who declared that henceforth 1 yard would be equal to the distance from his nose to his forefinger. How long would a yard be, on the basis of *your* body?

15. Following are some famous quotes that include units from the U.S. customary system. Replace the bold word in each quote with the most appropriate unit from the metric system.

    a. "The journey of a thousand **miles** begins with one step." —Lao Tzu

    b. "An **ounce** of performance is worth **pounds** of promise." —Mae West

    c. "Tart words make no friends; a spoonful of honey will catch more flies than a **gallon** of vinegar."—Benjamin Franklin

16. Say the United States did go metric.

    a. What would you propose as the size of a standard sheet of paper? Justify your choice.

    b. What would you propose as the two standard sizes for soft drinks to replace 12 ounces and 16 ounces?

    c. Instead of buying a pound of butter, what would we buy?

17. a. Why might the United States want to go metric?

b. What are some challenges to the United States in going metric?

18. The average human heart pumps about 60 milliliters of blood for each beat.

a. About how many liters of blood does your heart pump each day?

b. Do you think a reasonable degree of precision for this question would be to report the answer to the nearest 100 liters, 10 liters, 1 liter, or 1 milliliter? Explain your reasoning.

19. Suppose we measured the passage of time during the day as the Babylonians did. Compare the length of a daytime hour on the longest day of the year to the length of a daytime hour on the shortest day of the year where you live. Part of this question has to do with resourcefulness, an important trait for teachers to develop. Where can you go to get the information needed to answer this question?

20. Which is greater, an increase of 5° Fahrenheit or an increase of 5° Celsius? Justify your answer.

21. A train is traveling 60 miles per hour and is about to enter a tunnel that is 1200 feet long. How long will it take the train to pass completely through the tunnel? What additional information do you need in order to solve this problem?

22. a. One way to determine the speed of the current of a river is to measure how long it takes a leaf to pass under a bridge. For example, let's say it takes a leaf 9 seconds to pass under a bridge that is 35 feet wide. What is the speed of the current in miles per hour?

b. How might we determine the speed of the current if we were in a boat?

23. Two cups are half full. Cup A holds coffee. Cup B holds milk. One teaspoon of coffee is taken from cup A to cup B. Then one teaspoon of the mixture is taken from cup B to cup A. The two cups now have the same amount of liquid. Is there more coffee in cup A than milk in cup B, the same amount of coffee in A as there is milk in B, or less coffee in A than milk in B? Explain your reasoning.

24. Let's say that a 12-ounce cup of coffee is 20 percent milk. If you add 6 ounces of coffee, what is the percentage of milk in the new mixture?

25. a. Determine the following dimensions for your body using centimeters: height, wrist, neck, leg, arm, foot.

b. Determine the following ratios: height to foot, height to leg, neck to wrist, height to arm.

c. Before determining the ratios for the entire class, predict which ratio will have the most variation and which the least variation. Explain your reasoning.

d. Gather the class data. Decide on a way to determine the variation and to compare the variation.

26. Will the ratio of the heights of an adult and a child be equivalent to the ratio of the lengths of their feet? If it will be, explain why you think so. If not, predict which ratio will be greater and why. Gather some data. Do the data support your predictions? Explain.

27. You will need a scale for these problems.

a. Buy several bananas. Compare the price per pound that you paid and the price per pound of what you actually eat. If there are enough data, graph the actual price for the entire class. Before you do so, predict the shape of the distribution: normal, uniform, and so on.

b. Do the same for peanuts.

c. Buy several oranges of two types, one type with a thick skin and one type with a thin skin. Compare the ratios for both kinds of oranges.

28. Before they wash their faces, most people run the tap until the water is hot. How much water is wasted waiting for the water to get hot?

a. With a partner, determine a plan for answering this question.

b. Determine the amount of water (in U.S. customary units) wasted at the place where you live.

c. Describe the relative precision of your answer. For example, if you say 3 quarts, is your answer to the nearest quart?

d. Graph the class data.

e. Let's say the average for the United States is close to the average in this class and that 200 million people in the United States wash their faces every day. How many gallons of water are wasted?

29. You will need a dropper for this problem. Let's say you have a leaky faucet that drips every 6 seconds. How much water is wasted in one day? Explain the assumptions that you made and describe your solution path.

30. Take a sheet of blank paper. Tear it in half. Place one half on top of the other. Tear these two sheets in half. Place one half (two sheets) on top of the other. Assume for the purposes of this problem that you were able to continue this process for a total of 20 times. What would be the height of the stack of paper?

31. Select a trip in your part of the country for which there is not one obvious shortest way.

a. Determine which way is shortest by using a map. Explain your work.

b. Have at least one person try more than one way to determine the actual distance and the actual time it took in each case.

c. Is the shorter way faster?

d. Describe the possible sources of error in part (a).

e. Determine the average speed for each route.

f. On the basis of the data you have, which way would you go, or do you need more data?

32. a. Measure your classroom in meters and then make a scale model of your classroom so that 1 centimeter on your model represents 1 meter.

b. Measure your classroom in feet and inches, and then make a scale model of your classroom so that 1 inch on your model represents 1 foot.

c. Are the computations in both problems about the same, or are the computations for the metric-scale map easier? Explain.

FROM STANDARDIZED ASSESSMENTS

2008 NECAP, Grade 6

33. Jeff bought 6 quarts of juice for a party. One glass holds 8 fluid ounces of juice. What is the total number of glasses Jeff can fill with the juice he bought? [1 quart = 32 fluid ounces]

2008 NECAP, Grade 4

34. Look at this toy car.

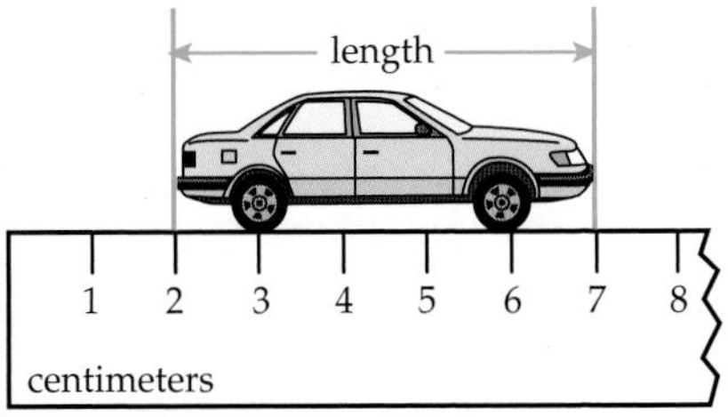

What is the length of the toy car in centimeters?

a. 4 centimeters **b.** 5 centimeters
c. 6 centimeters **d.** 7 centimeters

2006 NECAP, Grade 4

35. The height of a tree is 368 centimeters. Which measurement also shows the height of the tree?

a. 3 meters 68 centimeters **b.** 36 meters 8 centimeters
c. 36 meters 80 centimeters **d.** 300 meters 68 centimeters

2008 NECAP, Grade 5

36. Mr. Randall's class is making puppets. Each puppet needs 60 centimeters of string. How many puppets can Mr. Randall's class make from 12 meters of string? [1 meter = 100 centimeters]

a. 20 **b.** 50
c. 200 **d.** 500

2006 NECAP, Grade 6

37. Nadia is putting beads on a piece of string, as shown below.

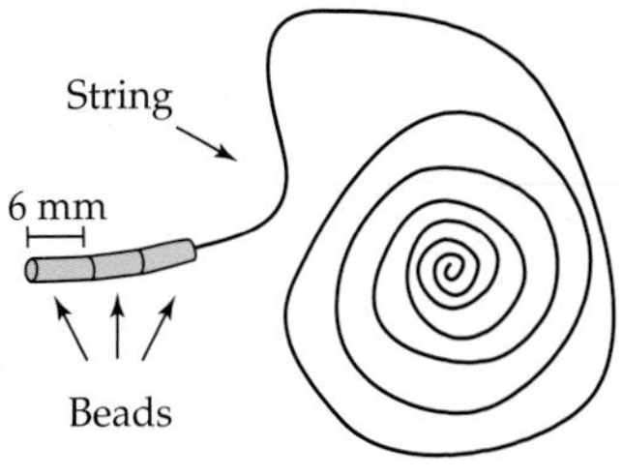

Each bead is 6 mm long. What is the greatest number of beads Nadia can put on a 60-cm string? [1 cm = 10 mm]

a. 10 **b.** 36
c. 100 **d.** 3600

SECTION 9.2 Perimeter and Area

What do you think?

- Can two shapes with different perimeters have the same area? Why or why not?
- What does *pi* mean?

In this section, we will investigate the concepts of perimeter and area, both for geometric figures for which there are formulas and for irregularly shaped objects. It may surprise you that many students do not understand these concepts well. For example, consider the data in Table 9.9 concerning questions reported in Results *from the Fourth Mathematics Assessment.*[6]

TABLE 9.9

Question		PERCENT CORRECT Grade 3	Grade 7
What is the perimeter of this rectangle?	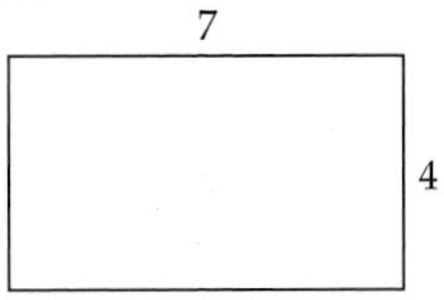	17	46

(continued)

TABLE 9.9

Question	PERCENT CORRECT Grade 3	PERCENT CORRECT Grade 7
What is the area of this rectangle?	20	56
What is the area of this rectangle? 6 5	5	46

Both the NCTM Standards and the Common Core State Standards stress that teachers should not just help students better understand how to perform the procedures for determining perimeters and areas. Rather, if the students develop these formulas through problem solving and reasoning, if they are required to express their understanding in words also (not just in formulas), and if they can see connections among the formulas, *then* performance on these and more challenging problems will increase dramatically. The Common Core State Standards place perimeter of polygons and area of rectangles in third grade. Circumference and area of other figures, as well as surface area and volume, are learned in sixth and seventh grades. Eighth grade is where the Pythagorean Theorem is first learned.

INVESTIGATION 9.2a Perimeter or Area?

Your rectangular bedroom is 12 feet by 14 feet and you decide to get carpet and new baseboard. How much carpet and how much baseboard are needed? What units would we use to measure these? What are the similarities and differences between these two measurements?

DISCUSSION

Visuals help, so let's first draw a model of the room.

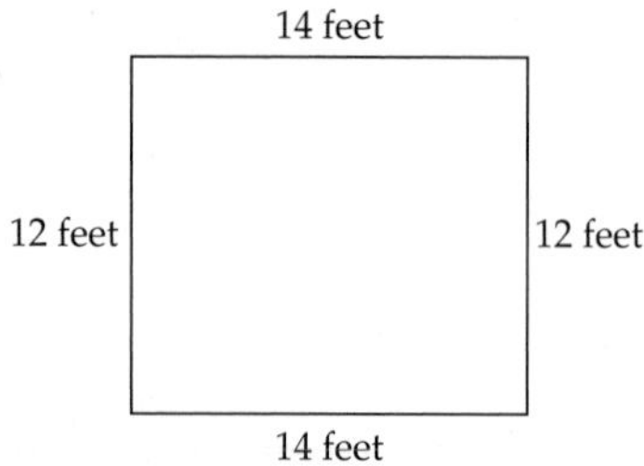

Let's start with the baseboard. Because these are boards going around the perimeter of the room, you will need 12 feet plus 12 feet plus 14 feet plus 14 feet, which equals 52 feet. However, we also need to consider the fact that there will be a doorway that will not have baseboard, so let's subtract 3 feet for the doorway, so we need 49 feet of baseboard.

This is a linear measurement because we would use 49 linear feet of board; in other words, think about "unfolding" all four sides into a straight line and it would be 49 feet long. Also, because we are adding feet plus feet we end up with feet, just like $x + x = 2x$.

How is covering the room with carpet different? Can we use feet to measure the space inside the rectangle? No, here we have to fill in the space with squares.

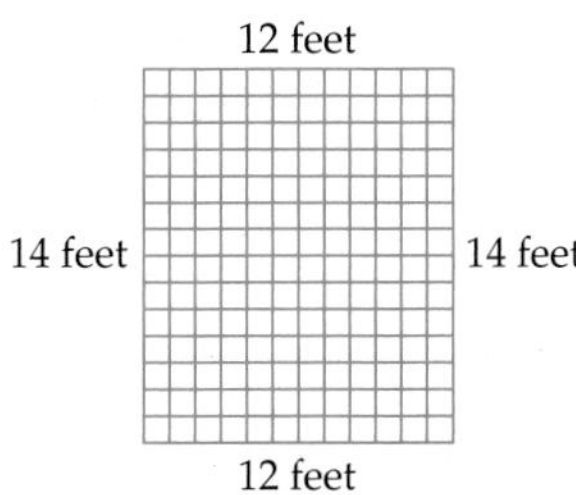

As the picture models, a length of 14 feet times a width of 12 feet means you will need 168 square feet to cover the room in carpet. Also, if we look at the units, we are multiplying feet times feet so we get square feet just like x times x is x^2.

PERIMETER

Let us begin with the concept of **perimeter**, essentially the distance around an object. Many practical applications of perimeter involve surrounding an object—for example, fencing a yard or running a baseboard around the base of a room. As explored in the first investigation, perimeter of a rectangle (or any polygon) is the distance around the outside of a figure and therefore is measured in linear units. The sum of the lengths of the sides gives us the perimeter of a polygon. Now let's learn about distances around circles.

CIRCUMFERENCE AND π

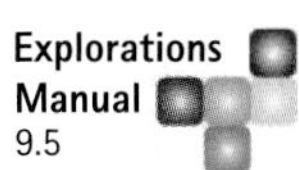

When we determine the distances around figures and objects, sometimes the path is not a straight line but rather is a circle. You may remember a formula involving the distance around a circle, or the **circumference**, and that it involves π. You may have constructed a definition of π in Exploration 9.6. Before reading on, take a few moments to think about what π means. That is, the value of π is 3.14 (to two decimal places), but what does π *mean*? For example, where did the number 3.14 come from? How is this number related to the circumference or diameter or radius of a circle (not the formula, but a description of the relationship)? Write your thoughts before reading on. . . .

In one sense, π is a ratio—that is, π is the ratio of the circumference to the diameter of *any* circle. If we could precisely measure the circumference and diameter of any circle and then divide the circumference of the circle by its diameter, we would *always* get π. If we call the circumference C and the diameter d, we have $\pi = \frac{C}{d}$.

Thus we have the formulas

$$C = \pi d \quad \text{and} \quad C = 2\pi r$$

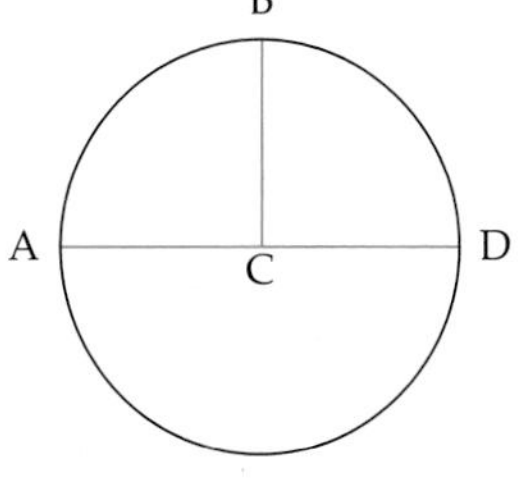

Figure 9.4

In Figure 9.4, C is the center of the circle, $\overline{AC}$, $\overline{BC}$, and $\overline{DC}$ are radii (*radii* is the plural of *radius*), and $\overline{AD}$ is a diameter.

There is another way to think about π. Imagine placing a string around the circumference of the circle in Figure 9.4 and then straightening out that string. If the diameter of the circle is d, then the length of the string is about $3.14d$. That is, if we unwrap the circumference, its length will always be about 3.14 times the length of the diameter. Thinking about unwrapping this string helps us to see why circumference is also measured in linear units (feet, meters, centimeters, etc.).

AREA

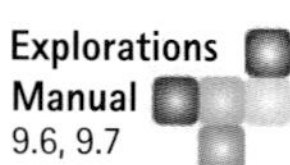

Explorations Manual 9.6, 9.7

Questions about area generally deal with "how much" it takes to *cover* an object—for example, how much fertilizer to cover a lawn, how much material to cover a bed. In order to answer area questions, we have to select an appropriate unit, and thus the answer takes the form of how many of those units. However, the units for perimeter and area are not the same, as seen in Investigation 9.2a. For example, if we have a 20-foot by 10-foot garden, we say that we need 60 feet of fence to surround the garden, but we would say that the area of the garden is 200 *square feet*. The need for units and the difference between units for perimeter, area, and volume are generally not well understood by students and therefore are worth emphasizing more than once.

Many people remember or recognize the basic formulas for determining areas, but few people understand why they work or where they came from. Let us investigate them for a bit. Understanding them will pay dividends when we attempt to solve more complex problems. The simplest case for area involves rectangles and squares.

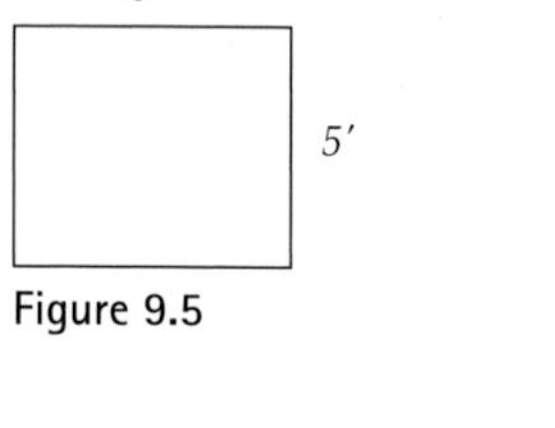

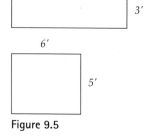

Figure 9.5

Thinking about area by looking at rectangles Let us focus on some more subtle aspects of the concept of area.

What does it mean to say that one rectangle (or other figure) covers more area than another? Consider the diagrams in Figure 9.5 of two pieces of plastic sheeting that I use to cover two different woodpiles in my back yard. One piece is 10 feet by 3 feet, and the other is 6 feet by 5 feet. Which one would you say is bigger and why?

When I ask young children if the sheets are the same size or if one is bigger, they generally say that the 10-foot by 3-foot sheet is bigger. However, the areas of the sheets are equal, 30 square feet. We can demonstrate this by covering each figure with 30 squares, each of which is 1 foot by 1 foot (Figure 9.6).

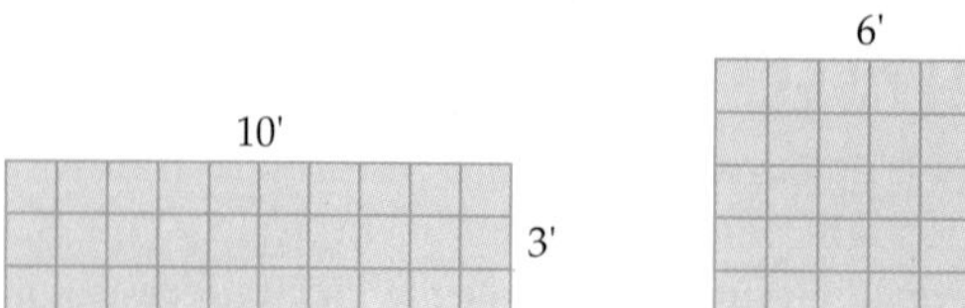

Figure 9.6

One of the models we used for multiplication in Chapter 3 was called the "area model." Just as we can model $6 \times 5 = 30$ with the rectangle, also the area of a 6 by 5 rectangle is 30 squares. This is a nice connection between geometry and understanding operations.

Although this may seem very simple and straightforward, many people are at least initially stumped when they try to explain what *area* means when the dimensions of a rectangle are not whole numbers—for example, when a rectangle is 3.5 centimeters by 2.5 centimeters. We say that the area of the rectangle is 8.75 square centimeters, because we know that the area of a rectangle is obtained by multiplying its length by its width. Can you explain what that 8.75 means, as if you were talking to a child who understands that in the previous problem, both sheets of plastic have an area of 30 square feet? Do this before reading on. . . .

Figure 9.7 shows one way of explaining what it means to say that the area is 8.75 square centimeters: cutting up the rectangle and reassembling the pieces in a certain way. Do you understand this diagram? Did you explain the meaning of the 8.75 square centimeters in another way that also makes sense? Take some time to see if you can understand the diagrams below before you read my explanation.

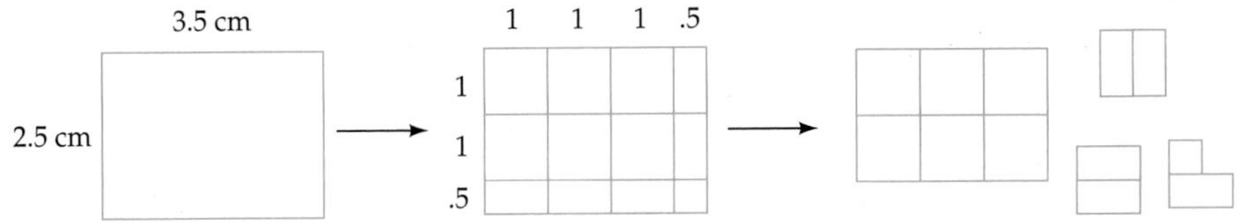

Figure 9.7

We can partition the rectangle in a manner similar to the way we used partitioning in the multiplication of fractions and decimals. That is, we have 6 one-centimeter by one-centimeter squares, and we can combine the remaining pieces to make another 2.75 square centimeters (Figure 9.7).

Base and height Most students use the more common terms *length* and *width*, so let's take a moment to examine what *base* and *height* mean, because they will become important when we investigate other figures for which the terms *length* and *width* are not appropriate.

Any polygon can be rotated so that its bottom side will be parallel to the bottom of the paper. Thus any side can be taken as the **base** of the polygon. For example, we can rotate the triangle in Figure 9.8 so that any one of its sides is the base. Which one we choose is generally an arbitrary decision or may be based on what measurements we know.

The **height** of a polygon is the distance from the side chosen as the base to the point farthest away, measured along a line perpendicular to the base. See Figure 9.8 for an illustration of this notion with triangles. What is the relationship between the height and the base in each triangle? They are always perpendicular.

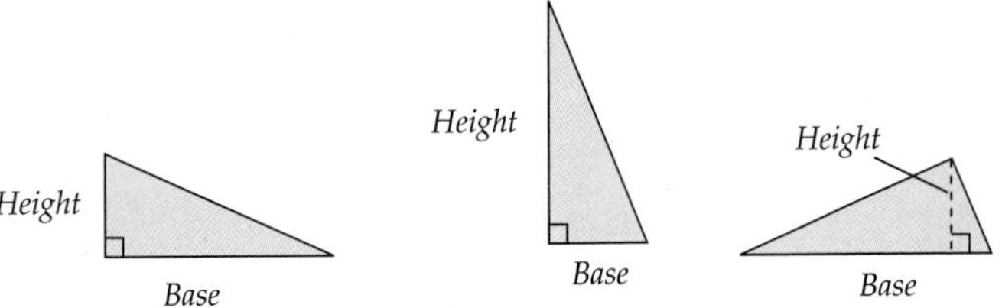

Figure 9.8

We can say that the area of a rectangle equals length times width, or base times height.

Understanding the area formula for parallelograms The formulas for many common figures are striking, both in their simplicity and in their connection to one another. If we learn the meaning of the formulas, rather than memorizing a list of formulas, then we gain a deeper understanding and do not have to memorize formulas with no understanding. The formula for determining the area of a parallelogram is connected to the formula for determining the area of a rectangle. The diagram at the right in Figure 9.9 illustrates the connection. Can you see it? How would you explain it?

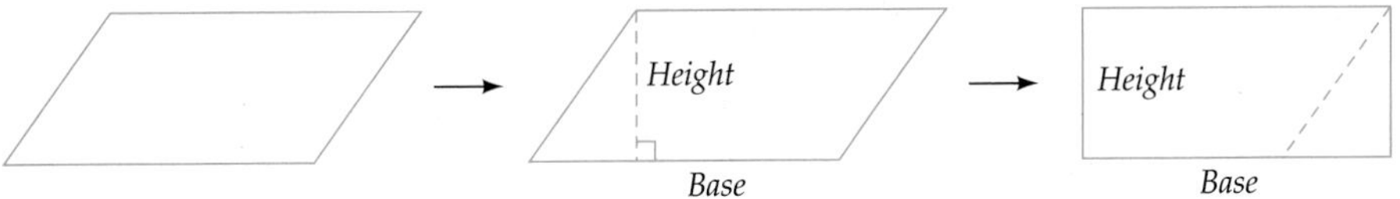

Figure 9.9

If we cut the triangle from the parallelogram and reconnect it at the right-hand side, we have transformed the parallelogram into a rectangle. We have not added or subtracted any area, so the areas of the parallelogram and the rectangle must be equal. From Figure 9.9, you can see that the base of the parallelogram is congruent to the base of the rectangle and that the height of the parallelogram is congruent to the height of the

CLASSROOM  CONNECTION

Grade 4

While some of these are easy for college students, you might find #7 challenging. Try it.

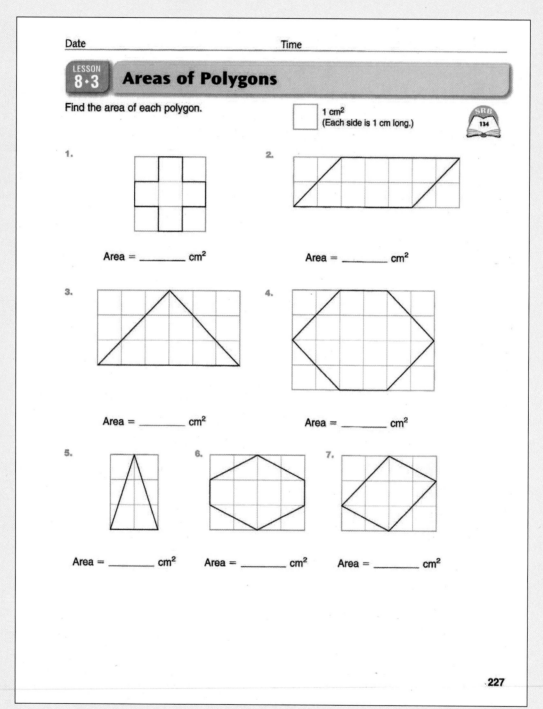

Date Time

LESSON 8·3 **Areas of Polygons**

Find the area of each polygon.

1 cm^2 (Each side is 1 cm long.)

1. 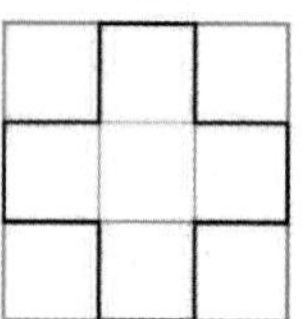

Area = ________ cm^2

2. 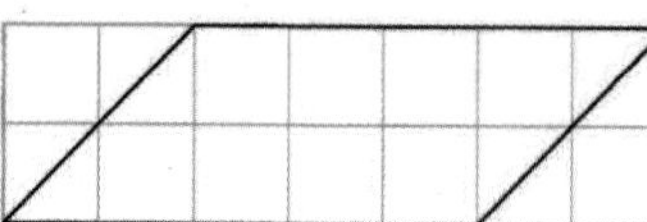

Area = ________ cm^2

3. 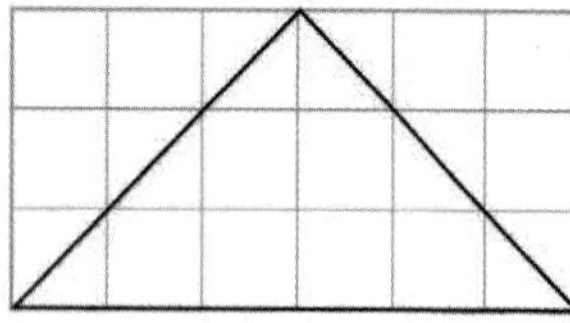

Area = ________ cm^2

4. 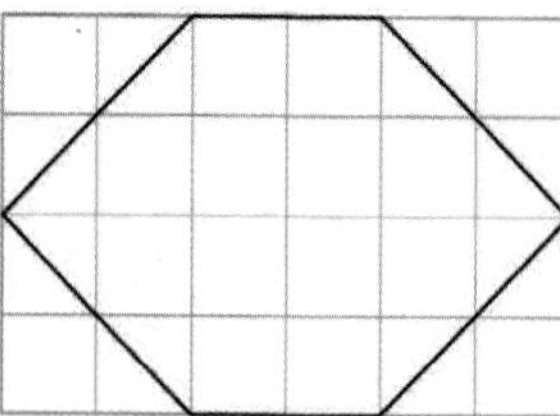

Area = ________ cm^2

5. 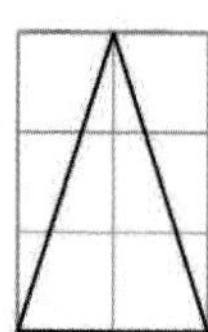

Area = ________ cm^2

6. 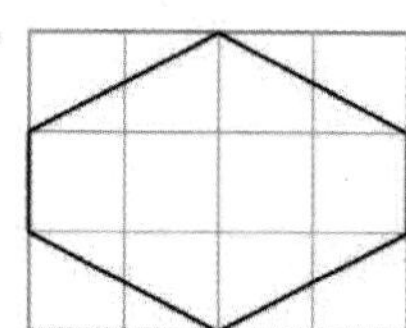

Area = ________ cm^2

7. 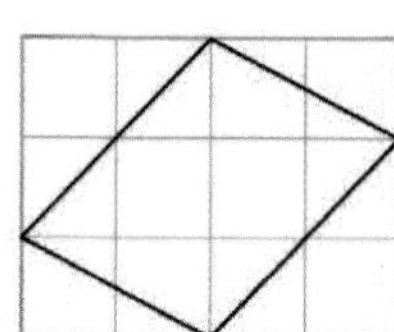

Area = ________ cm^2

227

From *Everyday Mathematics, Grade 4:* The University of Chicago School Mathematics Project: Student Math Journal, Volume 2, by Max Bell et al., Lesson 8-3, p. 227. Reprinted by permission of The McGraw-Hill Companies, Inc.

rectangle. Because we know that the areas of the two are equal, we can now state that *the area of any parallelogram is equal to the product of its base and its height;* that is, $\boldsymbol{A = bh}$ (Figure 9.10). A key aspect here is that the base and the height are perpendicular (they form a right angle).

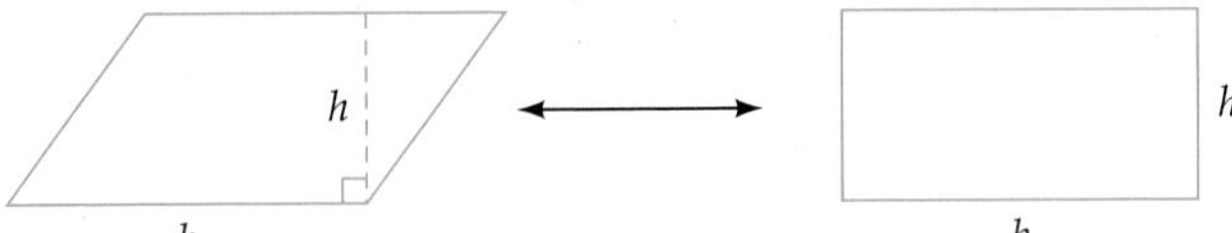

Figure 9.10

Understanding the area formula for triangles We will use what we have learned about parallelograms to develop the formula for the area of triangles. Just as we connected the area of any parallelogram to a rectangle, take a few minutes to see if you can adapt this idea to develop formulas for the areas of triangles. Then read on. . . .

CLASSROOM CONNECTION

You can see children's thinking about trapezoids in *Measuring Space in One, Two, and Three Dimensions: Casebook,* by D. Schifter, V. Bastable, and S. J. Russell with K. R. Woleck (Parsippany, NJ: Dale Seymour Publications, 2002), pp. 130–135.

Consider triangle MAN (Figure 9.11). If we make a congruent copy of this triangle and move that triangle into place as shown (that is, by rotating it 180 degrees), we form a parallelogram. Do you see how we can use the formula for parallelograms to determine the formula for triangles now? See if you can do it on your own before reading on. . . .

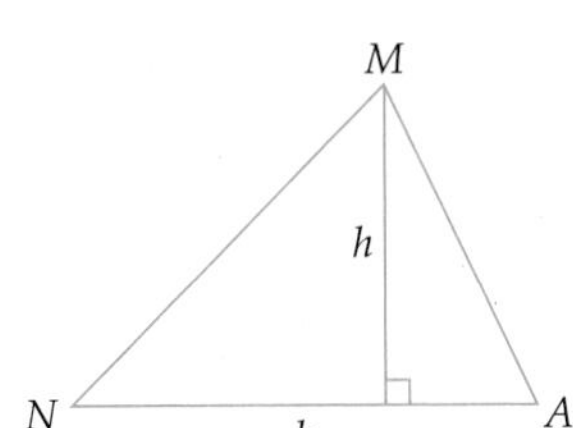

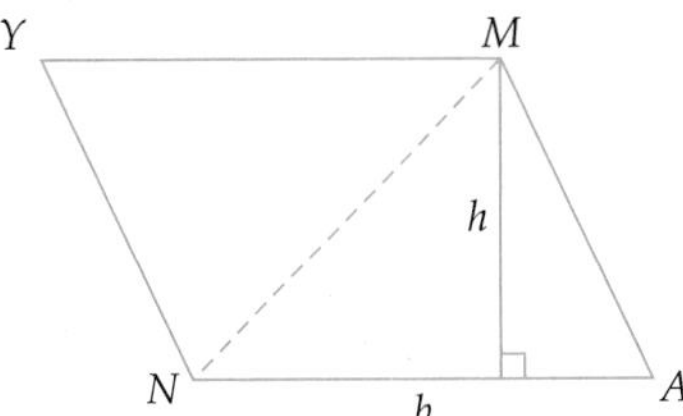

Figure 9.11

The base of the parallelogram and the base of the triangle are identical, and so are their heights.

The area of a triangle is $\frac{1}{2}$ of the area of the whole parallelogram. So, area of triangle = $\frac{1}{2}$ (area of parallelogram) = $\frac{1}{2}$ (base)(height). In other words, the formula we just found for the area of a triangle is $\boldsymbol{A = \frac{1}{2}bh}$.

Understanding the area formula for trapezoids Can you now apply what you have learned about the areas of these figures to determine the formula for the area of a trapezoid (Figure 9.12)? Try to do so before reading on. . . .

MATHEMATICS

Did you realize that a trapezoid can be thought of as a triangle with its head cut off? Actually, this is not just a silly thought, because this way of thinking about trapezoids helps us connect the area formula for the trapezoid with the triangle. We could have determined the formula for the area of a triangle by thinking of a triangle as a trapezoid where the top base = 0. In this case, the area is $\frac{1}{2}(a + b)h$, and since $a = 0$, this becomes $\frac{1}{2}bh$.

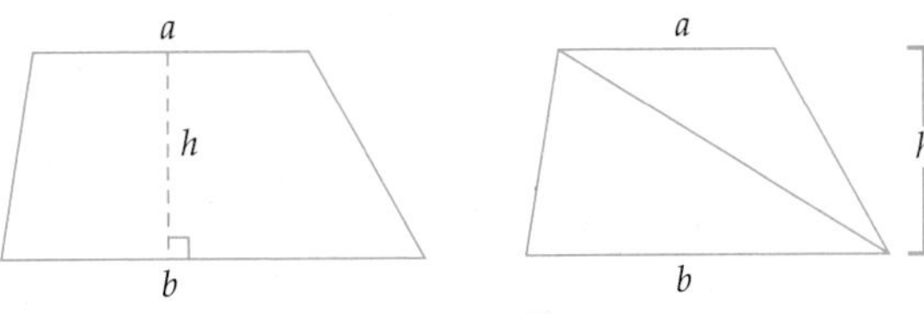

Figure 9.12

Figure 9.13

There are several ways to derive the formula for the area of a trapezoid. We will discuss two that connect to our preceding discussions. First, we can draw a diagonal of the trapezoid, which cuts (decomposes) it into two triangles. We know how to find the areas of these triangles: The areas are $\frac{1}{2}ah$ and $\frac{1}{2}bh$. The sum of the areas of the two triangles is equal to the area of the trapezoid, so *the area A of the trapezoid is* $\frac{1}{2}ah + \frac{1}{2}bh$, or, if we factor out the $\frac{1}{2}$ and the h, we get or $\boldsymbol{A = \frac{1}{2}(a + b)h}$ (Figure 9.13).

Figure 9.14 shows another method for finding the area of a trapezoid. If we construct a congruent trapezoid and connect it to the original trapezoid, we have a parallelogram, and the area of the parallelogram is equal to the product of its base and its height; that is, the area is equal to $(a + b)h$. Because the area of the trapezoid is equal to one-half the area of this parallelogram, the area of the trapezoid $= \frac{1}{2}(a + b)h$.

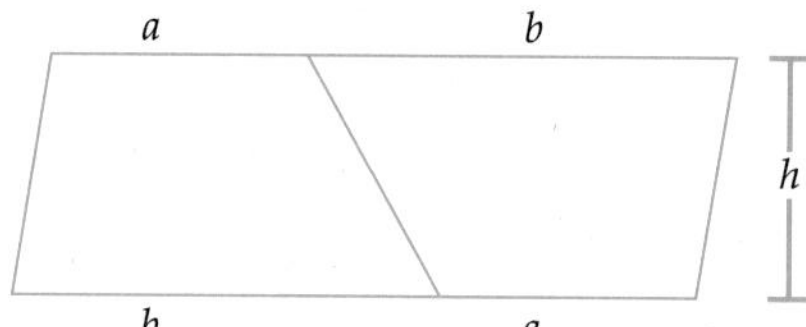

Figure 9.14

INVESTIGATION 9.2b Converting Units of Area

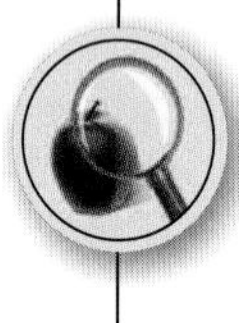

Converting area units is different than converting linear units, and thus an investigation will help you to own this idea. Malik and Lee have decided to pull up their old, shabby carpet and buy new carpet. The room measures 12 feet by 15 feet, so the area is 180 square feet. However, when they go to the carpet store, they find that the prices are in square yards. How many square yards is their floor?

DISCUSSION

A common answer is 60 square yards—that is, 180 square feet divided by 3, because there are 3 yards in 1 foot. However, that is wrong. Do you see why?

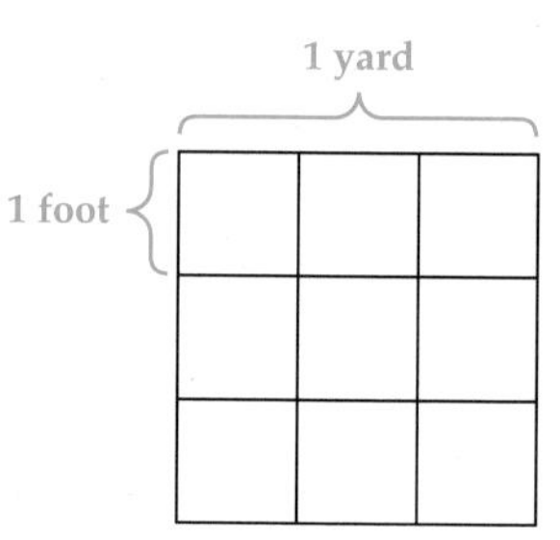

The diagram at the right illustrates why you must divide by 9. Even though there are 3 feet in 1 yard, there are 3^2 or 9 square feet in 1 square yard. So, the correct answer is 20 square yards (180 square feet divided by 9 square feet per square yard).

PYTHAGOREAN THEOREM

Although the relationship between the lengths of the sides of a right triangle had been known long before Pythagoras, it was Pythagoras who proved that *for any right triangle, the sum of the squares of the lengths of the two sides (say, a and b) is equal to the square of the length of the hypotenuse (say, c);* that is, $a^2 + b^2 = c^2$ (Figure 9.15[a]).

It is important to note that for the Greeks, this was not an algebraic relationship but rather a geometric relationship. That is, if you make a square whose sides are congruent to the length of side a, a square whose sides are congruent to the length of side b, and a square whose sides are congruent to the length of the hypotenuse c, then the sum of the areas of the two smaller squares will be equal to the area of the larger square (Figure 9.15[b]).

At the last count, there were over 370 different proofs of the Pythagorean theorem, including a proof by one of the U.S. presidents, James Garfield. The Pythagorean theorem has many practical applications, such as the one below, where it is used to determine the length of something that we cannot measure directly.

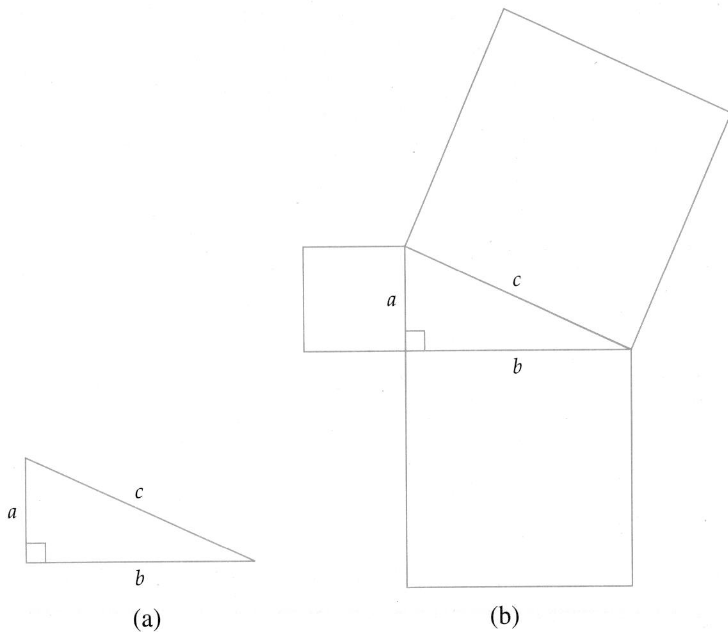

Figure 9.15

INVESTIGATION 9.2c Using the Pythagorean Theorem

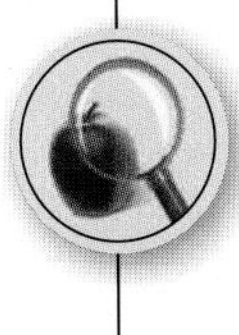

A child and his grandfather are flying a kite on a very windy day, so the kite string is straight. There are 80 feet of kite string out and the child wants to know how high off the ground the kite is. How can the grandfather (who happens to have a measuring tape with him) figure this out?

DISCUSSION

The grandfather can measure from where the child stands holding the kite to a point as directly underneath the kite as he can get. Suppose he measures 42 feet as shown here.

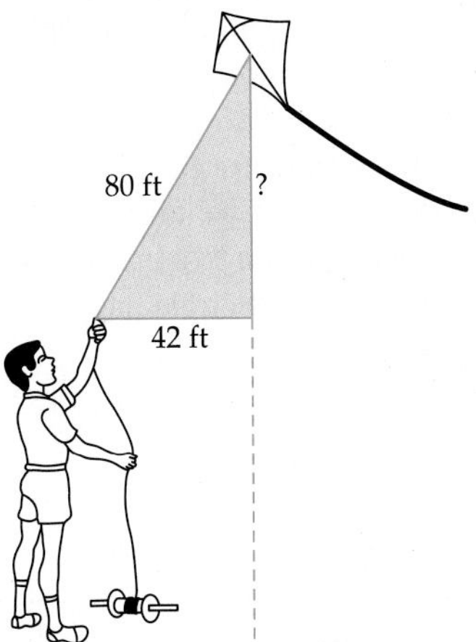

The height of the kite (from where the child is holding it) then can be found with the Pythagorean Theorem.

$$\text{height}^2 + 42^2 = 80^2$$

$$h^2 + 1764 = 6400$$

$$h^2 = 6400 - 1764 = 4636$$

$$h = \sqrt{4636} = 68.09 \text{ feet}$$

Note in the picture that the height we just found is actually not all the way to the ground, since the child is holding the kite string above the ground. So, the grandfather could add on the distance from where the child holds the kite string to the ground to more accurately answer his grandson's question.

INVESTIGATION 9.2d Understanding the Area Formula for Circles

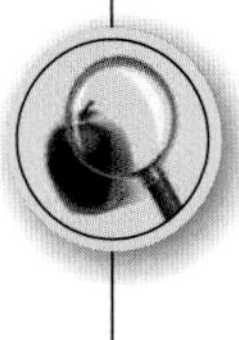

The last area formula that we will consider now is that for the area of a circle. The purpose of this investigation is in the realm of "number sense," helping you to see why $A \approx 3.14r^2$ (and therefore $\boldsymbol{A = \pi r^2}$) makes sense. Consider the circle in Figure 9.16(a), whose radius is r inches. In Figure 9.16(b), we have circumscribed a square around the circle. What is the area of the square? Determine this and then read on. . . .

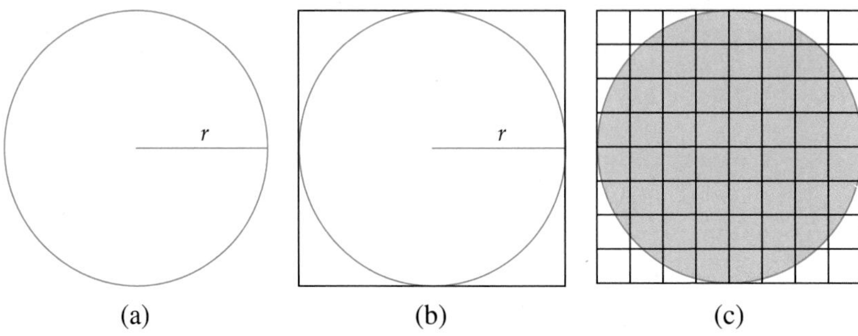

Figure 9.16

DISCUSSION

If the radius of the circle is r inches, then the length of each side of the square is $2r$, and thus the area of the square is $(2r)(2r) = 4r^2$.

Thus, the area of the circle is less than 4 times r^2. We can use our spatial sense to estimate that the circle covers about $\frac{3}{4}$ as much space as the square and thus approximate the area of the circle as $3r^2$, or we can place a grid over the figure and determine what fraction of the square is covered by the circle. In this case, we get a more accurate estimate of the area of the circle (see Figure 9.16[c]).

We can also derive the formula for the area of a circle by turning it into a shape that closely resembles a parallelogram. If we take a circle, cut it into sectors, and rearrange them, we get a shape that looks similar to a parallelogram. In the following figure, we cut the circle into 16 sectors. If we cut the circle into more sectors, we would get a shape that looks even more like a parallelogram.

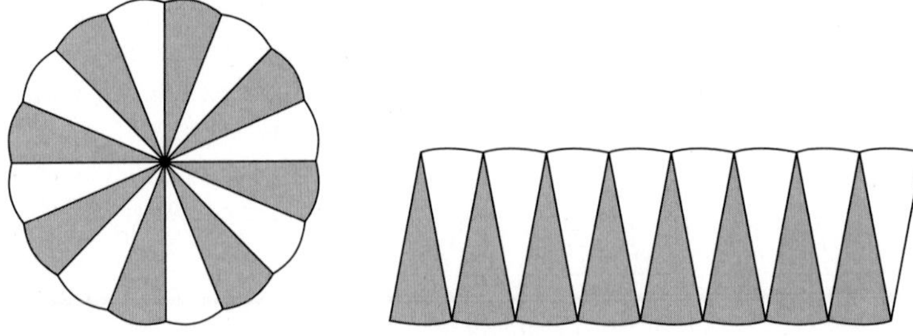

We already know that the area of a parallelogram is $A = \text{base} \times \text{height}$. What are the base and the height in relation to the original circle? Think about this before reading on. . . .

When we step back and consider, we can see that the height is the radius of the circle. If we look at the base, the length of the base is equivalent to the width of eight sectors, but eight sectors is equivalent to half of the total distance around the circle, or $\frac{1}{2}$ of the circumference. Since the circumference is $2\pi r$, then half the circumference is πr. The next figure shows this.

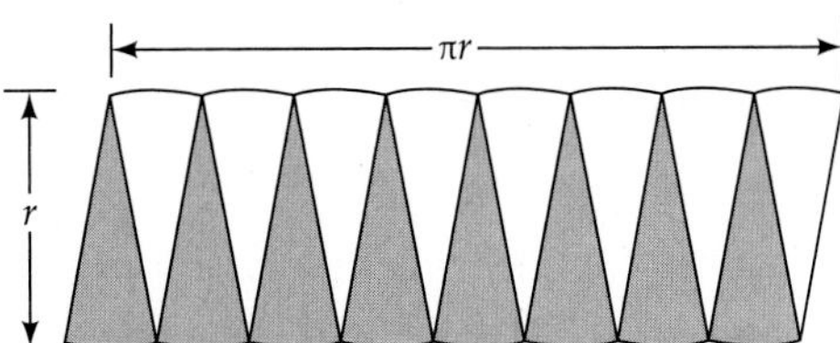

Therefore, since to find the area of a parallelogram we multiply base × height, the area of the figure is

$$A = (\pi r)(r) = \pi r^2$$

HISTORY

We know that the Egyptians and the Babylonians knew not only the formulas for the area of the square and the rectangle, but also the formula for the area of a trapezoid. They knew the formula for finding the area of a right triangle. There were correct and incorrect formulas for the areas of isosceles triangles. For example, there is an example in the Rhind papyrus (from ancient Egypt) of finding the area of an isosceles triangle by finding half the product of the base and the length of one of the equal sides, instead of the altitude. For the circle, the Babylonians developed the formula $A = \frac{C^2}{12}$ and the Egyptians developed the formula $A = \left(\frac{8}{9}d\right)^2$. It will be left as an exercise to see how close their formulas were to the actual formula.

INVESTIGATION 9.2e A 16-Inch Pizza versus an 8-Inch Pizza

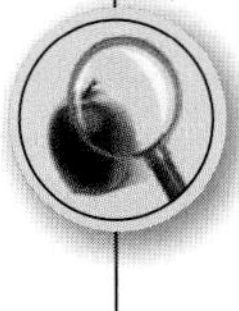

Let's say you are going out to have pizza with several friends. You are thinking of getting one large pizza—let's say its diameter is 16 inches. However, some people are vegetarians, and so you decide to get two little pizzas, each of which is 8 inches in diameter. Suddenly someone asks, "Are we getting the same amount of pizza if we get two little pizzas instead of one large one?" What do you think? Work on this before reading on. . . .

DISCUSSION

This is generally one of the most amazing investigations in this book because so many people are so surprised. To determine the area, we can use the area formulas: If the large pizza has a diameter of 16 inches, it has a radius of 8 inches. Therefore, its area is $\pi(8)^2 = 64\pi \approx 201$ square inches.

If the small pizza has a diameter of 8 inches, it has a radius of 4 inches. Therefore, its area is $\pi(4)^2 = 16\pi$, and the area of two small pizzas is 32π, or approximately 100.5 square inches.

Amazing, isn't it? Not only are two 8-inch pizzas not equal to one 16-inch pizza, but it takes *four* 8-inch pizzas to have the same area as *one* 16-inch pizza. Some people still don't believe their eyes. If you find this result amazing, draw a circle to represent the large pizza and then figure out how to place the two smaller pizzas inside the circle.

INVESTIGATION 9.2f How Big Is the Footprint?

The area of most objects cannot be determined by a formula. For example, biologists and environmentalists want to know the total surface area of the leaves of a tree in order to determine the amount of oxygen the tree produces. What if we wanted to measure the area of another irregularly shaped object, such as the footprint in Figure 9.17?

Take some time to think about how you might determine the area and to try out your ideas. Then read on. . . .

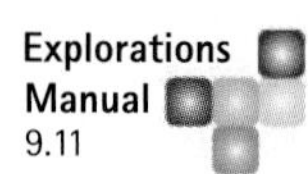

CLASSROOM CONNECTION

You can see children's thinking about determining the area of a hand in *Measuring Space in One, Two, and Three Dimensions: Casebook,* by D. Schifter, V. Bastable, and S. J. Russell with K. R. Woleck (Parsippany, NJ: Dale Seymour Publications, 2002), pp. 109–115.

Figure 9.17

Explorations Manual 9.11

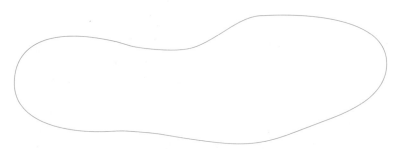

DISCUSSION

There are many possible strategies. Several are briefly described below.

STRATEGY 1 Use graph paper

This is one of the simpler strategies. There are two issues worth pursuing: (1) what to do with the partially filled squares, and (2) the advantages and disadvantages of graph paper with smaller squares.

STRATEGY 2 Draw rectangles and triangles

We can partition the footprint into rectangles and triangles. One advantage of this over the previous strategy is that in this case, one rectangle will account for a large portion of the print. One disadvantage is that there is room for error, depending on how the rectangles and triangles are made. This procedure could also become a bit tedious.

STRATEGY 3 Draw trapezoids

We can partition the footprint into trapezoids. This strategy arises from the realization that trapezoids are easily broken into rectangles and triangles; also, this method involves less computation than the previous method. We can approximate the area of the footprint with four trapezoids.

STRATEGY 4 Weigh it

We can trace this print onto a piece of thick posterboard and cut it out. We can then cut out another piece of posterboard in the shape of a rectangle. The ratio of the area of the footprint to the area of the rectangle will be equal to the ratio of the weight of the footprint to the weight of the rectangle. Therefore, we can use the following proportion:

$$\frac{\text{Area of footprint}}{\text{Area of rectangle}} = \frac{\text{weight of footprint}}{\text{weight of rectangle}}$$

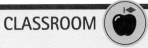

CLASSROOM CONNECTION

Rearranging a 16 by 12 Rectangle

Can you phrase your response to #8 in two different ways? 

Name Date

Measuring Polygons

Rearranging a 16 by 12 Rectangle

Here is a 16-inch by 12-inch rectangle:

Record its perimeter and area in the table below. Imagine cutting the rectangle in half, and attaching the two pieces together to make a new rectangle. Record the dimensions, perimeter, and area of the new rectangle in the table below. Do the same process at least three more times, and record the information in the table below.

Dimensions	Perimeter	Area
1. 16 inches by 12 inches		
2.		
3.		
4.		
5.		

6. What is happening to the area of each rectangle? Why?

7. What is happening to the perimeter of each rectangle? Why?

8. What do you notice about how the shape of the rectangle changes?

44 Unit 5 Sessions 2.5, 2.6

INVESTIGATION 9.2g Making a Fence with Maximum Area

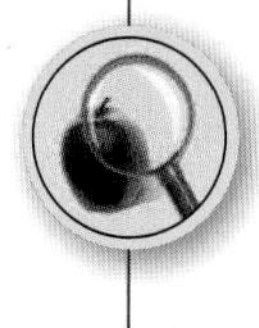

Joshua has decided to build a fence around his garden, and he buys 200 feet of chicken wire. After he gets home and unrolls the wire, he finds to his surprise that there are many different size gardens that he can make with 200 feet of fencing. He decides that he wants to get the largest garden from 200 feet of fencing. If he wants a garden that is rectangular in shape, what are the dimensions of that garden? Work on this before reading on. . . .

Explorations Manual 9.12

DISCUSSION

STRATEGY 1 Use guess–check–revise

TABLE 9.10

Length	Width	Area	Reasoning
80 feet	20 feet	1600 square feet	If the perimeter is 200, then $l + w = 100$. Reduce the length by 10 and increase the width by 10.
70 feet	30 feet	2100 square feet	This has more area, so reduce the length by 10 and increase the width by 10.
60 feet	40 feet	2400 square feet	This has more area, so reduce the length by 10 and increase the width by 10.
50 feet	50 feet	2500 square feet	Why does this feel like it "should" be the biggest garden?

STRATEGY 2 Reason and make a model

A very tactile way to "feel" this investigation is to cut a piece of string and tie the ends together. Now, using your two thumbs and two forefingers, make as skinny a rectangle as you can (Figure 9.18).

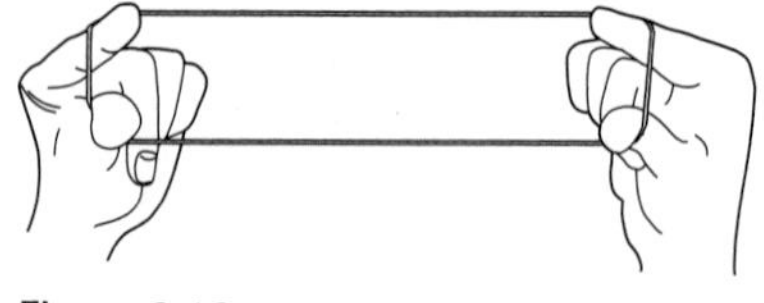

Figure 9.18

Now slowly move your thumbs and forefingers apart—that is, decrease the length and increase the width. What seems to be happening to the area? Why must it stop increasing when the length and width are equal?

STRATEGY 3 Be creative and adventurous

What if we made the garden in the shape of a circle? Do this before reading on. . . .

Because the circumference of the circle is 200 feet, the radius of the circle will be $r = \frac{C}{2\pi} \approx \frac{200}{2 \cdot 3.14} = 31.8$ feet. Now that we know the radius, we can find the area:

$$A = \pi r^2 \approx 3.14(31.8)^2 = 3175 \text{ square feet}$$

Compare the area of the circular garden to that of the square garden: The circular shape will give Joshua over 25 percent more area.

SUMMARY 9.2

In this section, we examined some measurement ideas and procedures related to perimeter and area. In one sense, perimeter concerns how much is needed to surround an object, and area involves how much is needed to cover an object. We have learned that π can be viewed both as how many times a diameter can wrap around a circle and as the ratio between a circle's circumference and its diameter. We have explored area formulas (and why they work) for triangles, for certain quadrilaterals, and for circles. We explored the Pythagorean Theorem. We have moved beyond more routine applications of perimeter and area to examine how areas of irregularly shaped figures are determined. Finally, we have found that the relationship between perimeter and area is not a simple one. For example, if we double the area of a square, we do not double its perimeter. Similarly, two objects can have the same perimeter but different areas, and two objects can have the same area but different perimeters.

9.2 Exercises

1. Determine the area and perimeter of the figures below.
   a. Acute scalene triangle

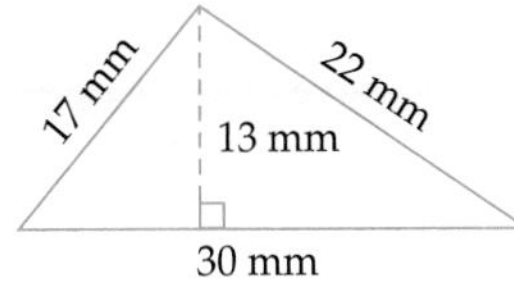

   b. Obtuse scalene triangle

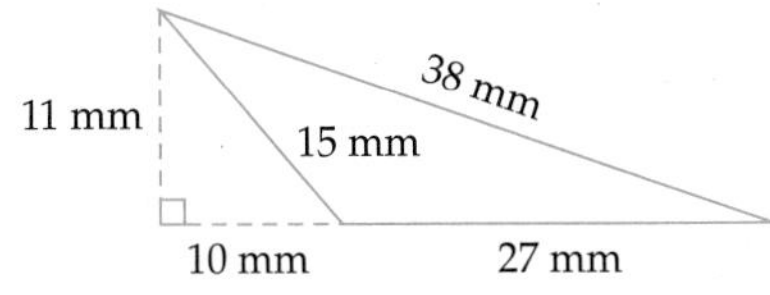

   c. Equilateral triangle

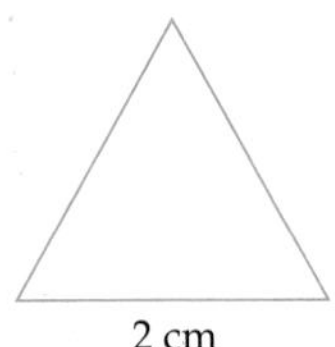

   d. Circle

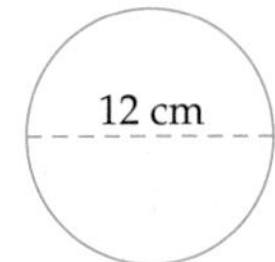

   e. Parallelogram

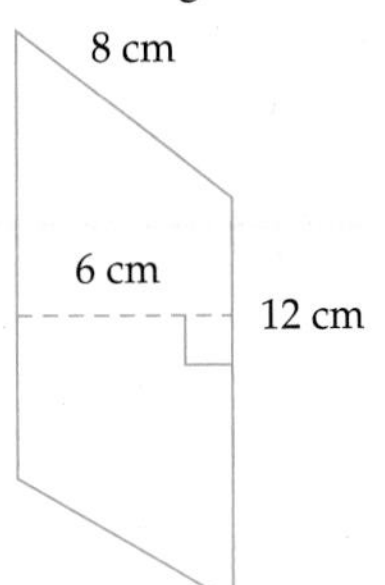

   f. Trapezoid

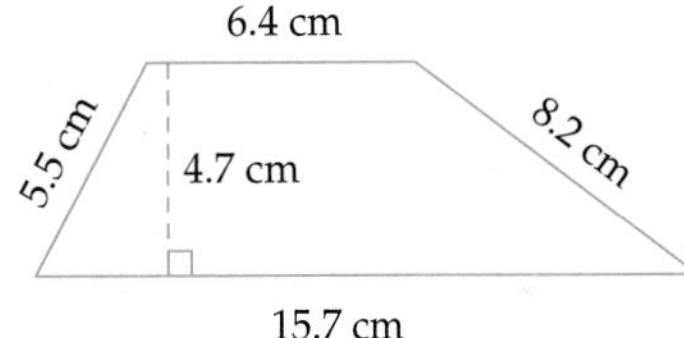

2. Determine the area and perimeter of the figures below.
   a.

   b.

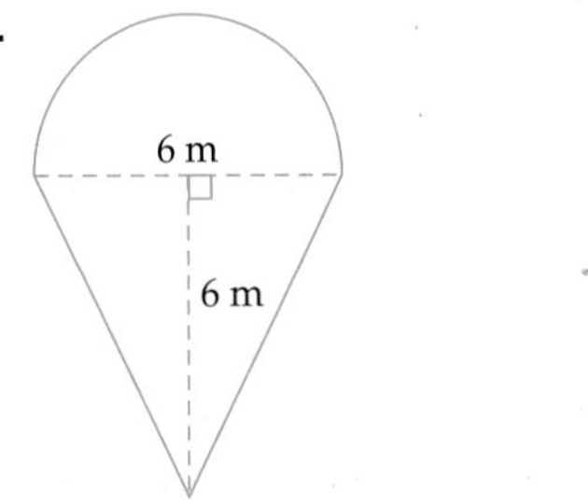

3. Determine the area of each of the following polygons on the Geoboard dot paper.

a. b. c.

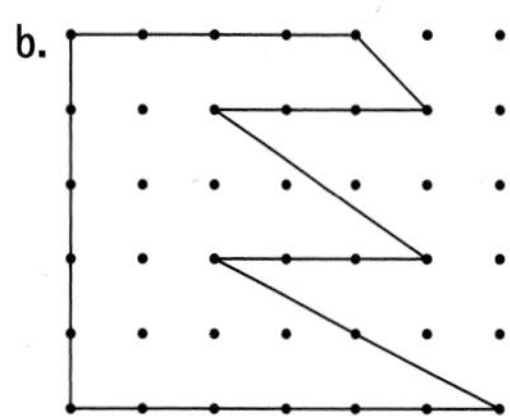

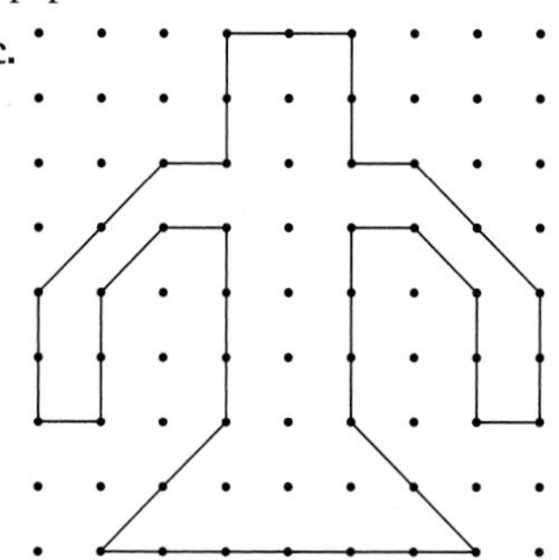

4. Find the area of the triangles below in square centimeters, using a ruler to measure the appropriate lengths.

a. b.

5. Determine the length of the arc $\widehat{AB}$ if the diameter of the circle is 10 feet and the angle is 128 degrees.

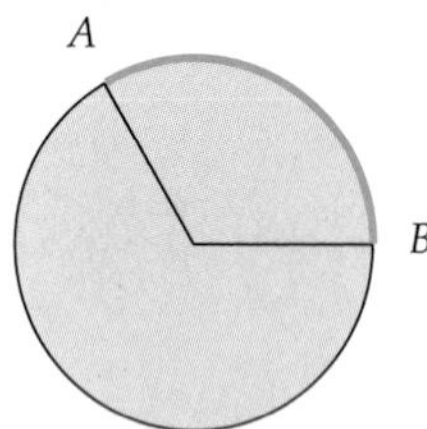

6. Find the length of x, to the nearest tenth of a foot. *Note:* The figure is not drawn to scale.

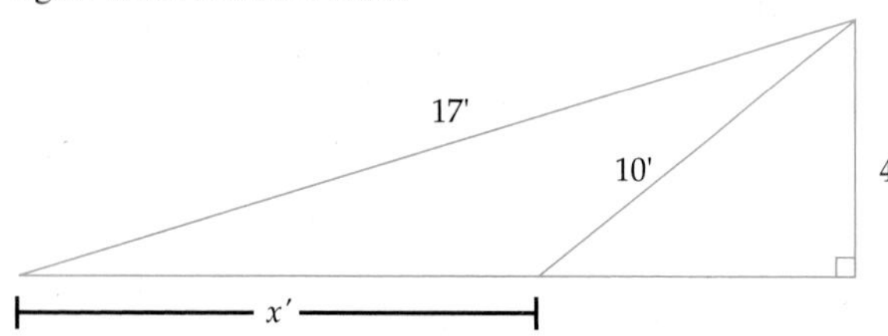

7. Determine the area of each figure below in two different ways.

a.

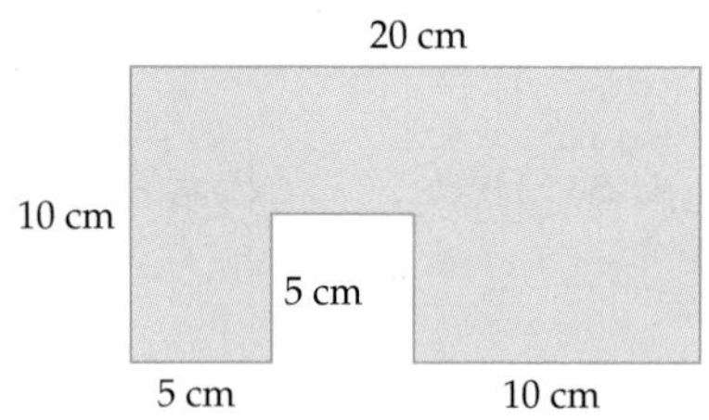

b.

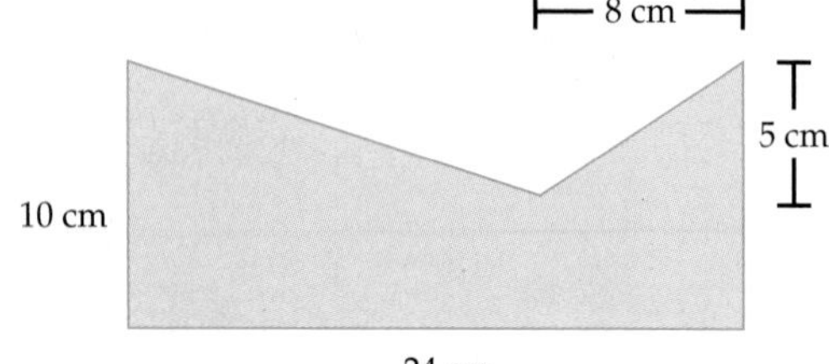

8. Let's say we have a square whose sides all measure 10 inches. Determine to two decimal places the dimensions of the square that has twice the area of this square.

9. How many square inches are there in 1 square foot?

DEEPENING YOUR UNDERSTANDING

10. Bernice wants to seed her large back yard with grass. Her back yard is in the shape of a rectangle with dimensions 120 feet by 80 feet. The seed costs \$3.99 for a 1-pound bag, and each bag covers up to 1050 square feet. How much will the seed cost?

11. One bag of bark mulch will cover 18 square feet. If a landscape architect wants to lay mulch on a border of plants around three sides of a 6-foot by 10-foot shed (as shown), how wide can the border be?

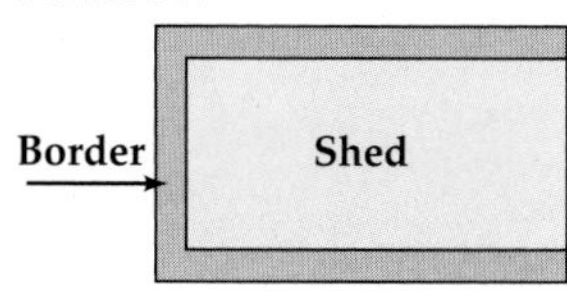

12. The indoor sod for the 1994 World Cup match in the Silverdome was grown in hexagonal units. Each hexagon was 7.5 feet on a side. How many hexagons were needed to cover the field? The soccer field was about 240 by 360 feet.

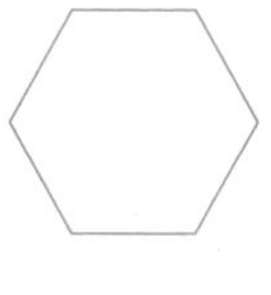

13. A farmer has a fence that encloses a square plot with an area of 36 square meters. If the farmer uses this fence to enclose a circular flower garden, what will the area of the garden be?

14. A Little Leaguer asks her mother to make her a home plate for a sandlot baseball game. According to her encyclopedia, an official plate is made from a square by making two 12-inch diagonal cuts that remove isosceles triangles as shown below. What is the length of the side of the original square?

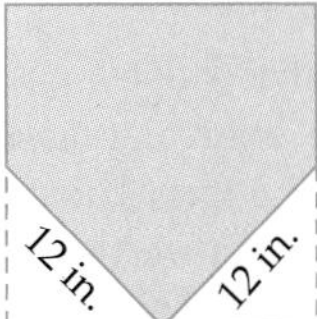

15. What percent of the quilt square below is dark blue?

16. If each of the circles has a diameter of 10 meters, what is the area of the region inside the four circles?

17. The figure below is a common representation of the famous yin-yang relationship, which originated in China.

a. Can you explain how the figure is made?

b. If the radius of the circle is r, what is the length of the curved line in the interior of the circle that separates the white region from the dark region?

18. A circular flower bed is 6 meters in diameter and has a circular sidewalk 1 meter wide around it. Find the area of the sidewalk in square meters.

19. How many 2-inch by 3-inch by 8-inch bricks will you need to build a (uniformly wide) brick walk with the shape shown in the figure below? (Lay the bricks so that the largest face is up.)

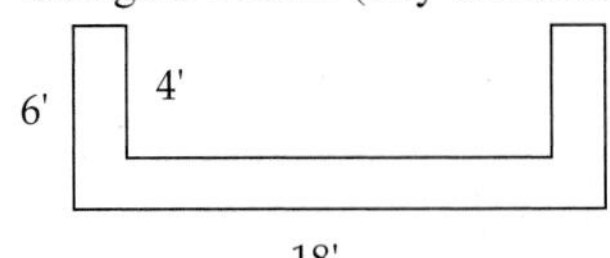

20. a. ***Classroom Connection*** Suppose a goat was tethered at the middle of a 100-foot-long fence and the length of the rope was 50 feet, as shown in the diagram below. Over how much area could the goat graze? Describe any assumptions you make in order to solve the problem.

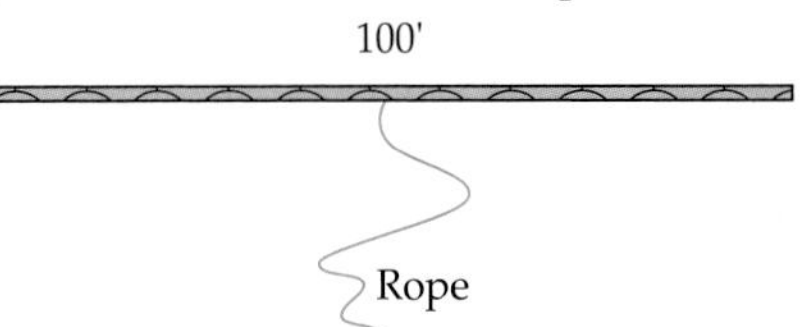

b. ***Classroom Connection*** What if the point at which the goat was tethered was at the middle of a 50-foot-long fence, as shown in the diagram below? Over how much area could the goat graze? Describe any assumptions you make in order to solve the problem.

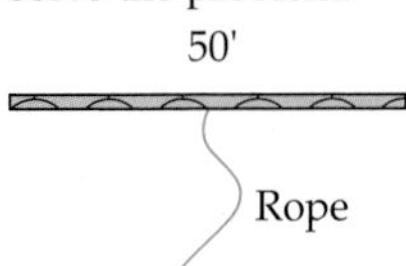

You can find a discussion of variations of the goat problem in "Reflections on Mathematics and the Connected Mathematics Project" in the February 1999 issue of *Mathematics Teaching in the Middle School,* pp. 324–330.

21. When I was a child, we visited Sequoia National Park. At that time, there was a redwood that we could actually drive through. Impossible you say? If the circumference of the tree was 83 feet, what was the diameter of the tree?

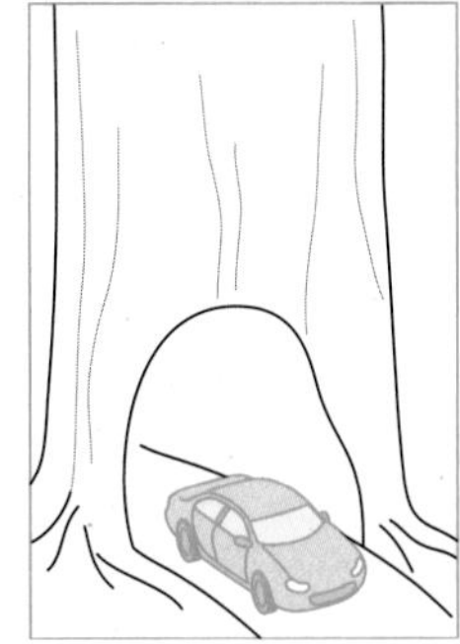

22. a. Make two rectangles that have the same perimeter but differ in area.

b. Make two rectangles that have the same area but differ in perimeter.

c. Make two parallelograms that have the same perimeter but differ in area.

d. Make two parallelograms that have the same area but differ in perimeter.

23. Archaeologists need to know geometry! Suppose the following piece of a wheel has been found. The archaeologist wants to know how big the actual wheel was. How would you determine this?

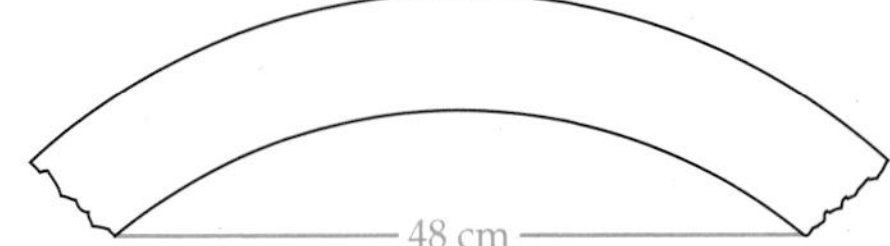

24. ***Classroom Connection*** This problem comes from "Figure This!" (http://www.figurethis.org/). The diagrams below show two kinds of windshield wipers. Which wiper cleans the greater area? Both wipers sweep a portion of a circle, although some thought will be required to figure out the shape that the second wiper covers.

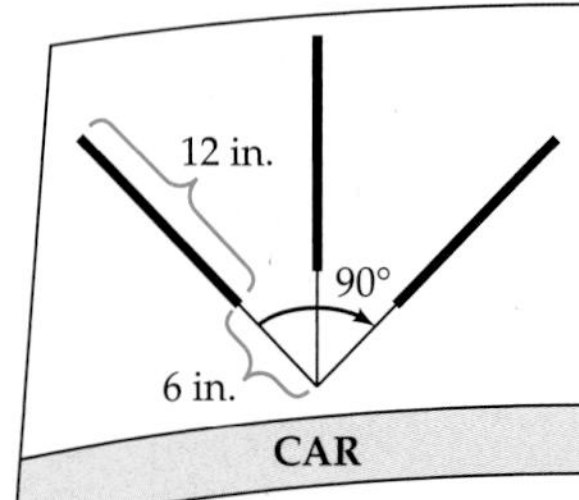

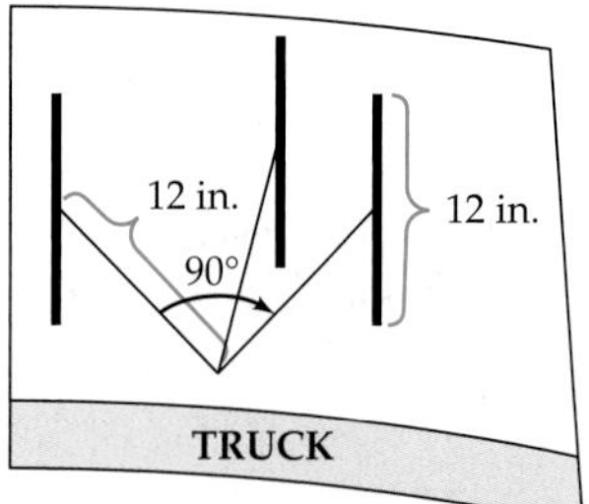

25. a. ***Classroom Connection*** Cut a 5 × 5 Geoboard into eight regions with equal areas such that none of the regions is rectangular in shape.

b. Below is one second-grader's solution, which is shown in "Exploring Area with Geoboards" in the October 1997 issue of *Teaching Children Mathematics,* pp. 72–75. Verify that the solution is correct.

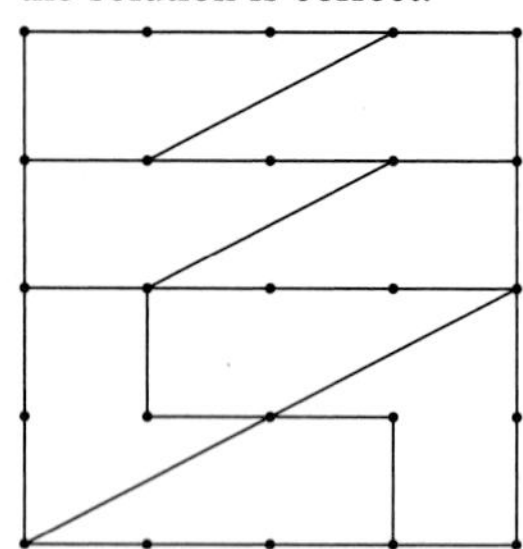

Source: "Exploring Area with Geoboards" in *Teaching Children Mathematics,* October 1997, p. 72–75. National Council of Teachers of Mathematics.

26. ***Classroom Connection*** The unit Geoboard consists of a rubber band that encloses the smallest square you can make on a Geoboard.

a. Make a square that has twice the area of the unit square. You can read children's solutions in "Responses to the Patterns in Squares Problem" in the November 2001 issue of *Teaching Children Mathematics.*

b. Make a square that has five times the area of the unit square.

27. ***Classroom Connection*** From "Solutions to the Height in Coins Problems" in the October 2008 issue of *Teaching Children Mathematics:* Which would be greater—the value of quarters arranged end to end on a flat surface to represent your height, or the value of nickels stacked vertically to represent your height?

28. Suppose a wire that is stretched tightly around the Earth is elongated by 5 meters and then placed back around the Earth. Could a person crawl under the elongated wire?

29. How many right triangles with sides 3 inches and 6 inches can be cut from a sheet of paper that is 18 inches by 40 inches?

30. The steps to the entrance of a school are 40 inches high. The school has decided to make the building wheelchair-accessible. If a city ordinance states that the ramp cannot be steeper than 5 degrees, how long must the ramp be? Solve this without using trigonometry.

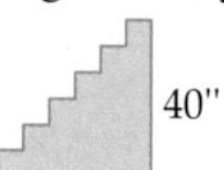

31. Some people have an intuitive sense of whether or not things will fit. For example, let's say you have bought a circular table top with a diameter of 7 feet and a thickness of 3 inches, and your doorway is 78 inches by 30 inches. First, predict whether or not the table top will fit. Then determine whether or not it will fit.

32. Compare the Babylonian formula for the area of the circle, $A = \frac{C^2}{12}$ and the Egyptian formula, $A = (\frac{8}{9}d)^2$. Which is more accurate? Justify your choice.

33. ***Classroom Connection*** *The Librarian Who Measured the Earth* by Kathryn Lasky tells the story of Eratosthenes (New York: Little, Brown, 1994). Around 240 B.C., Eratosthenes, a Greek mathematician, estimated the circumference of the Earth. His estimate was off by only 1 percent! He knew that on a certain day, the sun would be directly overhead at a town called Syene. However, the sun would not be exactly overhead in Alexandria, which was about 490 miles north of Syene. Using the figure below, determine how Eratosthenes measured the circumference of the Earth.

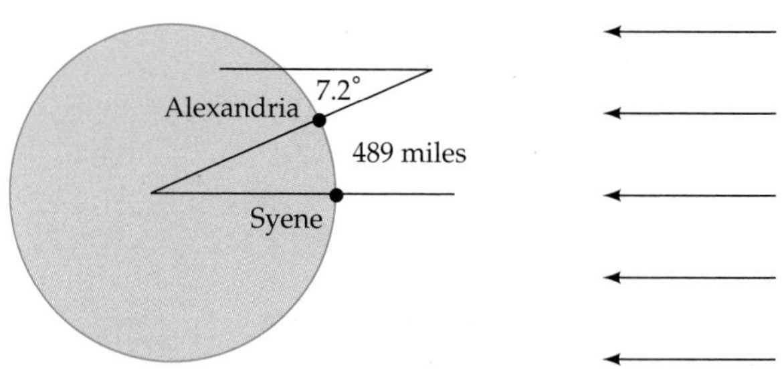

34. On a blank sheet of paper, draw from memory:

- A rectangle the size of a dollar bill
- A rectangle the size of a floppy disk
- A circle the size of a penny
- A circle approximating the top view of a soda can

a. Before you check your estimates, predict the relative accuracy of your estimates, from 1 to 4, with 1 denoting the drawing that you predict will be "closest" to the actual shape.

b. Determine the actual area of each of your shapes and of the actual objects. Describe how you determined which of your measurements was "closest" to the actual measurement.

35. Describe how you would find the surface area of the top of each of the following:

d. If the leaves in parts a and b were about the same size (in terms of area), which would have the larger perimeter? Why?

36. ***Classroom Connection*** The April 1999 issue of *The Mathematics Teacher* (pp. 294–298) describes a project in which high school students used Pick's theorem to determine the area of leaves, which biologists need to do. The formula is $A = \frac{b}{2} + I - 1$, where b is the number of boundary points and I is the number of interior points in a figure that has been traced on Geoboard dot paper.

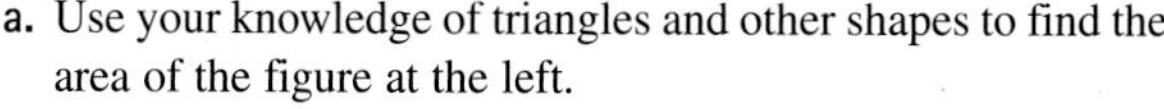

a. Use your knowledge of triangles and other shapes to find the area of the figure at the left.

b. Calculate the area of the figure at the left using Pick's theorem.

c. Determine the area of the leaf at the right using Pick's theorem.

37. Make a photocopy of a map of your state and determine the area of your state.

a. Show and explain your work.

b. Justify the precision in your answer—for example, the choice of unit and the number of decimal places, if any.

c. Describe and explain your degree of confidence in your result.

d. Describe any difficulties you had and how you overcame them.

38. The Treasury Department designed a new \$100 bill and printed \$80 billion in new \$100 bills. How large a room would be needed to contain all these \$100 bills?

39. One way of buying tennis balls is in a can that contains three tennis balls. Is the tin taller or bigger around?

40. Draw on Geoboard Dot Paper a polygon whose perimeter is 20 units and whose area is less than 10 square units. Then draw a figure whose perimeter is 20 units and whose area is greater than 20 square units. Describe your solution process. This description should include your reasoning, guesses that didn't work, and what you learned from those guesses.

41. Let's say you have two similar polygons and the ratio of their perimeters is 2 : 1. What is the ratio of corresponding sides? What is the ratio of their areas? Justify your answer.

42. How is the size of a TV measured? If we are comparing a 15-inch TV and a 30-inch TV, what is twice as big about the bigger TV?

43. If possible, determine the perimeter of the figures below. If there is not enough information, explain why.

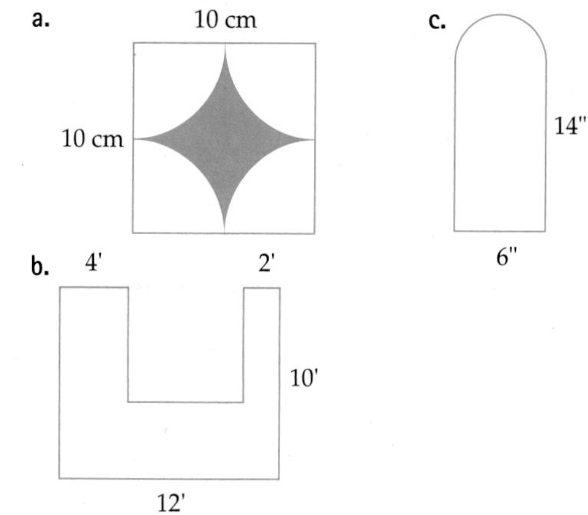

44. If possible, determine the area of the figures below. If there is not enough information, explain why.

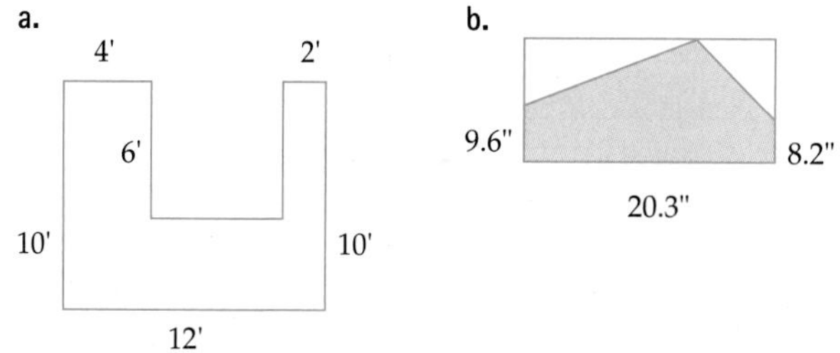

45. For each of the problems below, if there is not enough information given for there to be exactly one answer, explain why and provide one more piece of information that would enable us to determine the solution. If there is enough information, solve the problem.

a. The hypotenuse of a right isosceles triangle is 10 inches. Find the area.

b. The perimeter of one rectangle is twice the perimeter of another rectangle. The dimensions of the rectangle with the greater perimeter are 8 inches by 6 inches. Find the dimensions of the smaller rectangle.

c. The area of a rectangle is 20 centimeters. Find its perimeter.

d. A farmer is fencing his pasture. The fence posts will be 10 feet apart. The pasture is in the shape of a rectangle whose length is double its width. The farmer needs 54 fence posts. What are the dimensions of the pasture?

46. Anne used a grid to determine the area of an irregularly shaped figure, and she determined that the area was 23 square centimeters. Then she realized that she wanted square inches. How can she convert from square centimeters to square inches? There are 2.5 centimeters in 1 inch.

47. Adam is also confused. He wonders why, if you take a rectangle and multiply the length of each side by 1.5, the area of the new rectangle isn't 1.5 times as big as the area of the old rectangle. How would you explain this to him?

48. If a circle has a radius of 5 inches, we determine that the area of the circle is approximately 78.5 square inches. A fellow student has come to you and asked why we use *square inches* to represent the area of the circle. What does 78.5 square inches mean?

49. Cattle ranching is a primary cause of deforestation in Latin America. Since 1960, more than one quarter of all Central American forests have been razed to make pasture for cattle. Just one quarter-pound hamburger imported from Latin America requires the clearing of 6 square yards of rain forest. How much land would be required to provide one hamburger per year for each person in the United States?

50. How much land is used by all the streets and highways in the United States? As of 1990 there were 3,880,000 miles of roadway.

51. The Reverend Robert Walsh served aboard a ship designed to intercept slave ships in 1829. His account of the capture of the Feloz reveals the incredible conditions. He reported that 336 male slaves had been crammed into a space that was 3 feet 3 inches high and 40 feet by 21 feet. How much room did each man have?

52. A kitchen gadget called Perfect Pasta Portions enables people to determine the appropriate amount of spaghetti to boil for 1, 2, 3, or 4 portions. If the diameter of the hole for 1 serving is $\frac{7}{8}$ inch, what is the diameter of the hole for 4 servings?

53. Why is perimeter measured in linear units like inches, feet, or meters, but area is measured in square units like square inches, square feet, or square meters?

54. Explain what π means. Why is it such a special number?

55. ***Classroom Connection*** How many blades of grass are there on a soccer field? Look in the September 1999 issue of *Mathematics Teaching in the Middle School,* pp. 7–10, to see the children's work.

56. ***Classroom Connection*** *Counting on Frank* by Rod Clement (Milwaukee: Gareth Stevens Publishing, 1991) examines the kinds of questions that kids ask. Let us consider two here.

    a. How long would it take to fill a bathroom if both faucets are running?

    b. If ten humpback whales would fit in the narrator's house, how big is the narrator's house?

57. ***Classroom Connection*** *If You Hopped Like a Frog* by David Schwartz (New York: Scholastic Press, 1999) tells the readers that if you ate like a shrew, you would eat three times your weight every day.

    a. How many hamburgers would that be?

    b. How many boxes of cereal?

58. a. ***Classroom Connection*** What is the greatest number of blank squares you could enclose if you colored in 20 squares on a sheet of graph paper with the condition that each square had to have at least one entire side connected to an entire side of another square?

    b. ***Classroom Connection*** What if the condition was changed so that two squares could touch just at a vertex? This problem was posted in *Teaching Children Mathematics* in a more child-friendly way using cubes to make a pen to hold animals who took up the space of one cube. The children's solutions are shown in the December 1997 issue, pp. 212–214.

FROM STANDARDIZED ASSESSMENTS

2006 NECAP, Grade 6

59. Jasmine drew a rectangle with the following properties.

    - The area is 32 square centimeters.
    - The length is twice the width.

    What is the perimeter of Jasmine's rectangle? Show your work or explain how you know.

60. 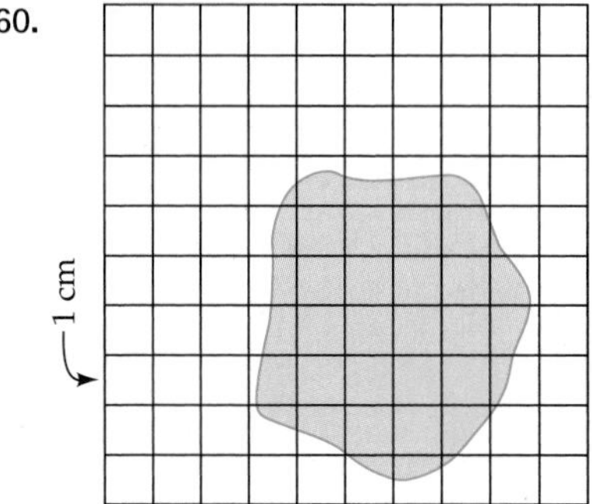

    About how many square cm is the area of the shaded region?

    a. 16   b. 26

    c. 36   d. 52

Source: From the *Fourth NAEP Mathematics Assessment,* p. 41. U.S. Department of Education, National Center for Education Statistics.

2007 NECAP, Grade 6

61. Look at this picture.

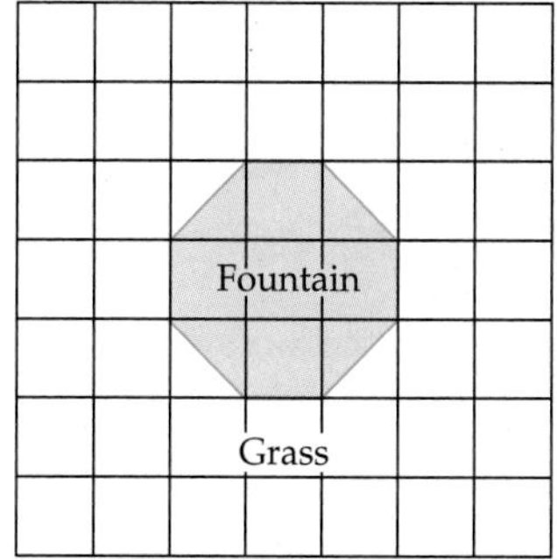

1 square represents 1 square foot.

What is the area of the grass around the fountain?

a. 46 square feet
b. 44 square feet
c. 42 square feet
d. 40 square feet

2008 NECAP, Grade 4

62. Look at these figures.

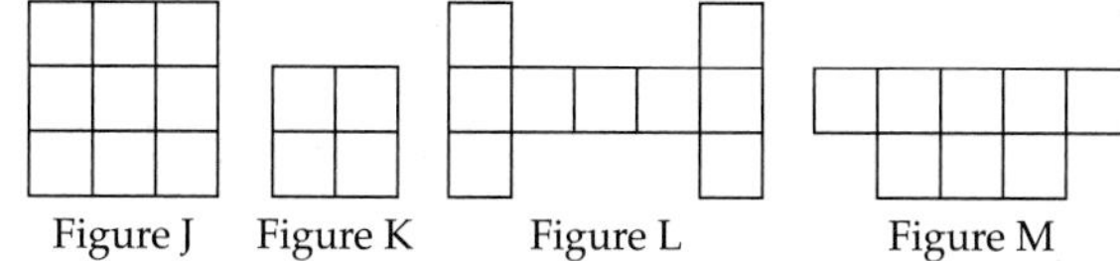

Which two figures have the same area?

a. Figures J and K
b. Figures J and L
c. Figures K and M
d. Figures L and M

63. 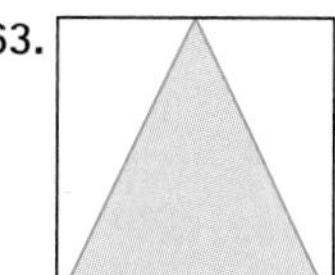

If the area of the shaded triangle shown above is 4 square inches, what is the area of the entire square?

a. 4 square inches
b. 8 square inches
c. 12 square inches
d. 16 square inches
e. Not enough information given

Source: *NAEP Mathematics Assessment*, 1992, p. 253. U.S. Department of Education, National Center for Education Statistics.

SECTION 9.3 Surface Area and Volume

What do you think?

- Is the relationship between surface area and volume a functional relationship? Why or why not?
- If we doubled the volume of a room, would its surface area double too?

In Section 9.2, we examined two attributes of two-dimensional objects: perimeter and area. In this section, we will examine two attributes of three-dimensional objects: surface area and volume. We can think of **surface area** as the area needed to cover all the faces of a three-dimensional object, and we can think of **volume** as the amount of space contained within that three-dimensional object.

According to the Common Core State Standards, the concept of volume and finding the volume of rectangular prisms is introduced in fifth grade. The concept of surface area begins in sixth grade. By the end of eighth grade, students should have learned volume of cones, cylinders, and spheres.

INVESTIGATION 9.3a Building a Box and Filling a Box

The Sweet Sugar Cube Company wants to redesign their boxes in which sugar cubes are sold. What will they need to do to figure out how much cardboard they need for the package? What will they need to do to determine how many sugar cubes each box will hold?

DISCUSSION

How would we calculate the amount of cardboard for one side of the box? We will find the area of one side. To make all sides, we would add the area of each side to get the total "surface area." Surface area is what the name implies, the areas of all the surfaces (or faces) of a three-dimensional figure. Because we are adding area (which is measured in square units), surface area is also measured in square units.

For the second question, we want to fill the inside space with cubes. This is the volume, the quantity that the three-dimensional object holds. Suppose we can fit 10 sugar cubes across the length of the bottom of the box, and 4 cubes along the width of the bottom of the box and 6 cubes high. Then we could fit a total of $10 \times 4 \times 6 = 240$ cubes. This sugar cube example helps us to see why volume is measured in cubic units (because we would fill it with cubes).

UNDERSTANDING THE SURFACE AREA OF PRISMS AND PYRAMIDS

Very simply, the *surface area of a prism or pyramid is equal to the sum of the areas of all of its faces.* Thus, to determine the surface area, we need to be able to determine the dimensions of each face. Let us consider a rectangular prism first. Determine the surface area of the rectangular prism in Figure 9.19 and then read on. It may help first to draw a net of that prism, which is shown at the right.The Common Core State Standards recommends using nets extensively in sixth grade to help students understand surface area. These visuals help students of all ages to better understand the concept.

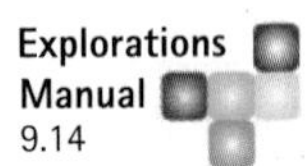

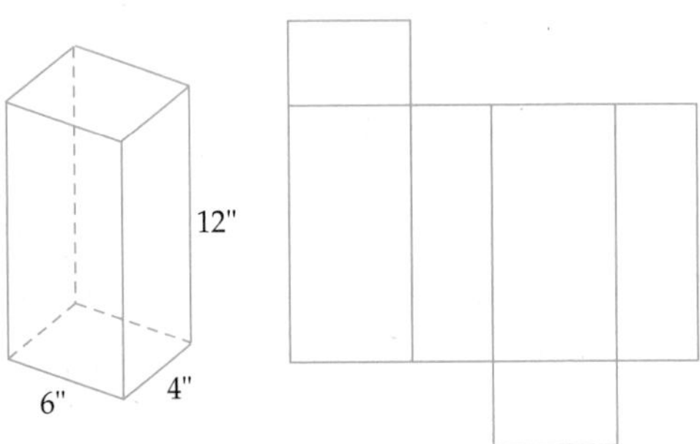

Figure 9.19

There are patterns in this prism that make its surface area easier to determine. You can see that the bottom and top bases are congruent; similarly, the front and back faces are congruent, and the right and left faces are congruent. Thus,

$$\text{Surface area of the prism} = 2(24 \text{ in}^2) + 2(72 \text{ in}^2) + 2(48 \text{ in}^2) = 288 \text{ in}^2$$

With a concrete example under our belts, let us now investigate the surface area of prisms and pyramids whose bases are regular polygons. What do we know about the different surfaces of prisms and pyramids that would make the task easier? Look at the diagrams in Figure 9.20, showing three prisms, one pyramid, and their nets. What do you see? Think before reading on. . . .

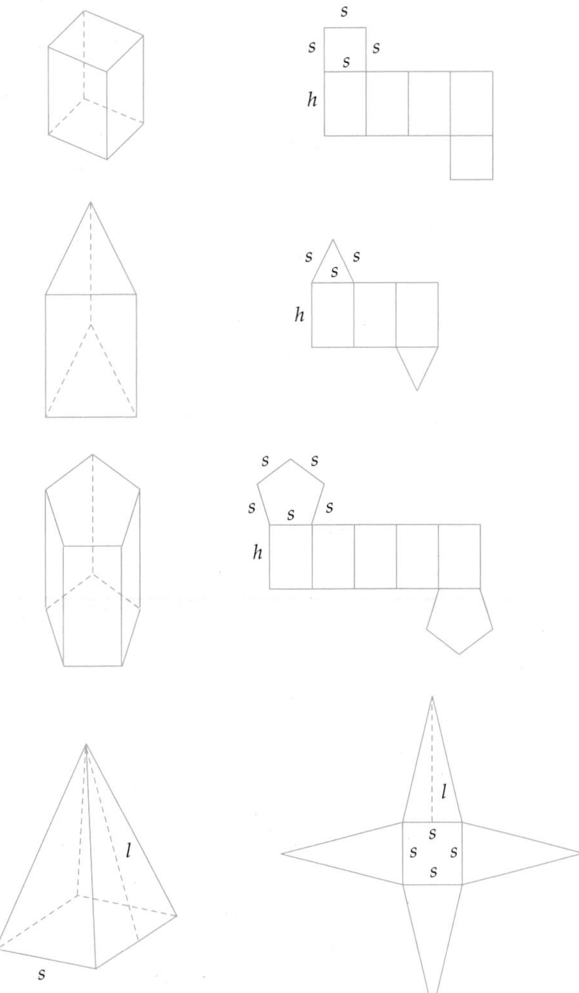

Figure 9.20

Unfolding these three-dimensional shapes to create the two-dimensional nets takes practice. With many sixth graders, as well as adults, physically unfolding paper models helps develop understanding. Another option is to draw out each face separately, find the areas of each, and add them up. Either way, to find the surface area of a prism or pyramid, we find the area of each face and add them to find the complete surface area. Because we are still talking about area, we are still measuring in square units—in other words we are covering the surface with squares.

UNDERSTANDING THE SURFACE AREA OF CYLINDERS

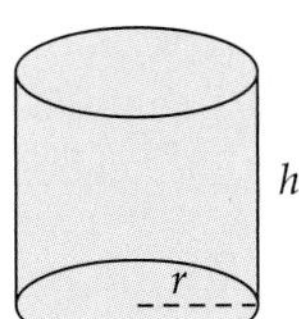

Figure 9.21

How do you think we might determine the surface area of a cylinder with a circular base (Figure 9.21)? Then read on. . . .

To find the surface area of a cylinder, we can begin by finding the area of one base and multiplying that by 2. What about the surface area of the rounded part (the part you would hold if this was a soda can)? How do you think we might find that? If you are not sure, find a cylinder. (Use an empty toilet paper tube or paper towel tube and cut it open, or wrap a piece of paper around a soup can or soda can.) What is the shape of the lateral face? Do this yourself before reading on. . . .

Yes, it is a rectangle! The height of the rectangle is equal to the height of the cylinder. What about the length of the base of the rectangle? How is it connected to the cylinder's base? Think and then read on. . . .

The length of the base of the rectangle is equal to the circumference of the circle (Figure 9.22)! With this knowledge, see if you can now discover the formula for the surface area of any cylinder in terms of the radius of the cylinder and the height of the cylinder. Then read on. . . .

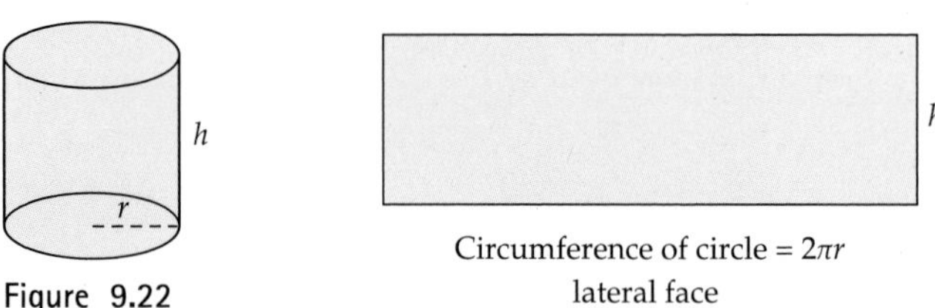

Figure 9.22

Since the Common Core State Standards stress thinking about nets, what would the net of a cylinder look like? Try sketching it and then read on.

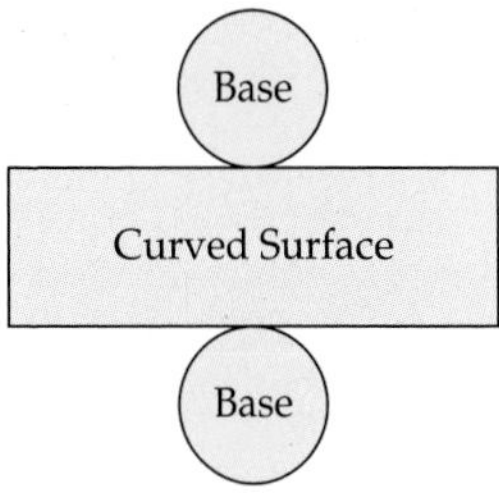

- The area of the base is πr^2, and because we have two bases, the area of the two bases is $2\pi r^2$.
- The area of the rectangle $=$ (the circumference of the circle) times (the height of the cylinder); that is, the area of the rectangle $= 2\pi rh$.

Thus,

$$\text{Surface area of a cylinder} = 2\pi r^2 + 2\pi rh$$

VOLUME

In Section 9.2, we connected the terms *perimeter* and *surround*, and we connected the terms *area* and *cover*. With respect to volume, we can say that **volume** reflects how much it will take to *fill* an object.

UNDERSTANDING THE VOLUMES OF PRISMS

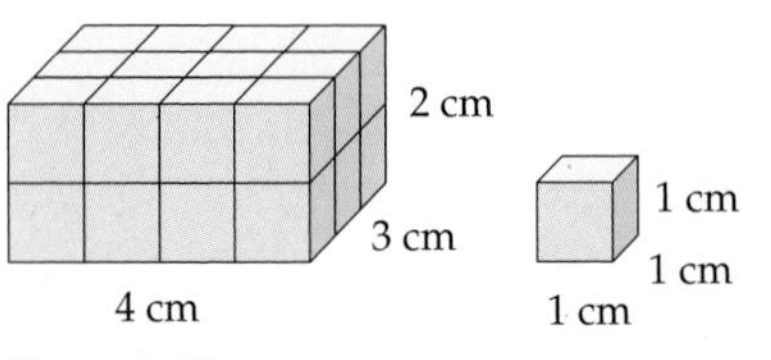

Figure 9.23

Let us first develop the formula for the volume of a rectangular prism. Imagine a box whose base is 4 centimeters by 3 centimeters and whose height is 2 centimeters (Figure 9.23). If we were to fill that box with little cubes, each with dimensions 1 centimeter by 1 centimeter by 1 centimeter—that is, with a volume of 1 cubic centimeter—how many little cubes would it take?

It would take 24 little cubes—that is, $4 \cdot 3 \cdot 2$. Therefore, it makes sense to say that *we can determine the volume of a rectangular prism by multiplying its length, its width, and its height:*

$$V = l \cdot w \cdot h$$

To fill up the bottom layer of the prism, we multiply 4 times 3, which is also finding the area of that rectangular base. Then we can think of it as multiplying by 2 since we have a height of 2, or two layers of cubes. So, another way to look at finding the volume of this

Picture It

How could you answer the question without using the formula $V = l \cdot w \cdot h$?

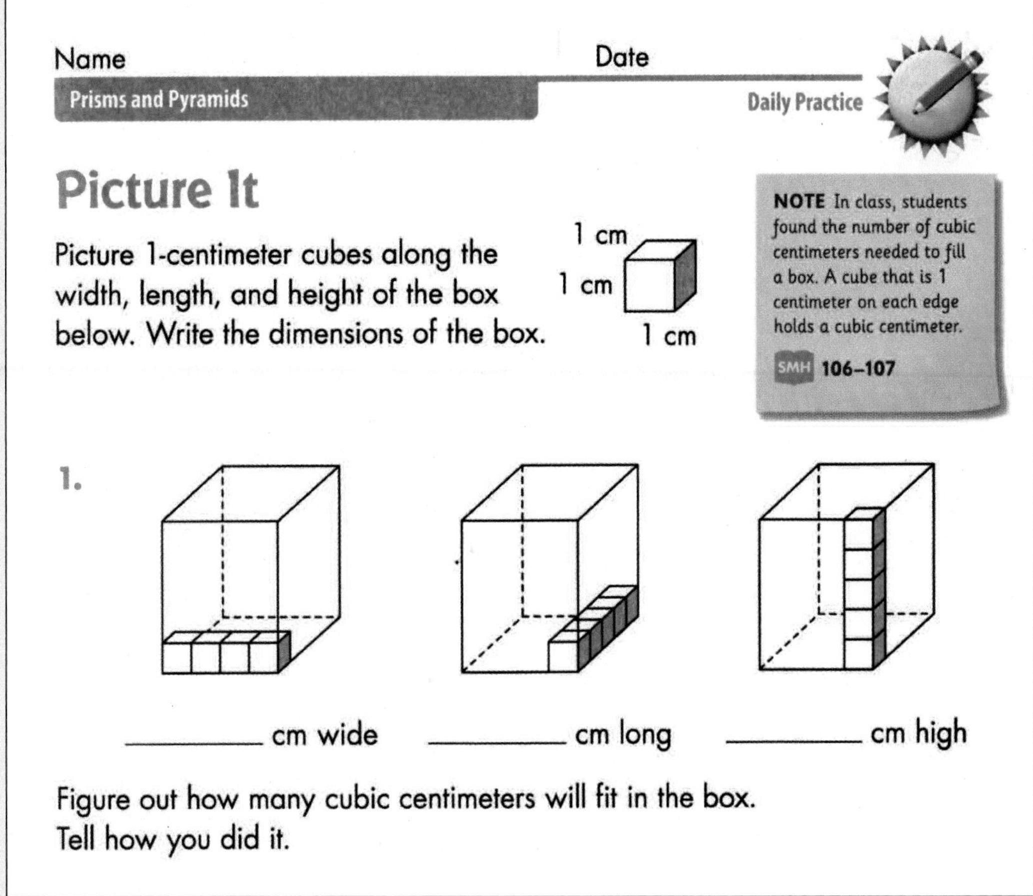

Name Date

Prisms and Pyramids — Daily Practice

Picture It

Picture 1-centimeter cubes along the width, length, and height of the box below. Write the dimensions of the box.

NOTE In class, students found the number of cubic centimeters needed to fill a box. A cube that is 1 centimeter on each edge holds a cubic centimeter.

SMH 106–107

1\.

______ cm wide ______ cm long ______ cm high

Figure out how many cubic centimeters will fit in the box. Tell how you did it.

rectangular prism, which will help us with other prisms and cylinders, is to consider that we find the area of the base (to fill up the bottom layer of cubes) and multiply it by the height. In other words,

Volume of prisms and cylinders = (Area of the Base)(Height)

INVESTIGATION 9.3b Finding Volume of Cylinders and Prisms

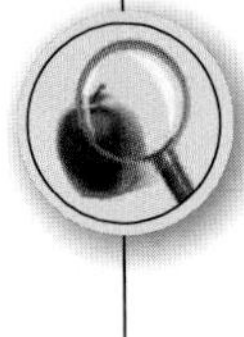

Using the idea that volume of a prism or cylinder = (Area of Base)(Height), determine the volume of the following figures:

A.

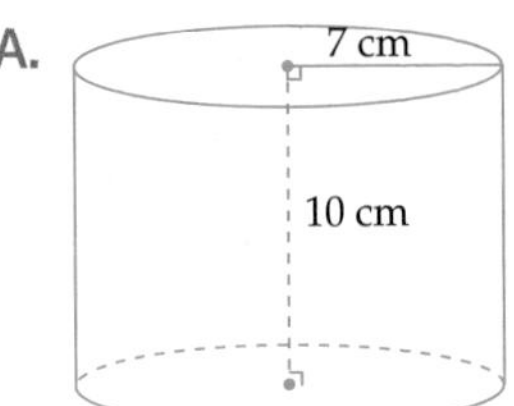

B.

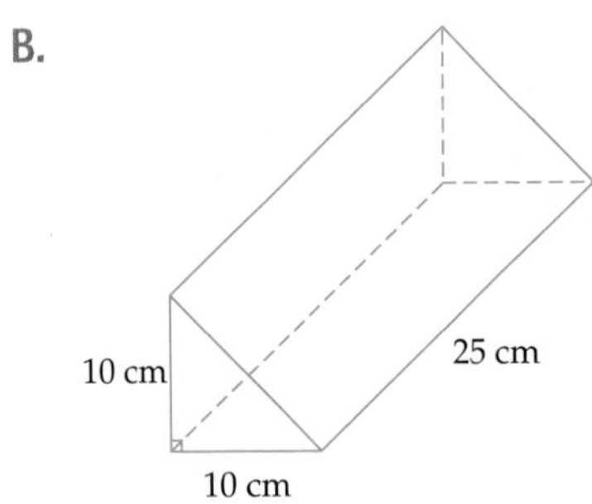

DISCUSSION

A. It is a little more difficult to visualize filling the base of a cylinder with cubes, but the concept is the same. We would have to cut some cubes to make them fit, as shown below.

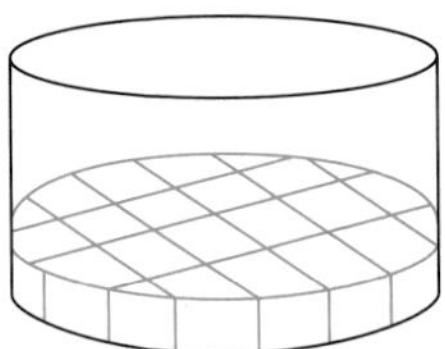

How does this help us to find the volume of the cylinder?

The area of the circular base would tell us how many cubes would be on the bottom layer, and then we multiply by the height of the cylinder to find out how many cubes would fit.

$$\begin{aligned}\text{Volume of cylinder} &= \text{(Area of Base)(Height)}\\ &= (\pi r^2)(h)\\ &= \pi(7)^2(10)\\ &= 490\pi \text{ cubic centimeters}\end{aligned}$$

The result of the previous equation is approximately 1539.38 cubic centimeters. In other words, if we had a cylinder this size, we could fit 1539 cubic centimeters (cubes that are 1 centimeter by 1 centimeter by 1 centimeter) inside the cylinder.

B. First we need to decide what the base is. Remember that the base is not always on the bottom. In this triangular prism, the base is the triangle and the height is the distance between the triangular bases (25 cm).

Volume = (Area of Base)(Height)

Since we have a triangular base, we use the formula Area of Base = $\frac{1}{2}$(base)(height). Note here that we are using the words "base" and "height" to mean different things. We have the Base and Height of the prism, and the base and height of the triangle. Using capital letters when we are talking about the three-dimensional figure (the prism) and lowercase letters when we are talking about the two-dimensional figures (the triangle) can help us differentiate them.

$$\begin{aligned}\text{Volume} &= \text{(Area of Base)(Height)}\\ &= (\tfrac{1}{2}bh)\text{(Height)}\\ &= (\tfrac{1}{2} \times 10 \times 10)(25)\\ &= 1250 \text{ cubic centimeters}\end{aligned}$$

UNDERSTANDING THE VOLUMES OF PYRAMIDS AND CONES

We will illustrate the formula of a pyramid using a cube. It so happens that three congruent pyramids will fit inside the cube. Most people need to see this to believe it, and Figure 9.24 provides the necessary information to make three congruent pyramids that can be joined together to make a cube. Two-dimensional drawings of three-dimensional objects can be difficult to visualize. Geosolids are manipulatives that enable you to demonstrate concepts related to volume. You can fill a geosolid pyramid with water and see that it takes three of them to fill up a geosolid cube with the same height and same size base.

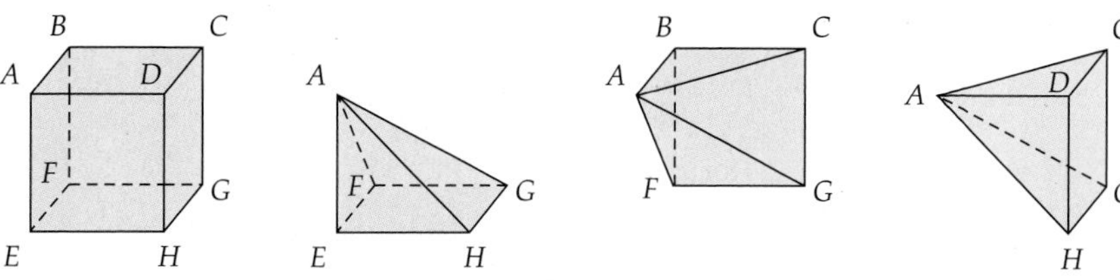

Figure 9.24

At this moment, let us examine the implications of this relationship for determining the volume of a pyramid. On the basis of this information, what do you predict will be the formula for the volume of a pyramid? Think before reading on. . . .

If three congruent pyramids can be fit into the cube, and the bases of the pyramids are congruent to the base of the cube, this means that the volume of the pyramid is one-third the volume of the cube and *that the formula for finding the volume of a pyramid is*

$$\text{Volume} = \frac{1}{3}\,(\textit{Area of Base})(\textit{Height})$$

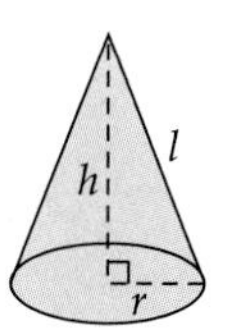

Figure 9.25

What about cones? Thinking of the similarities between pyramids and cones, can you hypothesize a formula for the volume of a cone with radius r, height h, and slant height l (Figure 9.25)? Try to do so before reading on. . . .

Geosolids can help us see this as well. You can fill a geosolid cone with water and see that it takes three of them to fill up a geosolid cylinder with the same height and same size base.

We can deduce that just as the volume of a pyramid is one-third the volume of a prism with the same base and height, the volume of a cone must be one-third the volume of a cylinder with the same base and height:

$$\text{Volume of a cone} = \frac{1}{3}\,(\textit{Area of Base})(\textit{Height})$$

Just as the base and height of triangles had to be perpendicular, so do the height and base of a pyramid or cone.

UNDERSTANDING THE VOLUME OF SPHERES

The derivation of the formula for the volume of a sphere is rather sophisticated.

$$\text{Volume of a sphere} = \frac{1}{3}\,\pi r^3 \textit{ where r is the radius of the sphere}$$

There is an interesting connection between the volume of a sphere and the volume of a cylinder that you can verify if your instructor has a hollow cylinder and sphere of the same radius, like the ones in the Geosolids set. The height of the cylinder must be equal to the diameter of the cylinder. Recall that the volume of a cylinder is *(Area of Base) (Height)*; if the base of the cylinder is a circle of radius r and the height of the cylinder is $2r$, then the volume of that cylinder is $2\pi r^3$.

First imagine placing the sphere inside the cylinder (Figure 9.26). Next, imagine filling the cylinder with water. Finally, we take out the sphere and put the water back in the cylinder (Figure 9.27). If this experiment is done carefully, the height of the water will be $\frac{1}{3}$ the height of the cylinder. It logically follows that the volume of the sphere is $\frac{2}{3}$ the volume of the cylinder, so we have

$$\text{Volume of sphere} = \frac{2}{3}(2\pi r^3) = \frac{4}{3}\pi r^3$$

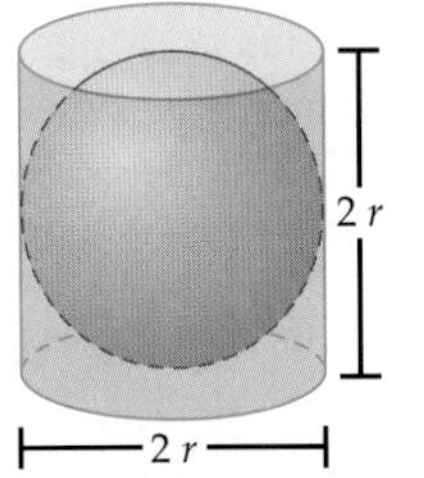

Figure 9.26

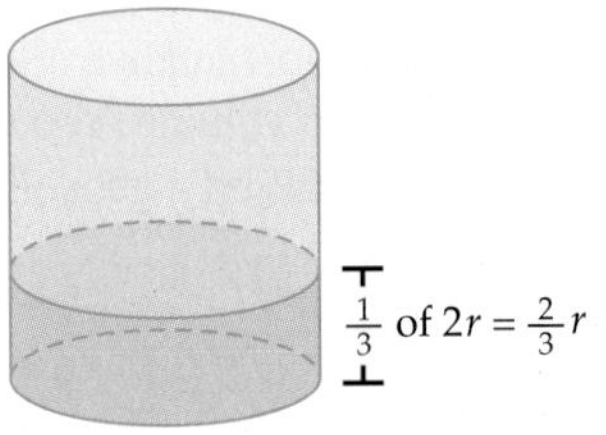

Figure 9.27

Archimedes was the first person to prove the formula for the volume of a sphere.

OUTSIDE THE CLASSROOM

With respect to understanding spheres, one of the more interesting mathematical problems is determining how to represent that sphere in two dimensions—for example, making a map of the planet Earth.

If you look at the traditional (Mercator) map of the world (Figure 9.28), you can see that this map misrepresents the relative sizes of the continents. For example, the area of Africa is $1\frac{1}{2}$ times the area of North America. However, North America appears to be larger than Africa in the Mercator projection. In 1974, the Peters projection (Figure 9.39) was created to represent accurately the relative sizes of Earth's land masses. The mathematical problem here is representing a three-dimensional object in two dimensions. If you make a map of the planet that is rectangular in shape (there are other possibilities), either you can have the shapes accurate (that is, this is what the continents would look like viewed from space) or you can have the relative sizes accurate, but you cannot have both! This is but one of many examples where the mathematical answer to a problem is "There is no perfect solution." There are many websites where you can explore this map-making issue in more depth, and there are many websites with lesson plans for elementary schools.

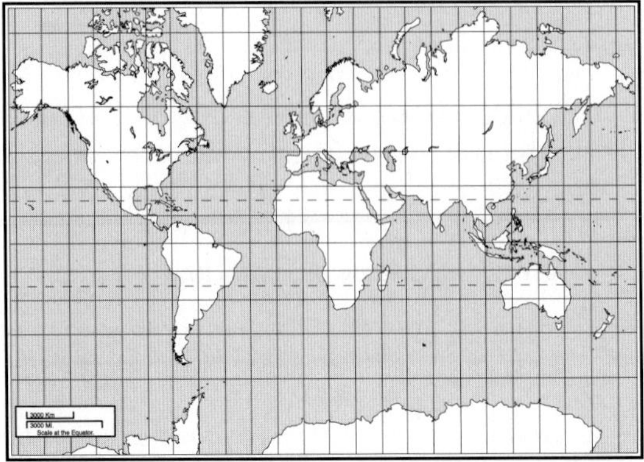

Figure 9.28

(continued)

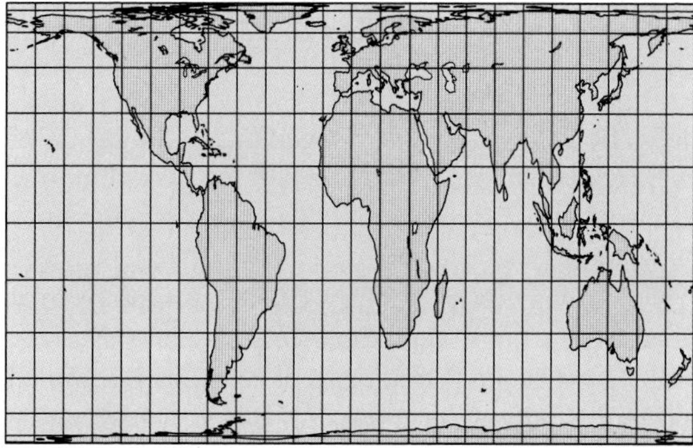

Figure 9.29

Source: From Which Map Is Best? Projections for World Maps (1986). Courtesy of American Congress on Surveying and Mapping. Used with permission.

INVESTIGATION 9.3c Are Their Pictures Misleading?

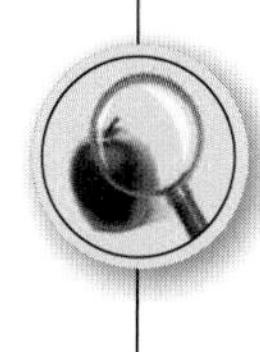

More and more newspapers, magazines, and brochures use graphs to display data in a way that will catch the reader's eye. The hypothetical example below illustrates the dangers. Let's say that Yummy Soda sold twice as much soda as Good Soda last year. The graph in Figure 9.30 is a pictorial representation of the sales of the two sodas that Yummy included in an ad. Why do you think Good Soda might object to the graph and actually file a lawsuit for unfair advertising? Write down your thoughts before reading on. . . .

Sales (millions of cans)
30
20
10
0
Good Soda
Yummy Soda

Figure 9.30

DISCUSSION

If the reader looks at this as simply a "cute" bar graph, the height of the Yummy bar is twice the height of the Good bar. However, the diameter of the base of the Yummy can is twice the diameter of the base of the Good can, causing the Yummy can to look more than twice as big as the Good can. What is the ratio of the volumes of the two cans? Determine this before reading on. . . .

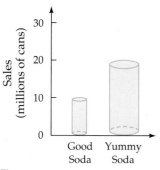

One solution is to take the actual dimensions:

- The Good Soda can has a radius of 2 millimeters and a height of 10 millimeters.
- The Yummy Soda can has a radius of 4 millimeters and a height of 20 millimeters.

Using the volume formula for cylinders, $V = \pi r^2 h$ we have

- Volume of Good Soda can $= 40\pi$
- Volume of Yummy Soda can $= 320\pi$

That is, the Yummy Soda can has 8 times the volume of the Good Soda can. *Note:* Do you see why we used π instead of using a decimal approximation? We also could have used the ratios—that is,

$$\text{Volume of Good Soda can} = \pi r^2 h$$

$$\text{Volume of Yummy Soda can} = \pi(2r)^2(2h) = 8\pi r^2 h$$

In either case, we find that the graph gives a visual impression that Yummy Soda is selling much more than twice as much as Good Soda. (It also gives the impression that Yummy cans of soda are bigger!)

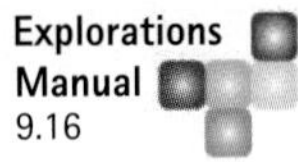
Explorations Manual 9.16

DETERMINING THE VOLUMES OF IRREGULARLY SHAPED OBJECTS

In Section 9.2, we acknowledged that formulas have limited usefulness in real life. Similarly, we often need to find the volume of a solid to which the formulas do not apply. How can we do this? Think about this before reading on. . . .

This is actually a very challenging problem that stumped ancient mathematician-scientists for quite some time. One way of addressing this question was developed by one of the greatest mathematicians of all times, Archimedes, who lived in the city of Syracuse, located in modern Italy, over 2000 years ago.

Archimedes was summoned by the king, who suspected that the goldsmith who had made his crown had not made it of solid gold but had mixed some silver in it. He wanted Archimedes to determine whether the crown was pure gold without having to cut into it.

Archimedes knew that silver was only about half as dense as gold. (We determine the density of an object by dividing its mass by its volume—for example, grams per cubic centimeter.) Archimedes was aware of the formula volume $\times$ density $=$ mass (for example, $\text{cm}^3 \times \text{g/cm}^3 = \text{g}$). Thus, if he could determine the volume of the crown, he would be able to solve the problem. But how was he to find the volume of the crown accurately?

Some time later, Archimedes was taking a bath when the solution suddenly came to him: He could determine the volume of the crown by submerging it in water, because two objects with the same volume will displace the same amount of water. Legend has it that he became so excited that he ran outside, still naked, shouting "Eureka," which means, in Greek, "I have found it!"

How did this discovery enable him to determine whether the crown was counterfeit?

Now that he knew a way to determine the volume with some precision, he could use this number and the density of gold in the formula given above to determine how much the crown would weigh if it were pure gold. (It turns out that the goldsmith had indeed cheated the king!)

INVESTIGATION 9.3d Finding the Volume of a Hollow Box

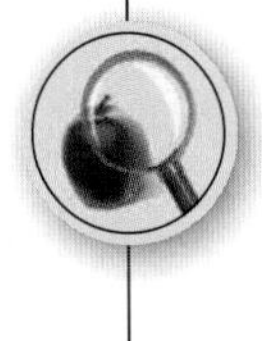

Imagine a large object that will be used to hold water or other materials (Figure 9.31). The outer dimensions are 12 meters by 10 meters by 8 meters. The walls are 1 meter thick, and so is the floor. What is the volume of the container in cubic meters? For example, if we were to make this object from concrete, we would need to know how much concrete to order. Work on this. Then read on. . . .

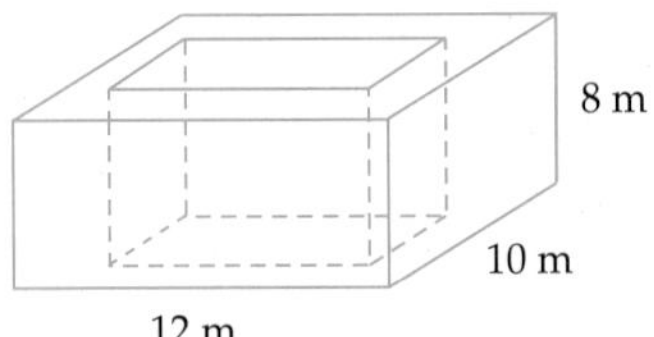

Figure 9.31

DISCUSSION

STRATEGY 1 Break the problem into parts

Just as a rectangular prism has six surfaces, a prism with no top will have five surfaces. Let us identify them and then determine the dimensions of each of them: bottom, front, right side, left side, back. If you did not think of this strategy, seeing if you can use it to determine the volume will be a useful exercise of your problem-solving tools. If you used a different strategy, do you get the same answer using this strategy? Work on this before reading on. . . .

Explorations Manual 9.15

The volume of the bottom will be

$$12\text{ m} \cdot 10\text{ m} \cdot 1\text{ m} = 120\text{ m}^3$$

The front and back sides are congruent, and the volume of each will be

$$12 \text{ m} \cdot 7 \text{ m} \cdot 1 \text{ m} = 84 \text{ m}^3$$

(Do you see why it is not 8 m? If not, make a model of the container.)
The right and left sides are congruent, and the volume of each will be

$$8 \text{ m} \cdot 7 \text{ m} \cdot 1 \text{ m} = 56 \text{ m}^3$$

(Do you understand these dimensions? If not, make a model.)
Thus, the volume $= 120 \text{ m}^3 + 2(84 \text{ m}^3) + 2(56 \text{ m}^3) = 400 \text{ m}^3$

STRATEGY 2 See the problem from a different perspective
For example, determine what the volume would be if the container were not hollow. Then determine the volume of the hollow region. Then subtract! Do you understand this strategy?

Volume of the whole region:

$$12 \text{ m} \cdot 10 \text{ m} \cdot 8 \text{ m} = 960 \text{ m}^3$$

Volume of the hollow:

$$10 \text{ m} \cdot 8 \text{ m} \cdot 7 \text{ m} = 560 \text{ m}^3$$

What is the difference between the two? Does this answer match the answer from Strategy 1? Exploration 9.16 provides more opportunities to apply your understanding of volume concepts.

Again, it is important to note that this discussion illustrated only two of several possible ways to solve this problem.

INVESTIGATION 9.3e Surface Area and Volume

In Section 9.2, we discovered that perimeter and area are not related in a simple way; for example, if you double the perimeter, that doesn't necessarily mean that you double the area. Let us investigate the relationship between surface area and volume. As you can imagine, there are many applications of this relationship in fields outside mathematics!

Consider a set of eight small cubes, each of whose dimensions are 1 centimeter by 1 centimeter by 1 centimeter (Figure 9.32).

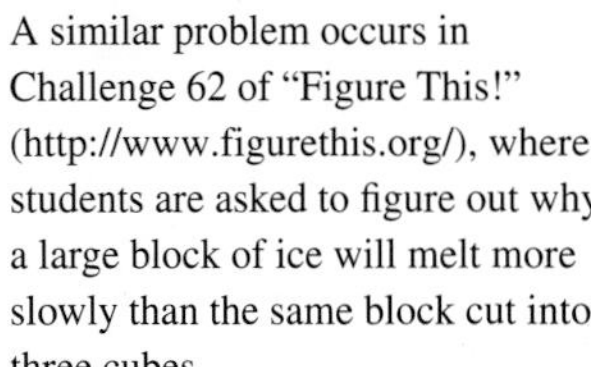

CLASSROOM CONNECTION

A similar problem occurs in Challenge 62 of "Figure This!" (http://www.figurethis.org/), where students are asked to figure out why a large block of ice will melt more slowly than the same block cut into three cubes.

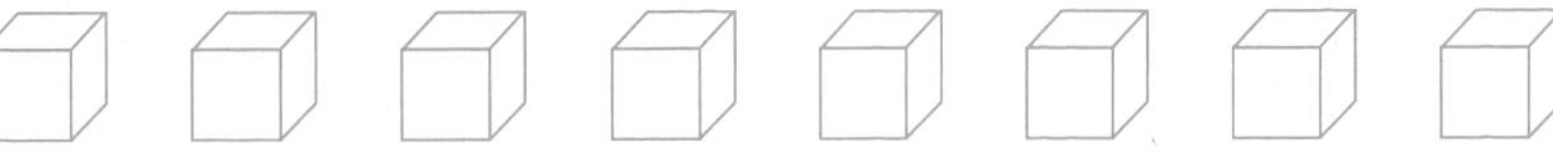

Figure 9.32

What arrangement of those cubes will have the smallest surface area?
What arrangement of those cubes will have the largest surface area?

Work on these questions before reading on. . . . 

DISCUSSION

There are two aspects of this question that are not well defined. First, some students, applying this idea to real-life phenomena such as melting ice, consider the "surface area" to be the amount of surface exposed. Thus they do not count the base. Other students imagine having to paint the whole arrangement, so they do count the base. In this investigation, when we say "surface area," we mean the latter.

The second part of the question that is not well defined is "arrangement." In this investigation, in order for something to count as an arrangement, at least one entire face of each cube will have to cover an entire face of another cube. Thus, simply putting the six cubes down on the table so that they are all disjoined does not count as an "arrangement" as we are defining this problem. With these two aspects of the problem now well defined, solve the problem before reading on. . . .

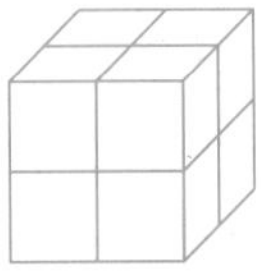

Figure 9.33

Arranging the cubes into a large cube will create the smallest surface area (Figure 9.33). If you didn't find this, make them into a cube (or draw a diagram) and determine the surface area.

Because all six faces of the large cube are congruent, we can multiply the area of one face (4 cm^2) by 6 (the number of faces), to get 24 cm^2.

The largest surface area is obtained by arranging the eight cubes in a line (Figure 9.34). If you didn't find this, please determine the surface area. If you did, you may want to compare your solution path with that of another student. Then read on. . . .

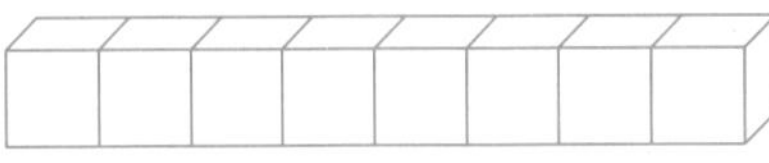

Figure 9.34

STRATEGY 1 See this arrangement as a prism with six sides
The surface area of the front, top, back, and bottom is 32 cm^2.
The surface area of the two bases (the two sides in this diagram) is 2 cm^2.
Thus the total surface area is 34 cm^2.

STRATEGY 2 See this arrangement as two end cubes and six cubes in the middle
The surface area of the six cubes in the middle is 24 cm^2.
The surface area of the two cubes on the end is 10 cm^2.
Thus the total surface area is 34 cm^2.

SUMMARY 9.3

If this section has been successful, the various formulas for surface area and volume make sense. In order to make sense of these formulas, we used some strategies like the ones we used to make sense of area formulas in Section 9.2 (breaking a figure into parts) and built on previous knowledge. Connections were an important aspect of making sense of these formulas. That is, the formulas for finding the surface area and volume of prisms are fairly straightforward. By seeing how prisms and cylinders are similar, we could then use the prism formulas to understand the cylinder formulas. Similarly, we connected the volume of a pyramid to the volume of a cube in order to understand the formula for the volume of a pyramid. Understanding the formulas and having a good problem-solving toolbox then enabled us to solve nonroutine problems. Finally, we examined the relationship between surface area and volume, realizing as we did so that this relationship, like that between perimeter and area, is not simple.

9.3 Exercises

1. Determine the surface area and volume of each of the following:

   a. $\measuredangle ABC$ is a right angle.
   $AB = 10$ inches, $BC = 12$ inches
   $AD = 3$ feet

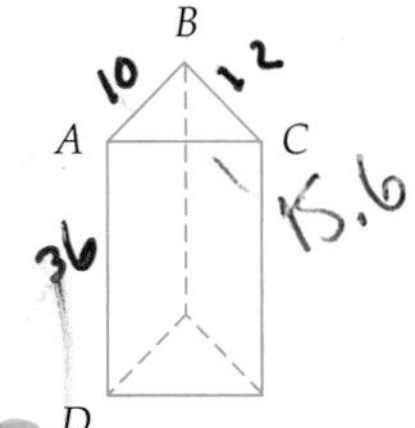

   b. The base is a square.

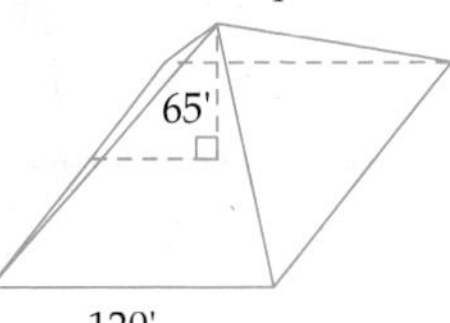

   c. The diameter is 5 feet 6 inches.
   The height is 11 feet 3 inches.

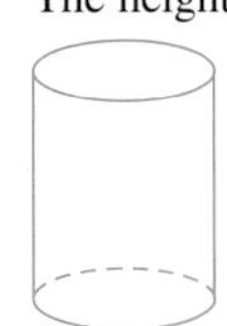

2. Find the missing numbers for the dimensions and measures of a rectangular prism.

	Length	Width	Height	Surface area	Volume
a.	10 m	5 m			900 m^3
b.		7 m	5 m	214 m^2	
c.	5 m			94 m^2	60 m^3
d.		2.5 m	4.6 m		172.5 m^3

3. a. If a sphere has a radius of 6 inches, what is its volume?

   b. If a cylinder has a height of 12 feet and a radius of 2 feet, what is its surface area? volume?

   c. If a cylinder has a volume of 400 m^3 and a radius of 5 m, what is its height?

4. a. Look at the triangular prism shown at the top of the right column. If you were to make a net for this polyhedron, what would be the dimensions of each face? (The triangle is a right triangle.)

   b. Look at the figure at the far right at the top of the right column. If you were to make a net for this polyhedron, what would be the exact dimensions of the "roof"?

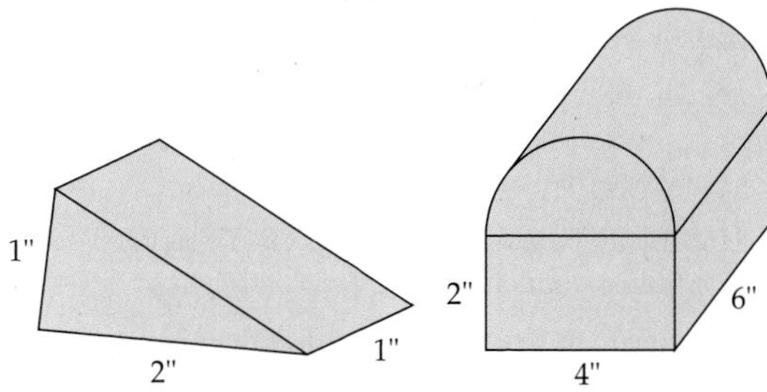

5. a. Determine the surface area and volume of the watering trough shown below. The length of the trough is 16 feet, each of the two ends of the trough is an isosceles triangle whose base is 2 feet, and the height is 1 foot.

   b. Determine the surface area of the inside of the room in the center below and the volume of the room. The base of the room is in the shape of an isosceles trapezoid. The longer base is 12 feet 4 inches, and the shorter base is 8 feet 9 inches. Both slant sides are 13 feet. The height of the room is 9 feet.

   c. The base of the pyramid at the right below is a square. The slant height of the pyramid is 18 feet. Determine the surface area and volume.

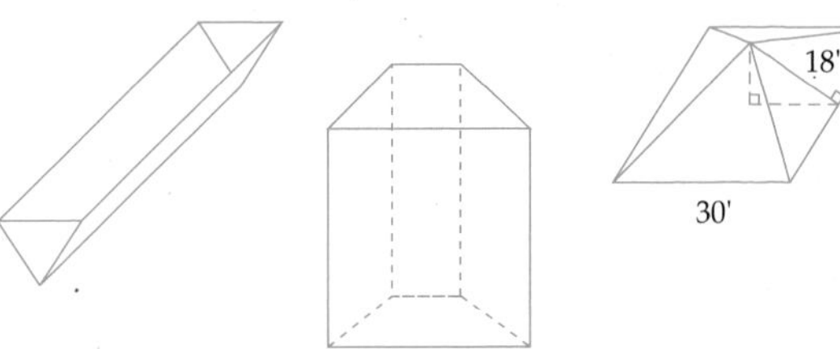

6. a. Determine the surface area and volume of the swimming pool below.

   b. Determine how many gallons of water the pool will hold.

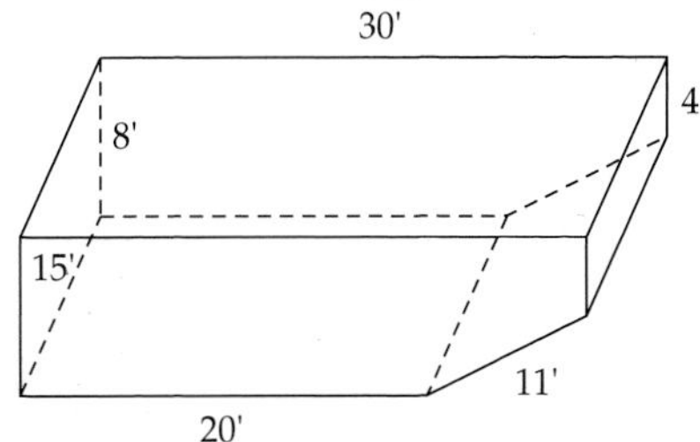

7. The dimensions of a cereal box are $b = 6$ inches, $h = 12$ inches and $w = 2$ inches.

   a. Determine its volume and surface area.

   b. Keeping the dimensions as whole numbers, modify the dimensions (but keep them as integers) to keep the volume the same but decrease the surface area and thus the amount of packaging.

   c. What is the maximum volume you can get with the same surface area in part a?

8. Imagine a piece of square paper that measures 20 by 20 cm. You can make a box (with no lid) by cutting a square of the same size from each corner and folding up what's left to make a box. Keeping the lengths of each sides integers, what is the maximum volume box that can be made?

9. A regular sheet of paper ($8\frac{1}{2}$ inches by 11 inches) can be rolled into a cylinder in two different ways. Which of these cylinders has the greater volume?

10. a. How many boxes that are 1 foot by 1 foot by 1 foot could you fit into your classroom?

b. How many boxes that are 6 inches by 6 inches by 6 inches could you fit into your classroom?

c. How many boxes of cereal could you fit into your classroom?

11. Each edge of the cube is 20 feet. Calculate the distance from point A to point D on each path.

a. From vertex A to vertex B to vertex C to vertex D

b. From vertex A to a point X halfway between vertex B and C, then from that point to vertex D

c. A straight line from vertex A to vertex D

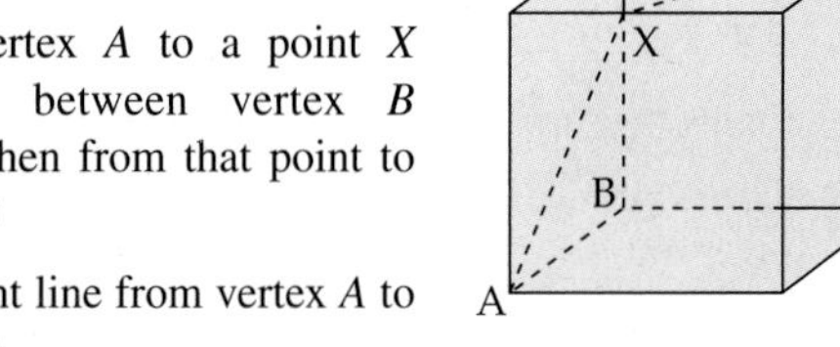

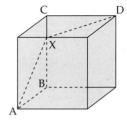

12. Let's say a natural disaster (hurricane, tornado, ice storm, or earthquake) has struck your area and thousands of people have been made homeless. The National Guard has a large supply of cots, measuring 2 feet by 6 feet.

a. Let's say the National Guard has tents that measure 20 feet by 25 feet. How many cots could they reasonably get in each tent?

b. Let's say the gym in a high school measures 120 feet by 80 feet and has two entrances. How many cots could they reasonably get in the gym?

DEEPENING YOUR UNDERSTANDING

13. Before we had children, my wife and I went camping in a pup tent, shown at the left in the figure that follows. After having two children, we bought a much bigger tent, like the one in the center. Determining the volume of that tent is rather complex, so let us assume that the big tent was actually a square pyramid, like the figure at the right, with a height of 8 feet.

a. How much bigger in volume is the new tent than the old tent? First predict the answer and explain the reasoning behind your prediction. Then do the computations.

b. Compare the amounts of material needed to make these tents. First predict the answer and explain the reasoning behind your prediction. Then do the computations.

c. Using what you have learned about two- and three-dimensional figures in the last three chapters, describe your hypotheses for how you might determine the actual volume of the middle tent. The height of the truncated pyramid is 6 feet, and the height of the small pyramid at the top is 2 feet.

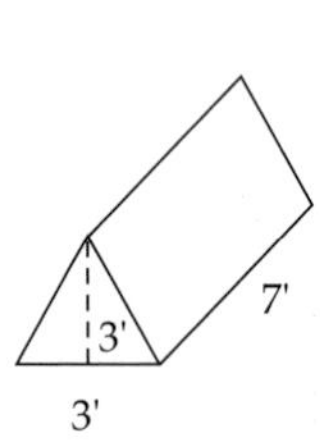

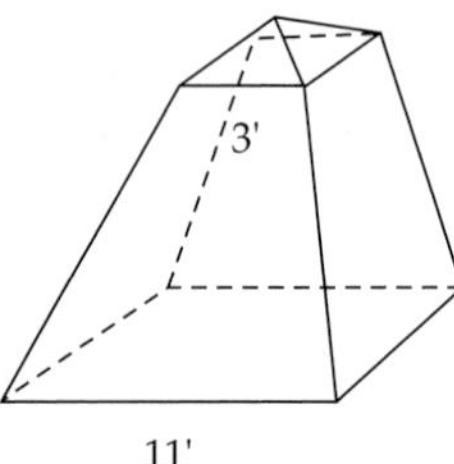

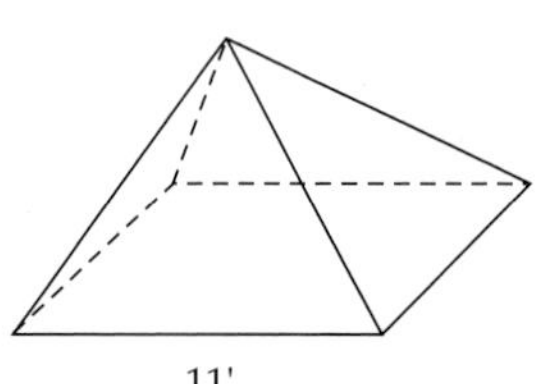

14. The napkin ring pictured is to be resilvered. How many square centimeters of surface area must be covered?

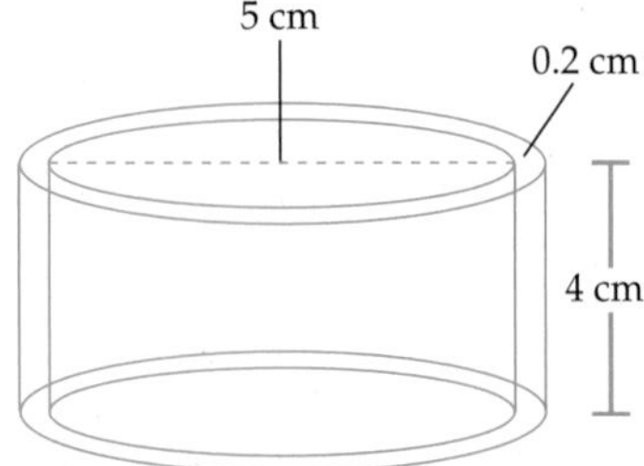

15. This past year the local elementary school has bought 50 boxes of paper, each box containing 10 reams. Each ream contains 500 sheets of paper. There are two shelves in the copy room, each 12 inches wide and 5 feet long. The second shelf is 2 feet above the first shelf and 3 feet below the ceiling. Assuming that the shelves are sturdy, how many sheets of paper can be stored on the shelves? What percent of the yearly purchase is this?

16. The U.S. federal debt, at the time of the writing of this exercise, was \$17 trillion. One way to get a sense of the enormousness of that number is to imagine how much space 17 trillion one-dollar bills would occupy. Determine the space. Then recommend a denomination that would enable a semitrailer to haul that much money.

17. The Three Gorges Dam in China required approximately 950 million cubic feet of concrete. Do something with this number to make it more understandable. For example, it would be equivalent to a block 1 mile long, 1 mile wide, and x feet high.

18. The Great Pyramid was built around 2600 B.C. and is considered to be one of the "Seven Wonders of the World." It was built for Pharaoh Khufu. The pyramid was made from about 2.3 million stone blocks, whose total weight was about 6 million tons! It has been estimated that it took 100,000 workers about 30 years to make it, and we believe they hauled the blocks only during the three months of the year when the Nile was flooded—that is, when the farmers weren't farming. The pyramid has a square base, and each side is 768 feet. The height of the pyramid is 481 feet.

    a. Compare the volume of the pyramid to the volume of the building in which you are taking this course.

    b. Compare the volume of the pyramid to the volume of a building 100 feet high covering a football field (120 yards long and 55 yards wide).

    c. Compare the Great Pyramid to the Transamerica Pyramid in San Francisco. The height and base of the Transamerica Pyramid are 870 feet and 117 feet, respectively.

    d. Many books have been written about many of the "coincidental" measurements of the pyramid. For example, it is said that the area of one of the triangular faces of the pyramid is equal to the square of the height of the pyramid. Are these two dimensions equal?

19. Joe has eight blocks of ice, each measuring 30 centimeters by 30 centimeters by 30 centimeters.

    a. How should he stack them if he wants them to melt as slowly as possible?

    b. How should he stack them if he wants them to melt as quickly as possible?

20. Let's say a company is manufacturing juice boxes that are rectangular prisms with dimensions 4 inches by 3 inches by 4 inches. The company has a warehouse whose dimensions are 40 feet by 30 feet by 20 feet. If the company makes 200,000 juice boxes, will it be able to store all the juice boxes in the warehouse?

21. Sharon has 150 CDs and has decided to make a wooden shelf to hold them. The space between CDs need only be 2 millimeters. Design a shelf that will hold at least 200 CDs.

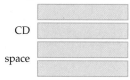

22. Given eight cubes, each with dimensions of 1 centimeter by 1 centimeter by 1 centimeter, can you find arrangements that will have surface areas for every whole number between the minimum and maximum that were found in Investigation 9.3C—that is, between 24 square centimeters and 34 square centimeters?

23. How many cubic centimeters would it take to fill 1 cubic meter?

24. Consider two cubes. The smaller cube measures 4 inches on a side, and the larger cube measures 10 inches on a side. What is the ratio of the surface areas of the two cubes? What is the ratio of the volumes of the two cubes?

25. Consider the two similar rectangular prisms below (not drawn to scale). The dimensions of the smaller prism are 3 inches by 2 inches by 4 inches, and the dimensions of the larger prism are 9 inches by 6 inches by 12 inches. That is, the ratio of their sides is 1:3

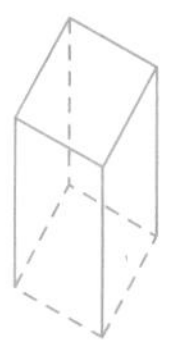

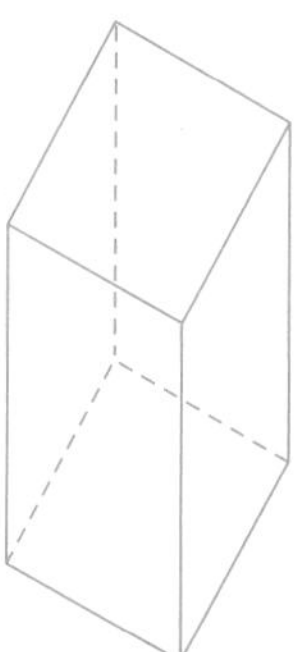

    a. Predict the ratio of their surface areas and explain your reasoning. Then determine the actual ratio of the surface areas. If your prediction was off, explain the error in your prediction.

    b. Predict the ratio of their volumes and explain your reasoning. Then determine the actual ratio of the volumes. If your prediction was off, explain the error in your prediction.

26. When my children were younger, we bought a set of dominoes. A complete set had every possible combination of numbers from 0 (represented by a blank) to 6. For example:

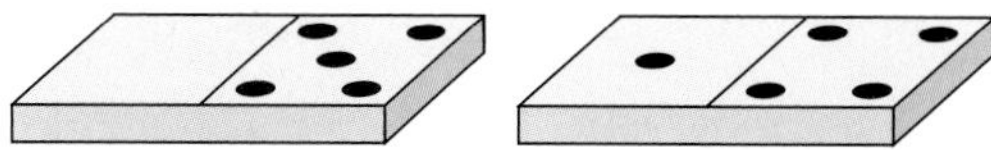

    a. How many dominoes are in a complete set?

    b. If each domino is 1 inch $\times$ 2 inches $\times$ $\frac{3}{8}$ inch, design a carrying case that would hold all of them.

27. This problem is taken from a book called *Mathematical Investigations: Book One.*[7] I have noticed that "flats" of berries and "flats" of young plants are not cubical in shape but, rather, are rectangular prisms.

Suppose you wanted a flat that would hold 4000 cubic centimeters of strawberries. What would be the dimensions of the prism that would require the least amount of cardboard to make?

Source: From *Mathematical Investigations: Book One*, by Randall Souviney, Murray Britt, Salvi Garguilo, and Peter Hughes (Dale Seymour, 1988), p. 10, 14.

28. a. The cake pictured has been cut into three equal pieces. One more person comes. How can you make one cut in the cake so that each of four persons will get the same amount of cake?

b. The cake pictured below has been cut into N equal pieces. One more person comes. How can you make one cut in the cake so that each of $N + 1$ persons will get the same amount of cake?

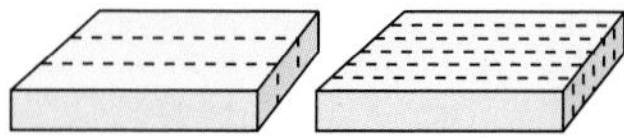

29. Below are three possible floor plans for a new office building. All the plans have almost exactly the same amount of floor space.

a. Analyze the advantages and disadvantages of each of these designs.

b. If each building were to be made of bricks, compare the relative amounts of bricks needed for each.

c. If the outside walls of each building were 10 feet tall and 1 gallon of paint covered approximately 500 square feet, determine the amount of paint that would be needed to paint each building.

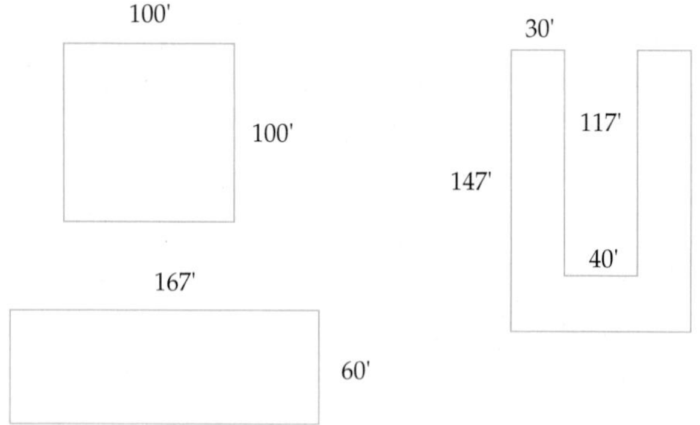

30. Each American produces about 3.5 kilograms of solid waste (garbage) per day. If 1 cubic meter of garbage weighs 90 kilograms, determine the size of the waste produced each day by:

a. The population of your college

b. The people in your state

c. The people in the United States

In each case, describe the assumptions you made in order to answer the question.

31. Many automobile commercials have stressed that the new model being advertised has more carrying capacity.

a. How might you determine the carrying capacity of two similar automobiles?

b. Select two comparable automobiles and determine the carrying capacity of each. Look up the carrying capacity given by the manufacturers. Do you agree with their amounts? If not, do you think one of the figures is erroneous, or do you think the term was defined differently?

32. Analyze the accompanying graph for its accuracy. First, predict whether or not you think it is accurate and explain your reasoning. Next do the measuring and calculating and explain your interpretation of the numbers. Then assess your prediction. If the prediction and the reasoning behind it were accurate, great. If either the prediction or the reasoning behind it was not accurate, explain the errors in the prediction.

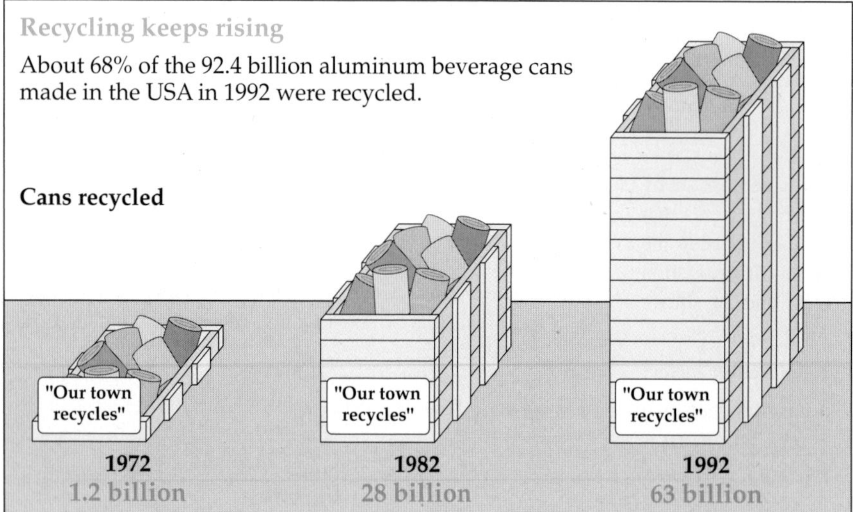

33. Make a net for a soccer ball. Describe your process. For example, how did you determine how many faces the soccer ball had and how to put them together? Describe any discoveries, conjectures, and questions that came from this exploration.

34. Predict how many square feet of floor space there are in the building in which you are taking this class. If it is a large building, your instructor may wish to have you determine the floor space in a specific subset of the whole building. Finally, as accurately as you can, determine the actual floor space.

35. How many ping-pong balls would it take to fill a classroom that measures 20 feet by 16 feet by 8 feet?

36. ***Classroom Connection*** The following questions can be done theoretically with rectangular prisms. However, give them a context, cereal boxes, and you will find these questions and similar ones in many elementary and middle school textbooks.

   a. Can two cereal boxes have same volume but differ in surface area?

   b. Pick a cereal box. Redesign it so that it has the same volume but less surface area.

37. Why do you think most products are not manufactured in the shape that requires the least amount of packaging material?

FROM STANDARDIZED ASSESSMENTS

2005 NECAP, Grade 6

38. Look at this structure.

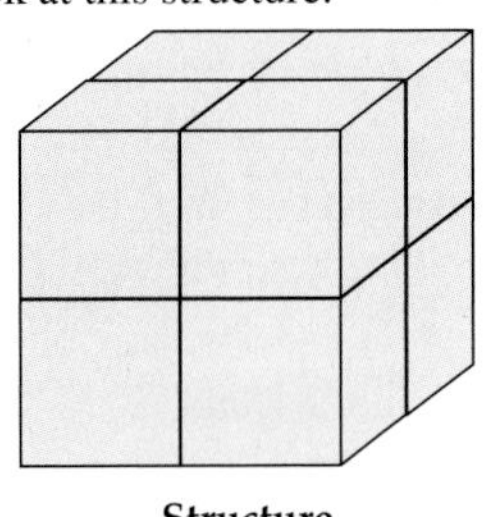

Structure

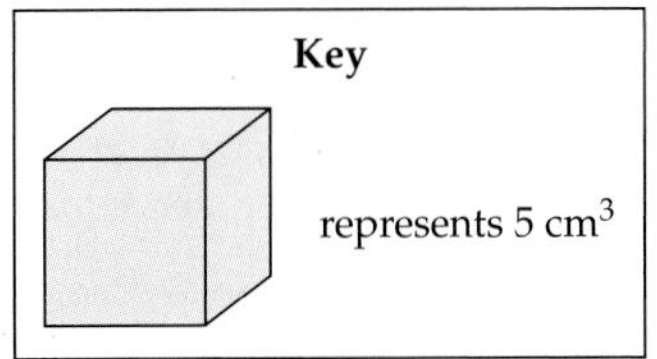

What is the volume of this structure?

a. 8 cm^3 b. 20 cm^3

c. 40 cm^3 d. 60 cm^3

39. A cereal company packs its oatmeal into cylindrical containers. The height of each container is 10 inches, and the radius of the bottom is 3 inches. What is the volume of the box to the nearest cubic inch?

Source: Results of the *Seventh NAEP Mathematics Assessment*, p. 230. U.S. Department of Education, National Center for Education Statistics.

LOOKING BACK on chapter nine

QUESTIONS TO SUMMARIZE BIG IDEAS

1. Why is measurement important?
2. What is the relationship between the metric system and the base ten number system?
3. What is the difference between perimeter and area?
4. What is the difference between surface area and volume?
5. How do we find the volume of a prism or cylinder? Explain the formula.
6. How do we find the volume of a pyramid or cone? Explain the formula.
7. What parts of this chapter are less clear to you at this point? What will you do to clarify those ideas?
8. Look back at the Mathematical Practices of the CCSS. In what ways did you engage in those practices during this chapter?

CHAPTER 9 SUMMARY

1. To measure an object means to assign a number to some attribute of the object.
2. Two objects do not have to be congruent to have the same measure. For example, two objects can have different shapes but have the same volume.
3. In order to measure, we must decide on a unit.
4. All units are conventions, as opposed to absolutes.
5. Virtually all measurements are approximations.
6. On some occasions, we cannot directly measure an attribute and must rely on indirect measurement.
7. Different attributes of an object—perimeter and area, and surface area and volume—are not related in simple ways.

BASIC CONCEPTS

Section 9.1: Systems of Measurement

metric system **510**	U.S. customary system **511**
meter **511**	liter **512**
gram **513**	weight **513**
mass **513**	precise **516**
accurate **516**	

Section 9.2: Perimeter and Area

Relationships between perimeter and area **520**

perimeter **521**	circumference **521**
π **521**	area **522**
base **523**	height **523**

Pythagorean theorem **526**

Understanding the area formulas for:

parallelograms **523**	triangles **525**
trapezoids **523**	circles **527**

Areas of irregularly shaped figures **530**

Section 9.3: Surface Area and Volume

Understanding the surface area formulas for:

prisms **540**	pyramids **540**
cylinders **541**	

Understanding the volume formulas for:

prisms **542**	cylinders **544**
pyramids **545**	cones **545**
spheres **546**	

REVIEW EXERCISES chapter nine

1. Describe the four-step process of measurement in your own words.
2. Which metric unit would you use to express each of the following? Briefly explain your choice. The choices for length are mm, cm, m, km; the choices for volume are mL, L; the choices for mass are mg, g, kg.
   a. The length of a pencil tip
   b. The volume of a bathtub
   c. The volume of a dose of cough medicine
   d. The mass of an ant
   e. The mass of an elephant
3. Fill in the blanks.
   a. 340 m______ cm b. 0.345 kg ______ g
   c. 2.75 L ______ mL
4. a. Describe a situation in which a measurement could be precise but not accurate.
   b. Describe a situation in which a measurement could be accurate but not precise.
5. Describe all the attributes of a puddle that we might measure—for example, the volume of the water in the puddle.
6. Find the exact area of the figure below in two different ways.

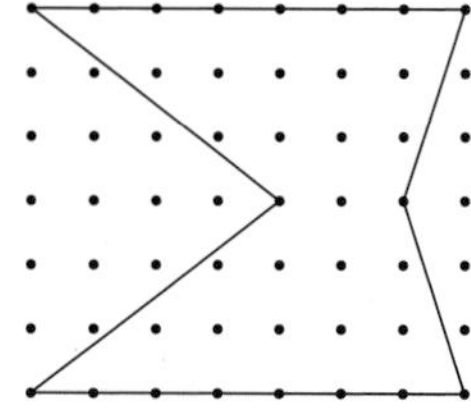

7. If possible, determine the area of the accompanying right triangle. If there is not enough information, explain why.

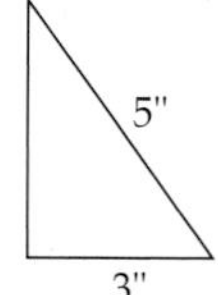

8. A swimming pool is shaped like the figure below. Each end is a semicircle, and the length and width of the rectangle are 40 feet and 20 feet, respectively. If there were a 1-foot-wide cement border around the pool, what would be the area of the border?

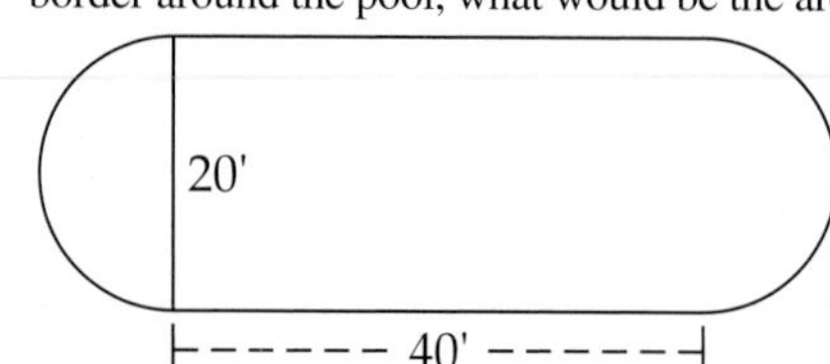

9. A meter trundle wheel is a convenient device for measuring distances along the ground. Every time the wheel makes one complete revolution, it has moved forward 1 meter. If you were to cut this wheel from a square piece of plywood, what would be the dimensions of the piece of wood?

10. Building codes generally require that the ventilating area be not less than $\frac{1}{150}$ of the area of the floor area of the attic space being ventilated. If the attic of a house is 50 feet by 25 feet, what is the minimum square inches of ventilation needed?

11. Below is a piece of wire mesh. Determine the dimensions of the square that would contain 1 million holes.

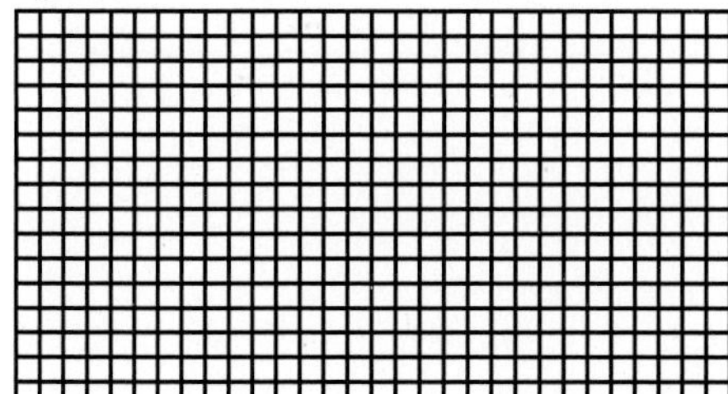

12. If you throw a dart at random and it lands on the dart board, what is the probability of its landing in the shaded region?

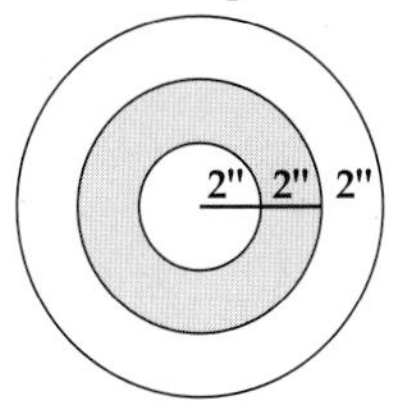

13. Kim decided he wanted to tile his bedroom with pennies. His bedroom measures 10 feet by 10 feet. He knows the area of a penny is πr^2, which is approximately 0.44 square inch. He computed the number of pennies by dividing the area of the room by the area of a penny. Here is his computation: (120 in. · 120 in.)/0.44 sq. in.

   a. However, this is not the right answer. What did he do wrong?

   b. Determine the right answer using an arrangement of laying the pennies in rows.

14. Randolph wants to know why you don't find the area of a parallelogram by multiplying the length times the width, as you do for rectangles. Explain why the formula is Area = (base)(height) instead of (length)(width).

15. Given that there are 12 inches in 1 foot, why can't we convert 1000 square inches to square feet by dividing by 12?

16. What happens to the area of a circle when you double the radius?

17. What percent of this quilt block is shaded?

18. The figure below is a map of a pond. Each centimeter on the map represents 10 meters. Determine the area, in square meters, of the pond by covering the pond with a grid and then finding the area of the pond by decomposing the pond into trapezoids. Show your work.

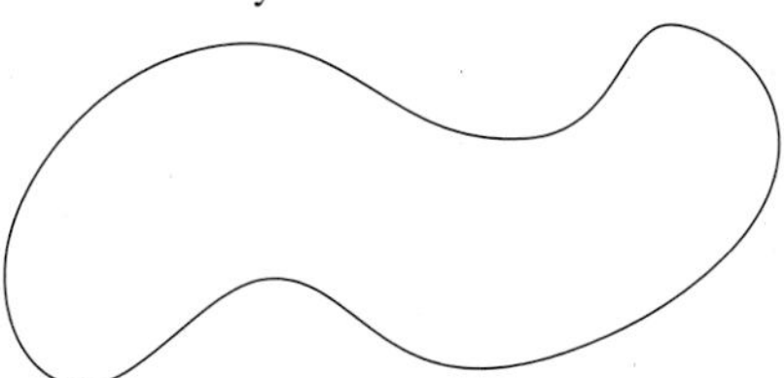

19. The federal government recently passed a law requiring providers of services to have all of their clients sign a privacy form. I filled out a form for the dentist, orthodontist, doctor, credit cards, and so on. Determine the volume of the space occupied by all of these forms if the average adult filled out 10 of them. Then convert this number to something that will be more readily grasped—for example, the dimensions of a building that would contain all the forms or how many filing cabinets that much paper would fill. Assume that there are 200 million adults in the United States.

20. Each side of the cube is 20 feet.

   a. If someone went from point A to point X (the midpoint of the side) and then from point X to point D, what would be the total distance traveled?

   b. Determine the direct distance between point A and point D.

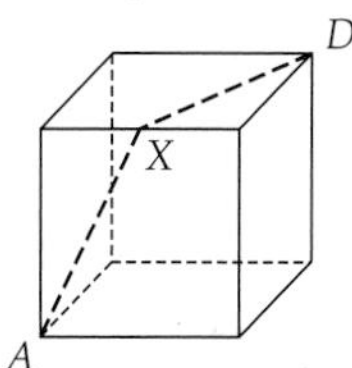

21. Let's say your college has decided to replace the worn cement paths between buildings. The width of each path is 6 feet, the depth of the concrete is 4 inches, and the total length of the new paths is 1200 feet. Concrete is normally ordered by the cubic yard.

   a. How many cubic yards should the company order?

   b. If customers want to allow 10% for spilling and overflow, how much should they order?

22. Determine the surface area (of the sides and roof) and the volume of the house-shaped figure below.

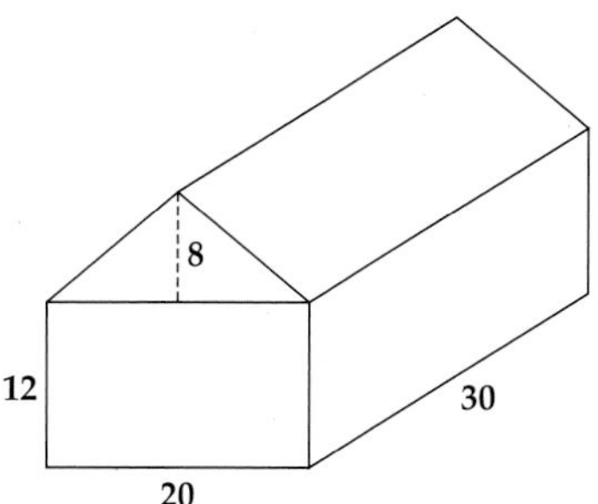

23. How many cubes that measure $\frac{1}{2}$ inch by $\frac{1}{2}$ inch by $\frac{1}{2}$ inch would fit into this container?

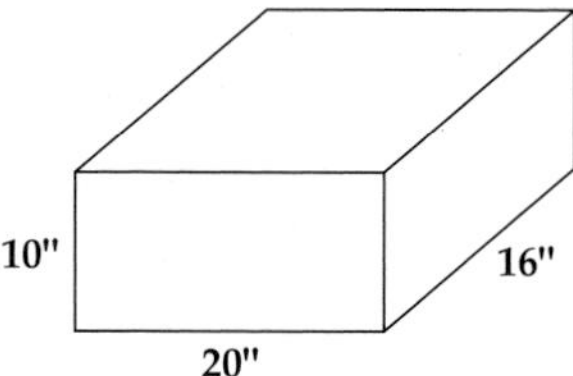

24. Design two different rectangular prisms that have the same surface area. Compare their volumes.

25. A company is manufacturing two kinds of juice boxes. One is in the shape of a rectangular prism with dimensions 10 cm by 6 cm by 4 cm. The other is a cylinder with radius 2.8 cm and height 10 cm. Compare the volumes and surface areas of these two juice boxes.

26. Our family bought a water-saving shower head. We determined that each shower now consumes 4 fewer gallons than before.

    a. If there are four of us and we each take one shower per day, how many gallons do we save per year?

    b. If we had a swimming pool in the form of a rectangular prism, and the pool measured 30 feet by 20 feet by 6 feet, would the amount saved in 1 year equal the volume of the swimming pool?

Geometry as Transforming Shapes

10

In Chapter 9 we focused on understanding measurement of shapes, and in this chapter, we are going to look at what happens when we *transform* shapes. This will help us better understand the patterns that emerge when shapes are put together, as in quilts, floor patterns, art, and other situations. The themes that are consistent through our investigations include: the notion of deconstructing—which leads to awareness of multiple attributes; classifying—which leads to understanding of structures; and looking for patterns and relationships—which leads to generalizations and, again, to understanding of structures. According to the Common Core State Standards, fourth graders learn about folding figures into matching parts over a line of symmetry. Then in eighth grade, congruence and transforming shapes are studied.

Look at the pictures in Figures 10.1 to 10.3. What do *you* see? Note your ideas before reading on.

Figure 10.1 shows an Islamic design composed entirely of two shapes: an 8-pointed star and a 15-sided polygon. Each star can be moved to the position of a nearby star by sliding it—horizontally, vertically, or diagonally. Each 15-gon can be moved to the position of a nearby 15-gon by turning it.

Figure 10.2 illustrates one of many parquet floor designs. Most people see pinwheels composed of 8 congruent isosceles trapezoids. Some people see these embedded in squares. Other people see congruent pinwheels of a different color. Others see white pinwheels composed of 4 white isosceles triangles. Each dark pinwheel can be moved to the position of a nearby light pinwheel by sliding it. Each trapezoid in a pinwheel can be moved to the position of an adjacent trapezoid by a turn.

In Figure 10.3, focus on the butterflies. Though the butterflies are in different positions and of different sizes, they all have the same basic shape. This is true for each of the animals pictured. The arrangement of the animals is not random but intentional. If you focus only on the butterflies, what do you notice? From one perspective, we can say that every butterfly has a twin, and each butterfly can be moved from its position to its twin's position by flipping, turning, or sliding or by a combination of these moves.

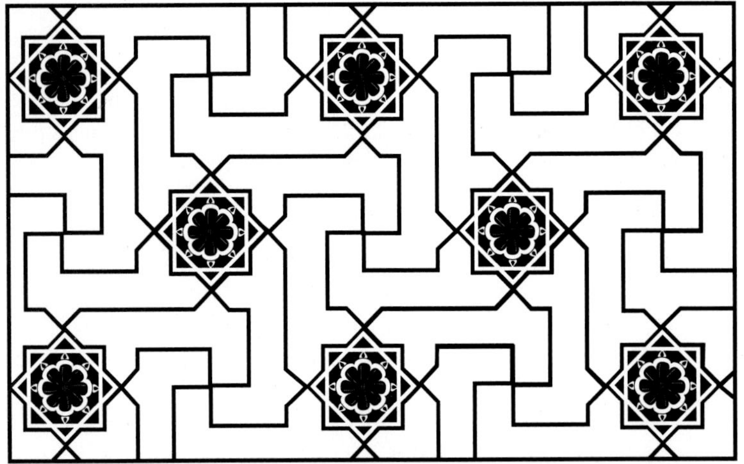

Figure 10.1

Figure 10.2

Figure 10.3

SECTION 10.1 Congruence Transformations

What do you think?

- How are translations, reflections, and rotations related?
- How might you describe the translation, reflection, or rotation of a three-dimensional figure in space?
- How are the operations translation, reflection, and rotation like the operations addition, subtraction, multiplication, and division?

TRANSFORMATIONS

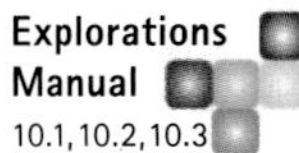
Explorations Manual 10.1, 10.2, 10.3

Place an object on the table in front of you (book, cell phone, isosceles triangle, etc.). Move the object in different ways so that its ending position is on the table (plane). Describe different motions that allow you to do this. These motions that take this object from one place to another are called transformations. Before we get into the mathematical aspect of transformations, what do you think when you hear this word?

The interesting examples offered by my students include the transformation of a frog into a prince in fairy tales and the transformation of a caterpillar to a butterfly. You are in the process of transforming from a student to a teacher. Virtually all examples of transformation give a sense of movement and change. In mathematics, there are many kinds of transformations. Figure 10.4 illustrates several. Which of these did you discover as you moved the object at the beginning of this section? Look at the various transformations of the letter P. How would you describe the transformation? Write your thoughts before reading on. . . .

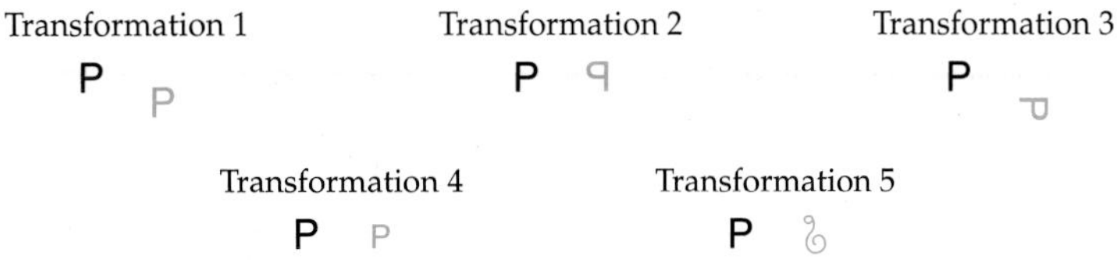

Transformation 1 P P

Transformation 2 P

Transformation 3 P

Transformation 4 P P

Transformation 5 P

Figure 10.4

Most people find the first four transformations fairly straightforward but find the last one very difficult. Rather than give the answer at this point, let us ask a more refined question: What is the same and what is different about each figure and its image after the transformation? Think about it, and then read on. . . .

Table 10.1 shows the informal names for the transformations, the formal names, what is changed, and what is unchanged.

TABLE 10.1

Informal name	Formal name	What is changed?	What is unchanged?
1. Slide	Translation	Position	Size, shape
2. Flip	Reflection	Position	Size, shape
3. Turn	Rotation	Position	Size, shape
4. Shrink	Similarity transformation	Position and size	Shape
5. Distortion	Topological transformation	Position, size, and shape	Neighborhood

In a moment, we will zoom in on the first three transformations. In Section 10.3, we will focus on transformations in which a figure is reduced or enlarged so that the shape is similar but the size has changed. You can explore topological transformations (stretching and bending) on the website. Ironically, young children's initial understanding of spatial ideas is more topological than Euclidean.

The following investigations will all be driven by this question: How can we describe what we have to do to move an object from here to there on a plane?

TRANSLATIONS

When you see the word *translation,* you might think of translating from one language to another: The translator replaces English words and phrases with Spanish words and phrases, for example, but the meaning of the phrase remains the same. In geometry, a translation involves moving an object along a straight line and not turning it, but the size and shape of the object remain the same.

Each of the designs in Figure 10.5 shows translations. Figure 10.5(a) is from the Native American Yuchi tribe and depicts storm clouds. Figure 10.5(b) shows a quilt border called Orange Peel. Figure 10.5(c) shows a papercutting, which is a common elementary school activity. As noted above, a translation is a special kind of congruence transformation. In each of the designs, we have congruent figures in the design, and we can imagine picking up part of the figure, sliding it in a straight line, and then putting it back down on top of another congruent part of the figure.

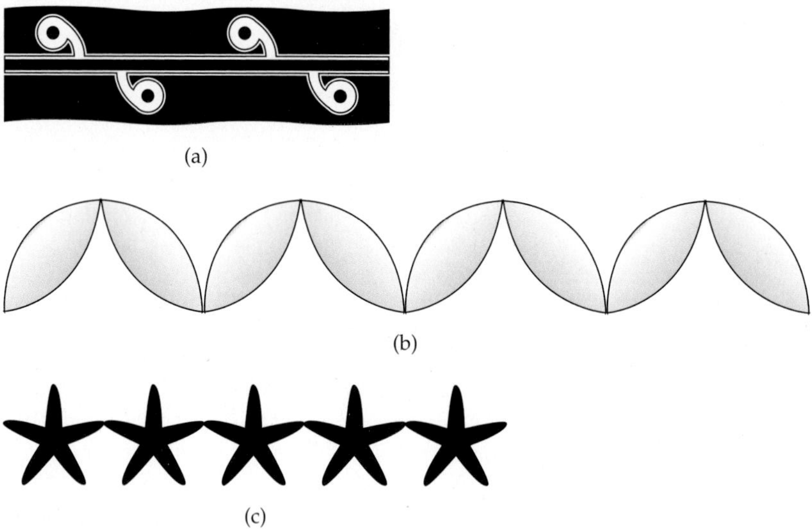

Figure 10.5

Sources: (a) Le Roy H. Appleton, *American Indian Design and Decoration* (New York: Dover Publications, 1971), Plate 15 (in insert). (b) From Jinny Beyer, *The Quilter's Album of Blocks & Borders* (Delaplane, VA: EPM Publications, Inc., 1986), p. 185. Reprinted by permission of the author.

INVESTIGATION 10.1a Understanding Translations

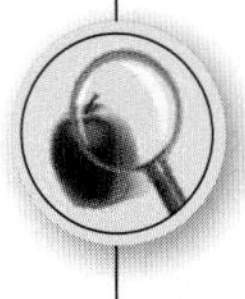

In order to make intricate designs, we often need to be precise in describing the translation. Let's say you were talking on the phone to a friend and you wanted to describe the translation of the $\triangle CAT$ in Figure 10.6 from its original location to its new location. The original triangle is $\triangle CAT$; its translation image is $\triangle C'A'T'$. How would you do this? You may assume that you and your friend have any resources you need: scissors, ruler, compass, graph paper, etc. Think and then read on. . . .

DISCUSSION

One common response is to say that we slide the triangle from one location to the other. The mathematical word for slide is *translation,* and we talk about a figure and its **image**. There are many different ways to describe a translation. Each of these descriptions represents different perspectives and preferences concerning how to communicate what has happened to $\triangle CAT$. Some of the perspectives are easier to describe if we draw the

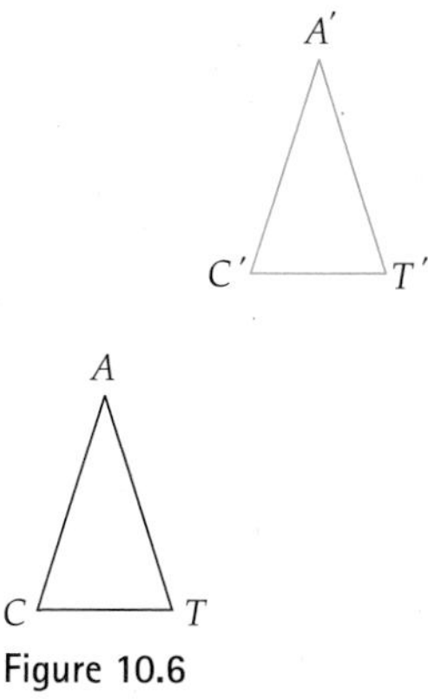

Figure 10.6

two triangles on graph paper (Figure 10.7). What they all have in common is the realization that when we translate a figure, every point on the figure moves the same distance and in the same direction.

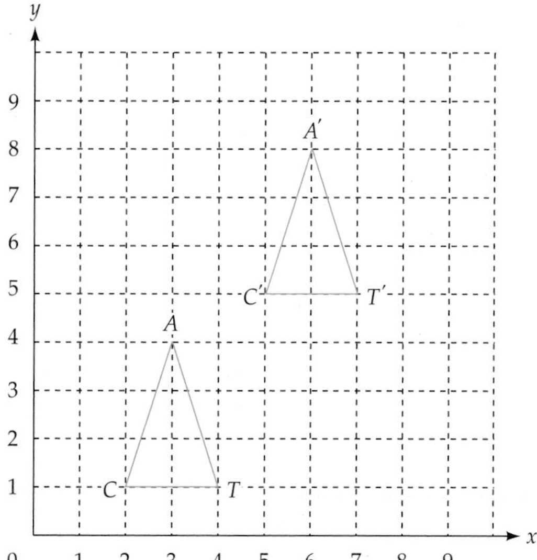

Figure 10.7

DESCRIPTION 1 Use "taxicab language"

Just as we could ask a cabbie to "go three blocks east and four blocks north," we could say that each point has been moved 3 units to the right and 4 units up. For example, copy $\triangle CAT$ onto another sheet of paper and cut it out. Then place it on the paper above and move the whole triangle, first 3 units to the right and then 4 units toward the top of the paper.

DESCRIPTION 2 Invent new notation

We can look at the x and y coordinates of the points to help us describe the translation. Point C moves from (2, 1) to (5, 5), point A moves from (3, 4) to (6, 8) and point T moves from (4, 1) to (7, 5). In each case, we add 3 to the x value and 4 to the y value. We can use notation to express this same idea:

$$(x, y) \rightarrow (x + 3, y + 4)$$

This notation gives the directions "go 3 units to the right and 4 units toward the top of the page" very succinctly.

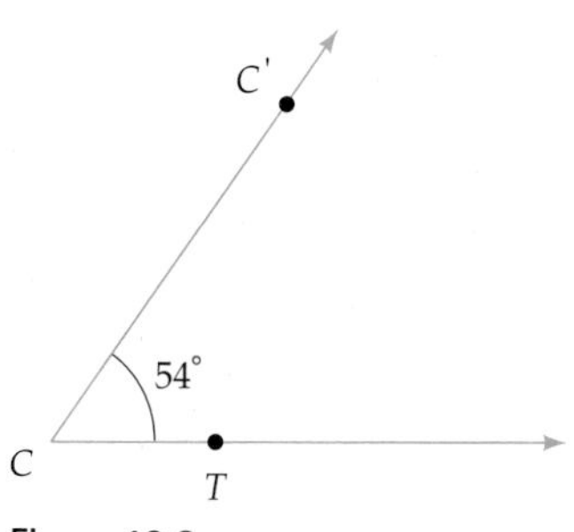

Figure 10.8

DESCRIPTION 3 Specify the distance and the angle

We could say that the figure has been moved a distance of 5 units at a 54-degree angle from the x-axis (Figure 10.8). That is, C and C' are 5 units apart; the distance between any point on $\triangle CAT$ and the corresponding point on $\triangle C'A'T'$ is 5 units. Similarly, the angle formed by the rays $\overrightarrow{CC'}$ and $\overrightarrow{CT}$ is equal to approximately 54 degrees. I encourage you to check this for yourself with a ruler and a protractor.

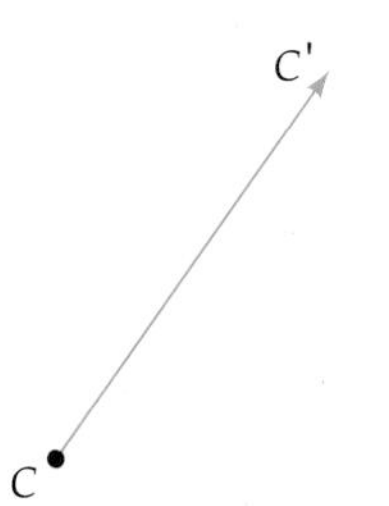

Figure 10.9

DESCRIPTION 4 Use vectors

Another alternative is to use a vector. By definition, a **vector** has a length and a direction. For example, the instructions for the translation could be shown by drawing one vector, as in Figure 10.9.

We will formally define **translation** as a transformation on a plane determined by moving each point in the figure the same distance in the same direction.

PROPERTIES OF TRANSLATIONS

Now let us determine some of the properties of translations. We know from Chapter 8 that two points determine a line. Therefore, if we connect each vertex in the $\triangle TAR$ to its image, $\triangle T'A'R'$, we have three line segments: TT', AA', and RR' (Figure 10.10). What relationship do you notice among these line segments?

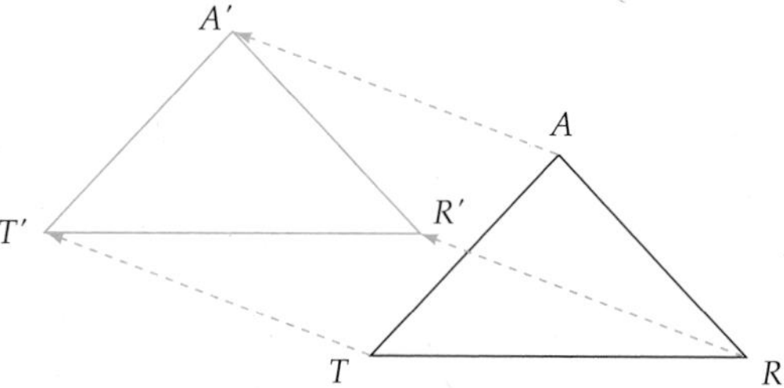

Figure 10.10

The three line segments are all congruent (same length) and all parallel (same direction). How might this make the task of drawing the translation image easier? Think and then read on. . . .

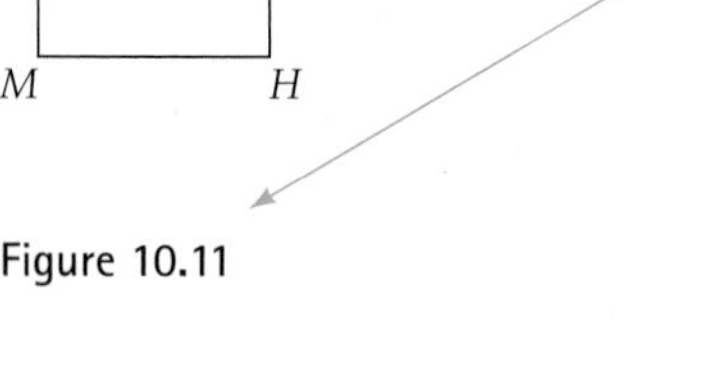

Figure 10.11

Translate the rectangle in Figure 10.11, using the translation vector shown.

We can draw four lines, each going through a vertex, that are the same length and parallel to the translation vector. If we copy the length of the vector using a compass, we can quickly mark off the same lengths on the lines to determine the vertices of rectangle $M'A'T'H'$ (Figure 10.12).

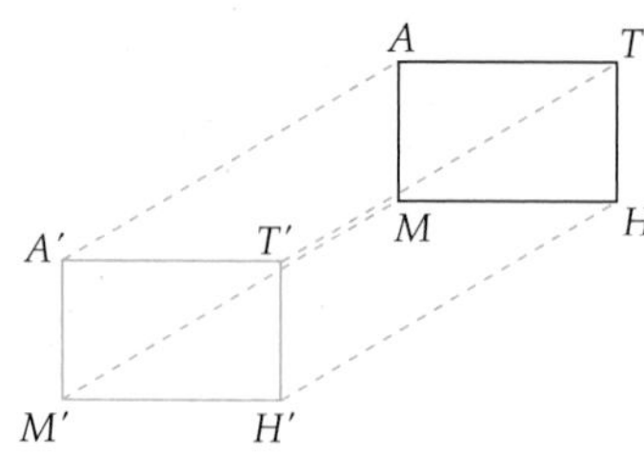

Figure 10.12

REFLECTIONS

Like translations, reflections can be accomplished in a variety of ways. One of the simplest is by folding paper. Trace the $\triangle FOX$ and the line below from Figure 10.13 onto a blank sheet of paper, and then fold the paper at the line m.

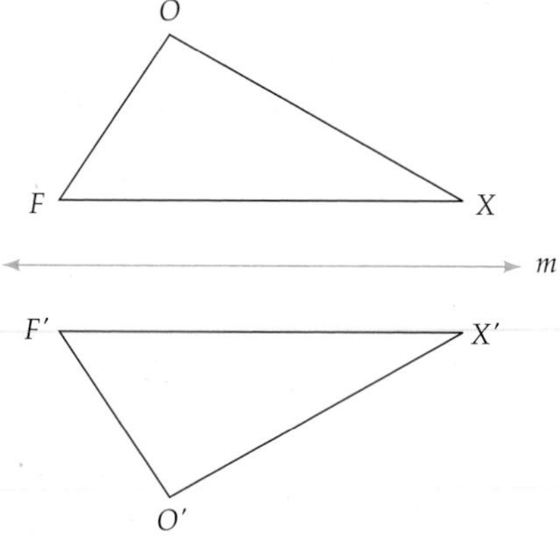

Figure 10.13

Now trace over the figure. When you fold the paper on the line, each point on the triangle coincides with its image on the reflected triangle. Do you think we need to trace the whole figure? What do you think is the minimum tracing needed?

If you trace the points F, O, and X, you will have the points F' O' and X' and we now need connect only these three points because we know that three noncollinear points determine a triangle.

INVESTIGATION 10.1b Understanding Reflections

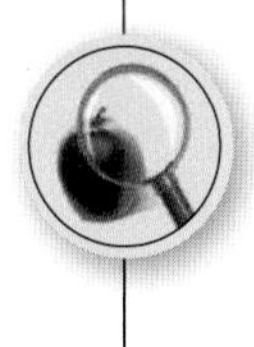

Figure 10.14 shows a trapezoid that has been reflected (flipped) in three different ways. In each case, bold lines denote the original figure and dotted lines denote the flipped image.

A. Find the line of reflection in each case.

B. Can you discover a rule that would allow us to find the line of reflection in all cases? Stop, think, and write.

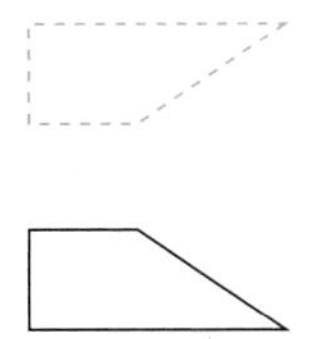

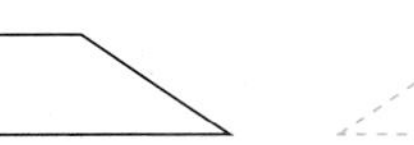

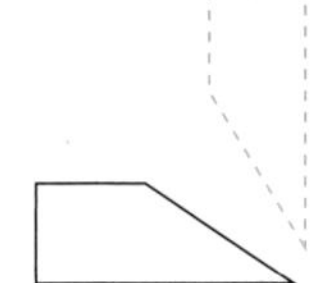

Figure 10.14

CLASSROOM CONNECTION

Although many elementary classrooms still use paper folding to help children understand reflections, many classrooms use other means—computer software, acetate sheets called Patty Paper, and Miras, which are made of Plexiglas. The bottom edge of the Mira acts as the reflection line. When the Mira is placed on the paper, the image of the object can be seen and then traced on the other side of the Mira.

DISCUSSION

A. You may find it helpful to start by tracing these figures on tracing paper and (by folding the paper) finding the line that makes the original figure coincide with its reflected image. Do you see what is happening? Try making figures of your own on paper; fold the paper and then trace the reflection image of the figure you drew. Do this several times. Play the "what-if" and "is it possible" games: What if the line of reflection went through the figure? Is it possible to have the line of reflection go right through the figure?

Paper folding is one of many ways to draw reflections. What other ways can you think of?

B. *Properties of reflections.* What is true for all reflections is that if we connect any point on the original figure with the corresponding point on the reflected figure, the line of reflection is the perpendicular bisector of that line segment (Figure 10.15). That is, $A'X = XA$, and $\overline{AA'}$ and line l are perpendicular.

This realization leads to a more formal definition of reflection. A **reflection** is a transformation that maps a figure so that a line, called the **line of reflection**, is the perpendicular bisector of every line segment joining a point on the figure and the corresponding point on the reflected figure.

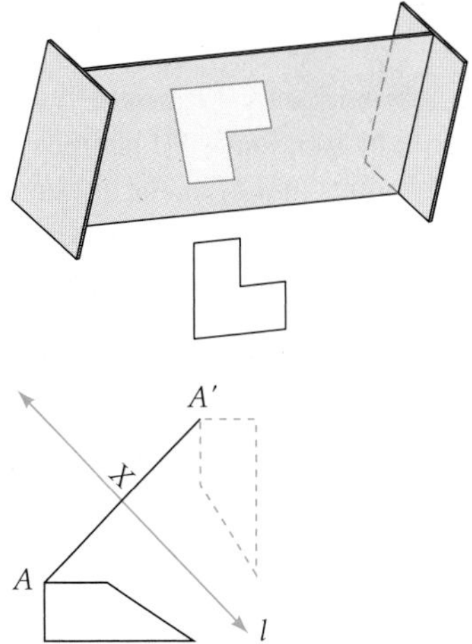

Figure 10.15

ROTATIONS

Consider the word "rotation." What does it mean? In what other contexts do we use the word? The rotation of a car wheel, hands on a clock, or of Earth on its axis are all familiar to us.

INVESTIGATION 10.1c Understanding Rotations

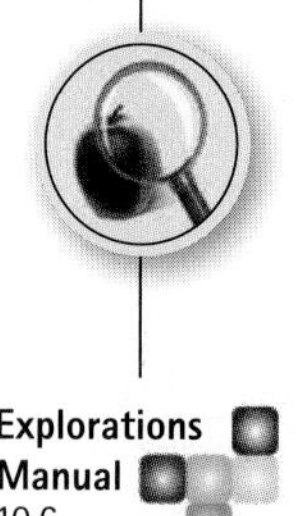

Explorations Manual 10.6

This investigation uncovers ways in which we can describe rotations. In each of the diagrams, the letter P has been rotated. However, as with translations and reflections, not all rotations are the same. In order to describe any rotation precisely, what do we need? You can play with this on your own by tracing the figures and playing around with them. . . .

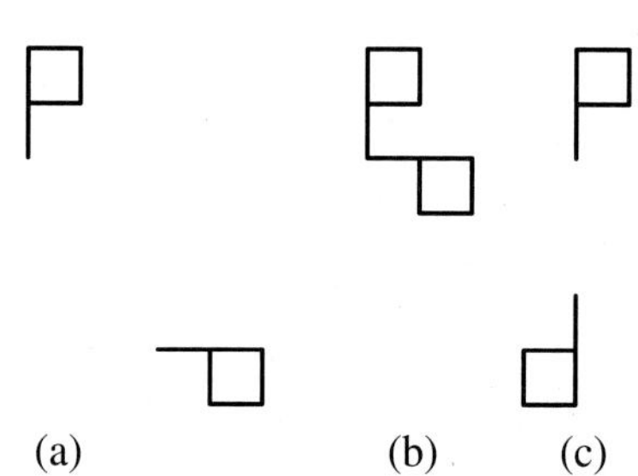

As with the concept of reflection, you might want to explore this yourself. You can do this by drawing the figure on a piece of paper and then tracing it on another piece of paper. Line up the two pieces of paper so that the top figure lies on the bottom figure, place the tip of your pencil or pen on the paper at some point, and then rotate the bottom sheet of paper. Draw the rotation image by tracing over the image you see from the bottom sheet. This is easier if the top sheet is a sheet of tracing paper, but it works with regular blank paper, too.

DISCUSSION

In order to rotate any figure (in a plane), we need to select a *center of rotation* (that is, the point that does not move), and we need to decide how much to rotate the figure. "How much" is determined by specifying an angle.

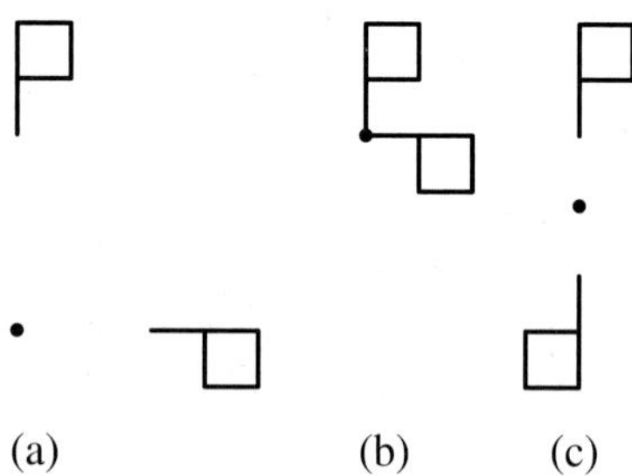

In part (a), the center of rotation is the point that is approximately 17 mm below the bottom of the figure and 2 mm to the left, and the degree of rotation is 90 degrees clockwise. If you trace the figure, place your pencil on the point, and turn the paper 90 degrees, the figure will now be in the new position.

In part (b), the center of rotation is the bottom of the figure, and the degree of rotation is 90 degrees clockwise.

In part (c), the location of the center of rotation is 5 mm below the figure, and the rotation is 180 degrees. Note that in the first and second figures above, the letter P was rotated 90 degrees clockwise. What both of these rotations have in common is that the orientation of the rotated figures is the same; in this case, both rotated images are lying on their sides. How, since the center of rotation was different, the distance and direction from the original figure are not the same.

Formally, we say that a **rotation** is a transformation on a plane determined by holding one point fixed and rotating the plane (in our case, the paper) about this point by a certain number of degrees in a certain direction. The fixed point is called the **center of rotation**.

These three transformations—translation, reflection, and rotation—are known as **congruence transformations** because the images are congruent to the original figure.

Before we go on, it is important to note one important attribute common to each of these three transformations. Recall that we found that when we translate a figure, each point on the figure is translated the same distance. Similarly, when we reflect a figure

MATHEMATICS

Technically, the center of rotation is always on the perpendicular bisector of the line joining two corresponding points. For example, if you connect the bottom points on the two figures and then draw a perpendicular line through the midpoint of that line, the center of rotation would be on that line.

Grade 4

Do this activity if your instructor has transparent mirrors or Miras. 

Date ______________ Time ______________

LESSON 10·1 Basic Use of a Transparent Mirror

SRB 106

A **transparent mirror** is shown at the right.

Notice that the mirror has a **recessed** drawing edge, along which lines are drawn. Some transparent mirrors have a drawing edge both on the top and on the bottom.

Place your transparent mirror on this page so that its drawing edge lies along line *MK* below. Then look through the transparent mirror to read the "backward" message.

back face
end
front face
end
drawing edge

M *K*

If you have followed the directions correctly, you are now able to read this message. Here are a few things to remember when using your transparent mirror:

- ◆ Always look into the front of the transparent mirror. In this position, the drawing edge will be facing you.
- ◆ Use your transparent mirror on flat surfaces like your desk or a tabletop.
- ◆ Use a sharp pencil when tracing along the drawing edge.
- ◆ Experiment and have fun!

274

From *Everyday Mathematics, Grade 4:* The University of Chicago School Mathematics Project: Student Math Journal, Volume 2, by Max Bell et al., Lesson 10-1, p. 274. Reprinted by permission of The McGraw-Hill Companies, Inc.

across a line, any point and its image are the same distance from the line of reflection. What do you think is equidistant with rotations?

When we rotate a figure, any point and its image are the same distance from the center of rotation. Thus the sense of "equidistance" is something all congruence transformations share.

Let us now examine some of the relationships among these three transformations.

INVESTIGATION 10.1d Understanding Translations, Reflections, and Rotations

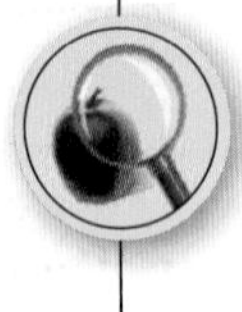

Now that we have investigated translations, reflections, and rotations, let us stop to check for understanding. In each case below, describe the transformation that has moved the dark pentomino to the position of the light pentomino.

a. b. c.

d. e. f.

DISCUSSION

a. The pentomino was reflected across a horizontal line below the pentomino.

b. The pentomino was rotated 75 degrees counterclockwise.

c. The pentomino was translated in a diagonal direction.

d. The pentomino was rotated 180 degrees.

e. The pentomino was reflected across a line making a 45-degree angle with a horizontal line.

f. There is no single translation, reflection, or rotation that will accomplish this move. You could accomplish it in two moves by first translating the figure to the right and then reflecting across a horizontal line between the two figures. Or you could reverse the order: First reflect the pentomino across the horizontal line and then translate the image to the right.

The combination of translation and reflection described in part (f) is called a glide reflection and will be discussed shortly. The key to understanding this transformation is to realize that the translation and the reflection line must be parallel.

COMBINING SLIDES, FLIPS, AND TURNS

Now that we understand these three basic transformations of the plane, let us examine combinations of these transformations. There are many ways in which we can combine these transformations. For example, we could reflect a figure across a vertical line and then translate it 4 units to the right. Or we could reflect a figure across a vertical line and then reflect the image across another vertical line. We call any combination of transformations a **composite transformation**.

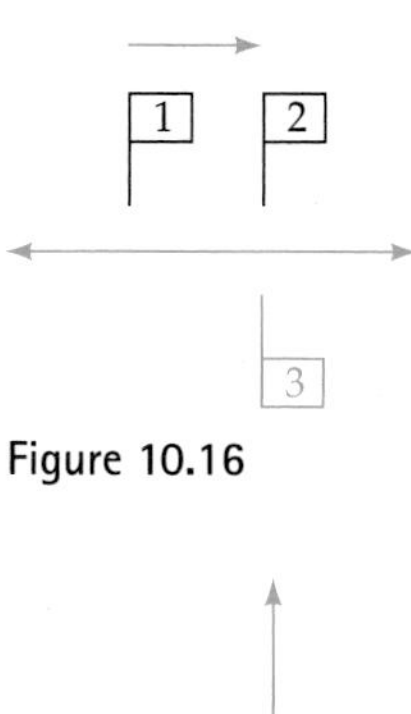

Figure 10.16

Figure 10.17

There is a special name for the composite transformation of a translation followed by a reflection with the condition that the translation vector and the line of reflection are parallel (that is, point in the same direction). We call such a composition a **glide reflection**. Recall the last problem in the previous investigation. I remember finding this transformation a bit elusive in terms of *really* understanding it, so let's take a minute to examine it. Figure 10.16 shows a flag being translated to position 2 and then being reflected across a horizontal line (which is parallel to the translation vector). The transformation of the flag from position 1 to position 3 is a glide reflection. Note that the translation vector does not need to be horizontal—it can be in any direction. But whatever the direction of the translation vector, the reflection line must be in the same direction.

Let me give you an example of a translation followed by a reflection that is not a glide reflection. Figure 10.17 shows a flag being translated to position 2 and then being reflected across a vertical line that is *not* parallel to the translation vector. The composite transformation of the flag is not a glide reflection, because the translation vector and the reflection line are not parallel.

Probably the most famous example of glide reflection—and one that has helped many students get the sense of glide reflections—is a diagram of footprints on the sand (Figure 10.18).

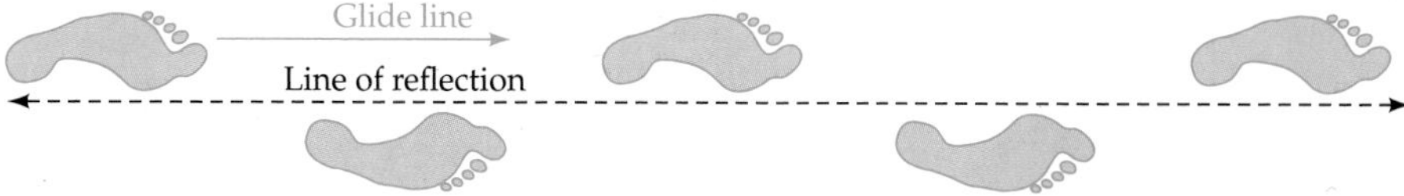

Figure 10.18

You can see the reflection (the left and right feet), and you know (from walking on a beach on the ocean or on a lake) the feeling of gliding along the beach. Also, you can connect your experience to the mathematical idea—that the line of reflection and the glide line are parallel (in the same direction), which is a *requirement* for all glide reflections.

Look at Figure 10.19, which is taken from a woodcut by M. C. Escher. Do you see the glide reflections? How would you explain them?

Any black bird can be mapped onto any white bird by translating the bird and then reflecting the translated image across a vertical line. Of course, to do it exactly, we would need to specify the translation vector (direction and distance) and the reflection line (direction and distance from the translated image of the black bird).

At this point, there are many possible "what-if" and "is it possible" questions about these transformations. Which ones come to your mind? Think before reading on. . . .

Such questions include the following:

1. Does it matter which one you do first when you do a composite transformation?
2. Is there a least number of steps that will enable you to map one figure onto a congruent image?
3. How many different congruence transformations are there?

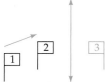

Figure 10.19

Rather than answer these questions directly, let us use the following investigation to examine them.

INVESTIGATION 10.1e Connecting Transformations

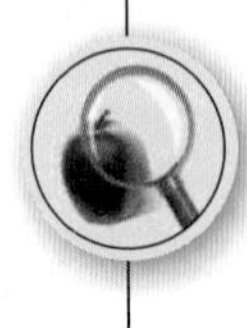

In each of the three pictures below, describe how you would get the flag in position 1 to position 2. Then read on. . . .

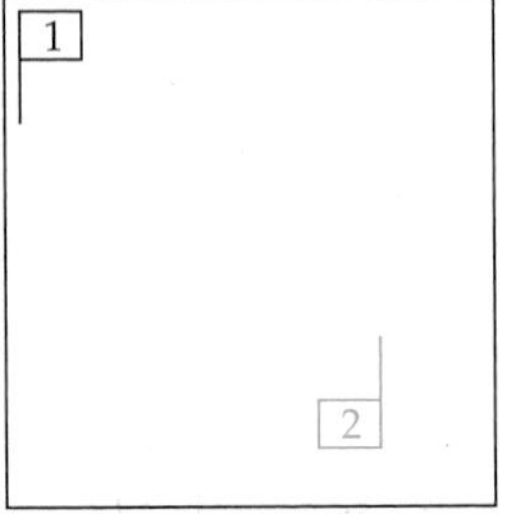

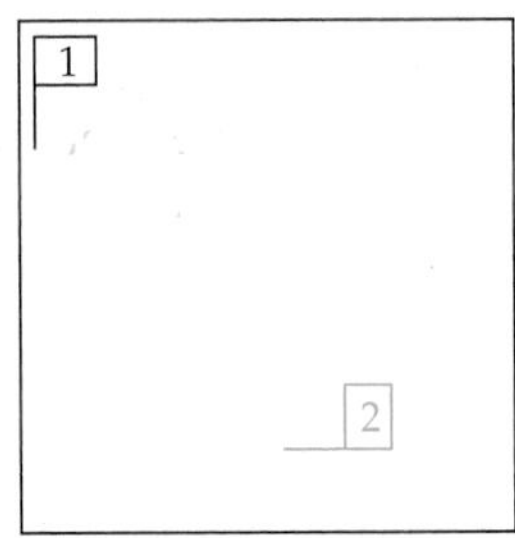

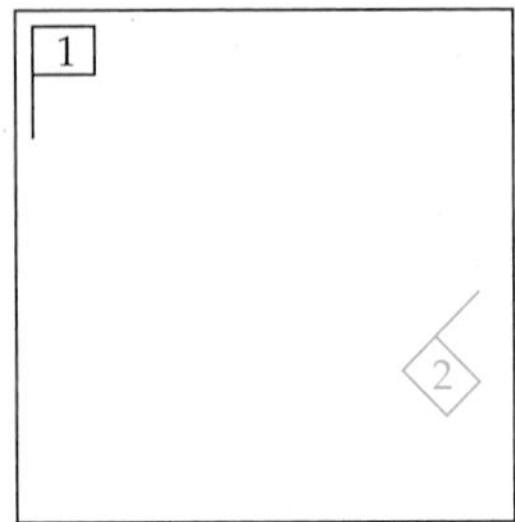

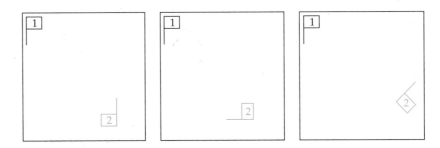

DISCUSSION

The curious reader may have realized that there is more than one right answer for each of these! For example, in the first picture, you could get from the first to the second position by first reflecting the flag across a vertical line and then reflecting the image across a horizontal line. Or you could reverse the order, doing the horizontal reflection first and the vertical reflection second. Or you could get from the first to the second position by a single rotation. Can you find the center of rotation?

Let us use this example to answer the first question above. In this case, it did not matter which reflection (vertical or horizontal) we did first. This leads most people to conclude that the operation of reflection is commutative. This is connected to our discovery, in Chapter 3, of the commutative properties of addition and multiplication. However, if you conclude that there is a commutative property of reflection, you are wrong! Two reflections produce the same image only if the lines intersect at right angles. If the two reflection lines are parallel, we do not get the same image when we reflect across line m and then across line n as we do when we reflect across n and then m. Thus the operation of reflection is commutative only if the lines of reflection intersect at right angles. This is just one of many examples in mathematics where we must think carefully before generalizing!

In the second picture, you could translate the flag to the right, then rotate the image 90 degrees clockwise, and then flip the image across a horizontal line. Or you could rotate the figure 90 degrees clockwise and then do a glide reflection by translating the image so that it is directly above the flag in position 2 and then reflecting this image. Or you could move from the first to the second position by a single reflection. (Can you find the line of reflection?)

The third situation is more complicated. You could reflect the flag across a horizontal or vertical line, then rotate the image the appropriate amount, and then translate this image the appropriate amount. However, there is a glide reflection that enables us to move the flag from position 1 to position 2.

Let us now formally answer the three questions posed above.

1. Sometimes the order in which a composite transformation is done does matter. Recall the operations on whole numbers where the order does not matter for addition and multiplication but does for subtraction and division.
2. Imagine any figure on a plane. Now move that figure to any other position on the plane in any orientation. No matter how complicated the shape, and no matter how complicated the relationship between the two, we will always be able to map the figure from the first to the second position with one of the four congruence transformations in one step. This conclusion has been proven.
3. This leads to the answer to the third question: These are *the* four congruence transformations. Mathematically, we use the term *sufficient*. That is, these four transformations are sufficient to describe the movement of any figure on a plane to any other place on the plane. This idea is further explored in the exercises.

OPERATIONS ON SHAPES AND OPERATIONS ON FIGURES

It is beyond the scope of this book to go into more detail about other properties of operations on shapes, but I did want to give you a glimpse of the deeper mathematical structures here. At this point, let us summarize some important connections between operations on numbers and operations on shapes. In both cases, there are many important subsets. In both cases, there are many operations that we can perform. In both cases, we can make tables for those operations (Exploration 3.8, Exploration 10.8). In both cases, we can learn a lot by taking apart (decomposing) numbers and shapes in various ways. For example, we can decompose 45 into $40 + 5$ or $9 \cdot 5$. In Chapter 8, you found that we can decompose any polygon into a number of triangles. Finally, just as we talked about properties of operations with numbers (such as closure, commutativity, associativity, identity, and inverse), we encounter the same properties when we do operations on shapes. Table 10.2 summarizes these connections between numbers and shapes, with examples for each.

TABLE 10.2

	Numbers	Shapes
Subsets	N, W, I, Q, R natural numbers, whole numbers, integers, rational numbers, real numbers	Triangles, quadrilaterals, regular polygons, concave, etc.
Operations	$+, -, \times, \div$, exponentiation, averaging	Translation, reflection, rotation, glide reflection, similarity transformation, topological transformation
Tables and patterns in tables	Addition and multiplication tables	Multiplication tables for symmetries of various shapes
Decompose/compose	Numbers can be decomposed additively and multiplicatively.	Any polygon can be decomposed into triangles.
Properties of operations	Closure, commutativity, associativity, identity, inverse	Closure, commutativity, associativity, identity, inverse

CONGRUENCE

In Chapter 8 we discussed and worked with different concepts of congruence. Now that we have moved (transformed) objects and have seen that moving them does not change the figure, we can give a more general definition of congruence: Two figures are **congruent** if one figure can be obtained from another by a sequence of rotations, reflections, translations, or glide reflections.

INVESTIGATION 10.1f Transformations and Art

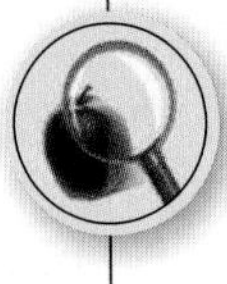

Transformations have a lot to do with the patterns we see in tiles, quilts, and other art, especially Islamic art and the work of M. C. Escher. Look at the two patterns in Figure 10.20, which we have seen before. Do you see translations, reflections, and/or rotations in these figures? Write your thoughts and then read on. . . .

Parquet

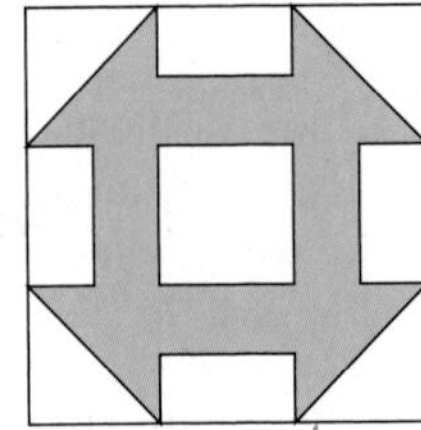

Churn Dash

Figure 10.20

DISCUSSION

In the parquet pattern, we can see that each light trapezoid can be mapped onto a dark trapezoid through a translation along a diagonal line. Each triangle can also be mapped onto another triangle through a 90-degree rotation. Furthermore, if we rotate the figure 90 degrees (clockwise or counterclockwise), the whole figure maps onto itself. If you don't see this, make a copy of the figure, superimpose the figure on the figure in the book, and rotate it.

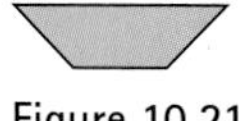

Figure 10.21

From another perspective, you can see that the "pinwheel" part of the parquet pattern can be generated by taking only one-eighth of the pinwheel (that is, one trapezoid) and rotating it 45 degrees seven times (Figure 10.21).

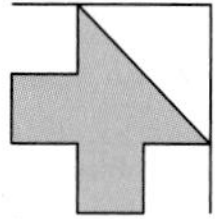

Figure 10.22

In the churn dash pattern, each of the white triangles can be mapped onto each of the other white triangles across a line of reflection. If we rotate the churn dash 90 degrees, with the center of rotation being the center of the figure, it maps onto itself. Furthermore, the churn dash can be generated by taking one-quarter of the figure and rotating it 90 degrees (clockwise or counterclockwise) three times (Figure 10.22) or by reflecting it across a horizontal line and then reflecting the original figure and its image together across a vertical line.

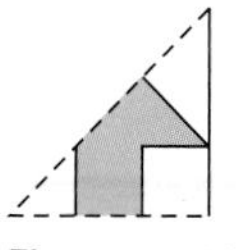

Figure 10.23

Actually, an even smaller piece of the figure (half of the piece of the churn dash in Figure 10.22) will generate the whole churn dash design (Figure 10.23). Don't just take my word for it. Do you see how?. . .

If we reflect Figure 10.23 across the diagonal dotted line, we get Figure 10.22. One fun way to verify that Figure 10.23 will generate the churn dash is to tape two small mirrors together and place the two mirrors along the dotted lines of Figure 10.23. Like magic, you will see the whole churn dash! From one perspective, we can call this region the unit of the churn dash; from another perspective, mathematicians call this region a **fundamental domain** or **fundamental region**—that is, a region that under some combination of transformations will produce the whole pattern. This piece of the churn dash is the smallest part of the figure that will generate the whole figure. This looking for the smallest piece of a figure or a design that will generate the whole figure or design occurs in many parts of mathematics, not just with shapes; it is called finding the fundamental region of a pattern.

SUMMARY 10.1

We have examined four fundamental ways in which we can transform a geometric figure: translation, reflection, rotation, and glide reflection. We have uncovered connections among these transformations. For example, performing two specific reflections is equivalent to performing a specific rotation. We have discovered that there are many similarities between numbers and shapes: They have important subsets, we do operations on them, we can make tables for the operations, we can decompose numbers and shapes, and there are important properties that help us better understand our operations on numbers and shapes.

These investigations of geometric transformations of two-dimensional figures in a plane can be extended. What if we examined transformations of three-dimensional objects? These investigations occupy the attention of many mathematicians and have applications in other fields. For example, the way in which atoms are packed helps to determine the properties of a compound.

10.1 Exercises

1. Below are a figure and a translation vector. Determine the translation. Explain at least two ways to describe the translation other than providing the vector.

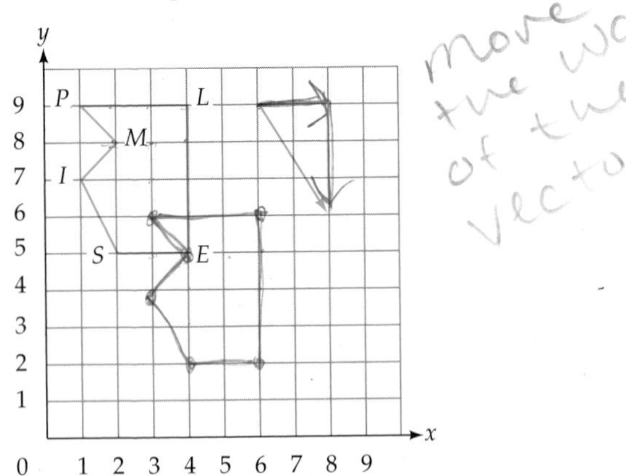

2. Below are a figure and a translation vector. Without using pencil or pen, determine whether the image overlaps the original figure or not. Explain your reasoning.

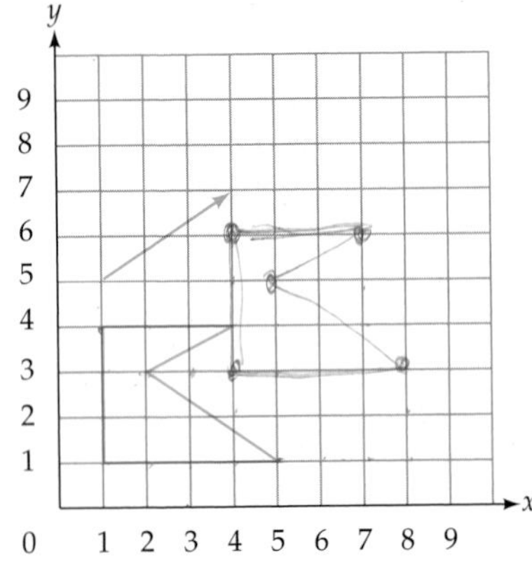

3. Find the image of the kite reflected across line r.

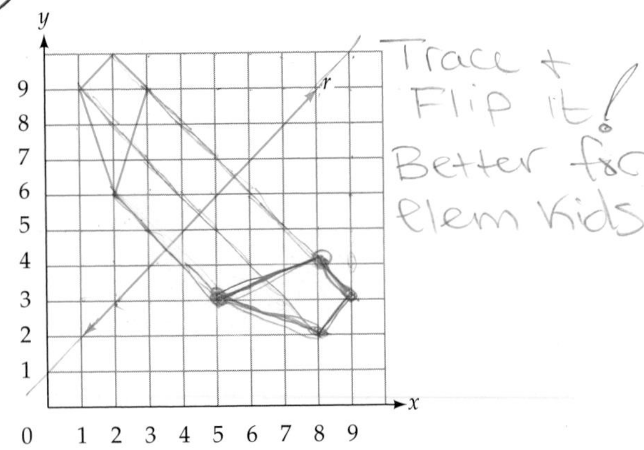

4. The figure below shows quadrilateral $ABCD$ and a line of reflection. Determine the coordinates of A', B', C', and D' using only reasoning (that is, without folding). Explain your reasoning.

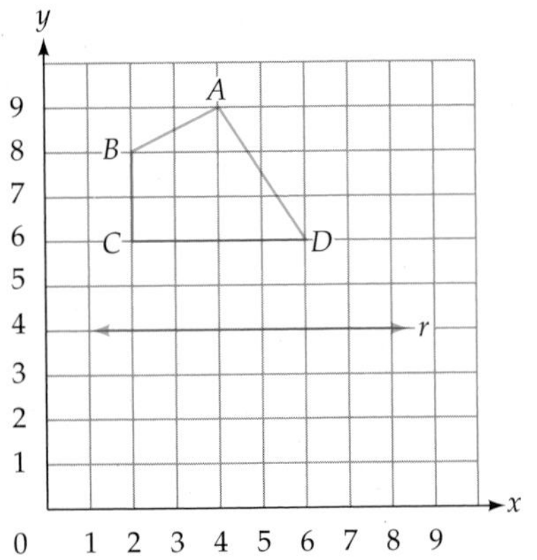

5. Below are a figure and its image. Without folding, sketch the approximate location of the line of reflection. Explain your reasoning.

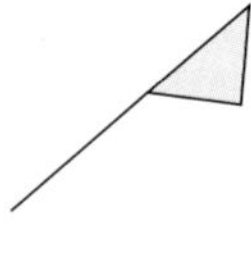

6. Without actually doing the reflection, predict the cases in which the following two images are reflections of each other across the line. Explain your reasoning.

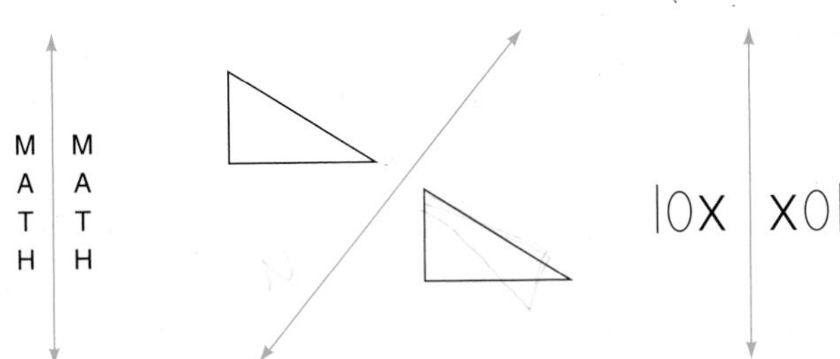

7. Determine the rotation image of ΔMAT if the triangle is rotated 90 degrees clockwise about point A.

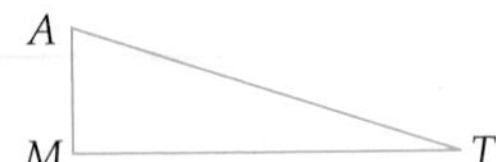

8. Determine the rotation image of trapezoid *FARM* if the figure is rotated 90 degrees counterclockwise about point *F*.

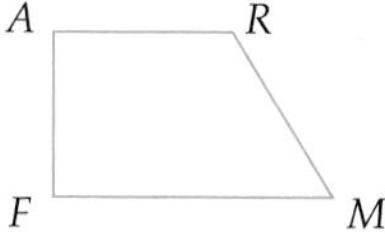

9. Determine the rotation image of the figure below if it is rotated 180 degrees clockwise about point *A*. Explain your reasoning.

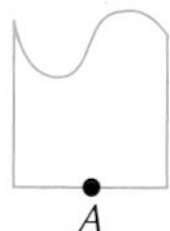

10. The following two exercises appeared in *Geometry and Spatial Sense,* which was one of the books in the Grades K–6 Addenda Series published in the 1990s. Both were suggested for second-graders to help them explore the idea of reflection. The additional part for you in both cases is first to predict where the mirror line is and to explain why you predict that mirror line will produce the number of dots shown.

(1) Place a mirror on or near the three circles shown below in such a way that you can see the numbers of circles shown:

a. Six circles
b. Five circles
c. Four circles
d. Three circles
e. Two circles
f. One circle

(2) For second-graders: Place a mirror on the Pattern Block design shown at the right to get the following figures.

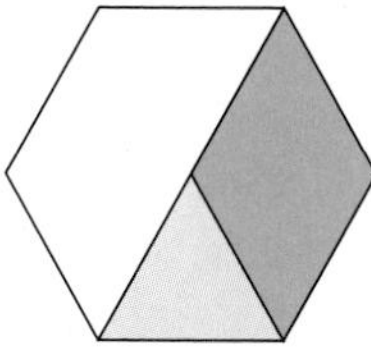

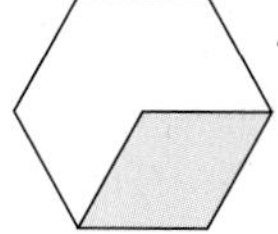
a.

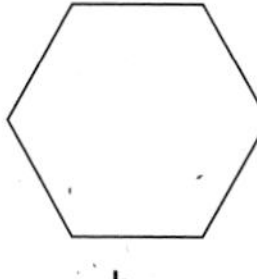
b.

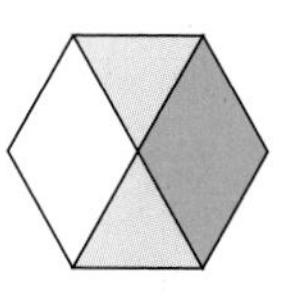
c.

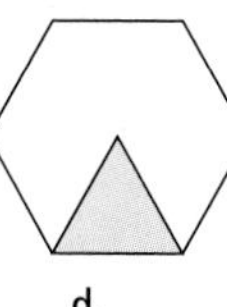
d.

Source: Geometry and Spatial Sense, by the National Council of Teachers of Mathematics.

11. *Navigating Through Geometry in Kindergarten–Grade 2* has some fascinating explorations for young children using either mirrors or Miras. One is called Monster Molly. The children explore how the placement of a mirror can produce different pictures. Molly is shown at the left. Where would you place the mirror to produce the image on the right?

Source: Reprinted with permission from *Navigating Through Geometry in Kindergarten–Grade 2*, copyright © by the National Council of Teachers of Mathematics.

12. a. Determine the coordinates of the vertices of the rotation image of trapezoid *ABCD* (below) if it is rotated 90 degrees clockwise about vertex *A*. Explain your reasoning.

b. Determine the coordinates of the vertices of the rotation image of trapezoid *ABCD* (below) if it is rotated 90 degrees clockwise about point *X*. Explain your reasoning.

c. Determine the coordinates of the vertices of the rotation image of trapezoid *ABCD* (below) if it is rotated 90 degrees clockwise about point *Y*. Explain your reasoning.

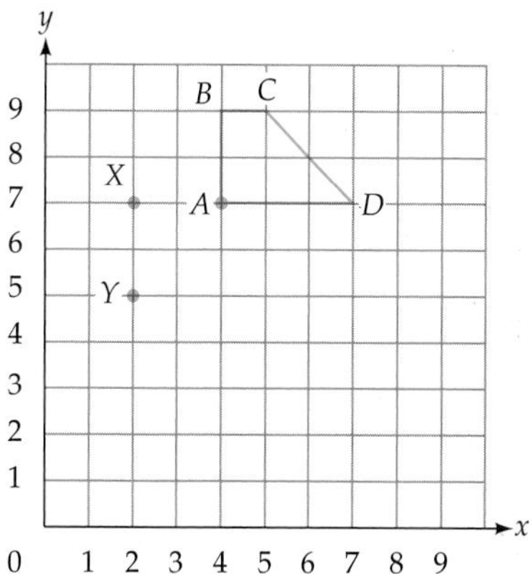

13. The diagram shows two triangles, *CAT* and *DOG*. It appears that they are reflections of each other. Prove that they are congruent.

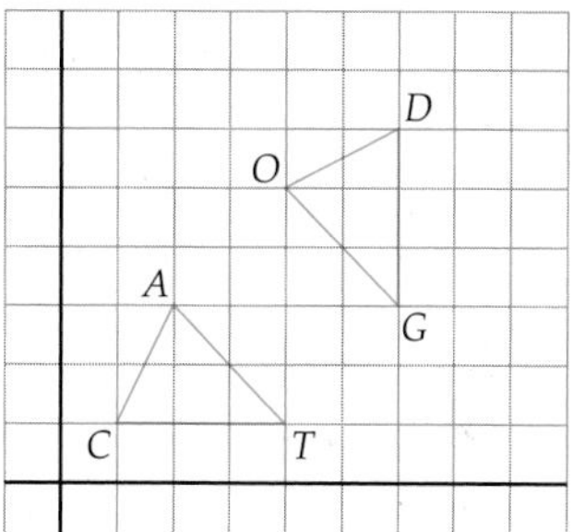

14. Describe how to get the shape from position A to position B in as few moves as possible, using translations, rotations, reflections, and/or glide reflections.

DEEPENING YOUR UNDERSTANDING

Exercises 15 and 16 involve various composite transformations. In each case:

a. First predict the relative location and sketch the image of the figure after the transformations. Explain your reasoning.

b. Perform the two transformations. If your prediction was correct, great. If not, explain the error in your reasoning. Explain also your plans for "correcting" that error.

c. Predict whether the doubly transformed image will be the same if you do the transformations in the opposite order. Then do the two transformations. If your prediction was correct, great. If not, explain the error in your reasoning. Explain also your plans for "correcting" that error.

15. First transformation: translation

Second transformation: reflect the figure across line *l*.

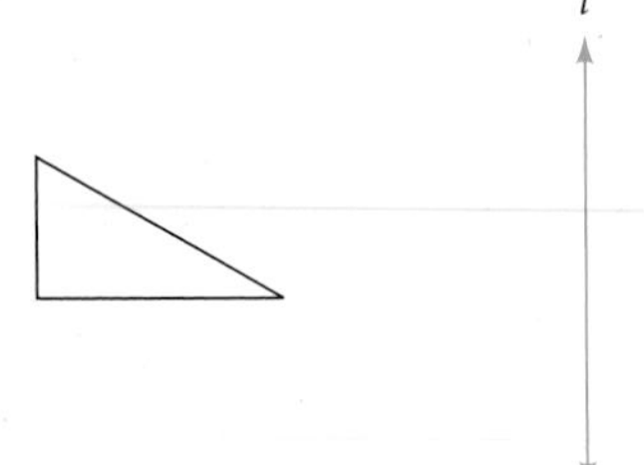

16.

First transformation: translation

Second transformation: reflection across line *l*

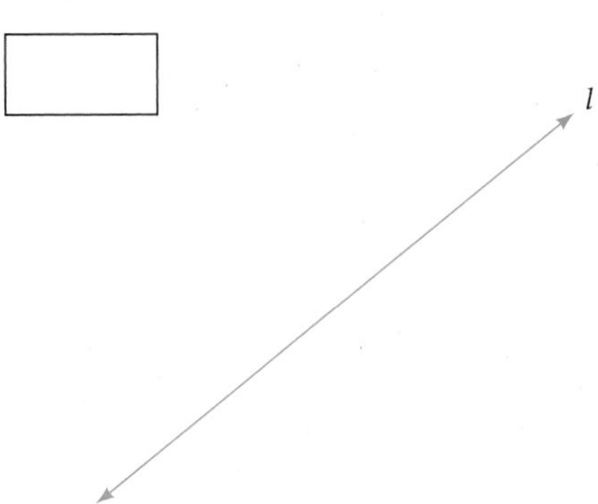

17. Under what circumstances do a translation then reflection commute? That is, when will the translation followed by a reflection produce an image that coincides with the reflection followed by the translation? Explain your reasoning.

18. **a.** Determine the location of the image of the trapezoid (below) after it is reflected across line *l* and its reflection image is then reflected across line *m*.

b. Predict whether the reflection of the trapezoid across line *m* and then across line *l* will give you the same image. Explain your reasoning.

c. Test your conjecture, and correct it if necessary.

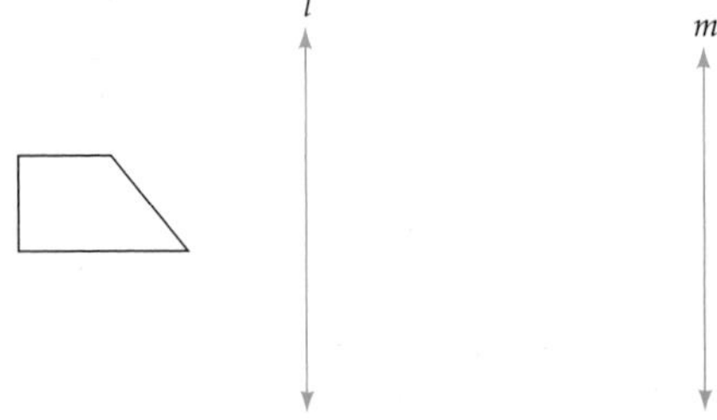

19. **a.** Determine the location of the image of the rectangle on the next page after it is reflected across line *l* and its reflection image is then reflected across line *m*.

b. Predict whether the reflection of the rectangle across line m and then across line *l* will give you the same image. Explain your reasoning.

c. Test your conjecture, and correct it if necessary.

20. a. Determine the location of the image of the trapezoid after it is reflected across line l and its reflection image is then reflected across line m.

b. Predict whether the reflection of the trapezoid across line m and then across line l will give you the same image. Explain your reasoning.

c. Test your conjecture, and correct it if necessary.

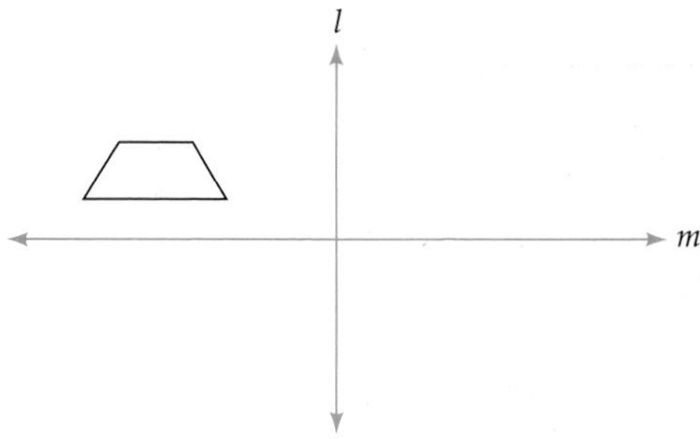

21. a. Rotate $\triangle DIG$ 90 degrees clockwise about point D. Then reflect the rotated image across line l.

b. Is there another transformation or composition mapping that will produce the same effect?

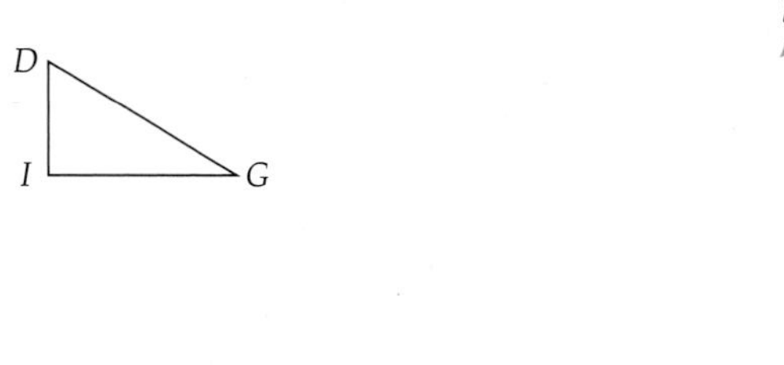

22. Consider parallelogram $STRU$, whose diagonals intersect at M.

a. Describe a transformation that would map $\overline{ST}$ onto $\overline{RU}$.

b. Describe a transformation that would map $\measuredangle SRU$ onto $\measuredangle TSR$.

c. If we reflect $\overline{UM}$ across line $\overleftrightarrow{SR}$, will the image be $\overline{MT}$? If so, explain why. If not, explain why not.

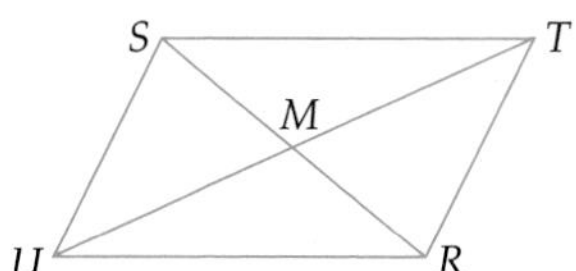

23. Consider isosceles $\triangle PRO$, where M is the midpoint of $\overline{PO}$.

a. Describe a transformation that would map $\overline{PR}$ onto $\overline{OR}$.

b. What transformation would map $\triangle PMR$ onto $\triangle OMR$?

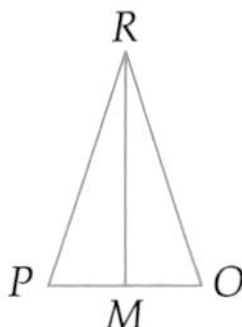

24. The figure below shows $\triangle CAT$, which has been reflected across line m.

a. What relationships do you see between the coordinates of the vertices in $\triangle CAT$ and those of its reflection image $\triangle C'A'T'$?

b. What generalizations can you make from this exercise? You may want to gather more evidence on your own—that is, to make and reflect different figures.

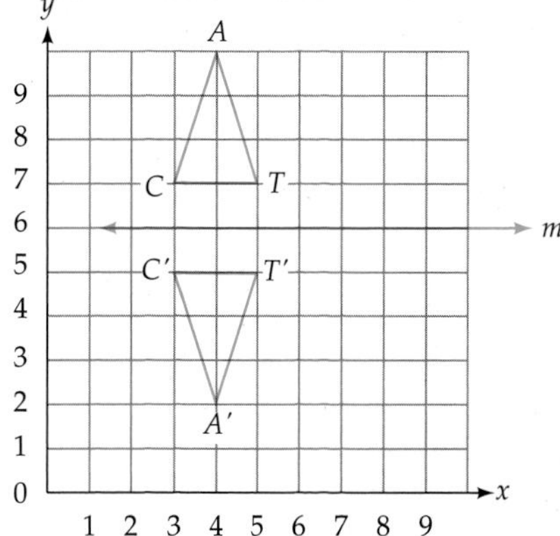

25. Who says geometry isn't practical? Describe where to aim the ball (white circle) to get into the hole (black circle) in a miniature golf course.

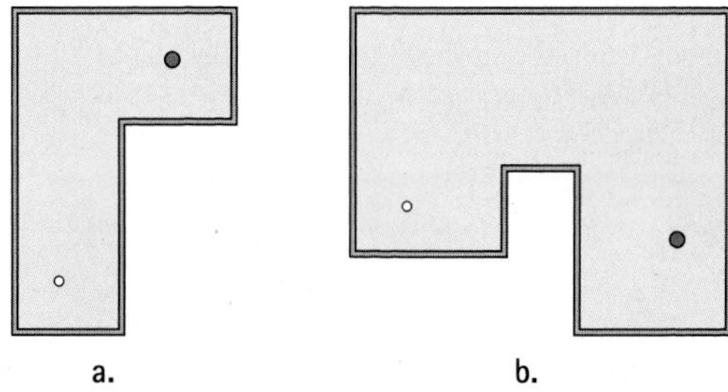

26. Look back to Figure 10.1. Describe any translations, reflections, rotations, and glide reflections that will move part of the figure on top of another part of the figure. For translations, supply a vector or other means of communicating the translation; for reflections, describe the line(s) of reflection; for rotations, describe the angle of rotation.

27. Follow the directions in Exercise 26 for the quilt patterns below.

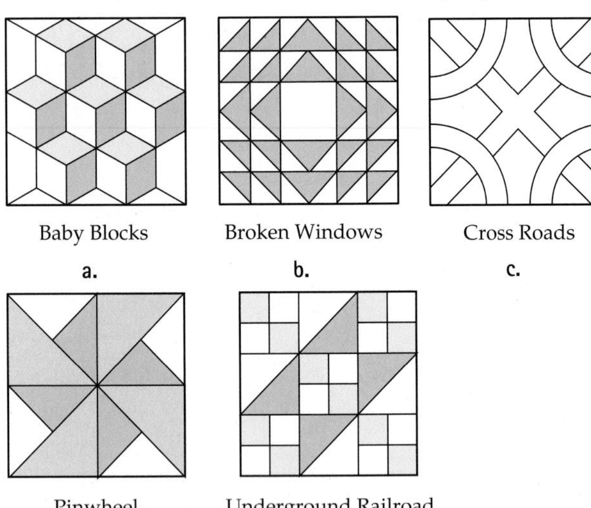

28. Follow the directions in Exercise 26 for the Islamic designs below.

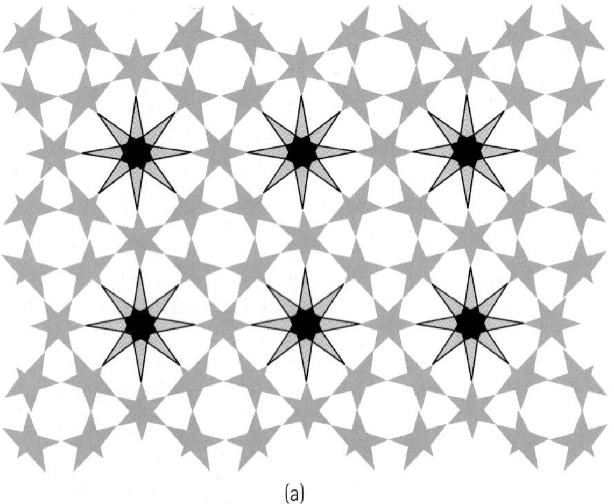

(a)

Source: From Introduction to Tessellations by Dale Seymour and Jill Britton. Copyright © 1989 Pearson Education, Inc., or its affiliates. Used by permission. All Rights Reserved.

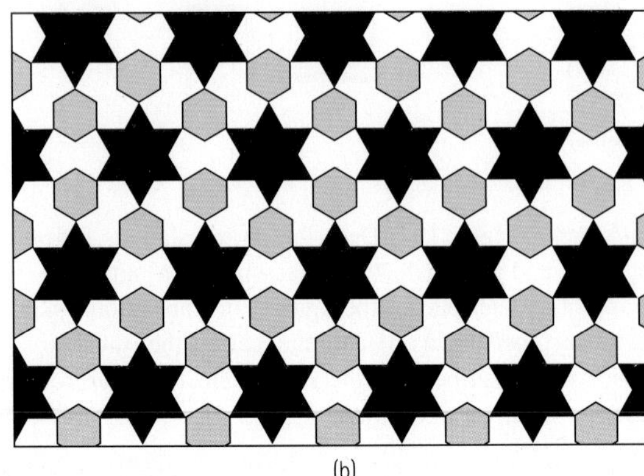

(b)

Source: From Introduction to Tessellations by Dale Seymour and Jill Britton. Copyright © 1989 Pearson Education, Inc., or its affiliates. Used by permission. All Rights Reserved.

29. Make the quilt blocks below using a computer software program. Describe how you used transformations to do so. In each case, you do not have to draw the whole block. That is, you can make a part of the block and then finish the block by translating, rotating, and/or reflecting part(s) of the pattern.

a. Baby Blocks (see Exercise 27a)

b. Broken Windows (see Exercise 27b)

c. Pinwheel (see Exercise 27d)

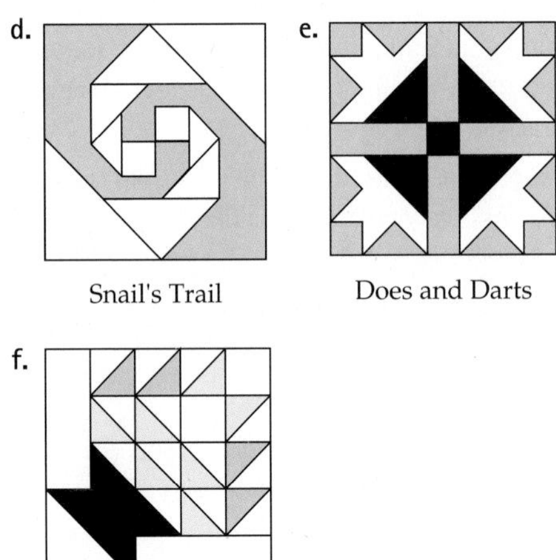

30. In Chapter 1, we discussed palindromes; for example, 1331 is a palindrome. There are word palindromes too. Some close friends of mine, math educators too, named their children HANNAH and AVIVA. We can make a new subset of palindromes, called reflection palindromes, that are legible when reflected. For example, if we reflect 1331 across a vertical line, it will read IƐƐI, and if we reflect 1331 across a horizontal line, it will read IƐƐI. But if we reflect AVIVA across a vertical line, we still have AVIVA. Explore different palindromes and reflections. Describe the characteristics of palindromes (both number and letter) that will be legible after reflections across horizontal lines, those that will be legible after reflections across vertical lines, and those that will be legible after reflections across either line.

31. a. If we reflect the angle formed by a clock's hands at 3 o'clock through a vertical line through the center of the clock, what will be the time?

b. If we rotate the angle formed by a clock's hands at 3 o'clock 90 degrees clockwise, what will be the time?

c. Make up and solve a similar problem.

32. The table is a base ten multiplication table in which only the ones digit is given. For example, because 8 times 2 = 16, only the 6 is shown.

1	**1**	**2**	**3**	**4**	**5**	**6**	**7**	**8**	**9**	**10**
1	1	2	3	4	5	6	7	8	9	0
2	2	4	6	8	0	2	4	6	8	0
3	3	6	9	2	5	8	1	4	7	0
4	4	8	2	6	0	4	8	2	6	0
5	5	0	5	0	5	0	5	0	5	0
6	6	2	8	4	0	6	2	8	4	0
7	7	4	1	8	5	2	9	6	3	0
8	8	6	4	2	0	8	6	4	2	0
9	9	8	7	6	5	4	3	2	1	0
10	0	0	0	0	0	0	0	0	0	0

a. Describe translations, reflections, and rotations within this table. For example, the first half of the 2s row (2, 4, 6, 8, 0) can be translated (5 boxes horizontally). Without transformation language, we would say the first half of the 2s row is the same as the second half of the 2s row. For the purposes of this exercise, consider only the location of the digit, not its spatial orientation.

b. Make a multiplication table for base twelve in which only the ones digit of the product is given. Describe the transformations that can map one row or column onto another row or column. What similarities and differences do you see between the two tables?

33. The object of the game below is to move the arrow from the top-left corner to the bottom-right corner, using transformations. The only restriction is that you cannot go back to a square once occupied.

a. Describe how to move the arrow from the top-left corner to the bottom-right corner in as few moves as possible.

b. Describe how to move the arrow from the top-left corner to the bottom-right corner in as few moves as possible, with the additional restriction that each move must be to a square that has at least one point in common with the square in which the figure resides.

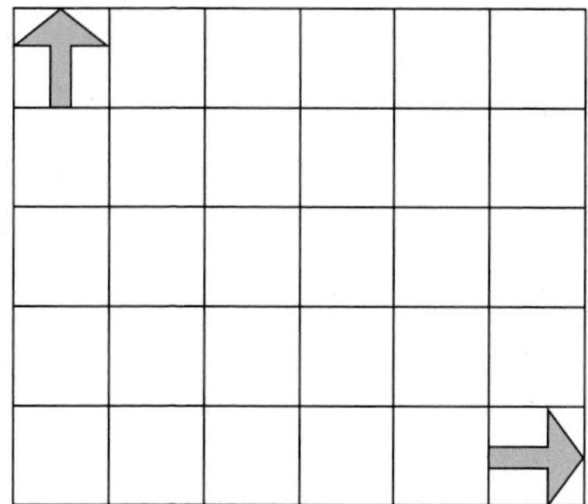

34. Go to http://www.shodor.org/interactivate/activities/Transmographer/ and explore the different transformations. Write about what you found interesting and what challenges you had.

FROM STANDARDIZED ASSESSMENTS

NECAP, 2008, Grade 7

35. Look at Figure P and Figure Q.

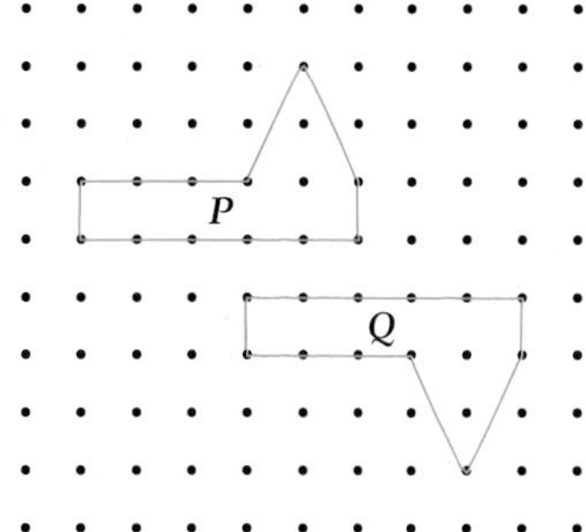

Which motion or motions will result in Figure P exactly covering Figure Q?

a. Slides only

b. Turns only

c. Flips and turns only

d. Flips and slides only

SECTION 10.2 Symmetry and Tessellations

What do you think?

- Does a figure have to be closed to have reflection symmetry? Why or why not?
- What are some practical applications of symmetry?

"Symmetry, as wide or as narrow as you may define its meaning, is one idea by which humanity through the ages has tried to comprehend and create order, beauty, and perfection."[1]

Symmetry pervades our lives, and many people find symmetric figures appealing. We see symmetry in both natural objects and human-made objects—snowflakes, starfish, flags, logos, quilts, and many kinds of art, to name but a few. Look at Figure 10.24. Each of its elements is a well-known symbol or figure. For each of them, write what it is about that figure that is visually appealing or that attracts your attention. Describe the symmetry within each figure. Then read on. . . .

CLASSROOM CONNECTION

Snowflake Bentley is the story of William Bentley, who was fascinated by snowflakes and who worked for many years to develop a way to photograph actual snowflakes. His book *Snow Crystals*, which has over 2000 photographs of snowflakes and other snow crystals, is widely available and is fascinating to children and adults alike.

Chinese Yin - Yang "Unity of opposites"

Canadian flag

Butterfly

Snowflake

Figure 10.24

I think it is very possible that the first mathematical acts of human beings had to do not with numbers but with symmetry. Archaeological records of virtually every culture show symmetry in some of that culture's artifacts: its baskets, its pottery, the artifacts created for celebrations and rituals.

Symmetry also abounds at the microscopic level. For example, many chemical molecules are symmetric, as are many viruses. In fact, the most common shape that viruses take is the icosahedron (remember Chapter 8?); it characterizes the viruses that cause herpes, chicken pox, and warts.

Symmetry abounds at the macroscopic level as well: Volcanoes are cone-shaped, stars are spherical, honeycombs are hexagonal, and galaxies tend to be spiral or elliptical; all of these shapes are distinguished by their symmetry. "Any understanding of nature must include an understanding of these patterns."[2] For example, the hexagonal shape of the honeycomb is the solution to the problem of how to get the maximum storage capacity from the minimum amount of building materials!

Symmetry also adds to our understanding of patterns made by humans. For example, understanding symmetry can make it easier to make many very exotic-looking prints on skirts and quilts and can also increase our appreciation of their beauty.

According to the Common Core State Standards, line symmetry (which we will call reflection symmetry to connect with the language used in Section 10.1) is learned in fourth grade. The other symmetries in this chapter are not in the K–8 CCSS. However, because some people find these ideas so interesting, we are including more types of symmetry as well as the concept of tessellations.

Let us then investigate this aspect of mathematics more deeply, to better understand what we mean by the term *symmetry* and what kinds of symmetries there are. We will investigate four kinds of symmetry: translation, reflection, rotation, and glide reflection.

Let us begin with a definition of **symmetry**: a transformation that places the object directly on top of itself.

REFLECTION AND ROTATION SYMMETRY

On the basis of our work with reflections in the previous section, what do you think reflection symmetry is? How would you explain reflection symmetry in your own words? Answer these questions before reading on. . . .

A figure has *reflection symmetry* if there is a line, called the **line of symmetry**, that can be drawn through the figure such that when we fold the paper on the line, the part of the figure on one side will lie directly on top of the other part.

Now let us take a look at rotation symmetry. Thinking of the definition of reflection symmetry and what you know about a rotation transformation, how would you describe rotation symmetry? Try to do so before reading on. . . .

CLASSROOM CONNECTION

Grade 2

This activity is similar to Exploration 10.1 in the *Explorations* manual. 

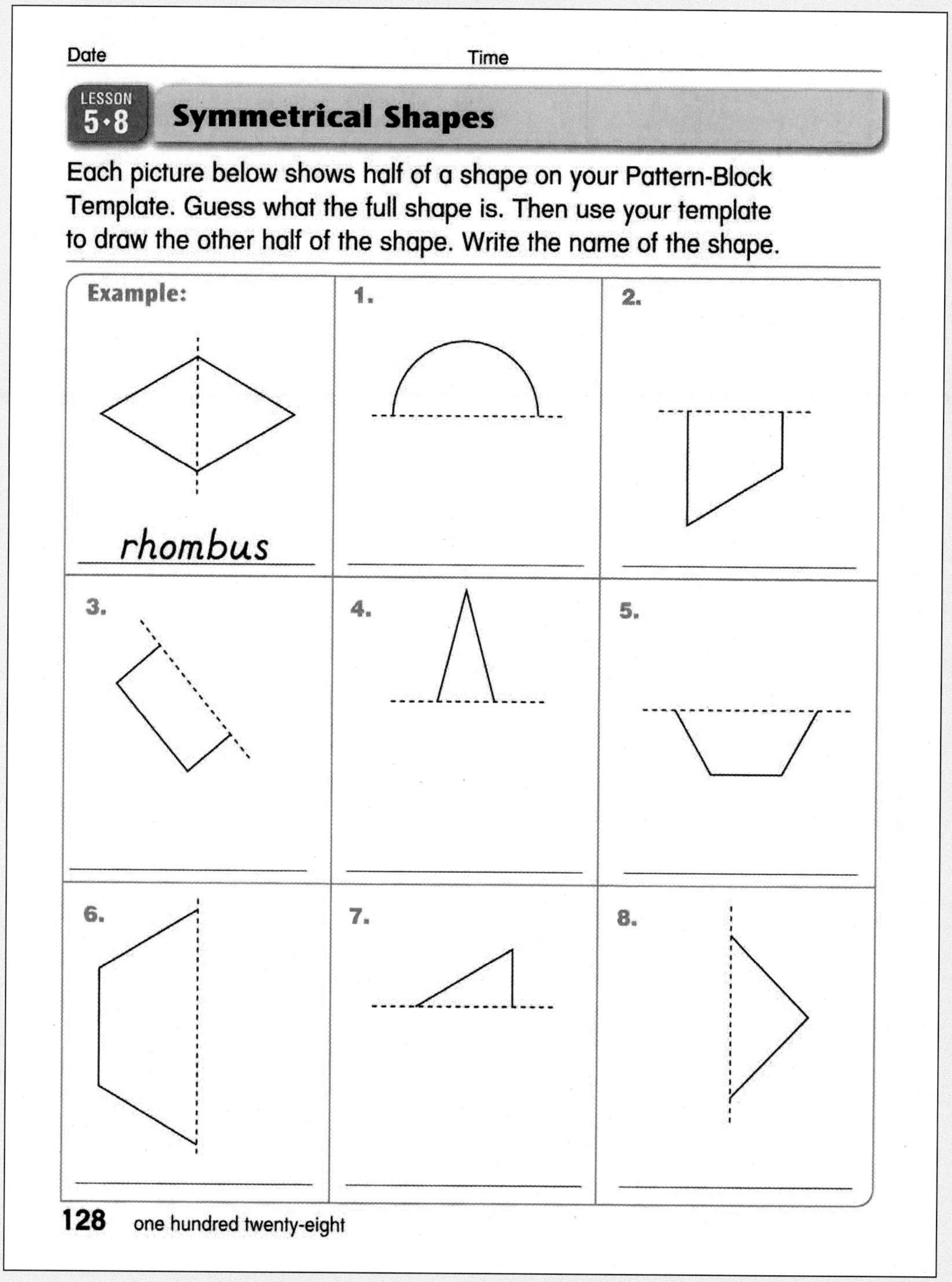

Date Time

LESSON 5·8 **Symmetrical Shapes**

Each picture below shows half of a shape on your Pattern-Block Template. Guess what the full shape is. Then use your template to draw the other half of the shape. Write the name of the shape.

Example: rhombus	**1.**	**2.**
3.	**4.**	**5.**
6.	**7.**	**8.**

128 one hundred twenty-eight

From *Everyday Mathematics, Grade 2:* The University of Chicago School Mathematics Project: Student Math Journal, Volume 1, by Max Bell et al., Lesson 5-8, p. 128. Reprinted by permission of The McGraw-Hill Companies, Inc.

Imagine tracing a figure and putting the traced image directly on top of the figure. Informally, we say that the figure has *rotation symmetry* if we can pick up the tracing, turn it by some amount less than a full turn, and place it down so that it lies directly on the original figure.

INVESTIGATION 10.2a Reflection and Rotation Symmetry in Triangles

Many of the geometric shapes we investigated in the previous chapter have reflection and/or rotation symmetry. Describe the reflection and rotation symmetries you see in the equilateral triangle, the isosceles triangle, and the scalene triangle below. Then read on. . . .

Explorations Manual 10.7

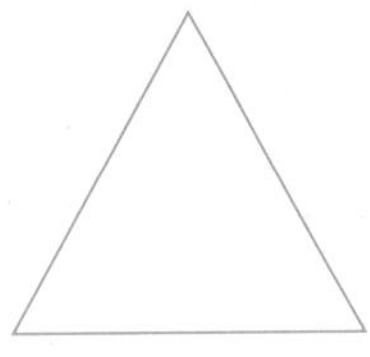

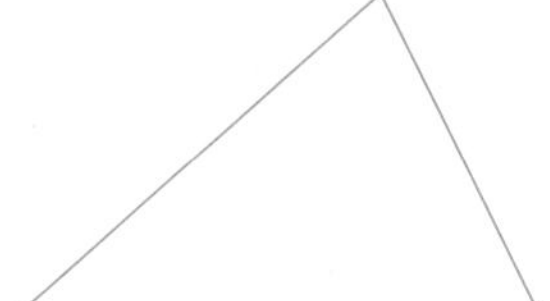

DISCUSSION

We find that the equilateral triangle has three lines of symmetry, the isosceles triangle has one line, and the scalene triangle has none.

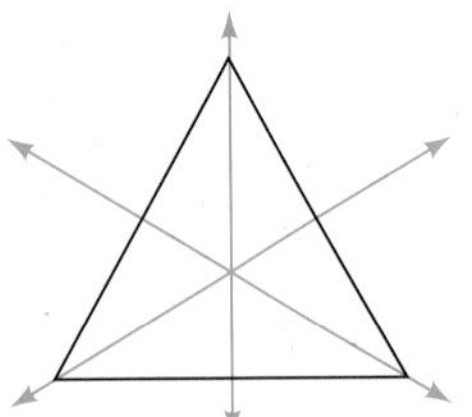

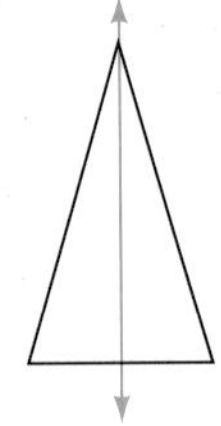

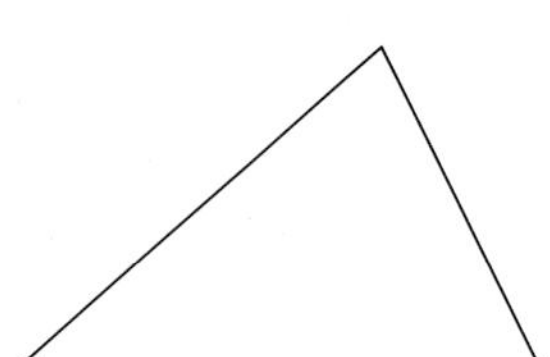

With respect to rotation symmetry, only the equilateral triangle has rotation symmetry. How would you describe that symmetry? Think before reading on. . . .

FINDING THE AMOUNT OF ROTATION IN ROTATION SYMMETRY

Some people think of rotation symmetry in terms of how much we have to turn the figure in order for its image to lie directly on the figure. If we turn the equilateral triangle $\frac{1}{3}$ of a whole turn (the center of rotation being the center of the triangle), the image will fit on top of the triangle. If we turn it another $\frac{1}{3}$ turn, the image will fit on top of the triangle again. If we turn it another $\frac{1}{3}$ turn, then we are back to its original position. Thus we can say that the equilateral triangle has $\frac{1}{3}$-turn symmetry.

Other people prefer to specify rotation symmetry by the number of degrees needed to make the rotated image fit on top of the figure. In this case, if we rotate the equilateral triangle 120 degrees, the image will fit on top of the figure. If we rotate the equilateral triangle another 120 degrees, the image will fit on top of the figure again. If we rotate the equilateral triangle another 120 degrees, the figure will be back to its original position.

Some of my students are very puzzled at this point and ask, "How did you figure out that it is 120 degrees?" Could you explain why, as though to a ten-year-old? If not, think before reading on. . . .

LANGUAGE

Some readers may be wondering why I didn't say that the equilateral triangle has $\frac{1}{3}$-turn rotation symmetry. Actually, I could have said so, but the word *rotation* is unnecessary.

Many students find the following illustration useful. The triangle at the left in the figure below is in its original position, with the vertex angles labeled 1, 2, and 3, respectively. Imagine turning the triangle clockwise until the image is in the same position as the original triangle. The second figure represents the new position. Rotate the triangle again until the image is in the same position as the original triangle. The third figure represents the new position. Rotate the triangle again until the image is in the same position as the original triangle. Because it took three rotations to get back to the original position, each rotation has to equal $\frac{360}{3} = 120$ degrees.

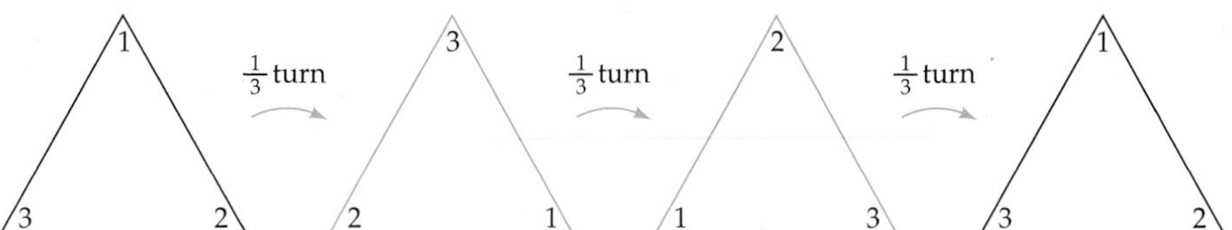

INVESTIGATION 10.2b Reflection and Rotation Symmetry in Quadrilaterals

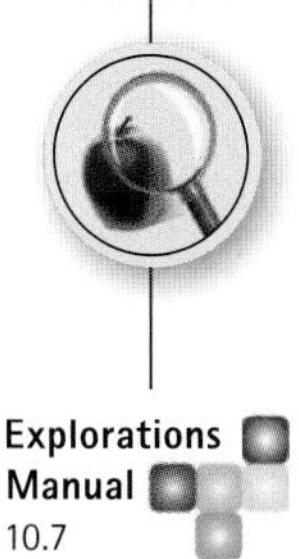

Explorations Manual 10.7

What about quadrilaterals? Determine the reflection and rotation symmetry of the kite, parallelogram, rhombus, rectangle, and square. I encourage you to use whatever means you need in order to understand this investigation thoroughly. You might trace the figures on a blank sheet of paper, on tracing paper, or on an overhead transparency. Work on this before reading on. . . .

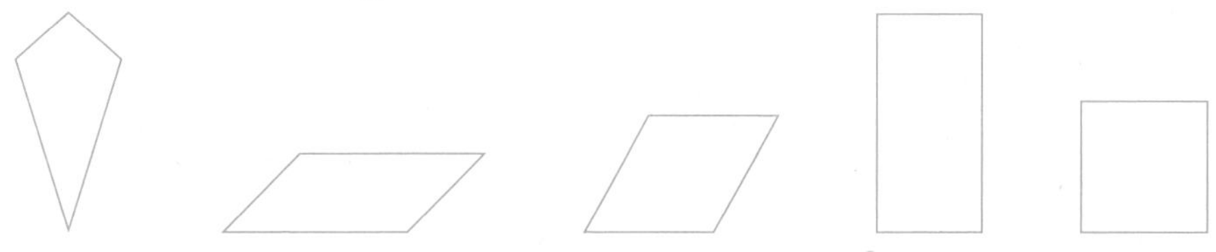

DISCUSSION

Kites have one line of symmetry, parallelograms have no lines of symmetry, rhombi have two, rectangles have two, and squares have four. The most common mistake my students make is to think that parallelograms have reflection symmetry. If you got this one wrong, make several parallelograms on a blank sheet of paper (tracing paper is even better), fold the parallelogram on either diagonal, and hold the paper up to the light. You will see clearly that the parallelogram does not fold onto itself in either case.

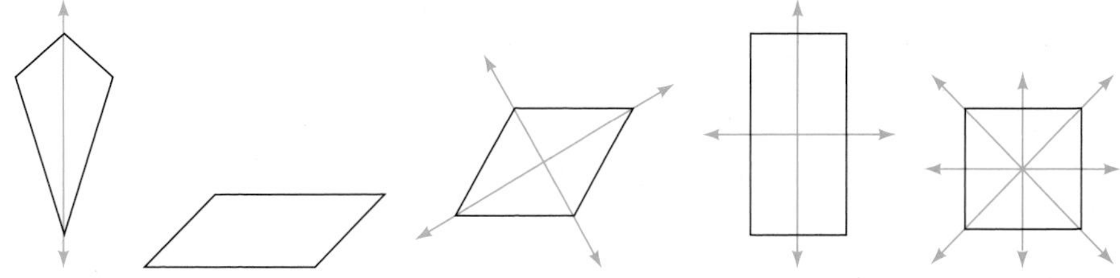

With respect to rotation symmetry, a kite has no rotation symmetry, a parallelogram has $\frac{1}{2}$-turn (or 180-degree rotation) symmetry, a rhombus has $\frac{1}{2}$-turn (or 180-degree rotation) symmetry, a rectangle has $\frac{1}{2}$-turn (or 180-degree rotation) symmetry, and a square has $\frac{1}{4}$-turn (or 90-degree) rotation symmetry.

As you may have already observed, many figures have 180-degree rotation symmetry but no other rotation symmetry. A figure has 180-degree rotation symmetry if it looks the

MATHEMATICS

It is important to note that when we say that a rectangle has two lines of symmetry, we mean that subset of rectangles that are not squares. Technically, squares are rectangles, so technically, some rectangles (those that are also squares) have four lines of symmetry.

In Section 8.2, we created a family tree for quadrilaterals. By definition, each descendant had the same properties as its immediate ancestor plus at least one new characteristic, thus creating a new class of shapes. This is also true for symmetry. Every quadrilateral has at least as many reflection and rotation symmetries as any quadrilateral above it in the "family tree."

same when turned halfway around. Because this particular kind of rotation symmetry is so common, it has its own special name: **point symmetry**. These three terms are synonymous: $\frac{1}{2}$-turn symmetry, 180-degree rotation symmetry, and point symmetry.

One last note on rotation symmetry. When I ask my students to tell me whether or not a figure has rotation symmetry, many of them will say that a figure has 360-degree rotation symmetry. What does it mean to say that a figure has 360-degree rotation symmetry? Think and read on. . . .

To say that a figure has 360-degree rotation symmetry means that if we turn it completely around, it will look the same. However, this is true for *any* figure. Because any figure has 360-degree rotation symmetry, we do not use this term when describing the rotation symmetries of a figure.

INVESTIGATION 10.2c Reflection and Rotation Symmetry in Other Figures

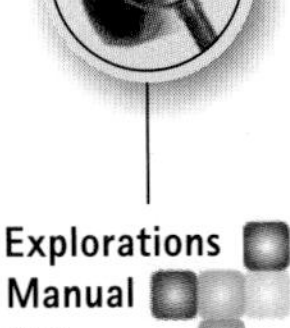

Explorations Manual 10.7

Below are regular polygons with 3, 4, 5, 6, 7, and 8 sides. Draw the lines of symmetry for each figure. What do you see? . . .

 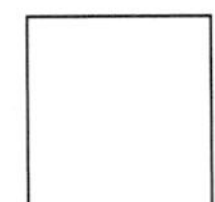 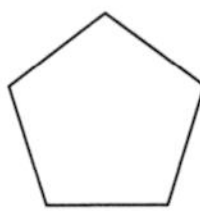 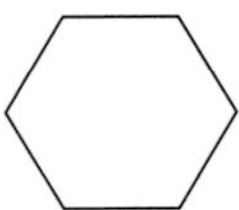 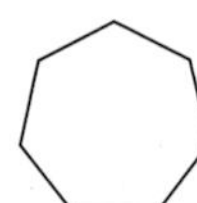 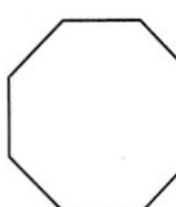

DISCUSSION

Many students realize that the lines of symmetry in the figures with 3, 5, and 7 sides (that is, with an odd number of sides) are somehow "different" from those in the figures with an even number of sides. How would you describe this difference? . . . .

When the polygon has an odd number of sides, the lines of symmetry all connect a vertex of the polygon to the middle of the opposite side. When the polygon has an even number of sides, half of the lines of symmetry connect two opposite vertices and half of the lines of symmetry connect the midpoints of two opposite sides (Figure 10.25). However, in all cases, the number of lines of symmetry is equal to the number of sides in the figure.

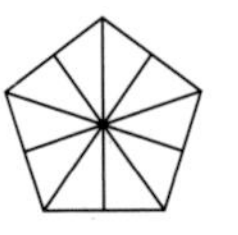 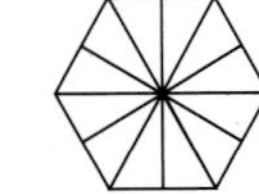

Figure 10.25

INVESTIGATION 10.2d Letters of the Alphabet and Symmetry

Examine the letters of the alphabet. Which letters have rotation symmetry? What kind(s) of rotation symmetry? Which letters have reflection symmetry? What kind(s) of reflection symmetry? Work on this and then read on. . . .

A B C D E F G H I J K L M N O P Q R S T U V W X Y Z

DISCUSSION

When we examine the letters in terms of reflection symmetry, we find that there are two main kinds of lines of reflection: vertical and horizontal.

The following letters have **vertical line symmetry**: A, H, I, M, O, T, U, V, W, X, Y. I think it is interesting that over half of the letters with vertical line symmetry occur at the end of the alphabet.

The following letters have **horizontal line symmetry**: B, C, D, E, H, I, O, X. Here we have the opposite phenomenon: Four consecutive letters occur near the beginning of the alphabet.

The following letters have point symmetry: H, I, N, O, S, X, Z.

Note: The symmetry of some letters depends on how they are drawn. For example, B and K do not always have horizontal line symmetry.

Above, I represented the solution to the question in a list form. In Figure 10.26 are two Venn diagrams. The first shows just the letters that have line symmetry, and the second shows all the letters. Seeing the letters grouped in this diagram can help us to see and better understand these symmetries.

Symmetries of the letters of the alphabet

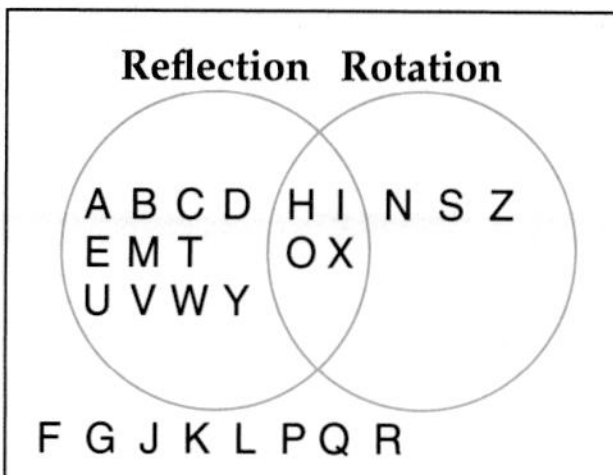

Reflection symmetries of the letters of the alphabet

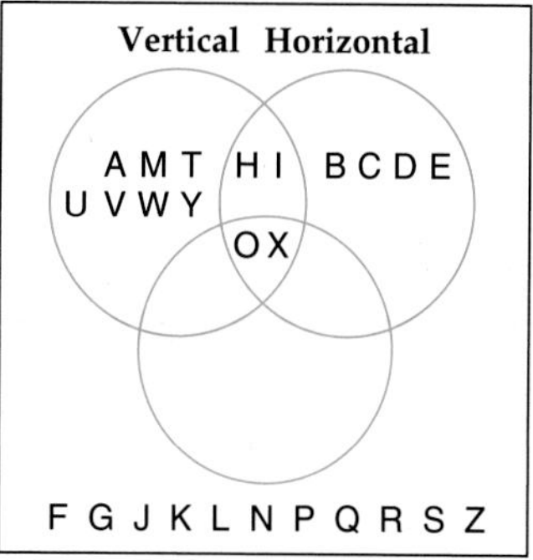

Figure 10.26

WHAT ARE PATTERNS?

We find patterns in our clothing, on quilts, on ceilings, on floors, on walls, and on the borders of many objects—in short, all over the place. To define the word *pattern* is very difficult, so let us take some time to develop this idea. . . .

INVESTIGATION 10.2e Patterns

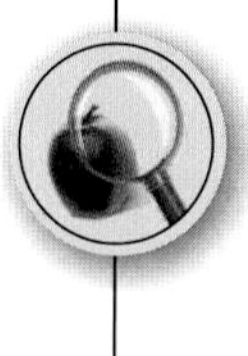

Patterns have been a recurring theme in this book. Looking for patterns is a tool for solving problems, and we have seen patterns in addition and multiplication tables, as well as other numerical patterns. When we talk about geometry, we almost inevitably encounter the word *pattern*. Let us explore this idea of pattern.

A. Below are three "brick" patterns. Look at them and think what they all have in common, and then try to come up with a definition of pattern. Then read on. . . .

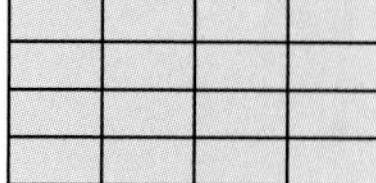

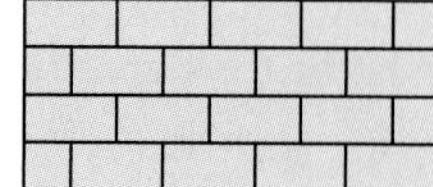

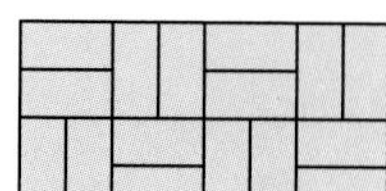

DISCUSSION

There are three attributes in all patterns. First, there is a unit, and, in some cases, multiple units. Second, there is a repetition of that unit, and third, the unit is repeated in some organized way.

In the first case, the unit is a rectangle, and it is repeated (translated) in a simple way. In the second case, the unit is still a rectangle; in this case, each row is staggered (translated) slightly—this pattern is used in brick walls and buildings because it is stronger. In the third case, we *can* see one brick as the unit; however, most people will think of the unit as two bricks side by side. The principle of organization here is rotation. Each pair of bricks is laid side by side, but rotated 90 degrees.

B. Now look at the three patterns below (Figures 10.27[a], [b], [c]). In each case, identify the unit and describe the way in which the unit is repeated to make the pattern. It may help to think of writing directions for someone who doesn't understand this idea of unit of a pattern. . . .

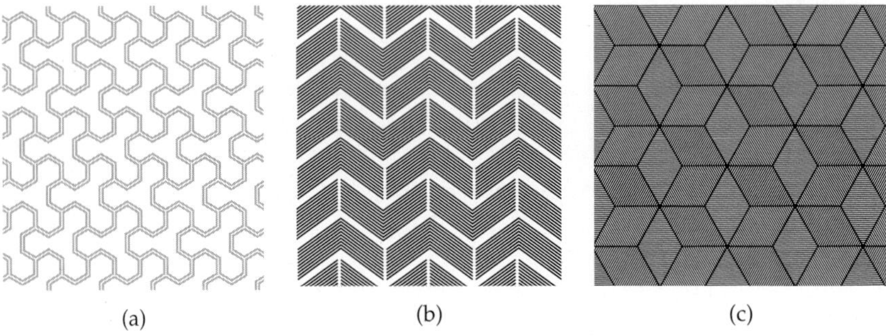

(a) (b) (c)

Figure 10.27

Sources: (a) ©RODINA OLENA/Shutterstock.com. (b) ©wanpatsorn/Shutterstock.com. (c) ©Svetlana Prikhnenko/Shutterstock.com.

DISCUSSION

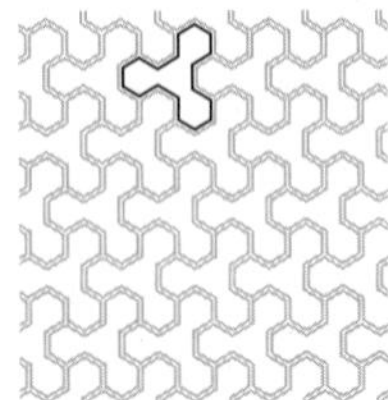

Figure 10.28

Although Figure 10.27a may appear complex at first, there is really only one shape (see Figure 10.28) that is simply repeated through translations. To use computer language, if you copy and paste the image at the right, you could "fill the plane" with this pattern. That is, you could fill as large a surface as you wished by repeating this unit.

In Figure 10.27b, there is also one shape—a particular kind of hexagon called a chevron (Figure 10.29a). However, in some cases it is "right side up" and in some cases, it is upside down. So if we take one chevron and reflect it across a horizontal line and then translate it so that the bottom right side of the original chevron meets the top left side of the second chevron (Figure 10.29b) we translate this unit horizontally to fill the plane.

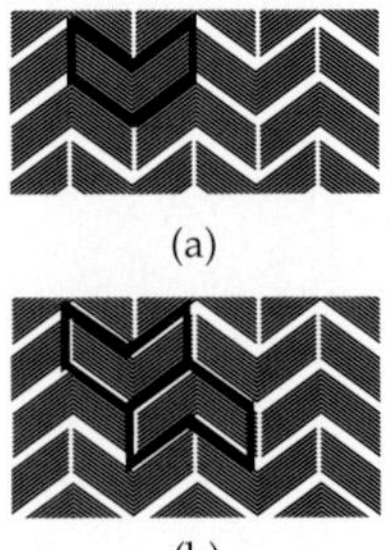

(a)

(b)

Figure 10.29

Figure 10.27c is more complex because we have two different quadrilaterals, one of which has two different orientations and there is different shading. Many people looking at this pattern will see a larger hexagon formed from five of the figures with the diamond in the center of the pattern, almost like the eye of the pattern (Figure 10.30a). This pattern is not quite the unit though. If you imagine translating it, you will not create the whole pattern. However if you take this composite hexagon and attach one of the diamonds to the right-bottom side of the hexagon (Figure 10.30b), you will be able to fill the plane by translating the figure.

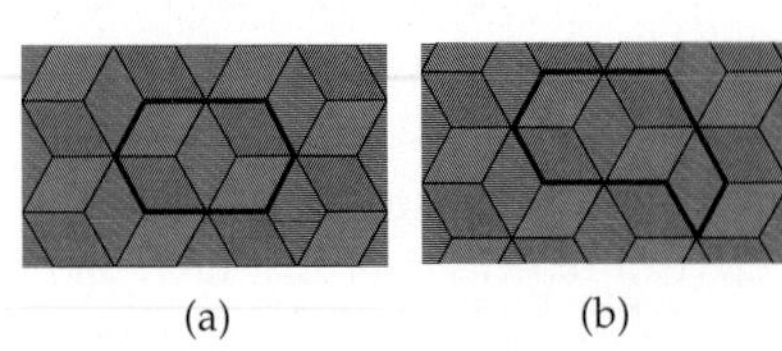

(a) (b)

Figure 10.30

This analysis of patterns works fine for certain kinds of patterns, such as checkerboard, herringbone, and quilt patterns. However, this definition always provokes a response in some of my students who consider the definition very carefully. They ask questions like this: "What about patterns you see on a sand dune, or patterns of light on the surface of a lake, or patterns you can see when you ride a canoe on a pond and there is no wind, or the patterns of ripples of sand along the seashore? Aren't these also patterns?" What a wonderful question!! Although these examples don't fit this crisp characterization we have described, there is some sense of repetition and organization in each case. The last thing I want to do is perpetuate the belief many students bring to mathematics class that mathematics is fundamentally "different" from the real world. Thus, even though we will focus here on that subset of patterns where we can find the unit and analyze the organization, I don't want to exclude this larger sense of patterns.

MATHEMATICS

The careful reader may have realized that we also speak of numerical patterns. Recall our work in Chapters 1 and 2. Do those numerical patterns have repetition? Is there a unit for those patterns? Is there some organization of those numbers? Or do we have to have two definitions—one for shape patterns and one for number patterns? What do you think?

Patterns appear everywhere in the natural world. The examples below are taken from a fascinating article by Marvin Harrell and Linda Fosnaugh in the May 1997 issue of *Mathematics Teaching in the Middle School* (pp. 380–389).

Bumblebee's eye at over 100 magnification

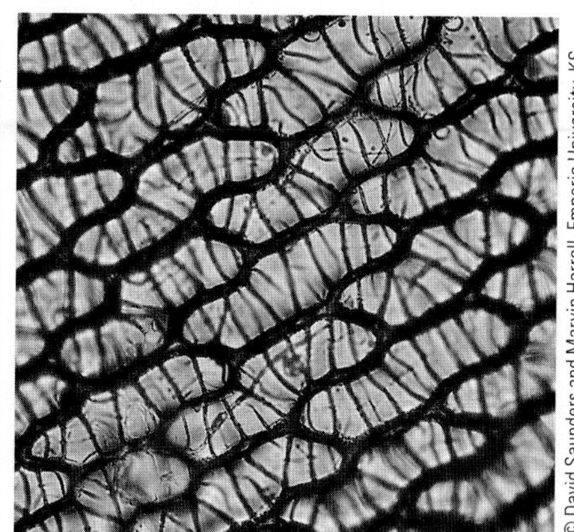

Sphagnum moss at over 200 magnification

MATHEMATICS

Just as our crisp characterization of the concept of pattern left out a lot of interesting patterns and was restricted to "perfect" patterns, so too did traditional geometry "leave out" a lot of shapes that aren't perfect or nearly perfect—shapes like clouds, smoke, ocean currents, and weather patterns. A whole new field of geometry has been born since I graduated from college in the 1970s. This field, called fractal geometry, came into being in part because classical Euclidean geometry is very limited in its ability to help us to understand such shapes.

We are almost ready to examine symmetries of patterns. However, before we do so, we need to revisit a concept that was presented in Chapter 8—infinity. Recall our discussion in Chapter 8 of the concept of a plane. That is, a mathematical plane is a flat surface that extends infinitely in all directions. Obviously, all examples of planes in a book must be finite, and thus any diagram of a plane is actually a piece (technically, a subset) of a plane. Similarly, mathematical patterns are theoretically infinite. Thus, when you see patterns below and are asked to translate, rotate, and reflect them, you need to realize that, theoretically, there are no edges to the pattern.

SYMMETRIES AND PATTERNS

LANGUAGE

Many people use the words *design* and *pattern* interchangeably. I have learned otherwise from friends who do design. A nice definition of design is "a set of shapes, generally contained in a larger shape where the shapes are arranged for a specific purpose—functional or aesthetic." Many designs have patterns in the design, but the design itself is generally not a pattern in the sense in which we use the term here.

We have now laid the groundwork for understanding and describing symmetries of patterns. Recall our definition of symmetry: a transformation that places the object directly on top of itself. Thus a pattern has **translation symmetry** if we can lift the pattern, translate it some distance and in some direction, and set it down on top of itself. A pattern has **reflection symmetry** if we can find a line such that if we reflect the pattern across that line, the pattern will fit directly on top of itself. A pattern has **rotation symmetry** if we can find a point such that if we turn the pattern around that point, the pattern will fit directly on top of itself. A pattern has **glide reflection symmetry** if we can find a translation and a line such that if we translate the pattern some distance and in some direction, and then reflect the pattern across a line that is parallel to the translation vector, the pattern will fit directly on top of itself.

Let us now focus on the subset of patterns whose characteristics can be clearly described. When an artist or a manufacturer wants to create a new design, the possibilities are virtually limitless. However, as soon as we replace *design* by *pattern,* the designers are limited by the mathematical laws that underlie the formation of patterns. Let us see what this means with one kind of pattern.

Now that we have the idea of translation symmetry, we can give a definition of **pattern**: a figure with translation symmetry. Do you see the connection between this definition and the earlier discussion of characteristics of patterns: having a unit that is repeated in an organized manner?

INVESTIGATION 10.2f Symmetries of Strip Patterns

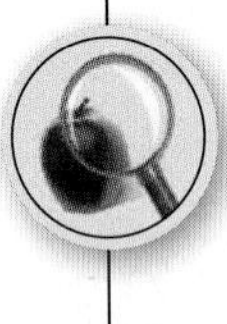

We will begin by looking at the mathematically simplest kinds of patterns, which are often called border or strip patterns because the motif (unit) is repeated in only one direction. With your understanding of translations, reflections, rotations, and glide reflections, determine the translation, reflection, rotation, and glide reflection symmetry for each strip pattern below.

A. What kind(s) of symmetry does the following strip pattern have? Think before reading on. . . .

P P P P P P P P P P

DISCUSSION

This pattern has translation symmetry because we can translate this pattern so that we can place it on top of itself. The translation vector is parallel to the bottom of the page, and the length of the vector is equal to the distance between any two shapes.

One way to verify this symmetry is to trace the flags on a blank sheet of paper or on an overhead transparency. You can then see that you can move the pattern over 1 unit and it fits onto itself. Remember that, just as the plane is considered to be infinite, extending in all directions, so too are patterns considered to be infinite. Thus, although my figure and your tracing will have a left-most and a right-most flag, the "pattern" is considered to have no ends.

B. What kind(s) of symmetry does the following strip pattern have? Think before reading on. . . .

DISCUSSION

This pattern has translation symmetry. This pattern also has two vertical lines of symmetry. There are actually two "different" sets of vertical lines that will enable you to put the figure onto itself if you trace this pattern. One of the lines has the flags facing the line, whereas the other line has two flags pointing away from the lines. If you don't see the vertical lines, turn to Appendix B, where the symmetries for these strips are shown.

Although learning why there are two "different" sets of vertical lines of symmetry for this pattern is not one of the big ideas of this section, coming to understand and appreciate mathematical language is one of the big ideas of the book. Thus, let us take just a few moments to pursue why we say there are two different vertical lines of symmetry. Actually, mathematicians say not that the two lines of symmetry are different but rather that they are inequivalent. Two lines of symmetry are equivalent if a symmetry of the pattern can move one reflection line to the other. For example, if you draw one vertical reflection line on the pattern above and then trace the figure and draw another vertical reflection line on the top sheet, these two lines of symmetry are equivalent if you can move the figure so that when the one mirror line is placed on top of the other, the figure fits on top of itself.

C. What kind(s) of symmetry does each strip pattern have? Think before reading on. . . .

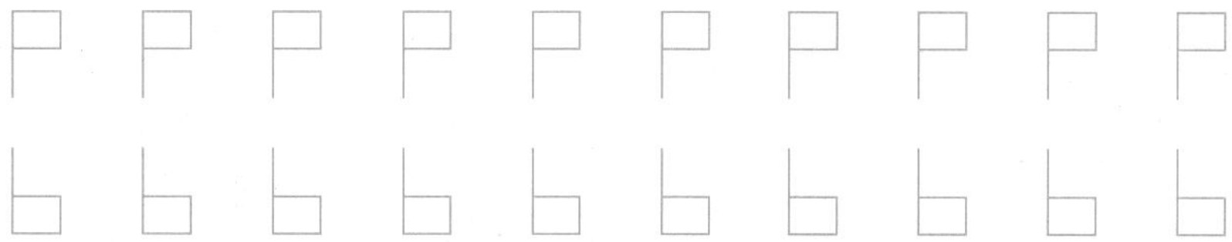

D.

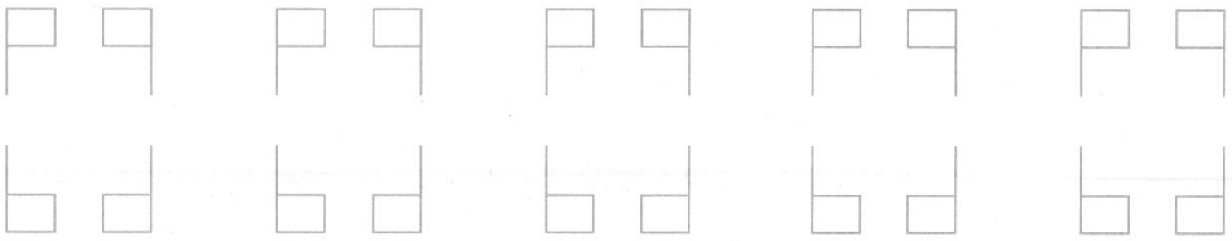

E.

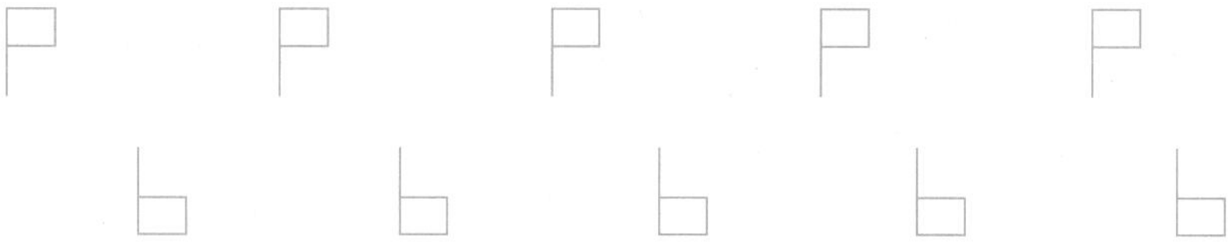

F.

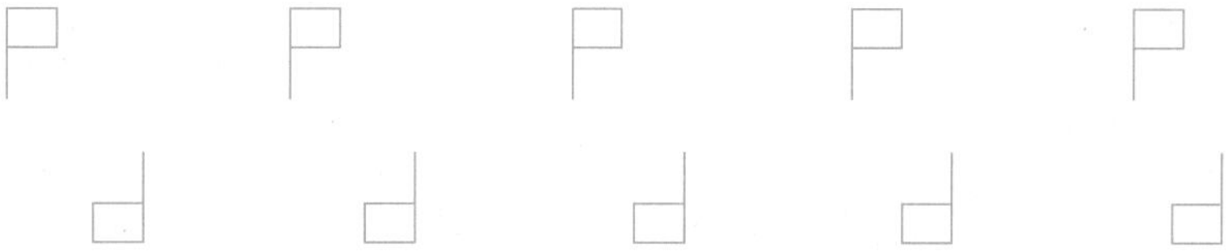

G.

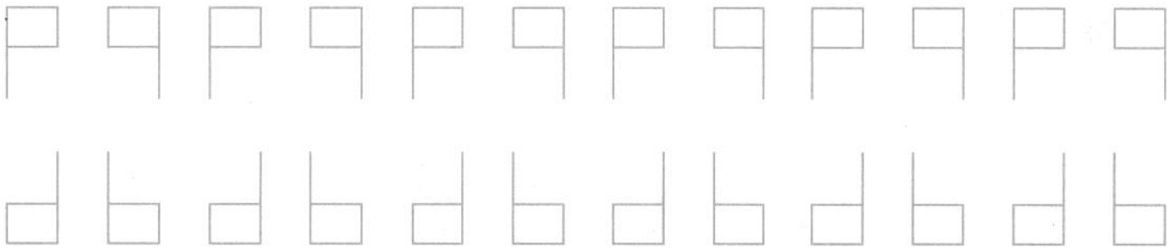

DISCUSSION

C. This pattern has translation symmetry. This pattern has horizontal line symmetry. If you trace this pattern and reflect the pattern across a horizontal line, it will fit on top of itself.

D. This pattern has translation symmetry. This pattern has both vertical and horizontal line symmetry. This pattern has 180-degree rotation symmetry.

E. This pattern has translation symmetry. This pattern has glide symmetry. If you reflect the pattern across a horizontal line and then move the pattern in the horizontal direction, it will fit on top of itself.

F. This pattern has translation symmetry. This pattern has $\frac{1}{2}$-turn symmetry or 180-degree rotation symmetry or point symmetry.

G. This pattern has translation symmetry. This pattern has vertical line symmetry and glide symmetry. If you reflect the pattern across a vertical line, it will fit on top of itself. If you reflect the pattern across a horizontal line and then move the pattern in the horizontal direction, it will fit on top of itself.

You can look on Appendix B to see the actual lines of reflection, centers of rotation, and glide vectors.

THE SEVEN SYMMETRY TYPES OF STRIP PATTERNS

I find it amazing that any strip or border pattern will fit into one of the seven symmetry types illustrated in A through G above. That is, just as the four congruence transformations are complete in that no more are needed to describe how to map any two congruent figures in a plane onto each other, these seven symmetry types are all that are needed to describe any strip pattern.

Let us use a table and introduce some notation to increase the number of readers who understand this. First, you will notice that every one of the strip patterns had translation symmetry. Therefore, we are technically looking at that subset of strip patterns that have translation symmetry. Actually, you will be hard pressed to find a quilt or building or other strip that does not have translation symmetry. As Ian Stewart put it, "Something in the human mind is attracted to symmetry."[3]

A few words about notation in the far right column of Table 10.3: We will use the letter m to indicate a line symmetry (m representing mirror), the letter g to represent a glide symmetry, the number 2 to indicate 2-turn symmetry, and the letter l to indicate a lack of line symmetry.

TABLE 10.3

	TYPE OF SYMMETRY				
Pattern	**Vertical line**	**Horizontal line**	**$\frac{1}{2}$-turn**	**Glide**	**Notation to describe**
A					ll
B	Yes				ml
C		Yes			lm
D	Yes	Yes	Yes		mm
E				Yes	lg
F			Yes		l2
G	Yes		Yes	Yes	mg

Some active readers look at this table and argue that there should be more than seven types of symmetry. For example, what about gm? What about m2? These are wonderful questions.

WALLPAPER PATTERNS

Let us examine another kind of planar pattern called wallpaper patterns, ones in which the motif (unit of repetition) can be repeated in two directions. It is important to note that the term *wallpaper pattern* refers to any pattern that can fill the plane if extended. Thus the three patterns that follow, which are often seen with bricks, are called wallpaper patterns by mathematicians.

Extending patterns in two dimensions leads to more complexity; consider, for example, all the wallpaper patterns that have been created—thousands, perhaps millions of different ones. However, when we classify them by the kinds of symmetry they possess, any wallpaper pattern falls into one of exactly 17 categories. I find that fascinating! It is a relatively recent discovery, though this issue puzzled mathematicians and scientists for centuries. Interestingly enough, the same 17 categories can be used to classify all crystals. Examining each of these 17 categories is beyond the scope of this book, but the interested reader can search for wallpaper patterns on the Web. You will find many descriptions.

INVESTIGATION 10.2g Analyzing Brick Patterns

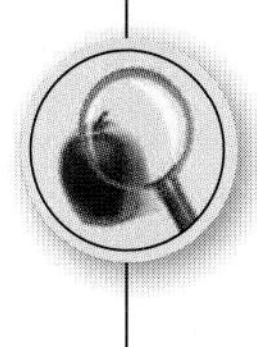

A. What kinds of symmetries does this simple brick pattern have? Think before reading on. . . .

DISCUSSION

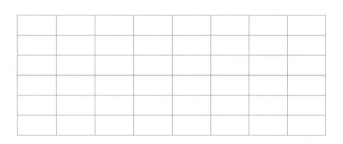

It might not surprise you to find that this figure has lots of symmetry!

It has translation symmetry. In fact, the figure can be translated horizontally, vertically, diagonally down to the right, and diagonally up to the right. Thus this figure has four different lines of translation symmetry.

This figure has two vertical lines of symmetry. *Note:* Technically, you could say it has infinitely many vertical lines of symmetry. Recalling the discussion of inequivalence in the previous investigation, we say that this pattern has two inequivalent vertical lines of symmetry.

Similarly, the figure has two inequivalent lines of horizontal symmetry.

The pattern has $\frac{1}{2}$-turn rotation symmetry, and there are three inequivalent centers of rotation. That is, there are three different centers of rotation that cannot be placed on top of one another by any symmetry of the pattern. See Appendix B to see the three centers of rotation. (The reader who finds the notion of inequivalent reflection lines and centers of rotation difficult to understand completely should not be discouraged: These are relatively advanced ideas.)

Finally, when looking for glide symmetry, we find that there are no glide lines that are not already mirror lines. Thus we say that this figure has no nontrivial glide symmetry.

B. What about this slightly more complex brick pattern? What symmetries does it have?

DISCUSSION

This pattern has translation symmetry. Like the previous pattern, this one can be translated horizontally, vertically, diagonally down to the right, and diagonally up to the right. Thus this figure has four different lines of translational symmetry. However, if you draw the translation vectors, you will find that they are not identical to the four in part A. You can check your thinking on this page.

This figure has one line of vertical symmetry.

The figure also has one line of horizontal symmetry.

The pattern has $\frac{1}{2}$-turn rotation symmetry, and there are three inequivalent centers of rotation. That is, there are three different kinds of points on which you can place your pen, rotate the figure 180 degrees, and have the pattern fit on top of itself.

Finally, this figure has one nontrivial glide symmetry. Did you find it? See Appendix B to check your thinking.

C. Finally, what about this more complex brick pattern? What symmetries does it have?

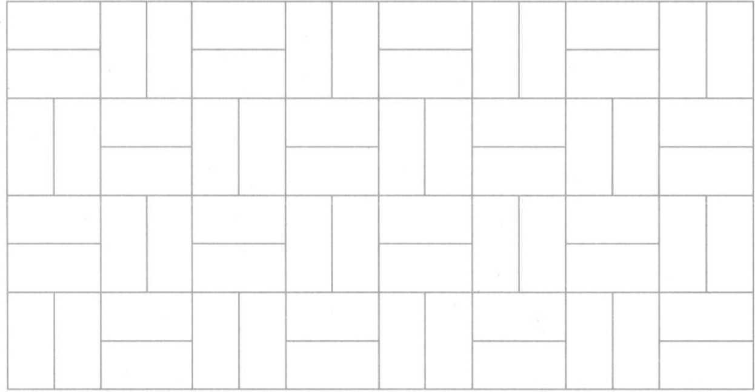

DISCUSSION

This pattern has translation symmetry. Like the two preceding patterns, this pattern can be translated diagonally down to the right and diagonally up to the right, although the four translation vectors are not identical to either of the four above.

This figure has one line of vertical symmetry.

Similarly, the figure has one line of horizontal symmetry.

The pattern has $\frac{1}{2}$-turn rotation symmetry, and there are three inequivalent centers of rotation.

Finally, this figure has two inequivalent glide lines. Can you specify the reflection and slide lines? See Appendix B to check your thinking.

SYMMETRY BREAKING

Our discussion of symmetry would be incomplete if we did not also talk about symmetry breaking. Although there is indeed "something in the human mind that is attracted to symmetry" and symmetry is one important factor in most people's sense of beauty, at the same time, perfect symmetry is repetitive and predictable. I will not attempt a rigorous definition of symmetry breaking but, rather, will give some examples and descriptions. From one perspective, symmetry breaking occurs when we take a pattern that has "lots" of symmetry and do something to reduce the symmetry. Consider Figure 10.31(a). This figure has four lines of symmetry and fourfold symmetry. If we add two colors [Figure 10.31(b)], we still have fourfold symmetry, but now we have no lines of symmetry. Thus we have broken the symmetry. One basic quilt design can be modified in many ways that break the symmetry. If we color in the center [Figure 10.31(c)], we now have only twofold symmetry and no lines of symmetry. Finally, we can redraw the lines in the middle to give a pinwheel effect. This design now has $\frac{1}{4}$-turn symmetry [Figure 10.31(d)].

OUTSIDE THE CLASSROOM

The notion that symmetry is a mathematical concept as well as an aesthetic one was brought home to me several years ago. I was asked to facilitate a four-day workshop for college professors who were invited to Dartmouth College to get a taste of several different courses designed to bridge mathematics and other disciplines. In my workshop, there were 24 college professors: about half were Ph.D. mathematicians and half were art professors. My job was to keep them from killing each other. Seriously, part of my job was to help them recognize what they had in common. One of the many things they had in common was a love of patterns and a love of breaking patterns.

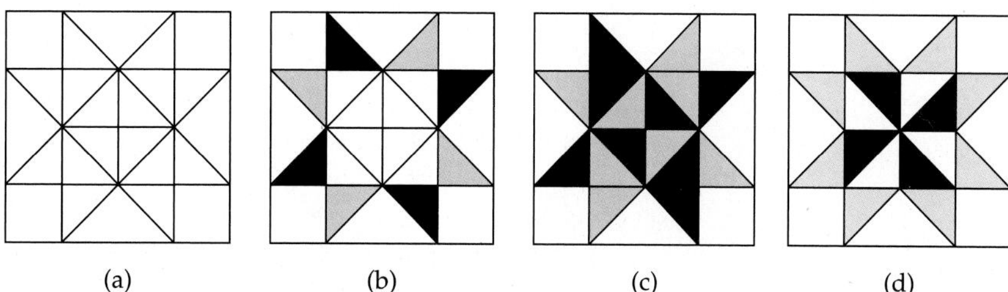

Figure 10.31

These figures illustrate another way of describing symmetry breaking: A symmetry is expected, but that expectation is not met. If you look closely at Oriental carpets, which often have very intricate designs and plenty of symmetry, you will see that there is a playfulness with symmetry that results in many intriguing patterns within the larger, more obvious patterns. "In art, too, it seems that the approximation of symmetry, rather than its precision, teases the mind as it pleases the eye."[4]

In nature, symmetry is always imperfect, just as you will find in the next chapter that measurements, by definition, are always approximations. However, just as a mathematical model of a problem will nicely represent the important or "big" ideas of a problem or phenomenon, while at the same time idealizing or simplifying other parts, so too do physicists, chemists, biologists, and other scientists often speak in terms of symmetries in their work as though the symmetries were perfect. Did you ever wonder why our faces are not perfectly symmetric but rather "almost" symmetric? Figure 10.32(a) is a photograph of Edgar Allen Poe. Figure 10.32(b) shows what he would look like if you reflected his right side, and Figure 10.32(c) shows what he would look like if you reflected his left side.

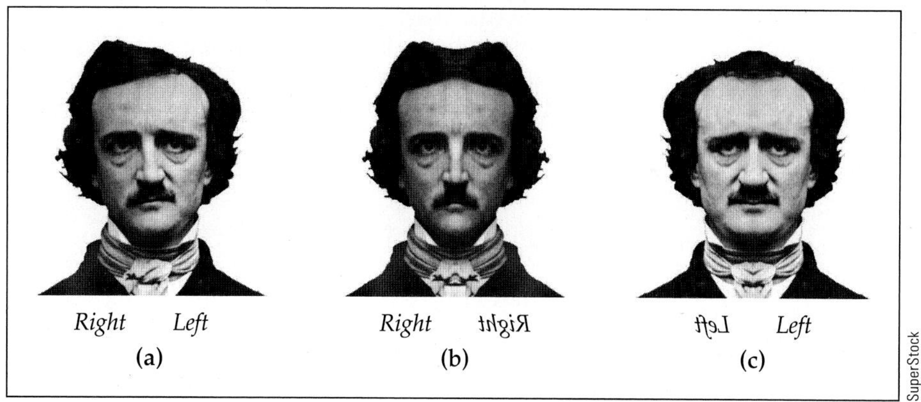

Figure 10.32

SYMMETRY FOR THREE-DIMENSIONAL OBJECTS

Thus far we have examined symmetry with two-dimensional figures. However, we see symmetry in many three-dimensional figures. Buildings often have symmetry; think of the Taj Mahal or the U.S. Capitol building. Everyday objects such as forks, nuts and bolts, and windows have symmetry. Many natural forms, including starfish, honeycombs, pyrite, and crystals, have symmetry.

Let us briefly examine symmetry of three-dimensional objects. We will focus on reflection and rotation symmetry.

When we defined reflection symmetry with two-dimensional figures, we used a line. Therefore, it should make sense that when we define reflection symmetry with three-dimensional figures, we will use a plane. Because we have added one more dimension to the figure (two to three), we add one more dimension to the instrument of reflection (one to two).

If there is a plane that can be drawn through a three-dimensional figure so that one half of the figure is a mirror image of the other, then the figure has reflection symmetry. The plane is called the **plane of symmetry**.

Similarly, when we defined rotation symmetry with two-dimensional figures, we used a point (a one-dimensional object). We define rotation symmetry with three-dimensional figures using a line (a two-dimensional object).

If there is a line that can be drawn, around which a three-dimensional figure can be rotated so that it coincides with itself, then the figure has rotation symmetry. The line about which the figure is rotated is called the **axis of symmetry**.

Figure 10.33 shows two three-dimensional objects that have reflection and/or rotation symmetry. In each case, determine and try to describe the plane(s) of symmetry and the axis (or axes) of symmetry. Then read on. . . .

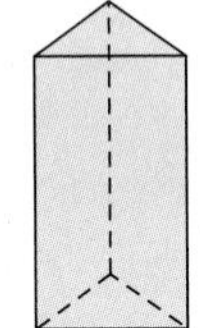

Right isosceles triangle prism

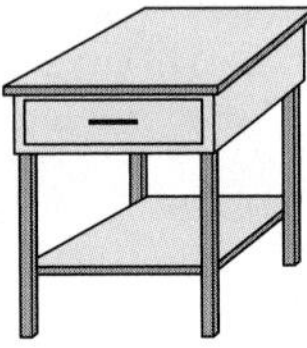

Nightstand

Figure 10.33

The prism has 180-degree rotation symmetry; the axis of symmetry is a line that is parallel to and midway between these bases and which bisects the vertex of the right dihedral angle. The prism has two planes of symmetry: a horizontal plane parallel to and midway between the bases and a vertical plane that contains the lines of symmetry of the two bases. The nightstand has one plane of symmetry, a vertical plane that is parallel to the left and right sides.

I have talked with scientists about the usefulness of symmetry in their fields. A chemist told me that molecules of similar symmetry give similar spectra and that advances in our mathematical understanding of symmetry have helped organic chemists to understand the structure of organic compounds. At a molecular level, the atomic structure of various elements and compounds determines how they will be packed together. As Marjorie Senechal has written, "if all we learn about symmetry is to identify it, we miss the whole point. Symmetry is an effect, not a cause."[5]

TESSELLATIONS

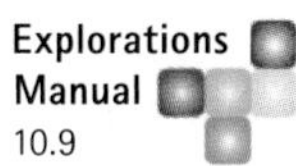
Explorations Manual 10.9

For thousands of years, people have chosen to cover surfaces (for example, floors, walls, tables, beds, or roads) with patterns. We will use transformational geometry, including our knowledge of symmetries of patterns, to examine one kind of covering called tessellations. Look at the pictures in Figure 10.34, all of which are examples of tessellations. From these pictures, what do you think *tessellation* means? That is, a pattern that tessellates is one in which . . .

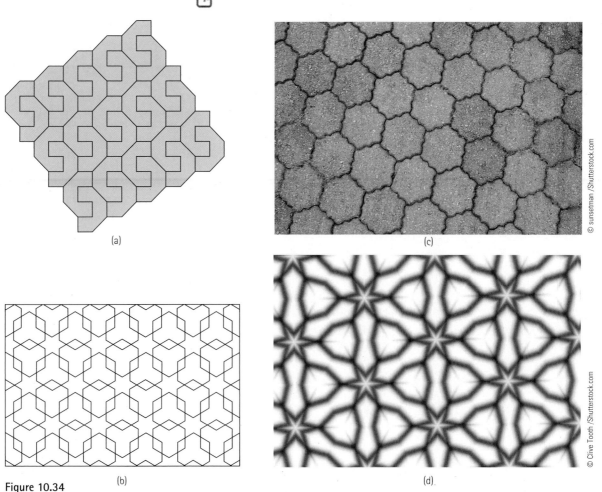

(a)

(c)

© sunsetman /Shutterstock.com

(b)

(d)

© Clive Tooth /Shutterstock.com

Figure 10.34

LANGUAGE

The word *tessellate* comes from the Latin word *tessella*, which refers to small tiles that the Romans used for pavement and to decorate buildings.

MATHEMATICS

Where else have you seen tessellations in this book? Figures 10.1 and 10.2 are tessellations, and the three brick designs in Investigation 10.2g are tessellations.

A figure or a combination of figures **tessellates** the plane if a regular repetition of the figure or figures covers the plane so that there are no gaps and no overlapping of figures. Look at the four designs in Figure 10.34. Do you see that all of these designs qualify as tessellations?

One difficulty that people commonly encounter when first making sense of this definition has to do with the borders. When discussing tessellations, we will not be concerned with whether the borders are smooth or straight. For example, if we were covering a floor with square tiles, unless both the length and the width of the floor were equal to a multiple of the length of the tile, the person laying the floor would have to cut the square to fit the edge of the room. Thus, when we say that a pattern tessellates, we are talking not about the edges of the design, but about the design itself.

We will investigate some basic ideas related to two-dimensional tessellations here. Exploration 10.9 carries these ideas further.

INVESTIGATION 10.2h Which Triangles Tessellate?

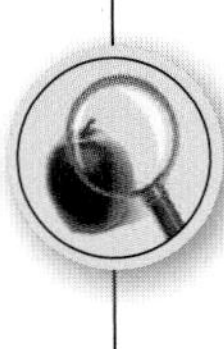

One starting point for investigating tessellations is to ask which figures will tessellate. We will begin with the triangle (Figure 10.35). What triangles tessellate? Just a reminder that you will learn more if you really do think before reading on. . . .

Note your first thoughts as predictions. Then make copies of some triangles and test your ideas. One simple way to do so is to fold a piece of paper in half, then in half again, and then in half again. Now if you make a triangle and then cut the figure, you will have 8 copies of the triangle.

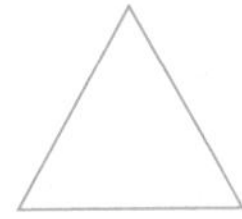
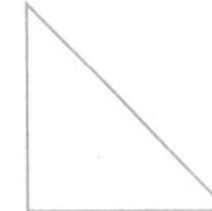

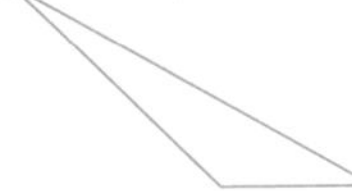

Figure 10.35

DISCUSSION

Many students are surprised to find that all triangles tessellate! Did you predict this? If not, make some copies of different triangles and confirm this fact for yourself. Then come back to the text. Now that you realize that all triangles tessellate, can you justify this—that is, explain why? Work on this question before reading on. . . .

Let us begin with a scalene triangle, shown in Figure 10.36(a). I have labeled the three angles of the triangle. If we rotate the triangle 180 degrees (or reflect it vertically and then horizontally), we can join the triangle and its image together (Figure 10.36[b]). The combining of the triangles creates a parallelogram. If we take this parallelogram and translate it, we now have four triangles (Figure 10.36[c]). Figure 10.37 shows that we can extend this pattern in all directions infinitely.

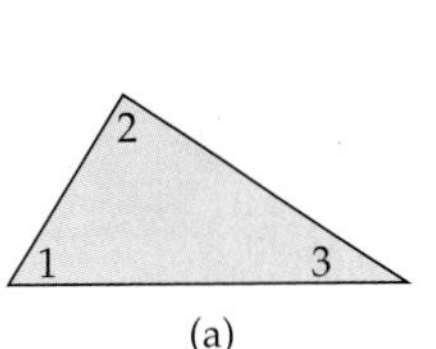

(a)

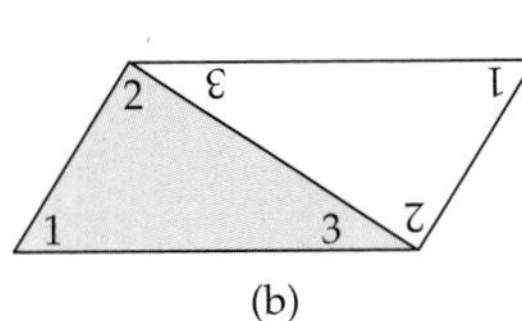

(b)

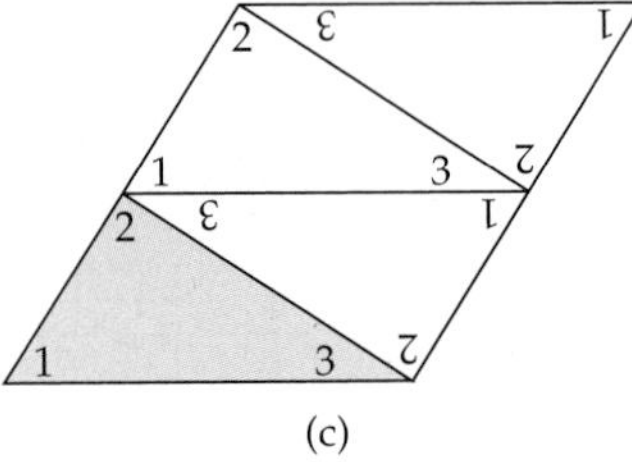

(c)

Figure 10.36

Our eyes tell us that it "looks as if" the triangle tessellates. We will now take some time to confirm what our eyes tell us. Again, this is not a formal geometry course, so we will give not a two-column proof but a more informal justification. Figure 10.38 repeats Figure 10.36(c), but with some vertices labeled to make communication easier. We can see that we have angles $\measuredangle 1$, $\measuredangle 3$, and $\measuredangle 2$ meeting at vertex *B*.

We know from Chapter 8 that the sum of the angles of any triangle is 180 degrees, and therefore $\measuredangle ABC = 180°$. However, this means that $\overleftrightarrow{AC}$ is a straight line. Looking back at Figure 10.37, we can see that there will be six angles at *any* vertex ($\measuredangle 1$, $\measuredangle 2$, $\measuredangle 3$, $\measuredangle 1$, $\measuredangle 2$, and $\measuredangle 3$), and the sum of these six angles is 360°. As you might be thinking, 360 is an important number; that is, one complete revolution about a point is 360 degrees. Thus we can repeat this triangle in such a way that there are no gaps or overlap, and so the triangle tessellates.

CLASSROOM CONNECTION

Recall the discussion of the van Hiele levels in Chapter 8. Where is your level of understanding of the justification that all triangles tessellate?

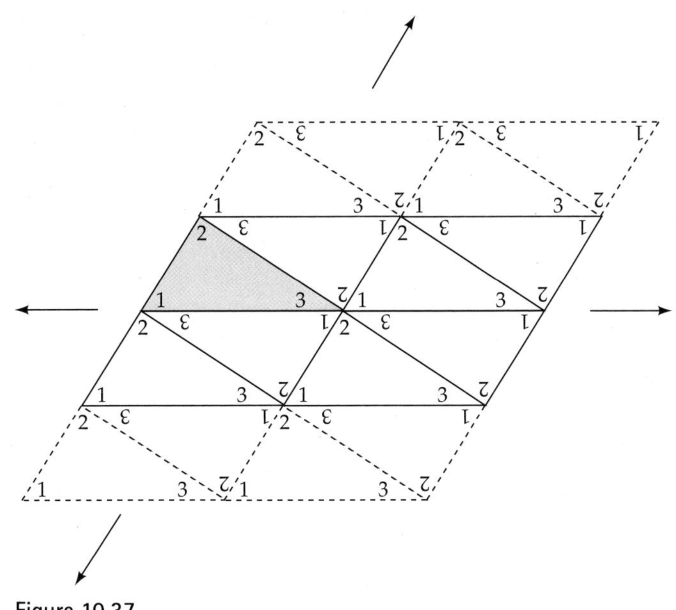

Figure 10.37

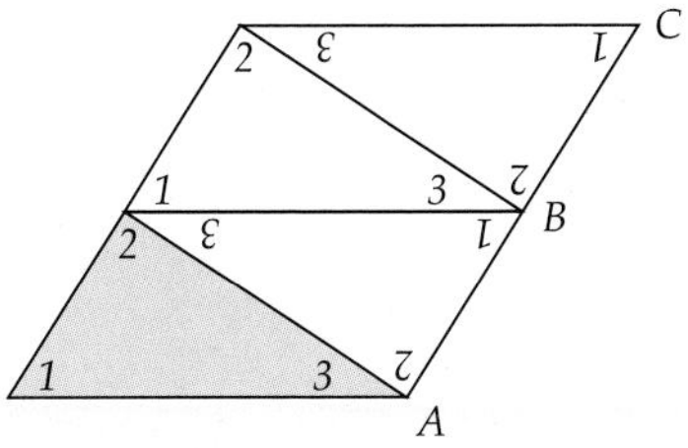

Figure 10.38

In a similar fashion, we can show that any quadrilateral also tessellates the plane. Does this surprise you? If so, make a number of different quadrilaterals to convince yourself that, indeed, *any* quadrilateral does tessellate. Justification of this will be left as an exercise.

INVESTIGATION 10.2i Which Regular Polygons Tessellate?

Another common starting point for understanding tessellations is to examine that subset of all polygons called regular polygons—that is, polygons in which all sides are the same length and all angles have the same measure. Which regular polygons will tessellate? Work on this before reading on. You might want to use Figure 10.39 as a template to make copies, or you might choose to apply your understanding of the properties of regular polygons. In either case, do some thinking on your own before reading on. . . .

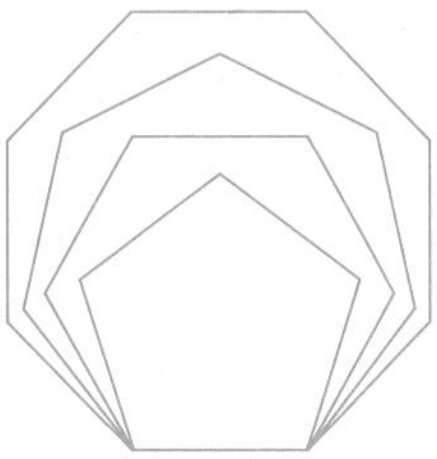

Figure 10.39

DISCUSSION

It turns out that only three regular polygons tessellate: the equilateral triangle, the square, and the regular hexagon. (Note that an equilateral triangle is a regular triangle and that a square is a regular quadrilateral.) The reason has to do with the sum of the angles.

Whether or not you concluded that these three are the only regular polygons that tessellate, take a few minutes to see whether you can justify this conclusion—that is, explain why it is true. Then read on. . . .

If you recall from Chapter 8, the sum of the angles of any polygon is $(n - 2)180$, where n is the number of sides in the polygon. From this formula, we can find the measure of each interior angle of regular polygons, shown in Table 10.4. Can you see how these numbers help us to understand why only three regular polygons tessellate? Think before reading on. . . .

TABLE 10.4

Figure	Sum of angles (degrees)	Measure of each interior angle (degrees)
Equilateral triangle	180	60
Square	360	90
Regular pentagon	540	108
Regular hexagon	720	120
Regular octagon	1080	135

Table 10.5 helps us to understand why the equilateral triangle, square, and regular hexagon tessellate, but no other regular polygons will. If we place three pentagons at a common vertex, the sum of those angles is 324. This is not quite 360; however, there is not enough room to place a fourth pentagon at that vertex (Figure 10.40). Similarly, if we place two octagons at a common vertex, the sum of the two angles is 270, and there is not enough room to place a third octagon at that vertex.

TABLE 10.5

Figure	Number of angles at common vertex	Sum of those angles (degrees)
Equilateral triangle	6	360
Square	4	360
Regular pentagon	3	324
Regular pentagon	4	432
Regular hexagon	3	360
Regular octagon	2	270
Regular octagon	3	405

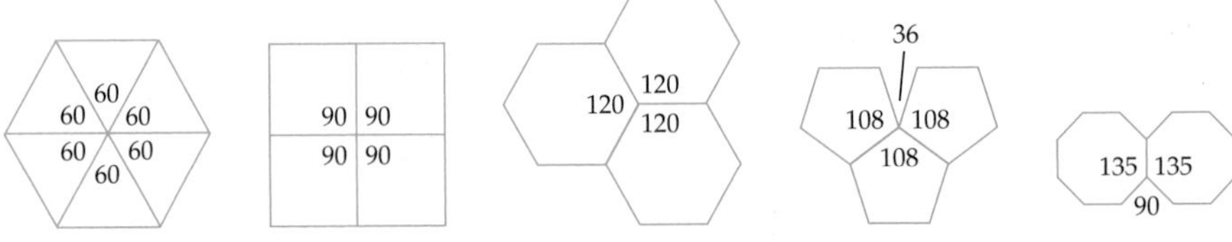

Figure 10.40

CLASSROOM CONNECTION

In *A Cloak for the Dreamer,* three sons make quilts for the Archduke. The first two sons use rectangles that tessellate. The third son makes circles that don't tessellate, so there are holes in the coat. They salvage his work by turning the circles into hexagons, which do tessellate. How could you cut a hexagon out of a circle?

Figure 10.41 shows what these three tessellations look like when multiple copies are joined together. These will look familiar to you from common ceiling or floor patterns, and the hexagon tessellation is constructed by honeybees. We call these three tessellations *regular tessellations,* which are defined as tessellations composed of congruent regular polygons.

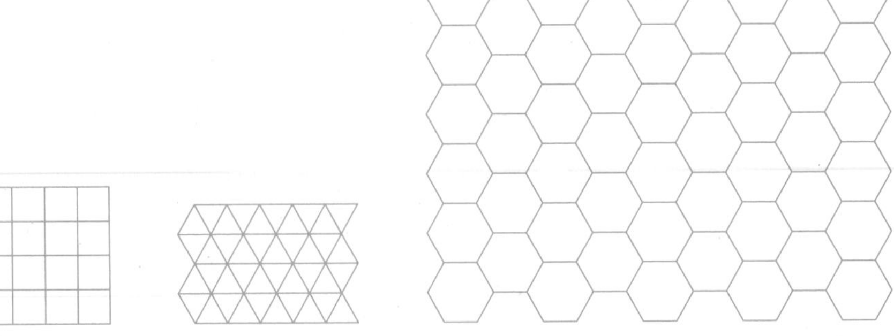
Figure 10.41

If you are playing the "what-if" game in your mind, you may have realized that 270 + 90 = 360 and conjectured that a combination of octagons and squares *will* tessellate. This is true, as Figure 10.42 illustrates. There are many combinations of regular polygons that will tessellate. Mathematicians have names for certain subsets. One that we will consider here is the **semiregular tessellation**—a tessellation of two or more regular polygons that are arranged so that the same polygons appear in the same order around each vertex point. The tessellation at the left in Figure 10.42 is a semiregular tessellation because the two figures are a square and a regular octagon. The tessellation at the right in Figure 10.42 is not a semiregular tessellation because the two figures are a square and a nonregular octagon. As you can see, by varying the lengths of the sides of the octagons (that is, the ratio of the noncongruent sides), we can change the appearance of the tessellation.

OUTSIDE THE CLASSROOM

This semiregular tessellation is a common kitchen floor design.

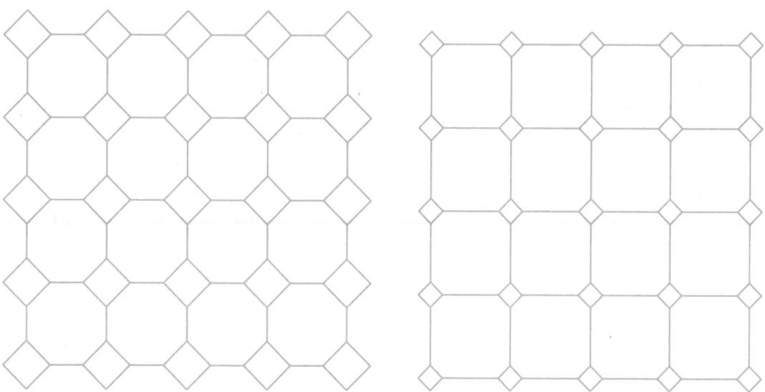

Figure 10.42

There are only eight possible semiregular tessellations. There is another whole class of tessellations made with regular polygons called demiregular tessellations. Here, however, we will move from making tessellations with regular polygons to making tessellations with nonregular polygons. This is where the artists and designers come in!

INVESTIGATION 10.2j Tessellating Trapezoids

In this investigation, you will deepen your knowledge about why certain figures tessellate. First, though, cover Figure 10.43 before reading on. This will better enable you to own the knowledge you will construct here. Now, recall the "brick" rectangles in Investigation 10.2g and our discussions about multiple attributes in Chapter 8. One reason why the brick rectangle has multiple tessellation patterns is that the length of the rectangle is exactly twice the width. Using a protractor and a ruler, measure the trapezoid to the left and describe its attributes. Then read on. . . .

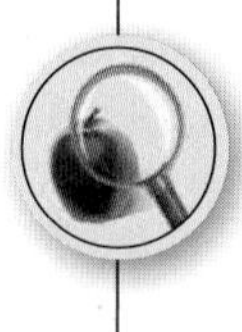

With respect to angles: the trapezoid has two right angles, one acute angle, and one obtuse angle.

The measures of the acute and obtuse angles are 45 degrees and 135 degrees, respectively.

The right angles are adjacent (as opposed to opposite).

With respect to sides: the trapezoid's left side and top base are congruent, and the bottom base is twice the length of the top base.

Now make about 12 or more copies of this trapezoid. You can use Geoboard Dot Paper, graph paper, or blank paper and a ruler. How many different ways can you get this trapezoid to tessellate? In other words, how many different tessellation patterns can you make?

DISCUSSION

The specific attributes of this trapezoid open the possibility of multiple tessellation patterns Figure 10.43 shows four possibilities.

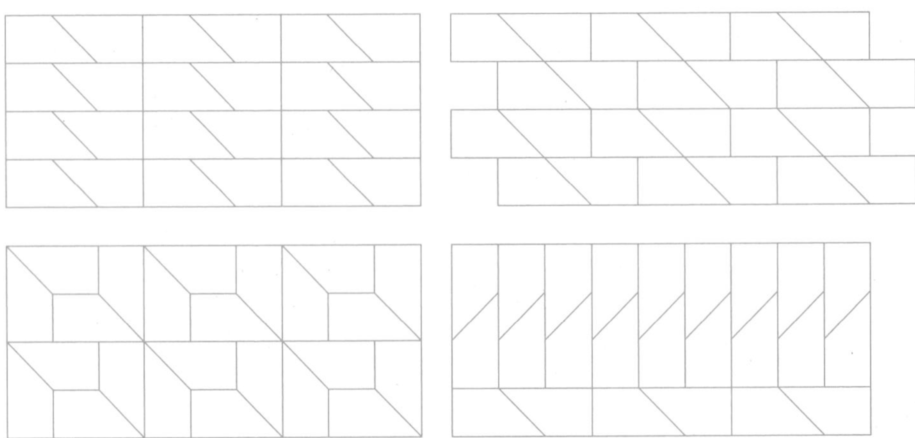

Figure 10.43

Once we move out of regular polygons, verifying that a pattern tessellates becomes more complicated. One way to make the task easier is to make sure that the sum of the angles at any point in the pattern is 360 degrees. To make *this* task easier, we use the term **vertex point**, which we will define as any point on the figure where polygons share a common vertex. Mathematicians have proved that all vertex points in regular and semiregular polygons are identical. In those cases, we need only verify that the sum of the angles of one vertex point is 360 degrees to prove that the figure tessellates. Look at the figures below and determine how many different vertex points there are in each figure. Determine the sum of the angles at those points. . . .

The figures below show the vertex points for two of the figures. In the figure at the left, you can see that A and B are equivalent vertex points. The sum of the angles at these vertices is $45 + 135 + 180 = 360$ degrees. The sum of the angles at vertex C is easily seen to be 360 because we have four right angles. In the figure at the right, there are two different vertex points. The sum of the angles at point A is $45 + 135 + 45 + 135 =$ 360 degrees, and the sum of the angles at point B is $90 + 90 + 180 = 360$ degrees.

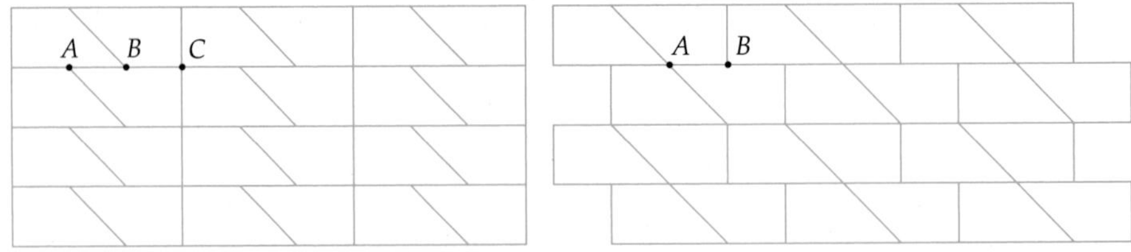

Now, we revisit this notion of attributes. *Any* trapezoid with two adjacent right angles can be tessellated in the fashion shown by the first two figures below. Do you see why? Since the sum of the angles of a trapezoid is 360 degrees and two of the angles are 90 degrees, the other two angles must be supplementary. This knowledge, coupled with a straight angle measuring 180 degrees, enables us to verify that the sum of the angles at any vertex point is 360 degrees.

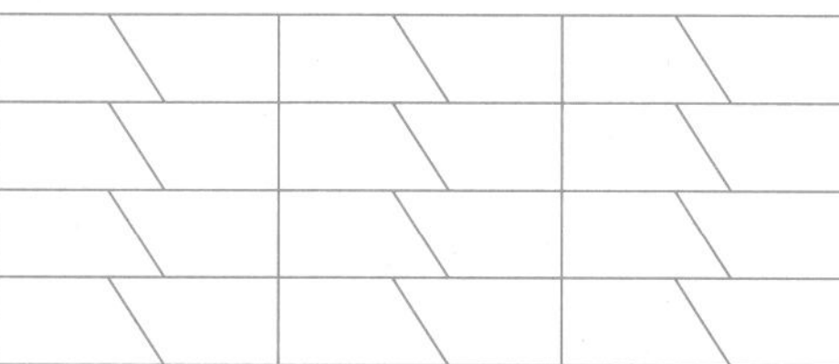

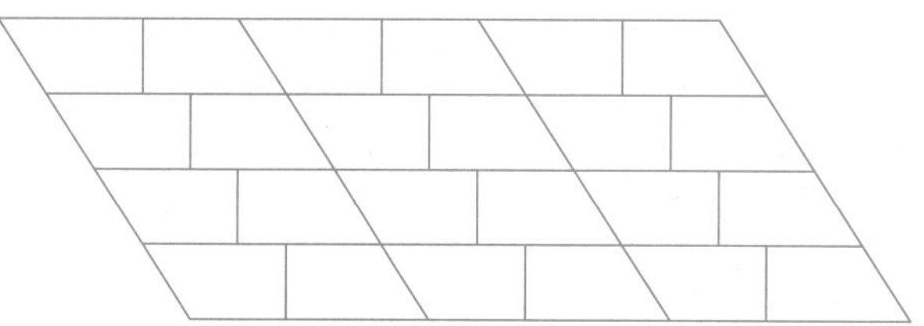

However, the other two designs made with the trapezoid at the beginning of this investigation will not work for every trapezoid, as the figures below illustrate.

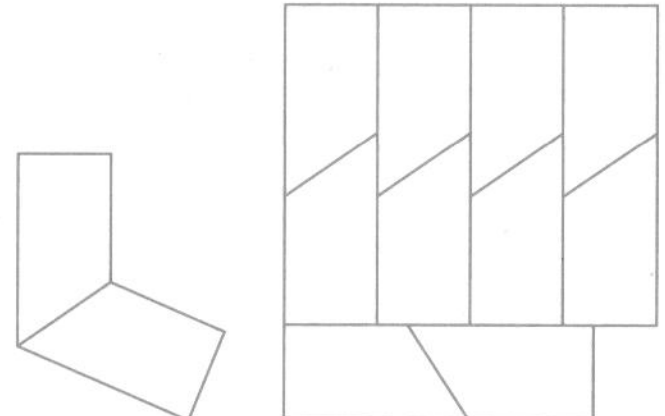

INVESTIGATION 10.2k More Tessellating Polygons

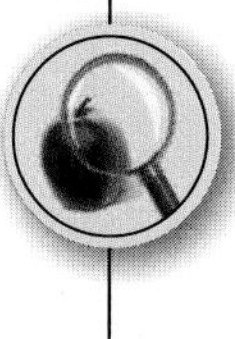

What about other shapes? Although the regular pentagon does not tessellate, many pentagons do. Similarly, many arrows tessellate. Can you figure out why the pentagon and the arrow below tessellate? What attributes do they have? Think before reading on. . . .

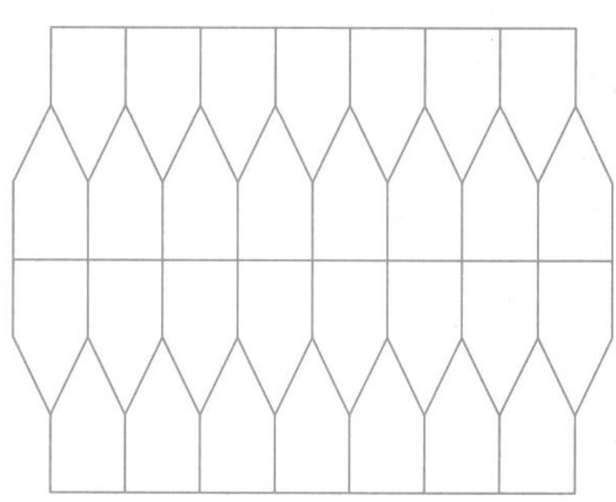

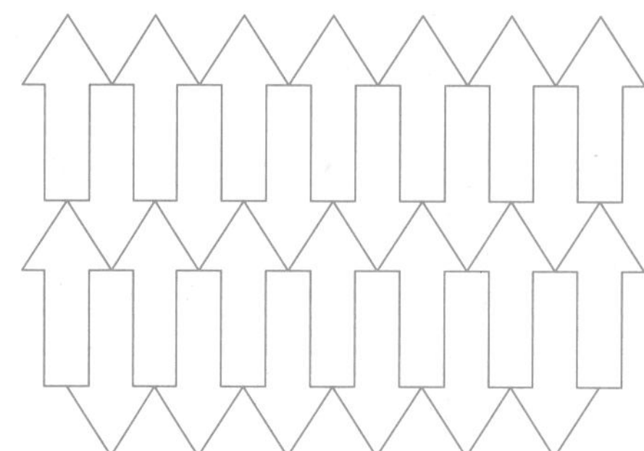

DISCUSSION

Some readers think that I should have told the angles, but in this case, other attributes of the figures—symmetry and perpendicularity—eliminate the need to know specific angles.

For the pentagons, there are two distinct vertex points. Make sure that you see them. If you are not sure, check another vertex point that you think might not be equivalent and examine its attributes—what angles and sides meet at that point.

The first vertex point is simply the meeting of four right angles, so it is easy to verify that the sum of the angles at that vertex point is 360 degrees. For the other vertex point, we can use our knowledge of geometry! First, we can break the pentagon into a square and a triangle (Figure 10.44). Because the pentagon is symmetric, we know that the triangle is isosceles, which means the base angles are congruent. Thus, we can use algebra and call the angles of the triangle a, a, and b.

The sum of the angle at this vertex point is

$$90 + 90 + a + b + a$$

However, we know that $a + b + a = 180$ since the sum of the angles of any triangle is 180 degrees. Therefore, the sum of the angles of this and every vertex point is 360 degrees.

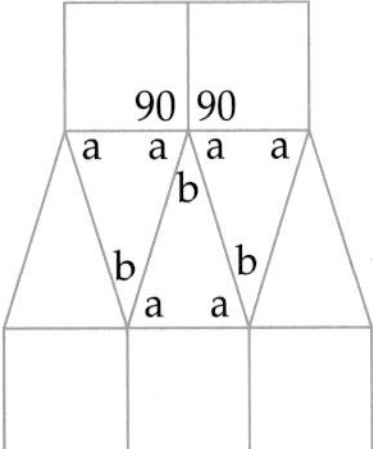

Figure 10.44

Because the arrows can also be broken apart into rectangles and isosceles triangles, the verification that it tessellates is similar and will be left as an exercise.

I need to sound a cautionary note about making tessellating figures. The shape below has some nice rotational symmetry and all its angles are 90 degrees and multiples of 90 degrees. At first glance (see the figure at the right below), it appears that it will tessellate.

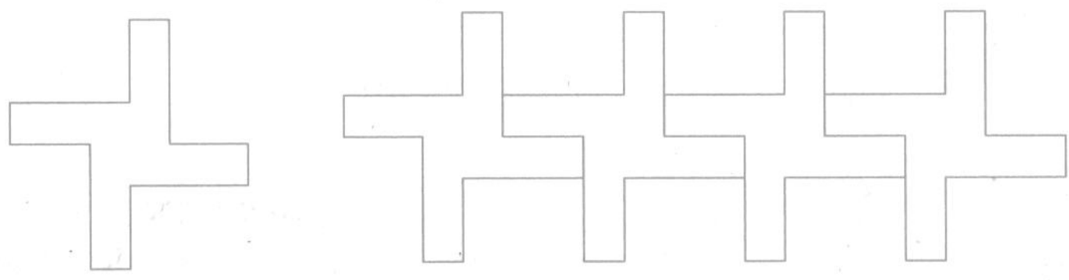

However, when we try to make a second row, to continue the tessellation, we find that there is simply no way to place copies of that figure—by translating, rotating, or reflecting—so that there will be no gaps, as the figure at the left below shows. However, if we continue with this idea, we do make a *pattern* that does tessellate; in this case, the unit of the tessellation is the original figure and a square.

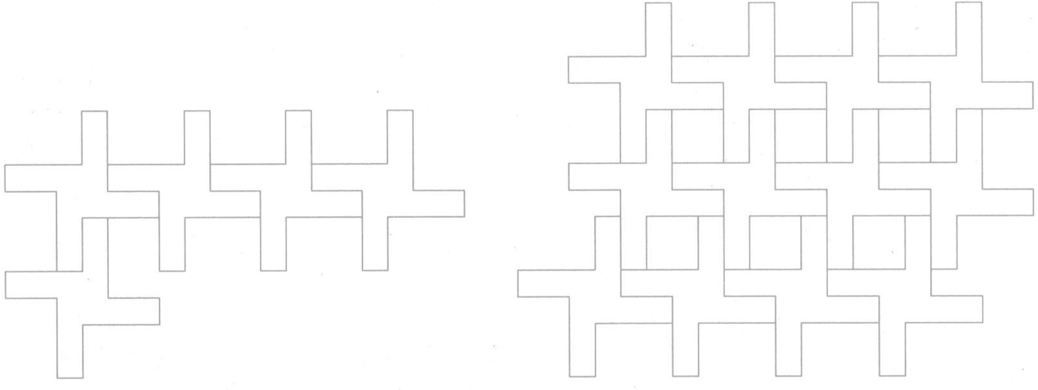

Thus far, we have restricted our investigation of tessellations to polygons. However, many nonpolygonal figures will tessellate. Recall our discussion of Escher and Figure 10.19 with the tessellating swans. M. C. Escher popularized tessellations that move from mathematics into art. How did he do it? Making interesting tessellation figures is not as hard as you think. One kind is shown here. Begin with a square (although you could begin with any parallelogram). Make a shape inside the square (it need not be a polygon) that has one side on the edge of the square. Translate this shape to the opposite side. You can do this more than once. The example below connects to jigsaw puzzles, in that I have cut out a circular shape from two sides and then translated that side.

LANGUAGE

Here is where the language of Chapter 8 makes communication easier. When I said, "you could begin with any parallelogram," this tells you, because you know that rectangles and rhombuses are also parallelograms, that I am including them too.

Exploration 10.9 offers many interesting ways to extend your understanding of tessellations.

INVESTIGATION 10.2I Generating Pictures Through Transformations

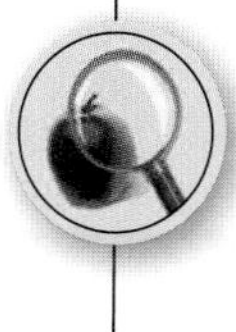

Let us apply our understanding of translations, reflections, rotations, and glide reflections to see how a tessellation figure is composed. Figure 10.45 shows one of my favorite tessellation figures, one that appears in many Islamic designs. Let's say this was a jigsaw puzzle and all the pieces were congruent. How would you put the puzzle together? I have numbered some of the pieces to make communication simpler. Make your own plan before reading on. . . .

MATHEMATICS

Recall our discussion of unit in Chapter 4 and Chapter 9. One of the goals of this course is for you to see that unit is not just "inch, foot, ounce, etc." What do you think *unit* means?

DISCUSSION

We have learned that if we have two congruent figures in the same plane, they can be mapped onto each other by some combination of translation, reflection, rotation, and glide reflection. In this pattern, each of the figures has line symmetry. As we look for patterns within this tessellation, we notice that each figure has many reflection neighbors and many rotation neighbors. For example, figures 1 and 7 are reflection images of each other; so are figures 2 and 8, figures 3 and 9, and figures 4 and 5, to name but a few. Similarly, the members of these pairs are also rotation images of each other. If we consider figures 2, 4, 5, and 8 as a unit, then we can generate the whole pattern by translating this unit in a diagonal direction. If you turn the paper 45 degrees, it is much easier to see that these four figures can easily generate the whole pattern!

Figure 10.45

Source: From Introduction to Tessellations by Dale Seymour and Jill Britton. Copyright © 1989 Pearson Education, Inc., or its affiliates. Used by permission. All Rights Reserved.

SUMMARY 10.2

We have found that each of the transformations from Section 10.1 leads to a type of symmetry—translation, reflection, rotation, or glide reflection. One of the more important goals is for you to see symmetry as a movement—as a verb, rather than a noun. When we see symmetry in an object, we do so by moving that object or parts of that object in our minds. We have seen that symmetry has many applications in the world of art (quilts, mosaics), in everyday life (wallpaper and tiles), and in science (symmetry at the atomic level). We have seen that symmetry helps us to understand better how and why tessellations work. Finally, we have introduced the notion of three-dimensional symmetry, which has many important applications in everyday life and in science.

10.2 Exercises

1. Describe the rotation and reflection symmetries in the figures below.

a.
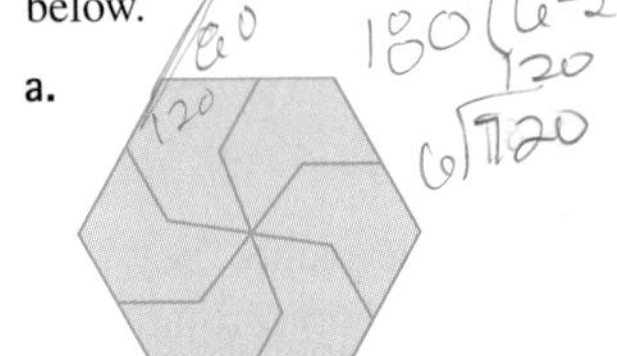

b.

c.

d.

2. Describe the rotation and reflection symmetries in the figures below.

a.

Maple leaf
Canada

b.

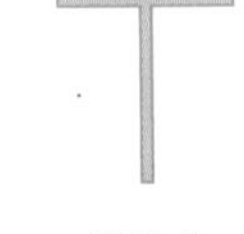
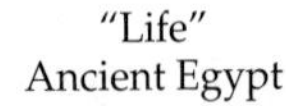

"Life"
Ancient Egypt

c.

"Unity is Strength"
Ghana

d.

Inca design

3. Describe the rotation and reflection symmetries in the flags below.

a.

Barbados

b.
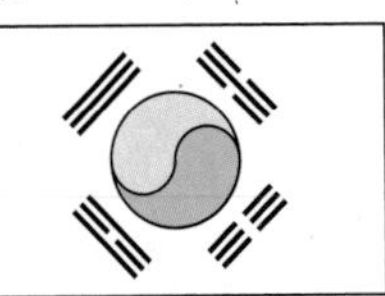

Switzerland

c.

Korea

4. Describe the rotation and reflection symmetries in the logos below.

a.
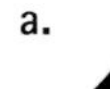

Chevrolet

b.

Volkswagen

Sources: (a) Courtesy of Chevrolet Motor Division, General Motors Corporation. (b) Volkswagen logo courtesy of Volkswagen of America.

c.
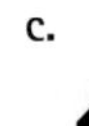

d.
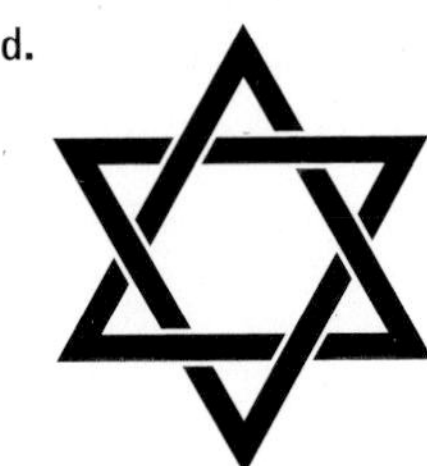

Source: The Octagon logo is a Registered trademark of JPMorgan Chase & Co. and is used here with its expressed permission.

5. Describe the rotation and reflection symmetries in the quilt patterns below.

a.
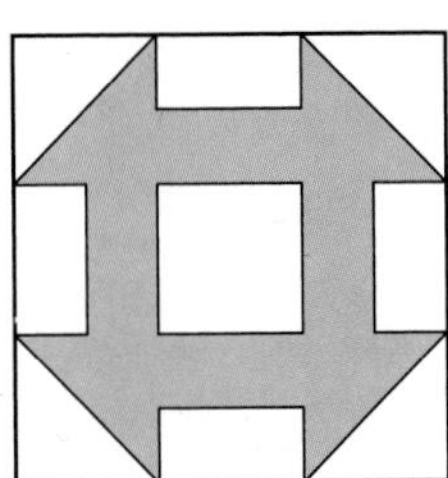

Churn Dash

b.
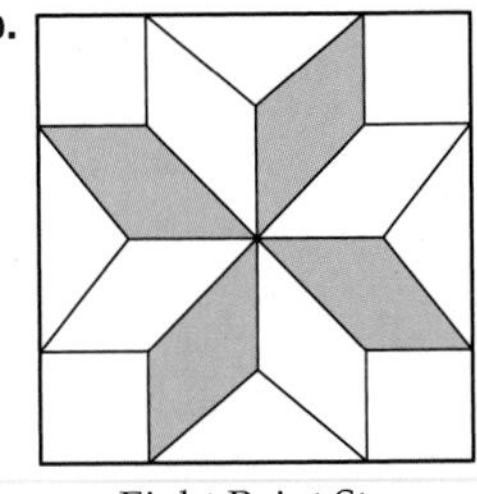

Eight Point Star

c.

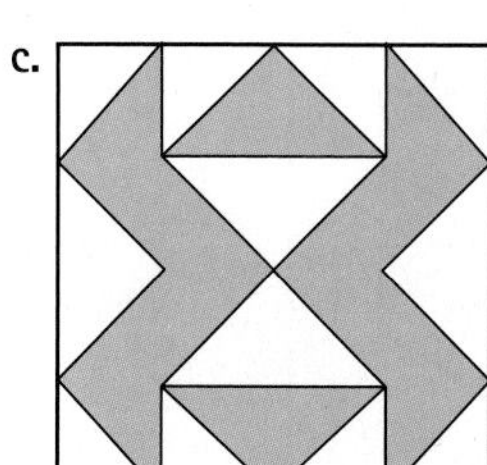

Fool's Puzzle

d.

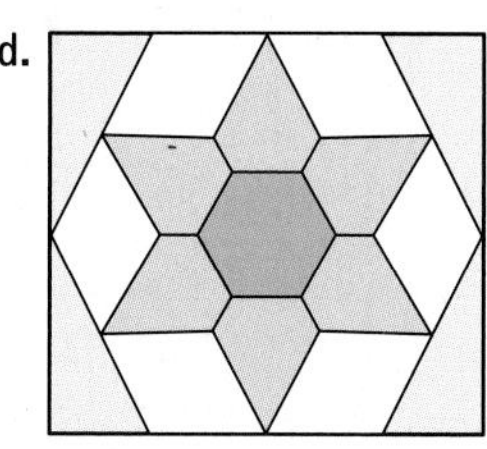

Diamonds in the Sky

e.

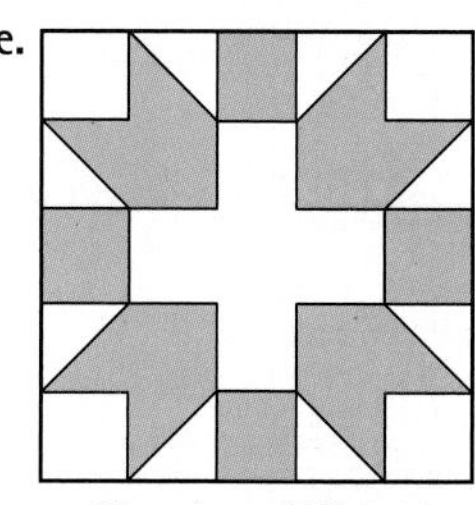

Hearth and Home

6. a. Describe the rotation and reflection symmetries of the tangram pieces. You can see images of these online simply by googling tangrams, your instructor may have some, and they are in the *Explorations* manual.

   b. Make a design that has symmetry with some or all of the tangram pieces. Explain the symmetry.

7. Classify the pentominoes by the kind(s) of symmetry that they have. Explain the different subsets.

8. Describe the rotation and reflection symmetries of the ten digits in base ten.

9. Describe the symmetries of the following strip patterns.

10. Copy the following shapes onto Geoboard dot paper and determine whether each shape does or does not tessellate.

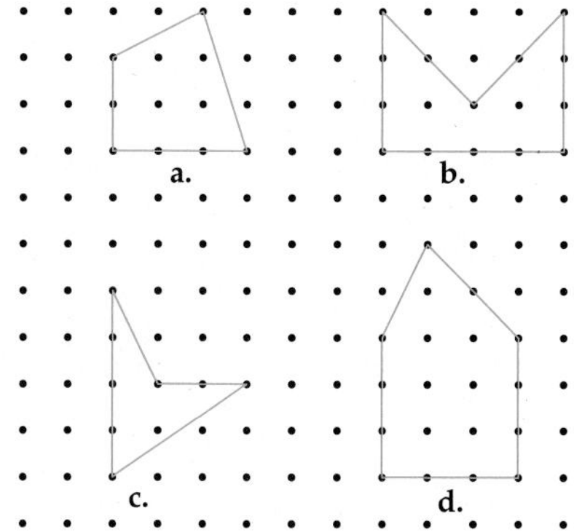

11. Verify that the arrow pattern in Investigation 10.2k does indeed tessellate.

12. Look at the tessellation patterns in Figure 10.34. Describe the symmetries of specified shapes in the patterns.

    a. Figure 10.34a

    b. Figure 10.34b

    c. Figure 10.34c

    d. Figure 10.34d

13. Which of the 12 pentominoes will tessellate? Your instructor may have a set of these, or you can view images of them on the Internet.

14. Which of the following figures tessellate? If you believe a figure tessellates, you need to show why. If you believe a figure does not tessellate, you need to explain why not.

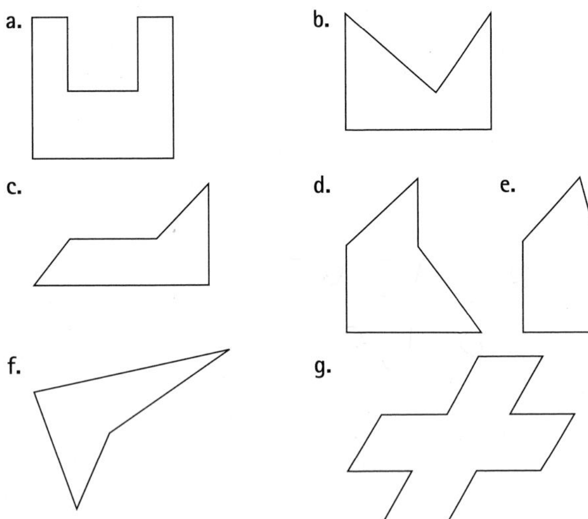

15. Describe each unit and the means by which it is repeated to form the patterns in Figure 10.34.

a. Figure 10.34a

b. Figure 10.34b

c. Figure 10.34c

d. Figure 10.34d

16. Describe each unit and the means by which it is repeated to form the following patterns.

a.

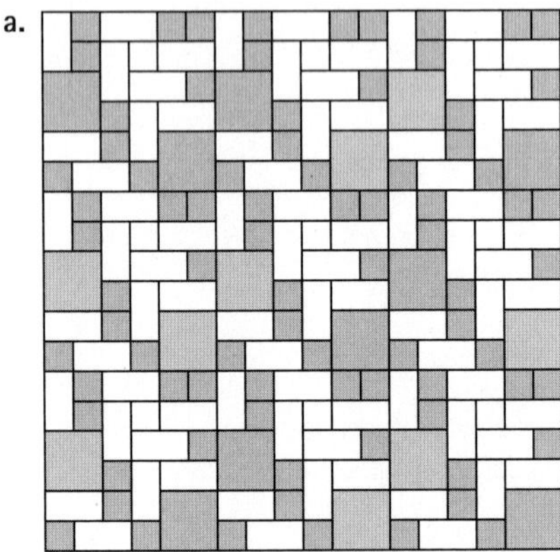

Bathroom tile floor design

b.

Ancient Mesopotamian motif

c.

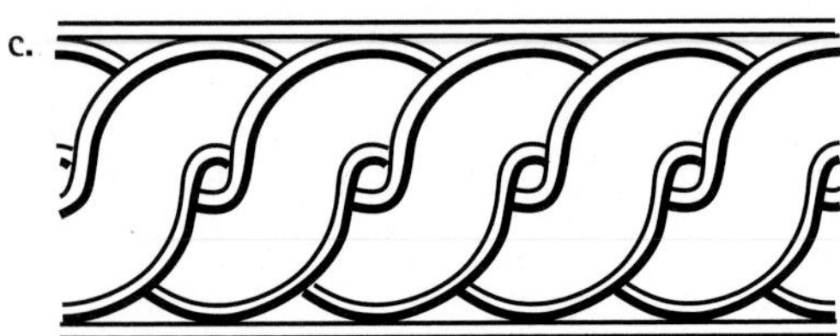

Ancient Greek and Persian motif

d.

Arabian design

e.

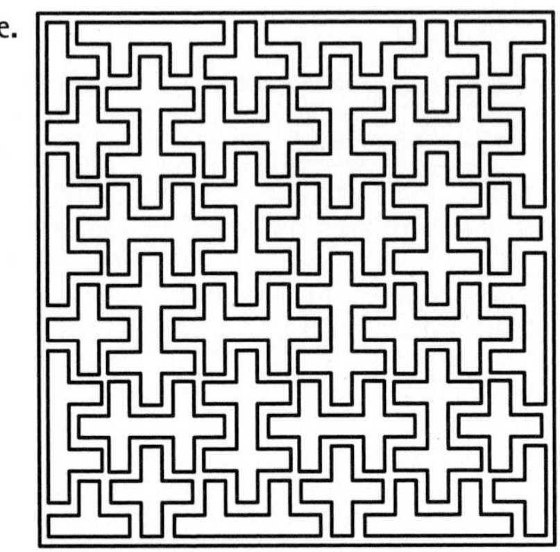

Chinese lattice design

Source: Parts b–e from Peter Stevens, *Handbook of Regular Patterns: An Introduction to Symmetry in Two Dimensions* (Fig. 14.4a, p. 121; Fig. 16.9a, p. 144; Fig. 33.3e, p. 295; Fig. 31.5c, p. 276). Copyright © 1981 Massachusetts Institute of Technology. Reprinted by permission of The MIT Press.

DEEPENING YOUR UNDERSTANDING

17. Recall the family tree for the properties of quadrilaterals in Chapter 8. When we look at the symmetries of the quadrilaterals in the family tree, will that same family tree still hold true? That is, does each descendant quadrilateral have the same symmetries as its ancestor plus at least one more symmetry?

18. Tell whether the statements below are true or false. If a statement is true, explain why. If it is false, provide a counterexample.

a. If a figure has rotation symmetry, it must also have reflection symmetry.

b. If a figure has point symmetry, it must also have rotation symmetry.

c. If a figure does not have point symmetry, it cannot have reflection symmetry.

19. Create a figure (not one copied from this book) that has the symmetry described below. In each case, describe your thought processes, including ideas that did not pan out.

a. 60-degree rotation symmetry

b. 120-degree rotation symmetry

c. 90-degree rotation symmetry

d. Two lines of symmetry

e. Three lines of symmetry

f. At least one line of symmetry and some kind of rotation symmetry

g. Reflection symmetry but not rotation symmetry

h. Rotation symmetry and exactly one line of reflection

20. a. Complete the figure following so that it has point symmetry but no other rotation symmetry.

b. Complete the figure following so that it has two lines of symmetry.

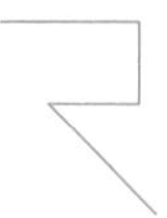

c. Complete the figure following so that it has two lines of symmetry.

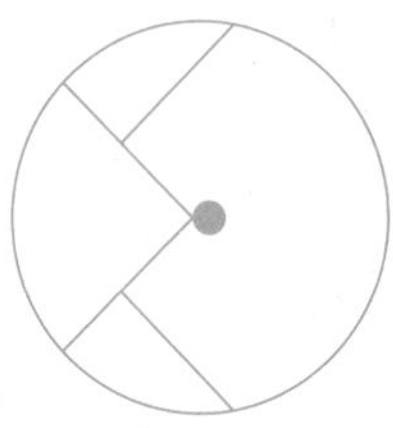

d. Complete the figure following so that it has point symmetry and no reflection symmetry.

21. In each part, the square piece of paper shown at the right is folded along the dashed lines and then cut as shown at the left. Predict what the paper will look like when unfolded.

a.

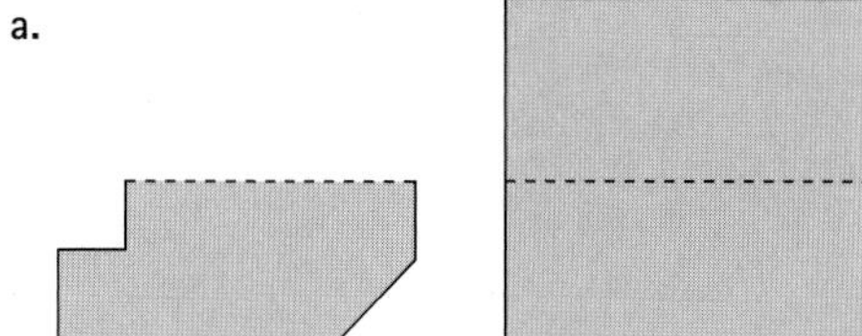

b.

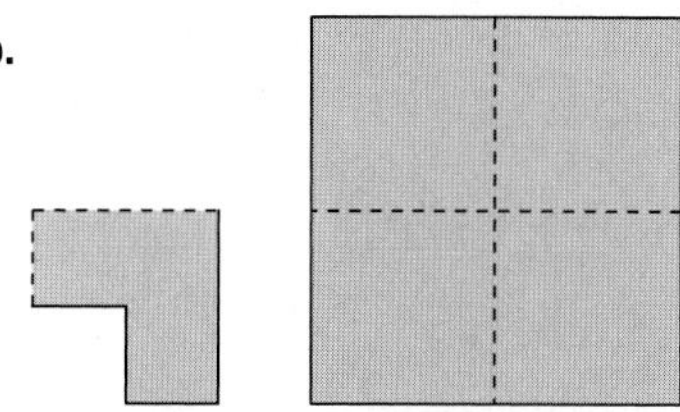

22. Sketch where you would cut the folded paper shown at the left in order to make the shape shown at the right when you unfold the paper. The dashed line represents the fold line.

a.

b.

c.

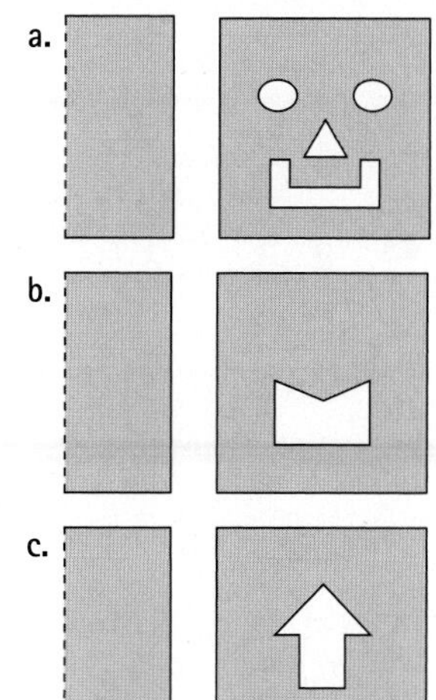

23. Sketch where you would cut the folded paper shown at the left in order to make the shape shown at the right when you unfold the paper. The dashed lines represent the two fold lines.

a.

b.

c.

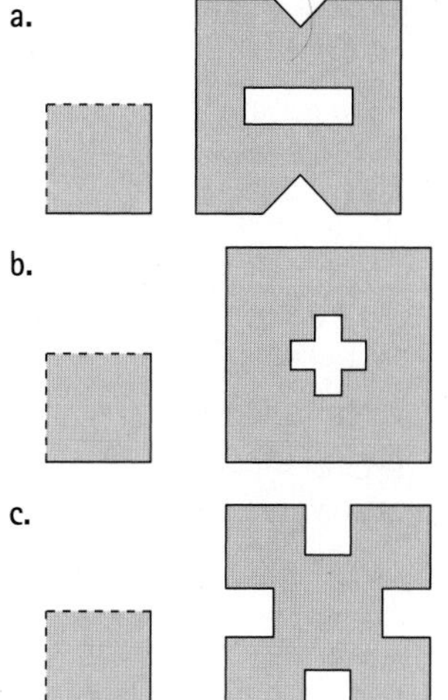

24. Describe the symmetries of the following pictures, which were taken at the Taj Mahal.

a.

© Tom Bassarear

b.

© Tom Bassarear

c.

© Tom Bassarear

d.

© Tom Bassarear

25. In each case, make your own original strip pattern with the symmetry indicated, and justify that it does indeed have this symmetry.

a. ll **b.** lm **c.** ml **d.** mm **e.** 12 **f.** mg **g.** lg

26. Make the following strip pattern using grid paper or using a computer graphics program. Explain your solution path.

27. Describe the rotation and reflection symmetries in the Navajo designs as they are drawn below. Then, imagining them to extend to the right and left indefinitely, describe any translation symmetries.

a.

b.

c.

28. Make the designs shown in Exercise 27 using computer software. Describe how you made the designs.

29. a. Describe the symmetries of the quilt at the left below. Ignore the border.
    b. Describe the symmetries of the quilt at the right below.

© Christie's Images/CORBIS

© bobby oksa 2010/Shutterstock.com

c. Describe the symmetries of the quilt blocks selected by your instructor from the sampler quilt at the right.

© Vespasian/Alamy

30. Look at the pattern below.
    a. Find a shape and an image of that shape after a translation. Describe the shape and specify the translation.
    b. Find a shape and an image of that shape after a reflection. Describe the shape and specify the reflection.
    c. Find a shape and an image of that shape after a rotation. Describe the shape and specify the rotation.
    d. Find a shape and an image of that shape after a glide reflection. Describe the shape and specify the glide reflection.

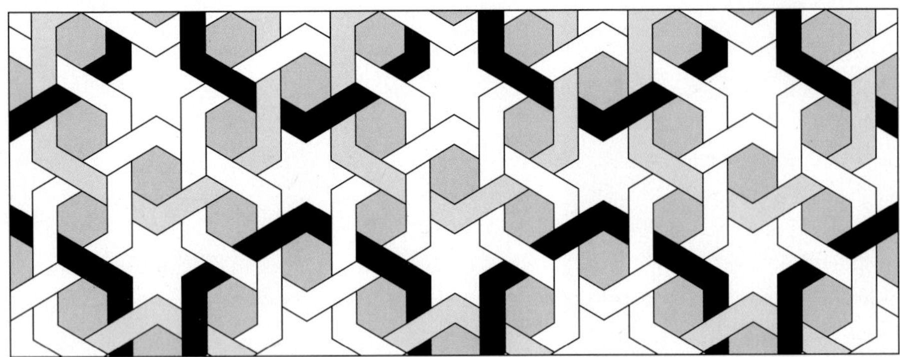

Source: From Introduction to Tessellations by Dale Seymour and Jill Britton. Copyright © 1989 Pearson Education, Inc., or its affiliates. Used by permission. All Rights Reserved.

31. Give instructions for making the patterns below as though you were talking to someone on the phone. Use symmetries when possible.

a.
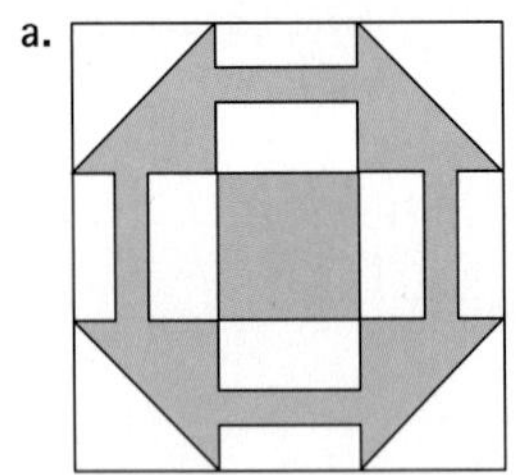
Shoo fly

b.
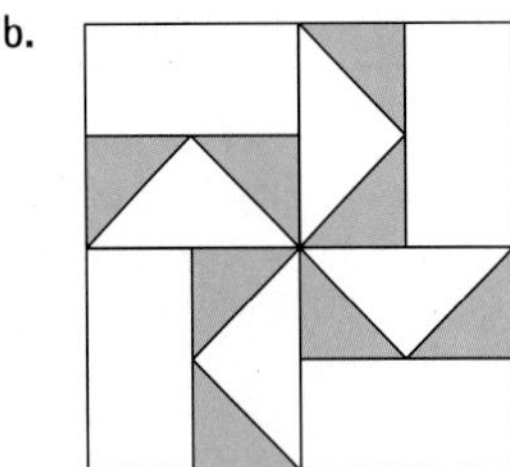
Amish design

c.
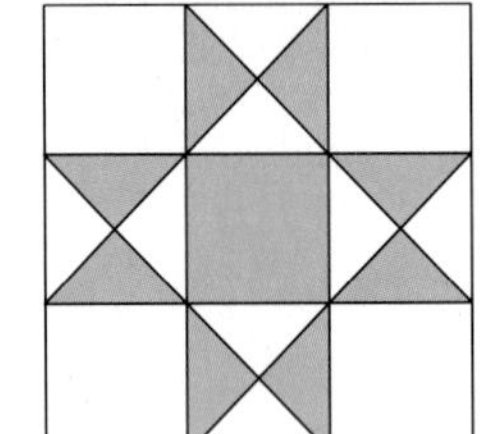
Star

32. Design a flag that has
    a. Rotation symmetry
    b. Reflection symmetry
    c. Rotation symmetry but no reflection symmetry
    d. Reflection symmetry but no rotation symmetry

33. Examine a deck of cards and separate them into subsets that have the same kinds of symmetry.

34. The word "HIDE" has one (horizontal) line of symmetry. This means that if we flip the word over, it will still spell HIDE. Find another word that has horizontal symmetry.

35. The word "WITH" has one (vertical) line of symmetry if we write it vertically. Find another word that has vertical symmetry.

W

I

T

H

36. Find a two-dimensional or three-dimensional figure on campus or from your home that has some symmetries. Sketch the figure and describe the symmetries.

37. Find examples of children's toys that have symmetries.

38. Describe the symmetries in a nut and explain why a nut might have these symmetries.

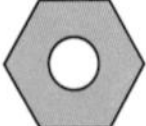

39. Copy the following shapes onto isometric dot paper (dots arranged in a grid of equilateral triangles) and determine whether each shape does or does not tessellate.

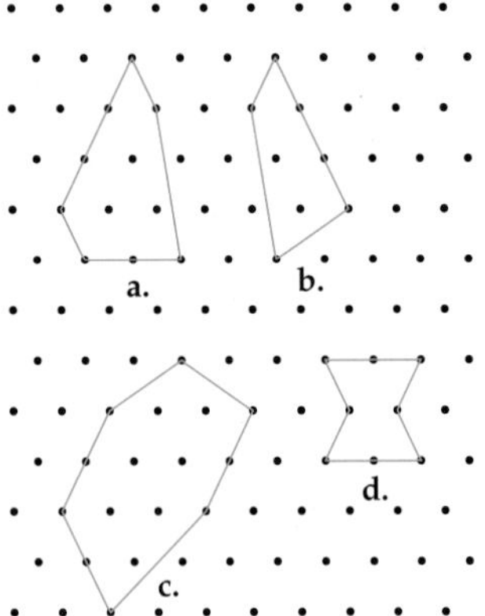

40. Select one pair of figures from Figure 10.45 Investigation 10.21.
    a. Describe a translation that will map that pair onto another pair.
    b. Describe a reflection that will map that pair onto another pair.
    c. Describe a rotation that will map that pair onto another pair.
    d. Describe a glide reflection that will map that pair onto another pair.

41. You have learned that a regular hexagon will tessellate. Below are two conjectures. There are three parts to what you turn in: your response: true or false; your justification of your response; and your work—that is, the hexagons that you made up and tested to see whether they tessellated or not.

   a. If a hexagon has line or rotation symmetry, then it must tessellate.

   b. All convex hexagons will tessellate.

42. Look at the tessellation (right), a hexagon that has been decomposed into six congruent hexagons called chevrons.

   a. Can you identify two chevrons in which the one is a translation of the other? Describe the translation.

   b. Can you identify two chevrons in which the one is a reflection of the other? Describe the reflection.

   c. Can you identify two chevrons in which the one is a rotation of the other? Describe the rotation.

   d. Can you identify two chevrons in which the one is a glide reflection of the other? Describe the glide reflection.

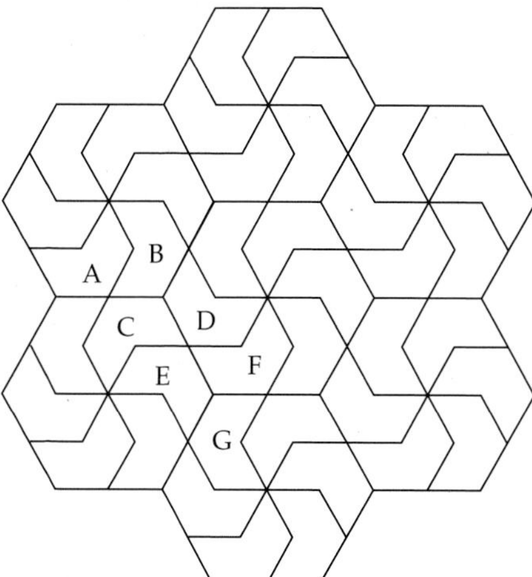

SECTION  10.3 Similarity

What do you think?

- What does it mean to say that two figures or two objects are similar?
- Are an object and its shadow similar? Why or why not?

In Section 10.1, we explored several transformations that let us change the position of an object but preserve its size and shape. That is, all of the images were congruent to the original figure. In this section, we will investigate a transformation in which the size of the figure is changed but the shape is not; we will call these transformations **similarity transformations**. According to the Common Core State Standards, these concepts appear in the eighth grade. Because some states define an elementary teaching license as grades K–8, we are including these ideas here.

Most people are comfortable with the idea of saying that two objects are similar if they have the same shape but different size. We encounter the notion of similarity frequently in daily life: When we use maps, when we use a photocopier to shrink or enlarge a picture or drawing, and in scale models. In each case, there is a scale: The map might say 1 inch = 50 miles. The photocopier will use percentages—for example, make a copy 75%. And scale models may include the information that they exhibit a 1 : 100 scale—that is, the actual object is 100 times the size of the scale model (Figure 10.46).

We will investigate the mathematical idea of similarity, which simply makes the notion of "same shape, different size" more precise from three different perspectives.

First, let us use the ideas developed in Chapter 8 about the multiple attributes of shapes.

CLASSROOM CONNECTION

You have used the concept "similar" since you were young. "Similarity seems to be a very fundamental concept. Preschoolers understand that miniature animals, doll clothes, and play houses are all small versions of familiar things. The fact that even such young children know what these tiny objects are supposed to represent shows that they intuitively understand change of scale. Building and taking apart scale models of towers, bridges, houses, and shapes of any kind give the child—of any age—a firm grasp of this idea."[6]

Lynton Gardiner/Getty Images

© Zina Seletskaya/Shutterstock.com

Figure 10.46

INVESTIGATION 10.3a Understanding Similarity

Explorations Manual 10.12

Figure 10.47 includes four pairs of shapes that are mathematically similar. Look at each pair. What is it that makes the larger polygon similar to the smaller one? Look at angles and sides. Measure them with a protractor and a ruler.

DISCUSSION

Angles In Chapter 8, we defined two polygons as being congruent if all corresponding angles were congruent and all corresponding sides were congruent. What relationships do you notice between angles and sides of the similar figures that might enable us to define two polygons as being similar? Write down your present hypothesis before reading on. . . .

It can be determined that in each of the four cases in Figure 10.47, the corresponding angles of the similar figures are congruent. If you worked with the pattern blocks in Exploration 10.11, you may have come to the same conclusion. You can confirm this relationship either with a protractor or by placing the smaller figure on top of the larger one and verifying by inspection.

Sides What about the sides? Obviously, the sides are not congruent. What relationship do they have? Think about this before reading on. . . .

One observation is that if both figures are aligned as shown, then corresponding sides are parallel. Another possible observation involves the lengths. Measure the lengths of corresponding sides. Do you notice anything? Do this before reading on. . . .

If we measure the sides of each figure and compare the lengths of corresponding sides, we find that the ratios are equal. For example, the lengths of the smaller right

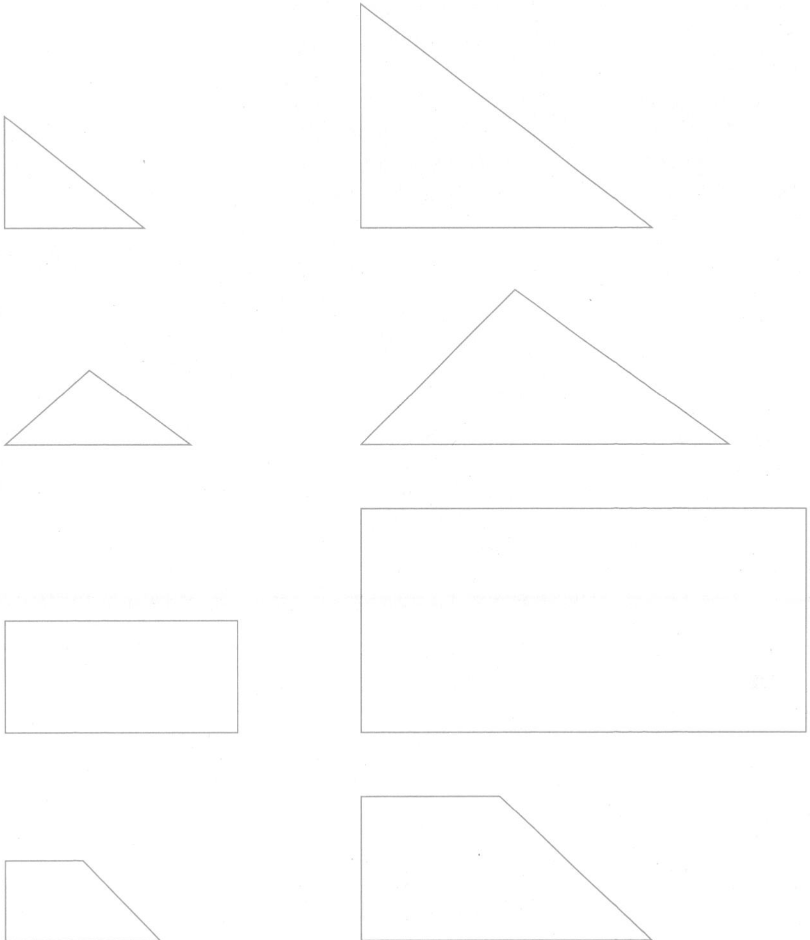

Figure 10.47

triangle are 15 mm, 20 mm, and 25 mm. The lengths of the larger right triangle are 30 mm, 40 mm, and 50 mm. That is, each of the sides of the large triangle is twice the length of the corresponding side of the smaller triangle. Another way of expressing this is to say that the ratios of corresponding sides are equal.

This common ratio of the lengths of corresponding sides is called the **scale factor**.

From Chapter 5 we know that when two or more ratios are equal, we can use the term *proportional.* Thus, when two figures are similar, the corresponding sides are proportional. It is this language that is generally used to define similarity, because it takes fewer words.

Two polygons are **similar** if corresponding angles are congruent and corresponding sides are proportional.

We use the symbol $\sim$ to denote similarity. For example, in Figure 10.48, $\triangle ABC \sim \triangle XYZ$.

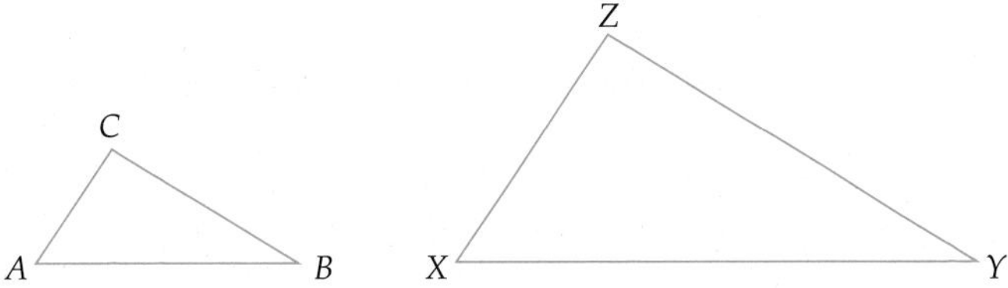

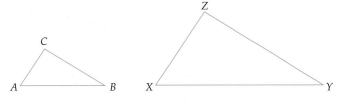

Figure 10.48

In the following investigation, we will use a technique developed many centuries ago by artists.

INVESTIGATION 10.3b Similarity Using an Artistic Perspective

We will develop the concept with a simple shape. The figure below shows how an artist might enlarge a figure. The small rectangle has been enlarged, using a scale factor of 3. Do you understand how this works? Could you use this method to enlarge another figure? Investigate this method before reading on. . . .

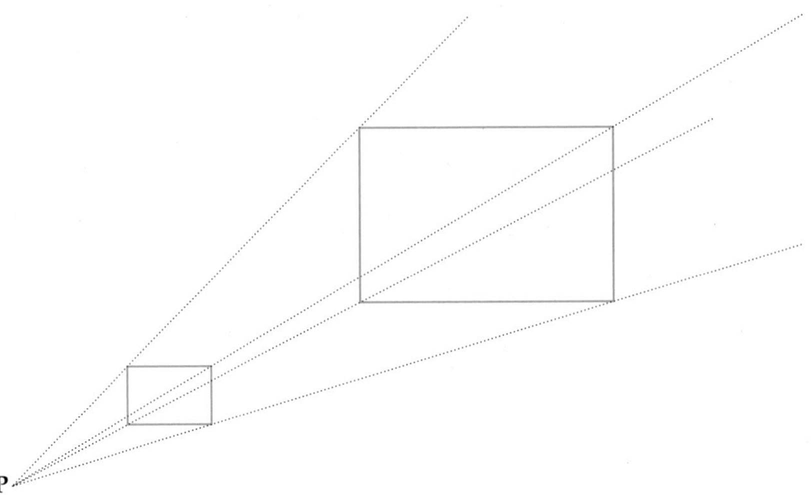

DISCUSSION

We draw the rectangle and a point *P* on a sheet of paper and draw lines that go through each of the four vertices of the rectangle. We measure the distance from the point *P* to the top left vertex of the rectangle, and then we mark a point on that line that is three times the distance from point *P*. If we do this for the other three vertices of the rectangle and connect the points, we have a new rectangle whose height is three times the height of the original rectangle.

Our last investigation adds the tool of coordinate geometry to the notion of perspective lines to bring out more relationships.

INVESTIGATION 10.3c Using Coordinate Geometry to Understand Similarity

Look at the two similar rectangles that have been drawn using perspective lines that have now been placed on the coordinate plane. What do you notice about the relationships among the coordinates? Think before reading on. . . .

DISCUSSION

There are two ways to illustrate the connections. One is to notice that the relationship between the *x*- and *y*-coordinates of each ordered pair is the same as that between the coordinates of the corresponding ordered pair of the other polygon. That is, the ratios 2 : 4 and 8 : 16 are the same—that is, 1 : 2. By the same token, both 6 : 2 and 24 : 8 equal 3 : 1.

Another way of showing the relationship is to multiply the coordinates of the first rectangle by 4:

$$4(2, 2) = (8, 8)$$
$$4(2, 4) = (8, 16)$$
$$4(6, 2) = (24, 8)$$
$$4(6, 4) = (24, 16)$$

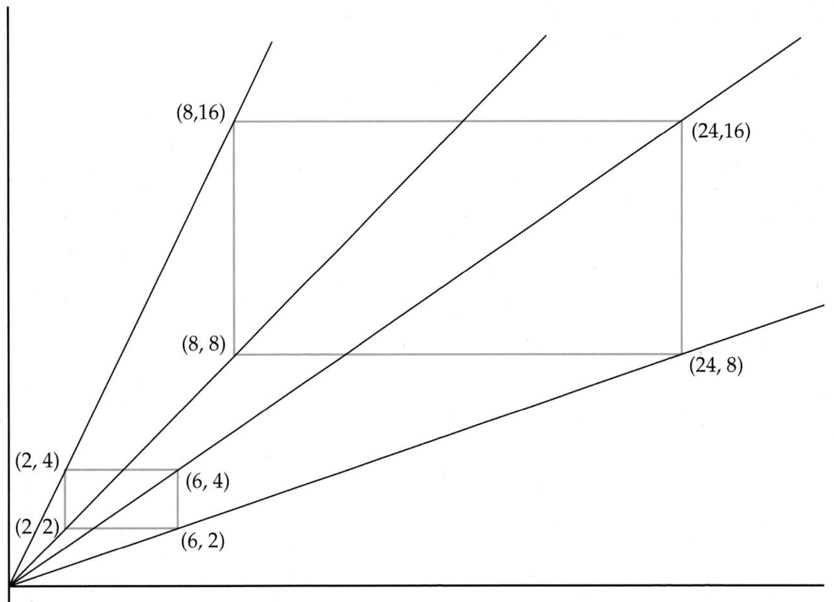

This observation enables us to enlarge any object by any scale factor. For example, if we want to make it 2.5 times as tall, we place the original object on the coordinate plane, find the coordinates of each vertex, and then multiply them by 2.5.

SUMMARY 10.3

In this section, we have investigated the idea of similarity so that your thinking about this concept could move up the van Hiele scale—that is, beyond "similarity means they look alike." We have also looked at how some ways to draw similar figures connect to the concept of similarity. In the explorations, you will find other ways to look at the idea of similarity.

10.3 Exercises

1. In each case below, the two polygons are similar. Find the length of the side labeled x.

a. 6 cm

6 cm

9 cm

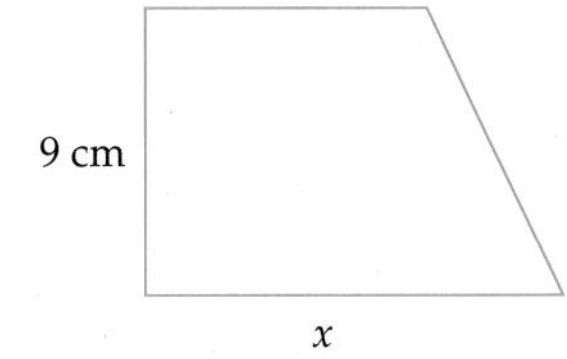

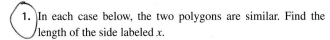

b.

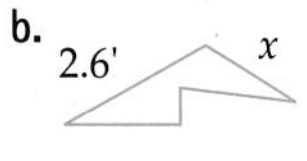

2. The two triangles below are similar, but they are not oriented on the page the same way. Determine the corresponding parts and explain how you did so.

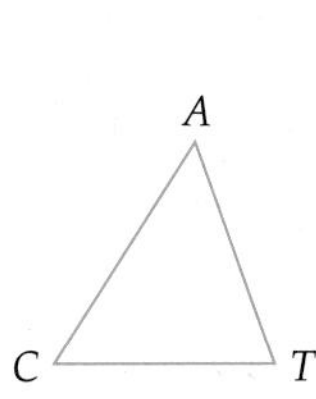

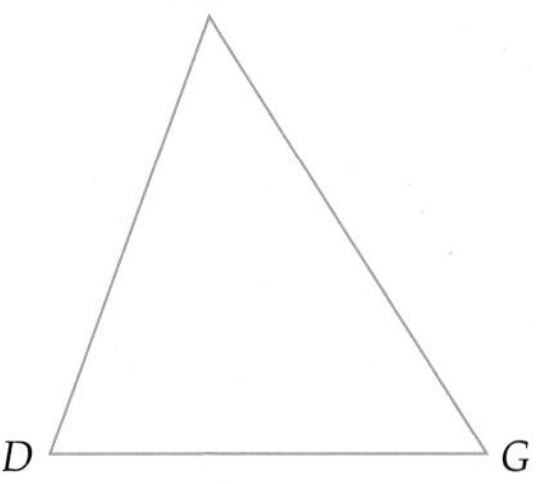

3. The figure at the right below has been created from four triangles that are congruent to the triangle at the left. Are the two triangles similar? Explain why or why not.

4. For each of the questions below, answer yes or no. In either case, justify your response.

   a. Are all isosceles triangles similar?

   b. Are all equilateral triangles similar?

   c. Are all squares similar?

   d. Are all rhombuses similar?

   e. Are congruent polygons similar?

DEEPENING YOUR UNDERSTANDING

5. Gerald has a photo that measures 8 inches × 11 inches and a frame that is 4 inches × 6 inches.

   a. Explain to him why we cannot reduce the photo so that it will fit perfectly in the frame.

   b. Explain to him his options for cutting so that the reduced photo will fit.

6. When we use an overhead projector in a classroom, is the image on the screen similar to the object on the overhead projector?

7. Below is a map of Utah. If the actual distance across the bottom of Utah is 275 miles, determine the scale of the map. That is, 1 inch = x miles.

8. The diagram below (not drawn to scale) shows how we can find the distance across a lake using the principle of similar triangles.

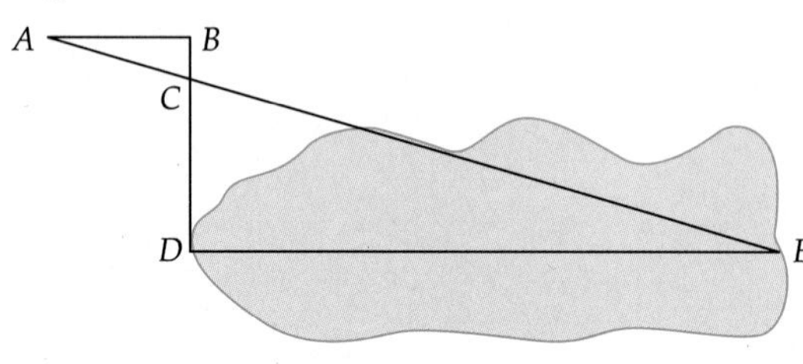

   a. Explain how points C, B, and A would be determined, as though you were talking to someone who understands the idea of similar triangles but has never heard of this method before.

   b. Explain why $\triangle ABC \sim \triangle EDC$.

   c. If $m\overline{AB} = 324$ feet, $m\overline{BC} = 103$ feet, and $m\overline{CD} = 212$ feet, what is the distance across the pond?

9. Explain how you could use shadows and similarity to find the height of a building.

10. A common standard setting for enlarging on a copy machine is 121 percent. What does this mean?

11. Find a quilt block that contains similar figures. Explain why the figures are similar.

12. Find a tessellation pattern that contains similar figures. Explain why the figures are similar.

13. Solomon Golumb invented reptiles, figures that both tessellate and also replicate themselves.

    Consider the trapezoid below.

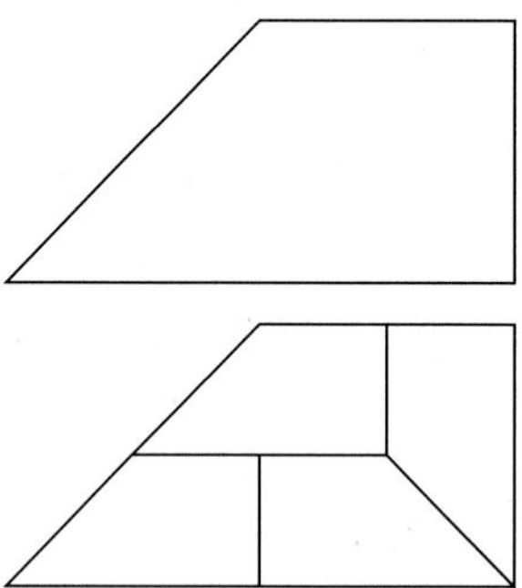

This trapezoid tessellates. Now we will see how it "replicates itself." We can divide this trapezoid into four smaller congruent trapezoids, each of which is similar to the original—sort of like having children!

   a. Why is "reptile" an appropriate name for such figures?

   b. Prove that the four little trapezoids are similar to the larger trapezoid.

   c. Make another trapezoid that is a reptile.

   d. Describe the attributes that a trapezoid must have in order to be a reptile.

   e. Find at least two ways in which this trapezoid will tile the plane.

14. The hexagon below is another reptile.

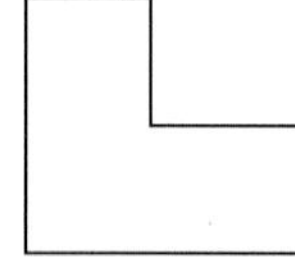

   a. Determine how to divide this hexagon into several little congruent hexagons, all of which are similar to the original.

   b. Find at least two ways in which this hexagon will tile the plane.

FROM STANDARDIZED ASSESSMENTS

NECAP 2007, Grade 5

15. Look at these triangles.

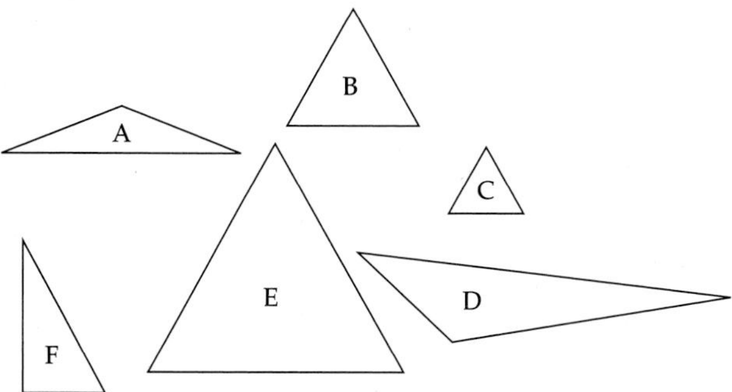

Which triangles are similar?

LOOKING BACK on chapter ten

QUESTIONS TO SUMMARIZE BIG IDEAS

1. What are the types of congruence transformations? Describe each one in as much detail as you can.
2. What are the types of symmetry? Describe each one in as much detail as you can.
3. What is a tessellation? What figures will tessellate a plane?
4. What does it mean to say that two figures are similar?
5. What parts of this chapter are less clear to you at this point? What will you do to clarify those ideas?
6. Look back at the Mathematical Practices of the CCSS. In what ways did you engage in those practices during this chapter?

CHAPTER 10 SUMMARY

1. Both congruence and similarity are kinds of transformations.
2. Congruence can be defined in Euclidean terms (the classical definition) and in terms of congruence mappings.
3. There are different ways to perform and describe different transformations. Which one(s) we use depends partly on preference and partly on purpose.
4. Translations, reflections, rotations, and glide reflections are congruence transformations—the figure is moved, but its shape and size are not changed.
5. The geometric transformations of translation, reflection, rotation, and glide reflection are operations, with many similarities to the operations of addition, subtraction, multiplication, and division.
6. Symmetry has to do with congruence within a figure.
7. The many symmetries in geometric figures include translation, reflection, rotation, and glide reflection symmetry. Many figures have more than one kind of symmetry.

BASIC CONCEPTS

Section 10.1: Congruence Transformations

Section 10.2: Symmetry and Tessellations

symmetry **580** line of symmetry **580**
point symmetry **584** vertical line symmetry **584**
horizontal line symmetry **585**
diagonal line symmetry **585**

Symmetry for two-dimensional figures:

translation symmetry **587** reflection symmetry **587**
rotation symmetry **587** glide reflection symmetry **587**
pattern **588**

Symmetry for three-dimensional objects:

plane of symmetry **594** axis of symmetry **594**

Tessellations:

tessellate **595** semiregular tessellations **599**
vertex point **600**

Section 10.3: Similarity

similarity transformation **611** scale factor **613**
similar **613**

REVIEW EXERCISES chapter ten

1. Perform the following transformation on the given figure, and give its new coordinates:
   a. Translation $(x, y) \rightarrow (x + 4, y + 0)$
   b. Rotate 90 degrees counterclockwise about the point (4, 3).
   c. Reflect the figure across the x-axis.

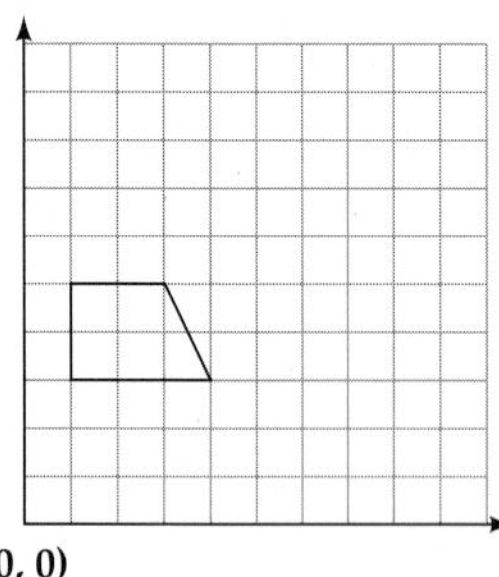

2. Without folding, determine the reflection image of the triangle at the right. Briefly justify your method.

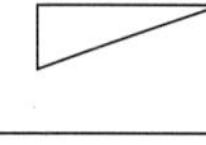

3. Sketch the image by rotating the figure 180 degrees around point P.

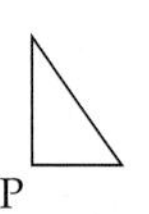

4. a. Find the image of the figure below reflected across line a and then across line b.
   b. Find another way to get the same outcome.
   c. Find the image of the figure reflected across line c and then across line d.
   d. Find another way to get the same outcome.

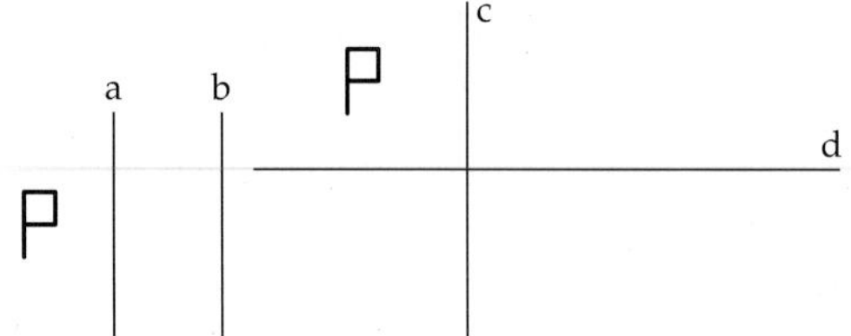

5. Describe how to get the shape from position A to position B in as few moves as possible, using translations, rotations, reflections, and/or glide reflections.

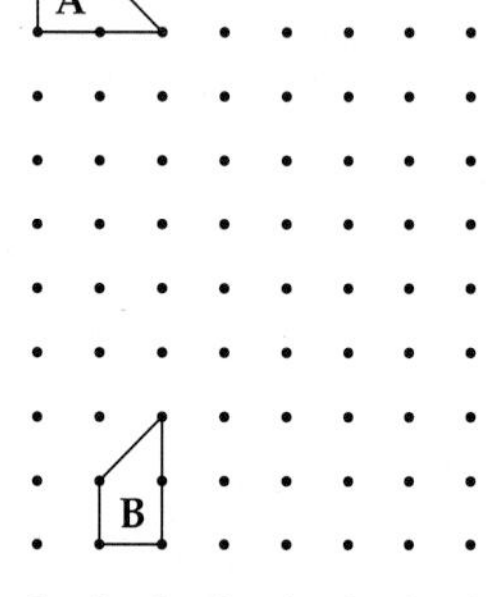

6. How are rotations and reflections alike? How are they different?
7. If a glide reflection is a composite transformation, why is it included as one of the basic four congruence transformations? That is, why not call translation, reflection, and rotation the basic three?
8. Describe the reflection and rotation symmetries of the following figures.

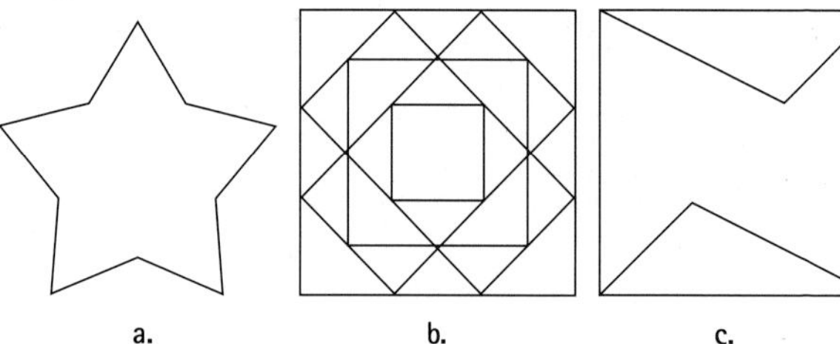

9. a. Complete the first figure so that it has reflection but not rotation symmetry.
   b. Complete the second figure so that it has reflection and rotation symmetry.
   c. Complete the third figure so that it has rotation but not reflection symmetry.

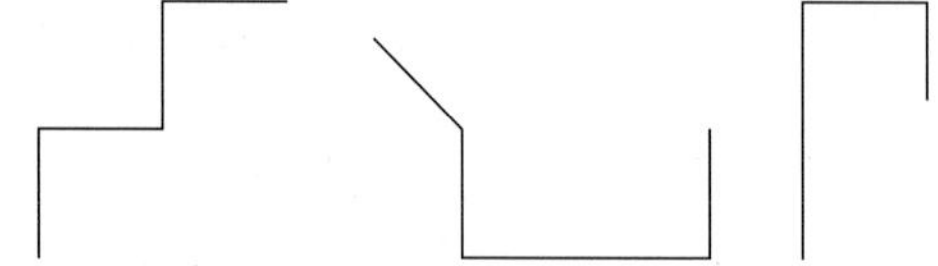

10. a. Make an original figure that has reflection but not rotation symmetry.

    b. Make an original figure that has reflection and rotation symmetry.

    c. Make an original figure that has rotation but not reflection symmetry.

11. a. What does it mean to say that a figure has rotation symmetry?

    b. What does it mean to say that a figure has reflection symmetry?

12. Describe the symmetries of the following strip patterns.

a.

b.

c.

d.

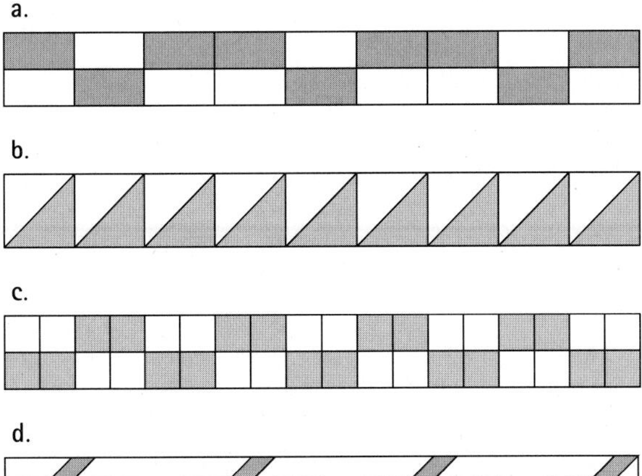

13. In the tessellations below, describe the unit and the means by which it is repeated.

a.

b.

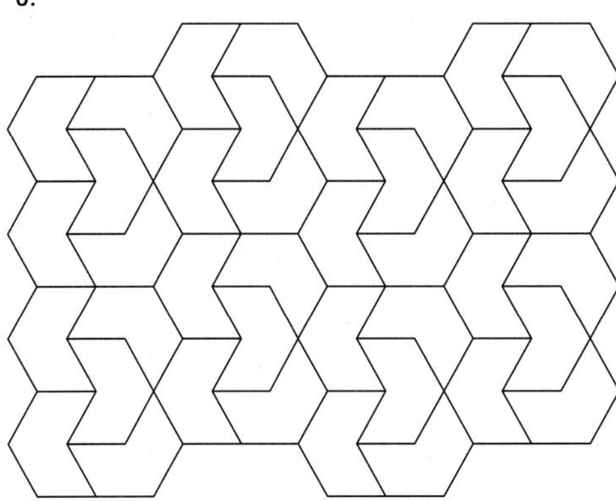

14. Determine whether each figure below tessellates or not. If it does tessellate, show the tessellation and explain why it tessellates.

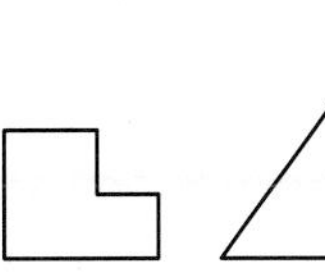

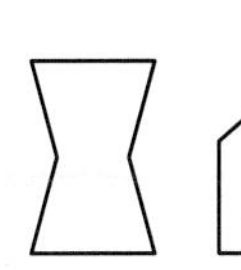
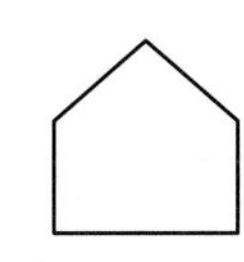

a. b. c. d. e.

15. The following two pentagons are similar. Find the length of the side marked x.

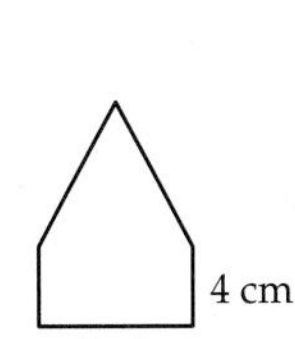

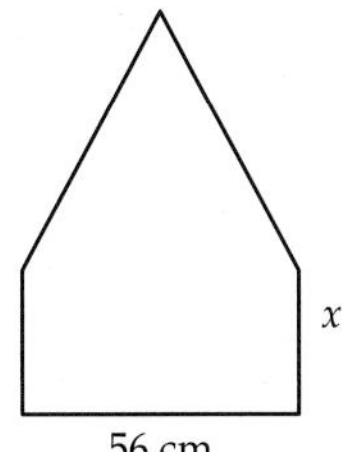

16. Are all right isosceles triangles similar? Explain why or why not.

17. Using the lines-of-perspective method, make a larger copy of the given figure so that each side of the copy is twice as long as each side of the given figure.

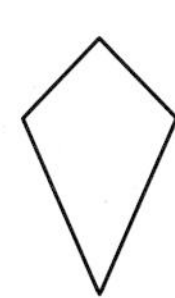

18. If the coordinates of a triangle are (3, 4), (8, 4), and (4, 7) and the projection point P is at the origin, predict the coordinates of a similar triangle whose sides will each be twice the length of the corresponding sides of that triangle. Explain your reasoning.

APPENDIX A

Selected Answers

CHAPTER 1 EXERCISES

Section 1.1

1. Answers vary.
3. Answers vary.
5. Student A: / Student B:

Student A	Student B
86	86
× 47	× 47
602	42
344	560
4042	240
	3200
	4042

7. You pay \$16 for 5 movies, so you pay \$3.20 per movie.
9. Sally makes \$4.98 per hour.
11. **a.** If one penny has a diameter of 0.75 inches, then we have 216.525 billion inches; 18.04375 billion feet; 3,417,400 miles.
    b. 347 pennies per second; 10,950,000,000 pennies per year.
13. **a.** Need to make assumptions for the distance driven and the average cost of gasoline.
    b. Answers will vary. If the distance is 2400 miles and the cost of gas averages \$3.00/gallon, then it will take \$175 extra to drive the van.
    c. Answers will vary. Using the assumptions in part b if the price of gas rises 40¢, then it is now averaging \$3.40/gallon. Using the same procedure as in part b, it costs \$53.33 more for the van and \$30 more for the sedan.
15. **a.** 3600 beats per minute
    b. 216,000 beats per hour
17. 36 toothpicks

Section 1.2

1. Answers vary.
3. 67 games
5. We don't know, because an answer of 3:15 assumes that the power was off only a matter of seconds. It is possible that the power went off at 1 A.M. and came back on at 3:15 A.M. Thus we can say only that the power came back on at 3:15 because electric clocks begin at 12 A.M. when the power comes back on. If it is now 6:45 A.M., and the clock says 3:30, that means the power came back on $3\frac{1}{2}$ hours ago, that is, at 3:15 A.M.
7. **a.** $\frac{104 - (8 \times 6)}{7} = $ 8-inch space between each plate and between the end plates and the wall.

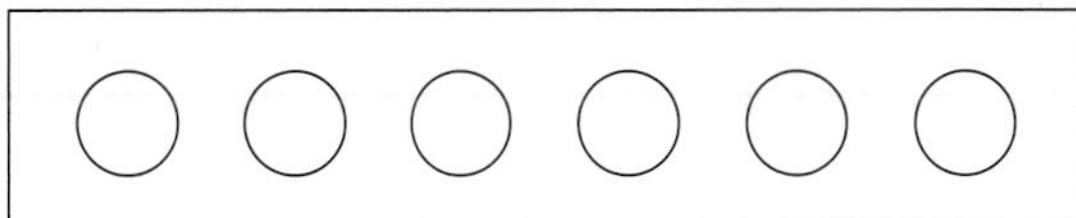

    b. $\frac{104 - (8 \times 5)}{6} = 10\frac{2}{3}$ inches
    c. $\frac{104 - (12 \times 2) - (8 \times 6)}{5} = 6\frac{2}{5}$ inches
9. 8 tricycles and 24 bicycles
11. The three possibilities are:

Bicycles	Tricycles
6	0
3	2
0	4

13. 81 tribbles and 16 chalkas
15. **a.** The possible scores are 4 through 22, 24, 25, 26, 28, and 32.
    b. Two ways: 1 + 1 + 2 + 8 and 2 + 2 + 4 + 4
    c. There must be an even number of ones to have the result be even. There are three ways to get an even number: no 1s, two 1s, or four 1s. Of course, the last possibility will not give us 12. If we use two 1s, then the remaining two throws must result in a total of 10. The only way to do this is 2 + 8. Finally, using no 1s, suppose we have one 8-point throw, then we must score the remaining 4 points in three throws, which is impossible. Thus an 8-point dart cannot be used. If we have a 4-point throw, then the remaining three throws must result in 8 points. The only way to do this is 2 + 2 + 4.
17. **a.** 17 bottles
    b. $1\frac{2}{3}$ (or about 1.7) ounces of leftover perfume/jug
    c. jug size of 11 ounces or multiples of 11 ounces
19. **a.** Fill the 5-gallon pail and use it to fill the 3-gallon pail (leaving 2 gallons in the 5-gallon pail). Empty the 3-gallon pail and put the 2 gallons left in the 5-gallon pail into the 3-gallon pail. Fill the 5-gallon pail again and use it to finish filling the 3-gallon pail. This leaves 4 gallons in the 5-gallon pail.

b. Fill the 3-gallon jug from the 8-gallon jug and empty it into the 5-gallon jug. Fill the 3-gallon jug again (from the 8-gallon jug).

21. The ball bounced 4 times.

23. 380 valentines. This is not the same as the handshakes problem. If students *A* and *B* meet and shake hands, then the handshake counts for both *A* and *B*. However, if *A* gives *B* a valentine, it only counts for *A*. *B* still has to give a valentine to *A*.

25. a. 1474

b. 1628

c. 5115

27. 8712

29. Answers will vary.

31. I have simply given the completed squares for questions 31 and 32 rather than just the missing numbers.

a.

9	19	5
7	11	15
17	3	13

b.

4	18	8
14	10	6
12	2	16

c.

1	26	0
8	9	10
18	−8	17

33. a. Answers will vary.

b. Answers will vary.

35. a.

50	25	10	5	1
1	1	1	3	0
1	0	5	0	0
0	3	2	1	0

b.

50	25	10	5	1
1	0	3	3	0
0	3	0	4	0
0	2	4	1	0

c.

50	25	10	5	1
1		2	3	15
0	1	3	7	10
0	2	3	1	15
0	0	7	4	10
0	0	3	13	5

37. Answers will vary.

39. a. The sum of the first seven terms is 33, which is one less than the ninth term. The sum of the first eight terms is 54, which is one less than the tenth term. So, the sum of the first *n* terms is equal to one less than the $(n + 2)$nd term, or $A_1 + A_2 + A_3 + \cdots + A_n = A_{n+2} - 1$.

b. $A_n \times A_{n+2} = (A_{n+1})^2 \pm 1$

41. Many possible answers

43. 325

45. Yes. Use three 15¢ stamps and four 33¢ stamps.

47. 56, 62

49. 6 tables: 4 round and 2 square

51. 21 votes. The other dogs get 20, 20, and 19.

CHAPTER 1 REVIEW EXERCISES

1. 316 student tickets

2. 16 stools

3. 15 coins equaling 92¢

50	25	10	5	1
1	1	0	1	12
1	0	0	7	7
0	1	1	11	2
0	1	5	2	7
0	2	1	5	7
0	0	5	8	2

4. $216

5. Nine ways

6. 40 posts

7. $1.17

8. 43,200

9. There were two 10¢ stamps and ten 5¢ stamps.

10. The product is a four-digit number. Starting from the left, the first two digits have a value one less than the number you are multiplying by 99. The second two digits equal what you get if you subtract that number from 100. Using symbols, if we let the number $= ab$, then the first two digits are $(ab - 1)$ and the second two digits are $(100 - ab)$.

11. P

12. There are so many possibilities. Here are several:
The magic sum is 34.
The sum of the 4 numbers in the center is also 34.
If you partition the 4×4 square into four 2×2 squares, the sum of the numbers in each of the 2×2 squares is also 34.
There are 2 odd and 2 even numbers in each row and column.
The two middle rows and/or columns can be switched without affecting the square.
The top-left to bottom-right diagonal has a constant difference of 3.
If you look at the middle two columns, each pair of numbers contains consecutive numbers.

13. a. The sum of the numbers along the length of the stick equals the number at the end of the stick.

b. Since the largest number on the chart is 924, work backwards: 1, 6, 21, 56, 126, 252, 462.

c. The handle would be the diagonal row of 13 ones along the left edge of the chart, and at the end of the stick would be 12.

d. Answers will vary.

14. a. Three ways

b. 23, 27, 29, 30, and 31 are impossible.

15. a. Fill the 9-gallon pail and use it to fill the 4-gallon pail. Empty that pail and fill it again from the 9-gallon pail. You now have 1 gallon left in the 9-gallon pail.

b. Fill the 4-gallon pail and empty it into the 9-gallon pail. Do it again. Fill the 4-gallon pail a third time and now fill the

9-gallon pail—it takes 1 more gallon to do so. You now have exactly 3 gallons left in the 4-gallon pail.

16. 11 hours

17. 26 packages can be made. 2 ounces will remain.

CHAPTER 2 EXERCISES

SECTION 2.1

1. a. $0 \notin \varnothing$ or $0 \notin \{\ \}$

b. $3 \notin B$

3. a. {e, l, m, n, t, a, r, y} and $\{x \mid x$ is a letter in the word "elementary"$\}$ *or* $\{x \mid x$ is one of these letters: e, l, m, n, t, a, r, y$\}$

b. $\{x \mid x$ is a country in Europe$\}$. It would not be practical to name all the countries in Europe because there are so many countries and the list today is not the same as the list 20 years ago.

c. {2, 3, 5, 7, 11, 13, 17, 19, 23, 29, 31, 37, 41, 43, 47, 53, 59, 61, 67, 71, 73, 79, 83, 89, 97}.
Also $\{x \mid x$ is a prime less than 100$\}$.

d. The set of fractions between 0 and 1 is infinite.
$\{x \mid x$ is a fraction between zero and one$\}$.

e. {name1, name2, name3, . . .}.
$\{x \mid x$ is a student in this class$\}$.

5. a. $\in$; 3 is an element of the set.

b. $\subset$; {3} is a subset of the set.

c. $\in$; {1} is an element of this set of sets.

d. $\subset$; $\{a\}$ is a subset of the set.

e. $\not\subset$ or $\notin$; $\{ab\}$ is neither a subset nor an element.

f. $\subset$; the null set is a subset of every set.

7. a.

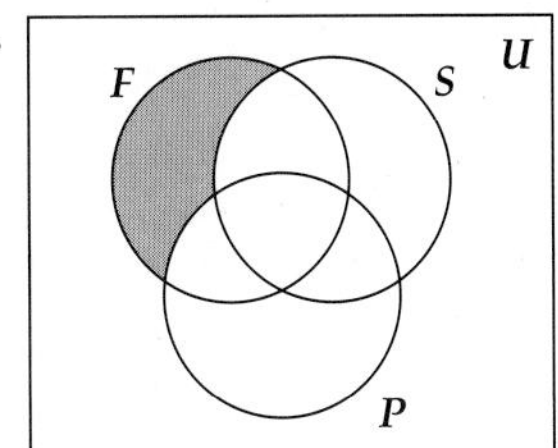

$F \cap (\overline{S \cup P})$ or $F \cap \overline{S} \cap \overline{P}$

b.

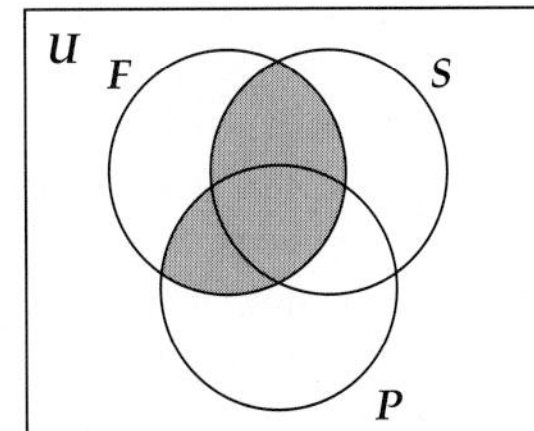

American females who smoke and/or have a health problem

c.

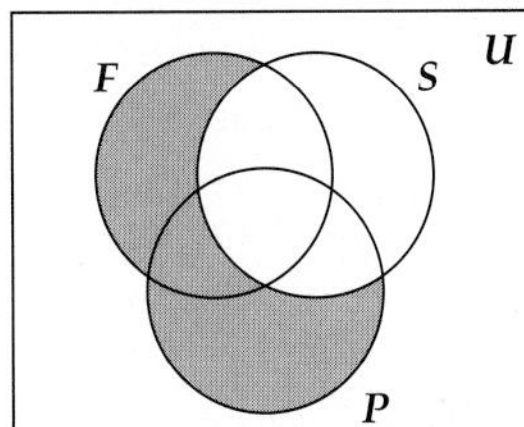

Nonsmokers who are female and/or have health problems

d. $F \cap S$
Females who smoke

e. $\overline{F} \cap (S \cap P)$
Males who smoke and have a health problem

9. a.

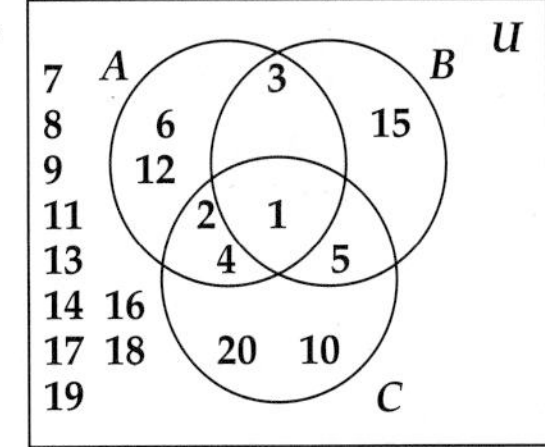

b. All numbers in U that don't evenly divide 12, 15, or 20; $\overline{A \cup B \cup C}$ or $\overline{A} \cap \overline{B} \cap \overline{C}$

c. All numbers in U that evenly divide 12 and 20, but not 15; $\overline{B} \cap (A \cap C)$

d. All numbers in U except 1 and 3

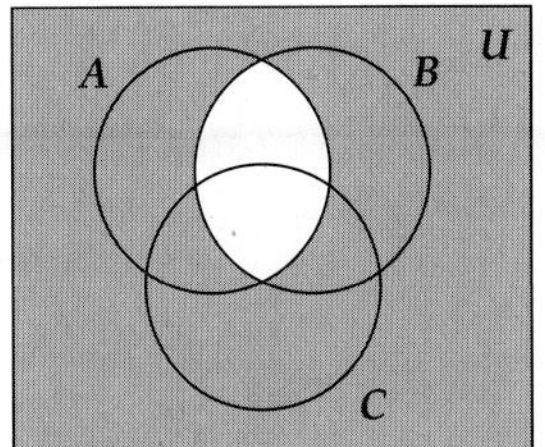

e. All numbers from 1 to 20, except those that divide 12 or 15 evenly

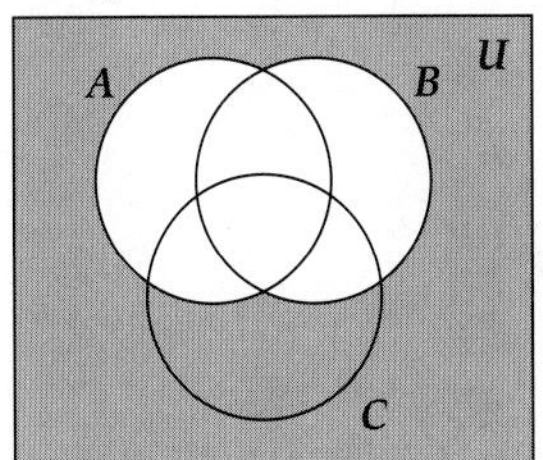

f. Note: This description is ambiguous; it depends on how one interprets "or." $A \cup B$

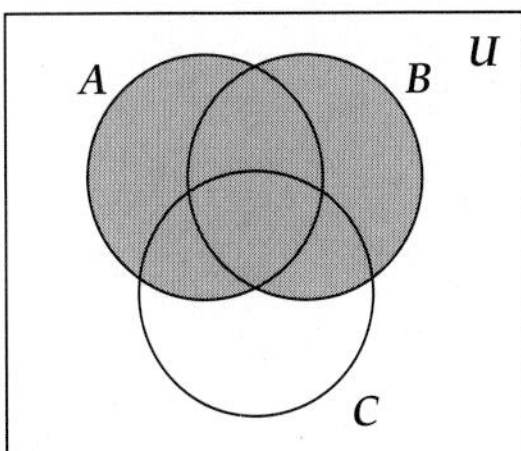

11. 15 possible committees. Label the members with *A, B, C, D, E,* and *F*.
The committees could be: *AB, AC, AD, AE, AF, BC, BD, BE, BF, CD, CE, CF, DE, DF, EF.*

13. Answers will vary.

15. The circles enable us to easily represent visually all the possible subsets.
The diagram is not equivalent because there is no region corresponding to elements that are in all three sets.

17. a. $6 + 8 + 12 + 3 = 29\%$

b. $6 + 25 + 15 = 46\%$

c. Those people who agree with his foreign policy and those people who agree with his economic and his social policy

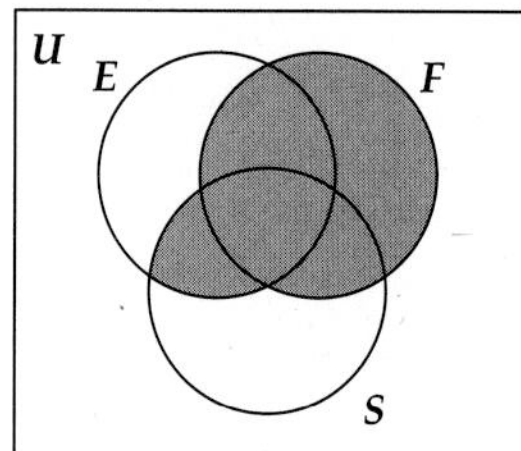

d. The center region, because they agree with all three of his policies

19. a. Construct a Venn diagram. $100 - 11 - 10 - 23 = 56\%$

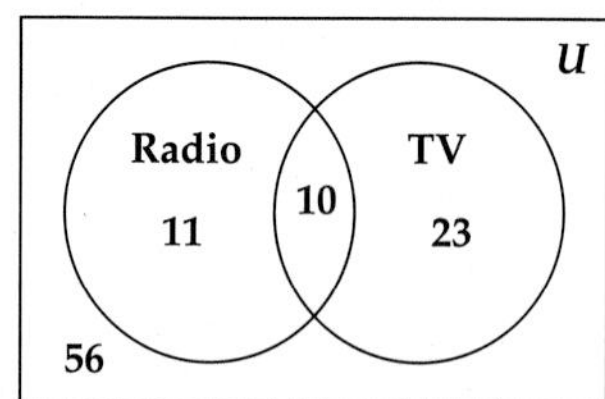

b. $11 + 23 = 34\%$

21. Answers will vary.

23. Answers will vary.

Section 2.2

1. a.

	Mayan	Luli	South American
7		lokep moile tanlip	
8			teyente toazumba
12			caya-ente-cayupa
13		is yaoum moile tamlip	caya-ente-toazumba
15		is yaoum is alapea	
16	ho buluc		toazumba-ente-tey
21	hun hunkal	is eln yaoum moile alapea	cajesa-ente-tey
22	ca huncal	is eln yaoum moile tamop	cajesa-ente-cayupa

b. Answers will vary.

c. Answers will vary.

3.		**Egyptian**	**Roman**	**Babylonian**
a.	312	𓍢𓍢𓍢𓎆𓏺𓏺	CCCXII	𒁹𒁹𒁹𒁹𒁹 𒌋𒁹𒁹
b.	1206	𓆼𓍢𓍢𓏺𓏺𓏺𓏺𓏺𓏺	MCCIIIIII or MCCVI	𒌋𒌋 𒁹𒁹𒁹𒁹𒁹𒁹
c.	6000	𓆼𓆼𓆼𓆼𓆼𓆼	MMMMMM	𒁹 𒌋𒌋𒌋𒌋 ◣◣
d.	10,000	𓂭	MMMMMMMMMM	𒁹𒁹 𒌋𒌋𒌋𒌋 𒁹𒁹𒁹𒁹𒁹𒁹 𒌋𒌋𒌋𒌋
e.	123,456	𓆐𓂭𓂭𓆼𓆼𓆼𓍢𓍢𓍢𓍢𓎆𓎆𓎆𓎆𓎆𓏺𓏺𓏺𓏺𓏺𓏺	Can't do	𒌋𒌋𒌋𒁹𒁹𒁹𒁹 𒌋𒁹𒁹𒁹𒁹𒁹𒁹𒁹 𒌋𒌋𒌋𒁹𒁹𒁹𒁹𒁹𒁹

5. a. 26 **b.** 540 **c.** 25 **d.** 450
e. Three thousand four hundred
f. 3450

7. a.

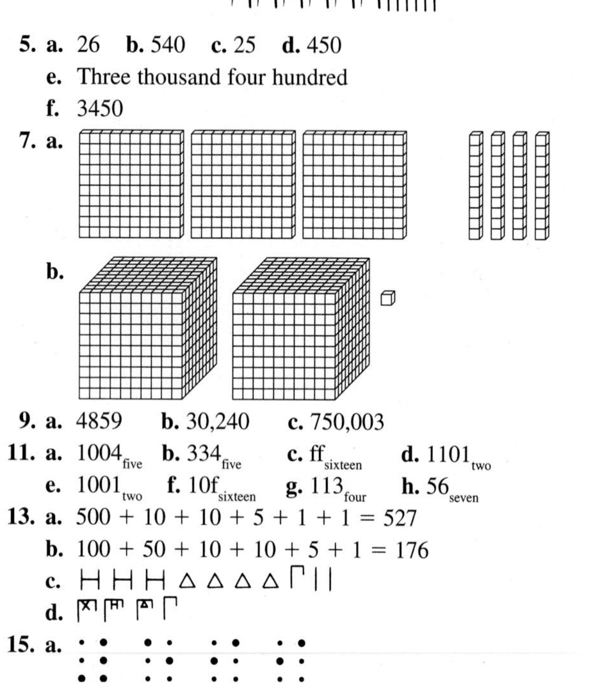

b.

9. a. 4859 **b.** 30,240 **c.** 750,003

11. a. 1004_{five} **b.** 334_{five} **c.** $\text{ff}_{\text{sixteen}}$ **d.** 1101_{two}
e. 1001_{two} **f.** $10\text{f}_{\text{sixteen}}$ **g.** 113_{four} **h.** 56_{seven}

13. a. $500 + 10 + 10 + 5 + 1 + 1 = 527$
b. $100 + 50 + 10 + 10 + 5 + 1 = 176$
c. ΗΗΗΔΔΔΔ𐅃ΙΙ
d. 𐅇𐅅𐅄𐅃

15. a. ⠼⠑⠊⠊
b. ⠼⠉⠂⠋⠚⠉⠂⠃⠚⠚
c. 460,859
d. 135,246

17. a. 585 cartons of milk

b. It has all 6 characteristics because this system is essentially base six. The places are called cartons, boxes, crates, flats, and pallets. The value of each place is 6 times that of the previous place.

19. The child does not realize that every ten numbers you need a new prefix. At "twenty-ten" the ones place is filled up, but the child does not realize this. Alternatively, the child does not realize the cycle so that a new prefix comes after nine.

21. Yes, 5 is the middle number between 0 and 10.

23. We mark our years, in retrospect, from the approximate birth year of Jesus Christ—this is why they are denoted 1996 A.D.; A.D. stands for Anno Domini, Latin for "in the year of our Lord." Because we are marking in retrospect from a fixed point, we call the first hundred years after that point the first century, the second hundred years the second century, and so on. The first hundred years are numbered zero (for the period less than a year after Jesus' birth) through ninety-nine. This continues until we find that the twentieth century is numbered 1900 A.D. through 1999 A.D.

25. Place value is the idea of assigning different *number values* to *digits* depending on their position in a numeral. This means that the numeral 4 (four) would have a different value in the "ones" place than in the "hundreds" place because 4 ones are very different from 4 hundreds.

27. a. 11.57 days
b. 11,570 days, or 31.7 years

29. a. 21 **b.** 35 **c.** 55 **d.** 279 **e.** 26 **f.** 259
g. 51 **h.** 300 **i.** 13 **j.** 17 **k.** 153 **l.** 2313

31. Base nine

33. x = nine

35. This has to do with dimensions. The base ten long is 2 times the length of the base five long. When we go to the next place, we now have a new dimension, so the value will be 2×2 as much. This links to measurement. If we compare two cubes, one of whose sides is double the length of the other, the ratio of lengths of sides is 2:1, the ratio of the surface area is 4:1, and the ratio of the volumes is 8:1.

37. a.

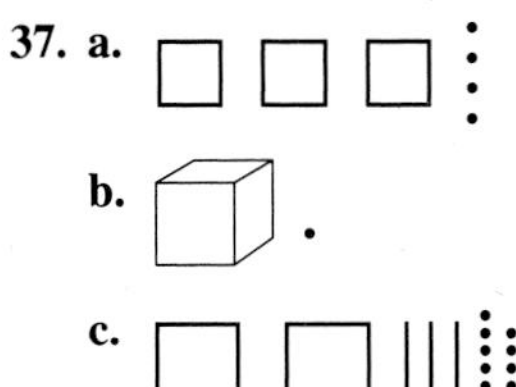

b.

c.

39. Answers will vary.

41. 835

43. b

45. b

CHAPTER 2 REVIEW EXERCISES

1. a. $\{x \mid x = 10^n, n = 0,1,2,3,...\}$

b. $\{1, 10, 100, 1000, 10{,}000,...\}$ or $\{10^0, 10^1, 10^2, 10^3, ...\}$

2. a. $\in$ **b.** $\subset$ **c.** $\subset$ **d.** $\not\subset$

3. a. $D \not\subset E$ **b.** $0 \notin \{\ \}$

4. a.

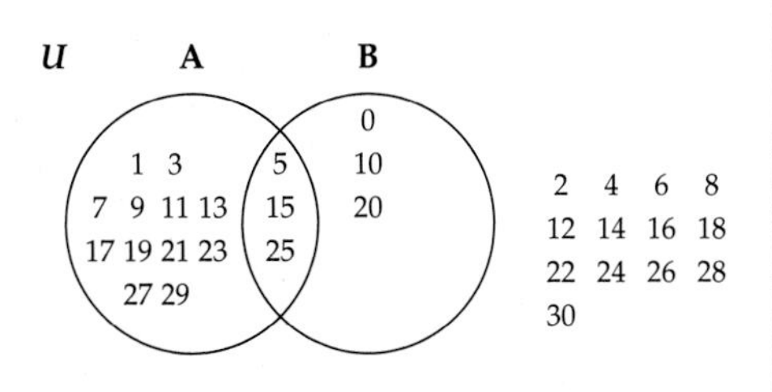

b. 5, 15, 25

c. The set of even numbers between 0 and 30

5. a.

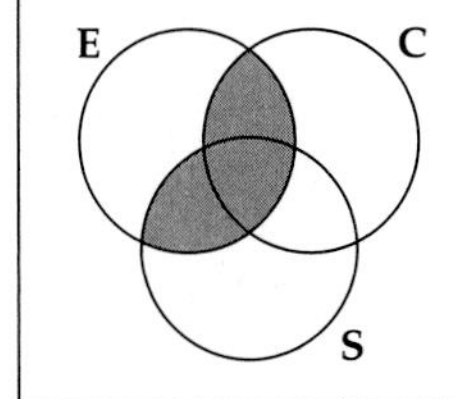

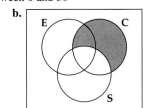

b.

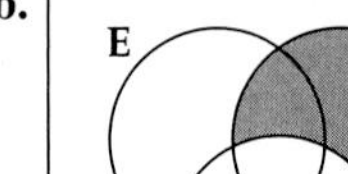

6. 50 have both.

7. The former means the same elements, and the latter means the same number of elements.

8.

	Egyptian	Roman	Babylonian
a. 47	𓎆𓎆𓎆𓎆 𓏺𓏺𓏺𓏺𓏺𓏺𓏺	XLVII	◀◀◀◀▼▼▼▼▼▼▼
b. 95	𓎆𓎆𓎆𓎆𓎆𓎆𓎆𓎆𓎆 𓏺𓏺𓏺𓏺𓏺	XCV	▼◀◀◀▼▼▼▼▼
c. 203	𓍢𓍢𓏺𓏺𓏺	CCIII	▼▼▼◀◀▼▼▼
d. 3210	𓆼𓆼𓆼 𓍢𓍢𓎆	MMMCCX	◀◀◀◀◀ ▼▼▼◀◀◀

9. a. 410_{five} **b.** 1300_{five} **c.** 1010_{two}

10. a. 4314_{five} **b.** 30034_{five} **c.** 1011_{two}

11. $25 \times 2 + 4 \times 5 + 3 = 73$

12. Because $1000_{five} = 125_{ten}$, I would rather have $\$200_{ten}$.

13. They both have the value of 3 flats, 2 longs, and 1 single. Because base six flats and longs have greater value than base five flats and longs, the two numbers do not have the same value.

14. Because we are dealing with powers. Thus, the value of a base ten flat is $2 \times 2 = 4$ times the value of a base five flat.

15. The value of the fifth place in base ten is $10^4 = 10{,}000$. The value of the fifth place in base five is $5^4 = 625$. $10000 \div 625 = 16$.

16. There are many possible responses. Here are three:
"One-zero" is the amount obtained when the first place is full.
It means you have used up all the single digits in your base.
It is the first two-digit number.

17. There are several equivalent representations:
$2000 + 60 + 8$
$2 \times 1000 + 0 \times 100 + 6 \times 10 + 8 \times 1$
$2 \times 1000 + 6 \times 10 + 8 \times 1$
$2 \times 10^3 + 0 \times 10^2 + 6 \times 10^1 + 8 \times 10^0$
$2 \times 10^3 + 6 \times 10^1 + 8 \times 10^0$

18. Answers will need to include all six characteristics described in the section.

CHAPTER 3 EXERCISES

Section 3.1

1. Answers will vary.

3. We know that each horizontal move to the right on the table increases the value by 1, and we know that any vertical move down on the table increases the value by 1. Each move on the diagonals described results from a horizontal move to the right and a vertical move down, so the value will increase by 2. Algebraically, we can find the location in the table of the sum of any two numbers, x and y. We can then use the structure of the addition table to determine the value of other cells in relation to $x + y$. The value of the cell that is diagonally down to the right must be 2 units greater than $x + y$. If we extend the table, we find that the sum of the next term on this diagonal is $x + y + 4$, that is $(x + y + 2) + 2$. Thus, the value of terms on this diagonal increases by 2 each time.

	x	$x + 1$
y	$x + y$	$x + y + 1$
$y + 1$	$x + y + 1$	$x + y + 2$

5. Answers will vary. One possibility for each is given below.

a. $47 + 53 = 100$ (compatible numbers). Now add 2 more to get 102.

b. $70 + 77 = 147$. Now subtract 1 to get 146.

c. $56 + 20 = 76$. Now take away 1 to get 75.

d. $575 + 125 = 700$. Now add 3 to get 703.

e. $900 + 70 + 13 = 970 + 13 = 983$

f. $70 + 30 = 100$, $100 + 180 = 280$, $280 + 60 = 340$, $340 + (5 + 5) = 350$, $350 + 13 = 363$

g. $387 + 24 = 400 + 11 = 411$. Then $411 + 53 = 464$.

h. $300 + 400 - 5 - 6 = 700 - 11 = 689$

i. $295 + 400 - 6 = 695 - 6 = 689$

j. $186 + 600 - 2 = 786 - 2 = 784$

7. a. The child is adding one place at a time, beginning with the ones place but using the break and bridge strategy.

b.

$48 + 27 = 48 + (2 + 25)$	subtraction
$= (48 + 2) + 25$	associative
$= 50 + 25$	addition
$= 50 + (20 + 5)$	substitution
$= (50 + 20) + 5$	associative
$= 70 + 5$	addition
$= 75$	addition

c. $36 + 4 = 40$ and then $40 + 80 = 120$.

d. $268 + 347 = (260 + 8) + (340 + 7) = (260 + 340) + (8 + 7) = 600 + 15 = 615$

9. One explanation: The digits in the ones place are added first: $6 + 8 = 14$. Then the digits in the tens place are added: $3 + 4 = 7$. Since the 7 is in the tens place, it has a value of 70. Finally, the two partial sums are added: $14 + 70 = 84$.

11. Answers will vary.

13. a. The student added the first digits of all the numbers to get $1 + 1 + 2 + 4 + 1 = 9$, then the second digits together to get $0 + 0 + 0 + 0 = 0$, and last the third digits $0 + 1 = 1$. The child does not understand the place values of the digits.

b. One possibility is that the student got 20 and then just added the other digits: $2 + 1 + 4 + 9$. The other possibility is that the student made a simple addition mistake (this happens frequently with 9s) and had $9 + 4 + 2 = 16$.

c. When adding the ones digits $8 + 3 = 11$, the child has placed 11 in the ones column instead of placing 1 in the ones and 1 in the tens column. The child does not understand why numbers greater than 9 are traded to the next column.

15. a. 10100_{two} **b.** 10000_{five}

c. 154_{six} **d.** $bf7_{sixteen}$

17.
$$\begin{array}{r} 4\,2\,0\,4_{six} \\ +\ 3\,5\,5_{six} \\ \hline 5\,0\,0\,3_{six} \end{array}$$

19. Answers will vary. One possibility for each is given below.

a. Leading digit: $1000 + 160 = 1160$

b. Using only leading digit we get: 16,000. Then, using compatible numbers ($900 + 100$), we can add another 1000 to get 17,000. Depending on one's short-term memory, one can go even further.

c. Leading digit and rounding: 146,000. Add leading digits: $8 + 3 + 2 = 13$ (representing 130,000). Then round each number to the nearest thousand. Then add second leading digit: $9 + 3 + 4 = 16$ (representing 16,000). $130{,}000 + 16{,}000 = 146{,}000$.

d. Rounding: $500{,}000 + 100{,}000 + 700{,}000 = 1{,}300{,}000$

e. Leading digit, compatible numbers, and rounding: 2800. $4 + 3 + 1 + 5 + 9 + 3 = 25$ (representing 2500). Then $73 + 34 \approx 100$; $45 + 55 = 100$; $65 + 43 \approx 100$. $2500 + 100 + 100 + 100 = 2800$.

f. Leading digit: 340,000

g. Leading digit and rounding: $2 + 3 + 6 + 1 + 2 = 14$ "hundred millions"; that is, 1,400,000,000.

21. a. $<$ goes in the circle. Reasoning: A quick estimation of the 1st three numbers is about 1100. We quickly see that this is less than $624 + 736$.

b. $<$ goes in the circle. Reasoning: Using leading digit, the sum of the first three numbers is 1100 plus less than 100. The sum of the next three numbers is 1000 plus more than 200.

23. a.
$$\begin{array}{r} 256 \\ +588 \\ \hline 844 \end{array}$$
b.
$$\begin{array}{r} 653 \\ 5246 \\ +\ 465 \\ \hline 6364 \end{array}$$
c.
$$\begin{array}{r} 5775 \\ +\ 7648 \\ \hline 13423 \end{array}$$

25. a. $862 + 731 = 1593$

b. $268 + 137 = 405$

Note: Other arrangements are possible for each problem.

27. This is one solution, but there are many more:
$$\begin{array}{r} 372 \\ 168 \\ +\ 459 \\ \hline 999 \end{array}$$

29. c

Section 3.2

1. Answers will vary.

3. a. $7 - 5 = 2$ **b.** $7 - 5 = 2$

c. $8 - 2 - 2 - 2 - 2 = 0$

5. a. $a - b \neq b - a$, because $b - a = -(a - b)$. For example, let $a = 7$ and $b = 4$. Then, $7 - 4 = 3$, and $4 - 7 = -3$.

b. $(a - b) - c \neq a - (b - c)$, because $(a - b) - c = a - b - c$ and $a - (b - c) = a - b + c$. For example, $(10 - 4) - 3 \neq 10 - (4 - 3)$, because $6 - 3 \neq 10 - 1$.

7. One possible way: $\$145{,}000 - \$116{,}000 = (\$145{,}000 - \$115{,}000) - \$1000 = \$30{,}000 - \$1000 = 29{,}000$

9. One possible way: $4132 - 2824$: $32 - 24 = 08$, $41 - 28 = 13$, $4132 - 2824 = 1308$ students

11. a. Samantha is using rectangles to represent tens and dots to represent ones. When she writes out the part to be subtracted, she crosses those pieces off the first list, trading from the tens symbols as needed.

b. Alice is basically doing the same thing as Samantha only she writes one of the tens rectangles as a sum giving 10 (4 + 6), so she can trade the 6 and leave behind the 4.

13. a.

$904 - 300 = 604$	The method works because
$604 - 60 = 544$	$904 - 367 = 904 - (300 + 60 + 7)$
$544 - 7 = 537$	$= 904 - 300 - 60 - 7$

b.

$367 + 7 = 374$	$904 - 367 = abc \Rightarrow 904 = 367 + abc$
$374 + 30 = 404$	First he adds what is needed to get
$404 + 500 = 904$	the c digit correct, trading the
$500 + 30 + 7 = 537$	extra to the tens column. Then he follows by working on "balancing" the tens and hundreds column.

c.

$904 - 360 = 544$	This method is like the method in
$544 - 7 = 537$	part (a), only the first two steps have been combined.

15. a. Step 1: Regroup

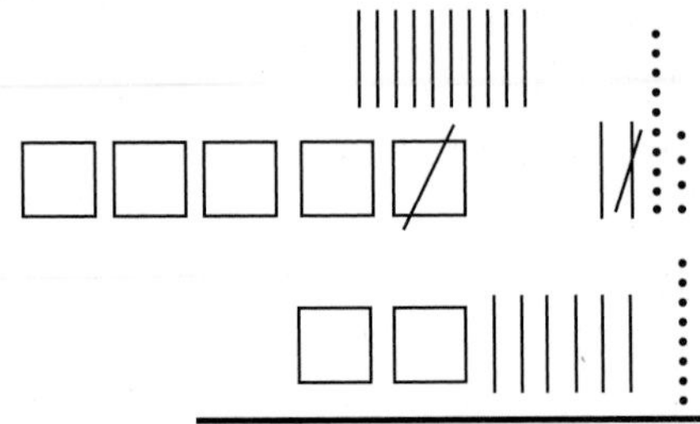

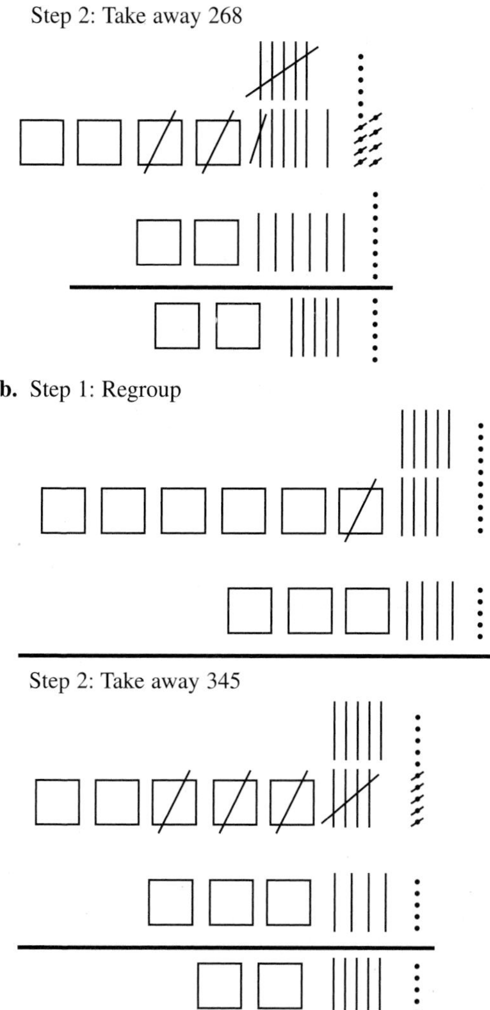

17. a. The child subtracted, $8 - 6$. Some children automatically subtract the smaller number from the larger number, even if it changes the order of the numbers in the problem.

b. The child wrote $6 - 8 = 2$, knew that 8 is bigger, and so traded from the tens place.

c. The child renamed the 6 in the ones column as 16 before subtracting, but didn't change the 7 in the tens place to 6.

d. The child probably reversed the numbers in the ones column in order to subtract, $8 - 0 = 8$, instead of renaming the 7 in order to subtract, $10 - 8$. Alternatively, the student could have reasoned that when you have a 0 in the minuend, you just bring down the digit in the subtrahend.

e. The child changed the zero in the ones place to 10, but didn't rename the 7 as 6 in the tens place.

f. The child renamed the 7 as 6 and each of the zeros as 10. The zero in the tens place should have been renamed as 9.

g. The child does not realize he is trading the 10 from the larger place-value column. He knows to do $14 - 9$, but does not change the 6 to 5 to compensate. Thus the next two column subtractions are wrong.

h. The child is looking at the column $2 - 9$ and rearranging to actually do $9 - 2 = 7$. Since $3 - 2$ does not need rearranging in the child's eyes, she does $3 - 2 = 1$. The child does not understand that there is a difference between $a - b$ and $b - a$.

19. Answers will vary. One possibility for each is given below.

a. Leading digit: 2100

b. Round both up to next thousand: $66{,}000 - 30{,}000 = 36{,}000$

c. 28 to 30 is 2, 30 to 72 is 42, so an estimate is 44,000.

d. 16 to 26 is 10, to 36 is thus 20, to 43 is thus 27, that is, 27,000.

e. Think of $413 - 285$. 285 to 300 is 15, to 413 is 113 more. $113 + 15 = 128$. Thus, estimate is 128,000.

f. $18 + 14 = 32$. So 180,000 to 320,000 is 140,000. Can refine the estimate by taking off 4000 to say 136,000.

g. Rounding: just over 1 million

h. Rounding: about 40 million

21. a. There is more than one solution. b can be any digit, but $a = 0$, $c = 4$, and $d = 2$.

b. There are seven solutions: in each case $y = x + 2$.

23. Not reasonable. By using leading digit and looking at the hundreds place, we can quickly see that the answer will be less than 6000.

25. $3020_{\text{five}} - 441_{\text{five}} = 2024_{\text{five}}$

27. Some possibilities: $968 - 734$, $946 - 712$. There are many!

29. The mule was carrying 5 bales of cloth and the horse was carrying 7 bales.

31. The primary reason for the lack of tables is the inverse relationship between the two operations. If you know your addition and multiplication "facts," then you already know the corresponding subtraction and division "facts." For example, if you know that $4 \times 8 = 32$, then you know that $32 \div 8 = 4$, etc. Also, a subtraction or division table is unnecessary—if you know your addition or multiplication facts, you don't need a subtraction or division table. Thus, a subtraction or division table would send a mixed message that this is something new and important. A second reason is that a subtraction or division table would imply more complexity. Since subtraction and division are not commutative, there would be more "facts" to learn.

33. 132 miles between Fresno and Bakersfield

35. a

37. b

Note: 64% of fourth-graders got this correct.

Section 3.3

1. Answers will vary.

3. 4×3 because we are adding 3 four times

5. No. Assign numerical values to a, b, and c to show that $a + (b \times c) \neq (a + b)(a + c)$.

7. a. The upper and lower halves are symmetric, which illustrates the commutative property of multiplication.

b. There are many ways to express the patterns. The most basic is that if you look at the two ends of the diagonal, you find the same number. As you move from the ends to the center of the diagonal, you continue to find matching numbers. Some diagonals have a center number with no partner and some do not. The diagonals with a center number have odd numbers at the ends.

c. Both products of the diagonals have the same factors. If the factors of the top left term are x and y, the factors of the products of the diagonals are x, y, $x + 1$, and $y + 1$. Using algebraic notation, we can say:

	x	$x + 1$
y	xy	$y(x + 1)$
$y + 1$	$(y + 1)x$	$(y + 1)(x + 1)$

In each case, we are adding 9 to the previous amount. Adding 9 is equivalent to adding a long and then taking away one single. Thus, we have +1 in the 10s place and −1 in the ones place.

9. When 9 is multiplied by a number n, the product can be thought of as 10 times the number minus the number, $9n = 10n - n$. In this case, $9 \times 7 = (10 \times 7) - 7$. When n is 10 or less, the value of the digit in the tens place of the product is $10(n - 1)$, and the value of the digit in the ones place is $9 - (n - 1)$. Thus the sum of the digits is equal to 9. Since the bent finger represents n, the $n - 1$ fingers to the left of n accurately represent the tens place, and the remaining fingers to the right of the bent finger represent the ones place.

11. 900 people

13. a.

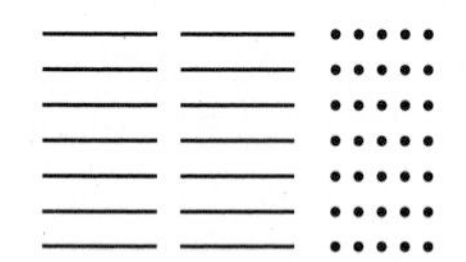

14 tens + 35 = 175

b. 10 10 10 10 111

10
1
1

$$\begin{array}{r} 400 \\ 30 \\ 80 \\ +\ \ 6 \\ \hline 516 \end{array}$$

c. 10 10 10 1

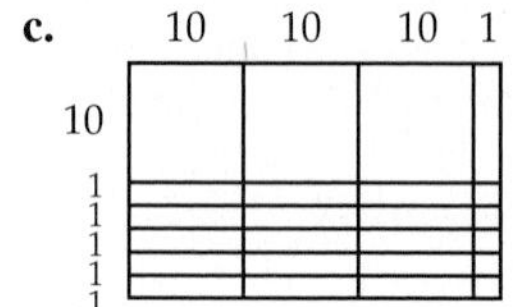

$30 \times 15 = 300 + 150 = 450$

$29 \times 15 = (30 \times 15) - 15 = 435$

15.

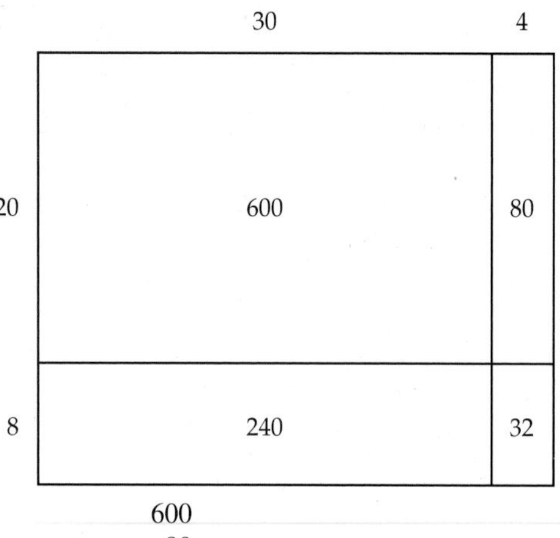

$$\begin{array}{r} 600 \\ 80 \\ 240 \\ +\ 32 \\ \hline 952 \end{array}$$

17. a.

$$\begin{array}{rl} 72 & \\ \times\ 34 & \\ \hline 8 & \leftarrow 4 \times 2 \\ 280 & \leftarrow 4 \times 70 \\ 60 & \leftarrow 30 \times 2 \\ 2100 & \leftarrow 30 \times 70 \\ \hline 2448 & \end{array}$$

This representation is nice because you don't have to do any trading, like using FOIL on $(70 + 2)(30 + 4)$.

b. This method is almost the same as the method in part (a); the order of multiplication is different.

c.

$$\begin{array}{rl} 72 & \\ \times\ 34 & \\ \hline 8 & \leftarrow 4 \times 2 \\ 28\ & \leftarrow 4 \times 7 \text{ (shifted because the 7 is in the tens place)} \\ 6\ \ & \leftarrow 3 \times 2 \text{ (shifted because the 3 is in the tens place)} \\ 21\ \ \ & \leftarrow 3 \times 7 \text{ (shifted because the 3 and 7 are both in the tens place)} \\ \hline 2448 & \end{array}$$

19. a. Set up the numbers as shown at the right. Find the four partial products and place them in the appropriate places: the ones digit in the lower spot and the tens digit in the higher spot of each cell.

4 6
1 2 | 1 8 | 3
2 8 | 4 2 | 7
1 7 0 2

Note: If the product of the two numbers is less than 10, you can either leave the top cell blank or place a 0 in that spot. To find the product, find the sum of the numbers on each of the diagonal rows. If the sum of any diagonal row is above ten, trade the digit in the tens place to the next diagonal.

b. One way to understand why the lattice method works is to compare it to obtaining the product using expanded form (shown at the right). Looking at both algorithms, you can see the placement of the partial products: 42, 28, 18, and 12. The lattice algorithm somehow enables each digit to be in its proper place. You can verify the correct placement of each digit by examining the actual value represented by each partial product. That is, $6 \times 7 = 42$ and the 2 is in the ones place and the 4 is in the 10s place. Similarly, 6×3 represents 6×30, which is 180. In the traditional algorithm, we don't even write the 0 (we just move over). Similarly, in the lattice algorithm, we deal only with "significant" digits: The 8 must be in the tens place and the 1 must be in the hundreds place.

$$\begin{array}{r} 40 + 6 \\ 30 + 7 \\ \hline 42 \\ 280 \\ 180 \\ 1200 \end{array}$$

c. You must know the multiplication table (how to multiply one-digit numbers) and how to add a list of one-digit numbers and trade if necessary. You must also be familiar with the ones and tens digit places in base ten.

d. Answers will vary.

21. Rather than trading the regroupings above the multiplication (where they are added in immediately following the multiplication step), the student places them in the addition rows (where they are added in during the addition step). She does this so that the multiplication and addition are completely separated.

23. The child put the 8 from 28 below and then also traded the 8.

25. Answers will vary.

27. b is wrong. Estimate it as $20 \times 900 = 18{,}000$, which is not even close to the answer, 30,102.

29. 3720 miles. Answers on cost will vary depending on price of gas and miles per gallon that the car gets.

31. The value of the largest place contributes more to the final product than all of the other places combined.

33. **a.** A novice would say reasonable because $60 \times 80 = 4800$ and thus is under 4800, but a more sophisticated student would say that, thinking of the four partial products, it will be clearly less than 4798.

b. No. This answer essentially takes the smallest and largest of the partial products: 8×4 and 4×3. The answer is also off by one full place: $400 \times 300 = 120{,}000$—and that error is the larger of the two.

35. **a.** 62×49. Reasoning 72×49 would be close to $70 \times 50 = 3500$, which is too big.

b. There are several possible ways to get 5 in the ones place. If we look at partial products and begin with a "middle possibility," $(90 + 5) \times (60 + 5)$, we can see that we have $5400 + 450 + 300 + 25$ which is going to be about 6200. So we can try a bit smaller, $(90 + 3) \times (60 + 5)$. This gives us $5400 + 450 + 180 + 15 = 6045$.

c. A quick estimate enables to know that the value of the tens place of the second number is 4. Thus we have 50 something times $43 = 2193$. We can get the answer by dividing 2193 by 43.

d. A quick glance tells us that we are looking at 80 something $\times$ 50 something and the answer has a 1 in the ones place and we need 611 from the partial products. What numbers can give us a 1 in the ones place? So we try 83×57, 87×53, 81×51, and 89×59.

37. A ream of paper is typically 500 sheets of paper.

a. 600,000 sheets of paper

b. \$18,000

c. \$12,000 savings

39. Answers will vary.

41. **a.** About 1 per hour means about 24 per day; round to 25 since it is slightly more than 1 per hour. $25 \times 365 = 9125$ per year.

b. Answers will vary. Some possibilities: How did they find out? Do they test both driver's and passenger's blood for alcohol when there is a fatal accident? Does this imply the drinker is the driver of the car or in the car? Does this count all kinds of traffic accidents: motorcycle, car hits pedestrian?

43. **a.** 3,796,416

45. **a.** This shortcut will work for any two-digit numbers that begin with the same tens digit where the sum of the ones digits is 10.

b. Label the digits of the two numbers to be multiplied: $ab \times ac$. The last two digits of the result come from multiplying $b \times c$. The digits that precede this come from $a \times (a + 1)$.

c.
$$\begin{aligned} ab \times ac &= (10a + b)(10a + c) \\ &= 100a^2 + 10ab + 10ac + bc \\ &= 10a(10a + b + c) + bc \\ &= 10a(10a + 10) + bc \\ &= 100a(a + 1) + bc \end{aligned}$$

The 100 in the product $100a(a + 1)$ simply moves the term two places to the left; thus, the product $a(a + 1)$ gives the first part of the answer and bc gives the second part.

47. **a.** The ones place of the missing multiplier is an 8. Without knowing the product, that is all we can say.

b. 68×83 will be above 5000 and 78×83 will be above 6000, so the multiplier is 68.

c. 8×83 and 18×83 will both be less than 2000, so the missing multiplier is either 8 or 18.

49. **a.** $530 \times 71 = 37{,}630$

b. $357 \times 01 = 357$

51. **a.** 425 **b.** 1872 **c.** 5040

53. **a.** $(30 + 4)(20 + 3) = 600 + 90 + 80 + 12$

$(3x + 4)(2x + 3) = 6x^2 + 9x + 8x + 12$

If x represents the number 10, then $6x^2 + 9x + 8x + 12 = 600 + 90 + 80 + 12$.

b. The FOIL algorithm works because it specifies the four partial products.

55. Alike: sum (product) of any row, any column, both diagonals are identical.

Difference: multiplication magic square, don't seem to need all numbers in the square to be different.

57. This is a game to be played by students so there are no answers.

59. 7

Section 3.4

1. 42 jelly beans. The partitioning model.

3. **a.** $3 \times 2 = 6$ (repeated addition)

b. $8 \div 2 = 4$ (repeated subtraction) or $8 \div 4 = 2$ (partitioning)

5. **a.** 300 **b.** 80 **c.** 50 **d.** 300 **e.** 40 **f.** 50

7. **a–b.** Answers will vary.

9. There are multiple ways to get answers.

a. $6 \times 10 = 60$ and $6 \times 5 = 30$, which gives an estimate of 15.

b. $\frac{450}{15} = 30$ and $\frac{45}{15} = 3$, so a quick estimate is 30; a refined estimate is 33.

c. $\frac{180}{6} = 30$ and $\frac{18}{6} = 3$, so a quick estimate is 30; a refined estimate is 33.

d. Cancel the zeros and we have $\frac{26}{2}$, which is 13. Here the exact answer is obtained.

e. $\frac{500}{25} = 20$ and $\frac{75}{25} = 3$, so a quick estimate is 20; a refined estimate is 23.

f. $\frac{750}{25} = 30$ and $\frac{100}{25} = 4$, so a quick estimate is 30; the exact answer is 34.

g. Cancel the zeros and we have $\frac{240}{5}$. Since $\frac{240}{10} = 24$, $\frac{240}{5} = 48$.

11. 4 cans

13. \$700 per month

15. **a.** 3023_{five} **b.** $203_{\text{five}}\text{r}4_{\text{five}}$ or $(203\frac{1}{2})_{\text{five}}$

17. 675 cases per week and 2700 cases per month, for a 5-day week and a 4-week month

19. Round up to get 53 weeks

21. ≈89 days

23. (a) The quotient must be greater than 25.

25. Not reasonable because $60 \times 60 = 3600$ and if you think about the partial products (shown below), you can see that their value is clearly more than 400.

$$\begin{array}{r} 60 + 7 \\ \times\ 60 + 8 \end{array}$$

27. Melanie needs to sell 160 tapes.

29. Suppose each glass of juice contains 8 ounces; then each guest will get 16 ounces of juice. Wei needs to buy 60 containers of juice. The juice will cost \$53.40.

31. One. As I was going to St. Ives, . . . everyone else was going the other way!

33. Answers will vary. Assumptions that must be made include the size of the page, the size of the type, and the number of columns per page.

35. 3 million crimes per year would mean about 17,000 crimes per school day (dividing by 180 school days per year). If we divide 17,000 by 50 states, that's about 340 crimes per day per state. That still seems high, so the term "crime" must be broad: vandalism, minor assault. Does it mean "crimes reported to police"?

37. Not quite. At this rate, she will burn exactly 370 calories.

39. A carton contains 4 packs $\times \frac{6 \text{ sodas}}{\text{pack}} = 24$ sodas. 247 students divided by 24 sodas per carton is 10 remainder 7. Round up to get 11 cartons total.

41. 1961, 30, 48, 20,000, 20

43. **a.** They will make the trip in less than 30 days.
b. Exact answer: 5010 miles

45. Exact answer: $238.19

47. (c) is wrong; the actual quotient is 78.

49. Francie is correct. There are many ways to arrive at this conclusion. Thinking of partitioning, if you increase the whole (dividend) and decrease the number of groups (divisor), each group will get more. If you increase the whole, then you've got to increase the divisor too.

51. Each child gets 4_{five} gumdrops.

53. **a.**

a	b	$a + b$	$a \cdot b$
65	66	131	4290
1208	72	1280	86,976

b.

a	b	$a + b$	$a \div b$
153	9	162	17
19	1	20	19

c.

a	b	$a - b$	$a \cdot b$
44	24	20	1056

d.

a	b	$a + b$	$a \div b$
10	2	20	5
34,225	37	1,266,325	925

55. **a.** $2\overline{)974} = 487$
b. $9\overline{)247} = 27.4$

57. **a.** Divisible by 3, 4, 6, 8
b. None of these numbers divide 2,345,678.

59. **a.** No **b.** No **c.** Yes **d.** Yes

61. 42,857

63. **a.** One solution is $66 + 80 - 3$.
b. One solution is $23 \times 60 + 23 \times 10$.
c. One solution is $(260 \div 10) \div 2$.
d. One solution is $25 \times 44 + 1 \times 44$.
e. One solution is $653 + 150 = 803$ and $803 + 31 = 834$, so the answer is 181.

65. **a.** $[(A + B + C) \times D]$ cents
b. $(A + B) \times C$ dollars
c. $A + BC - D - E$ dollars

67. d

69. a

71. b

73. 752; he accidentally multiplied.

CHAPTER 3 REVIEW EXERCISES

1. **a.** $7 + 2$ **b.** $7 - 4$ **c.** 2×3 **d.** $12 \div 4$

2.

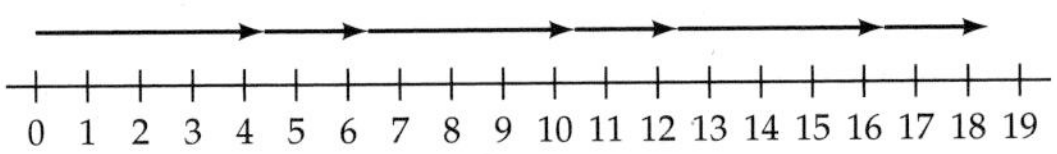

3. Responses will vary. This algorithm shows the sum of each place, beginning with the largest place. Once the sum for each place is written down, it is relatively easy to add those numbers mentally to write down the sum.

4. $345 - 97$

5. We have simply changed the representation of the minuend from 8 hundreds to 7 hundreds + 9 tens + 10 ones.

6. Responses will vary.

7. Responses will vary. At the heart of the matter is that, regardless of the places involved, 10 of something is being exchanged for 1 of something else.

8. Responses will vary. A common solution path is to see the four partial sums and add them: 6 hundreds, 14 tens, 12 tens, and 28 ones.

9. The digits being multiplied to get the second row are 37×2. However, this represents 37×20, which is 740. We commonly omit the zero and just write 74; however, because its real value is 740, we must place the 7 and the 4 in the correct places.

10. Responses will vary. At the heart of the matter is repeated addition. Essentially, we have 13 groups of 25. We can more easily get the answer by finding 10 groups of 25 (250) and 3 groups of 25 (75) and adding these amounts together.

11. 541×82. Responses will vary.

12. Responses will vary. One response is to note that a visual representation of the problem involves four partial products, only two of which are shown here. Pete's method gives us only two of the four products needed.

13. Responses will vary. The heart of the response is to note that if we are adding, when we take 1 from 17 and give it to 29, it is literally 1. However, if we are multiplying, when we take 1 from 17 and give it to 29, we are really taking one group of 29, rather than just 1.

14. Responses will vary. At a concrete level, using the example cited, the origins of 12×20 are (3×4) and (4×5). The origins of 15 and 16 are (3×5) and (4×4). Using the commutative and associative properties, we can show the equivalence of (3×4) times (4×5) and (3×5) times (4×4).

15. $a = 16$ and $b = 30$ or vice versa.

16. Using a calculator, $73932500 \div 97665 \approx 757.00097$. Multiplying 757×97665 gives us 73932405 , which we can subtract from 73932500 to get 95. Thus $x = 757$ and $y = 95$.

17. The answer is an infinite sequence beginning with 91 and increasing by 75 each time. That is, 91, 166, 241, 316, and so on.

18. 12

19. $79 \times 9 \times 11 = 7821$

20. In each case, we are determining how much is being used up by the digit in the divisor. That is, when we say 5 goes into 43 eight times, we are saying that we are using up $80 \times 5 = 400$.

21. Responses will vary.

22. Responses will vary.

23. a. Commutative and associative properties transform the problem into (36 + 64) + 82, which equals 100 + 82.
b. The distributive property transforms the problem into 20×19.
c. Representing 1592 as 1600 − 8 , and then using the distributive property of division over subtraction transforms the problem into $\frac{1600}{8} - \frac{8}{8}$.

24–28. Responses will vary. One possible response for each is presented here.

24. a. $3684 + 8312 \approx 3700 + 8300 = 12{,}000$
$2853 + 6241 \approx 2900 + 6200 = 9100$
$12000 + 9100 = 21{,}100$ for an estimate.
b. $44268 - 28843 \approx 44000 - 29000 = 45000 - 30000 = 15{,}000$ for an estimate.
c. $468 \times 9 \approx 468 \times 10 = 4680$, which is an overestimate by 468. So $4680 - 470 = 4210$ is an estimate.
d. $48 \div 14 \approx 3\frac{1}{2}$, so multiplying by 100, I would estimate 3500.

25. \$345,300 − \$216,250 ≈ \$345,000 − \$215,000 = \$130,000

26. She makes approximately 30 trips per semester and drives about 125 miles each trip. $30 \times 125 = 3750$ miles.

27. $489 \div 19 \approx 500 \div 20 = 25$ miles/gallon

28. Round each price to the nearest 50¢: \$1.50 + \$2.50 + \$2.00 + \$0.50 + \$3.50 + \$1.00 + \$1.00 + \$1.00 = \$13.00

29. When we change 638×42 to 638×40, we are decreasing the answer not by 2 but by 638×2, roughly 1200. Thus, when we round up 638, we need to make up that 1200. If we round 638 to 640, we are increasing by 2×42; if we round 638 to 660, we are increasing by 22×42.

30. 338 days

31. 3 times

32. a. 600,000 pages
b. \$18,000
c. \$12,000

33. a. 11330_{five} **b.** 2211_{five} **c.** 3113_{five} **d.** 3214_{five}

34. Base seven

35. Base eight

36. Base four

37. a.
$$\begin{array}{r} {\scriptstyle 1\,1} \\ 3684 \\ +\,4248 \\ \hline 7932 \end{array}$$
b.
$$\begin{array}{r} {\scriptstyle 5} \\ \not{6}\not{0}\not{0}3 \\ -\,3284 \\ \hline 2719 \end{array}$$
c. $a = 4$; $b = 6$

CHAPTER 4 EXERCISES

Section 4.1

1. a. −218 **b.** −2 **c.** 19 **d.** −78 **e.** 14
f. −18 **g.** 7 **h.** 2 **i.** 221 **j.** −6

3. −5° Fahrenheit

5. \$864

7. −\$19

9. −40° Celsius

11. *One way:* $-19 + (-6) = -25$; $-25 + (-22) = -47$; $-47 + 8 = -39$; $-39 + (-4) = -43$; $-43 + 7 = -36$; and $-36 + 1 = -35$
Another way: $(-19 + (-6) + (-22) + (-4)) + (8 + 7 + 1) = -51 + 16 = -35$

13. $2\frac{2}{3}$ hours (2 hours, 40 minutes) difference in flight time

15. Answers will vary. Here are some possibilities:

a.

18	7	11
5	2	3
13	5	8

b.

−5	−3	−2
−2	−1	−1
−3	−2	−1

c.

3	5	−2
8	8	0
−5	−3	−2

d.

a	b	$a - b$
c	d	$c - d$
$a - c$	$b - d$	$a - b - c + d$

17. Answers will vary.

19. a. Always positive. The absolute value of a nonzero expression is always a positive.
b. It depends. If $x > y$, then it will be positive. If $x < y$, then it will be negative.
c. Always positive. $x^2 + y^2 > xy$.
d. It depends. If $x^2 + 2xy > y^2$, then it will be positive.

21. 176 pounds

23. b

Section 4.2

1. Answers will vary. Some possibilities include:

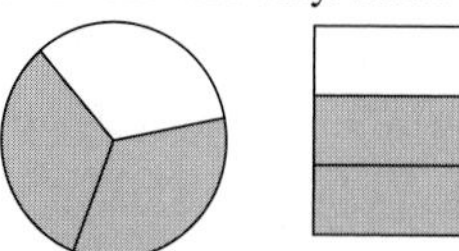

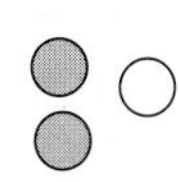

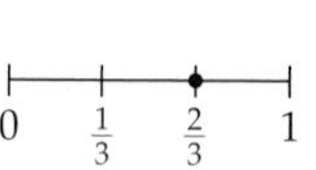

$\frac{4}{6}$ $.\overline{6}$

3. Larger piece $= \frac{1}{3}$, smaller piece $= \frac{1}{15}$

5. a. $\frac{3}{4}$ **b.** $\frac{1}{4}$ **c.** $\frac{2}{5}$ **d.** $\frac{1}{10}$ **e.** $\frac{1}{4}$ **f.** $\frac{1}{16}$

7. Answers will vary.
a. $\frac{2}{10}, \frac{3}{15}, \frac{4}{20}$ **b.** $\frac{6}{8}, \frac{9}{12}, \frac{12}{16}$
c. $\frac{4}{6}, \frac{6}{9}, \frac{8}{12}$ **d.** $\frac{10}{12}, \frac{15}{18}, \frac{20}{24}$

9. a. $\frac{5}{8}$ **b.** $\frac{2}{3}$ **c.** $\frac{9}{10}$ **d.** $\frac{2}{5}$ **e.** $\frac{7}{9}$ **f.** $\frac{7}{11}$
g. $\frac{29}{30}$ **h.** $\frac{1}{5}$

11. a. The value of the regions from greatest to least is $\frac{1}{2}, \frac{1}{3}, \frac{1}{6}$.
b. The value of the regions from greatest to least is $\frac{3}{8}, \frac{1}{4}, \frac{1}{6}, \frac{1}{18}, \frac{1}{12}$.
c. Answers will vary.

13.
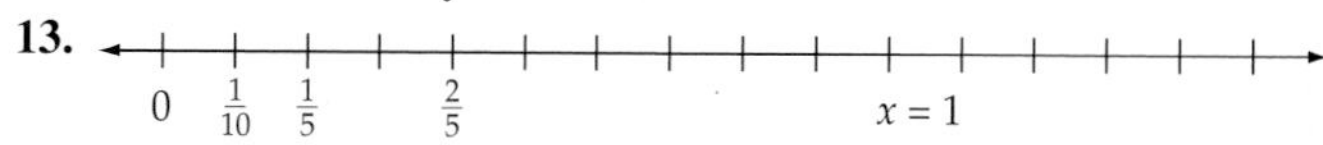

15. a. **b.** **c.** 16

17. a. $\frac{3}{7} < \frac{1}{2} < \frac{5}{8}$

b. $\frac{5}{6} < \frac{9}{10}$ Each 1 piece away from 1

c. $\frac{2}{7} < \frac{1}{3} < \frac{4}{11}$

d. $\frac{7}{9} < \frac{15}{17}$ Each 2 pieces away from 1

e. $\frac{2}{9} < \frac{1}{4} < \frac{5}{16}$

f. $\frac{3}{4} = \frac{75}{100} < \frac{79}{100}$

19. If you round both numbers down slightly, you have $\frac{35{,}000}{50{,}000} \approx \frac{7}{10}$.

21. a. $\frac{5}{8}$

b. $\frac{5}{6}$

c. $\frac{37}{158}$ is between $\frac{1}{5}$ and $\frac{1}{4}$. The middle thermometer is in that range.

23. $\frac{1}{6}$. Even though there are five sections, they are not equal in size. The shaded portion is $\frac{1}{3}$ of $\frac{1}{2}$ of the box.

25. This is a valid, though unconventional, response. If you take a whole and divide it into 2 equal parts and then shade in $1\frac{1}{2}$ of them, you have the same area as if you had divided the whole into 4 parts and shaded in 3 of them.

Alternatively, the ratio of $1\frac{1}{2}$:2 is equivalent to 3:4.

27. a.

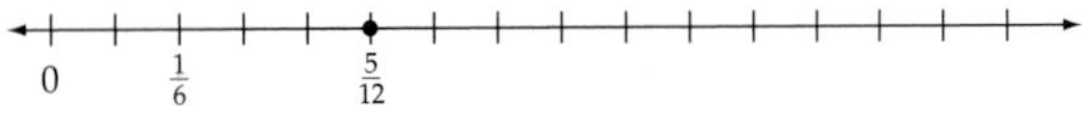

b.

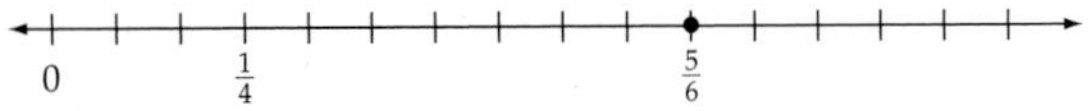

29. a. About $\frac{3}{4}$ of $\frac{1}{4}$ of a tank, so $\frac{3}{4} \cdot \frac{1}{4} = \frac{3}{16}$. Given that this is an approximation, one could also say $\frac{2}{3}$ of $\frac{1}{4}$, which equals $\frac{1}{6}$.

b. About $\frac{1}{2}$ of a tank plus another $\frac{1}{2}$ of $\frac{1}{4}$ of a tank, so $\frac{1}{2} + \frac{1}{2} \cdot \frac{1}{4} = \frac{1}{2} + \frac{1}{8} = \frac{5}{8}$.

31. $\frac{10}{13}$ is closer to 1 than to $\frac{1}{2}$. $\frac{1}{2}$ of 13 is $6\frac{1}{2}$, so $\frac{1}{2} = \frac{6\frac{1}{2}}{13}$. $\frac{10}{13}$ is $\frac{3\frac{1}{2}}{13}$ more than $\frac{1}{2}$ and $\frac{3}{13}$ less than 1.

33. a. $\frac{5}{15}, \frac{3}{4}, \frac{4}{5}$. $\frac{5}{15}$ is less than $\frac{1}{2}$; $\frac{3}{4}$ and $\frac{4}{5}$ are greater than $\frac{1}{2}$. $\frac{3}{4}$ is $\frac{1}{4}$ less than 1; $\frac{4}{5}$ is $\frac{1}{5}$ less than 1. Since $\frac{1}{4}$ is greater than $\frac{1}{5}$, $\frac{3}{4}$ is less than $\frac{4}{5}$.

b. $\frac{5}{11}, \frac{2}{3}, \frac{6}{7}$. $\frac{5}{11}$ is less than $\frac{1}{2}$; $\frac{2}{3}$ and $\frac{6}{7}$ are greater than $\frac{1}{2}$. $\frac{2}{3}$ is $\frac{1}{3}$ less than 1; $\frac{6}{7}$ is $\frac{1}{7}$ less than 1. Since $\frac{1}{3}$ is greater than $\frac{1}{7}$, $\frac{2}{3}$ is less than $\frac{6}{7}$.

c. $\frac{3}{50}, \frac{1}{3}, \frac{2}{5}, \frac{5}{8}, \frac{3}{4}$. $\frac{3}{50} < \frac{5}{50} = \frac{1}{10}$ and $\frac{1}{10} < \frac{1}{3}$, so $\frac{3}{50} < \frac{1}{3}$. $\frac{1}{3} = \frac{2}{6}$. Since $\frac{1}{6}$ is less than $\frac{1}{5}$, $\frac{2}{6}$ is less than $\frac{2}{5}$, so $\frac{1}{3}$ is less than $\frac{2}{5}$. $\frac{5}{8}$ is $\frac{1}{8}$ more than $\frac{1}{2}$; $\frac{3}{4}$ is $\frac{1}{4}$ more than $\frac{1}{2}$. Since $\frac{1}{4}$ is greater than $\frac{1}{8}$, $\frac{3}{4}$ is greater than $\frac{5}{8}$.

d. $\frac{3}{10}, \frac{2}{5}, \frac{4}{7}, \frac{5}{6}, \frac{7}{8}$. $\frac{3}{10}$ and $\frac{2}{5}$ are the only ones less than $\frac{1}{2}$. $\frac{3}{10} < \frac{1}{3} < \frac{2}{5}$. $\frac{4}{7}$ is closer to $\frac{1}{2}$ than to 1. $\frac{5}{6}$ and $\frac{7}{8}$ are closer to 1 than to $\frac{1}{2}$. $\frac{5}{6} = 1 - \frac{1}{6}$ and $\frac{7}{8} = 1 - \frac{1}{8}$ and $\frac{1}{8} < \frac{1}{6}$, so $\frac{7}{8}$ is closer to 1 than $\frac{5}{6}$ is.

35. a. $\frac{2}{3}$ and $\frac{3}{4}$ are reasonable answers.

b. $\frac{3}{10}$ is closest to the actual fraction, but $\frac{1}{4}$ and $\frac{1}{3}$ are reasonable approximations.

c. $\frac{3}{4}$

d. $\frac{3}{5}$

37. Approximately $\frac{1}{5}$ of the trip is left. $\frac{24}{115} \approx \frac{22}{115}$, $22 \times 5 = 110$, and $22 \times 6 = 132$, so $\frac{1}{5}$ is the best answer.

39. Fuller. There are several ways to justify this. One way: How much larger (multiplicatively) is the denominator? That is, 32 times what = 264 and 58 times what = 402? Mentally, we can determine that $32 \cdot 8 = 256$ and $58 \cdot 7 = 406$.

That is, $\frac{32}{264}$ is slightly less than $\frac{1}{8}$ and $\frac{58}{402}$ is slightly greater than $\frac{1}{7}$. So $\frac{58}{402}$ is larger.

41. a. There are 48 students (30 girls and 18 boys) in the chorus.

b. $\frac{1}{4}$ and $\frac{1}{5}$ are both reasonable.

43. a. He was 12 and I was 48. In 6 years, he will be 18 and I will be 54; in 24 years, he will be 36 and I will be 72.

b. There are many solutions: … (10, 40), (12, 48), (14, 56) …

c. There are many answers. One is that both numbers are multiples of 10.

45. If the original fraction is less than one, the new fraction is larger. The difference between numerator and denominator becomes less significant as they increase, so they will be proportionately closer together, making for a larger fraction.

47. a.

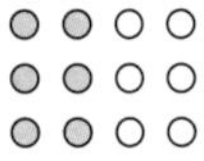

b. 

c. 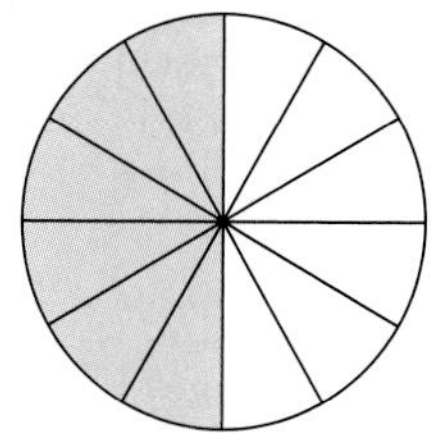

49. The student is trying to compare two fractions that have a different unit. The model needs to have the same unit.

51. a. 3 **b.** 3

53. $\frac{3}{12}$ or $\frac{1}{4}$

55. a

57. c

Section 4.3

1. a. $80\frac{17}{24}$ **b.** $-2\frac{23}{120}$ **c.** $23\frac{13}{20}$ **d.** $191\frac{11}{16}$ **e.** $-3\frac{7}{12}$

f. $66\frac{7}{24}$ **g.** 99 **h.** 135 **i.** $13\frac{1}{2}$ **j.** $\frac{1}{8}$ **k.** $\frac{3}{8}$

3. a. They added the numerators, but instead of first getting a common denominator, they just multiplied the denominators.

b. They added the denominators.

c. They added the denominators and changed to a mixed number using a base ten idea, 13 = 1 whole and 3 parts, instead of 1 whole $\left(\frac{8}{8}\right)$ and 5 parts.

d. They subtracted the larger fraction from the smaller fraction, ignoring the order.

e. They simply subtracted the numerators and denominators.

f. $7\frac{1}{8}$ should equal $6\frac{9}{8}$ not $6\frac{11}{8}$.

g. They multiplied 2×4 and 3×3, then put the digits side by side.

h. They multiplied both the numerator and the denominator by 5.

i. They cross-multiplied, one numerator by the other denominator.

j. They divided the numerators and divided the denominators.

k. They took the reciprocal of the first fraction instead of the second.

l. They divided each part separately; $8 \div 2 = 4$ and $\frac{1}{8} \div \frac{1}{4} = \frac{1}{2}$.

5. a. $\frac{23}{6}$ **b.** $\frac{23}{4}$

7. Repeated subtraction. Since $\frac{5}{5} = 1$, we repeatedly subtract $\frac{5}{5}$ from $\frac{13}{5}$ until our remainder is less than $\frac{5}{5}$.

9. Diagrams will vary.

a. $\frac{2}{3} \times 2\frac{3}{4}$:

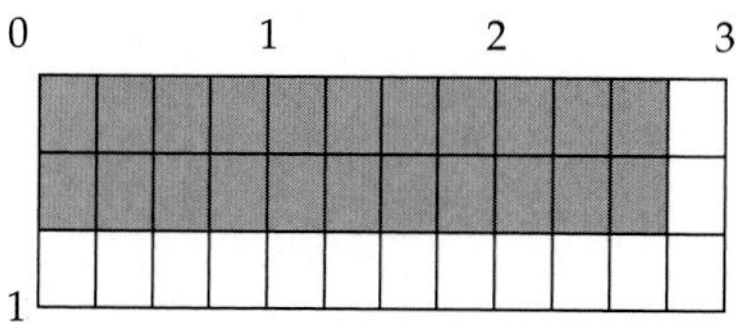

Rearranging the shaded area:

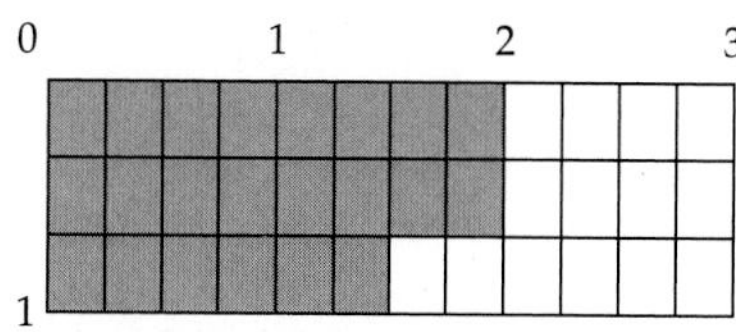

Total area $= 1 + \frac{2}{3} + \frac{2}{12} = 1\frac{5}{6}$

b. $2\frac{2}{3} \times 3\frac{1}{2}$:

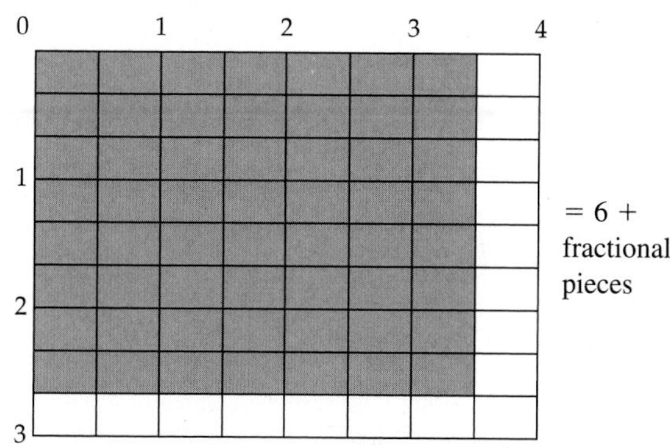

We have 6 whole pieces. Rearranging the fractional pieces:

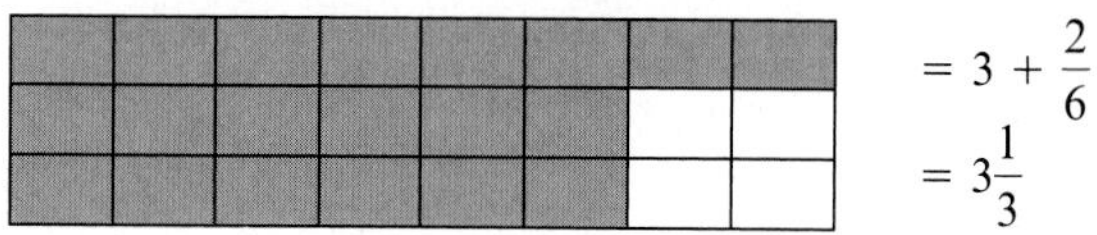

$= 3 + \frac{2}{6}$

$= 3\frac{1}{3}$

Total area $= 6 + 3\frac{1}{3} = 9\frac{1}{3}$

11. a. Less than 10. $\frac{5}{8} + \frac{3}{8} = 1$ and $\frac{3}{42} < \frac{3}{8}$.

b. Less than 2. One possible way: $\frac{3}{4} + \frac{1}{16} < 1$, so the sum of the three is less than 2.

c. Greater than 2. Since $\frac{1}{4} > \frac{1}{10}, \frac{1}{4} + 1\frac{9}{10} > 2$.

d. Between $5\frac{1}{2}$ and 6. $2\frac{2}{3} < 3$. Thus, $8\frac{1}{2} - 2\frac{2}{3} > 8\frac{1}{2} - 3 = 5\frac{1}{2}$.

e. Greater than 20. $8\frac{1}{2} \times 3$ (round up, round down) $= 25\frac{1}{2}$.

Alternatively $8 \times 2 = 16$, $8 \times \frac{7}{8} = 7$; we are already past 20.

f. Greater than $\frac{1}{2}$. If we double $4\frac{7}{8}$, we will clearly be over 9.

13. 114 teachers

15. $\frac{5}{42}$ of the weight is additives. Let 1 = total weight. Subtract the weight of the water and juice:

$$1 - \left(\frac{5}{7} + \frac{1}{6}\right) = 1 - \left(\frac{30}{42} + \frac{7}{42}\right) = 1 - \frac{37}{42} = \frac{5}{42}.$$

17. a. $213\frac{1}{3}$ ounces left **b.** $17\frac{7}{9}$ glasses

19. a. The child is adding the fractions as if they are pieces of a pie. The child took $\frac{1}{2}$ from $2\frac{1}{2}$ and added it to $\frac{1}{4}$. This gave 2 whole plus $1\frac{1}{4}$. The child took apart $2\frac{1}{2}$ into $2 + \frac{1}{2}$ and then used the associative property to connect $\frac{3}{4}$ and $\frac{1}{4}$.

b. The child is taking fractions apart and putting them back together and using the associative and commutative properties: $2\frac{1}{2} + \frac{3}{4} = \frac{1}{2} + 2 + \frac{3}{4} = \frac{1}{2} + 2\frac{3}{4} = \frac{1}{4} + \frac{1}{4} + 2\frac{3}{4} = \frac{1}{4} + 3 = 3\frac{3}{4}$.

21. a. $\frac{3}{4}$ is 3 pieces of a 4-piece pie. Take $\frac{1}{2}$ of each of the three pieces. The half pieces are each $\frac{1}{8}$ of the pie, so three of them would be $\frac{3}{8}$.

b. $\frac{3}{4}$ is 3 pieces of a 4-piece pie. Divide two of the three pieces so that you keep one and give one away. Divide the last of the original three pieces in half, keep one and give one away. What you have left is $\frac{1}{4} + \frac{1}{8} = \frac{3}{8}$.

23. It is not just repeated addition. The most general model of multiplication is that it represents the area of the rectangle formed by the two numbers. If you make a 1×1 square and find $\frac{1}{4}$ of that square and then find $\frac{1}{2}$ of $\frac{1}{4}$ of that rectangle, then you have a $\frac{1}{2}$ by $\frac{1}{4}$ rectangle, which is $\frac{1}{8}$ of the square.

25. a. *One way:* $26 \times 11 \div 12$. *Another way:* $11 \div 12 \times 26$.

b. $26 \times 12 \div 11$

27. a. If $4\frac{1}{2} \times 60 = 270$, then $4\frac{1}{2} \times 15 = 67\frac{1}{2}$ and $4\frac{1}{2} \times \frac{1}{2} = 2\frac{1}{4}$. So the answer is $67\frac{1}{2} + 2\frac{1}{4} = 69\frac{3}{4}$.

b. If we double both numbers, we have 9×31. The answer to $4\frac{1}{2} \times 15\frac{1}{2}$ will be $\frac{1}{4}$ of 9×31. If $9 \times 30 = 270$ and $9 \times 1 = 9$, then $9 \times 31 = 279$. One-half of $279 = 139\frac{1}{2}$ and one-half of $139\frac{1}{2} = 69\frac{3}{4}$.

c. $15 \times 36 = 540$, so $15 \times 18 = 270$ and $15 \times 9 = 135$ and $15 \times 4\frac{1}{2} = 67\frac{1}{2}$. Now we need one-half of $4\frac{1}{2}$, which is $2\frac{1}{4}$. So $15 \times 4\frac{1}{2} + \frac{1}{2} \times 4\frac{1}{2} = 69\frac{3}{4}$.

29. The conceptual error is that the student does not realize $\frac{a+b}{c+d} \neq \frac{a}{c} + \frac{b}{d}$.

To demonstrate: $\dfrac{9\frac{1}{4}}{3\frac{3}{4}} = \dfrac{9+\frac{1}{4}}{3+\frac{3}{4}} \neq \frac{9}{3} + \dfrac{\frac{1}{4}}{\frac{3}{4}}$, rather $\dfrac{9\frac{1}{4}}{3\frac{3}{4}} = \dfrac{9}{3\frac{3}{4}} + \dfrac{\frac{1}{4}}{3\frac{3}{4}}$.

31. a. $6 \div \frac{3}{4} = 6 \times \frac{4}{3} = \frac{24}{3} = 8$ guests can be served.

b. Thinking of each pint in 4 parts, there are $6 \times 4 = 24$ parts to split up. Give 3 to each person, $24 \div 3 = 8$ guests.

c. $\frac{3}{4} \times ? = 6 \Rightarrow \frac{3}{4} \times ? = \frac{24}{4} \Rightarrow \frac{3}{4} \times \frac{8}{1} = \frac{24}{4}$, thus $\frac{3}{4} \times 8 = 6$.

d. Giving $\frac{3}{4}$ to each guest, we'll add $\frac{3}{4} + \frac{3}{4} + \frac{3}{4} + \ldots$ until we get 8.

e. We have 6 pints and we'll give away $\frac{3}{4}$ pint repeatedly until we run out.

33. a. D **b.** A **c.** B **d.** E

35. 20 packages with 1 ounce left over

37. 13 boxes

39. The assumption is of monogamy and no same-sex marriages. That is, the number of married women = the number of married men. Thus, there are more men than women in this community. One solution path: since $\frac{2}{3}$ of the women = $\frac{1}{2}$ of the men, draw a diagram to represent this fact. (Shown at the right.) Now, all the parts are the same size. Since there are seven parts in the whole, $\frac{3}{7}$ of the population is single.

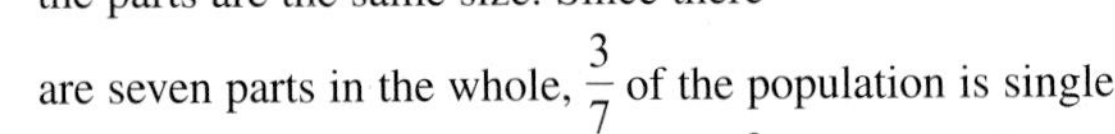

41. It will take 4 pressings to get at least $\frac{3}{4}$ of the juice.

It will take 6 pressings to get at least $\frac{9}{10}$ of the juice.

43. a. \$48

b. Answers will vary.

45. $\frac{2}{9} - \frac{1}{5}$

47. $\frac{7}{8} \div \frac{1}{9}$ if you assumed proper fractions; $\frac{8}{2} \div \frac{1}{9}$ if you did not

49. a. 110_{five}

b. 20_{five}

51. The student is "adding straight across." Either a set, linear or area model could be used to help with understanding. Using one of these models the need and process for finding a common denominator becomes clear.

53. a. 300 **b.** 120

55. In 60 seconds, when Ann has run 5 laps and Bob has run 4 laps (60 is the LCM of 12 and 15).

57. d

59. 4 gallons

61. b

63. Between 10 pounds and 11 pounds (3 bottles) and between 14 pounds and 15 pounds (4 bottles)

Section 4.4

1. a.

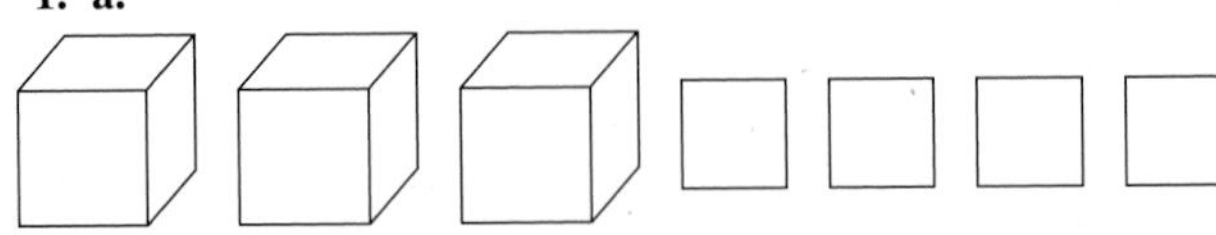

b. 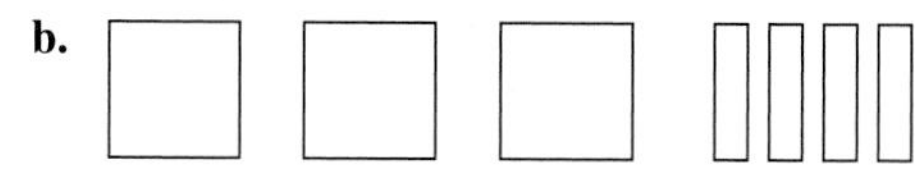

c. 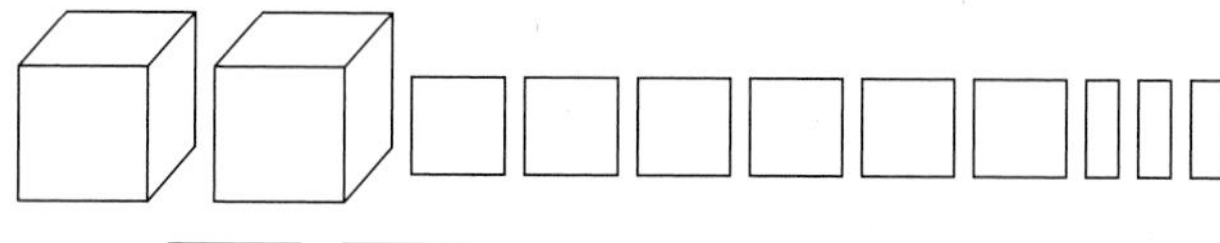

d. 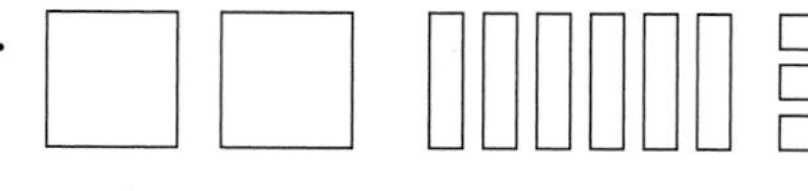

e. 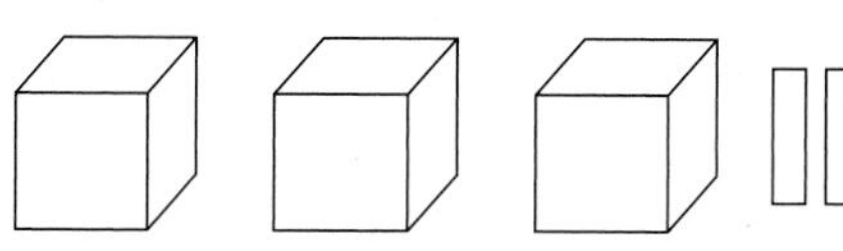

f. 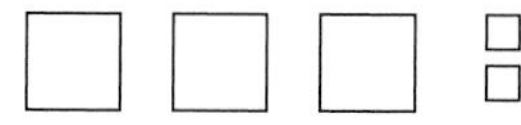

g. 1.64 **h.** 0.164 **i.** 2.05 **j.** .205

3. a. $4 + \frac{6}{10} = 4\frac{6}{10} = 4\frac{3}{5}$ **b.** $\frac{7}{10} + \frac{5}{100} = \frac{75}{100} = \frac{3}{4}$

c. $1 + \frac{2}{10} + \frac{3}{100} + \frac{4}{1000} = 1\frac{234}{1000} = 1\frac{117}{500}$

d. $4 + \frac{6}{100} = 4\frac{6}{100} = 4\frac{3}{50}$

5. a. $\frac{2}{3}$ **b.** $\frac{5}{9}$ **c.** $\frac{1}{11}$ **d.** $\frac{1}{7}$

7. a. 0.4 **b.** 0.98899 **c.** 0.05 **d.** 0.087

9. a. 0.0084, 0.058, 0.56, 0.6 **b.** 0.0086, 0.065, 0.9, 1.04

11. \$24,440,000 or \$24.44 million

13. a. 0.067 **b.** 4060.034

15. a. 8.24 **b.** 16.804 **c.** 1.85 **d.** 1.928

e. 11.18 **f.** 16.5435 **g.** $26.\overline{6}$ **h.** 24

17. a. 4.1 **b.** .64 **c.** .027 **d.** −.135 **e.** .03

f. 42 **g.** 5.6 **h.** 16 **i.** 31.2 **j.** .5

k. 2 **l.** 22.1 **m.** .609 **n.** .0452

19. a. Answers will vary. One possibility: 0.9995

b. Answers will vary. One possibility: $3\frac{445}{1000}$

21. Since the value of the digit in the 10ths place is the digit ÷ 10, the oneths place would be the digit ÷ 1. But this is the ones place. Thus the ones place and oneths place are the same place.

23. There are many answers.

a. 3276 = 36 × 91, so 32.76 = 3.6 × 9.1

b. 476 = 2 × 238, so 4.76 = 0.2 × 23.8

c. .72 ÷ 2 = 36 = .072 ÷ .2

25. a. We want 0.23 of 0.8 m^3. The word "of" is a good indication of multiplication: 0.23 × 0.8 = 0.184 m^3.

b. $\frac{40 \text{ miles}}{\cancel{\text{gallons}}} \times 0.75 \cancel{\text{ gallons}} = 30$ miles. Using multiplication, the units cancel properly to give us miles.

c. $5 \text{ liters} \times \frac{0.2 \text{ liters}}{\text{cup}}$ cannot be correct because the units do not cancel. Therefore this a division problem:

$5 \cancel{\text{ liters}} \times \frac{\text{cup}}{0.2 \cancel{\text{ liters}}} = 5 \div 0.2 = 25$ cups.

d. $7.2 \cancel{\text{ pounds}} \times \frac{\text{box}}{0.8 \cancel{\text{ pounds}}} = 7.2 \div 0.8 = 9$ boxes

e. 75 roses ÷ 5 bouquets = 15 roses per bouquet

f. $\frac{1 \text{ yard}}{\$15.00} \times \frac{1}{0.65 \text{ yards}}$ will not work because the \$ units are left in the denominator instead of the numerator. Thus, we try $\frac{\$15.00}{\cancel{\text{yard}}} \times 0.65 \cancel{\text{ yard}} = \$9.75$.

g. We want the answer to be in pounds per person: 5 pounds ÷ 12 people = $\frac{5}{12}$ pound of cookies per person.

h. $\frac{16 \text{ miles}}{\cancel{\text{second}}} \times 0.85 \cancel{\text{ second}} = 13.6$ miles

i. $13.9 \cancel{\text{ meters stretched}} \times \frac{1 \text{ meter original}}{3.3 \cancel{\text{ meters stretched}}} =$

13.9 ÷ 3.3 ≈ 4.2 meters original length

27. Answers will vary.

29. 102 is the number of vials that can be completely filled. 0.4 is the fraction of another vial that can be filled. 102 vials will use 127.5 ounces. The remaining 0.5 ounce is 0.4 of a 1.25-ounce vial.

31. \$632

33. \$70.64

35. a. 1.23456789×10^8 **b.** 3×10^{15}

c. 5.6×10^{-10} **d.** 3.02×10^{-7}

37. 12,090,000,000,000, which is approximately \$40,000 for every person in the United States (based on a population of 300 million).

39. Not much. Assume the average citizen drives 12,500 miles a year and has a car that gets 25 miles per gallon. The person would buy 500 gallons. \$0.05/gallon × 500 gallons = \$25.

41. a. The length of each candle

b. Answers will vary.

43. a. Light travels at a speed of 186,000 miles per second. To find out how far it travels in a year, you would do the following calculation: (186,000 miles/second)(60 seconds/minute)(60 minutes/hour)(24 hours/day)(365.25 days/year).

b. (14 × 3356 × 789) × 10,000,000,000 = 37,070,376 × 10,000,000,000 = 370,703,760,000,000,000

45. I guess they figure readers will be confused if they report it accurately, for example, $8\frac{1}{3}$ as 8.3 and $8\frac{2}{3}$ as 8.7.

47. $\frac{21}{60} = 0.35$, so it is 7.35 minutes.

49. a. 3.75 **b.** 6.67 **c.** 7:12 P.M. **d.** 8:00 A.M.

51. a. Diagram will be equivalent to $\frac{1}{5}$.

b. A0.A

53. a. 4.8 **b.** 4.05

55. e

57. d

CHAPTER 4 REVIEW EXERCISES

1. a. −4 **b.** $\frac{1}{4}$

2. a. −74 **b.** 19 **c.** −6 **d.** −18

3. −2

4. It means $(-3) + (-3) + (-3) + (-3) + (-3) + (-3)$.
5. The first symbol represents an operation. The second symbol signifies the value of the number. To mix them up is to become sloppy, which is not a good habit to get into. The number sentence given reads, "Negative 3 minus 4 is equal to negative 7."
6. Yes. The whole has been divided into four regions of equivalent value, and three of those four regions are shaded. The horizontal line is extraneous.
7. Because $\frac{5}{6} = \frac{35}{42}$ and $\frac{6}{7} = \frac{36}{42}$, you can go to 84ths and see that $\frac{71}{84}$ is between the two.
8. There are three spaces between 0 and $\frac{1}{5}$, thus each space is $\frac{1}{5} \div 3 = \frac{1}{5} \times \frac{1}{3} = \frac{1}{15}$. Counting by 15ths from 0 to the x, we can see that x is $\frac{11}{15}$.
9.

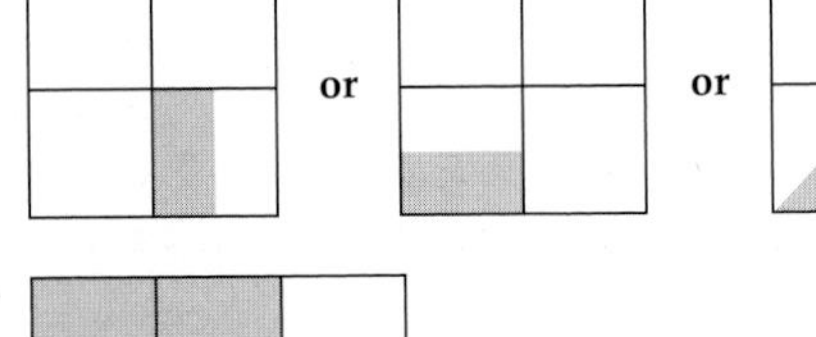

10.

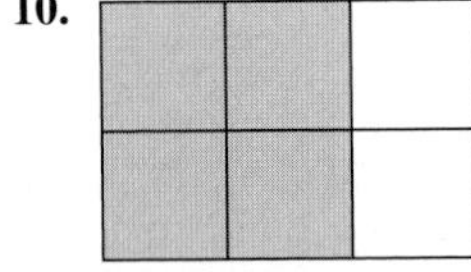

11. Answers will vary. Figure should have four dots.
12. **a.** Both are two parts from 1. Because 11ths are greater than 13ths, $\frac{2}{11}$ will be farther from 1, so $\frac{13}{15}$ is greater.

    b. $\frac{7}{12}$ is greater. Using $\frac{1}{2}$ as a benchmark, we see that $\frac{7}{12} > \frac{1}{2}$ and $\frac{13}{28} < \frac{1}{2}$.
13. When we divide 1 by 2 in base five we get $0.2222\ldots_{\text{five}}$.
14. It means they have the same value.
15. There are many ways to express the reason. One is to say that, by definition, the numerator of a fraction means how many "equal" pieces one has. If the denominators are not identical, then the numerators represent pieces of different size.
16. There are many ways to express why the hypothesis is invalid. One is to liken it to multiplication of two-digit numbers, such as 73×21. If you just multiply 70×20 and 3×1, you are missing the "two longs" regions.
17. We can solve it without the algorithm by finding $7\frac{3}{4}$, $2\frac{1}{3}$ times, that is:

$$7\frac{3}{4} + 7\frac{3}{4} + \frac{1}{3}\left(6\frac{3}{4} + 1\right) = 7\frac{3}{4} + 7\frac{3}{4} + 2\frac{1}{4} + \frac{1}{3} = 18\frac{1}{12}$$

18. It will be in region B because the answer is clearly positive, but it will be less than either of the two fractions because of the operator model of fractions.
19. She can make 15 cakes and will have $1\frac{1}{4}$ cups of flour left over.
20. You could do it in either of two ways: $4\frac{3}{4}$ is $\frac{1}{2}$ of $9\frac{1}{2}$, so this problem is more than $\frac{1}{2}$. Or $\frac{1}{2}$ of $8\frac{7}{8}$ would be $4\frac{3.5}{8}$, and because $\frac{3}{4}$ is greater than $\frac{3.5}{8}$, the answer is greater than $\frac{1}{2}$.
21. Answers will vary.
22. **a.** 1200 is a reasonable rough estimate, because $\frac{2}{3}$ of 18 is 12, and 1754 is close to 1800. Using the distributive property, we can get closer: $\frac{2}{3}(1800 - 46) = 1200 - 30 - 1170$.

    b. She has about $9 \times 7 = 63\ \text{cm}^2$.
23. **a.** 0.625 **b.** 0.004 **c.** $\frac{6}{25}$
24. The number of valid answers is infinite. They include 0.44441, 0.44442, and so on.
25. -0.54 0.045 0.454 $\frac{5}{11}$
26. 23.3. The two nearest tenths are 23.2 and 23.3.
27. $4,306,000,000
28. **a.** $3.4 \times 10^{10} \times 5.6 \times 10^{7} = 3.4 \times 5.6 \times 10^{10} \times 10^{7} = 19.04 \times 10^{17} \times 1.904 \times 10^{18}$

    b. $5.82 \times 10^{5} \div 2.3 \times 10^{10} = \frac{5.82 \times 10^{5}}{2.3 \times 10^{10}} = \frac{5.82}{2.3} \times 10^{-5} = 2.5 \times 10^{-5}$
29. **a.** 0.112 cubic meters. We multiply because we have the tank fraction to use as the operator.

    b. 53 cups. We divide because the problem is repeated subtraction–repeatedly taking away 0.15 liter.

    c. 8 boxes. Same reasoning as in part (b).

    d. $6.06. This can be seen as division as a proportion: $\frac{4.85}{0.8} = \frac{x}{1}$.

CHAPTER 5 EXERCISES

Section 5.1

Notes: (1) All of these problems can be solved in more than one way; some of the alternative solution paths are noted. (2) This is not an algebra course; most of these problems can be solved with proportions, but there are no problems for which solving a proportion is the only solution path.

1. 143.75 calories
3. 165 dentists
5. **a.** $750 **b.** $1250 **c.** $977.08
7. 860,000 people
9. Approximately $51\frac{1}{2}$ feet
11. Yes
13. $11\frac{3}{4}$ inches (approximate answer)
15. **a.** 48 cents **b.** 96 cents **c.** $4.41
17. Answers will vary.
19. **a.** 0.15 pound **b.** 14 ounces or .875 pound **c.** 49 million
21. Amy's red blood cell count is high.
23. 10 questions
25. 3.5 feet

27. 37 handshakes per minute, which is one handshake in less than 2 seconds

29. 2520 students

31. $1\frac{1}{2}$ hours more

33. More prison inmates; about 800,000 physicians; about 1,500,000 prison inmates

35. The ratio between boys and girls becomes greater, because the ratio of boys to girls added to the class, 1:1, is greater than the original ratio, 3:8.

37. a. The proportions of the ingredients vary with the quantity, due to the cooking process.

b. 8 cups liquid, 5 cups cereal, $\frac{1}{2}$ teaspoon salt. Answers may vary.

39. $\frac{199}{325} = 61.2\%$ and $\frac{5}{8} = 62.5\%$. These percentages are close enough to say that the advertisement is accurate. Also, $\frac{199}{325} \approx \frac{200}{325}$, which simplifies to $\frac{8}{13}$, but $\frac{5}{8}$ sounds better than $\frac{8}{13}$ for advertising purposes.

41. The person who gets paid twice a month will have a larger paycheck.

43. If you assume that the truck slowed her down by 20 miles per hour and Sonja claims that she would have arrived at class 20 minutes earlier, then she followed the truck for $6\frac{2}{3}$ miles. It seems unlikely that she would not have been able to pass the truck in this distance, unless she was driving on a narrow, winding road.

45. a. Answers will vary. For example, there are more than 100,000 more black males in prison than white males.

b. Answers will vary. For example, the rate of incarceration for black males is more than 7 times the rate of incarceration of white males.

c. Answers will vary. For example, from the raw numbers, the numbers for white females and black females are close. However, the rates for black females is almost 5 times the rate for white females. The disparity or inequity becomes more visible with rates.

47. a. Answers will vary. For example, the numbers in South Africa are slightly more than the number for India.

b. Answers will vary. For example, while the numbers for South Africa and India are similar, the rate in South Africa is more than 20 times the rate in India.

c. Answers will vary.

49. Additive decline 5.2 vs 8.5
Multiplicative decline 48% vs 38%

51. d

Section 5.2

1.

a. 18	**b.** 19.55	**c.** 1211.8	**d.** 54.4	**e.** 0.8625
f. 240	**g.** 80%	**h.** 85.7%	**i.** 8.52%	**j.** 26.1%
k. 18.75%	**l.** 145.8%	**m.** 46.35	**n.** 480	**o.** 7.192
p. 70	**q.** 32.4	**r.** 43.65	**s.** 1185.6	**t.** 160

3. 35%. Using compatible numbers, 23 × 3 = 69, so 23 is just over 33%.

5. 700 patients. One way to estimate is to use guess–check–revise. 30% of 1000 is 300, so try smaller. Using this method, you can get the actual answer.

7. a. 131%. Estimating: 42,000 to 84,000 is 100% increase. Answer is more than 100%.

b. 26%. Estimating: Using compatible numbers, 15,000/60,000 $= \frac{1}{4}$.

c. 43%. Estimating: Increase is about 31,000, or just under $\frac{1}{2}$.

d. 132%. Estimating: Similar strategy as in (a).

9. Approximately 6%. The baby lost $\frac{1}{2}$ pound. $\frac{1}{2} \div 8 = \frac{1}{16} \approx \frac{6}{100}$. Adult would have lost approximately 10 pounds. If you see $6\% \approx \frac{1}{16}$, then $\frac{1}{16} \times 160 = 10$.

11. About 12.8 million. 12,812,800

13. Approximately 41% of eligible voters aged 18–24 voted in 2008

15. $2200

17. $963.64

19. $30,941.04 after 2 years

21. $22,373,000

23. About 1.2 miles

25. The "whole" each year is not the same. Let's take someone making $40,000. If they get a 6% raise per year, their salary after 4 years would be $50,499.08. If they got 12% in the first year and then 4% for the next three years, their salary after 4 years would be $50,393.91. However, and here is where math is not simple, the total 4 years' salary under the first plan would be $185,483.70 compared to $190,241.60 under the second plan. So after 4 years, they would have more total salary under the second plan but their base salary would be slightly higher under the first plan.

At a simpler level, think of $40,000 with a 50% raise the first year and then no raise the second year vs a 25% raise each year. In the first plan, your salary jumps to $60,000 and then stays at $60,000 the second year. Under the second plan, your salary jumps to $50,000 the first year and then it goes up 25% of $50,000 the second year and so it goes up to $62,500.

27. a. The high school principals make 91% more than the teachers; the junior high school principals make 79% more than the teachers; and the elementary school principals make 68% more than the teachers.

b. The junior high school principals make 6% less than the high school principals. The elementary school principals make 12% less than the high school principals. The teachers make 48% less than the high school principals.

29. a. 5 out of 1000, or 1 out of 200, children will have a severe reaction.

b. Answers will vary. Possible questions include the following: What are the risks of not having the child vaccinated? How does the reaction compare with the illness itself? Are some children more likely than others to have a serious reaction?

31. 7% grade means that a hill changes 7 vertical feet for every 100 horizontal feet, or 70 vertical feet for every 1000 horizontal feet, or a proportionate amount of change. The percent grade gives drivers an indication of the steepness of a hill.

33. a. $348.40 per month

b. The raise was less than the rate of inflation.
$\frac{\$1200}{\$23,400} \times 100 \approx 5\%$ raise

35. If the cost of the item is less than $100, take $10 off. If the cost of the item is greater than $100, take 10% off.

37. a. $10.29

b. \$20,580
c. This assumes that she works 40 hours a week and 50 weeks a year.
d. She would be making \$10.33 an hour, which is about \$80 more a year.

39. \$121,885.98

41. $16\frac{2}{3}$

43. $2\frac{1}{2}$ hours

45. a. Using leading digit to get the total number of cars, we get about 140 million and $\frac{13}{140}$ is approximately $\frac{1}{10}$ or 10%.
b. 9.8%
c. The data are probably available from the Registry of Motor Vehicles, which every state has. I would guess these data are pretty reliable.

47. Impossible to say without knowing the total number of people who attended each game.

49. a. 48 inches $\div 3\frac{3}{4} = 12$ with 3 inches remaining
b. $\frac{3}{48} \times 100 = 6.25\%$ waste
c. 6.25% of \$40,000 = \$2,500 wasted
d. A 45-inch coil would result in no waste (although there are certainly other correct answers).
e. Both coils result in less waste than the 48-inch coil.
f. Answers will vary. It is essential to explain that since $48 \div 3.75 = 12.8$, the 0.8 represents $\frac{8}{10}$ of the 3.75 inches that is needed.

51. b

CHAPTER 5 REVIEW EXERCISES

1. \$69.29
2. \$1227.50
3. The ratio will increase.
4. They would need to add 7 teachers.
5. 559 miles
6. 50 more males
7. The birth rate is a ratio. Thus, to keep the ratio the same, if the population increases, the number of births also needs to increase. If the rate is down, it means that the number of births did not increase at the same rate as the population did.
8. We use rates to enable us to compare "apples and oranges." For example, let's say two cities had the same number of cases of a disease, but the second city had five times as many people as the first. If we get just raw numbers, we don't see the difference. However, if we convert the numbers to rates, the rate for the first city is five times the rate for the second, which tells us that the disease is much more prevalent in the first city.
9. Still need to raise \$240,000
10. To keep the same taste, she has to add both juices according to the ratio, which, in simplest terms, is 2 cups of grape juice for every 3 cups of orange juice. Thus, if she added 4 cups of grape juice, she would need to add 6 cups of orange juice to keep the ratios the same.
11. \$22,373,000
12. 67% weight gain
13. The task will be complete in one more hour.
14. $\approx$ 252 sec/mile = 4.2 min/mile or 4 min 12 sec/mile
15. \$1,575,000
16. 66% of the students have pets.
17. $\approx 0.046 = 4.6\%$ of the world's population lived in the United States in 2008.
18. 30 students
19. $\approx 0.80 = 80\%$
20. $\approx 0.023 = 2.3\%$
21. I had punched 0.0055 instead of 0.055.
22. Technically, the answer is \$125.64, but it is more likely that the original price was \$126. When you take 30% off \$126, you get \$88.20, which would probably be rounded down to \$88.
23. If the original price is greater than \$100.
24. **a.** Between 70 and 75 **b.** $\frac{123}{185} \approx \frac{12}{18} = \frac{2}{3} \approx 67\%$
25. Technically it comes to \$305,300, but the more reasonable answer is \$300,000, because 115% is an average.
26. His salary increase has outpaced the increase in the car price. $\frac{3200}{8400} = 0.38$ and $\frac{20{,}400}{59{,}000} = 0.35$, so the 1976 purchase took a bigger portion of his income than the 2006 Honda did.
27. An increase of 71%
28. Previous year $= \frac{72.0}{1.092} \approx 65.9$ ppm

CHAPTER 6 EXERCISES

Section 6.1

1. Answers will vary.

3. a. 15 **b.** 78 **c.** $\frac{n(n+1)}{n}$

5. a. 7 buses
b.

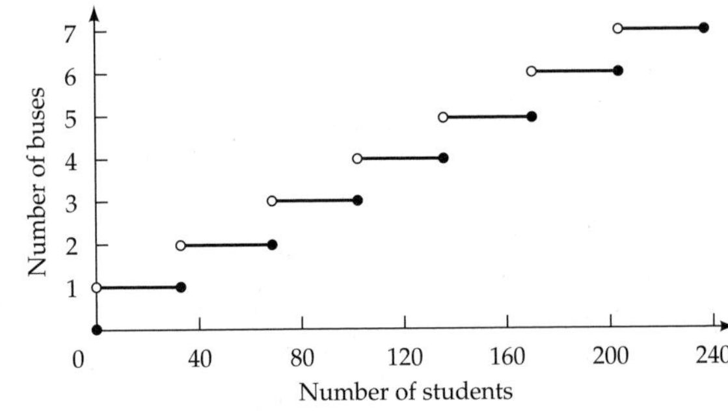

7. $2n^2 + 3n$

9. $12n - 3$

11. a. $T_n^2 - T_{n-1}^2$
b. It is a function. Justifications will vary.

13. $3n + 2$

15. a. If we assume the surface area of one square on a block is 1 square unit, then a stack of n blocks has a surface area of $4n + 2$ square units.
b. $6n + 4$

17. a. 8 hours
b. 5 hours

19. Answers will vary.

21. a. Answers will vary. Some patterns may be that the y values are increasing by 3 each time, or that they are 3(1), 3(2), 3(3), etc.
b. $y = 300$

c. The y value is 3 times the x value.

d. $f(x) = 3x$

23. a. $4^2 + 5^2 + 20^2 = 21^2$

b. $1^3 + 2^3 + 3^3 + 4^3 + 5^3 = 15^2$

c. $16 + 17 + 18 + 19 + 20 = 21 + 22 + 23 + 24$

Section 6.2

1. $\frac{3x + 6}{3} - x = x + 2 - x = 2$

3. $\frac{2x + 10}{2} - 5 = x + 5 - 5 = x$

5. Answers will vary.

7. The models are rectangles for even numbers, and rectangles with the "one left over" added on for odd numbers. If we take an even rectangle and multiply it an even or odd number of times, it will still be a rectangle (even). If we take an odd rectangle with the one left over and multiply it an odd number of times, there will still be one left over.
Algebraically, even times even = $2k \times 2n = 2(k2n)$ = even (2 times a number)
odd times even = $(2k + 1)(2n) = 2(2k + 1)(n)$ = even (2 times a number)
odd times odd = $(2k + 1)(2n + 1) = 4kn + 2k + 2n + 1$ = odd (1 left over)

9. a. $x = 8$. The balance scale below shows the first setup. Then we would subtract four +1s from each side to keep it balanced and are left with x = 8.

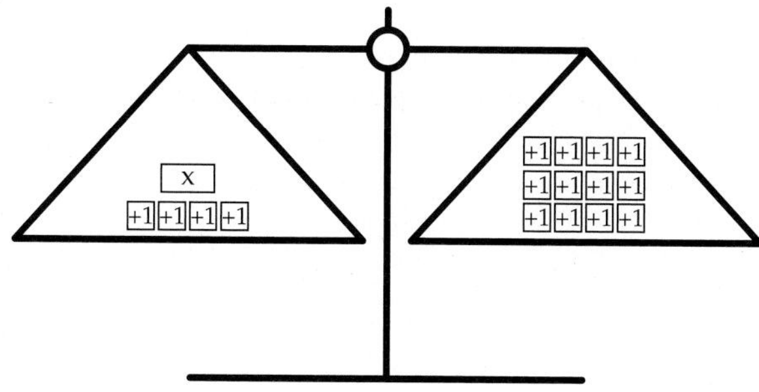

b. $n = 10$. From the setup shown here, we would then add four +1s to each side to keep it balanced, and which would leave the x only on the left side.

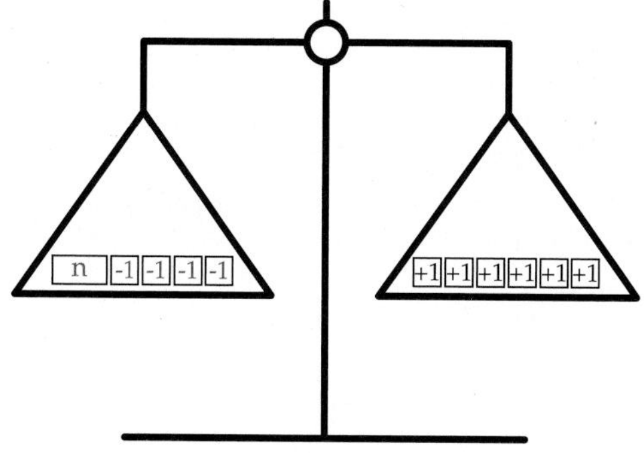

c. $x = 4$. After we subtract the two +1s from both sides, we can see that each x would equal +4.

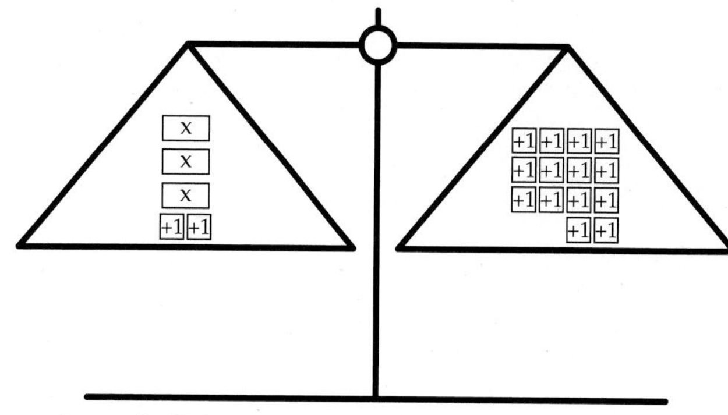

d. $x = 5$. We would add four +1s to each side which would give the two x's on the left side and ten +1s on the right side, so each x would equal 5.

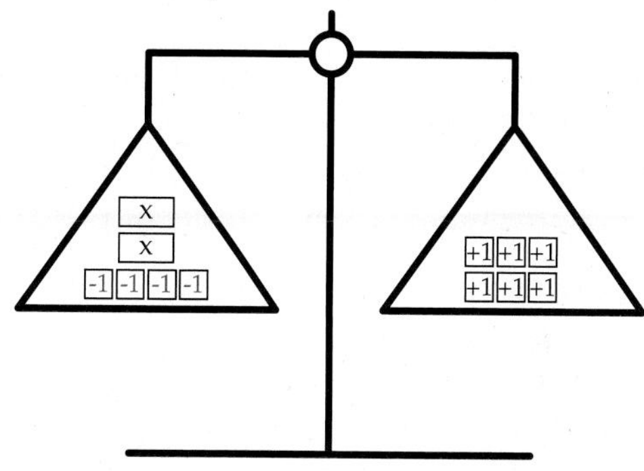

11. a. Yes

b. When the 4 is subtracted from both sides, the inequality stays the same.

13. a. $x > -3$

b. $x > 8/5$

15. Answers will vary.

17. a. 28 candies

b. $\frac{x}{2} + 8 = 22$

19. b

21. c

Section 6.3

1. Answers will vary.

3. a. Values in table may vary.

Number of People	Fee
10	190
20	230
30	270
40	310
50	350

b. $C = 150 + 4P$, where C is the cost, or fee, and P is the number of people.

c.

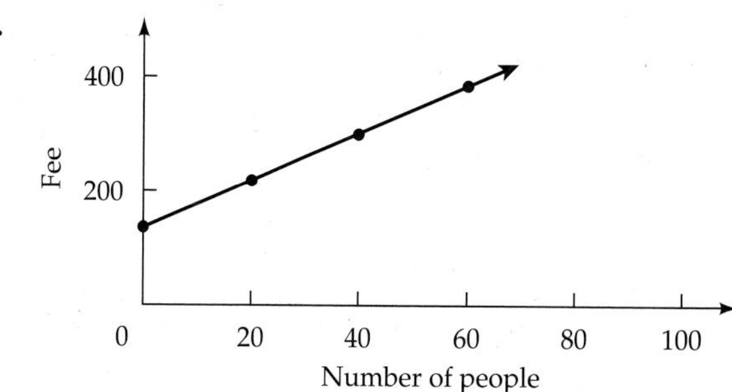

d. 78 people attended.

e. Answers will vary.

5. a. $x = 20$

b. Methods will vary.

c. Both are functions. In both plans, for a given number of checks, there is only one possible charge.

7. 164.9 cm or 1.65 m

9. 1.04 miles, or approximately 1 mile

11. book = \$4; notepad = \$7; pen = \$3

13. oval = 5 ; triangle = 4 ; rhombus = 6

15. a.

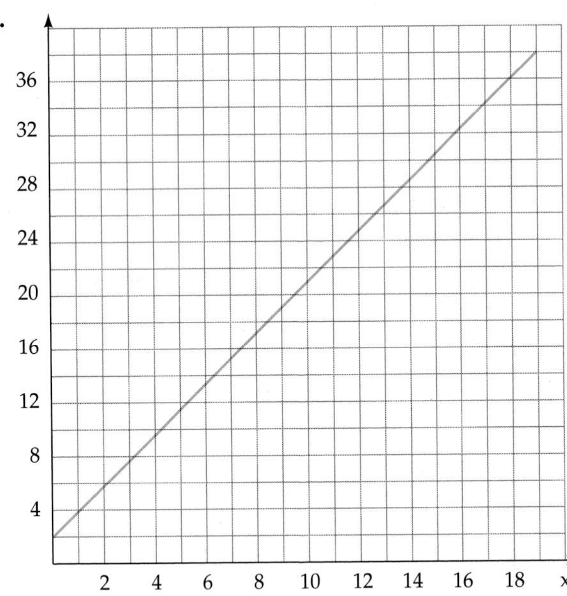

b. 200

17. a.

1200

600	T	T

A — T

b. $A + T = 1200$; $A = 600 + T$. Thomas earned \$300 and Adam earned \$900.

Section 6.4

1. a. \$3400

b. A decrease of \$800 per year

c. In $6\frac{1}{4}$ years

3. The first graph

5. Answers will vary.

7. Answers will vary.

9. Answers will vary.

11. a. The ramp rises 1 foot for every 12-foot increase horizontally.

b. 72 feet

13. a. Geometric, times 2

b. Arithmetic, plus 4

c. Square numbers; neither

15. Answers will vary.

17. No, it is not a constant rate of change, because the slopes of the lines change.

19. The population increased by 174,000 people during those 2 weeks.

CHAPTER 6 REVIEW EXERCISES

1. 316 student tickets

2. 16 stools

3. 43,200

4. There were two 10-cent stamps and ten 5-cent stamps.

5. a. 31; 121; $6n + 1$

b. 32; 524, 288; $2^{(n-1)}$

c. 242; 3,486,784,400; $3^n - 1$

d. 95; 3,145,727; $6 \cdot 2^{(n-1)} - 1$

6. "Equal" means to have the same value.

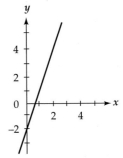

7. a.

x	$f(x)$
0	−2
1	1
2	4
3	7

$0 \to -2$
$1 \to 1$
$2 \to 4$
$3 \to 7$

b. 16

8. 28 rubles

9. $2(n + 2) + 1 = 2n + 5$

10. $2(2n + 1) + 1 = 4n + 3$

11. a. 110

b. $n(n + 1)$ gives the nth number.

12. a. 46

b. $4n + 6$

13. If he averages 400 or more minutes per month, then use the second plan. Otherwise, use the first plan.

14. Graph (a) is not a match because the temperature of the coffee will not go below room temperature. Graph (b) is not a good match because it implies that the coffee cools at a steady rate. The coffee will cool more rapidly at the beginning. As it gets closer to room temperature, the cooling will get slower. Graph (c) is the best match because the rate of cooling slows down over time.

15.

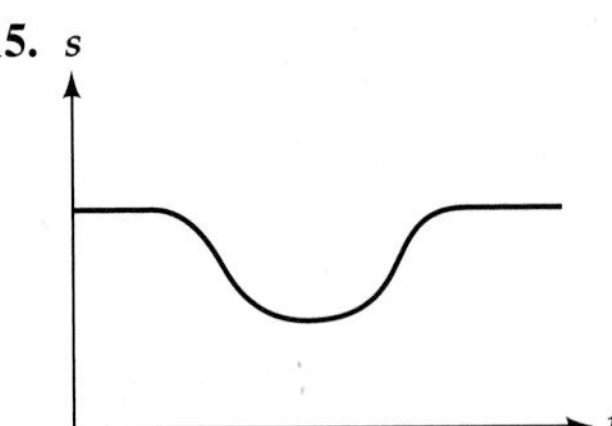

16. Answers will vary. **One story:** Time is independent, distance is dependent. Jackie walked steadily for a period of time and then sat down.

CHAPTER 7 EXERCISES

Section 7.1

Note: In cases where students are asked to explain the graph or to describe questions about reliability, validity, etc., there are many possible valid ways to answer those questions. Because of space limitations, only one response is given here. However, it should be interpreted as "one of many valid responses," as opposed to the right response or even the best response.

1. Answers will vary. Possibilities are given here:

a. How many times can a third-grader dribble a ball in one minute—only one hand can be used, the ball must visibly bounce off the floor each time, it can only touch the floor (i.e., not bounce off the wall).

b. How long can you hop on one foot—person can select one foot but cannot alternate from one foot to the other, everyone must hop with arms at sides or hop with arms out straight, all hop with same footwear, the other foot should be in same position, e.g., just off floor or by the knee, etc., can't touch anything with hands, for each hop the foot must visibly come off the ground.

c. How many concerts have you attended in the past year—can include free or paying, just focused on music, a street festival counts as one concert.

d. How much time do you study in a week—include all time spent on your courses outside class, reading, studying, working on projects. Round to the nearest hour.

3.

```
  X X
  X X
  X X
X X X X
X X X X   X
X X X X  X X  X X                X     X X   X
1 1 2 2 2 2 2 2 2 2 2 2 3 3 3 3 3 3 3 3 3 3 4 4 4 4 4 4 4
8 9 0 1 2 3 4 5 6 7 8 9 0 1 2 3 4 5 6 7 8 9 0 1 2 3 4 5 6
```

a. There were 27 students whose ages range from 18–46. The cluster, from 18–21, contains 18 students or $\frac{2}{3}$ of the class. Only 4 students are older than 27.

b. 24. That feels like the center of gravity, as if all these x's were on a see-saw.

c. Mean = 23.9, median = 21, modes = 19 and 21

5. a. Mean = 123, median = 122, mode = 121

b. They do not tell us anything about the shape of the data—the range, the spread, gaps, clusters, or outliers.

c.

```
                           X
                           X
                           X X
                         X X X       X
X                      X X X X  X X  X X
X     X  X       X X X X X X X X X X X X X X   X X    X    X
107 109 111 113 115 117 119 121 123 125 127 129 131 133 135 137 139
```

d. The number of raisins range from 107 to 138; most lie between 117 and 129.

e. It does not tell you the average.

f.

100–109	2
110–119	6
120–129	25
130–139	4

g.

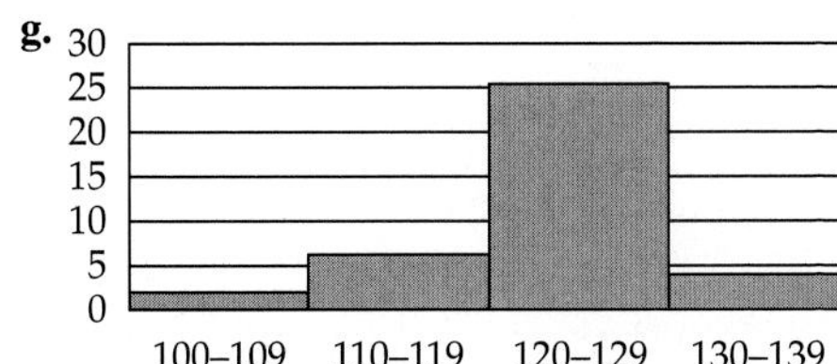

h. The histogram tells us that most of the boxes had between 120 and 129 raisins.

i.

100–104	0
105–109	2
110–114	2
115–119	4
120–124	16
125–129	9
130–134	2
135–139	2

j.

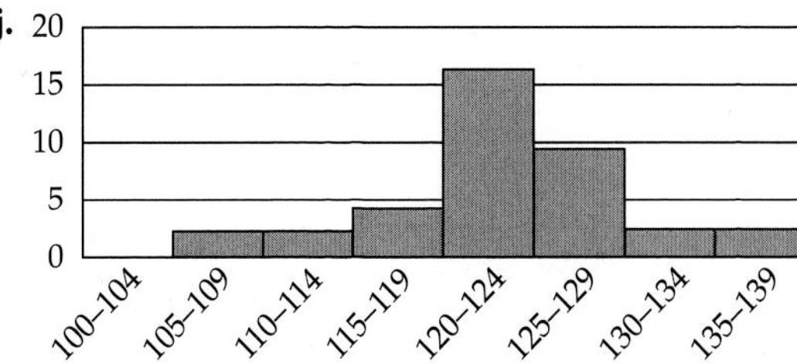

k. The histogram tells us that most of the boxes had between 120 and 129 raisins.

l. They are alike in that they both show a spike in the 120s. They are different in that the second histogram shows that the largest concentration is in the 120–124 range and that, other than the 120s, the number of raisins in each interval is relatively constant.

m. Answers will vary.

n. They are packaged by weight, not by number.

7. a. Answers will vary.

b. Mean = 4.2 years, median = 4 years

9. 93

11. Let x be the number that is removed.

$$\frac{4(7) + x}{5} = 6 \Rightarrow 28 + x = 30 \Rightarrow x = 2$$

13. a. Mean = 72.75, median = 77.5

b. Delete one below and one above the median.

c. Delete two below the median.

d. Delete two scores whose sum is greater than 146.

e. Delete one score above and one below the median; the sum of the scores must be greater than 146.

15. a.

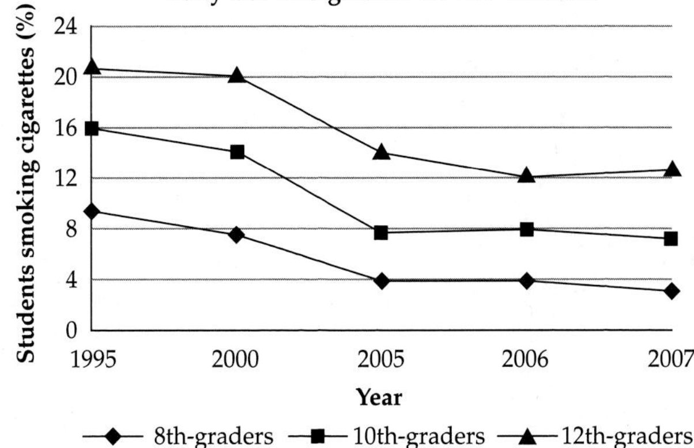

b. For all grade levels, the percentage of students that are smoking cigarettes has decreased over the last 12 years.

c. 8th grade: decrease of ≈ 68%
10th grade: decrease of ≈ 56%
12th grade: decrease of ≈ 43%

17. a.

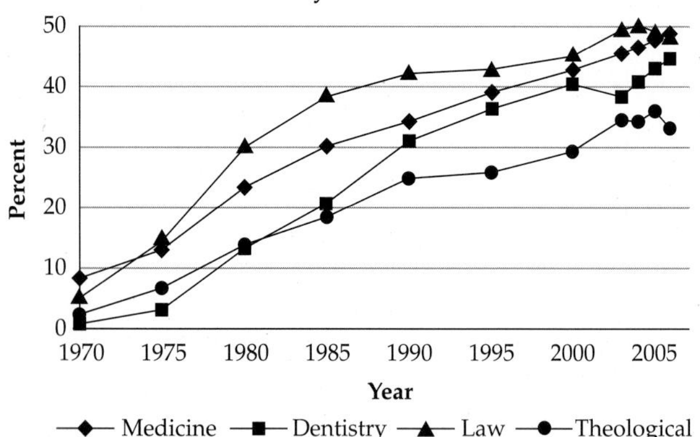

b. Answers will vary.

c. The percentages of women receiving degrees in these fields have risen dramatically in the past 36 years. By 2006 almost 5 out of every 10 degrees in medicine and law went to women, and about 3 in 10 theological degrees went to women.

Answers to 19–23 will vary. Examples of possible answers are given.

19. a. Almost $\frac{3}{4}$ of Americans drink at least 1 cup of coffee a day.

b. There is no information about how they got the data. Did they ask people of different ages, incomes, ethnicities?

c. No problems with the graph

21. a. The percentage of employees given a day off for Martin Luther King Day seemed to increase from 1986 to 2006, but the percentage in 2010 was less than in 2006.

b. We don't know that the numbers increased steadily between 1986 and 2006. We also don't know what kinds of employees were surveyed. All federal employees get that day off.

c. The x-axis is not labeled, though it is pretty clear that it represents years.

23. a. About $\frac{2}{3}$ of Americans dying of AIDS are between 25 and 44.

b. What year is this for?

c. Each wedge is a different shade of blue except for the largest wedge. It is hard at first to know where Under 15 is.

25. a. The mode is New Hampshire. The mode is the datum that occurs the most.

b. There is no mean. Although you and computer programs can compute the numbers, the number 4.6 is meaningless.

27. a. Mean $= \frac{44}{12} \approx 3.7$, median $= 2$

b. There should be spaces between the bars representing 2, 4, 6, 9, and 11 siblings. Without those spaces, one could interpret the data as being clustered together.

29. Answers will vary.

31. There are at most a few employees making substantially more than \$10 per hour, but they are outnumbered by those making less than \$7 per hour.

33. Think of a set of data where the mean and median are close and add one more datum that is an outlier on the high end. The median will either stay the same or shift to the next highest datum, that the mean will jump appreciably. If you have outliers, you can have a situation where the mean is not really close to the center of the data. See the example below for wages (in thousands of dollars) in a small company.

12, 12, 12, 15, 15, 15, 18, 18, 18, 18, 18, 18, 20, 20, 20, 30, 30, 30, 75, 100

In this case, the median is 18,000. The mean is 26,000 and $\frac{3}{4}$ of the employees make less than 26,000. That is, $\frac{3}{4}$ of the data are below average.

35. Answers will vary.

37. Rates allow us to compare different wholes. For example, say there were 20 murders last year in a city of 800,000 and 30 murders in a city of 1,800,000. If we just show the raw data, and the reader did not know the populations of the city, one could conclude that the second city was more dangerous. However, if we use 100,000 as our unit, we would say that the first city has a murder rate of 2.5 per 100,000 people and the second city has a murder rate of 1.7 per 100,000 people.

39. What is crucial is that there is a whole. Many data can be in percentages but there is no whole. For example, see problem 21, where a circle graph would be meaningless.

41. a.

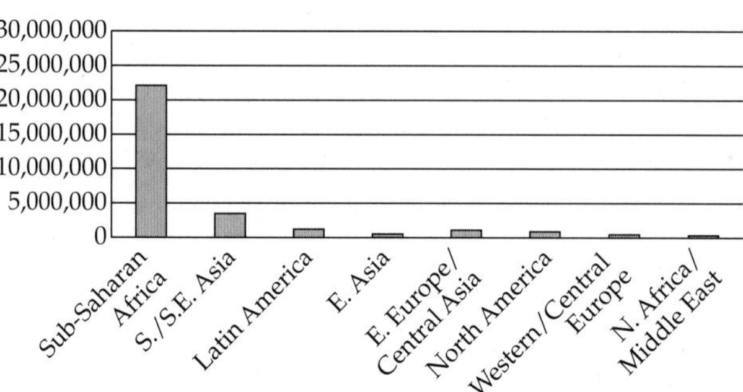

b. The number of HIV/AIDS cases in Sub-Saharan Africa clearly stands out on this graph.

c. Answers will vary.

43. a. Answers will vary.

b.

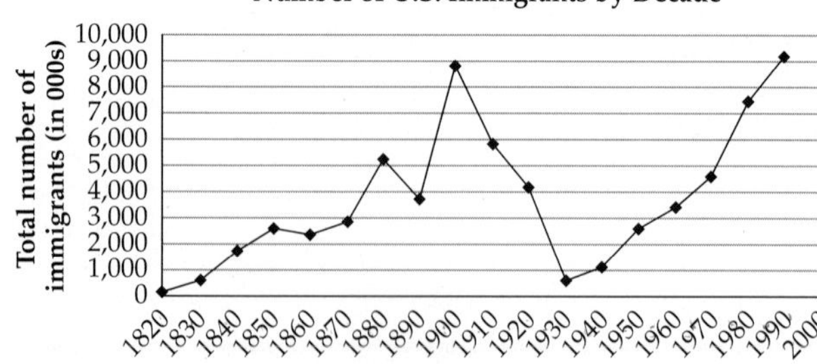

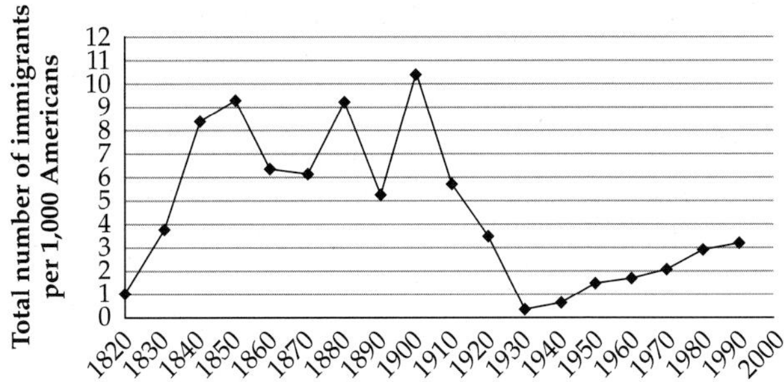

The first graph makes it appear that there has been a rapid increase in the number of immigrants in the last few decades. The second graph gives a more accurate impression that the rate of increase is actually not as high as it has been in the past.

c. Answers will vary.

d. Answers will vary.

45. a. The number of AIDS diagnoses for Blacks is equal to that for all other groups combined.

b. Since the data came from the Centers for Disease Control and Prevention, the data are probably fairly accurate. However, as with problem 44, I would wonder about illegal immigrants and homeless people. Are the numbers for those populations underreported?

c. A bar graph makes sense. A circle graph would make it easier to realize that Blacks represent $\frac{1}{2}$ of the total number of cases.

47. a. In Germany and in the United States, the overwhelming majority of seatwork time is on practicing procedures, whereas it is less than $\frac{1}{2}$ the time in Japanese classrooms.

b. Same as for problem 46

c. The choice of a bar graph is fine.

49. a. About $\frac{5}{6}$ of all immigrants to the United States in 1900 were from Europe, as opposed to about $\frac{1}{6}$ in 2000. Latin Americans represented less than 2% of the immigrants in 1900 compared to almost $\frac{1}{2}$ in 2000.

b.

	1900		2000
Europe	84.9%	Europe	15.3%
Latin America	1.3%	Latin America	51.0%
Asia	1.2%	Asia	25.5%
Other regions	12.6%	Other regions	8.1%

c. With the table, you lose the breakdown of the sections of Latin America in 2000 unless you make another table, but that would also be cumbersome for some readers. With the graph, you can more visually see the tremendous change in demographics.

51. a. If you enclose each column of x's in a line plot with a bar, you have a histogram.

b. If you put a dot at the center of the top of each bar and connect the dots, you have a line graph.

53. Answers will vary.

55. Between 1970 and 1980

57. c

59. I would use the median. In both cases, more than half of the data are very close to the median. In the first case, four of the five data values are substantially above the mean.

Section 7.2

1. a. It tells us the range: 52 – 81 and that there is a cluster in the mid to high 60s. We can say that about $\frac{1}{2}$ of the class has a pulse rate between 65 and 69.

X X X
X X X X X X X X X X X X X X X

5 5 5 5 5 5 5 5 6 6 6 6 6 6 6 6 6 6 7 7 7 7 7 7 7 7 7 7 8 8
2 3 4 5 6 7 8 9 0 1 2 3 4 5 6 7 8 9 0 1 2 3 4 5 6 7 8 9 0 1

b. It tells us the range and that most of the pulses are in the 60s.

5 | 289
6 | 236778899
7 | 145
8 | 1

c. It tells us the range, that the median is 67, and that about half the class's pulse is between 62 and 70.

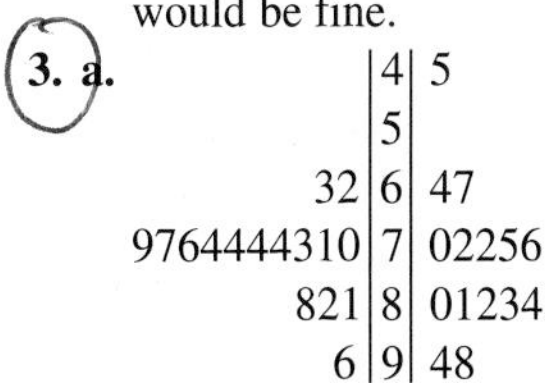

5 5 5 5 5 5 5 5 6 6 6 6 6 6 6 6 6 6 7 7 7 7 7 7 7 7 7 7 8 8
2 3 4 5 6 7 8 9 0 1 2 3 4 5 6 7 8 9 0 1 2 3 4 5 6 7 8 9 0 1

d. I would have everyone sitting down. I would make sure that everyone got their pulse in the same way, e.g., finger on the wrist. I would say "1, 2, 3, start" and then "stop" after 30 seconds. Some people don't have the best concentration, so getting the number for 30 seconds and then doubling it would be fine.

3. a.

	4	5
	5	
32	6	47
9764444310	7	02256
821	8	012348
6	9	48

b. The first class has a much larger range (45 to 98) compared to (62 to 96). The first class has about the same number of scores in the 70s as in the 80s, while the second class has a large cluster in the 70s.

c.

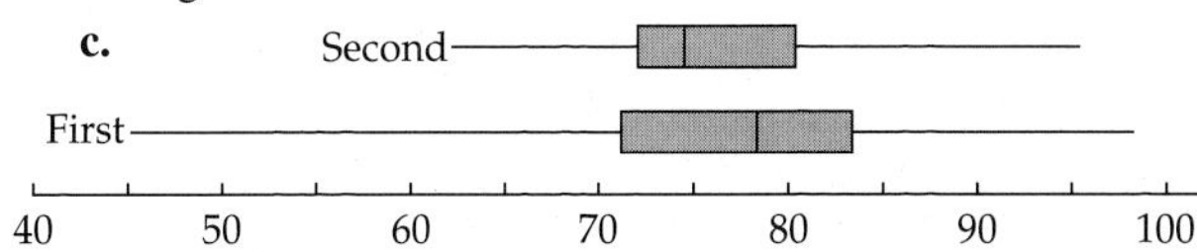

d. We can still see that the range for the first class is much greater. We can see that the median for the first class is higher (78 vs. 74). We can also see that the middle half of the scores are similar (about 71 to 83 in the first class and about 72 to 80 in the second class).

5. The second class had a median of 79 compared with 77 in the first class. The mean for the first class is 80.8. The mean for the second class is 80.0. The second class was bimodal at 77 and 85, whereas the mode for the first class was 76. The second class had a smaller range of 65 to 96 compared with 58 to 100 in the first class.

7. a. Grouped frequency bar graph, histogram, or circle graph is appropriate. Also appropriate are box-and-whisker and line plots.

b. Since the distribution is skewed, the mean, median, and mode are not convergent. The median is 21. The data are bimodal at 19 and 21. The mean is 23.9.

c. Data is skewed to the right.
Range: 18-46
Clusters: 18-21
Biggest gap: between 27 and 37
Outliers: 37, 42, 43, 47
Standard deviation: 8.0

9. a.

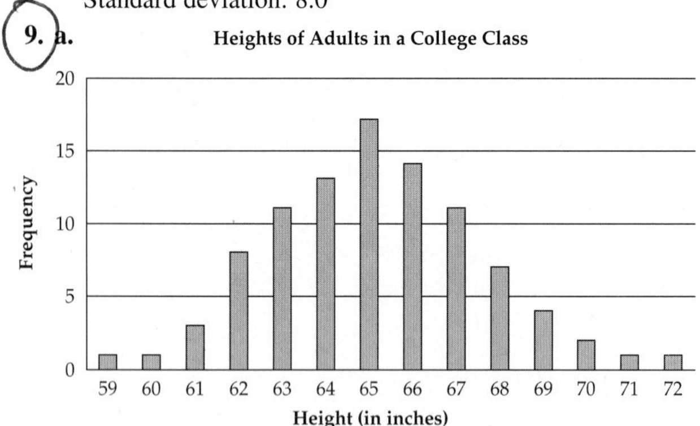

b. Mean = 65.15, standard deviation 2.43

c. 70%

11. 95% are between 61 and 71 inches. About 1% are less than 5 feet.

13. a. 50 tires

b. Without z-scores, one can approximate the area under the curve. By various means, one can conclude that approximately 6% will wear out before 55,000 miles.

15. a. Answers will vary depending on the grouped frequency

b.

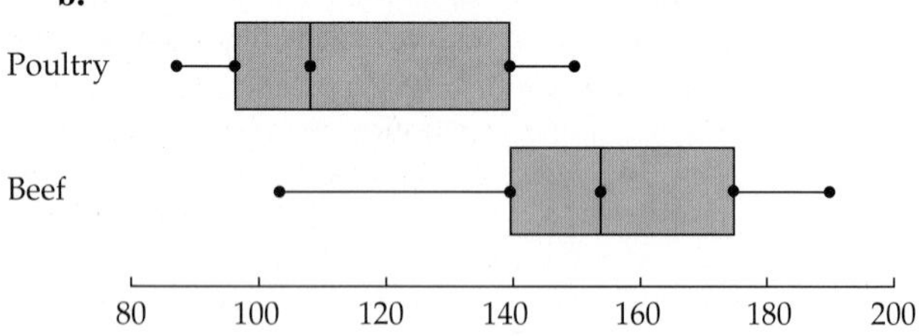

c. Answers will vary.

d. The relationship between calories and sodium content is pretty strong. That is, in general, the more the calories the more the sodium.

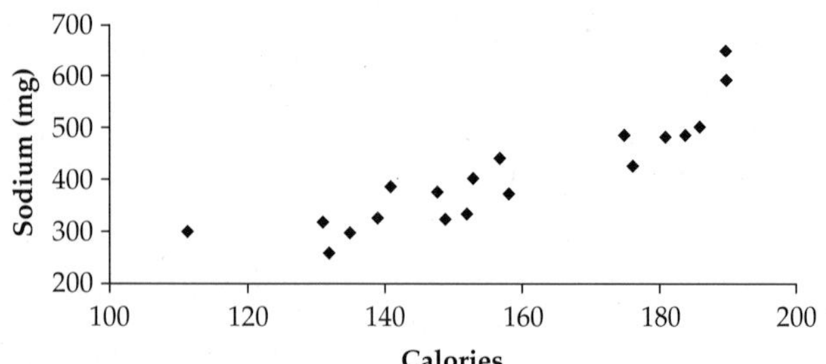

17 a. Positive correlation

b. No correlation

c. Negative correlation

d. Positive correlation

e. Positive correlation

f. Positive correlation

g. Positive correlation

h. Positive correlation

19. a. Whom did you survey? I would not say that over half of the families I know eat dinner together 5 or more days per week. Does it count if only part of the family is there? Were these data gathered from two-parent families or from one- and two-parent families?

b. How often does your family eat dinner together in an average week during the school year? (I would give them the categories in the table below or I would ask for a specific number. For example, a response of "2 or 3" would create problems in comparing to the data given.)

c.

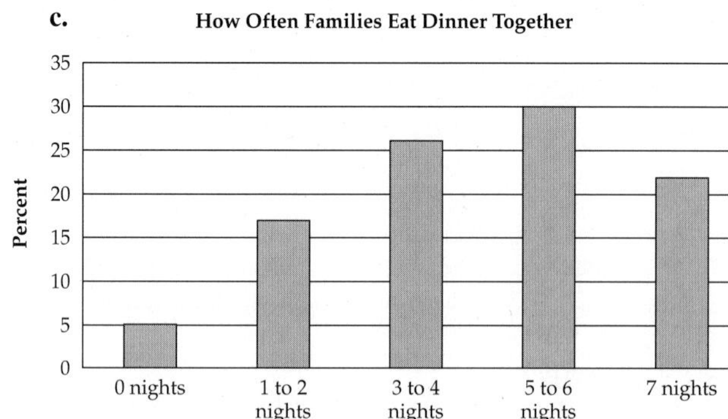

d. A circle graph would have been okay here, but there is a numerical progression (from 0 to 1 or 2 to 3 or 4 to 5 or 6 to 7), and it is easier to follow this progression with a bar graph. One main advantage of a circle graph is that it gives you the part of the whole; in this case, the data are in percentages, so you get that from the bar graph also.

e. The bar graph does help you see that the percentages increase as the number of days per week eating together increases, up to eating dinner together every day.

21. I would want to know if they gave these categories (0-4, etc.) or asked parents to estimate and then grouped them into these categories. I would want to know if the researchers took a representative sample; for example, kids get more colds in the Northeast than in the Southwest because of the severity of the winters.

23. 76.7

25. Answers will vary.

27. a. For the students: mean = 2.1, median = 1.5 siblings, and mode = 1 sibling
For the mothers: mean = 3.8; median = 3 siblings, and modes = 2 and 3 siblings

b. In this case, the difference between the two means and the two medians is about $1\frac{1}{2}$. So I might say that the students in the class have, on average, $1\frac{1}{2}$ fewer siblings than their mothers did.

c. We really can't. While we know from the Census that the size of families has been steadily decreasing, the demographics (characteristics) of the students in the class is by no means representative of the overall population.

d. The line plots show that the data for the students are more tightly clustered around 0 to 2 and then steadily decrease after that. The 9 is clearly an outlier.

```
    X X
    X X   X
  X X X   X
X X X X X     X   X     X
0 1 2 3 4 5 6 7 8 9 10 11
        parents
```

```
X
X
X X
X X X
X X X X
X X X X X X       X
0 1 2 3 4 5 6 7 8 9 10 11
        students
```

e. The boxplots nicely show the "shift" in data from the mothers to the children. The mothers' median is equal to the students' upper quartile. That is, half of the mothers have 3 or more siblings; only $\frac{1}{4}$ of the students have 3 or more siblings. The ranges in the data are comparable.

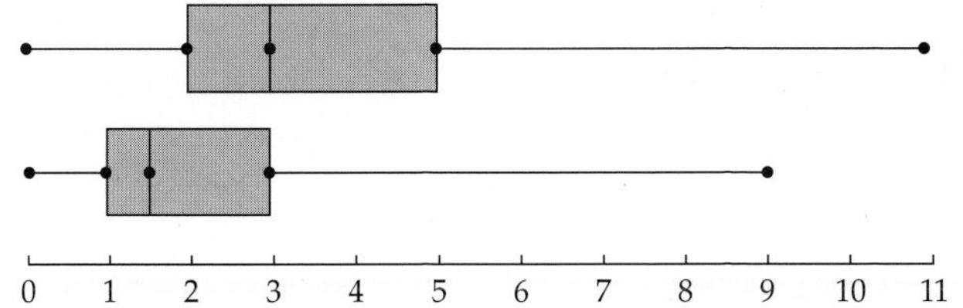

f. The standard deviations are 2.1 and 2.8, respectively. The mean absolute deviations are 2.11 and 2.17, respectively.

g. We would need to know the ages of each student and the student's mother.

h. It would tell us if the size of a student's family is related to the size of the mother's family. That is, do students with small families tend to have mothers from small families and do students with larger families tend to have mothers from larger families?

i. Answers will vary.

29. a. Make sure all of the intervals have the same number of years in them.

b. Answers will vary.

c. Answers will vary.

d. The graph from part (a) clearly shows that there are many brand-new teachers in the district—which is not evident from looking at the average.

31. a. The football payrolls range from about \$62 to \$120 million with a median of about \$82 million. The middle half of the teams has payrolls between \$77 and \$91 million. There appears to be a cluster of payrolls between \$77 and \$82 million.

The baseball payrolls range from about \$30 to \$210 million with a median of about \$66 million.

The middle half of the teams has payrolls between \$48 and \$87 million. The longest whisker is longer than the other whisker and boxes combined, which indicates that there are probably gaps between \$87 and \$210 million and that \$210 million might be an outlier.

b. The football distribution is probably slightly skewed to the right. The baseball distribution is strongly skewed to the right.

c. The baseball payrolls have a range (\$175 million) that is more than the range of the football payrolls (\$70 million). About $\frac{1}{2}$ of the baseball teams have a payroll smaller than all of the football teams.

33. a. A cluster

b. A high probability of gaps or, in the case of whiskers, the possibility of outliers

c. 

Normal distribution is symmetric.

d. 

Skewed to the right means more clustered at the left and more spread out at the right.

35. Without z-scores, one can approximate the area under the curve. By various means, one can conclude that approximately 35%, or about 700 tires, will wear out before 45,000 miles.

37. Her math score was the highest in terms of standard deviations above the mean.

39. Answers will vary.

41. There is a mild positive correlation.

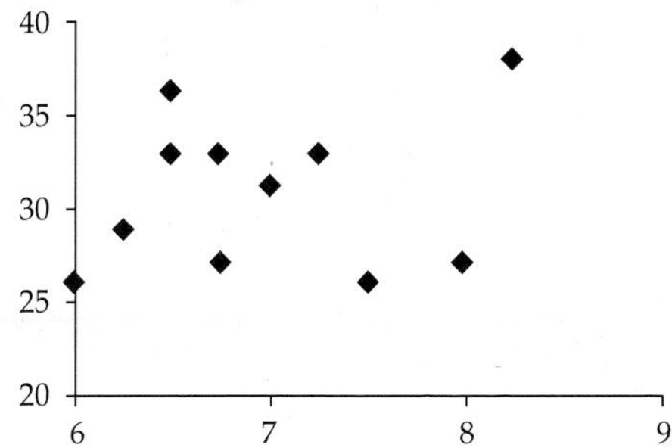

43. a. If we are going to make generalizations about the population called "teenage boys," the driving records of the boys surveyed in these places are likely to be worse than if we had a representative sample.

b. If we are going to make generalizations about the population called "citizens," then parents is not a random sample. Even if we are going to make generalizations about the population called "parents," then parents at a PTA meeting is still not representative. Only a fraction of parents attend PTA meetings.

c. If we are going to make generalizations about the population called "residents," this will not be a representative sample: Not all people have telephones, and certain people are home less than others, for example, young single people.

d. You would probably get different results if you asked their opinion on December 23!

45. Her mean speed is about 17 miles per hour.

47. Julie is right. Although 148.12 grade points/46 credits = 3.22 GPA, one's total grade points is either a whole number or a mixed number.

Section 7.3

1. $\frac{1}{3}$

3. .14 or 1 in 7

5. a. .1 **b.** .3 **c.** .6 **d.** .4 **e.** .9

7. a. $\frac{1}{4}$ **b.** $\frac{3}{8}$

9. a. $\frac{1}{36}$

b. $\frac{18}{36} = \frac{1}{2}$

c. $\frac{15}{36} = \frac{5}{12}$

d. $\frac{6}{36} = \frac{1}{6}$

11. $\frac{2}{3}$ chance of landing in room A

13. $\frac{2}{9}$

15. $\frac{1}{6}$

17. No, it is not a fair game.

19. The expected value from the spinner is \$1.625. Take the \$2 and run!

21. $\frac{2}{3}$

23. \$375

25. a. It is fair.

b. It is not fair. Give 1 point to player A if the number is even and 3 points to player B if the number is odd.

c. It is fair.

27. a. $\frac{1}{2}$

b. $\frac{23}{94}$

29. $\frac{1}{12}$

31. $\frac{53}{80}$

33. $\frac{53}{64} \approx 0.83$

35. a. $\frac{5}{16} = 0.3125$

b. $\frac{35}{128} \approx 0.27$

c. $\frac{63}{256} \approx 0.25$

d. Answers will vary.

e. $\binom{50}{25} \div 2^{50} \approx 11\%$

37. The most likely sum is 5. The probability of rolling a 5 is $\frac{4}{16} = \frac{1}{4}$.

39. Answers will vary. P(3 doubles in a row) = $\frac{1}{6} \times \frac{1}{6} \times \frac{1}{6} = \frac{1}{216} \approx 0.00463.$

41. Answers will vary. Here is one set of possibilities: One die must have the same odd number on all its faces. The other die could have 1, 1, 3, 3, 5, 5, or three of one odd number and three of another.

43. The probabilities of winning are: player 1, $\frac{1}{9}$; player 2, $\frac{2}{9}$; and player 3, $\frac{6}{9}$.

Thus, give each player this number of points when they win: player 1: 6 points, player 2: 3 points, and player 3: 1 point.

45. No

47. No. We need to remember the law of large numbers. If you play many times, you are more likely to win about $\frac{1}{4}$ of the times. But with a small sample size, the unlikely is possible though not probable, like rolling doubles 4 times in a row.

49. Facts: There are at least one red, one blue, and one green ball, and there are at least three different colors of balls in the bag. Inferences will vary—there are probably more red than blue (or green) balls; there are probably at least twice as many red as blue (or green) balls; there are probably less than 10 colors, etc.

51. Answers will vary.

Section 7.4

1. a. $\frac{1}{52}$ **b.** $\frac{1}{13}$ **c.** $\frac{3}{13}$

3. a. $4 \times 3 \times 2 = 4! = 24$

b. 16, determined by making the arrangements

5. 10,000

7. $\frac{3}{51} = \frac{1}{17}$

9. ${}_{12}C_3 = 220$ ways

11. If the flavors were scooped in any order, there would be ${}_9C_3 = 84$ possibilities. If you specified the order of the flavors, there would be ${}_9P_3 = 504$ possibilities.

13. P(4 of a kind) − 0.0002. P(3 of a kind) = 0.0211. P(2 of a kind) = 0.423.

15. $\frac{1}{120}$

17. The probability that each child will get the popsicle he or she wants is $\frac{26}{27}$.

19. a. 21 days

b. 7 (SM, MT, TW, WT, TF, FS, SS)

c. 6 days

21. 39,916,800 (11!) possible words from *mathematics.* Assuming you use all the letters.

23. $n!$ can be written as follows: $n(n-1)(n-2)\ldots(n-r+1)(n-r)!$ When ${}_nP_r$ is expressed as a fraction, $(n-r)!$ in the numerator and the denominator cancel each other, leaving $n(n-1)(n-2)\ldots(n-r+1)$, which is the other expression for ${}_nP_r$.

25. Answers will vary.

CHAPTER 7 REVIEW EXERCISES

1. a.

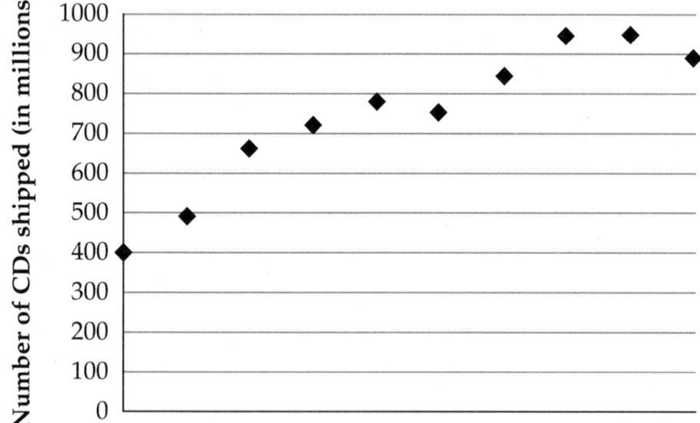

b. The number of CDs shipped rose steadily between 1992 and 2001, with a dip in 1997 and 2001. The number of units shipped in 2001 was more than double the number shipped 9 years earlier.

2. a. See graph. Note that because the years are not evenly spaced, we need to leave blank spots for missing years. And we should not connect the dots, because we can't be sure what the data are for the missing years.

b. The percentage of adults who smoke declined pretty steadily between 1965 and 1990. Since 1990, the percentage has declined only slightly, from 25% to 23%.

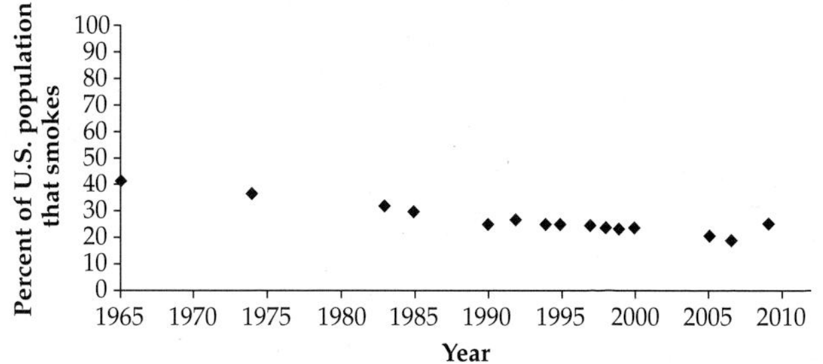

3. a.

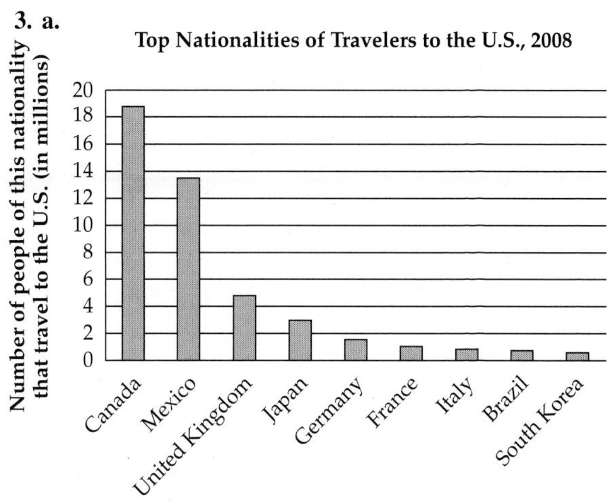

b. Answers will vary.

4. a.

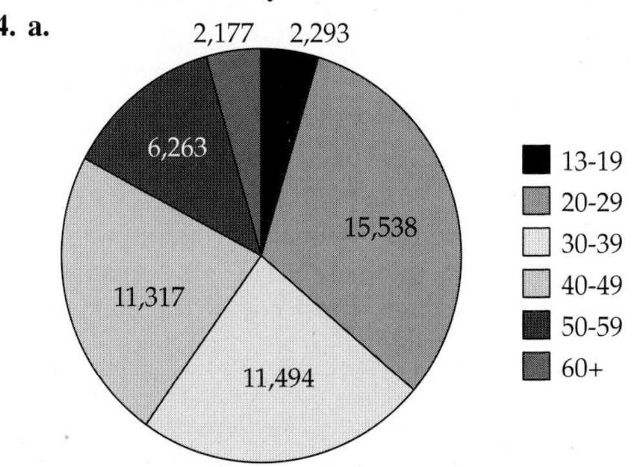

b. Answers will vary.

5–6. Responses for these questions will vary.

7. a. One possible graph is below.

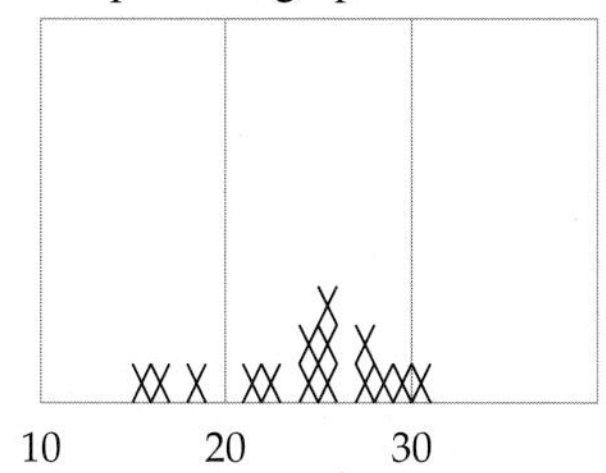

b. The mean is 23.7, the median is 25, and the mode is 25.

c. The number of drops recorded varied from 15 to 30. Over half of the data were between 21 and 27 drops.

8. a. 4.1 pounds

b. 2.8 pounds

c. Were the dogs all approximately the same size? For example, comparing the weight loss of a 100-pound dog to the weight loss of a 10-pound dog does not make sense.

9. a.

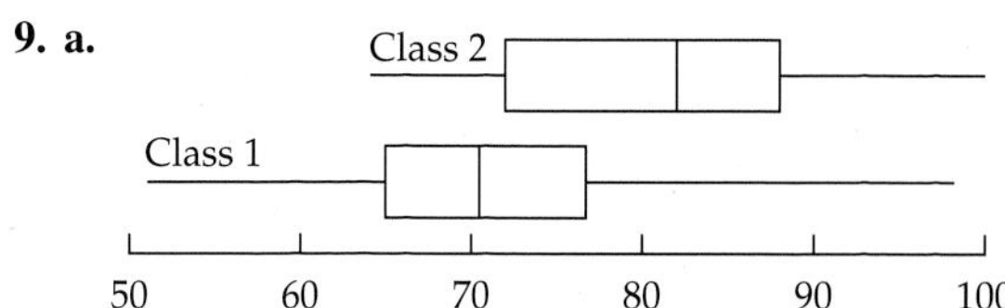

The boxplot gives a quick snapshot. It tells us that the first class had a significantly higher range; that the second class did better overall—it had a higher median; and that $\frac{3}{4}$ of the second class is above 70 compared to only $\frac{1}{2}$ of the first class.

b.

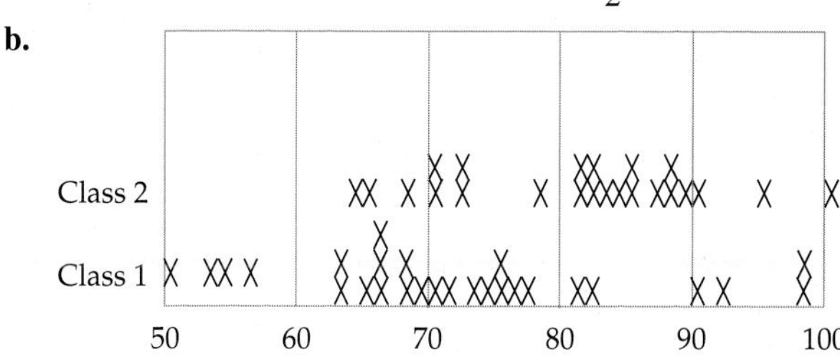

The line plot lets us see the range; it also lets us see the clusters. The first class is relatively spread out; the second class has a cluster in the 80s.

c. *Box-and-whisker:* pros—quick snapshot; cons—you don't have all the data.

Line: pros—you have all the data; you can see the distribution (spread, range, clusters, gaps); cons—you don't have the quartiles.

d. Means are 72.1 and 80.8; medians are 70.5 and 82; modes are not useful here.

e. Ranges are 47 vs. 36.

f. Standard deviations are 11.9 vs. 9.3.

g. 71% vs. 70%

10. He needs at least 83.

11. 48 students

12. a. 21 **b.** 1.9 **c.** 2

13. a. A situation with outliers

b. A situation where you would want to know standard deviation; mean and standard deviation go together well; grades

14. a.

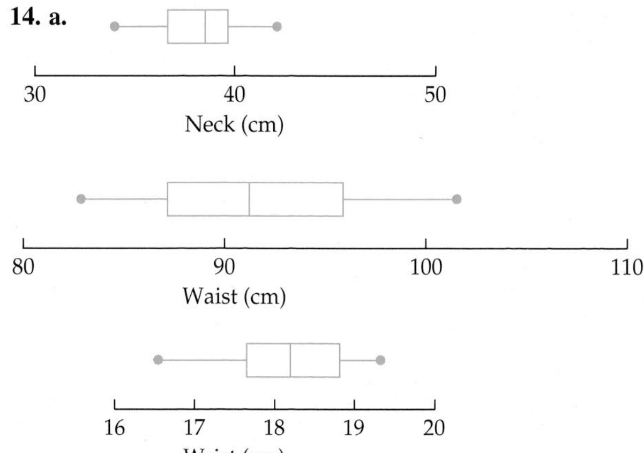

b. The correlation between neck and waist is a weak positive relationship. The correlation between neck and wrist is a strong positive relationship.

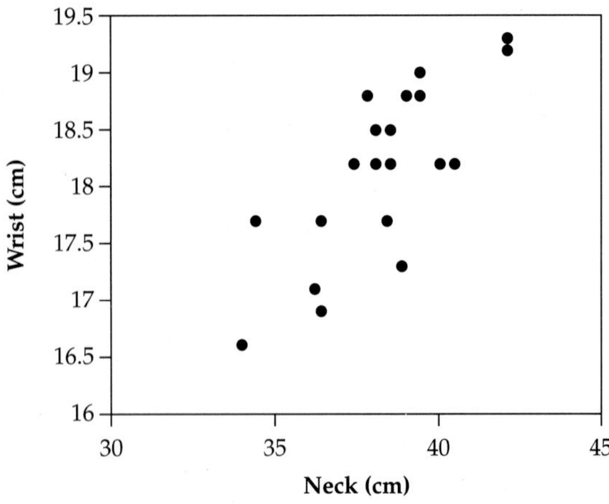

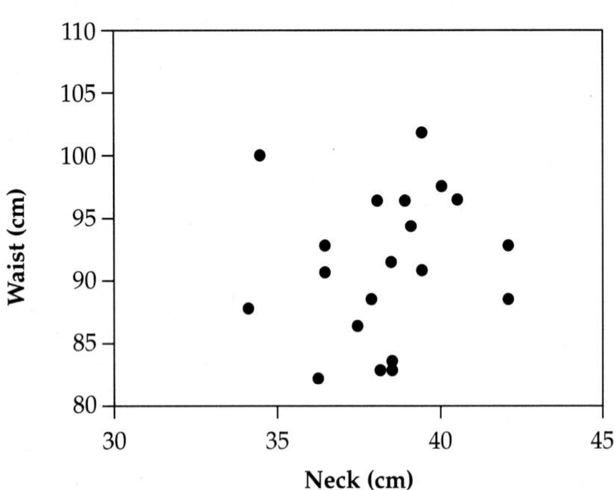

15. $\frac{1}{4} \times \frac{2}{8} = \frac{1}{16}$ is the probability of drawing two white circles.

16. a. $\frac{1}{2}$

b. The probability is 1 because there are only two colors.

c. $\frac{5}{42}$ (We can only consider the possibility that there are red socks.)

17. Most likely sum is 9, which can be arrived at with 1&8, 2&7, 3&6, 4&5, 5&4, 6&3, 7&2, or 8&1. There are 64 ways to roll the dice, so the probability of rolling a sum of 9 is $\frac{8}{64} = \frac{1}{8}$.

18. $\frac{1}{4}$

19. The probability of randomly choosing a number with a zero is $\frac{9}{90} = \frac{1}{10}$. The probability of randomly choosing a number with a 5 is $\frac{18}{90} = \frac{1}{5}$.

20. It is not a fair game; $\frac{3}{4}$ chance of even; $\frac{1}{4}$ chance of odd. To make it fair, give 3 points to player A.

21. Not fair. Probability of a match is $\frac{1}{3}$. You could make it fair by giving player A 2 points if the colors match and player B 1 point if they don't match.

22. $5! = 120$ ways to display the sports on the poster

23. It doesn't matter what the first card is, there are 12 out of 51 ways to get a matching suit with the second card, thus $\frac{12}{51} = \frac{4}{17}$.

24. 24

25. 20%

26. a. 45

b. 120

27. a. ${}_6C_2 = \frac{6!}{4!2!} = 15$ ways to choose a schedule

b. 3 **c.** 5

28. $26 \times 10 \times 10 \times 10 \times 26 = 676{,}000$ different license plates

CHAPTER 8 EXERCISES

Section 8.1

1. a. A tetromino is the mathematical name for a Tetris piece.

b. *Faulty definition:* Four squares that touch each other.

Fixed definition: Four squares on a sheet of graph paper where each square intersects at least one of the others at a whole edge.

3. a. The lines are the same length.

b. The circles are the same size.

c. Yes

d. The segment on the far right.

e. Answers will vary. Most likely, a triangle with circles at its vertices will be seen.

f. It is a 2-dimensional drawing of an impossible 3-dimensional figure.

5. Twelve different rays

7. a. False. The lines could be skewed.

b. True. Given any two parallel lines, there is a plane that will contain both lines.

c. True. The lines contain three distinct sets of points: those exclusively on one line, those exclusively on the other line, and the point of intersection. These points determine the plane in which the lines lie.

d. False. It could be skewed.

e. False. Think of two lines on the plane determined by this sheet of paper and a third line perpendicular to this plane.

f. False. Two planes cannot intersect at a point.

9. Several answers are possible. Examples are given.

a. $\angle ABG$ and $\angle GBE$; $\angle AEB$ and $\angle EAB$

b. $\angle FEG$ and $\angle GEB$; $\angle AGB$ and $\angle BGE$

c. $\angle ABC$ and $\angle EBD$

d. $\angle ABC$ and $\angle ABG$

11. a. 22 times **b.** 9:00 and 3:00 **c.** 150° and (210°)

d. 75° **e.** 27° **f.** Answers will vary.

15. a. **b.**

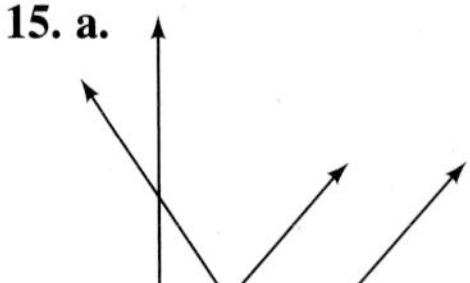

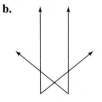

c.

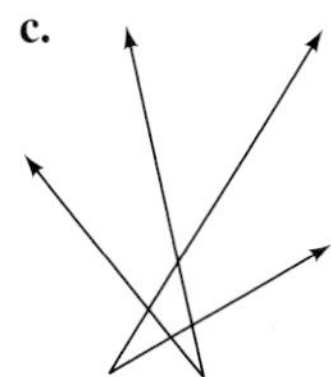

17. Answers may vary. A possible letter "e" for each font is given below.

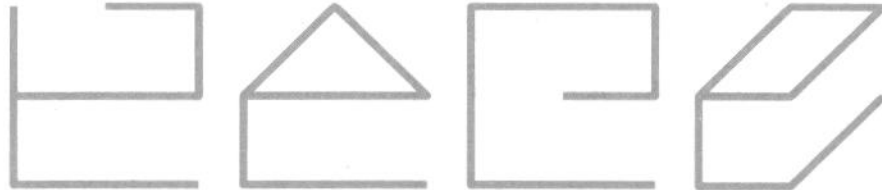

19. These students are seeing that the rays are longer, but not looking at the opening between the rays.

Section 8.2

1. Answers will vary.
3. **a.** I would say the triangle and kite are most alike, but each pair has similarities.
   b. There are several properties that are common to all three figures. You could make an argument for similarities between the second and third figures or for similarities between the first and third figures.
   c. The first and third figure are the most alike: regular polygons; multiple pairs of parallel sides; all angles are obtuse.
5. **a.** 20 triangles
   b. 11 rectangles
   c. There are many, for example, trapezoids of different size, parallelograms of different size, pentagons (convex and concave), hexagons (convex and concave), and polygons with more sides.
   d. Answers will vary.
   e. Answers will vary.
7. **a.**

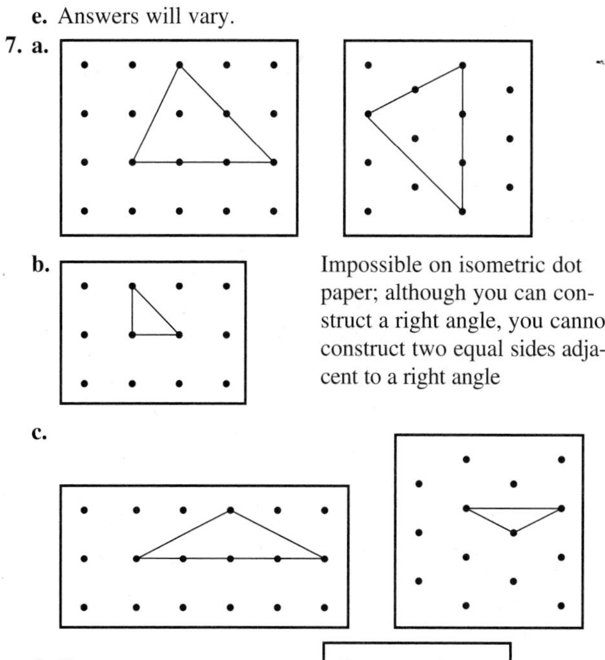

b. Impossible on isometric dot paper; although you can construct a right angle, you cannot construct two equal sides adjacent to a right angle

c.

d. Cannot construct an equilateral triangle on geoboard paper

e.

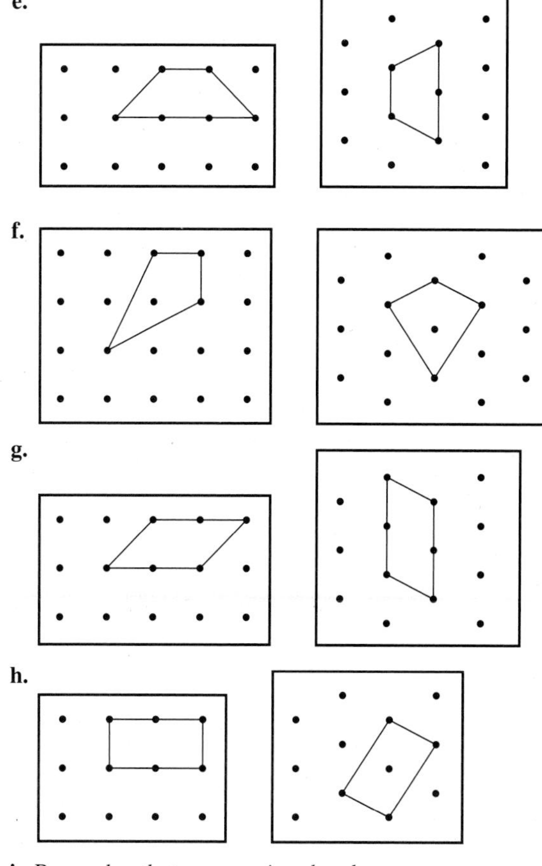

f.

g.

h.

i. Remember that a square *is* a rhombus.

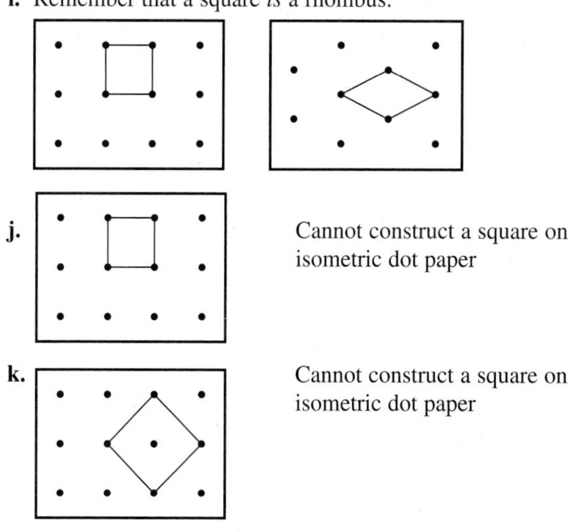

j. Cannot construct a square on isometric dot paper

k. Cannot construct a square on isometric dot paper

9. **a.** Concave polygon **b.** Not a polygon
   c. Convex polygon **d.** Not a polygon
   e. Concave polygon
11. **a.** Parallelogram, rhombus, rectangle, square
    b. Rectangle, square
    c. Square, rhombus
    d. Parallelogram, rhombus, rectangle, square
    e. Rectangle, square, isosceles trapezoid
    f. Rectangle, square, and an infinite number of unnamed quadrilaterals
    g. Parallelogram, rhombus, rectangle, square

h. Kite, unless it is a rhombus

i. Rhombus

j. Rectangle

13. a.

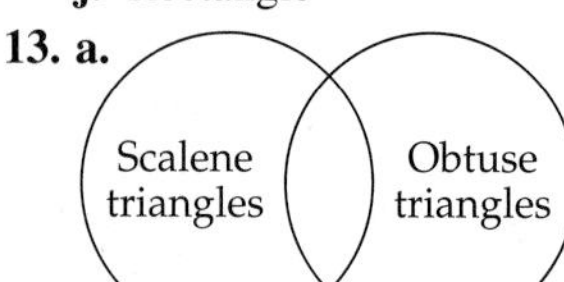

b.

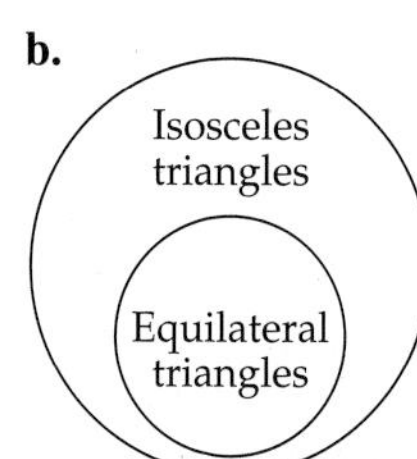

c.

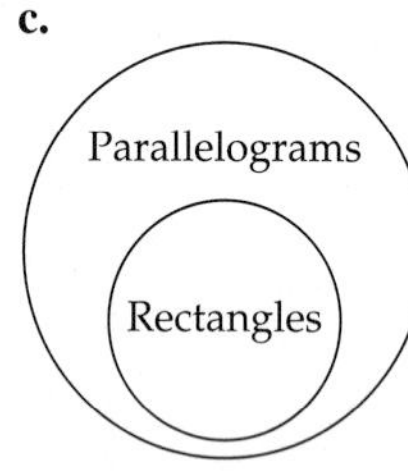

d.

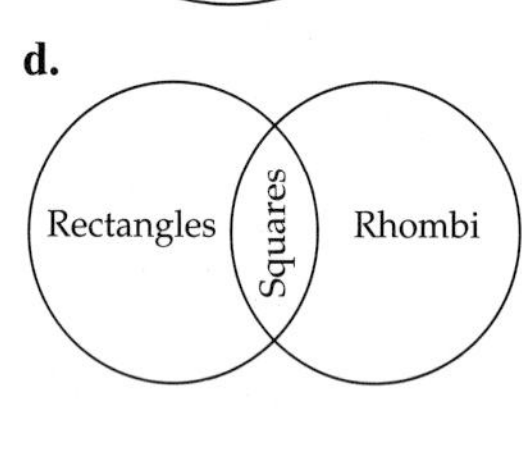

15. A letter "U" (assuming you connect *H* with *A*)

17. a. $\left(\frac{0+4}{2}, \frac{4+12}{2}\right) = (2, 7)$

b. $\left(\frac{3+7}{2}, \frac{4+12}{2}\right) = (5, 8)$

19. a. (0, 0), (6, 0), (6, 6), and (0, 6)

b. The other two vertices are on the horizontal line going through (7, −2) and are equidistant from (7, −2); for example, (5, −2) and (9, −2).

c. The coordinates are (10, 0) and (0, 10), or (10, 0) and (0, −10).

d. Answers will vary.

21. If the definition does not include both parts, then there could be more than one kind of regular *n*-gon for a given length of sides. For example, a regular four-sided figure could be a square or any variety of rhombus. A hexagon could be convex or concave.

23. Answers will vary.

25. a. You can make many different hexagons, both concave and convex, that have all sides equal but not all angles equal. Some have symmetry, some don't.

b. Many possibilities

c. Many possibilities, all are concave

d. Many shapes are possible. All must have four consecutive right angles.

e. Two possibilities: both have five 90° angles and one 270° angle. One kind has reflection symmetry and the other doesn't.

f. Yes, squares and rectangles are trapezoids.

g. None

h. No

i. Yes, many possibilities

27. A square is derived from a rhombus and a rectangle, because it has four congruent sides that meet at right angles.

29. b. Only one way

c. Label the angles *a*, *b*, and *c*. At each point where three triangles meet, the angles are *a*, *b*, and *c*. Since we know that the sum of these is 180°, each of those points is a straight line.

d. Large triangle is similar to the small triangle.

31. Note that $n - 2$ triangles can be inscribed in any regular *n*-gon originating from the same vertex. All the triangles have angles summing to 180°, so the *n*-gon has angles summing to $180(n - 2)°$. For example:

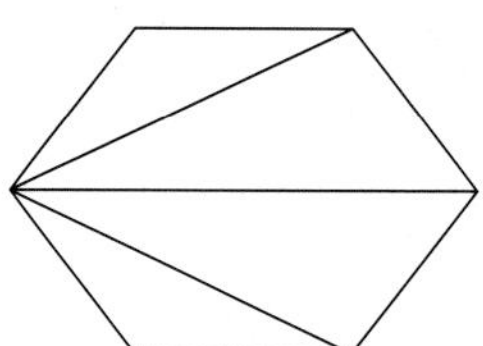

This 6-gon (hexagon) has 4 inscribed triangles and $180 \times 4 = 720°$.

33. *One way:* Use a straight edge and compass. Draw any line through the circle, call the points where it crosses the circle *A* and *B*. Using the compass, find the perpendicular bisector of *AB*, and call the points where this meets the circle *C* and *D* (notice that *CD* is a diameter of the circle). Now bisect *CD*, and the midpoint of *CD* will be the center of the circle.

35. a. A 3RIT is a figure made from 3 right isosceles triangles joined together so that when a side meets a side, there is no overlap.

b. Here are some examples:

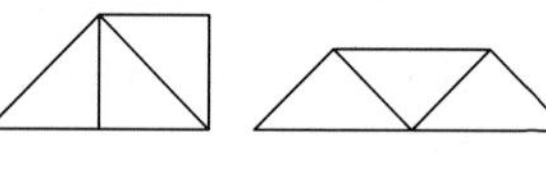
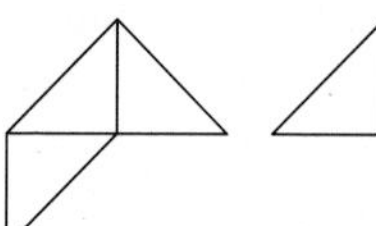
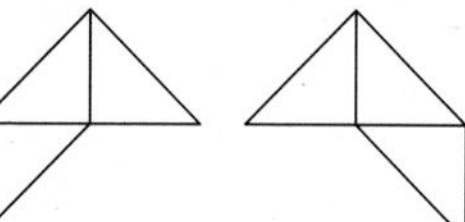

c. They are congruent. You can flip and rotate one to have it in the same orientation as the other.

d. Here are some examples:

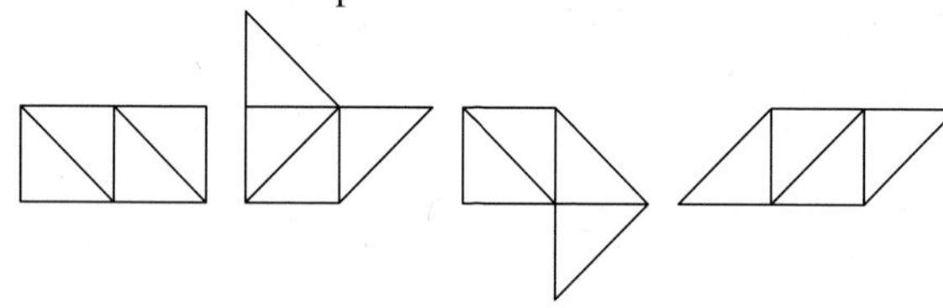

37. *A*(0, 0) *F*(8, 4)
B(8, 0) *G*(8, 8)
C(4, 4) *H*(4, 8)
D(6, 2) *I*(2, 6)
E(6, 6) *J*(0, 8)

39. a. Orient the triangle so that its noncongruent side is on the *x*-axis with one vertex at the origin. Let the other two vertices be called (4*a*, 0) and (2*a*, 2*b*). The midpoints of the isosceles sides are (*a*, *b*) and (3*a*, *b*). The base measures $d = \sqrt{(4a - 0)^2 + (0 - 0)^2} = \sqrt{16a^2} = 4a$ and the distance of the segment connecting the midpoints is $d = \sqrt{(3a - a)^2 + (b - b)^2} = \sqrt{4a^2} = 2a$. Therefore the segment is one-half the length of the base.

b. True for all triangles. Can prove the same way as in part (a).

41. a. There are many ways to show this. One way is to make use of the Pythagorean theorem.

b. One diagonal has slope 1 and the other has slope -1. $(1)(-1) = -1$, so the lines are perpendicular. (Note that this assumes that the square is oriented with its sides running vertically and horizontally.)

43. Answers will vary, but might include it has four sides, it has more than one right angle, all the sides are congruent, all the angles are congruent, and opposite sides are parallel.

45. d

47. a. They don't have the same number of sides and angles.
b. All the sides and angles are congruent.

Section 8.3

1. a. A cube and a rectangular prism are two possibilities.
b. Pentagonal pyramid

3. Answers will vary. There are several correct answers.
a. $\triangle ABC$ **b.** Point A **c.** $\overline{AB}$

5. a. Octagonal prism **b.** Triangular prism
c. Rectangular prism **d.** Right cylinder

7. a. They have at least one pair of parallel sides.
b. Prisms have two parallel bases that are congruent polygons.
c. At least one simple closed curve base

9. Figures on isometric dot paper should look the same as the figures in the book.

11. a. Rectangle **b.** Rectangle **c.** Rectangle
d. Rectangle **e.** Rectangle

13. $2n$ vertices, $n + 2$ faces, $3n$ edges

15. Answers will vary.

17. Yes. If all the edges of the base are of different lengths, then the triangular faces will not be congruent.

19.

Base of Prism	Number of Diagonals
Triangular	6
Square	16
Pentagonal	30
n-gon	$\frac{v(v-4)}{2}$

21. The sides will always be isosceles triangles.

23. There are 9.

25. a

27. Answers will vary.

29. a

31. d

33. c

CHAPTER 8 REVIEW EXERCISES

1. Answers will vary.

2. There are multiple answers for each question. One answer is given for each.
a. $\angle ABG$ and $\angle GBF$
b. $\angle CBG$ and $\angle GBF$
c. $\angle ABG$ and $\angle DBE$
d. $\angle DBE$ and $\angle EBG$

3. a. False. Consider two lines that lie on this paper and one that is perpendicular to the paper.
b. False. They could be skewed.

4. If you do not start with some undefined terms, some of the definitions will be circular, similar to dictionary definitions. For example, A is defined in terms of B, and B is defined in terms of C, and C is defined in terms of A.

5. Answers will vary. One option is to say that a triangle is a shape made by three line segments such that the endpoints of each segment touch the endpoints of another segment.

6. 6

7. Answers will vary. One option is to draw a quadrilateral and one diagonal, thus making two triangles. Knowing that the sum of the angles of a triangle is 180° leads to the conclusion that the sum of the angles of the quadrilateral is 360°.

8. They are convex polygons, with four sides, four right angles, opposite sides congruent and parallel, and the diagonals congruent and bisecting each other.

9.
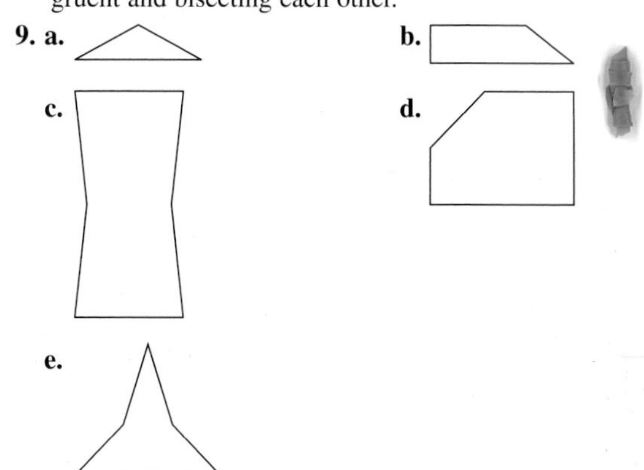

10. The case of five right angles is the most challenging. In a hexagon, there are six angles whose sum is 720°. If five of the angles are right angles, then the sixth angle would be 270°, which is possible if the hexagon is convex.

11.

The First Figure	The Second Figure
Hexagon	Hexagon
Concave	Convex
Two sides parallel	Three pairs of parallel sides
Two pairs of congruent sides	
4 acute angles	0 acute angles
2 reflex angles	0 reflex angles
0 right angles	2 right angles
0 obtuse angles	4 obtuse angles
3 pairs of congruent angles	3 pairs of congruent angles
1 line of symmetry	2 lines of symmetry
No rotation symmetry	180° rotation symmetry

12. Answers will vary.

13. a. (1) Kiwis are composed of a pentagon with a line segment protruding from one vertex.
(2) The protruding segment is perpendicular to the side the segment would intersect between the endpoints of the side if it were extended.
(3) Kiwis have 2 adjacent right angles, which is equivalent to saying that each kiwi has 2 parallel sides that meet a third side at 90° angles. The first not-kiwi has 6 sides. The second not-kiwi has the protruding segment not perpendicular to the opposite side. The third not-kiwi has the line segment inside the pentagon. The fourth not-kiwi does not have 2 adjacent right angles.

b. Figure (a) is not a kiwi—the protruding segment would not intersect the opposite side if extended. Figure (b) is a kiwi. Figure (c) is not a kiwi—the line segment is partially inside and partially outside the pentagon. Figure (d) is not a kiwi—it does not have 2 adjacent right angles.

14. Answers will vary. One description: Draw a square with one side parallel to the bottom edge of the paper and then draw the diagonal that begins at the bottom left corner. Extend the base of the square to the right, about one-third the length of the square. Connect this endpoint to the top right vertex of the square.

15. No, because it implies that there are some rectangles that are not parallelograms.

16.

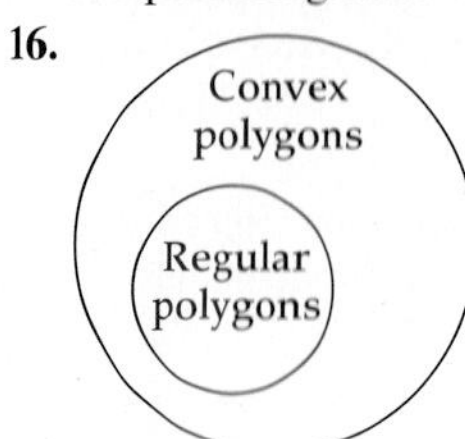

17. = (2.5, 7.5)

18. (2, 3)

19. (6, 6) and (12, 6), (−2, 6) and (4, 6), (6, 0) and (12, 0), (−2, 0) and (4, 0)

20. 6 vertices, 9 edges, and 5 faces

21. They have a base and an apex.

22. a. Answers will vary. The top, front, and side views will not work here because two different buildings have the same top, front, and side views. Descriptions can build from the ground up, from one side to another, or from front to back. Also valid is the top view with numbers to indicate the number of cubes in each spot. The key in most cases is to orient the reader correctly.

b. Answers will vary.

23.

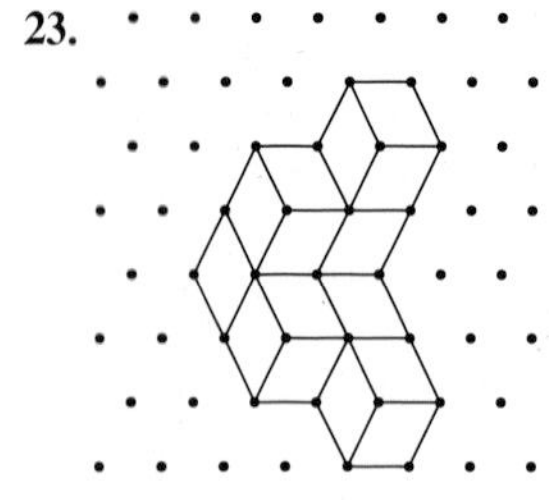

24.

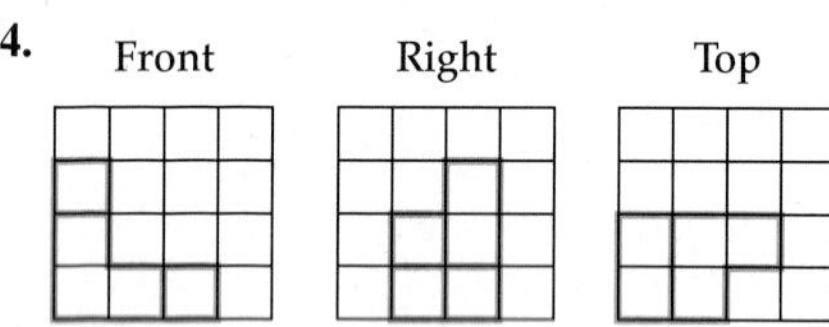

25. a. 7 cubes

b. It will be a mirror image of the right-side view.

c.

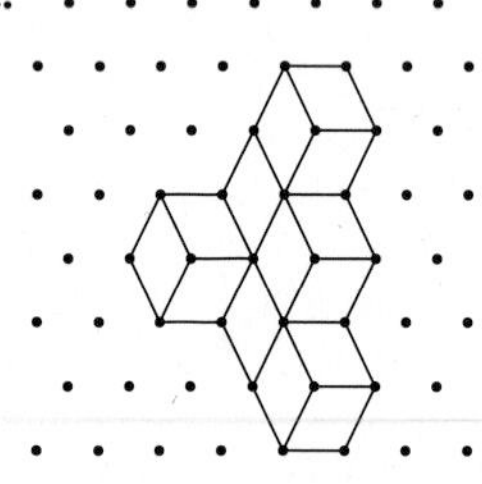

26. There are many possible nets. To be valid, it would have to fold up to make the prism. Two are shown below.

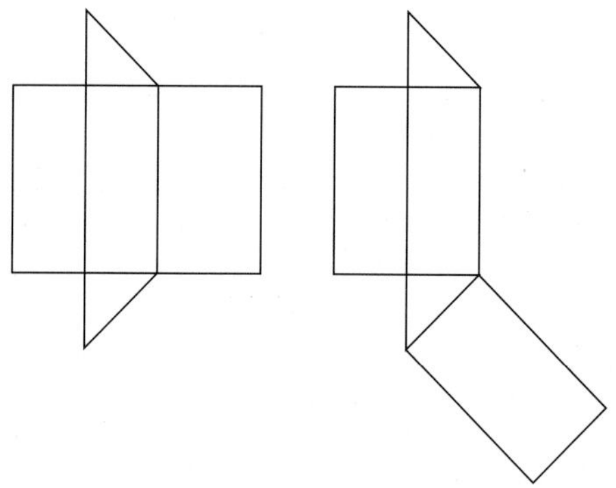

27. Answers will vary. One response: a line segment connecting two nonadjacent vertices

28. a. Trapezoid

b. Rectangle

29.

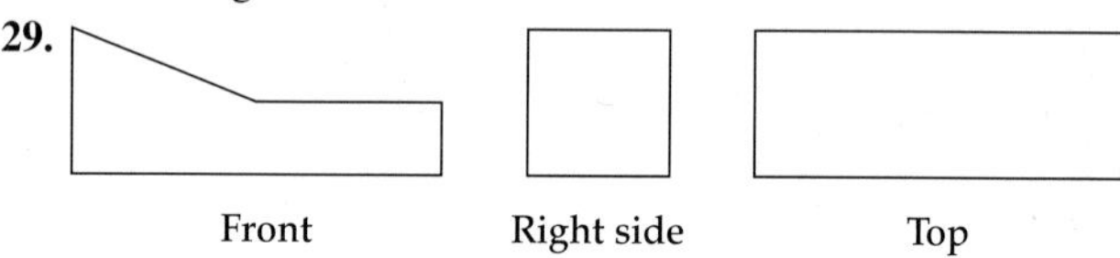

CHAPTER 9 EXERCISES

Section 9.1

1. Answers may vary. Following are examples for two of the given objects.

a. Surface area, volume, amount of pollutants, temperature of the water (at various levels), depth

c. Height, surface area of walls (for painting), surface area of roof (for shingles), surface area of windows, surface area of floors, ratio of area of windows to area of floors, ratio of area of windows to area of floors (to determine adequacy of ventilation)

3. a. 0.5 **b.** 45 **c.** 0.670 **d.** 3600 **e.** 0.450 **f.** 35,000 **g.** 0.024

5. *Mental math:* Answers will vary.
One possible response: Each yard is a little less than $3\frac{1}{2}$ inches less than a meter, so 400 yards would be about $3\frac{1}{2}$ inches × 400 = 1400 inches less than 400 meters. Then notice that 1400 inches ≈ 1440 inches. Since 1440 inches = 120 feet = 40 yards, we would expect the answer to be a few yards less than 40 yards.
Actual calculation: ≈ 33 yards
Calculation: $400\text{m} \times \frac{39\text{ m}}{1\text{ m}} \times \frac{1\text{ yd}}{36} \approx 433$ yards

7. The first estimate is "off" by 5 feet out of 105 feet, which is a little less than 5%. The second estimate is "off " by 5 feet out of 15 feet, which is about a 33% difference.

9. 1 liter is the volume in a cube that is 10 cm by 10 cm by 10 cm (or 1000 cm^3). A kilogram is the mass of 1 liter of water at 4 degrees Celsius.

11. 30 days

13. Answers will vary.

15. a. Kilometers

b. Gram; kilogram

c. Liter

17. Answers will vary.

19. Answers will vary.

21. The answer cannot be determined unless you know the length of the train.

23. Same amount
25–29. Answers will vary.
31. Answers will vary.
33. 24 glasses
35. a
37. c

Section 9.2

1. a. $P = 69$ mm, $A = 195$ mm^2
b. $P = 80$ mm, $A = 148.5$ mm^2
c. $P = 6$ cm, $A = \sqrt{3} \approx 1.7$ cm^2
d. $P \approx 37.68$ cm, $A \approx 113.04$ cm^2
e. $P = 40$ cm, $A = 72$ cm^2
f. $P = 35.8$ cm, $A \approx 51.9$ cm^2
3. a. 15 square units **b.** $19\frac{1}{2}$ square units
c. 29 square units
5. Approximately 11.2 feet. Arc length $= \frac{128}{360}\pi(10)$.
7. Strategies will vary.
a. 175 square centimeters
b. 180 square centimeters
9. 144 square inches
11. About 8–9 inches
13. ≈ 45.84 square meters
15. $44.4\% = \frac{4}{9}$ of the square
17. a. Answers will vary. The curved line traces the top half of one circle and the bottom half of another circle. These circles have their centers along a diameter of the larger circle and radii that are half the radius of the larger circle.
b. πr
19. 312 bricks
21. The diameter of the tree is approximately 16.4 feet.
23. Answers will vary.
25. a. Answers will vary.
b. The area of each piece is 2 square units.
27. Answers will vary according to the person's height. If you assume 1 quarter is 2.5 cm long, the value of a line of quarters 170 cm long is \$17. If you assume 1 nickel is 2 mm tall, then the value of a stack of nickels 170 cm high is \$42.50.
29. 78 triangles
31. No, it will not fit.
33. 24,450 miles. 489 miles is $\frac{7.2}{360}$ of Earth's circumference.
35. Answers will vary.
37–38. Answers will vary.
39. This is counterintuitive: Let d be the diameter of the ball. The height of the can is $3d$, but the distance around the can is $3.14d$!
41. Ratio of corresponding sides is 2:1. Ratio of areas is 4:1.
43. a. 10π cm
b. Not possible, because we don't know the height of the part that is cut out.
c. $34 + 3\pi$ inches
45. a. 50 square inches **b.** Not possible
c. Answers will vary. If the rectangle is 4 centimeters × 5 centimeters, $P = 18$ centimeters. If the rectangle is 10 centimeters × 2 centimeters, $P = 24$ centimeters.
d. 180 feet by 90 feet
47. The rectangle is getting bigger in its length and width. The area of the new rectangle is $1.5^2 = 2.25$ times the area of the old rectangle.
49. A population of 300 million would give an area of 581 square miles, which is over $\frac{1}{3}$ the size of Rhode Island!
51. The computation yields the almost unbelieveable answer of 2.5 square feet per person. If we make a rectangle that is 5 feet long, it would have to be 6 inches wide. This number helps us to understand why as many as $\frac{1}{3}$ of the people died during the voyage. Other accounts of slave ships report that each male had a space about 16 inches wide.
53. For the perimeter we are measuring linear distances (the perimeter could be stretched into a line), and for the area we are filling in the space with squares.
55. Answers will vary depending on the density of the grass.
57. a. If we assume that each hamburger is $\frac{1}{4}$ pound and if a person weighed 100 pounds, that person would eat 400 hamburgers a day.
b. If we take a cereal box that has 12 ounces of cereal and divide 100 pounds by $\frac{3}{4}$ of a pound, we get 133 boxes of cereal!
59. 8 cm by 4 cm
61. c
63. b

Section 9.3

1. a. $S.A. = 1474.3$ square inches
$V = 2160$ cubic inches
b. $S.A. = 35{,}630$ square feet
$V = 312{,}000$ cubic feet
c. $S.A. \approx 242$ square feet
$V \approx 267$ cubic feet
3. Using 3.14 for π,
a. $S.A. = 144\pi$ or 452.16 square inches, $V = 288\pi$ or 904.32 cubic inches
b. $S.A. = 48\pi$ or 150.72 square inches, $V = 47\pi$ or 150.72 cubic inches
c. $h = \frac{16}{\pi}$ or 5.1 m
5. a. $S.A. \approx 47$ square feet
$V = 16$ cubic feet
b. $S.A. = 695.2$ square feet, $V = 1221$ cubic feet. One strategy is to first change all lengths to inches.
$V = 1222$ cubic feet
c. $S.A. = 1980$ square feet
$V = 2985$ cubic feet
7. a. $S.A. = 216$ square inches, $V = 144$ cubic inches
b. Make it 6 inches by 6 inches by 4 inches
$S.A. = 168$ square inches
c. Make it a cube. Length of each side would be 5.24 inches.
$V \approx 144$ cubic inches
9. If you roll it so the height is 11 inches, the volume is 62.9 cubic inches.
If you roll it so the height is 8.5 inches, the volume is 81.7 cubic inches.
11. a. 60 feet
b. 44.7 feet
c. 34.6 feet
13. a. Predictions and explanations will vary.
The new tent is about 291 cubic feet larger than the old tent.
b. Predictions and explanations will vary.
The new tent uses almost 258 square feet of additional material.
c. Answers will vary.
15. About 105,000 sheets. The shelves can hold about 42% of the yearly purchase.

17. ≈ 34 feet

19. a. Stack them in two layers with each layer a square, 2 × 2.

b. Stack them end to end in one layer, with space between each block.

21. Answers will vary.

23. Because 1 cubic meter is a cube that is 1 meter by 1 meter by 1 meter, or 100 cm by 100 cm by 100 cm, we find the answer by: 100 × 100 × 100 = 1,000,000 cubic cm.

25. a. *S.A.* of small prism : *S.A.* of large prism :: 52 : 468 = 1 : 9

b. Volume of small prism : volume of large prism :: 24 : 648 = 1 : 27

27. 20 centimeters × 20 centimeters × 10 centimeters

29. a. Answers will vary.

b. The 167′ × 60′ building would require about 8% more bricks than the square building. The U-shaped building would require 82% more bricks.

c. 8 gallons, 9.1 gallons, 14.6 gallons

31–33. Answers will vary.

35. Answers will vary. Depends on measured dimensions of a ping-pong ball and how you intend to pack the balls.

37. There are other considerations. A cylindrical shape might require less packing material, but there will be wasted space when many of the products are packed into boxes for shipping to stores.

39. 283 cubic inches

CHAPTER 9 REVIEW EXERCISES

1. Answers will vary.

2. a. mm **b.** L **c.** mL **d.** mg **e.** kg

3. a. 34,000 cm **b.** 345 g **c.** 2750 mL

4. Answers will vary. Two situations are given here.

a. Measuring the length of a figure with a ruler but beginning at 1 instead of 0, a mistake commonly made by children

b. Reporting the distance between Los Angeles and San Francisco to the nearest 100 miles

5. Attributes include surface area, perimeter, temperature, depth, and average depth.

6. Here are two ways: (1) Find the area of the rectangle that encloses the figure and subtract the area of the two triangles: 42 − 12 − 3 = 27 square units. (2) Use the symmetry of the figure—the bottom half is a trapezoid with area 13.5 square units.

7. Area = 6 square inches

8. The area of the border is 146 square feet.

9. The dimensions are 31.83 centimeters × 31.83 centimeters.

10. 1200 square inches

11. 200 centimeters on a side, which is equivalent to a square that is 2 meters on a side

12. The total area of the dartboard is $\pi \cdot 6^2 = 36\pi$. The shaded region is $\pi \cdot 4^2 - \pi \cdot 2^2 = 16\pi - 4\pi = 12\pi$. So the probability of landing in the shaded region is $\frac{12\pi}{36\pi} = \frac{1}{3} =$ 0.33 or 33%.

13. a. There is empty space between the pennies because circles do not tessellate.

b. 25,600 pennies, using 3.14 for π

14. Answers will vary. One response: We can take any parallelogram, slice off a right triangle from one side, and translate that triangle to the opposite side. The parallelogram becomes a rectangle, whose area we know to be $l \times w$. We know first that the rectangle and parallelogram have the same area, that the length of the rectangle is the same as the base of the parallelogram, and that the width of the rectangle is the same as the height of the parallelogram. Therefore, $l \times w$ (of the rectangle) $= b \cdot h$ (of the parallelogram).

15. Answers will vary. Two responses: (1) Because 12 inches = 1 foot, 144 square inches = 1 square foot. (2) One could also draw a square that measures 1 foot by 1 foot; thus the area is 1 square foot. When we convert to inches, the same square is 12 inches × 12 inches = 144 square inches.

16. It quadruples.

17. $37\frac{1}{2}\%$

18. The area of the figure will be about 8.1 square centimeters, or 810 square millimeters, which means that the actual area of the pond is 810 square meters.

19. One ream of paper measures 8.5 inches by 11 inches by 2 inches. Thus all the forms will take up about 433,000 cubic feet. A cube that measures 76 feet on a side would be able to contain all the paper. In more conventional terms, a building 233 feet by 233 feet by 8 feet would be needed to contain all the paper.

20. a. 44.7 feet

b. 34.6 feet

21. a. 88.9 cubic yards

b. 97.8 cubic yards

22. The surface area is approximately 2128 square feet. For the volume, break the figure into a rectangular prism and a triangular prism: 9600 cubic feet.

23. 25,600 cubes

24. There are many examples. Two are 15 × 1 × 1, with a surface area of 62 square units and a volume of 15 cubic units; and 5 × 3 × 2, which also has a surface area of 62 square units but a volume of 30 cubic units.

25. Surface area of the prism is 248 square centimeters.
Surface area of the cylinder is 225 square centimeters.
Volume of the cylinder is 246 cubic centimeters.
Volume of the prism is 240 cubic centimeters.

26. a. 5840 gallons

b. No. 1 gallon = 0.1337 cubic foot. The family saves 5840 gallons, which converts to 781 cubic feet. Because the swimming pool contains 3600 cubic feet, we would fill about $\frac{1}{5}$ of the pool.

CHAPTER 10 EXERCISES

Section 10.1

1. Each point of the figure has been moved 2 units to the right and 3 units down. Also, the figure has been moved approximately 3.6 units along a vector that makes a 55-degree angle with the *x*-axis.

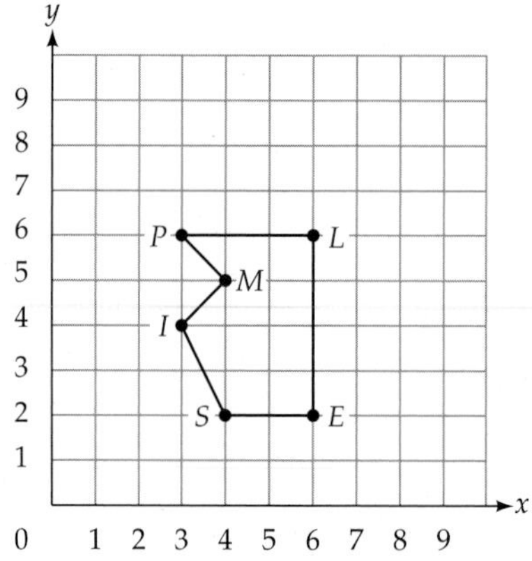

3.

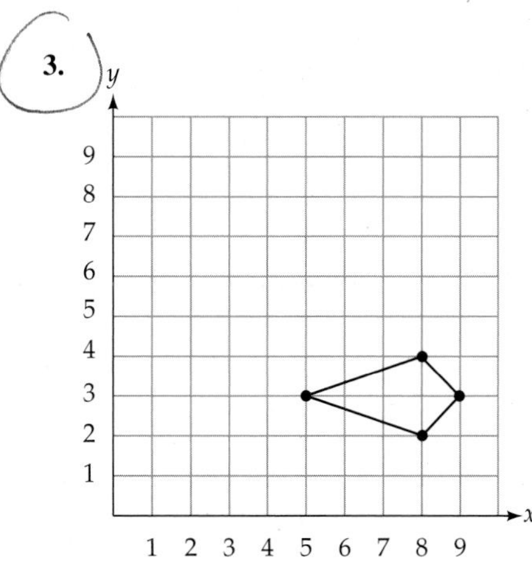

5. Each point is the same distance from the line as its reflected image.

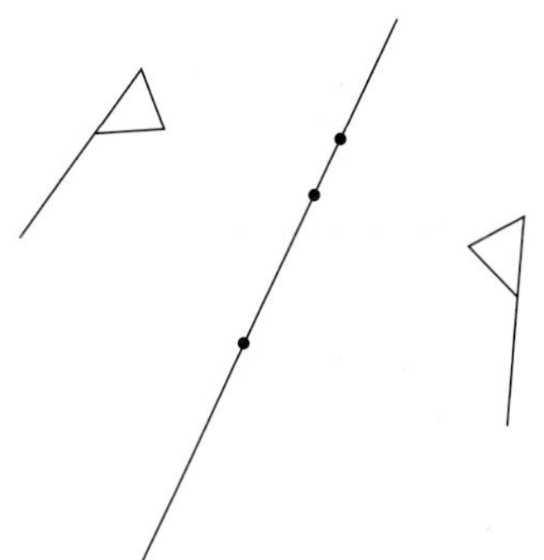

7.

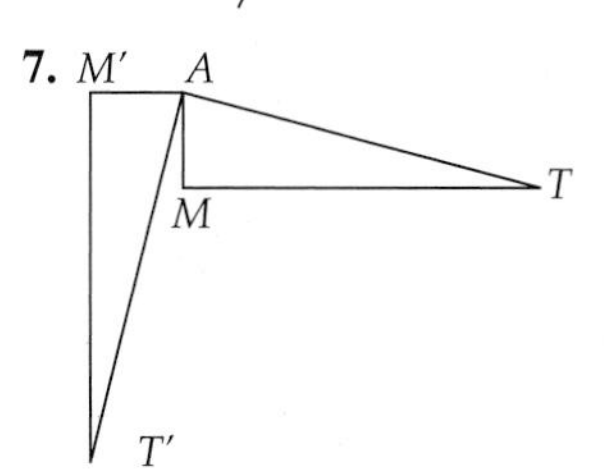

(The original figure is shown for reference purposes.)

9.

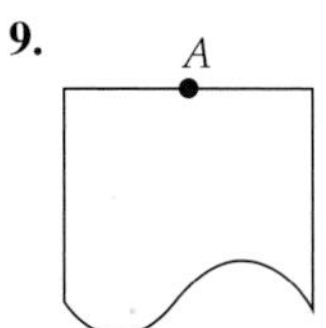

11. Place the mirror so that it bisects Molly's right eye and right leg (facing the left side of Molly).

13. There are many ways to do this. One way is to draw a parallel line between $\overline{AT}$ and $\overline{OG}$ and show that the figures are reflections over this line.

15. a. Answers will vary.

b. (The original figure is shown for reference purposes.)

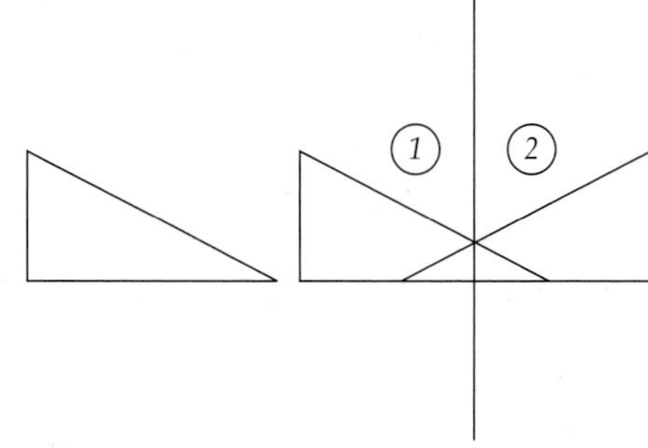

c.

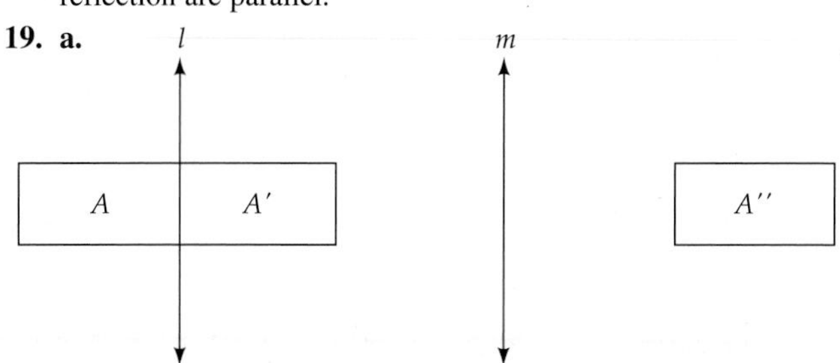

The image from translating and then reflecting does not coincide with the image from reflecting and then translating.

17. They will commute when the translation vector and the line of reflection are parallel.

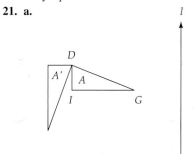

19. a.

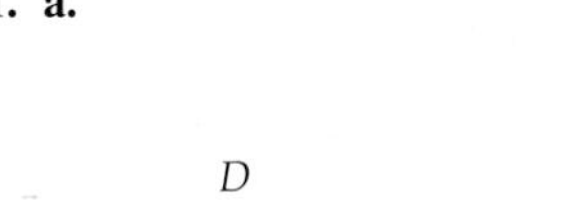

(The original figure is shown for reference purposes.)

b. The images are not the same. The image is located to the left of the original figure by a distance that is approximately equal to the distance between lines *l* and *m*.

21. a.

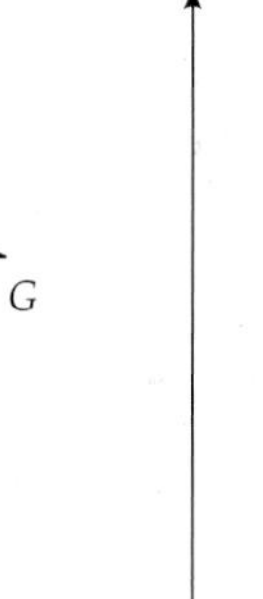

(The original figure is shown for reference purposes.)

b. Reflect the figure across line *l*, then rotate the image 90 degrees counterclockwise about point *D*.

23. a. Reflect $\overline{PR}$ across $\overline{RM}$.

b. Reflect $\triangle PMR$ across $\overline{RM}$.

25. a.

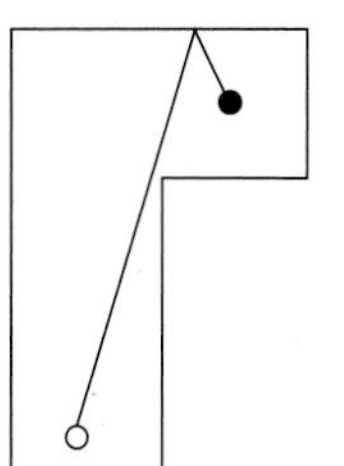

b.

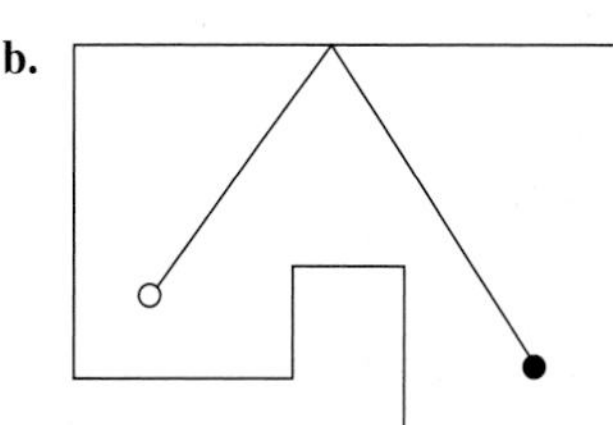

27. Answers will vary. Some possibilities are given.
 a. Baby Blocks: If we see the figure as composed of rhombuses, each rhombus can be mapped onto a neighboring rhombus by a 60-degree rotation.
 b. Broken Windows: The top row can be translated onto the second row. A similar translation can be described with respect to columns.
 c. Cross Roads: There are horizontal, vertical, and diagonal lines of reflection through the center of the design, and it can be rotated 90 degrees, 180 degrees, or 270 degrees onto itself.
 d. Pinwheel: The design can be rotated 90 degrees, 180 degrees, or 270 degrees onto itself. Each white triangle can be rotated 90 degrees and then translated onto another white triangle.
 e. Underground Railroad: The design contains two alternating unit squares that are translated diagonally. One unit square consists of four smaller squares, two light and two dark. The whole design can be reflected across diagonal lines that pass through the center and can be rotated 180 degrees about the center.

29. Answers will vary.

31. **a.** 9 o'clock
 b. Approximately 6:15
 c. Answers will vary.

33. Answers will vary.

35. c and d

Section 10.2

1. **a.** 60-degree rotation symmetry
 b. Five lines of symmetry through the vertices; 72-degree rotation symmetry
 c. Horizontal and vertical line symmetry; point symmetry
 d. Translation symmetry if seen as a pattern extending to the right and left

3. **a.** Vertical line symmetry
 b. Horizontal, vertical, and diagonal line symmetry; 90-degree rotation symmetry
 c. No symmetry because of the line segments

5. **a.** Horizontal, vertical, and diagonal line symmetry; 90-degree rotation symmetry
 b. 90-degree rotation symmetry
 c. Horizontal and vertical line symmetry; point symmetry
 d. Horizontal and vertical lines of symmetry
 e. Horizontal, vertical, and two diagonal lines of symmetry; 90-degree rotation symmetry

7.

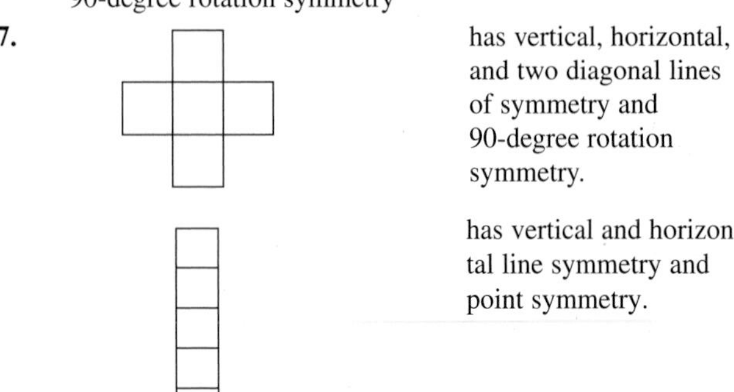

has vertical, horizontal, and two diagonal lines of symmetry and 90-degree rotation symmetry.

has vertical and horizontal line symmetry and point symmetry.

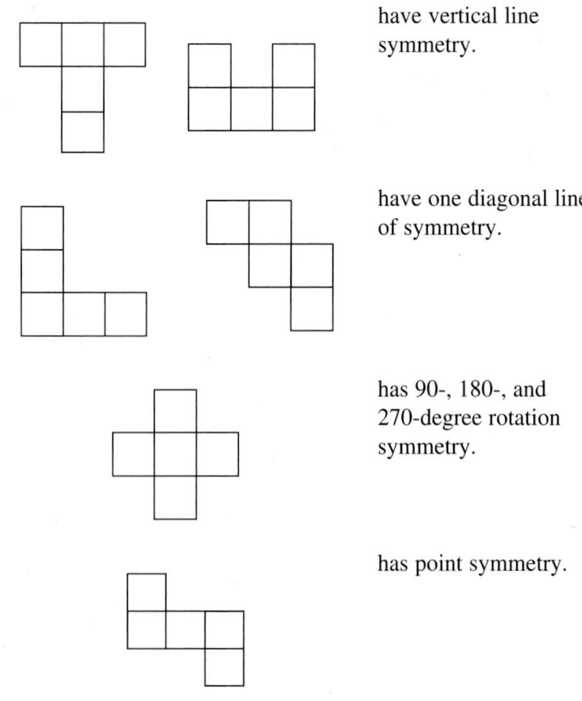

have vertical line symmetry.

have one diagonal line of symmetry.

has 90-, 180-, and 270-degree rotation symmetry.

has point symmetry.

The remaining pentominoes have no symmetry.

9. Notation after semicolon is from page 590.
 a. rotation; 12
 b. translation, rotation; 12
 c. translation, glide; 1g
 d. translation, vertical reflection, horizontal reflection, rotation; mm
 e. translation; 11
 f. translation, vertical reflection, glide; mg

11. You need to verify that the sum of the angles at each noncongruent vertex point is indeed 360 degrees.

13. All 12 pentominoes tessellate.

15. **a.** If the unit is one of the C shapes, then the means of tesselation is a diagonal tranlation to make one strip of C's, then rotate that entire strip 180 degrees and translate. Then repeat these two steps. If the unit is two C's nested into each other, then the means is to translate those two C's diagonally to make a long strip. Then repeat.
 b. The unit will be six interlocking hexagons (which form a hexagonal kind of shape). Translate this unit diagonally. Then translate this whole pattern in the other diagonal direction.
 c. The unit is an individual brick. Translate the brick diagonally. Then translate this whole pattern in the other diagonal direction.
 d. This figure is composed of three shapes: a star, a hexagon, and an octagon. One unit consists of the star, two hexagons, and three octagons. The two hexagons are beneath the star and have one common vertex each. The three octagons surround the hexagon that lies to the lower left of the star.

17. It holds.

19. **a–g.** Answers will vary.
 h. Impossible

21. a. **b.**

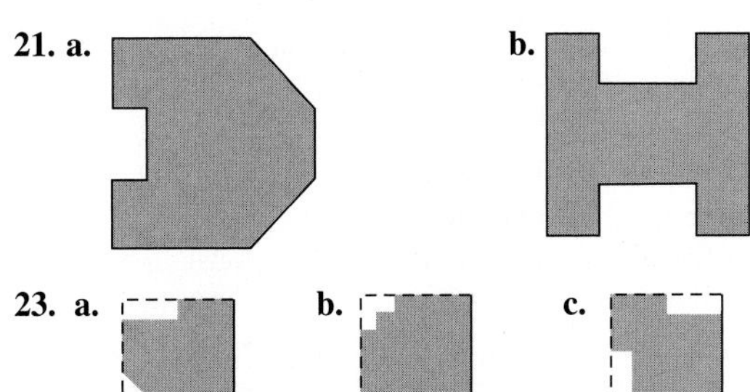

23. a. **b.** **c.**

25. Answers will vary.

27. All three have the same symmetry: translation, vertical reflection, horizontal reflection, 180-degree rotation; mm

29. a. 90-degree rotation symmetry and 4 lines of reflection symmetry

b. 4 lines of symmetry, 90-degree rotation symmetry

c. Answers will vary.

31–33. Answers will vary.

35. Answers will vary. Possible responses: MATH, WAIT.

37. Answers will vary.

39. All the shapes tesselate.

41. a. False. Justifications may vary.

b. False. Justifications may vary.

Section 10.3

1. a. 13.5 cm

b. 2.2 cm

3. Yes, the angles are the same and the sides of the figure to the right are twice as long as the sides of the smaller triangle.

5. a. Corresponding sides of 8×11 and 4×6 rectangles are not proportional.

b. Answers will vary.

7. 1 inch = 275 miles

9. Find the height of an object near the building, the length of its shadow, and the length of the shadow of the building. Then write and solve the proportion:

$$\frac{\text{Height of object}}{\text{Length of shadow}} = \frac{\text{Unknown height of building, } x}{\text{Length of building's shadow}}$$

11. Answers will vary.

13. a. The smaller congruent shapes look like the scales on a reptile's body. Another possibility: they are repeating tiles.

b. Answers will vary.

c. Answers will vary.

d. Must be either a parallelogram or have one side perpendicular to the bases. Also, it needs two congruent sides.

e. There are at least two rectangular tilings and one tiling with a composite unit that looks like a "+" sign.

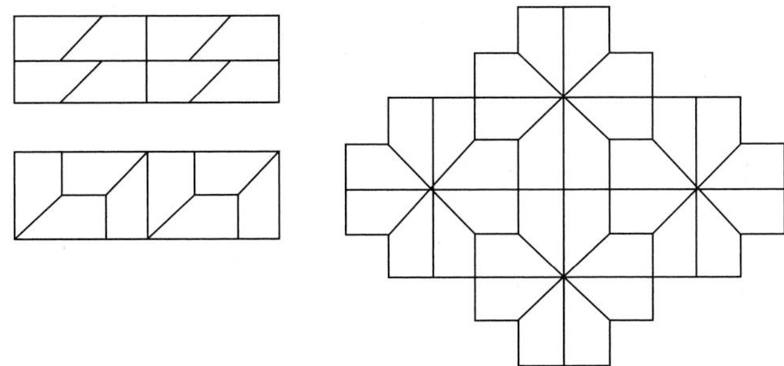

15. It appears that triangles B, C, and E are all similar.

CHAPTER 10 REVIEW EXERCISES

1.

2. Answers will vary.

3.

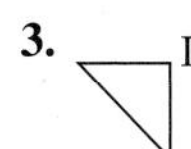

4.

a.

b. Translate the figure to the right.

c.

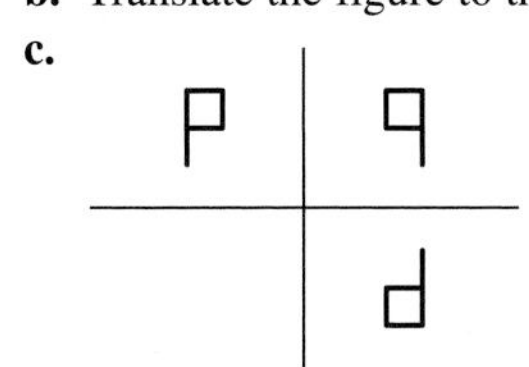

d. Rotate the figure 180 degrees about the intersection of the two lines.

5. The only one-step solution is to rotate the figure 90 degrees counterclockwise, about the point that is three units to the right and three units below the bottom right vertex of the top trapezoid. There are several two-step solutions; for example, translate six units down and then rotate 90 degrees counterclockwise.

6. Alike—They both move the object to a different position, keep the figure the same size and the same shape. Different—The orientation is changed in a reflection but not in a rotation. For example, imagine rotating and reflecting the letter P. To see the rotated P, you could just turn the page until it was "right" again. However, no matter how you turned the page, the reflected P just wouldn't look right—it would feel backward. The curved part is now facing left instead of right.

7. By including glide reflection, we have the theorem that any figure can be moved to any other spot on a plane in *exactly* one move.

8. a. $\frac{1}{5}$-turn symmetry, or 72-degree rotation symmetry; five lines of reflection

b. $\frac{1}{4}$-turn symmetry, or 90-degree rotation symmetry; four lines of reflection

c. $\frac{1}{2}$-turn symmetry, or 180-degree rotation symmetry; no reflection symmetry

9. There are many possibilities. One valid figure is given for each case.

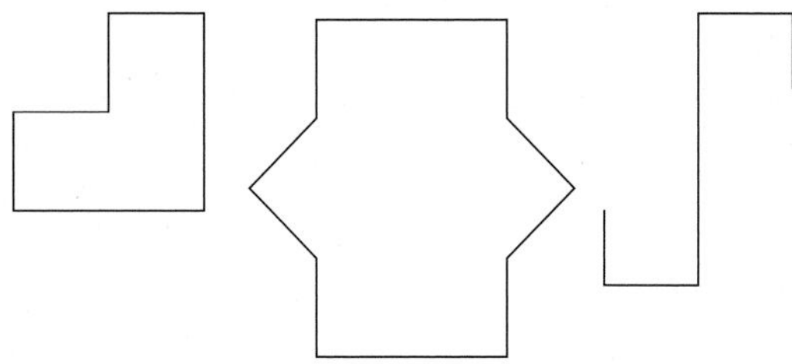

10. Answers will vary.

11. a. It means that there is at least one way to pick up the figure, turn it some amount less than 360 degrees, and lay it down so it fits back exactly on itself.

b. It means that there is at least one line that you can draw through the figure such that if you fold the figure along that line, the half of the figure on one side of the line will fit exactly on the other half.

12. The symmetries can be described with notation or visually. The notation is given.

a. ml

b. ll

c. mg

d. mg

13. a. There is more than one possibility. For example, the figure at the left can be seen as the unit. In this case, the means by which it is repeated is a translation to the right. On the other hand, the figure on the right can also be seen as the unit. In this case, the means by which it is repeated is also a translation to the right.

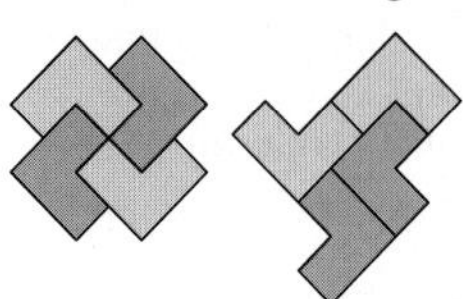

b. In this case, the unit consists of the five chevrons shown below. The means by which it is repeated is translation in a diagonal direction.

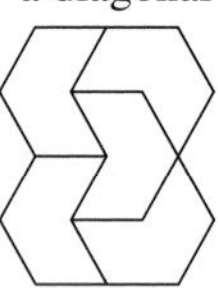

14. a. This figure tessellates. The sum of all angles at every vertex point is 360 degrees; this is because all angles are 90 degrees or 270 degrees.

b. This figure tessellates because all quadrilaterals tessellate.

c. This figure tessellates because it is made by beginning with a square, modifying the bottom, and then translating that modification to the top side.

d. This figure tessellates. Because of the symmetry of the hexagon, the four acute angles are congruent and the two reflex angles are congruent. If we label the acute angles a and the reflex angles b, we have $4a + 2b = 720$, which simplifies to $2a + b = 360$. Thus the sum of the angles at each vertex point is 360 degrees.

e. This figure tessellates. The reasoning here is virtually identical to the reasoning for part (d), except that the sum of the angles of a pentagon is 540 degrees.

15. 28 cm

16. Yes. No matter what the size, all three corresponding angles are congruent.

17.

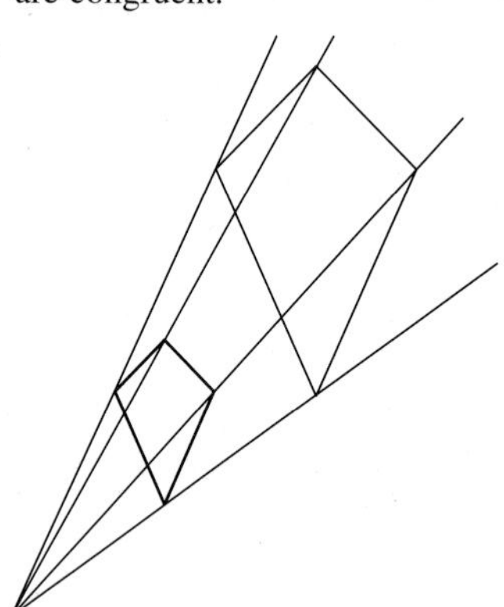

18. (6, 8), (16, 8), and (8, 14). The coordinates of each point of a similar triangle will be double the coordinates of the corresponding points on the original triangle.

19. Answers will vary. For example, any simple, closed curve will suffice for 0; any simple open curve will suffice for 1, 2, 3. Any simple, closed curve with two straight or curved segments protruding will suffice for 4.

APPENDIX B

Answers to Questions in Text

Chapter 1

Investigation 1.2a **p. 13**

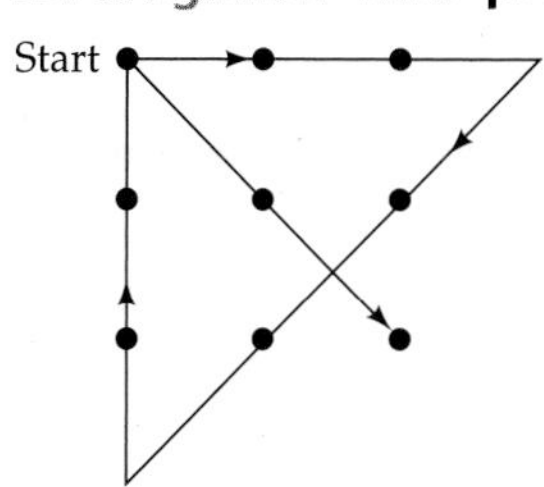

Chapter 2

Patterns in counting: missing words in Table 2.3 **p. 56**

Number	Greenland Eskimos	Aztecs	Luo of Kenya
8	achfineq-pinasut		ab-adek
9			ab-angwen
12	achqaneq-machdlug	matlacti-on-ome	
13	achqaneq-pinasut	matlacti-on-yey	apar-adek
15			apar-abich
16	achfechsaneq-atauseq	caxtulli-on-ce	apar-ab-chiel

The Egyptian numeration system: symbols **p. 56**

1. 1323 **2.** 143,022 **3.** 𓆼𓍢𓍢II **4.** 𓍢𓍢𓍢IIII

The Roman numeration system: symbols **p. 57**

1. 1011 **2.** 1597 **3.** MCII **4.** CCCXIX

Working with the system **p. 58**

The Later Roman symbol for 444 was CDXLIV.

The Babylonian numeration system: places **p. 59**

1. $120 + 12 = 132$

2. $5 \times 60 + 42 = 342$

3. $1 \times 60^2 + 11 \times 60 + 21 = 3600 + 660 + 21 = 4281$

4. 1202 would be seen as $20 \times 60 + 2 =$

5. 304 would be seen as five 60s plus 4, or

The Mayan numeration system: symbols **p. 60**

1. $(2 \times 20) + 6 = 46$
2. $(15 \times 20) + 0 = 300$
3. $(1 \times 360) + 0 + 13 = 373$
4. $= (6 \times 20) + 3$
5. $= (12 \times 20) + 16$
6. $= (1 \times 7200) + (0 \times 360) + (5 \times 20) + 0$

Chapter 7

Investigation 7.3f **p. 399**

TABLE 7.18

6 6 1	6 5 2	6 4 3	5 5 3	4 4 5
6 1 6	6 2 5	6 3 4	5 3 5	4 5 4
1 6 6	5 6 2	4 6 3	3 5 5	5 4 4
	5 2 6	4 3 6		
	2 5 6	3 6 4		
	2 6 5	3 4 6		

TABLE 7.19

6 6 1	5 6 2	4 6 3	3 6 4	2 6 5	1 6 6
6 5 2	5 5 3	4 5 4	3 5 5	2 5 6	
6 4 3	5 4 4	4 4 5	3 4 6		
6 3 4	5 3 5	4 3 6			
6 2 5	5 2 6				
6 1 6					

Sample Space of Rolling 3 Dice **p. 399**

Note: The first column is simply to make communication easier; for example, we can talk about patterns in rows 19 to 24.

TABLE 7.23

1	1 1 1	2 1 1	3 1 1	4 1 1	5 1 1	6 1 1
2	1 1 2	2 1 2	3 1 2	4 1 2	5 1 2	6 1 2
3	1 1 3	2 1 3	3 1 3	4 1 3	5 1 3	6 1 3
4	1 1 4	2 1 4	3 1 4	4 1 4	5 1 4	6 1 4
5	1 1 5	2 1 5	3 1 5	4 1 5	5 1 5	6 1 5
6	1 1 6	2 1 6	3 1 6	4 1 6	5 1 6	6 1 6
7	1 2 1	2 2 1	3 2 1	4 2 1	5 2 1	6 2 1
8	1 2 2	2 2 2	3 2 2	4 2 2	5 2 2	6 2 2
9	1 2 3	2 2 3	3 2 3	4 2 3	5 2 3	6 2 3
10	1 2 4	2 2 4	3 2 4	4 2 4	5 2 4	6 2 4
11	1 2 5	2 2 5	3 2 5	4 2 5	5 2 5	6 2 5
12	1 2 6	2 2 6	3 2 6	4 2 6	5 2 6	6 2 6
13	1 3 1	2 3 1	3 3 1	4 3 1	5 3 1	6 3 1
14	1 3 2	2 3 2	3 3 2	4 3 2	5 3 2	6 3 2
15	1 3 3	2 3 3	3 33	4 3 3	5 3 3	6 3 3
16	1 3 4	2 3 4	3 3 4	4 3 4	5 3 4	6 3 4
17	1 3 5	2 3 5	3 3 5	4 3 5	5 3 5	6 3 5
18	1 3 6	2 3 6	3 3 6	4 3 6	5 3 6	6 3 6
19	1 4 1	2 4 1	3 4 1	4 4 1	5 4 1	6 4 1
20	1 4 2	2 4 2	3 4 2	4 4 2	5 4 2	6 4 2
21	1 4 3	2 4 3	3 4 3	4 4 3	5 4 3	6 4 3
22	1 4 4	2 4 4	3 4 4	4 4 4	5 4 4	6 4 4
23	1 4 5	2 4 5	3 4 5	4 4 5	5 4 5	6 4 5
24	1 4 6	2 4 6	3 4 6	4 4 6	5 4 6	6 4 6
25	1 5 1	2 5 1	3 5 1	4 5 1	5 5 1	6 5 1
26	1 5 2	2 5 2	3 5 2	4 5 2	5 5 2	6 5 2
27	1 5 3	2 5 3	3 5 3	4 5 3	5 5 3	6 5 3
28	1 5 4	2 5 4	3 5 4	4 5 4	5 5 4	6 5 4
29	1 5 5	2 5 5	3 5 5	4 5 5	5 55	6 5 5
30	1 5 6	2 5 6	3 5 6	4 5 6	5 5 6	6 5 6
31	1 6 1	2 6 1	3 6 1	4 6 1	5 6 1	6 6 1
32	1 6 2	2 6 2	3 6 2	4 6 2	5 6 2	6 6 2
33	1 6 3	2 6 3	3 6 3	4 6 3	5 6 3	6 6 3
34	1 6 4	2 6 4	3 6 4	4 6 4	5 6 4	6 6 4
35	1 6 5	2 6 5	3 6 5	4 6 5	5 6 5	6 6 5
36	1 6 6	2 6 6	3 6 6	4 6 6	5 6 6	6 6 6

TABLE 7.24

	3	4	5	6	7	8	9	10	11	12	13	14	15	16	17	18
1	111	112	113	114	115	116	126	136	146	156	166	266	366	466	566	666
2		121	122	123	124	125	135	145	155	165	256	356	456	556	656	
3		211	131	132	133	134	144	154	164	246	265	365	465	565	665	
4			212	141	142	143	163	163	236	255	346	446	546	646		
5			221	213	151	152	162	226	245	264	355	455	555	655		
6			311	222	214	161	216	235	254	336	364	464	564	664		
7				231	223	215	225	244	263	345	436	536	636			
8				312	232	224	234	253	326	354	445	545	645			
9				321	241	233	243	262	335	363	454	554	654			
10				411	313	242	252	316	344	426	463	563	663			
11					322	251	261	325	353	435	526	626				
12					331	314	315	334	362	444	535	635				
13					412	323	324	343	416	453	544	644				
14					421	332	333	352	425	462	553	653				
15					511	341	342	361	434	516	562	662				
16						413	351	415	443	525	616					
17						422	414	424	452	534	625					
18						431	423	433	461	543	634					
19						512	432	442	515	552	643					
20						521	441	451	524	561	652					
21						611	513	514	533	615	661					
22							522	523	542	624						
23							531	532	551	633						
24							612	541	614	642						
25							621	613	623	651						
26								622	632							
27								631	541							

Chapter 10

Investigation 10.2f pp. 588–590

v = translation vector

m = mirror line

· = center of rotation

g = glide vector

m_g = mirror line for glide reflection

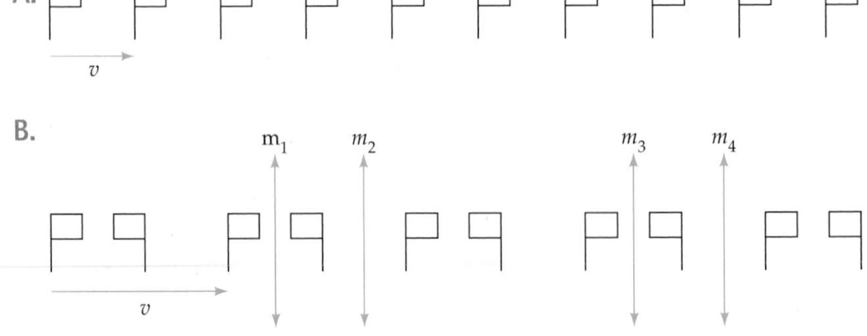

m_1 and m_3 are equivalent.
m_2 and m_4 are equivalent.

C.

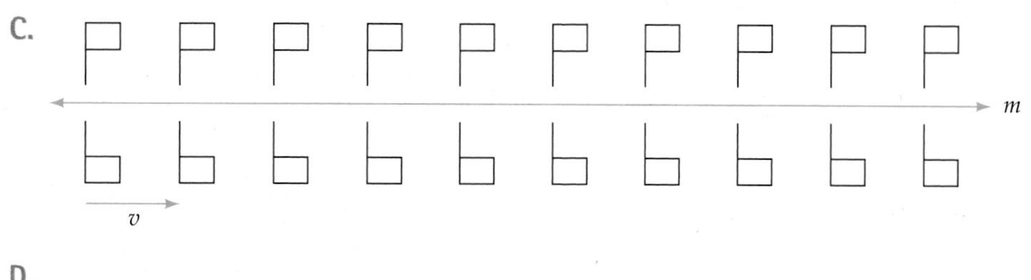

D.

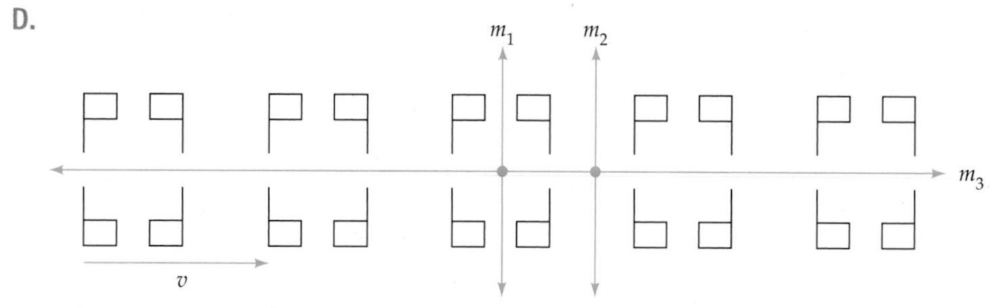

E.

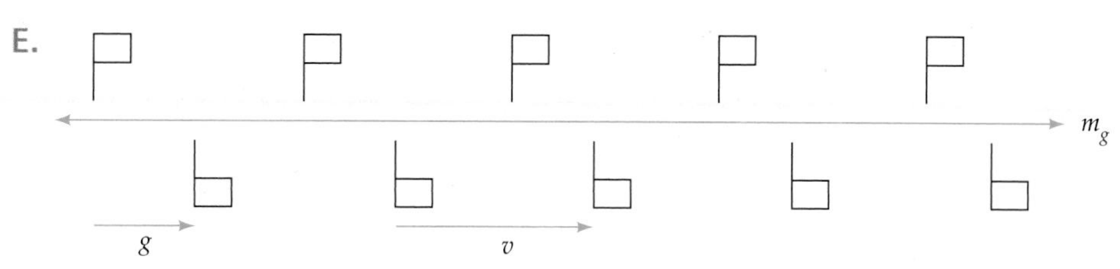

F.

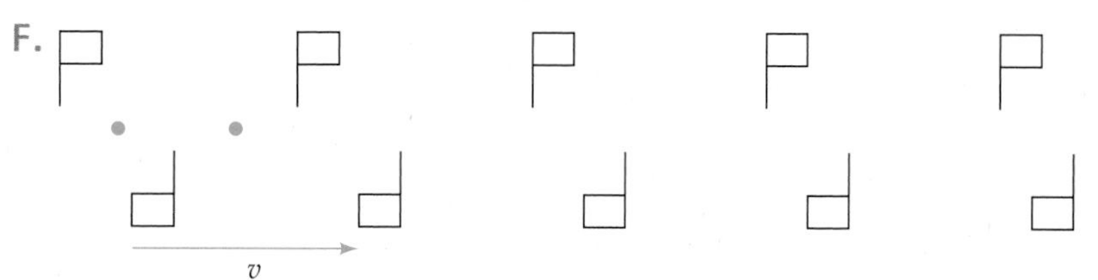

G.

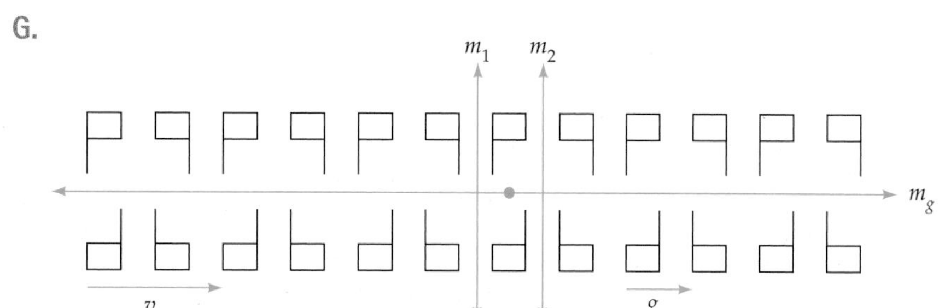

Investigation 10.2g **pp. 591–592**

m = mirror line

· = center of rotation

g = glide vector

m_g = mirror line for glide reflection

A.

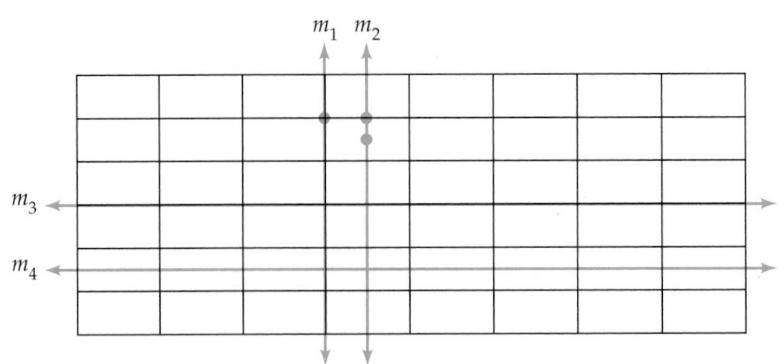

Translation vectors

B.

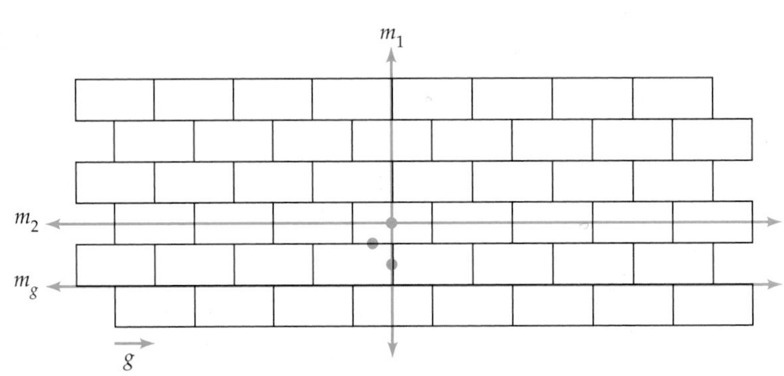

Translation vectors

C.

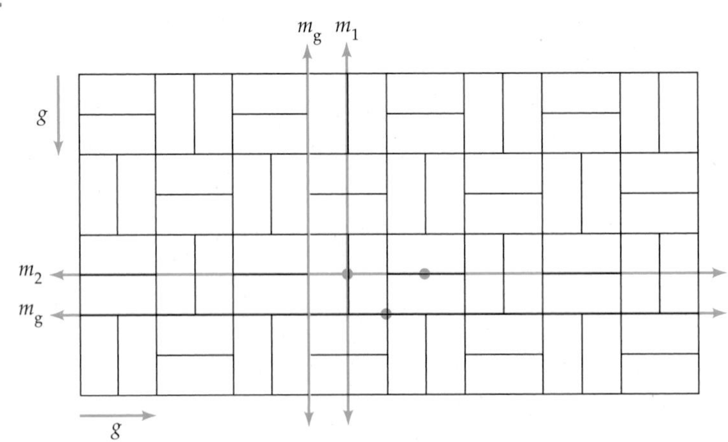

Translation vectors

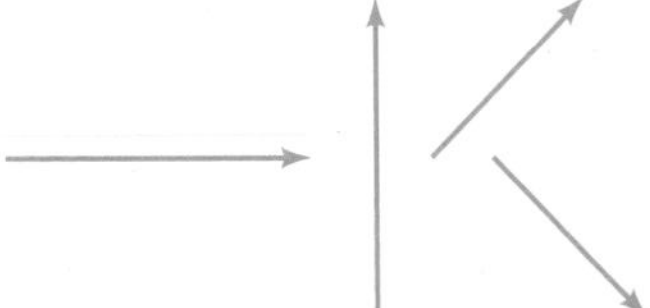

ENDNOTES

CHAPTER 1

p. 1 1. NCTM, *Curriculum and Evaluation Standards for School Mathematics: Executive Summary* (Reston, VA: NCTM, 1989), p. 5.

p. 3 2. At appropriate places, you will see the boxes indicating a specific connection that I will be making—connections called MATHEMATICS, HISTORY, OUTSIDE THE CLASSROOM, LANGUAGE, and CLASSROOM CONNECTION. These notes will help develop the notion of the connectedness of mathematics.

p. 3 3. Lynn Arthur Steen, ed., *On the Shoulders of Giants: New Approaches to Numeracy* (Washington, DC: National Academy Press, 1990).

p. 3 4. *Mathematics: The Science of Patterns* (New York: W. H. Freeman, 1996); *The Language of Mathematics: Making the Invisible Visible* (New York: W. H. Freeman, 1998); and *Life by the Numbers* (New York: Wiley, 1998).

p. 3 5. American Association for the Advancement of Science, *Benchmarks for Scientific Literacy* (New York: Oxford University Press, 1993), p. 25.

p. 4 6. Deborah Meier, *The Power of Their Ideas: Lessons for America from a Small School in Harlem* (Boston: Beacon Press, 1995).

CHAPTER 2

p. 40 1. *Introduction to Mathematics* (New York, 1911), pp. 59–69, cited in Robert Moritz, *On Mathematics* (New York: Dover Publications, 1914), p. 199.

p. 60 2. Georges Ifrah, *From One to Zero: A Universal History of Numbers,* p. 373. Translated from the French, 1985.

p. 60 3. Robert Kalpan, *The Nothing That Is* (New York: Oxford University Press, 1999), p. 102.

p. 60 4. Karl Menninger, *Number Words and Number Symbols: A Cultural History of Numbers,* trans. Paul Broneer (Cambridge, MA: M.I.T. Press, 1969), p. 418.

p. 62 5. H.A. Freebury, *A History of Mathematics* (New York: Macmillan, 1958), p. 170.

p. 62 6. Karl Menninger, *Number Words and Number Symbols: A Cultural History of Numbers* (Cambridge, MA: The MIT Press, 1969), p. 422.

p. 68 7. David M. Schwartz, *How Much Is a Million?* (New York: Lothrop, Lee, & Shepard Books, 1985).

p. 68 8. Edward Packard, *Big Numbers: And Pictures That Show Just How Big They Are!* (Brookfield, CT: Millbrook Press, 2000).

CHAPTER 3

p. 75 1. Frank J. Swetz, *Capitalism and Arithmetic: The New Math of the 15th Century* (La Salle, IL: Open Court, 1987), p. 181.

p. 77 2. Lauren Resnick, "A Developmental Theory of Number Understanding," in *The Development of Mathematical Thinking,* ed. Herbert Ginsburg (New York: Academic Press, 1983), p. 114.

p. 88 3. Louis Charles Karpinski, *The History of Arithmetic* (New York: Russell & Russell, 1965), p. 81.

p. 105 4. Karl Menninger, *Number Words and Number Symbols: A Cultural History of Numbers* (Cambridge, MA: MIT Press, 1969), p. 413.

CHAPTER 4

p. 163 1. Although there are still numbers beyond irrational numbers, we will limit ourselves to the sets of numbers mentioned here because these are the sets that we focus on in K–8 mathematics.

p. 175 2. Louis Karpinski, *The History of Arithmetic* (New York: Russell & Russell, 1965), p. 127.

p. 180 3. Susan J. Lamon, *Teaching Fractions and Ratios with Understanding* (Mahwah, NJ: Lawrence Erlbaum Associates, 1999).

p. 191 4. These two problems are discussed in the March 2008 issue of *Teaching Children Mathematics,* pp. 421–425.

p. 213 5. Louis Karpinski, *The History of Arithmetic* (New York: Russell & Russell, 1965), p. 128.

p. 223 6. Louis Karpinski, *The History of Arithmetic* (New York: Russell & Russell, 1965), p. 131.

p. 236 7. H.A. Freebury, *A History of Mathematics* (New York: Macmillan, 1961), p. 92.

CHAPTER 5

p. 250 1. Judah Schwartz, "Intensive Quantity and Referent Transforming Arithmetic Operations," in *Number Concepts and Operations in the Middle Grades*, ed. James Hiebert and Merylyn Behr (Reston, VA: NCTM, 1988), p. 43.

p. 250 2. James Hiebert and Merylyn Behr, "Introduction: Capturing the Major Themes," in *Number Concepts and Operations in the Middle Grades*, ed. James Hiebert and Merylyn Behr (Reston, VA: NCTM, 1988), p. 2.

p. 254 3. Kathleen Cramer, Thomas Post, and Sarah Currier, "Learning and Teaching Ratio and Proportion: Research Implications," in *Research Ideas for the Classroom: Middle Grades Mathematics,* ed. Douglas T. Owens (New York: Macmillan, 1993), pp. 159–178.

p. 261 4. A kilowatt is 1000 watts. For example, if you run a lamp with a 100-watt light bulb for 10 hours, you have used 1000 watt-hours or 1 kilowatt-hour.

p. 267 5. *Historical Topics for the Mathematics Classroom: 31st Yearbook* (Reston, VA: NCTM, 1969), p. 147.

CHAPTER 6

p. 301 1. Thomas Carpenter, Megan Frank, and Linda Levi, *Thinking Mathematically: Integrating Arithmetic and Algebra in Elementary School* (Portsmouth, NH: Heinemann, 2003), p. 12.

CHAPTER 7

p. 328 1. *Information Please* (Boston: Houghton Mifflin, 1993), p. 392.

p. 330 2. *New York Times,* July 8, 1994.

p. 336 3. *The Time Almanac 2003* (Boston: Information Please, 2002), p. 343.

p. 337 4. *A Statistical Abstract of the United States 2002* (Washington, DC: U.S. Department of Commerce, Bureau of the Census, 2001), p. 18.

p. 352 5. Center for Statistical Education and the American Statistical Association, *Teaching Statistics: Guidelines for Elementary Through High School* (Palo Alto, CA: Dale Seymour Publications), p. 23.

p. 370 6. Technically, because we are finding the standard deviation of a sample as opposed to the whole population, we should determine the standard deviation by dividing by $n - 1$ instead of n, as we would do in an ordinary average. A full explanation of why one should divide by $n - 1$ is beyond the scope of this book. Our focus is on understanding the concept, so I have chosen to divide by n, as have the authors of most elementary mathematics textbooks.

p. 379 7. Harold W. Stevenson, Max Lummis, Shinying Lee, and James W. Stigler, *Making the Grade in Mathematics: Elementary School Mathematics in the United States, Taiwan, and Japan* (Reston, VA: NCTM, 1990), p. 2.

p. 388 8. From *Instructor's Course Planner* to accompany *Children* 4e, by John Santrock (Dubuque, IA: Brown & Benchmark, 1995), p. 36. © 1995. Reprinted by permission of The McGraw-Hill Companies.

p. 389 9. Marjorie Senechal, *On the Shoulders of Giants: New Approaches to Numeracy*, National Academy of Sciences, 1990. Reprinted with permission from *On the Shoulders of Giants: New Approaches to Numeracy.* Copyright © 1990 by the National Academy of Sciences. Courtesy of the National Academy Press, Washington, DC.

p. 409 10. Lois Lowry, *The Giver* (Boston: Houghton Mifflin Company/Walter Lorraine Book, 1993).

p. 417 11. *Utne Reader,* November/December 1993, p. 19.

CHAPTER 8

p. 432 1. Douglas H. Clements and Michael T. Battista, "Geometry and Spatial Reasoning," in *Handbook of Research on Mathematics Teaching and Learning,* ed. Douglas A. Grouws (New York: Macmillan, 1992), p. 442.

p. 447 2. My writing of this section has been informed by Thomas Fox's "Implications of Research on Children's Understanding of Geometry" in the May 2000 issue of *Teaching Children Mathematics,* pp. 572–576.

p. 456 3. From "Shape Up!" by Christine Oberdorf and Jennifer Taylor-Cox, published in the February 1999 issue of *Teaching Children Mathematics,* pp. 340–345.

p. 459 4. From *The American Heritage Dictionary of the English Language,* Third Edition.

p. 483 5. Marjorie Senechal, *On the Shoulders of Giants,* p. 173. Reprinted with permission from *On the Shoulders of Giants: New Approaches to Numeracy.* Copyright © 1990 by the National Academy of Sciences. Courtesy of the National Academy Press, Washington, D.C.

p. 491 6. H.A. Freebury, *A History of Mathematics* (New York: Macmillan, 1961), p. 36; H.A. Eves, *An Introduction to the History of Mathematics* (New York: Holt, Rinehart and Winston, 1969), p. 68.

p. 491 7. *Historical Topics for the Mathematics Classroom: Thirty-first Yearbook* (Reston, VA: NCTM, 1969), p. 220.

p. 497 8. From *Examining Features of Shape: Casebook* by Deborah Schifter, Virginia Bastable, and Susan Jo Russell, with Danielle Harrington and Marion Reynolds (Parsippany, NJ: Dale Seymour, 2002), p. 26.

CHAPTER 9

p. 507 1. Daniel Boorstin, *The Discoverers* (New York: Random House, 1983), p. 34.

p. 507 2. H. Arthur Klein, *The World of Measurements* (New York: Simon and Schuster, 1974), p. 163.

p. 507 3. Terry A. Richardson, *A Guide to Metrics* (Ann Arbor, MI: Prakken Publications, 1978), p. 2.

p. 510 4. S. Carl Hirsch, *Meter Means Measure: The Story of the Metric System* (New York: Viking Press, 1973), p. 29.

p. 519 5. Mary M. Lindquist, ed., *Results from the Fourth Mathematics Assessment* (Reston, VA: NCTM, 1989), p. 40.

p. 533 6. Randall Souviney, Murray Britt, Salvi Garguilo, and Peter Hughes, *Mathematical Investigations: Book One* (Palo Alto, CA: Dale Seymour Publications, 1988), pp. 10, 14. Reprinted by permission.

CHAPTER 10

p. 579 1. Hermann Weyl, *Symmetry* (Princeton, NJ: Princeton University Press, 1952), p. 5.

p. 580 2. Ian Stewart, *Nature's Numbers* (New York: HarperCollins, 1995), p. 83. Excerpt from *Nature's Numbers* by Ian Stewart. Copyright © 1995 by Ian Stewart. Basic Books/Perseus Books Group.

p. 590 3. Ian Stewart, *Nature's Numbers* (New York: HarperCollins, 1995), p. 73. Excerpt from *Nature's Numbers* by Ian Stewart. Copyright © 1995 by Ian Stewart. Basic Books/Perseus Books Group.

p. 593 4. http://forum.swarthmore.edu/geometry/rugs/symmetry/breaking.html.

p. 594 5. Marjorie Senechal, *On the Shoulders of Giants,* pp. 141–142. Reprinted with permission from *On the Shoulders of Giants: New Approaches to Numeracy.* Copyright © 1990 by the National Academy of Sciences. Courtesy of the National Academy Press, Washington, DC.

p. 612 6. Marjorie Senechal, *On the Shoulders of Giants,* pp. 141–142. Reprinted with permission from *On the Shoulders of Giants: New Approaches to Numeracy.* Copyright © 1990 by the National Academy of Sciences. Courtesy of the National Academy Press, Washington, DC.

INDEX

B

C

D

E

F

N

O

Q

R

S